Multiples and Prefixes for Metric Units*

Multiple	Prefix (and Abbreviation)	Pronunciation
10^{24}	yotta- (Y)	yot'ta (*a* as in *a*bout)
10^{21}	zetta- (Z)	zet'ta (*a* as in *a*bout)
10^{18}	exa- (E)	ex'a (*a* as in *a*bout)
10^{15}	peta- (P)	pet'a (as in *pet*al)
10^{12}	tera- (T)	ter'a (as in *terr*ace)
10^{9}	giga- (G)	ji'ga (*ji* as in *jig*gle, *a* as in *a*bout)
10^{6}	mega- (M)	meg'a (as in *mega*phone)
10^{3}	kilo- (k)	kil'o (as in *kilo*watt)
10^{2}	hecto- (h)	hek'to (*heck-toe*)
10	deka- (da)	dek'a (*deck* plus *a* as in *a*bout)
10^{-1}	deci- (d)	des'i (as in *deci*mal)
10^{-2}	centi- (c)	sen'ti (as in *senti*mental)
10^{-3}	milli- (m)	mil'li (as in *mili*tary)
10^{-6}	micro- (μ)	mi'kro (as in *micro*phone)
10^{-9}	nano- (n)	nan'oh (*an* as in *ann*ual)
10^{-12}	pico- (p)	pe'ko (*peek-oh*)
10^{-15}	femto- (f)	fem'toe (*fem* as in *fem*inine)
10^{-18}	atto- (a)	at'toe (as in *an*atomy)
10^{-21}	zepto- (z)	zep'toe (as in *ze*ppelin)
10^{-24}	yocto- (y)	yock'toe (as in *sock*)

*For example, 1 gram (g) multiplied by 1000 (10^{3}) is 1 kilogram (kg); 1 gram multiplied by 1/1000 (10^{-3}) is 1 milligram (mg).

SI Base Units

Physical Quantity		
Length		
Mass		
Time	second	
Electric current	ampere	A
Temperature	kelvin	K
Amount of substance	mole	mol
Luminous intensity	candela	cd

Some SI Derived Units

Physical Quantity	Name of Unit	Symbol	SI Unit
Frequency	hertz	Hz	s^{-1}
Energy	joule	J	$kg \cdot m^2/s^2$
Force	newton	N	$kg \cdot m/s^2$
Pressure	pascal	Pa	$kg/(m \cdot s^2)$
Power	watt	W	$kg \cdot m^2/s^3$
Electric charge	coulomb	C	$A \cdot s$
Electric potential	volt	V	$kg \cdot m^2/(A \cdot s^3)$
Electric resistance	ohm	Ω	$kg \cdot m^2/(A^2 \cdot s^3)$
Capacitance	farad	F	$A^2 \cdot s^4/(kg \cdot m^2)$
Inductance	henry	H	$kg \cdot m^2/(A^2 \cdot s^2)$
Magnetic field	tesla	T	$kg/(A \cdot s^2)$

Pythagorean Theorem (right triangle)

$$r = \sqrt{x^2 + y^2}$$

Quadratic Formula

If $ax^2 + bx + c = 0$, then

$$x = \frac{-b \pm \sqrt{b^2 - 4ac}}{2a}$$

Trigonometric Relationships

Definitions of Trigonometric Functions

$$\sin\theta = \frac{y}{r} \qquad \cos\theta = \frac{x}{r} \qquad \tan\theta = \frac{\sin\theta}{\cos\theta} = \frac{y}{x}$$

$\theta°$ (rad)	$\sin\theta$	$\cos\theta$	$\tan\theta$
0° (0)	0	1	0
30° ($\pi/6$)	0.500	$\sqrt{3}/2 \approx 0.866$	$\sqrt{3}/3 \approx 0.577$
45° ($\pi/4$)	$\sqrt{2}/2 \approx 0.707$	$\sqrt{2}/2 \approx 0.707$	1.00
60° ($\pi/3$)	$\sqrt{3}/2 \approx 0.866$	0.500	$\sqrt{3} \approx 1.73$
90° ($\pi/2$)	1	0	∞

Law of Cosines

$$a^2 = b^2 + c^2 - 2bc \cos A$$

Law of Sines

$$\frac{a}{\sin A} = \frac{b}{\sin B} = \frac{c}{\sin C}$$

(*a*, *b*, *c* sides, *A*, *B*, *C* angles opposite sides for any plane triangle)

Conversion Factors

Mass

$1\ g = 10^{-3}\ kg$
$1\ kg = 10^3\ g$
$1\ u = 1.66 \times 10^{-24}\ g = 1.66 \times 10^{-27}\ kg$
1 metric ton $= 1000\ kg$

Length

$1\ nm = 10^{-9}\ m$
$1\ cm = 10^{-2}\ m = 0.394\ in.$
$1\ m = 10^{-3}\ km = 3.28\ ft = 39.4\ in.$
$1\ km = 10^3\ m = 0.621\ mi$
$1\ in. = 2.54\ cm = 2.54 \times 10^{-2}\ m$
$1\ ft = 0.305\ m = 30.5\ cm$
$1\ mi = 5280\ ft = 1609\ m = 1.609\ km$

Area

$1\ cm^2 = 10^{-4}\ m^2 = 0.155\,0\ in^2$
$\quad = 1.08 \times 10^{-3}\ ft^2$
$1\ m^2 = 10^4\ cm^2 = 10.76\ ft^2 = 1550\ in^2$
$1\ in^2 = 6.94 \times 10^{-3}\ ft^2 = 6.45\ cm^2$
$\quad = 6.45 \times 10^{-4}\ m^2$
$1\ ft^2 = 144\ in^2 = 9.29 \times 10^{-2}\ m^2 = 929\ cm^2$

Volume

$1\ cm^3 = 10^{-6}\ m^3 = 3.53 \times 10^{-5}\ ft^3$
$\quad = 6.10 \times 10^{-2}\ in^3$
$1\ m^3 = 10^6\ cm^3 = 10^3\ L = 35.3\ ft^3$
$\quad = 6.10 \times 10^4\ in^3 = 264\ gal$
$1\ liter = 10^3\ cm^3 = 10^{-3}\ m^3 = 1.056\ qt$
$\quad = 0.264\ gal = 0.035\,3\ ft^3$
$1\ in^3 = 5.79 \times 10^{-4}\ ft^3 = 16.4\ cm^3$
$\quad = 1.64 \times 10^{-5}\ m^3$
$1\ ft^3 = 1728\ in^3 = 7.48\ gal = 0.028\,3\ m^3$
$\quad = 28.3\ L$
$1\ qt = 2\ pt = 946\ cm^3 = 0.946\ L$
$1\ gal = 4\ qt = 231\ in^3 = 0.134\ ft^3 = 3.785\ L$

Time

$1\ h = 60\ min = 3600\ s$
$1\ day = 24\ h = 1440\ min = 8.64 \times 10^4\ s$
$1\ y = 365\ days = 8.76 \times 10^3\ h$
$\quad = 5.26 \times 10^5\ min = 3.16 \times 10^7\ s$

Angle

$1\ rad = 57.3°$

$1° = 0.0175\ rad$	$60° = \pi/3\ rad$
$15° = \pi/12\ rad$	$90° = \pi/2\ rad$
$30° = \pi/6\ rad$	$180° = \pi\ rad$
$45° = \pi/4\ rad$	$360° = 2\pi\ rad$

$1\ rev/min = (\pi/30)\ rad/s = 0.104\,7\ rad/s$

Speed

$1\ m/s = 3.60\ km/h = 3.28\ ft/s$
$\quad = 2.24\ mi/h$
$1\ km/h = 0.278\ m/s = 0.621\ mi/h$
$\quad = 0.911\ ft/s$
$1\ ft/s = 0.682\ mi/h = 0.305\ m/s$
$\quad = 1.10\ km/h$
$1\ mi/h = 1.467\ ft/s = 1.609\ km/h$
$\quad = 0.447\ m/s$
$60\ mi/h = 88\ ft/s$

Force

$1\ N = 0.225\ lb$
$1\ lb = 4.45\ N$
Equivalent weight of a mass of 1 kg
$\quad$ on Earth's surface $= 2.2\ lb = 9.8\ N$

Pressure

$1\ Pa\ (N/m^2) = 1.45 \times 10^{-4}\ lb/in^2$
$\quad = 7.5 \times 10^{-3}\ torr\ (mm\ Hg)$
$1\ torr\ (mm\ Hg) = 133\ Pa\ (N/m^2)$
$\quad = 0.02\ lb/in^2$
$1\ atm = 14.7\ lb/in^2 = 1.013 \times 10^5\ N/m^2$
$\quad = 30\ in.\ Hg = 76\ cm\ Hg$
$1\ lb/in^2 = 6.90 \times 10^3\ Pa\ (N/m^2)$
$1\ bar = 10^5\ Pa$
$1\ millibar = 10^2\ Pa$

Energy

$1\ J = 0.738\ ft \cdot lb = 0.239\ cal$
$\quad = 9.48 \times 10^{-4}\ Btu = 6.24 \times 10^{18}\ eV$
$1\ kcal = 4186\ J = 3.968\ Btu$
$1\ Btu = 1055\ J = 778\ ft \cdot lb = 0.252\ kcal$
$1\ cal = 4.186\ J = 3.97 \times 10^{-3}\ Btu$
$\quad = 3.09\ ft \cdot lb$
$1\ ft \cdot lb = 1.36\ J = 1.29 \times 10^{-3}\ Btu$
$1\ eV = 1.60 \times 10^{-19}\ J$
$1\ kWh = 3.6 \times 10^6\ J$

Power

$1\ W = 0.738\ ft \cdot lb/s = 1.34 \times 10^{-3}\ hp$
$\quad = 3.41\ Btu/h$
$1\ ft \cdot lb/s = 1.36\ W = 1.82 \times 10^{-3}\ hp$
$1\ hp = 550\ ft \cdot lb/s = 745.7\ W$
$\quad = 2545\ Btu/h$

Mass–Energy Equivalents

$1\ u = 1.66 \times 10^{-27}\ kg \leftrightarrow 931.5\ MeV$
1 electron mass $= 9.11 \times 10^{-31}\ kg$
$\quad = 5.49 \times 10^{-4}\ u \leftrightarrow 0.511\ MeV$
1 proton mass $= 1.672\,62 \times 10^{-27}\ kg$
$\quad = 1.007\,276\ u \leftrightarrow 938.27\ MeV$
1 neutron mass $= 1.674\,93 \times 10^{-27}\ kg$
$\quad = 1.008\,665\ u \leftrightarrow 939.57\ MeV$

Temperature

$T_F = \frac{9}{5} T_C + 32$
$T_C = \frac{5}{9}(T_F - 32)$
$T_K = T_C + 273$

cgs Force

$1\ dyne = 10^{-5}\ N = 2.25 \times 10^{-6}\ lb$

cgs Energy

$1\ erg = 10^{-7}\ J = 7.38 \times 10^{-6}\ ft \cdot lb$

VOLUME 1

College Physics

SEVENTH EDITION

Jerry D. Wilson
LANDER UNIVERSITY
GREENWOOD, SC

Anthony J. Buffa
CALIFORNIA POLYTECHNIC STATE UNIVERSITY
SAN LUIS OBISPO, CA

Bo Lou
FERRIS STATE UNIVERSITY
BIG RAPIDS, MI

ADDISON-WESLEY

SAN FRANCISCO BOSTON NEW YORK
CAPE TOWN HONG KONG LONDON MADRID MEXICO CITY
MONTREAL MUNICH PARIS SINGAPORE SYDNEY TOKYO TORONTO

_____ Smith
___ _e Editor_: Nancy Whilton
___ rial Manager_: Laura Kenney
___oject Editor_: Chandrika Madhavan
Editorial Assistant: Dyan Menezes
Media Producer: David Huth
Director of Marketing: Christy Lawrence
Executive Marketing Manager: Scott Dustan
Managing Editor: Corinne Benson
Sr. Production Supervisor: Nancy Tabor
Production Service: Prepare, Inc.
Project Manager: Simone Lukashov
Illustrations: ArtWorks and Prepare, Inc.
Text Design: Seventeeth Street Studios
Cover Design: Derek Bacchus
Manufacturing Buyer: Jeff Sargent
Photo Research: Cypress Integrated Systems
Manager, Rights and Permissions: Zina Arabia
Image Permission Coordinator: Richard Rodrigues
Cover Printer: Courier, Kendallville
Text Printer and Binder: Courier, Kendallville
Cover Image: Adam Petty/Corbis
Photo Credits: See page C-1.

Library of Congress Cataloging-in-Publication Data

Wilson, Jerry D.
  College physics / Jerry D. Wilson, Anthony J. Buffa, Bo Lou. -- 7th ed.
    p. cm.
  ISBN 978-0-321-60183-4 -- ISBN 978-0-321-59277-4
  1. Physics--Textbooks. I. Buffa, Anthony J. II. Lou, Bo. III. Title.
  QC21.3.W35 2010
  530--dc22

                                              2008051063

ISBN: 978-0-321-59270-5

Addison-Wesley
is an imprint of

www.pearsonhighered.com

1 2 3 4 5 6 7 8 9 10—CRK—13 12 11 10 09

About the Authors

JERRY D. WILSON, a native of Ohio, is Emeritus Professor of Physics and former chair of the Division of Biological and Physical Sciences at Lander University in Greenwood, South Carolina. He received a B.S. degree from Ohio University, an M.S. degree from Union College, and, in 1970, a Ph.D. from Ohio University. He earned his M.S. degree while employed as a Materials Behavior physicist.

As a doctoral graduate student, Professor Wilson held the faculty rank of Instructor and began teaching physical science courses. During this time, he coauthored a physical science text that is now in its twelfth edition. In conjunction with his teaching career, Professor Wilson continued his writing and has authored or coauthored six titles. Having retired from full-time teaching, he continues to write, producing, among other works, *The Curiosity Corner*, a weekly column for local newspapers that can also be found on the Internet. He and his wife Sandy are lake dwellers in Greenwood, SC.

ANTHONY J. BUFFA received his B.S. degree in physics from Rensselaer Polytechnic Institute (RPI) in Troy, New York, and M.S. and Ph.D. degrees in physics from the University of Illinois, Urbana–Champaign. In 1970, Professor Buffa joined the faculty at California Polytechnic State University, San Luis Obispo. Recently retired, he teaches at Cal Poly as an Emeritus Professor. During his career he was involved in nuclear physics research at several national laboratories. On campus, he was a research associate in the department's radioanalytical facility. During his tenure at Cal Poly, he has taught courses from introductory physical science to quantum mechanics, developed and revised many laboratory experiments, and taught physics to local K-12 teachers at NSF workshops. Combining physics with interests in art and architecture, Dr. Buffa develops his own artwork and sketches, which he uses to increase his effectiveness in teaching. In addition to teaching, during his (partial) retirement, he and his wife intend to travel more and enjoy their future grandkids.

BO LOU is currently Professor of Physics at Ferris State University in Michigan. His primary teaching responsibilities are undergraduate introductory physics lectures and laboratories. Professor Lou emphasizes the importance of conceptual understanding of the basic laws and principles of physics and their practical applications to the real world. He is also an enthusiastic advocate of using technology in teaching and learning.

Professor Lou received his B.S. and M.S. degrees in optical engineering from Zhejiang University (China) in 1982 and 1985, respectively, and a Ph.D. in condensed matter physics from Emory University in 1989.

Dr. Lou, his wife Lingfei, and their daughter Alina reside in Big Rapids, Michigan. The family enjoys travel, nature, and tennis.

ActivPhysics™ OnLine Activities

Activ ONLINE Physics   www.masteringphysics.com

Brief Contents

Learn By Drawing

Applications

(Insights appear in **boldface**, and **"(bio)"** indicates a biomedical application)

Preface

We believe there are two basic goals in any introductory physics course: (1) to impart an understanding of the basic concepts of physics and (2) to enable students to use these concepts to solve a variety of problems.

These goals are linked. We want students to apply their conceptual understanding as they solve problems. Unfortunately, students often begin the problem-solving process by searching for an equation. There is the temptation to try to plug numbers into equations before visualizing the situation or considering the physical concepts that could be used to solve the problem. In addition, students often do not check their numerical answer to see if it matches their understanding of the relevant physical concept.

We feel, and users agree, that the strengths of this textbook are as follows:

Conceptual Basis. Giving students a secure grasp of physical principles will almost invariably enhance their problem-solving abilities. We have organized discussions and incorporated pedagogical tools to ensure that conceptual insight drives the development of practical skills.

Concise Coverage. To maintain a sharp focus on the essentials, we have avoided topics of marginal interest. We do not derive relationships when they shed no additional light on the principle involved. It is usually more important for students in this course to understand what a relationship means and how it can be used than to understand the mathematical or analytical techniques employed to derive it.

Applications. *College Physics* is known for the strong mix of applications related to medicine, science, technology, and everyday life in its text narrative and Insight boxes. The seventh edition continues to have a wide range of applications we have also increased the number of biological and biomedical applications in recognition of the high percentage of premed and allied health majors who take this course. A complete list of applications, with page references, is found on pages viii–xi.

A complete list of applications, with page references, is found on pages viii–xi.

We have added new material to further student understanding and make physics more relevant, interesting, and memorable for students.

Learning Path To provide students with a clear overview of the key concepts that they will be expected to learn as the chapter progresses, we have incorporated a flow chart that shows the learning path that students will take. It is reinforced throughout the chapter to keep students focused on the key concepts as they proceed.

Physics Facts. Each chapter begins with four to six interesting facts about discoveries or everyday phenomena applicable to the chapter.

Learning Path Review. Each end-of-chapter summary includes visual representations of the key concepts from the chapter to serve as a reminder for students as they review.

Biological Applications. The number and scope of biological and biomedical applications make them a popular feature. Examples of biological applications include "*g*'s of Force and Effects on the Human Body," "People Power: Using Body Energy," "Osteoporosis and Bone

---

INSIGHT 9.1 Osteoporosis and Bone Mineral Density (BMD)

Bone is a living, growing tissue. Your body is continuously taking up old bone (resorption) and making new bone tissue. In the early years of life, bone growth is greater than bone loss. This continues until a peak bone mass is reached as a young adult. After this, bone growth is slowly outpaced by bone loss. Bones naturally become less dense and weaker with age. Osteoporosis ("porous bone") occurs when bones deteriorate to the point where they are easily fractured (Fig. 1).

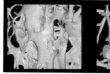

▲ FIGURE 1 **Bone mass loss** An X-ray micrograph of the bone structure of the vertebrae of a 50-year-old (left) and a 70-year-old (right). Osteoporosis, a condition characterized by bone weakening caused by loss of bone mass, is evident for the vertebrae on the right.

Osteoporosis and low bone mass affect an estimated 24 million Americans, most of whom are women. Osteoporosis results in an increased risk of bone fractures, particularly of the hip and the spine. Many women take calcium supplements to help prevent this.

To understand how bone density is measured, let's first distinguish between *bone* and *bone tissue*. Bone is the solid material composed of a protein matrix, most of which has calcified. Bone tissue includes the marrow spaces within the matrix. (Marrow is the soft, fatty, vascular tissue in the interior cavities of bones and is a major site of blood cell production.) The marrow volume varies with the bone type.

If the volume of an intact bone is measured (for example, by water displacement), then the *bone tissue density* can be com-

puted, commonly in grams per cubic centimeter, after the bone is weighed to determine mass. If you burn the bone, weigh the remaining ash, and divide by the volume of the overall bone (bone tissue), you get the *bone tissue mineral density*, which is commonly called the **bone mineral density (BMD)**.

To measure the BMD of bones *in vivo*, types of radiation transmission through the bone are measured, which is related to the amount of bone mineral present. Also, a "projected" area of the bone is measured. Using these measurements, a projected BMD is computed in units of mg>cm². Figure 2 illustrates the magnitude of the effect of bone density loss with aging.

The diagnosis of osteoporosis relies primarily on the measurement of BMD. The mass of a bone, measured by a BMD test (also called a *bone densitometry test*), generally correlates to the bone strength. It is possible to predict fracture risk, much as blood pressure measurements can help predict stroke risk. Bone density testing is recommended for all women age 65 and older, and for younger women at an increased risk of osteoporosis. This testing also applies to men. Osteoporosis is often thought to be a woman's disease, but 20% of osteoporosis cases occur in men. A BMD test cannot predict the certainty of developing a fracture, but only predicts the degree of risk.

So how is BMD measured? This is where the physics comes in. Various instruments, divided into *central devices* and *peripheral devices*, are used. Central devices are used primarily to measure the bone density of the hip and spine. Peripheral devices are smaller, portable machines that are used to measure the bone density in such places as the heel or finger.

The most widely used central device relies on *dual energy X-ray absorptiometry* (DXA), which uses X-ray imaging to measure bone density. (See Section 20.4 for a discussion of X-rays.) The DXA scanner produces two X-ray beams of different energy levels. The amount of X-rays that pass through a bone is measured for each beam; the amounts vary with the density of bone. The calculated bone density is based on the difference between the two beams. The procedure is nonintrusive and takes 10–20 min, and the X-ray exposure is usually about one-tenth of that of a chest X-ray (Fig. 3).

▶ FIGURE 2 **Bone density loss with aging** An illustration of how normal bone density loss for a female hip bone increases with age (scale on right). Osteopenia refers to decreased calcification or bone density. A person with osteopenia is at risk for developing osteoporosis, a condition that causes bones to become brittle and prone to fracture.

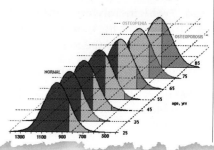

Mineral Density (BMD)," and "The Magnetic Force in Future Medicine."

We have enhanced the following pedagogical features in the seventh edition:

Learn by Drawing Boxes. Visualization is one of the most important steps in problem solving. In many cases, if students can make a sketch of a problem, they can solve it. "Learn by Drawing" features offer students specific help on making certain types of sketches and graphs that will provide key insights into a variety of physical situations.

Learning Path Questions (LPQs) and Did You Learn? (DYL). The usual section objectives have been replaced by LPQs at the beginning of each chapter section. The LPQs are two to three general questions on important section topics. They alert the student to important concepts covered in the section. The DYL at the end of each section is a new feature. They are general statements of items that should have been learned in the section.

Demonstrations. Six new demonstrations with photos have been added to the seventh edition. Examples include "Tension in a String: Action and Reaction Forces" and "Buoyancy and Density."

DEMONSTRATION 4 | Simple Harmonic Motion (SHM) and Sinusoidal Oscillation

A demostration to show that SHM can be represented by a sinusoidal function. A "graph" of the function is generated with an analogue of a strip chart recorder.

A salt-filled funnel oscillates, suspended from two strings.

Away we go. The salt falls on a black-painted poster board that will be pulled in a direction perpendicular to the plane of the funnel's oscillation.

The salt trail traces out a plot of displacement versus time, or $y = A \sin vt + d2$. Note that in this case the phase constant is about $d = 90°$ and $y = A \cos vt$. (Why?)

Suggested Problem-Solving Procedure. Section 1.7 provides a framework for thinking about problem solving. This section includes the following:

- An overview of problem-solving strategies

- A six-step procedure that is general enough to apply to most problems in physics, but is easily used in specific situations

- Examples that illustrate the detailed problem-solving process, showing how the general procedure is applied in practice

Problem-Solving Strategies and Hints. The initial treatment of problem solving is followed throughout with an abundance of suggestions, tips, cautions, shortcuts, and useful techniques for solving specific kinds of problems. These strategies and hints help students apply general principles to specific contexts as well as avoid common pitfalls and misunderstandings.

Conceptual Examples. These Examples ask students to think about a physical situation and conceptually solve a question or choose the correct prediction out of a set of possible outcomes on the basis of an understanding of relevant principles. The discussion that follows ("Reasoning and Answer") explains clearly how the correct answer can be identified, as well as why the other answers are wrong.

Worked Examples. We have tried to make in-text Examples as clear and detailed as possible. The aim is not merely to show students which equations to use, but to explain the strategy being employed and the role of each step in the overall plan. Students are encouraged to learn the "why" of each step along with the "how." Our goal is to provide a model for students to use as they solve problems. Each worked Example includes the following:

- *Thinking It Through* focuses students on the critical thinking and analysis they should undertake before beginning to use equations.

- *Given* and *Find* are provided as the first part of every *Solution* to remind students the importance of identifying what is known and what needs to be solved.

- *Follow-Up Exercises* at the end of each Conceptual Example and each worked Example further reinforce the importance of conceptual understanding and offer additional practice. (Answers to Follow-Up Exercises are given at the back of the text.)

Integrated Examples. In order to further emphasize the connection between conceptual understanding and quantitative problem solving, we have developed Integrated Examples for each chapter. These Examples work through a physical situation both qualitatively and quantitatively. The qualitative portion is solved by conceptually choosing the correct answer from a set of possible answers. The quantitative portion involves a mathematical solution related to the conceptual part, demonstrating how conceptual understanding and numerical calculations go hand in hand.

Pulling It Together Examples. These worked examples show students how to work problems that involve multiple concepts from the chapter and provide a bridge from the individual worked examples in the chapter to the end-of-chapter comprehensive Pulling It Together problems.

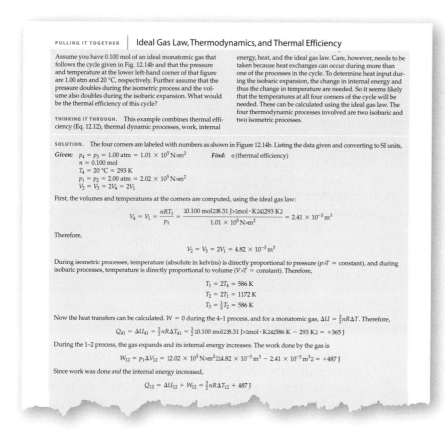

PULLING IT TOGETHER | Ideal Gas Law, Thermodynamics, and Thermal Efficiency

Assume you have 0.100 mol of an ideal monatomic gas that follows the cycle given in Fig. 12.14b and that the pressure and temperature at the lower left-hand corner of that figure are 1.00 atm and 20 °C, respectively. Further assume that the pressure doubles during the isometric process and the volume also doubles during the isobaric expansion. What would be the thermal efficiency of this cycle?

THINKING IT THROUGH. This example combines thermal efficiency (Eq. 12.12), thermal dynamic processes, work, internal energy, heat, and the ideal gas law. Care, however, needs to be taken because heat exchanges can occur during more than one of the processes in the cycle. To determine heat input during the isobaric expansion, the change in internal energy and thus the change in temperature are needed. So it seems likely that the temperatures at all four corners of the cycle will be needed. These can be calculated using the ideal gas law. The four thermodynamic processes involved are two isobaric and two isometric processes.

SOLUTION. The four corners are labeled with numbers as shown in Figure 12.14b. Listing the data given and converting to SI units,

Given: $p_4 = p_3 = 1.00$ atm $= 1.01 \times 10^5$ N>m^2 Find: e (thermal efficiency)
$n = 0.100$ mol
$T_4 = 20$ °C $= 293$ K
$p_1 = p_2 = 2.00$ atm $= 2.02 \times 10^5$ N>m^2
$V_2 = V_3 = 2V_4 = 2V_1$

First, the volumes and temperatures at the corners are computed, using the ideal gas law:

$$V_4 = V_1 = \frac{nRT_1}{p_1} = \frac{10.100 \text{ mol}238.31 \text{ J>1mol} \cdot \text{K}241293 \text{ K2}}{1.01 \times 10^5 \text{ N>m}^2} = 2.41 \times 10^{-3} \text{ m}^3$$

Therefore,

$$V_2 = V_3 = 2V_1 = 4.82 \times 10^{-3} \text{ m}^3$$

During isometric processes, temperature (absolute in kelvins) is directly proportional to pressure ($p>T$ = constant), and during isobaric processes, temperature is directly proportional to volume ($V>T$ = constant). Therefore,

$$T_1 = 2T_4 = 586 \text{ K}$$
$$T_2 = 2T_1 = 1172 \text{ K}$$
$$T_3 = \tfrac{1}{2}T_2 = 586 \text{ K}$$

Now the heat transfers can be calculated. $W = 0$ during the 4–1 process, and for a monatomic gas, $\Delta U = \tfrac{3}{2}nR\Delta T$. Therefore,

$$Q_{41} = \Delta U_{41} = \tfrac{3}{2}nR\Delta T_{41} = \tfrac{3}{2} 10.100 \text{ mol}238.31 \text{ J>1mol} \cdot \text{K}241586 \text{ K} - 293 \text{ K2} = +365 \text{ J}$$

During the 1–2 process, the gas expands and its internal energy increases. The work done by the gas is

$$W_{12} = p_1\Delta V_{12} = 12.02 \times 10^5 \text{ N>m}^2214.82 \times 10^{-3} \text{ m}^3 - 2.41 \times 10^{-3} \text{ m}^32 = +487 \text{ J}$$

Since work was done *and* the internal energy increased,

$$Q_{12} = \Delta U_{12} + W_{12} = \tfrac{3}{2}nR\Delta T_{12} + 487 \text{ J}$$

End-of-Chapter Exercises. Each section of the end-of-chapter material begins with multiple-choice questions (**MC**) to allow students a quick self-test for that section. These are followed by short-answer conceptual questions (**CQ**) that test students' conceptual understanding and ask students to reason from principles. Quantitative problems round out the Exercises in each section. *College Physics* provides short answers to all odd-numbered Exercises (quantitative *and* conceptual) at the back of the text, so students can check their understanding.

Paired Exercises. To encourage students to work problems on their own, most sections include at least one set of Paired Exercises that deal with similar situations. The first problem in a pair is solved in the *Student Study Guide and Solutions Manuals*; the second problem, which explores a situation similar to that presented in the first problem, has only an answer at the back of the book.

Integrated Exercises. Like the Integrated Examples in the chapter, Integrated Exercises (IE) ask students to solve a problem quantitatively as well as answer a conceptual question dealing with the exercise. By answering both parts, students can see if their numerical answer matches their conceptual understanding.

Pulling It Together: Multiconcept Exercises. To ensure that students can synthesize concepts, each chapter concludes with a section of comprehensive exercises drawn from all sections of the chapter and perhaps basic principles from previous chapters.

ABSOLUTELY ZERO TOLERANCE FOR ERRORS CLUB (AZTEC)

We have continued to ensure accuracy through the Absolutely Zero Tolerance for Errors Club (AZTEC). Tony Buffa of Cal Poly San Luis Obispo headed the AZTEC team and was supported by the text's co-authors as well as by *Wayne Anderson* and *Sen-Ben Liao*. Every end-of-chapter Exercise was worked by three of the five team members individually and independently. The results were collected and discrepancies were resolved by a team discussion. While there has never been a text that was absolutely free of errors, that was our goal; we worked very hard to make the book error-free.

The seventh edition is supplemented by a media and print ancillary package developed to address the needs of both students and instructors.

FOR THE INSTRUCTOR

Instructor Solutions Manual, prepared by Wayne Anderson, this manual is available online at the Instructor Resource Center: www.pearsonhighered.com/educator. It includes complete, worked-out solutions to all end-of-chapter exercises.

Instructor Resource Manual with Notes on ConcepTest Questions Available at the Instructor Resource Center, www.pearsonhighered.com/educator, this online manual has two parts. The first part, prepared by Katherine Whatley and Judith Beck (both of University of North Carolina, Asheville), contains sample syllabi, lecture outlines, notes, demonstration suggestions, readings, and additional references and resources. The second part, prepared by Cornelius Bennhold and Gerald Feldman (both of George Washington University), contains an overview of the development and implementation of ConcepTests, as well as instructor notes for each

ConcepTest found on the Instructor Resource Center and available on the Instructor Resource DVD.

Test Bank Available at the Instructor Resource Center, www.pearsonhighered.com/educator, the test bank was fully revised by Delena Gatch (University of North Alabama). This online, cross-platform test bank offers more than 2800 multiple-choice, true/false, and short-answer/essay questions, approximately 50% conceptual. The questions are organized and referenced by chapter section and by question type.

Instructor Resource DVD (ISBN 0-321-59273-5) This cross-platform DVD provides virtually every electronic asset you'll need in and out of the classroom. The DVD is organized by chapter and includes all text illustrations and tables from *College Physics*, seventh edition in jpeg format. The IRDVD also contains ConcepTest "Clicker" Questions in PowerPoint, the eleven Physics You Can See demonstration videos, and pdf files of the *Instructor Resource Manual with Notes on ConcepTest Questions.*

(MP) **MasteringPhysics™** Available at www.masteringphysics.com, this homework, tutorial, and assessment system is designed to assign, assess, and track each student's progress using a wide diversity of tutorials and extensively pre-tested problems. Half of the end-of-chapter problems from the text are available in MasteringPhysics. This system provides instructors with a fast and effective way to assign uncompromising, wide-ranging online homework assignments of just the right difficulty and duration. The tutorials coach 90% of students to the correct answer with specific wrong-answer feedback. The powerful post-assignment diagnostics allow instructors to assess the progress of their class as a whole or to quickly identify individual students' areas of difficulty.

myeBook The interactive myebook is available through MasteringPhysics either automatically when MasteringPhysics is packaged with new books, or as a purchased upgrade online. Allowing students access to the text wherever they have acces to the Internet, myeBook comprises the full text, including figures that can be enlarged for better viewing. Within myeBook, students are able to pop up definitions and terms to help with vocabulary and the reading of the material. Students can also take notes in myeBook using the annotation feature at the top of each page.

ActivPhysics Online™ Accessed through the Self Study area of www.masteringphysics.com, ActivPhysics Online provides a comprehensive library of more than 420 tried and tested ActivPhysics applets. In addition, it provides a suite of applet-based tutorials developed by education pioneers Alan Van Heuvelen and Paul D'Allessandris. The online exercises are designed to encourage students to confront misconceptions, reason qualitatively, and learn to think critically. They cover all topics from mechanics to electricity and magnetism and from optics to modern physics. The ActivePhysics OnLine companion workbooks help students work through complex concepts and understand them more clearly.

FOR THE STUDENT

Student Study Guide and Selected Solutions Manual by Bo Lou (Ferris State University); Volume 1: 0-321-59274-3; Volume 2: 0-321-59278-6 This guide presents chapter-by-chapter reviews, chapter summaries and discussions, mathematical summary, additional worked examples, practice quizzes, and solutions to paired and selected exercises.

(MP) **MasteringPhysics™** Available at www.masteringphysics.com, this homework, tutorial, and assessment system is based on years of research into how students work physics problems and precisely where they need help. Studies show that students who use MasteringPhysics significantly increase their final scores compared to those who use handwritten homework. MasteringPhysics achieves this improvement by providing students with instantaneous feedback specific to their wrong answers, simpler sub-problems upon request when they get stuck, and partial credit for their method(s) used. This individualized, 24/7 Socratic tutoring is recommended by nine out of ten students to their peers as the most effective and time-efficient way to study.

myeBook is available through MasteringPhysics either automatically when MasteringPhysics is packaged with new books, or as a purchased upgrade online. Allowing students access to the text wherever they have access to the Internet, myeBook comprises the full text, including figures that can be enlarged for better viewing. Within myeBook, students are able to pop up definitions and terms to help with vocabulary and the reading of the material. Students can also take notes in myeBook using the annotation feature at the top of each page.

ActivPhysics Online™ Accessed through the Self Study area of www.masteringphysics.com, ActivPhysics Online provides students with a suite of highly regarded applet-based self-study tutorials (see description on previous page). The following workbooks provide a range of tutorial problems designed to use the *ActivPhysics OnLine* simulations, helping students work through complex concepts and understand them more clearly:

- **ActivPhysics OnLine Workbook** Volume 1: Mechanics—Thermal Physics—Oscillations & Waves (ISBN 0-8053-9060-X)
- **ActivPhysics OnLine Workbook** Volume 2: Electricity & Magnetism—Optics—Modern Physics (ISBN 0-8053-9061-8)

Pearson Tutor Services (www.pearsontutorservices. com) Each student's subscription to MasteringPhysics also contains complimentary access to Pearson Tutor Services, powered by Smarthinking, Inc. By logging in with their MasteringPhysics ID and password, they will be connected to highly qualified e-structors who provide additional, interactive online tutoring on the major concepts of physics. Some restrictions apply; offer subject to change.

Acknowledgments

The members of AZTEC—Wayne Anderson and Sen-Ben Liao—as well as accuracy reviewer Todd Pedlar deserve more than a special thanks for their tireless, timely, and extremely thorough review of this book.

Dozens of other colleagues, listed in the upcoming section, helped us identify ways to make the seventh edition a better learning tool for students. We are indebted to them, as their thoughtful and constructive suggestions benefited the book greatly.

We owe many thanks to the editorial and production team at Addison-Wesley, including Nancy Whilton, Executive Editor, and Chandrika Madhavan, Project Editor. In particular, the authors wish to acknowledge the outstanding performance of Simone Lukashov, Production Editor. His courteous, conscientious, and cheerful manner made for an efficient and enjoyable production process.

In addition, I (Tony Buffa) once again extend many thanks to my co-authors, Jerry Wilson and Bo Lou, for their cheerful helpfulness and professional approach to the work on this edition. As always, several colleagues of mine at Cal Poly gave of their time for fruitful discussions. Among them are Professors Joseph Boone, Ronald Brown, and Theodore Foster. My family—my wife, Connie, and daughters, Jeanne and Julie—was, as always, a continuous and welcomed source of support. I also acknowledge the support of my father, Anthony Buffa, Sr. Lastly, I thank the students in my classes who contributed excellent ideas over the past few years.

Finally, we would like to urge anyone using the book—student or instructor—to pass on to us any suggestions that you have for its improvement. We look forward to hearing from you.

—Jerry D. Wilson
jwilson@greenwood.net

—Anthony J. Buffa
abuffa@calpoly.edu

—Bo Lou
loub@ferris.edu

REVIEWERS OF PREVIOUS EDITIONS

David Aaron
South Dakota State University

William Achor
Western Maryland College

E. Daniel Akpanumoh
Houston Community College, Southwest

Alice Hawthorne Allen
Virginia Tech

Arthur Alt
College of Great Falls

Zaven Altounian
McGill University

Frederick Anderson
University of Vermont

Ifran Azeem
Embry-Riddle Aeronautical University

Charles Bacon
Ferris State College

Ali Badakhshan
University of Northern Iowa

Anand Batra
Howard University

Raymond D. Benge
Tarrant County College

Michael Berger
Indiana University

William Berres
Wayne State University

Frederick Bingham
University of North Carolina, Wilmington

Timothy C. Black
University of North Carolina, Wilmington

Mary Boleware
Jones County Junior College

James Borgardt
Juniata College

Hugo Borja
Macomb Community College

Bennet Brabson
Indiana University

Jeffrey Braun
University of Evansville

Art Braundmeier
Southern Illinois University, Edwardsville

Michael Browne
University of Idaho

Michael L. Broyles
Collin County Community College

Debra L. Burris
Oklahoma City Community College

David Bushnell
Northern Illinois University

Lyle Campbell
Oklahoma Christian University

James Carroll
Eastern Michigan State University

Aaron Chesir
Lucent Technologies

Lowell Christensen
American River College

Philip A. Chute
University of Wisconsin–Eau Claire

Robert Coakley
University of Southern Maine

Lawrence Coleman
University of California–Davis

Lattie F. Collins
East Tennessee State University

Sergio Conetti
University of Virginia, Charlottesville

James Cook
Middle Tennessee State University

David M. Cordes
Belleville Area Community College

James R. Crawford
Southwest Texas State University

William Dabby
Edison Community College

Purna Das
Purdue University

J. P. Davidson
University of Kansas

Donald Day
Montgomery College

Richard Delaney
College of Aeronautics

Jason Donav
University of Puget Sound

Robert M. Drosd
Portland Community College

James Ellingson
College of DuPage

Donald Elliott
Carroll College

Bruce Emerson
Central Oregon Community College

Arnold Feldman
University of Hawaii

Milton W. Ferguson
Norfolk State University

John Flaherty
Yuba College

Rober J. Foley
University of Wisconsin–Stout

Lewis Ford
Texas A&M University

Donald Foster
Wichita State University

Donald R. Franceschetti
Memphis State University

Frank Gaev
ITT Technical Institute–Ft. Lauderdale

Rex Gandy
Auburn University

Simon George
California State–Long Beach

Barry Gilbert
Rhode Island College

Phillip Gilmour
Tri-County Technical College

Richard Grahm
Ricks College

Tom J. Gray
University of Nebraska

Allen Grommet
East Arkansas Community College

Douglas Al Harrington
Northeastern State University

Gary Hastings
Georgia State University

Xiaochun He
Georgia State University

J. Erik Hendrickson
University of Wisconsin–Eau Claire

Al Hilgendorf
University of Wisconsin–Stout

Brian Hinderliter
North Dakota State University

Joseph M. Hoffman
Frostburg State University

Andy Hollerman
University of Louisiana, Layfayette

Ben Yu-Kuang Hu
University of Akron

Jacob W. Huang
Towson University

Porter Johnson
Illinois Institute of Technology

Randall Jones
Loyola University

Omar Ahmad Karim
University of North Carolina–Wilmington

S. D. Kaviani
El Camino College

Victor Keh
ITT Technical Institute–Norwalk, California

John Kenny
Bradley University

Andrew W. Kerr
University of Findlay

Jim Ketter
Linn-Benton Community College

James Kettler
Ohio University, Eastern Campus

Dana Klinck
Hillsborough Community College

Chantana Lane
University of Tennessee–Chattanooga

Phillip Laroe
Carroll College

Rubin Laudan
Oregon State University

Bruce A. Layton
Mississippi Gulf Coast Community College

R. Gary Layton
Northern Arizona University

Kevin Lee
University of Nebraska

Paul Lee
California State University, Northridge

Federic Liebrand
Walla Walla College

Mark Lindsay
University of Louisville

Bryan Long
Columbia State Community College

Michael LoPresto
Henry Ford Community College

Dan MacIsaac
Northern Arizona University

Terrence Maher
Alamance Community College

Robert March
University of Wisconsin

Trecia Markes
University of Nebraska–Kearney

Aaron McAlexander
Central Piedmont Community College

William McCorkle
West Liberty State University

John D. McCullen
University of Arizona

Michael McGie
California State University–Chico

Kevin McKone
Copiah Lincoln Community College

Kenneth L. Menningen
University of Wisconsin, Stevens Point

Michael Mikhaiel
Passaic County Community College

Ramesh C. Misra
Minnesota State University, Mankato

Sandra Moffet
Linn Benton Community College

Paul Morris
Abilene Christian University

Gary Motta
Lassen College

J. Ronald Mowrey
Harrisburg Area Community College

Gerhard Muller
University of Rhode Island

K. W. Nicholson
Central Alabama Community College

Erin O'Connor
Allan Hancock College

Michael Ottinger
Missouri Western State College

James Palmer
University of Toledo

Anthony Pitucco
Glendale Community College

William Pollard
Valdosta State University

R. Daryl Pedigo
Austin Community College

T. A. K. Pillai
University of Wisconsin–La Crosse

Darden Powers
Baylor University

Donald S. Presel
University of Massachusetts–Dartmouth

Kent J. Price
Morehead State University

E. W. Prohofsky
Purdue University

Dan R. Quisenberry
Mercer University

W. Steve Quon
Ventura College

David Rafaelle
Glendale Community College

George Rainey
California State Polytechnic University

Michael Ram
SUNY–Buffalo

William Riley
Ohio State University

Salvatore J. Rodano
Harford Community College

William Rolnick
Wayne State University

John B. Ross
Indiana University-Purdue University, Indianapolis

Robert Ross
University of Detroit–Mercy

Craig Rottman
North Dakota State University

Gerald Royce
Mary Washington College

Roy Rubins
University of Texas, Arlington

Sid Rudolph
University of Utah

Om Rustgi
Buffalo State College

Anne Schmiedekamp
Pennsylvania State University–Ogontz

Cindy Schwarz
Vassar College

Terry Scott
University of Northern Colorado

Ray Sears
University of North Texas

Mark Semon
Bates College

Rahim Setoodeh
Milwaukee Area Technical College

Bartlett Sheinberg
Houston Community College

Jerry Shi
Pasadena City College

Martin Shingler
Lakeland Community College

Peter Shull
Oklahoma State University

Thomas Sills
Wilbur Wright College

Larry Silva
Appalachian State University

Michael Simon
Housatonic Community Technical College

Christopher Sirola
Tri-County Technical College

Gene Skluzacek
St. Petersburg College

Soren P. Sorensen
University of Tennessee–Knoxville

Ross Spencer
Brigham Young University

Mark Sprague
East Carolina State University

Steven M. Stinnett
McNeese State University

Dennis W. Suchecki
San Diego Mesa College

Frederick J. Thomas
Sinclair Community College

Jacqueline Thornton
St. Petersburg Junior College

Anthony Trippe
ITT Technical Institute–San Diego

Gabriel Umerah
Florida Community College–Jacksonville

John Underwood
Austin Community College

Tristan T. Utschig
Lewis-Clark State College

Lorin Vant-Hull
University of Houston

Pieter B. Visscher
University of Alabama

Karl Vogler
Northern Kentucky University

John Walkup
California Polytechnic State University

Arthur J. Ward
Nashville State Technical Institute

Larry Weinstein
Old Dominion University

John C. Wells
Tennessee Technical University

Steven P. Wells
Louisiana Technical University

Christopher White
Illinois Institute of Technology

Arthur Wiggins
Oakland Community College

Kevin Williams
ITT Technical Institute–Earth City

Linda Winkler
Appalachian State University

Jeffery Wragg
College of Charleston

Rob Wylie
Carl Albert State University

Anthony Zable
Portland Community College

John Zelinsky
Southern Illinois University

John Zelinsky
Community College of Baltimore County, Essex

Dean Zollman
Kansas State University

Contents

6   Linear Momentum and Collisions   180

7   Circular Motion and Gravitation   222

8   Rotational Motion and Equilibrium   266

9   Solids and Fluids   311

10   Temperature and Kinetic Theory   355

College Physics

1 Measurement and Problem Solving†

PHYSICS FACTS

✦ Tradition holds that in the twelfth century King Henry I of England decreed that the yard should be the distance from the tip of his royal nose to the thumb of his outstretched arm. (Had King Henry's arm been 3.37 inches longer, the yard and the meter would be equal in length.)

✦ The abbreviation for the pound, *lb*, comes from the Latin word *libra*, which was a Roman unit of weight approximately equal to a pound. The word *pound* comes from the Latin *pondero*, "to weigh." Libra is also a sign of the zodiac and is symbolized by a set of scales (used for weight measurement).

✦ Is the old saying "A pint's a pound the world around" true? It depends on what you are talking about. The saying is a good approximation for water and other similar liquids. Water weighs 8.3 pounds per gallon, so one-eighth of that, or a pint, weighs 1.04 lb.

I s it first and ten in the chapter-opening photo? A measurement is needed, as with many other things in our lives. Length measurements tell us how far it is between cities, how tall you are, and as in the photo, if it's first and ten (yards to go). Time measurements tell you how long it is until the class ends, when the semester or quarter begins, and how old you are. Drugs taken because of illnesses are given in measured doses. Lives depend on various measurements made by doctors, medical technologists, and pharmacists in the diagnosis and treatment of disease.

†The mathematics needed in this chapter involves scientific (powers-of-10) notation and trigonometry relationships. You may want to review these topics in Appendix I.

Measurements enable us to compute quantities and solve problems. Units of measurement are also important in measurements and problem solving. For example, in finding the volume of a rectangular box, if you measure its dimensions in inches, the volume would have units of in^3 (cubic inches); if measured in centimeters, then the units would be cm^3 (cubic centimeters). Measurement and problem solving are part of our lives. They play a particularly central role in our attempts to describe and understand the physical world, as will be seen in this chapter. For some reasons why one should study physics, see Insight 1.1.

1.1   Why and How We Measure

LEARNING PATH QUESTIONS

➡ What does physics attempt to do?

➡ How does a unit become a standard?

➡ What is a system of units?

Imagine that someone is giving you directions to her house. Would you find it helpful to be told, "Drive along Elm Street for a little while, and turn right at one of the lights. Then keep going for quite a long way"? Or would you want to deal with a bank that sent you a statement at the end of the month saying, "You still have some money left in your account. Not a great deal, though."

Measurement is important to all of us. It is one of the concrete ways in which we deal with our world. This concept is particularly true in physics. *Physics is concerned with the description and understanding of nature*, and measurement is one of its most important tools.

INSIGHT 1.1 | Why Study Physics?

The question "Why study physics?" occurs to many students at some time during their college careers. The truth is that there are probably as many answers as there are students, much as with any other subject. However, the answers can usually be arranged into several general groups, as follows.

You are probably not a *physics major*, but for these students, the answer is obvious. Introductory physics provides the foundation of their careers. The fundamental goal of physics is to discover and understand the rules ("laws") that govern observed phenomena. These students will use their knowledge of physics continually in various fields. As an example of an application, consider the development of the laser in the 1960s. It currently plays an important role in various fields—medicine, industry, music (CD-DVD players), and so on.

You are also probably not an *engineering "applied physics" major*. For these students, physics provides the basis of the engineering principles used to solve technological (applied and practical) problems. Some of these students may not use physics directly, but a good understanding of physics is crucial to the problem solving needed in technological advances. For example, after the discovery of the transistor by physicists, engineers then developed uses for it. Decades later it evolved into the modern computer chip, which is an electrical computing network containing millions of tiny transistor elements.

More than likely you are a *life or biological science major* (such as biology, premedicine, preveterinary, medical technology, or physical therapy). In this case, physics can provide a background understanding of the principles involved in your work. Although the applications of the laws of physics may not be immediately obvious, understanding them can be a valuable tool. For example, if you are a medical professional, it may be necessary to evaluate MRI (magnetic resonance imaging) results, a procedure that is now commonplace. Would you be surprised to know that MRI scans are based on a physical phenomenon called *nuclear magnetic resonance*, first discovered by physicists and still used for measuring nuclear and solid-state properties?

If you are a student in a *nonscience major*, the physics requirement is intended to provide a well-rounded education, that is, the ability to evaluate technology in the context of societal needs. For example, you may be called on to vote on tax benefits for an energy production source, and you may want to evaluate the pros and cons of the process. Or you may be tempted to vote for an official who has strong views on nuclear waste disposal. Are these views scientifically justified? To fully evaluate them, a knowledge of physics is necessary.

So as you can see, there is no one answer to the question "Why study physics?" However, there is one overriding theme: Knowledge of the laws of physics can provide an excellent background for understanding of the world around you, or it can simply help make you a better and more well-rounded citizen.

There are ways of describing the physical world that do not involve measurement. For instance, we might talk about the color of a flower or a dress. But the perception of color is subjective; it may vary from one person to another. Indeed, many people are color-blind and cannot tell certain colors apart. Light received by our eyes can be described in terms of wavelengths and frequencies. Different wavelengths are associated with different colors because of the physiological response of our eyes to light. But unlike the sensations or perceptions of color, wavelengths can be measured. They are the same for everyone. In other words, measurements are objective. *Physics attempts to describe and understand nature in an objective way through measurement.*

STANDARD UNITS

Measurements are expressed in terms of unit values, or units. As you are probably aware, a large variety of units are used to express measured values. Some of the earliest units of measurement, such as the foot, were originally referenced to parts of the human body. Even today, the hand is still used as a unit to measure the height of horses. One hand is equal to 4 inches (in.). If a unit becomes officially accepted, it is called a **standard unit**. Traditionally, a government or international body establishes standard units.

A group of standard units and their combinations is called a **system of units**. Two major systems of units are in use today—the metric system and the British system. The latter is still widely used in the United States, but has virtually disappeared in the rest of the world, having been replaced by the metric system.

Different units in the same system or units of different systems can be used to describe the same thing. For example, your height can be expressed in inches, feet, centimeters, meters—or even miles, for that matter (although this unit would not be very convenient). It is always possible to convert from one unit to another, and such conversions are sometimes necessary. However, it is best, and certainly most practical, to work consistently within the same system of units, as will be seen.

DID YOU LEARN?
➡ Physics attempts to describe and understand nature through measurement.
➡ A government or international body establishes measurement standards.
➡ A group of standard units and their combinations form a system of units. The two major systems of units in use today are the metric system and the British system.

1.2 SI Units of Length, Mass, and Time

LEARNING PATH QUESTIONS
➡ What is the difference between base and derived units?
➡ How are the meter (m), the kilogram (kg), and the second (s) currently defined?

Length, mass, and time are fundamental physical quantities that are used to describe a great many quantities and phenomena. In fact, the topics of mechanics (the study of motion and force) covered in the first part of this book require *only* these physical quantities. The system of units used by scientists to represent these and other quantities is based on the metric system.

Historically, the metric system was the outgrowth of proposals for a more uniform system of weights and measures in France during the seventeenth and eighteenth centuries. The modern version of the metric system is called the **International System of Units**, officially abbreviated as **SI** (from the French *Système International des Unités*).

The SI includes *base quantities* and *derived quantities*, which are described by base units and derived units, respectively. **Base units**, such as the meter (m), the kilogram (kg), and the second (s) are defined by standards. Other quantities that are expressed in terms of combinations of base units are called **derived units**.

▶ FIGURE 1.1 **The SI length standard: the meter** (a) The meter was
originally defined as 1/10 000 000 of
the distance from the North Pole to
the equator along a meridian running through Paris, of which a portion was measured between
Dunkirk and Barcelona. A metal bar
(called the Meter of the Archives)
was constructed as a standard.
(b) The meter is currently defined in
terms of the speed of light. See the
text for description.

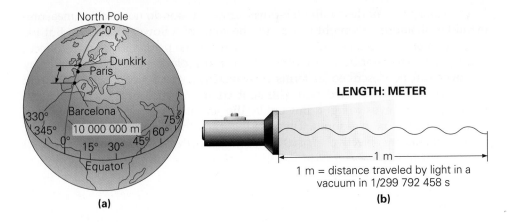

(Think of how we commonly measure the length of a trip in miles (mi) and the amount of time the trip takes in hours (h). To express how fast, or the rate we travel, the derived unit of miles per hour (mi/h) is used, which represents distance traveled per unit of time, or length per time.)

LENGTH

Length is the base quantity used to measure distances or dimensions in space. We commonly say that length is the distance between two points. But the distance between any two points depends on how the space between them is traversed, which may be in a straight or a curved path.

The SI unit of length is the **meter (m)**. The meter was originally defined as 1/10 000 000 of the distance from the North Pole to the equator along a meridian running through Paris (▲Fig. 1.1a).* A portion of this meridian between Dunkirk, France, and Barcelona, Spain, was surveyed to establish the standard length, which was assigned the name *metre*, from the Greek word *metron*, meaning "a measure." (The American spelling is *meter*.) A meter is 39.37 in.—slightly longer than a yard (3.37 in. longer).

The length of the meter was initially preserved in the form of a material standard: the distance between two marks on a metal bar (made of a platinum–iridium alloy) that was stored under controlled conditions in France and called the Meter of the Archives. However, it is not desirable to have a reference standard that changes with external conditions, such as temperature. In 1983, the meter was redefined in terms of a more accurate standard, an unvarying property of light: the length of the path traveled by light in a vacuum during an interval of 1/299 792 458 of a second (Fig. 1.1b). Light travels 299 792 458 m in a second, and the speed of light in a vacuum is $c = 299\ 792\ 458$ m/s. (c is the common symbol for the speed of light.) Thus, light travels 1 m in 1/299 792 458 s. Note that the length standard is referenced to time, which can be measured with great accuracy.

MASS

Mass is the base quantity used to describe amounts of matter. The more massive an object, the more matter it contains. The SI unit of mass is the **kilogram (kg)**. The kilogram was originally defined in terms of a specific volume of water, that is, a cube 0.10 m (10 cm) on a side (thereby associating the mass standard with the

---

*Note that this book and most physicists have adopted the practice of writing large numbers with a thin space for three-digit groups—for example, 10 000 000 (not 10,000,000). This is done to avoid confusion with the European practice of using a comma as a decimal point. For instance, 3.141 in the United States would be written as 3,141 in Europe. Large decimal numbers, such as 0.537 84, may also be separated, for consistency. Spaces are generally used for numbers with more than four digits on either side of the decimal point.

length standard). However, the kilogram is now referenced to a specific material standard: the mass of a prototype platinum–iridium cylinder kept at the International Bureau of Weights and Measures in Sèvres, France (►Fig. 1.2). The United States has a duplicate of the prototype cylinder. The duplicate serves as a reference for secondary standards that are used in everyday life and commerce. It is hoped that the kilogram may eventually be referenced to something other than a material standard.

You may have noticed that the phrase *weights and measures* is generally used instead of *masses and measures*. In the SI, mass is a base quantity, but in the British system, weight is used to describe amounts of mass—for example, weight in pounds instead of mass in kilograms. The weight of an object is the gravitational attraction that the Earth exerts on the object. For example, when you weigh yourself on a scale, your weight is a measure of the downward gravitational force exerted on your mass by the Earth. Weight is a measure of mass in this way near the Earth's surface, because weight and mass are directly proportional to each other.

But treating weight as a base quantity creates some problems. A base quantity should have the same value everywhere. This is the case with mass—an object has the same mass, or amount of matter, regardless of its location. *But this is not true of weight.* For example, the weight of an object on the Moon is less than its weight on the Earth (one-sixth as much). This is because the Moon is less massive than the Earth and the gravitational attraction exerted on an object by the Moon (the object's weight) is less than that exerted by the Earth. That is, an object with a given amount of mass has a particular weight on the Earth, but on the Moon, the same amount of mass will weigh only about one-sixth as much. Similarly, the weight of an object would vary for different planets.

For now, keep in mind that in a given location, such as on the Earth's surface, *weight is related to mass, but they are not the same.* Since the weight of an object of a certain mass can vary with location, it is much more practical to take mass as the base quantity, as the SI does. Base quantities should remain the same regardless of where they are measured, under normal or standard conditions. The distinction between mass and weight will be more fully explained in a later chapter. Our discussion until then will be chiefly concerned with mass.

TIME

Time is a difficult concept to define. A common definition is that time is the continuous, forward flow of events. This statement is not so much a definition as an observation that time has never been known to run backward, as it might appear to do when you view a film run backward in a projector. Time is sometimes said to be a fourth dimension, accompanying the three dimensions of space (x, y, z, t). That is, if something exists in space, it also exists in time. In any case, events can be used to mark time measurements. The events are analogous to the marks on a meterstick used for measurements of length. [An old view: Time does not exist in itself, but only through the perceived object, from which the concepts of past, of present, and of future ensue. *Lucretis* (c. 99 BC–c. 55 BC)]

The SI unit of time is the **second (s)**. The solar "clock" was originally used to define the second. A solar day is the interval of time that elapses between two successive crossings of the same longitude line (meridian) by the Sun. A second was fixed as 1/86 400 of this apparent solar day (1 day = 24 h = 1440 min = 86 400 s). However, the elliptical path of the Earth's motion around the Sun causes apparent solar days to vary in length.

As a more precise standard, an average, or mean, solar day was computed from the lengths of the apparent solar days during a solar year. In 1956, the second was referenced to this mean solar day. But the mean solar day is not exactly the same for each yearly period because of minor variations in the Earth's motions and a very small, but steady, slowing of its rate of rotation due to tidal friction. So scientists kept looking for something better.

MASS: KILOGRAM

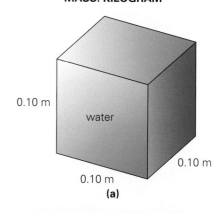

(a)

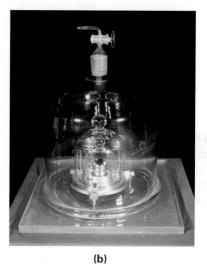

(b)

▲ **FIGURE 1.2 The SI mass standard: the kilogram** (a) The kilogram was originally defined in terms of a specific volume of water, that of a cube 0.10 m (10 cm) on a side, thereby associating the mass standard with the length standard. (b) The standard kilogram is now defined by a metal cylinder. The international prototype of the kilogram is kept at the French Bureau of Weights and Measures. It was manufactured in the 1880s of an alloy of 90% platinum and 10% iridium. Copies have been made for use as 1-kg national prototypes, one of which is the mass standard for the United States. (Shown in the photo.) It is kept at the National Institute of Standards and Technology (NIST) in Gaithersburg, MD. (Notice that the bell jar can be evacuated so the cylinder can be stored under partial vacuum.)

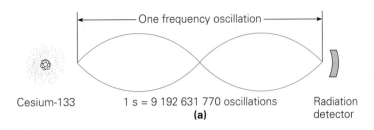

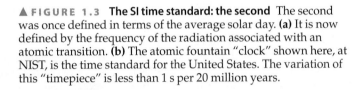

Cesium-133   1 s = 9 192 631 770 oscillations   Radiation detector
(a)

(b)

▲ **FIGURE 1.3   The SI time standard: the second** The second was once defined in terms of the average solar day. **(a)** It is now defined by the frequency of the radiation associated with an atomic transition. **(b)** The atomic fountain "clock" shown here, at NIST, is the time standard for the United States. The variation of this "timepiece" is less than 1 s per 20 million years.

In 1967, an atomic standard was adopted as a better reference. The second was defined by the radiation frequency of the cesium-133 atom. This "atomic clock" used a beam of cesium atoms to maintain our time standard, with a variation of about 1 s in 300 years. In 1999, another cesium-133 atomic clock was adopted, the atomic fountain clock, which, as the name implies, is based on the radiation frequency of a fountain of cesium atoms rather than a beam (▲ Fig. 1.3). The variation of this "timepiece" is less than 1 s per 20 million years!*

A modern practical application involving length and time in designating a position or location on the Earth is the GPS. See Insight 1.2, Global Positioning System (GPS)

*An even more precise clock, the all-optical atomic clock, is under development. It is so named because it uses laser technology and measures a time interval of 0.000 01 s. This new clock does not use cesium atoms, but rather a single cooled ion of liquid mercury linked to a laser oscillator. The frequency of the mercury ion is 100 000 times the frequency of cesium atoms, hence the shorter, more precise time interval.

INSIGHT 1.2 | ## Global Positioning System (GPS)

The GPS consists of a network of two dozen satellites. These solar-powered satellites circle the Earth at altitudes of about 20 000 km (12 400 mi), making two complete orbits every day. The orbits are arranged so that there are at least four satellites observable at any time from anywhere on the Earth (Fig. 1).

Originally developed for the Department of Defense as a military navigation system, the GPS is now available to everyone. All you need is a GPS receiver to find your location anywhere on Earth, except where the satellite radio signals cannot be received such as in caves or underwater.

GPS receivers are becoming increasingly commonplace for finding locations in navigation and other applications. They are used by hunters, hikers, and boaters. GPSs are found in automobiles to provide locations for roadside assistance,

FIGURE 1   Global Positioning System (GPS) An artist's conception of GPS satellites.

along with sophisticated systems that can look up addresses and give directions to a particular location. The accuracy of a receiver depends on how much you want to spend. High-end receivers have accuracies down to 1 m. Really expensive units can come within 1 cm!

So how does the GPS determine a position on the Earth (latitude and longitude)? The electronics and so on are quite complicated, but the basic principles of locating a position can be understood. The process involves triangulation. You have probably seen one form of this on TV or in a movie where police are trying to locate a radio transmitter. One receiver gets a "fix" or direction of the transmitter and a straight line is drawn on a map. Another receiver at another location does the same, and where the two directional lines cross is the location of the transmitter. Just to make sure, a third receiver is used for a three-line intersection.

However in the case of the GPS, it is distance rather than direction that is used. Let's consider a two-dimensional example of finding a location. Suppose you are at a big university and want to find your location on a campus map. Stopping a

passing student, you ask how far it is to the bell tower over there; the answer is one block. Drawing a circle with a one-block radius with the bell tower at the center, you know that you are somewhere on the circle (Fig. 2a).

That doesn't help much, so you ask another student how far it is to the gym; the answer is two blocks. Drawing a circle with a two-block radius with the gym at the center, you know you are at either point A or B where the circles intersect (Fig. 2b). Doing the same for the campus gate, which you are told is three blocks away, you now know your location is at point A where the three circles intersect (Fig. 2c).

The same idea works in three dimensions on spheres. The satellites send time radio signals to the receiver and its electronics interprets these in terms of the satellites' distance. The satellites carry highly accurate "atomic clocks" for time measurements. For GPS to work, the clocks in orbit must be "in sync" with the corresponding clocks on the Earth. If not, the travel time will be incorrect and the distances will be wrong. Due to the satillites' orbital speeds (several kilometers per second), there are special relativity time dilations to account for, along with general relativity effects. (See the Chapter 26 Insight 26.1, Relativity in Everyday Living.)

The distance to a satellite is computed by a simple equation, $d = vt$ (distance = speed × time, Section 2.1). Here, radio waves, which travel at the speed of light, are used, so $d = ct$, where $c = 3.0 \times 10^8$ m/s (186 000 mi/s). Then, analogous to the previous two-dimensional example, the positions and distances provide three circles on the globe, the intersection of which is the receiver's location (Fig. 3).*

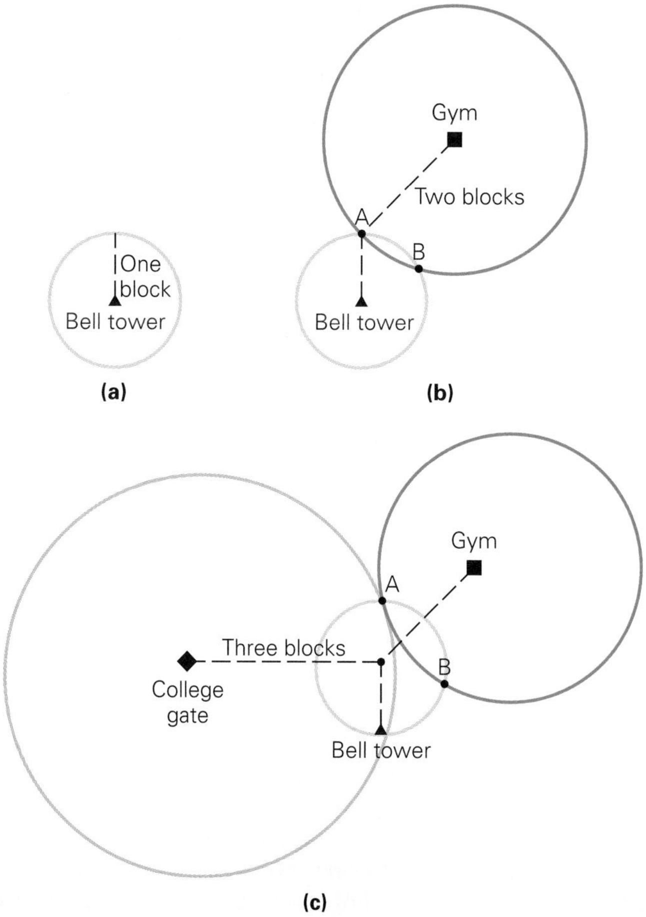

FIGURE 2 Finding a location Triangulation can be used to find a location. **(a)** You are somewhere on the circle. **(b)** You are at either point A or point B. **(c)** You are at point A, where all three circles intersect. See text for detailed description.

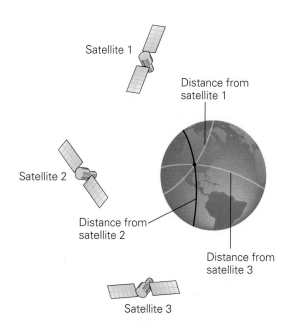

FIGURE 3 Location on Earth Satellite data provide three circles on the globe, the intersection of which is the receiver's location.

---

*Actually, the receiver's altitude should also be supplied. By adding a fourth satellite, the receiver's latitude, longitude, and altitude can be determined.

TABLE 1.1   The Seven Base Units of the SI

Name of Unit (abbreviation)	Property Measured
meter (m)	length
kilogram (kg)	mass
second (s)	time
ampere (A)	electric current
kelvin (K)	temperature
mole (mol)	amount of substance
candela (cd)	luminous intensity

SI BASE UNITS

The SI has seven *base units* for seven base quantities, which are assumed to be mutually independent. In addition to the meter, kilogram, and second for (1) length, (2) mass, and (3) time, SI units include (4) electric current (charge/second) in amperes (A), (5) temperature in kelvins (K), (6) amount of substance in moles (mol), and (7) luminous intensity in candelas (cd). See Table 1.1.

    The foregoing quantities are thought to compose the smallest number of base quantities needed for a full description of everything observed or measured in nature.

DID YOU LEARN?
- ➡ Base units are defined by standards. Derived units are combinations of base units.
- ➡ The meter is defined in terms of the speed of light, the standard mass of 1 kg is associated with a platinum–iridium cylinder, (the only SI standard unit referenced to a material artifact), and the second is defined by the radiation frequency of the cesium-133 atom in an "atomic clock."

1.3   More about the Metric System

LEARNING PATH QUESTIONS
- ➡ What is the difference between the mks and cgs systems of units?
- ➡ What is the proper order, from smallest to largest, of the metric prefixes kilo-, milli-, mega-, micro-, and centi- ?
- ➡ Why does 1 L of water have a mass of 1 kg?

The metric system involving the standard units of length, mass, and time, now incorporated into the SI, was once called the **mks system** (for *meter–kilogram–second*). Another metric system that has been used in dealing with relatively small quantities is the **cgs system** (for *centimeter–gram–second*). In the United States, the system still generally in use is the British (or English) engineering system, in which the standard units of length, mass, and time are foot, slug, and second, respectively. You may not have heard of the slug, because as mentioned earlier, gravitational force (weight) is commonly used instead of mass—pounds instead of slugs—to describe quantities of matter. As a result, the British system is sometimes called the **fps system** (for *foot–pound–second*).

    The metric system is predominant throughout the world and is coming into increasing use in the United States. Because it is simpler mathematically, the SI is the preferred system of units for science and technology. SI units are used throughout most of this book. All quantities can be expressed in SI units. However, some units from other systems are accepted for limited use as a matter of practicality—for example, the time unit of hour and the temperature unit of

TABLE 1.2 Some Multiples and Prefixes for Metric Units*

Multiple†	Prefix (and Abbreviation)	Pronunciation
10^{12}	tera- (T)	ter'a (as in *terra*ce)
10^9	giga- (G)	jig'a (*jig* as in *jig*gle, *a* as in *a*bout)
10^6	mega- (M)	meg'a (as in *mega*phone)
10^3	kilo- (k)	kil'o (as in *kilo*watt)
10^2	hecto- (h)	hek'to (*heck-toe*)
10	deka- (da)	dek'a (*deck* plus *a* as in *a*bout)
10^{-1}	deci- (d)	des'i (as in *deci*mal)
10^{-2}	centi- (c)	sen'ti (as in *senti*mental)
10^{-3}	milli- (m)	mil'li (as in *mili*tary)
10^{-6}	micro- (μ)	mi'kro (as in *micro*phone)
10^{-9}	nano- (n)	nan'o (*an* as in *an*nual)
10^{-12}	pico- (p)	pe'ko (*peek-oh*)
10^{-15}	femto- (f)	fem'to (*fem* as in *fem*inine)
10^{-18}	atto- (a)	at'toe (as in *a*natomy)

*For example, 1 gram (g) multiplied by 1000, or 10^3, is 1 kilogram (kg); 1 gram multiplied by 1/1000, or 10^{-3}, is 1 milligram (mg).

†The most commonly used prefixes are printed in blue. Note that the abbreviations for the multiples 10^6 and greater are capitalized, whereas the abbreviations for the smaller multiples are lowercased.

degree Celsius. British units will sometimes be used in the early chapters for comparison purposes, since these units are still employed in everyday activities and many practical applications.

The increasing worldwide use of the metric system means that you should be familiar with it. One of the greatest advantages of the metric system is that it is a decimal, or base-10, system. This means that larger or smaller units may be obtained by multiplying or dividing by powers of 10. A list of some multiples and corresponding prefixes for metric units is given in Table 1.2.

For metric measurements, the prefixes *micro-, milli-, centi-, kilo-,* and *mega-* are the ones most commonly used—for example, microsecond (μs), millimeter (mm), centimeter (cm), kilogram (kg), and megabyte (MB) as for computer disk or CD storage sizes. The decimal characteristics of the metric system make it convenient to change measurements from one size of metric unit to another. In the British system, different conversion factors must be used, such as 16 for converting pounds to ounces and 12 for converting feet to inches, whereas in the metric system, the conversion factors are multiples of 10. For example, 100 (10^2) to convert meters to centimeters (1 m = 100 cm) and 1000 (10^3) to convert meters to millimeters (1 m = 1000 mm).

You are already familiar with one base-10 system—U.S. currency. Just as a meter can be divided into 10 decimeters, 100 centimeters, or 1000 millimeters, the "base unit" of the dollar can be broken down into 10 "decidollars" (dimes), 100 "centidollars" (cents), or 1000 "millidollars" (tenths of a cent, or mills, used in figuring property taxes and bond levies). Since all the metric prefixes are powers of 10, there are no metric analogues for quarters or nickels.

The official metric prefixes help eliminate confusion. For example, in the United States, a billion is a thousand million (10^9); in Great Britain, a billion is a million million (10^{12}). The use of metric prefixes eliminates any confusion, since *giga-* indicates 10^9 and *tera-* stands for 10^{12}. You will probably be hearing more about *nano-*, the prefix that indicates 10^{-9}, with respect to nanotechnology (*nanotech* for short). In general, nanotechnology is any technology done on the

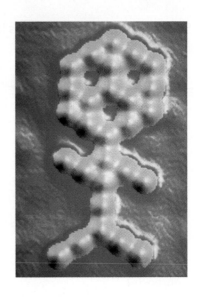

► **FIGURE 1.4** **Molecular Man**
This figure was crafted by moving
28 molecules, one at a time. Each of
the gold-colored peaks is the image
of a carbon monoxide molecule. The
molecules rest on a single crystal
platinum surface. "Molecular Man"
measures 5 nm tall and 2.5 nm wide
(hand to hand). It would take about
16 000 such figures, linked hand to
hand, to span a single human hair.
The molecules in the figure were
positioned using a special microscope at very low temperatures.

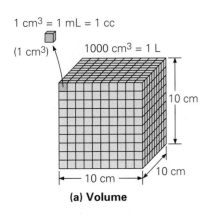

(a) Volume

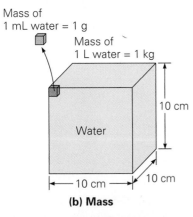

(b) Mass

▲ **FIGURE 1.5** **The liter and the
kilogram** Other metric units are
derived from the meter. **(a)** A unit of
volume (capacity) was taken to be
the volume of a cube 10 cm, or
0.10 m, on a side and was given the
name *liter* (L). **(b)** The mass of a liter
of water was defined to be 1 kg.
Note that the decimeter cube contains 1000 cm³, or 1000 mL. Thus,
1 cm³, or 1 mL, of water has a mass
of 1 g.

nanometer scale. A nanometer (nm) is one billionth (10^{-9}) of a meter, about the
width of three to four atoms. Basically, nanotechnology involves the manufacture
or building of things one atom or molecule at a time, so the nanometer is the
appropriate scale. One atom or molecule at a time? That may sound a bit far-
fetched, but it's not (see ▲Fig. 1.4).

The chemical properties of atoms and molecules are well understood. For
example, rearranging the atoms in coal can produce a diamond. (This is already
done without nanotechnology using heat and pressure.) Nanotechnology presents
the possibility of constructing novel molecular devices or "machines" with extraordinary properties and abilities, for example, in medicine. Nanostructures might
be injected into the body to go to a particular site, such as a cancerous growth, and
deliver a drug directly. Other organs of the body would then be spared any effects
of the drug. (This process might be considered nanochemotherapy.)

It is difficult for us to grasp or visualize the new concept of nanotechnology.
Even so, keep in mind that a nanometer is one billionth of a meter. The diameter of
an average human hair is about 40 000 nm—huge compared with the new
nanoapplications. The future should be an exciting nanotime.

VOLUME

In the SI, the standard unit of volume is the cubic meter (m³)–the three-dimensional
derived unit of the meter base unit. Because this unit is rather large, it is often more
convenient to use the nonstandard unit of volume (or capacity) of a cube 10 cm on a
side. This volume was given the name *litre*, which is spelled **liter (L)** in the United
States. The volume of a liter is 1000 cm³ (10 cm × 10 cm × 10 cm). Since
1 L = 1000 mL (milliliters, mL) it follows that 1 mL = 1 cm³. See ◄Fig. 1.5a. [The
cubic centimeter is sometimes abbreviated as cc, particularly in chemistry and biology. Also, the milliliter is sometimes abbreviated as ml, but the capital L is preferred
(mL) so as not to be confused with the numeral one, 1.]

Recall from Fig. 1.2 that the standard unit of mass, the kilogram, was originally
defined to be the mass of a cubic volume of water 10 cm, or 0.10 m, on a side, or
the mass of one liter (1 L) of water*. That is, *1 L of water has a mass of 1 kg* (Fig. 1.5b).
Also, since 1 kg = 1000 g and 1 L = 1000 cm³ (= 1000 mL), then *1 cm³ (or 1 mL)
of water has a mass of 1 g.*

---

*This is specified at 4 °C. A volume of water changes slightly with temperature (thermal expansion,
Section 10.4). For our purposes here, a volume of water will be considered to remain constant under
normal temperature conditions.

| EXAMPLE 1.1 | The Metric Ton (or Tonne): Another Unit of Mass |

As discussed, the metric unit of mass was originally related to length, with a liter (1000 cm³) of water having a mass of 1 kg. The standard metric unit of volume is the cubic meter (m³) and this volume of water was used to define a larger unit of mass called the *metric ton* (or *tonne*, as it is sometimes spelled). A metric ton is equivalent to how many kilograms?

THINKING IT THROUGH. A cubic meter is a relatively large volume and holds a large amount of water (more than a cubic yard; why?). The key is to find how many cubic volumes measuring 10 cm on a side (liters) are in a cubic meter. A large number would be expected.

SOLUTION. Each liter of water has a mass of 1 kg, so we need to find out how many liters are in 1 m³. Since there are 100 cm in a meter, a cubic meter is simply a cube with sides 100 cm in length. Therefore, a cubic meter (1 m³) has a volume of 10^2 cm $\times$ 10^2 cm $\times$ 10^2 cm $= 10^6$ cm³. Since 1 L has a volume of 10^3 cm³, there must be $(10^6$ cm³$)/(10^3$ cm³$/$L$) = 1000$ L in 1 m³. Thus, 1 metric ton is equivalent to 1000 kg.

Note that this line of reasoning can be expressed very concisely in a single ratio:

$$\frac{1 \text{ m}^3}{1 \text{ L}} = \frac{100 \text{ cm} \times 100 \text{ cm} \times 100 \text{ cm}}{10 \text{ cm} \times 10 \text{ cm} \times 10 \text{ cm}} = 1000 \quad \text{or} \quad 1 \text{ m}^3 = 1000 \text{ L}$$

FOLLOW-UP EXERCISE. What would be the length of the sides of a cube that contained a metric kiloton of water? (*Answers to all Follow-Up Exercises are given in Appendix VI at the back of the book.*)

You are probably more familiar with the liter than you think. The use of the liter is becoming quite common in the United States, as ▸ Fig. 1.6 indicates.

Because the metric system is coming into increasing use in the United States, you may find it helpful to have an idea of how metric and British units compare. The relative sizes of some units are illustrated in ▾ Fig. 1.7. The mathematical conversion from one unit to another will be discussed shortly.

DID YOU LEARN?

➡ The mks (meter-kilogram-second) system has standard units. The cgs (centimeter-gram-second) system, although not standard, is useful in measuring relatively small quantities.

➡ The values of commonly used metric prefixes in ascending order are micro- (10^{-6}), milli- (10^{-3}), centi- (10^{-2}), kilo- (10^3), and mega- (10^6).

➡ The kilogram was defined to be the mass of a cube of water 10 cm on a side. This volume was taken to be a liter, so 1 L of water has a mass of 1 kg.

▲ **FIGURE 1.6 Two, three, one, and one-half liters** The liter is now a common volume unit for soft drinks.

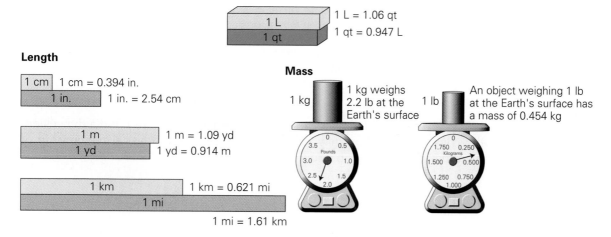

Volume

1 L = 1.06 qt
1 qt = 0.947 L

Length

1 cm = 0.394 in.
1 in. = 2.54 cm

1 m = 1.09 yd
1 yd = 0.914 m

1 km = 0.621 mi

1 mi = 1.61 km

Mass

1 kg weighs 2.2 lb at the Earth's surface

An object weighing 1 lb at the Earth's surface has a mass of 0.454 kg

▲ **FIGURE 1.7 Comparison of some SI and British units** The bars illustrate the relative magnitudes of each pair of units. (*Note*: The comparison scales are different in each case.)

1.4 Unit Analysis

LEARNING PATH QUESTIONS

➡ How is unit analysis useful?

➡ What units should be used in working a problem?

➡ What does density represent?

TABLE 1.3   Some Units of Common Quantities

Quantity	Unit
mass	kg
time	s
length	m
area	m^2
volume	m^3
velocity (v)	m/s
acceleration (a or g)	m/s^2

The fundamental or base quantities used in physical descriptions are called *dimensions*. For example, length, mass, and time are dimensions. You could measure the distance between two points and express it in units of meters, centimeters, or feet, but the quantity would still have the dimension of length.

Dimensions provide a procedure by which the consistency of equations may be checked. In practice, it is convenient to use specific units, such as m, kg, and s. (See Table 1.3.) Such units can be treated as algebraic quantities and be canceled. Using units to check equations is called **unit analysis**, which shows the consistency of units and whether an equation is dimensionally correct.

You have used equations and know that an equation is a mathematical equality. Since physical quantities used in equations have units, *the two sides of an equation must be equal not only in numerical value (magnitude), but also in units (dimensions)*. For example, suppose you had the length quantities $a = 3.0$ m and $b = 4.0$ m. Inserting these values into the equation $a \times b = c$ gives $3.0 \text{ m} \times 4.0 \text{ m} = 12 \text{ m}^2$. Both sides of the equation are numerically equal ($3 \times 4 = 12$), and both sides have the same units, $\text{m} \times \text{m} = \text{m}^2 = (\text{length})^2$. If an equation is correct by unit analysis, it must be dimensionally correct. Example 1.2 demonstrates the further use of unit analysis.

EXAMPLE 1.2 | **Checking Dimensions: Unit Analysis**

A professor puts two equations on the board: (a) $v = v_0 + at$ and (b) $x = v/2a$, where x is distance in meters (m); v and v_0 are velocities in meters/second (m/s); a is acceleration in (meters/second)/second, or meters/second^2 (m/s^2); and t is time in seconds (s). Are the equations dimensionally correct? Use unit analysis to find out.

THINKING IT THROUGH. Simply insert the units for the quantities in each equation, cancel, and check the units on both sides.

SOLUTION.

(a) The equation is

$$v = v_0 + at$$

Inserting units for the physical quantities gives (Table 1.3)

$$\frac{m}{s} = \frac{m}{s} + \left(\frac{m}{s^2} \times s\right) \quad \text{or} \quad \frac{m}{s} = \frac{m}{s} + \left(\frac{m}{s \times s} \times s\right)$$

Notice that units cancel like numbers in a fraction. Then, simplifying,

$$\frac{m}{s} = \frac{m}{s} + \frac{m}{s} \quad \text{\textit{(dimensionally correct)}}$$

The equation is dimensionally correct, since the units on each side are meters per second. (The equation is also a correct relationship, as will be seen in Chapter 2.)

(b) Using unit analysis, the equation

$$x = \frac{v}{2a}$$

is

$$m = \frac{\left(\dfrac{m}{s}\right)}{\left(\dfrac{m}{s^2}\right)} = \frac{m}{s} \times \frac{s^2}{m} \quad \text{or} \quad m = s \quad \begin{array}{l}\text{\textit{(not dimensionally}}\\\text{\textit{correct)}}\end{array}$$

The meter (m) is not the same unit as the second (s), so in this case, the equation is not dimensionally correct (length ≠ time), and therefore is also not physically correct.

FOLLOW-UP EXERCISE. Is the equation $ax = v^2$ dimensionally correct? *(Answers to all Follow-Up Exercises are given in Appendix VI at the back of the book.)*

Unit analysis will tell if an equation is dimensionally correct, but a dimensionally consistent equation may not correctly express the physical relationship of quantities. For example, in terms of units, the equation

$$x = at^2$$

is

$$m = (m/s^2)(s^2) = m$$

This equation is dimensionally correct (length = length). But, as will be learned in Chapter 2, it is not *physically* correct. The correct form of the equation—both dimensionally and physically—is $x = \frac{1}{2}at^2$. (The fraction $\frac{1}{2}$ has no dimensions; it is an exact dimensionless number.) Unit analysis *cannot* tell you if an equation is physically correct, only whether or not it is dimensionally consistent.

MIXED UNITS

Unit analysis also allows you to check for mixed units. In general, when working problems, *you should always use the same system of units and the same unit for a given dimension throughout an exercise.*

Suppose you wanted to buy a rug to fit a rectangular floor area and you measure the sides to be 4.0 yd $\times$ 3.0 m. The area of the rug would then be $A = l \times w = 4.0$ yd $\times$ 3.0 m $= 12$ yd $\cdot$ m, which might cause a problem at the carpet store. Note that this equation is dimensionally correct, $(length)^2 = (length)^2$, but the units are inconsistent or mixed. So, unit analysis will point out *mixed units*. Note that it is possible for an equation to be dimensionally correct, even if the units are mixed.

Let's look at mixed units in an equation. Suppose that you used centimeters (cm) as the unit for x in the equation

$$v^2 = v_0^2 + 2ax$$

and the units for the other quantities as in Example 1.2. In terms of units, this equation would give

$$\left(\frac{m}{s}\right)^2 = \left(\frac{m}{s}\right)^2 + \left(\frac{m \times cm}{s^2}\right)$$

or

$$\frac{m^2}{s^2} = \frac{m^2}{s^2} + \frac{m \times cm}{s^2}$$

which is dimensionally correct, $(length)^2/(time)^2$, on both sides of the equation. But the units are mixed (m and cm). The value of x in centimeters needs to be converted to meters to be used in the equation.

DETERMINING THE UNITS OF QUANTITIES

Another aspect of unit analysis that is very important in physics is the determination of the units of quantities from defining equations. For example, the **density (ρ)** of an object (represented by the Greek letter rho, ρ) is defined by the equation

$$\rho = \frac{m}{V} \quad \left(\frac{kg}{m^3}\right) \tag{1.1}$$

where m is the object's mass and V its volume. (Density is the mass per unit volume and is a measure of the compactness of the mass of an object or substance.) In SI units, mass is measured in kilograms and volume in cubic meters, which gives the derived SI unit for density as kilograms per cubic meter (kg/m^3).

How about the units of π? The relationship between the circumference (c) and the diameter (d) of a circle is given by the equation $c = \pi d$, so $\pi = c/d$. If the lengths are measured in meters, then unitwise,

$$\pi = \frac{c}{d}\left(\frac{\cancel{m}}{\cancel{m}}\right)$$

Thus, π has no units. It is unitless, or a dimensionless constant.

DID YOU LEARN?
  ➥ Unit analysis can tell if an equation is dimensionally correct, but not physically correct.
  ➥ In working problems, the same system of units should be used throughout.
  ➥ Density is a measure of the compactness of the mass of an object (mass/volume).

1.5 Unit Conversions

LEARNING PATH QUESTIONS

➥ What is an equivalence statement?

➥ How are conversion factors written?

Because units in different systems, or even different units in the same system, can be used to express the same quantity, it is sometimes necessary to convert the units of a quantity from one unit to another. For example, we may need to convert feet to yards or convert inches to centimeters. You already know how to do many unit conversions. If a sidewalk is 12 ft long, what is its length in yards? Your immediate answer is 4 yd.

How did you do this conversion? Well, you must have known a relationship between the units of foot and yard. That is, you know that 3 ft = 1 yd. This is what is called an *equivalence statement*. As was seen in Section 1.4, the numerical values and units on both sides of an equation must be the same. In equivalence statements, we commonly use an equal sign to indicate that 1 yd and 3 ft stand for the *same*, or *equivalent, length*. The numbers are different because they stand for different *units* of length.

Mathematically, to change units **conversion factors** are used, which are simply equivalence statements expressed in the form of ratios—for example, 1 yd/3 ft or 3 ft/1 yd. (The "1" is often omitted in the denominators of such ratios for convenience—for example, 3 ft/yd.) To understand why such ratios are useful, note the expression 1 yd = 3 ft in ratio form:

$$\frac{1\ yd}{3\ ft} = \frac{3\ ft}{3\ ft} = 1 \qquad \text{or} \qquad \frac{3\ ft}{1\ yd} = \frac{1\ yd}{1\ yd} = 1$$

As can be seen, a conversion factor has an actual value of unity or one—and you can multiply any quantity by one without changing its value or size. Thus, *a conversion factor simply lets you express a quantity in terms of other units without changing its physical value or size.*

The manner in which 12 ft is converted to yards may be expressed mathematically as follows:

$$12\ \cancel{ft} \times \frac{1\ yd}{3\ \cancel{ft}} = 4\ yd \quad \text{(units cancel)}$$

Using the appropriate conversion factor form, the units cancel, as shown by the slash marks, giving the correct unit analysis, yd = yd.

Suppose you are asked to convert 12.0 in. to centimeters. You may not know the conversion factor in this case, but it can be obtained from a table (such as the one that appears inside the front cover of this book). The needed relationships are 1 in. = 2.54 cm or 1 cm = 0.394 in. It makes no difference which of these equivalence statements you use. The question, once you have expressed the equivalence statement as a ratio conversion factor, is whether to multiply or divide by that factor to make the conversion. *In doing unit conversions, take advantage of unit analysis*—that is, let the units determine the appropriate form of conversion factor.

Note that the equivalence statements can give rise to two forms of the conversion factors: 1 in./2.54 cm and 2.54 cm/in. When changing inches to centimeters, the appropriate form for multiplying is either 2.54 cm/in. or 1 cm/0.394 in. When changing centimeters to inches, use the form 1 in./2.54 cm or 0.394 in./cm. For example,

$$12.0\ \cancel{in.} \times \frac{2.54\ cm}{\cancel{in.}} = 30.5\ cm$$

$$15.0\ \cancel{cm} \times \frac{0.394\ in.}{\cancel{cm}} = 5.91\ in.$$

The multiplication of conversion factors in cancelling units is usually more convenient than division. In general, the multiplication form of conversion factors will be used throughout this book.*

A few commonly used equivalence statements are not dimensionally or physically correct; for example, consider 1 kg = 2.2 lb, which is used for quickly determining the weight of an object near the Earth's surface given its mass. The kilogram is a unit of mass, and the pound is a unit of weight. This means that 1 kg is *equivalent* to 2.2 lb; that is, a 1-kg *mass* has a *weight* of 2.2 lb. Since mass and weight are directly proportional, the dimensionally incorrect conversion factor 1 kg/2.2 lb may be used (but *only* near the Earth's surface).

(a)

EXAMPLE 1.3 | Converting Units: Use of Conversion Factors

A championship male pole vaulter goes over a bar set at 6.14 m. A championship female vaulter clears a height of 4.82 m. What is the difference in these heights in feet?

THINKING IT THROUGH. After using the correct conversion factor, the rest is arithmetic.

SOLUTION.

From the conversion table, 1 m = 3.28 ft, so converting the heights to feet:

$$6.14 \text{ m} \times \frac{3.28 \text{ ft}}{\text{m}} = 20.1 \text{ ft}$$

$$4.82 \text{ m} \times \frac{3.28 \text{ ft}}{\text{m}} = 15.8 \text{ ft}$$

And the difference in height is $\Delta h = 20.1 \text{ ft} - 15.8 \text{ ft} = 4.3 \text{ ft}$.

Another approach would be to subtract the heights in meters and have only a single conversion:

$$6.14 \text{ m} - 4.82 \text{ m} = 1.32 \text{ m} \times \frac{3.28 \text{ ft}}{\text{m}} = 4.33 \text{ ft}$$

The answers aren't the same. Is there something wrong? No, as will be discussed in the Section 1.6 Problem-Solving Hint, *The "Correct" Answer*, the difference is usually due to rounding differences. This may occur when working a problem by another method.

Another foot–meter conversion is shown in ▶ Fig. 1.8a. Is it correct?

FOLLOW-UP EXERCISE. Rather than use a single conversion factor from the table, use commonly known factors to convert a 30-day month to seconds. *(Answers to all Follow-Up Exercises are given in Appendix VI at the back of the book.)*

(b)

▲ FIGURE 1.8 **Unit conversion** Signs sometimes list both the British and metric units, as shown here for elevation **(a)** and speed **(b)**. Note the highlighted km.

EXAMPLE 1.4 | More Conversions: A Really Long Capillary System

Capillaries, the smallest blood vessels of the body, connect the arterial system with the venous system and supply our tissues with oxygen and nutrients (▶Fig. 1.9). It is estimated that if all of the capillaries of an average adult were unwound and spread out end to end, they would extend to a length of about 64 000 km. (a) How many miles is this length? (b) Compare this length with the circumference of the Earth.

THINKING IT THROUGH. (a) This conversion is straightforward—just use the appropriate conversion factor. (b) How is the circumference of a circle or sphere calculated? There is an equation to do so, but the radius or diameter of the Earth must be known. (If you do not remember one of these values, see the solar system data table inside the back cover of this book.)

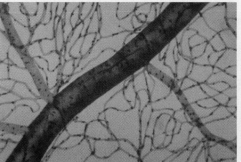

◀FIGURE 1.9 **Capillary system** Capillaries connect the arterial and venous systems in our bodies. They are the smallest blood vessels, but their total length is impressive.

(continued on next page)

---

*Quantities may also be divided by conversion factors. For example,

$$12 \text{ in.} \Big/ \left(\frac{1 \text{ in.}}{2.54 \text{ cm}} \right) = 12 \text{ in.} \times (2.54 \text{ cm/in.}) = 30 \text{ cm}$$

Using the multiplication form saves the step of inverting the ratio.

SOLUTION.

(a) From the conversion table, 1 km = 0.621 mi, so

$$\frac{64\ 000\ \cancel{km} \times 0.621\ mi}{1\ \cancel{km}} = 40\ 000\ mi \quad \text{(rounded off)}$$

(b) A length of 40 000 mi is substantial. To see how this length compares with the circumference (c) of the Earth, recall that the radius of the Earth is approximately 4000 mi, so the diameter (d) is 8000 mi. The circumference of a circle is given by $c = \pi d$ (Appendix I-C), and

$$c = \pi d \approx 3 \times 8000\ mi \approx 24\ 000\ mi \quad \text{(rounded off)}$$

[To make a general comparison, π ($= 3.14\ldots$) is rounded off to 3. The $\approx$ symbol means "approximately equal to."]
So,

$$\frac{\text{capillary length}}{\text{Earth's circumference}} \approx \frac{40\ 000\ mi}{24\ 000\ mi} = 1.7$$

The capillaries of your body have a total length that would extend about 1.7 times around the world. Wow!

FOLLOW-UP EXERCISE. Taking the average distance between the East Coast and West Coast of the continental United States to be 4800 km, how many times would the total length of your body's capillaries cross the country? (*Answers to all Follow-Up Exercises are given in Appendix VI at the back of the book.*)

EXAMPLE 1.5 | ## Converting Units of Area: Choosing the Correct Conversion Factor

A hall bulletin board has an area of 2.5 m². What is this area (a) in square centimeters (cm²), and (b) square inches (in²)?

THINKING IT THROUGH. This problem is a conversion of area units, and we know that 1 m = 100 cm. So, some squaring must be done to get square meters related to square centimeters.

SOLUTION. A common error in such conversions is the use of incorrect conversion factors. Because 1 m = 100 cm, it is sometimes assumed that 1 m² = 100 cm², which is *wrong*. The correct area conversion factor may be obtained directly from the correct linear conversion factor, 100 cm/1 m, or 10^2 cm/1 m, by *squaring* the linear conversion factor:

$$\left(\frac{10^2\ cm}{1\ m}\right)^2 = \frac{10^4\ cm^2}{1\ m^2}$$

Hence, 1 m² = 10^4 cm² (= 10 000 cm²).

(a) Then using the conversion factor explicitly squared:

$$2.5\ m^2 \times \left(\frac{10^2\ cm}{1\ m}\right)^2 = 2.5\ \cancel{m^2} \times \frac{10^4\ cm^2}{1\ \cancel{m^2}} = 2.5 \times 10^4\ cm^2$$

(b) Using the cm² result found in (a), by a similar procedure,

$$2.54 \times 10^4\ cm^2 \left(\frac{0.394\ in.}{cm}\right)^2 =$$

$$2.54 \times 10^4\ \cancel{cm^2} \times \left(\frac{0.155\ in^2}{\cancel{cm^2}}\right) = 3.94 \times 10^3\ in^2$$

FOLLOW-UP EXERCISE. How many cubic centimeters are in 1 m³? (*Answers to all Follow-Up Exercises are given in Appendix VI at the back of the book.*)

EXAMPLE 1.6 | ## The Better Deal

A grocery store has a sale on sodas. A 2-L bottle sells for $1.35, and the price of a half-gallon bottle is $1.32. Which is the better buy?

THINKING IT THROUGH. The answer is obtained by knowing the price per common volume. This means that liters must be converted to quarts or vice versa. (The cancellation slashes will now be omitted as being understood.)

SOLUTION. To get a common volume, let's convert liters to quarts using the conversion factor given inside the front cover of the book:

$$2.0\ L \left(\frac{1.056\ qt}{L}\right) = 2.1\ qt$$

Then in terms of quarts, the base price of the liquid in the 2.0-L bottle is

$$\frac{\$1.35}{2.1\ qt} = \frac{\$0.64}{qt}$$

Similarly for the half-gallon volume:

$$\frac{\$1.32}{2.0\ qt} = \frac{\$0.66}{qt}$$

So, the better buy is the 2-L bottle, even though its price for 2 L is higher. (Keep in mind a liter is larger than a quart.)

FOLLOW-UP EXERCISE. Work the Example the other way—changing quarts to liters—to see if the result is the same. (*Answers to all Follow-up Exercies are given in Appendix VI at the back of the book.*)

Some examples of the importance of unit conversion are given in the accompanying Insight 1.3, Is Unit Conversion Important?

DID YOU LEARN?

➡ 3 ft = 1 yd is not an equation, but an equivalence statement representing equivalent lengths: 3 ft is the equivalent length of 1 yd.

➡ The conversion factor for foot and yard may be written 3 ft/1 yd or 1 yd/3 ft.

INSIGHT 1.3 | **Is Unit Conversion Important?**

The answer to this question is, you bet! Here are a couple of cases in point. In 1999, the $125 million Mars Climate Orbiter was making a trip to the Red Planet to investigate its atmosphere (Fig. 1). The spacecraft approached the planet in September, but suddenly contact between the Orbiter and personnel on Earth was lost, and the Orbiter was never heard from again. Investigations showed that the orbiter had approached Mars at a far lower altitude than planned. Instead of the Orbiter passing 147 km (91 mi) above the Martian surface, tracking data showed that it was on a trajectory that would have taken it as close as 57 km (35 mi) from the surface. As a result, the spacecraft either burned up in the Martian atmosphere or crashed into the surface.

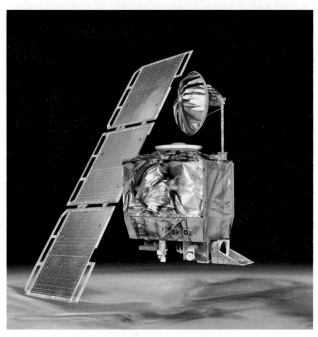

FIGURE 1   Mars Climate Orbiter An artist's conception of the orbiter near the surface of Mars. The actual orbiter either burned up in the Martian atmosphere or crashed into the surface. The cause was attributed to a mix-up in units, resulting in the loss of a $125 million spacecraft.

How could this have happened? Investigations showed that the failure of the Orbiter was primarily a problem of a lack of unit conversion. At Lockheed Martin Astronautics, which built the spacecraft, the engineers calculated the navigational information in British units. When scientists at NASA's Jet Propulsion Laboratory received the data, they assumed that the information was in metric units, as was called for in the mission specifications. The unit conversions weren't made, and a $125 million spacecraft was lost on the Red Planet—causing more than a few red faces.

Closer to Earth, in 1983 Air Canada Flight 143 was on course from Montréal to Edmonton, Canada, with sixty-one passengers in a new Boeing 767, at the time the most advanced jetliner in the world. Almost halfway into the flight, a warning light came on for a fuel pump, then for another, and finally for all four pumps. The engines quit, and this advanced plane was now a glider, about 100 mi from the nearest major airport, at Winnipeg. Without engines, Flight 143's descent would bring it down 10 mi short of the airport, so it was diverted to an old Royal Canadian Air Force landing field at Gimli. The pilot maneuvered the powerless plane to a landing, stopping just short of a barrier. Did the plane, which was dubbed "The Gimli Glider," have bad fuel pumps? No—it had run out of fuel!

This near-disaster was caused by another conversion problem. The fuel computers weren't working properly so the mechanics had used the old procedure of measuring the fuel in the tanks with a dipstick. In this method, the length of the stick that is wet is used to determine the volume of fuel by means of conversion values in tables. Air Canada had for years computed the amount of fuel in pounds, but the new 767's fuel consumption was expressed in kilograms. Even worse, the dipstick procedure gave the amount of fuel onboard in liters instead of pounds or kilograms. The result was that the aircraft was loaded with 22 300 lb of fuel instead of the required 22 300 kg. Since 1 lb has a mass of 0.45 kg, the plane had less than half the required fuel.

These incidents underscore the importance of using appropriate units, making correct unit conversions, and working consistently in the same system of units. Several exercises at the end of the chapter will challenge you to develop your skills in accurate unit conversions.

1.6   Significant Figures

LEARNING PATH QUESTIONS

➡ Are there some numbers without uncertainty or error?
➡ What numbers generally have uncertainty and error?
➡ What determines the degree of accuracy or number of significant figures of a measured quantity?

Most of the time, you will be given numerical data when asked to solve a problem. In general, such data are either exact numbers or measured numbers (quantities). **Exact numbers** are numbers without any uncertainty or error. This category includes numbers such as the 100 used to calculate a percentage and the 2 in the equation $r = d/2$ relating the radius and diameter of a circle. **Measured numbers** are numbers obtained from measurement processes and thus generally have some degree of uncertainty or error.

▲ **FIGURE 1.10 Significant figures and insignificant figures** For the division operation 5.3/1.67, a calculator with a floating decimal point gives many digits. A calculated quantity can be no more accurate than the least accurate quantity involved in the calculation, so this result should be rounded off to two significant figures—that is, 3.2.

When calculations are done with measured numbers, the uncertainty and/or error of measurement is *propagated*, or carried along, by the mathematical operations. The question of how to report a result arises. For example, suppose that you are asked to find time (t) from the equation $x = vt$ and are given that $x = 5.3$ m and $v = 1.67$ m/s. That is,

$$t = \frac{x}{v} = \frac{5.3 \text{ m}}{1.67 \text{ m/ s}} = ?$$

Doing the division operation on a calculator yields a result such as 3.173 652 695 s (◄Fig. 1.10). How many figures, or digits, should you report in the answer?

The uncertainty of the result of a mathematical operation may be computed by statistical methods.* However, a simpler and more widely used procedure for estimating uncertainty involves the use of **significant figures (sf)**, sometimes called *significant digits*. The degree of accuracy of a measured quantity depends on how finely divided the measuring scale of the instrument is. For example, you might measure the length of an object as 2.5 cm with one instrument and 2.54 cm with another. The second instrument with a finer scale provides more significant figures and thus a greater degree of accuracy.

Basically, *the significant figures in any measurement are the digits that are known with certainty, plus one digit that is uncertain.* This set of digits is usually defined as all of the digits that can be read directly from the instrument used to make the measurement, plus one uncertain digit that is obtained by estimating the fraction of the smallest division of the instrument's scale.

The quantities 2.5 cm and 2.54 cm have two and three significant figures, respectively. This is rather evident. However, some confusion may arise when a quantity contains one or more zeros. For example, how many significant figures does the quantity 0.0254 m have? What about 104.6 m, or 2705.0 m? In such cases, the following rules will be used to determine significant figures:

1. Zeros at the beginning of a number are not significant. They merely locate the decimal point. For example,

   0.0254 m has three significant figures (2, 5, 4)

2. Zeros within a number are significant. For example,

   104.6 m has four significant figures (1, 0, 4, 6)

3. Zeros at the end of a number after the decimal point are significant. For example,

   2705.0 m has five significant figures (2, 7, 0, 5, 0)

4. In whole numbers without a decimal point that end in one or more zeros (trailing zeros)—for example, 500 kg—the zeros may or may not be significant. In such cases, it is not clear which zeros serve only to locate the decimal point and which are actually part of the measurement. For example, if the first zero after the 5 in 5̲00 kg is the estimated digit in the measurement, then there are only two significant figures. Similarly, if the last zero is the estimated digit (50̲0 kg), then there are three significant figures. This ambiguity may be removed by using scientific (powers-of-10) notation:

   5.0×10^2 kg has two significant figures

   5.00×10^2 kg has three significant figures

This notation is helpful in expressing the results of calculations with the proper numbers of significant figures, as will be seen shortly. (Appendix I includes a review of scientific notation.)

(*Note*: To avoid confusion regarding numbers having trailing zeros used as given quantities in text examples and exercises, the trailing zeros will be

*Measurement error can arise because of a miscalibrated instrument and/or a personal error in reading the instrument.

considered significant. For example, assume that a time of 20 s has two significant figures, even if it is not written out as 2.0×10^1 s.)

It is important to report the results of mathematical operations with the proper number of significant figures. This is accomplished by using rules for (1) multiplication and division and (2) addition and subtraction. To obtain the proper number of significant figures, the results are rounded off. Here are some general rules that will be used for mathematical operations and rounding.

SIGNIFICANT FIGURES IN CALCULATIONS

1. When multiplying and dividing quantities, leave as many significant figures in the answer as there are in the quantity with the least number of significant figures.
2. When adding or subtracting quantities, leave the same number of decimal places (rounded) in the answer as there are in the quantity with the least number of decimal places.

RULES FOR ROUNDING*

1. If the first digit to be dropped is less than 5, leave the preceding digit as is.
2. If the first digit to be dropped is 5 or greater, increase the preceding digit by one.

The rules for significant figures mean that the result of a calculation can be no more accurate than the least accurate quantity used. That is, accuracy cannot be gained performing mathematical operations. Thus, the result that should be reported for the division operation discussed at the beginning of this section is

$$\frac{\overset{(2\ sf)}{5.3 \text{ m}}}{\underset{(3\ sf)}{1.67 \text{ m/s}}} = 3.2 \text{ s} \quad (2\ sf)$$

The result is rounded off to two significant figures. (See Fig. 1.10.)

Applications of these rules are shown in the following Examples.

EXAMPLE 1.7 | ## Using Significant Figures in Multiplication and Division: Rounding Applications

The following operations are performed and the results rounded off to the proper number of significant figures:

Multiplication:

$$\underset{(2\ sf)}{2.4 \text{ m}} \times \underset{(3\ sf)}{3.65 \text{ m}} = 8.76 \text{ m}^2 = 8.8 \text{ m}^2 \quad \textit{(rounded to two sf)}$$

Division:

$$\frac{\overset{(4\ sf)}{725.0 \text{ m}}}{\underset{(3\ sf)}{0.125 \text{ s}}} = 5800 \text{ m/s} = 5.80 \times 10^3 \text{ m/s} \quad \textit{(represented with three sf; why?)}$$

FOLLOW-UP EXERCISE. Perform the following operations, and express the answers in the standard powers-of-10 notation (one digit to the left of the decimal point) with the proper number of significant figures: (a) $(2.0 \times 10^5 \text{ kg})(0.035 \times 10^2 \text{ kg})$ and (b) $(148 \times 10^{-6} \text{ m})/(0.4906 \times 10^{-6} \text{ m})$. *(Answers to all Follow-Up Exercises are given in Appendix VI at the back of the book.)*

*It should be noted that these rounding rules give an approximation of accuracy, as opposed to the results provided by more advanced statistical methods.

EXAMPLE 1.8 | ## Using Significant Figures in Addition and Subtraction: Application of Rules

The following operations are performed by finding the number that has the least number of decimal places. (Units have been omitted for convenience.)

Addition:

In the numbers to be added, note that 23.1 has the least number of decimal places (one):

$$
\begin{array}{r}
23.1 \\
0.546 \\
+1.45 \\
\hline
25.096
\end{array}
\xrightarrow[\text{(rounding off)}]{} 25.1
$$

Subtraction:

The same rounding procedure is used. Here, 157 has the least number of decimal places (none).

$$
\begin{array}{r}
157 \\
-5.5 \\
\hline
151.5
\end{array}
\xrightarrow[\text{(rounding off)}]{} 152
$$

FOLLOW-UP EXERCISE. Given the numbers 23.15, 0.546, and 1.058, (a) add the first two numbers and (b) subtract the last number from the first. *(Answers to all Follow-Up Exercises are given in Appendix VI at the back of the book.)*

Suppose that you have to deal with mixed operations—multiplication and/or division *and* addition and/or subtraction. What do you do in this case? Just follow the regular rules for order of algebraic operations, and observe significant figures as you go.*

The number of digits reported in a result depends on the number of digits in the given data. The rules for rounding will generally be observed in this book. However, there will be exceptions that may make a difference, as explained in the following Problem-Solving Hint.

PROBLEM-SOLVING HINT: THE "CORRECT" ANSWER

When working problems, you naturally strive to get the correct answer and will probably want to check your answers against those listed in the Answers to Odd-Numbered Exercises section in the back of the book. However, on occasion, your answer may differ slightly from that given, even though you have solved the problem correctly. There are several reasons why this could occur.

It is best to round off only the final result of a multipart calculation, but this practice is not always convenient in elaborate calculations. Sometimes, the results of intermediate steps are important in themselves and need to be rounded off to the appropriate number of digits as if each were a final answer. Similarly, Examples in this book are often worked in steps to show the stages in the *reasoning* of the solution. The results obtained when the results of intermediate steps are rounded off may differ slightly from those obtained when only the final answer is rounded.

Rounding differences may also occur when using conversion factors. For example, in changing 5.0 mi to kilometers using the conversion factor listed inside the front cover of this book in different forms,

$$5.0 \text{ mi}\left(\frac{1.609 \text{ km}}{1 \text{ mi}}\right) = (8.045 \text{ km}) = 8.0 \text{ km} \quad \textit{(two significant figures)}$$

and

$$5.0 \text{ mi}\left(\frac{1 \text{ km}}{0.621 \text{ mi}}\right) = (8.051 \text{ km}) = 8.1 \text{ km} \quad \textit{(two significant figures)}$$

The difference arises because of rounding of the conversion factors. Actually, 1 km = 0.6214 mi, so 1 mi = (1/0.6214) km = 1.609 269 km ≈ 1.609 km. (Try repeating these conversions with the unrounded factors, and see what you get.) To avoid rounding differences in conversions, the multiplication form of a conversion factor will generally be used, as in the first of the foregoing equations, unless there is a convenient exact factor, such as 1 min/60 s.

Slight differences in answers may also occur when different methods are used to solve a problem. Keep in mind that when solving a problem, *if your answer differs from that in the text in only the last digit, the disparity is most likely the result of a rounding difference for an alternative method of solution being used.*

---

*Order of operations: (1) calculations done from left to right, (2) calculations inside parentheses, (3) multiplication and division, (4) addition and subtraction.

1.7 Problem Solving

LEARNING PATH QUESTIONS

➥ What is the first step in problem solving?

➥ What is a final step in problem solving?

➥ What is meant by an order of magnitude?

An important aspect of physics is problem solving. In general, this involves the application of physical principles and equations to data from a particular situation in order to find some unknown or wanted quantity. There is no universal method for approaching problem solving that will automatically produce a solution. However, although there is no magic formula for problem solving, there are some sound practices that can be very useful. The steps in the following procedure are intended to provide you with a framework that can be applied to solving most of the problems you will encounter during your course of study. (Modifications may be made to suit your own style.)

These steps will generally be used in dealing with the Example problems throughout the text. Additional problem-solving hints will be given where appropriate.

GENERAL PROBLEM-SOLVING STEPS

1. *Read the problem carefully, and analyze it.* What is given, and what is wanted?

2. *Where appropriate, draw a diagram as an aid in visualizing and analyzing the physical situation of the problem.* This step may not be necessary in every case, but it is often useful.

3. *Write down the given data and what is to be found. Make sure the data are expressed in the same system of units (usually SI).* If necessary, use the unit conversion procedure learned earlier in the chapter. Some data may not be given explicitly. For example, if a car "starts from rest," its initial speed is zero ($v_0 = 0$); in some instances, you may be expected to know certain quantities, such as the acceleration due to gravity, g, or can look them up in tables.

4. *Determine which principle(s) and equation(s) are applicable to the situation, and how they can be used to get from the information given to what is to be found.* You may have to devise a strategy that involves several steps. Also, try to simplify equations as much as possible through algebraic manipulation. The fewer calculations you do, the less likely you are to make a mistake—*so don't put in numbers until you have to.*

5. *Substitute the given quantities (data) into the equation(s) and perform calculations.* Report the result with the proper units and proper number of significant figures.

6. *Consider whether the results are reasonable.* Does the answer have an appropriate magnitude? (This means, is it in the right ballpark?) For example, if a person's calculated mass turns out to be 4.60×10^2 kg, the result should be questioned, since a mass of 460 kg has a weight of 1010 lb. (Also, in motion problems, direction may be important.) ▶Fig. 1.11 summarizes the main steps in the form of a flowchart.

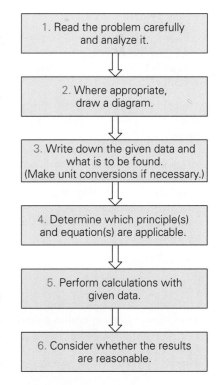

▲ FIGURE 1.11 **A flowchart for the suggested problem-solving procedure**

TABLE 1.4   Types of Examples

Example—primarily mathematical in nature

    Sections: **Thinking It Through**
                     Solution

Integrated Example—(a) conceptual multiple choice, (b) mathematical follow-up

    Sections: **(a) Conceptual Reasoning**
                     (b) Quantitative Reasoning and Solution

Conceptual Example—in general, needs only reasoning to obtain the answer, although some simple math may be required at times to justify the reasoning

    Sections: **Reasoning and Answer**

Pulling It Together—at the end of chapter, these examples demonstrate the use of several different concepts. Their purpose is to create a Learning Bridge from the chapter Learning Path to the End of Chapter Exercises, particularly the multiconcept type.

    Sections: **Same as Example or Integrated Example depending on the question(s) asked**

In general, there are four types of examples in this book, as listed in Table 1.4. The preceding steps are applicable to the three types, because they include calculations. Conceptual Examples, in general, do not follow these steps, being primarily conceptual in nature. The chapter Putting It Together is a multiconcept example.

In reading the worked Examples and Integrated Examples, you should be able to recognize the general application or flow of the preceding steps. This format will be used throughout the text. Let's take an Example and an Integrated Example as illustrations. Comments will be made in these examples to point out the problem-solving approach and steps that will not be made in the text Examples, but should be understood. Since no physical principles have really been covered, math and trig problems will be used, which should serve as a good review.

EXAMPLE 1.9 | Finding the Outside Surface Area of a Cylindrical Container

A closed cylindrical container used to store material from a manufacturing process has an outside radius of 50.0 cm and a height of 1.30 m. What is the total outside surface area of the container?

THINKING IT THROUGH. (In this type of Example, the Thinking It Through section generally combines problem-solving steps 1 and 2 given previously.)

   It should be noted immediately that the length units are given in mixed units, so a unit conversion will be in order. To visualize and analyze the cylinder, drawing a diagram is helpful (▶ Fig. 1.12). With this information in mind, proceed to finding the solution, using the expression for the area of a cylinder (the combined areas of the circular ends and the cylinder's side).

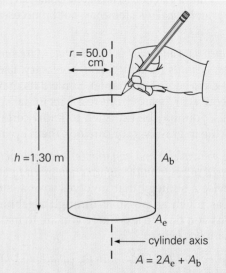

◀ **FIGURE 1.12**
A helpful step in problem solving Drawing a diagram helps you visualize and better understand the situation.

SOLUTION. Writing what is given and what is to be found (step 3 in our procedure):

Given: $r = 50.0$ cm   *Find:* A (the total outside surface
          $h = 1.30$ m   area of the cylinder)

First, let's tend to the mixed units. You should be able in this case to immediately write $r = 50.0$ cm $= 0.500$ m. But often

conversions are not obvious, so going through the unit conversion for illustration:

$$r = 50.0 \text{ cm}\left(\frac{1 \text{ m}}{100 \text{ cm}}\right) = 0.500 \text{ m}$$

There are general equations for areas (and volumes) of commonly shaped objects. The area of a cylinder can be easily looked up (given in Appendix I-C), but suppose you didn't have such a source. In this case, you may be able to figure it out for yourself. Looking at Fig. 1.12, note that the outside surface area of a cylinder consists of that of two circular ends and that of a rectangle (the body of the cylinder laid out flat). Equations for the areas of these common shapes are generally remembered. So the area of the two ends would be

$$2A_e = 2 \times \pi r^2 \quad (2 \text{ times the area of the circular end;}$$
$$\text{area of a cicle} = \pi r^2)$$

and the area of the body of the cylinder is

$$A_b = 2\pi r \times h \quad (\text{circumference of circular end}$$
$$\text{times height})$$

Then the total area A is

$$A = 2A_e + A_b = 2\pi r^2 + 2\pi rh$$

The data could be put into the equation, but sometimes an equation may be simplified to save some calculation steps.

$$A = 2\pi r(r + h) = 2\pi(0.500 \text{ m})(0.500 \text{ m} + 1.30 \text{ m})$$
$$= \pi(1.80 \text{ m}^2) = 5.65 \text{ m}^2$$

The result appears reasonable considering the cylinder's dimensions.

FOLLOW-UP EXERCISE. If the wall thickness of the cylinder's side and ends is 1.00 cm, what is the inside volume of the cylinder? *(Answers to all Follow-Up Exercises are given in Appendix VI at the back of the book.)*

INTEGRATED EXAMPLE 1.10 | ## Sides and Angles*

(a) A gardener has a rectangular plot measuring 3.0 m × 4.0 m. She wishes to use half of this area to make a triangular flower bed. Of the two types of triangles shown in ► Fig. 1.13, which should she use to do this: (1) the right triangle, (2) the isosceles triangle—two sides equal, or (3) either one? (b) In laying out the flower bed, the gardener decides to use a right triangle. Wishing to line the sides with rows of stone, she wants to know the total length (L) of the triangle sides. She would also like to know the values of the acute angles of the triangle. Can you help her so she doesn't have to do physical measurements? (Appendix I includes a review of trigonometric relationships as well as the marginal note on the next page.)

(A) CONCEPTUAL REASONING. The rectangular plot has a total area of 3.0 m × 4.0 m = 12 m². It is obvious that the right triangle divides the plot in half (Fig. 1.13). This is not as obvious for the isosceles triangle. But with a little study you should see that the outside areas could be arranged such that their combined area would be the same as that of the shaded isosceles triangle. So the isosceles triangle also divides the plot in half and the answer is (3). [This could be proven mathematically by computing the areas of the triangles. Area = $\frac{1}{2}$(altitude × base).]

(B) QUANTITATIVE REASONING AND SOLUTION. To find the total length of the sides, the length of the hypotenuse of the triangle is needed. This can be done using the Pythagorean theorem, $x^2 + y^2 = r^2$, and

$$r = \sqrt{x^2 + y^2} = \sqrt{(3.0 \text{ m})^2 + (4.0 \text{ m})^2} = \sqrt{25 \text{ m}^2} = 5.0 \text{ m}$$

(Or directly, you may have noticed that this is a 3–4–5 right triangle.) Then,

$$L = 3.0 \text{ m} + 4.0 \text{ m} + 5.0 \text{ m} = 12.0 \text{ m}$$

The acute angles of the triangle can be found by using trigonometry. Referring to the angles in Fig. 1.13,

$$\tan \theta_1 = \frac{\text{side opposite}}{\text{side adjacent}} = \frac{4.0 \text{ m}}{3.0 \text{ m}}$$

and

$$\theta_1 = \tan^{-1}\left(\frac{4.0 \text{ m}}{3.0 \text{ m}}\right) = 53°$$

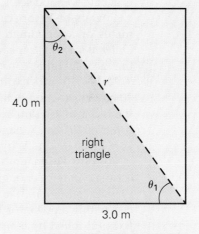

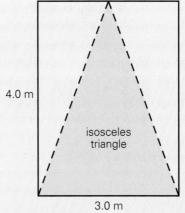

◄ **FIGURE 1.13**
A flower bed project
Two types of triangles for a new flower bed.

Similarly,

$$\theta_2 = \tan^{-1}\left(\frac{3.0 \text{ m}}{4.0 \text{ m}}\right) = 37°$$

The two angles add to 90° as would be expected with the right angle (53° + 37° = 90°).

FOLLOW-UP EXERCISE. What are the total length of the sides and the interior angles for the isosceles triangle in Fig. 1.13? *(Answers to all Follow-Up Exercises are given in Appendix VI at the back of the book.)*

---

*Here and throughout the text, angles will be considered exact, that is, they do not determine the number of significant figures.

Basic trigonometric functions:

$$\cos \theta = \frac{x}{r} \left(\frac{\text{side adjacent}}{\text{hypotenuse}} \right)$$

$$\sin \theta = \frac{y}{r} \left(\frac{\text{side opposite}}{\text{hypotenuse}} \right)$$

$$\tan \theta = \frac{\sin \theta}{\cos \theta} = \frac{y}{x} \left(\frac{\text{side opposite}}{\text{side adjacent}} \right)$$

These examples illustrate how the problem-solving steps are woven into finding the solution of a problem. You will see this pattern throughout the solved examples in the text, although not as explicitly explained. Try to develop your problem-solving skills in a similar manner.

APPROXIMATION AND ORDER-OF-MAGNITUDE CALCULATIONS

At times when solving a problem, you may not be interested in an exact answer, but want only an estimate or a "ballpark" figure. Approximations can be made by rounding off quantities so as to make the calculations easier and, perhaps, obtainable without the use of a calculator. For example, suppose you want to get an idea of the area of a circle with radius $r = 9.5$ cm. Then, rounding 9.5 cm ≈ 10 cm, and $\pi \approx 3$ instead of 3.14,

$$A = \pi r^2 \approx 3(10 \text{ cm})^2 = 300 \text{ cm}^2$$

(Note that significant figures are not a concern in calculations involving approximations.) The answer is not exact, but it is a good approximation. Compute the exact answer and see.

Powers-of-10, or scientific, notation is particularly convenient in making estimates or approximations in what are called **order-of-magnitude calculations**. *Order of magnitude* means that a quantity is expressed to the power of 10 closest to the actual value. For example, in the foregoing calculation, approximating 9.5 cm ≈ 10 cm is expressing 9.5 as 10^1, and we say that the radius is *on the order of* 10 cm. Expressing a distance of 75 km $\approx 10^2$ km indicates that the distance is on the order of 10^2 km. The radius of the Earth is 6.4×10^3 km $\approx 10^4$ km, or on the order of 10^4 km. A nanostructure with a width of 8.2×10^{-9} m is on the order of 10^{-8} m, or 10 nm. (Why an exponent of -8?)

An order-of-magnitude calculation gives only an estimate, of course. But this estimate may be enough to provide you with a better grasp or understanding of a physical situation. Usually, the result of an order-of-magnitude calculation is precise within a power of 10, or *within an order of magnitude*. That is, the number (prefix) multiplied by the power of 10 is somewhere between 1 and 10. For example, if a length result of 10^5 km were obtained, it would be expected that the exact answer was somewhere between 1×10^5 km and 10×10^5 km.

EXAMPLE 1.11 | ## Order-of-Magnitude Calculation: Drawing Blood

A medical technologist draws 15 cc of blood from a patient's vein. Back in the lab, it is determined that this volume of blood has a mass of 16 g. Estimate the density of the blood in SI units.

THINKING IT THROUGH. The data are given in cgs (centimeter-gram-second) units, which are often used for

practicality when dealing with small, whole-number quantities in some situations. The cc abbreviation is commonly used in the medical and chemistry fields for cm^3. Density (ρ) is mass per unit volume, where $\rho = m/V$ (Section 1.4).

SOLUTION.
First, changing to SI standard units:

> **Given:** $m = 16 \text{ g} \left(\dfrac{1 \text{ kg}}{1000 \text{ g}} \right) = 1.6 \times 10^{-2} \text{ kg} \approx 10^{-2} \text{ kg}$ **Find:**   estimate of ρ (density)

> $V = 15 \text{ cm}^3 \left(\dfrac{1 \text{ m}}{10^2 \text{ cm}} \right)^3 = 1.5 \times 10^{-5} \text{ m}^3 \approx 10^{-5} \text{ m}^3$

So, we have

$$\rho = \frac{m}{V} \approx \frac{10^{-2} \text{ kg}}{10^{-5} \text{ m}^3} = 10^3 \text{ kg/m}^3$$

This result is quite close to the average density of whole blood, $1.05 \times 10^3 \text{ kg/m}^3$.

FOLLOW-UP EXERCISE. A patient receives 750 cc of whole blood in a transfusion. Estimate the mass of the blood, in standard units. (*Answers to all Follow-Up Exercises are given in Appendix VI at the back of the book.*)

EXAMPLE 1.12 | How Many Red Cells Are in Your Blood?

The blood volume in the human body varies with a person's age, body size, and sex. On average, this volume is about 5.0 L. A typical value of red blood cells (erythrocytes) per volume is 5 000 000 (5.0×10^6) cells per cubic millimeter. Estimate how many red blood cells you have in your body.

THINKING IT THROUGH. The red blood cell count in cells per cubic millimeter is sort of a red blood cell "number density." Multiplying this figure by the total volume of blood [(cells/volume) $\times$ total volume] will give the total number of cells. But note that the volumes must have the same units. First let's start by converting 5.0 L to cubic meters (m^3): $1 \, L = 10^{-3} \, m^3$. (See inside front cover.)

SOLUTION.

Given: $V = 5.0 \, L$

$$= 5.0 \, L \left(10^{-3} \frac{m^3}{L} \right)$$

$$= 5.0 \times 10^{-3} \, m^3 \simeq 10^{-2} \, m^3$$

Find: the approximate number of red blood cells in the body

$$\text{cells/volume} = 5.0 \times 10^6 \frac{\text{cells}}{\text{mm}^3} \approx 10^7 \frac{\text{cells}}{\text{mm}^3}$$

Then, changing to cubic meters,

$$\frac{\text{cells}}{\text{volume}} \simeq 10^7 \frac{\text{cells}}{\text{mm}^3} \left(\frac{10^3 \, \text{mm}}{1 \, \text{m}} \right)^3 \approx 10^{16} \frac{\text{cells}}{\text{m}^3}$$

(*Note*: The conversion factor for liters to cubic meters was obtained directly from the conversion tables, but there is no conversion factor given for converting cubic millimeters to cubic meters, so a known conversion factor is cubed.) Then,

$$\left(\frac{\text{cells}}{\text{volume}} \right) (\text{total volume}) \approx \left(10^{16} \frac{\text{cells}}{\text{m}^3} \right) (10^{-2} \, m^3)$$

$$= 10^{14} \text{ red blood cells}$$

That's a bunch of cells. Red blood cells (erythrocytes) are one of the most abundant cells in the human body.

FOLLOW-UP EXERCISE. The average number of white blood cells (leukocytes) in human blood is normally 5000 to 10 000 cells per cubic millimeter. Estimate the number of white blood cells you have in your body. (*Answers to all Follow-Up Exercises are given in Appendix VI at the back of the book.*)

DID YOU LEARN?
➥ It is essential to initially understand a problem.
➥ Answers should be checked to ensure that they are reasonable in magnitude, and in some cases, direction.
➥ For an order of magnitude or a "ball park" figure, a number may be approximated by expressing it to the closest power of ten.

PULLING IT TOGETHER | Painting by Pythagoras

A painter uses a 10.0 ft ladder to reach the top area of a wall. The ladder is to be placed so that its top supports are at a height of 8.50 ft above the floor. (a) How far out from the wall are the bottom ladder feet? (b) What angle does the ladder make with respect to the vertical? (c) If the painter wants to adjust the ladder so that the top supports are 60.0 cm below the initial 8.50 ft height, how much further (in cm) must the floor feet be moved from the wall?

THINKING IT THROUGH. In part (a), if you make a quick sketch, you can see that the third side of a right triangle is being asked for, so the Pythagorean theorem applies. In (b), a trig function is clearly appropriate. (c) This is a repeat of (a) but a conversion of units is needed.

SOLUTION.

Given: $L = 10.0 \, \text{ft}$
(triangle hypotenuse)

$h = 8.50 \, \text{ft}$
(initial height of top ladder supports on wall)

$\Delta h = 60.0 \, \text{cm}$
(distance top ladder supports move down)

Find: (a) d (distance ladder feet from wall)

(b) angle of ladder from vertical

(c) Δd (distance ladder feet move out)

(a) From the Pythagorean theorem, $L^2 = h^2 + d^2$. Thus
$$d = \sqrt{L^2 - h^2} = \sqrt{(10.0 \, \text{ft})^2 - (8.50 \, \text{ft})^2} = 5.27 \, \text{ft}$$

(b) From a sketch the angle is given by
$$\theta = \tan^{-1}\left(\frac{d}{h} \right) = \tan^{-1}\left(\frac{5.27 \, \text{ft}}{8.50 \, \text{ft}} \right) = 31.8°$$

(c) Since the answer is asked for in centimeters, let's convert all the dimensions into cm:

$$L = 10.0 \, \text{ft} \times \frac{12.0 \, \text{in.}}{\text{ft}} \times \frac{2.54 \, \text{cm}}{\text{in.}} = 305 \, \text{cm}$$

$$h = 8.50 \, \text{ft} \times \frac{12.0 \, \text{in.}}{\text{ft}} \times \frac{2.54 \, \text{cm}}{\text{in.}} = 259 \, \text{cm}$$

$$d = 5.27 \, \text{ft} \times \frac{12.0 \, \text{in.}}{\text{ft}} \times \frac{2.54 \, \text{cm}}{\text{in.}} = 161 \, \text{cm}$$

Then the new distance of the top supports up the wall is
$$h' = h - \Delta h = 259 \, \text{cm} - 60.0 \, \text{cm} = 199 \, \text{cm}$$

Applying the Pythagorean theorem one more time to find the new distance d' out from the wall and the change in that distance:
$$d' = \sqrt{L^2 - (h')^2} = \sqrt{(305 \, \text{cm})^2 - (199 \, \text{cm})^2} = 231 \, \text{cm}$$

and
$$\Delta d = d' - d = 231 \, \text{cm} - 161 \, \text{cm} = 70 \, \text{cm}$$

Learning Path Review

- **SI units of length, mass, and time.** The meter (m), the kilogram (kg), and the second (s), respectively.

LENGTH: METER

1 m = distance traveled by light in a
vacuum in 1/299 792 458 s

MASS: KILOGRAM

0.10 m
water
0.10 m
0.10 m

One frequency oscillation

Cesium-133   1 s = 9 192 631 770 oscillations   Radiation detector

- **Liter (L).** A volume of $10 \text{ cm} \times 10 \text{ cm} \times 10 \text{ cm} = 1000 \text{ cm}^3$ or 1000 mL. A liter of water has a mass of 1 kg or 1000 g. Therefore, 1 cm^3 or 1 mL has a mass of 1 gram.

Volume

1 L
1 qt

1 L = 1.06 qt
1 qt = 0.947 L

- **Unit analysis.** Unit analysis can be used to determine the consistency of an equation, that is, if the equation is dimensionally correct, but not if physically correct. Unit analysis can also be used to find the unit of a quantity.

- **Significant figures (digits).** The digits that are known with certainty, plus one digit that is uncertain, in a measured value.

- **Problem solving.** Problems should be worked using a consistent procedure. Order-of-magnitude calculations may be done when an estimated value is desired.

  Suggested Procedure for Problem Solving:
  1. Read the problem carefully and analyze it.
  2. Where appropriate draw a diagram.
  3. Write down the given data and what is to be found. (Make unit conversions if necessary.)
  4. Determine which principle(s) and equation(s) are applicable.
  5. Perform calculations with given data.
  6. Consider whether the results are reasonable.

- **Density (ρ).** The mass per unit volume of an object or substance, which is a measure of the compactness of the material it contains:

$$\rho = \frac{m}{V} \left(\frac{\text{mass}}{\text{volume}} \right)$$

Learning Path Questions and Exercises   For instructor-assigned homework, go to www.masteringphysics.com

MULTIPLE CHOICE QUESTIONS

1.2 SI UNITS OF LENGTH, MASS, AND TIME

1. How many base units are there in the SI: (a) 3, (b) 5, (c) 7, or (d) 9?

2. The only SI standard represented by material standard or artifact is the (a) meter, (b) kilogram, (c) second, (d) electric charge.

3. Which of the following is the SI base unit for mass: (a) pound, (b) gram, (c) kilogram, or (d) ton?

4. Which of the following is not related to a volume of water: (a) kilogram, (b) pound, (c) gram, or (d) tonne?

1.3 MORE ABOUT THE METRIC SYSTEM

5. The prefix *giga-* means (a) 10^{-9}, (b) 10^9, (c) 10^{-6}, (d) 10^6.

6. The prefix *micro-* means (a) 10^6, (b) 10^{-6}, (c) 10^3, (d) 10^{-3}.

7. A new technology is concerned with objects the size of what metric prefix: (a) *nano-*, (b) *micro-*, (c) *mega-*, or (d) *giga-*?

8. Which of the following has the greatest volume: (a) 1 L, (b) 1 qt, (c) 2000 μL, or (d) 2000 mL?

9. Which of the following metric prefixes is the smallest: (a) micro-, (b) centi-, (c) nano-, or (d) milli-?

1.4 UNIT ANALYSIS

10. Both sides of an equation are equal in (a) numerical value, (b) units, (c) dimensions, (d) all of the preceding.

11. Unit analysis of an equation cannot tell you if (a) the equation is dimensionally correct, (b) the equation is physically correct, (c) the numerical value is correct, (d) both b and c.

1.5 UNIT CONVERSIONS

12. A good way to ensure proper unit conversion is to (a) use another measurement instrument, (b) always work in the same system of units, (c) use unit analysis, (d) have someone check your math.

13. You often see 1 kg = 2.2 lb. This expression means that (a) 1 kg is equivalent to 2.2 lb, (b) this is a true equation, (c) 1 lb = 2.2 kg, (d) none of the preceding.

14. You have a quantity of water and wish to express this in volume units that give the largest number. Which of the following units should be used: (a) in^3, (b) mL, (c) μL, or (d) cm^3?

1.6 SIGNIFICANT FIGURES

15. Which of the following has the greatest number of significant figures: (a) 103.07, (b) 124.5, (c) 0.09916, or (d) 5.408×10^5?

16. Which of the following numbers has four significant figures: (a) 140.05, (b) 276.02, (c) 0.004 006, or (d) 0.073 004?

17. In a multiplication and/or division operation involving the numbers 15 437, 201.08, and 408.0×10^5, the result should have how many significant figures: (a) 3, (b) 4, (c) 5, or (d) any number?

1.7 PROBLEM SOLVING

18. An important step in problem solving before mathematically solving an equation is (a) checking units, (b) checking significant figures, (c) checking with a friend, (d) checking to see if the result will be reasonable.

19. An important final step in problem solving before reporting an answer is (a) saving your calculations, (b) reading the problem again, (c) seeing if the answer is reasonable, (d) checking your results with another student.

20. In order-of-magnitude calculations, you should (a) pay close attention to significant figures, (b) work primarily in the British system, (c) get results within a factor of 100, (d) express a quantity to the power of 10 closest to the actual value.

CONCEPTUAL QUESTIONS

1.2 SI UNITS OF LENGTH, MASS, AND TIME

1. Why are there not more SI base units?

2. Why is weight not a base quantity?

3. What replaced the original definition of the second and why? Is the replacement still used?

4. Give a couple of major differences between the SI and the British system.

1.3 MORE ABOUT THE METRIC SYSTEM

5. If a fellow student tells you he saw a 3-cm-long ladybug, would you believe him? How about another student saying she caught a 10-kg salmon?

6. Explain why 1 mL is equivalent to 1 cm^3.

7. Explain why a metric ton is equivalent to 1000 kg.

1.4 UNIT ANALYSIS

8. Can unit analysis tell you whether you have used the correct equation in solving a problem? Explain.

9. The equation for the area of a circle from two sources is given as $A = \pi r^2$ and $A = \pi d^2/2$. Can unit analysis tell you which is correct? Explain.

10. How might unit analysis help determine the units of a quantity?

11. Why is π unitless?

1.5 UNIT CONVERSIONS

12. Are an equation and an equivalence statement the same? Explain.

13. Does it make any difference whether you multiply or divide by a conversion factor? Explain.

14. A popular saying is "Give him an inch and he'll take a mile." What would be the equivalent numerical values and units in the metric system?

1.6 SIGNIFICANT FIGURES

15. What is the purpose of significant figures?

16. Are all the significant figures reported for a measured value accurately known? Explain.

17. How are the number of significant figures determined for the results of calculations involving (a) multiplication, (b) division, (c) addition, and (d) subtraction?

18. Why is 5 chosen to be the major digit for rounding?

1.7 PROBLEM SOLVING

19. What are the main steps in the problem-solving procedure suggested in this chapter?

20. When you do order-of-magnitude calculations, should you be concerned about significant figures? Explain.

21. When doing an order-of-magnitude calculation, how accurate can you expect the answer to be? Explain.

22. The largest organ of the human body is the skin. The total external skin area of the average human covers an area of approximately 2.0 m^2. If you were asked to compute the approximate skin area, how would you go about it? (*Hint*: see Example 8.18.)

23. Is the following statement reasonable? It took 300 L of gasoline to fill the car's tank. (Justify your answer.)

24. Is the following statement reasonable? A car traveling 30 km/h through a school speed zone exceeds the speed limit of 25 mi/h. (Justify your answer.)

EXERCISES*

Integrated Exercises (IEs) are two-part exercises. The first part typically requires a conceptual answer choice based on physical thinking and basic principles. The following part is quantitative calculations associated with the conceptual choice made in the first part of the exercise.

Many exercise sections include "paired" exercises. These exercise pairs, identified with **red numbers,** *are intended to assist you in problem solving and learning. In a pair, the first exercise (even numbered) is worked out in the Study Guide so that you can consult it should you need assistance in solving it. The second exercise (odd numbered) is similar in nature, and its answer is given in Appendix VII at the back of the book.*

1.3 MORE ABOUT THE METRIC SYSTEM

1. • The metric system is a decimal (base-10) system, and the British system is, in part, a duodecimal (base-12) system. Discuss the ramifications if our monetary system had a duodecimal base. What would be the possible values of our coins if this were the case?

2. • (a) In the British system, 16 oz = 1 pt and 16 oz = 1 lb. Is something wrong here? Explain. Here's an old one: A pound of feathers weighs more than a pound of gold. How can that be? [*Hint:* Look up *ounce* in the dictionary.]

3. • Convert the following: (a) 40 000 000 bytes to MB, (b) 0.5722 mL to L, (c) 2.684 m to cm, and (d) 5 500 bucks to kilobucks.

4. •• A sailor tells you that if his ship is traveling at 25 knots (nautical miles per hour), it is moving faster than the 25 mi/h your car travels. How can that be?

5. •• A rectangular container measuring 25 cm × 35 cm × 55 cm is filled with water. What is the mass of this volume of water?

6. •• What size cube (in centimeters) would have a volume equal to that of a quart?

7. •• (a) What volume in liters is a cube 20 cm on a side? (b) If the cube is filled with water, what is the mass of the water?

1.4 UNIT ANALYSIS

8. • Show that the equation $x = x_o + vt$, where v is velocity, x and x_o are lengths, and t is time, is dimensionally correct.

9. • If x refers to distance, v_o and v to velocities, a to acceleration, and t to time, which of the following equations is dimensionally correct: (a) $x = v_o t + at^3$, (b) $v^2 = v_o^2 + 2at$, (c) $x = at + vt^2$, or (d) $v^2 = v_o^2 + 2ax$?

10. •• Use SI unit analysis to show that the equation $A = 4\pi r^2$, where A is the area and r is the radius of a sphere, is dimensionally correct.

11. •• The general equation for a parabola is $y = ax^2 + bx + c$, where a, b, and c are constants. What are the units of each constant if y and x are in meters?

12. •• You are told that the volume of a sphere is given by $V = \pi d^3/4$, where V is the volume and d is the diameter of the sphere. Is this equation dimensionally correct? (Use SI unit analysis to find out.)

13. •• The correct equation for the volume of a sphere is $V = 4\pi r^3/3$, where r is the radius of the sphere. Is the equation in Exercise 12 correct? If not, what should it be when expressed in terms of d?

14. •• The units for pressure (p) in terms of SI base units are known to be $\dfrac{kg}{m \cdot s^2}$. For a physics class assignment, a student derives an expression for the pressure exerted by the wind on a wall in terms of the air density (ρ) and wind speed (v) and her result is $p = \rho v^2$. Use SI unit analysis to show that her result is dimensionally consistent. Does this prove that this relationship is physically correct?

15. •• Is the equation for the area of a trapezoid, $A = \frac{1}{2}a(b_1 + b_2)$, where a is the height and b_1 and b_2 are the bases, dimensionally correct? (▼Fig. 1.14.)

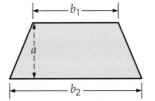

◀ **FIGURE 1.14 The area of a trapezoid** See Exercise 15.

16. ••• Newton's second law of motion (Section 4.3) is expressed by the equation $F = ma$, where F represents force, m is mass, and a is acceleration. (a) The SI unit of force is, appropriately, called the newton (N). What are the units of the newton in terms of base quantities? (b) An equation for force associated with uniform circular motion (Section 7.3) is $F = mv^2/r$, where v is speed and r is the radius of the circular path. Does this equation give the same units for the newton?

17. ••• The angular momentum (L) of a particle of mass m moving at a constant speed v in a circle of radius r is given by $L = mvr$ (Section 8.5). (a) What are the units of angular momentum in terms of SI base units? (b) The units of kinetic energy in terms of SI base units are $\dfrac{kg \cdot m^2}{s^2}$. Using SI unit analysis, show that the expression for the kinetic energy of this particle in terms of its angular momentum, $K = \dfrac{L^2}{2mr^2}$, is dimensionally correct. (c) In the previous equation, the term mr^2 is called the *moment of inertia* of the particle in the circle. What are the units of moment of inertia in terms of SI base units?

18. ••• Einstein's famous mass–energy equivalence is expressed by the equation $E = mc^2$, where E is energy, m is mass, and c is the speed of light. (a) What are the SI base units of energy? (b) Another equation for energy is $E = mgh$, where m is mass, g is the acceleration due to gravity, and h is height. Does this equation give the same units as in part (a)?

*Keep in mind here and throughout the text that your answer to an odd-numbered exercise may differ slightly from that given in Appendix VII at the back of the book because of rounding. See the Problem-Solving Hint: The "Correct" Answer in this chapter.

1.5 UNIT CONVERSION

19. • Figure 1.8 (top) shows the elevation of a location in both feet and meters. Is the conversion correct?

20. IE • (a) If you wanted to express your height with the largest number, which units would you use: (1) meters, (2) feet, (3) inches, or (4) centimeters? Why? (b) If you are 6.00 ft tall, what is your height in centimeters?

21. • If the capillaries of an average adult were unwound and spread out end to end, they would extend to a length over 40 000 mi (Fig. 1.9). If you are 1.75 m tall, how many times your height would the capillary length equal?

22. IE • (a) Compared with a 2-L soda bottle, a half-gallon soda bottle holds (1) more, (2) the same amount of, (3) less soda. (b) Verify your answer for part (a).

23. • (a) A football field is 300 ft long and 160 ft wide. What are the field's dimensions in meters? (b) A football is 11.0 to $11\frac{1}{4}$ in. long. What is its length in centimeters?

24. • Suppose that when the United States goes completely metric, the dimensions of a football field are established as 100 m by 54 m. Which would be larger, the metric football field or a current football field (see Exercise 23a), and what would be the difference between the areas?

25. • Water is sold in pint bottles. What is the mass of the water in a full bottle?

26. • How many (a) quarts and (b) gallons are there in 10.0 L?

27. • A submarine is submerged 175 fathoms below the surface. What is its depth in meters? (A *fathom* is an old nautical measurement equal to 2 yd.)

28. •• Driving a jet-powered car, Royal Air Force pilot Andy Green broke the sound barrier on land for the first time and achieved a record land speed of more than 763 mi/h in Black Rock Desert, Nevada, on October 15, 1997 (▼Fig. 1.15). (a) What is this speed expressed in m/s? (b) How long would it take the jet-powered car to travel the length of a 300-ft football field at this speed? (*Hint*: $v = d/t$.)

▲ FIGURE 1.15   **Record run**   See Exercise 28.

29. IE •• (a) Which of the following represents the greatest speed: (1) 1 m/s, (2) 1 km/h, (3) 1 ft/s, or (4) 1 mi/h? (b) Express the speed 15.0 m/s in mi/h.

30. •• An automobile speedometer is shown in ▼Fig. 1.16. (a) What would be the equivalent scale readings (for each empty box) in kilometers per hour? (b) What would be the 70-mi/h speed limit in kilometers per hour?

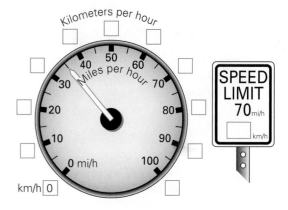

▲ FIGURE 1.16   **Speedometer readings**   See Exercise 30.

31. •• A person weighs 170 lb. (a) What is his mass in kilograms? (b) Assuming the density of the average human body is about that of water (which is true), estimate his body's volume in both cubic meters and liters. Explain why the smaller unit of the liter is more appropriate (convenient) for describing a volume of this size.

32. •• If the components of the human circulatory system (arteries, veins, and capillaries) were completely extended and placed end to end, the length would be on the order of 100 000 km. Would the length of the circulatory system reach around the circumference of the Moon? If so, how many times?

33. •• The human heartbeat, as determined by the pulse rate, is normally about 60 beats/min. If the heart pumps 75 mL of blood per beat, what volume of blood is pumped in one day in liters?

34. •• Some common product labels are shown in ▼Fig. 1.17. From the units on the labels, find (a) the number of milliliters in 2 fl. oz and (b) the number of ounces in 100 g.

▲ FIGURE 1.17   **Conversion factors**   See Exercise 34.

35. •• ▼Fig. 1.18 is a picture of red blood cells seen under a scanning electron microscope. Normally, women possess about 4.5 million of these cells in each cubic millimeter of blood. If the blood flow to the heart is 250 mL/min, how many red blood cells does a woman's heart receive each second?

◀ FIGURE 1.18
Red blood cells
See Exercise 35.

36. •• A student was 18 in. long when she was born. She is now 5 ft 6 in. tall and 20 years old. How many centimeters a year did she grow on average?

37. ••• How many minutes of arc does the Earth rotate in 1 min of time?

38. ••• The density of metal mercury is 13.6 g/cm³. (a) What is this density as expressed in kilograms per cubic meter? (b) How many kilograms of mercury would be required to fill a 0.250-L container?

39. ••• The Roman Coliseum used to be flooded with water to re-create ancient naval battles. Assuming the circular floor be 250 m in diameter and the water to have a depth of 10 ft, (a) how many cubic meters of water are required? (b) How much mass would this water have in kilograms? (c) How much would the water weigh in pounds?

40. ••• In the Bible, Noah is instructed to build an ark 300 cubits long, 50.0 cubits wide, and 30.0 cubits high (▼Fig. 1.19). Historical records indicate a cubit is equal to half a yard. (a) What would be the dimensions of the ark in meters? (b) What would be the ark's volume in cubic meters? To approximate, assume that the ark is to be rectangular.

▲ FIGURE 1.19   **Noah and his ark** See Exercise 40.

1.6 SIGNIFICANT FIGURES

41. • Express the length 50 500 μm (micrometers) in centimeters, decimeters, and meters, to three significant figures.

42. • Using a meterstick, a student measures a length and reports it to be 0.8755 m. What is the smallest division on the meterstick scale?

43. • Determine the number of significant figures in the following measured numbers: (a) 1.007 m, (b) 8.03 cm, (c) 16.272 kg, (d) 0.015 μs (microseconds).

44. • Express each of the numbers in Exercise 43 with two significant figures.

45. • Round the following numbers to two significant figures: (a) 95.61, (b) 0.00208, (c) 9438, (d) 0.000344

46. • Which of the following quantities has three significant figures: (a) 305.0 cm, (b) 0.0500 mm, (c) 1.000 81 kg, (d) 8.06 × 10⁴ m²?

47. •• The cover of your physics book measures 0.274 m long and 0.222 m wide. What is its area in square meters?

48. •• The interior storage compartment of a restaurant refrigerator measures 1.3 m high, 1.05 m wide, and 67 cm deep. Determine its volume in cubic feet.

49. IE •• The top of a rectangular table measures 1.245 m by 0.760 m. (a) The smallest division on the scale of the measurement instrument is (1) m, (2) cm, (3) mm. Why? (b) What is the area of the tabletop?

50. IE •• The outside dimensions of a cylindrical soda can are reported as 12.559 cm for the diameter and 5.62 cm for the height. (a) How many significant figures will the total outside area have: (1) two, (2) three, (3) four, or (4) five? Why? (b) What is the total outside surface area of the can in square centimeters?

51. •• Express the following calculations using the proper number of significant figures: (a) 12.634 + 2.1, (b) 13.5 − 2.134, (c) π(0.25 m)², (d) √2.37/3.5

52. IE ••• In doing a problem, a student adds 46.9 m and 5.72 m and then subtracts 38 m from the result. (a) How many decimal places will the final answer have: (1) zero, (2) one, or (3) two? Why? (b) What is the final answer?

53. ••• Work this exercise by the two given procedures as directed, commenting on and explaining any difference in the answers. Use your calculator for the calculations. Compute p = mv, where v = x/t, given x = 8.5 m, t = 2.7 s, and m = 0.66 kg. (a) First compute v and then p. (b) Compute p = mx/t without an intermediate step. (c) Are the results the same? If not, why?

1.7 PROBLEM SOLVING

54. • A corner construction lot has the shape of a right triangle. If the two sides perpendicular to each other are 37 m long and 42.3 m long, what is the length of the hypotenuse?

55. •• The lightest solid material is silica aerogel, which has a typical density of only about 0.10 g/cm³. The molecular structure of silica aerogel is typically 95% empty space. What is the mass of 1 m³ of silica aerogel?

56. •• A *cord* of wood is a volume of cut wood equal to a stack 8.0 ft long, 4.0 ft wide, and 4.0 ft high. How many cords are there in 3.0 m³?

57. •• Nutrition Facts labels now appear on most foods. An abbreviated label concerned with fat is shown in ▼Fig. 1.20. When burned in the body, each gram of fat supplies 9 Calories. (A food Calorie is really a kilocalorie, as will be learned in Chapter 11.) (a) What percentage of the Calories in one serving is supplied by fat? (b) You may notice that our answer doesn't agree with the listed Total Fat percentage in Fig. 1.20. This is because the given Percent Daily Values are the percentages of the maximum recommended amounts of nutrients (in grams) contained in a 2000-Calorie diet. What are the maximum recommended amounts of total fat and saturated fat for a 2000-Calorie diet?

Nutrition Facts
Serving Size: 1 can
Calories: 310

Amount Per Serving	% Daily Value*
Total Fat 18 g	28%
Saturated Fat 7g	35%

* Percent Daily Values are based on a 2,000 Calorie diet.

◀ **FIGURE 1.20**
Nutrition Facts
See Exercise 57.

58. •• The thickness of the numbered pages of a textbook is measured to be 3.75 cm. (a) If the last page of the book is numbered 860, what is the average thickness of a page? (b) Repeat the calculation by using order-of-magnitude calculations.

59. •• The mass of the Earth is 5.98×10^{24} kg. What is the average density of the Earth in standard units?

60. IE •• To go to a football stadium from your house, you first drive 1000 m north, then 500 m west, and finally 1500 m south. (a) Relative to your home, the football stadium is (1) north of west, (2) south of east, (3) north of east, (4) south of west. (b) What is the straight-line distance from your house to the stadium?

61. •• Two chains of length 1.0 m are used to support a lamp, as shown in ▼Fig. 1.21. The distance between the two chains along the ceiling is 1.0 m. What is the vertical distance from the lamp to the ceiling?

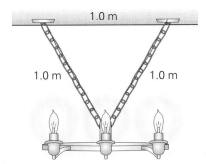

▲ **FIGURE 1.21** **Support the lamp** See Exercise 61.

62. •• Tony's Pizza Palace sells a medium 9.0-in. (diameter) pizza for $7.95, and a large 12-in. pizza for $13.50. Which pizza is the better buy?

63. •• Two students go into Tony's Pizza Palace and order a 12-in. (diameter) pizza. Shortly thereafter, the waitress brings an 8-in. pizza special. She explains that the 12-in. pizza was given to someone else by mistake and they could have the 8-in. now and she would bring another 8-in. shortly to make up for the missing 12-in. pizza. Was this a good deal?

64. •• In ▼Fig. 1.22, which black region has the greater area, the center circle or the outer ring?

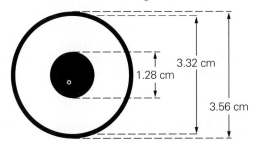

▲ **FIGURE 1.22** **Which black area is greater?** See Exercise 64.

65. •• The Channel Tunnel, or "Chunnel," which runs under the English Channel between Great Britain and France, is 31 mi long. (There are actually three separate tunnels.) A shuttle train that carries passengers through the tunnel travels with an average speed of 75 mi/h. On average, how long, in minutes, does the shuttle take to make a one-way trip through the Chunnel?

66. •• Human adult blood contains, on average, $7000/mm^3$ white blood cells (leukocytes) and $250\,000/mm^3$ platelets (thrombocytes). If a person has a blood volume of 5.0 L, estimate the total number of white cells and platelets in the blood.

67. •• The average number of hairs on the normal human scalp is 125 000. A healthy person loses about 65 hairs per day. (New hair from the hair follicle pushes the old hair out.) (a) How many hairs are lost in one month? (b) Pattern baldness (top-of-the-head hair loss) affects about 35 million men in the United States. If an average of 15% of the scalp is bald, how many hairs are lost per year by one of these "bald is beautiful" people?

68. IE •• A car is driven 13 mi east and then a certain distance due north, ending up at a position 25° north of east of its initial position. (a) The distance traveled by the car due north is (1) less than, (2) equal to, (3) greater than 13 mi. Why? (b) What distance due north does the car travel?

69. IE ••• At the Indianapolis 500 time trials, each car makes four consecutive laps, with its overall or average speed determining that car's place on race day. Each lap covers 2.5 mi (exact). During a practice run, cautiously and gradually taking his car faster and faster, a driver records the following average speeds for each successive lap: 160 mi/h, 180 mi/h, 200 mi/h, and 220 mi/h. (a) Will his average speed be (1) exactly the average of these speeds (190 mi/h), (2) greater than 190 mi/h, or (3) less than 190 mi/h? Explain. (b) To corroborate your conceptual reasoning, calculate the car's average speed.

70. ••• Approximately 118 mi wide, 307 mi long, and averaging 279 ft in depth, Lake Michigan is the second-largest Great Lake by volume. Estimate its volume of water in cubic meters.

71. **IE ●●●** In the Tour de France, a bicyclist races up two successive (straight) hills of different slope and length. The first is 2.00 km long at an angle of 5° above the horizontal. This is immediately followed by one 3.00 km long at 7°. (a) What will be the overall (net) angle from start to finish: (1) smaller than 5°, (2) between 5° and 7°, or (3) greater than 7°? (b) Calculate the actual overall (net) angle of rise experienced by this racer from start to finish, to corroborate your reasoning in part (a).

72. **●●●** A student wants to determine the distance from the lakeshore to a small island (▶Fig. 1.23). He first draws a 50-m line parallel to the shore. Then, he goes to the ends of the line and measures the angles of the lines of sight from the island relative to the line he has drawn. The angles are 30° and 40°. How far is the island from the shore?

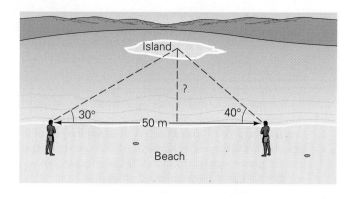

▲ **FIGURE 1.23   Measuring with lines of sight** See Exercise 72

PULLING IT TOGETHER: MULTICONCEPT EXERCISES

The Multiconcept Exercises require the use of more than one fundamental concept for their understanding and solution from this chapter.

73. A farmer owns a piece of land in the shape of an equilateral triangle, 200 m on a side, which is totally fenced in. He wishes to construct an additional fence parallel to the side fronting the road (▼Fig. 1.24) so that the area fronting the road takes up one-third of the total area. This area will be for his horses. On the remaining two-thirds he plans to construct his dream home. How far back from the road (shown as the distance *h*) should the fence be located?

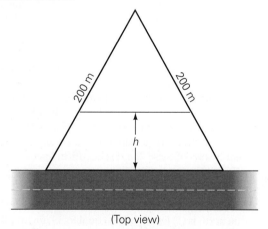

▲ **FIGURE 1.24   Don't fence me in** See Exercise 73.

74. In a radioactivity experiment, a solid lead brick (with same measurements as a patio brick, 2.00 in. × 4.00 in. × 8.00 in., except with a density that is 11.4 times that of water) is to be modified to hold a solid cylindrical piece of plastic. To accomplish this, the machinists are told to drill a cylindrical hole 2.0 cm in diameter through the center of the brick parallel to the longest side of the brick. (a) What is the mass of lead (in kilograms) removed from the brick? (b) What percentage of the original lead remains in the brick? (c) Assuming the

cylindrical hole is completely filled with plastic (with a density twice that of water), determine the overall (average) density of the brick/plastic combination after fabrication is complete.

75. A spherical shell is formed by taking a solid sphere of radius 20.0 cm and hollowing out a spherical section from the shell's interior. Assume the hollow section and the sphere itself have the same center location (that is, they are concentric). (a) If the hollow section takes up 90.0 percent of the total volume, what is its radius? (b) What is the ratio of the outer area to the inner area of the shell?

76. Two separate seismograph stations receive indication of an earthquake in the form of a wave traveling to them in a straight line from the epicenter and shaking the ground at their locations. Station B is 50 km due east of station A. The epicenter is located due north of station A and 30° north of due west from station B. (a) Draw a sketch and use it to determine the distance from the epicenter to A. (b) Determine the distance from the epicenter to B. (c) Station C is located an additional 20 km east of B. At what angle does C report the direction of the epicenter to be?

77. You are sailing a radio-controlled model powerboat on a perfectly circular pool of water. The boat travels at a constant 0.500 m/s. It takes 30.0 s to make the trip from one side of the pool, through the center, to the other side. (a) How long would it take the boat to travel completely around the edge of the pool? (b) If the pool is uniformly 1.50 m deep, how many gallons of water does it hold?

78. A certain material has a density of 9.0 g/cm³. It is formed into a solid rectangular brick with dimensions 1.0 cm × 2.0 cm × 4.0 cm. (a) What is its mass in kilograms? (b) If you wanted to make a cube of this same material containing twice the mass of this brick, what would be the length of one side of the cube?

2 Kinematics: Description of Motion†

PHYSICS FACTS

✦ "Give me matter and motion and I will construct the universe." Rene Descartes (1640)

✦ Nothing can exceed the speed of light (in vacuum), 3.0×10^8 m/s (186 000 mi/s).

✦ A bullet from a high-powered rifle travels at a speed of about 2900 km/h (1800 mi/h).

✦ NASA's X-43A uncrewed jet flew at a speed of 7700 km/h (4800 mi/h) —faster than a speeding bullet.

✦ Electrical signals between your brain and muscles travel at about 435 km/h (270 mi/h).

✦ A person at the equator is traveling at a speed of 1600 km/h (1000 mi/h) due to the Earth's rotation.

✦ Aristotle thought heavy objects fall faster than lighter ones. Galileo wrote, "Aristotle says that an iron ball falling from a height of one hundred cubits reaches the ground before a one-pound ball has fallen a single cubit. I say they arrive at the same time."

The cheetah is running at full stride in the chapter-opening photo. This fastest of all land animals is capable of attaining speeds up to 113 km/h, or 70 mi/h. The sense of motion in the photograph is so strong that you can almost feel the air rushing by. And yet this sense of motion is an illusion. Motion takes place in time, but the photo can "freeze" only a single instant. You'll find that without the dimension of time, motion cannot be described at all.

The description of motion involves the representation of a restless world. Nothing is ever

†The mathematics needed in this chapter involves general algebraic equation manipulation. You may want to review this in Appendix I.

perfectly still. You may sit, apparently at rest, but your blood flows, and air moves into and out of your lungs. The air is composed of gas molecules moving at different speeds and in different directions. And while experiencing stillness, you, your chair, the building you are in, and the air you breathe are all rotating and revolving through space with the Earth, part of a solar system in a spiraling galaxy in an expanding universe.

The branch of physics concerned with the study of motion and what produces and affects motion is called **mechanics**. The roots of mechanics and of human interest in motion go back to early civilizations. The study of the motions of heavenly bodies, or *celestial mechanics*, grew out of measuring time and location. Several early Greek scientists, notably Aristotle, put forth theories of motion that were useful descriptions, but were later proved to be incomplete or incorrect. Our currently accepted concepts of motion were formulated in large part by Galileo (1564–1642) and Isaac Newton (1642–1727).

Mechanics is usually divided into two parts: (1) kinematics and (2) dynamics. **Kinematics** deals with the *description* of the motion of objects, without consideration of what causes the motion. **Dynamics** analyzes the *causes* of motion. This chapter covers kinematics and reduces the description of motion to its simplest terms by considering linear motion, that is, motion in a straight line. You'll learn to analyze changes in motion—speeding up, slowing down, and stopping. Along the way, a particularly interesting case of accelerated motion will be presented: free fall (motion under the influence of gravity only).

2.1   Distance and Speed: Scalar Quantities

LEARNING PATH QUESTIONS

➥ Why is distance a scalar quantity?
➥ What is the difference between average speed and instantaneous speed?
➥ When are the average and instantaneous speeds equal?

DISTANCE

Motion is observed all around us. But what is motion? This question seems simple; however, you might have some difficulty giving an immediate answer. (And, it's not fair to use forms of the verb *to move* to describe motion.) After a little thought, you should be able to conclude that **motion** (or moving) *involves changing position*. Motion can be described in part by specifying *how far* something travels in changing position—that is, the distance it travels. **Distance** is simply the *total path length* traversed in moving from one location to another. For example, you may drive to school from your hometown and express the distance traveled in miles or kilometers. In general, the distance between two points depends on the path traveled (◄Fig. 2.1).

Along with many other quantities in physics, distance is a scalar quantity. A **scalar quantity** is one with only magnitude or size. That is, a *scalar* has only a numerical value, such as 160 km or 100 mi. (Note that the magnitude includes units.) Distance tells you the magnitude only—how far, but not the direction. Other examples of scalars are quantities such as 10 s (time), 3.0 kg (mass), and 20 °C (temperature). Some scalars may have negative values, for example, −10 °F.

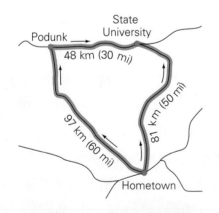

▲ **FIGURE 2.1   Distance—total path length** In driving to State University from Hometown, one student may take the shortest route and travel a distance of 81 km (50 mi). Another student takes a longer route in order to visit a friend in Podunk before returning to school. The longer trip is in two segments, but the distance traveled is the total length, 97 km + 48 km = 145 km (90 mi).

SPEED

When something is in motion, its position changes with time. That is, it moves a certain distance in a given amount of time. Both *length* and *time* are therefore important quantities in describing motion. For example, imagine a car and a pedestrian moving down a street and traveling a distance of one block. You would expect the car to travel faster and thus to cover the same distance in a shorter time than the person. A length–time relationship can be expressed by using the *rate* at which distance is traveled, or **speed**.

Average speed ($\bar{s}$) is the distance d traveled, that is, the actual path length, divided by the total time Δt elapsed in traveling that distance:

$$\text{average speed} = \frac{\text{distance traveled}}{\text{total time to travel that distance}} \tag{2.1}$$

$$\bar{s} = \frac{d}{\Delta t} = \frac{d}{t_2 - t_1}$$

SI unit of speed: meters per second (m/s)

A symbol with a bar over it is commonly used to denote an average. The Greek letter delta, Δ, is used to represent a change or difference in a quantity, in this case the time difference between the beginning (t_1) and end (t_2) of a trip, or the elapsed total time.

The SI standard unit of speed is meters per second (m/s, length/time), although kilometers per hour (km/h) is used in many everyday applications. The British standard unit is feet per second (ft/s), but a commonly used unit is miles per hour (mi/h). Often, the initial time is taken to be zero, $t_1 = 0$, as in resetting a stopwatch, and thus the equation is written $\bar{s} = d/t$, where it is understood that t is the total time.

Since distance is a scalar (as is time), speed is also a scalar. The distance does *not* have to be in a straight line (see Fig. 2.1). For example, you probably have computed the average speed of an automobile trip by using the distance obtained from the starting and ending odometer readings. Suppose these readings were 17 455 km and 17 775 km, respectively, for a 4.0-h trip. Subtracting the readings gives a total traveled distance d of 320 km, so the average speed of the trip is $d/t = 320 \text{ km}/4.0 \text{ h} = 80 \text{ km/h}$ (or about 50 mi/h).

Average speed gives a general description of motion over a time interval Δt. In the case of the auto trip with an average speed of 80 km/h, the car's speed wasn't *always* 80 km/h. With various stops and starts on the trip, the car must have been moving more slowly than the average speed at various times. It therefore had to be moving more rapidly than the average speed another part of the time. With an average speed, you don't know how fast the car was moving at any particular instant of time during the trip. By analogy, the average test score of a class doesn't tell you the score of any particular student.

On the other hand, **instantaneous speed** tells how fast something is moving *at a particular instant of time*. That is, when $\Delta t \rightarrow 0$ (the time interval approaches zero), which represents an instant of time. The speedometer of a car gives an approximate instantaneous speed. For example, the speedometer shown in ►Fig. 2.2 indicates a speed of about 44 mi/h, or 70 km/h. If the car travels with constant speed (so the speedometer reading does not change), then the average and instantaneous speeds will be equal. (Do you agree? Think of the previous average test score analogy. What if all of the students in the class got the same score?)

◄ **FIGURE 2.2 Instantaneous speed** The speedometer of a car gives the speed over a very short interval of time, so its reading approaches the instantaneous speed. Note the speeds are given in mi/h and km/h. (MPH is a nonstandard abbreviation.)

EXAMPLE 2.1 | Slow Motion: Rover Moves Along

In January 2004, a Mars Exploration Rover touched down on the surface of Mars and rolled out for exploration (▼Fig. 2.3). The average speed of the Rover on flat, hard ground is 5.0 cm/s. (a) Assuming the Rover traveled continuously over this terrain at its average speed, how much time would it take to travel 2.0 m nonstop in a straight line? (b) However, in order to ensure a safe drive, the Rover was equipped with hazard avoidance software that caused it to stop and assess its location every few seconds. It was programmed to drive at its average speed for 10 s, then stop and observe the terrain for 20 s before moving onward for another 10 s and repeating the cycle. Taking its programming into account, what would be the Rover's average speed in traveling the 2.0 m? (There were actually two Rovers on this mission, named Spirit and Opportunity. At the time of this writing, Spring 2008, both Rovers are still functioning after over four years on the Red Planet.)

▲ **FIGURE 2.3 A Mars exploration rover** Twin Rovers landed on opposite sides of the Martian planet in search of answers about the history of water on Mars.

THINKING IT THROUGH.   (a) Because the average speed and the distance are known, the time can be computed from the equation for average speed (Eq. 2.1). (b) Here, to calculate the average speed, the *total* time, which includes stops, must be used.

SOLUTION.
Listing the data in symbol form (cm/s is converted directly to m/s):

Given:
(a) $\bar{s} = 5.0$ cm/s $= 0.050$ m/s
    $d = 2.0$ m
(b) cycles of 10-s travel, 20-s stops

Find:
(a) Δt (time to travel distance)
(b) $\bar{s}$ (average speed)

(a) Rearranging Eq. 2.1, $\bar{s} = \dfrac{d}{\Delta t}$, to solve for time,

$$\Delta t = \frac{d}{\bar{s}} = \frac{2.0 \text{ m}}{0.050 \text{ m/s}} = 40 \text{ s}$$

So it takes the Rover 40 s to travel a path length of 2.0 m.

(b) Here, the total time for the 2.0-m distance is needed. In each 10-s interval, a distance of 0.050 m/s $\times$ 10 s $= 0.50$ m would be traveled. So, the total time would be four 10-s intervals for actual travel and three 20-s intervals of stopping, giving $\Delta t = 4 \times 10$ s $+ 3 \times 20$ s $= 100$ s. Then

$$\bar{s} = \frac{d}{\Delta t} = \frac{d}{t_2 - t_1} = \frac{2.0 \text{ m}}{100 \text{ s}} = 0.020 \text{ m/s}$$

FOLLOW-UP EXERCISE.   Suppose the Rover's programming was for 5.0 s of travel and for 10-s stops. How long would it take to travel the 2.0 m in this case? (*Answers to all Follow-Up Exercises are given in Appendix VI at the back of the book.*)

DID YOU LEARN?
➥ Distance is the total path length and has only magnitude.
➥ Average speed is speed over a period of time; instantaneous speed is speed at a particular instant of time.
➥ If the speed is constant, the average and instantaneous speeds are equal.

 2.2 ## One-Dimensional Displacement and Velocity: Vector Quantities

LEARNING PATH QUESTIONS
➥ How are displacement and velocity related?
➥ When are average velocity and instantaneous velocity related for linear motion?

DISPLACEMENT

For straight-line, or linear, motion, it is convenient to specify position by using the familiar two-dimensional Cartesian coordinate system with *x*- and *y*-axes at right angles. A straight-line path can be in any direction relative to the axes, but for convenience, the coordinate axes are usually oriented so that the motion is along one of them. (See the accompanying Learn by Drawing 2.1, Cartesian Coordinates and One-Dimensional Displacement.)

As was discussed in the previous section, distance is a scalar quantity with only magnitude (and units). However, to more completely describe motion, more information can be given by adding a *direction*. This information is particularly easy to convey for a change of position in a straight line. **Displacement** is defined as the straight-line distance between two points, along with the *direction* directly from the starting point to the final position. Unlike distance (a scalar), displacement can have either positive or negative values, with the signs indicating the directions along a coordinate axis.

As such, displacement is a **vector quantity**. In other words, a *vector* has both magnitude and direction. For example, when describing the displacement of an airplane as 25 km north, this is a *vector* description (magnitude and direction). Other vector quantities include velocity and acceleration, as will be learned later in the chapter.

There is an algebra that applies to vectors, which involves how to specify and deal with the direction part of the vector. This is done relatively easily in one dimension by using + and − signs to indicate directions. To illustrate this with displacements, consider the situation shown in ▼Fig. 2.4, where x_1 and x_2 indicate the initial and final positions, respectively, on the x-axis as a student moves in a straight line from his locker to the physics lab. As can be seen in Fig. 2.4a, the scalar distance traveled is 8.0 m. To specify displacement (a vector) between x_1 and x_2, we use the expression

$$\Delta x = x_2 - x_1 \qquad (2.2)$$

where Δ is again used to represent a change in a quantity. Then, as in Fig. 2.4b,

$$\Delta x = x_2 - x_1 = +9.0 \text{ m} - (+1.0 \text{ m}) = +8.0 \text{ m}$$

where the + signs indicate the positions on the positive x-axis. Hence, the student's displacement (magnitude and direction) is 8.0 m in the positive x-direction, as indicated by the positive (+) result in Fig. 2.4b. (As in "regular" mathematics, the plus sign is often omitted, as being understood, so this displacement can be written as $\Delta x = 8.0$ m instead of $\Delta x = +8.0$ m.)

Cartesian coordinates and one-dimensional displacement

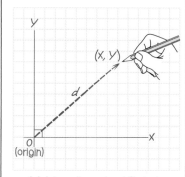

(a) A two-dimensional Cartesian coordinate system. A displacement vector d locates a point (x, y).

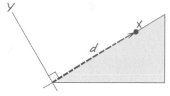

(b) For one-dimensional, or straight-line, motion, it is convenient to orient one of the coordinate axes along the direction of motion.

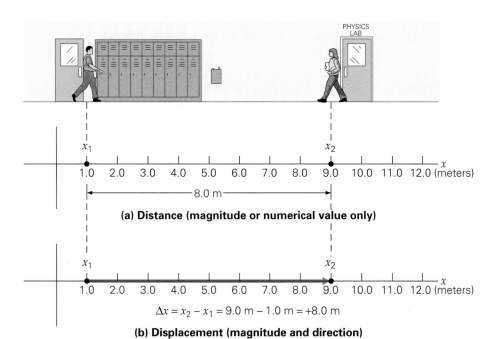

▲ **FIGURE 2.4 Distance (scalar) and displacement (vector)** **(a)** The distance (straight-line path) between the student on the left and the physics lab is 8.0 m and is a scalar quantity. **(b)** To indicate displacement, x_1 and x_2 specify the initial and final positions, respectively. The displacement is then $\Delta x = x_2 - x_1 = 9.0 \text{ m} - 1.0 \text{ m} = +8.0$ m—that is, 8.0 m in the positive x-direction.

Vector quantities in this book are usually indicated by boldface type with an over-arrow; for example, a displacement vector is indicated by $\vec{d}$ or $\vec{x}$, and a velocity vector is indicated by $\vec{v}$. However, when working in one dimension, this notation is not needed. Instead, plus and minus signs can be used to indicate the only two possible directions. The x-axis is commonly used for horizontal motions, and a plus $(+)$ sign is taken to indicate the direction to the right, or in the "positive x-direction," and a minus $(-)$ sign indicates the direction to the left, or in the "negative x-direction."

Keep in mind that these signs only "point" in *particular directions*. An object moving along the negative x-axis toward the origin would be moving in the positive x-direction. How about an object moving along the positive x-axis toward the origin? If you said in the negative x-direction, you are correct.

Suppose the other student in Fig. 2.4 walks from the physics lab (her initial position is different, $x_1 = +9.0$ m) to the end of the lockers (the final position is now $x_2 = +1.0$ m). Her displacement would be

$$\Delta x = x_2 - x_1 = +1.0 \text{ m} - (+9.0 \text{ m}) = -8.0 \text{ m}$$

The minus sign indicates that the direction of the displacement was in the negative x-direction or to the left in the figure. In this case, we say that the two students' displacements are equal (in magnitude) and opposite (in direction).

VELOCITY

As has been learned, speed, like the distance it incorporates, is a scalar quantity—it has magnitude only. Another more descriptive quantity used to describe motion is *velocity*. Speed and velocity are often used synonymously in everyday conversation, but the terms have different meanings in physics. Speed is a scalar, and velocity is a vector—velocity has both magnitude and direction. Unlike speed, one-dimensional velocities can have both positive and negative values, indicating the only two possible directions (as with displacement).

Velocity tells how fast something is moving *and* in which direction it is moving. And just as there are average and instantaneous speeds, there are average and instantaneous velocities involving vector displacements. The **average velocity** is the displacement divided by the total travel time. In one dimension, this involves just motion along one axis, which is taken to be the x-axis. In this case,

$$\text{average velocity} = \frac{\text{displacement}}{\text{total travel time}} \qquad (2.3)^*$$

$$\bar{v} = \frac{\Delta x}{\Delta t} = \frac{x_2 - x_1}{t_2 - t_1}$$

SI unit of velocity meters per second (m/s)

In the case of more than one displacement (such as for successive displacements), the average velocity is equal to the *total or net* displacement divided by the total time. The total displacement is found by adding the displacements algebraically according to the directional signs.

You might be wondering whether there is a relationship between average speed and average velocity. A quick look at Fig. 2.4 will show that if all the motion

*Another common form of this equation is

$$\bar{v} = \frac{\Delta x}{\Delta t} = \frac{(x_2 - x_1)}{(t_2 - t_1)} = \frac{(x - x_0)}{(t - t_0)} = \frac{(x - x_0)}{t}$$

or, after rearranging,

$$x = x_0 + \bar{v}t \qquad (2.3)$$

where x_0 is the initial position, x is the final position, and $\Delta t = t$ with $t_0 = 0$. See Section 2.3 for more on this notation.

is in one direction, that is, there is no reversal of direction, the distance is equal to the magnitude of the displacement. Then the average speed is equal to the magnitude of the average velocity. *However, be careful.* This set of relationships is not true if there is a reversal of direction, as Example 2.2 shows.

EXAMPLE 2.2 | ## There and Back: Average Velocities

A jogger jogs from one end to the other of a straight 300-m track in 2.50 min and then jogs back to the starting point in 3.30 min. What was the jogger's average velocity (a) in jogging to the far end of the track, (b) coming back to the starting point, and (c) for the total jog?

THINKING IT THROUGH. The average velocities are computed from the defining equation. Note that the times given are the Δt's associated with the particular displacements.

SOLUTION.

Given: $\Delta x_1 = +300$ m (taking the initial direction as positive)
$\Delta x_2 = -300$ m (taking the direction of the return trip as negative)
$\Delta t_1 = 2.50$ min$(60$ s/min$) = 150$ s
$\Delta t_2 = 3.30$ min$(60$ s/min$) = 198$ s

Find: Average velocities for
(a) the first leg of the jog,
(b) the return jog,
(c) the total jog

(a) The jogger's average velocity for the trip down the track is found using Eq. 2.3:

$$\bar{v}_1 = \frac{\Delta x_1}{\Delta t_1} = \frac{+300 \text{ m}}{150 \text{ s}} = +2.00 \text{ m/s}$$

(b) Similarly, for the return trip,

$$\bar{v}_2 = \frac{\Delta x_2}{\Delta t_2} = \frac{-300 \text{ m}}{198 \text{ s}} = -1.52 \text{ m/s}$$

(c) For the total trip, there are two displacements to consider, down and back, so these are added together to get the total displacement, and then divided by the total time:

$$\bar{v}_3 = \frac{\Delta x_1 + \Delta x_2}{\Delta t_1 + \Delta t_2} = \frac{300 \text{ m} + (-300 \text{ m})}{150 \text{ s} + 198 \text{ s}} = 0 \text{ m/s}$$

The average velocity for the total trip is zero! Do you see why? Recall from the definition of displacement that the magnitude of displacement is the straight-line distance between two points. The displacement from one point back to the same point is zero; hence the average velocity is zero. (See ▶ Fig. 2.5.)

The total or net displacement for this case could have been found by simply taking $\Delta x = x_{\text{final}} - x_{\text{initial}} = 0 - 0 = 0$, where the initial and final positions are taken to be the origin, but it was done in parts here for illustration purposes.

▲ **FIGURE 2.5 Back home again!** Despite having covered nearly 110 m on the base paths, at the moment the runner slides through the batter's box (his original position) into home plate, his displacement is zero—at least, if he is a right-handed batter. No matter how fast he ran the bases, his average velocity for the round trip is zero.

FOLLOW-UP EXERCISE. Find the jogger's average speed for each of the cases in this Example, and compare these with the magnitudes of the respective average velocities. [Will the average speed for part (c) be zero?] *(Answers to all Follow-Up Exercises are given in Appendix VI at the back of the book.)*

As Example 2.2 shows, average velocity provides only an overall description of motion. One way to take a closer look at motion is to take smaller time intervals, that is, to let the observation time interval (Δt) become smaller and smaller. As with speed, when Δt approaches zero, an **instantaneous velocity** is obtained, which describes how fast something is moving and in which direction *at a particular instant of time.*

Δx (km)	Δt (h)	$\Delta x/\Delta t$
50	1.0	50 km/1.0 h = 50 km/h
100	2.0	100 km/2.0 h = 50 km/h
150	3.0	150 km/3.0 h = 50 km/h

(a)

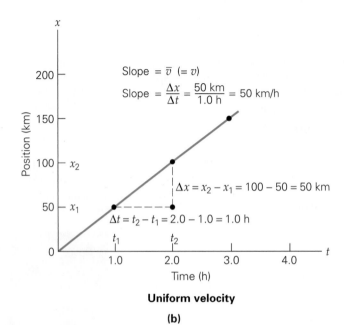

Uniform velocity

(b)

Instantaneous velocity is defined mathematically as

$$v = \lim_{\Delta t \to 0} \frac{\Delta x}{\Delta t} \tag{2.4}$$

This expression is read as "the instantaneous velocity is equal to the limit of $\Delta x/\Delta t$ as Δt goes to zero." The time interval does not ever equal zero (why?), but *approaches* zero. Instantaneous velocity is technically still an average velocity, but over such a very small Δt that it is essentially an average "at an instant in time," which is why it is called the *instantaneous* velocity.

Uniform motion means motion with a constant or uniform velocity (constant magnitude *and* constant direction). As a one-dimensional example of this, the car in ▲Fig. 2.6 has a uniform velocity. It travels the same distance and experiences the same displacement in equal time intervals (50 km each hour) and the direction of its motion does not change. Hence, the magnitudes of the average velocity and instantaneous velocity are equal in this case. The average of a constant is equal to that constant.

GRAPHICAL ANALYSIS

Graphical analysis is often helpful in understanding motion and its related quantities. For example, the motion of the car in Fig. 2.6a may be represented on a plot of position versus time, or x versus t. As can be seen from Fig. 2.6b, a straight line is obtained for a uniform, or constant, velocity on such a graph.

Recall from Cartesian graphs of y versus x that the slope of a straight line is given by $\Delta y/\Delta x$. Here, with a plot of x versus t, the slope of the line, $\Delta x/\Delta t$, is

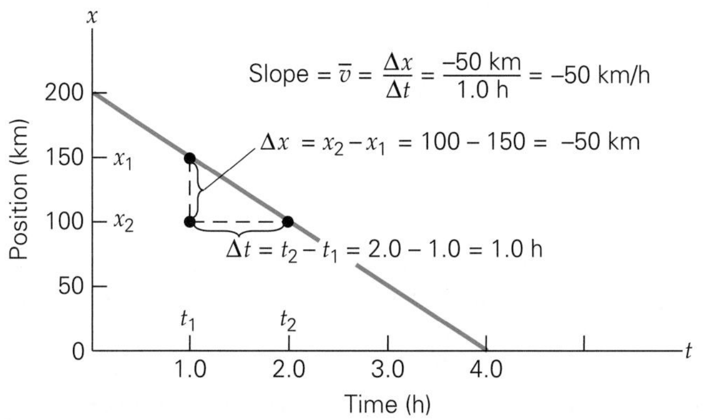

▲ **FIGURE 2.7 Position-versus-time graph for an object in uniform motion in the negative *x*-direction** A straight line on an *x*-versus-*t* plot with a negative slope indicates uniform motion in the negative *x*-direction. Note that the object's location changes at a constant rate. At $t = 4.0$ h, the object is at $x = 0$. How would the graph look if the motion continues for $t > 4.0$ h?

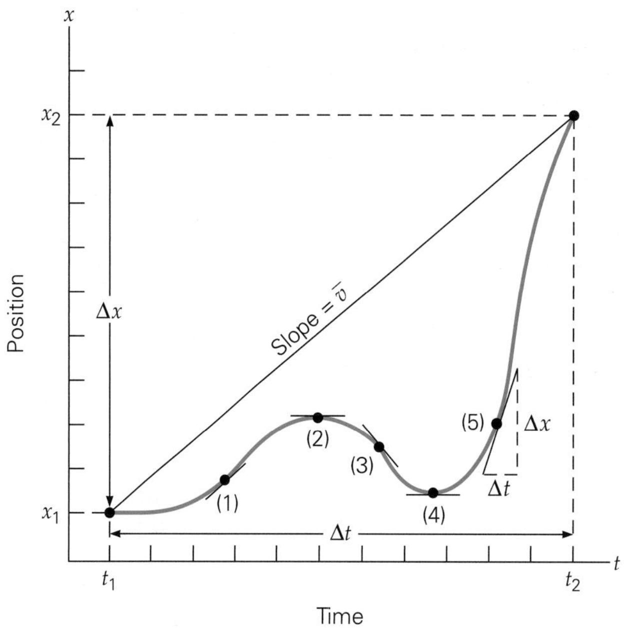

▲ **FIGURE 2.8 Position-versus-time graph for an object in nonuniform linear motion** For a nonuniform velocity, an *x*-versus-*t* plot is a curved line. The slope of the line between two points is the average velocity between those positions, and the instantaneous velocity at any time *t* is the slope of a line tangent to the curve at that point. Five tangent lines are shown, with the intervals for $\Delta x/\Delta t$ in the fifth. Can you describe the object's motion in words?

therefore equal to the average velocity $\bar{v} = \Delta x/\Delta t$. For uniform motion, this value is equal to the instantaneous velocity. That is, $\bar{v} = v$. (Why?) The numerical value of the slope is the magnitude of the velocity, and the sign of the slope gives the direction. A positive slope indicates that *x* increases with time, so the motion is in the positive *x*-direction. (The plus sign is often omitted as being understood, which will be done in general from here on.)

Suppose that a plot of position versus time for a car's motion is a straight line with a negative slope, as in ▲Fig. 2.7. What does this indicate? As the figure shows, the position (*x*) values get smaller with time at a constant rate, indicating that the car is traveling in uniform motion, but now in the negative *x*-direction which correlates with the negative value of the slope.

In most instances, the motion of an object is *nonuniform*, meaning that different distances are covered in equal intervals of time. An *x*-versus-*t* plot for such motion in one dimension is a curved line, as illustrated in ▲Fig. 2.8. The average velocity of the object during any interval of time is the slope of a straight line between the two points on the curve that correspond to the starting and ending times of the interval. In the figure, since $\bar{v} = \Delta x/\Delta t$, the average velocity of the total trip is the slope of the straight line joining the beginning and ending points of the curve.

The instantaneous velocity is equal to the slope of the tangent line to the curve at the time of interest. Five typical tangent lines are shown in Fig. 2.8. At (1), the slope is positive, and the motion is therefore in the positive *x*-direction. At (2), the slope of a horizontal tangent line is zero, so there is no motion for an instant. That is, the object has instantaneously stopped ($v = 0$) at that time. At (3), the slope is negative, so the object is moving in the negative *x*-direction. Thus, the object stopped in the process of changing direction at point (2). What is happening at points (4) and (5)?

By drawing various tangent lines along the curve, it can be seen that their slopes vary, in magnitude and direction (sign), indicating that the instantaneous velocity is changing with time. An object in nonuniform motion can speed up, slow down, or change direction. How nonuniform motion is described in the topic of Section 2.3.

DID YOU LEARN?

➥ Velocity is the time rate of change of displacement.

➥ For linear motion in one direction, average speed is equal to the magnitude of the average velocity. (This is not true if there is a reversal in direction.)

2.3   Acceleration

LEARNING PATH QUESTIONS
➥ What is evidence of an acceleration?
➥ What is required for a deceleration?
➥ Is a negative acceleration always a deceleration?

The basic description of motion involving the time rate of change of position (and direction) is called *velocity*. Going one step further, we can consider how this *rate of change* itself changes. Suppose an object is moving at a constant velocity and then the velocity changes. Such a change in velocity is called an *acceleration*. The gas pedal on an automobile is commonly called the *accelerator*. When you press down on the accelerator, the car speeds up; when you let up on the accelerator, the car slows down. In either case, there is a change in velocity with time. **Acceleration** is defined as the time rate of change of velocity.

Analogous to average velocity, the **average acceleration** is defined as the change in velocity divided by the time taken to make the change:

$$\text{average acceleration} = \frac{\text{change in velocity}}{\text{time to make the change}} \tag{2.5}$$

$$\bar{a} = \frac{\Delta v}{\Delta t}$$

$$= \frac{v_2 - v_1}{t_2 - t_1} = \frac{v - v_0}{t - t_0}$$

SI unit of acceleration: meters per second squared (m/s^2)

Note that the initial and final variables have been changed to a more commonly used notation. That is, v_0 and t_0 are the initial or original velocity and time, respectively, and v and t are the general velocity and time at some point in the future, such as when you want to know the velocity v at a particular time t. (This may or may not be the final velocity of a particular situation. There may be an acceleration after this time.)

From $\Delta v/\Delta t$, the SI units of acceleration can be seen to be meters per second (Δv) per second (Δt), that is, $(\text{m/s})/\text{s}$ or $\text{m/(s} \cdot \text{s)}$, which is commonly expressed as meters per second squared (m/s^2). In the British system, the units are feet per second squared (ft/s^2).

Because velocity is a vector quantity, so is acceleration, as acceleration represents a change in velocity. Since velocity has both magnitude and direction, a change in velocity may involve changes in either or both of these factors. Thus an acceleration may result from a change in *speed* (magnitude), a change in *direction*, or a change in *both*, as illustrated in ▸Fig. 2.9.

For straight-line, linear motion, plus and minus signs will be used to indicate the directions of velocity and acceleration, as was done for linear displacements. Eq. 2.5 is commonly simplified and written as

$$\bar{a} = \frac{v - v_0}{t} \tag{2.6}$$

where t_0 is taken to be zero. (v_0 may not be zero, so it cannot generally be omitted.)

Analogous to instantaneous velocity, **instantaneous acceleration** is the acceleration at a particular instant of time. This quantity is expressed mathematically as

$$a = \lim_{\Delta t \to 0} \frac{\Delta v}{\Delta t} \tag{2.7}$$

The conditions of the time interval approaching zero are the same here as described for instantaneous velocity.

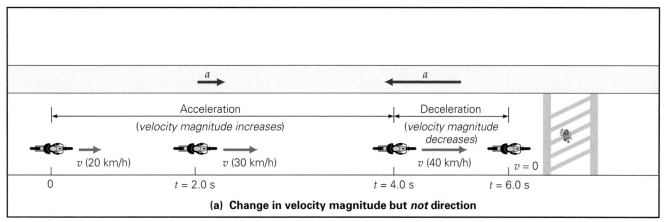

(a) Change in velocity magnitude but *not* direction

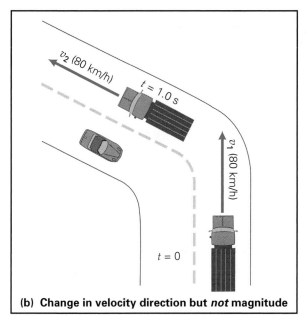

(b) Change in velocity direction but *not* magnitude

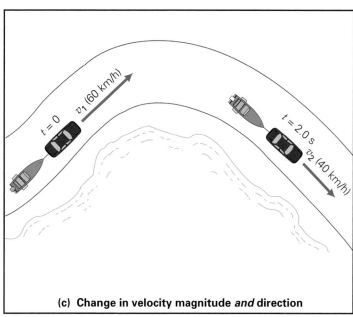

(c) Change in velocity magnitude *and* direction

▲ **FIGURE 2.9 Acceleration—the time rate of change of velocity** Since velocity is a vector quantity, with magnitude and direction, an acceleration can occur when there is **(a)** a change in magnitude, but not direction; **(b)** a change in direction, but not magnitude; or **(c)** a change in both magnitude and direction.

EXAMPLE 2.3 | ## Slowing It Down: Average Acceleration

A couple in a sport-utility vehicle (SUV) are traveling at 90 km/h on a straight highway. The driver sees an accident in the distance and slows down to 40 km/h in 5.0 s. What is the average acceleration of the SUV?

THINKING IT THROUGH. To find the average acceleration, the variables as defined in Eq. 2.6 must be given, and they are.

SOLUTION. Listing the data and converting units,

Given: $v_o = (90 \text{ km/h})\left(\dfrac{0.278 \text{ m/s}}{1 \text{ km/h}}\right)$
$\qquad = 25 \text{ m/s}$
$\qquad v = (40 \text{ km/h})\left(\dfrac{0.278 \text{ m/s}}{1 \text{ km/h}}\right)$
$\qquad = 11 \text{ m/s}$
$\qquad t = 5.0 \text{ s}$

Find: $\bar{a}$ (average acceleration)

[Here, the instantaneous velocities are assumed to be in the positive direction, and conversions to standard units (meters per second) are made right away, since it is noted that the speed is given in km/h. In general, standard units should be used.]

Given the initial and final velocities and the time interval, the average acceleration can be found by using Eq. 2.6:

$$\bar{a} = \frac{v - v_o}{t} = \frac{11 \text{ m/s} - (25 \text{ m/s})}{5.0 \text{ s}} = -2.8 \text{ m/s}^2$$

The minus sign indicates the direction of the (vector) acceleration. In this case, the acceleration is opposite to the direction of the motion and the car slows. Such an acceleration is sometimes called a *deceleration*, since the car is slowing. (*Note:* This is why v_o cannot arbitrarily be set to zero, because as shown here there may be motion, and $v_o \neq 0$ at $t_o = 0$.)

FOLLOW-UP EXERCISE. Does a negative acceleration necessarily mean that a moving object is slowing down (decelerating) or that its speed is decreasing? [*Hint:* See the accompanying Learn by Drawing 2.2, Signs of Velocity and Acceleration.] (*Answers to all Follow-Up Exercises are given in Appendix VI at the back of the book.*)

CONSTANT ACCELERATION

Although acceleration can vary with time, our study of motion will generally be restricted to constant accelerations for simplicity. (An important constant acceleration is the acceleration due to gravity near the Earth's surface, which will be considered in the next section.) Since for a constant acceleration, the average acceleration is equal to the constant value ($\bar{a} = a$), the bar over the acceleration in Eq. 2.6 may be omitted. Thus, for a constant acceleration, the equation relating velocity, acceleration, and time is commonly written (rearranging Eq. 2.6) as follows:

$$v = v_0 + at \quad \textit{(constant acceleration only)} \tag{2.8}$$

(Note that the at term represents the *change* in velocity, since $at = v - v_0 = \Delta v$.)

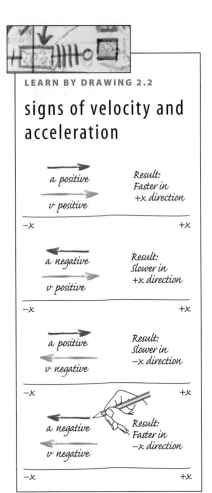

EXAMPLE 2.4

Fast Start, Slow Stop: Motion with Constant Acceleration

A drag racer starting from rest accelerates in a straight line at a constant rate of 5.5 m/s^2 for 6.0 s. (a) What is the racer's velocity at the end of this time? (b) If a parachute deployed at this time causes the racer to slow down uniformly at a rate of 2.4 m/s^2, how long will the racer take to come to a stop?

THINKING IT THROUGH. The racer first speeds up and then slows down, so close attention must be given to the directional signs of the vector quantities. Choose a coordinate system with the positive direction in the direction of the initial velocity. (Draw a sketch of the situation for yourself.) The answers can then be found by using the appropriate equations. Note that there are two different parts to the motion, with two different accelerations. Let's distinguish these phases with subscripts of 1 and 2.

SOLUTION. Taking the initial motion to be in the positive direction, we have the following data:

Given: (a) $v_0 = 0$ (at rest)
$a_1 = 5.5$ m/s^2
$t_1 = 6.0$ s

(b) $v_0 = v_1$ [from part (a)]
$v_2 = 0$ (comes to stop)
$a_2 = -2.4$ m/s^2 (opposite direction of v_0)

Find: (a) v_1 (final velocity for first part of the motion)

(b) t_2 (time for second part of the motion)

The data have been listed in two parts. This practice helps avoid confusion with symbols. Note that the final velocity v_1 that is to be found in part (a) becomes the initial velocity v_0 for part (b).

(a) To find the final velocity v_1, Eq. 2.8 may be used directly:

$$v_1 = v_0 + a_1t_1 = 0 + (5.5 \text{ m/s}^2)(6.0 \text{ s}) = 33 \text{ m/s}$$

(b) Here we want to find time, so solving Eq. 2.6 for t_2 and using $v_0 = v_1 = 33$ m/s from part (a),

$$t_2 = \frac{v_2 - v_0}{a_2} = \frac{0 - (33 \text{ m/s})}{-2.4 \text{ m/s}^2} = 14 \text{ s}$$

Note that the time comes out positive, as it should. Why?

FOLLOW-UP EXERCISE. What is the racer's instantaneous velocity 10 s after the parachute is deployed? *(Answers to all Follow-Up Exercises are given in Appendix VI at the back of the book.)*

Motions with constant accelerations are easy to represent graphically by plotting instantaneous velocity versus time. In this case, the *v*-versus-*t* plot is a straight line, the slope of which is equal to the acceleration, as illustrated in ▶Fig. 2.10. Note that Eq. 2.8 can be written as $v = at + v_0$, which, as you may recognize, has the form of an equation of a straight line, $y = mx + b$ (slope m and intercept b).

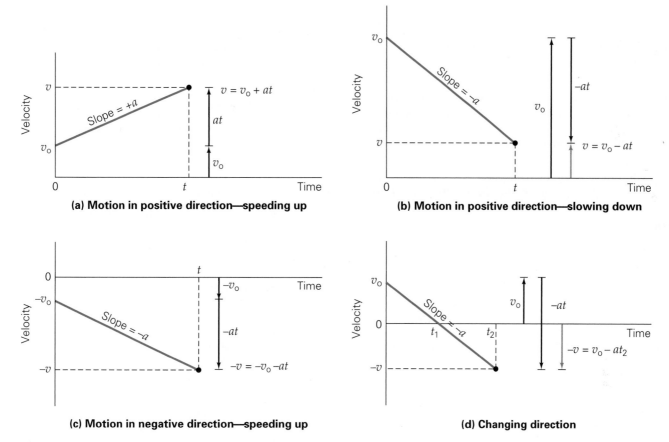

▲ FIGURE 2.10 Velocity-versus-time graphs for motions with constant accelerations The slope of a v-versus-t plot is the acceleration. **(a)** A positive slope indicates an increase in the velocity in the positive direction. The vertical arrows to the right indicate how the acceleration adds velocity to the initial velocity, v_0. **(b)** A negative slope indicates a decrease in the initial velocity, v_0, or a deceleration. **(c)** Here a negative slope indicates a negative acceleration, but the initial velocity is in the negative direction, $-v_0$, so the speed of the object increases in that direction. **(d)** The situation here is initially similar to that of part (b) but ends up resembling that in part (c). Can you explain what happened at time t_1?

In Fig. 2.10a, the motion is in the positive direction, and the acceleration term adds to the velocity after $t = 0$, as illustrated by the vertical arrows at the right of the graph. Here, the slope is positive ($a > 0$). In Fig. 2.10b, the negative slope ($a < 0$) indicates a negative acceleration that produces a slowing down, or deceleration. However, Fig. 2.10c illustrates how a negative acceleration can speed things up (for motion in the negative direction). The situation in Fig. 2.10d is slightly more complex. Can you explain what is happening there?

When an object moves at a constant acceleration, its velocity changes by the same amount in each equal time interval. For example, if the acceleration is 10 m/s^2 in the same direction as that of the initial velocity, the object's velocity increases by 10 m/s each second. As an example of this, suppose that the object has an initial velocity v_0 of 20 m/s in a particular direction at $t_0 = 0$. Then, for $t = 0, 1.0, 2.0, 3.0,$ and 4.0 s, the magnitudes of the velocities are 20, 30, 40, 50, and 60 m/s, respectively.

The average velocity may be computed in the regular manner (Eq. 2.3), or you may recognize that the uniformly increasing series of numbers 20, 30, 40, 50, and 60 has an average value of 40 (the midway value of the series) and $\bar{v} = 40$ m/s. Note that the average of the initial and final values also gives the average of the series— that is, $(20 + 60)/2 = 40$. Only when the velocity changes at a uniform rate because of a constant acceleration is $\bar{v}$ then the average of the initial and final velocities:

$$\bar{v} = \frac{v + v_0}{2} \quad \textit{(constant acceleration only)} \qquad (2.9)$$

EXAMPLE 2.5 | On the Water: Using Multiple Equations

A motorboat starting from rest on a lake accelerates in a straight line at a constant rate of 3.0 m/s² for 8.0 s. How far does the boat travel during this time?

THINKING IT THROUGH.   We have only one equation for distance (Eq. 2.3, $x = x_o + \bar{v}t$), but this equation cannot be used directly. The average velocity must first be found, so multiple equations and steps are involved.

SOLUTION.
Reading the problem, summarizing the given data, and identifying what is to be found (assuming the boat to accelerate in the +x-direction) gives the following:

Given: $x_o = 0$ *Find:* x (distance)
           $v_o = 0$
           $a = 3.0$ m/s²
           $t = 8.0$ s

(Note that all of the units are standard.)
    In analyzing the problem, one might reason: To find x, Eq. 2.3 needs to be used in the form $x = x_o + \bar{v}t$. (The average velocity $\bar{v}$ must be used because the velocity is changing and thus not constant.) With time given, the solution to the prob-

lem then involves finding $\bar{v}$. Then by Eq. 2.9, $\bar{v} = (v + v_o)/2$, and with $v_o = 0$, only the final velocity v is needed to solve the problem. Equation 2.8, $v = v_o + at$, can be used to calculate v from the given data. So it follows that:
    The velocity of the boat at the end of 8.0 s is

$$v = v_o + at = 0 + (3.0 \text{ m/s}^2)(8.0 \text{ s}) = 24 \text{ m/s}$$

The average velocity over that time interval is

$$\bar{v} = \frac{v + v_o}{2} = \frac{24 \text{ m/s} + 0}{2} = 12 \text{ m/s}$$

Finally, the magnitude of the displacement, which in this case is the same as the distance traveled, is given by Eq. 2.3 (choosing the boat's initial location at the origin, so $x_o = 0$):

$$x = \bar{v}t = (12 \text{ m/s})(8.0 \text{ s}) = 96 \text{ m}$$

FOLLOW-UP EXERCISE.   (Sneak preview.) In Section 2.4, the following equation will be derived: $x = v_o t + \frac{1}{2}at^2$. Use the data in this Example to see if this equation gives the distance traveled. (*Answers to all Follow-Up Exercises are given in Appendix VI at the back of the book.*)

DID YOU LEARN?
➡ A change in velocity is evidence of an acceleration.
➡ A deceleration requires a velocity in the opposite direction of the acceleration.
➡ A negative acceleration for motion in the negative direction increases the velocity.

2.4 Kinematic Equations (Constant Acceleration)

LEARNING PATH QUESTIONS
➡ How is the velocity affected for a constant acceleration?
➡ What is necessary for a moving object to come to a stop?

The description of motion in one dimension with constant acceleration requires only three basic equations. From previous sections, these equations are

$$x = x_o + \bar{v}t \tag{2.3}$$

$$\bar{v} = \frac{v + v_o}{2} \quad \textit{(constant acceleration only)} \tag{2.9}$$

$$v = v_o + at \quad \textit{(constant acceleration only)} \tag{2.8}$$

(Keep in mind that the first equation, Eq. 2.3, is general and is not limited to situations in which there is constant acceleration, as are the latter two equations.)
    However, as Example 2.5 showed, the description of motion in some instances requires multiple applications of these equations, which may not be obvious at first. It would be helpful if there were a way to reduce the number of operations in solving kinematic problems, and there is—by combining equations algebraically.
    For instance, suppose an expression that gives location x in terms of time and acceleration is wanted, rather than one in terms of time and average velocity (as in Eq. 2.3). Eliminating $\bar{v}$ from Eq. 2.3 by substituting for $\bar{v}$ from Eq. 2.9 into Eq. 2.3,

$$x = x_o + \bar{v}t \ \ (2.3) \qquad \text{and} \qquad \bar{v} = \frac{v + v_o}{2} \ \ (2.9)$$

and on substituting,

$$x = x_o + \tfrac{1}{2}(v + v_o)t \quad \textit{(constant acceleration only)} \tag{2.10}$$

Then, substituting for v from Eq. 2.8 ($v = v_o + at$) gives

$$x = x_o + \tfrac{1}{2}(v_o + at + v_o)t$$

Simplifying,

$$x = x_o + v_o t + \tfrac{1}{2}at^2 \quad \textit{(constant acceleration only)} \tag{2.11}$$

Essentially, this series of steps was done in Example 2.5. The combined equation allows the displacement of the motorboat in that Example to be computed directly:

$$x - x_o = \Delta x = v_o t + \tfrac{1}{2}at^2 = 0 + \tfrac{1}{2}(3.0 \text{ m/s}^2)(8.0 \text{ s})^2 = 96 \text{ m}$$

Much easier, isn't it?

We may want an expression that gives velocity as a function of position x rather than time (as in Eq. 2.8). In this case, t can be eliminated from Eq. 2.8 by using Eq. 2.10 in the form

$$v + v_o = 2\frac{(x - x_o)}{t}$$

Then, multiplying this equation by Eq. 2.8 in the form $(v - v_o) = at$ gives

$$(v + v_o)(v - v_o) = 2a(x - x_o)$$

and using the relationship $v^2 - v_o^2 = (v + v_o)(v - v_o)$,

$$v^2 = v_o^2 + 2a(x - x_o) \quad \textit{(constant acceleration only)} \tag{2.12}$$

PROBLEM-SOLVING HINT

Students in introductory physics courses are sometimes overwhelmed by the various kinematic equations. Keep in mind that equations and mathematics are the tools of physics. As any mechanic or carpenter will tell you, tools make your work easier as long as you are familiar with them and know how to use them. The same is true for physics tools. Summarizing the equations for linear motion with *constant* acceleration:

$$v = v_o + at \tag{2.8}$$
$$x = x_o + \tfrac{1}{2}(v + v_o)t \tag{2.10}$$
$$x = x_o + v_o t + \tfrac{1}{2}at^2 \tag{2.11}$$
$$v^2 = v_o^2 + 2a(x - x_o) \tag{2.12}$$

This set of equations is used to solve the majority of kinematic problems. (Occasionally there may be interest in average speed or velocity, and for that Eq. 2.3 can be used.)

Note that each of the equations in the list has four or five variables. All but one of the variables in an equation must be known in order to be able to solve for what you are trying to find. Generally, an equation with the unknown or wanted quantity is chosen. But, as pointed out, the other variables in the equation must be known. If they are not, then the wrong equation was chosen or another equation must be used to find the variables. (Another possibility is that not enough data are given to solve the problem, but that is not the case in this textbook.)

Always try to understand and visualize a problem. Listing the data as described in the suggested problem-solving procedure in Section 1.7 may help you decide which equation to use, by determining the known and unknown variables. Remember this approach as you work through the remaining Examples in the chapter. Also, don't overlook any *implied data or restrictive conditions*, as illustrated in the following examples.

CONCEPTUAL EXAMPLE 2.6 | ## Something Is Wrong!

A student working a problem with a constantly accelerating object wants to find v, and is given that $v_o = 0$ and $t = 3.0$ s, but is not given the acceleration a. He observes the kinematic equations and decides, using $v = at$ and $x = \frac{1}{2}at^2$ (with $x_o = v_o = 0$), that the unknown a can be eliminated. With $a = v/t$ and $a = 2x/t^2$ and equating,

$$\frac{v}{t} = \frac{2x}{t^2}$$

but x is not known, so he decides to use $x = vt$ to eliminate it, and

$$\frac{v}{t} = \frac{2vt}{t^2}$$

Simplifying,

$$v = 2v \quad \text{or} \quad 1 = 2!$$

What's wrong here?

REASONING AND ANSWER. Obviously something is big-time wrong, and it goes back to the problem-solving procedure given in Section 1.7. Step 4 there states: Determine which principle(s) and equation(s) are applicable to the situation. Since only equations were used, one equation must not apply to the situation. On inspection and analyzing, this can be seen to be $x = vt$, which applies only to *nonaccelerated* motion, and hence doesn't apply to the problem.

FOLLOW-UP EXERCISE. Given only v_o and t, is there any way to find v using the given kinematic equations? Explain. *(Answers to all Follow-Up Exercises are given in Appendix VI at the back of the book.)*

EXAMPLE 2.7 | ## Moving Apart: Where Are They Now?

Two riders on dune buggies sit 10 m apart on a long, straight track, facing in opposite directions. Starting at the same time, both riders accelerate at a constant rate of 2.0 m/s². How far apart will the dune buggies be at the end of 3.0 s?

THINKING IT THROUGH. The dune buggies are initially 10 m apart, so they can be positioned anywhere on the x-axis. It is convenient to place one at the origin so that one initial position (x_o) is zero. A sketch of the situation is shown in ▾Fig. 2.11.

SOLUTION. Listing the data:

Given: $x_{o_A} = 0$
$a_A = -2.0$ m/s²
$t = 3.0$ s
$x_{o_B} = 10$ m
$a_B = 2.0$ m/s²

Find: separation distance at $t = 3.0$ s

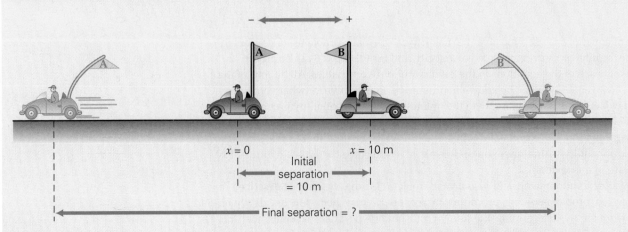

▲ FIGURE 2.11 Away they go! Two dune buggies accelerate away from each other. How far apart are they at a later time?

The displacement of each vehicle is given by Eq. 2.11 [the only displacement (Δx) equation with acceleration (a)]: $x = x_\mathrm{o} + v_\mathrm{o}t + \frac{1}{2}at^2$. But there is no v_o in the Given list. Some implied data must have been missed. It should be quickly noted that $v_\mathrm{o} = 0$ for both vehicles, so

$$x_\mathrm{A} = x_{\mathrm{o}_\mathrm{A}} + v_{\mathrm{o}_\mathrm{A}}t + \tfrac{1}{2}a_\mathrm{A}t^2$$
$$= 0 + 0 + \tfrac{1}{2}(-2.0 \text{ m/s}^2)(3.0 \text{ s})^2 = -9.0 \text{ m}$$

And for buggie B with a nonzero x_o,

$$x_\mathrm{B} = x_{\mathrm{o}_\mathrm{B}} + v_{\mathrm{o}_\mathrm{B}}t + \tfrac{1}{2}a_\mathrm{B}t^2$$
$$= 10 \text{ m} + 0 + \tfrac{1}{2}(2.0 \text{ m/s}^2)(3.0 \text{ s})^2 = 19 \text{ m}$$

Hence vehicle A is 9.0 m to the left of the origin on the $-x$-axis, whereas vehicle B is at a position of 19 m to the right of the origin on the $+x$-axis. And so, the separation distance between the two dune buggies is 19 m + 9 m = 28 m.

FOLLOW-UP EXERCISE.   Would it make any difference in the separation distance if vehicle B had been initially put at the origin instead of vehicle A? Try it and find out. (*Answers to all Follow-Up Exercises are given in Appendix VI at the back of the text.*)

EXAMPLE 2.8 | ## Putting On the Brakes: Vehicle Stopping Distance

The stopping distance of a vehicle is an important factor in road safety. This distance depends on the initial speed (v_o) and the braking capacity, which produces the deceleration, a, assumed to be constant. Express the stopping distance x in terms of these quantities.

THINKING IT THROUGH.   The signs of the velocity and acceleration are taken to be plus and minus respectively, indicating they are in opposite directions so the car comes to a stop. Again, a kinematic equation is required, and the appropriate one may be better determined by listing what is given and what is to be found. Notice that the distance x is wanted and time is not involved.

SOLUTION.   Here the quantities are variables and represented in symbol form:

Given: $+v_\mathrm{o}$
$-a$
$v = 0$ (car comes to stop)
$x_\mathrm{o} = 0$ (car taken to be initially at the origin)

Find: stopping distance x (in terms of the given variables)

Again, it is helpful to make a sketch of the situation, particularly when vector quantities are involved (▼Fig. 2.12). Since Eq. 2.12 has the variables we want, it should allow us to find the stopping distance x. Expressing the negative acceleration explicitly and assuming $x_\mathrm{o} = 0$ gives

$$v^2 = v_\mathrm{o}^2 - 2ax$$

Since the vehicle comes to a stop ($v = 0$), solving for x:

$$x = \frac{v_\mathrm{o}^2}{2a}$$

This equation gives x expressed in terms of the vehicle's initial speed and stopping acceleration. Notice that the

stopping distance x is proportional to the *square* of the initial speed. Doubling the initial speed therefore increases the stopping distance by a factor of 4 (for the same deceleration). That is, if the stopping distance is x_1 for an initial speed of v_1, then for a twofold increase in the initial speed ($v_2 = 2v_1$), the stopping distance would increase fourfold:

$$x_1 = \frac{v_1^2}{2a}$$

$$x_2 = \frac{v_2^2}{2a} = \frac{(2v_1)^2}{2a} = 4\left(\frac{v_1^2}{2a}\right) = 4x_1$$

The same result can be obtained by using ratios:

$$\frac{x_2}{x_1} = \frac{v_2^2}{v_1^2} = \left(\frac{v_2}{v_1}\right)^2 = 2^2 = 4$$

Do you think this consideration is important in setting speed limits, for example, in school zones? (The driver's reaction time should also be considered. A method for approximating a person's reaction time is given in Section 2.5.)

FOLLOW-UP EXERCISE.   Tests have shown that the Chevy Blazer has an average braking deceleration of 7.5 m/s^2, while that of a Toyota Celica is 9.2 m/s^2. Suppose these two vehicles are being driven down a straight, level road at 97 km/h (60 mi/h), with the Celica in front of the Blazer. A cat runs across the road ahead of them, and both drivers apply their brakes at the same time and come to safe stops (not hitting the cat). Assuming constant acceleration and the same reaction times for both drivers, what is the minimum safe tailgating distance for the Blazer so that there won't be a rear-end collision with the Celica when the two vehicles come to a stop? (*Answers to all Follow-Up Exercises are given in Appendix VI at the back of the book.*)

◀**FIGURE 2.12 Vehicle stopping distance** A sketch to help visualize the situation.

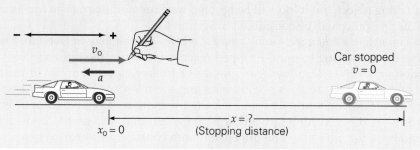

Car stopped
$v = 0$

$x = ?$
$x_\mathrm{o} = 0$ (Stopping distance)

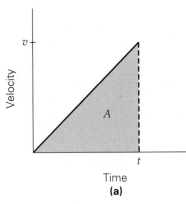

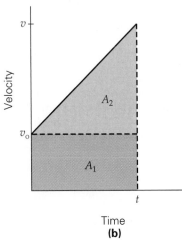

▲ **FIGURE 2.13** *v*-versus-*t* **graphs, one more time** **(a)** In the straight-line plot for a constant acceleration, the area under the curve is equal to Δx, the distance covered. **(b)** If v_0 is not zero, the distance is still given by the area under the curve Δx, but here divided into two parts, areas A_1 and A_2.

GRAPHICAL ANALYSIS OF KINEMATIC EQUATIONS

As was shown in Fig. 2.10, plots of v versus t give straight-line graphs where the slopes are values of the constant accelerations. There is another interesting aspect of *v*-versus-*t* graphs. Consider the one shown in ◄Fig. 2.13a, particularly the shaded area under the curve. Suppose we calculate the area of the shaded triangle, where, in general, $A = \frac{1}{2}ab \left[\text{Area} = \frac{1}{2}(\text{altitude})(\text{base}) \right]$.

For the graph in Fig. 2.13a, the altitude is v and the base is t, so $A = \frac{1}{2}vt$. But from the equation $v = v_0 + at$, we have $v = at$, where $v_0 = 0$ (zero intercept on graph). Therefore,

$$A = \tfrac{1}{2}vt = \tfrac{1}{2}(at)t = \tfrac{1}{2}at^2 = \Delta x$$

Hence, Δx is equal to the area under a *v*-versus-*t* curve.

Now look at Fig. 2.13b. Here, there is a nonzero value of v_0 at $t = 0$, so the object is initially moving. Consider the two shaded areas. We know that the area of the triangle is $A_2 = \frac{1}{2}at^2$, and the area of the rectangle can be seen (with $x_0 = 0$) to be $A_1 = v_0 t$. Adding these areas to get the total area yields

$$A_1 + A_2 = v_0 t + \tfrac{1}{2}at^2 = \Delta x$$

This is just Eq. 2.11, which is equal to the area under the *v*-versus-*t* curve.

> **DID YOU LEARN?**
> ➡ The velocity changes linearly as a function of time for a constant acceleration and gives a straight line on a *v* versus-*t* graph.
> ➡ An acceleration in the opposite direction of the motion is needed for a moving object to come to a stop.

2.5 Free Fall

LEARNING PATH QUESTIONS

➡ What is required for an object to be in free fall?
➡ What is different about free fall on the Moon compared to that on the Earth?
➡ A heavy object and a light object are in free fall, having been dropped from equal heights. Which object strikes the ground first?

One of the more common cases of constant acceleration is the acceleration due to gravity near the Earth's surface. When an object is dropped, its initial velocity (at the instant it is released) is zero. At a later time while falling, it has a nonzero velocity. There has been a change in velocity and thus, by definition, an acceleration. This **acceleration due to gravity (g)** near the Earth's surface has an approximate magnitude of

$$g = 9.80 \text{ m/s}^2 \quad \textit{(acceleration due to gravity)}$$

(or 980 cm/s^2) and is directed downward (toward the center of the Earth). In British units, the value of g is about 32.2 ft/s^2.

The values given here for g are only approximate because the acceleration due to gravity varies slightly at different locations as a result of differences in elevation and regional average mass densities of the Earth. These small variations will be ignored in this book unless otherwise noted. (Gravitation is studied in more detail in Section 7.5.) Air resistance is another factor that affects (reduces) the acceleration of a falling object, but it too will be ignored here for simplicity. (The frictional effect of air resistance will be considered in Section 4.6.)

Objects in motion solely under the influence of gravity are said to be in **free fall**. The words *free fall* may bring to mind dropped objects. However, the term applies to any motion under the *sole* influence of gravity. Objects released from rest, thrown upward or downward, are all in free fall once they are released. That is, after $t = 0$ (the time of release), only gravity is influencing the motion. (Even when an object projected upward is traveling upward, *it is still accelerating downward*.) Thus, the set of equations for motion in one dimension with constant acceleration given in the last section can be used to describe free fall.

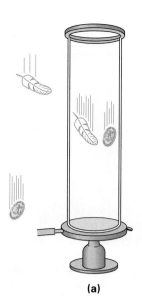
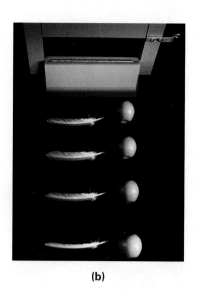

(a)   (b)

◀**FIGURE 2.14 Free fall and air resistance (a)** When dropped simultaneously from the same height, a feather falls more slowly than a coin, because of air resistance. But when both objects are dropped in an evacuated container with a good partial vacuum, where air resistance is negligible, the feather and the coin both have the same constant acceleration. **(b)** An actual demonstration with multiflash photography: An apple and a feather are released simultaneously through a trap door into a large vacuum chamber, and they fall together—almost. Because the chamber has only a partial vacuum, there is still some air resistance. (Can you tell?)

The acceleration due to gravity, g, has the same value for all free-falling objects, regardless of their mass or weight. It was once thought that heavier bodies accelerate faster than lighter bodies. This concept was part of Aristotle's theory of motion. You can easily observe that a coin accelerates faster than a sheet of paper when dropped simultaneously from the same height. But in this case, air resistance plays a noticeable role. If the paper is crumpled into a compact ball, it gives the coin a much better race. Similarly, a feather "floats" down much more slowly than a coin falls. However, in a near-vacuum, where there is negligible air resistance, the feather and the coin have the same acceleration—the acceleration due to gravity (▲Fig. 2.14).

Astronaut David Scott performed a similar experiment on the Moon in 1971 by simultaneously dropping a feather and a hammer from the same height. He did not need a vacuum pump. The Moon has no atmosphere and therefore no air resistance. The hammer and the feather reached the lunar surface together, but both had a smaller acceleration and fell at a slower rate than on Earth. The acceleration due to gravity near the Moon's surface is only about one-sixth of that near the Earth's surface ($g_M \approx g/6$).

Currently accepted ideas about the motion of falling bodies are due in large part to Galileo. He challenged Aristotle's theory and experimentally investigated the motion of such objects. Legend has it that Galileo studied the accelerations of falling bodies by dropping objects of different weights from the top of the Leaning Tower of Pisa. (See the accompanying Insight 2.1, Galileo Galilei and the Leaning Tower of Pisa.)

It is customary to use y to represent the vertical direction and to take upward as positive (as with the vertical y-axis of Cartesian coordinates). Because the acceleration due to gravity is always downward, it is in the negative y-direction. This negative acceleration, $a = -g = -9.80 \text{ m/s}^2$, should be substituted into the equations of motion. However, the relationship $a = -g$ may be expressed explicitly in the equations for linear motion for convenience:

$$v = v_0 - gt \tag{2.8'}$$

$$y = y_0 + v_0 t - \tfrac{1}{2}gt^2 \quad \begin{array}{l}\textit{(free-fall equations with}\\ a_y = -g \textit{ expressed explicity)}\end{array} \tag{2.11'}$$

$$v^2 = v_0^2 - 2g(y - y_0) \tag{2.12'}$$

Equation 2.10 applies to free fall as well, but it does not contain g:

$$y = y_0 + \tfrac{1}{2}(v + v_0)t \tag{2.10'}$$

INSIGHT 2.1   Galileo Galilei and the Leaning Tower of Pisa

Galileo Galilei (▶ Fig. 1) was born in Pisa, Italy, in 1564 during the Renaissance. Today, he is known throughout the world by his first name and is often referred to as the father of modern science and experimental physics, which attests to the magnitude of his scientific contributions.

One of Galileo's greatest contributions to science was the establishment of the scientific method—that is, investigation through experiment. In contrast, Aristotle's approach was based on deduction. By the scientific method, for a theory to be valid, it must correctly predict, or agree with, experimental results. If it doesn't, it is invalid or requires modification. Galileo said, "I think that in the discussion of natural problems we ought not to begin at the authority of places of Scripture, but at sensible experiments and necessary demonstrations."*

Probably the most popular and well-known legend about Galileo is that he performed experiments with falling bodies by dropping objects from the Leaning Tower of Pisa (▼ Fig. 2).

FIGURE 1 Galileo Galilei is alleged to have performed free-fall experiments by dropping objects off the Leaning Tower of Pisa.

There is some debate as to whether Galileo actually did this, but there is little doubt that he questioned Aristotle's view on the motion of falling objects. In 1638, Galileo wrote,

> Aristotle says that an iron ball of one hundred pounds falling from a height of one hundred cubits reaches the ground before a one-pound ball has fallen a single cubit. I say that they arrive at the same time. You find, on making the experiment, that the larger outstrips the smaller by two finger-breadths, that is, when the larger has reached the ground, the other is short of it by two finger-breadths; now you would not hide behind these two fingers the ninety-nine cubits of Aristotle.[†]

The experiments at the Tower of Pisa supposedly took place around 1590. In his writings of about that time, Galileo mentions dropping objects from a high tower, but never specifically names the Tower of Pisa. A letter written to Galileo from another scientist in 1641 describes the dropping of a cannonball and a musket ball from the Tower of Pisa. The first account of Galileo doing a similar experiment was written a dozen years after his death by Vincenzo Viviani, his last pupil and first biographer. It is not known whether Galileo told this story to Viviani in his declining years or Viviani created this picture of his former teacher.

The important point is that Galileo recognized (and probably experimentally showed) that free-falling objects fall with the same acceleration regardless of their mass or weight. (See Fig. 2.14.) Galileo gave no reason as to why all objects in free fall have the same acceleration, but Newton did, as will be learned in a later chapter.

FIGURE 2 The Leaning Tower of Pisa The tower, constructed as a belfry for a nearby cathedral, was built on shifting subsoil. Construction began in 1173, and the tower started to shift one way and then the other before inclining to its present direction. Today, the tower leans about 5 m (16 ft) from the vertical at the top. It was closed in 1990, and efforts were made to stabilize and correct the leaning. Some improvement was made and the tower is now open to the public.

*From *Growth of Biological Thought: Diversity, Evolution & Inheritance*, by F. Meyr (Cambridge, MA: Harvard University Press, 1982).

[†]From *Aristotle, Galileo, and the Tower of Pisa*, by L. Cooper (Ithaca, NY: Cornell University Press, 1935).

The origin ($y = 0$) of the frame of reference is usually taken to be at the initial position of the object. Writing $-g$ explicitly in the equations is a reminder of its direction. Then the value of g is simply inserted as 9.80 m/s^2.

The equations can also be written with $a = g$, for example, $v = v_0 + gt$, with the directional minus sign associated directly with g. In this case, a value of $g = -9.80$ m/s^2 must be substituted for g each time. However, either method works, and *the choice is arbitrary*. Your instructor may prefer one method over the other.

Note that you must be explicit about the directions of vector quantities. The location y and the velocities v and v_o may be positive (up) or negative (down), but the acceleration due to gravity is always downward.

The use of these equations and the sign convention (with $-g$ explicitly expressed in the equations) are illustrated in the following Examples. (This convention will be used throughout the text.)

EXAMPLE 2.9 | **A Stone Thrown Downward: The Kinematic Equations Revisited**

A boy on a bridge throws a stone vertically downward with an initial speed of 14.7 m/s toward the river below. If the stone hits the water 2.00 s later, what is the height of the bridge above the water?

THINKING IT THROUGH. This is a free-fall problem, but note that the initial velocity is downward, which is taken as the negative direction. It is important to express this factor explicitly. Draw a sketch to help you analyze the situation if needed.

SOLUTION. As usual, first writing what is given and what is to be found:

> *Given:* $v_o = -14.7$ m/s   *Find:* y (bridge height
> $t = 2.00$ s   above water)
> $g\ (= 9.80$ m/s$^2)$

Notice that g is listed as a positive number, since by our convention the directional minus sign has already been put into the previous equations of motion.

Which equation(s) will provide the solution using the given data? It should be evident that the distance the stone travels in an amount of time t is given directly by Eq. 2.11'. Taking $y_o = 0$:

$$y = v_o t - \tfrac{1}{2}gt^2$$
$$= (-14.7 \text{ m/s})(2.00 \text{ s}) - \tfrac{1}{2}(9.80 \text{ m/s}^2)(2.00 \text{ s})^2$$
$$= -29.4 \text{ m} - 19.6 \text{ m} = -49.0 \text{ m}$$

The minus sign indicates that the rock's displacement is downward, as it should be. Thus the height bridge is 49.0 m.

FOLLOW-UP EXERCISE. How much longer would it take for the stone to reach the river if the boy in this Example had dropped the ball rather than thrown it? *(Answers to all Follow-Up Exercises are given in Appendix VI at the back of the book.)*

Reaction time is the time it takes a person to notice, think, and act in response to a situation—for example, the time between first observing and then responding to an obstruction on the road ahead by applying the brakes. Reaction time varies with the complexity of the situation (and with the individual). In general, the largest part of a person's reaction time is spent thinking, but practice in dealing with a given situation can reduce this time. The following Example gives a simple method for measuring reaction time.

EXAMPLE 2.10 | **Measuring Reaction Time: Free Fall**

A person's reaction time can be measured by having another person drop a ruler (without warning) even with and through the first person's thumb and forefinger, as shown in ▸ Fig. 2.15. After observing the unexpected release, the first person grasps the falling ruler as quickly as possible, and the length of the ruler below the top of the finger is noted. Suppose the ruler descends 18.0 cm before it is caught. What is the person's reaction time?

THINKING IT THROUGH. Both distance and time are involved. This observation indicates which kinematic equation should be used.

▸ **FIGURE 2.15 Reaction time**
A person's reaction time can be measured by having the person grasp a dropped ruler.

(continued on next page)

SOLUTION. Notice that only the distance of fall is given. However, a couple of other things are known, such as v_o and g. So, taking $y_o = 0$:

Given: $y = -18.0$ cm $= -0.180$ m   *Find:* t (reaction time)

        $v_o = 0$

        $g \, (= 9.80 \text{ m/s}^2)$

(Note that the distance y has been converted to meters. Why?) It can be seen that Eq. 2.11'; applies here (with $v_o = 0$), giving

$$y = -\tfrac{1}{2}gt^2$$

Then solving for t,

$$t = \sqrt{\frac{2y}{-g}} = \sqrt{\frac{2(-0.180 \text{ m})}{-9.80 \text{ m/s}^2}} = 0.192 \text{ s}$$

Try this experiment with a fellow student and measure your reaction time. Why do you think another person besides you should drop the ruler?

FOLLOW-UP EXERCISE. A popular trick is to substitute a crisp dollar bill lengthwise for the ruler in Fig. 2.15, telling the person that he or she can have the dollar if able to catch it. Is this proposal a good deal? (The length of a dollar is 15.7 cm.) *(Answers to all Follow-Up Exercises are given in Appendix VI at the back of the book.)*

Here are some interesting facts about free-fall motion of an object thrown upward in the absence of air resistance. First, if the object returns to its launch elevation, the times of flight upward and downward are the same. Similarly, note that at the very top of the trajectory, the object's velocity is zero for an instant, but the acceleration (even at the top) remains a constant 9.8 m/s^2 downward. It is a common misconception that at the top of the trajectory the acceleration is zero. If this were the case, the object would remain there, as if gravity had been turned off!

Finally, the object returns to the starting point with the same speed as that at which it was launched. (The velocities have the same magnitude, but are opposite in direction.)

EXAMPLE 2.11 | ## Free Fall Up and Down: Using Implicit Data

A worker on a scaffold in front of a billboard throws a ball straight up. The ball has an initial speed of 11.2 m/s when it leaves the worker's hand at the top of the billboard (▶ Fig. 2.16). (a) What is the maximum height the ball reaches relative to the top of the billboard? (b) How long does it take the ball to reach this height? (c) What is the position of the ball at $t = 2.00$ s?

THINKING IT THROUGH. In part (a), only the upward part of the motion has to be considered. Note that the ball stops (zero instantaneous velocity) at the maximum height, which allows this height to be determined. In part (b), knowing the maximum height allows the determination of the upward time of flight. In part (c), the distance–time equation (Eq.2.11') applies for any time and therefore allows calculation of the position (y) of the ball relative to the launch point at $t = 2.00$ s.

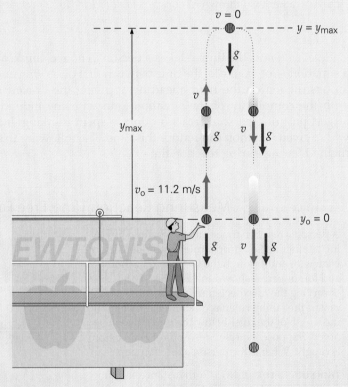

▶ **FIGURE 2.16 Free fall up and down** Note the lengths of the velocity and acceleration vectors at different times. (The upward and downward paths of the ball are horizontally displaced for illustration purposes.)

SOLUTION. It might appear that all that is given is the initial velocity v_o at time t_o. However, a couple of other pieces of information are implied that should be recognized. One, of course, is the acceleration g, and the other is the velocity at the maximum height where the ball stops. Here, in changing direction, the velocity of the ball is momentarily zero, so (again taking $y_o = 0$):

Given: $v_o = 11.2$ m/s   *Find:*   (a) y_{max} (maximum height above launch point)
         $g \, (= 9.80 \text{ m/s}^2)$ (b) t_u (time upward)
         $v = 0$ (at y_{max})   (c) y (at $t = 2.00$ s)
         $t = 2.00$ s [for part (c)]

(a) Notice that the height $(y_o = 0)$ is referenced to the top of the billboard. For this part of the problem, we need be concerned with only the upward motion—a ball is thrown upward and stops at its maximum height y_{max}. With $v = 0$ at this height, y_{max} may be found directly from Eq. 2.12′,

$$v^2 = 0 = v_o^2 - 2gy_{max}$$

So,

$$y_{max} = \frac{v_o^2}{2g} = \frac{(11.2 \text{ m/s})^2}{2(9.80 \text{ m/s}^2)} = 6.40 \text{ m}$$

relative to the top of the billboard $(y_o = 0$; see Fig. 2.16).

(b) The time the ball travels upward to its maximum height is designated t_u. This is the time it takes for the ball to reach y_{max}, where $v = 0$. Since v_o and v are known, the time t_u can be found directly from Eq. 2.8′,

$$v = 0 = v_o - gt_u$$

So,

$$t_u = \frac{v_o}{g} = \frac{11.2 \text{ m/s}}{9.80 \text{ m/s}^2} = 1.14 \text{ s}$$

(c) The height of the ball at $t = 2.00$ s is given directly by Eq. 2.11′:

$$y = v_o t - \tfrac{1}{2}gt^2$$
$$= (11.2 \text{ m/s})(2.00 \text{ s}) - \tfrac{1}{2}(9.80 \text{ m/s}^2)(2.00 \text{ s})^2$$
$$= 22.4 \text{ m} - 19.6 \text{ m} = 2.8 \text{ m}$$

Note that this height is 2.8 m above, or measured upward from, the reference point $(y_o = 0)$. The ball has reached its maximum height in 1.14 s and is on the way back down.

Considered from another reference point, the situation in part (c) can be analyzed by imagining dropping a ball from a height of y_{max} above the top of the billboard with $v_o = 0$ and asking how far it falls in a time $t = 2.00$ s $- t_u = 2.00$ s $-$ 1.14 s = 0.86 s. The answer is (this time with $y_o = 0$ at the maximum height)

$$y = v_o t - \tfrac{1}{2}gt^2 = 0 - \tfrac{1}{2}(9.80 \text{ m/s}^2)(0.86 \text{ s})^2 = -3.6 \text{ m}$$

This height is the same as the position found previously, but is measured with respect to the maximum height as the reference point; that is,

$$y_{max} - 3.6 \text{ m} = 6.4 \text{ m} - 3.6 \text{ m} = 2.8 \text{ m}$$

above the starting point.

FOLLOW-UP EXERCISE. At what height does the ball in this Example have a speed of 5.00 m/s? [*Hint*: The ball attains this height twice—once on the way up, and once on the way down.] (*Answers to all Follow-Up Exercises are given in Appendix VI at the back of the book.*)

PROBLEM-SOLVING HINT

When working vertical projectile problems involving motions up and down, it is often convenient to divide the problem into two parts and consider each part separately. As seen in Example 2.11, for the upward part of the motion, the velocity is zero at the maximum height. A quantity of zero usually simplifies the calculations. Similarly, the downward part of the motion is analogous to that of an object dropped from a height where the initial velocity can be taken as zero.

However, as Example 2.11 shows, the appropriate equations may be used directly for any position or time of the motion. For instance, note in part (c) that the height was found directly for a time *after* the ball had reached the maximum height. The velocity of the ball at that time could also have been found directly from Eq. 2.8′, $v = v_o - gt$.

Also, note that the initial position was consistently taken as $y_o = 0$. This assumption is taken for convenience when the situation involves only one object (then $y_o = 0$ at $t_o = 0$). Using this convention can save a lot of time in writing and solving equations.

The same is true with only one object in horizontal motion: You can usually take $x_o = 0$ at $t_o = 0$. There are a couple of exceptions to this case, however. The first is if the problem specifies the object to be initially located at a position other than $x_o = 0$, and the second is if the problem involves two objects, as in Example 2.7. In the latter case, if one object is taken to be initially at the origin, the other's initial position is not zero.

EXAMPLE 2.12 | Lunar Landing

A Lunar Lander makes a descent toward a level plain on the Moon. It descends slowly by using retro (braking) rockets. At a height of 6.0 m above the surface, the rockets are shut down with the Lander having a downward speed of 1.5 m/s. What is the speed of the Lander just before touching down?

THINKING IT THROUGH This appears to be analogous to a simple free-fall problem of throwing an object downward—and it is, but the situation takes place on the Moon. It was noted previously that the acceleration due to gravity on the Moon, g_M, is one-sixth of that on the Earth, g_E. (No problem with air resistance on the Moon—it has no atmosphere.)

(continued on next page)

SOLUTION.

Given: $y = -6.0$ m   *Find:* v (just before
  $v_o = -1.5$ m/s   touching down)
  $g_M = g_E/6$
  $= (9.8$ m/s$^2)/6$
  $= 1.6$ m/s^2

Then Eq. 2.12' can be used:
$$v^2 = v_o^2 - 2g_M\, y = (-1.5 \text{ m/s})^2 - 2(1.6 \text{ m/s}^2)(-6.0 \text{ m})$$
$$= 21 \text{ m}^2/\text{s}^2$$

So,
$$v^2 = 21 \text{ m}^2/\text{s}^2 \quad \text{and} \quad v = \sqrt{21 \text{ m}^2/\text{s}^2} = \pm 4.6 \text{ m/s}$$

This is the velocity, which we know is downward, so the negative root is selected, and $v = -4.6$ m/s, and the speed is 4.6 m/s.

FOLLOW-UP EXERCISE.   From the 6.0-m height, how long did the Lander's descent take? (*Answers to all Follow-Up Exercises are given in Appendix VI at the back of the book.*)

DID YOU LEARN?

➥ When gravitational attraction is the sole influence on an object, it is in free fall. This includes objects dropped from rest, projected upward, or thrown downward.

➥ The gravitational attraction on the Moon is one-sixth of that on the Earth, hence objects in free fall on the Moon fall more slowly than on the Earth.

➥ Heavy and light objects in free fall dropped from equal heights strike the ground at the same time because the acceleration due to gravity is the same for both.

PULLING IT TOGETHER | ## Learning Kinematics by Playing

A student drops a tennis ball from a dormitory window 13.0 m above the ground. At that instant, a student on the ground launches another ball straight up directly toward the dropped ball with an initial speed of 15.0 m/s in order to hit the dropped ball. The upward ball is launched from shoulder height of 1.0 m above the ground. (a) Sketch the location of each ball as a function of time. You should show two curves, labeled 1 for the dropped ball and 2 for the thrown ball. (b) How long does it take before they hit? (c) How far above the ground are the balls when they hit? (d) Which way is the thrown ball moving up or down when they hit, or is it at rest at that instant?

THINKING IT THROUGH.   This example uses kinematic equations in free-fall situations—down and up. (a) Since both balls experience constant acceleration, the curves will be parabolas, intersecting at some height and time to be determined from the mathematics. Sign conventions are crucial. (b) When the balls collide, their heights above the launch point are the same. Setting these equal should enable the determination of the time. (c) Once the time is known, the height of the balls can be determined from kinematics. (d) The sign of the thrown ball's velocity at this time will tell which way it is moving on collision.

SOLUTION.

Given: 13.0 m − 1.0 m = 12.0 m. (Initial distance between the balls.) The height of 1.0 m will be taken as $y_o = 0$ for convenience, making the window at $y = 12.0$ m, relative to this reference point.

  $v_{o_1} = 0$ (dropped ball initial velocity)
  $v_{o_2} = 15.0$ m/s (thrown ball initial velocity)
  $y_{o_1} = 12.0$ m (dropped ball initial location)
  $y_{o_2} = 0$ m (thrown ball initial location)

Find: (a) each ball's location variation with time
  (b) t (time to hit)
  (c) y (distance above ground when they hit)
  (d) direction of thrown ball at collision

(a) The dropped ball's trajectory has zero slope to start since its initial velocity is zero, and it is a downward curving parabola starting at $y = 12.0$ m (▸Fig. 2.17). The thrown ball also will be a downward-curving parabola, but due to its initial upward velocity, it starts with an upward slope starting at $y = 0$ m. The intersection point represents the balls' common location and the time when they collide.

(b) Each ball's location follows the general one-dimensional free fall equation $y = y_o + v_o t - \frac{1}{2}gt^2$. By putting in the numbers for each ball, an equation for each ball's location as a function of time is obtained. [All times are in seconds (s), velocities in meters per second (m/s) and accelerations in meters per second squared (m/s^2), but units are omitted here for convenience.]

$$y_1 = y_{o_1} + v_{o_1}t - \tfrac{1}{2}gt^2 = 12.0 + 0 - 4.90t^2$$

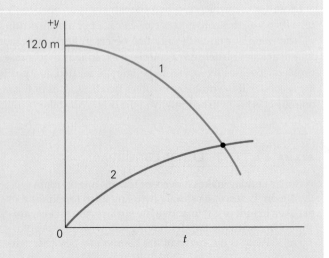

▲ **FIGURE 2.17   Ball location versus time.**

and
$$y_2 = y_{o_2} + v_{o_2}t - \tfrac{1}{2}gt^2 = 0 + 15.0t - 4.90t^2$$
Setting $y_1 = y_2$ yields $12.0 - 4.90t^2 = 15.0t - 4.90t^2$, which can be solved for t since it becomes $12.0 = 15.0t$. Thus the collision time is $t = 0.80$ s.

(c) Either location equation can be used to determine the height at collision. Ball 1 has a simpler equation, hence
$$y_1 = 12.0 + 0 - 4.90(0.80)^2 = 8.86 \text{ m}$$
(You should check to see that $y_2 = 8.86$ m also.)

Then the height above the ground is
$1.0 \text{ m} + y_1 = 1.0 \text{ m} + 8.86 \text{ m} = 9.9 \text{ m}$.

(d) The general equation for velocity in one-dimensional free fall is $v = v_o - gt$. Applying this to ball 2:
$$v_2 = v_{o_2} - gt = 15.0 - 9.80(0.80) = +7.16 \text{ m/s}$$
So from the plus sign for the velocity, it is clear that ball 2 is rising when the balls collide.

Learning Path Review

- **Motion** involves a change of position; it can be described in terms of the distance moved (a scalar) or the displacement (a vector).

- A **scalar** quantity has magnitude (value and units) only; a **vector** quantity has magnitude *and* direction.

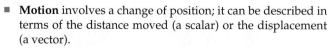
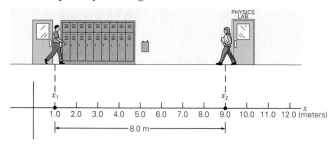

- **Average speed** ($\bar{s}$) (a scalar) is the distance traveled divided by the total time:
$$\text{average speed} = \frac{\text{distance traveled}}{\text{total time to travel that distance}}$$
$$\bar{s} = \frac{d}{\Delta t} = \frac{d}{t_2 - t_1} \tag{2.1}$$

- **Average velocity** (a vector) is the displacement divided by the total travel time:
$$\text{average velocity} = \frac{\text{displacement}}{\text{total travel time}}$$
$$\bar{v} = \frac{\Delta x}{\Delta t} = \frac{x_2 - x_1}{t_2 - t_1} \quad \text{or} \quad x = x_o + \bar{v}t \tag{2.3}$$

- **Instantaneous velocity** (a vector) describes how fast something is moving and in what direction at a particular instant of time.

- **Acceleration** is the time rate of change of velocity and hence is a vector quantity:
$$\text{average acceleration} = \frac{\text{change in velocity}}{\text{time to make the change}}$$

$$\bar{a} = \frac{\Delta v}{\Delta t} = \frac{v_2 - v_1}{t_2 - t_1} \tag{2.5}$$

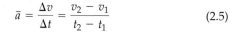

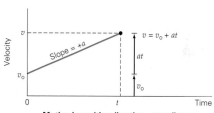

Motion in positive direction—speeding up

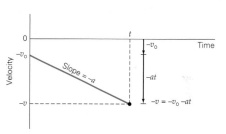

Motion in negative direction—speeding up

- **The kinematic equations for *constant* acceleration:**

$$\bar{v} = \frac{v + v_o}{2} \tag{2.9}$$

$$v = v_o + at \tag{2.8}$$

$$x = x_o + \tfrac{1}{2}(v + v_o)t \tag{2.10}$$

$$x = x_o + v_o t + \tfrac{1}{2}at^2 \tag{2.11}$$

$$v^2 = v_o^2 + 2a(x - x_o) \tag{2.12}$$

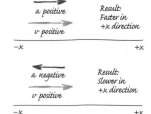

■ An object in **free fall** has a constant acceleration of magnitude $g = 9.80 \text{ m/s}^2$ (acceleration due to gravity) near the surface of the Earth.

■ Expressing $a = -g$ in the kinematic equations for constant acceleration in the y-direction yields the following:

$$v = v_\text{o} - gt \tag{2.8'}$$

$$y = y_\text{o} + \tfrac{1}{2}(v + v_\text{o})t \tag{2.10'}$$

$$y = y_\text{o} + v_\text{o}t - \tfrac{1}{2}gt^2 \tag{2.11'}$$

$$v^2 = v_\text{o}^2 - 2g(y - y_\text{o}) \tag{2.12'}$$

Learning Path Questions and Exercises
For instructor-assigned homework, go to www.masteringphysics.com

MULTIPLE CHOICE QUESTIONS

2.1 DISTANCE AND SPEED: SCALAR QUANTITIES
AND
2.2 ONE-DIMENSIONAL DISPLACEMENT AND VELOCITY: VECTOR QUANTITIES

1. A scalar quantity has (a) only magnitude, (b) only direction, (c) both magnitude and dirrection.

2. Which of the following is always true about the magnitude of the displacement: (a) It is greater than the distance traveled; (b) it is equal to the distance traveled; (c) it is less than the distance traveled; or (d) it is less than or equal to the distance traveled?

3. A vector quantity has (a) only magnitude, (b) only direction, (c) both direction and magnitude.

4. What can be said about average speed relative to the magnitude of the average velocity? (a) greater than, (b) equal to, (c) both a and b.

5. Distance is to displacement as (a) centimeters is to meters, (b) a vector is to a scalar, (c) speed is to velocity, (d) distance is to time.

2.3 ACCELERATION

6. On a position-versus-time plot for an object that has a constant acceleration, the graph is (a) a horizontal line, (b) a nonhorizontal and nonvertical straight line, (c) a vertical line, (d) a curve.

7. An acceleration may result from (a) an increase in speed, (b) a decrease in speed, (c) a change of direction, (d) all of the preceding.

8. A negative acceleration can cause (a) an increase in speed, (b) a decrease in speed, (c) either a or b.

9. The gas pedal of an automobile is commonly referred to as the *accelerator*. Which of the following might also be called an accelerator: (a) the brakes, (b) the steering wheel, (c) the gear shift, or (d) all of the preceding? Explain.

10. For a constant acceleration, what changes uniformly? (a) acceleration, (b) velocity, (c) displacement, (d) distance.

11. Which one of the following is true for a deceleration? (a) The velocity remains constant. (b) The acceleration is negative. (c) The acceleration is in the direction opposite to the velocity. (d) The acceleration is zero.

12. A car accelerates from 80 km/h to 90 km/h, while a moped accelerates from 0 to 20 km/h in twice the time. Which of the following is true: (a) The car has the greater acceleration; (b) the moped has the greater acceleration; or (c) they both have the same magnitude of acceleration?

2.4 KINEMATIC EQUATIONS (CONSTANT ACCELERATION)

13. For a constant linear acceleration, the velocity-versus-time graph is (a) a horizontal line, (b) a vertical line, (c) a nonhorizontal and nonvertical straight line, (d) a curved line.

14. For a constant linear acceleration, the position-versus-time graph would be (a) a horizontal line, (b) a vertical line, (c) a nonhorizontal and nonvertical straight line, (d) a curve.

15. An object accelerates uniformly from rest for t seconds. The object's average speed for this time interval is (a) $\tfrac{1}{2}at$, (b) $\tfrac{1}{2}at^2$, (c) $2at$, (d) $2at^2$.

2.5 FREE FALL

16. An object is thrown vertically upward. Which of the following statements is true: (a) Its velocity changes nonuniformly; (b) its maximum height is independent of the initial velocity; (c) its travel time upward is slightly greater than its travel time downward; or (d) its speed on returning to its starting point is the same as its initial speed?

17. The free-fall motion described in this section applies to (a) an object dropped from rest, (b) an object thrown vertically downward, (c) an object thrown vertically upward, (d) all of the preceding.

18. A dropped object in free fall (a) falls 9.8 m each second, (b) falls 9.8 m during the first second, (c) has an increase in speed of 9.8 m/s each second, (d) has an increase in acceleration of 9.8 m/s^2 each second.

19. An object is thrown straight upward. At its maximum height, (a) its velocity is zero, (b) its acceleration is zero, (c) both a and b.

20. When an object is thrown vertically upward, it is accelerating on (a) the way up, (b) the way down, (c) both a and b.

CONCEPTUAL QUESTIONS

2.1 DISTANCE AND SPEED: SCALAR QUANTITIES
AND
2.2 ONE-DIMENSIONAL DISPLACEMENT AND VELOCITY: VECTOR QUANTITIES

1. Can the displacement of a person's trip be zero, yet the distance involved in the trip be nonzero? How about the reverse situation? Explain.

2. You are told that a person has walked 750 m. What can you safely say about the person's final position relative to the starting point?

3. If the displacement of an object is 300 m north, what can you say about the distance traveled by the object?

4. Speed is the magnitude of velocity. Is average speed the magnitude of average velocity? Explain.

5. The average velocity of a jogger on a straight track is computed to be +5 km/h. Is it possible for the jogger's instantaneous velocity to be negative at any time during the jog? Explain.

2.3 ACCELERATION

6. A car is traveling at a constant speed of 60 mi/h on a circular track. Is the car accelerating? Explain.

7. Does a fast-moving object always have higher acceleration than a slower object? Give a few examples, and explain.

8. A classmate states that a negative acceleration always means that a moving object is decelerating. Is this statement true? Explain.

9. Describe the motions of the two objects that have the velocity-versus-time plots shown in ▼ Fig. 2.18.

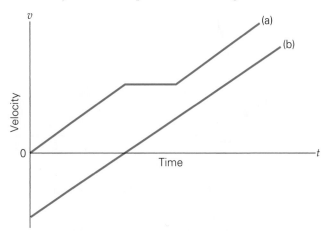

10. An object traveling at a constant velocity v_o experiences a constant acceleration in the same direction for a period of time t. Then an acceleration of equal magnitude is experienced in the opposite direction of v_o for the same period of time t. What is the object's final velocity?

11. Car A is in a straight-line distance d from a starting line, and Car B is a distance of $2d$ from the line. Accelerating uniformly from rest, it is desired that both cars cross the starting line at the same speed. If so, which car has the greater acceleration, and how much greater?

2.4 KINEMATIC EQUATIONS (CONSTANT ACCELERATION)

12. If an object's velocity-versus-time graph is a horizontal line, what can you say about the object's acceleration?

13. In solving a kinematic equation for x, which has a negative acceleration, is x necessarily negative?

14. How many variables must be known to solve a kinematic equation?

15. Consider Eq. 2.12, $v^2 = v_o^2 + 2a(x - x_o)$. An object starts from rest ($v_o = 0$) and accelerates. Since v is squared and therefore always positive, can the acceleration be negative? Explain.

2.5 FREE FALL

16. When a ball is thrown upward, what are its velocity and acceleration at its highest point?

17. If the instantaneous velocity of an object is zero, is the acceleration necessarily zero?

18. Imagine you are in space far away from any planet, and you throw a ball as you would on the Earth. Describe the ball's motion.

19. A person drops a stone from the window of a building. One second later, she drops another stone. How does the distance between the stones vary with time?

20. How would free fall on the Moon differ from that on the Earth?

▲ FIGURE 2.18 **Description of motion** See Conceptual Question 9.

EXERCISES

Integrated Exercises (IEs) are two-part exercises. The first part typically requires a conceptual answer choice based on physical thinking and basic principles. The following part is quantitative calculations associated with the conceptual choice made in the first part of the exercise.

Many exercise sections will include "paired" exercises. These exercise pairs, identified with **red numbers**, *are intended to assist you in problem solving and learning. In a pair, the first exercise (even numbered) is worked out in the Study Guide so that you can consult it should you need assistance in solving it. The second exercise (odd numbered) is similar in nature, and its answer is given at in Appendix VII the back of the book.*

2.1 DISTANCE AND SPEED: SCALAR QUANTITIES
AND
2.2 ONE-DIMENSIONAL DISPLACEMENT AND VELOCITY: VECTOR QUANTITIES

1. • What is the magnitude of the displacement of a car that travels half a lap along a circle that has a radius of 150 m? How about when the car travels a full lap?

2. • A motorist travels 80 km at 100 km/h, and 50 km at 75 km/h. What is the average speed for the trip?

3. • An Olympic sprinter can run 100 yd in 9.0 s. At the same rate, how long would it take the sprinter to run 100 m?

4. • A senior citizen walks 0.30 km in 10 min, going around a shopping mall. (a) What is her average speed in meters per second? (b) If she wants to increase her average speed by 20% when walking a second lap, what would her travel time in minutes have to be?

5. •• A hospital patient is given 500 cc of saline by IV. If the saline is received at a rate of 4.0 mL/min, how long will it take for the half liter to run out?

6. •• A hospital nurse walks 25 m to a patient's room at the end of the hall in 0.50 min. She talks with the patient for 4.0 min, and then walks back to the nursing station at the same rate she came. What was the nurse's average speed?

7. •• A train makes a round trip on a straight, level track. The first half of the trip is 300 km and is traveled at a speed of 75 km/h. After a 0.50 h layover, the train returns the 300 km at a speed of 85 km/h. What is the train's (a) average speed and (b) average velocity?

8. IE •• A car travels three-quarters of a lap on a circular track of radius R. (a) The magnitude of the displacement is (1) less than R, (2) greater than R, but less than 2R, (3) greater than 2R. (b) If R = 50 m, what is the magnitude of the displacement?

9. •• The interstate distance between two cities is 150 km. (a) If you drive the distance at the legal speed limit of 65 mi/h, how long would the trip take? (b) Suppose on the return trip you pushed it up to 80 mi/h (and didn't get caught). How much time would you save?

10. IE •• A race car travels a complete lap on a circular track of radius 500 m in 50 s. (a) The average velocity of the race car is (1) zero, (2) 100 m/s, (3) 200 m/s, (4) none of the preceding. Why? (b) What is the average speed of the race car?

11. IE •• A student runs 30 m east, 40 m north, and 50 m west. (a) The magnitude of the student's net displacement is (1) between 0 and 20 m, (2) between 20 m and 40 m, (3) between 40 m and 60 m. (b) What is his net displacement?

12. •• A student throws a ball vertically upward such that it travels 7.1 m to its maximum height. If the ball is caught at the initial height 2.4 s after being thrown, (a) what is the ball's average speed, and (b) what is its average velocity?

13. •• An insect crawls along the edge of a rectangular swimming pool of length 27 m and width 21 m (▼ Fig. 2.19). If it crawls from corner A to corner B in 30 min, (a) what is its average speed, and (b) what is the magnitude of its average velocity?

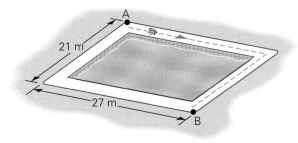

▲ **FIGURE 2.19 Speed versus velocity** See Exercise 13. (Not drawn to scale; insect is displaced for clarity.)

14. •• A plot of position versus time is shown in ▼ Fig. 2.20 for an object in linear motion. (a) What are the average velocities for the segments AB, BC, CD, DE, EF, FG, and BG? (b) State whether the motion is uniform or nonuniform in each case. (c) What is the instantaneous velocity at point D?

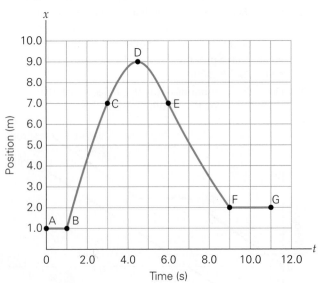

▲ **FIGURE 2.20 Position versus time** See Exercise 14.

15. •• In demonstrating a dance step, a person moves in one dimension, as shown in ▶ Fig. 2.21. What are (a) the average speed and (b) the average velocity for each phase of the motion? (c) What are the instantaneous

velocities at $t = 1.0$ s, 2.5 s, 4.5 s, and 6.0 s? (d) What is the average velocity for the interval between $t = 4.5$ s and $t = 9.0$ s? [*Hint:* Recall that the overall displacement is the displacement between the starting point and the ending point.]

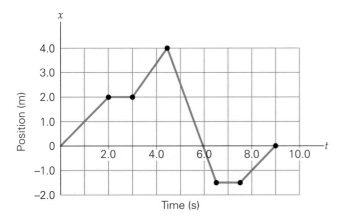

▲ **FIGURE 2.21 Position versus time** See Exercise 15.

16. •• A high school kicker makes a 30.0-yd field goal attempt (in American football) and hits the crossbar at a height of 10.0 ft. (a) What is the net displacement of the football from the time it leaves the ground until it hits the crossbar? (b) Assuming the football took 2.50 s to hit the crossbar, what was its average velocity? (c) Explain why you *cannot* determine its average speed from these data.

17. •• The location of a moving particle at a particular time is given by $x = at - bt^2$, where $a = 10$ m/s and $b = 0.50$ m/s^2. (a) Where is the particle at $t = 0$? (b) What is the particle's displacement for the time interval $t_1 = 2.0$ s and $t_2 = 4.0$ s?

18. •• The displacement of an object is given as a function of time by $x = 3t^2$ m. What is the magnitude of the average velocity for (a) $\Delta t = 2.0$ s $- 0$ s, and (b) $\Delta t = 4.0$ s $- 2.0$ s?

19. •• Short hair grows at a rate of about 2.0 cm/month. A college student has his hair cut to a length of 1.5 cm. He will have it cut again when the length is 3.5 cm. How long will it be until his next trip to the barber shop?

20. ••• A student driving home for the holidays starts at 8:00 AM to make the 675-km trip, practically all of which is on nonurban interstate highways. If she wants to arrive home no later than 3:00 PM, what must be her minimum average speed? Will she have to exceed the 65-mi/h speed limit?

21. ••• A regional airline flight consists of two legs with an intermediate stop. The airplane flies 400 km due north from airport A to airport B. From there, it flies 300 km due east to its final destination at airport C. (a) What is the plane's displacement from its starting point? (b) If the first leg takes 45 min and the second leg 30 min, what is the average velocity for the trip? (c) What is the average speed for the trip? (d) Why is the average speed not the same as the magnitude for the average velocity?

22. ••• Two runners approaching each other on a straight track have constant speeds of 4.50 m/s and 3.50 m/s, respectively, when they are 100 m apart (▶Fig. 2.22).

How long will it take for the runners to meet, and at what position will they meet if they maintain these speeds?

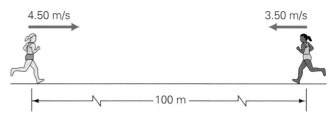

▲ **FIGURE 2.22 When and where do they meet?** See Exercise 22.

2.3 ACCELERATION

23. • An automobile traveling at 15.0 km/h along a straight, level road accelerates to 65.0 km/h in 6.00 s. What is the magnitude of the auto's average acceleration?

24. • A sports car can accelerate from 0 to 60 mi/h in 3.9 s. What is the magnitude of the average acceleration of the car in meters per second squared?

25. • If the sports car in Exercise 24 can accelerate at a rate of 7.2 m/s^2, how long does the car take to accelerate from 0 to 60 mi/h?

26. IE •• A couple is traveling by car down a straight highway at 40 km/h. They see an accident in the distance, so the driver applies the brakes, and in 5.0 s the car uniformly slows down to rest. (a) The direction of the acceleration vector is (1) in the same direction as, (2) opposite to, (3) at 90° relative to the velocity vector. Why? (b) By how much must the velocity change each second from the start of braking to the car's complete stop?

27. •• A paramedic drives an ambulance at a constant speed of 75 km/h on a straight street for ten city blocks. Because of heavy traffic, the driver slows to 30 km/h in 6.0 s and travels two more blocks. What was the average acceleration of the vehicle?

28. •• During liftoff, a hot-air balloon accelerates upward at a rate of 3.0 m/s^2. The balloonist drops an object over the side of the gondola when the speed is 15 m/s. (a) What is the object's acceleration after it is released (relative to the ground)? (b) How long does it take to hit the ground?

29. •• A new-car owner wants to show a friend how fast her sports car is. The friend gets in his car and drives down a straight, level highway at a constant speed of 60 km/h to a point where the sports car is waiting. As the friend's car just passes, the sports car accelerates at a rate of 2.0 m/s^2. (a) How long does it take for the sports car to catch up to the friend's car? (b) How far down the road does the sports car catch up to the friend's car? (c) How fast is the sports car going at this time?

30. •• After landing, a jetliner on a straight runway taxis to a stop at an average velocity of -35.0 km/h. If the plane takes 7.00 s to come to rest, what are the plane's initial velocity and acceleration?

31. •• A train on a straight, level track has an initial speed of 35.0 km/h. A uniform acceleration of 1.50 m/s^2 is applied while the train travels 200 m. (a) What is the speed of the train at the end of this distance? (b) How long did it take for the train to travel the 200 m?

32. ●● A hockey puck sliding along the ice to the left hits the boards head-on with a speed of 35 m/s. As it reverses direction, it is in contact with the boards for 0.095 s, before rebounding at a slower speed of 11 m/s. Determine the average acceleration the puck experienced while hitting the boards. Typical car accelerations are 5.0 m/s^2. Comment on the size of your answer, and why it is so different from this value, especially when the puck speeds are similar to car speeds.

33. ●● What is the acceleration for each graph segment in ▼Fig. 2.23? Describe the motion of the object over the total time interval.

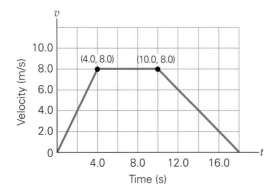

▲ **FIGURE 2.23 Velocity versus time** See Exercises 33 and 51.

34. ●● ▼Figure 2.24 shows a plot of velocity versus time for an object in linear motion. (a) Compute the acceleration for each phase of motion. (b) Describe how the object moves during the last time segment.

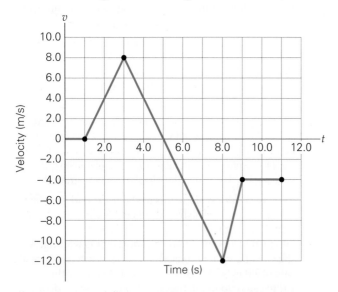

▲ **FIGURE 2.24 Velocity versus time** See Exercises 34 and 55.

35. ●● A car initially traveling to the right at a steady speed of 25 m/s for 5.0 s applies its brakes and slows at a constant rate of 5.0 m/s^2 for 3.0 s. It then continues traveling to the right at a steady but slower speed with no additional braking for another 6.0 s. (a) To help with the calculations, make a sketch of the car's velocity versus time, being sure to show all three time intervals. (b) What is its velocity after the 3.0 s of braking? (c) What was its displacement during the total 14.0 s of its motion? (d) What was its average speed for the 14.0 s?

36. ●●● A train normally travels at a uniform speed of 72 km/h on a long stretch of straight, level track. On a particular day, the train must make a 2.0-min stop at a station along this track. If the train decelerates at a uniform rate of 1.0 m/s^2 and, after the stop, accelerates at a rate of 0.50 m/s^2, how much time is lost because of stopping at the station?

2.4 KINEMATIC EQUATIONS (CONSTANT ACCELERATION)

37. ● At a sports car rally, a car starting from rest accelerates uniformly at a rate of 9.0 m/s^2 over a straight-line distance of 100 m. The time to beat in this event is 4.5 s. Does the driver beat this time? If not, what must the minimum acceleration be to do so?

38. ● A car accelerates from rest at a constant rate of 2.0 m/s^2 for 5.0 s. (a) What is the speed of the car at the end of that time? (b) How far does the car travel in this time?

39. ● A car traveling at 25 mi/h is to stop on a 35-m-long shoulder of the road. (a) What is the required magnitude of the minimum acceleration? (b) How much time will elapse during this minimum deceleration until the car stops?

40. ● A motorboat traveling on a straight course slows uniformly from 60 km/h to 40 km/h in a distance of 50 m. What is the boat's acceleration?

41. ●● The driver of a pickup truck going 100 km/h applies the brakes, giving the truck a uniform deceleration of 6.50 m/s^2 while it travels 20.0 m. (a) What is the speed of the truck in kilometers per hour at the end of this distance? (b) How much time has elapsed?

42. ●● A roller coaster car traveling at a constant speed of 20.0 m/s on a level track comes to a straight incline with a constant slope. While going up the incline, the car has a constant acceleration of 0.750 m/s^2 in magnitude. (a) What is the speed of the car at 10.0 s on the incline? (b) How far has the car traveled up the incline at this time?

43. ●● A rocket car is traveling at a constant speed of 250 km/h on a salt flat. The driver gives the car a reverse thrust, and the car experiences a continuous and constant deceleration of 8.25 m/s^2. How much time elapses until the car is 175 m from the point where the reverse thrust is applied? Describe the situation for your answer.

44. ●● Two identical cars capable of accelerating at 3.00 m/s^2 are racing on a straight track with running starts. Car A has an initial speed of 2.50 m/s; car B starts with speed of 5.00 m/s. (a) What is the separation of the two cars after 10 s? (b) Which car is moving faster after 10 s?

45. ●● According to Newton's laws of motion (which will be studied in Chapter 4), a frictionless 30° incline should provide an acceleration of 4.90 m/s^2 down the incline. A student with a stopwatch finds that an object, starting from rest, slides down a 15.00-m very smooth incline in exactly 3.00 s. Is the incline frictionless?

46. IE ●● An object moves in the $+x$-direction at a speed of 40 m/s. As it passes through the origin, it starts to experience a constant acceleration of 3.5 m/s^2 in the $-x$-direction. (a) What will happen next? (1) The object will reverse its direction of travel at the origin; (2) the object will keep traveling in the $+x$-direction; (3) the object will travel in the $+x$-direction and then reverses its direction. Why? (b) How much time elapses before the object returns to the origin? (c) What is the velocity of the object when it returns to the origin?

47. ●● A rifle bullet with a muzzle speed of 330 m/s is fired directly into a special dense material that stops the bullet in 25.0 cm. Assuming the bullet's deceleration to be constant, what is its magnitude?

48. ●● The speed limit in a school zone is 40 km/h (about 25 mi/h). A driver traveling at this speed sees a child run onto the road 13 m ahead of his car. He applies the brakes, and the car decelerates at a uniform rate of 8.0 m/s^2. If the driver's reaction time is 0.25 s, will the car stop before hitting the child?

49. ●● Assuming a reaction time of 0.50 s for the driver in Exercise 48, will the car stop before hitting the child?

50. ●● A bullet traveling horizontally at a speed of 350 m/s hits a board perpendicular to its surface, passes through and emerges on the other side at a speed of 210 m/s. If the board is 4.00 cm thick, how long does the bullet take to pass through it?

51. ●● (a) Show that the area under the curve of a velocity-versus-time plot for a constant acceleration is equal to the displacement. [*Hint*: The area of a triangle is $ab/2$, or one-half the altitude times the base.] (b) Compute the distance traveled for the motion represented by Fig. 2.23.

52. IE ●● An object initially at rest experiences an acceleration of 2.00 m/s^2 on a level surface. Under these conditions, it travels 6.00 m. Let's designate the first 3.00 m as phase 1 with a subscript of 1 for those quantities, and the second 3.00 m as phase 2 with a subscript of 2. (a) The times for traveling each phase should be related by which condition: (1) $t_1 < t_2$, (2) $t_1 = t_2$, or (3) $t_1 > t_2$? (b) Now calculate the two travel times and compare them quantitatively.

53. IE ●● A car initially at rest experiences loss of its parking brake and rolls down a straight hill with a constant acceleration of 0.850 m/s^2, traveling a total of 100 m. Let's designate the first half of the distance as phase 1 with a subscript of 1 for those quantities, and the second half as phase 2 with a subscript of 2. (a) The car's speeds at the end of each phase should be related by which condition (1) $v_1 < \frac{1}{2}v_2$, (2) $v_1 = \frac{1}{2}v_2$, or (3) $v_1 > \frac{1}{2}v_2$? (b) Now calculate the two speeds and compare them quantitatively.

54. ●● An object initially at rest experiences an acceleration of 1.5 m/s^2 for 6.0 s and then travels at that constant velocity for another 8.0 s. What is the object's average velocity over the 14-s interval?

55. ●●● Figure 2.24 shows a plot of velocity versus time for an object in linear motion. (a) What are the instantaneous velocities at $t = 8.0$ s and $t = 11.0$ s? (b) Compute the final displacement of the object. (c) Compute the total distance the object travels.

56. IE ●●● (a) A car traveling at a speed of v can brake to an emergency stop in a distance x. Assuming all other driving conditions are similar, if the traveling speed of the car doubles, the stopping distance will be (1) $\sqrt{2}x$, (2) $2x$, (3) $4x$. (b) A driver traveling at 40.0 km/h in a school zone can brake to an emergency stop in 3.00 m. What would be the braking distance if the car were traveling at 60.0 km/h?

57. ●●● A car accelerates horizontally from rest on a level road at a constant acceleration of 3.00 m/s^2. Down the road, it passes through two photocells ("electric eyes") designated by 1 for the first one and 2 for the second one) that are separated by 20.0 m. The time interval to travel this 20.0-m distance as measured by the electric eyes is 1.40 s. (a) Calculate the speed of the car as it passes *each* electric eye. (b) How far is it from the start to the first electric eye? (c) How long did it take the car to get to the first electric eye?

58. ●●● An automobile is traveling on a long, straight highway at a steady 75.0 mi/h when the driver sees a wreck 150 m ahead. At that instant, she applies the brakes (ignore reaction time). Between her and the wreck are two different surfaces. First there is 100 m of ice, where the deceleration is only 1.00 m/s^2. From then on, it is dry concrete, where the deceleration is a more normal 7.00 m/s^2. (a) What was the car's speed just after leaving the icy portion of the road? (b) What is the total distance her car travels before it comes to a stop? (c) What is the total time it took the car to stop?

2.5 FREE FALL

59. ● A student drops a ball from the top of a tall building; the ball takes 2.8 s to reach the ground. (a) What was the ball's speed just before hitting the ground? (b) What is the height of the building?

60. IE ● The time it takes for an object dropped from the top of cliff A to hit the water in the lake below is twice the time it takes for another object dropped from the top of cliff B to reach the lake. (a) The height of cliff A is (1) one-half, (2) two times, (3) four times that of cliff B. (b) If it takes 1.80 s for the object to fall from cliff A to the water, what are the heights of cliffs A and B?

61. ● For the motion of a dropped object in free fall, sketch the general forms of the graphs of (a) v versus t and (b) y versus t.

62. ● You can perform a popular trick by dropping a dollar bill (lengthwise) through the thumb and forefinger of a fellow student. Tell your fellow student to grab the dollar bill as fast as possible, and he or she can have the dollar if able to catch it. (The length of a dollar is 15.7 cm, and the average human reaction time is about 0.20 s. See Fig. 2.15.) Is this proposal a good deal? Justify your answer.

63. ● A juggler tosses a ball vertically a certain distance. How much higher must the ball be tossed so as to spend twice as much time in the air?

64. ● A boy throws a stone straight upward with an initial speed of 15.0 m/s. What maximum height will the stone reach before falling back down?

65. ● In Exercise 64, what would be the maximum height of the stone if the boy and the stone were on the surface of the Moon, where the acceleration due to gravity is only one-sixth of that of the Earth's?

66. ●● The Petronas Twin Towers in Malaysia and the Chicago Sears Tower have heights of about 452 m and 443 m, respectively. If objects were dropped from the top of each, what would be the difference in the time it takes the objects to reach the ground?

67. ●● In an air bag test, a car traveling at 100 km/h is remotely driven into a brick wall. Suppose an identical car is dropped onto a hard surface. From what height would the car have to be dropped to have the same impact as that with the brick wall?

68. ●● You throw a stone vertically upward with an initial speed of 6.0 m/s from a third-story office window. If the window is 12 m above the ground, find (a) the time the stone is in flight and (b) the speed of the stone just before it hits the ground.

69. IE ●● A Super Ball is dropped from a height of 4.00 m. Assuming the ball rebounds with 95% of its impact speed, (a) the ball would bounce to (1) less than 95%, (2) equal to 95%, or (3) more than 95% of the initial height? (b) How high will the ball go?

70. ●● In ▼Fig. 2.25, a student at a window on the second floor of a dorm sees his math professor walking on the sidewalk beside the building. He drops a water balloon from 18.0 m above the ground when the professor is 1.00 m from the point directly beneath the window. If the professor is 1.70 m tall and walks at a rate of 0.450 m/s, does the balloon hit her? If not, how close does it come?

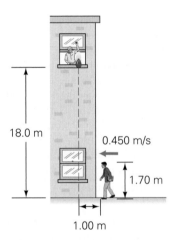

▲ **FIGURE 2.25 Hit the professor** See Exercise 70. (This figure is not drawn to scale.)

71. ●● A photographer in a helicopter ascending vertically at a constant rate of 12.5 m/s accidentally drops a camera out the window when the helicopter is 60.0 m above the ground. (a) How long will the camera take to reach the ground? (b) What will its speed be when it hits?

72. IE ●● The acceleration due to gravity on the Moon is about one-sixth of that on the Earth. (a) If an object were dropped from the same height on the Moon and on the Earth, the time it would take to reach the surface on the Moon is (1) $\sqrt{6}$, (2) 6, or (3) 36 times the time it would take on the Earth. (b) For a projectile with an initial velocity of 18.0 m/s upward, what would be the maximum height and the total time of flight on the Moon and on the Earth?

73. ●●● It takes 0.210 s for a dropped object to pass a window that is 1.35 m tall. From what height above the top of the window was the object released? (See ▶Fig. 2.26.)

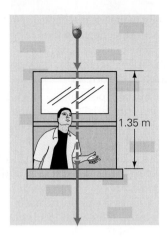

◀ **FIGURE 2.26**
From where did it come? See Exercise 73.

74. ●●● A tennis ball is dropped from a height of 10.0 m. It rebounds off the floor and comes up to a height of only 4.00 m on its first rebound. (Ignore the small amount of time the ball is in contact with the floor.) (a) Determine the ball's speed just before it hits the floor on the way down. (b) Determine the ball's speed as it leaves the floor on its way up to its first rebound height. (c) How long is the ball in the air from the time it is dropped until the time it reaches its maximum height on the first rebound?

75. ●●● A pollution-sampling rocket is launched straight upward with rockets providing a constant acceleration of 12.0 m/s² for the first 1000 m of flight. At that point the rocket motors cut off and the rocket itself is in free fall. Ignore air resistance. (a) What is the rocket's speed when the engines cut off? (b) What is the maximum altitude reached by this rocket? (c) What is the time it takes to get to its maximum altitude?

76. ●●● A test rocket containing a probe to determine the composition of the upper atmosphere is fired vertically upward from an initial position at ground level. During the time t while its fuel lasts, the rocket ascends with a constant upward acceleration of magnitude $2g$. Assume that the rocket travels to a small enough height that the Earth's gravitational force can be considered constant. (a) What are the speed and height, in terms of g and t, when the rocket's fuel runs out? (b) What is the maximum height the rocket reaches in terms of g and t? (c) If $t = 30.0$ s, calculate the rocket's maximum height.

77. ●●● A car and a motorcycle start from rest at the same time on a straight track, but the motorcycle is 25.0 m behind the car (▼Fig. 2.27). The car accelerates at a uniform rate of 3.70 m/s² and the motorcycle at a uniform rate of 4.40 m/s². (a) How much time elapses before the motorcycle overtakes the car? (b) How far will each have traveled during that time? (c) How far ahead of the car will the motorcycle be 2.00 s later? (Both vehicles are still accelerating.)

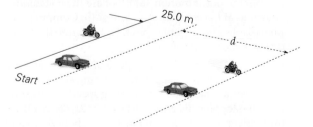

▲ **FIGURE 2.27 A tie race** See Exercise 77. (This figure is not drawn to scale.)

PULLING IT TOGETHER: MULTICONCEPT EXERCISES

The Multiconcept Exercises require the use of more than one fundamental concept for their understanding and solution. The concepts may be from this chapter, but may include those from previous chapters.

78. Two joggers run at the same average speed. Jogger A cuts directly north across the diameter of the circular track, while jogger B takes the full semicircle to meet his partner on the opposite side of the track. Assume their common average speed is 2.70 m/s and the track has a diameter of 150 m. (a) How many seconds ahead of jogger B does jogger A arrive? (b) How do their travel distances compare? (c) How do their displacements compare? (d) How do their average velocities compare?

79. Many highways with steep downhill areas have "runaway truck" inclined paths just off the main roadbed. These paths are designed so that if a vehicle's braking system gives out, the driver can steer it onto this incline (usually composed of loose gravel or sand). The idea is that the vehicle can then roll up the incline and come permanently and safely to rest with no need of a braking system. In one region of Hawaii the incline distance is 300 m and provides a (constant) deceleration of 2.50 m/s^2. (a) What is the maximum speed that a runaway vehicle can have as it enters the incline? (b) How long would such a vehicle take to come to rest? (c) Suppose another vehicle moving 10 mi/h (4.47 m/s) faster than the maximum value enters the incline. What speed will it have as it leaves the gravel-filled area?

80. The Taipei 101 Tower in Taipei, Taiwan is a 509-m (1667-ft), 101-story building (▼Fig. 2.28). The outdoor observation deck is on the 89th floor, and two high-speed elevators that service it reach a peak speed of 1008 m/min on the way up and 610 m/min on the way down. Assuming these peak speeds are reached at the midpoint of the run and that the accelerations are constant for each leg of the runs, (a) what are the accelerations for the up and down runs? (b) How much longer is the trip down than the trip up?

▲ FIGURE 2.28 **A tall one** The Taipei 101 Tower in Taipei, Taiwan, is a tall building with 101 stories. It has a height of 509 m (1671 ft). See Exercise 80.

81. From street level, Superman spots Lois Lane in trouble—the evil villain, Lex Luthor, is dropping her from near the top of the Empire State Building. At that very instant, the Man of Steel starts upward at a constant acceleration to attempt a midair rescue of Lois. Assuming she was dropped from a height of 300 m and that Superman can accelerate straight upward at 15.0 m/s^2, determine (a) how far Lois falls before he catches her, (b) how long Superman takes to reach her, and (c) their speeds at the instant he reaches her. Comment on whether these speeds might be a danger to Lois, who, being a mere mortal, might get hurt running into the impervious Man of Steel if the speeds are too great.

82. In the 1960s there was a contest to find the car that could do the following two maneuvers (one right after the other) in the shortest *total* time: First, accelerate from rest to 100 mi/h (45.0 m/s), and then brake to a complete stop. (Ignore the reaction time correction that occurs between the speeding-up and slowing-down phases and assume that all accelerations are constant.) For several years, the winner was the "James Bond car," the Aston Martin. One year it won the contest when it took a *total* of only 15.0 seconds to perform these two tasks! Its braking acceleration (deceleration) was known to be an excellent 9.00 m/s^2. (a) Calculate the time it took during the braking phase. (b) Calculate the distance it traveled during the braking phase. (c) Calculate the car's acceleration during the speeding-up phase. (d) Calculate the distance it took to reach 100 mi/h.

83. Let's investigate a possible vertical landing on Mars that includes two segments: free fall followed by a parachute deployment. Assume the probe is close to the surface, so the Martian acceleration due to gravity is constant at 3.00 m/s^2. Suppose the lander is initially moving vertically downward at 200 m/s at a height of 20 000 m above the surface. Neglect air resistance during the free-fall phase. Assume it first free falls for 8000 m. (The parachute doesn't open until the lander is 12 000 m from the surface. See ▼Fig. 2.29.) (a) Determine the lander's speed at the end of the 8000-m free-fall drop. (b) At 12 000 m above the surface, the parachute deploys and the lander *immediately* begins to slow. If it can survive hitting the surface at speeds of up to 20.0 m/s, determine the minimum constant deceleration needed during this phase. (c) What is the total time taken to land from the original height of 20 000 m?

▶ FIGURE 2.29
Down she comes
See Exercise 83.

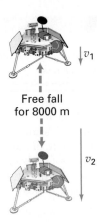

Free fall
for 8000 m

Parachute
slowdown for
the last 12 000 m

Just above the Martian surface

84. You are driving slowly in the right lane of a straight country road. For a while, a car to your left has lagged 50.0 m *behind you* at the same speed of 25.0 mi/h.

Suddenly that car speeds up and passes you, traveling at a constant acceleration until it is 40.0 m in front of you 7.00 s later. (a) Qualitatively sketch the location-versus-time graphs for both cars on the same axes, letting $t = 0$ be the start of the acceleration, and $x = 0$ be the location of the other car at that time. (b) Determine the other car's acceleration. (c) How far did each of you travel during the passing procedure? (d) What is the other car's speed at the end of the passing procedure?

85. A car is traveling on a straight, level road under wintry conditions. Seeing a patch of ice ahead of her, the driver of the car slams on her brakes and skids on dry pavement for 50 m, decelerating at 7.5 m/s². Then she hits the icy patch and skids another 80 m before coming to rest. If her initial speed was 70 mi/h, what was the deceleration on the ice?

86. On a water slide ride, you start from rest at the top of a 45.0-m-long incline (filled with running water) and accelerate down at 4.00 m/s². You then enter a pool of water and skid along the surface for 20.0 m before stopping. (a) What is your speed at the bottom of the incline? (b) What is the deceleration caused by the water in the pool? (c) What was the total time for you to stop? (d) How fast were you moving after skidding the first 10.0 m on the water surface?

87. A toy rocket is launched (from the ground) vertically upward with a constant acceleration of 30.0 m/s². After traveling 1000 m, its engines stop. When it reaches the very top of its motion, it falls for 0.500 s before a parachute deploys and it descends safely to the ground at the speed it has at that time. (a) What is the maximum altitude reached by the rocket? (b) How long does the rocket take to get to its maximum altitude? (c) How long does the total trip, from launch to ground impact, take?

88. A Superball is dropped from a height of 2.5 m and rebounds off the floor to a height of 2.1 m. If the ball is in contact with the floor for 0.70 ms, determine (a) the direction and (b) magnitude of the ball's average acceleration due to the floor.

3 Motion in Two Dimensions†

PHYSICS FACTS

◆ Word origins:
  – *kinematics*: from the Greek *kinema*, meaning "motion."
  – *velocity*: from the Latin *velocitas*, meaning "swiftness."
  – *acceleration*: from the Latin *accelerare*, meaning "hasten."

◆ Projectiles:
  – "Big Bertha," a gun used by the Germans in World War I, with a barrel length of 6.7 m (22 ft), could project an 820-kg (1800-lb) shell 15 km (9.3 mi).
  – The "Paris Gun," also used by the Germans in World War I, with a barrel length of 34 m (112 ft), could project a 120-kg (264-lb) shell 131 km (81 mi). Designed to bombard Paris, France, the shell reached a maximum height of 40 km (25 mi) during its 170-s trajectory.
  – A bullet fired from a high-powered rifle has a muzzle speed on the order of 2900 km/h (1800 mi/h).

You *can* get there from here! It's a matter of knowing which way to head at the crossroads (chapter-opening photo). But did you ever wonder why so many streets and roads meet at right angles? There's a good reason. Living on the Earth's surface, we are used to describing locations mainly in two dimensions, and one of the easiest ways to do this is by referring to a pair of mutually perpendicular axes. When you want to tell someone how to get to a particular place in the city, you might say, "Go four blocks on Main St., turn right onto Oak St. and go three more

†The mathematics needed in this chapter involves trigonometric functions. You may want to review these in Appendix I.

blocks." In the country, it might be "Go south for five miles and then another half mile east." In each case, one needs to know how far to go in each of *two* directions that are 90° apart.

The same approach can be used to describe motion—and the motion doesn't have to be in a straight line. As will be learned shortly, a generalized version of vectors introduced in Section 2.2 can be used to describe motion in curved paths as well. Such analysis of *curvilinear* motion will eventually allow you to analyze the behavior of batted balls, planets circling the Sun, and even the motions of electrons in atoms.

Two-dimensional curvilinear motion can be analyzed by using rectangular components of motion. Essentially, the curved motion is broken down or resolved into rectangular (*x* and *y*) components so the motion can be considered linearly in both dimensions. The kinematic equations introduced in Chapter 2 can be applied to these components. For an object moving in a curved path, for example, the *x*- and *y*-coordinates of its motion will give the object's position at any time.

3.1   Components of Motion

LEARNING PATH QUESTIONS

➡ How is motion in two dimensions described?

➡ What are the magnitudes of the components of velocity (*v*) in two dimensions?

➡ What is the major restriction for using the kinematic equations for components of motion?

In Section 2.1, an object moving in a straight line was considered to be moving along one of the Cartesian axes (*x* or *y*). But what if the motion is not along an axis? For example, consider the situation illustrated in ▶Fig. 3.1. Here, three balls are moving uniformly across a tabletop. The ball rolling in a straight line along the side of the table, designated as the *x*-direction, is moving in one dimension. That is, its motion can be described with a single coordinate, *x*. Similarly, the motion of the ball rolling along the end of the table in the *y*-direction can be described by a single *y*-coordinate. However, for this coordinate choice, both *x*- and *y*-coordinates are needed to describe the motion of the ball rolling diagonally across the table; that is, the motion is described *in two dimensions*.

You might observe that if the diagonally moving ball were the only object to consider, the *x*-axis could be chosen to be in the direction of that ball's motion, and the motion would thereby be reduced to one dimension. This observation is true, but once the coordinate axes are fixed, motions not along the axes must be described with two coordinates (*x*, *y*), or in two dimensions. Also, keep in mind that not all motions in a plane (two dimensions) are in straight lines. Think about the path of a ball you toss to another person. The path is curved for such projectile motion. (This motion will be considered in Section 3.3.) In general, both coordinates are needed.

In considering the motion of the ball moving diagonally across the table in Fig. 3.1a, it can be thought of as moving in the *x*- and *y*-directions simultaneously. That is, it has a velocity in the *x*-direction (v_x) *and* a velocity in the *y*-direction (v_y) at the same time. The combined velocity components describe the actual motion of the ball. If the ball has a constant velocity *v* in a direction at an angle θ relative to the *x*-axis, then the velocities in the *x*- and *y*-directions are obtained by resolving, or breaking down, the velocity vector into **components of motion** in these direc-

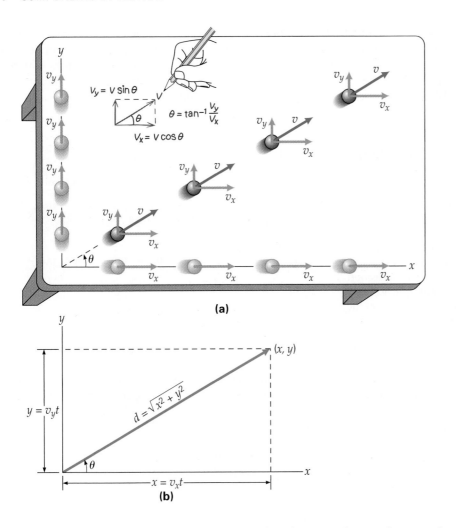

(a)

(b)

◀**FIGURE 3.1 Components of motion** (a) The velocity (and displacement) for uniform, straight-line motion—that of the dark purple ball—may have x- and y-components (v_x and v_y as shown in the pencil drawing), because of the chosen orientation of the coordinate axes. Note that the velocity and displacement of the ball in the x-direction are exactly the same as those that a ball rolling along the x-axis with a uniform velocity of v_x would have. A comparable relationship holds true for the ball's motion in the y-direction. Since the motion is uniform, the ratio v_y/v_x (and therefore θ) is constant. (b) The coordinates (x, y) of the ball's position and the distance d the ball has traveled from the origin can be found at any time t.

tions (see the pencil drawing in Fig. 3.1a). As this drawing shows, the v_x and v_y components have magnitudes of

$$v_x = v \cos \theta \qquad (3.1a)$$

and

$$v_y = v \sin \theta \qquad (3.1b)$$

(Notice that $v = \sqrt{v_x^2 + v_y^2}$, so v is a combination of the velocities in the x- and y-directions.)

You are familiar with the use of two-dimensional length components in finding the x- and y-coordinates in a Cartesian system. For the ball rolling on the table, its position (x, y), or the distance traveled from the origin in each of the component directions at time t, is given by (Eq. 2.11 with $a = 0$)

$$x = x_0 + v_x t \quad \textit{(magnitudes of displacement components} \\ \textit{under condition of constant velocity and} \qquad (3.2a) \\ \textit{zero acceleration)}$$

$$y = y_0 + v_y t \qquad (3.2b)$$

respectively. (Here, the x_0 and y_0 are the ball's coordinates at $t = 0$, which may be other than zero.) The ball's straight-line distance from the origin at any given time is then $d = \sqrt{x^2 + y^2}$ (Fig. 3.1b).

Note that $\tan \theta = v_y/v_x$ (see Fig. 3.1a.). So the direction of the motion relative to the x-axis is given by $\theta = \tan^{-1}(v_y/v_x)$. Also, $\theta = \tan^{-1}(y/x)$.

In this introduction to components of motion, the velocity vector has been taken to be in the first quadrant ($0 < \theta < 90°$), where both the x- and y-components are positive. But, as will be shown in more detail in the next section, vectors may be in any quadrant, and one or both of their components can be negative. Can you tell in which quadrants the v_x and v_y components would both be negative?

EXAMPLE 3.1 | On a Roll: Using Components of Motion

If the diagonally moving ball in Fig. 3.1a has a constant velocity of 0.50 m/s at an angle of 37° relative to the x-axis, find how far it travels in 3.0 s by using x- and y-components of its motion.

THINKING IT THROUGH. Given the magnitude and direction (angle) of the velocity of the ball, the x- and y-components of the velocity can be found. Then the distance in each direction can be computed. Since the x- and y-axes are at right angles to each other, the Pythagorean theorem gives the distance of the straight-line path of the ball, as shown in Fig. 3.1b. (Note the procedure: Separate the motion into components, calculate what is needed in each direction, and recombine if necessary.)

SOLUTION.   Listing the data,

Given: $v = 0.50$ m/s   *Find:* d (distance traveled)

$\theta = 37°$

$t = 3.0$ s

The distance traveled by the ball in terms of its x- and y-components is given by $d = \sqrt{x^2 + y^2}$. To find x and y as given by Eq. 3.2, we first need to compute the velocity components v_x and v_y (Eq. 3.1):

$$v_x = v \cos 37° = (0.50 \text{ m/s})(0.80) = 0.40 \text{ m/s}$$
$$v_y = v \sin 37° = (0.50 \text{ m/s})(0.60) = 0.30 \text{ m/s}$$

Then, taking $x_0 = 0$ and $y_0 = 0$, the component distances are

$$x = v_x t = (0.40 \text{ m/s})(3.0 \text{ s}) = 1.2 \text{ m}$$

and

$$y = v_y t = (0.30 \text{ m/s})(3.0 \text{ s}) = 0.90 \text{ m}$$

and the distance of the path is

$$d = \sqrt{x^2 + y^2} = \sqrt{(1.2 \text{ m})^2 + (0.90 \text{ m})^2} = 1.5 \text{ m}$$

FOLLOW-UP EXERCISE.   Suppose that a ball is rolling diagonally across a table with the same speed as in this Example, but from the lower right corner, which is taken as the origin of the coordinate system, toward the upper left corner at an angle of 37° relative to the $-x$-axis. What would be the velocity components in this case? (Would the distance change?) *(Answers to all Follow-Up Exercises are given in Appendix VI at the back of the book.)*

PROBLEM-SOLVING HINT

Note that for this simple case, the distance can also be obtained directly from $d = vt = (0.50 \text{ m/s})(3.0 \text{ s}) = 1.5 \text{ m}$. However, this Example was solved in a more general way to illustrate the use of components of motion. The direct solution would have been evident if the equations had been combined algebraically before calculation, that is, as

$$x = v_x t = (v \cos \theta)t$$

and

$$y = v_y t = (v \sin \theta)t$$

from which it follows that

$$d = \sqrt{x^2 + y^2} = \sqrt{(v \cos \theta)^2 t^2 + (v \sin \theta)^2 t^2} = \sqrt{v^2 t^2 (\cos^2 \theta + \sin^2 \theta)} = vt$$

Before embarking on the first solution strategy that occurs to you, pause for a moment to see whether there might be an easier or more direct way of approaching the problem.

KINEMATIC EQUATIONS FOR COMPONENTS OF MOTION

Example 3.1 involved two-dimensional motion in a plane. With a constant velocity (constant components v_x and v_y), the motion is in a straight line. The motion may also be accelerated. For motion in a plane with a *constant acceleration* that has components a_x and a_y, the displacement and velocity components are given by the kinematic equations of Section 2.4 written separately for the x- and y-directions:

$$x = x_o + v_{x_o}t + \tfrac{1}{2}a_xt^2 \qquad (3.3a)$$
$$y = y_o + v_{y_o}t + \tfrac{1}{2}a_yt^2 \qquad (3.3b)$$
$$v_x = v_{x_o} + a_xt \qquad (3.3c)$$
$$v_y = v_{y_o} + a_yt \qquad (3.3d)$$

(constant acceleration only)

If an object is initially moving with a constant velocity and suddenly experiences an acceleration in the direction of the velocity or opposite to it, it will continue in a straight-line path, either speeding up or slowing down, respectively.

If, however, the acceleration is at some angle other than 0° or 180° to the velocity vector, the motion will be along a curved path. For the motion of an object to be *curvilinear*—that is, to vary from a straight-line path—an acceleration not parallel to the velocity is required. For such a curved path, the ratio of the velocity components varies with time. That is, the direction of the motion, $\theta = \tan^{-1}(v_y/v_x)$, varies with time, because one or both of the velocity components do.

Consider a ball initially moving along the x-axis, as illustrated in ▼Fig. 3.2. Assume that, starting at a time $t_o = 0$, the ball receives a constant acceleration a_y in the y-direction. The magnitude of the x-component of the ball's displacement is given by $x = v_xt$, where the $\tfrac{1}{2}a_xt^2$ term of Eq. 3.3a drops out because there is no acceleration in the x-direction ($a_x = 0$). Prior to t_o, the motion was in a straight line along the x-axis. But at any time after t_o, the y-coordinate is not zero, but is given by $y = \tfrac{1}{2}a_yt^2$ (Eq. 3.3b with $y_o = 0$ and $v_{y_o} = 0$). The result is a *curved* path for the ball.

Note that the length (magnitude) of the velocity component v_y changes with time, while that of the v_x component remains constant. The total velocity vector *at any time* is tangent to the curved path of the ball. It is at an angle θ relative to the positive x-axis, given by $\theta = \tan^{-1}(v_y/v_x)$, which now changes with time, as can be seen in Fig. 3.2 and in Example 3.2.

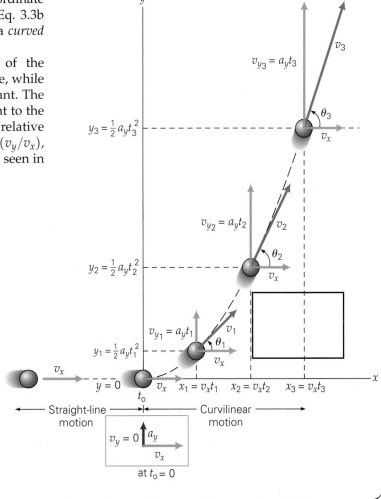

▶ **FIGURE 3.2 Curvilinear motion** An acceleration not parallel to the instantaneous velocity produces a curved path. Here, an acceleration a_y is applied at $t_o = 0$ to a ball initially moving with a constant velocity v_x. The result is a curved path with the velocity components as shown. Notice how v_y increases with time, while v_x remains constant.

EXAMPLE 3.2 | A Curving Path: Vector Components

Suppose that the ball in Fig. 3.2 has an initial velocity of 1.50 m/s along the x-axis. Starting at $t_o = 0$, the ball receives an acceleration of 2.80 m/s² in the y-direction. (a) What is the position of the ball 3.00 s after t_o? (b) What is the velocity of the ball at that time?

THINKING IT THROUGH. Keep in mind that the motions in the x- and y-directions can be analyzed independently. For part (a), simply compute the x- and y-positions at the given time, taking into account the acceleration in the y-direction. For part (b), find the component velocities, and vectorially combine them to get the total velocity.

SOLUTION. Referring to Fig. 3.2,

Given: $v_{x_o} = v_x = 1.50$ m/s   *Find:* (a) (x, y) (position coordinates)
  $v_{y_o} = 0$ (b) v (velocity, magnitude and direction)
  $a_x = 0$
  $a_y = 2.80$ m/s²
  $t = 3.00$ s

(a) At 3.00 s after $t_o = 0$, Eqs. 3.3a and 3.3b tell us that the ball has traveled the following distances from the origin $(x_o = y_o = 0)$ in the x- and y-directions:

$$x = v_{x_o}t + \tfrac{1}{2}a_x t^2 = (1.50 \text{ m/s})(3.00 \text{ s}) + 0 = 4.50 \text{ m}$$

$$y = v_{y_o}t + \tfrac{1}{2}a_y t^2 = 0 + \tfrac{1}{2}(2.80 \text{ m/s}^2)(3.00 \text{ s})^2 = 12.6 \text{ m}$$

Thus, the position of the ball is $(x, y) = (4.50 \text{ m}, 12.6 \text{ m})$. If you had computed the distance $d = \sqrt{x^2 + y^2}$, what would have been obtained? [This quantity is the magnitude of the *displacement*, or straight-line distance, from the origin to the $(x, y) = (4.50 \text{ m}, 12.6 \text{ m})$ position.]

(b) The x-component of the velocity is given by Eq. 3.3c:

$$v_x = v_{x_o} + a_x t = 1.50 \text{ m/s} + 0 = 1.50 \text{ m/s}$$

(This component is constant, since there is no acceleration in the x-direction.) Similarly, the y-component of the velocity is given by Eq. 3.3d:

$$v_y = v_{y_o} + a_y t = 0 + (2.80 \text{ m/s}^2)(3.00 \text{ s}) = 8.40 \text{ m/s}$$

The velocity therefore has a magnitude of

$$v = \sqrt{v_x^2 + v_y^2} = \sqrt{(1.50 \text{ m/s})^2 + (8.40 \text{ m/s})^2} = 8.53 \text{ m/s}$$

and its direction relative to the +x-axis is

$$\theta = \tan^{-1}\left(\frac{v_y}{v_x}\right) = \tan^{-1}\left(\frac{8.40 \text{ m/s}}{1.50 \text{ m/s}}\right) = 79.9°$$

FOLLOW-UP EXERCISE. Suppose that the ball in this Example also received an acceleration of 1.00 m/s² in the +x-direction starting at t_o. What would be the position of the ball 3.00 s after t_o in this case?

PROBLEM-SOLVING HINT

When using the kinematic equations, it is important to note that motion in the x- and y-directions can be analyzed independently—the factor connecting them being time *t*. That is, you can find (x, y) and/or (v_x, v_y) at a given time *t*. Also, keep in mind that the initial positions are often set $x_o = 0$ and $y_o = 0$, which means that the object is located at the origin at $t_o = 0$. If the object is actually elsewhere at $t_o = 0$, then the values of x_o and/or y_o would have to be used in the appropriate equations. (See Eqs. 3.3a and b.)

DID YOU LEARN?
➥ Two-dimensional motion is described by considering an object to be moving in the x- and y- directions simultaneously.
➥ The magnitudes of velocity components are $v_x = v \cos \theta$ and $v_y = v \sin \theta$, where θ is the angle of the x- and y-components.
➥ The kinetic equations for the components of motion are for constant acceleration only.

3.2   **Vector Addition and Subtraction**

LEARNING PATH QUESTIONS
➥ What is a resultant vector?
➥ Given the rectangular vector components, $C_x = C \cos \theta$ and $C_y = C \sin \theta$, how is the angle determined?
➥ What is a unit vector?

Many physical quantities, including those describing motion, have a direction associated with them—that is, they are vectors. You have already worked with a few such quantities related to motion (displacement, velocity, and acceleration) and will encounter more during the course of study. A very important technique in the analysis of many physical situations is the addition (and subtraction) of vectors. By adding or combining such quantities (**vector addition**), the resultant, or net, vector is obtained. This *resultant* vector is the *vector sum*.

You have already been adding vectors. In Section 2.2, displacements in one dimension were added to get the net displacement. In this chapter, vector components of motion in two dimensions will be added to get net effects. Notice that in Example 3.2, the velocity components v_x and v_y were combined to get the resultant velocity.

In this section, vector addition and subtraction in general, along with common vector notation, will be considered. As will be learned, these operations are not the same as scalar or numerical addition and subtraction, with which you are already familiar. Vectors have magnitudes *and* directions, so different rules apply.

In general, there are geometrical (graphical) methods and analytical (computational) methods of vector addition. The geometrical methods are useful in helping you visualize the concepts of vector addition, particularly with a quick sketch. Analytical methods are more commonly used, however, because they are faster and more precise.

In Section 3.1, the concern was chiefly about vector components. The notation for the magnitudes of components was, for example, v_x and v_y. To represent vectors, the notation $\vec{A}$ and $\vec{B}$—a boldface symbol with an overarrow—will be used.

VECTOR ADDITION: GEOMETRIC METHODS

Triangle Method   To add two vectors—say, to add $\vec{B}$ to $\vec{A}$ (that is, to find $\vec{A} + \vec{B}$) by the **triangle method**—you first draw $\vec{A}$ on a sheet of graph paper to some scale (▼Fig. 3.3a). For example, if $\vec{A}$ represents a displacement in meters, a convenient

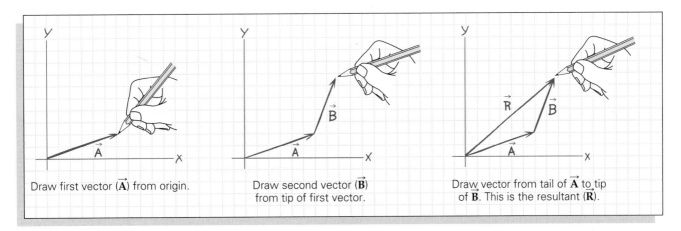

Draw first vector ($\vec{A}$) from origin.

Draw second vector ($\vec{B}$) from tip of first vector.

Draw vector from tail of $\vec{A}$ to tip of $\vec{B}$. This is the resultant ($\vec{R}$).

(a)

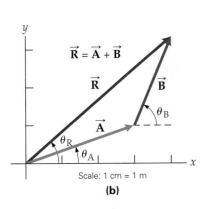

Scale: 1 cm = 1 m

(b)

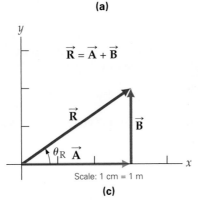

Scale: 1 cm = 1 m

(c)

▲ **FIGURE 3.3 Triangle method of vector addition (a)** The vectors $\vec{A}$ and $\vec{B}$ are placed tip to tail. The vector that extends from the tail of $\vec{A}$ to the tip of $\vec{B}$, forming the third side of the triangle, is the resultant or sum $\vec{R} = \vec{A} + \vec{B}$. **(b)** When the vectors are drawn to scale, the magnitude of $\vec{R}$ can be found by measuring the length of $\vec{R}$ and using the scale conversion, and the direction angle θ_R can be measured with a protractor. Analytical methods can also be used. For a nonright triangle, as in part (b), the laws of sines and cosines can be used to determine the magnitude of $\vec{R}$ and θ_R (Appendix I). **(c)** If the vector triangle is a right triangle, $\vec{R}$ is easily obtained via the Pythagorean theorem, and the direction angle is given by an inverse trigonometric function.

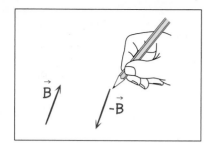

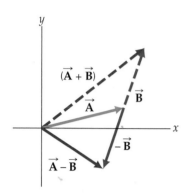

▲ **FIGURE 3.4 Vector subtraction** Vector subtraction is a special case of vector addition; that is, $\vec{A} - \vec{B} = \vec{A} + (-\vec{B})$, where $-\vec{B}$ has the same magnitude as $\vec{B}$, but is in the opposite direction. (See the sketch.) Thus, $\vec{A} + \vec{B}$ is not the same as $\vec{B} - \vec{A}$, in either length or direction. Can you show that $\vec{B} - \vec{A} = -(-\vec{A} - \vec{B})$ geometrically?

scale is 1 cm : 1 m, or 1 cm of vector length on the graph corresponds to 1 m of displacement. As shown in Fig. 3.3b, the direction of the $\vec{A}$ vector is specified as being at an angle θ_A relative to a coordinate axis, usually the *x*-axis.

Next, draw $\vec{B}$ with its tail starting at the tip of $\vec{A}$. (Thus, this method is also called the *tip-to-tail method*.) The vector from the tail of $\vec{A}$ to the tip of $\vec{B}$ is then the vector sum $\vec{R}$, or the resultant of the two vectors: $\vec{R} = \vec{A} + \vec{B}$.

If the vectors are drawn to scale, the magnitude of $\vec{R}$ can be found by measuring its length and using the scale conversion. In such a graphical approach, the direction angle θ_R is measured with a protractor. If the magnitudes and directions (angles θ) of $\vec{A}$ and $\vec{B}$ are known, the magnitude and direction of $\vec{R}$ can be found analytically by using trigonometric methods. For the nonright triangle in Fig. 3.3b, the laws of sines and cosines can be used. (See Appendix I.)

This tip-to-tail method can be extended to any number of vectors. The vector from the tail of the first vector to the tip of the last vector is the resultant or vector sum. For more than two vectors, it is called the *polygon method*.

The resultant of the vector right triangle in Fig. 3.3c would be much easier to find using the Pythagorean theorem for the magnitude and an inverse trigonometric function to find the direction angle. Notice that $\vec{R}$ is made up of *x*- and *y*-vectors $\vec{A}$ and $\vec{B}$. Such *x*- and *y*-components are the basis of the convenient analytical component method, which will be discussed shortly.

Vector Subtraction   Vector subtraction is a special case of vector addition:

$$\vec{A} - \vec{B} = \vec{A} + (-\vec{B})$$

That is, to subtract $\vec{B}$ from $\vec{A}$, a *negative* $\vec{B}$ is added to $\vec{A}$. In Section 2.2, you learned that a minus sign simply means that the direction of a vector is opposite that of one with a plus sign (for example, $+x$ and $-x$). The same is true with vectors represented by boldface notation. The vector $-\vec{B}$ has the same magnitude as the vector $\vec{B}$, but is in the opposite direction (◄Fig. 3.4). The vector diagram in Fig. 3.4 provides a graphical representation of $\vec{A} - \vec{B}$.

VECTOR COMPONENTS AND THE ANALYTICAL COMPONENT METHOD

Probably the most widely used analytical method for adding multiple vectors is the **component method**. It will be used again and again throughout the course of our study, so a basic understanding of the method is *essential*. Learn this section well.

Adding Rectangular Vector Components   *Rectangular components* means that vector components are at right 90° angles to each other, usually taken in the rectangular coordinate *x*- and *y*-directions. You have already had an introduction to the addition of such components in the discussion of the velocity components of motion in Section 3.1. For the general case, suppose that $\vec{A}$ and $\vec{B}$, two vectors at right angles, are added, as illustrated in ▸Fig. 3.5a. The right angle makes the math easy. The magnitude of $\vec{C}$ is given by the Pythagorean theorem:

$$C = \sqrt{A^2 + B^2} \tag{3.4a}$$

The orientation of $\vec{C}$ relative to the *x*-axis is given by the angle

$$\theta = \tan^{-1}\left(\frac{B}{A}\right) \tag{3.4b}$$

This notation is how a resultant is expressed in **magnitude–angle form**.

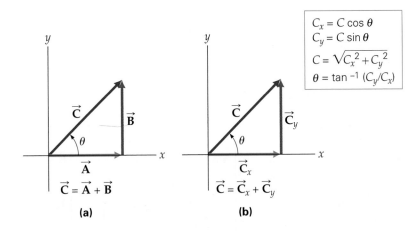

$$C_x = C \cos \theta$$
$$C_y = C \sin \theta$$
$$C = \sqrt{C_x{}^2 + C_y{}^2}$$
$$\theta = \tan^{-1}(C_y/C_x)$$

$\vec{C} = \vec{A} + \vec{B}$

(a)

$\vec{C} = \vec{C}_x + \vec{C}_y$

(b)

◀ **FIGURE 3.5 Vector compo-
nents (a)** The vectors $\vec{A}$ and $\vec{B}$ along
the x- and y-axes, respectively, add
to give $\vec{C}$. **(b)** A vector $\vec{C}$ may be
resolved into rectangular compo-
nents $\vec{C}_x$ and $\vec{C}_y$.

Resolving a Vector into Rectangular Components; Unit Vectors   Resolving a
vector into rectangular components is essentially the reverse of adding the rectangular components of the vector. Given a vector $\vec{C}$, Fig. 3.5b illustrates how it may
be resolved into x and y vector components $\vec{C}_x$ and $\vec{C}_y$. Simply complete the vector
triangle with x- and y-components. As the diagram shows, the magnitudes, or
vector lengths, of these components are given by

$$C_x = C \cos \theta \qquad (3.5a)$$
(vector components)
$$C_y = C \sin \theta \qquad (3.5b)$$

respectively (similar to $v_x = v \cos \theta$ and $v_y = v \sin \theta$ in Example 3.1).* The angle
of direction of $\vec{C}$ can also be expressed in terms of the components, since
$\tan \theta = C_y/C_x$, or

$$\theta = \tan^{-1}\left(\frac{C_y}{C_x}\right) \qquad \begin{array}{l}\textit{(direction of vector}\\ \textit{from magnitudes of components)}\end{array} \qquad (3.6)$$

Another way of expressing the magnitude and direction of a vector involves the
use of unit vectors. For example, as illustrated in ▶Fig. 3.6, a vector $\vec{A}$ can be written
as $\vec{A} = A\hat{a}$. The numerical magnitude is represented by A, and $\hat{a}$ is a **unit vector**,
which indicates direction. That is, $\hat{a}$ has a magnitude of unity, or one, with no units,
and simply indicates a vector's direction. For example, a velocity along the x-axis can
be written $\vec{v} = (4.0 \text{ m/s})\hat{x}$ (that is, 4.0 m/s magnitude in the $+x$-direction).

Note in Fig. 3.6 how $-\vec{A}$ would be represented in this notation. Although the
minus sign is sometimes put in front of the numerical magnitude, this quantity is
an absolute number or value. The minus actually goes with the unit vector:
$-\vec{A} = -A\hat{a} = A(-\hat{a})$.† That is, the unit vector is in the $-\hat{a}$ direction (opposite $\hat{a}$).
A velocity of $\vec{v} = (-4.0 \text{ m/s})\hat{x}$ has a magnitude of 4.0 m/s in the $-x$ direction, that
is, $\vec{v} = (4.0 \text{ m/s})(-\hat{x})$.

This notation can be used to express explicitly the rectangular components of a
vector. For example, the ball's displacement from the origin in Example 3.2 could
be written $\vec{d} = (4.50 \text{ m})\hat{x} + (12.6 \text{ m})\hat{y}$, where $\hat{x}$ and $\hat{y}$ are unit vectors in the x-
and y-directions. In some instances, it may be more convenient to express a gen-
eral vector in this unit vector **component form**:

$$\vec{C} = C_x\hat{x} + C_y\hat{y} \qquad (3.7)$$

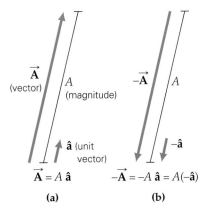

▲ **FIGURE 3.6 Unit vectors
(a)** A unit vector $\hat{a}$ has a magnitude of
unity, or one, and thereby simply
indicates a vector's direction. Written
with the magnitude A, it represents
the vector $\vec{A}$, and $\vec{A} = A\hat{a}$.
(b) For the vector $-\vec{A}$, the unit vector is $-\hat{a}$, and $-\vec{A} = -A\hat{a} = A(-\hat{a})$.

*Figure 3.5b illustrates only a vector in the first quadrant, but the equations hold for all quadrants
when vectors are referenced to either the positive or negative x-axis. The directions of the components
are indicated by $+$ and $-$ signs, as will be shown shortly.

†The notation is sometimes written with an absolute value, $\vec{A} = |A|\hat{a}$, or $-\vec{A} = -|A|\hat{a}$, so as to
clearly show that the magnitude of $\vec{A}$ is a positive quantity.

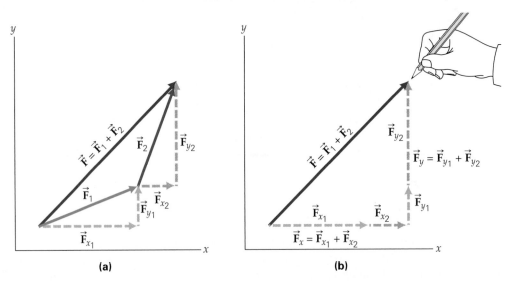

▲ **FIGURE 3.7   Component addition** **(a)** In adding vectors by the component method, each vector is first resolved into its x- and y-component vectors. **(b)** The sums of the x- and y-components of vectors $\vec{F}_1$ and $\vec{F}_2$ are $\vec{F}_x = \vec{F}_{x_1} + \vec{F}_{x_2}$ and $\vec{F}_y = \vec{F}_{y_1} + \vec{F}_{y_2}$, respectively.

VECTOR ADDITION USING COMPONENTS

The **analytical component method of vector addition** involves resolving the vectors into rectangular vector components and adding the components for each axis independently. This method is illustrated graphically in ▲Fig. 3.7 for two vectors $\vec{F}_1$ and $\vec{F}_2$.* *The sums of the x- and y-component vectors being added are equal to the corresponding vector components of the resultant vector.*

The same principle applies if you are given three (or more) vectors to add. You could find the resultant by applying the graphical tip-to-tail method. However, this technique involves drawing the vectors to scale and using a protractor to measure angles, which can be time consuming. But in using the component method, you do not have to draw the vectors tip to tail. In fact, it is usually more convenient to put all of the tails together at the origin, as shown in ▶Fig. 3.8a. Also, the vectors do not have to be drawn to scale, since the approximate sketch is just a visual aid in applying the analytical method.

Basically, in the component method, the vectors to be added are resolved into their x- and y-components, and the respective components added and then recombined to find the resultant. The resultant of the three vectors in Fig. 3.8a is shown in Fig. 3.8b. By looking at the x-components, it can be seen that the vector sum of these components is in the $-x$-direction. Similarly, the sum of the y-components is in the $+y$-direction. (Note that $\vec{v}_2$ is in the y-direction and has a zero x-component, just as a vector in the x-direction would have a zero y-component.)

The x- and y-components of the resultant are $\vec{v}_x = \vec{v}_{x_1} + \vec{v}_{x_3}$ and $\vec{v}_y = \vec{v}_{y_1} + \vec{v}_{y_2} + \vec{v}_{y_3}$. When the numerical values of the vector components are computed and put into these equations, you will have values for $v_x < 0$ (negative) and $v_y > 0$ (positive) as shown in Fig. 3.8b.

Notice also in Fig. 3.8b that the directional angle θ of the resultant is referenced to the x-axis, as are the individual vectors in Fig. 3.8a. *In adding vectors by the component method, all vectors will be referenced to the nearest x-axis—that is, the +x-axis or $-x$-axis.* This policy eliminates angles greater than 90° (as occurs when customarily measuring angles counterclockwise from the $+x$-axis) and the use of double-angle formulas, such as $\cos(\theta + 90°)$. This greatly simplifies calculations. The

*The symbol $\vec{F}$ is commonly used to denote force, a very important vector quantity that will be studied in Chapter 4. Here, $\vec{F}$ is employed as a general vector, but its use provides familiarity with the notation used in the next chapter, where knowledge of the addition of forces is essential.

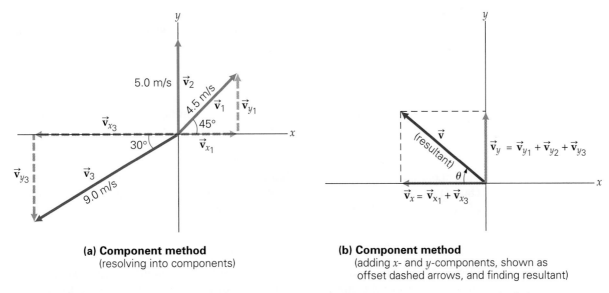

(a) Component method
(resolving into components)

(b) Component method
(adding x- and y-components, shown as
offset dashed arrows, and finding resultant)

▲ FIGURE 3.8 **Component method of vector addition** **(a)** In the analytical component method, all the vectors to be added ($\vec{v}_1$, $\vec{v}_2$ and $\vec{v}_3$) are first placed with their tails at the origin so that they may be easily resolved into rectangular components. **(b)** The respective summations of all the x-components and all the y-components are then added to give the components of the resultant $\vec{v}$.

recommended procedures for adding vectors analytically by the component method can be summarized as follows:

PROCEDURES FOR ADDING VECTORS BY THE COMPONENT METHOD

1. Resolve the vectors to be added into their x- and y-components. Use the acute angles (angles less than 90°) between the vectors and the x-axis to find the magnitudes, and indicate the directions of the components by plus and minus signs (▼Fig. 3.9).

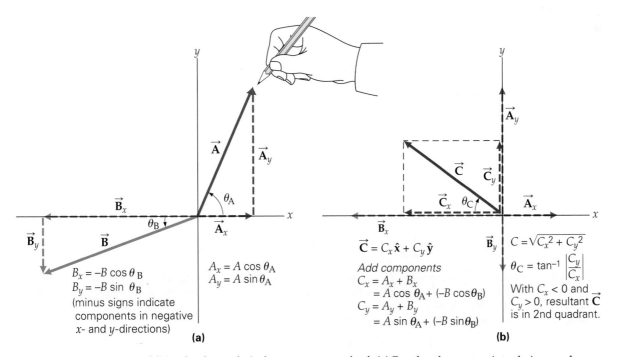

$B_x = -B \cos \theta_B$
$B_y = -B \sin \theta_B$
(minus signs indicate components in negative x- and y-directions)
(a)

$A_x = A \cos \theta_A$
$A_y = A \sin \theta_A$

$\vec{C} = C_x \hat{x} + C_y \hat{y}$

Add components
$C_x = A_x + B_x$
$\quad = A \cos \theta_A + (-B \cos \theta_B)$
$C_y = A_y + B_y$
$\quad = A \sin \theta_A + (-B \sin \theta_B)$
(b)

$C = \sqrt{C_x^2 + C_y^2}$
$\theta_C = \tan^{-1} \left| \dfrac{C_y}{C_x} \right|$

With $C_x < 0$ and $C_y > 0$, resultant $\vec{C}$ is in 2nd quadrant.

▲ FIGURE 3.9 **Vector addition by the analytical component method** **(a)** Resolve the vectors into their x- and y-components. **(b)** Add all of the x-components and all of the y-components together vectorially to obtain the x- and y-components $\vec{C}_x$ and $\vec{C}_y$, respectively, of the resultant. Express the resultant in either component form or magnitude–angle form. All angles are referenced to the $+x$- or $-x$-axis to keep them less than 90°.

2. Add all of the x-components together, and all of the y-components together vectorially to obtain the x- and y-components of the resultant, or vector sum.

3. Express the resultant vector, using:
   (a) the unit vector component form—for example, $\vec{\mathbf{C}} = C_x\hat{\mathbf{x}} + C_y\hat{\mathbf{y}}$—or
   (b) the magnitude–angle form.

For the latter notation, find the magnitude of the resultant by using the summed x- and y-components and the Pythagorean theorem:

$$C = \sqrt{C_x^2 + C_y^2}$$

Find the angle of direction (relative to the x-axis) by taking the inverse tangent ($\tan^{-1}$) of the *absolute value* (that is, the positive value, ignoring any minus signs) of the ratio of the magnitudes of y- and x-components:

$$\theta = \tan^{-1}\left|\frac{C_y}{C_x}\right|$$

Designate the quadrant in which the resultant lies. This information is obtained from the signs of the summed components or from a sketch of their addition via the triangle method. (See Fig. 3.9.) The angle θ is the angle between the resultant and the x-axis in that quadrant.

EXAMPLE 3.3

Applying the Analytical Component Method: Separating and Combining x- and y-Components

Let's apply the procedural steps of the component method to the addition of the vectors in Fig. 3.8a. The vectors with units of meters per second represent velocities.

THINKING IT THROUGH.   Follow and learn the steps of the procedure. Basically, the vectors are resolved into components and the respective components are added to get the components of the resultant, which then may be expressed in (unit vector) component form or magnitude–angle form.

SOLUTION.   The rectangular components of the vectors are shown in Fig. 3.8b. Summing these components and taking the values from Fig. 3.8a,

$$\vec{\mathbf{v}} = v_x\hat{\mathbf{x}} + v_y\hat{\mathbf{y}} = (v_{x_1} + v_{x_2} + v_{x_3})\hat{\mathbf{x}} + (v_{y_1} + v_{y_2} + v_{y_3})\hat{\mathbf{y}}$$

where

$$v_x = v_{x_1} + v_{x_2} + v_{x_3} = v_1\cos 45° + 0 - v_3\cos 30°$$
$$= (4.5\text{ m/s})(0.707) - (9.0\text{ m/s})(0.866) = -4.6\text{ m/s}$$

and

$$v_y = v_{y_1} + v_{y_2} + v_{y_3} = v_1\sin 45° + v_2 - v_3\sin 30°$$
$$= (4.5\text{ m/s})(0.707) + (5.0\text{ m/s}) - (9.0\text{ m/s})(0.50) = 3.7\text{ m/s}$$

Expressed in tabular form, the components are as follows:

	x-Components		*y*-Components
v_{x_1}	$+v_1\cos 45° = +3.2$ m/s	v_{y_1}	$+v_1\sin 45° = +3.2$ m/s
v_{x_2}	$= 0$ m/s	v_{y_2}	$= +5.0$ m/s
v_{x_3}	$-v_3\cos 30° = -7.8$ m/s	v_{y_3}	$-v_3\sin 30° = -4.5$ m/s
Sums:	$v_x = -4.6$ m/s		$v_y = +3.7$ m/s

The directions of the components are indicated by signs. (The + sign is sometimes omitted as being understood.) Here, $\vec{\mathbf{v}}_2$ has no x-component. *Note that in general, for the analytical component method, the x-components are cosine functions and the y-components are sine functions, as long as they are referenced to the nearest part of the x-axis.*

In component form, the resultant vector is

$$\vec{\mathbf{v}} = (-4.6\text{ m/s})\hat{\mathbf{x}} + (3.7\text{ m/s})\hat{\mathbf{y}}$$

In magnitude–angle form, the resultant velocity has a magnitude of

$$v = \sqrt{v_x^2 + v_y^2} = \sqrt{(-4.6\text{ m/s})^2 + (3.7\text{ m/s})^2} = 5.9\text{ m/s}$$

Since the x-component is negative and the y-component is positive, the resultant lies in the *second quadrant* at an angle of

$$\theta = \tan^{-1}\left|\frac{v_y}{v_x}\right| = \tan^{-1}\left(\frac{3.7 \text{ m/s}}{4.6 \text{ m/s}}\right) = 39°$$

above the $-x$-axis because of the negative x component (see Fig. 3.8b).

FOLLOW-UP EXERCISE. Suppose in this Example that there were an additional velocity vector $\vec{v}_4 = (+4.6 \text{ m/s})\hat{x}$. What would be the resultant of all four vectors in this case?

Although our discussion is limited to motion in two dimensions (in a plane), the component method is easily extended to three dimensions. For a velocity in three dimensions, the vector has x-, y-, and z-components: $\vec{v} = v_x\hat{x} + v_y\hat{y} + v_z\hat{z}$ and magnitude $v = \sqrt{v_x^2 + v_y^2 + v_z^2}$.

LEARN BY DRAWING 3.1

make a sketch and add them up

(a) A sketch is made for the vectors $\vec{A}$ and $\vec{B}$. In a vector drawing, the vector lengths are usually set to some scale—for example, 1 cm : 1 m—but in a quick sketch, the vector lengths are estimated. **(b)** By shifting $\vec{B}$ to the tip of $\vec{A}$ and putting in $\vec{D}$, the vector $\vec{C}$ can be found from $\vec{A} + \vec{B} + \vec{C} = \vec{D}$.

EXAMPLE 3.4 | **Find the Vector: Add Them Up**

Given two displacement vectors, $\vec{A}$, with a magnitude of 8.0 m in a direction 45° below the $+x$-axis, and $\vec{B}$, which has an x-component of $+2.0$ m and a y-component of $+4.0$ m. Find a vector $\vec{C}$ so that $\vec{A} + \vec{B} + \vec{C}$ equals a vector $\vec{D}$ that has a magnitude of 6.0 m in the $+y$-direction.

THINKING IT THROUGH. Here again, a sketch helps to understand the situation and gives a general idea of the attributes of $\vec{C}$. This would be something like the accompanying Learn By Drawing 3.1, Make a Sketch and Add Them Up. Note that in part (a) both $\vec{A}$ and $\vec{B}$ have $+x$-components, so $\vec{C}$ would have to have a $-x$-component to cancel these components. (It is given that the resultant $\vec{D}$ points only in the $+y$-direction.) $\vec{B}_y$ and $\vec{D}$ are in the $+y$-direction, but the $\vec{A}_y$-component is larger in the $-y$-direction, so $\vec{C}$ would have to have a $+y$-component. With this information, it can be seen that $\vec{C}$ lies in the second quadrant. A polygon sketch [shown in part (b) of the Learn by Drawing] confirms this observation.

So $\vec{C}$ has second-quadrant components and it has a relatively large magnitude (from the lengths of the vectors in the polygon drawing). This information gives an idea of what we are looking for, making it easier to see if the results from the analytic solution are reasonable.

SOLUTION.

Given: $\vec{A}$: 8.0 m, 45° below the $+x$-axis (fourth quadrant)

$\vec{B}_x = (2.0 \text{ m})\hat{x}$

$\vec{B}_y = (4.0 \text{ m})\hat{y}$

Find: $\vec{C}$ such that

$\vec{A} + \vec{B} + \vec{C} = \vec{D} = (+6.0 \text{ m})\hat{y}$

Setting up the components in tabular form again so they can be easily seen:

x-Components	*y*-Components
$A_x = A\cos 45° = (8.0 \text{ m})(0.707) = +5.7 \text{ m}$	$A_y = -A\sin 45° = -(8.0 \text{ m})(0.707)$ $= -5.7 \text{ m}$
$B_x = +2.0 \text{ m}$	$B_y = +4.0 \text{ m}$
$C_x = ?$	$C_y = ?$
$D_x = 0$	$D_y = +6.0 \text{ m}$

To find the components of $\vec{C}$, where $\vec{A} + \vec{B} + \vec{C} = \vec{D}$, the x- and y-components are summed separately:

x: $\quad \vec{A}_x + \vec{B}_x + \vec{C}_x = \vec{D}_x$

or

$+5.7 \text{ m} + 2.0 \text{ m} + C_x = 0 \quad$ and $\quad C_x = -7.7 \text{ m}$

y: $\quad \vec{A}_y + \vec{B}_y + \vec{C}_y = \vec{D}_y$

or

$-5.7 \text{ m} + 4.0 \text{ m} + C_y = 6.0 \text{ m} \quad$ and $\quad C_y = +7.7 \text{ m}$

(continued on next page)

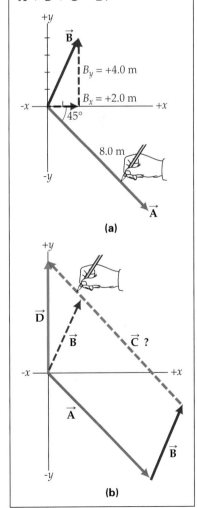

So,

$$\vec{C} = (-7.7 \text{ m})\hat{x} + (7.7 \text{ m})\hat{y}$$

The result can also be expressed in magnitude–angle form:

$$C = \sqrt{C_x^2 + C_y^2} = \sqrt{(-7.7 \text{ m})^2 + (7.7 \text{ m})^2} = 11 \text{ m}$$

and

$$\theta = \tan^{-1}\left|\frac{C_y}{C_x}\right| = \tan^{-1}\left|\frac{7.7 \text{ m}}{-7.7 \text{ m}}\right| = 45° \quad \text{(above the } -x \text{ axis; why?)}$$

FOLLOW-UP EXERCISE.   Suppose $\vec{D}$ pointed in the opposite direction [that is, $\vec{D} = (-6.0 \text{ m})\hat{y}$]. What would $\vec{C}$ be in this case?

DID YOU LEARN?
➥ A resultant is the vector sum of added (or subtracted) vectors.
➥ θ is given by the component ratio $\dfrac{C_y}{C_x} = \dfrac{C \sin\theta}{C \cos\theta} = \tan\theta$, and $\theta = \tan^{-1}\left|\dfrac{C_y}{C_x}\right|$
➥ A unit vector has a magnitude of unity, or one, with no units, and simply indicates a vector's direction.

3.3   Projectile Motion

LEARNING PATH QUESTIONS
➥ Why do two balls, one dropped and one horizontally projected from the same height, strike the ground at the same time?
➥ For projections with v_x and v_y components, which is constant and why?

A familiar example of two-dimensional, curvilinear motion is that of an object that is thrown or projected by some means. The motion of a stone thrown across a stream or a golf ball driven off a tee are examples of **projectile motion**. A special case of projectile motion in one dimension occurs when an object is projected vertically upward (or downward or dropped). This case was treated in Section 2.5 in terms of free fall. General two-dimensional projectile motion is in free fall too, because the only acceleration of a projectile is that due to gravity (air resistance neglected). Vector components can be used to analyze projectile motion by simply breaking up the motion into its x- and y-components and treating them separately.

HORIZONTAL PROJECTIONS

It is instructive to first analyze the motion of an object projected horizontally, or parallel to a level surface. Suppose that you throw an object horizontally with an initial velocity v_{x_o} as in ▸Fig. 3.10. Projectile motion is analyzed beginning at the instant of release ($t = 0$). Once the object is released, there is no longer a horizontal acceleration ($a_x = 0$), so throughout the object's path, the horizontal velocity remains constant: $v_x = v_{x_o}$.

According to the equation $x = x_o + v_x t$ (Eq. 3.2a), the projected object would continue to travel in the horizontal direction indefinitely. However, you know that this is not what happens. As soon as the object is projected, it is in free fall in the vertical direction, with $v_{y_o} = 0$ (vertically it behaves as though it had been dropped) and $a_y = -g$. In other words, the projected object travels at a uniform velocity in the horizontal direction, while *at the same time* undergoing acceleration in the downward direction under the influence of gravity. The result is a curved path, as illustrated in Fig. 3.10. (Compare the motions in Fig. 3.10 and Fig. 3.2. Do you see any similarities?) If there were no horizontal motion, the object would simply drop to the ground in a straight line. In fact, the time of flight of the horizontally projected object is *exactly the same as if it were a dropped object falling vertically*.

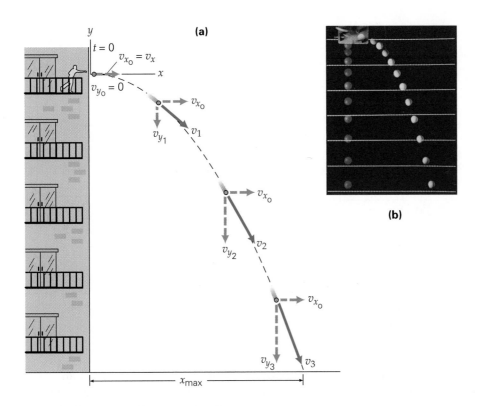

(a)

(b)

◀**FIGURE 3.10 Horizontal projection (a)** The velocity components of a projectile launched horizontally show that the projectile travels to the right as it falls downward. Note the increase in v_y. **(b)** A multiflash photograph shows the paths of two golf balls. One was projected horizontally at the same time that the other was dropped straight down. The horizontal lines are 15 cm apart, and the interval between flashes was $\frac{1}{30}$ s. The vertical motions of the balls are the same. Why? Can you describe the horizontal motion of the yellow ball?

Note the components of the velocity vector in Fig. 3.10a. The length of the horizontal component of the velocity vector remains the same, but the length of the vertical component increases with time. What is the instantaneous velocity at any point along the path? (Think in terms of vector addition, covered in Section 3.2.) The photo in Fig. 3.10b shows the actual motions of a horizontally projected golf ball and one that is simultaneously dropped from rest. The horizontal reference lines show that the balls fall vertically at the same rate. The only difference is that the horizontally projected ball also travels to the right as it falls.

EXAMPLE 3.5 | **Starting at the Top: Horizontal Projection**

Suppose that the ball in Fig. 3.10a is projected from a height of 25.0 m above the ground and is thrown with an initial horizontal velocity of 8.25 m/s. (a) How long is the ball in flight before striking the ground? (b) How far from the building does the ball strike the ground?

THINKING IT THROUGH. In looking at the components of motion, we see that part (a) involves the time it takes the ball to fall vertically, analogous to a ball dropped from that height. This time is also the time the ball travels in the horizontal direction. The horizontal speed is constant, so the horizontal distance requested in part (b) can be found.

SOLUTION. Writing the data with the origin chosen as the point from which the ball is thrown and downward taken as the negative direction:

Given: $y = -25.0$ m
$v_{x_o} = 8.25$ m/s
$a_x = 0$
$v_{y_o} = 0$
$a_y = -g$
($x_o = 0$ and $y_o = 0$ because of our choice of axes location.)

Find: (a) t (time of flight)
(b) x (horizontal distance)

(a) As noted previously, the time of flight is the same as the time it takes for the ball to fall vertically to the ground. To find this time, the equation $y = y_o + v_{y_o}t - \frac{1}{2}gt^2$ can be used, in which the negative direction of g is expressed explicitly, as was done in Section 2.4. With $v_{y_o} = 0$,

$$y = -\frac{1}{2}gt^2$$

So,

$$t = \sqrt{\frac{2y}{-g}} = \sqrt{\frac{2(-25.0 \text{ m})}{-9.80 \text{ m/s}^2}} = 2.26 \text{ s}$$

(b) The ball travels in the x-direction for the same amount of time it travels in the y-direction (that is, 2.26 s). Since there is no acceleration in the horizontal direction, the ball travels in this direction with a uniform velocity. Thus, with $x_o = 0$ and $a_x = 0$,

$$x = v_{x_o}t = (8.25 \text{ m/s})(2.26 \text{ s}) = 18.6 \text{ m}$$

FOLLOW-UP EXERCISE. (a) Choose the axes to be at the base of the building, and show that the resulting equation is the same as in the Example. (b) What is the velocity (in component form) of the ball just before it strikes the ground?

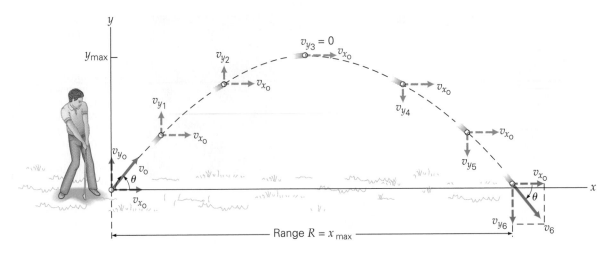

▲ **FIGURE 3.11** **Projection at an angle** The velocity components of the ball are shown for various times. (Directions are indicated by signs, with the + sign being omitted as conventionally understood.) Note that $v_y = 0$ at the top of the arc, or at y_{max}. The range R is the maximum horizontal distance, or x_{max}. (Notice that $v_o = v_6$ in magnitude. Why?)

PROJECTIONS AT ARBITRARY ANGLES

The general case of projectile motion involves an object projected at an arbitrary angle θ relative to the horizontal—for example, a golf ball hit by a club (▲Fig. 3.11). During projectile motion, the object travels up and down while traveling horizontally with a constant velocity. (Does the ball have acceleration? Yes. At each point of the motion, gravity acts, and $\vec{a} = -g\hat{y}$.)

This motion is also analyzed by using its components. As before, upward is taken as the positive direction and downward as the negative direction. The initial velocity v_o is first resolved into rectangular components:

$$v_{x_o} = v_o \cos \theta \tag{3.8a}$$
$$v_{y_o} = v_o \sin \theta \qquad \textit{(initial velocity components)} \tag{3.8b}$$

There is no horizontal acceleration and the acceleration due to gravity acts in the negative y-direction. Thus, the x-component of the velocity is constant and the y-component varies with time (see Eq. 3.3d):

$$v_x = v_{x_o} = v_o \cos \theta \tag{3.9a}$$
$$v_y = v_{y_o} - gt = v_o \sin \theta - gt \qquad \textit{(projectile motion velocity components)} \tag{3.9b}$$

The components of the instantaneous velocity at various times are illustrated in Fig. 3.11. The instantaneous velocity is the sum of these components and is tangent to the curved path of the ball at any point. Notice that the ball strikes the ground at the same speed (but with $-v_{y_o}$) and at the same angle below the horizontal as it was launched.

Similarly, the displacement components are given by ($x_o = y_o = 0$):

$$x = v_{x_o}t = (v_o \cos \theta)t \tag{3.10a}$$
$$y = v_{y_o}t - \tfrac{1}{2}gt^2 = (v_o \sin \theta)t - \tfrac{1}{2}gt^2 \qquad \textit{(projectile motion locations)} \tag{3.10b}$$

The curve described by these equations, or the path of motion (trajectory) of the projectile, is called a **parabola**. The path of projectile motion is often referred to as a *parabolic arc*. Such arcs are commonly observed (◄Fig. 3.12).

Note that, as in the case of horizontal projection, *time is the common feature shared by the components of motion.* Aspects of projectile motion that may be of interest in various situations include the time of flight, the maximum height reached, and the **range (R)**, which is the maximum horizontal distance traveled.

▲ **FIGURE 3.12** **Parabolic arcs** Sparks of hot metal projectiles from welding describe parabolic arcs.

EXAMPLE 3.6 | Teeing Off: Projection at an Angle

Suppose a golf ball is hit off the tee with an initial velocity of 30.0 m/s at an angle of 35° to the horizontal, as in Fig. 3.11. (a) What is the maximum height reached by the ball? (b) What is its range?

THINKING IT THROUGH. The maximum height involves the y-component; the procedure for finding this is like that for finding the maximum height of a ball projected vertically upward. The ball travels in the x-direction for the same amount of time it would take for the ball to go up and down.

SOLUTION.

Given: $v_o = 30.0$ m/s
$\theta = 35°$
$a_y = -g$
(x_o and $y_o = 0$ and final $y = 0$)

Find: (a) y_{max}
(b) $R = x_{max}$

Let us compute v_{x_o} and v_{y_o} explicitly so simplified kinematic equations can be used:

$$v_{x_o} = v_o \cos 35° = (30.0 \text{ m/s})(0.819) = 24.6 \text{ m/s}$$
$$v_{y_o} = v_o \sin 35° = (30.0 \text{ m/s})(0.574) = 17.2 \text{ m/s}$$

(a) Just as for an object thrown vertically upward, $v_y = 0$ at the maximum height (y_{max}). Thus, the time to reach the maximum height (t_u) can be found by using Eq. 3.3b, with v_y set equal to zero:

$$v_y = 0 = v_{y_o} - gt_u$$

Solving for t_u,

$$t_u = \frac{v_{y_o}}{g} = \frac{17.2 \text{ m/s}}{9.80 \text{ m/s}^2} = 1.76 \text{ s}$$

(Note that t_u represents the amount of time the ball moves upward.)

The maximum height y_{max} is then obtained by substituting t_u into Eq. 3.10b:

$$y_{max} = v_{y_o}t_u - \tfrac{1}{2}gt_u^2$$
$$= (17.2 \text{ m/s})(1.76 \text{ s}) - \tfrac{1}{2}(9.80 \text{ m/s}^2)(1.76 \text{ s})^2 = 15.1 \text{ m}$$

The maximum height could also be obtained directly from Eq. 2.11′, $v_y^2 = v_{y_o}^2 - 2gy$, with $y = y_{max}$ and $v_y = 0$. However, the method of solution used here illustrates how the time of flight is obtained.

(b) As in the case of vertical projection, the time in going up is equal to the time in coming down, so the total time of flight is $t = 2t_u$ (to return to the elevation from which the object was projected, $y = y_o = 0$, as can be seen from $y - y_o = v_{y_o}t - \tfrac{1}{2}gt^2 = 0$, and $t = 2v_{y_o}/g = 2t_u$.)

The range R is equal to the horizontal distance traveled (x_{max}), which is easily found by substituting the total time of flight $t = 2t_u = 2(1.76 \text{ s}) = 3.52 \text{ s}$ into Eq. 3.10a:

$$R = x_{max} = v_x t = v_{x_o}(2t_u) = (24.6 \text{ m/s})(3.52 \text{ s}) = 86.6 \text{ m}$$

FOLLOW-UP EXERCISE. How would the values of maximum height (y_{max}) and the range (x_{max}) compare with those found in this Example if the golf ball had been similarly teed off on the surface of the Moon? [*Hint:* $g_M = g/6$; that is, acceleration due to gravity on the Moon is one-sixth of that on the Earth.] Do not do any numerical calculations. Find the answers by "sight reading" the equations.

The range of a projectile is an important consideration in various applications. This factor is particularly important in sports in which a maximum range is desired, such as golf and javelin throwing.

In general, what is the range of a projectile launched with velocity v_o at an angle θ? In order to answer this question, consider the equation used in Example 3.6 to calculate the range, $R = v_x t$. First let's look at the expressions for v_x and t. Since there is no acceleration in the horizontal direction,

$$v_x = v_{x_o} = v_o \cos \theta$$

and the total time t (as shown in Example 3.6) is

$$t = \frac{2v_{y_o}}{g} = \frac{2v_o \sin \theta}{g}$$

and R is given by,

$$R = v_x t = (v_o \cos \theta)\left(\frac{2v_o \sin \theta}{g}\right) = \frac{2v_o^2 \sin \theta \cos \theta}{g}$$

Using the trigonometric identity $\sin 2\theta = 2 \cos \theta \sin \theta$ (see Appendix I),

$$R = \frac{v_o^2 \sin 2\theta}{g} \qquad \begin{array}{l}(\textit{projectile range } x_{max}, \\ \textit{only for } y_{initial} = y_{final})\end{array} \qquad (3.11)$$

Note that the range depends on the magnitude of the initial velocity (or speed), v_o, and that the angle of projection, θ, and g are assumed to be constant. Keep in mind that this equation applies only to the *special*, but common, case of $y_{initial} = y_{final}$, that is, when the landing point is at the same height as the launch point.

EXAMPLE 3.7 | ## A Throw from the Bridge

A young girl standing on a bridge throws a stone with an initial velocity of 12 m/s at a downward angle of 45° to the horizontal, in an attempt to hit a block of wood floating in the river below (▶ Fig. 3.13). If the stone is thrown from a height of 20 m and it just reaches the water when the block is 13 m from the bridge, does the stone hit the block? (Assume that the block does not move appreciably and that it is in the plane of the throw.)

THINKING IT THROUGH. The question is, what is the range of the stone? If this range is the same as the distance between the block and the bridge, then the stone hits the block. To find the range of the stone, we need to find the time of descent (from the y-component of motion) and then use this time to find the distance x_{max}. (Time is the connecting factor.)

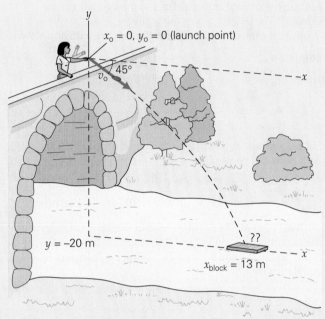

▲ **FIGURE 3.13 A throw from the bridge—hit or miss?** See Example text for description.

SOLUTION.

Given: $v_o = 12$ m/s
$\theta = 45°$ $v_{x_o} = v_o \cos 45° = 8.5$ m/s
$y = -20$ m $v_{y_o} = -v_o \sin 45° = -8.5$ m/s
$x_{block} = 13$ m
$(x_o = y_o = 0)$

Find: Range or x_{max} of stone from bridge. (Is it the same as the block's distance from the bridge?)

To find the time for upward travel, $v_y = v_{y_o} - gt$ was used in Example 3.6, where $v_y = 0$ at the top of the arc. However, in this case, v_y is not zero when the stone reaches the river, so to use this equation, v_y is needed. This may be found from the kinematic equation Eq. 2.11',

$$v_y^2 = v_{y_o}^2 - 2gy$$

as

$$v_y = \sqrt{(-8.5 \text{ m/s})^2 - 2(9.8 \text{ m/s}^2)(-20 \text{ m})} = -22 \text{ m/s}$$

(negative root because v_y is downward).
Then solving $v_y = v_{y_o} - gt$ for t,

$$t = \frac{v_{y_o} - v_y}{g} = \frac{-8.5 \text{ m/s} - (-22 \text{ m/s})}{9.8 \text{ m/s}^2} = 1.4 \text{ s}$$

The stone's horizontal distance from the bridge at this time is

$$x_{max} = v_{x_o}t = (8.5 \text{ m/s})(1.4 \text{ s}) = 12 \text{ m}$$

So the girl's throw falls short by a meter (the block is at 13 m).
Note that Eq. 3.10b, $y = y_o + v_{y_o}t - \frac{1}{2}gt^2$, could have been used to find the time, but this calculation would have involved solving a quadratic equation.

FOLLOW-UP EXERCISE. (a) Why was it assumed that the block was in the plane of the throw? (b) Why wasn't Eq. 3.11 used in this Example to find the range? Show that Eq. 3.11 works in Example 3.6, but not in Example 3.7, by computing the range in each case and comparing your results with the answers found in the Examples.

CONCEPTUAL EXAMPLE 3.8 | Which Has the Greater Speed?

Consider two balls, both thrown with the same initial speed v_0, but one at an angle of 45° above the horizontal and the other at an angle of 45° below the horizontal (▼Fig. 3.14). Determine whether, upon reaching the ground, (a) the ball projected upward will have the greater speed, (b) the ball projected downward will have the greater speed, or (c) both balls will have the same speed. *Clearly establish the reasoning and physical principle(s) used in determining your answer before checking it. That is, **why** did you select your answer?*

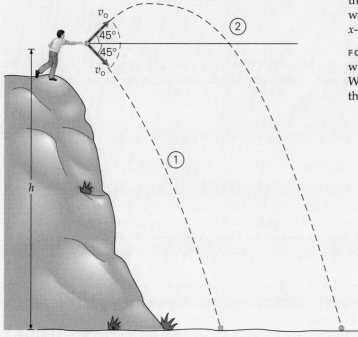

◀FIGURE 3.14 Which has the greater speed? See Example text for description.

REASONING AND ANSWER. At first, you might think the answer is (b), because this ball is projected downward. But the ball projected upward falls from a greater maximum height, so perhaps the answer is (a). To solve this dilemma, look at the horizontal line in Fig. 3.14 between the two velocity vectors that extends beyond the upper trajectory. From this diagram, it can be seen that the trajectories for both balls are the same below this line. Moreover, the downward velocity of the upper ball on reaching this line is v_0 at an angle of 45° below the horizontal. (See Fig. 3.11.) Therefore, relative to the horizontal line and below, the conditions are identical, with the same y-component and the same constant x-component. So the answer is (c).

FOLLOW-UP EXERCISE. Suppose the ball thrown downward was thrown at an angle of −40° and the upward ball at 45°. Which ball would hit the ground with the greater speed in this case?

PROBLEM-SOLVING HINT

The range of a projectile projected downward, as in Fig. 3.14, is found as illustrated in Example 3.7. But what about the range of a projectile projected upward? This case might be thought of as an "extended range" problem. One way to solve it is to divide the trajectory into two parts—(1) the arc above the horizontal line and (2) the downward part below the horizontal line—such that $x_{max} = x_1 + x_2$. You know how to find x_1 (Example 3.6) and x_2 (Example 3.7). Another way to solve the problem is to use $y = y_0 + v_{y_0}t - \frac{1}{2}gt^2$, where y is the final position of the projectile, and solve for t, the total time of flight. You would then use that value in the equation $x = v_{x_0}t$.

Equation 3.11, $R = \dfrac{v_0^2 \sin 2\theta}{g}$, allows the range to be computed for a particular projection angle and initial velocity on a level surface. However, we are sometimes interested in the maximum range for a given initial velocity—for example, the maximum range of an artillery piece that fires a projectile with a particular muzzle velocity. Is there an optimum angle that gives the maximum range? Under ideal conditions, the answer is yes.

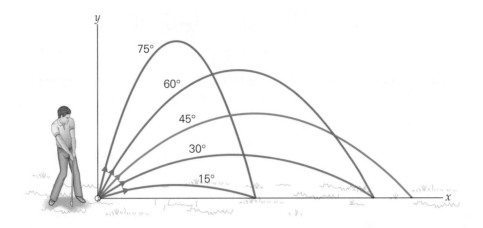

▶ **FIGURE 3.15 Range** For a projectile with a given initial speed, the maximum range is ideally attained with a projection of 45° (no air resistance). For projection angles above and below 45°, the range is shorter, and it is equal for angles equally different from 45° (for example, 30° and 60°).

For a particular v_o, the range is a maximum (R_{max}) when $\sin 2\theta = 1$, since this value of θ yields the maximum value of the sine function (which varies from 0 to 1). Thus,

$$R_{max} = \frac{v_o^2}{g} \qquad (y_{initial} = y_{final}) \qquad (3.12)$$

Because this maximum range is obtained when $\sin 2\theta = 1$ and because $\sin 90° = 1$,

$$2\theta = 90° \quad \text{or} \quad \theta = 45°$$

for the maximum range for a given initial speed when the projectile returns to the elevation from which it was projected. At a greater or smaller angle, for a projectile with the same initial speed, the range will be less, as illustrated in ▲Fig. 3.15. Also, the range is the same for angles equally above and below 45°, such as 30° and 60°.

Thus, to get the maximum range, a projectile *ideally* should be projected at an angle of 45°. However, up to now, air resistance has been neglected. In actual situations, such as when a baseball is thrown or hit, this factor may have a significant effect. Air resistance reduces the speed of the projectile, thereby reducing the range. As a result, when air resistance is a factor, the angle of projection for maximum range is less than 45°, which gives a greater initial horizontal velocity (▼Fig. 3.16). Other factors, such as spin and wind, may also affect the range of a projectile. For example, backspin on a driven golf ball provides lift and the projection angle for the maximum range may be considerably less than 45°.

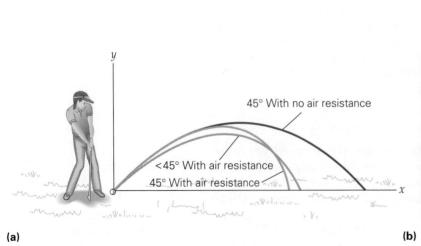

(a)

(b)

▲ **FIGURE 3.16 Air resistance and range** **(a)** When air resistance is a factor, the angle of projection for maximum range is less than 45°. **(b)** Javelin throw. Because of air resistance, the javelin is thrown at an angle less than 45° in order to achieve maximum range.

Keep in mind that for the maximum range to occur at a projection angle of 45°, the components of initial velocity must be equal—that is, $\tan^{-1}(v_{y_o}/v_{x_o}) = 45°$ and $\tan 45° = 1$, so that $v_{y_o} = v_{x_o}$. However, this condition may not always be physically possible, as Conceptual Example 3.9 shows.

CONCEPTUAL EXAMPLE 3.9 | The Longest Jump: Theory and Practice

In a long-jump event, does the jumper normally have a launch angle of (a) less than 45°, (b) exactly 45°, or (c) greater than 45°? *Clearly establish the reasoning and physical principle(s) used in determining your answer before checking it. That is, **why** did you select your answer?*

REASONING AND ANSWER. Air resistance is not a major factor here (although wind speed is taken into account for record setting in track-and-field events). Therefore, it would seem that in order to achieve maximum range, the jumper would take off at an angle of 45°. But there is another physical consideration. Let's look more closely at the jumper's initial velocity components (▼Fig. 3.17a).

To maximize a long jump, the jumper runs as fast as possible and then pushes upward as strongly as possible to maximize both velocity components. The initial vertical velocity component v_{y_o} depends on the upward push of the jumper's legs, whereas the initial horizontal velocity component v_{x_o} depends mostly on the running speed toward the jump point. In general, a greater velocity can be achieved by running than by jumping, so $v_{x_o} > v_{y_o}$. Then, since $\theta = \tan^{-1}(v_{y_o}/v_{x_o})$, then $\theta < 45°$, where $v_{y_o}/v_{x_o} < 1$ in this case. Hence, the answer is (a)—it certainly could not be (c). A typical launch angle for a long jump is 20° to 25°. (If a jumper increased the launch angle to be closer to the ideal 45°, then the running speed would have to decrease, resulting in a decrease in range.)

FOLLOW-UP EXERCISE. When driving in and jumping to score, basketball players seem to be suspended momentarily, or to "hang" in the air (Fig. 3.17b). Explain the physics of this effect.

(a) **(b)**

◀**FIGURE 3.17 Athletes in action (a)** To maximize a long jump, a jumper runs as fast as possible and then pushes upward as strongly as he can to maximize the velocity components (v_x and v_y). **(b)** When driving in toward the basket and jumping to score, basketball players seem to be suspended momentarily, or "hang" in the air.

EXAMPLE 3.10 | A "Slap Shot": Is It Good?

A hockey player hits a "slap shot" in practice (with no goalie present) when he is 15.0 m directly in front of the net. The net is 1.20 m high, and the puck is initially hit at an angle of 5.00° above the ice with a speed of 35.0 m/s. (a) Determine whether the puck makes it into the net. (b) If it does, determine whether the puck is rising or falling vertically as it crosses the front plane of the net.

THINKING IT THROUGH. First let's make a sketch of the situation using x–y coordinates, assuming that the puck is at the origin at the time it is hit and showing the net and its height as in ▼Fig. 3.18. Note that the launch angle is exaggerated. An angle of 5.00° is quite small, but then again, the top of the net is not overly high (1.20 m).

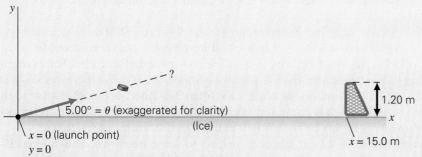

◀**FIGURE 3.18 Slap shot** Is it a goal? See Example text for description.

To determine whether the shot is of goal quality, we need to know whether the puck's trajectory takes it above the net or into the net. That is, what is the puck's height (y) when its horizontal distance is $x = 15.0$ m? Whether the puck is rising or falling at this horizontal distance depends on when the puck reaches its maximum height. The appropriate equation(s) should provide this information; but keep mind that time is the connecting factor between the x- and y-components.

(continued on next page)

SOLUTION.　Listing the data as usual,

Given:　$x = 15.0 \text{ m}, x_o = 0$　　　　*Find:*　(a) Whether the puck goes into the net
　　　　$y_{net} = 1.20 \text{ m}, y_o = 0$　　　　　　　(b) If so, is it rising or falling?
　　　　$\theta = 5.00°$
　　　　$v_o = 35.0 \text{ m/s}$
　　　　$v_{x_o} = v_o \cos 5.00° = 34.9 \text{ m/s}$
　　　　$v_{y_o} = v_o \sin 5.00° = 3.05 \text{ m/s}$

The vertical location of the puck at any time t is given by $y = v_{y_o}t - \frac{1}{2}gt^2$, so we need to know how long the puck takes to travel the 15.0 m to the net. The connecting factor of the components is time, so this time can be found from the x motion:

$$x = v_{x_o}t \quad \text{or} \quad t = \frac{x}{v_{x_o}} = \frac{15.0 \text{ m}}{34.9 \text{ m/s}} = 0.430 \text{ s}$$

So on reaching the front of the net, the puck is at a height of

$$y = v_{y_o}t - \frac{1}{2}gt^2 = (3.05 \text{ m/s})(0.430 \text{ s}) - \frac{1}{2}(9.80 \text{ m/s}^2)(0.430 \text{ s})^2$$
$$= 1.31 \text{ m} - 0.906 \text{ m} = 0.40 \text{ m}$$

Goal!

The time (t_u) for the puck to reach its maximum height is given by $v_y = v_{y_o} - gt_u$, where $v_y = 0$ and

$$t_u = \frac{v_{y_o}}{g} = \frac{3.05 \text{ m/s}}{9.80 \text{ m/s}^2} = 0.311 \text{ s}$$

and with the puck reaching the net in 0.430 s, it is descending.

FOLLOW-UP EXERCISE.　At what distance from the net did the puck start to descend?

DID YOU LEARN?

➡ The vertical components of a dropped ball and a ball horizontally projected from the same height are the same, and the balls fall at the same rate hitting the ground at the same time.

➡ The v_x component is constant in projectile motion because there is no acceleration in that direction, whereas the v_y component changes with time because of the acceleration due to gravity.

*3.4　Relative Velocity

LEARNING PATH QUESTIONS

➡ What is relative in relative velocity?

➡ When adding velocity vectors, such as $\vec{v}_{ac} = \vec{v}_{ab} + \vec{v}_{bc}$, what do the subscripts stand for and how is it known that the equation is set up correctly?

Velocity is not absolute, but is dependent on the observer. That is, its description is *relative* to the observer's state of motion. If an object is observed moving with a certain velocity, then that velocity must be relative to something else. For example, a bowling ball moves down the alley with a certain velocity, and its velocity is relative to the alley. The motions of objects are often described as being relative to the Earth or ground, which is commonly thought of as a *stationary* frame of reference. In other instances it may be convenient to use a *moving* frame of reference.

Measurements must be made with respect to some reference. This reference is usually taken to be the origin of a coordinate system. The point you designate as the origin of a set of coordinate axes is arbitrary and entirely a matter of choice. For example, you may "attach" the coordinate system to the road or the ground and then measure the displacement or velocity of a car relative to these axes. For a "moving" frame of reference, the coordinate axes may be attached to a car moving along a highway. In analyzing motion from another reference frame, you do not change the physical situation or what is taking place, only the point of view from which you describe it. Hence, motion is *relative* (to some reference frame), and is referred to as **relative velocity**. Since velocity is a vector, vector addition and subtraction are helpful in determining relative velocities.

RELATIVE VELOCITIES IN ONE DIMENSION

When the velocities are linear (along a straight line) in the same or opposite directions and all have the same reference (such as the ground), the relative velocities can be found by using vector subtraction. As an illustration, consider cars moving with constant velocities along a straight, level highway, as in ▼ Fig. 3.19. The velocities of the cars shown in the figure are *relative to the Earth, or the ground*, as indicated by the reference set of coordinate axes in Fig. 3.19a, with motions along the *x*-axis. They are also relative to the stationary observers standing by the highway and sitting in the parked car A. That is, these observers see the cars as moving with velocities $\vec{v}_B = +90$ km/h and $\vec{v}_C = -60$ km/h. The relative velocity of two objects is given by the velocity (vector) difference between them. For example, the velocity of car B *relative to car A* is given by

$$\vec{v}_{BA} = \vec{v}_B - \vec{v}_A = (+90 \text{ km/h})\hat{x} - 0 = (+90 \text{ km/h})\hat{x}$$

Thus, a person sitting in car A would see car B move away (in the positive *x*-direction) with a speed of 90 km/h. For this linear case, the directions of the velocities are indicated by plus and minus signs (in addition to the minus sign in the equation).

Similarly, the velocity of car C relative to an observer in car A is

$$\vec{v}_{CA} = \vec{v}_C - \vec{v}_A = (-60 \text{ km/h})\hat{x} - 0 = (-60 \text{ km/h})\hat{x}$$

The person in car A would see car C approaching (in the negative *x*-direction) with a speed of 60 km/h.

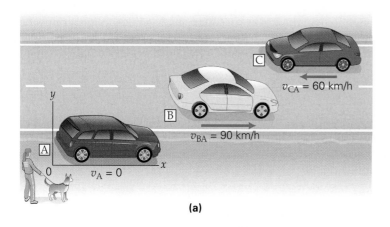

(a)

◀**FIGURE 3.19 Relative velocity** The observed velocity of a car depends on, or is relative to, the frame of reference. The velocities shown in **(a)** are relative to the ground or to the parked car. In **(b)**, the frame of reference is with respect to car B, and the velocities are those that a driver of car B would observe. (See text for description.) **(c)** These aircraft, performing air-to-air refueling, are normally described as traveling at hundreds of kilometers per hour. To what frame of reference do these velocities refer? What is their velocity relative to each other?

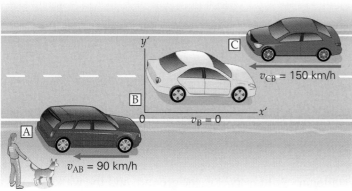

(b)

(c)

But suppose that you want to know the velocities of the other cars *relative to car B* (that is, from the point of view of an observer in car B) or relative to a set of coordinate axes with the origin fixed on car B (Fig. 3.19b). Relative to these axes, car B is not moving; it acts as the fixed reference point. The other cars are moving relative to car B. The velocity of car C relative to car B is

$$\vec{v}_{CB} = \vec{v}_C - \vec{v}_B = (-60 \text{ km/h})\hat{x} - (+90 \text{ km/h})\hat{x} = (-150 \text{ km/h})\hat{x}$$

Similarly, car A has a velocity relative to car B of

$$\vec{v}_{AB} = \vec{v}_A - \vec{v}_B = 0 - (+90 \text{ km/h})\hat{x} = (-90 \text{ km/h})\hat{x}$$

Notice that relative to B, the other cars are both moving in the negative *x*-direction. That is, car C is approaching car B with a velocity of 150 km/h in the negative *x*-direction, and car A appears to be receding from car B with a velocity of 90 km/h in the negative *x*-direction. (Imagine yourself in car B, and take that position as stationary. Car C would appear to be coming toward you at a high speed, and car A would be getting farther and farther away, as though it were moving backward relative to you.) Note that in general,

$$\vec{v}_{AB} = -\vec{v}_{BA}$$

(Prove this for yourself.)

What about the velocities of cars A and B relative to car C? From the point of view (or reference point) of car C, both cars A and B would appear to be approaching or moving in the positive *x*-direction. For the velocity of car B relative to car C,

$$\vec{v}_{BC} = \vec{v}_B - \vec{v}_C = (90 \text{ km/h})\hat{x} - (-60 \text{ km/h})\hat{x} = (+150 \text{ km/h})\hat{x}$$

Can you show that $\vec{v}_{AC} = (+60 \text{ km/h})\hat{x}$? Also note the situation in Fig. 3.19c.

In some instances, velocities do not all have the same reference point. In such cases, relative velocities can be found by means of vector addition. To solve problems of this kind, *it is essential to identify the velocity references with care.*

Let's look first at a one-dimensional (linear) example. Suppose that a straight moving walkway in a major airport moves with a velocity of $\vec{v}_{wg} = (+1.0 \text{ m/s})\hat{x}$, where the subscripts indicate the velocity of the walkway (w) relative to the ground (g). A passenger (p) on the walkway (w) trying to make a flight connection walks with a velocity of $\vec{v}_{pw} = (+2.0 \text{ m/s})\hat{x}$ relative to the walkway. What is the passenger's velocity relative to an observer standing next to the walkway (that is, relative to the ground)?

This velocity, $\vec{v}_{pg}$, is given by

$$\vec{v}_{pg} = \vec{v}_{pw} + \vec{v}_{wg} = (2.0 \text{ m/s})\hat{x} + (1.0 \text{ m/s})\hat{x} = (3.0 \text{ m/s})\hat{x}$$

Thus, the stationary observer sees the passenger as traveling with speed of 3.0 m/s down the walkway. (Make a sketch, and show how the vectors add.) An explanation of the indicator line on the w symbols follows.

PROBLEM-SOLVING HINT

Notice the pattern of the subscripts in this example. On the right side of the equation, the two inner subscripts out of the four total subscripts are the same (w). Basically, the walkway (w) is used as an intermediate reference frame. The outer subscripts (p and g) are sequentially the same as those for the relative velocity on the left side of the equation. When adding relative velocities, always check to make sure that the subscripts have this relationship—it indicates that you have set up the equation correctly.

What if a passenger got on the walkway going in the opposite direction and walked with the same speed as that of the walkway? Now it is essential to indicate the direction in which the passenger is walking by means of a minus sign: $\vec{v}_{pw} = (-1.0 \text{ m/s})\hat{x}$. In this case, relative to the stationary observer,

$$\vec{v}_{pg} = \vec{v}_{pw} + \vec{v}_{wg} = (-1.0 \text{ m/s})\hat{x} + (1.0 \text{ m/s})\hat{x} = 0$$

so the passenger is stationary with respect to the ground, and the walkway acts as a treadmill. (Good physical exercise.)

RELATIVE VELOCITIES IN TWO DIMENSIONS

Of course, velocities are not always in the same or opposite directions. However, with the knowledge of how to use rectangular components to add or subtract vectors, problems involving relative velocities in two dimensions can be solved, as Examples 3.11 and 3.12 show.

EXAMPLE 3.11 | **Across and Down the River: Relative Velocity and Components of Motion**

The current of a 500-m-wide straight river has a flow rate of 2.55 km/h. A motorboat that travels with a constant speed of 8.00 km/h in still water crosses the river (▼Fig. 3.20). (a) If the boat's bow points directly across the river toward the opposite shore, what is the velocity of the boat relative to the stationary observer sitting at the corner of the bridge? (b) How far downstream will the boat's landing point be from the point directly opposite its starting point?

THINKING IT THROUGH. Careful designation of the given quantities is very important—the velocity of what, relative to what? Once this is done, part (a) should be straightforward. (See the previous Problem-Solving Hint.) For part (b), kinematics is used, where the time it takes the boat to cross the river is the key.

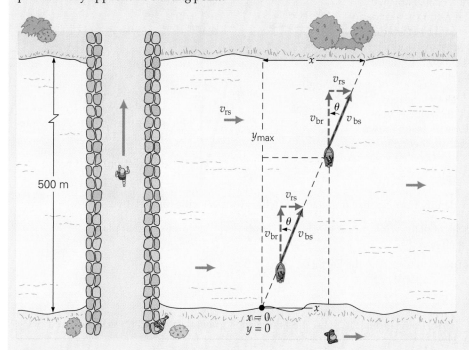

◀**FIGURE 3.20 Relative velocity and components of motion**
As the boat moves across the river, it is carried downstream by the current.

SOLUTION. As indicated in Fig. 3.20, the river's flow velocity ($\vec{v}_{rs}$, river to shore) is taken to be in the x-direction and the boat's velocity ($\vec{v}_{br}$, boat to river) to be in the y-direction. Note that the river's flow velocity is *relative to the shore* and that the boat's velocity is *relative to the river*, as indicated by the subscripts. Listing the data,

Given: $y_{max} = 500$ m   (river width)

$\vec{v}_{rs} = (2.55$ km/h $)\hat{x}$ (velocity of river *relative to shore*)

$= (0.709$ m/s$)\hat{x}$

$\vec{v}_{br} = (8.00$ km/h$)\hat{y}$

$= (2.22$ m/s$)\hat{y}$ (velocity of boat *relative to river*)

Find: (a) $\vec{v}_{bs}$ (velocity of boat *relative to shore*)
(b) x (distance downstream)

Notice that as the boat moves toward the opposite shore, it is also carried downstream by the current. These velocity components would be clearly apparent to the jogger crossing the bridge and to the person sauntering downstream in Fig. 3.20. If both observers stay even with the boat, the velocity of each will match one of the components of the boat's velocity. Since the velocity components are constant, the boat travels in a straight line diagonally across the river (much like the ball rolling across the table in Example 3.1).

(a) The velocity of the boat relative to the shore ($\vec{v}_{bs}$) is given by vector addition. In this case,

$$\vec{v}_{bs} = \vec{v}_{br} + \vec{v}_{rs}$$

(continued on next page)

Since the velocities are not in the same direction and not along one axis, their magnitudes cannot be added directly. Notice in Fig. 3.20 that the vectors form a right triangle, so the Pythagorean theorem can be applied to find the magnitude of v_{bs}:

$$v_{bs} = \sqrt{v_{br}^2 + v_{rs}^2} = \sqrt{(2.22 \text{ m/s})^2 + (0.709 \text{ m/s})^2}$$
$$= 2.33 \text{ m/s}$$

The direction of this velocity is defined by

$$\theta = \tan^{-1}\left(\frac{v_{rs}}{v_{br}}\right) = \tan^{-1}\left(\frac{0.709 \text{ m/s}}{2.22 \text{ m/s}}\right) = 17.7°$$

(b) To find the distance x that the current carries the boat downstream, we use components. Note that in the y-direction, $y_{max} = v_{br}t$, and

$$t = \frac{y_{max}}{v_{br}} = \frac{500 \text{ m}}{2.22 \text{ m/s}} = 225 \text{ s}$$

which is the time it takes the boat to cross the river.

During this time, the boat is carried downstream by the current a distance of

$$x = v_{rs}t = (0.709 \text{ m/s})(225 \text{ s}) = 160 \text{ m}$$

FOLLOW-UP EXERCISE.   What is the distance traveled by the boat in crossing the river?

EXAMPLE 3.12 | ## Flying into the Wind: Relative Velocity

An airplane with an airspeed of 200 km/h (its speed in still air) flies in a direction such that with a west wind of 50.0 km/h, it travels in a straight line northward. (Wind direction is specified by the direction *from* which the wind blows, so a west wind blows from west to east.) To maintain its course due north, the plane must fly at an angle, as illustrated in ▶Fig. 3.21. What is the speed of the plane along its northward path?

THINKING IT THROUGH.   Here again, the velocity designations are important, but as Fig. 3.21 shows, the velocity vectors form a right triangle, and the magnitude of the unknown velocity can be found by using the Pythagorean theorem.

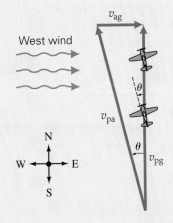

◀**FIGURE 3.21   Flying into the wind**   To fly directly north, the plane's heading (θ-direction) must be west of north.

SOLUTION.   As always, it is important to identify the reference frame to which the given velocities are relative.

Given: $\vec{v}_{pa}$ = 200 km/h at angle θ (velocity of plane with respect to still air = air speed)

$\vec{v}_{ag}$ = 50.0 km/h east (velocity of air with respect to the Earth, or ground = wind speed)

Plane flies due north with velocity $\vec{v}_{pg}$

Find: v_{pg} (ground speed of plane)

The speed of the plane with respect to the Earth, or the ground, v_{pg}, is called the plane's *ground speed*, and v_{pa} is its airspeed. Vectorially, the respective velocities are related by

$$\vec{v}_{pg} = \vec{v}_{pa} + \vec{v}_{ag}$$

If no wind were blowing ($v_{ag} = 0$), the ground speed and airspeed would be equal. However, a headwind (a wind blowing directly toward the plane) would cause a slower ground speed, and a tailwind would cause a faster ground speed. The situation is analogous to that of a boat going upstream versus downstream.

Here, $\vec{v}_{pg}$ is the resultant of the other two vectors, which can be added by the triangle method. Using the Pythagorean theorem to find v_{pg}, noting that v_{pa} is the hypotenuse of the triangle:

$$v_{pg} = \sqrt{v_{pa}^2 - v_{ag}^2} = \sqrt{(200 \text{ km/h})^2 - (50.0 \text{ km/h})^2} = 194 \text{ km/h}$$

(Note that it was convenient to use the units of kilometers per hour, since the calculation did not involve any other units.)

FOLLOW-UP EXERCISE.   What must be the plane's heading (θ-direction) in this Example for the plane to fly directly north?

DID YOU LEARN?

➥ Relative velocity describes the relative motion between different frames of reference.

➥ The subscripts in an equation to compute relative velocity stand for the different velocities, and for the overall motion, the innermost two subscripts must be the same.

PULLING IT TOGETHER | Soccer, Kinematics, and Vectors

A soccer player kicks a ball at an angle of 50° above the horizontal as shown in ▶ Fig. 3.22. The player is 10.0 m from the base of a rectangular building that is 5.0 m high (with a flat roof). The roof is 10.0 m wide. The ball is kicked with an initial speed of 15.0 m/s.

(a) Show that the ball *will* clear the front wall of the building. (b) Does the maximum height of the ball occur while it is over the roof, before it passes over the roof, or after it passes over the roof? (c) Determine whether the ball lands on the roof or beyond the building. (d) Wherever the ball lands, determine its location relative to the back wall of the building. (e) What is the ball's velocity, in unit vector notation, just before it lands?

THINKING IT THROUGH.   This example demonstrates the use of kinematic equation in 2-dimensions, components, and unit vectors. (a) This involves determining the ball's height above the ground at the time it is in the vertical plane of the front wall. The key here is to determine the time from the constant

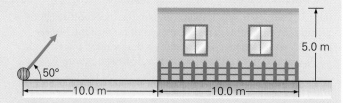

▲ **FIGURE 3.22 Where does the ball go?** See Exampe text for description.

horizontal velocity. Hence, determining the initial velocity components is crucial. (b) Maximum height occurs when the vertical velocity is zero. This enables the determination of the time and hence the horizontal distance. (c) This involves finding the time when the ball is 5.0 m above the ground level. (d) Determining where the ball lands means finding its x-coordinate. Part (c) gives the time. (e) The horizontal velocity does not change, so all that is needed is the vertical velocity at the time of landing.

SOLUTION.

Given: 10.0 m (distance of player from building)   *Find:*   (a) ball's height when it has
        $v_o = 15.0$ m/s at 50° (ball initial velocity)   traveled horizontally 10.0 m
        $h = 5.0$ m (wall height)   (b) where y_{max} occurs
        $L = 100$ m (roof width)   (c) if ball lands on roof
                                                       (d) landing location
                                                       (e) $\vec{v}$ (landing velocity)

(a) Locating the kick at the origin of the x–y coordinate system, the ball's coordinates vary with time according to

$$x = v_{x_o}t$$
$$y = v_{y_o}t - \tfrac{1}{2}gt^2$$

where $v_{x_o} = v_o \cos 50° = 9.64$ m/s and $v_{y_o} = v_o \sin 50° = 11.5$ m/s. To find the time to reach the vertical plane of the near wall, set $x = 10.0$ m in the x equation and solve for t:

$$t = \frac{x}{v_{x_o}} = \frac{10.0 \text{ m}}{9.64 \text{ m/s}} = 1.04 \text{ s}$$

The ball's height at this time is found by substituting this time into the y equation:

$$y = (11.5 \text{ m/s})(1.04 \text{ s}) - \tfrac{1}{2}(9.80 \text{ m/s}^2)(1.04 \text{ s})^2 = 6.35 \text{ m}$$

Since this is greater than the wall height of 5.0 m, the ball does clear the wall.

(b) To find the time (t_{max}) for the ball to get to its maximum height (y_{max}), set the vertical velocity component to zero and solve. The general equation for the vertical velocity is $v_y = v_{y_o} - gt$, hence $0 = v_{y_o} - gt_{max}$, and

$$t_{max} = \frac{v_{y_o}}{g} = \frac{11.5 \text{ m/s}}{9.80 \text{ m/s}^2} = 1.17 \text{ s}$$

Use this time to determine the ball's horizontal location (x) when it is at its maximum height:

$$x = v_{x_o}t_{max} = (9.64 \text{ m/s})(1.17 \text{ s}) = 11.3 \text{ m}$$

Since x is less than the 20 m horizontal distance to the back wall but greater than the 10 m horizontal distance to the near wall, it reaches its maximum height while over the roof.

(c) To determine whether the ball lands on the roof, the time for it to reach the roof height ($y = 5.0$ m) is needed. There will be two such times (why?). The general equation for the ball's height above the ground is $y = v_{y_o}t - \tfrac{1}{2}gt^2$, which, after substituting, becomes $5.00 = 11.5t - 4.90t^2$, expressed in meters and seconds. (Units are omitted for convenience.) In standard quadratic form this is $4.90t^2 - 11.5t + 5.00 = 0$. The quadratic formula yields the two roots:

$$t = \frac{11.5 \pm \sqrt{(-11.5)^2 - 4(4.90)(5.00)}}{2(4.90)} = \frac{11.5 \pm 5.85}{9.80}$$

$$\text{and}\quad t = 0.58 \text{ s or } 1.77 \text{ s}$$

Since it takes the ball 1.04 s to reach the front wall's plane, the 0.58 s is the time when the ball is 5.0 m high as it rises on its way to the building. The 1.77 s refers to the time at which the ball is 5.0 m high on the way down after clearing the front wall. This latter time needs to be compared to the time to reach the vertical plane of the back wall, where $x = 20.0$ m. To get to the plane of the back wall requires a time of

$$t = \frac{x}{v_{x_o}} = \frac{20.0 \text{ m}}{9.64 \text{ m/s}} = 2.08 \text{ s}$$

Since this time is longer than 1.77 s, the ball lands on the roof.

(d) To determine where on the roof the ball lands, find x for the 1.77 s time in part (c).

$$x = v_{x_o}t = (9.64 \text{ m/s})(1.77 \text{ s}) = 17.1 \text{ m}$$

(continued on next page)

Since the back wall is at $x = 20.0$ m, the ball lands $20.0 - 17.1$ or 2.9 m before the rear edge of the roof.

(e) The velocity just before landing on the roof requires both x- and y-components. The x-component is constant, hence $v_x = +9.64$ m/s. The y-component is determined from the 1.77 s elapsed time and the vertical velocity equation,

$$v_y = v_{y_0} - gt = 11.5 \text{ m/s} - (9.80 \text{ m/s}^2)(1.77 \text{ s}) = -5.85 \text{ m/s}$$

As expected, the vertical component is negative and smaller in magnitude than the initial vertical component (why?). Finally, the velocity just before hitting the rooftop in unit vector notation is

$$\vec{v} = v_x\hat{x} + v_y\hat{y} = (9.64\hat{x} - 5.85\hat{y}) \text{ m/s}$$

Learning Path Review

- Motion in two dimensions is analyzed by considering the motion of linear components. The connecting factor between components is time.

   Components of Initial Velocity:

   $$v_{x_0} = v_0 \cos\theta \qquad (3.1a)$$
   $$v_{y_0} = v_0 \sin\theta \qquad (3.1b)$$

   Components of Displacement *(constant acceleration only)*:

   $$x = x_0 + v_{x_0}t + \tfrac{1}{2}a_x t^2 \qquad (3.3a)$$
   $$y = y_0 + v_{y_0}t + \tfrac{1}{2}a_y t^2 \qquad (3.3b)$$

   Components of Velocity *(constant acceleration only)*:

   $$v_x = v_{x_0} + a_x t \qquad (3.3c)$$
   $$v_y = v_{y_0} + a_y t \qquad (3.3d)$$

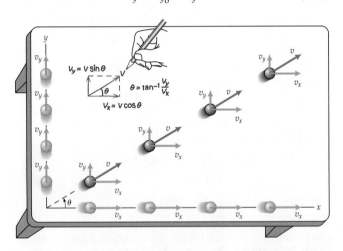

- Of the various methods of vector addition, the component method is most useful. A resultant vector can be expressed in **magnitude–angle form** or in **unit vector component form**.

- **Vector Representation:**

$$\left. \begin{array}{l} C = \sqrt{C_x^2 + C_y^2} \\[2mm] \theta = \tan^{-1}\left|\dfrac{C_y}{C_x}\right| \end{array} \right\} \quad \textit{(magnitude–angle form)} \qquad (3.4a)$$

$$\vec{C} = C_x\hat{x} + C_y\hat{y} \qquad \textit{(component form)} \qquad (3.7)$$

- **Projectile motion** is analyzed by considering horizontal and vertical components separately—constant velocity in the horizontal direction and an acceleration due to gravity, g, in the downward vertical direction. (The foregoing equations for constant acceleration then have an acceleration of $a = -g$ instead of a.)

- **Range (R)** is the maximum horizontal distance traveled.

$$R = \frac{v_0^2 \sin 2\theta}{g} \qquad \begin{array}{l}\textit{(projectile range } x_{\max} \\ \textit{only for } y_{\text{initial}} = y_{\text{final}}) \end{array} \qquad (3.11)$$

$$R_{\max} = \frac{v_0^2}{g} \qquad (y_{\text{initial}} = y_{\text{final}}) \qquad (3.12)$$

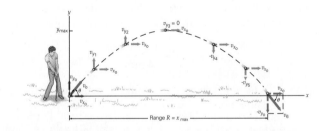

- **Relative velocity** is expressed *relative* to a particular reference frame.

Learning Path Questions and Exercises   For instructor-assigned homework, go to www.masteringphysics.com

MULTIPLE CHOICE QUESTIONS

3.1 COMPONENTS OF MOTION

1. On Cartesian axes, the x-component of a vector is generally associated with a (a) cosine, (b) sine, (c) tangent, (d) none of the foregoing.

2. The equation $x = x_0 + v_{x_0}t + \tfrac{1}{2}a_x t^2$ applies (a) to all kinematic problems, (b) only if v_{y_0} is zero, (c) to constant accelerations, (d) to negative times.

3. For an object in curvilinear motion, (a) the object's velocity components are constant, (b) the y-velocity component is necessarily greater than the x-velocity component, (c) there is an acceleration nonparallel to the object's path, (d) the velocity and acceleration vectors must be at right angles (90°).

4. Which one of the following *cannot* be a true statement about an object: (a) It has zero velocity and a nonzero acceleration; (b) it has velocity in the x-direction and acceleration in the y-direction; (c) it has velocity in the y-direction and acceleration in the y-direction; of (d) it has constant velocity and changing acceleration?

3.2 VECTOR ADDITION AND SUBTRACTION

5. Two linear vectors of magnitudes 3 and 4 are added. The magnitude of the resultant vector is (a) 1, (b) 7, (c) between 1 and 7.

6. The resultant of $\vec{A} - \vec{B}$ is the same as (a) $\vec{B} - \vec{A}$, (b) $-\vec{A} + \vec{B}$, (c) $-(\vec{A} + \vec{B})$, (d) $-(\vec{B} - \vec{A})$.

7. A unit vector has (a) magnitude, (b) direction, (c) neither of these, (d) both of these.

3.3 PROJECTILE MOTION

8. If air resistance is neglected, the motion of an object projected at an angle consists of a uniform downward acceleration combined with (a) an equal horizontal acceleration, (b) a uniform horizontal velocity, (c) a constant upward velocity, (d) an acceleration that is always perpendicular to the path of motion.

9. A football is thrown on a long pass. Compared to the ball's initial horizontal velocity component, the velocity at the highest point is (a) greater, (b) less, (c) the same.

10. A football is thrown on a long pass. Compared to the ball's initial vertical velocity, the vertical component of its velocity at the highest point is (a) greater, (b) less, (c) the same.

*3.4 RELATIVE VELOCITY

11. You are traveling in a car on a straight, level road going 70 km/h. A car coming toward you appears to be traveling 130 km/h. How fast is the other car going: (a) 130 km/h, (b) 60 km/h, (c) 70 km/h, or (d) 80 km/h?

12. Two cars approach each other on a straight, level highway. Car A travels at 60 km/h and car B at 80 km/h. The driver of car B sees car A approaching at a speed of (a) 60 km/h, (b) 80 km/h, (c) 20 km/h, (d) greater than 100 km/h.

13. For the situation in Exercise 12, at what speed does the driver of car A see car B approaching: (a) 60 km/h, (b) 80 km/h, (c) 20 km/h, or (d) greater than 100 km/h?

CONCEPTUAL QUESTIONS

3.1 COMPONENTS OF MOTION

1. Can the x-component of a vector be greater than the magnitude of the vector? How about the y-component? Explain.

2. Is it possible for an object's velocity to be perpendicular to the object's acceleration? If so, describe the motion.

3. Describe the motion of an object that is initially traveling with a constant velocity and then receives an acceleration of constant magnitude (a) in a direction parallel to the initial velocity, (b) in a direction perpendicular to the initial velocity, and (c) that is always perpendicular to the instantaneous velocity or direction of motion.

3.2 VECTOR ADDITION AND SUBTRACTION

4. What are the conditions for two vectors to add to zero?

5. (a) Can a vector be less than one of its components? (b) How about equal to one of its components?

6. Can a nonzero vector have a zero x-component? Explain.

7. Is it possible to add a vector quantity to a scalar quantity?

8. Can $\vec{A} + \vec{B}$ equal zero, when $\vec{A}$ and $\vec{B}$ have nonzero magnitudes? Explain.

9. Are any of the vectors in ▸ Fig. 3.23 equal?

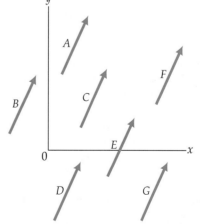

◀ FIGURE 3.23
Different vectors?
See Conceptual
Question 9.

3.3 PROJECTILE MOTION

10. A golf ball is hit on a level fairway. When it lands, its velocity vector has rotated through an angle of 90°. What was the launch angle of the golf ball? [*Hint*: See Fig. 3.11.]

11. Figure 3.10b shows a multiflash photograph of one ball dropping from rest and, at the same time, another ball projected horizontally from the same height. The two balls hit the ground at the same time. Explain.

12. In ▼Fig. 3.24, a spring-loaded "cannon" on a wheeled car fires a metal ball vertically. The car is given a push and set in motion horizontally with constant velocity. A pin is pulled with a string to launch the ball, which travels upward and then falls back into the moving cannon every time. Why does the ball always fall back into the cannon? Explain.

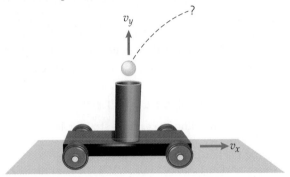

▲ **FIGURE 3.24   A ballistics car** See Conceptual Question 12 and Exercise 52.

13. A rifle is sighted-in so that a bullet hits the bull's-eye of a target 1000 m away on the same level. (a) If the same sighting is used to shoot a target uphill, should one aim above, below, or right at the bull's-eye? (b) How about shooting downhill?

*3.4 RELATIVE VELOCITY

14. Sitting in a parked bus, you suddenly look up at a bus moving alongside and it appears that you are moving. Why is this? How about with both buses moving in opposite directions?

15. A student walks on a treadmill moving at 4.0 m/s and remains at the same place in the gym. (a) What is the student's velocity relative to the gym floor? (b) What is the student's speed relative to the treadmill?

16. You are running in the rain along a straight sidewalk to your dorm. If the rain is falling vertically downward relative to the ground, how should you hold your umbrella so as to minimize the rain landing on you? Explain.

17. When driving to the basket for a layup, a basketball player usually tosses the ball gently upward relative to herself. Explain why.

18. When you are riding in a fast-moving car, in what direction would you throw an object up so it will return to your hand? Explain.

EXERCISES

Integrated Exercises (IEs) are two-part exercises. The first part typically requires a conceptual answer choice based on physical thinking and basic principles. The following part is quantitative calculations associated with the conceptual choice made in the first part of the exercise.

Many exercise sections will include "paired" exercises. These exercise pairs, identified with red numbers, *are intended to assist you in problem solving and learning. In a pair, the first exercise (even numbered) is worked out in the Study Guide so that you can consult it should you need assistance in solving it. The second exercise (odd numbered) is similar in nature, and its answer is given in Appendix VII at the back of the book.*

3.1 COMPONENTS OF MOTION

1. ● An airplane climbs at an angle of 15° with a horizontal component of speed of 200 km/h. (a) What is the plane's actual speed? (b) What is the magnitude of the vertical component of its velocity?

2. IE ● A golf ball is hit with an initial speed of 35 m/s at an angle less than 45° above the horizontal. (a) The horizontal velocity component is (1) greater than, (2) equal to, (3) less than the vertical velocity component. Why? (b) If the ball is hit at an angle of 37°, what are the initial horizontal and vertical velocity components?

3. IE ● The x- and y-components of an acceleration vector are 3.0 m/s² and 4.0 m/s², respectively. (a) The magnitude of the acceleration vector is (1) less than 3.0 m/s², (2) between 3.0 m/s² and 4.0 m/s², (3) between 4.0 m/s² and 7.0 m/s², (4) equal to 7.0 m/s². (b) What are the magnitude and direction of the acceleration vector?

4. ● If the magnitude of a velocity vector is 7.0 m/s and the x-component is 3.0 m/s, what is the y-component?

5. ●● The x-component of a velocity vector that has an angle of 37° to the +x-axis has a magnitude of 4.8 m/s. (a) What is the magnitude of the velocity? (b) What is the magnitude of the y-component of the velocity?

6. IE ●● A student walks 100 m west and 50 m south. (a) To get back to the starting point, the student must walk in a general direction of (1) south of west, (2) north of east, (3) south of east, (4) north of west. (b) What displacement will bring the student back to the starting point?

7. ●● A student strolls diagonally across a level rectangular campus plaza, covering the 50-m distance in 1.0 min (▼Fig. 3.25). (a) If the diagonal route makes a 37° angle with the long side of the plaza, what would be the distance traveled if the student had walked halfway around the outside of the plaza instead of along the diagonal route? (b) If the student had walked the outside route in 1.0 min at a constant speed, how much time would she have spent on each side?

▲ **FIGURE 3.25   Which way?** See Exercise 7.

8. •• A ball rolls at a constant velocity of 1.50 m/s at an angle of 45° below the +x-axis in the fourth quadrant. If we take the ball to be at the origin at $t = 0$ what are its coordinates (x, y) 1.65 s later?

9. •• A ball rolling on a table has a velocity with rectangular components $v_x = 0.60$ m/s and $v_y = 0.80$ m/s. What is the displacement of the ball in an interval of 2.5 s?

10. •• A hot air balloon rises vertically with a speed of 1.5 m/s. At the same time, there is a horizontal 10 km/h wind blowing. In which direction is the balloon moving?

11. IE •• During part of its trajectory (which lasts exactly 1 min) a missile travels at a constant speed of 2000 mi/h while maintaining a constant orientation angle of 20° from the vertical. (a) During this phase, what is true about its velocity components: (1) $v_y > v_x$, (2) $v_y = v_x$, or (3) $v_y < v_x$? [*Hint*: Make a sketch and be careful of the angle.] (b) Determine the two velocity components analytically to confirm your choice in part (a) and also calculate how far the missile will rise during this time.

12. •• At the instant a ball rolls off a rooftop it has a horizontal velocity component of +10.0 m/s and a vertical component (downward) of 15.0 m/s. (a) Determine the angle of the roof. (b) What is the ball's speed as it leaves the roof?

13. •• A particle moves at a speed of 3.0 m/s in the +x-direction. Upon reaching the origin, the particle receives a continuous constant acceleration of 0.75 m/s² in the −y-direction. What is the position of the particle 4.0 s later?

14. •• At a constant speed of 60 km/h, an automobile travels 700 m along a straight highway that is inclined 4.0° to the horizontal. An observer notes only the vertical motion of the car. What is the car's (a) vertical velocity magnitude and (b) vertical travel distance?

15. ••• A baseball player hits a home run into the right field upper deck. The ball lands in a row that is 135 m horizontally from home plate and 25.0 m above the playing field. An avid fan measures its time of flight to be 4.10 s. (a) Determine the ball's average velocity components. (b) Determine the magnitude and angle of its average velocity. (c) Explain why you cannot determine its average *speed* from the data given.

3.2. VECTOR ADDITION AND SUBTRACTION

16. • Using the triangle method, show graphically that (a) $\vec{A} + \vec{B} = \vec{B} + \vec{A}$ and (b) if $\vec{A} - \vec{B} = \vec{C}$, then $\vec{A} = \vec{B} + \vec{C}$

17. IE • (a) Is vector addition associative? That is, does $(\vec{A} + \vec{B}) + \vec{C} = \vec{A} + (\vec{B} + \vec{C})$? (b) Justify your answer graphically.

18. • The vectors $\vec{A}$ and $\vec{B}$ are perpendicular to one another and in the same plane. Prove that $\vec{A}_x\vec{B}_x + \vec{A}_y\vec{B}_y = 0$.

19. • (a) What is the sum of $\vec{A} = 3.0\,\hat{x} + 5.0\,\hat{y}$ and $\vec{B} = 1.0\,\hat{x} - 3.0\,\hat{y}$? (b) What are the magnitude and direction of $\vec{A} + \vec{B}$?

20. • For the two vectors $\vec{x}_1 = (20\text{ m})\hat{x}$ and $\vec{x}_2 = (15\text{ m})\hat{x}$, compute and show graphically (a) $\vec{x}_1 + \vec{x}_2$, (b) $\vec{x}_1 - \vec{x}_2$, and (c) $\vec{x}_2 - \vec{x}_1$.

21. •• If the vector is added to vector , the result is . If is subtracted from , the result is . What is the magnitude of $\vec{A}$?

22. •• Two boys are pulling a box across a horizontal floor as shown in ▼Fig 3.26. If $F_1 = 50.0$ N and $F_2 = 100$ N, find the resultant (or sum) force by (a) the graphical method and (b) the component method.

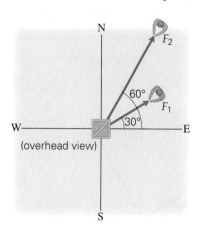

◀ FIGURE 3.26
Adding force vectors
See Exercises 22 and 43.

23. •• For each of the given vectors, give a vector that, when added to it, yields a *null vector* (a vector with a magnitude of zero). Express the vector in the form other than that in which it is given (component or magnitude–angle): (a) $\vec{A} = 4.5$ cm, 40° above the +x-axis; (b) $\vec{B} = (2.0\text{ cm})\hat{x} - (4.0\text{ cm})\hat{y}$; (c) $\vec{C} = 8.0$ cm at an angle of 60° above the −x-axis.

24. IE •• (a) If each of the two components (x and y) of a vector are doubled, (1) the vector's magnitude doubles, but the direction remains unchanged; (2) the vector's magnitude remains unchanged, but the direction angle doubles; or (3) both the vector's magnitude and direction angle double. (b) If the x- and y-components of a vector of 10 m at 45° are tripled, what is the new vector?

25. •• Two vectors are given by $\vec{A} = 4.0\,\hat{x} - 2.0\,\hat{y}$ and $\vec{B} = 1.0\,\hat{x} + 5.0\,\hat{y}$. What is (a) $\vec{A} + \vec{B}$, (b) $\vec{B} - \vec{A}$, and (c) a vector $\vec{C}$ such that $\vec{A} + \vec{B} + \vec{C} = 0$?

26. •• Two brothers are pulling their other brother on a sled (▼Fig 3.27). (a) Find the resultant (or sum) of the vectors $\vec{F}_1$ and $\vec{F}_2$ (b) If $\vec{F}_1$ in the figure were at an angle of 27° instead of 37° with the +x-axis, what would be the resultant (or sum) of $\vec{F}_1$ and $\vec{F}_2$?

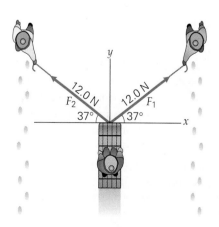

◀ FIGURE 3.27
Vector addition
See Exercise 26.

27. •• Given two vectors, $\vec{A}$ which has a length of 10.0 and makes an angle of 45° below the $-x$-axis, and $\vec{B}$ which has an x-component of $+2.0$ and a y-component of $+4.0$, (a) sketch the vectors on x–y axes, with all their "tails" starting at the origin, and (b) calculate $\vec{A} + \vec{B}$.

28. •• The velocity of object 1 in component form is $\vec{v}_1 = (+2.0 \text{ m/s})\hat{x} + (-4.0 \text{ m/s})\hat{y}$. Object 2 has twice the speed of object 1 but moves in the opposite direction. (a) Determine the velocity of object 2 in component notation. (b) What is the speed of object 2?

29. •• For the vectors shown in ▼Fig. 3.28, determine $\vec{A} + \vec{B} + \vec{C}$.

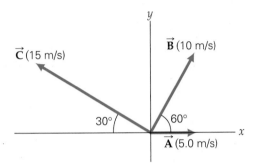

▲ **FIGURE 3.28 Adding vectors**
See Exercises 29 and 30.

30. •• For the velocity vectors shown in Fig. 3.28, determine $\vec{A} - \vec{B} - \vec{C}$.

31. •• Given two vectors $\vec{A}$ and $\vec{B}$ with magnitudes A and B, respectively, you can subtract $\vec{B}$ from $\vec{A}$ to get a third vector $\vec{C} = \vec{A} - \vec{B}$. If the magnitude of $\vec{C}$ is equal to $C = A + B$, what is the relative orientation of vectors $\vec{A}$ and $\vec{B}$?

32. •• In two successive chess moves, a player first moves his queen two squares forward, then moves the queen three steps to the left (from the player's view). Assume each square is 3.0 cm on a side. (a) Using forward (toward the player's opponent) as the positive y-axis and right as the positive x-axis, write the queen's net displacement in component form. (b) At what net angle was the queen moved relative to the leftward direction?

33. •• Referring to the parallelogram in ▼Fig. 3.29, express $\vec{C}, \vec{C} - \vec{B}$, and $(\vec{E} - \vec{D} + \vec{C})$ in terms of $\vec{A}$ and $\vec{B}$.

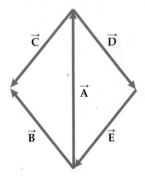

◀ **FIGURE 3.29**
Vector combos
See Exercise 33.

34. •• Two force vectors, $\vec{F}_1 = (3.0 \text{ N})\hat{x} - (4.0 \text{ N})\hat{y}$ and $\vec{F}_2 = (-6.0 \text{ N})\hat{x} + (4.5 \text{ N})\hat{y}$, are applied to a particle. What third force $\vec{F}_3$ would make the net, or resultant, force on the particle zero?

35. •• A student works three problems involving the addition of two different vectors $\vec{F}_1$ and $\vec{F}_2$. He states that the magnitudes of the three resultants are given by (a) $F_1 + F_2$, (b) $F_1 - F_2$, and (c) $\sqrt{F_1^2 + F_2^2}$. Are these results possible? If so, describe the vectors in each case.

36. •• A block weighing 50 N rests on an inclined plane. Its weight is a force directed vertically downward, as illustrated in ▼Fig. 3.30. Find the components of the force parallel to the surface of the plane and perpendicular to it.

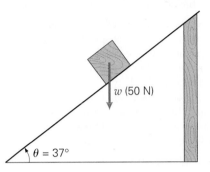

◀ **FIGURE 3.30**
Block on an inclined plane See Exercise 36.

37. •• Two displacements, one with a magnitude of 15.0 m and a second with a magnitude of 20.0 m, can have any angle you want. (a) How would you create the sum of these two vectors so it has the largest magnitude possible? What is that magnitude? (b) How would you orient them so the magnitude of the sum was at its minimum? What value would that be? (c) Generalize the result to any two vectors.

38. ••• A person walks from point A to point B as shown in ▼Fig. 3.31. What is the person's displacement relative to point A?

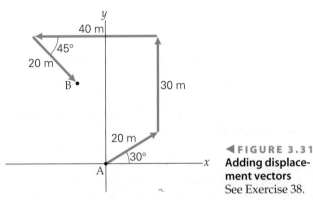

◀ FIGURE 3.31
Adding displacement vectors
See Exercise 38.

39. IE ••• A meteorologist tracks the movement of a thunderstorm with Doppler radar. At 8:00 PM, the storm was 60 mi northeast of her station. At 10:00 PM, the storm is at 75 mi north. (a) The general direction of the thunderstorm's velocity is (1) south of east, (2) north of west, (3) north of east, (4) south of west. (b) What is the average velocity of the storm?

40. IE ••• A flight controller determines that an airplane is 20.0 mi south of him. Half an hour later, the same plane is 35.0 mi northwest of him. (a) The general direction of the airplane's velocity is (1) east of south, (2) north of west, (3) north of east, (4) west of south. (b) If the plane is flying with constant velocity, what is its velocity during this time?

41. IE ●●● ▼Fig. 3.32 depicts a decorative window (the thick inner square) weighing 100 N suspended in a patio opening (the thin outer square). The upper two corner cables are each at 45° and the left one exerts a force (F_1) of 100 N on the window. (a) How does the magnitude of the force exerted by the upper right cable (F_2) compare to that exerted by the upper left cable: (1) $F_2 > F_1$, (2) $F_2 = F_1$, or $F_2 < F_1$? (b) Use your result from part (a) to help determine the force exerted by the bottom cable (F_3).

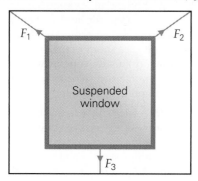

▲ **FIGURE 3.32 A suspended patio window** See Exercise 41.

42. ●●● A golfer lines up for her first putt at a hole that is 10.5 m exactly northwest of her ball's location. She hits the ball 10.5 m and straight, but at the wrong angle, 40° from due north. In order for the golfer to have a "two-putt green," determine (a) the angle of the second putt and (b) the magnitude of the second putt's displacement. (c) Determine why you cannot determine the length of travel of the second putt.

43. ●●● Two students are pulling a box as shown in Fig. 3.26, where $F_1 = 100$ N and $F_2 = 150$ N. What third force would cause the box to be stationary when all three forces are applied?

3.3 PROJECTILE MOTION

(Assume angles to be exact for significant figure purposes.)

44. ● A ball with a horizontal speed of 1.0 m/s rolls off a bench 2.0 m high. (a) How long will the ball take to reach the floor? (b) How far from a point on the floor directly below the edge of the bench will the ball land?

45. ● An electron is ejected horizontally at a speed of 1.5×10^6 m/s from the electron gun of a computer monitor. If the viewing screen is 35 cm from the end of the gun, how far will the electron travel in the vertical direction before hitting the screen? Based on your answer, do you think designers need to worry about this gravitational effect?

46. ● A ball rolls horizontally with a speed of 7.6 m/s off the edge of a tall platform. If the ball lands 8.7 m from the point on the ground directly below the edge of the platform, what is the height of the platform?

47. ● A ball is projected horizontally with an initial speed of 5.0 m/s. Find its (a) position and (b) velocity at $t = 2.5$ s.

48. ● An artillery crew wants to shell a position on level ground 35 km away. If the gun has a muzzle velocity of 770 m/s, to what angle of elevation should the gun be raised?

49. ●● A pitcher throws a fastball horizontally at a speed of 140 km/h toward home plate, 18.4 m away. (a) If the batter's combined reaction and swing times total 0.350 s, how long can the batter watch the ball after it has left the pitcher's hand before swinging? (b) In traveling to the plate, how far does the ball drop from its original horizontal line?

50. IE ●● Ball A rolls at a constant speed of 0.25 m/s on a table 0.95 m above the floor, and ball B rolls on the floor directly under the first ball with the same speed and direction. (a) When ball A rolls off the table and hits the floor, (1) ball B is ahead of ball A, (2) ball B collides with ball A, (3) ball A is ahead of ball B. Why? (b) When ball A hits the floor, how far from the point directly below the edge of the table will each ball be?

51. ●● The pilot of a cargo plane flying 300 km/h at an altitude of 1.5 km wants to drop a load of supplies to campers at a particular location on level ground. Having the designated point in sight, the pilot prepares to drop the supplies. (a) What should the angle be between the horizontal and the pilot's line of sight when the package is released? (b) What is the location of the plane when the supplies hit the ground?

52. ●● A wheeled car with a spring-loaded cannon fires a metal ball vertically (Fig. 3.24). If the vertical initial speed of the ball is 5.0 m/s as the cannon moves horizontally at a speed of 0.75 m/s, (a) how far from the launch point does the ball fall back into the cannon, and (b) what would happen if the cannon were accelerating?

53. ●● A convertible travels down a straight, level road at a slow speed of 13 km/h. A person in the car throws a ball with a speed of 3.6 m/s forward at an angle of 30° to the horizontal. Where is the car when the ball lands?

54. ●● A good-guy stuntman is being chased by bad guys on a building's level roof. He comes to the edge and is to jump to the level roof of a lower building 4.0 m below and 5.0 m away. What is the minimum launch speed the stuntman needs to complete the jump? (Landing on the edge is assumed complete.)

55. ●● An astronaut on the Moon fires a projectile from a launcher on a level surface so as to get the maximum range. If the launcher gives the projectile a muzzle velocity of 25 m/s, what is the range of the projectile? [*Hint*: The acceleration due to gravity on the Moon is only one-sixth of that on the Earth.]

56. ●● In 2004 two Martian probes successfully landed on the Red Planet. The final phase of the landing involved bouncing the probes until they came to rest (they were surrounded by protective inflated "balloons"). During one of the bounces, the telemetry (electronic data sent back to Earth) indicated that the probe took off at 25.0 m/s at an angle of 20° and landed 110 m away (and then bounced again). Assuming the landing region was level, determine the acceleration due to gravity near the Martian surface.

57. ●● In laboratory situations, a projectile's range can be used to determine its speed. To see how this is done, suppose a ball rolls off a horizontal table and lands 1.5 m out from the edge of the table. If the tabletop is 90 cm above the floor, determine (a) the time the ball is in the air, and (b) the ball's speed as it left the table top.

58. ●● A stone thrown off a bridge 20 m above a river has an initial velocity of 12 m/s at an angle of 45° above the horizontal (▼Fig. 3.33). (a) What is the range of the stone? (b) At what velocity does the stone strike the water?

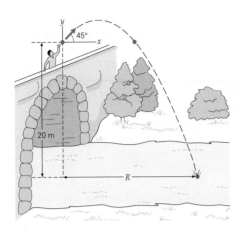

▲ **FIGURE 3.33** **A view from the bridge**
See Exercise 58.

59. ●● If the maximum height reached by a projectile launched on level ground is equal to half the projectile's range, what is the launch angle?

60. ●● William Tell is said to have shot an apple off his son's head with an arrow. If the arrow was shot with an initial speed of 55 m/s and the boy was 15 m away, at what launch angle did Bill aim the arrow? (Assume that the arrow and apple are initially at the same height above the ground.)

61. ●●● This time, William Tell is shooting at an apple that hangs on a tree (▼ Fig. 3.34). The apple is a horizontal distance of 20.0 m away and at a height of 4.00 m above the ground. If the arrow is released from a height of 1.00 m above the ground and hits the apple 0.500 s later, what is the arrow's initial velocity?

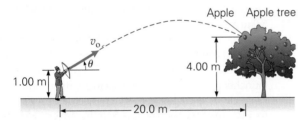

▲ **FIGURE 3.34** **Hit the apple** See Exercise 61.
(Not drawn to scale.)

62. ●●● The apparatus for a popular lecture demonstration is shown in ▼ Fig. 3.35. A gun is aimed directly at a can,

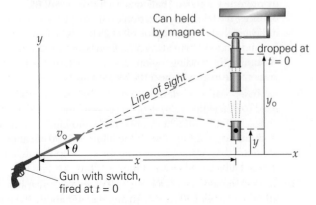

▲ **FIGURE 3.35** **A sure shot** See Exercise 62.
(Not drawn to scale.)

which is released at the same time that the gun is fired. This gun won't miss as long as the initial speed of the bullet is sufficient to reach the falling target before the target hits the floor. Verify this statement, using the figure. [*Hint*: Note that $y_o = x \tan \theta$.]

63. **IE** ●●● A shot-putter launches the shot from a vertical distance of 2.0 m off the ground (from just above her ear) at a speed of 12.0 m/s. The initial velocity is at an angle of 20° above the horizontal. Assume the ground is flat. (a) Compared to a projectile launched at the same angle and speed at ground level, would the shot be in the air (1) a longer time, (2) a shorter time, or (3) the same amount of time? (b) Justify your answer explicitly; determine the shot's range and velocity just before impact in unit vector (component) notation.

64. ●●● A ditch 2.5 m wide crosses a trail bike path (▼Fig 3.36). An upward incline of 15° has been built on the approach so that the top of the incline is level with the top of the ditch. What is the minimum speed a trail bike must be moving to clear the ditch? (Add 1.4 m to the range for the back of the bike to clear the ditch safely.)

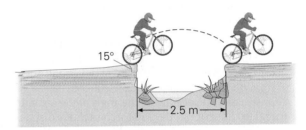

▲ **FIGURE 3.36** **Clear the ditch** See Exercise 64.

65. ●●● A ball rolls down a roof that makes an angle of 30° to the horizontal (▼ Fig. 3.37). It rolls off the edge with a speed of 5.00 m/s. The distance to the ground from that point is two stories or 7.00 m. (a) How long is the ball in the air? (b) How far from the base of the house does it land? (c) What is its speed just before landing?

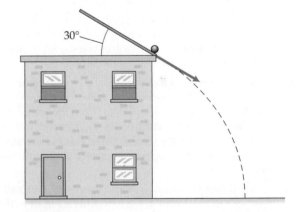

▲ **FIGURE 3.37** **There she rolls** See Exercise 65.
(Not drawn to scale.)

66. ●●● A quarterback passes a football—at a velocity of 50 ft/s at an angle of 40° to the horizontal—toward an intended receiver 30 yd downfield. The pass is released 5.0 ft above the ground. Assume that the receiver is stationary and that he will catch the ball if it comes to him.

Will the pass be completed? If not, will the throw be long or short?

67. ••• A 2.05-m-tall basketball player takes a shot when he is 6.02 m from the basket (at the three-point line). If the launch angle is 25° and the ball was launched at the level of the player's head, what must be the release speed of the ball for the player to make the shot? The basket is 3.05 m above the floor.

*3.4 RELATIVE VELOCITY

68. • While you are traveling in a car on a straight, level interstate highway at 90 km/h, another car passes you in the same direction; its speedometer reads 120 km/h. (a) What is your velocity relative to the other driver? (b) What is the other car's velocity relative to you?

69. • A shopper is in a hurry to catch a bargain in a department store. She walks up the escalator, rather than letting it carry her, at a speed of 1.0 m/s relative to the escalator. If the escalator is 10 m long and moves at a speed of 0.50 m/s, how long does it take for the shopper to get to the next floor?

70. • A motorboat's speed in still water is 2.0 m/s. The driver wants to go directly across a river with a current speed of 1.5 m/s. At what angle upstream should the boat be steered?

71. • A person riding in the back of a pickup truck traveling at 70 km/h on a straight, level road throws a ball with a speed of 15 km/h relative to the truck in the direction opposite to the truck's motion. What is the velocity of the ball (a) relative to a stationary observer by the side of the road, and (b) relative to the driver of a car moving in the same direction as the truck at a speed of 90 km/h?

72. • In Exercise 71, what are the relative velocities if the ball is thrown in the direction of the truck?

73. •• In a 500-m stretch of a river, the speed of the current is a steady 5.0 m/s. How long does a boat take to finish a round trip (upstream and downstream) if the speed of the boat is 7.5 m/s relative to still water?

74. •• A moving walkway in an airport is 75 m long and moves at a speed of 0.30 m/s. A passenger, after traveling 25 m while standing on the walkway, starts to walk at a speed of 0.50 m/s relative to the surface of the walkway. How long does she take to travel the total distance of the walkway?

75. IE •• A swimmer swims north at 0.15 m/s relative to still water across a river that flows at a rate of 0.20 m/s from west to east. (a) The general direction of the swimmer's velocity, relative to the riverbank, is (1) north of east, (2) south of west, (3) north of west, (4) south of east. (b) Calculate the swimmer's velocity relative to the riverbank.

76. •• A boat that travels at a speed of 6.75 m/s in still water is to go directly across a river and back (►Fig. 3.38). The current flows at 0.50 m/s. (a) At what angle(s) must the boat be steered? (b) How long does it take to make the round trip? (Assume that the boat's speed is constant at all times, and neglect turnaround time.)

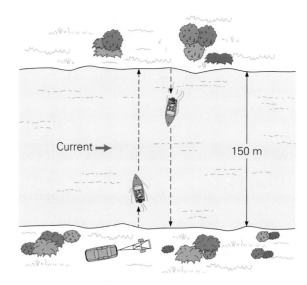

▲ FIGURE 3.38 **Over and back** See Exercise 76. (Not drawn to scale.)

77. •• A pouring rain comes straight down with a raindrop speed of 6.0 m/s. A woman with an umbrella walks eastward at a brisk clip of 1.5 m/s to get home. At what angle should she tilt her umbrella to get the maximum protection from the rain?

78. IE •• It is raining, and there is no wind. When you are sitting in a stationary car, the rain falls straight down relative to the car and the ground. But when you're driving, the rain appears to hit the windshield at an angle. (a) As the velocity of the car increases, this angle (1) also increases, (2) remains the same, (3) decreases. Why? (b) If the raindrops fall straight down at a speed of 10 m/s, but appear to make an angle of 25° to the vertical, what is the speed of the car?

79. •• If the flow rate of the current in a straight river is greater than the speed of a boat in the water, the boat cannot make a trip *directly across* the river. Prove this statement.

80. IE •• You are in a fast powerboat that is capable of a sustained steady speed of 20.0 m/s in still water. On a swift, straight section of a river you travel parallel to the bank of the river. You note that you take 15.0 s to go between two trees on the riverbank that are 400 m apart. (a) (1) Are you traveling with the current, (2) are you traveling against the current, or (3) is there no current? (b) If there is a current [reasoned in part (a)], determine its speed.

81. •• An observer by the side of a straight, level, north-south road watches a car (A) moving south at a rate of 75 km/h. A driver in another car (B) going north at 50 km/h also observes car A. (a) What is car A's velocity as observed from car B? (Take north to be positive.) (b) If the roadside observer sees car A brake to a stop in 6.0 s, what constant acceleration would be measured? (c) What constant acceleration would the driver in car B measure for the braking car A?

82. ••• An airplane flies due north with an air speed of 250 km/h. A steady wind at 75 km/h blows eastward. (Air speed is the speed relative to the air.) (a) What is the plane's ground speed (v_{pg})? (b) If the pilot wants to fly due north, what should his heading be?

83. ••• A shopper in a mall is on an escalator that is moving downward at an angle of 41.8° below the horizontal at a constant speed of 0.75 m/s. At the same time a little boy drops a toy parachute from a floor above the escalator and it descends at a steady vertical speed of 0.50 m/s. Determine the speed of the parachute toy as observed from the moving escalator.

84. ••• An airplane is flying at 150 mi/h (its speed in still air) in a direction such that with a wind of 60.0 mi/h blowing from east to west, the airplane travels in a straight line southward. (a) What must be the plane's heading (direction) for it to fly directly south? (b) If the plane has to go 200 mi in the southward direction, how long does it take?

PULLING IT TOGETHER: MULTICONCEPT EXERCISES

The Multiconcept Exercises require the use of more than one fundamental concept for their understanding and solution. The concepts may be from this chapter, but may include those from previous chapters.

85. A hockey puck slides along a horizontal ice surface at 20.0 m/s, hits a flat vertical wall, and bounces off. Its initial velocity vector makes an angle of 35° with the wall and it comes off at an angle of 25° moving at 10.0 m/s. Choose the +x-axis to be along the wall in the direction of motion and the y-axis to be perpendicular (into) to the wall. (a) Write each velocity in unit vector notation. (b) Determine the change in velocity in unit vector notation. (c) Determine the magnitude and direction, relative to the wall, of this velocity change.

86. A football is kicked off the flat ground at 25.0 m/s at an angle of 30° relative to the ground. (a) Determine the total time it is in the air. (b) Find the angle of its velocity with respect to the ground after it has been in the air for one-fourth of this time. (c) Repeat for one-half and three-fourths of the total time. (d) For each of these times, determine its speed. Comment on the speed changes as it follows its parabolic arc. Do they make sense physically?

87. A railroad flatbed car is set up for a physics demonstration. It is set to roll horizontally on its straight rails at a constant speed of 12.0 m/s. On it is rigged a small launcher capable of launching a small lead ball vertically upward, relative to the bed of the car, at a speed of 25.0 m/s. (a) Compare [by making two sketches] the description of the motion from the point of view of two different observers: one riding on the car and one at rest on the ground next to the tracks. (b) How long does it take the ball to return to its launch location? (c) Compare the ball's velocity at the top of its motion from the viewpoint of each of the two observers and explain any differences. (d) What is the launch angle (relative to the ground) and speed of the ball according to the ground observer? (e) How far down the rails has the car moved when the ball lands back at the launcher? Compare this distance to how far the ball has moved relative to the car.

88. A sailboat is traveling due north at 2.40 m/s on a calm lake with no noticeable water currents. From the crow's nest at the top of its 10.0-m-high mast, one of the passengers drops her digital camera. (a) Make a sketch of the camera's trajectory from the point of view of the passengers on the deck below and from the point of view of passengers on a nearby boat at rest relative to the lake. (b) Determine the camera's initial velocity relative to the ship and relative to the lake surface. (c) How long does the camera take to hit the deck? (d) What is its total travel distance as determined by the boat passengers? (e) Compare this to the magnitude of the net displacement, as determined by the passengers on the nearby boat and comment on why they are different.

89. At a merging on-ramp of a busy Los Angeles freeway, car A is moving directly east on the freeway at a steady speed of 35.0 m/s. Car B is merging onto the freeway from the on-ramp, which points 10° north of due east, moving at 30.0 m/s. (See ▼Fig. 3.39.) If the two cars collide, it will be at the point marked **x** in the figure, which is 350 m down the road from the position of car A. Use the x–y coordinate system to signify E–W versus N–S directions. (a) What is the velocity of car B relative to car A? (b) Show that they do *not* collide at point **x**. (c) Determine how far apart the cars are (and which car is ahead) when car B reaches point **x**.

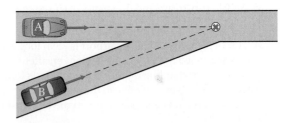

▲ **FIGURE 3.39 Los Angeles freeway**
See Exercise 89.

4 Force and Motion†

PHYSICS FACTS

✦ Isaac Newton was born on Christmas Day, 1642, the same day that Galileo died. (By our current Gregorian calendar, Newton's birth date is January 4, 1643. England did not begin using the Gregorian calendar until 1752.)

✦ Issac Newton
- demonstrated white light to be a mixture of colors and theorized that light was made up of particles, which he called corpuscles, rather than waves. We now have the dual nature of light, with light both behaving as a wave and made up of particles called photons.
- developed the fundamentals of calculus. Gottfried Leibniz, a German mathematician, independently developed a similar version of calculus. There was a lifelong, bitter dispute between Newton and Leibniz over who should receive the credit for doing so first.
- built the first reflecting telescope with a power of 40x.

You don't have to understand any physics to know that what's needed to get the car in the chapter-opening photo moving is a push or a pull. If the frustrated men (or the tow truck that may soon be called) can apply enough *force*, the car will move. But what's keeping the car stuck in the snow? A car's engine can generate plenty of force—so why doesn't the driver just put the car into reverse and back out? For a car to move, another force is needed besides that exerted by the engine—*friction*. Here, the problem is most likely that there is not enough friction between the tires and the snowy surface.

†The mathematics needed in this chapter involves trigonometric functions. You may want to review them in Appendix I.

Chapters 2 and 3 covered how to analyze motion in terms of kinematics. Now our attention turns to the study of *dynamics*—that is, what *causes* motion and changes in motion. This leads to the concepts of force and inertia. The study of force and motion occupied many early scientists. The English scientist Isaac Newton (1642–1727; ◄Fig. 4.1) summarized the various relationships and principles of those early scientists into three statements, or laws, which not surprisingly are known as *Newton's laws of motion*. These laws sum up the concepts of dynamics. In this chapter, you'll learn what Newton had to say about force and motion.

▲ FIGURE 4.1 Isaac Newton
Newton (1642–1727), one of the greatest scientific minds of all time, made fundamental contributions to mathematics, astronomy, and several branches of physics, including optics and mechanics. He formulated the laws of motion and universal gravitation (Section 7.5) and was one of the inventors of calculus. He did some of his most profound work when he was in his mid-twenties.

4.1 The Concepts of Force and Net Force

LEARNING PATH QUESTIONS
➡ What is meant by a force?
➡ What is meant by a net force?

Let's first take a closer look at the meaning of force. It is easy to give examples of forces, but how would you generally define this concept? An operational definition of force is based on observed effects. That is, a force is recognized and described in terms of what it does. From your own experience, you know that *forces can produce changes in motion*. A force can set a stationary object into motion. It can also speed up or slow down a moving object and/or change the direction of its motion. In other words, *a force can produce a change in velocity (speed and/or direction)—that is, an acceleration*. Therefore, an observed *change* in motion, including motion starting from rest, is evidence of a force. This concept leads to a common definition of **force**:

> A force is something that is capable of changing an object's state of motion, that is, changing its velocity or producing an acceleration.

The word *capable* is very significant here. It takes into account the fact that a force may be acting on an object, but its capability to produce a change in motion may be balanced, or canceled, by one or more other forces. The net effect is then zero. Thus, a force may not necessarily produce a change in motion. However, it follows that if a force acts *alone*, the object on which it acts *will* have a change in velocity or an acceleration.

Since a force can produce an acceleration—a vector quantity—force must itself be a vector quantity, with both magnitude and direction. When several forces act on an object, the interest is often in their combined effect—the net force. The **net force**, $\vec{F}_{net}$, is the vector sum $\sum \vec{F}_i$, or resultant, of all the forces acting on an object or system.* Consider the opposite forces illustrated in ►Fig. 4.2a. The net force is zero when forces of equal magnitude act in opposite directions (Fig. 4.2b, where signs are used to indicate directions). Such forces are said to be balanced. A nonzero net force is referred to as an unbalanced force (Fig. 4.2c). In this case, the situation can be analyzed as though only one force equal to the net force were acting. An unbalanced, or nonzero, net force always produces an acceleration. In some instances, an applied unbalanced force may also deform an object, that is, change its size and/or shape (as will be seen in Section 9.1). A deformation involves a change in motion for some part of an object; hence, there is an acceleration.

Forces are sometimes divided into two types or classes. The more familiar of these classes is *contact forces*. Such forces arise because of physical contact between objects.

---

*In the notation $\sum \vec{F}_i$ the Greek letter sigma means the "sum of" the individual forces, as indicated by the *i* subscript: $\sum \vec{F}_i = \vec{F}_1 + \vec{F}_2 + \vec{F}_3 + \cdots$, that is, a vector sum. The *i* subscript is sometimes omitted as being understood, $\sum \vec{F}$.

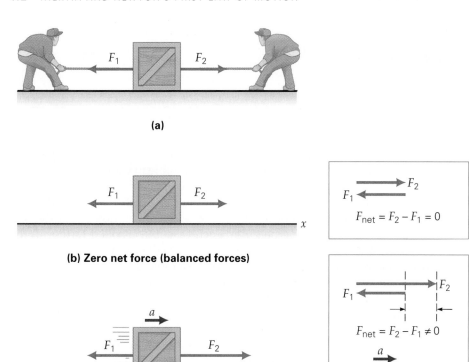

(a)

(b) Zero net force (balanced forces)

$$F_{net} = F_2 - F_1 = 0$$

(c) Nonzero net force (unbalanced forces)

$$F_{net} = F_2 - F_1 \neq 0$$

◀ FIGURE 4.2 **Net force**
(a) Opposite forces are applied to a crate. **(b)** If the forces are of equal magnitude, the vector resultant, or the net force acting on the crate is zero. The forces acting on the crate are said to be balanced. **(c)** If the forces are unequal in magnitude, the resultant is not zero. A nonzero net force (F_{net}), or unbalanced force, then acts on the crate, producing an acceleration (for example, setting the crate in motion if it was initially at rest).

For example, when you push on a door to open it or throw or kick a ball, you exert a contact force on the door or ball.

The other class of forces is called *action-at-a-distance forces*. Examples of these forces include the gravitational force, the electrical force between two charges, and the magnetic force between two magnets. The Moon is attracted to the Earth and maintained in orbit by a gravitational force, but there seems to be nothing physically transmitting that force. (In Chapter 30, the modern view of how such action-at-a-distance forces are thought to be transmitted is given.)

Now, with a better understanding of the concept of force, let's see how force and motion are related through Newton's laws.

DID YOU LEARN?
➥ A force is something that is capable of changing an object's velocity, or producing an acceleration.
➥ A net force is the vector sum of the forces acting on an object.

4.2 Inertia and Newton's First Law of Motion

LEARNING PATH QUESTIONS
➥ What is inertia?
➥ How is inertia related to mass?
➥ In the absence of an unbalanced force, what can be said about an object's motion?

The groundwork for Newton's first law of motion was laid by Galileo. In his experimental investigations, Galileo dropped objects to observe motion under the influence of gravity. (See the related Chapter 2 Insight 2.1, Galileo Galilei and the Leaning Tower of Pisa.) However, the relatively large acceleration due to gravity causes dropped objects to move quite fast and quite far in a short time. From the kinematic equations in Section 2.4, it can be seen that 3.0 s after being dropped, an object in free fall has a speed of about 29 m/s (64 mi/h) and has fallen a distance of 44 m (about 48 yd, or almost half the length of a football field).

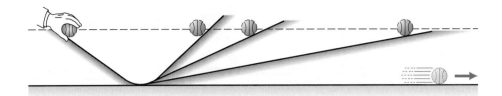

▶ FIGURE 4.3 **Galileo's
experiment** A ball rolls farther
along the upward incline as the
angle of incline is decreased. On a
smooth, horizontal surface, the ball
rolls a greater distance before coming to rest. How far would the ball
travel on an ideal, perfectly smooth
surface? (The ball would slide,
rather than roll, in this case because
of the absence of friction.)

Thus, experimental measurements of free-fall distance versus time were particu-larly difficult to make with the instrumentation available in Galileo's time.

To slow things down so he could study motion, Galileo used balls rolling on inclined planes. He allowed a ball to roll down one inclined plane and then up another with a different degree of incline (▲Fig. 4.3). Galileo noted that the ball rolled to approximately the same height in each case, but it rolled farther in the horizontal direction when the angle of incline was smaller. When allowed to roll onto a horizontal surface, the ball traveled a considerable distance and went even farther when the surface was made smoother. Galileo wondered how far the ball would travel if the horizontal surface could be made perfectly smooth (friction-less). Although this situation is impossible to attain experimentally, Galileo rea-soned that in this ideal case with an infinitely long surface, the ball would continue to travel indefinitely with straight-line, uniform motion, since there would be nothing (no net force) to cause its motion to change. (The ball would actually slide, not roll, in this ideal case of the absence of friction.)

According to Aristotle's theory of motion, which had been accepted for about 1900 years prior to Galileo's time, the normal state of a body was to be at rest (with the exception of celestial bodies, which were thought to be naturally in motion). Aristotle no doubt observed that objects moving on a surface tend to slow down and come to rest, so this conclusion would have seemed logical to him. However, from his experiments, Galileo concluded that bodies in motion exhibit the behav-ior of maintaining that motion, and if an object were initially at rest, it would remain so unless something caused it to move.

Galileo called this tendency of an object to maintain its initial state of motion inertia. That is,

▲ FIGURE 4.4 **A difference in
inertia** The larger punching bag has
more mass and hence more inertia,
or resistance to a change in motion.

> **Inertia** is the natural tendency of an object to maintain a state of rest or to remain in uniform motion in a straight line (constant velocity).

For example, if you've ever tried to stop a slowly rolling automobile by pushing on it, you felt its resistance to a change in motion. Physicists describe the property of inertia in terms of observed behavior. A comparative example of inertia is illus-trated in ◀Fig. 4.4. If the two punching bags have the same density (mass per unit volume; see Section 1.4), the larger one has more mass and therefore more inertia, as you would quickly notice when punching each bag.

Newton related the concept of inertia to mass. Originally, he called mass a quantity of matter, but later redefined it as follows:

> **Mass** is a quantitative measure of inertia.

That is, a massive object has more inertia, or more resistance to a change in motion, than does a less massive object. For example, a car has more inertia than a bicycle.

Newton's first law of motion, sometimes called the *law of inertia*, summarizes these observations:

> In the absence of an unbalanced applied force ($\vec{\mathbf{F}}_{net} = 0$), a body at rest remains at rest, and a body in motion remains in motion with a constant velocity (constant speed and direction).

That is, if the net force acting on an object is zero, then its acceleration is zero. It may be moving with a constant velocity, or be at rest—in both cases, $\Delta\vec{\mathbf{v}} = 0$ or $\vec{\mathbf{v}} = $ constant.

4.3   Newton's Second Law of Motion

LEARNING PATH QUESTIONS

➡ In the expression $F = ma$, what do F and m represent?

➡ What is the difference between mass and weight?

➡ What is the SI unit of weight?

A change in motion, or an acceleration (that is, a change in velocity—speed and/or direction), is evidence of a net force. All experiments indicate that the acceleration of an object is directly proportional to, and in the direction of, the applied net force; that is, in vector notation,

$$\vec{a} \propto \vec{F}_{net}$$

For example, suppose you separately cued two identical billiard balls. If you hit the second ball twice as hard as the first (that is, you applied twice as much force), you would expect the acceleration of the second ball to be twice as great as that of the first ball (and still in the direction of the force).

However, as Newton recognized, the inertia or mass of the object also plays a role. For a given net force, the more massive the object, the less its acceleration will be. For example, if you hit two balls of different masses with the same force, the less massive ball would experience a greater acceleration. Specifically, the acceleration is inversely proportional to mass:

$$\vec{a} \propto \frac{\vec{F}_{net}}{m}$$

or in words,

The acceleration of an object is directly proportional to the net force acting on it and inversely proportional to its mass. The direction of the acceleration is in the direction of the applied net force.

▾ Figure 4.5 presents some illustrations of this principle.

▾ **FIGURE 4.5 Newton's second law** The relationships among force, acceleration, and mass shown here are expressed by Newton's second law of motion (assuming no friction on the cart wheels, which would slide).

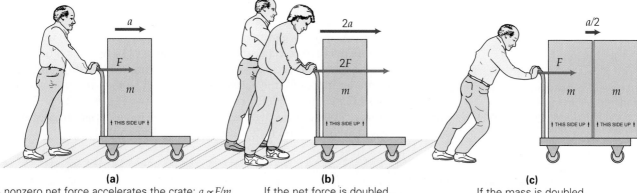

(a)
A nonzero net force accelerates the crate: $a \propto F/m$

(b)
If the net force is doubled, the acceleration is doubled.

(c)
If the mass is doubled, the acceleration is halved.

Rewritten as $\vec{F}_{net} \propto m\vec{a}$. **Newton's second law of motion** is commonly expressed in equation form as

$$\vec{F}_{net} = m\vec{a} \quad \text{(Newton's second law)} \tag{4.1}$$

SI unit of force: newton (N) or
kilogram-meter per second squared (kg·m/s²)

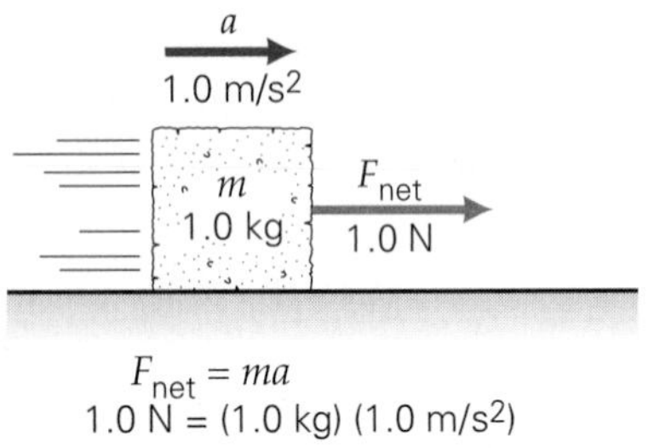

where $\vec{F}_{net} = \Sigma\vec{F}_i$. Equation 4.1 defines the SI unit of force, which is appropriately called the **newton (N)**.

By unit analysis, Eq. 4.1 shows that a newton in base units is defined as $1\,N = 1\,kg \cdot m/s^2$. That is, a net force of 1 N gives a mass of 1 kg an acceleration of $1\,m/s^2$ (◀Fig. 4.6). The British system unit of force is the pound (lb). One pound is equivalent to about 4.5 N (actually, 4.448 N). An average apple weighs about 1 N.

Newton's second law, $\vec{F}_{net} = m\vec{a}$, allows the quantitative analysis of force and motion. It might be thought of as a cause-and-effect relationship, with the force being the cause and acceleration being the motional effect. Notice that if the net force acting on an object is zero, the object's acceleration is zero, and it remains at rest or in uniform motion, which is consistent with Newton's first law. For a nonzero net force (an unbalanced force), the resulting acceleration is in the same direction as the net force.*

▲ FIGURE 4.6 The newton (N) A net force of 1.0 N acting on a mass of 1.0 kg produces an acceleration of 1.0 m/s² (on a frictionless surface).

MASS AND WEIGHT

Equation 4.1 can be used to relate mass and weight. Recall from Section 1.2 that weight is the gravitational force of attraction that a celestial body exerts on an object. For us, this is mainly the force of the gravitational attraction of the Earth. Its effects are easily demonstrated: When you drop an object, it falls (accelerates) toward the Earth. Since there is only one force is acting on the object (air resistance neglected), its **weight ($\vec{w}$)** is the net force $\vec{F}_{net}$, and the acceleration due to gravity ($\vec{g}$) can be substituted for $\vec{a}$ in Eq. 4.1. Therefore in terms of magnitudes,

$$w = mg \tag{4.2}$$

$$(F_{net} = ma)$$

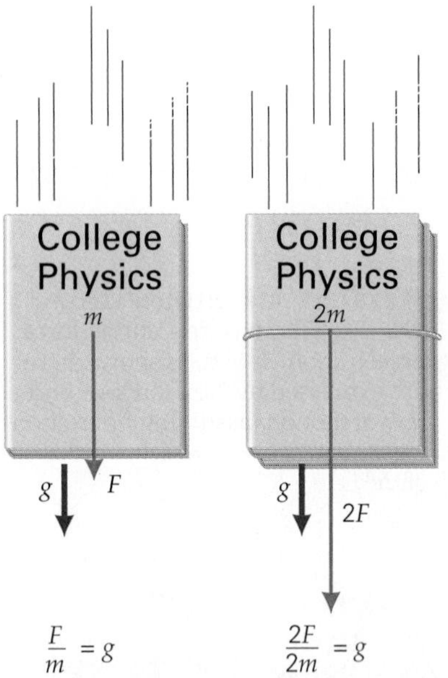

Thus the weight of an object with 1.0 kg of mass is $w = mg = (1.0\,kg)(9.8\,m/s^2) = 9.8\,N$.

That is, 1.0 kg of mass has a weight of approximately 9.8 N, or 2.2 lb, near the Earth's surface. Although weight and mass are simply related through Eq. 4.2, keep in mind that *mass is the fundamental property*. Mass doesn't depend on the value of g, but weight does. As pointed out previously, the acceleration due to gravity on the Moon is about one-sixth that on the Earth. The weight of an object on the Moon would thus be one-sixth of its weight on the Earth, but its mass, which reflects the quantity of matter it contains and its inertia, would be the same in both places.

Newton's second law, along with the fact that $w \propto m$, explains why all objects in free fall have the same acceleration (Section 2.5). Consider, for example, two falling objects, one with twice the mass of the other. The object with twice as much mass would have twice as much weight, or two times as much gravitational force acting on it. But the more massive object also has twice the inertia, so twice as much force is needed to give it the same acceleration. Expressing this relationship mathematically, for the smaller mass (m), the acceleration is $a = F_{net}/m = mg/m = g$, and for the larger mass ($2m$), the acceleration is the same: $a = F_{net}/m = 2mg/(2m) = g$ (◀Fig. 4.7). Some other effects of g, which you may have experienced, are discussed in Insight 4.1, g's of Force and Effects on the Human Body.

▲ FIGURE 4.7 Newton's second law and free fall In free fall, all objects fall with the same constant acceleration g. An object with twice the mass of another has twice as much gravitational force acting on it. But with twice the mass, the object also has twice the inertia, so twice as much force is needed to give it the same acceleration.

*It may appear that Newton's first law of motion is a special case of Newton's second law, but not so. The first law defines what is called an inertial reference frame (Section 26.1): a frame in which Newton's first law holds. That is, in an inertial frame, an object on which there is no net force does not accelerate. Since Newton's first law holds in this frame, the second law of motion ($\vec{F}_{net} = m\vec{a}$) also holds.

INSIGHT 4.1 | *g*'s of Force and Effects on the Human Body

The value of g at the Earth's surface is referred to as the *standard acceleration* and is used as a nonstandard unit. For example, when a spacecraft lifts off, astronauts are said to experience an acceleration of "several g's." This expression means that the astronauts' acceleration is several times the standard acceleration g. Since $g = w/m$, it can be seen that g is the (weight) *force per unit mass*. Thus, the term **g's of force** is used to express force in terms of multiples of the standard acceleration.

To help understand this nonstandard unit of force, let's look at some examples. During the takeoff of a jet airliner, passengers experience an average horizontal force of about $0.20g$. This means that as the plane accelerates down the runway, the seat back exerts a horizontal force on you of about one-fifth of your weight (to accelerate you along with the plane), but you experience a feeling of being pushed back into the seat. On takeoff at an angle of 30°, the force increases to about $0.70g$.

When a person is subjected to several g's vertically, blood can begin to pool in the lower extremities, which may cause blood vessels to distend or capillaries to rupture. Under such conditions, the heart has a difficult time pumping blood throughout the body. At about $4g$, the pooling of blood in the lower body deprives the head of sufficient oxygen. Lack of blood circulation to the eyes can cause temporary blindness, and if the brain is deprived of oxygen, a person becomes disoriented and quickly "blacks out" or loses consciousness. The average person can withstand several g's for only a short period of time.

The maximum force on astronauts in a space shuttle on blastoff is about $3g$. But jet fighter pilots are subjected to as much as $9g$ when pulling out of a downward dive. These pilots wear "g-suits," which are designed to prevent blood pooling. The common g-suit is inflated by compressed air and applies pressure to the pilot's lower body to prevent blood from accumulating there. Work is being done on the development of a hydrostatic g-suit that contains liquid, which is less restrictive than air. When the number of g's increases, the liquid, like the blood in the body, flows into the lower part of the suit and applies pressure to the legs.

On the Earth, where only $1g$ is experienced, a partial "g-suit" of sorts is used to prevent blood clots in patients who have undergone hip replacement surgery. Each year 400 to 800 people die in the first three months after such surgery, primarily because blood clots form in a leg, break off into the bloodstream, and lodge in the lungs—giving rise to a condition called *pulmonary embolism*. In other cases, a blood clot in the leg may slow the flow of blood to the heart. These complications arise more often after hip replacement surgery than after almost any other surgery and occur after the patient has left the hospital.

Studies have shown that pneumatic (operated by air) compression of the legs during the hospital stay reduces these risks. A plastic leg cuff inflates every few minutes, forcing blood from the lower leg (Fig. 1). This mechanical massaging prevents blood from pooling in the veins and clotting. By using both this technique and anticlotting drug therapy, many of the postoperative deaths can be prevented.

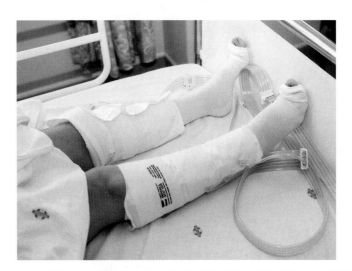

FIGURE 1 Pneumatic massage
The leg cuffs inflate periodically, forcing the blood from the lower legs and preventing it from pooling in the veins, particularly after hip surgery.

Newton's second law, $\vec{\mathbf{F}}_{net} = m\vec{\mathbf{a}}$, allows us to analyze dynamic situations. In using this law, keep in mind that F_{net} is the *magnitude of the net force* and m is the *total mass of the system*. The boundaries defining a system may be real or imaginary. For example, a system might consist of all the gas molecules in a particular sealed vessel. But you might also define a system to be all the gas molecules in an arbitrary cubic meter of air. In dynamics, there are often occasions to work with systems made up of two or more discrete masses—the Earth and Moon, for instance, or a series of blocks on a tabletop, or a tractor and wagon, as in Example 4.1.

EXAMPLE 4.1 | Newton's Second Law: Finding Acceleration

A tractor pulls a loaded wagon on a level road with a constant horizontal force of 440 N (▶Fig. 4.8). If the mass of the wagon is 200 kg and that of the load is 75 kg, what is the magnitude of the wagon's acceleration? (Ignore frictional forces.)

THINKING IT THROUGH. This problem is a direct application of Newton's second law. The two separate masses (wagon and contents) make up the system.

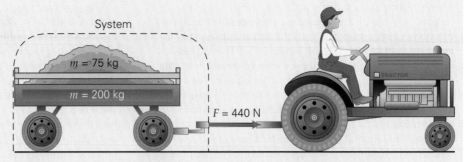

▲**FIGURE 4.8** **Force and acceleration** See Example text for description.

SOLUTION. Listing the given data and what is to be found,

Given: $F = 440$ N
$m_1 = 200$ kg (wagon)
$m_2 = 75$ kg (load)

Find: a (acceleration)

In this case, F is the net force, and the acceleration is given by Eq. 4.2, $F_{net} = ma$, where m is the total mass. Solving for the magnitude of a,

$$a = \frac{F_{net}}{m} = \frac{F_{net}}{m_1 + m_2} = \frac{440 \text{ N}}{200 \text{ kg} + 75 \text{ kg}} = 1.60 \text{ m/s}^2$$

and the direction of a is in the direction of F_{net} or the direction in which the tractor is pulling. Note that m is the *total* mass of the wagon and its contents. In reality, there would be a total opposing force of friction, $-f$. Suppose there were an effective frictional force of magnitude $f = 140$ N. In this case, the net force would be the vector sum of the force exerted by the tractor and the frictional force. Then the acceleration would be (using directional signs)

$$a = \frac{F_{net}}{m} = \frac{F - f}{m_1 + m_2} = \frac{440 \text{ N} - 140 \text{ N}}{275 \text{ kg}} = 1.09 \text{ m/s}^2$$

Again, the direction of a is in the direction of F_{net}.

With a constant net force, the acceleration is also constant, so the kinematic equations of Section 2.4 can be applied. Suppose the wagon started from rest ($v_0 = 0$). Could you find how far it traveled in 4.00 s? Using the appropriate kinematic equation (Eq. 2.11, with $x_0 = 0$) for the case with friction,

$$x = v_0 t + \tfrac{1}{2}at^2 = 0 + \tfrac{1}{2}(1.09 \text{ m/s}^2)(4.00 \text{ s})^2 = 8.72 \text{ m}$$

FOLLOW-UP EXERCISE. Suppose the applied force on the wagon is 550 N. With the same frictional force, what would be the wagon's velocity 4.00 s after starting from rest? (*Answers to all Follow-Up Exercises are given in Appendix VI at the back of the book.*)

EXAMPLE 4.2 | Newton's Second Law: Finding Mass

A student weighs 588 N. What is her mass?

THINKING IT THROUGH. Newton's second law allows us to determine an object's mass if we know the object's weight (force), since g is known.

Given: $w = 588$ N

Find: m (mass)

SOLUTION. Recall that weight is a (gravitational) force and it is related to the mass of an object by $w = mg$ (Eq. 4.2), where g is the acceleration due to gravity (9.80 m/s²). Rearranging the equation,

$$m = \frac{w}{g} = \frac{588 \text{ N}}{9.80 \text{ m/s}^2} = 60.0 \text{ kg}$$

In countries that use the metric system, the kilogram unit of mass is used to express "weight" rather than a force unit. It would be said that this student weighs 60.0 "kilos."

Recall that 1 kg of mass has a weight of 2.2 lb on the Earth's surface. Then in British units, she would weigh
60.0 kg (2.2 lb/kg) = 132 lb.

FOLLOW-UP EXERCISE. (a) A person in Europe is a bit overweight and would like to lose 5.0 "kilos." What would be the equivalent loss in pounds? (b) What is your "weight" in kilos?

As has been learned, a dynamic system may consist of more than one object. In applications of Newton's second law, it is often advantageous, and sometimes necessary, to isolate a given object within a system. This isolation is possible because *the motion of any part of a system is also described by Newton's second law,* as Example 4.3 shows.

EXAMPLE 4.3 | Newton's Second Law: All or Part of the System?

Two blocks with masses $m_1 = 2.5$ kg and $m_2 = 3.5$ kg rest on a frictionless surface and are connected by a light string (►Fig. 4.9).* A horizontal force (F) of 12.0 N is applied to m_1, as shown in the figure. (a) What is the magnitude of the acceleration of the masses (that is, of the total system)? (b) What is the magnitude of the force (T) in the string? [When a rope or string is stretched taut, it is said to be under tension. For a very light string, the force at the right end of the string has the same magnitude (T) as the force at the left end.]

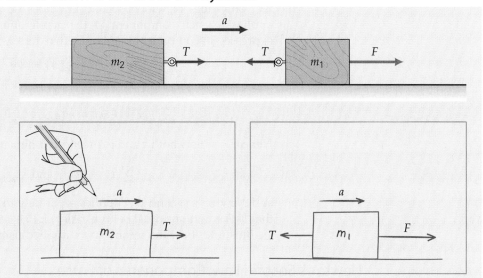

Isolating the masses

▲ **FIGURE 4.9 An accelerated system** See Example text for description.

THINKING IT THROUGH. It is important to remember that Newton's second law may be applied to a total system or any part of it (a subsystem, so to speak). This capability allows for the analysis of a particular component of a system, if desired. Identification of all of the acting forces is critical, as this Example shows. Then $F_{net} = ma$ is applied to each subsystem or component.

SOLUTION. Carefully listing the data and what is to be found:

Given: $m_1 = 2.5$ kg
$m_2 = 3.5$ kg
$F = 12.0$ N

Find: (a) a (acceleration)
(b) T (tension, a force)

Given an applied force, the acceleration of the masses can be found from Newton's second law. It is important to keep in mind that Newton's second law applies to the total system *or to any part of it*—that is, to the total mass ($m_1 + m_2$) or individually to m_1 or m_2. However, *you must be sure to correctly identify the appropriate force or forces in each case.* The net force acting on the combined masses, for example, is not the same as the magnitude of the net force acting on m_2 considered separately, as will be seen.

(a) First, taking the system as a whole (that is, considering both m_1 and m_2), the net force acting on this system is F. Note that in considering the total system, we are concerned only about the net external force acting on it. The *internal* equal and opposite T forces are not a consideration in this case, since they cancel. Then, using Newton's second law:

$$a = \frac{F_{net}}{m} = \frac{F}{m_1 + m_2} = \frac{12.0 \text{ N}}{2.5 \text{ kg} + 3.5 \text{ kg}} = 2.0 \text{ m/s}^2$$

The acceleration of both blocks is in the direction of the applied force, as indicated in the figure.

(b) Under tension, a force is exerted on an object by a string is directed along the string. Note in the figure that it is assumed the tension is transmitted *undiminished* through the string. That is, the tension is the same everywhere in the string. Thus, the magnitude of T acting on m_2 is the same as that acting on m_1. This is actually true only if the string has zero mass. Only such idealized *light* (that is, of negligible mass) strings or ropes will be considered in this book.

So there is a force of magnitude T on each of the masses, because of tension in the connecting string. To find the value of T, a *part* of the system that is affected by this force is considered. Each block may be considered as a separate system to which Newton's second law applies. In these subsystems, the tension comes into play explicitly. Note in the sketch of the isolated m_2 in Fig. 4.9 that the only force acting to accelerate this mass is T. From the values of m_2 and a, the magnitude of this force is given directly by

$$F_{net} = T = m_2a = (3.5 \text{ kg})(2.0 \text{ m/s}^2) = 7.0 \text{ N}$$

An isolated sketch of m_1 is also shown in Fig. 4.9, and Newton's second law can equally well be applied to this block to find T. The forces must be added vectorially to get the net force on m_1 that produces its acceleration. Recalling that vectors in one dimension can be written with directional signs and magnitudes,

$$F_{net} = F - T = m_1a \quad \text{(direction of F taken to be positive)}$$

Then, solving for T,

$$T = F - m_1a$$
$$= 12.0 \text{ N} - (2.5 \text{ kg})(2.0 \text{ m/s}^2) = 12.0 \text{ N} - 5.0 \text{ N} = 7.0 \text{ N}$$

FOLLOW-UP EXERCISE. Suppose that an additional horizontal force to the left of 3.0 N is applied to m_2 in Fig. 4.9. What would be the tension in the connecting string in this case?

*When an object is described as being "light," its mass can be ignored in analyzing the situation given in the problem. That is, here the mass of the string is negligible relative to the other masses.

THE SECOND LAW IN COMPONENT FORM

Not only does Newton's second law hold for any part of a system, but it also applies to each component of the acceleration. For example, a force may be expressed in component notation in two dimensions as follows:

$$\sum \vec{F}_i = m\vec{a}$$

and

$$\sum (F_x\hat{\mathbf{x}} + F_y\hat{\mathbf{y}}) = m(a_x\hat{\mathbf{x}} + a_y\hat{\mathbf{y}}) = ma_x\hat{\mathbf{x}} + ma_y\hat{\mathbf{y}} \qquad (4.3a)$$

Hence, to satisfy both x and y directions independently,

$$\sum F_x = ma_x \quad \text{and} \quad \sum F_y = ma_y \qquad (4.3b)$$

and Newton's second law applies separately to each component of motion. Note that *both* equations must be true. (Also, $\sum F_z = ma_z$ in three dimensions.) Example 4.4 illustrates how the second law is applied using components.

EXAMPLE 4.4 | # Newton's Second Law: Components of Force

A block of mass 0.50 kg travels with a speed of 2.0 m/s in the positive x-direction on a flat, frictionless surface. On passing through the origin, the block experiences a constant force of 3.0 N at an angle of 60° relative to the x-axis for 1.5 s (▶ Fig. 4.10). What is the velocity of the block at the end of this time?

THINKING IT THROUGH.   With the force at an angle to the initial motion, it would appear that the solution is complicated. But note in the insert in Fig. 4.10 that the force can be resolved into components. The motion can then be analyzed in each component direction.

SOLUTION.   Listing the given data and what is to be found:

Given: $m = 0.50$ kg   **Find:** $\vec{v}$ (velocity at the
   $v_{x_0} = 2.0$ m/s   end of 1.5 s)
   $v_{y_0} = 0$
   $F = 3.0$ N, $\theta = 60°$
   $t = 1.5$ s

First let's find the magnitudes of the force components:

$$F_x = F\cos 60° = (3.0\ \text{N})(0.500) = 1.5\ \text{N}$$
$$F_y = F\sin 60° = (3.0\ \text{N})(0.866) = 2.6\ \text{N}$$

Then, applying Newton's second law to each direction to find the components of acceleration,

$$a_x = \frac{F_x}{m} = \frac{1.5\ \text{N}}{0.50\ \text{kg}} = 3.0\ \text{m/s}^2$$

$$a_y = \frac{F_y}{m} = \frac{2.6\ \text{N}}{0.50\ \text{kg}} = 5.2\ \text{m/s}^2$$

Next, from the kinematic equation relating velocity and acceleration (Eq. 2.8), the magnitudes of the velocity components of the block are given by

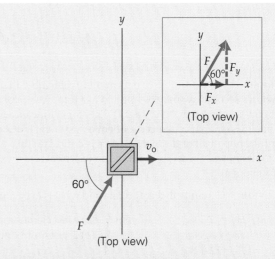

▲ FIGURE 4.10 **Off the straight and narrow** A force is applied to a moving block when it reaches the origin, and the block then begins to deviate from its straight-line path. The vector components are shown in the box.

$$v_x = v_{x_0} + a_x t = 2.0\ \text{m/s} + (3.0\ \text{m/s}^2)(1.5\ \text{s}) = 6.5\ \text{m/s}$$
$$v_y = v_{y_0} + a_y t = 0 + (5.2\ \text{m/s}^2)(1.5\ \text{s}) = 7.8\ \text{m/s}$$

And, at the end of the 1.5 s, the velocity of the block is

$$\vec{v} = v_x\hat{\mathbf{x}} + v_y\hat{\mathbf{y}} = (6.5\ \text{m/s})\hat{\mathbf{x}} + (7.8\ \text{m/s})\hat{\mathbf{y}}$$

FOLLOW-UP EXERCISE.   (a) What is the direction of the velocity at the end of the 1.5 s? (b) If the force were applied at an angle of 30° (rather than 60°) relative to the x-axis, how would the results of this Example be different?

DID YOU LEARN?
➥ In $F = ma$, the symbol F is the net force (F_{net}) and m is the *total* mass of the system.
➥ Mass (m) is an invariant fundamental property. Weight is related to mass ($w = mg$), but can vary with variations in g.
➥ Since weight is a force ($w = mg$), it has units of kg · m/s^2, which is taken to be a newton (N).

4.4 Newton's Third Law of Motion

LEARNING PATH QUESTIONS

➥ Is it possible to have a single force?

➥ For a third law force pair, which force is the action and which is the reaction?

➥ What is a normal force?

Newton formulated a third law that is as far-reaching in its physical significance as the first two laws. For a simple introduction to the third law, consider the forces involved in seatbelt safety. When the brakes are suddenly applied when you are riding in a moving car, because of your inertia you continue to move forward as the car slows. (The frictional force on the seat of your pants is not enough to stop you.) In doing so, you exert forward forces on the seatbelt and shoulder strap. The belt and strap exert corresponding backward reaction forces on you, causing you to slow down with the car. If you hadn't buckled up, you would keep going (Newton's first law) until another backward force, such as that applied by the dashboard or windshield, slowed you down. (In an abrupt collision stop, hopefully the air bags would come into effect; see Chapter 6 Insight 6.1, The Automobile Air Bag and Martian Air Bags.)

We commonly think of forces as occuring singly. However, Newton recognized that it is impossible to have a single force. He observed that in any force application, there is always a mutual interaction; therefore, forces occur in pairs. An example given by Newton was the following: If you press on a stone with a finger, then the finger is also pressed by, or receives a force from, the stone.

Newton termed the paired forces *action* and *reaction*, and **Newton's third law of motion** is as follows:

| For every force (action), there is an equal and opposite force (reaction). |

In symbol notation, Newton's third law may be expressed:

$$\vec{\mathbf{F}}_{12} = -\vec{\mathbf{F}}_{21}$$

That is, $\vec{\mathbf{F}}_{12}$ is the force exerted *on* object 1 *by* object 2, and $-\vec{\mathbf{F}}_{21}$ is the equal and opposite force exerted *on* object 2 *by* object 1. (The minus sign indicates the opposite direction.) *Which force is considered the action or the reaction is arbitrary;* $\vec{\mathbf{F}}_{21}$ may be the reaction to $\vec{\mathbf{F}}_{12}$ or vice versa.

At a glance, Newton's third law may seem to contradict Newton's second law: If there are always equal and opposite forces, how can there be a nonzero net force? An important thing to remember about the force pair of the third law is that *the action–reaction forces do not act on the same object*. The second law is concerned with a force (or forces) acting on a particular object (or system). The opposing forces of the third law act on *different* objects. Hence, these forces cannot cancel each other nor have a vector sum of zero when the second law is applied to the individual objects.

To illustrate this distinction, consider the situations shown in ▶ Fig. 4.11. We often tend to forget reaction forces. For example, in the left portion of Fig. 4.11a, the obvious force that acts on a block sitting on a table is the Earth's gravitational attraction, which is expressed by the weight mg. But, *there has to be another force* acting on the block. For the block not to accelerate, the table must exert an upward force $\vec{\mathbf{N}}$ of which the magnitude is equal to the block's weight. Thus, $\Sigma F_y = +N - mg = ma_y = 0$ where the directions of the vectors are indicated by plus and minus signs.

In reaction to $\vec{\mathbf{N}}$, the block exerts a downward force on the table, $-\vec{\mathbf{N}}$, whose magnitude is the same as the block's weight, mg. However, $-\vec{\mathbf{N}}$ is *not* the object's weight. Weight and $-\vec{\mathbf{N}}$ have two different origins: Weight is the action-at-a-distance gravitational force, and $-\vec{\mathbf{N}}$ is a contact force between the two surfaces.

You can easily demonstrate that this upward force on the block is there by placing the block on your hand and holding it stationary—you exert an upward force on the block and you would feel a reaction force of $-\vec{\mathbf{N}}$ on your hand. If you applied a greater force, that is, $N > mg$, then the block would accelerate upward.

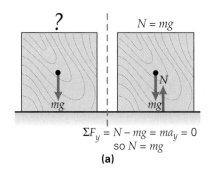

$\Sigma F_y = N - mg = ma_y = 0$
so $N = mg$

(a)

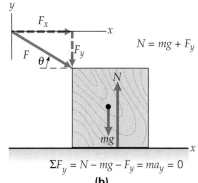

$\Sigma F_y = N - mg - F_y = ma_y = 0$

(b)

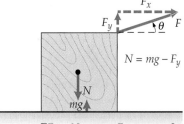

$\Sigma F_y = N - mg + F_y = ma_y = 0$

(c)

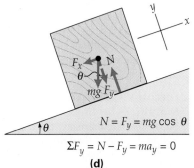

$N = F_y = mg \cos \theta$

$\Sigma F_y = N - F_y = ma_y = 0$

(d)

▲ **FIGURE 4.11 Distinctions between Newton's second and third laws** Newton's second law deals with the forces acting on a particular object (or system). Newton's third law deals with the force pair that acts on different objects. See text for description.

The force that a surface exerts on an object is called a *normal* force and the symbol N is used to denote the force. *Normal* means *perpendicular*. The **normal force** that a surface exerts on an object is always perpendicular to the surface. In Fig. 4.11a, the normal force is equal and opposite to the weight of the block (but not a third law pair. Why?). However, the normal force is not always equal and opposite to an object's weight. The normal force is a "reaction" force; it reacts to the situation. Examples of this given in Figs. 4.11b, c, and d, are described here with the summation of the vertical components ($\sum F_y$).

(Fig. 4.11b) Applied force at a downward angle.

$$\sum F_y: \quad N - mg - F_y = ma_y = 0, \quad \text{and} \quad N = mg + F_y \quad (N > mg)$$

(Fig. 4.11c) Applied force at an upward angle.

$$\sum F_y: \quad N - mg + F_y = ma_y = 0, \quad \text{and} \quad N = mg - F_y \quad (N < mg)$$

(Fig. 4.11d) Block on an inclined plane. (Normal force perpendicular to the surface of the plane.)

$$\sum F_y: \quad N - F_y = ma_y = 0, \quad \text{and} \quad N = F_y = mg \cos \theta$$

where $mg \cos \theta$ is the weight component perpendicular to the plane.

In the case of Fig. 4.11d, the weight component down the plane, F_x, would accelerate the block down the plane in the absence of an equal opposing frictional force between the block and surface of the plane.

CONCEPTUAL EXAMPLE 4.5 | ## Where Are the Newton's Third Law Force Pairs?

A woman waiting to cross the street holds a briefcase in her hand as shown in ▶ Fig. 4.12a. Identify all of the third law force pairs involving the briefcase in this situation.

REASONING AND ANSWER. The briefcase is being held motionless, so its acceleration is zero, and $\sum F_y = 0$. Focusing only on the case, two equal and opposite forces acting on it can be identified—the downward weight of the case and the upward applied force by the hand. However, these two forces *cannot* be a third law force pair because they act on the *same* object.

On an overall inspection, you should realize that the reaction force to the upward force of the hand on the briefcase is a downward force on the hand. Then how about the reaction force to the weight of the case? Since weight is the attractive gravitational force on the case by the Earth, the corresponding force on the Earth by the case makes up the third law force pair.

FOLLOW-UP EXERCISE. The woman inadvertently drops her briefcase as illustrated in Fig. 4.12b. Are there any third law force pairs in this situation? Explain.

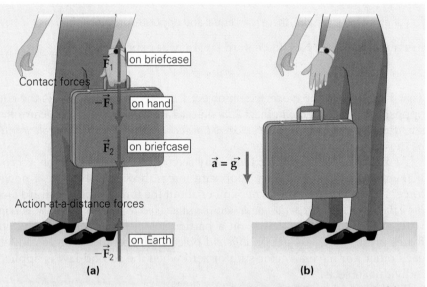

▲ **FIGURE 4.12** **Force pairs of Newton's third law** **(a)** When a person holds a briefcase, there are two force pairs: a contact pair ($\vec{F}_1$ and $-\vec{F}_1$) and an action-at-a-distance (gravity) pair ($\vec{F}_2$ and $-\vec{F}_2$). The net force acting on the briefcase is zero: The upward contact force ($\vec{F}_1$ on the briefcase) balances the downward weight force. Note, however, that the upward contact force and downward weight force are *not* a third law pair. **(b)** Any third law force pairs? See the Follow-Up Exercise.

Jet propulsion is yet another example of Newton's third law in action. In the case of a rocket, the rocket and exhaust gases exert equal and opposite forces on each other. As a result, the exhaust gases are accelerated away from the rocket, and the rocket is accelerated in the opposite direction. When a big rocket "blasts off," as in a Space Shuttle launch, it produces a fiery release of exhaust. A common misconception is that the exhaust gases "push against" the launch pad to accelerate the rocket. If this interpretation were true, there would be no space travel, since there is nothing to "push against" in space. The correct explanation is one of action (the expanding gases exert a force on the rocket) and reaction (the rocket exert a force on the gas).

Another action–reaction pair is given in Insight 4.2, Sailing into the Wind—Tacking.

INSIGHT 4.2 | ## Sailing into The Wind—Tacking

A sailboat can easily sail in the direction of the wind (which fills the sails). However, after sailing some distance in the windward direction, the skipper usually wants to return to home port—which involves somehow "sailing into the wind." This may sound rather impossible, but it isn't. The process is called *tacking*, and can be explained and understood by using force vectors and Newton's laws.

A sailboat cannot sail directly upwind, since the wind force on the sail would accelerate the boat backward, or opposite the desired direction. The wind filling the sail exerts a force F_s perpendicular to the sail (Fig. 1a). If the boat is steered at an angle relative to the wind direction, there is a component of force parallel to the boat's heading ($F_{||}$). On this course, some distance upwind is gained, but it would never get the boat back to port. The perpendicular component ($F_\perp$) acts sideways and would put the boat way off course.

So, being an old salt, the skipper "tacks" or maneuvers the boat so that the parallel force component is changed by 90°

(Fig. 1b). The skipper continually repeats the maneuver, and using this zigzag course, the boat gets back to port (Fig. 2a).

What about the perpendicular force component? You might think that this would take the boat way off course. It would, and does a little, but most of the perpendicular force is balanced by the force of water on the keel of the boat, which is underneath (Fig. 2b). The water resistance exerts an opposite force on the keel, which cancels out most of the sideways perpendicular force, producing little, if any, acceleration in that direction.

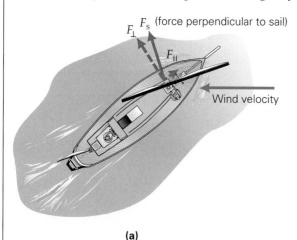

(a)

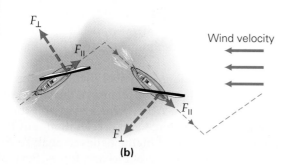

(b)

FIGURE 1 Let's go tacking (a) The wind filling the sail exerts a force perpendicular to the sail (F_s) We can resolve this force vector into components. The one parallel to the motion of the boat ($F_{||}$) has an upwind component. **(b)** By changing the direction of the sail, the skipper can "tack" the boat upwind.

(a)

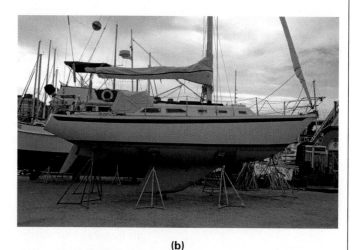

(b)

FIGURE 2 Into the wind (a) As the skipper turns the boat into the wind, the tacking begins. **(b)** The perpendicular force component in tacking would take the boat off course sideways. But the water resistance on the keel under the boat exerts an opposite force and cancels out most of the sideways force.

| ## Tension in a String: Action and Reaction Forces

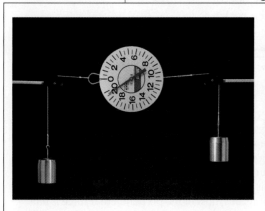 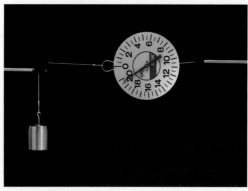 

(a) Two suspended 2-kilogram masses are attached to opposite sides of a scale (calibrated in newtons). The total suspended weight is $w = mg = (4.00 \text{ kg})(9.80 \text{ m/s}^2) = 39.2 \text{ N}$, yet the scale reads about 20 N. Is something wrong with the scale?

(b) No, think of it in this manner. The effect of the weight of the mass on the right is replaced by fixing the end of the string. The other mass stretches the scale spring, giving a reading of about 20 N [or $w = mg = (2.00 \text{ kg})(9.80 \text{ m/s}^2) = 19.6 \text{ N}$].

(c) Similarly, the other end of the scale can be fixed. (A fixed pulley merely changes the direction of the force, and the scale can be hung vertically with the same effect.) In all cases, the tension in the string is 19.6 N, as the scale shows.

DID YOU LEARN?

➥ Forces always occur in pairs, but act on *different* objects.

➥ If $\vec{F}_1$ is the action, then $-\vec{F}_2$ is the reaction, and vice versa.

➥ A normal force is the force a surface exerts on an object and is always perpendicular to the surface.

4.5   More on Newton's Laws: Free-Body Diagrams and Translational Equilibrium

LEARNING PATH QUESTIONS

➥ What is the difference between a space diagram and a free-body diagram?

➥ What is the condition for translational equilibrium, and does equilibrium mean that an object is at rest?

Now that you have been introduced to Newton's laws and some applications in analyzing motion, the importance of these laws should be evident. They are so simply stated, yet so far-reaching. The second law is probably the most often applied, because of its mathematical relationship. However, the first and third laws are often used in qualitative analysis, as our continuing study of the different areas of physics will reveal.

In general, we will be concerned with applications that involve constant forces. Constant forces result in constant accelerations and allow the use of the kinematic equations from Section 2.4 in analyzing the motion. When there is a variable force, Newton's second law holds for the *instantaneous* force and acceleration, but the acceleration will vary with time, requiring advanced mathematics to analyze. So in general, our study will be limited to constant accelerations and forces. Several examples of applications of Newton's second law are presented in this section so that you can become familiar with its use. This small but powerful equation will be used again and again throughout the book.

There is still one more item in the problem-solving arsenal that is a great help with force applications—free-body diagrams. These are explained in the following Problem-Solving Strategy.

PROBLEM-SOLVING STRATEGY: FREE-BODY DIAGRAMS

In illustrations of physical situations, sometimes called *space diagrams*, force vectors are drawn at different locations to indicate their points of application. However, presently being concerned with only linear motions, vectors in *free-body diagrams* (FBDs) may be shown as emanating from a common point, which is usually chosen as the origin of the x–y axes. One of the axes is generally chosen along the direction of the net force acting on an object, since that is the direction in which the object will accelerate. Also, it is often important to resolve force vectors into components, and properly chosen x–y axes simplify this task.

In a free-body diagram, the vector arrows do not have to be drawn exactly to scale. However, the diagram should clearly show whether there is a net force and whether forces balance each other in a particular direction. When the forces aren't balanced, by Newton's second law, there must be an acceleration.

In summary, the general steps in constructing and using free-body diagrams are as follows. (Refer to the accompanying Learn by Drawing 4.1, Forces on an Object on an Inclined Plane and Free-Body Diagrams as you read.)

1. Make a sketch, or space diagram, of the situation (if one is not already available) and identify the forces acting on each body of the system. A space diagram is an illustration of the physical situation that identifies the force vectors.
2. Isolate the body for which the free-body diagram is to be constructed. Draw a set of Cartesian axes, with the origin at a point through which the forces act and with one of the axes along the direction of the body's acceleration. (The acceleration will be in the direction of the net force, if there is one.)
3. Draw properly oriented force vectors (including angles) on the diagram, emanating from the origin of the axes. If there is an unbalanced force, assume a direction of acceleration and indicate it with an acceleration vector. Be sure to include only those forces that act on the isolated body of interest.
4. Resolve any forces that are not directed along the x- or y-axis into x- or y-components (use plus and minus signs to indicate direction). Use the free-body diagram and force components to analyze the situation in terms of Newton's second law of motion. (*Note*: If you assume that the acceleration is in one direction, and in the solution it comes out with the opposite sign, then the acceleration is actually in the opposite direction from that assumed. For example, if you assume that $\vec{a}$ is in the $+x$-direction, but you get a negative answer, then $\vec{a}$ is in the $-x$-direction.)

Free-body diagrams are a particularly useful way of following one of the suggested problem-solving procedures in Section 1.7: Draw a diagram as an aid in visualizing and analyzing the physical situation of the problem. *Make it a practice to draw free-body diagrams for force problems, as done in the following Examples.*

EXAMPLE 4.6 | Up or Down? Motion on a Frictionless Inclined Plane

Two masses are connected by a light string running over a light pulley of negligible friction, as illustrated in the Learn by Drawing (LBD) 4.1 space diagram. One mass ($m_1 = 5.0$ kg) is on a frictionless $20°$ inclined plane, and the other ($m_2 = 1.5$ kg) is freely suspended. What is the acceleration of the masses?

THINKING IT THROUGH. Apply the preceding Problem-Solving Strategy.

SOLUTION. Following the usual procedure of listing the data and what is to be found:

Given: $m_1 = 5.0$ kg   *Find:* $\vec{a}$ (acceleration)
$m_2 = 1.5$ kg
$\theta = 20°$

(To help visualize the forces involved, isolate m_1 and m_2 and draw free-body diagrams for each mass.) For mass m_1, there are three concurrent forces (forces acting through a common point). These forces are T, its weight m_1g, and N, where T is the tension force of the string on m_1 and N is the normal force of the plane on the block. (See 3 in the LBD 4.1.) The forces are shown as emanating from their common point of action. (Recall that a vector can be moved as long as its direction and magnitude are not changed.)

(continued on next page)

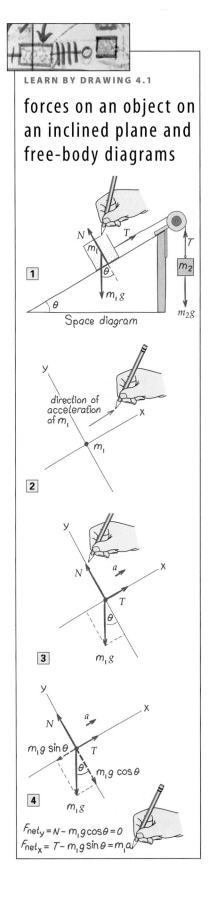

LEARN BY DRAWING 4.1

forces on an object on an inclined plane and free-body diagrams

1   Space diagram

2   direction of acceleration of m_1

3   m_1g

4   m_1g
$F_{net_y} = N - m_1g\cos\theta = 0$
$F_{net_x} = T - m_1g\sin\theta = m_1a$

Start by assuming that m_1 accelerates up the plane, which is taken to be in the $+x$-direction. (It makes no difference whether it is assumed that m_1 accelerates up or down the plane, as will be seen shortly.) Notice that $m_1 g$ (the weight) is broken down into components. The x-component is opposite to the assumed direction of acceleration, and the y-component acts perpendicularly to the plane and is balanced by the normal force N. (There is no acceleration in the y-direction, so there is no net force in this direction.)

Then, applying Newton's second law in component form (Eq. 4.3b) to m_1,

$$\sum F_{x_1} = T - m_1 g \sin \theta = m_1 a$$

$$\sum F_{y_1} = N - m_1 g \cos \theta = m_1 a_y = 0 \quad \text{(} a_y = 0, \text{ no net force, so the forces cancel)}$$

And for m_2,

$$\sum F_{y_2} = m_2 g - T = m_2 a_y = m_2 a$$

where the masses of the string and pulley have been neglected. Since they are connected by a string, the accelerations of m_1 and m_2 have the same magnitudes, so $a_x = a_y = a$.

Then adding the first and last equations to eliminate T,

$$m_2 g - m_1 g \sin \theta = (m_1 + m_2)a$$

(net force = *total* mass × acceleration)

(Note that this is the equation that would be obtained by applying Newton's second law to the system as a whole, because in the system of both blocks, the T forces are internal forces and cancel.)

Then, solving for a:

$$a = \frac{m_2 g - m_1 g \sin 20°}{m_1 + m_2}$$

$$= \frac{(1.5 \text{ kg})(9.8 \text{ m/s}^2) - (5.0 \text{ kg})(9.8 \text{ m/s}^2)(0.342)}{5.0 \text{ kg} + 1.5 \text{ kg}}$$

$$= -0.32 \text{ m/s}^2$$

The minus sign indicates that the acceleration is opposite to the assumed direction. That is, m_1 actually accelerates down the plane, and m_2 accelerates upward. As this example shows, if you assume the acceleration to be in the wrong direction, the sign on the result will give the correct direction anyway.

Could you find the tension force T in the string if asked to do so? How this task could be done should be quite evident from the free-body diagram.

FOLLOW-UP EXERCISE.   (a) In this Example, what is the minimum amount of mass for m_2 that would cause m_1 not to accelerate up or down the plane? (b) Keeping the masses the same as in the Example, how should the angle of incline be adjusted so that m_1 would not accelerate up or down the plane?

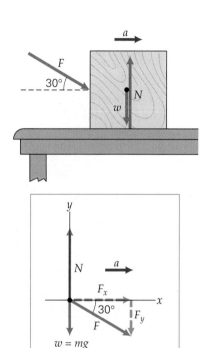

▲ **FIGURE 4.13 Finding force from motional effects** See Example 4.7.

EXAMPLE 4.7 | ## Components of Force and Free-Body Diagrams

A force of 10.0 N is applied at an angle of 30° to the horizontal on a 1.25-kg block initially at rest on a frictionless surface, as illustrated in ▲ Fig. 4.13. (a) What is the magnitude of the block's acceleration? (b) What is the magnitude of the normal force?

THINKING IT THROUGH.   The applied force may be resolved into components. The horizontal component accelerates the block. The vertical component affects the normal force. (Review Fig. 4.11 if necessary.) And drawing a free-body diagram for the block, as in Fig. 4.13, is helpful.

SOLUTION.

Given: $F = 10.0 \text{ N}$ *Find:*   (a) a (acceleration)
             $m = 1.25 \text{ kg}$ (b) N (normal force)
             $\theta = 30°$
             $v_o = 0$

(a) The acceleration of the block can be calculated using Newton's second law, and the axes are chosen so that a is in the $+x$-direction. As the free-body diagram shows, only a component (F_x) of the applied force F acts in this direction. The component of F in the direction of motion is $F_x = F \cos \theta$. Applying Newton's second law in the x-direction to calculate the acceleration:

$$F_x = F \cos 30° = ma_x$$

and

$$a_x = \frac{F \cos 30°}{m} = \frac{(10.0 \text{ N})(0.866)}{1.25 \text{ kg}} = 6.93 \text{ m/s}^2$$

(b) The acceleration found in part (a) is the acceleration of the block, since the block accelerates only in the x-direction. Since $a_y = 0$, the sum of the forces in the y-direction must be zero.

That is, the downward component of F acting on the block, F_y, and its downward weight force, w, must be balanced by the upward normal force N that the surface exerts on the block. If this were not the case, then there would be a net force and an acceleration in the y-direction.

Summing the forces in the y-direction with upward taken as positive,

$$\sum F_y = N - F_y - w = 0$$

or

$$N - F \sin 30° - mg = 0$$

and

$$N = F \sin 30° + mg = (10.0 \text{ N})(0.500) + (1.25 \text{ kg})(9.80 \text{ m/s}^2)$$
$$= 17.3 \text{ N}$$

The surface then exerts a force of 17.3 N upward on the block, which balances the sum of the downward forces acting on it.

FOLLOW-UP EXERCISE.   (a) Suppose the applied force on the block is applied for only a short time. What is the magnitude of the normal force after the applied force is removed? (b) If the block slides off the edge of the table, what would be the net force on the block just after it leaves the table (with the applied force removed)?

PROBLEM-SOLVING HINT

There is no single fixed way to go about solving a problem. However, some general strategies or procedures are helpful in solving problems involving Newton's second law. When using the suggested problem-solving procedures introduced in Section 1.7, you might include the following steps when solving problems involving force applications:

■ Draw a free-body diagram for each individual body, showing all of the forces acting on that body.
■ Depending on what is to be found, apply Newton's second law either to the system as a whole (in which case internal forces cancel) or to a part of the system. Basically, *you want to obtain an equation (or set of equations) containing the quantity for which you want to solve.* Review Example 4.3. (If there are two unknown quantities, application of Newton's second law to two parts of the system may give you two equations and two unknowns. See Example 4.6.)
■ Keep in mind that Newton's second law may be applied to components of acceleration and that forces may have to be resolved into components to do this. Review Example 4.7.

TRANSLATIONAL EQUILIBRIUM

Several forces may act on an object without producing an acceleration. In such a case, with $\vec{a} = 0$, from Newton's second law,

$$\sum \vec{F}_i = 0 \qquad (4.4)$$

That is, the vector sum of the forces, or the net force, is zero, so the object either remains at rest (as in ▸Fig. 4.14) *or* moves with a constant velocity (Newton's first law). In such cases, objects are said to be in **translational equilibrium**. When remaining at rest, an object is said to be in *static translational equilibrium.*

It follows that the sums of the rectangular components of the forces for an object in translational equilibrium are also zero (why?):

$$\sum \vec{F}_x = 0$$
$$\sum \vec{F}_y = 0 \qquad (4.5)$$

(translational equilibrium only)

For three-dimensional problems, $\sum \vec{F}_z = 0$ also applies. However, our discussion will be restricted to forces in two dimensions.

Equations 4.5 give what is often referred to as the **condition for translational equilibrium**. Let's apply this translational equilibrium condition to a case involving static equilibrium.

▶ **FIGURE 4.14 Many forces, no acceleration (a)** At least five different external forces act on this physics professor. (Here, f is the force of friction.) Nevertheless, she experiences no acceleration. Why? **(b)** Adding the force vectors by the polygon method reveals that the vector sum of the forces is zero. The professor is in static translational equilibrium.

EXAMPLE 4.8 | Keep It Straight: In Static Equilibrium

Keeping a broken leg bone straight while it is healing some-times requires *traction*, which is the procedure in which the bone is held under stretching tension forces at both ends to keep it aligned. Consider a leg under tractional tension as shown in ▼ Fig. 4.15. The cord is attached to a suspended mass of 5.0 kg and runs over a pulley. The attached cord above the pulley makes an angle of $\theta = 40°$ with the vertical. Neglect-ing the mass of the lower leg and the pulley and assuming all the strings are ideal, determine the magnitude of the tension in the horizontal cord.

THINKING IT THROUGH.   The pulley is in a static equilibrium and thus has no net force on it. If the forces are summed both vertically and horizontally, they independently should add to zero. This should allow the tension in the horizontal string to be found.

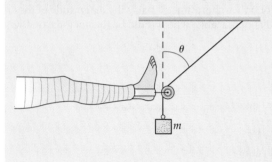

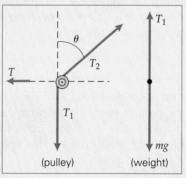

(pulley)   (weight)

◄ **FIGURE 4.15 Static transla-tional equilibrium** See Example text for description.

SOLUTION.

Given: Listing the data: *Find:* T (tension) in the
    $m = 5.0$ kg   horizontal cord
    $\theta = 40°$

Draw free-body diagrams for the pulley and suspended mass (shown in Fig. 4.15). It should be clear that the horizon-tal string must exert a force to the left on the pulley as shown. Summing the vertical forces on m, it can be seen that $T_1 = mg$. Then, summing the vertical forces on the pulley,

$$\sum \vec{F}_y = +T_2 \cos \theta - T_1 = 0$$

and summing the horizontal forces:

$$\sum \vec{F}_x = +T_2 \sin \theta - T = 0$$

Solving the latter equation for T, and substituting T_2 from the first:

$$T = T_2 \sin \theta = \frac{T_1}{\cos \theta} \sin \theta = mg \tan \theta$$

where $T_1 = mg$. Putting in the numbers,

$$T = mg \tan \theta = (5.0 \text{ kg})(9.8 \text{ m/s}^2) \tan 40° = 41 \text{ N}$$

FOLLOW-UP EXERCISE.   Suppose the attending physician requires a tractional force on the bottom of the foot of 55 N. If the sus-pended mass was kept the same, would you increase or decrease the angle of the upper string? Prove your answer by calculat-ing the required angle.

EXAMPLE 4.9 | On Your Toes: In Static Equilibrium

An 80-kg person stands on one foot with the heel elevated (► Fig. 4.16a). This gives rise to a tibia force F_1 and an Achilles tendon "pull" force F_2 as illustrated in Fig. 4.16b. Typical angles are $\theta_1 = 15°$ and $\theta_2 = 21°$, respectively. (a) Find gen-eral equations for F_1 and F_2, and show that θ_2 must be greater than θ_1 to prevent damage to the Achilles tendon. (b) Compare the force applied by the Achilles tendon with the weight of the person.

THINKING IT THROUGH.   This is a case of static translational equilibrium, so the x- and y-compo-nents can be summed to get equations for F_1 and F_2.

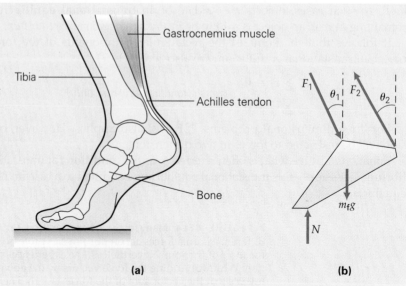

► **FIGURE 4.16 On your toes (a)** A per-son stands on one foot with the heel ele-vated. **(b)** The foot forces involved for this position (not to scale).

(a)   (b)

SOLUTION. Listing what is given and what is to be found,

Given: $m = 80$ kg
F_1 = tibia force
F_2 = tendon "pull"
$\theta_1 = 15°, \theta_2 = 21°$
(The mass of the foot m_f is not given.)

Find: (a) general equations for F_1 and F_2
(b) comparison of tendon force F_2 and the person's weight

(a) It is assumed that the person of mass m is at rest, standing on one foot. Then, summing the force components on the foot (Fig. 4.16b),

$$\sum \vec{F}_x = +F_1 \sin \theta_1 - F_2 \sin \theta_2 = 0$$

$$\sum \vec{F}_y = +N - F_1 \cos \theta_1 + F_2 \cos \theta_2 - m_f g = 0$$

where m_f is the mass of the foot. From the F_x equation,

$$F_1 = F_2 \left(\frac{\sin \theta_2}{\sin \theta_1} \right) \tag{1}$$

Substituting into the F_y equation,

$$N - F_2 \left(\frac{\sin \theta_2}{\sin \theta_1} \right) \cos \theta_1 + F_2 \cos \theta_2 - m_f g = 0$$

Solving for F_2 with $N = mg$ yields,

$$F_2 = \frac{N - m_f g}{\left(\frac{\sin \theta_2}{\tan \theta_1} \right) - \cos \theta_2} = \frac{mg - m_f g}{\cos \theta_2 \left(\frac{\tan \theta_2}{\tan \theta_1} - 1 \right)} \tag{2}$$

Examining the F_2 in Eq. 2, we see that if $\theta_2 = \theta_1$, or $\tan \theta_2 = \tan \theta_1$, then F_2 is very large. (Why?) So to have a finite force, we must have $\tan \theta_2 > \tan \theta_1$ or $\theta_2 > \theta_1$, and $21° > 15°$, so Nature obviously knows her physics.

Then, substituting F_2 into Eq. 1 to find F_1,

$$F_1 = F_2 \left(\frac{\sin \theta_2}{\sin \theta_1} \right) = \left[\frac{(m - m_f) g}{\cos \theta_2 \left(\frac{\tan \theta_2}{\tan \theta_1} \right) - 1} \right] \left(\frac{\sin \theta_2}{\sin \theta_1} \right)$$

$$= \frac{(m - m_f)g \tan \theta_2}{\left(\frac{\tan \theta_2}{\tan \theta_1} - 1 \right) \sin \theta_1} = \frac{\tan \theta_2 (m - m_f)g}{\cos \theta_1 \tan \theta_2 - \sin \theta_1}$$

(Check the trig manipulation on this last step.)

(b) The person's weight is $w = mg$, where m is the mass of the person's body. This is to be compared with F_2. Then, with $m \gg m_f$ (total body mass much greater than mass of the foot), to a good approximation, m_f may be assumed negligible compared to m, that is, $w - m_f g = mg - m_f g \approx w$. So for F_2,

$$F_2 = \frac{w - m_f g}{\cos \theta_2 \left(\frac{\tan \theta_2}{\tan \theta_1} - 1 \right)} \approx \frac{w}{\cos 21° \left(\frac{\tan 21°}{\tan 15°} - 1 \right)} = 2.5w$$

The Achilles tendon force is thus approximately 2.5 times the person's weight. No wonder folks stretch or tear this tendon, even without jumping.

FOLLOW-UP EXERCISE. (a) Compare the tibia force with the weight of the person. (b) Suppose the person jumped upward from the one-foot toe position (as in taking a running jump shot in basketball). How would this jump affect F_1 and F_2?

DID YOU LEARN?

➡ A space diagram shows force vectors drawn at the points of application; a free-body diagram shows the force vectors emanating from a common point, usually the origin of the x–y axes.

➡ The condition for translational equilibrium is $\sum \vec{F}_i = 0$. This does not mean the object is at rest. By Newton's first law, it could be moving with a constant velocity.

4.6 Friction

LEARNING PATH QUESTIONS

➡ What is the difference between static friction and kinetic friction?

➡ What is a coefficient of friction?

➡ Why does air resistance depend on the size and shape of an object?

Friction refers to the ever-present resistance to motion that occurs whenever two materials, or media, are in contact with each other. This resistance occurs for all types of media—solids, liquids, and gases—and is characterized as the **force of friction ($\vec{f}$)**.

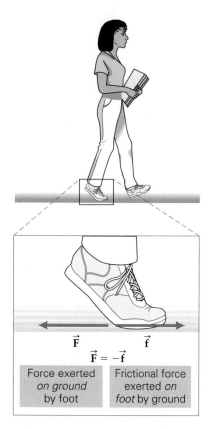

▲ **FIGURE 4.17 Friction and walking** The force of friction, $\vec{\mathbf{f}}$, is shown in the direction of the walking motion. The force of friction prevents the foot from slipping backward while the other foot is brought forward. If you walk on a deep-pile rug, $\vec{\mathbf{F}}$ is evident in that the pile will be bent backward.

For simplicity, up to now various kinds of friction (including air resistance) have been generally ignored in examples and exercises. Now knowing how to describe motion, you are ready to consider situations that are more realistic, in that the effects of friction are included.

In some situations, an increase in friction is desired—for example, when putting sand on an icy road or sidewalk to improve traction. This might seem contradictory, since an increase in friction presumably would increase the resistance to motion. We commonly say that friction opposes motion, and think that the force of friction is in the opposite direction of motion. However, consider the forces involved in walking, as illustrated in ◄Fig. 4.17. The force of friction does resist motion (that of the foot), but is in the direction of the (walking) motion. Without friction, the foot would slip backward, as when walking on a slippery surface.

As another example, consider a worker standing in the center of the bed of a flatbed truck that is accelerating in the forward direction. If there were no friction between the worker's shoes and the truck bed, the truck would slide out from under him. Obviously, there is a frictional force between the shoes and the bed, and it is in the forward direction. This is necessary for the worker to accelerate with the truck.

So there are situations where friction is desired (▼Fig. 4.18a), and situations where reduced friction is needed (Fig. 4.18b). Another situation where reduced friction is promoted is in the lubrication of moving machine parts. This allows the parts to move more freely, thereby lessening wear and reducing the expenditure of energy. Automobiles would not run without friction-reducing oils and greases.

This section is concerned chiefly with friction between solid surfaces. All surfaces are microscopically rough, no matter how smooth they appear or feel. It was originally thought that friction was due primarily to the mechanical interlocking of surface irregularities, or *asperities* (high spots). However, research has shown that friction between the contacting surfaces of ordinary solids (metals in particular) is due mostly to local adhesion. When surfaces are pressed together, local welding or bonding occurs in a few small patches where the largest asperities make contact. To overcome this local adhesion, a force great enough to pull apart the bonded regions must be applied.

Friction between solids is generally classified into three types: static, kinetic (sliding), and rolling. **Static friction** includes all cases in which the frictional force is sufficient to prevent relative motion between surfaces. Suppose you want to move a large desk. You push on it, but the desk doesn't move. The force of static friction between the desk's legs and the floor opposes and equals the horizontal force you are applying, so there is no motion—a static condition.

Kinetic friction (or **sliding**) **friction**, occurs when there is relative (sliding) motion at the interface of the surfaces in contact. When pushing on the desk, you can eventually get it sliding, but there is still a great deal of resistance between the desk's legs and the floor—kinetic friction.

► **FIGURE 4.18 Increasing and decreasing friction (a)** To get a fast start, drag racers need to make sure that their wheels don't slip when the starting light goes on. Just before the start of the race, they floor the accelerator to maximize the friction between their tires and the track by "burning in" the tires. This "burn in" is done by spinning the wheels with the brakes on until the tires are extremely hot. The rubber becomes so sticky that it almost welds itself to the surface of the road. **(b)** Water serves as a good lubricant to reduce friction in rides such as this one.

(a)

(b)

Rolling friction occurs when one surface rotates as it moves over another surface, but does not slip or slide at the point or area of contact. Rolling friction, such as that occurring between a train wheel and a rail, is attributed to small, local deformations in the contact region. This type of friction is difficult to analyze and will not be considered.

FRICTIONAL FORCES AND COEFFICIENTS OF FRICTION

Now let's look at the forces of friction on stationary and sliding objects. These forces are called the *force of static friction* and the *force of kinetic (sliding) friction*, respectively. Experimentally, it has been found that the force of friction ($\vec{f}$) depends on both the nature of the two surfaces, and, to a good approximation, the normal force ($\vec{N}$) that a surface exerts on an object, that is, $\vec{f} \propto \vec{N}$. For an object on a horizontal surface, and with no other vertical forces, this force is equal in magnitude to the object's weight. (Why?) However, as was shown in the previous Learn By Drawing 4.1, on an inclined plane the normal force is in response to only a component of the weight force.

The force of static friction $\vec{f}_s$ between surfaces in contact acts in the direction that opposes the initiation of relative motion between the surfaces. The magnitude takes on a range of values given by

$$f_s \leq \mu_s N \quad \text{(static conditions)} \tag{4.6}$$

where μ_s is a constant of proportionality called the **coefficient of static friction**. ("μ_s" is the Greek letter mu. Note that it is dimensionless. How do you know this from the equation?)

The less-than-or-equal-to sign ($\leq$) indicates that the force of static friction may have different values from zero up to some maximum value. To understand this concept, look at ▼Fig. 4.19. In Fig. 4.19a, one person pushes on a file cabinet, but it doesn't move. With no acceleration, the net force on the cabinet is zero, and $F - f_s = 0$, or $F = f_s$. Suppose that a second person also pushes, and the file cabinet still doesn't budge. Then f_s must now be larger, since the applied force has

▼ **FIGURE 4.19 Force of friction versus applied force (a)** In the static region of the graph, as the applied force F increases, so does f_s; that is, $f_s = F$ and $f_s < \mu_s N$ **(b)** When the applied force F exceeds $f_{s_{max}} = \mu_s N$ the heavy file cabinet is set into motion. **(c)** Once the cabinet is moving, the frictional force decreases, since kinetic friction is less than static friction ($f_k < f_{s_{max}}$) Thus, if the applied force is maintained, there is a net force, and the cabinet is accelerated. For the cabinet to move with constant velocity, the applied force must be reduced to equal the kinetic friction force: $f_k = \mu_k N$.

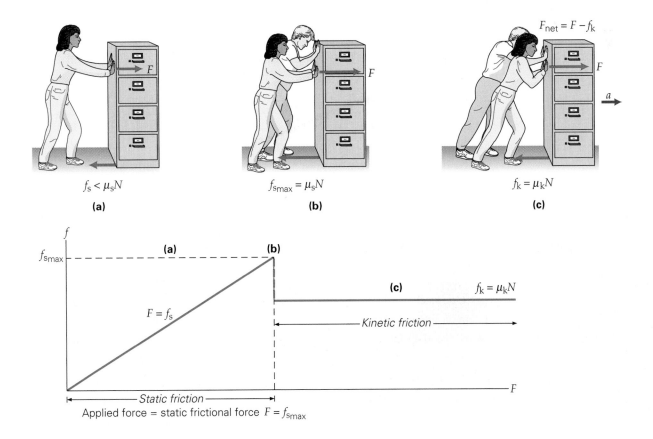

$f_s < \mu_s N$

(a)

$f_{s_{max}} = \mu_s N$

(b)

$F_{net} = F - f_k$

$f_k = \mu_k N$

(c)

Applied force = static frictional force $F = f_{s_{max}}$

been increased. Finally, if the applied force is made large enough to overcome the static friction, motion occurs (Fig. 4.19c). The greatest, or maximum, force of static friction is exerted just before the cabinet starts to slide (Fig. 4.19b), and for this case, Eq. 4.7 gives the maximum value of static friction:

$$f_{s_{max}} = \mu_s N \quad \textit{(maximum value of static friction)} \tag{4.7}$$

Once an object is sliding, the force of friction changes to kinetic friction ($\vec{f}_k$) This force acts in the direction opposite to the direction of the object's motion and has a magnitude of

$$f_k = \mu_k N \quad \textit{(sliding conditions)} \tag{4.8}$$

where μ_k is the **coefficient of kinetic friction** (sometimes called the *coefficient of sliding friction*). Note that Eqs. 4.7 and 4.8 are *not* vector equations, since f and N are in different directions. Generally, the coefficient of kinetic friction is less than the coefficient of static friction ($\mu_k < \mu_s$), which means that the force of kinetic friction is less than $f_{s_{max}}$. The coefficients of friction between some common materials are listed in ▼Table 4.1.

Note that the force of static friction (f_s) exists in response to an applied force. The magnitude of f_s and its direction depend on the magnitude and direction of the applied force. Up to its maximum value, the force of static friction is equal in magnitude and opposite in direction to the applied force (F), since there is no acceleration ($F - f_s = ma = 0$). Thus, if the person in Fig. 4.19a were to push on the cabinet in the opposite direction, f_s would also change direction to oppose the new push. If there were no applied force F, then f_s would be zero. When the magnitude of F exceeds that of $f_{s_{max}}$, the cabinet begins moving (accelerates), and kinetic friction comes into play, with $f_k = \mu_k N$. If the magnitude of F is reduced to that of f_k, the cabinet will slide with a constant velocity; if the magnitude of F is maintained greater than that of f_k, the cabinet will continue to accelerate.

It has been experimentally determined that the coefficients of friction (and therefore the forces of friction) are nearly independent of the contact area between

TABLE 4.1   Approximate Values for Coefficients of Static and Kinetic Friction between Certain Surfaces

Friction between Materials	μ_s	μ_k
Aluminum on aluminum	1.90	1.40
Glass on glass	0.94	0.35
Rubber on concrete		
dry	1.20	0.85
wet	0.80	0.60
Steel on aluminum	0.61	0.47
Steel on steel		
dry	0.75	0.48
lubricated	0.12	0.07
Teflon on steel	0.04	0.04
Teflon on Teflon	0.04	0.04
Waxed wood on snow	0.05	0.03
Wood on wood	0.58	0.40
Lubricated ball bearings	<0.01	<0.01
Synovial joints (at the ends of most long bones—for example, elbows and hips)	0.01	0.01

metal surfaces. This means that the force of friction between a brick-shaped metal block and a metal surface is the same regardless of whether the block is lying on a larger side or a smaller side.

Finally, keep in mind that although the equation $f = \mu N$ holds in general for frictional forces, it may not remain linear. That is, μ is not always constant. For example, the coefficient of kinetic friction varies somewhat with the relative speed of the surfaces. However, for speeds up to several meters per second, the coefficients are relatively constant. For simplicity, our discussion will neglect any variations due to speed (or area), and the forces of static and kinetic friction will be assumed to depend only on the load (N) and the nature of the two surfaces as expressed by the given coefficients of friction.

EXAMPLE 4.10 | Pulling a Crate: Static and Kinetic Forces of Friction

(a) In ▼Fig. 4.20, if the coefficient of static friction between the 40.0-kg crate and the floor is 0.650, what is the magnitude of the minimum horizontal force the worker must pull to get the crate moving? (b) If the worker maintains that force once the crate starts to move and the coefficient of kinetic friction between the surfaces is 0.500, what is the magnitude of the acceleration of the crate?

THINKING IT THROUGH. This situation involves applications of the forces of friction. In (a), the maximum force of static friction must be calculated. In (b), if the worker maintains an applied force of this magnitude after the crate is in motion, there will be an acceleration, since $f_k < f_{s_{max}}$

SOLUTION. As usual, listing the given data and what is to be found:

Given: $m = 40.0$ kg   *Find:* (a) F (minimum force neces-
$\quad\quad\ \mu_s = 0.650$ $\quad\quad\quad\quad\quad\quad\quad$ sary to move crate)
$\quad\quad\ \mu_k = 0.500$ $\quad\quad\quad\quad\quad\quad$ (b) a (acceleration)

(a) The crate will not move until the magnitude of the applied force F slightly exceeds that of the maximum static frictional force $f_{s_{max}}$. So $f_{s_{max}}$ must be found to see what force

the worker needs to apply. The weight of the crate and the normal force are equal in magnitude in this case (see the free-body diagram in Fig. 4.20), so the magnitude of the maximum force of static friction is

$$f_{s_{max}} = \mu_s N = \mu_s(mg)$$
$$= (0.650)(40.0 \text{ kg})(9.80 \text{ m/s}^2) = 255 \text{ N}$$

So the crate will begin to move when the applied force F exceeds 255 N.

(b) Now with the crate in motion, the kinetic friction f_k acts on the crate. However, this force is smaller than the applied force $F = f_{s_{max}} = 255$ N, because $\mu_k < \mu_s$. Hence, there is a net force on the crate and the acceleration of the crate can be found by using Newton's second law in the x-direction:

$$\sum F_x = +F - f_k = F - \mu_k N = ma_x$$

Solving for a_x,

$$a_x = \frac{F - \mu_k N}{m} = \frac{F - \mu_k(mg)}{m}$$
$$= \frac{255 \text{ N} - (0.500)(40.0 \text{ kg})(9.80 \text{ m/s}^2)}{40.0 \text{ kg}} = 1.48 \text{ m/s}^2$$

▶ **FIGURE 4.20**
Forces of static and kinetic friction See Example text for description.

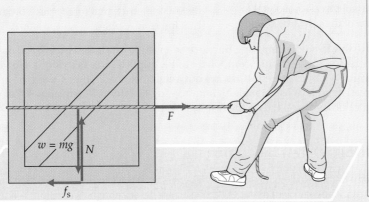

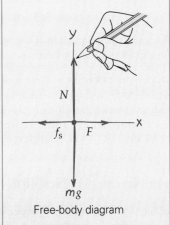

Free-body diagram

FOLLOW-UP EXERCISE. On the average, by what factor does μ_s exceed μ_k for nonlubricated, metal-on-metal surfaces? (See Table 4.1.)

Let's look at another worker with the same crate, but this time with the worker applying the force at an angle (▼Fig. 4.21).

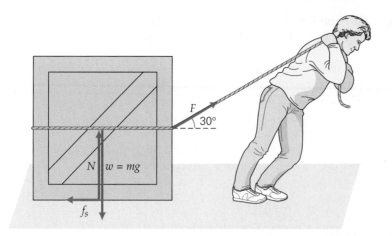

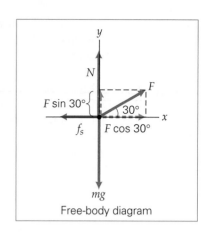

Free-body diagram

▲ **FIGURE 4.21 Pulling at an angle: a closer look at the normal force** See Example 4.11.

EXAMPLE 4.11 | ## Pulling at an Angle: A Closer Look at the Normal Force

A worker pulling a crate applies a force at an angle of 30° to the horizontal, as shown in Fig. 4.21. What is the magnitude of the minimum force he must apply to move the crate? (Before looking at the solution, would you expect that the force needed in this case would be greater or less than that in Example 4.10?)

THINKING IT THROUGH.   Since the applied force is at an angle to the horizontal surface, the vertical component will affect the normal force. (See Fig. 4.11.) This change in the normal force will, in turn, affect the maximum force of static friction.

SOLUTION.   The data are the same as in Example 4.10, except that the force is applied at an angle.

Given: $\theta = 30°$ *Find:* F (minimum force necessary to move the crate)

In this case, the crate will begin to move when the *horizontal component* of the applied force, $F \cos 30°$, slightly exceeds the maximum static friction force. So for the maximum friction:

$$F \cos 30° = f_{s_{max}} = \mu_s N$$

However, the magnitude of the normal force is not equal to the weight of the crate here, because of the upward compo-

nent of the applied force. (See the free-body diagram in Fig. 4.21.) Then by Newton's second law, since $a_y = 0$.

$$\sum F_y = +N + F \sin 30° - mg = 0$$

and

$$N = mg - F \sin 30°$$

In effect, the applied force here partially supports the weight of the crate. Substituting this expression for N into the first equation gives

$$F \cos 30° = \mu_s(mg - F \sin 30°)$$

Solving for F,

$$F = \frac{mg}{(\cos 30°/\mu_s) + \sin 30°}$$

$$= \frac{(40.0 \text{ kg})(9.80 \text{ m/s}^2)}{(0.866/0.650) + 0.500} = 214 \text{ N}$$

Thus, less applied force is needed in this case, reflecting the fact that the frictional force is less, because of the reduced normal force.

FOLLOW-UP EXERCISE.   Note that in this Example, applying the force at an angle produces two effects. As the angle between the applied force and the horizontal increases, the horizontal component of the applied force is reduced. However, the normal force also gets smaller, resulting in a lower $f_{s_{max}}$. Does one effect always outweigh the other? That is, does the applied force F necessary to move the crate always decrease with increasing angle? [*Hint:* Investigate F for different angles. For example, compute F for 20° and 50°. You already have a value for 30°. What do the results tell you?]

EXAMPLE 4.12 | ## No Slip, No Slide: Static Friction

A crate sits in the middle of the bed on a flatbed truck that is traveling at 80 km/h on a straight, level road. The coefficient of static friction between the crate and the truck bed is 0.40. When the truck comes uniformly to a stop, the crate does not slide, but remains stationary on the truck. What is the minimum stopping distance for the truck so the crate does not slide on the truck bed?

THINKING IT THROUGH.   There are three forces on the crate, as shown in the free-body diagram in ▶ Fig. 4.22 (assuming that the truck is initially traveling in the +x-direction). But wait.

There is a net force in the $-x$-direction, and hence there should be an acceleration in that direction. What does this mean? It means that relative to the ground, the crate is decelerating at the same rate as the truck, which is necessary for the crate not to slide—the crate and the truck slow down uniformly together.

The force creating this acceleration for the crate is the static force of friction. The acceleration is found using Newton's second law, and then is used in one of the kinematic equations to find the distance.

SOLUTION.

Given: $v_{x_0} = 80$ km/h $= 22$ m/s   *Find:* x (minimum stopping distance)

$\mu_s = 0.40$

Applying Newton's second law to the crate using the maximum f_s to find the minimum stopping distance,

$$\sum F_x = -f_{s_{max}} = -\mu_s N = -\mu_s mg = ma_x$$

Solving for a_x,

$$a_x = -\mu_s g = -(0.40)(9.8 \text{ m/s}^2) = -3.9 \text{ m/s}^2$$

which is the maximum deceleration of the truck so the crate does not slide.

Hence, the minimum stopping distance (x) for the truck is based on this acceleration and given by Eq. 2.12, where $v_x = 0$ and x_0 is taken to be zero. So,

$$v_x^2 = 0 = v_{x_0}^2 + 2(a_x)x$$

Solving for x,

$$x = \frac{v_{x_0}^2}{-2a_x} = \frac{(22 \text{ m/s})^2}{-2(-3.9 \text{ m/s}^2)} = 62 \text{ m}$$

Is the answer reasonable? This distance is about two-thirds the length of a football field.

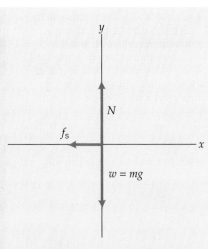

▲ **FIGURE 4.22 Free-body diagram** See Example text for description.

FOLLOW-UP EXERCISE.   Draw a free-body diagram and describe what happens in terms of accelerations and coefficients of friction if the crate starts to slide forward on the truck bed when the truck is braking to a stop (in other words, if a_x exceeds -3.9 m/s^2).

AIR RESISTANCE

Air resistance refers to the resistance force acting on an object as it moves through air. In other words, air resistance is a type of frictional force. In analyses of falling objects, you can usually ignore the effect of air resistance and still get good approximations for those falling relatively short distances. However, for longer distances, air resistance cannot be ignored.

Air resistance occurs when a moving object collides with air molecules. Therefore, air resistance depends on the object's shape and size (which determine the area of the object that is exposed to collisions) as well as its speed. The larger the object and the faster it moves, the more collisions there will be with air molecules. (Air density is also a factor, but this quantity can be assumed to be constant near the Earth's surface.) To reduce air resistance (and fuel consumption), automobiles are made more "streamlined," and airfoils are used on trucks and campers (▶ Fig. 4.23).

Consider a falling object. Since air resistance depends on speed, as a falling object accelerates under the influence of gravity, the retarding force of air resistance increases (▼ Fig. 4.24a). Air resistance for human-sized objects as a general rule is proportional to the square of the speed, v^2, so the resistance builds up rather

▲ **FIGURE 4.23 Airfoil** The airfoil at the top of the truck's cab makes the truck more streamlined and therefore reduces air resistance.

◀ **FIGURE 4.24 Air resistance and terminal velocity (a)** As the speed of a falling object increases, so does the frictional force of air resistance. **(b)** When this force of friction equals the weight of the object, the net force is zero, and the object falls with a constant (terminal) velocity. **(c)** A plot of speed versus time, showing these relationships.

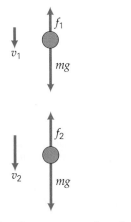

(a) As v increases, so does f.

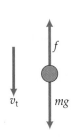

(b) When $f = mg$, the object falls with a constant (terminal) velocity.

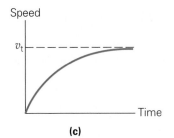

(c)

▶ **FIGURE 4.25 Terminal velocity**
Skydivers assume a spread-eagle
position to maximize air resistance.
This causes them to reach terminal
velocity more quickly and prolongs
the time of fall. Shown here is a formation of sky divers viewed from
below.

rapidly. Thus when the speed doubles, the air resistance increases by a factor of 4.
Eventually, the magnitude of the retarding force equals that of the object's weight
force (Fig. 4.24b), so the net force on it is zero. The object then falls with a maximum
constant velocity, which is called the **terminal velocity**, with magnitude v_t.

This can be easily seen from Newton's second law. For the falling object,

$$F_{net} = ma$$

or

$$mg - f = ma$$

where f is the air resistance (friction) and downward has been taken as positive for
convenience. Solving for a,

$$a = g - \frac{f}{m}$$

where a is the magnitude of the instantaneous downward acceleration.

Notice that the acceleration for a falling object when air resistance is included
is less than g; that is, $a < g$. As the object continues to fall, its speed increases, and
the force of air resistance, f, increases (since it is speed dependent) until $a = 0$ when
$f = mg$ and $f - mg = 0$. The object then falls at its constant terminal velocity.

For a skydiver with an unopened parachute, terminal velocity is about
200 km/h (about 125 mi/h). To reduce the terminal velocity so that it can
be reached sooner and the time of fall extended, a skydiver will try to increase
exposed body area to a maximum by assuming a spread-eagle position (▲Fig. 4.25).
This position takes advantage of the dependence of air resistance on the size and
shape of the falling object. Once the parachute is open (giving a larger exposed area
and a shape that catches the air), the additional air resistance slows the diver down
to about 40 km/h (25 mi/h), which is more preferable for landing.

CONCEPTUAL EXAMPLE 4.13   Race You Down: Air Resistance and Terminal Velocity

From a high altitude, a balloonist simultaneously drops two
balls of identical size, but appreciably different in mass.
Assuming that both balls reach terminal velocity during the
fall, which of the following is true? (a) The heavier ball
reaches terminal velocity first; (b) the balls reach terminal
velocity at the same time; (c) the heavier ball hits the ground
first; (d) the balls hit the ground at the same time. *Clearly*

*establish the reasoning and physical principle(s) used in determin-
ing your answer before checking it next. That is,* **why** *did you
select your answer?*

REASONING AND ANSWER.   Terminal velocity is reached when
the weight of a ball is balanced by the frictional air resistance.
Both balls initially experience the same acceleration, g, and their

speeds and the retarding forces of air resistance increase at the same rate. The weight of the lighter ball will be balanced first, so (a) and (b) are incorrect with the lighter ball reaching terminal velocity ($a = 0$) first, the heavier ball continues to accelerate and pulls ahead of the lighter ball. Hence, the heavier ball hits the ground first, and the answer is (c), and (d) is incorrect.

FOLLOW-UP EXERCISE.   Suppose the heavier ball were much larger in size than the lighter ball. How might this difference affect the outcome?

You see an example of terminal velocity quite often. Why do clouds stay seemingly suspended in the sky? Certainly the water droplets or ice crystals (high clouds) should fall—and they do. However, they are so small that their terminal velocity is reached quickly, and the very slow rate of their descent goes unnoticed. In addition, there may be some helpful updrafts that keep the water droplets and ice crystals from reaching the ground.

An extraterrestrial use of "air" resistance is called *aerobraking*. This spaceflight technique uses a planetary atmosphere to slow down a spacecraft. As the craft passes through the top layer of the planetary atmosphere, the atmospheric "drag" slows and lowers the craft's speed so as to put it in the desired orbit. Many passes may be needed, with the spacecraft passing in and out of the atmosphere to achieve the proper final orbit.

Aerobraking is a worthwhile technique because it eliminates the need for a heavy load of chemical propellants that would otherwise be needed to place the spacecraft in orbit. This allows a greater payload of scientific instruments for investigations.

DID YOU LEARN?
- ➥ Static friction prevents motion and has a maximum value of $f_{s_{max}} = \mu_s N$. Kinetic friction acts on a sliding body.
- ➥ The coefficient of friction is a constant of proportionality between frictional and normal forces.
- ➥ Air resistance depends on the area of an object exposed to air molecule collisions.

PULLING IT TOGETHER | ## Newton Helps Superman: Kinematics and Forces

Traveling at 90.0 mi/h, a driver applies the brakes to his fast-moving car and skids out of control on a wet, concrete horizontal road. The 2000-kg car is headed directly toward a student waiting to catch a bus to campus who is standing 58.0 m down the road. Fortunately, Superman is flying overhead and surveys the situation. Remembering from his physics class that the coefficient of kinetic friction between rubber and rough wet concrete is 0.800, he quickly determines that friction alone will not stop the car in time.

So, he flies down and exerts a constant force of $F = 13\,000$ N (2925 lb) on the car's hood at a downward angle of 30° (▶ Fig. 4.26).

(a) Show that Superman was correct in his determination that the force of friction (assumed constant) alone would not stop the car in time. (b) Show that Superman's applied force saves the day. (c) How close did the car come to the student before stopping?

THINKING IT THROUGH.   Involved in this example are Newton's laws, friction, free-body diagrams, the summing of forces in two dimensions, and constant acceleration kinematics. (a) Stopping the car before a collision requires a minimum deceleration that can be calculated from the initial speed and distance. Comparing this to the deceleration provided by friction will tell whether friction alone can do the job—if it can, the frictional acceleration will be greater than the required minimum acceleration. The force of friction can be determined from the coefficient of friction and the normal force. From this, the deceleration due

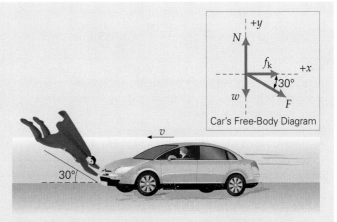

▲ **FIGURE 4.26 Superman to the rescue.** Superman applies a force to the skidding car. Will it be enough to save the student? See Example text for description.

to friction alone can be obtained. (b) This involves using the net force (which includes Superman's applied force) to find the deceleration using Newton's second law and ensuring that it matches or exceeds the required minimum deceleration. See the free-body diagram of the forces in Fig. 4.26. (c) Using the deceleration and initial speed, the stopping distance can be computed from the appropriate kinematic equation.

(continued on next page)

SOLUTION.

Given: initial car speed, $v_o = 90.0 \text{ mi/h} \left(\dfrac{0.447 \text{ m/s}}{\text{mi/h}} \right) = 40.2 \text{ m/s}$

        $m = 2000 \text{ kg}$
        $\mu_k = 0.800$
        Superman's force and angle: $F = 13\,000 \text{ N}$, at an angle of $\theta = 30°$
        $d = 58.0 \text{ m}$

Find: (a) Show that friction alone isn't enough to stop the car to avoid a tragedy.
      (b) Show that Superman's applied force is enough to stop the car in time.
      (c) How close the car comes to the student.

(a) Assuming a constant frictional force, the minimum deceleration can be determined from kinematics, using $v^2 = v_o^2 + 2a(x - x_o)$. Setting the final velocity to zero gives the minimum deceleration a_{min} as follows:

$$0 = v_o^2 + 2a_{min}(x - x_o)$$

and

$$a_{min} = -\frac{v_o^2}{2(x - x_o)} = -\frac{(-40.2 \text{ m/s})^2}{2(-58.0 \text{ m})} = +13.9 \text{ m/s}^2$$

Notice how the signs work out: the displacement is negative and the acceleration comes out positive, opposite the velocity, exactly what is needed for slowing down. (That is, x and v are in the negative direction in this case, and a is in the positive direction.)

The net force in this friction-only case is the backward-pointing force of kinetic friction. The normal force on the car is the same magnitude as the car's weight (why?), hence

$$f_k = \mu_k N = \mu_k w = \mu_k mg = (0.800)(2000 \text{ kg})(9.80 \text{ m/s}^2)$$
$$= 1.57 \times 10^4 \text{ N}$$

Then, being the only force in the x (horizontal) direction, by Newton's second law,

$$F_x = f_k = 1.57 \times 10^4 \text{ N} = ma$$

and

$$a = \frac{F_x}{m} = \frac{1.57 \times 10^4 \text{ N}}{2000 \text{ kg}} = +7.85 \text{ m/s}^2$$

Note that the positive sign means that the deceleration is opposite the car's velocity, consistent with slowing down. Since this is less than the minimum required deceleration, a_{min}, Superman's push is needed.

(b) To find the deceleration with Superman's applied force, first the normal force N is needed, which involves the vertical component of the applied force. (There are four forces acting on the car: weight, normal, frictional, and Superman forces.

See the free-body diagram in Fig. 4.26.) Summing the forces in the vertical direction, which add up to zero (why?):

$$F_y = N - w - F \sin 30° = 0$$

Solving for N, which is needed to determine the force of friction:

$$N = w + F \sin 30°$$
$$= mg + F \sin 30°$$
$$= (2000 \text{ kg})(9.80 \text{ m/s}^2) + (1.30 \times 10^4 \text{ N})(0.500)$$
$$= 2.61 \times 10^4 \text{ N}$$

Then the force of friction is

$$f_k = \mu_k N = (0.800)(2.61 \times 10^4 \text{ N}) = 2.09 \times 10^4 \text{ N}$$

To determine the deceleration with Superman in the picture, the forces in the horizontal direction are summed for Newton's second law:

$$F_x = f_k + F \cos 30°$$
$$= 2.09 \times 10^4 \text{ N} + (1.30 \times 10^4 \text{ N})(0.866)$$
$$= 3.22 \times 10^4 \text{ N} = ma$$

Solving for the deceleration:

$$a = \frac{F_x}{m} = \frac{3.22 \times 10^4 \text{ N}}{2000 \text{ kg}} = +16.1 \text{ m/s}^2$$

This exceeds the minimum value of $a_{min} = +13.9 \text{ m/s}^2$, so Superman does save the day.

(c) Since the acceleration, $a = 16.1 \text{ m/s}^2$, and the initial and final velocities, $v_o = -40.2 \text{ m/s}$ and $v = 0$, are known, the stopping distance can be found using the appropriate kinematic equation:

$$v^2 = v_o^2 + 2ax$$

Then,

$$x = \frac{v^2 - v_o^2}{2a} = \frac{0 - (-40.2 \text{ m/s})^2}{2(16.1 \text{ m/s}^2)} = -50.2 \text{ m}$$

So the car stops a distance of $58.0 \text{ m} - 50.2 \text{ m} = 7.8 \text{ m}$ from the student.

Learning Path Review

- A **force** is something that is capable of changing an object's state of motion. To produce a change in motion, there must be a nonzero net, or unbalanced, force:

  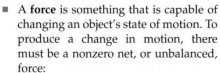
  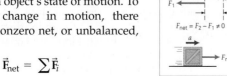

  $$\vec{F}_{net} = \sum \vec{F}_i$$

- **Newton's first law of motion** is also called the *law of inertia*, where inertia is the natural tendency of an object to maintain its state of motion. It states that in the absence of a net applied force, a body at rest remains at rest, and a body in motion remains in motion with constant velocity.

■ **Newton's second law** relates the net force acting on an object or system to the (total) mass and the resulting acceleration. It defines the cause-and-effect relationship between force and acceleration:

$$\sum \vec{F}_i = \vec{F}_{net} = m\vec{a} \qquad (4.1)$$

A nonzero net force accelerates the crate: $a \propto F/m$

The equation for **weight** in terms of mass is a form of Newton's second law:

$$w = mg \qquad (4.2)$$

The component form of Newton's second law:

$$\sum (F_x\hat{\mathbf{x}} + F_y\hat{\mathbf{y}}) = m(a_x\hat{\mathbf{x}} + a_y\hat{\mathbf{y}}) = ma_x\hat{\mathbf{x}} + ma_y\hat{\mathbf{y}} \qquad (4.3a)$$

and

$$\sum F_x = ma_x \quad \text{and} \quad \sum F_y = ma_y \qquad (4.3b)$$

■ **Newton's third law** states that for every force, there is an equal and opposite reaction force. The opposing forces of a third law force pair always act on different objects.

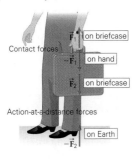

■ An object is said to be in **translational equilibrium** when it either is at rest or moves with a constant velocity. When remaining at rest, an object is said to be in *static translational equilibrium*. The condition for translational equilibrium is represented as

$$\sum \vec{F}_i = 0 \qquad (4.4)$$

or

$$\sum \vec{F}_x = 0 \quad \text{and} \quad \sum \vec{F}_y = 0 \qquad (4.5)$$

■ **Friction** is the resistance to motion that occurs between contacting surfaces. (In general, friction occurs for all types of media—solids, liquids, and gases.)

■ The frictional force between surfaces is characterized by coefficients of friction (μ), one for the static case and one for the kinetic (moving) case. In many cases, $f = \mu N$ where N is the normal force—the force perpendicular to the surface (that is, the force exerted *by* the surface *on* the object). As a ratio of forces (f/N), μ is unitless.

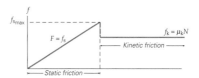

Force of Static Friction:

$$f_s \leq \mu_s N \qquad (4.6)$$

$$f_{s_{max}} = \mu_s N \quad \text{(maximum value of static friction)} \qquad (4.7)$$

Force of Kinetic (Sliding) Friction:

$$f_k = \mu_k N \qquad (4.8)$$

■ The force of air resistance on a falling object increases with increasing speed. It eventually attains a constant velocity, called the *terminal velocity*.

Learning Path Questions and Exercises

For instructor-assigned homework, go to www.masteringphysics.com

MULTIPLE CHOICE QUESTIONS

4.1 THE CONCEPTS OF FORCE AND NET FORCE
AND
4.2 INERTIA AND NEWTON'S FIRST LAW OF MOTION

1. Mass is related to an object's (a) weight, (b) inertia, (c) density, (d) all of the preceding.

2. A force (a) always produces motion, (b) is a scalar quantity, (c) is capable of producing a change in motion, (d) both a and b.

3. If an object is moving at constant velocity, (a) there must be a force in the direction of the velocity, (b) there must be no force in the direction of the velocity, (c) there must be no net force, (d) there must be a net force in the direction of the velocity.

4. If the net force on an object is zero, the object could (a) be at rest, (b) be in motion at a constant velocity, (c) have zero acceleration, (d) all of the preceding.

5. The force required to keep a rocket ship moving at a constant velocity in deep space is (a) equal to the weight of

the ship, (b) dependent on how fast the ship is moving, (c) equal to that generated by the rocket's engines at half power, (d) zero.

4.3 NEWTON'S SECOND LAW OF MOTION

6. The newton unit of force is equivalent to (a) kg · m/s, (b) kg · m/s^2, (c) kg · m^2/s, (d) none of the preceding.

7. The acceleration of an object is (a) inversely proportional to the acting net force, (b) directly proportional to its mass, (c) directly proportional to the net force and inversely proportional to its mass, (d) none of these.

8. The weight of an object is directly proportional to (a) its mass, (b) its inertia, (c) the acceleration due to gravity, (d) all of the preceding.

4.4 NEWTON'S THIRD LAW OF MOTION

9. The action and reaction forces of Newton's third law (a) are in the same direction, (b) have different magnitudes, (c) act on different objects, (d) can be the same force.

10. A brick hits a glass window. The brick breaks the glass, so (a) the magnitude of the force of the brick on the glass is greater than the magnitude of the force of the glass on the brick, (b) the magnitude of the force of the brick on the glass is smaller than the magnitude of the force of the glass on the brick, (c) the magnitude of the force of the brick on the glass is equal to the magnitude of the force of the glass on the brick, (d) none of the preceding.

11. A freight truck collides head-on with a passenger car, causing a lot more damage to the car than to the truck. From this condition, we can say that (a) the magnitude of the force of the truck on the car is greater than the magnitude of the force of the car on the truck, (b) the magnitude of the force of the truck on the car is smaller than the magnitude of the force of the car on the truck, (c) the magnitude of the force of the truck on the car is equal to the magnitude of the force of the car on the truck, (d) none of the preceding.

4.5 MORE ON NEWTON'S LAWS: FREE-BODY DIAGRAMS AND TRANSLATIONAL EQUILIBRIUM

12. The kinematic equations of Chapter 2 can be used (a) only with constant forces, (b) only with constant velocities, (c) with variable accelerations, (d) all of the preceding.

13. The condition(s) for translational equilibrium is (are) (a) $\sum F_x = 0$, (b) $\sum F_y = 0$, (c) $\sum \vec{F}_i = 0$, (d) all of the preceding.

4.6 FRICTION

14. In general, the frictional force (a) is greater for smooth than rough surfaces, (b) depends significantly on sliding speeds, (c) is proportional to the normal force, (d) depends significantly on the surface area of contact.

15. The coefficient of kinetic friction, μ_k, (a) is usually greater than the coefficient of static friction, μ_s, (b) usually equals μ_s, (c) is usually smaller than μ_s, (d) equals the applied force that exceeds the maximum static force.

16. A crate sits in the middle of the bed of a flatbed truck. The driver accelerates the truck gradually from rest to a normal speed, but then has to make a sudden stop to avoid hitting a car. If the crate slides as the truck stops, the frictional force would be (a) in the forward direction, (b) in the backward direction, (c) zero.

17. Two people, a 100-kg man and a 50-kg woman, jump out of a plane together and open their identical parachutes at the same time. Who will strike the ground first: (a) the man, (b) the woman, or (c) both together?

CONCEPTUAL QUESTIONS

4.1 THE CONCEPTS OF FORCE AND NET FORCE
AND
4.2 INERTIA AND NEWTON'S FIRST LAW OF MOTION

1. (a) If an object is at rest, there must be no forces acting on it. Is this statement correct? Explain. (b) If the net force on an object is zero, can you conclude that the object is at rest? Explain.

2. When on a jet airliner that is taking off, you feel that you are being "pushed" back into the seat. Use Newton's first law to explain why.

3. An object weighs 300 N on Earth and 50 N on the Moon. Does the object also have less inertia on the Moon?

4. Consider an air-bubble level that is sitting on a horizontal surface (▶ Fig. 4.27). Initially, the air bubble is in the middle of the horizontal glass tube. (a) If the level is pushed and a force is applied to accelerate it, which way would the bubble move? Which way would the bubble move if the force is then removed and the level slows down, due to friction? (b) Such a level is sometimes used as an "accelerometer" to indicate the direction of the acceleration. Explain the principle involved. [*Hint:* Think about pushing a pan of water.]

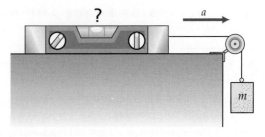

▲ **FIGURE 4.27 An air-bubble level/accelerometer** See Conceptual Question 4.

5. As a follow-up to Conceptual Question 4, consider a child holding a helium balloon in a closed car at rest. What would the child observe when the car (a) accelerates from rest and (b) brakes to a stop? (The balloon does not touch the roof of the car.)

6. The following is an old trick (▼Fig. 4.28). If a tablecloth is yanked out very quickly, the dishes on it will barely move. Why?

▲ FIGURE 4.28 **Magic or physics?** See Conceptual Question 6.

7. Another old one: Referring to ▶Fig. 4.29, (a) how would you pull to get the upper string to break? (b) How would you pull to get the lower string to break?

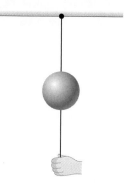

▲ FIGURE 4.29 **Give it a pull.** See Conceptual Question 7.

8. A student weighing 600 N crouches on a scale and suddenly springs vertically upward. Will the scale read more or less than 600 N just before the student leaves the scale?

4.3 NEWTON'S SECOND LAW OF MOTION

9. An astronaut has a mass of 70 kg when measured on Earth. What is her weight in deep space, far from any celestial body? What is her mass there?

10. In general, this chapter has considered forces that are applied to objects of constant mass. What would be the situation if mass were added to or lost from a system while a constant force was being applied to the system? Give examples of situations in which this set of events might happen.

11. The engines of most rockets produce a constant thrust (forward force). However, when a rocket is fired, its acceleration increases with time as the engine continues to operate. Is this situation a violation of Newton's second law? Explain.

12. In football, good wide receivers usually have "soft" hands for catching balls (▼Fig. 4.30). How would you interpret this description on the basis of Newton's second law?

▲ FIGURE 4.30 **Soft hands** See Conceptual Question 12.

4.4 NEWTON'S THIRD LAW OF MOTION

13. Here is a story of a horse and a farmer: One day, the farmer attaches a heavy cart to the horse and demands that the horse pull the cart. "Well," says the horse, "I cannot pull the cart, because, according to Newton's third law, if I apply a force to the cart, the cart will apply an equal and opposite force on me. The net result will be that I cannot pull the cart, since all the forces will cancel. Therefore, it is impossible for me to pull this cart." The farmer was very upset! What could he say to persuade the horse to move?

14. Is something wrong with the following statement? When a baseball is hit with a bat, there are equal and opposite forces on the bat and baseball. The forces then cancel, and there is no motion.

4.5 MORE ON NEWTON'S LAWS: FREE-BODY DIAGRAMS AND TRANSLATIONAL EQUILIBRIUM

15. Draw the free-body diagram for a person sitting in the seat of an aircraft (a) that is accelerating down the runway for takeoff, and (b) after takeoff at a 20° angle to the ground.

16. A person pushes perpendicularly on a block of wood that has been placed against a wall. Draw a free-body diagram of the block and identify the reaction forces to all the forces on the block.

17. A person on a bathroom scale (not the digital type) stands on the scale with his arms at his side. He then quickly raises his arms over his head, and notices that the scale reading increases as he brings his arms upward. Similarly, there is a decrease as he brings his arms downward. Why does the scale reading change? (Try this yourself.)

4.6 FRICTION

18. Identify the direction of the friction force in the following cases: (a) a book sitting on a table; (b) a box sliding on a horizontal surface; (c) a car making a turn on a flat road; (d) the initial motion of a machine part delivered on a conveyor belt in an assembly line.

19. The purpose of a car's antilock brakes is to prevent the wheels from locking up so as to keep the car rolling rather than sliding. Why would rolling decrease the stopping distance as compared with sliding?

20. Shown in ▼ Fig. 4.31 are the front and rear wings of an Indy racing car. These wings generate *down force*, which is the vertical downward force produced by the air moving over the car. Why is such a down force desired? An Indy car can create a down force equal to twice its weight. Why not simply make the cars heavier?

▲ FIGURE 4.31   **Down force** See Conceptual Question 20.

21. (a) We commonly say that friction opposes motion. Yet when we walk, the frictional force is in the direction of our motion (Fig. 4.17). Is there an inconsistency in terms of Newton's second law? Explain. (b) What effects would wind have on air resistance? [*Hint:* The wind can blow in different directions.]

22. Why are drag-racing tires wide and smooth, whereas passenger-car tires are narrower and have tread (▼ Fig. 4.32)? Are there frictional and/or safety considerations? Does this difference between the tires contradict the fact that friction is independent of surface area?

▲ FIGURE 4.32   **Racing tires versus passenger-car tires: safety** See Conceptual Question 22.

23. How could you approximately determine the coefficient of kinetic friction between your shoes and a fairly smooth floor? [*Hint:* See Exercise 70.]

Integrated Exercises (IEs) are two-part exercises. The first part typically requires a conceptual answer choice based on physical thinking and basic principles. The following part is quantitative calculations associated with the conceptual choice made in the first part of the exercise.

Throughout the text, many exercise sections will include "paired" exercises. These exercise pairs, identified with red numbers, *are intended to assist you in problem solving and learning. In a pair, the first exercise (even numbered) is worked out in the Study Guide so that you can consult it should you need assistance in solving it. The second exercise (odd numbered) is similar in nature, and its answer is given in Appendix VII at the back of the book.*

4.1 THE CONCEPTS OF FORCE AND NET FORCE
AND
4.2 INERTIA AND NEWTON'S FIRST LAW OF MOTION

1. ● Which has more inertia, 20 cm³ of water or 10 cm³ of aluminum, and how many times more? (See Table 9.2.)

2. ● Two forces act on a 5.0-kg object sitting on a frictionless horizontal surface. One force is 30 N in the $+x$-direction, and the other is 35 N in the $-x$-direction. What is the acceleration of the object?

3. ● In Exercise 2, if the 35-N force acted downward at an angle of 40° relative to the horizontal, what would be the acceleration in this case?

4. ● A net force of 4.0 N gives an object an acceleration of 10 m/s². What is the mass of the object?

5. ● Consider a 2.0-kg ball and a 6.0-kg ball in free fall. (a) What is the net force acting on each? (b) What is the acceleration of each?

6. **IE** ●● A hockey puck with a weight of 0.50 lb is sliding freely across a section of very smooth (frictionless) horizontal ice. (a) When it is sliding freely, how does the upward force of the ice on the puck (the normal force) compare with the upward force when the puck is sitting permanently at rest: (1) The upward force is greater when the puck is sliding; (2) the upward force is less when it is sliding; (3) the upward force is the same in both situations? (b) Calculate the upward force on the puck in both situations.

---

*Unless otherwise stated, all objects are located near the Earth's surface, where $g = 9.80$ m/s².

7. •• A 5.0-kg block at rest on a frictionless surface is acted on by forces $F_1 = 5.5$ N and $F_2 = 3.5$ N as illustrated in ▼ Fig. 4.33. What additional force will keep the block at rest?

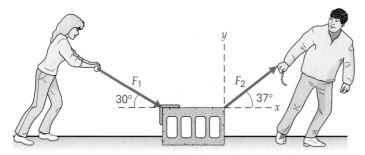

▲ FIGURE 4.33 **Two applied forces** See Exercise 7.

8. IE •• (a) You are told that an object has zero acceleration. Which of the following is true: (1) The object is at rest; (2) the object is moving with constant velocity; (3) either (1) or (2) is possible; or (4) neither 1 nor 2 is possible. (b) Two forces on the object are $F_1 = 3.6$ N at 74° below the +x-axis and $F_2 = 3.6$ N at 34° above the −x-axis. Is there a third force on the object? Why or why not? If there is a third force, what is it?

9. IE •• A fish weighing 25 lb is caught and hauled onto the boat. (a) Compare the tension in the fishing line when the fish is brought up vertically at a constant speed to the tension when the fish is held vertically at rest for the picture-taking ceremony on the wharf. In which case is the tension largest: (1) When the fish is moving up; (2) when the fish is being held steady; or (3) the tension is the same in both situations? (b) Calculate the tension in the fishing line.

10. ••• A 1.5-kg object moves up the y-axis at a constant speed. When it reaches the origin, the forces $F_1 = 5.0$ N at 37° above the +x-axis, $F_2 = 2.5$ N in the +x-direction, $F_3 = 3.5$ N at 45° below the −x-axis, and $F_4 = 1.5$ N in the −y-direction are applied to it. (a) Will the object continue to move along the y-axis? (b) If not, what simultaneously applied force will keep it moving along the y-axis at a constant speed?

11. IE ••• Three horizontal forces (the only horizontal ones) act on a box sitting on a floor. One (call it F_1) acts due east and has a magnitude of 150 lb. A second force (call it F_2) has an easterly component of 30.0 lb and a southerly component of 40.0 lb. The box remains at rest. (Neglect friction.) (a) Sketch the two known forces on the box. In which quadrant is the unknown third force: (1) the first quadrant; (2) the second quadrant; (3) the third quadrant; or (4) the fourth quadrant? (b) Find the unknown third force in newtons and compare your answer to the sketched estimate.

4.3 NEWTON'S SECOND LAW OF MOTION

12. • A 6.0-N net force is applied to a 1.5-kg mass. What is the object's acceleration?

13. • A force acts on a 1.5-kg, mass, giving it an acceleration of 3.0 m/s². (a) If the same force acts on a 2.5-kg mass, what acceleration would be produced? (b) What is the magnitude of the force?

14. • A loaded Boeing 747 jumbo jet has a mass of 2.0×10^5 kg. What net force is required to give the plane an acceleration of 3.5 m/s² down the runway for take-offs?

15. IE • A 6.0-kg object is brought to the Moon, where the acceleration due to gravity is only one-sixth of that on the Earth. (a) The mass of the object on the Moon is (1) zero, (2) 1.0 kg, (3) 6.0 kg, (4) 36 kg. Why? (b) What is the weight of the object on the Moon?

16. •• A gun is fired and a 50-g bullet is accelerated to a muzzle speed of 100 m/s. If the length of the gun barrel is 0.90 m, what is the magnitude of the accelerating force? (Assume the acceleration to be constant.)

17. IE ••• ▼Fig. 4.34 shows a product label. (a) This label is correct (1) on the Earth; (2) on the Moon, where the acceleration due to gravity is only one-sixth of that on the Earth; (3) in deep space, where there is little gravity; (4) all of the preceding. (b) What mass of lasagne would a label show for an amount that weighs 2 lb on the Moon?

▲ FIGURE 4.34 **Correct label?** See Exercise 17.

18. •• In a college homecoming competition, eighteen students lift a sports car. While holding the car off the ground, each student exerts an upward force of 400 N. (a) What is the mass of the car in kilograms? (b) What is its weight in pounds?

19. IE •• (a) A horizontal force acts on an object on a frictionless horizontal surface. If the force is halved and the mass of the object is doubled, the acceleration will be (1) four times, (2) two times, (3) one-half, (4) one-fourth as great. (b) If the acceleration of the object is 1.0 m/s², and the force on it is doubled and its mass is halved, what is the new acceleration?

20. •• A force of 50 N acts on a mass m_1, giving it an acceleration of 4.0 m/s². The same force acts on a mass m_2 and produces an acceleration of 12 m/s². What acceleration will this force produce if the total system is $m_1 + m_2$?

21. •• A student weighing 800 N crouches on a scale and suddenly springs vertically upward. His roommate notices that the scale reads 900 N momentarily just as he leaves the scale. With what acceleration does he leave the scale?

22. •• The engine of a 1.0-kg toy plane exerts a 15-N forward force. If the air exerts an 8.0-N resistive force on the plane, what is the magnitude of the acceleration of the plane?

23. •• When a horizontal force of 300 N is applied to a 75.0-kg box, the box slides on a level floor, opposed by a force of kinetic friction of 120 N. What is the magnitude of the acceleration of the box?

24. IE •• A rocket is far away from all planets and stars, so gravity is not a consideration. It is using its rocket engines to accelerate upward with an acceleration $a = 9.80$ m/s². On the floor of the main deck is a crate (object with brick pattern) with a mass of 75.0 kg (▼Fig. 4.35). (a) How many forces are acting on the crate: (1) zero; (2) one; (3) two; (4) three? (b) Determine the normal force on the crate and compare it to the normal force the crate would experience if it were at rest on the surface of the Earth.

◄ **FIGURE 4.35 Away we go** See Exercise 24.

25. •• An object (mass 10.0 kg) slides *upward* on a slippery vertical wall. A force F of 60 N acts at an angle of 60° as shown in ▼Fig. 4.36. (a) Determine the normal force exerted on the object by the wall. (b) Determine the object's acceleration.

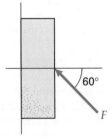

◄ **FIGURE 4.36 Up a wall** See Exercise 25. (Drawing not to scale.)

26. •• In an emergency stop to avoid an accident, a shoulder-strap seatbelt holds a 60-kg passenger in place. If the car was initially traveling at 90 km/h and came to a stop in 5.5 s along a straight, level road, what was the average force applied to the passenger by the seatbelt?

27. IE •• A student is assigned the task of measuring the startup acceleration of a large RV (recreational vehicle) using an iron ball suspended from the ceiling by a long string. In accelerating from rest, the ball no longer hangs vertically, but at an angle to the vertical. (a) Is the angle of the ball forward or backward from the vertical? (b) If the string makes an angle of 3.0 degrees from the vertical, what is the initial acceleration of the RV?

28. •• A force of 10 N acts on two blocks on a frictionless surface (▼Fig. 4.37). (a) What is the acceleration of the system? (b) What force does block A exert on block B? (c) What force does block B exert on block A?

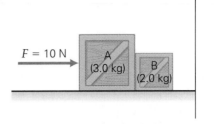

◄ **FIGURE 4.37 Forces: inside and out** See Example 28.

29. •• A 2.0-kg object has an acceleration of 1.5 m/s² at 30° above the −x-axis. Write the force vector producing this acceleration in component form.

30. ••• In a pole-sliding game among friends, a 90-kg man makes a total vertical drop of 7.0 m while gripping the pole which exerts and upward force (call it F_p) on him. Starting from rest and sliding with a constant acceleration, his slide takes 2.5 s. (a) Draw the man's free body diagram being sure to label all the forces. (b) What is the magnitude of the upward force exerted on the man by the pole? (c) A friend whose mass is only 75 kg, slides down the same distance, but the pole force is only 80% of the force on his buddy. How long did the second person's slide take?

4.4 NEWTON'S THIRD LAW OF MOTION

31. IE • A book is sitting on a horizontal surface. (a) There is (are) (1) one, (2) two, or (3) three force(s) acting on the book. (b) Identify the reaction force to each force on the book.

32. •• In an Olympic figure-skating event, a 65-kg male skater pushes a 45-kg female skater, causing her to accelerate at a rate of 2.0 m/s². At what rate will the male skater accelerate? What is the direction of his acceleration?

33. IE •• A sprinter of mass 65.0 kg starts his race by pushing horizontally backward on the starting blocks with a force of 200 N. (a) What force causes him to accelerate out of the blocks: (1) his push on the blocks; (2) the downward force of gravity; or (3) the force the blocks exert forward on him? (b) Determine his initial acceleration as he leaves the blocks.

34. •• Jane and John, with masses of 50 kg and 60 kg, respectively, stand on a frictionless surface 10 m apart. John pulls on a rope that connects him to Jane, giving Jane an acceleration of 0.92 m/s² toward him. (a) What is John's acceleration? (b) If the pulling force is applied constantly, where will Jane and John meet?

35. IE ••• During a daring rescue, a helicopter rescue squad initially accelerates a little girl (mass 25.0 kg) vertically off the roof of a burning building. They do this by dropping a rope down to her, which she holds on to as they pull her up. Neglect the mass of the rope. (a) What force causes the girl to accelerate vertically upward: (1) her weight; (2) the pull of the helicopter on the rope; (3) the pull of the girl on the rope; or (4) the pull of the rope on the girl? (b) Determine the pull of the rope (the tension) if she initially accelerates upward at 0.750 m/s².

4.5 MORE ON NEWTON'S LAWS: FREE-BODY DIAGRAMS AND TRANSLATIONAL EQUILIBRIUM

36. •• A 75.0-kg person is standing on a scale in an elevator. What is the reading of the scale in newtons if the elevator is (a) at rest, (b) moving up at a constant velocity of 2.00 m/s, and (c) accelerating up at 2.00 m/s²?

37. •• In Exercise 36, what if the elevator is accelerating down at 2.00 m/s²?

38. **IE** • (a) When an object is on an inclined plane, the normal force exerted by the inclined plane on the object is (1) less than, (2) equal to, (3) more than the weight of the object. Why? (b) For a 10-kg object on a 30° inclined plane, what are the object's weight and the normal force exerted on the object by the inclined place?

39. **IE** •• The weight of a 500-kg object is 4900 N. (a) When the object is on a moving elevator, its measured weight could be (1) zero, (2) between zero and 4900 N, (3) more than 4900 N, (4) all of the preceding. Why? (b) Describe the motion if the object's measured weight is only 4000 N in a moving elevator.

40. •• A boy pulls a box of mass 30 kg with a force of 25 N in the direction shown in ▼Fig. 4.38. (a) Ignoring friction, what is the acceleration of the box? (b) What is the normal force exerted on the box by the ground?

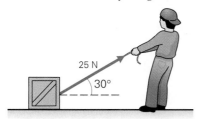

25 N

30°

▲ **FIGURE 4.38** **Pulling a box** See Exercise 40.

41. •• A girl pushes a 25-kg lawn mower as shown in ▼Fig. 4.39. If $F = 30$ N and $\theta = 37°$ (a) what is the acceleration of the mower, and (b) what is the normal force exerted on the mower by the lawn? Ignore friction.

F

θ

▲ **FIGURE 4.39** **Mowing the lawn** See Exercise 41.

42. •• A 3000-kg truck tows a 1500-kg car by a chain. If the net forward force on the truck by the ground is 3200 N, (a) what is the acceleration of the car, and (b) what is the tension in the connecting chain?

43. •• A block of mass 25.0 kg slides down a frictionless surface inclined at 30°. To ensure that the block does not accelerate, what is the smallest force that you must exert on it and what is its direction?

44. **IE** •• (a) An Olympic skier coasts down a slope with an angle of inclination of 37°. Neglecting friction, there is (are) (1), one, (2) two, (3) three force(s) acting on the skier. (b) What is the acceleration of the skier? (c) If the skier has a speed of 5.0 m/s at the top of the slope, what is his speed when he reaches the bottom of the 35-m-long slope?

45. •• A car coasts (engine off) up a 30° grade. If the speed of the car is 25 m/s at the bottom of the grade, what is the distance traveled by the car before it comes to rest?

46. •• Assuming ideal frictionless conditions for the apparatus shown in ▼Fig. 4.40, what is the acceleration of the system if (a) $m_1 = 0.25$ kg, $m_2 = 0.50$ kg, and $m_3 = 0.25$ kg, and (b) $m_1 = 0.35$ kg, $m_2 = 0.15$ kg, and $m_3 = 0.50$ kg?

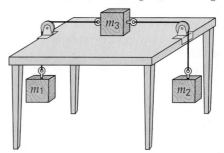

m_3

m_1

m_2

◀ **FIGURE 4.40** **Which way will they accelerate?** See Exercises 46, 80, and 81.

47. **IE** •• A rope is fixed at both ends on two trees and a bag is hung in the middle of the rope (causing the rope to sag vertically). (a) The tension in the rope depends on (1) only the tree separation, (2) only the sag, (3) both the tree separation and sag, (4) neither the tree separation nor the sag. (b) If the tree separation is 10 m, the mass of the bag is 5.0 kg, and the sag is 0.20 m, what is the tension in the line?

48. •• A 55-kg gymnast hangs vertically from a pair of parallel rings. (a) If the ropes supporting the rings are attached to the ceiling directly above, what is the tension in each rope? (b) If the ropes are supported so that they make an angle of 45° with the ceiling, what is the tension in each rope?

49. •• A physicist's car has a small lead weight suspended from a string attached to the interior ceiling. Starting from rest, after a fraction of a second the car accelerates at a steady rate for about 10 s. During that time, the string (with the weight on the end of it) makes a backward (opposite the acceleration) angle of 15.0° from the vertical. Determine the car's (and the weight's) acceleration during the 10-s interval.

50. •• A 10-kg mass is suspended as shown in ▼Fig. 4.41. What is the tension in the cord between points A and B?

51. •• Referring to Fig. 4.41, what are the tensions in all the cords?

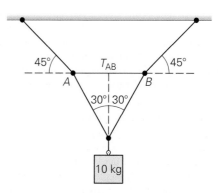

45° T_{AB} 45°
A B
30° 30°

10 kg

▲ **FIGURE 4.41** **Under tension** See Exercises 50 and 51.

52. •• At the end of most landing runways in airports, an extension of the runway is constructed using a special substance called *formcrete*. Formcrete can support the weight of cars, but crumbles under the weight of airplanes to slow them down if they run off the end of a runway. If a plane of mass 2.00×10^5 kg is to stop from a speed of 25.0 m/s on a 100-m-long stretch of formcrete, what is the average force exerted on the plane by the formcrete?

53. •• A rifle weighs 50.0 N and its barrel is 0.750 m long. It shoots a 25.0-g bullet, which leaves the barrel at a speed (muzzle velocity) of 300 m/s after being uniformly accelerated. What is the magnitude of the force exerted on the rifle by the bullet?

54. •• A horizontal force of 40 N acting on a block on a frictionless, level surface produces an acceleration of 2.5 m/s². A second block, with a mass of 4.0 kg, is dropped onto the first. What is the magnitude of the acceleration of the combination of blocks if the same force continues to act? (Assume that the second block does not slide on the first block.)

55. •• The *Atwood machine* consists of two masses suspended from a fixed pulley, as shown in ▼Fig. 4.42. It is named after the British scientist George Atwood (1746–1807), who used it to study motion and to measure the value of g. If $m_1 = 0.55$ kg and $m_2 = 0.80$ kg, (a) what is the acceleration of the system, and (b) what is the magnitude of the tension in the string?

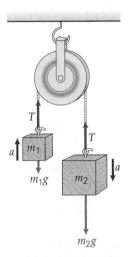

◀ **FIGURE 4.42 Atwood machine** See Exercises 55, 56, and 57.

56. •• An Atwood machine (see Fig. 4.42) has suspended masses of 0.25 kg and 0.20 kg. Under ideal conditions, what will be the acceleration of the smaller mass?

57. ••• One mass, $m_1 = 0.215$ kg, of an ideal Atwood machine (see Fig. 4.42) rests on the floor 1.10 m below the other mass, $m_2 = 0.255$ kg, (a) If the masses are released from rest, how long does it take m_2 to reach the floor? (b) How high will mass m_1 ascend from the floor? (*Hint*: When m_2 hits the floor, m_1 continues to move upward.)

58. IE ••• Two blocks are connected by a light string and accelerated upward by a pulling force F. The mass of the upper block is 50.0 kg and that of the lower block is 100 kg. The upward acceleration of the system as a whole is 1.50 m/s². Neglect the mass of the string. (a) Draw the free-body diagram of each block. Use the diagrams to determine which of the following is true

for the magnitude of the string tension T compared to other forces: (1) $T > w_2$ and $T < F$; (2) $T > w_2$ and $T > F$; (3) $T < w_2$ and $T < F$; or (4) $T = w_2$ and $T < F$? (b) Apply Newton's laws to find the required pull, F. (c) Find the tension in the string, T.

59. ••• Two blocks on a level, frictionless table are in contact. The mass of the left block is 5.00 kg and the mass of the right block is 10.0 kg, and they accelerate to the left at 1.50 m/s². A person on the left exerts a force (F_1) of 75.0 N to the right. Another person exerts an unknown force (F_2) to the left. (a) Determine the force F_2. (b) Calculate the force of contact N between the two blocks (that is, the normal force at their vertical touching surfaces).

60. ••• In the frictionless apparatus shown in ▼Fig. 4.43, $m_1 = 2.0$ kg. What is m_2 if both masses are at rest? How about if both masses are moving at constant velocity?

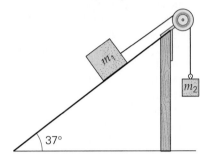

▲ **FIGURE 4.43 Inclined Atwood machine** See Exercises 60, 61, and 79.

61. ••• In the ideal setup shown in Fig. 4.43, $m_1 = 3.0$ kg and $m_2 = 2.5$ kg. (a) What is the acceleration of the masses? (b) What is the tension in the string?

4.6   FRICTION

62. IE • A 20-kg box sits on a rough horizontal surface. When a horizontal force of 120 N is applied, the object accelerates at 1.0 m/s². (a) If the applied force is doubled, the acceleration will (1) increase, but less than double; (2) also double; (3) increase, but more than double. Why? (b) Calculate the acceleration to prove your answer to part (a).

63. • The coefficients of static and kinetic friction between a 50.0-kg box and a horizontal surface are 0.500 and 0.400 respectively. (a) What is the acceleration of the object if a 250-N horizontal force is applied to the box? (b) What is the acceleration if the applied force is 235 N?

64. • In moving a 35.0-kg desk from one side of a classroom to the other, a professor finds that a horizontal force of 275 N is necessary to set the desk in motion, and a force of 195 N is necessary to keep it in motion at a constant speed. What are the coefficients of (a) static and (b) kinetic friction between the desk and the floor?

65. • A 40-kg crate is at rest on a level surface. If the coefficient of static friction between the crate and the surface is 0.69, what horizontal force is required to get the crate moving?

66. •• A packing crate is placed on a 20° inclined plane. If the coefficient of static friction between the crate and the plane is 0.65, will the crate slide down the plane if released from rest? Justify your answer.

67. •• A 1500-kg automobile travels at 90 km/h along a straight concrete highway. Faced with an emergency situation, the driver jams on the brakes, and the car skids to a stop. What is the car's stopping distance for (a) dry pavement and (b) wet pavement?

68. •• A hockey player hits a puck with his stick, giving the puck an initial speed of 5.0 m/s. If the puck slows uniformly and comes to rest in a distance of 20 m, what is the coefficient of kinetic friction between the ice and the puck?

69. •• A crate sits on a flat-bed truck that is traveling with a speed of 50 km/h on a straight, level road. If the coefficient of static friction between the crate and the truck bed is 0.30, in how short a distance can the truck stop with a constant acceleration without the crate sliding?

70. •• A block is projected with a speed of 2.5 m/s on a horizontal surface. If the block comes to rest in 1.5 m, what is the coefficient of kinetic friction between the block and the surface?

71. •• A block is projected with a speed of 3.0 m/s on a horizontal surface. If the coefficient of kinetic friction between the block and the surface is 0.60, how far does the block slide before coming to rest?

72. IE •• A person has a choice while trying to push a crate across a horizontal pad of concrete: push it at a downward angle of 30°, or pull it at an upward angle of 30°. (a) Which choice is most likely to require less force on the part of the person: (1) pushing at a downward angle; (2) pulling at the same angle, but upward; or (3) pushing or pulling shouldn't matter? (b) If the crate has a mass of 50.0 kg and the coefficient of kinetic friction between it and the concrete is 0.750, calculate the required force to move it across the concrete at a steady speed for both situations.

73. •• Suppose the slope conditions for the skier shown in ▼Fig. 4.44 are such that the skier travels at a constant velocity. From the photo, could you find the coefficient of kinetic friction between the snowy surface and the skis? If so, describe how this would be done.

▲ FIGURE 4.44 **A down slope run** See Exercise 73.

74. •• A 5.0-kg wooden block is placed on an adjustable wooden inclined plane. (a) What is the angle of incline above which the block will *start* to slide down the plane? (b) At what angle of incline will the block then slide down the plane at a constant speed?

75. •• A block that has a mass of 2.0 kg and is 10 cm wide on each side just begins to slide down an inclined plane with a 30° angle of incline (▼Fig. 4.45). Another block of the same height and same material has base dimensions of 20 cm × 10 cm and thus a mass of 4.0 kg. (a) At what critical angle will the more massive block start to slide down the plane? Why? (b) Estimate the coefficient of static friction between the block and the plane.

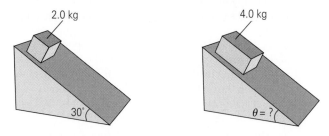

▲ FIGURE 4.45 **At what angle will it begin to slide?** See Exercise 75.

76. •• In the apparatus shown in ▼Fig. 4.46, $m_1 = 10$ kg and the coefficients of static and kinetic friction between m_1 and the table are 0.60 and 0.40, respectively. (a) What mass of m_2 will just barely set the system in motion? (b) After the system begins to move, what is the acceleration?

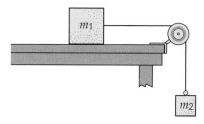

▲ FIGURE 4.46 **Friction and motion** See Exercise 76.

77. •• In loading a fish delivery truck, a person pushes a block of ice up a 20° incline at constant speed. The push is 150 N in magnitude and parallel to the incline. The block has a mass of 35.0 kg. (a) Is the incline frictionless? (b) If not, what is the force of kinetic friction on the block of ice?

78. ••• An object (mass 3.0 kg) slides *upward* on a vertical wall *at constant velocity* when a force F of 60 N acts on it at an angle of 60° to the horizontal. (a) Draw the free-body diagram of the object. (b) Using Newton's laws find the normal force on the object. (c) Determine the force of kinetic friction on the object.

79. ••• In the apparatus shown in Fig. 4.43, $m_1 = 2.0$ kg and the coefficients of static and kinetic friction between m_1 and the inclined plane are 0.30 and 0.20, respectively. (a) What is m_2 if both masses are at rest? (b) What is m_2 if both masses are moving at constant velocity?

80. ••• For the apparatus shown in Fig. 4.40, what is the minimum value of the coefficient of static friction between the block (m_3) and the table that would keep the system at rest if $m_1 = 0.25$ kg, $m_2 = 0.50$ kg, and $m_3 = 0.75$ kg?

81. ••• If the coefficient of kinetic friction between the block and the table in Fig. 4.40 is 0.560, and $m_1 = 0.150$ kg and $m_2 = 0.250$ kg, (a) what should m_3 be if the system is to move with a constant speed? (b) If $m_3 = 0.100$ kg, what is the magnitude of the acceleration of the system?

The Multiconcept Exercises require the use of more than one fundamental concept for their understanding and solution. The concepts may be from this chapter, but may include those from previous chapters.

82. **IE** One block (A, mass 2.00 kg) rests atop another (B, mass 5.00 kg) on a horizontal surface. The surface is a powered walkway accelerating to the right at 2.50 m/s^2. B does not slip on the walkway surface, nor does A slip on B's top surface. (a) Sketch the free-body diagram of each block. Use these to determine the force responsible for A's acceleration. Is it (1) the pull of the walkway, (2) the normal force on A by the top surface of B, (3) the force of static friction on the bottom surface of B, or (4) the force of static friction acting on A due to the top surface of B? (b) Determine the forces of static friction on each block.

83. Two blocks (A and B) remain stuck together as they are pulled to the right by a force $F = 200 \text{ N}$ (▼Fig. 4.47). B is on a rough horizontal tabletop (coefficient of kinetic friction of 0.800). (a) What is the acceleration of the system? (b) What is the force of friction between the two objects?

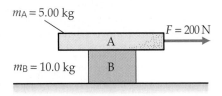

$m_A = 5.00 \text{ kg}$

A $F = 200 \text{ N}$

$m_B = 10.0 \text{ kg}$ B

▲ **FIGURE 4.47 Take it away** See Exercise 83.

84. **IE** To haul a boat out of the water for the winter, a worker at the storage facility uses a wide strap with cables operating at the same angle (measured from the horizontal) on either side of the boat (▼Fig. 4.48). (a) As the boat comes up vertically and θ decreases, the tension in the cables (1) increases, (2) decreases, (3) stays the same. (b) Determine the tension in each cable if the boat has a mass of 500 kg and the angle of each cable is 45°

T_2 T_1

θ θ

▲ **FIGURE 4.48 Hoist it up** See Exercise 84.

from the horizontal and the boat is being momentarily held at rest. Compare this to the tension when the boat is raised and held at rest so the angle becomes 30°.

85. You are in charge of an accident reconstruction case for the local police department. In order to determine car speeds, skid mark lengths are measured. To determine the coefficient of kinetic friction, you get into an identical car, and at a speed of 65.2 mi/h, you lock its brakes and skid 51.5 m to rest. (a) Determine the car's deceleration. (b) What is the coefficient of kinetic friction between the tires and road surface? (c) The car in the accident actually skidded 57.3 m. What was its initial speed?

86. **IE** Compare two different situations in which a ball and hard surface exert forces on one another. First, a putty ball is placed gently on the floor and left at rest. Then it is dropped from a height of 2.00 m and comes to rest without a bounce, leaving a 1.15-cm-deep dent in the putty. (a) In which case does the ball exert more force on the floor? In which case is it most likely to dent the floor? Explain. (b) Calculate the force exerted by the ball on the floor (in terms of its weight w) in the first case. (c) Determine the average acceleration of the ball and the average force exerted by the ball on the floor (in terms of the ball's weight w) in the second case.

87. A hockey puck impacts a goalie's plastic mask horizontally at 122 mi/h and rebounds horizontally off the mask at 47 mi/h. If the puck has a mass of 170 g and it is in contact with the mask for 25 ms, (a) what is the average force (including direction) that the puck exerts on the mask? (b) Assuming that this average force accelerates the goalie (neglect friction with the ice), with what speed will the goalie move, assuming she was at rest initially and has a total mass of 85 kg?

88. A 2.50-kg block is placed on a rough surface inclined at 30°. The block is propelled and launched at a speed of 1.60 m/s *down* the incline and comes to rest after sliding 1.10 m. (a) Draw the free-body diagram of the block while it is sliding. Also indicate your coordinate system axes. (b) Starting with Newton's second law applied along both axes of your coordinate system, use your free-body diagram to generate two equations. (c) Solve these equations for the coefficient of kinetic friction between the block and the incline surface. [*Hint:* You will need to first determine the block's acceleration.]

5 Work and Energy†

PHYSICS FACTS

✦ *Kinetic* comes from the Greek *kinein*, meaning "to move."

✦ *Energy* comes from the Greek *energeia*, meaning "activity."

✦ The United States has 5% of the world's population, yet consumes about 26% of its energy supply.

✦ Recycling aluminum takes 95% less energy than making aluminum from raw materials.

✦ The human body uses muscles to propel itself, turning stored energy into motion. There are 630 active muscles in your body and they act in groups.

✦ The human body operates within the limits imposed by the law of conservation of total energy, needing dietary energy equal to the energy expended in the overall work of daily activities, internal activities, and system heat losses.

✦ Energy is neither created nor destroyed. The amount of energy in the universe is constant or conserved.

A description of pole vaulting, as shown in the chapter-opening photo, might be as follows: The athlete runs with a pole, plants it into the ground, and tries to vault his body over a bar set at a certain height. However, a physicist might give a different description: The athlete has chemical potential energy stored in his body. He uses this potential energy to do work in running down the path to gain speed, or kinetic energy. When he plants the pole, most of his kinetic energy goes into elastic potential energy of the bent pole.

†The mathematics in this chapter involves trigonometric functions. You may want to review these in Appendix I.

This potential energy is used to lift the vaulter in doing work against gravity, and is partially converted into gravitational potential energy. At the top, there is just enough kinetic energy left to carry the vaulter over the bar. On the way down, the gravitational potential energy is converted back to kinetic energy, which is absorbed by the mat in doing work to stop the fall. The pole vaulter participates in a game of work–energy, a game of give and take.

This chapter centers on two concepts that are important in both science and everyday life—*work* and *energy*. We commonly think of work as being associated with doing or accomplishing something. Because work makes us physically (and sometimes mentally) tired, machines have been invented to decrease the amount of effort expended personally. Thinking about energy tends to bring to mind the cost of fuel for transportation and heating, or perhaps the food that supplies the energy our bodies need to sustain life processes and to do work.

Although these notions do not really define work and energy, they point in the right direction. As you may have surmised, work and energy are closely related. In physics, as in everyday life, when something possesses energy, it has the ability to do work. For example, water rushing through the sluices of a dam has energy of motion, and this energy allows the water to do the work of driving a turbine or dynamo to generate electricity. Conversely, no work can be performed without energy.

Energy exists in various forms: mechanical energy, chemical energy, electrical energy, heat energy, nuclear energy, and so on. A transformation from one form to another may take place, but the total amount of energy is *conserved*, meaning there is always the same amount. This point makes the concept of energy very useful. When a physically measurable quantity is conserved, it not only gives us an insight that leads to a better understanding of nature, but also usually provides another approach to practical problems. (You will be introduced to other conserved quantities and conservation laws during the course of our study of physics.)

5.1   Work Done by a Constant Force

LEARNING PATH QUESTIONS

➥ What is the work done by a constant force?
➥ How does negative work arise?
➥ What is meant by total, or net, work?

The word *work* is commonly used in a variety of ways: We go to work; work on projects; work at our desks or on computers; work on problems. In physics, however, *work* has a very specific meaning. Mechanically, work involves force and displacement, and the word *work* is used to describe quantitatively what is accomplished when a force acts on an object as it moves through a distance. In the simplest case of a *constant* force acting on an object, work that the force does is defined as follows:

> The **work** done by a constant force acting on an object is equal to the product of the magnitudes of the displacement and the force, or component of the force, parallel to that displacement.

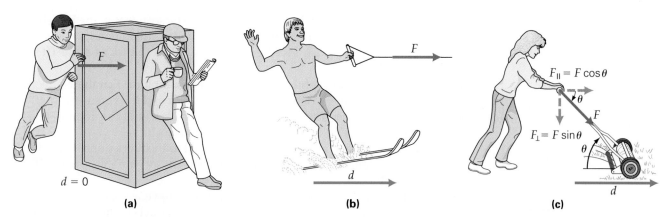

▲ FIGURE 5.1 Work done by a constant force—the product of the magnitudes of the parallel component of force and the displacement **(a)** If there is no displacement, no work is done: $W = 0$. **(b)** For a constant force in the same direction as the displacement, $W = Fd$. **(c)** For a constant force at an angle to the displacement, $W = (F \cos \theta)d$.

Work then involves a force acting on an object and moving it through a distance. A force may be applied, as in ▲Fig. 5.1a, but *if there is no motion (no displacement), then no work is done*. However when there is motion, a constant force F acting *in the same direction* as the displacement d does work (Fig. 5.1b). The work (W) done in this case is defined as the product of their magnitudes:

$$W = Fd \qquad (5.1)$$

and work is a scalar quantity. (As you might expect, when work is done as in Fig. 5.1b, energy is expended. The relationship between work and energy is discussed in Section 5.3.)

In general, work is done on an object by a force, or force *component*, parallel to the line of motion or displacement of the object (Fig. 5.1c). That is, if the force acts at an angle θ to the object's displacement, then $F_{\|} = F \cos \theta$ is the component of the force parallel to the displacement. Thus, a more general equation for work done by a constant force is*

$$W = F_{\|} d = (F \cos \theta)d \quad \textit{(work done by a constant force)} \qquad (5.2)$$

Notice that θ is the angle *between* the force and the displacement vectors. As a reminder of this factor, $\cos \theta$ may be written between the magnitudes of the force and displacement, $W = F(\cos \theta)d$. If $\theta = 0°$ (that is, force and displacement are in the same direction, as in Fig. 5.1b), then $W = F(\cos 0°)d = Fd$, so Eq. 5.2 reduces to Eq. 5.1. The perpendicular component of the force, $F_{\perp} = F \sin \theta$, does no work, since there is no displacement in this direction.

The units of work can be determined from the equation $W = Fd$. With force in newtons and displacement in meters, work has the SI unit of newton-meter (N·m). This unit is called a **joule (J)**:[†]

$$Fd = W$$

$$1 \, \text{N·m} = 1 \, \text{J}$$

For example, the work done by a force of 25 N on an object as the object moves through a parallel displacement of 2.0 m is $W = Fd = (25 \, \text{N})(2.0 \, \text{m}) = 50 \, \text{N·m}$, or 50 J.

---

*The product of two vectors (force and displacement) is a special type of vector multiplication and yields a scalar quantity equal to $(F \cos \theta)d$. Thus, work is a scalar—it does not have direction. It can, however, be positive, zero, or negative, depending on the angle.

[†]The joule (J), pronounced "jool," was named in honor of James Prescott Joule (1818–1889), a British scientist who investigated work and energy.

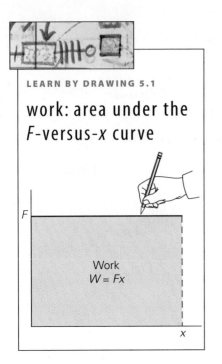

LEARN BY DRAWING 5.1

work: area under the F-versus-x curve

From the previous displayed equation, it can also be seen that in the British system, work would have the unit pound-foot. However, this name is commonly written in reverse. The British standard unit of work is the **foot-pound (ft · lb)**. One ft · lb is equal to 1.36 J.

Work can be analyzed graphically. Suppose a constant force F in the x-direction acts on an object as it moves a distance x. Then $W = Fx$ and if F versus x is plotted, a horizontal straight-line graph is obtained such as shown in the accompanying Learn by Drawing 5.1, Work: Area under the F-versus-x Curve.

The area under the line is Fx, so this area is equal to the work done by the force over the given distance. Work done by a nonconstant, or variable, force will be considered later.*

Remember that *work is a scalar quantity* and may have a positive or negative value. In Fig. 5.1b, the work is positive, because the force acts in the same direction as the displacement (and cos 0° is positive). The work is also positive in Fig. 5.1c, because a force component acts in the direction of the displacement (and cos θ is positive).

However, if the force, or a force component, acts in the opposite direction of the displacement, the work is negative, since the cosine term is negative. For example, for $\theta = 180°$ (force opposite to the displacement), cos 180° = −1, so the work is negative: $W = F_{\parallel}d = F(\cos 180°)d = -Fd$. An example is a braking force that slows down or decelerates an object. See the associated Learn by Drawing 5.2, Determining the Sign of Work.

EXAMPLE 5.1 | ## Applied Psychology: Mechanical Work

A student holds her 1.5-kg psychology textbook out a second-story dormitory window until her arm is tired; then she releases it (▶ Fig. 5.2). (a) How much work is done on the book by the student in simply holding it out the window? (b) How much work is done by the force of gravity during the time in which the book falls 3.0 m?

THINKING IT THROUGH.   Analyze the situations in terms of the definition of work, keeping in mind that force and displacement are the key factors.

SOLUTION.   Listing the data,

Given: $v_o = 0$ (initially at rest)   *Find:*   (a) W (work done by student in holding)
$m = 1.5$ kg   (b) W (work done by gravity in falling)
$d = 3.0$ m

(a) Even though the student gets tired (because work is performed within the body to maintain muscles in a state of tension), she does *no work on the book* in merely holding it stationary. She exerts an upward force on the book (equal in magnitude to its weight), but the displacement is zero in this case ($d = 0$). Thus, $W = Fd = F \times 0 = 0$ J.

(b) While the book is falling, the only force acting on it is the force of gravity (neglecting air resistance), which is equal in magnitude to the weight of the book: $F = w = mg$. The displacement is in the same direction as the force ($\theta = 0°$) and has a magnitude of $d = 3.0$ m so the work done by gravity is

$$W = F(\cos 0°)d = (mg)d = (1.5 \text{ kg})(9.8 \text{ m/s}^2)(3.0 \text{ m}) = +44 \text{ J}$$

(+ because the force and displacement are in the same direction.)

FOLLOW-UP EXERCISE.   A 0.20-kg ball is thrown upward. How much work is done on the ball by gravity as the ball rises between heights of 2.0 m and 3.0 m? *(Answers to all Follow-Up Exercises are given in Appendix VI at the back of the book.)*

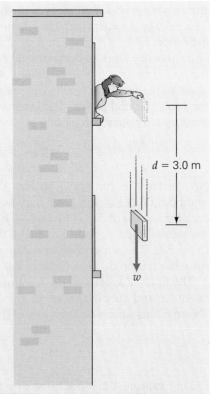

▲ **FIGURE 5.2 Mechanical work requires motion** See Example text for description.

---

*Work is the area under the F-versus-x curve even if the curve is not a straight line. Finding the work in such cases generally requires advanced mathematics.

EXAMPLE 5.2 | Hard Work

A worker pulls a 40.0-kg crate with a rope, as illustrated in ▾Fig. 5.3. The coefficient of kinetic (sliding) friction between the crate and the floor is 0.550. If he moves the crate with a constant velocity a distance of 7.00 m, how much work is done?

THINKING IT THROUGH. A good thing to do first in problems such as this is to draw a free-body diagram. This is shown in the figure. To find the work, the force F must be known. As usual in such cases, this is done by summing the forces.

Given: $m = 40.0 \text{ kg}$ **Find:** W (work done in moving the crate
$\mu_k = 0.550$ 7.00 m)
$d = 7.00 \text{ m}$
$\theta = 30°$ (from figure)
v (constant)

SOLUTION. Then, summing the forces in the x- and y-directions and setting these equal to zero (with a constant velocity $F_{net} = 0$):

$$\sum F_x = F \cos 30° - f_k = F \cos 30° - \mu_k N = ma_x = 0$$

$$\sum F_y = N + F \sin 30° - mg = ma_y = 0$$

To find F, the second equation may be solved for N, which is then substituted in the first equation.

$$N = mg - F \sin 30°$$

(Notice that N is not equal to the weight of the crate. Why?) And, substituting N into the first equation,

$$F \cos 30° - \mu_k(mg - F \sin 30°) = 0$$

Solving for F and putting in values:

$$F = \frac{\mu_k \, mg}{(\cos 30° + \mu_k \sin 30°)} = \frac{(0.550)(40.0 \text{ kg})(9.80 \text{ m/s}^2)}{(0.866) + (0.550)(0.500)]} = 189 \text{ N}$$

Then,

$$W = F(\cos 30°)d = (189 \text{ N})(0.866)(7.00 \text{ m}) = 1.15 \times 10^3 \text{ J}$$

FOLLOW-UP EXERCISE. It takes about 3.80×10^4 J of work to lose 1.00 g of body fat. What distance would the worker have to pull the crate to lose 1 g of fat? (Assume all the work goes into fat reduction.) Make an estimate before solving and see how close you come.

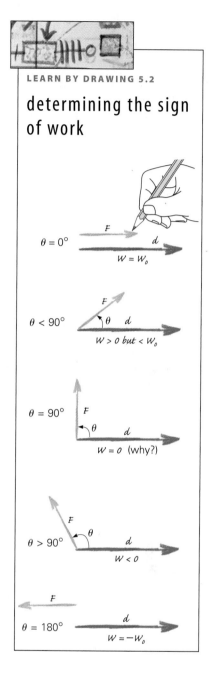

LEARN BY DRAWING 5.2

determining the sign of work

$\theta = 0°$
$W = W_o$

$\theta < 90°$
$W > 0$ but $< W_o$

$\theta = 90°$
$W = 0$ (why?)

$\theta > 90°$
$W < 0$

$\theta = 180°$
$W = -W_o$

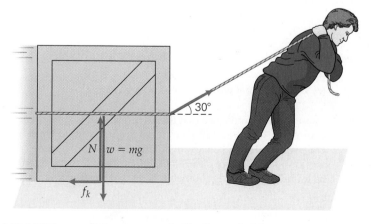

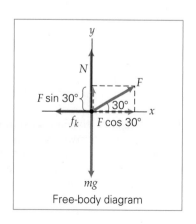

Free-body diagram

▲ **FIGURE 5.3 Doing some work** See Example 5.2.

It is commonly said that a force does work *on* an object. For example, the force of gravity does work on a falling object, such as the book in Example 5.1. Also, when you lift an object, *you* do work *on* the object. This is sometimes described as doing work *against* gravity, because the force of gravity acts in the direction opposite that of the applied lift force and opposes it. For example, an average-sized apple has a weight of about 1 N. So, when lifting such an apple a distance of 1 m with a force equal to its weight, 1 J of work is done against gravity $[W = Fd = (1\text{ N})(1\text{ m}) = 1\text{ J}]$. This gives an idea of how much work 1 J represents.

In both Examples 5.1 and 5.2, work was done by a single constant force. If more than one force acts on an object, the work done by each can be calculated separately. That is:

> The *total*, or *net, work* is defined as the work done by all the forces acting on an object, or the scalar sum of the work done by each force.

This concept is illustrated in Example 5.3.

EXAMPLE 5.3 | ## Total or Net Work

A 0.75-kg block slides with a uniform velocity down a 20° inclined plane (▼ Fig. 5.4). (a) How much work is done by the force of friction on the block as it slides the total length of the plane? (b) What is the net work done on the block? (c) Discuss the net work done if the angle of incline is adjusted so that the block accelerates down the plane.

THINKING IT THROUGH.   (a) The length of the plane can be found using trigonometry, so this part boils down to finding the force of friction. (b) The net work is the sum of all the work done by the individual forces. (*Note*: Since the block has a uniform, or constant, velocity, the net force on it is zero. This observation should tell you the answer, but it will be shown explicitly in the solution.) (c) If there is acceleration, Newton's second law applies, which involves a net force, so there may be net work.

▶ **FIGURE 5.4 Total or net work** See Example text for description.

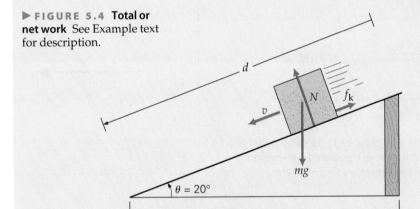

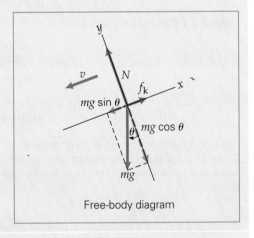

Free-body diagram

SOLUTION.   Listing the data given, and specifically what is to be found:

Given: $m = 0.75$ kg
$\theta = 20°$
$L = 1.2$ m (from Fig. 5.4)

Find: (a) W_f (work done on the block by friction)
(b) W_{net} (net work on the block)
(c) W (discuss net work with block accelerating)

(a) Note from the Fig. 5.4 free-body diagram that only two forces do work, because there are only two forces parallel to the motion: f_k, the force of kinetic friction, and $mg \sin \theta$, the component of the block's weight acting down the plane. The normal force N and $mg \cos \theta$, the component of the block's weight, act perpendicular to the plane and do no work. (Why?)

First finding the work done by the frictional force:

$$W_f = f_k(\cos 180°)d = -f_k d = -\mu_k N d$$

The angle 180° indicates that the force and displacement are in opposite directions. (It is common in such cases to write $W_f = -f_k d$ directly, since kinetic friction typically opposes motion.) The distance d the block slides down the plane can be found by using trigonometry. Note that $\cos \theta = L/d$, so

$$d = \frac{L}{\cos \theta}$$

We know that $N = mg \cos \theta$, but what is μ_k? It would appear that some information is lacking. When this situation occurs, look for another approach to solve the problem. As noted earlier, there are only two forces parallel to the motion, and they are opposite, so with a constant velocity their magnitudes are equal, $f_k = mg \sin \theta$. Thus,

$$W_f = -f_k d = -(mg \sin \theta)\left(\frac{L}{\cos \theta}\right) = -mg\,L \tan 20°$$

$$= -(0.75 \text{ kg})(9.8 \text{ m/s}^2)(1.2 \text{ m})(0.364) = -3.2 \text{ J}$$

(b) To find the net work, the work done by gravity needs to be calculated and then added to the result in part (a). Since $F_\parallel$ for gravity is just $mg \sin \theta$,

$$W_g = F_\parallel d = (mg \sin \theta)\left(\frac{L}{\cos \theta}\right) = mgL \tan 20° = +3.2 \text{ J}$$

where the calculation is the same as in part (a) except for the sign. Then, the net work is

$$W_{net} = W_g + W_f = +3.2 \text{ J} + (-3.2 \text{ J}) = 0$$

(constant velocity, zero net force, zero net work). Remember that work is a scalar quantity, so scalar addition is used to find net work.

(c) If the block accelerates down the plane, then from Newton's second law, $F_{net} = mg \sin \theta - f_k = ma$. The component of the gravitational force ($mg \sin \theta$) is greater than the opposing frictional force (f_k), so net work is done on the block, because now $|W_g| > |W_f|$. You may be wondering what the effect of nonzero net work is. As will be shown shortly, nonzero net work causes a change in the amount of kinetic energy an object has.

FOLLOW-UP EXERCISE. In part (c) of this Example, is it possible for the frictional work to be greater in magnitude than the gravitational work? What would this condition mean in terms of the block's speed?

PROBLEM-SOLVING HINT

Note that in part (a) of Example 5.3, the equation for W_f was simplified by using algebraic expressions for N and d instead of by computing these quantities initially. It is a good rule of thumb not to plug numbers into an equation until you have to. Simplifying an equation through cancellation is easier with symbols and saves computation time.

DID YOU LEARN?

➡ The product of the magnitudes of the displacement and the force, or component of force, parallel to the displacement gives the work done by a constant force.

➡ If a force or a force component acts in the opposite direction of the displacement, the work done by the force is negative.

➡ The work done by all the forces acting on an object, or the scalar sum of all the work, gives the total, or net, work.

5.2 Work Done by a Variable Force

LEARNING PATH QUESTIONS

➡ What is meant by "a spring force is a function of position"?

➡ If a spring, or force, constant of one spring is greater than that of another, what does this imply?

The discussion in the preceding section was limited to work done by constant forces. In general, however, forces are variable; that is, they change in magnitude and/or angle with time and/or position.

An example of a variable force that does work is illustrated in ▼Fig. 5.5, which depicts a spring being stretched by an applied force F_a. As the spring is stretched (or compressed) farther and farther, its restoring force (the spring force that opposes the stretching or compression) becomes greater, and an increased applied force is required. For most springs, the spring force (F_s) is directly proportional to the change in length of the spring from its unstretched length. In equation form, this relationship is expressed

$$F_s = -k\Delta x = -k(x - x_o)$$

or, if $x_o = 0$,

$$F_s = -kx \quad \text{(ideal spring force)} \quad (5.3)$$

▶ **FIGURE 5.5 Spring force**
(a) An applied force F_a stretches the spring, and the spring exerts an equal and opposite force F_s on the hand. **(b)** The magnitude of the force depends on the change Δx in the spring's length. This change is measured from to the end of the unstretched spring at x_o.

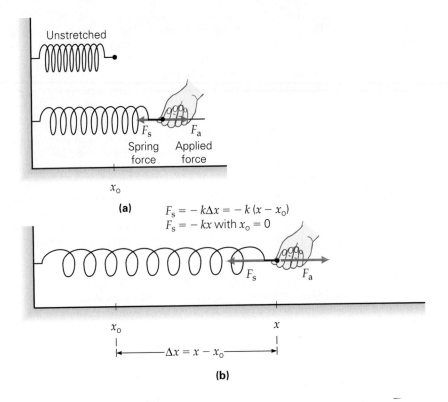

$$F_s = -k\Delta x = -k(x - x_o)$$
$$F_s = -kx \text{ with } x_o = 0$$

where x now represents the distance the spring is stretched (or compressed) from its unstretched length. As can be seen, the force varies with x. This is described by saying that the *force is a function of position.*

The k in this equation is a constant of proportionality and is commonly called the **spring constant**, or **force constant**. The greater the value of k, the stiffer or stronger the spring. As you should be able to prove to yourself, the SI unit of k is newtons per meter (N/m). The minus sign in Eq. 5.3 indicates that the spring force acts in the direction opposite to the displacement when the spring is either stretched or compressed. Equation 5.3 is a form of what is known as *Hooke's law,* named after Robert Hooke, a contemporary of Newton.

The relationship expressed by the spring force equation holds only for ideal springs. Real springs approximate this linear relationship between force and displacement within certain limits. If a spring is stretched beyond a certain point, called its *elastic limit,* the spring will be permanently deformed, and the linear relationship will no longer apply.

Computing the work done by variable forces generally requires calculus. But it is fortunate that the spring force is a special case that can be computed graphically. A plot of F (the applied force) versus x is shown in ◀Fig. 5.6. The graph has a straight-line slope of k, with $F = kx$, where F is the applied force doing work in stretching the spring.

As described earlier, work is the area under an F-versus-x curve, and here it is in the form of a triangle, as indicated by the shaded area in the figure. Then, computing this area,

$$\text{area} = W = \tfrac{1}{2}(\text{altitude} \times \text{base})$$

or

$$W = \tfrac{1}{2}Fx = \tfrac{1}{2}(kx)x = \tfrac{1}{2}kx^2$$

where $F = kx$. Thus,

$$W = \tfrac{1}{2}kx^2 \qquad \begin{array}{l}\textit{(work done in stretching or compressing}\\ \textit{a spring from } x_o = 0)\end{array} \qquad (5.4)$$

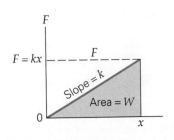

▲ **FIGURE 5.6 Work done by a uniformly variable spring force** A graph of F versus x, where F is the applied force doing work in stretching a spring, is a straight line with a slope of k. The work is equal to the area under the line, which is that of a triangle with area = $\tfrac{1}{2}$(altitude × base). Then $W = \tfrac{1}{2}Fx = \tfrac{1}{2}(kx)x = \tfrac{1}{2}kx^2$.

EXAMPLE 5.4 | Determining the Spring Constant

A 0.15-kg mass is attached to a vertical spring and hangs at rest a distance of 4.6 cm below its original position (▶ Fig. 5.7). An additional 0.50-kg mass is then suspended from the first mass and the system is allowed to descend to a new equilibrium position. What is the total extension of the spring? (Neglect the mass of the spring.)

THINKING IT THROUGH. The spring constant k appears in Eq. 5.3. Therefore, to find the value of k for a particular instance, the spring force and distance the spring is stretched (or compressed) must be known.

SOLUTION. The data given are as follows:

Given: $m_1 = 0.15$ kg   *Find:* x (total stretch distance)
$\qquad x_1 = 4.6$ cm $= 0.046$ m
$\qquad m_2 = 0.50$ kg

The total stretch distance is given by $x = F/k$, where F is the applied force, which in this case is the weight of the mass suspended on the spring. (The minus sign in Eq. 5.3 is ignored here for convenience.) However, the spring constant k is not given. But, k may be found from the data pertaining to the suspension of m_1 and resulting displacement x_1. (This method is commonly used to determine spring constants.) As seen in Fig. 5.7a, the magnitudes of the weight force and the restoring spring force are equal, since $a = 0$, their magnitudes may be equated:

$$F_s = kx_1 = m_1 g$$

Solving for k,

$$k = \frac{m_1 g}{x_1} = \frac{(0.15 \text{ kg})(9.8 \text{ m/s}^2)}{0.046 \text{ m}} = 32 \text{ N/m}$$

Then, knowing k, the total extension of the spring can be found from the balanced-force situation shown in Fig. 5.7b:

$$F_s = (m_1 + m_2)g = kx$$

Thus,

$$x = \frac{(m_1 + m_2)g}{k} = \frac{(0.15 \text{ kg} + 0.50 \text{ kg})(9.8 \text{ m/s}^2)}{32 \text{ N/m}} = 0.20 \text{ m (or 20 cm)}$$

FOLLOW-UP EXERCISE. How much work is done by gravity in stretching the spring through both displacements in Example 5.4?

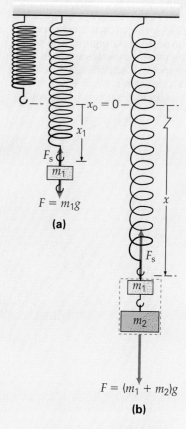

▲ **FIGURE 5.7 Determining the spring constant and the work done in stretching a spring** See Example text for description.

PROBLEM-SOLVING HINT

The reference position x_o used to determine the change in length of a spring is arbitrary but is usually chosen as $x_o = 0$ for convenience. *The important quantity in computing work is the difference in position, Δx, or the net change in the length of the spring from its unstretched length.* As shown in ▼ Fig. 5.8 for a mass suspended on a spring, x_o can be referenced to the unloaded length of the spring or to the loaded position, which may be taken as the zero position for convenience. In Example 5.4, x_o was referenced to the end of the unloaded spring.

When the net force on the suspended mass is zero, the mass is said to be at its *equilibrium position* (as in Fig. 5.7a with m_1 suspended). This position, rather than the unloaded length, may be taken as a zero reference ($x_o = 0$; see Fig. 5.8b). The equilibrium position is a convenient reference point for cases in which the mass oscillates up and down on the spring. Also, since the displacement is in the vertical direction, the x's are often replaced by y's.

DID YOU LEARN?
➥ The spring force depends on the length of the spring from its unstretched position, for either an extension or a compression.
➥ A spring with a greater spring constant would apply a greater spring force, or is a stiffer spring.

► **FIGURE 5.8 Displacement reference** The reference position x_o is arbitrary and is usually chosen for convenience. It may be **(a)** at the end of the spring at its unloaded position or **(b)** at the equilibrium position when a mass is suspended on the spring. The latter is particularly convenient in cases in which the mass oscillates up and down on the spring.

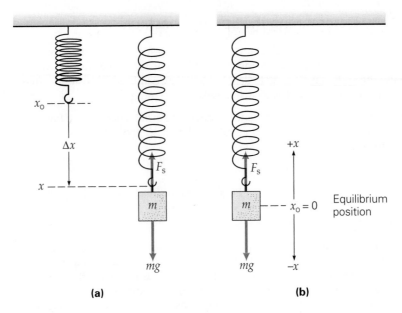

(a)   (b)

5.3 The Work–Energy Theorem: Kinetic Energy

LEARNING PATH QUESTIONS

➥ Why is kinetic energy called "the energy of motion"?

➥ How does the work–energy theorem relate work and energy?

➥ How is a change in kinetic energy computed?

Now that we have an operational definition of work, let's take a look at how work is related to energy. Energy is one of the most important concepts in science. It is described as something that objects or systems possess. Basically, work is something that is *done on* objects, whereas energy is something that objects *have*, which is the ability to do work.

One form of energy that is closely associated with work is *kinetic energy*. (Another basic form of energy, *potential energy*, will be discussed in Section 5.4.) Consider an object at rest on a frictionless surface. Let a horizontal force act on the object and set it in motion. Work is done *on* the object, but where does the work "go," so to speak? It goes into setting the object into motion, or changing its *kinetic* conditions. Because of its motion, we say the object has gained energy—kinetic energy, which gives it the capability to do work.

For a constant force doing work on a moving object parallel to the direction of motion, as illustrated in ▼Fig. 5.9, the force does an amount of work $W = Fx$. But what are the kinematic effects? The force gives the object a constant acceleration, and from Eq. 2.12, $v^2 = v_0^2 + 2ax$ (with $x_o = 0$),

$$a = \frac{v^2 - v_0^2}{2x}$$

$$\boxed{W = K - K_o = \Delta K}$$

$$K_o = \tfrac{1}{2}mv_0{}^2 \qquad\qquad K = \tfrac{1}{2}mv^2$$

$$\xrightarrow{\ v_0\ } \qquad\qquad \xrightarrow{\ v\ }$$

$F \longrightarrow \boxed{m}$ $F \longrightarrow \boxed{m}$

(Frictionless)

◄── x ──►

$$\boxed{W = Fx}$$

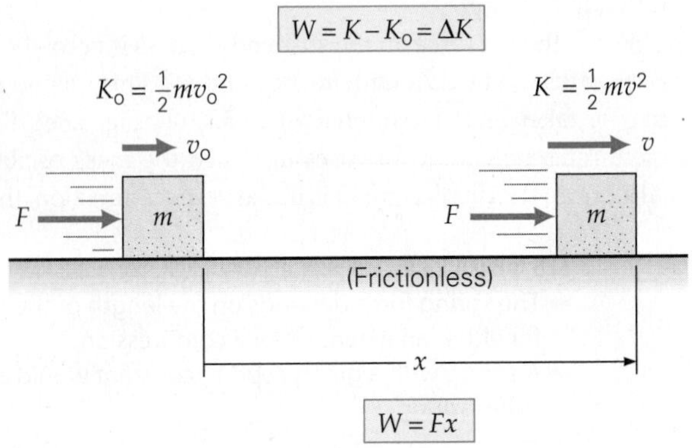

► **FIGURE 5.9 The relationship of work and kinetic energy** The work done on a block by a constant force in moving it along a horizontal frictionless surface is equal to the change in the block's kinetic energy: $W = \Delta K$.

where v_o may or may not be zero. Writing the magnitude of the force in the form of Newton's second law and substituting in the expression for a from the previous equation gives

$$F = ma = m\left(\frac{v^2 - v_o^2}{2x}\right)$$

Using this expression in the equation for work,

$$W = Fx = m\left(\frac{v^2 - v_o^2}{2x}\right)x$$

$$= \tfrac{1}{2}mv^2 - \tfrac{1}{2}mv_o^2$$

The term $\tfrac{1}{2}mv^2$ is defined as the **kinetic energy (K)** of the moving object:

$$K = \tfrac{1}{2}mv^2 \quad \textit{(kinetic energy)} \qquad (5.5)$$

SI unit of energy: joule (J)

Kinetic energy is often called the *energy of motion*. Note that it is directly proportional to the square of the (instantaneous) speed of a moving object, and therefore cannot be negative.

Then, in terms of kinetic energy, the previous expression for work may be written as

$$W = \tfrac{1}{2}mv^2 - \tfrac{1}{2}mv_o^2 = K - K_o = \Delta K$$

or

$$W = \Delta K \qquad (5.6)$$

where it is understood that W *is the net work if more than one force acts on the object,* as shown in Example 5.3. This equation is called the **work–energy theorem**, and it relates the work done on an object to the change in the object's kinetic energy. That is, *the net work done on a body by all the forces acting on it is equal to the change in kinetic energy of the body.* Both work and energy have units of joules, and both are *scalar* quantities. The work–energy theorem is true in general for variable forces and not just for the special case considered in deriving Eq. 5.6.

To illustrate that net work is equal to the change in kinetic energy, recall that in Example 5.1 the force of gravity did $+44$ J of work on a book that fell from rest through a distance of $y = 3.0$ m. At that position and instant, the falling book had 44 J of kinetic energy. Since $v_o = 0$ in this case, $\tfrac{1}{2}mv^2 = mgy$. Substituting this expression into the equation for the work done on the falling book by gravity,

$$W = Fd = mgy = \frac{mv^2}{2} = K = \Delta K$$

where $K_o = 0$. Thus the kinetic energy gained by the book is equal to the net work done on it: 44 J in this case. (As an exercise, confirm this fact by calculating the speed of the book and computing its kinetic energy.)

The work–energy theorem tells us that when work is done on an object, there is a change in or a transfer of energy. In general, then, it might be said that *work is a measure of the transfer of kinetic energy* to the object. For example, a force doing work on an object that causes the object to speed up gives rise to an increase in the object's kinetic energy. Conversely, (negative) work done by the force of kinetic friction may cause a moving object to slow down and decrease its kinetic energy. So for an object to have a change in its kinetic energy, net work must be done on the object, as Eq. 5.6 indicates.

When an object is in motion, it possesses kinetic energy and thus has the capability to do work. For example, a moving automobile has kinetic energy and can do work in crumpling a fender in a fenderbender—not *useful* work in that case, but still work. Another example of work done by kinetic energy is shown in ▶ Fig. 5.10.

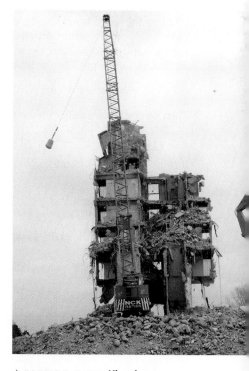

▲ **FIGURE 5.10 Kinetic energy and work** A moving object, such as a wrecking ball, processes kinetic energy and can do work. A massive ball is used in demolishing buildings.

EXAMPLE 5.5 | A Game of Shuffleboard: The Work–Energy Theorem

A shuffleboard player (▼ Fig. 5.11) pushes a 0.25-kg puck that is initially at rest such that a constant horizontal force of 6.0 N acts on it through a distance of 0.50 m. (Neglect friction.) (a) What are the kinetic energy and the speed of the puck when the force is removed? (b) How much work would be required to bring the puck to rest?

THINKING IT THROUGH.   Apply the work–energy theorem. If the amount of work done can be found, then this gives the change in kinetic energy.

▲ **FIGURE 5.11   Work and kinetic energy**   See Example text for description.

SOLUTION.   Listing the given data as usual,

Given: $m = 0.25$ kg   *Find:*   (a) K (kinetic energy)
$F = 6.0$ N v (speed)
$d = 0.50$ m   (b) W (work done in
$v_o = 0$ stopping puck)

(a) Since the speed is not known, the kinetic energy ($K = \frac{1}{2}mv^2$) cannot be computed directly. However, kinetic energy is related to work by the work–energy theorem. The work done on the puck by the player's applied force F is

$$W = Fd = (6.0 \text{ N})(0.50 \text{ m}) = 3.0 \text{ J}$$

Then, by the work–energy theorem,

$$W = \Delta K = K - K_o = 3.0 \text{ J}$$

But $K_o = \frac{1}{2}mv_o^2 = 0$, because $v_o = 0$, so

$$K = 3.0 \text{ J}$$

The speed can be found from the kinetic energy. Since $K = \frac{1}{2}mv^2$,

$$v = \sqrt{\frac{2K}{m}} = \sqrt{\frac{2(3.0 \text{ J})}{0.25 \text{ kg}}} = 4.9 \text{ m/s}$$

(b) As you might guess, the work required to bring the puck to rest is equal to the puck's kinetic energy (that is, the amount of energy that the puck must lose to come to a stop). To confirm this equality, the previous calculation is essentially performed in reverse, with $v_o = 4.9$ m/s and $v = 0$:

$$W = K - K_o = 0 - K_o = -\frac{1}{2}mv_o^2 = -\frac{1}{2}(0.25 \text{kg})(4.9 \text{ m/s})^2$$
$$= -3.0 \text{ J}$$

The minus sign indicates that the puck loses energy as it slows down. The work is done *against* the motion of the puck; that is, the opposing force is in a direction opposite that of the motion. (In the real-life situation, the opposing force could be friction.)

FOLLOW-UP EXERCISE.   Suppose the puck in this Example had twice the final speed when released. Would it then take twice as much work to stop the puck? Justify your answer numerically.

PROBLEM-SOLVING HINT

Notice how work–energy considerations were used to find speed in Example 5.5. This operation can be done in another way as well. First, the acceleration could be found from $a = F/m$, and then the kinematic equation $v^2 = v_o^2 + 2ax$ could be used to find v (where $x = d = 0.50$ m). The point is that many problems can be solved in different ways, and finding the fastest and most efficient way is often the key to success. As our discussion of energy progresses, it will be seen how useful and powerful the notions of work and energy are, both as theoretical concepts and as practical tools for solving many kinds of problems.

CONCEPTUAL EXAMPLE 5.6 | Kinetic Energy: Mass versus Speed

In a football game, a 140-kg guard runs at a speed of 4.0 m/s, and a 70-kg free safety moves at 8.0 m/s. Which of the following is a correct statement? (a) The players have the same kinetic energy. (b) The safety has twice as much kinetic energy as the guard. (c) The guard has twice as much kinetic energy as the safety. (d) The safety has four times as much kinetic energy as the guard.

REASONING AND ANSWER.   The kinetic energy of a body depends on both its mass and speed. You might think that, with half the mass but twice the speed, the safety would have the same kinetic energy as the guard, but this is not the case. As observed from the relationship $K = \frac{1}{2}mv^2$, kinetic energy is directly proportional to the mass, but is proportional to the *square* of the speed. Thus, having half the mass decreases the kinetic energy by a factor of 2. So if the two athletes had equal speeds, the safety would have half as much kinetic energy as the guard.

However, doubling the speed increases the kinetic energy, not by a factor of 2 but by a factor of 2^2, or 4. Thus, the safety, with half the mass but twice the speed, would have $\frac{1}{2} \times 4 = 2$ times as much kinetic energy as the guard, and so the answer is (b).

Note that to answer this question, it was not necessary to calculate the kinetic energy of each player. But this can be done to verify the answer:

$$K_{\text{safety}} = \tfrac{1}{2}m_s v_s^2 = \tfrac{1}{2}(70 \text{ kg})(8.0 \text{ m/s})^2 = 2.2 \times 10^3 \text{ J}$$
$$K_{\text{guard}} = \tfrac{1}{2}m_g v_g^2 = \tfrac{1}{2}(140 \text{ kg})(4.0 \text{ m/s})^2 = 1.1 \times 10^3 \text{ J}$$

which explicitly shows the answer to be correct.

FOLLOW-UP EXERCISE.   Suppose that the safety's speed were only 50% greater than the guard's, or 6.0 m/s. Which athlete would then have the greater kinetic energy, and how much greater?

PROBLEM-SOLVING HINT

Note that the work–energy theorem relates the work done to the *change* in the kinetic energy. Often, $v_o = 0$ and $K_o = 0$, so $W = \Delta K = K$. But take care! You *cannot* simply use the square of the change or difference in speed, $(v - v_o)^2 = (\Delta v)^2$, to calculate ΔK, as you might at first think. In terms of speed,

$$W = \Delta K = K - K_o = \tfrac{1}{2}mv^2 - \tfrac{1}{2}mv_o^2 = \tfrac{1}{2}m(v^2 - v_o^2)$$

Note that $(v^2 - v_o^2)$ is not the same as $(v - v_o)^2 = (\Delta v)^2$ because $(v - v_o)^2 = v^2 - 2vv_o + v_o^2$. Hence, the change in kinetic energy is *not* equal to $\tfrac{1}{2}m(v - v_o)^2 = \tfrac{1}{2}m(\Delta v)^2 \neq \Delta K$.

This observation means that to calculate work, or the change in kinetic energy, you must compute the kinetic energy of an object at one point or time (using the instantaneous speed to get the instantaneous kinetic energy) and also at another location or time. Then subtract the quantities to find the change in kinetic energy, or the work. Alternatively, you can find the difference of the *squares* of the speeds $(v^2 - v_o^2)$ first in computing the change, but remember never to use the square of the difference of the speeds. To see this hint in action, look at Conceptual Example 5.7.

CONCEPTUAL EXAMPLE 5.7 | An Accelerating Car: Speed and Kinetic Energy

A car traveling at 5.0 m/s speeds up to 10 m/s, with an increase in kinetic energy that requires work W_1. Then the car's speed increases from 10 m/s to 15 m/s, requiring additional work W_2. Which of the following relationships accurately compares the two amounts of work: (a) $W_1 > W_2$, (b) $W_1 = W_2$, or (c) $W_2 > W_1$?

REASONING AND ANSWER.   As noted previously, the work–energy theorem relates the work done on the car to the *change* in its kinetic energy. Since the speeds have the same increment in each case ($\Delta v = 5.0$ m/s), it might appear that (b) would be the answer. However, keep in mind that the work is equal to the *change* in kinetic energy and involves $v_2^2 - v_1^2$, *not* $(\Delta v)^2 = (v_2 - v_1)^2$.

So the greater the speed of an object, the greater its kinetic energy. The *difference* in kinetic energy in changing speeds (or the work required to change speed) would then be greater for higher speeds for the same Δv. Therefore, (c) is the answer.

The main point is that the Δv values are the same, but more work is required to increase the kinetic energy of an object at higher speeds.

FOLLOW-UP EXERCISE.   Suppose the car speeds up a third time, from 15 m/s to 20 m/s, a change requiring work W_3. How does the work done in this increment compare with W_2? Justify your answer numerically. [*Hint*: Use a ratio.]

5.4 Potential Energy

LEARNING PATH QUESTIONS

➥ How may an object's potential energy be changed?

➥ How do different reference points affect the difference in the gravitational potential energy of two positions?

(a)

(b)

▲ **FIGURE 5.12 Potential energy**
Potential energy has many forms.
(a) Work must be done to bend the bow, giving it potential energy. That energy is converted into kinetic energy when the arrow is released.
(b) Gravitational potential energy is converted into kinetic energy when something falls. (Where did the gravitational potential energy of the water and the diver come from?)

An object in motion has kinetic energy. However, whether an object is in motion or not, it may have another form of energy—potential energy. As the name implies, an object having potential energy has the *potential* to do work. You can probably think of many examples: a compressed spring, a drawn bow, and water held back by a dam. In all such cases, the potential to do work derives from the *position* or *configuration* of bodies. A spring has energy because it is compressed, a bow because it is drawn, and the water because it has been lifted above the surface of the Earth (◄Fig. 5.12). Consequently, **potential energy (U)**, is often called *the energy of position* (and/or configuration).

Unlike kinetic energy, which is associated with motion, potential energy is a form of mechanical energy associated with the position of an object within a system (or configuration). Potential energy is a property of the system, rather than the object. If the configuration of a system of objects changes, so does the potential energy of a particular object within that system.

In a sense, potential energy can be thought of as stored work. You have already seen an example of potential energy in Section 5.2 when work was done in stretching a spring from its equilibrium position. Recall that the work done in such a case is $W = \frac{1}{2}kx^2$ (with $x_o = 0$). Note that the amount of work done depends on the amount of stretching (x). Because work is done, there is a *change* in the spring's potential energy (ΔU), which is equal to the work done *by the applied force* in stretching (or compressing) the spring:

$$W = \Delta U = U - U_o = \frac{1}{2}kx^2 - \frac{1}{2}kx_o^2$$

Thus, with $x_o = 0$ and $U_o = 0$, as they are commonly taken for convenience, the *potential energy of a spring* is

$$U = \frac{1}{2}kx^2 \quad \textit{(potential energy of a spring)} \tag{5.7}$$

SI unit of energy: joule (J)

[*Note*: Since the potential energy varies as x^2, the previous Problem-Solving Hint also applies, and when $x_o \neq 0$, then $x^2 - x_o^2 \neq (x - x_o)^2$. That is, the potential energy of a spring must be calculated at different positions and then subtracted to find ΔU.]

Perhaps the most well known type of potential energy is gravitational potential energy. In this case, position refers to the height of an object above some reference point, such as the floor or the ground. Suppose that an object of mass m is lifted a distance Δy (▶Fig. 5.13). Work is done against the force of gravity, and an applied force at least equal to the object's weight is necessary to lift the object: $F = w = mg$. The work done in lifting is then equal to the change in potential energy. Expressing this relationship in equation form, since there is no overall change in kinetic energy,

work done by external force = change in gravitational potential energy

or

$$W = F\Delta y = mg(y - y_o) = mgy - mgy_o = \Delta U = U - U_o$$

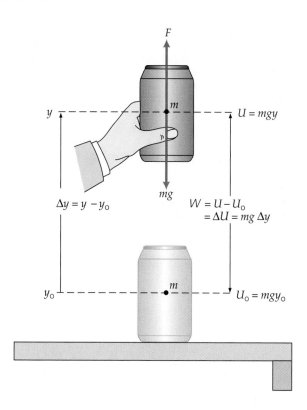

◀ FIGURE 5.13 **Gravitational potential energy** The work done in lifting an object is equal to the change in gravitational potential energy: $W = F\Delta y = mg(y - y_o)$.

where y is used as the vertical coordinate. With the common choice of $y_o = 0$, such that $U_o = 0$, the **gravitational potential energy** is

$$U = mgy \qquad (5.8)$$

SI unit of energy: joule (J)

(Eq. 5.8 represents the gravitational potential energy on or near the Earth's surface, where g is considered to be constant. A more general form of gravitational potential energy will be given in Chapter 7.5.)

EXAMPLE 5.8 | More Energy Needed

To walk 1000 m on level ground, a 60-kg person requires an expenditure of about 1.0×10^5 J of energy. What is the total energy required if the walk is extended another 1000 m along a 5.0° incline as shown in ▶ Fig. 5.14? (Neglect any frictional changes.)

THINKING IT THROUGH. To walk an additional 1000 m would require the 1.0×10^5 J *plus* the additional energy for doing work against gravity in walking up the incline. From the figure, the increase in height can be seen to be $h = d \sin \theta$, where d is 1000 m.

SOLUTION. Listing the given data:

Given: $m = 60$ kg   *Find:* E (total expended energy)
         $E_o = 1.0 \times 10^5$ J (for 1000 m)
         $\theta = 5.0°$
         $d = 1000$ m (for each part of the work)

The additional expended energy in going up the incline is equal to gravitational potential energy gained. So,

$$\Delta U = mgh = (60 \text{ kg})(9.8 \text{ m/s}^2)(1000 \text{ m})\sin 5.0° = 5.1 \times 10^4 \text{ J}$$

Then, the total energy expended for the 2000-m walk is

$$\text{Total } E = 2E_o + \Delta U = 2(1.0 \times 10^5 \text{J}) + 0.51 \times 10^5 \text{J} = 2.5 \times 10^5 \text{J}$$

(continued on next page)

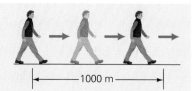

1000 m

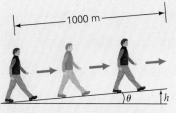

1000 m

▲ FIGURE 5.14 **Adding potential energy** See Example text for description.

Notice that the value of ΔU was expressed as a multiple of 10^5 in the last equation so it could be added to the E_0 term, and the result was rounded to two significant figures per the rules given in Chapter 1.6.

FOLLOW-UP EXERCISE.   If the angle of incline were doubled and the walk *just* up the incline is repeated, will the additional energy expended by the person in doing work against gravity be doubled? Justify your answer.

EXAMPLE 5.9 | ## A Thrown Ball: Kinetic Energy and Gravitational Potential Energy

A 0.50-kg ball is thrown vertically upward with an initial velocity of 10 m/s ($\blacktriangleright$ Fig. 5.15). (a) What is the change in the ball's kinetic energy between the starting point and the ball's maximum height? (b) What is the change in the ball's potential energy between the starting point and the ball's maximum height? (Neglect air resistance.)

THINKING IT THROUGH.   Kinetic energy is lost and gravitational potential energy is gained as the ball travels upward.

SOLUTION.   Studying Fig. 5.15 and listing the given data,

Given: $m = 0.50$ kg   *Find:*   (a) ΔK (the change in kinetic energy between y_0 and y_{max})
$v_0 = 10$ m/s   (b) ΔU (the change in potential energy between y_0
$a = g$ and y_{max})

(a) To find the *change* in kinetic energy, the kinetic energy is computed at each point. The initial velocity is v_0 and at the maximum height $v = 0$, so $K = 0$. Thus,

$$\Delta K = K - K_0 = 0 - K_0 = -\tfrac{1}{2}mv_0^2 = -\tfrac{1}{2}(0.50 \text{ kg})(10 \text{ m/s})^2 = -25 \text{ J}$$

That is, the ball loses 25 J of kinetic energy as negative work is done on it by the force of gravity. (The gravitational force and the ball's displacement are in opposite directions.)

(b) To find the change in potential energy, we need to know the ball's height above its starting point when $v = 0$. Using Eq. 2.11', $v^2 = v_0^2 - 2gy$ (with $y_0 = 0$ and $v = 0$), to find y_{max},

$$y_{max} = \frac{v_0^2}{2g} = \frac{(10 \text{ m/s})^2}{2(9.8 \text{ m/s}^2)} = 5.1 \text{ m}$$

Then, with $y_0 = 0$ and $U_0 = 0$

$$\Delta U = U = mgy_{max} = (0.50 \text{ kg})(9.8 \text{ m/s}^2)(5.1 \text{ m}) = +25 \text{ J}$$

The potential energy increases by 25 J, as might be expected. This is an example of the conservation of energy, as will be discussed shortly.

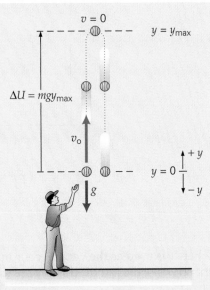

$\blacktriangle$ **FIGURE 5.15 Kinetic and potential energies** (The ball is displaced sideways for clarity.) See Example text for description.

FOLLOW-UP EXERCISE.   In this Example, what are the overall changes in the ball's kinetic and potential energies when the ball returns to the starting point?

ZERO REFERENCE POINT

An important point is illustrated in Example 5.9, namely, the choice of a zero reference point. Potential energy is the energy of *position*, and the potential energy at a particular position (U) is meaningful only when referenced to the potential energy at some other position (U_0). The reference position or point is arbitrary, as is the origin of a set of coordinate axes for analyzing a system. Reference points are usually chosen with convenience in mind—for example, $y_0 = 0$. The value of the potential energy at a particular position depends on the reference point used. However, *the difference, or change, in potential energy associated with two positions is the same regardless of the reference position.*

If, in Example 5.9, ground level had been taken as the zero reference point, then U_0 at the release point would not have been zero. However, U at the maximum height would have been greater, and $\Delta U = U - U_0$ would have been the same. This concept is illustrated in $\blacktriangleright$ Fig. 5.16. Note in Fig. 5.16a that the potential energy can be negative. When an object has a negative potential energy, it is said to be in a potential energy *well*, which is analogous to being in an actual well. Work is needed to raise the object to a higher position in the well or to get it out of the well.

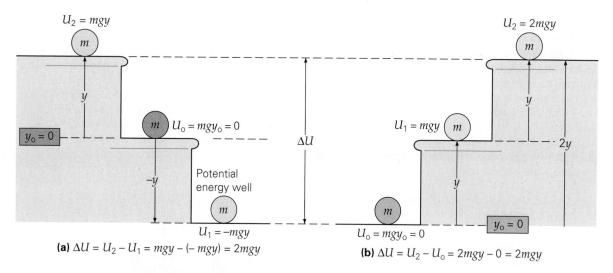

▲ **FIGURE 5.16** **Reference point and change in potential energy** **(a)** The choice of a reference point (zero height) is arbitrary and may give rise to a negative potential energy. An object is said to be in a potential energy well in this case. **(b)** The well may be avoided by selecting a new zero reference. Note that the difference, or change, in potential energy (ΔU) associated with the two positions is the same, regardless of the reference point. There is no physical difference, even though there are two coordinate systems and two different zero reference points.

It is also said that gravitational potential energy is *independent of path*. This means that only the change in height Δh (or Δy) is the consideration, not the path that leads to the change in height. An object could travel many paths leading to the same Δh.

DID YOU LEARN?
➥ A change in position (or configuration) may result in a change in potential energy for an object.
➥ The difference, or change, in potential energy associated with two positions is the same irrespective of reference points.

5.5 Conservation of Energy

LEARNING PATH QUESTIONS
➥ What is meant by "the total energy of the universe is conserved"?
➥ When is the total mechanical energy conserved?
➥ How does work done by nonconservative force affect the mechanical energy of a system?

Conservation laws are the cornerstones of physics, both theoretically and practically. Most scientists would probably name conservation of energy as the most profound and far-reaching of these important laws. Saying that a physical quantity is *conserved* means it is constant, or has a constant value. Because so many things continually change in physical processes, conserved quantities are extremely helpful in our attempts to understand and describe a situation. Keep in mind, though, that many quantities are conserved only under special conditions.

One of the most important conservation laws is that concerning conservation of energy. (You have seen this topic in Example 5.9.) A familiar statement is that *the total energy of the universe is conserved*. This statement is true, because the whole universe is taken to be a system. A system is defined as a definite quantity of matter enclosed by boundaries, either real or imaginary. In effect, the universe is the largest possible closed, or isolated, system we can imagine. On a smaller scale, a classroom might be considered a system, and so might an arbitrary cubic meter of air.

INSIGHT 5.1 | People Power: Using Body Energy

The human body is energy inefficient. That is, a lot of energy doesn't go into doing useful work and is wasted. It would be advantageous to convert some of this energy into useful work. Attempts are being made to do this through "energy harvesting" from the human body. Normal body activities produce motion, flexing and stretching, compression, and body heat—this is energy there for the taking. Harvesting the energy is a difficult job, but using advances in nanotechnology (Chapter 1.3) and materials science, the effort is being made.

One older example of using body energy is the self-winding wristwatch, which is wound mechanically from the wearer's wrist movements. (Today, batteries have all but taken over.) An ultimate goal in "energy harvesting" is to convert some of the body's energy into electricity—even if only a small amount. How might this be done? Here are a couple of ways:

- Piezoelectric devices. When mechanically stressed, piezoelectric substances, like some ceramics, can generate electrical energy.
- Thermoelectric materials, which convert heat resulting from some temperature difference into electrical energy.

These methods have severe limitations and produce only small amounts of electricity. But with miniaturization and nanotechnology, the results could be practical. Researchers have already developed boots that use the compression of walking on a compound to produce enough energy to power a radio.

A more recent application is the "backpack generator." The mounted backpack's load is suspended by springs (▶ Fig. 1). The up-and-down hip motion of a person wearing the backpack makes the suspended load bounce up and down. This motion turns a gear connected to a simple magnetic coil generator, similar to those used in flashlights that are energized by a rhythmic shaking (see Chapter 20 Insight 20.1, Electromagnetic Induction at Work; Flashlights and Antiterrorism). The body's mechanical energy with this device can generate up to 7 watts of electrical energy. A typical cell phone operates on about 1 watt. (The watt is a unit of power, J/s, energy/second; see Section 5.6).

Who knows what the future of technology may hold? Reflect on how many advances have occurred in your own lifetime.

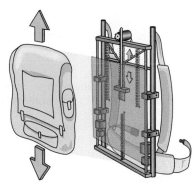

FIGURE 1   Backpack generator   The backpack frame straps to the body and the load is suspended from springs. The load moves up and down on the springs when the wearer walks and the toothed bar turns the gear on the electrical generator. With good strides, more than 7 watts of power is generated. That's enough to power a GPS (Global Positioning System) locator and a head-lamp.

Within a *closed system*, particles can interact with each other, but have absolutely no interaction with anything outside. In general, then, the amount of energy in a system remains constant when no work is done on or by the system, and no energy is transferred to or from the system (including thermal energy and radiation).

Thus, the **law of conservation of total energy** may be stated as follows:

The total energy of an isolated system is always conserved.

Within such a system, energy may be converted from one form to another, but the total amount of all forms of energy is constant, or unchanged. Energy can never be created or destroyed.

An application of energy in a nonconservative system is discussed in the accompanying Insight 5.1, People Power: Using Body Energy.

CONCEPTUAL EXAMPLE 5.10 | Violation of the Conservation of Energy?

A static, uniform liquid is in one side of a double container as shown in ▶ Fig. 5.17a. If the valve is open, the level will fall, because the liquid has (gravitational) potential energy. This may be computed by assuming all the mass of the liquid to be concentrated at its center of mass, which is at a height $h/2$. (More on the center of mass in Chapter 6.5.) When the valve is open, the liquid flows into the container on the right, and when static equilibrium is reached, each container has liquid to a height of $h/2$, with centers of mass at $h/4$. This being the case, the potential energy of the liquid before opening the valve was $U_o = (mg)h/2$, and afterward, with half the total mass in each container (Fig. 5.16b),

$$U = (m/2)g(h/4) + (m/2)g(h/4) = 2(m/2)g(h/4) = (mg)h/4.$$
Whoa. Was half of the energy lost?

REASONING AND ANSWER.   No; by the conservation of total energy, it must be around somewhere. Where might it have gone? When the liquid flows from one container to the other, because of internal friction and friction against the walls, half of the potential energy is first converted to kinetic energy (flow of liquid), then to heat (thermal energy), which is transferred to the surroundings as the liquid comes to equilibrium. (This means a constant temperature and no internal fluctuations.)

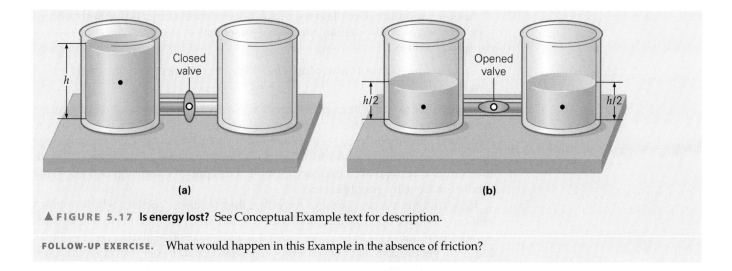

▲ **FIGURE 5.17 Is energy lost?** See Conceptual Example text for description.

FOLLOW-UP EXERCISE.   What would happen in this Example in the absence of friction?

CONSERVATIVE AND NONCONSERVATIVE FORCES

A general distinction can be made among systems by considering two categories of forces that may act within or on them: conservative and nonconservative forces. You have already been introduced to a couple of conservative forces: the force due to gravity and the spring force. A classic nonconservative force, friction, was considered in Chapter 4.6.

A **conservative force** is defined as follows:

> A force is said to be conservative if the work done by it in moving an object is independent of the object's path.

This definition means that the work done by a conservative force depends only on the initial and final positions of an object.

The concept of conservative and nonconservative forces is sometimes difficult to comprehend at first. Because this concept is so important in the conservation of energy, let's consider some illustrative examples to increase understanding.

First, what does *independent of path* mean? As an example of path independence, consider picking an object up from the floor and placing it on a table. This is doing work against the *conservative force of gravity*. The work done is equal to the potential energy gained, $mg\Delta h$, where Δh is the *vertical* distance between the object's position on the floor and its position on the table. This is the important point. You may have carried the object over to the sink before putting it on the table, or walked around to the other side of the table. But only the vertical displacement makes a difference in the work done because that is in the direction of the vertical force. (Note that it was said in the last section that gravitational potential energy is independent of path. Now you know why.)

For any horizontal displacement no work is done, since the displacement and force are at right angles. The magnitude of the work done is equal to the change in potential energy (under frictionless conditions only), and in fact, *the concept of potential energy is associated only with conservative forces*. A change in potential energy can be defined in terms of the work done by a conservative force.

Conversely, a **nonconservative force** *does depend on path*.

> A force is said to be nonconservative if the work done by it in moving an object depends on the object's path.

Friction is a nonconservative force. A longer path would produce more work done by friction than a shorter one, and more energy would be lost to heat on the longer path. So the work done against friction certainly depends on the path. Hence, in a sense, a conservative force allows you to conserve or store energy as potential energy, whereas a nonconservative force does not.

Another approach to explain the distinction between conservative and nonconservative forces is through an equivalent statement of the previous definition of conservative force:

> A force is conservative if the work done by it in moving an object through a round trip is zero.

Notice that for the *conservative* gravitational force, the force and displacement are sometimes in the same direction (in which case positive work is done by the force) and sometimes in opposite directions (in which case negative work is done by the force) during a round trip. Think of the simple case of the book falling to the floor and being placed back on the table. With positive and negative work, the total work done by gravity is zero.

However, for a *nonconservative* force like that of kinetic friction, which opposes motion or is in the opposite direction to the displacement, the total work done by such a force in a round trip can *never* be zero and is always negative (that is, energy is lost). But don't get the idea that nonconservative forces only take energy away from a system. On the contrary, nonconservative pushes and pulls (forces) that add to the energy of a system are often supplied, such as when you push a stalled car and get it moving.

CONSERVATION OF TOTAL MECHANICAL ENERGY

The idea of a conservative force allows us to extend the conservation of energy to the special case of mechanical energy, which greatly helps to better analyze many physical situations. *The sum of the kinetic and potential energies is called the* **total mechanical energy**:

$$\underset{\substack{\text{total} \\ \text{mechanical} \\ \text{energy}}}{E} = \underset{\substack{\text{kinetic} \\ \text{energy}}}{K} + \underset{\substack{\text{potential} \\ \text{energy}}}{U} \tag{5.9}$$

For a **conservative system** (that is, a system in which only conservative forces do work), the total mechanical energy is constant, or conserved:

$$E = E_0$$

Substituting for E and E_0 from Eq. 5.9,

$$K + U = K_0 + U_0 \tag{5.10a}$$

or

$$\tfrac{1}{2}mv^2 + U = \tfrac{1}{2}mv_0^2 + U_0 \tag{5.10b}$$

Equation 5.10b is a mathematical statement of the **law of the conservation of mechanical energy**:

> In a conservative system, the sum of all types of kinetic energy and potential energy is constant and equals the total mechanical energy of the system at any time.

Note: While the kinetic and potential energies in a conservative system may change, their sum is always constant. For a conservative system when work is done and energy is transferred within a system, Eq. 5.10a can be written as

$$(K - K_0) + (U - U_0) = 0 \tag{5.11a}$$

or

$$\Delta K + \Delta U = 0 \quad \textit{(for a conservative system)} \tag{5.11b}$$

This expression indicates that these quantities are related in a seesaw fashion: If there is a decrease in potential energy, then the kinetic energy must increase by an equal amount to keep the sum of the changes equal to zero. However, in a nonconservative system, mechanical energy is usually lost (for example, to the heat of friction), and thus $\Delta K + \Delta U < 0$. But as pointed out previously, a nonconservative force may instead add energy to a system (or have no effect at all).

EXAMPLE 5.11 | Look Out Below! Conservation of Mechanical Energy

A painter on a scaffold drops a 1.50-kg can of paint from a height of 6.00 m. (a) What is the kinetic energy of the can when the can is at a height of 4.00 m? (b) With what speed will the can hit the ground? (Neglect air resistance.) (c) Show that the expression for speed from energy considerations is the same as that from kinematics (Chapter 2.5).

THINKING IT THROUGH. Total mechanical energy is conserved, since only the conservative force of gravity acts on the system (the can). The initial total mechanical energy can be found, and the potential energy decreases as the kinetic energy (as well as speed) increases.

SOLUTION. Listing the given data and what is to be found:

Given: $m = 1.50$ kg
$y_o = 6.00$ m
$y = 4.00$ m
$v_o = 0$

Find: (a) K (kinetic energy at $y = 4.00$ m)
(b) v (speed just before hitting the ground)
(c) Compare speeds

(a) First, it is convenient to find the can's initial total mechanical energy, since this quantity is conserved while the can is falling. With $v_o = 0$, the can's total mechanical energy is initially all potential energy. Taking the ground as the zero reference point,

$$E = K_o + U_o = 0 + mgy_o = (1.50 \text{ kg})(9.80 \text{ m/s}^2)(6.00 \text{ m}) = 88.2 \text{ J}$$

The relation $E = K + U$ continues to hold while the can is falling, and now E is known. Rearranging the equation, $K = E - U$ and K can be found at $y = 4.00$ m

$$K = E - U = E - mgy = 88.2 \text{ J} - (1.50 \text{ kg})(9.80 \text{ m/s}^2)(4.00 \text{ m}) = 29.4 \text{ J}$$

Alternatively, the change in (in this case, the loss of) potential energy, ΔU, could have been computed. Whatever potential energy was lost must have been gained as kinetic energy (Eq. 5.11). Then,

$$\Delta K + \Delta U = 0$$

$$(K - K_o) + (U - U_o) = (K - K_o) + (mgy - mgy_o) = 0$$

With $K_o = 0$ (because $v_o = 0$),

$$K = mg(y_o - y) = (1.50 \text{ kg})(9.8 \text{ m/s}^2)(6.00 \text{ m} - 4.00 \text{ m}) = 29.4 \text{ J}$$

(b) Just before the can strikes the ground ($y = 0, U = 0$), the total mechanical energy is all kinetic energy,

$$E = K = \tfrac{1}{2}mv^2$$

Thus, the speed is,

$$v = \sqrt{\frac{2E}{m}} = \sqrt{\frac{2(88.2 \text{ J})}{1.50 \text{ kg}}} = 10.8 \text{ m/s}$$

(c) Basically, all of the potential energy of a free-falling object released from some height y is converted into kinetic energy just before the object hits the ground, so

$$|\Delta K| = |\Delta U|$$

(Why absolute values?) Thus,

$$\tfrac{1}{2}mv^2 = mgy$$

and

$$v = \sqrt{2gy} = \sqrt{2(9.8 \text{ m/s}^2)(6.00 \text{ m})} = 10.8 \text{ m/s}$$

Note that the mass cancels and is not a consideration. This result is also obtained from a kinematic equation (Eq. 2.12): $v^2 = v_o^2 - 2g(y - y_o)$. With $v_o = 0, y_o = 0$, and $-y$ (downward),

$$v = \sqrt{2gy}$$

FOLLOW-UP EXERCISE. A painter on the ground wishes to toss a paintbrush vertically upward a distance of 5.0 m to her partner on the scaffold. Use methods of conservation of mechanical energy to determine the minimum speed that she must give to the brush.

CONCEPTUAL EXAMPLE 5.12 | A Matter of Direction? Speed and Conservation of Energy

Three balls of equal mass m are projected with the same speed in different directions, as shown in ▼Fig. 5.18. If air resistance is neglected, which ball would you expect to strike the ground with the greatest speed: (a) ball 1, (b) ball 2, (c) ball 3, or (d) all balls strike with the same speed?

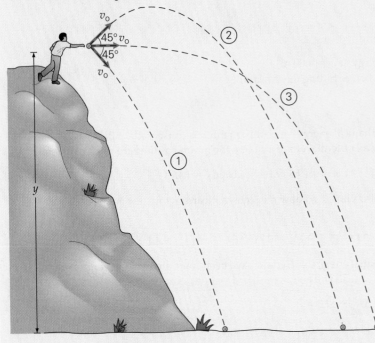

▲ FIGURE 5.18 **Speed and energy** See Example text for description.

REASONING AND ANSWER. All of the balls have the same initial kinetic energy, $K_o = \frac{1}{2}mv_o^2$. (Recall that energy is a scalar quantity, and the different directions of projection do not produce any difference in the kinetic energies.) Regardless of their trajectories, all of the balls ultimately descend a distance y relative to their common starting point, so they all lose the same amount of potential energy. (Recall that U is energy of *position* and is *independent* of path.)

By the law of conservation of mechanical energy, the amount of potential energy each ball loses is equal to the amount of kinetic energy it gains. Since all of the balls start with the same amount of kinetic energy and gain the same amount of kinetic energy, all three will have equal kinetic energies just before striking the ground. This means that their speeds must be equal, so the answer is (d).

Although balls 1 and 2 are projected at 45° angles, this factor is not relevant. Since the change in potential energy is independent of path, it is independent of the projection angle. The vertical distance between the starting point and the ground is the same (y) for projectiles at any angle. (*Note*: Although the strike speeds are equal, the *times* the balls take to reach the ground are different. Refer to Chapter 3 Conceptual Example 3.11 for another approach.)

FOLLOW-UP EXERCISE. Would the balls strike the ground with different speeds if their masses were different? (Neglect air resistance.)

EXAMPLE 5.13 | Conservative Forces: Mechanical Energy of a Spring

A 0.30-kg block sliding on a horizontal frictionless surface with a speed of 2.5 m/s, as depicted in ▼Fig. 5.19, strikes a light spring that has a spring constant of 3.0×10^3 N/m. (a) What is the total mechanical energy of the system? (b) What is the kinetic energy K_1 of the block when the spring is compressed a distance $x_1 = 1.0$ cm? (Assume that no energy is lost in the collision.)

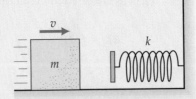

◀ FIGURE 5.19 **Conservative force and the mechanical energy of a spring** See Example text for description.

THINKING IT THROUGH. (a) Initially, the total mechanical energy is all kinetic energy. (b) The total energy is the same as in part (a), but it is now divided between kinetic energy and spring potential energy (assuming the spring is not fully compressed).

SOLUTION.

Given: $m = 0.30$ kg $\qquad$ *Find:* (a) E (total mechanical energy)
$\qquad v_o = 2.5$ m/s $\qquad\qquad\quad$ (b) K_1 (kinetic energy)
$\qquad k = 3.0 \times 10^3$ N/m
$\qquad x_1 = 1.0$ cm $= 0.010$ m

(a) Just before the block makes contact with the spring, the total mechanical energy of the system is all in the form of kinetic energy,

$$E = K_o = \tfrac{1}{2}mv_o^2 = \tfrac{1}{2}(0.30 \text{ kg})(2.5 \text{ m/s})^2 = 0.94 \text{ J}$$

Since the system is conservative (that is, no mechanical energy is lost), this quantity is the total mechanical energy at any time.

(b) When the spring is compressed a distance x_1, it has gained potential energy $U_1 = \frac{1}{2}kx_1^2$, and the block has kinetic energy K_1, so

$$E = K_1 + U_1 = K_1 + \frac{1}{2}kx_1^2$$

Solving for K_1,

$$
\begin{aligned}
K_1 &= E - \tfrac{1}{2}kx_1^2 \\
&= 0.94\,\text{J} - \tfrac{1}{2}(3.0 \times 10^3\,\text{N/m})(0.010\,\text{m})^2 \\
&= 0.94\,\text{J} - 0.15\,\text{J} = 0.79\,\text{J}
\end{aligned}
$$

FOLLOW-UP EXERCISE. How far will the spring in this Example be compressed when the block comes to a stop? (Solve using energy principles.)

See the accompanying Learn by Drawing 5.3, Energy Exchanges: A Falling Ball for another example of energy exchange.

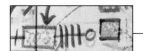

LEARN BY DRAWING 5.3

energy exchanges: a falling ball

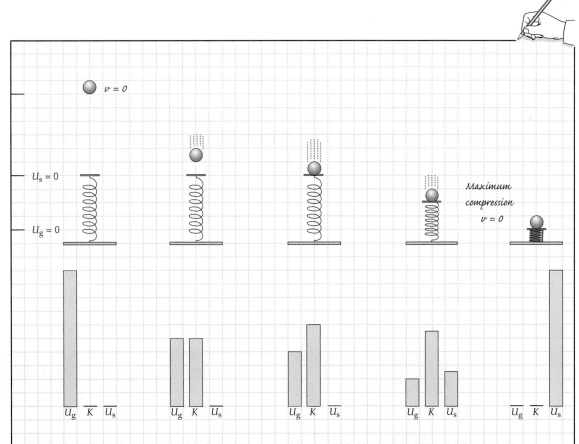

Both the physical situation and the graphs of gravitational potential energy (U_g), kinetic energy (K), and spring potential energy (U_s) are drawn to scale. (Air resistance, the mass of the spring, and any energy loss in the collision are assumed to be negligible.) Why is the spring energy only one-quarter of the total when the spring is halfway compressed?

▲ **FIGURE 5.20 Nonconservative force and energy loss** Friction is a nonconservative force—when friction is present and does work, mechanical energy is not conserved. Can you tell from the photo what is happening to the work being done by the motor on the grinding wheel after the work is converted into rotational kinetic energy? (Note that the worker is wisely wearing a face shield rather than just goggles as the sign in the background suggests.)

TOTAL ENERGY AND NONCONSERVATIVE FORCES

In the preceding examples, the force of friction was ignored; however, friction is probably the most common nonconservative force. In general, both conservative and nonconservative forces can do work on objects. But when nonconservative forces do work, the total mechanical energy is not conserved. Mechanical energy is "lost" through the work done by nonconservative forces, such as friction.

You might think that an energy approach can no longer be used to analyze problems involving such nonconservative forces, since mechanical energy can be lost or dissipated (◄Fig. 5.20). However, in some instances, the total energy can be used to find out how much energy was lost to the work done by a nonconservative force. Suppose an object initially has mechanical energy and that nonconservative forces do an amount of work W_{nc} on it. Starting with the work–energy theorem,

$$W = \Delta K = K - K_o$$

In general, the net work (W) may be done by both conservative forces (W_c) and nonconservative forces (W_{nc}), so

$$W_c + W_{nc} = K - K_o \tag{5.12}$$

But from Eq. 5.10a, the work done by conservative forces is equal to $-\Delta U = -(U - U_o)$, and $W_c = U_o - U$, so Eq. 5.12 then becomes

$$W_{nc} = K - K_o - (U_o - U)$$
$$= (K + U) - (K_o + U_o)$$

Therefore,

$$W_{nc} = E - E_o = \Delta E \tag{5.13}$$

Hence, the work done by the nonconservative forces acting on a system is equal to the change in mechanical energy. Notice that for dissipative forces, $E_o > E$ Thus, the change is negative, indicating a decrease in mechanical energy. This condition agrees in sign with W_{nc} which, for friction, would also be negative. Example 5.14 illustrates this concept.

EXAMPLE 5.14 | ## Nonconservative Force: Downhill Racer

A skier with a mass of 80 kg starts from rest at the top of a slope and skis down from an elevation of 110 m (► Fig. 5.21). The speed of the skier at the bottom of the slope is 20 m/s. (a) Show that the system is nonconservative. (b) How much work is done by the nonconservative force of friction?

THINKING IT THROUGH. (a) If the system is nonconservative, then $E_o \neq E$, and these quantities can be computed. (b) The work cannot be determined from force–distance considerations, but W_{nc} is equal to the difference in total energies (Eq. 5.13).

SOLUTION.

Given: $m = 80$ kg
$v_o = 0$
$v = 20$ m/s
$y_o = 110$ m

Find: (a) Show that E is not conserved.
(b) W_{nc} (work done by friction)

$v = 20$ m/s

110 m

▲ **FIGURE 5.21 Work done by a nonconservative force** See Example text for description.

(a) If the system is conservative, the total mechanical energy is constant. Taking $U_o = 0$ at the bottom of the hill, the initial energy at the top of the hill is

$$E_o = U = mgy_o = (80 \text{ kg})(9.8 \text{ m/s}^2)(110 \text{ m}) = 8.6 \times 10^4 \text{ J}$$

And the energy at the bottom of the slope is all kinetic, thus

$$E = K = \tfrac{1}{2}mv^2 = \tfrac{1}{2}(80 \text{ kg})(20 \text{ m/s})^2 = 1.6 \times 10^4 \text{ J}$$

Therefore, $E_o \neq E$, so this system is not conservative.

(b) The amount of work done by the nonconservative force of friction is equal to the change in the mechanical energy, or to the amount of mechanical energy lost (Eq. 5.13):

$$W_{nc} = E - E_o = (1.6 \times 10^4 \text{ J}) - (8.6 \times 10^4 \text{ J}) = -7.0 \times 10^4 \text{ J}$$

This quantity is more than 80% of the initial energy. (Where did this energy actually go?)

FOLLOW-UP EXERCISE. In free fall, air resistance is sometimes negligible, but for skydivers, air resistance has a very practical effect. Typically, a skydiver descends about 450 m before reaching a terminal velocity (Chapter 4.6) of 60 m/s. (a) What is the percentage of energy loss to nonconservative forces during this descent? (b) Show that after terminal velocity is reached, the rate of energy loss in J/s is given by $(60\ mg)$, where m is the mass of the skydiver.

EXAMPLE 5.15 | **Nonconservative Force: One More Time**

A 0.75-kg block slides on a frictionless surface with a speed of 20 m/s. It then slides over a rough area 1.0 m in length and onto another frictionless surface. The coefficient of kinetic friction between the block and the rough surface is 0.17. What is the speed of the block after it passes across the rough surface?

THINKING IT THROUGH. The task of finding the final speed implies that equations involving kinetic energy can be used, where the final kinetic energy can be found by using the conservation of *total* energy. Note that the initial and final energies are kinetic energies, since there is no change in gravitational potential energy. It is always good to make a sketch of the situation for clarity and understanding (▼Fig. 5.22).

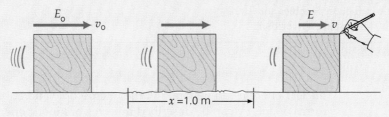

▲ **FIGURE 5.22 A nonconservative rough spot** See Example text for description.

SOLUTION. Listing the data,

Given: $m = 0.75$ kg   *Find:* v (final speed of block)
$x = 1.0$ m
$\mu_k = 0.17$
$v_o = 2.0$ m/s

For this nonconservative system, from Eq. 5.13

$$W_{nc} = E - E_o = K - K_o$$

In the rough area, the block loses energy, because of the work done by friction (W_{nc}) and thus

$$W_{nc} = -f_k x = -\mu_k N x = -\mu_k mgx$$

[negative because f_k and the displacement x are in opposite directions; that is, $[f_k(\cos 180°)x = -f_k x]$.

Then, rearranging the energy equation and writing the terms out in detail,

$$K = K_o + W_{nc}$$

or

$$\tfrac{1}{2}mv^2 = \tfrac{1}{2}mv_o^2 - \mu_k mgx$$

Solving for v yields,

$$v = \sqrt{v_o^2 - 2\mu_k gx}$$
$$= \sqrt{(2.0 \text{ m/s})^2 - 2(0.17)(9.8 \text{ m/s}^2)(1.0 \text{ m})}$$
$$= 0.82 \text{ m/s}$$

Note that the mass of the block was not needed. Also, it can be easily shown that the block lost more than 80% of its energy to friction.

FOLLOW-UP EXERCISE. Suppose the coefficient of kinetic friction between the block and the rough surface were 0.25. What would happen to the block in this case?

Note that in a closed nonconservative system, the *total energy* (*not* the total mechanical energy) is conserved (including nonmechanical forms of energy, such as thermal energy). But not all of the energy is available for mechanical work. For a conservative system, you get back what you put in, so to speak. That is, if you do work on the system, the transferred energy is available to do work. Conservative systems are idealizations, because all real systems are nonconservative to some degree. However, working with ideal conservative systems gives an insight into the conservation of energy.

Total energy is always conserved in a closed system. During the course of study, you will learn about other forms of energy, such as thermal, electrical, and

INSIGHT 5.2 | Hybrid Energy Conversion

As was learned, energy may be transformed from one form to another. An interesting example is the conversion that takes place in the new hybrid automobiles. A hybrid car has both a gasoline (internal combustion) engine and a battery-driven electric motor, both of which may be used to power the vehicle.

A moving car has kinetic energy, and when you step on the brake pedal to slow the car down, kinetic energy is lost. Normally, the brakes of a car accomplish this slowing by friction, and energy is dissipated as heat (conservation of energy). However, in the braking of a hybrid car, some of the energy is converted to electrical energy and stored in the battery of the electric motor. This is called *regenerative braking*. That is, instead of using regular friction brakes to slow the car, the electric motor is used. In this mode, the motor runs in reverse and acts as a generator, converting the lost kinetic energy into electrical energy. (See Chapter 20.2 for generator operation.) The energy is stored in the battery for later use (▶ Fig. 1).

Hybrid cars must also have regular friction brakes to be used when rapid braking is needed. (See the Chapter 20 Insight 20.2, Electromagnetic Induction: Hobbies and Transportation for a more detailed discussion on hybrids.)

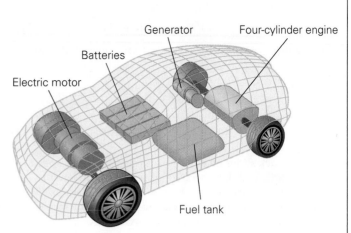

FIGURE 1 Hybrid car A diagram showing the major components. See text for description.

nuclear energies. In general, on the microscopic and submicroscopic levels, these forms of energy can be described in terms of kinetic energy and potential energy. Also, you will learn that mass is a form of energy and that the law of conservation of energy must take this form into account in order to be applied to the analysis of nuclear reactions.

A modern example of energy conversion is given in Insight 5.2, Hybrid Energy Conversion.

DID YOU LEARN?
- ➡ The total energy of a closed or isolated system is always conserved.
- ➡ In a conservative system (one in which only conservative forces act), the total mechanical energy, $E = K + U$, is conserved.
- ➡ When a nonconservative force, such as friction, does work, mechanical energy is not conserved.

5.6 Power

LEARNING PATH QUESTIONS
- ➡ What does power tell you?
- ➡ What does greater efficiency mean?

A particular task may require a certain amount of work, but that work might be done over different lengths of time or at different rates. For example, suppose that you have to mow a lawn. This task takes a certain amount of work, but you might do the job in a half hour, or you might take an hour. There's a practical distinction to be made here. That is, there is usually not only an interest in the amount of work done, but also an interest in how fast it is done—that is, the rate at which it is done. *The time rate of doing work* is called **power**.

The average power ($\overline{P}$) is the work done divided by the time it takes to do the work, or work per unit of time:

$$\overline{P} = \frac{W}{t}$$

(5.14)

The work (and power) done by a constant force of magnitude F acting while an object moves through a parallel displacement of magnitude d is

$$\overline{P} = \frac{W}{t} = \frac{Fd}{t} = F\left(\frac{d}{t}\right) = F\overline{v} \qquad (5.15)$$

SI unit of power: J/s or watt (W)

where it is assumed that the force is in the direction of the displacement. Here, $\overline{v}$ is the magnitude of the average velocity. If the velocity is constant, then $\overline{P} = P = Fv$. If the force and displacement are not in the same direction, then we can write

$$\overline{P} = \frac{F(\cos\theta)d}{t} = F\overline{v}\cos\theta \qquad (5.16)$$

where θ is the angle between the force and the displacement.

As can be seen from Eq. 5.15, the SI unit of power is joules per second (J/s), but this unit is given another name, the **watt (W)**:

$$1\,\text{J/s} = 1\,\text{watt (W)}$$

The SI unit of power is named in honor of James Watt (1736–1819), a Scottish engineer who developed one of the first practical steam engines. A common unit of electrical power is the *kilowatt* (kW).

The British unit of power is foot-pound per second (ft · lb/s). However, a larger unit coined by Watt, the **horsepower (hp),** is more commonly used:*

$$1\,\text{hp} = 550\,\text{ft} \cdot \text{lb/s} = 746\text{W}$$

Power tells how fast work is being done *or* how fast energy is transferred. For example, motors have power ratings commonly given in horsepower. A 2-hp motor can do a given amount of work in half the time that a 1-hp motor would take, or twice the work in the same amount of time. That is, a 2-hp motor is twice as "powerful" as a 1-hp motor.

EXAMPLE 5.16 | A Crane Hoist: Work and Power

A crane hoist like the one shown in ▼Fig. 5.23 lifts a load of 1.0 metric ton a vertical distance of 25 m in 9.0 s at a constant velocity. How much useful work is done by the hoist each second?

◄ **FIGURE 5.23**
Power delivery
See Example text for description.

SOLUTION.

Given: $m = 1.0$ metric ton
$\quad\quad = 1.0 \times 10^3\,\text{kg}$
$\quad y = 25\,\text{m}$
$\quad t = 9.0\,\text{s}$

Find: P (power, work per second)

Since the load moves with a constant velocity, $\overline{P} = P$. (Why?) The work is done against gravity, so $F = mg$, and

$$P = \frac{W}{t} = \frac{Fd}{t} = \frac{mgy}{t}$$
$$= \frac{(1.0 \times 10^3\,\text{kg})(9.8\,\text{m/s}^2)(25\,\text{m})}{9.0\,\text{s}}$$
$$= 2.7 \times 10^4\,\text{W (or 27 kW)}$$

Thus, since a watt (W) is a joule per second (J/s), the hoist did 2.7×10^4 J of work each second. Note that the velocity has a magnitude of $v = d/t = 25\,\text{m}/9.0\,\text{s} = 2.8\,\text{m/s}$, and the displacement is parallel to the applied force, therefore the power could be found using $P = Fv$.

THINKING IT THROUGH. The useful work done each second (that is, per second) is the power output, so this is what needs to be found (Eq. 5.15).

FOLLOW-UP EXERCISE. If the hoist motor of the crane in this Example is rated at 70 hp, what percentage of this power output goes into useful work?

*In Watt's time, steam engines were replacing horses for work in mines and mills. To characterize the performance of his new engine, which was more efficient than existing ones, Watt used the average rate at which a horse could do work as a unit—a horsepower.

EXAMPLE 5.17 | Cleaning Up: Work and Time

The motors of two vacuum cleaners have net power outputs of 1.00 hp and 0.500 hp, respectively. (a) How much work in joules can each motor do in 3.00 min? (b) How long does each motor take to do 97.0 kJ of work?

THINKING IT THROUGH.   (a) Since power is work/time ($P = W/t$), the work can be computed. Note that power is given in horsepower units, which is converted to watts. (b) This part of the problem is another application of Eq. 5.15.

SOLUTION.

Given: $P_1 = 1.00$ hp $= 746$ W   *Find:* (a) W (work for each)
$\quad\quad\quad P_2 = 0.500$ hp $= 373$ W $\quad\quad\quad$ (b) t (time for each)
$\quad\quad\quad$ (a) $t = 3.00$ min $= 180$ s
$\quad\quad\quad$ (b) $W = 97.0$ kJ $= 97.0 \times 10^3$ J

(a) Since $P = W/t$, for the 1.00-hp motor:

$$W_1 = P_1 t = (746\text{ W})(180\text{ s}) = 1.34 \times 10^5\text{ J}$$

And for the 0.500-hp motor:

$$W_2 = P_2 t = (373\text{ W})(180\text{ s}) = 0.67 \times 10^5\text{ J}$$

Note that in the same amount of time, the smaller motor does half as much work as the larger one, as would be expected.

(b) The times are given by $t = W/P$, and for the same amount of work,

$$t_1 = \frac{W}{P_1} = \frac{97.0 \times 10^3\text{ J}}{746\text{ W}} = 130\text{ s}$$

and

$$t_2 = \frac{W}{P_2} = \frac{97.0 \times 10^3\text{ J}}{373\text{ W}} = 260\text{ s}.$$

So, the smaller motor takes twice as long as the larger one to do the same amount of work.

FOLLOW-UP EXERCISE.   (a) A 10-hp motor breaks down and is temporarily replaced with a 5-hp motor. What can you say about the rate of work output? (b) Suppose the situation were reversed—a 5-hp motor is replaced with a 10-hp motor. What can you say about the rate of work output for this case?

EFFICIENCY

Machines and motors are commonly used items in our daily lives, and comments are made about their efficiencies—for example, one machine is more efficient than another. Efficiency involves work, energy, and/or power. Both simple and complex machines that do work have mechanical parts that move, so some input energy is always lost because of friction or some other cause (perhaps in the form of sound). Thus, not all of the input energy goes into doing useful work.

Mechanical efficiency is essentially a measure of what you get out for what you put in—that is, the *useful* work output compared with the energy input. **Efficiency, ε,** is given as a fraction (or percentage):

$$\varepsilon = \frac{\text{work output}}{\text{energy input}}\ (\times\ 100\%) = \frac{W_{\text{out}}}{E_{\text{in}}}\ (\times\ 100\%) \tag{5.17}$$

Efficiency is a unitless quantity

For example, if a machine has a 100-J (energy) input and a 40-J (work) output, then its efficiency is

$$\varepsilon = \frac{W_{\text{out}}}{E_{\text{in}}} = \frac{40\text{ J}}{100\text{ J}} = 0.40\ (\times\ 100\%) = 40\%$$

An efficiency of 0.40, or 40%, means that 60% of the energy input is lost because of friction or some other cause and doesn't serve its intended purpose. Note that if both terms of the ratio in Eq. 5.17 are divided by time t, we obtain $W_{out}/t = P_{out}$ and $E_{in}/t = P_{in}$. So efficiency can be written in terms of power, P:

$$\varepsilon = \frac{P_{out}}{P_{in}} \ (\times\ 100\%) \tag{5.18}$$

EXAMPLE 5.18 | **Home Improvement: Mechanical Efficiency and Work Output**

The motor of an electric drill with an efficiency of 80% has a power input of 600 W. How much useful work is done by the drill in 30 s?

THINKING IT THROUGH. Given the efficiency and power input, the power output P_{out} can readily be found from Eq. 5.18. This quantity is related to the work output $(P_{out} = W_{out}/t)$, from which W_{out} may be found.

SOLUTION.

Given: $\varepsilon = 80\% = 0.80$ $\qquad$ **Find:** W_{out} (work output)
$\qquad P_{in} = 600$ W
$\qquad t = 30$ s

First, rearranging Eq. 5.18 to find the power output:

$$P_{out} = \varepsilon P_{in} = (0.80)(600\ \text{W}) = 4.8 \times 10^2\ \text{W}$$

Then, substituting this value into the equation relating power output and work output,

$$W_{out} = P_{out}t = (4.8 \times 10^2\ \text{W})(30\ \text{s}) = 1.4 \times 10^4\ \text{J}$$

FOLLOW-UP EXERCISE. (a) Is it possible to have a mechanical efficiency of 100%? (b) What would an efficiency of greater than 100% imply?

▼Table 5.1 lists the typical efficiencies of some machines. You may be surprised by the relatively low efficiency of the automobile. Much of the energy input (from gasoline combustion) is lost as exhaust heat and through the cooling system (more than 60%), and friction accounts for a good deal more. About 20% of the input energy is converted to useful work that goes into propelling the vehicle. Air conditioning, power steering, radio, and MP3 and CD players are nice, but they also use energy and contribute to the car's decrease in efficiency.

DID YOU LEARN?
➡ How fast work is done or how fast energy is transferred is expressed by power.
➡ A machine with more useful work output for a given energy input has a greater efficiency.

TABLE 5.1 **Typical Efficiencies of Some Machines**

Machine	Efficiency (approximate %)
Compressor	85
Electric motor	70–95
Automobile (hybrid cars with an efficiency of 25%)	20
Human muscle*	20–25
Steam locomotive	5–10

*Technically not a machine, but used to perform work.

PULLING IT TOGETHER | ## Springs, Energy, and Friction

A spring with a spring constant of 2000 N/m, is in contact with a 1.00-kg block on a table (▶ Fig. 5.24). The spring with the block is compressed 10.0 cm and then released, and the block accelerates on a smooth (frictionless) table surface. Once the spring reaches its fully relaxed position, the block continues on without the spring, but the table surface is now rough, having a coefficient of kinetic friction of 0.500. The table is against an elastic wall 50.0 cm from where the block leaves the spring.

(a) Assume the block rebounds off the wall with no loss of speed. Using work-energy concepts, show how to determine whether the block makes it back to the spring or whether it stops before doing so. (b) If the block does make it back to the spring, how far is the spring compressed? If it doesn't, where does the block come to rest?

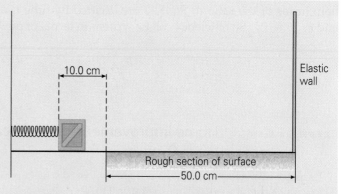

▲ **FIGURE 5.24 Does it make it back?** The block is propelled by a compressed spring over a rough surface and rebounds from an elastic wall. Does the block make it back to the spring?

THINKING IT THROUGH.   Energy, a nonconservative force (friction), and Newton's laws are involved in this example. (a) The spring's potential energy will be converted to the block's kinetic energy. Whether the block makes it back to the spring or not depends on the balance of the total initial mechanical energy and the work done by friction during the distance down the table and back (100 cm). If the latter is larger than the former, the block will stop short of the spring, that is, its kinetic energy will drop to zero before it hits the spring. Otherwise, its remaining energy will be used to recompress the spring. (b) Depending on the result in part (a), either the leftover kinetic energy will be stored in the spring, enabling the recompression to be determined, or, if the initial kinetic energy is all lost due to frictional (nonconservative) work, then the distance needed to do that and therefore the final resting location of the block can be determined.

SOLUTION.

Given: $k = 2.00 \times 10^3$ N/m (spring constant)
$m = 1.00$ kg (block mass)
$d = 50.0$ cm $= 0.500$ m (distance to wall)
$x_o = 10.0$ cm $= 0.100$ m (initial spring
                               compression)
$\mu_k = 0.500$ (coefficient of kinetic friction)

Find: (a) whether the block makes it back to the spring
(b) the final compression of the spring, or the
     final location of the block

(a) Because the first part of the table surface is smooth, the spring's potential energy (U_s) is completely converted into the block's kinetic energy (K_b). Thus the total initial mechanical energy is

$$K_b = U_s = \tfrac{1}{2}kx_o^2 = \tfrac{1}{2}(2.00 \times 10^3 \text{ N/m})(0.100 \text{ m})^2 = 10.0 \text{ J}$$

Now how much nonconservative work the force of kinetic friction (f_k) does must be determined. First the force of friction is found by setting equal the magnitudes of the normal force (N) and the block's weight (mg):

$$f_k = \mu_k N = \mu_k mg = (0.500)(1.00 \text{ kg})(9.80 \text{ m/s}^2) = 4.90 \text{ N}$$

Therefore, the maximum magnitude of work that could be done by this force over the 100 cm of rough surface is

$$W_f = f_k(2d) = (4.90 \text{ N})(1.00 \text{ m}) = 4.90 \text{ J}$$

This is less than the total mechanical energy, so there is some mechanical energy "left over" upon the block's return to the spring. Thus the spring will be (partially) recompressed. (Why partially?)

(b) When the block returns to the spring, it has $K = K_b - W_f = 10.0 \text{ J} - 4.90 \text{ J} = 5.10 \text{ J}$ of kinetic energy left. This energy will go into recompressing the spring a distance x according to the energy conversion:

$$K = U$$
$$5.10 \text{ J} = \tfrac{1}{2}kx^2$$

Solving for x,

$$x = \sqrt{\frac{2(5.10 \text{ J})}{2.00 \times 10^3 \text{ N/m}}} = 0.0714 \text{ m} = 7.14 \text{ cm}$$

Note that this distance is less than the original compression (10.0 cm), as expected.

Learning Path Review

- **Work done by a constant force** is the product of the magnitude of the displacement and the component of the force parallel to the displacement:

$$W = (F \cos \theta)d \tag{5.2}$$

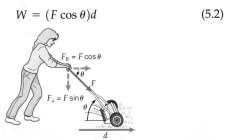

- Calculating work done by a variable force requires advanced mathematics. An example of a variable force is the **spring force**, given by *Hooke's law*:

$$F_s = -kx \tag{5.3}$$

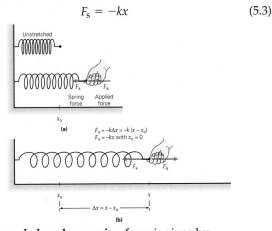

The **work done by a spring force** is given by

$$W = \tfrac{1}{2}kx^2 \tag{5.4}$$

- **Kinetic energy** is the energy of motion and is given by

$$K = \tfrac{1}{2}mv^2 \tag{5.5}$$

- By the **work–energy theorem**, the net work done on an object is equal to the change in the kinetic energy of the object:

$$W = K - K_0 = \Delta K \tag{5.6}$$

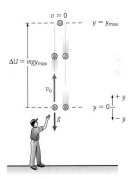

- **Potential energy** is the energy of position and/or configuration. The elastic **potential energy of a spring** is given by

$$U = \tfrac{1}{2}kx^2 \quad (\text{with } x_0 = 0) \tag{5.7}$$

The most common type of potential energy is **gravitational potential energy**, associated with the gravitational attraction near the Earth's surface.

$$U = mgy \quad (\text{with } y_0 = 0) \tag{5.8}$$

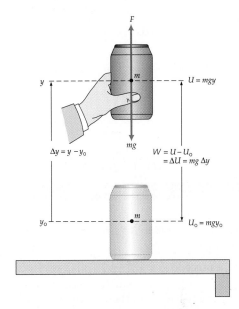

- **Conservation of energy**: The total energy of the universe or of an isolated system is always conserved.
  Conservation of mechanical energy: The total mechanical energy (kinetic plus potential) is constant in a conservative system:

$$\tfrac{1}{2}mv^2 + U = \tfrac{1}{2}mv_0^2 + U_0 \tag{5.10b}$$

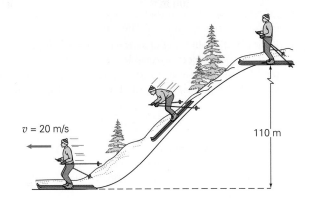

- In systems with **nonconservative forces**, where mechanical energy is lost, the work done by a nonconservative force is given by

$$W_{nc} = E - E_0 = \Delta E \tag{5.13}$$

■ **Power** is the time rate of doing work (or expending energy). **Average power** is given by

$$\overline{P} = \frac{W}{t} = \frac{Fd}{t} = F\overline{v} \qquad (5.15)$$

(constant force in direction of d and v)

$$\overline{P} = \frac{F(\cos\theta)d}{t} = F\overline{v}\cos\theta \qquad (5.16)$$

(constant force acts at an angle θ between d and v)

■ **Efficiency** relates work output to energy (work) input as a fraction or percent:

$$\varepsilon = \frac{W_{\text{out}}}{E_{\text{in}}} \, (\times 100\%) \qquad (5.17)$$

$$\varepsilon = \frac{P_{\text{out}}}{P_{\text{in}}} \, (\times 100\%) \qquad (5.18)$$

Learning Path Questions and Exercises

For instructor-assigned homework, go to www.masteringphysics.com

MULTIPLE CHOICE QUESTIONS

5.1 WORK DONE BY A CONSTANT FORCE

1. The units of work are (a) $N\cdot m$, (b) $kg\cdot m^2/s^2$, (c) J, (d) all of the preceding.

2. For a particular force and displacement, the most work is done when the angle between them is (a) 30°, (b) 60°, (c) 90°, (d) 180°.

3. A pitcher throws a fastball. When the catcher catches it, (a) positive work is done, (b) negative work is done, (c) the net work is zero.

4. Work done in free fall (a) is only positive, (b) is only negative, or (c) can be either positive or negative.

5. Which one of the following has units of work: (a) N, (b) N/s, (c) $J\cdot s$, or (d) $N\cdot m$?

5.2 WORK DONE BY A VARIABLE FORCE

6. The work done by a variable force of the form $F = kx$ is equal to (a) kx^2, (b) kx, (c) $\frac{1}{2}kx^2$, (d) none of the preceding.

5.3 THE WORK—ENERGY THEOREM: KINETIC ENERGY

7. Which of the following is a scalar quantity: (a) work, (b) force, (c) kinetic energy, or (d) both a and c?

8. If the angle between the net force and the displacement of an object is greater than 90°, (a) kinetic energy increases, (b) kinetic energy decreases, (c) kinetic energy remains the same, (d) the object stops.

9. Two identical cars, A and B, traveling at 55 mi/h collide head-on. A third identical car, C, crashes into a brick wall going 55 mi/h. Which car has the least damage: (a) car A, (b) car B, (c) car C, or (d) all the same?

10. Which of the following objects has the least kinetic energy: (a) an object of mass $4m$ and speed v, (b) an object of mass $3m$ and speed $2v$, (c) an object of mass $2m$ and speed $3v$, or (d) an object of mass m and speed $4v$?

5.4 POTENTIAL ENERGY

11. A change in gravitational potential energy (a) is always positive, (b) depends on the reference point, (c) depends on the path, (d) depends only on the initial and final positions.

12. The change in gravitational potential energy can be found by calculating $mg\Delta h$ and subtracting the reference point potential energy: (a) true, (b) false.

13. The reference point for gravitational potential energy may be (a) zero, (b) negative, (c) positive, (d) all of the preceding.

5.5 CONSERVATION OF ENERGY

14. Energy cannot be (a) transferred, (b) conserved, (c) created, (d) in different forms.

15. If a nonconservative force acts on an object, and does work, then (a) the object's kinetic energy is conserved, (b) the object's potential energy is conserved, (c) the mechanical energy is conserved, (d) the mechanical energy is not conserved.

16. The speed of a pendulum is greatest (a) when the pendulum's kinetic energy is a minimum, (b) when the pendulum's acceleration is a maximum, (c) when the pendulum's potential energy is a minimum, (d) none of the preceding.

17. Two springs are identical except for their force constants, $k_2 > k_1$. If the same force is used to stretch the springs, (a) spring 1 will be stretched farther than spring 2, (b) spring 2 will be stretched farther than spring 1, (c) both will be stretched the same distance.

18. If the two springs in Exercise 17 are compressed the same distance, on which spring is more work done: (a) spring 1, (b) spring 2, or (c) equal work on both?

19. Two identical stones are thrown from the top of a tall building. Stone 1 is thrown vertically downward with an initial speed v, and stone 2 is thrown vertically upward with the same initial speed. Neglecting air resistance, which stone hits the ground with a greater speed: (a) stone 1, (b) stone 2, or (c) both have the same speed?

20. In Exercise 19, if air resistance is taken into account, which stone hits the ground with a greater speed: (a) stone 1, (b) stone 2, or (c) both have the same speed?

5.6 POWER

21. Which of the following is not a unit of power: (a) J/s, (b) W · s, (c) W, or (d) hp?

22. Consider a 2.0-hp motor and a 1.0-hp motor. Compared to the 2.0-hp motor, for a given amount of work, the 1.0-hp motor can (a) do twice as much work in half the time, (b) half the work in the same time, (c) one quarter of the work in three quarters of the time, (d) none of the preceding.

CONCEPTUAL QUESTIONS

1. (a) As a weightlifter lifts a barbell from the floor in the "clean" procedure (▼Fig. 5.25a), has he done work? Why or why not? (b) In raising the barbell above his head in the "jerk" procedure, is he doing work? Explain. (c) In holding the barbell above his head (Fig. 5.25b), is he doing more work, less work, or the same amount of work as in lifting the barbell? Explain. (d) If the weightlifter drops the barbell, is work done on the barbell? Explain what happens in this situation.

(a)   (b)

▲ FIGURE 5.25 **Man at work?** See Conceptual Question 1.

2. You are carrying a backpack across campus. What is the work done by your vertical carrying force on the backpack? Explain.

3. A jet plane flies in a vertical circular loop. In what regions of the loop is the work done by the force of gravity on the plane positive and/or negative? Is the work constant? If not, are there maximum and minimum instantaneous values? Explain.

4. When walking up stairs, it is easier to do so in a zigzag path rather than going straight up. Why is this? (*Hint:* Think of an inclined plane.)

5. Can an object possess work?

5.2 WORK DONE BY A VARIABLE FORCE

6. Does it take twice the work to stretch a spring 2 cm from its equilibrium position as it does to stretch it 1 cm from its equilibrium position?

7. If a spring is compressed 2.0 cm from its equilibrium position and then compressed an additional 2.0 cm, how much more work is done in the second compression than in the first? Explain.

5.3 THE WORK—ENERGY THEOREM: KINETIC ENERGY

8. You want to decrease the kinetic energy of an object as much as you can. You can do so by either reducing the mass by half or reducing the speed by half. Which option should you pick, and why?

9. A certain amount of work W is required to accelerate a car from rest to a speed v. How much work is required to accelerate the car from rest to a speed of $v/2$?

10. A certain amount of work W is required to accelerate a car from rest to a speed v. If instead an amount of work equal to $2W$ is done on the car, what is the car's speed?

11. Car B is traveling twice as fast as car A, but car A has four times the mass of car B. Which car has the greater kinetic energy?

5.4 POTENTIAL ENERGY

12. If a spring changes its position from x_0 to x, what is the change in potential energy then proportional to? (Express the answer in terms of x_0 and x.)

13. A lab notebook sits on a table 0.75 m above the floor. Your lab partner tells you the book has zero potential energy, and another student says it has 8.0 J of potential energy. Who is correct?

14. An object is said to have a negative potential energy. Because you prefer not to work with negative numbers, how could you make the object to have a positive potential energy without moving it?

5.5 CONSERVATION OF ENERGY

15. For a classroom demonstration, a bowling ball suspended from a ceiling is displaced from the vertical position to one side and released from rest just in front of the nose of a student (▼Fig. 5.26). If the student doesn't move, why won't the bowling ball hit his nose?

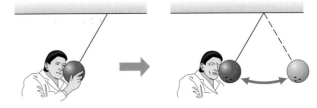

▲ FIGURE 5.26 **In your face?** See Conceptual Question 15.

16. When you throw an object into the air, is its initial speed the same as its speed just before it returns to your hand? Explain by applying the concept of conservation of mechanical energy.

17. A student throws a ball vertically upward so it just reaches the height of a window on the second floor of a dormitory. At the same time that the ball is thrown upward, a student at the window drops an identical ball. Are the mechanical energies of the balls the same at half the height of the window? Explain.

5.6 POWER

18. If you check your electricity bill, you will note that you are paying the power company for so many kilowatt-hours (kWh). Are you really paying for power? Explain.

19. (a) Does efficiency describe how fast work is done? Explain. (b) Does a more powerful machine always perform more work than a less powerful one? Explain.

20. Two students who weigh the same start at the same ground floor location at the same time to go to the same classroom on the third floor by different routes. If they arrive at different times, which student will have expended more power? Explain.

EXERCISES

*Integrated Exercises (**IE**s) are two-part exercises. The first part typically requires a conceptual answer choice based on physical thinking and basic principles. The following part is quantitative calculations associated with the conceptual choice made in the first part of the exercise.*

    Throughout the text, many exercise sections will include "paired" exercises. These exercise pairs, identified with red numbers, *are intended to assist you in problem solving and learning. In a pair, the first exercise (even numbered) is worked out in the Study Guide so that you can consult it should you need assistance in solving it. The second exercise (odd numbered) is similar in nature, and its answer is given in Appendix VII at the back of the book.*

1. • If a person does 50 J of work in moving a 30-kg box over a 10-m distance on a horizontal surface, what is the minimum force required?

2. • A 5.0-kg box slides a 10-m distance on ice. If the coefficient of kinetic friction is 0.20, what is the work done by the friction force?

3. • A passenger at an airport pulls a rolling suitcase by its handle. If the force used is 10 N and the handle makes an angle of 25° to the horizontal, what is the work done by the pulling force while the passenger walks 200 m?

4. •• A 3.00-kg block slides down a frictionless plane inclined 20° to the horizontal. If the length of the plane's surface is 1.50 m, how much work is done, and by what force?

5. •• Suppose the coefficient of kinetic friction between the block and the plane in Exercise 4 is 0.275. What would be the net work done in this case?

6. •• A father pulls his young daughter on a sled with a constant velocity on a level surface a distance of 10 m, as illustrated in ▼Fig. 5.27a. If the total mass of the sled and

the girl is 35 kg and the coefficient of kinetic friction between the sled runners and the snow is 0.20, how much work does the father do?

7. •• A father pushes horizontally on his daughter's sled to move it up a snowy incline, as illustrated in Fig. 5.27b. If the sled moves up the hill with a constant velocity, how much work is done by the father in moving it from the bottom to the top of the hill? (Some necessary data are given in Exercise 6.)

8. •• A block on a level frictionless surface has two horizontal forces applied, as shown in ▼Fig. 5.28. (a) What force F_2 would cause the block to move in a straight line to the right? (b) If the block moves 50 cm, how much work is done by each force? (c) What is the total work done by the two forces?

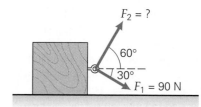

$F_2 = ?$

60°

30°

$F_1 = 90 N$

▲ **FIGURE 5.28** **Make it go straight** See Exercise 8.

9. •• A 0.50-kg shuffleboard puck slides a distance of 3.0 m on the board. If the coefficient of kinetic friction between the puck and the board is 0.15, what work is done by the force of friction?

10. •• A crate is dragged 3.0 m along a rough floor with a constant velocity by a worker applying a force of 500 N to a rope at an angle of 30° to the horizontal. (a) How many forces are acting on the crate? (b) How much work does each of these forces do? (c) What is the total work done on the crate?

11. **IE** •• A hot-air balloon ascends at a constant rate. (a) The weight of the balloon does (1) positive work, (2) negative work, (3) no work. Why? (b) A hot-air balloon with a mass of 500 kg ascends at a constant rate of 1.50 m/s for 20.0 s. How much work is done by the upward buoyant force? (Neglect air resistance.)

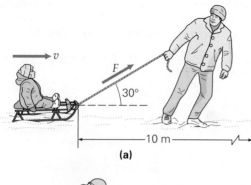

v

F

30°

—10 m—

(a)

v

F

3.6 m

15°

(b)

▲ **FIGURE 5.27** **Fun and work** See Exercises 6 and 7.

12. **IE** •• A hockey puck with a mass of 200 g and an initial speed of 25.0 m/s slides freely to rest in the space of 100 m on a sheet of horizontal ice. How many forces do nonzero work on it as it slows: (a) (1) none, (2) one, (3) two, or (4) three? Explain. (b) Determine the work done by all the individual forces on the puck as it slows.

13. **IE** •• An eraser with a mass of 100 g sits on a book at rest. The eraser is initially 10.0 cm from any edge of the book. The book is suddenly yanked very hard and slides out from under the eraser. In doing so, it partially drags the eraser with it, although not enough to stay on the book. The coefficient of kinetic friction between the book and the eraser is 0.150. (a) The sign of the work done by the force of kinetic friction of the book on the eraser is (1) positive, (2) negative, or (3) zero work is done by kinetic friction. Explain. (b) How much work is done by the book's frictional force on the eraser by the time it falls off the edge of the book?

14. ••• A 500-kg, light-weight helicopter ascends from the ground with an acceleration of 2.00 m/s². Over a 5.00-s interval, what is (a) the work done by the lifting force, (b) the work done by the gravitational force, and (c) the net work done on the helicopter?

15. ••• A man pushes horizontally on a desk that rests on a rough wooden floor. The coefficient of static friction between the desk and floor is 0.750 and the coefficient of kinetic friction is 0.600. The desk's mass is 100 kg. He pushes just hard enough to get the desk moving and continues pushing with that force for 5.00 s. How much work does he do on the desk?

16. **IE** ••• A student could either pull or push, at an angle of 30° from the horizontal, a 50-kg crate on a horizontal surface, where the coefficient of kinetic friction between the crate and surface is 0.20. The crate is to be moved a horizontal distance of 15 m. (a) Compared with pushing, pulling requires the student to do (1) less, (2) the same, or (3) more work. (b) Calculate the minimum work required for both pulling and pushing.

5.2 WORK DONE BY A VARIABLE FORCE

17. • To measure the spring constant of a certain spring, a student applies a 4.0-N force, and the spring stretches by 5.0 cm. What is the spring constant?

18. • A spring has a spring constant of 30 N/m. How much work is required to stretch the spring 2.0 cm from its equilibrium position?

19. • If it takes 400 J of work to stretch a spring 8.00 cm, what is the spring constant?

20. • If a 10-N force is used to compress a spring with a spring constant of 4.0×10^2 N/m, what is the resulting spring compression?

21. **IE** • A certain amount of work is required to stretch a spring from its equilibrium position. (a) If twice the work is performed on the spring, the spring will stretch more by a factor of (1) $\sqrt{2}$ (2) 2, (3) $1/\sqrt{2}$ (4) $\frac{1}{2}$. Why? (b) If 100 J of work is done to pull a spring 1.0 cm, what work is required to stretch it 3.0 cm?

22. •• Compute the work done by the variable force in the graph of F versus x in ▶ Fig. 5.29. [*Hint*: The area of a triangle is $A = \frac{1}{2}$ altitude × base]

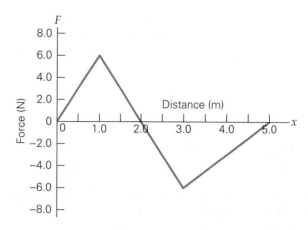

▲ **FIGURE 5.29 How much work is done?** See Exercise 22.

23. **IE** •• A spring with a force constant of 50 N/m is to be stretched from 0 to 20 cm. (a) The work required to stretch the spring from 10 cm to 20 cm is (1) more than, (2) the same as, (3) less than that required to stretch it from 0 to 10 cm. (b) Compare the two work values to prove your answer to part (a).

24. **IE** •• In gravity-free interstellar space, a spaceship fires its rockets to speed up. The rockets are programmed to increase thrust from zero to 1.00×10^4 N with a linear increase over the course of 18.0 km. Then the thrust decreases linearly back to zero over the next 18.0 km. Assuming the rocket was stationary to start, (a) during which segment will more work (magnitude) be done: (1) the first 60 s, (2) the second 60 s, or (3) the work done is the same in both segments? Explain your reasoning. (b) Determine quantitatively how much work is done in each segment.

25. •• A particular spring has a force constant of 2.5×10^3 N/m. (a) How much work is done in stretching the relaxed spring by 6.0 cm? (b) How much more work is done in stretching the spring an additional 2.0 cm?

26. •• For the spring in Exercise 25, how much mass would have to be suspended from the vertical spring to stretch it (a) the first 6.0 cm and (b) the additional 2.0 cm?

27. ••• In stretching a spring in an experiment, a student inadvertently stretches it past its elastic limit; the force-versus-stretch graph is shown in ▼ Fig. 5.30. Basically, after it reaches its limit, the spring begins to behave as if it were considerably stiffer. How much work was done on the spring? Assume that on the force axis, the tick marks are every 10 N, and on the x-axis, they are every 10 cm or 0.10 m.

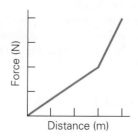

▲ **FIGURE 5.30 Past the limit** See Exercise 27.

28. ••• A spring (spring 1) with a spring constant of 500 N/m is attached to a wall and connected to another weaker spring (spring 2) with a spring constant of 250 N/m on a horizontal surface. Then an external force of 100 N is applied to the end of the weaker spring (#2). How much potential energy is stored in each spring?

5.3   THE WORK—ENERGY THEOREM: KINETIC ENERGY

29. IE • A 0.20-kg object with a horizontal speed of 10 m/s hits a wall and bounces directly back with only half the original speed. (a) What percentage of the object's initial kinetic energy is lost: (1) 25%, (2) 50%, or (3) 75%? (b) How much kinetic energy is lost in the ball's collision with the wall?

30. • A 1200-kg automobile travels at 90 km/h. (a) What is its kinetic energy? (b) What net work would be required to bring it to a stop?

31. • A constant net force of 75 N acts on an object initially at rest as it moves through a parallel distance of 0.60 m. (a) What is the final kinetic energy of the object? (b) If the object has a mass of 0.20 kg, what is its final speed?

32. IE •• A 2.00-kg mass is attached to a vertical spring with a spring constant of 250 N/m. A student pushes on the mass vertically upward with her hand while slowly lowering it to its equilibrium position. (a) How many forces do nonzero work on the object: (1) one, (2) two, or (3) three? Explain your reasoning. (b) Calculate the work done on the object by each of the forces acting on it as it is lowered into position.

33. •• The stopping distance of a vehicle is an important safety factor. Assuming a constant braking force, use the work–energy theorem to show that a vehicle's stopping distance is proportional to the square of its initial speed. If an automobile traveling at 45 km/h is brought to a stop in 50 m, what would be the stopping distance for an initial speed of 90 km/h?

34. IE •• A large car of mass $2m$ travels at speed v. A small car of mass m travels with a speed $2v$. Both skid to a stop with the same coefficient of friction. (a) The small car will have (1) a longer, (2) the same, (3) a shorter stopping distance. (b) Calculate the ratio of the stopping distance of the small car to that of the large car. (Use the work–energy theorem, not Newton's laws.)

35. ••• An out-of-control truck with a mass of 5000 kg is traveling at 35.0 m/s (about 80 mi/h) when it starts descending a steep (15°) incline. The incline is icy, so the coefficient of friction is only 0.30. Use the work–energy theorem to determine how far the truck will skid (assuming it locks its brakes and skids the whole way) before it comes to rest.

36. ••• If the work required to speed up a car from 10 km/h to 20 km/h is 5.0×10^3 J, what would be the work required to increase the car's speed from 20 km/h to 30 km/h?

5.4   POTENTIAL ENERGY

37. • How much more gravitational potential energy does a 1.0-kg hammer have when it is on a shelf 1.2 m high than when it is on a shelf 0.90 m high?

38. IE • You are told that the gravitational potential energy of a 2.0-kg object has decreased by 10 J. (a) With this information, you can determine (1) the object's initial height, (2) the object's final height, (3) both the initial and the final height, (4) only the difference between the two heights. Why? (b) What can you say has physically happened to the object?

39. •• Six identical books, 4.0 cm thick and each with a mass of 0.80 kg, lie individually on a flat table. How much work would be needed to stack the books one on top of the other?

40. IE •• The floor of the basement of a house is 3.0 m below ground level, and the floor of the attic is 4.5 m above ground level. (a) If an object in the attic were brought to the basement, the change in potential energy will be greatest relative to which floor: (1) attic, (2) ground, (3) basement, or (4) all the same? Why? (b) What are the respective potential energies of 1.5-kg objects in the basement and attic, relative to ground level? (c) What is the change in potential energy if the object in the attic is brought to the basement?

41. •• A 0.50-kg mass is placed on the end of a vertical spring that has a spring constant of 75 N/m and eased down into its equilibrium position. (a) Determine the change in spring (elastic) potential energy of the system. (b) Determine the system's change in gravitational potential energy.

42. •• A horizontal spring, resting on a frictionless tabletop, is stretched 15 cm from its unstretched configuration and a 1.00-kg mass is attached to it. The system is released from rest. A fraction of a second later, the spring finds itself compressed 3.0 cm from its unstretched configuration. How does its final potential energy compare to its initial potential energy? (Give your answer as a ratio, final to initial.)

43. ••• A student has six textbooks, each with a thickness of 4.0 cm and a weight of 30 N. What is the minimum work the student would have to do to place all the books in a single vertical stack, starting with all the books on the surface of the table?

44. ••• A 1.50-kg mass is placed on the end of a spring that has a spring constant of 175 N/m. The mass–spring system rests on a frictionless incline that is at an angle of 30° from the horizontal (▼Fig. 5.31). The system is eased into its equilibrium position, where it stays. (a) Determine the change in elastic potential energy of the system. (b) Determine the system's change in gravitational potential energy.

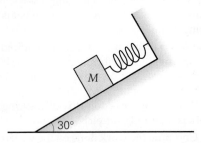

▲ FIGURE 5.31 **Changes in potential energy** See Exercise 44.

5.5 CONSERVATION OF ENERGY

45. • A 0.300-kg ball is thrown vertically upward with an initial speed of 10.0 m/s. If the initial potential energy is taken as zero, find the ball's kinetic, potential, and mechanical energies (a) at its initial position, (b) at 2.50 m above the initial position, and (c) at its maximum height.

46. • What is the maximum height reached by the ball in Exercise 45?

47. •• Referring to Fig. 3.13, find the speed with which the stone strikes the water using energy considerations.

48. IE •• A girl swings back and forth on a swing with ropes that are 4.00 m long. The maximum height she reaches is 2.00 m above the ground. At the lowest point of the swing, she is 0.500 m above the ground. (a) The girl attains the maximum speed (1) at the top, (2) in the middle, (3) at the bottom of the swing. Why? (b) What is the girl's maximum speed?

49. •• A 1.00-kg block (*M*) is on a flat frictionless surface (▼Fig. 5.32). This block is attached to a spring initially at its relaxed length (spring constant is 50.0 N/m). A light string is attached to the block and runs over a frictionless pulley to a 450-g dangling mass (*m*). If the dangling mass is released from rest, how far does it fall before stopping?

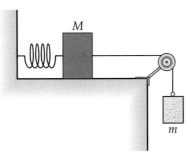

▲ FIGURE 5.32 **How far does it go?** See Exercise 49.

50. IE •• A 500-g (small) mass on the end of a 1.50-m-long string is pulled aside 15° from the vertical and shoved downward (toward the bottom of its motion) with a speed of 2.00 m/s. (a) Is the angle on the other side (1) greater than, (2) less than, or (3) the same as the angle on the initial side (15°)? Explain in terms of energy. (b) Calculate the angle it goes to on the other side, neglecting air resistance.

51. •• A 0.20-kg rubber ball is dropped from a height of 1.0 m above the floor and it bounces back to a height of 0.70 m. (a) What is the ball's speed just before hitting the floor? (b) What is the speed of the ball just as it leaves the ground? (c) How much energy was lost and where did it go?

52. •• A skier coasts down a very smooth, 10-m-high slope similar to the one shown in Fig. 5.21. If the speed of the skier on the top of the slope is 5.0 m/s, what is his speed at the bottom of the slope?

53. •• A roller coaster travels on a frictionless track as shown in ▶Fig. 5.33. (a) If the speed of the roller coaster at point A is 5.0 m/s, what is its speed at point B? (b) Will it reach point C? (c) What minimum speed at point A is required for the roller coaster to reach point C?

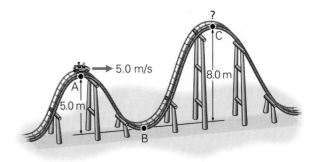

▲ FIGURE 5.33 **Energy conversion(s)** See Exercise 53.

54. •• A simple pendulum has a length of 0.75 m and a bob with a mass of 0.15 kg. The bob is released from an angle of 25° relative to a vertical reference line (▼Fig. 5.34). (a) Show that the vertical height of the bob when it is released is $h = L(1 - \cos 25°)$. (b) What is the kinetic energy of the bob when the string is at an angle of 9.0°? (c) What is the speed of the bob at the bottom of the swing? (Neglect friction and the mass of the string.)

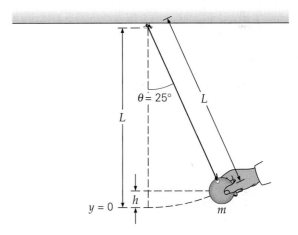

▲ FIGURE 5.34 **A pendulum swings** See Exercise 54.

55. •• Suppose the simple pendulum in Exercise 54 were released from an angle of 60°. (a) What would be the speed of the bob at the bottom of the swing? (b) To what height would the bob swing on the other side? (c) What angle of release would give half the speed of that for the 60° release angle at the bottom of the swing?

56. •• A 1.5-kg box that is sliding on a frictionless surface with a speed of 12 m/s approaches a horizontal spring. (See Fig. 5.19.) The spring has a spring constant of 2000 N/m. If one end of the spring is fixed and the other end changes its position, (a) how far will the spring be compressed in stopping the box? (b) How far will the spring be compressed when the box's speed is reduced to half of its initial speed?

57. •• A 0.50-kg mass is suspended on a spring that stretches 3.0 cm. (a) What is the spring constant? (b) What added mass would stretch the spring an additional 2.0 cm? (c) What is the change in potential energy when the mass is added?

58. ●● A vertical spring with a force constant of 300 N/m is compressed 6.0 cm and a 0.25-kg ball placed on top. The spring is released and the ball flies vertically upward. How high does the ball go?

59. ●● A block with a mass m_1 = 6.0 kg sitting on a frictionless table is connected to a suspended mass m_2 = 2.0 kg by a light string passing over a frictionless pulley. Using energy considerations, find the speed at which m_2 hits the floor after descending 0.75 m. (*Note:* A similar problem in Example 4.6 was solved using Newton's laws.)

60. ●●● A hiker plans to swing on a rope across a ravine in the mountains, as illustrated in ▼Fig. 5.35, and to drop when she is just above the far edge. (a) At what horizontal speed should she be moving when she starts to swing? (b) Below what speed would she be in danger of falling into the ravine? Explain.

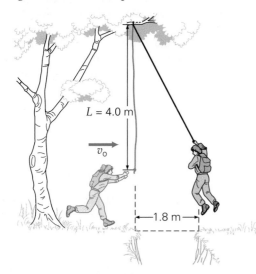

L = 4.0 m

v_0

1.8 m

▲ **FIGURE 5.35 Can she make it?** See Exercise 60.

61. ●●● In Exercise 52, if the skier has a mass of 60 kg and the force of friction retards his motion by doing 2500 J of work, what is his speed at the bottom of the slope?

62. ●●● A 1.00-kg block (*M*) is on a frictionless, 20° inclined plane. The block is attached to a spring (*k* = 25 N/m) that is fixed to a wall at the bottom of the incline. A light string attached to the block runs over a frictionless pulley to a 40.0-g suspended mass. The suspended mass is given an initial downward speed of 1.50 m/s. How far does it drop before coming to rest? (Assume the spring is unlimited in how far it can stretch.)

5.6 POWER

63. ● A girl consumes 8.4×10^6 J (2000 food calories) of energy per day while maintaining a constant weight. What is the average power she produces in a day?

64. ● A 1500-kg race car can go from 0 to 90 km/h in 5.0 s. What average power is required to do this?

65. ● The two 0.50-kg weights of a cuckoo clock descend 1.5 m in a three-day period. At what rate is their total gravitational potential energy decreased?

66. ●● A pump lifts 200 kg of water per hour a height of 5.0 m. What is the minimum necessary power output rating of the water pump in watts and horsepower?

67. ●● A race car is driven at a constant velocity of 200 km/h on a straight, level track. The power delivered to the wheels is 150 kW. What is the total resistive force on the car?

68. ●● An electric motor with a 2.0-hp output drives a machine with an efficiency of 40%. What is the energy output of the machine per second?

69. ●● Water is lifted out of a well 30.0 m deep by a motor rated at 1.00 hp. Assuming 90% efficiency, how many kilograms of water can be lifted in 1 min?

70. ●● How much power must you exert to horizontally drag a 25.0-kg table 10.0 m across a brick floor in 30.0 s at constant velocity, assuming the coefficient of kinetic friction between the table and floor is 0.550?

71. ●●● A 3250-kg aircraft takes 12.5 min to achieve its cruising altitude of 10.0 km and cruising speed of 850 km/h. If the plane's engines deliver, on average, 1500 hp during this time, what is the efficiency of the engines?

72. ●●● A sleigh and driver with a total mass of 120 kg are pulled up a hill with a 15° incline by a horse, as illustrated in ▼Fig. 5.36. (a) If the overall retarding frictional force is 950 N and the sled moves up the hill with a constant velocity of 5.0 km/h, what is the power output of the horse? (Express in horsepower, of course. Note the magnitude of your answer, and explain.) (b) Suppose that in a spurt of energy, the horse accelerates the sled uniformly from 5.0 km/h to 20 km/h in 5.0 s. What is the horse's maximum instantaneous power output? Assume the same force of friction.

f   15°

▲ **FIGURE 5.36 A one-horse open sleigh** See Exercise 72.

73. ●●● A construction hoist exerts an upward force of 500 N on an object with a mass of 50 kg. If the hoist started from rest, determine the power it expended to lift the object vertically for 10 s under these conditions.

The Multiconcept Exercises require the use of more than one fundamental concept for their understanding and solution. The concepts may be from this chapter, but may include those from previous chapters.

74. Two identical springs (neglect their masses) are used to "play catch" with a small block of mass 100 g (▼Fig. 5.37). Spring A is attached to the floor and compressed 10.0 cm with the mass on the end of it (loosely). Spring A is released from rest and the mass is accelerated upward. It impacts the spring attached to the ceiling, compresses it 2.00 cm, and stops after traveling a distance of 30.0 cm from the relaxed position of spring A to the relaxed position of spring B as shown. Determine the spring constant of the two springs (same since they are identical).

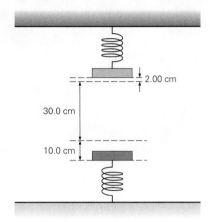

▲ **FIGURE 5.37 Playing catch** See Exercise 74.

75. A 200-g ball is launched from a height of 20.0 m above a lake. Its launch angle is 40° and it has an initial kinetic energy of 90.0 J. (a) Use energy methods to determine its maximum height above the lake surface. (b) Use projectile motion kinematics to repeat part (a). (c) Use energy methods to determine its speed just before impact with the water. (d) Repeat part (c) using projectile motion kinematics.

76. A 1.20-kg ball is projected straight upward with an initial speed of 18.5 m/s and reaches a maximum height of 14.7 m. (a) Show numerically that total mechanical energy is *not* conserved during this part of the ball's motion. (b) Determine the work done on the ball by the force of air resistance. (c) Calculate the average air resistance force on the ball and the ball's average acceleration.

77. An ideal spring of force constant k is hung vertically from the ceiling, and a held object of mass m is attached to the loose end. You carefully and slowly ease that mass down to its equilibrium position by keeping your hand under it until it reaches that position. (a) Show that the spring's change in length is given by $d = \dfrac{mg}{k}$. (b) Show that the work done by the spring is $W_{sp} = -\dfrac{m^2g^2}{2k}$. (c) Show that the work done by gravity is $W_g = \dfrac{m^2g^2}{k}$. Explain why these two works do *not* add to zero. Since the overall change in kinetic energy is zero, you might think they should, no? (d) Show that the work done by your hand is $W_{hand} = -\dfrac{m^2g^2}{2k}$ and that the hand exerted an average force of half the object's weight.

78. A winch is capable of hauling a ton of bricks vertically two stories (6.25 m) in 19.5 s. If the winch's motor is rated at 5.00 hp, determine its efficiency during raising the load.

79. **IE** A 0.455-kg soccer ball is kicked off level ground at an angle of 40° with an initial speed of 30.0 m/s. Neglecting air resistance, (a) at its maximum height off the ground, its kinetic energy will be (1) less than its value at launch, but not zero, (2) more than its value at launch, (3) zero. Explain. (b) Determine its kinetic energy when it is at its maximum height above the ground and compare it to the kinetic energy at launch. [*Hint:* What is its velocity at the top of its arc? Review projectile motion if necessary.]

6 Linear Momentum and Collisions

PHYSICS FACTS

- *Momentum* is the Latin word for "motion."
- Newton called momentum a "quantity of motion." From the *Principia*: *The quantity of motion is the measure of the same, arising from the velocity and quantity of matter, conjointly.*
- Newton called impulse a "motive force."
- A collision is the meeting or inter-action of particles or objects, resulting in an exchange of energy and/or momentum.
- There does not have to be physical contact for a collision. A spacecraft in a gravity-assisted fly-by of a planet is in a collision (energy and momentum transfer).
- It is a common misconception that on rocket blastoff, the fiery engine exhaust striking and "pushing" against the launch pad propels the rocket upward. If this were the case, how could rocket engines be used in space, where there is noth-ing to "push" against?

Tomorrow, sportscasters may say that the momentum of the entire game changed as a result of the clutch hit shown in the photo-graph. One team is said to have gained momentum and went on to win the game. But regardless of the effect on the team, it's clear that the momentum of the *ball* in the chapter-opening photograph must have changed dramatically. The ball was traveling toward the plate at a good rate of speed—with a lot of momentum. But a collision with a hardwood bat—with plenty of momentum of its own—changed the ball's direction in a fraction of a second. A fan might say that the batter turned the ball around. After studying Newton's second law in Section 4.3, you might say that the force the bat applied to the ball gave it a large acceleration, reversing its velocity vector. Yet if you summed the momenta (plural of momentum) of the ball and bat just before the collision and just afterward, you'd discover that although both the ball and the bat had momentum changes, the total momentum didn't change.

If you were bowling and the ball bounced off the pins and rolled back toward you, you would probably be very surprised. But why? What leads us to expect that the ball will send the pins flying and continue on its way, rather than rebounding? You might say that the momentum of the ball carries it forward even after the collision (and you would be right)—but what does that really mean? In this chapter, the concept of *momentum* will be studied and you will learn how it is particularly useful in analyzing motion and collisions.

6.1 Linear Momentum

LEARNING PATH QUESTIONS

➥ What is meant by the *total* linear momentum of a system?
➥ How is momentum related to Newton's second law?

The term *momentum* may bring to mind a football player running down the field, knocking down players who are trying to stop him. Or you might have heard someone say that a team lost its momentum (and so lost the game). Such everyday usages give some insight into the meaning of momentum. They suggest the idea of mass in motion and therefore inertia. We tend to think of heavy or massive objects in motion as having a great deal of momentum, even if they move very slowly. However, according to the technical definition of momentum, a light object can have just as much momentum as a heavier one, and sometimes more.

Newton referred to what modern physicists term **linear momentum (p)** as "the quantity of motion . . . arising from velocity and the quantity of matter conjointly." In other words, the momentum of a body is proportional to *both* its mass and velocity. By definition,

the linear momentum of an object is the product of its mass and velocity:

$$\vec{\mathbf{p}} = m\vec{\mathbf{v}} \tag{6.1}$$

SI unit of momentum: kilogram-meter per second (kg · m/s)

It is common to refer to linear momentum as simply *momentum*. Momentum is a vector quantity that has the same direction as the velocity, and x- and y-components with magnitudes of $p_x = mv_x$ and $p_y = mv_y$, respectively.

Equation 6.1 expresses the momentum of a single object or particle. For a system of more than one particle, the **total linear momentum ($\vec{\mathbf{P}}$)** of the system is the vector sum of the *momenta* of the individual particles:

$$\vec{\mathbf{P}} = \vec{\mathbf{p}}_1 + \vec{\mathbf{p}}_2 + \vec{\mathbf{p}}_3 + \cdots = \Sigma\vec{\mathbf{p}}_i \tag{6.2}$$

(*Note:* $\vec{\mathbf{P}}$ signifies the *total* momentum, while $\vec{\mathbf{p}}$ signifies an *individual* momentum.)

EXAMPLE 6.1 | Momentum: Mass and Velocity

A 100-kg football player runs with a velocity of 4.0 m/s straight down the field. A 1.0-kg artillery shell leaves the barrel of a gun with a muzzle velocity of 500 m/s. Which has the greater momentum (magnitude), the football player or the shell?

THINKING IT THROUGH. Given the mass and velocity of an object, the magnitude of its momentum can be calculated from Eq. 6.1.

SOLUTION. As usual, first listing the given data and using the subscripts p and s to refer to the player and shell, respectively;

Given: $m_p = 100$ kg   *Find:* p_p and p_s (magnitudes of
       $v_p = 4.0$ m/s   the momenta)
       $m_s = 1.0$ kg
       $v_s = 500$ m/s

The magnitude of the momentum of the football player is

$$p_p = m_p v_p = (100 \text{ kg})(4.0 \text{ m/s}) = 4.0 \times 10^2 \text{ kg} \cdot \text{m/s}$$

and that of the shell is

$$p_s = m_s v_s = (1.0 \text{ kg})(500 \text{ m/s}) = 5.0 \times 10^2 \text{ kg} \cdot \text{m/s}$$

Thus, the less massive shell has the greater momentum. Remember, the magnitude of momentum depends on *both* the mass *and* the magnitude of the velocity.

FOLLOW-UP EXERCISE. What would the football player's speed have to be for his momentum to have the same magnitude as the artillery shell's momentum? Would this speed be realistic? (*Answers to all Follow-Up Exercises are given in Appendix VI at the back of the book.*)

INTEGRATED EXAMPLE 6.2 | Linear Momentum: Some Ballpark Comparisons

Consider the three objects shown in ▼Fig. 6.1—a .22-caliber bullet, a cruise ship, and a glacier. Assuming each to be moving at its normal speed, (a) which would you expect to have the greatest linear momentum: (1) the bullet, (2) the ship, or (3) the glacier? (b) Estimate the masses and speeds and compute order-of-magnitude values of the linear momentum of the objects.

(A) CONCEPTUAL REASONING. Certainly the bullet travels the fastest and the glacier the slowest, with the cruise ship in between. But momentum, $p = mv$, is equally dependent on mass as well as speed. The fast bullet has a tiny mass compared with that of the ship and the glacier. The slow glacier has a huge mass that greatly overshadows that of the bullet, and to some extent that of the ship. The cruise ship weighs a great deal and has considerable mass. Which object has the greater momentum also depends on the relative speeds. The glacier "creeps" along compared with the ship, so the very slow speed

of the glacier counterbalances its huge mass to make its momentum less than might be expected. Assuming the speed difference to be greater than the mass difference for the ship and glacier, the ship would have the larger momentum. Similarly, because of the fast bullet's relatively tiny mass, it would be expected to have the least momentum. So with this reasoning, the largest momentum goes to the ship and the smallest momentum to the bullet, and the answer would be (2).

(B) QUANTITATIVE REASONING AND SOLUTION. With no physical data given, you are asked to estimate the masses and velocities (speeds) of the objects so as to be able to compute their momenta [which will verify the reasoning in part (a)]. As is often the case in real-life problems, it may be difficult to estimate the values, so you would try to look up approximate values for the various quantities. For this example, the estimates will be provided. (Note that the units given in references vary, and it is important to convert units correctly.)

(a)   (b)   (c)

▲ **FIGURE 6.1 Three moving objects: a comparison of momenta and kinetic energies** **(a)** A .22-caliber bullet shattering a ballpoint pen; **(b)** a cruise ship; **(c)** a glacier, Glacier Bay, Alaska. See Example text for description.

Given: Estimates of weight (mass) and speed for the bullet, cruise ship, and glacier.

Find: The approximate magnitudes of the momenta for the bullet (p_b), cruise ship (p_s), and glacier (p_g).

Bullet: A typical .22-caliber bullet has a weight of about 30 grains and a muzzle velocity of about 1300 ft/s. (A grain, abbreviated gr, is an old British unit. It was once commonly used for pharmaceuticals, such as 5-gr aspirin tablets; 1 lb = 7000 gr.)

Ship: A ship like the one shown in Fig. 6.1b would have a weight of about 70 000 tons and a speed of about 20 knots. (A knot is another old unit, still commonly used in nautical contexts; 1 knot = 1.15 mi/h.)

Glacier: The glacier might be 1 km wide, 10 km long, and 250 m deep, and move at a rate of about 1.0 m per day. (There is much variation among glaciers. Therefore, these figures must involve more assumptions and rougher estimates than those for the bullet or ship. For example, a uniform, rectangular cross-sectional area is assumed for the glacier. The depth is particularly difficult to estimate from a photograph; a minimum value is given by the fact that glaciers must be at least 50–60 m thick before they can "flow." Observed speeds range from a few centimeters to as much as 40 m a day for valley glaciers such as the one shown in Fig. 6.1c. The value chosen here is considered a typical one.)

Then, converting the data to metric units and giving orders of magnitude yield the following:

Bullet:

$$m_b = 30 \text{ gr}\left(\frac{1 \text{ lb}}{7000 \text{ gr}}\right)\left(\frac{1 \text{ kg}}{2.2 \text{ lb}}\right) = 0.0019 \text{ kg} \approx 10^{-3} \text{ kg}$$

$$v_b = (1.3 \times 10^3 \text{ ft/s})\left(\frac{0.305 \text{ m/s}}{\text{ft/s}}\right) = 4.0 \times 10^2 \text{ m/s} \approx 10^2 \text{ m/s}$$

Ship:

$$m_s = 7.0 \times 10^4 \text{ ton}\left(\frac{2.0 \times 10^3 \text{ lb}}{\text{ton}}\right)\left(\frac{1 \text{ kg}}{2.2 \text{ lb}}\right) = 6.4 \times 10^7 \text{ kg} \approx 10^8 \text{ kg}$$

$$v_s = 20 \text{ knots}\left(\frac{1.15 \text{ mi/h}}{\text{knot}}\right)\left(\frac{0.447 \text{ m/s}}{\text{mi/h}}\right) = 10 \text{ m/s} = 10^1 \text{ m/s}$$

Glacier:

$$\text{width } w \approx 10^3 \text{ m, length } l \approx 10^4 \text{ m, depth } d \approx 10^2 \text{ m}$$

$$v_g = (1.0 \text{ m/day})\left(\frac{1 \text{ day}}{86\ 400 \text{ s}}\right) = 1.2 \times 10^{-5} \text{ m/s} \approx 10^{-5} \text{ m/s}$$

We have all the speeds and masses except for m_g, the mass of the glacier. To compute this value, the density of ice is needed, since $m = \rho V$ (Eq. 1.1). The density of ice is less than that of water (ice floats in water), but the two are not very different, so the density of water, $1.0 \times 10^3 \text{ kg/m}^3$, will be used to simplify the calculations.

Thus, the mass of the glacier is approximated as

$$m_g = \rho V = \rho(l \times \omega \times d)$$
$$\approx (10^3 \text{ kg/m}^3)(10^4 \text{ m})(10^3 \text{ m})(10^2 \text{ m}) = 10^{12} \text{ kg}$$

Then, calculating the magnitudes of the momenta of the objects,

Bullet: $\qquad\qquad\qquad p_b = m_b v_b \approx (10^{-3} \text{ kg})(10^2 \text{ m/s}) = 10^{-1} \text{ kg} \cdot \text{m/s}$

Ship: $\qquad\qquad\qquad p_s = m_s v_s \approx (10^8 \text{ kg})(10^1 \text{ m/s}) = 10^9 \text{ kg} \cdot \text{m/s}$

Glacier: $\qquad\qquad\qquad p_g = m_g v_g \approx (10^{12} \text{ kg})(10^{-5} \text{ m/s}) = 10^7 \text{ kg} \cdot \text{m/s}$

So the ship does have the largest momentum, and the bullet has the smallest according to the estimates.

FOLLOW-UP EXERCISE. Which of the objects in this Example has (1) the greatest kinetic energy and (2) the least kinetic energy? Justify your choices using order-of-magnitude calculations. (Notice here that the dependence is on the square of the speed, $K = \frac{1}{2}mv^2$)

EXAMPLE 6.3 | Total Momentum: A Vector Sum

What is the total momentum for each of the systems of particles illustrated in ▼Fig. 6.2a and b?

THINKING IT THROUGH.   The total momentum is the vector sum of the individual momenta (Eq. 6.2). This quantity can be computed using the components of each vector.

SOLUTION.

Given: Magnitudes and directions of momenta from Fig. 6.2

Find: (a) Total momentum ($\vec{P}$) for Fig. 6.2a
(b) Total momentum ($\vec{P}$) for Fig. 6.2b

(a) The total momentum of a system is the vector sum of the momenta of the individual particles, so

$$\vec{P} = \vec{p}_1 + \vec{p}_2 = (2.0 \text{ kg} \cdot \text{m/s})\hat{x} + (3.0 \text{ kg} \cdot \text{m/s})\hat{x} = (5.0 \text{ kg} \cdot \text{m/s})\hat{x} \quad (+x\text{-direction})$$

(b) Computing the total momenta in the x- and y-directions gives

$$\vec{P}_x = \vec{p}_1 + \vec{p}_2 = (5.0 \text{ kg} \cdot \text{m/s})\hat{x} + (-8.0 \text{ kg} \cdot \text{m/s})\hat{x}$$

$$= -(3.0 \text{ kg} \cdot \text{m/s})\hat{x} \quad (-x\text{-direction})$$

$$\vec{P}_y = \vec{p}_3 = (4.0 \text{ kg} \cdot \text{m/s})\hat{y} \quad (+y\text{-direction})$$

Then

$$\vec{P} = \vec{P}_x + \vec{P}_y = (-3.0 \text{ kg} \cdot \text{m/s})\hat{x} + (4.0 \text{ kg} \cdot \text{m/s})\hat{y}$$

or

$$P = 5.0 \text{ kg} \cdot \text{m/s at } 53° \text{ relative to the negative } x\text{-axis}$$

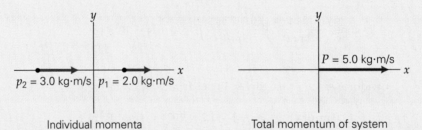

Individual momenta   Total momentum of system

(a) $\vec{P} = \vec{p}_1 + \vec{p}_2$

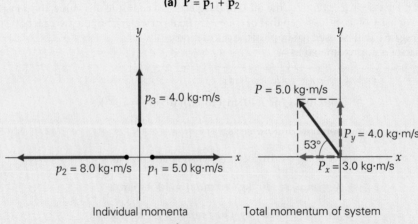

Individual momenta   Total momentum of system

(b) $\vec{P} = \vec{p}_1 + \vec{p}_2 + \vec{p}_3$

▲ **FIGURE 6.2 Total momentum** The total momentum of a system of particles is the vector sum of the particles' individual momenta. See Example text for description.

FOLLOW-UP EXERCISE.   In this Example, if $\vec{p}_1$ and $\vec{p}_2$ in part (a) were added to $\vec{p}_2$ and $\vec{p}_3$ in part (b), what would be the total momentum?

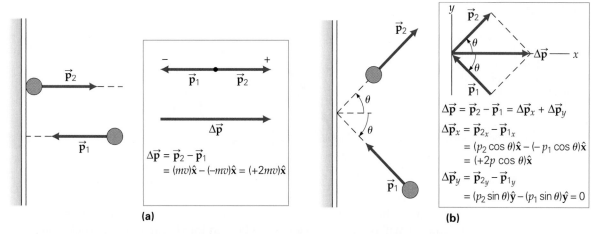

▲ FIGURE 6.3 Change in momentum The change in momentum is given by the *difference* in the momentum vectors. **(a)** Here, the vector sum is zero, but the vector *difference*, or change in momentum, is not. (The particles are displaced for convenience.) **(b)** The change in momentum is found by computing the change in the components.

In Example 6.3a, each of the momenta were along one of the coordinate axes and thus were added straightforwardly. If the motion of one (or more) of the particles is not along an axis, its momentum vector may be broken up, or resolved, into rectangular components, and then individual components can be added to find the components of the total momentum, just as you learned to do with force components in Section 4.3.

Since momentum is a vector, a change in momentum can result from a change in magnitude and/or direction. Examples of changes in the momenta of particles because of changes of direction on collision are illustrated in ▲ Fig. 6.3. In the figure, the magnitude of a particle's momentum is taken to be the same both before and after collision (as indicated by the arrows of equal length). Figure 6.3a illustrates a direct rebound—a 180° change in direction. Note that the change in momentum ($\Delta\vec{p}$) is the *vector* difference and that directional signs for the vectors are important. Figure 6.3b shows a glancing collision, for which the change in momentum is given by analyzing changes in the x- and y-components.

FORCE AND MOMENTUM

As you know from Section 4.3, if an object has a change in velocity, a net force must be acting on it. Similarly, since momentum is directly related to velocity, a change in momentum also requires a net force. In fact, Newton originally expressed his second law of motion in terms of momentum rather than acceleration. The force–momentum relationship may be seen by starting with $\vec{F}_{net} = m\vec{a}$ and using $\vec{a} = (\vec{v} - \vec{v}_0)/\Delta t$, where the mass is assumed to be constant. Thus,

$$\vec{F}_{net} = m\vec{a} = \frac{m(\vec{v} - \vec{v}_0)}{\Delta t} = \frac{m\vec{v} - m\vec{v}_0}{\Delta t} = \frac{\vec{p} - \vec{p}_0}{\Delta t} = \frac{\Delta\vec{p}}{\Delta t}$$

or

$$\vec{F}_{net} = \frac{\Delta\vec{p}}{\Delta t} \qquad \textit{(Newton's second law of motion in terms of momentum)} \qquad (6.3)$$

where $\vec{F}_{net}$ is the *average* net force on the object if the acceleration is not constant (or the *instantaneous* net force if Δt goes to zero).

Expressed in this form, Newton's second law states that *the net external force acting on an object is equal to the time rate of change of the object's momentum.* It is easily seen from the development of Eq. 6.3 that the equations $\vec{F}_{net} = m\vec{a}$ and $\vec{F}_{net} = \Delta\vec{p}/\Delta t$ are

▶ FIGURE 6.4 **Change in the momentum of a projectile** The total momentum vector of a projectile is tangential to the projectile's path (as is its velocity); this vector changes in magnitude and direction, because of the action of an external force (gravity). The x-component of the momentum is constant. (Why?)

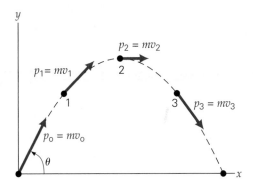

equivalent if the mass is constant. In some situations, however, the mass may vary. This factor will not be a consideration here in the discussion of particle collisions, but a special case will be given later in the chapter. The more general form of Newton's second law, Eq. 6.3, is true even if the mass varies.

Just as the equation $\vec{\mathbf{F}}_{net} = m\vec{\mathbf{a}}$ indicates that an acceleration is evidence of a net force, the equation $\vec{\mathbf{F}}_{net} = \Delta\vec{\mathbf{p}}/\Delta t$ indicates that *a change in momentum is evidence of a net force*. For example, as illustrated in ▲Fig. 6.4, the momentum of a projectile is tangential to the projectile's parabolic path and changes in both magnitude and direction. The change in momentum indicates that there is a net force acting on the projectile, which of course is the force of gravity. Changes in momentum were illustrated in Fig. 6.3. Can you identify the forces in these two cases? Think in terms of Newton's third law.

DID YOU LEARN?

➡ The vector sum of the momenta of all the individual particles or objects is the total linear momentum of a system.

➡ The time rate of change of momentum is equal to the net force, $\vec{\mathbf{F}}_{net} = \Delta\vec{\mathbf{p}}/\Delta t$ (equivalent to $\vec{\mathbf{F}}_{net} = m\vec{\mathbf{a}}$).

(a)

(b)

▲ FIGURE 6.5 **Collision impulse (a)** A collision impulse causes the football to be deformed. **(b)** The impulse is the area under the curve of an F-versus-t graph. Note that the impulse force on the ball is not constant, but rises to a maximum.

6.2 Impulse

LEARNING PATH QUESTIONS

➡ How is the impulse–momentum theorem analogous to the work–energy theorem?

➡ Can kinetic energy be expressed in terms of momentum?

➡ Do objects have to come into contact to have a "collision"?

When two objects—such as a hammer and a nail, a golf club and a golf ball, or even two cars—collide, they can exert a large force on one another for a short period of time, or an *impulse*. (◀Fig. 6.5a). The force is not constant in this situation. However, Newton's second law in momentum form is still useful for analyzing such situations by using average values. Written in this form, the law states that the *average* force is equal to the time rate of change of momentum: $\vec{\mathbf{F}}_{avg} = \Delta\vec{\mathbf{p}}/\Delta t$ (Eq. 6.3). Rewriting the equation to express the change in momentum (with only one force acting on the object),

$$\vec{\mathbf{F}}_{avg}\Delta t = \Delta\vec{\mathbf{p}} = \vec{\mathbf{p}} - \vec{\mathbf{p}}_o \qquad (6.4)$$

The term $\vec{\mathbf{F}}_{avg}\Delta t$ is known as the **impulse ($\vec{\mathbf{I}}$)** of the force:

$$\vec{\mathbf{I}} = \vec{\mathbf{F}}_{avg}\Delta t = \Delta\vec{\mathbf{p}} = m\vec{\mathbf{v}} - m\vec{\mathbf{v}}_o \qquad (6.5)$$

SI unit of impulse and momentum: newton-second (N·s)

TABLE 6.1 Some Typical Contact Times (Δt)

	Δt (milliseconds)
Golf ball (hit by a driver)	1.0
Baseball (hit off tee)	1.3
Tennis (forehand)	5.0
Football (kick)	8.0
Soccer (header)	23.0

Thus, *the impulse exerted on an object is equal to the change in the object's momentum.* This statement is referred to as the **impulse–momentum theorem**. Impulse has units of newton-second (N·s), which are also units of momentum ($1\ \text{N}\cdot\text{s} = 1\ \text{kg}\cdot\text{m/s}^2\cdot\text{s} = 1\ \text{kg}\cdot\text{m/s}$).

In Section 5.3, it was learned that by the work–energy theorem ($W_{net} = F_{net}\Delta x = \Delta K$), the area under an F_{net}-versus-x curve is equal to the net work, or change in kinetic energy. Similarly, the area under an F_{net}-versus-t curve is equal to the impulse, or the change in momentum (Fig. 6.5b). Forces between interacting objects usually varies with time and are therefore not constant forces. However, in general, it is convenient to talk about the equivalent *constant* average force $\vec{F}_{avg}$ acting over a time interval Δt to give the same impulse (same area under the force-versus-time curve), as shown in ►Fig. 6.6. Some typical contact times in sports are given in ▲Table 6.1.

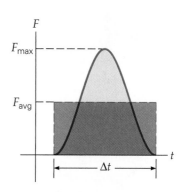

▲ **FIGURE 6.6 Average impulse force** The area under the average force curve ($F_{avg}\Delta t$, within the dashed red lines) is the same as the area under the F-versus-t curve, which is usually difficult to evaluate.

EXAMPLE 6.4 | Teeing Off: The Impulse–Momentum Theorem

A golfer drives a 0.046-kg ball from an elevated tee, giving the ball an initial horizontal speed of 40 m/s (about 90 mi/h). What is the magnitude of the average force exerted by the club on the ball during this time?

THINKING IT THROUGH. The average force on the ball is equal to the time rate of change of its momentum, and this can be computed (Eq. 6.5). The Δt of the collision is obtained from Table 6.1.

SOLUTION.

Given: $m = 0.046\ \text{kg}$ *Find:* F_{avg} (average force)
$v = 40\ \text{m/s}$
$v_0 = 0$
$\Delta t = 1.0\ \text{ms} = 1.0 \times 10^{-3}\ \text{s}$ (Table 6.1)

The mass and the initial and final velocities are given, so the change in momentum can be easily found. Then the magnitude of the average force can be computed from the impulse–momentum theorem:

$$F_{avg}\Delta t = p - p_0 = mv - mv_0$$

and

$$F_{avg} = \frac{mv - mv_0}{\Delta t} = \frac{(0.046\ \text{kg})(40\ \text{m/s}) - 0}{1.0 \times 10^{-3}\ \text{s}} = 1.8 \times 10^3\ \text{N (or about 410 lb)}$$

[This is a very large force compared with the weight of the ball, $w = mg = (0.046\ \text{kg})(9.8\ \text{m/s}^2) = 0.45\ \text{N}$ (or about 0.22 lb).] The force is in the direction of the acceleration and is the *average* force. The instantaneous force is even greater than this value near the midpoint of the time interval of the collision (Δt in Fig. 6.6).

FOLLOW-UP EXERCISE. Suppose the golfer in this Example drives the ball with the same average force, but "follows through" on the swing so as to increase the contact time to 1.5 ms. What effect would this change have on the initial horizontal speed of the drive?

$$F_{avg} \Delta t = mv_o$$

(a)

$$F_{avg} \Delta t = mv_o$$

(b)

▲ **FIGURE 6.7 Adjust the impulse**
(a) The change in momentum in catching the ball is a constant mv_o. If the ball is stopped quickly (small Δt), the impulse force is large (big F_{avg}) and stings the catcher's bare hands. **(b)** Increasing the contact time (large Δt) by moving the hands with the ball reduces the impulse force and makes catching more enjoyable.

Example 6.4 illustrates the large forces that colliding objects can exert on one another during short contact times. In some cases, the contact time may be shortened to maximize the impulse—for example, in a karate chop. However, in other instances, the Δt may be manipulated to reduce the force. Suppose there is a fixed change in momentum in a given situation. Then, with $\Delta p = F_{avg} \Delta t$, if Δt could be made longer, the average impulse F_{avg} would be reduced.

You have probably tried to minimize the impulse on occasion. For example, in catching a hard, fast-moving ball, you quickly learn not to catch it with your arms rigid, but rather to move your hands with the ball. This movement increases the contact time and reduces the impulse and the "sting" (◄Fig. 6.7).

When jumping from a height onto a hard surface, you should not land stiff-legged. The abrupt stop (small Δt) would apply a large impulse to your leg bones and joints and could cause injury. If you bend your knees as you land, the impulse is vertically upward, opposite your velocity ($F_{avg} \Delta t = \Delta p = -mv_o$ with the final velocity being zero). Thus, increasing the time interval Δt makes the impulse smaller. Another example in which the contact time is increased to decrease the impulse is given in Insight 6.1, The Automobile Air Bag and Martian Air Bags.

EXAMPLE 6.5 | **Impulse and Body Injury**

A 70.0-kg worker jumps stiff-legged from a height of 1.00 m onto a concrete floor. (a) What is the magnitude of the impulse he feels on landing, assuming a sudden stop in 8.00 ms? (b) What is the average force?

THINKING IT THROUGH.   The impulse is $F_{avg} \Delta t$ which cannot be calculated directly from the given data. But impulse is equal to the change in momentum, $F_{avg} \Delta t = \Delta p = mv - mv_o$. So the impulse can be calculated from the difference in momenta if v_o is found.

SOLUTION.

Given: $m = 70.0$ kg   *Find:*   (a) I (impulse on worker)
            $h = 1.00$ m   (b) F_{avg} (average force)
            $\Delta t = 8.00$ ms $= 8.00 \times 10^{-3}$ s

(a) There are two different parts here: (1) the worker descending after jumping and (2) the sudden stop after hitting the floor. So we must be careful with notation and distinguish between the two parts (a and b) with the subscripts of 1 and 2, respectively. (b) Knowing Δt, the average force can be calculated.

(a) Here, the initial velocity of the worker is $v_{1o} = 0$, and the final velocity just before hitting the floor may be found using $v^2 = v_o^2 - 2gh$ (Eq. 2.12'), with the result of

$$v_{1f} = -\sqrt{2gh}$$

where the minus sign indicates direction (downward).

(b) The v_{1f} in (a) is then the initial velocity with which the stiff-legged worker hits the floor, that is, $v_{2o} = v_{1f} = -\sqrt{2gh}$, and the final velocity at the second phase is $v_{2f} = 0$. Then,

$$I = F_{avg} \Delta t = \Delta p = mv_{2f} - mv_{2o} = 0 - m(-\sqrt{2gh}) = +m\sqrt{2gh}$$
$$= (70.0 \text{ kg})\sqrt{2(9.80 \text{ m/s}^2)(1.00 \text{ m})} = 310 \text{ kg} \cdot \text{m/s}$$

where the impulse is in the upward direction.

With a Δt of 8.00×10^{-3} s for the sudden stop on impact, this would give a force of

$$F_{avg} = \frac{\Delta p}{\Delta t} = \frac{310 \text{ kg} \cdot \text{m/s}}{8.00 \times 10^{-3} \text{ s}} = 3.88 \times 10^4 \text{ N}   \textit{(about 8730 lb of force!)}$$

and the force is upward on the stiff legs.

FOLLOW-UP EXERCISE.   Suppose the worker bent his knees and increased the contact time to 0.60 s on landing. What would be the average impulse force on him in this case?

In some instances, the impulse force may be relatively constant and the contact time (Δt) deliberately increased to produce a greater impulse, and thus a greater change in momentum ($F_{avg} \Delta t = \Delta p$). This is the principle of "following through" in sports, for example, when hitting a ball with a bat or racquet, or driving a golf ball.

▶ **FIGURE 6.8 Increasing the contact time** (a) A golfer follows through on a drive. One reason he does so is to increase the contact time so that the ball receives greater impulse and momentum. (b) The follow-through on a long putt increases the contact time for greater momentum, but the main reason here is for directional control.

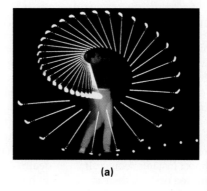

(a)

(b)

In the latter case (▶ Fig. 6.8a), assuming that the golfer supplies the same average force with each swing, the longer the contact time, the greater the impulse or change in momentum the ball receives. That is, with $F_{avg}\Delta t = mv$ (since $v_o = 0$), the greater the value of Δt, the greater the final speed of the ball. (This principle was illustrated in the Follow-Up Exercise in Example 6.4.) In some instances, a long follow-through may primarily be used to improve control of the ball's direction (Fig. 6.8b).

The word *impulse* implies that the impulse force acts only briefly (like an "impulsive" person), and this is true in many instances. However, the definition of *impulse* places no limit on the time interval of a collision over which the force may act. Technically, a comet at its closest approach to the Sun is involved in a collision, because in physics, collision forces do *not* have to be contact forces. Basically, a **collision** is any interaction between objects in which there is an exchange of momentum and/or energy.

As you might expect from the work–energy theorem and the impulse–momentum theorem, momentum and kinetic energy are directly related. A little algebraic manipulation of the equation for kinetic energy (Eq. 5.5) allows us to express kinetic energy (K) in terms of the *magnitude* of momentum (p):

$$K = \tfrac{1}{2}mv^2 = \frac{(mv)^2}{2m} = \frac{p^2}{2m} \qquad (6.6)$$

Thus, kinetic energy and momentum are intimately related, but they are different quantities.

DID YOU LEARN?

➥ Impulse is equal to the change in momentum. Work is equal to the change in kinetic energy.

➥ An interaction between objects in which there is an exchange of momentum and/or energy is a collision. Direct contact is not needed, for example, a comet making a pass around the Sun is in a collision.

➥ In terms of momentum, the kinetic energy is $K = p^2/2m$.

6.3 Conservation of Linear Momentum

LEARNING PATH QUESTIONS

➥ Why is the total momentum of a system conserved if the net force on the system is zero?

➥ Why is the net internal force of a system always zero?

Like total mechanical energy, the total momentum of a system is a conserved quantity under certain conditions. This fact allows us to analyze a wide range of situations and solve many problems readily. Conservation of momentum is one of the most important principles in physics. In particular, it is used to analyze collisions of objects ranging from subatomic particles to automobiles in traffic accidents.

A dark, rainy night—a car goes out of control and hits a big tree head-on! But the driver walks away with only minor injuries, because he had his seatbelt buckled and his car's air bags deployed. Air bags, along with seatbelts, are safety devices designed to prevent (or lessen) injuries to passengers in automobile collisions.

When a car collides with something basically immovable, such as a tree or a bridge abutment, or has a head-on collision with another vehicle, the car stops almost instantaneously. If the front-seat passengers have not buckled up (and there are no air bags), they keep moving until acted on by an external force (by Newton's first law). For the driver, this force is supplied by the steering wheel and column, and for the passenger, by the dashboard and/or windshield.

Even when everyone has buckled up, there can be injuries. Seatbelts absorb energy by stretching, and they widen the area over which the force is exerted. However, if a car is going fast enough and hits something truly immovable, there may be too much energy for the belts to absorb. This is where the air bag comes in (Fig. 1).

The bag inflates automatically on hard impact, cushioning the driver (and front-seat passenger if both sides are equipped with air bags). In terms of impulse, the air bag increases the stopping contact time—the fraction of a second it takes your head to sink into the inflated bag is many times longer than the instant in which you would have stopped otherwise by hitting a solid surface such as the windshield. A longer contact time means a reduced average impact force and thus less likelihood of an injury. (Because the bag is large, the total impact force is also spread over a greater area of the body, so the force on any one part of the body is also less.)

How does an air bag inflate during the little time that elapses between a front-end impact and the instant the driver would hit the steering column? An air bag is equipped with sensors that detect the sharp deceleration associated with a head-on collision the instant it begins. If the deceleration exceeds the sensors' threshold settings, a control unit sends an electric current to an igniter in the air bag, which sets off a chemical explosion that generates gas to inflate the bag at an explosive rate. The complete process from sensing to full inflation takes only on the order of 25 *thousandths* of a second (0.025 s).

Air bags have saved many lives. However, in some cases, the deployment of air bags has caused problems. An air bag is not a soft, fluffy pillow. When activated, it is ejected out of its compartment at speeds up to 320 km/h (200 mi/h) and can hit a person with enough force to cause severe injury and even death. Adults are advised to sit at least 13 cm (6 in.) from the air bag compartment and to buckle up—always. Children should sit in the rear seat, out of the reach of air bags.*

MARTIAN AIR BAGS

Air bags on Mars? They were there in 1997 when a robotic rover from the spacecraft *Pathfinder* landed on Mars. And in 2004,

▲ **FIGURE 1 Impulse and safety** An automobile air bag increases the contact time that a person in a crash would experience with the dashboard or windshield, thereby decreasing the impulse force that could cause injury.

*Guidelines from the National Highway Traffic Safety Administration (www.nhtsa.gov).

For the linear momentum of a single object to be conserved (that is, to remain constant with time), one condition must hold that is apparent from the momentum form of Newton's second law (Eq. 6.3). If the net force acting on a particle is zero, that is,

$$\vec{\mathbf{F}}_{net} = \frac{\Delta \vec{\mathbf{p}}}{\Delta t} = 0$$

then

$$\Delta \vec{\mathbf{p}} = 0 = \vec{\mathbf{p}} - \vec{\mathbf{p}}_o$$

where $\vec{\mathbf{p}}_o$ is the initial momentum and $\vec{\mathbf{p}}$ is the momentum at some later time. Since these two values are equal, the momentum is conserved, and

$$\vec{\mathbf{p}} = \vec{\mathbf{p}}_o \quad \text{or} \quad m\vec{\mathbf{v}} = m\vec{\mathbf{v}}_o$$

final momentum = initial momentum

Note that this conservation is consistent with Newton's first law: An object remains at rest ($\vec{\mathbf{p}} = 0$), or in motion with a *uniform* velocity (constant $\vec{\mathbf{p}} \neq 0$), unless acted on by a net external force.

(a)

(b)

(c)

more air bags arrived with the Mars Exploration Rover Mission and the touchdown of two Rovers. Spacecraft landings are usually softened by retrorockets fired intermittently toward the planet surface. However, firing retrorockets very near the Martian surface would have left trace amounts of foreign combustion chemicals on the surface. Since one objective of the Mars missions was to analyze the chemical composition of Martian rocks and soil, another method of landing had to be developed.

The solution? Probably the most expensive air bag system ever created, costing approximately $5 million to develop and install. The Rovers were surrounded by 4.6-m-(15-ft)-diameter "beach balls" for an air bag landing (Fig. 2a).

On entering the Martian atmosphere, the spacecraft was traveling at about 27 000 km/h (17 000 mi/h). A high-altitude rocket system and parachute slowed it down to about 80–100 km/h (50–60 mi/h). At an altitude of about 200 m (660 ft), gas generators inflated the air bags, which allowed the bag-covered Rovers to bounce and roll a bit on landing (Fig. 2b). The air bags then deflated, and out rolled the Rovers (Fig. 2c).

FIGURE 2 More bounce to the ounce (a) "Beach ball" air bags were used to protect *Pathfinder* and Mars Rovers. **(b)** Artist's conception of bouncing air bags of a Mars Rover. **(c)** A Rover coming out safely.

The conservation of momentum can be extended to a system of particles if Newton's second law is written in terms of the net force acting on the system and of the momenta of the particles: $\vec{\mathbf{F}}_{net} = \Sigma \vec{\mathbf{F}}_i$ and $\vec{\mathbf{P}} = \Sigma \vec{\mathbf{p}}_i = \Sigma m_i \vec{\mathbf{v}}_i$.

Because $\vec{\mathbf{F}}_{net} = \Delta \vec{\mathbf{P}}/\Delta t$, and if there is no net external force acting *on the system*, then $\vec{\mathbf{F}}_{net} = 0$, and $\Delta \vec{\mathbf{F}} = 0$; so $\vec{\mathbf{P}} = \vec{\mathbf{P}}_o$, and the *total* momentum is conserved. This generalized condition is referred to as the law of **conservation of linear momentum**:

$$\vec{\mathbf{P}} = \vec{\mathbf{P}}_o \qquad (6.7)$$

Thus, the total linear momentum of a system, $\vec{\mathbf{P}} = \Sigma \vec{\mathbf{p}}_i$, is conserved if the net external force acting on the system is zero.

There are various ways to achieve this condition. For example, recall from Section 5.5 that a *closed*, or *isolated*, system is one on which no net external force acts, so the total linear momentum of an isolated system is conserved.

Within a system, internal forces may act—for example, when particles collide. These are force pairs of Newton's third law, and there is a good reason that such forces are not explicitly referred to in the condition for the conservation of momentum.

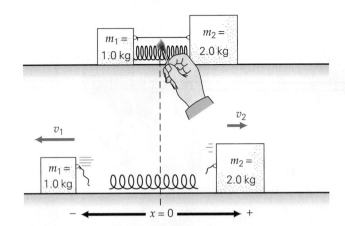

▲ **FIGURE 6.9   An internal force and the conservation of momentum** The spring force is an internal force, so the momentum of the system is conserved. See Example 6.6.

By Newton's third law, these internal forces are equal and opposite and vectorially cancel each other. Thus, *the net internal force of a system is always zero.*

An important point to understand, however, is that the momenta of *individual* particles or objects within a system may change. But in the absence of a net external force, the *vector sum* of all the momenta (the total system momentum $\vec{\mathbf{P}}$) remains the same. If the objects are initially at rest (that is, the total momentum is zero) and then are set in motion as the result of internal forces, the total momentum must still add to zero. This principle is illustrated in ▲ Fig. 6.9 and analyzed in Example 6.6. Objects in an isolated system may transfer momentum among themselves, but the total momentum after the changes must add up to the initial value, assuming the net external force on the system is zero.

The conservation of momentum is often a powerful and convenient tool for analyzing situations involving motion and collisions. Its application is illustrated in the following Examples. (Notice that conservation of momentum, in many cases, bypasses the need to know the forces involved.)

EXAMPLE 6.6 | ## Before and After: Conservation of Momentum

Two masses, $m_1 = 1.0$ kg and $m_2 = 2.0$ kg, are held on either side of a light compressed spring by a light string joining them, as shown in Fig. 6.9. The string is burned (negligible external force), and the masses move apart on the frictionless surface, with m_1 having a velocity of 1.8 m/s to the left. What is the velocity of m_2?

THINKING IT THROUGH. With no net external force (the weights are each canceled by a normal force), the total momentum of the system is conserved. It is initially zero, so after the string is burned, the momentum of m_2 must be *equal to and opposite* that of m_1. (Vector addition gives zero total momentum. Also, note that the term *light* indicates that the masses of the spring and string can be ignored.)

SOLUTION. Listing the data:

Given: $m_1 = 1.0$ kg   **Find:** v_2 (velocity—
          $m_2 = 2.0$ kg   speed and
          $v_1 = -1.8$ m/s (left)   direction)

Here, the system consists of the two masses and the spring. Since the spring force is internal to the system, the momentum

of the system is conserved. It should be apparent that the initial total momentum of the system ($\vec{\mathbf{P}}_o$) is zero, and therefore the final momentum must also be zero. Thus,

$$\vec{\mathbf{P}}_o = \vec{\mathbf{P}} = 0 \quad \text{and} \quad \vec{\mathbf{P}} = \vec{\mathbf{p}}_1 + \vec{\mathbf{p}}_2 = 0$$

(The momentum of the "light" spring does not come into the equations, because its mass is negligible.) Then,

$$\vec{\mathbf{p}}_2 = -\vec{\mathbf{p}}_1$$

which means that the momenta of m_1 and m_2 are equal and opposite. Using directional signs (with + indicating the direction to the right in the figure),

$$m_2 v_2 = -m_1 v_1$$

and

$$v_2 = -\left(\frac{m_1}{m_2}\right) v_1 = -\left(\frac{1.0 \text{ kg}}{2.0 \text{ kg}}\right)(-1.8 \text{ m/s}) = +0.90 \text{ m/s}$$

Thus, the velocity of m_2 is 0.90 m/s in the positive x-direction, or to the right in the figure. This value is half that of v_1 as you might have expected, since m_2 has twice the mass of m_1.

INTEGRATED EXAMPLE 6.7 | ## Conservation of Linear Momentum: Fragments and Components

A 30-g bullet with a speed of 400 m/s strikes a glancing blow to a target brick of mass 1.0 kg. The brick breaks into two fragments. The bullet deflects at an angle of 30° above the +x-axis and has a reduced speed of 100 m/s. One piece of the brick (with mass 0.75 kg) goes off to the right, or in the initial direction of the bullet, with a speed of 5.0 m/s. (a) Taking the +x-axis to the right, will the other piece of the brick move in the (1) second quadrant, (2) third quadrant, or (3) fourth quadrant? (b) Determine the speed and direction of the other piece of the brick immediately after collision (where gravity can be neglected).

(A) CONCEPTUAL REASONING. The conservation of linear momentum can be applied because there is no net external force on the system— the bullet and brick. Initially, all of the momentum is in the forward +x-direction, (▶ Fig. 6.10). Afterward, one piece of the brick flies off in the +x-direction, and the bullet at an angle of 30° to the x-axis. The bullet's momentum has a positive y-component, so the other piece of the brick must have a negative y-component because there was no initial momentum in the y-direction. Hence, with the total momentum in the +x-direction (before and after), the answer is (3) fourth quadrant.

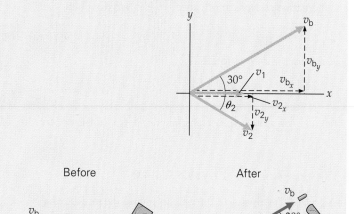

▲ **FIGURE 6.10 A glancing collision** Momentum is conserved in an isolated system. The motion in two dimensions may be analyzed in terms of the components of momentum, which are also conserved.

(B) QUANTITATIVE REASONING AND SOLUTION. There is one object with momentum before collision (the bullet), and three with momenta afterward (the bullet and the two fragments). By the conservation of linear momentum, the total (vector) momentum after collision equals that before collision. As is often the case, a sketch of the situation is helpful, with the vectors resolved in component form (Fig. 6.10). Applying the conservation of linear momentum should allow the velocity (speed and direction) of the second fragment to be determined.

Given: $m_b = 30$ g $= 0.030$ kg   *Find:* v_2 (speed of the smaller brick fragment)
$\qquad v_{b_o} = 400$ m/s (initial bullet speed) $\qquad\theta_2$ (direction of the fragment relative to
$\qquad v_b = 100$ m/s (final bullet speed) $\qquad\qquad$ the original direction of the bullet)
$\qquad \theta_b = 30°$ (final bullet angle)
$\qquad M = 1.0$ kg (brick mass)
$\qquad m_1 = 0.75$ kg and $\theta_1 = 0°$ (mass and angle of
$\qquad\qquad\qquad\qquad$ the large fragment)
$\qquad v_1 = 5.0$ m/s
$\qquad m_2 = 0.25$ kg (mass of small fragment)

With no external forces (gravity neglected), the total linear momentum is conserved. Therefore, both the x- and y-components of the total momentum can be equated before and after (see Fig. 6.10):

$$\qquad\qquad\qquad\text{\textit{before}}\qquad\qquad\qquad\text{\textit{after}}$$
$$x:\quad m_b v_{b_o} = m_b v_b \cos\theta_b + m_1 v_1 + m_2 v_2 \cos\theta_2$$
$$y:\qquad 0 = m_b v_b \sin\theta_b - m_2 v_2 \sin\theta_2$$

The x-equation can be rearranged to solve for the magnitude of the x-velocity of the smaller fragment:

$$v_2 \cos\theta_2 = \frac{m_b v_{b_o} - m_b v_b \cos\theta_b - m_1 v_1}{m_2}$$

$$= \frac{(0.030\text{ kg})(400\text{ m/s}) - (0.030\text{ kg})(100\text{ m/s})(0.886) - (0.75\text{ kg})(5.0\text{ m/s})}{0.25\text{ kg}}$$

$$= 22\text{ m/s}$$

(continued on next page)

Similarly, the y-equation can be solved for the magnitude of the y-velocity component of the smaller fragment:

$$v_2 \sin \theta_2 = \frac{m_b v_b \sin \theta_b}{m_2} = \frac{(0.030 \text{ kg})(100 \text{ m/s})(0.50)}{0.25 \text{ kg}} = 6.0 \text{ m/s}$$

Forming a ratio,

$$\frac{v_2 \sin \theta_2}{v_2 \cos \theta_2} = \frac{6.0 \text{ m/s}}{22 \text{ m/s}} = 0.27 = \tan \theta_2$$

(where the v_2 terms cancel, and $\frac{\sin \theta_2}{\cos \theta_2} = \tan \theta_2$). Then,

$$\theta_2 = \tan^{-1}(0.27) = 15°$$

and from the x-equation,

$$v_2 = \frac{22 \text{ m/s}}{\cos 15°} = \frac{22 \text{ m/s}}{0.97} = 23 \text{ m/s}$$

FOLLOW-UP EXERCISE.   Is the kinetic energy conserved for the collision in this Example? If not, where did the energy go?

EXAMPLE 6.8 | ## Physics on Ice

A physicist is lowered from a helicopter to the middle of a smooth, level, frozen lake, the surface of which has negligible friction, and challenged to make her way off the ice. Walking is out of the question. (Why?) As she stands there pondering her predicament, she decides to use the conservation of momentum by throwing her heavy, identical mittens, which will provide her with the momentum to get herself to shore. To get to the shore more quickly, which should this sly physicist do: throw both mittens at once or throw them separately, one after the other with the same speed?

THINKING IT THROUGH.   The initial momentum of the system (physicist and mittens) is zero. With no net external force, by the conservation of momentum, the total momentum *remains* zero. So if the physicist throws the mittens in one direction, she will go in the opposite direction (because momenta vectors in opposite directions add to zero). Then which way of throwing gives greater speed? If both the mittens were thrown together, the magnitude of their momentum would be

$2mv$, where v is relative to the ice and m is the mass of one mitten.

When thrown separately, the first mitten would have a momentum of mv. The physicist and the second mitten would then be in motion, and throwing the second mitten would add some more momentum to the physicist and increase her speed, but would the speed now be greater than that if both mittens were thrown simultaneously?

Let's analyze the conditions of the second throw. After throwing the first mitten, the physicist "system" would have less mass. With less mass, the second throw would produce a greater acceleration and speed things up. But on the other hand, after the first throw, the second mitten is moving with the person, and when thrown in the opposite direction, the mitten would have a velocity less than v relative to the ice (or to a stationary observer). So which effect would be greater? What do you think? Sometimes situations are difficult to analyze intuitively, and you must apply scientific principles to figure them out.

SOLUTION.

Given: m = mass of single mitten
M = mass of physicist
$-v$ = velocity of thrown mitten(s), in the negative direction
V_P = velocity of physicist in the positive direction

Find: Which method of mitten throwing gives the physicist the greater speed

When the mittens are thrown together, by the conservation of momentum,

$$0 = 2m(-v) + MV_p \quad \text{and} \quad V_p = \frac{2mv}{M} \quad \textit{(thrown together)} \tag{1}$$

When they are thrown separately,

$$\text{First throw:}\quad 0 = m(-v) + (M + m)V_{P_1} \quad \text{and} \quad V_{P_1} = \frac{mv}{M + m} \quad \textit{(thrown separately)} \tag{2}$$

$$\text{Second throw:}\quad (M + m)V_{P_1} = m(V_{P_1} - v) + MV_{P_2}$$

Note that in the last m term of the second throw, the quantities in the parentheses represent the velocity of the mitten relative to the ice. With an initial velocity of $+V_{P_1}$ when the first mitten is thrown in the negative direction, then $V_{P_1} - v$. (Recall relative velocities from Section 3.4.)

Solving for V_{P_2}:

$$V_{P_2} = V_{P_1} + \left(\frac{m}{M}\right)v = \frac{mv}{M + m} + \left(\frac{m}{M}\right)v = \left(\frac{m}{M + m} + \frac{m}{M}\right)v \tag{3}$$

where Eq. 2 was substituted for V_{P_1} after the first throw.

Now, when the mittens are thrown together (Eq. 1),

$$V_P = \left(\frac{2m}{M}\right)v$$

so the question is whether the result of Eq. 3 is greater or less than that of Eq. 1. Notice that with a greater denominator for the $m/(M + m)$ term in Eq. 3 it is less than the m/M term. So,

$$\left(\frac{m}{M + m} + \frac{m}{M}\right) < \frac{2m}{M}$$

and therefore, $V_P > V_{P_2}$, or (thrown together) > (thrown separately).

FOLLOW-UP EXERCISE. Suppose the second throw were in the direction of the physicist's velocity from the first throw. Would this throw bring her to a stop?

As mentioned previously, the conservation of momentum is used to analyze the collisions of objects ranging from subatomic particles to automobiles in traffic accidents. In many instances, however, external forces may be acting on the objects, which means that the momentum is not conserved.

But, as will be learned in the next section, the conservation of momentum often allows a good approximation *over the short time of a collision*, during which the internal forces (which conserve system momentum) are much greater than the external forces. For example, external forces such as gravity and friction also act on colliding objects, but are often relatively small compared with the internal forces of the collision. (This concept was implied in Example 6.7.) Therefore, if the objects interact for only a brief time, the effects of the external forces may be negligible compared with those of the large internal forces during that time and the conservation of linear momentum may be used.

DID YOU LEARN?
➡ If $\vec{F}_{net} = \Delta\vec{P}/\Delta t = 0$, then $\Delta\vec{P} = 0$ and $\vec{P}_{initial} = \vec{P}_{final}$, (conservation of linear momentum).
➡ By Newton's third law, the internal forces on particles are equal and opposite and cancel.

6.4 Elastic and Inelastic Collisions

LEARNING PATH QUESTIONS
➡ What are the conditions for elastic and inelastic collisions in an isolated system?
➡ How much energy is lost in an inelastic collision?

In general, a *collision* may be defined as a meeting or interaction of particles or objects that causes an exchange of energy and/or momentum. Taking a closer look at collisions in terms of the conservation of momentum is simpler for an isolated system, such as a system of particles (or balls) involved in head-on collisions. For simplicity, only collisions in one dimension will be considered, which can be analyzed in terms of the conservation of energy. On the basis of what happens to the total kinetic energy, two types of collisions are defined: *elastic* and *inelastic*.

(a)   (b)

▲ **FIGURE 6.11 Collisions (a)** Approximate elastic collisions. **(b)** An inelastic collision.

In an **elastic collision**, the total kinetic energy is conserved. That is, the *total kinetic energy* of all the objects of the system after the collision is the same as the *total* kinetic energy before the collision (▲Fig. 6.11a). Kinetic energy may be traded between objects of a system, but the total kinetic energy in the system remains constant. That is,

$$\text{total } K \text{ after } = \text{ total } K \text{ before}$$
$$K_f = K_i$$

(condition for an elastic collision) (6.8)

During such a collision, some or all of the initial kinetic energy is temporarily converted to potential energy as the objects are deformed. But after the maximum deformations occur, the objects *elastically* "spring" back to their original shapes, and the system regains all of its original kinetic energy. For example, two steel balls or two billiard balls may have a nearly elastic collision, with each ball having the same shape afterward as before; that is, there is no permanent deformation.

In an **inelastic collision**, total kinetic energy is *not* conserved (Fig. 6.11b). For example, one or more of the colliding objects may not regain the original shapes, and/or sound or frictional heat may be generated and some kinetic energy is lost. Then,

$$\text{total } K \text{ after } < \text{ total } K \text{ before}$$
$$K_f < K_i$$

(condition for an inelastic collision) (6.9)

For example, a hollow aluminum ball that collides with a solid steel ball may be dented. Permanent deformation of the ball takes work, and that work is done at the expense of the original kinetic energy of the system. Everyday collisions are inelastic.

For isolated systems, momentum is conserved in both elastic and inelastic collisions. *For an inelastic collision, only an amount of kinetic energy consistent with the conservation of momentum may be lost.* It may seem strange that kinetic energy can be lost and momentum still conserved, but this fact provides insight into the difference between scalar and vector quantities and the differences in their conservation requirements.

MOMENTUM AND ENERGY IN INELASTIC COLLISIONS

To see how momentum can remain constant while the kinetic energy changes (decreases) in inelastic collisions, consider the examples illustrated in ▶Fig. 6.12. In Fig. 6.12a, two balls of equal mass ($m_1 = m_2$) approach each other with equal

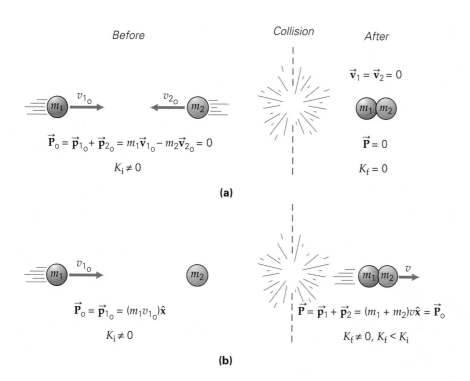

Before   Collision   After

$\vec{\mathbf{v}}_1 = \vec{\mathbf{v}}_2 = 0$

$\vec{\mathbf{P}}_o = \vec{\mathbf{p}}_{1_o} + \vec{\mathbf{p}}_{2_o} = m_1\vec{\mathbf{v}}_{1_o} - m_2\vec{\mathbf{v}}_{2_o} = 0$

$K_i \neq 0$

$\vec{\mathbf{P}} = 0$

$K_f = 0$

(a)

$\vec{\mathbf{P}}_o = \vec{\mathbf{p}}_{1_o} = (m_1 v_{1_o})\hat{\mathbf{x}}$

$K_i \neq 0$

$\vec{\mathbf{P}} = \vec{\mathbf{p}}_1 + \vec{\mathbf{p}}_2 = (m_1 + m_2)v\hat{\mathbf{x}} = \vec{\mathbf{P}}_o$

$K_f \neq 0, K_f < K_i$

(b)

◀ **FIGURE 6.12 Inelastic collisions** In inelastic collisions, momentum is conserved, but kinetic energy is not. Collisions like the ones shown here, in which the objects stick together, are called *completely* or *totally inelastic collisions*. The maximum amount of kinetic energy lost is consistent with the law of conservation of momentum.

and opposite velocities ($v_{1_o} = -v_{2_o}$). Hence, the total momentum before the collision is (vectorially) zero, but the (scalar) total kinetic energy is *not* zero. After the collision, the balls are stuck together and stationary, so the total momentum is unchanged—still zero.

Momentum is conserved because the forces of collision are internal to the system of the two balls—there is no net external force on the system. The total kinetic energy, however, has decreased to zero. In this case, some of the kinetic energy went into the work done in permanently deforming the balls. Some energy may also have gone into doing work against friction (producing heat) or may have been lost in some other way (for example, in producing sound).

It should be noted that the balls need not stick together after collision. In a less inelastic collision, the balls may recoil in opposite directions at reduced, but equal, speeds. The momentum would still be conserved (still equal to zero—why?), but the kinetic energy would again not be conserved. Under all conditions, the amount of kinetic energy that can be lost must be consistent with the conservation of momentum.

In Fig. 6.12b, one ball is initially at rest as the other approaches. The balls stick together after collision, but are still in motion. Both cases in Fig. 6.12 are examples of a **completely inelastic collision**, in which the objects stick together, and hence both objects have the same velocity after colliding. The coupling of colliding railroad cars is a practical example of a completely (or totally) inelastic collision.

Assume that the balls in Fig. 6.12b have different masses. Since the momentum is conserved even in inelastic collisions,

$$\overset{before}{m_1 v_{1_o}} = \overset{after}{(m_1 + m_2)v}$$

and

$$v = \left(\frac{m_1}{m_1 + m_2}\right)v_{1_o} \quad \begin{array}{l}\text{(m_2 initially at rest,}\\ \text{completely inelastic collision only)}\end{array} \quad (6.10)$$

Thus, v is less than v_{1_o}, since $m_1/(m_1 + m_2)$ must be less than 1. Now consider how much kinetic energy has been lost. Initially, $K_i = \frac{1}{2}m_1 v_o^2$, and after collision the final kinetic energy is:

$$K_f = \frac{1}{2}(m_1 + m_2)v^2$$

Substituting for v from Eq. 6.10 and simplifying the result,

$$K_f = \tfrac{1}{2}(m_1 + m_2)\left(\frac{m_1 v_{1_o}}{m_1 + m_2}\right)^2 = \frac{\tfrac{1}{2}m_1^2 v_{1_o}^2}{m_1 + m_2}$$

$$= \left(\frac{m_1}{m_1 + m_2}\right)\tfrac{1}{2}m_1 v_{1_o}^2 = \left(\frac{m_1}{m_1 + m_2}\right)K_i$$

and

$$\frac{K_f}{K_i} = \frac{m_1}{m_1 + m_2} \qquad \begin{array}{l}\text{(}m_2\text{ initially at rest,}\\ \text{completely inelastic collision only)}\end{array} \qquad (6.11)$$

Equation 6.11 gives the fractional amount of the initial kinetic energy that remains with the system after a completely inelastic collision. For example, if the masses of the balls are equal ($m_1 = m_2$), then $m_1/(m_1 + m_2) = \tfrac{1}{2}$, and $K_f/K_i = \tfrac{1}{2}$, or $K_f = K_i/2$. That is, half of the initial kinetic energy is lost.

Note that not all of the kinetic energy can be lost in this case, no matter what the masses of the balls are. The total momentum after collision cannot be zero, since it was not zero initially. Thus, after the collision, the balls must be moving and must have some kinetic energy ($K_f \neq 0$). *In a completely inelastic collision, the maximum amount of kinetic energy lost must be consistent with the conservation of momentum.*

EXAMPLE 6.9 | ## Stuck Together: Completely Inelastic Collision

A 1.0-kg ball with a speed of 4.5 m/s strikes a 2.0-kg stationary ball. If the collision is completely inelastic, (a) what are the speeds of the balls after the collision? (b) What percentage of the initial kinetic energy do the balls have after the collision? (c) What is the total momentum after the collision?

THINKING IT THROUGH.   Recall the definition of a *completely inelastic collision*. The balls stick together after collision; kinetic energy is *not* conserved, but total momentum is.

SOLUTION.   Listing the data:

Given: $m_1 = 1.0$ kg   **Find:** (a) v (speed after collision)
$\quad\quad\quad m_2 = 2.0$ kg $\quad\quad\quad$ (b) $\dfrac{K_f}{K_i}$ ($\times 100\%$)
$\quad\quad\quad v_o = 4.5$ m/s $\quad\quad\quad$ (c) $\vec{P}_f$ (total momentum after collision)

(a) The momentum is conserved and

$$\vec{P}_f = \vec{P}_o \qquad \text{or} \qquad (m_1 + m_2)v = m_1 v_o$$

The balls stick together and have the same speed after collision. This speed is then

$$v = \left(\frac{m_1}{m_1 + m_2}\right)v_o = \left(\frac{1.0\text{ kg}}{1.0\text{ kg} + 2.0\text{ kg}}\right)(4.5\text{ m/s}) = 1.5\text{ m/s}$$

(b) The fractional part of the initial kinetic energy that the balls have after the completely inelastic collision is given by Eq. 6.11. Notice that this fraction, as given by the masses, is the same as that for the speeds (Eq. 6.11) in this special case. By inspection,

$$\frac{K_f}{K_i} = \frac{m_1}{m_1 + m_2} = \frac{1.0\text{ kg}}{1.0\text{ kg} + 2.0\text{ kg}} = \frac{1}{3} = 0.33\ (\times 100\%) = 33\%$$

Let's show this relationship explicitly:

$$\frac{K_f}{K_i} = \frac{\tfrac{1}{2}(m_1 + m_2)v^2}{\tfrac{1}{2}m_1 v_0^2} = \frac{\tfrac{1}{2}(1.0\text{ kg} + 2.0\text{ kg})(1.5\text{ m/s})^2}{\tfrac{1}{2}(1.0\text{ kg})(4.5\text{ m/s})^2} = 0.33\ (= 33\%)$$

Keep in mind that Eq. 6.11 applies *only* to *completely* inelastic collisions in which m_2 is initially at rest. For other types of collisions, the initial and final values of the kinetic energy must be computed explicitly.

(c) The total momentum is conserved in all collisions (in the absence of external forces), so the total momentum after collision is the same as before collision. That value is the momentum of the incident ball, with a magnitude of

$$P_f = p_{1_o} = m_1 v_o = (1.0 \text{ kg})(4.5 \text{ m/s}) = 4.5 \text{ kg} \cdot \text{m/s}$$

and the same direction as that of the incoming ball. Also, as a double check,

$$P_f = (m_1 + m_2)v = 4.5 \text{ kg} \cdot \text{m/s}$$

FOLLOW-UP EXERCISE. A small hard-metal ball of mass m collides with a larger, stationary, soft-metal ball of mass M. A *minimum* amount of work W is required to make a dent in the larger ball. If the smaller ball initially has kinetic energy $K = W$, will the larger ball be dented in a completely inelastic collision between the two balls?

MOMENTUM AND ENERGY IN ELASTIC COLLISIONS

For *elastic* collisions, there are two conservation criteria: conservation of momentum (which holds for both elastic and inelastic collisions) and conservation of kinetic energy (for elastic collisions only). That is, for the elastic collision of two objects:

	before	*after*	

Conservation of momentum $\vec{\mathbf{P}}$: $m_1\vec{\mathbf{v}}_{1_o} + m_2\vec{\mathbf{v}}_{2_o} = m_1\vec{\mathbf{v}}_1 + m_2\vec{\mathbf{v}}_2$ (6.12)

Conservation of kinetic energy K: $\frac{1}{2}m_1 v_{1_o}^2 + \frac{1}{2}m_2 v_{2_o}^2 = \frac{1}{2}m_1 v_1^2 + \frac{1}{2}m_2 v_2^2$ (6.13)

▼Figure 6.13 illustrates two objects traveling prior to a one-dimensional, head-on collision with $v_{1_o} > v_{2_o}$ (both in the positive x-direction). For this two-object situation,

$$\text{Total momentum:} \quad m_1 v_{1_o} + m_2 v_{2_o} = m_1 v_1 + m_2 v_2 \tag{1}$$

(where signs are used to indicate directions and the v's indicate magnitudes).

$$\text{Kinetic energy:} \quad \frac{1}{2}m_1 v_{1_o}^2 + \frac{1}{2}m_2 v_{2_o}^2 = \frac{1}{2}m_1 v_1^2 + \frac{1}{2}m_2 v_2^2 \tag{2}$$

If the masses and the initial velocities of the objects are known (which they usually are), then there are two unknown quantities, the final velocities after collision. To find them, equations (1) and (2) are solved simultaneously. First the equation for momentum conservation is written as follows:

$$m_1(v_{1_o} - v_1) = -m_2(v_{2_o} - v_2) \tag{3}$$

Then, canceling the 1/2 terms in (2), rearranging, and factoring $[a^2 - b^2 = (a - b)(a + b)]$:

$$m_1(v_{1_o} - v_1)(v_{1_o} + v_1) = -m_2(v_{2_o} - v_2)(v_{2_o} + v_2) \tag{4}$$

Dividing equation (4) by (3) and rearranging yields

$$v_{1_o} - v_{2_o} = -(v_1 - v_2) \tag{5}$$

This equation shows that the magnitudes of the relative velocities before and after collision are equal. That is, the relative speed of approach of object m_1 to object m_2 before collision is the same as their relative speed of separation after collision. (See Section 3.4.) Notice that this relation is independent of the values of the masses of the objects, and holds for any mass combination as long as the collision is elastic and *one-dimensional*.

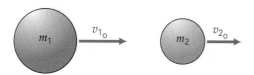

◀**FIGURE 6.13 Elastic collision coming up** Two objects traveling prior to collision with $v_{1_o} > v_{2_o}$. See text for description.

Then, combining equation (5) with (3) to eliminate v_2 and get v_1 in terms of the two initial velocities,

$$v_1 = \left(\frac{m_1 - m_2}{m_1 + m_2}\right)v_{1_o} + \left(\frac{2m_2}{m_1 + m_2}\right)v_{2_o} \tag{6.14}$$

Similarly, eliminating v_1 to find v_2,

$$v_2 = \left(\frac{2m_1}{m_1 + m_2}\right)v_{1_o} - \left(\frac{m_1 - m_2}{m_1 + m_2}\right)v_{2_o} \tag{6.15}$$

ONE OBJECT INITIALLY AT REST

For this common, special case, say with $v_{2_o} = 0$, there are only the first terms in Eqs. 6.14 and 6.15. In addition, if $m_1 = m_2$, then $v_1 = 0$ and $v_2 = v_{1_o}$. That is, the objects *completely* exchange momentum and kinetic energy. The incoming object is stopped on collision, and the originally stationary object moves off with the same velocity as the incoming ball, obviously conserving the system's momentum and kinetic energy. (A real-world example that comes close to these conditions is the head-on collision of billiard balls.)

You can also get some approximates for special cases from the equations for one object initially at rest (taken to be m_2):

For $m_1 \gg m_2$ (massive incoming ball): $v_1 \approx v_{1_o}$ and $v_2 \approx 2v_{1_o}$

That is, the massive incoming object is slowed down only slightly and the light (less massive) object is knocked away with a velocity almost twice that of the initial velocity of the massive object. (Think of a bowling ball hitting a pin.)

For $m_1 \ll m_2$ (light incoming ball): $v_1 \approx -v_{1_o}$ and $v_2 \approx 0$

That is, if a light (small mass) object elastically collides with a massive stationary one, the massive object remains *almost* stationary and the light object recoils backward with approximately the same speed that it had before collision.

EXAMPLE 6.10 | ## Elastic Collision: Conservation of Momentum and Kinetic Energy

A 0.30-kg ball with a speed of 2.0 m/s in the positive x-direction has a head-on elastic collision with a stationary 0.70-kg ball. What are the velocities of the balls after collision?

THINKING IT THROUGH.   The incoming ball is less massive than the stationary one, so it might expected that the objects separate in opposite directions after collision, with the less massive ball recoiling from the more massive one. Equations 6.14 and 6.15 can be used to find the velocities with $v_{2_o} = 0$.

SOLUTION.   Using the previous notation in listing the data,

Given: $m_1 = 0.30$ kg and $v_{1_o} = 2.0$ m/s   **Find:** v_1 and v_2
$m_2 = 0.70$ kg and $v_{2_o} = 0$

Directly from Eqs. 6.13 and Eq. 6.14, the velocities after collision are

$$v_1 = \left(\frac{m_1 - m_2}{m_1 + m_2}\right)v_{1_o} = \left(\frac{0.30 \text{ kg} - 0.70 \text{ kg}}{0.30 \text{ kg} + 0.70 \text{ kg}}\right)(2.0 \text{ m/s}) = -0.80 \text{ m/s}$$

$$v_2 = \left(\frac{2m_1}{m_1 + m_2}\right)v_{1_o} = \left[\frac{2(0.30 \text{ kg})}{0.30 \text{ kg} + 0.70 \text{ kg}}\right](2.0 \text{ m/s}) = 1.2 \text{ m/s}$$

FOLLOW-UP EXERCISE.   What would be the separation distance of the two objects 2.5 s after collision?

TWO COLLIDING OBJECTS, BOTH INITIALLY MOVING

Now let's look at some examples where both terms in Eqs. 6.14 and 6.15 are needed.

EXAMPLE 6.11 | Collisions: Overtaking and Coming Together

The precollision conditions for two elastic collisions are shown in ▸ Fig. 6.14. What are the final velocities in each case?

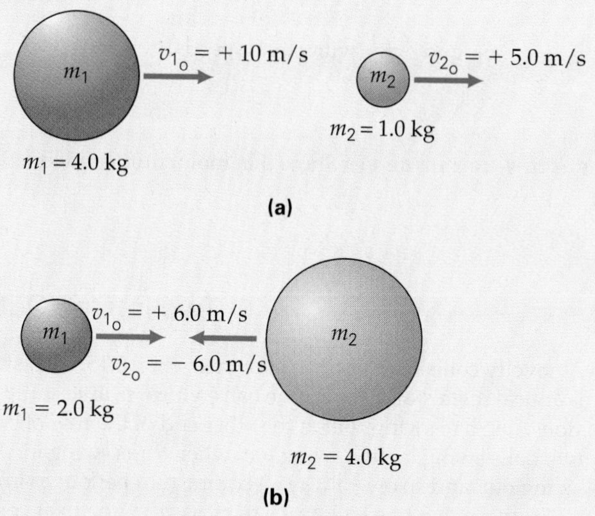

THINKING IT THROUGH. These collisions are direct applications of Eqs. 6.14 and 6.15. Notice that in (a) the 4.0-kg object will overtake and collide with the 1.0-kg object.

SOLUTION. Listing the data from the figure with the $+x$-direction taken to the right:

Given: (a) $m_1 = 4.0$ kg $\quad v_{1_0} = 10$ m/s $\qquad$ **Find:** v_1 and v_2
$\qquad m_2 = 1.0$ kg $\quad v_{2_0} = 5.0$ m/s $\qquad\qquad$ (velocities
$\qquad$ (b) $m_1 = 2.0$ kg $\quad v_{1_0} = 6.0$ m/s $\qquad\qquad$ after
$\qquad m_2 = 4.0$ kg $\quad v_{2_0} = -6.0$ m/s $\qquad\qquad$ collision)

Then, substituting into the collision equations,

(a) Eq. 6.14:

$$v_1 = \left(\frac{m_1 - m_2}{m_1 + m_2}\right)v_{1_0} + \left(\frac{2m_2}{m_1 + m_2}\right)v_{2_0}$$

$$= \left(\frac{4.0 \text{ kg} - 1.0 \text{ kg}}{4.0 \text{ kg} + 1.0 \text{ kg}}\right)10 \text{ m/s} + \left(\frac{2[1.0 \text{ kg}]}{4.0 \text{ kg} + 1.0 \text{ kg}}\right)5.0 \text{ m/s}$$

$$= \tfrac{3}{5}(10 \text{ m/s}) + \tfrac{2}{5}(5.0 \text{ m/s}) = 8.0 \text{ m/s}$$

Similarly, Eq. 6.15 gives:

$$v_2 = 13 \text{ m/s}$$

So the more massive object overtakes and collides with the less massive object, transferring momentum (increasing velocity).

(b) Applying the collision equations for this situation, (Eq. 6.14):

$$v_1 = \left(\frac{2.0 \text{ kg} - 4.0 \text{ kg}}{2.0 \text{ kg} + 4.0 \text{ kg}}\right)6.0 \text{ m/s} + \left(\frac{2[4.0 \text{ kg}]}{2.0 \text{ kg} + 4.0 \text{ kg}}\right)(-6.0 \text{ m/s})$$

$$= -\left(\tfrac{1}{3}\right)6.0 \text{ m/s} + \left(\tfrac{4}{3}\right)(-6.0 \text{ m/s}) = -10 \text{ m/s}$$

Similarly, Eq. 6.15 gives

$$v_2 = 2.0 \text{ m/s}$$

▲ FIGURE 6.14 Collisions (a) Overtaking and **(b)** coming together. See Example text for description.

Here, the less massive object goes in the opposite (negative) direction after collision, with a greater momentum obtained from the more massive object.

FOLLOW-UP EXERCISE. Show that in parts (a) and (b) of this Example, the amount of momentum gained by one object is the same as that lost by the other.

INTEGRATED EXAMPLE 6.12 | Equal and Opposite

Two balls of equal mass with equal but opposite velocities approach each other for a head-on elastic collision. (a) After collision, the balls will (1) move off stuck together, (2) both be at rest, (3) move off in the same direction, or (4) recoil in opposite directions. (b) Prove your answer explicitly.

(A) CONCEPTUAL REASONING. Make a sketch of the situation. Then, looking at the choices, (1) is eliminated because if they stuck together it would be an inelastic collision. If they come to rest after collision, momentum would be conserved

(why?), but not kinetic energy, so (2) is not applicable for an elastic collision. If they both moved off in the same direction after collision, (3), the momentum would not be conserved (zero before, nonzero after). The answer must be (4). This is the only option by which momentum and kinetic energy could be conserved. To maintain the zero momentum before collision, the objects would have to recoil in opposite directions with the same speeds as before collision.

(continued on next page)

(B) QUANTITATIVE REASONING AND SOLUTION. To explicitly show that (4) is correct, Eqs. 6.13 and 6.14 may be used. Since no numerical values are given, we work with symbols.

Given: $m_1 = m_2 = m$ (taking m_1 to be initially traveling in the $+x$ direction:)

v_{1_o} and $-v_{2_o}$ (with equal speeds)

Find: v_1 and v_2

Then, substituting into Eqs. 6.14 and 6.15, without writing the equations out [see part (a) in Example 6.11],

$$v_1 = \left(\frac{0}{2m}\right)v_{1_o} + \left(\frac{2m}{2m}\right)(-v_{2_o}) = -v_{2_o}$$

and

$$v_2 = \left(\frac{2m}{2m}\right)v_{1_o} + \left(\frac{0}{2m}\right)(-v_{2_o}) = v_{1_o}$$

From the results, it can be seen that after collision the balls recoil in opposite directions.

FOLLOW-UP EXERCISE.   Show that momentum and kinetic energy are conserved in this Example.

CONCEPTUAL EXAMPLE 6.13 | Two In, One Out?

A novelty collision device, as shown in ▸ Fig. 6.15, consists of five identical metal balls. When one ball swings in, after multiple collisions, one ball swings out at the other end of the row of balls. When two balls swing in, two swing out; when three swing in, three swing out, and so on—always the same number out as in.

Suppose that two balls, each of mass m, swing in at velocity v and collide with the next ball. Why doesn't one ball swing out at the other end with a velocity $2v$?

REASONING AND ANSWER.   The collisions along the horizontal row of balls are approximately elastic. The case of two balls swinging in and one ball swinging out with twice the velocity wouldn't violate the conservation of momentum: $(2m)v = m(2v)$. However, another condition applies if we assume elastic collisions—the conservation of kinetic energy. Let's check to see if this condition is upheld for this case:

$$\begin{array}{cc} before & after \\ K_i & = & K_f \\ \frac{1}{2}(2m)v^2 & \overset{?}{=} & \frac{1}{2}m(2v)^2 \\ mv^2 & \neq & 2mv^2 \end{array}$$

Hence, the kinetic energy would *not* be conserved if this happened, and the equation is telling us that this situation violates established physical principles and does not occur. Note that there's a big violation—more energy out than in.

FOLLOW-UP EXERCISE.   Suppose the first ball of mass m were replaced with a ball of mass $2m$. When this ball is pulled back and allowed to swing in, how many balls will swing out? [*Hint*: Think about the analogous situation for the first two balls as in Fig. 6.14a, and remember that the balls in the row are actually colliding. It may help to think of them as being separated.]

▲ **FIGURE 6.15 One in, one out** See Example text for description.

6.5 Center of Mass

LEARNING PATH QUESTIONS

➥ The center of mass concept applies to what type(s) of motion?

➥ How are the center of mass and the center of gravity related?

The conservation of total momentum gives a method of analyzing a "system of particles." Such a system may be virtually anything—for example, a volume of gas, water in a container, or a baseball. Another important concept, the center of mass, allows us to analyze the overall motion of a system of particles. It involves representing the whole system as a single particle or point mass. This concept will be introduced here and applied in more detail in the upcoming chapters.

It has been seen that if no net external force acts on a particle, the particle's linear momentum is constant. Similarly, if no net external force acts on a *system* of particles, the linear momentum of the *system* is constant. This similarity implies that a system of particles might be represented by an *equivalent* single particle. Moving rigid objects, such as balls, automobiles, and so forth, are essentially systems of particles and can be effectively represented by equivalent single particles when analyzing motion. Such representation is done through the concept of the **center of mass (CM)**:

> The center of mass is the point at which all of the mass of an object or system may be considered to be concentrated, for the purposes of describing its linear or translational motion only.

Even if a rigid object is rotating, an important result (beyond the scope of this text to derive) is that the center of mass still moves as though it were a particle (▼Fig. 6.16). The center of mass is sometimes described as the *balance point* of a solid object. For example, if you balance a meterstick on your finger, the center of mass of the stick is located directly above your finger, and all of the mass (or weight) seems to be concentrated there.

An expression similar to Newton's second law for a single particle applies to a *system* when the center of mass is used:

$$\vec{F}_{net} = M\vec{A}_{CM} \tag{6.16}$$

Here, $\vec{F}_{net}$ is the net *external* force on the system, M is the total mass of the system or the sum of the masses of the particles of the system ($M = m_1 + m_2 + m_3 + \cdots + m_n$, where the system has n particles), and $\vec{A}_{CM}$ is the acceleration of the center of mass of the system. In words, Eq. 6.16 says that the *center of mass* of a system of particles moves as though all the mass of the system were concentrated there and acted on by the resultant of the external forces. Note that the movement of the individual parts of the system is *not* predicted by Eq. 6.16.

It follows that *if the net external force on a system is zero*, the total linear momentum of the center of mass is conserved (that is, it stays constant), because

$$\vec{F}_{net} = M\vec{A}_{CM} = M\left(\frac{\Delta\vec{V}_{CM}}{\Delta t}\right) = \frac{\Delta(M\vec{V}_{CM})}{\Delta t} = \frac{\Delta\vec{P}_{CM}}{\Delta t} = 0 \tag{6.17}$$

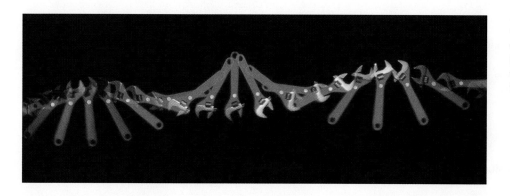

◀ **FIGURE 6.16 Center of mass** The center of mass of this sliding wrench moves in a straight line as though it were a particle. Note the white dot on the wrench that marks the center of mass.

▶ **FIGURE 6.17 System of parti-cles in one dimension** Where is the system's center of mass? See Example 6.14.

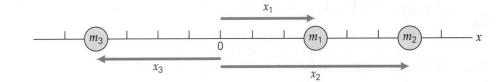

Then, $\Delta\vec{P}/\Delta t = 0$, which means that there is no change in $\vec{P}$ during a time Δt, or the total momentum of the system, $\vec{P} = M\vec{V}_{CM}$ is constant (but not necessarily zero). Since M is constant (why?), $\vec{V}_{CM}$ is a constant in this case. Thus, the center of mass either moves with a constant velocity or remains at rest.

Although you may more readily visualize the center of mass of a solid object, the concept of the center of mass applies to any system of particles or objects, even a quantity of gas. For a system of n particles arranged in one dimension, along the x-axis (▲Fig. 6.17), the location of the center of mass is given by

$$\vec{X}_{CM} = \frac{m_1\vec{x}_1 + m_2\vec{x}_2 + m_3\vec{x}_3 + \cdots + m_n\vec{x}_n}{m_1 + m_2 + m_3 + \cdots + m_n} \tag{6.18}$$

That is, X_{CM} is the x-coordinate of the center of mass (CM) of a system of particles. In shorthand notation (using signs to indicate vector directions in one dimension), this relationship is expressed as

$$X_{CM} = \frac{\sum m_i x_i}{M} \tag{6.19}$$

where $\sum$ is the summation of the products $m_i x_i$ for n particles ($i = 1, 2, 3, \ldots, n$). If $\sum m_i x_i = 0$, then $X_{CM} = 0$, and the center of mass of the one-dimensional system is located at the origin.

Other coordinates of the center of mass for systems of particles are similarly defined. For a two-dimensional distribution of masses, the coordinates of the center of mass are (X_{CM}, Y_{CM}).

EXAMPLE 6.14 | Finding the Center of Mass: A Summation Process

Three masses—2.0 kg, 3.0 kg, and 6.0 kg—are located at positions (3.0, 0), (6.0, 0), and (−4.0, 0), respectively, in meters from the origin (Fig. 6.17). Where is the center of mass of this system?

THINKING IT THROUGH. Since all $y_i = 0$, obviously $Y_{CM} = 0$, and the CM lies somewhere on the x-axis. The masses and the positions are given, so we can use Eq. 6.19 to calculate X_{CM} directly. However, keep in mind that the positions are located by vector displacements from the origin and are indicated in one dimension by the appropriate signs ($+$ or $−$).

SOLUTION. Listing the data,

Given: $m_1 = 2.0$ kg   *Find:* X_{CM} (CM coordinate)
        $m_2 = 3.0$ kg
        $m_3 = 6.0$ kg
        $x_1 = 3.0$ m
        $x_2 = 6.0$ m
        $x_3 = -4.0$ m

Then, performing the summation as indicated in Eq. 6.19 yields,

$$X_{CM} = \frac{\sum m_i x_i}{M}$$

$$= \frac{(2.0 \text{ kg})(3.0 \text{ m}) + (3.0 \text{ kg})(6.0 \text{ m}) + (6.0 \text{ kg})(-4.0 \text{ m})}{2.0 \text{ kg} + 3.0 \text{ kg} + 6.0 \text{ kg}} = 0$$

The center of mass is at the origin.

FOLLOW-UP EXERCISE. At what position should a fourth mass of 8.0 kg be added so the CM is at $x = +1.0$ m?

EXAMPLE 6.15 | A Dumbbell: Center of Mass Revisited

A dumbbell (▼Fig. 6.18) has a connecting bar of negligible mass. Find the location of the center of mass (a) if m_1 and m_2 are each 5.0 kg and (b) if m_1 is 5.0 kg and m_2 is 10.0 kg.

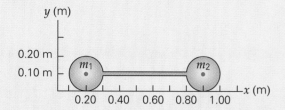

▲ FIGURE 6.18 **Location of the center of mass** See Example text for description.

THINKING IT THROUGH. This Example shows how the location of the center of mass depends on the distribution of mass. In part (b), you might expect the center of mass to be located closer to the more massive end of the dumbbell.

SOLUTION. Listing the data, with the coordinates to be used in Eq. 6.19,

Given: $x_1 = 0.20$ m
$x_2 = 0.90$ m
$y_1 = y_2 = 0.10$ m
(a) $m_1 = m_2 = 5.0$ kg
(b) $m_1 = 5.0$ kg
$m_2 = 10.0$ kg

Find: (a) (X_{CM}, Y_{CM}) (CM coordinates), with $m_1 = m_2$
(b) (X_{CM}, Y_{CM}), with $m_1 \neq m_2$

Note that each mass is considered to be a particle located at the center of the sphere (its center of mass).

(a) X_{CM} is given by a two-term sum.

$$X_{CM} = \frac{m_1 x_1 + m_2 x_2}{m_1 + m_2}$$

$$= \frac{(5.0\text{ kg})(0.20\text{ m}) + (5.0\text{ kg})(0.90\text{ m})}{5.0\text{ kg} + 5.0\text{ kg}} = 0.55\text{ m}$$

Similarly, $Y_{CM} = 0.10$ m, as you can prove for yourself. (You might have seen this right away, since each center of mass is at this height.) The center of mass of the dumbbell is then located at $(X_{CM}, Y_{CM}) = (0.55\text{ m}, 0.10\text{ m})$, or midway between the end masses.

(b) With $m_2 = 10.0$ kg,

$$X_{CM} = \frac{m_1 x_1 + m_2 x_2}{m_1 + m_2}$$

$$= \frac{(5.0\text{ kg})(0.20\text{ m}) + (10.0\text{ kg})(0.90\text{ m})}{5.0\text{ kg} + 10.0\text{ kg}} = 0.67\text{ m}$$

which is two-thirds of the way between the masses. (Note that the distance of the CM from the center of m_1 is $\Delta x = 0.67$ m $- 0.20$ m $= 0.47$ m. With the distance $L = 0.70$ m between the centers of the masses, $\Delta x/L = 0.47$ m/0.70 m $= 0.67$, or $\frac{2}{3}$.) You might expect the balance point of the dumbbell in this case to be closer to m_2 and it is. The y-coordinate of the center of mass is again $Y_{CM} = 0.10$ m.

FOLLOW-UP EXERCISE. In part (b) of this Example, take the origin of the coordinate axes to be at the point where m_1 touches the x-axis. What are the coordinates of the CM in this case, and how does its location compare with that found in the Example?

In Example 6.15, when the value of one of the masses changed, the x-coordinate of the center of mass changed. However, the centers of the end masses were still at the same height, and Y_{CM} remained the same. To increase Y_{CM}, one or both of the end masses would have to be in a higher position.

Now let's see how the concept of the center of mass can be applied to a realistic situation.

INTEGRATED EXAMPLE 6.16 | Internal Motion: Where's the Center of Mass and the Man?

A 75.0-kg man stands in the far end of a 50.0-kg boat 100 m from the shore, as illustrated in ▶Fig. 6.19. If he walks to the other end of the 6.00-m-long boat, (a) does the CM (1) move to the right, (2) move to the left, or (3) remain stationary? Neglect friction and assume the CM of the boat is at its midpoint. (b) After walking to the other end of the boat, how far is he from the shore?

(A) CONCEPTUAL REASONING. With no net external force, the acceleration of the center of mass of the man–boat system is zero (Eq. 6.18), and so is the total momentum by Eq. 6.17 $(\vec{P} = M\vec{V}_{CM} = 0)$. Hence, the velocity of the center of mass of the system is zero, or the center of mass is stationary and remains so to conserve system momentum; that is, X_{CM_i} (initial) $= X_{CM_f}$ (final), so the answer is (3).

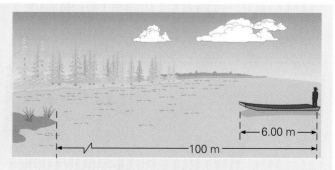

▲ FIGURE 6.19 **Walking toward shore** See Example text for description.

(continued on next page)

(B) QUANTITATIVE REASONING AND SOLUTION. The answer is *not* 100 m − 6.00 m = 94.0 m , because the boat moves as the man walks. Why? The positions of the masses of the man and the boat determine the location of the CM of the system, both before and after the man walks. Since the CM does not move, $X_{CM_i} = X_{CM_f}$. Using this fact and finding the value of X_{CM_i}, this value can be used in the calculation of X_{CM_f}, which will contain the unknown we are looking for.

Taking the shore as the origin ($x = 0$),

Given: $m_m = 75.0$ kg
$x_{m_i} = 100$ m
$m_b = 50.0$ kg
$x_{b_i} = 94.0$ m $+ 3.00$ m $= 97.0$ m
  (CM position of the boat)

Find: x_{m_f} (distance of man from shore)

Note that if we take the man's final position to be a distance x_{m_f} from the shore, then the final position of the boat's center of mass will be $x_{b_f} = x_{m_f} + 3.00$ m, since the man will be at the front of the boat, 3.00 m from its CM, but on the other side. Then initially,

$$X_{CM_i} = \frac{m_m x_{m_i} + m_b x_{b_i}}{m_m + m_b}$$

$$= \frac{(75.0 \text{ kg})(100 \text{ m}) + (50.0 \text{ kg})(97.0 \text{ m})}{75.0 \text{ kg} + 50.0 \text{ kg}} = 98.8 \text{ m}$$

Finally, the CM must be at the same location, since $V_{CM} = 0$. Then from Eq. 6.19,

$$X_{CM_f} = \frac{m_m x_{m_f} + m_b x_{b_f}}{m_m + m_b}$$

$$= \frac{(75.0 \text{ kg})x_{m_f} + (50.0 \text{ kg})(x_{m_f} + 3.00 \text{ m})}{75.0 \text{ kg} + 50.0 \text{ kg}} = 98.8 \text{ m}$$

Here, $X_{CM_f} = 98.8$ m $= X_{CM_i}$, since the CM does not move. Then, solving for x_{m_f},

$$(125 \text{ kg})(98.8 \text{ m}) = (125 \text{ kg})x_{m_f} + (50.0 \text{ kg})(3.00 \text{ m})$$

and

$$x_{m_f} = 97.6 \text{ m}$$

from the shore.

FOLLOW-UP EXERCISE.   Suppose the man then walks back to his original position at the opposite end of the boat. Would he then be 100 m from shore again?

CENTER OF GRAVITY

As you know, mass and weight are related. Closely associated with the concept of the center of mass is the concept of the **center of gravity (CG)**, the point where all of the *weight* of an object may be considered to be concentrated when the object is represented as a particle. If the acceleration due to gravity is constant in both magnitude and direction over the extent of the object, Eq. 6.20 can be rewritten as (with all $g_i = g$),

$$MgX_{CM} = \Sigma m_i g x_i \tag{6.20}$$

Then, the object's weight Mg acts as though its mass were concentrated at X_{CM}, and the center of mass and the center of gravity coincide. As you may have noticed, the location of the center of gravity was implied in some previous figures in Chapter 4, where the vector arrows for weight ($w = mg$) were drawn from a point at or near the center of an object.

For practical purposes, the center of gravity is usually considered to coincide with the center of mass. That is, the acceleration due to gravity is constant for all parts of the object. (Note the constant g in Eq. 6.20.) There would be a difference in the locations of the two points if an object were so large that the acceleration due to gravity was different at different parts of the object.

In some cases, the center of mass or the center of gravity of an object may be located by symmetry. For example, for a spherical object that is homogeneous (that is, the mass is distributed evenly throughout), the center of mass is at the geometrical center (or center of symmetry). In Example 6.15a, where the end masses of the dumbbell were equal, it was probably apparent that the center of mass was midway between them.

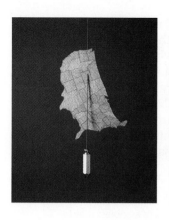  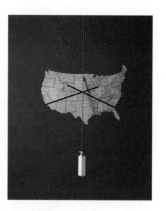

(b)

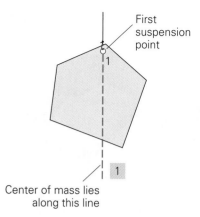

Center of mass lies
along this line

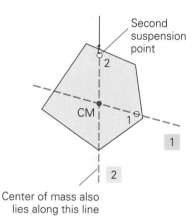

Center of mass also
lies along this line

(a)

▲ **FIGURE 6.20 Location of the center of mass by suspension** **(a, right)** The center of mass of a flat, irregularly shaped object can be found by suspending the object from two or more points. The CM (and CG) lies on a vertical line under any point of suspension, so the intersection of two such lines marks its location midway through the thickness of the body. The sheet could be balanced horizontally at this point. Why? **(b, above)** The process is illustrated with a cutout map of the United States. Note that a plumb line dropped from any other point (third photo) does in fact pass through the CM as located in the first two photos.

The location of the center of mass or center of gravity of an irregularly shaped object is not so evident and is usually difficult to calculate (even with advanced mathematical methods that are beyond the scope of this book). In some instances, the center of mass may be located experimentally. For example, the center of mass of a flat, irregularly shaped object can be determined experimentally by suspending it freely from different points (▲Fig. 6.20). A moment's thought should convince you that the center of mass (or center of gravity) always lies vertically below the point of suspension. Since the center of mass is defined as the point at which all the mass of a body can be considered to be concentrated, this is analogous to a particle of mass suspended from a string. Suspending the object from two or more points and marking the vertical lines on which the center of mass must lie locates the center of mass as the intersection of the lines.

The center of mass (or center of gravity) of an object may lie outside the body of the object (▼Fig. 6.21). For example, the center of mass of a homogeneous ring is at the ring's center. The mass in any section of the ring is compensated for by the mass in an equivalent section directly across the ring, and by symmetry, the center of mass is at the center of the ring. For an L-shaped object with uniform legs, equal in mass and length, the center of mass lies on a line that makes a 45° angle with both legs. Its location can easily be determined by suspending the L from a point on one of the legs and noting where a vertical line from that point intersects the diagonal line.

(a)

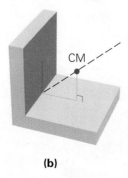

(b)

▶ **FIGURE 6.21 The center of mass may be located outside a body** The center of mass (and center of gravity) may lie either inside or outside a body, depending on the distribution of that object's mass. **(a)** For a uniform ring, the center of mass is at the center of the ring. **(b)** For an L-shaped object, if the mass distribution is uniform and the legs are of equal length, the center of mass lies on the diagonal between the legs.

In the high jump, the location of center of gravity (CG) is very important. Jumping raises the CG. It takes energy to do this, and the higher the jump, the more energy it takes. Therefore, a high jumper wants to clear the bar while keeping his CG low. A jumper will try to keep his CG as close to the bar as possible when passing over it. In the "Fosbury flop" style, made famous by Dick Fosbury in the 1968 Olympics, the jumper arches his body backward over the bar (◄Fig. 6.22). With the legs, head, and arms below the bar, the CG is lower than in the "layout" style, where the body is nearly parallel to the ground when going over the bar. With the "flop," a jumper may be able to make his CG (which is outside the body) pass underneath the bar while successfully clearing the bar.

▲ FIGURE 6.22 **Center of gravity** By arching his body backward over the bar, the high jumper lowers his center of gravity. See text for description.

DID YOU LEARN?
➥ For the center of mass, or the point at which all the mass of an object may be considered concentrated, only linear or translational motion applies (including at rest).
➥ The point at which all the weight of an object may be considered concentrated (the center of gravity) coincides with the center of mass when the acceleration due to gravity is constant.

6.6 Jet Propulsion and Rockets

LEARNING PATH QUESTIONS
➥ How is jet propulsion explained by the conservation of momentum?
➥ What is meant by "reverse thrust"?

The word *jet* is sometimes used to refer to a stream of liquid or gas emitted at a high speed—for example, a jet of water from a fountain or a jet of air from an automobile tire. **Jet propulsion** is the application of such jets to the production of motion. This concept usually brings to mind jet planes and rockets, but squid and octopi propel themselves by squirting jets of water (▼Fig. 6.23).

You have probably tried the simple application of blowing up a balloon and releasing it. Lacking any guidance or rigid exhaust system, the balloon zigzags around, driven by the escaping air. In terms of Newton's third law, the air is forced out by the contraction of the stretched balloon—that is, the balloon exerts a force on the air. Thus, there must be an equal and opposite reaction force exerted by the air on the balloon. It is this force that propels the balloon on its erratic path.

Jet propulsion is explained by Newton's third law, and in the absence of external forces, the conservation of momentum also applies. You may understand this concept better by considering the recoil of a rifle, taking the rifle and the bullet as an isolated system (►Fig. 6.24).

► FIGURE 6.23 **Jet propulsion** Squid and octopi propel themselves by squirting jets of water. Shown here is a Giant Octopus jetting away.

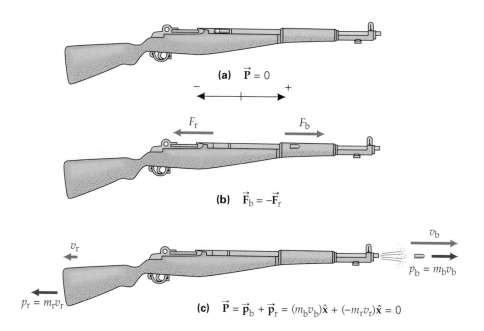

(a) $\vec{P} = 0$

(b) $\vec{F}_b = -\vec{F}_r$

(c) $\vec{P} = \vec{P}_b + \vec{P}_r = (m_b v_b)\hat{x} + (-m_r v_r)\hat{x} = 0$

◀ **FIGURE 6.24 Conservation of momentum** **(a)** Before the rifle is fired, the total momentum of the rifle and bullet (as an isolated system) is zero. **(b)** During firing, there are equal and opposite internal forces, and the instantaneous total momentum of the rifle–bullet system remains zero (neglecting external forces, such as those that arise when a rifle is being held). **(c)** When the bullet leaves the barrel, the total momentum of the system is still zero. [The vector equation is written in boldface (vector) notation and then in sign–magnitude notation so as to indicate directions.]

Initially, the total momentum of this system is zero. When the rifle is fired (by remote control to avoid external forces), the expansion of the gases from the exploding charge accelerates the bullet down the barrel. These gases push backward on the rifle as well, producing a recoil force (the "kick" experienced by a person firing a weapon). Since the initial momentum of the system is zero and the force of the expanding gas is an internal force, the momenta of the bullet and of the rifle must be equal and opposite at any instant. After the bullet leaves the barrel, there is no propelling force, so the bullet and the rifle move with constant velocities (unless acted on by a net external force such as gravity or air resistance).

Similarly, the thrust of a rocket is created by exhausting the gas from burning fuel out the rear of the rocket. The expanding gas exerts a force on the rocket that propels the rocket in the forward direction (▼Fig. 6.25). The rocket exerts a reaction force on the gas, so the gas is directed out the exhaust nozzle. If the rocket is at rest when the engines are turned on and there are no external forces (as in deep space, where friction is zero and gravitational forces are negligible), then the instantaneous momentum of the exhaust gas is equal and opposite to that of the rocket. The numerous exhaust gas molecules have small masses and high velocities, and the rocket has a much larger mass and a smaller velocity.

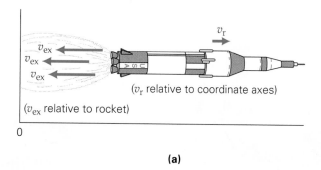

(a)

(b)

(c)

▲ **FIGURE 6.25 Jet propulsion and mass reduction** **(a)** A rocket burning fuel is continuously losing mass and thus becomes easier to accelerate. The resulting force on the rocket (the thrust) depends on the product of the rate of change of its mass with time and the velocity of the exhaust gases: $(\Delta m / \Delta t)\vec{v}_{ex}$. Since the mass is decreasing, $\Delta m / \Delta t$ is negative, and the thrust is opposite $\vec{v}_{ex}$. **(b)** The space shuttle uses a multistage rocket. Both of the two booster rockets and the huge external fuel tank are jettisoned in flight. **(c)** The first and second stages of a *Saturn V* rocket separating after 148 s of burn time.

Unlike a rifle firing a single shot, a rocket continuously loses mass when burning fuel. (The rocket is more like a machine gun.) Thus, the rocket is a system for which the mass is not constant. As the mass of the rocket decreases, it accelerates more easily. Multistage rockets take advantage of this fact. The hull of a burnt-out stage is jettisoned to give a further in-flight reduction in mass (Fig. 6.25c). The payload (cargo) is typically a very small part of the initial mass of rockets for space flights.

Suppose that the purpose of a spaceflight is to land a payload on the Moon. At some point on the journey, the gravitational attraction of the Moon will become greater than that of the Earth, and the spacecraft will accelerate toward the Moon. A soft landing is desirable, so the spacecraft must be slowed down enough to go into orbit around the Moon or land on it. This slowing down is accomplished by using the rocket engines to apply a *reverse thrust*, or braking thrust. The spacecraft is maneuvered through a 180° angle, or turned around, which is quite easy to do in space. The rocket engines are then fired, expelling the exhaust gas toward the Moon and supplying a braking action. That is, the force on the rocket is opposite its velocity.

You have experienced a reverse thrust effect if you have flown in a commercial jet. In this instance, however, the craft is not turned around. Instead, after touchdown, the jet engines are revved up, and a braking action can be felt. Ordinarily, revving up the engines accelerates the plane forward. The reverse thrust is accomplished by activating thrust reversers in the engines that deflect the exhaust gases forward (▾Fig. 6.26). The gas experiences an impulse force and a change in momentum in the forward direction (see Fig. 6.3b), and the engine and the aircraft have an equal and opposite momentum change, thus experiencing a braking impulse force.

Question: There are no end-of-chapter exercises on the material covered in this section, so test your knowledge with this one: Astronauts use handheld maneuvering devices (small rockets) to move around on space walks. Describe how these rockets are used. Is there any danger on an untethered space walk?

DID YOU LEARN?

➥ If the total momentum of a system is initially zero, an internal explosion will propel parts of the system in opposite directions (Newton's third law).

➥ A rocket or jet plane uses thrust to accelerate. By turning the propulsion thrust in the opposite direction (reverse thrust), the rocket or plane is slowed down or decelerated.

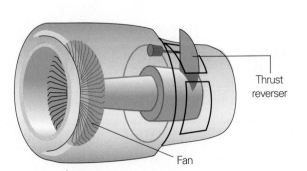

Thrust reverser

Fan

Normal operation

Thrust reverser activated

▲ **FIGURE 6.26 Reverse thrust** Thrust reversers are activated on jet engines during landing to help slow the plane. The gas experiences an impulse force and a change in momentum in the forward direction, and the plane experiences an equal and opposite momentum change and a braking impulse force.

| Pendulum and Putty Balls

A simple pendulum 1.50 m in length has a 500-g putty ball as an end bob. The pendulum is pulled aside 60° from the vertical and given an initial tangential speed of 1.20 m/s. At the bottom of the arc path, the bob hits and sticks to a 200-g putty ball sitting on a tee. (a) Determine the speed of the 500-g ball just before it hits the 200-g ball and the speed of the combined masses right after they stick together. (b) How much mechanical energy is lost in the collision? (c) What is the maximum angle to which the combined balls swing on the other side?

THINKING IT THROUGH. This example employs the use of energy, inelastic collisions, conservation of linear momentum,

and trigonometry. (a) The speed of the 500-g ball can be determined by applying the principle of mechanical energy conservation, but this cannot be applied to the inelastic collision. However, the principle of conservation of linear momentum can be applied horizontally at the very bottom of the arc path where the collision takes place and will give the balls' combined speed. (b) The mechanical energy loss during the collision is just the difference between the final and initial total kinetic energy, because where the collision takes place, the gravitational potential energy does not change. (c) After the collision, it is again valid to use mechanical energy conservation to determine the final height and angle of the combined system.

SOLUTION.

Given: $v_o = 1.20$ m/s (initial tangential speed of the ball)
$L = 1.50$ m (pendulum length)
$\theta = 60°$ (initial pendulum angle)
$m_1 = 500$ g $= 0.500$ kg (mass of initially moving ball)
$m_2 = 200$ g $= 0.200$ kg (mass of the target ball)

Find: (a) v and V (speed of original ball just before collision and that of the combined balls just after)
(b) ΔK mechanical (kinetic) energy lost
(c) maximum angle after collision

(a) The mechanical energy is conserved from just after the initial push of the descending ball to just before the collision. Initially the total mechanical energy is part kinetic energy and part gravitational potential energy. But at the bottom of the arc path, it is all kinetic energy, assuming that the zero point for gravitational potential energy is chosen to be at the bottom of the arc. The 500-g ball is initially at a height of $y_i = L(1 - \cos \theta_o) = (1.50$ m$)(1 - \cos 60°) = 0.75$ m. (Make a sketch to see this if it isn't clear.) Then, by the conservation of mechanical energy:

$$K_i + U_i = K_f + U_f$$

or

$$\tfrac{1}{2}m_1v_o^2 + m_1gy_i = \tfrac{1}{2}m_1v_1^2 + 0$$

where v is the ball's speed at the bottom of the arc. Solving for this speed,

$$v_1 = \sqrt{2gy_i + v_o^2}$$
$$= \sqrt{2(9.80 \text{ m/s}^2)(0.75 \text{ m}) + (1.20 \text{ m/s})^2} = 4.02 \text{ m/s}$$

By the conservation of linear momentum from just before to just after the collision, $m_1v_o = (m_1 + m_2)V$, where V represents the balls' combined speed. Solving for V,

$$V = \left(\frac{m_1}{m_1 + m_2}\right)v_1$$

$$= \left(\frac{0.500 \text{ kg}}{0.500 \text{ kg} + 0.200 \text{ kg}}\right)(4.02 \text{ m/s}) = 2.87 \text{ m/s}.$$

(b) From the speeds and masses, the kinetic energies can be found. Just before the collision,

$$K_1 = \tfrac{1}{2}m_1v_1^2 = \tfrac{1}{2}(0.500 \text{ kg})(4.02 \text{ m/s})^2 = 4.04 \text{ J}$$

Just after the collision, the final kinetic energy of the combined masses, K_2, is:

$$K_2 = \tfrac{1}{2}(m_1 + m_2)V^2 = \tfrac{1}{2}(0.700 \text{ kg})(2.87 \text{ m/s})^2 = 2.88 \text{ J}$$

Thus, $K_1 - K_2 = 4.04$ J $- 2.88$ J $= 1.16$ J of mechanical energy was lost (to heat and sound).

(c) To find the maximum angle on the other side of the arc, the principle of energy conservation is used from just after the collision to the location where the combined masses stop, that is, where the ball combination has no kinetic energy (using i and f to indicate initial and final, respectively):

$$K_i + U_i = K_f + U_f$$

The initial potential energy and the final kinetic energy are zero: $U_i = K_f = 0$. Then K_i is equal to $K_2 = 2.88$ J from part (b), and by the conservation of energy, $K_2 = U_f = (m_1 + m_2)gy_f$. Solving for the final height y_f,

$$y_f = \frac{K_2}{(m_1 + m_2)g}$$

$$= \frac{2.88 \text{ J}}{(0.700 \text{ kg})(9.80 \text{ m/s}^2)} = 0.420 \text{ m}$$

The angle is determined from the trigonometric relationship $y_f = L(1 - \cos \theta_f)$ solved for the final angle, $\cos \theta_f = 1 - \dfrac{y_f}{L}$,

or

$$\cos \theta_f = 1 - \frac{0.420 \text{ m}}{1.50 \text{ m}} = 0.72$$

and the angle is

$$\theta_f = \cos^{-1}(0.72) = 43.9°$$

which is less than the initial angle. (How would it be possible to make the final angle to be greater than the initial angle?)

Learning Path Review

■ The **linear momentum** ($\vec{\mathbf{p}}$) of a particle is a vector and is defined as the product of mass and velocity:

$$\vec{\mathbf{p}} = m\vec{\mathbf{v}} \tag{6.1}$$

■ The **total linear momentum** ($\vec{\mathbf{P}}$) of a system is the vector sum of the momenta of the individual particles:

$$\vec{\mathbf{P}} = \vec{\mathbf{p}}_1 + \vec{\mathbf{p}}_2 + \vec{\mathbf{p}}_3 + \cdots = \Sigma \vec{\mathbf{p}}_i \tag{6.2}$$

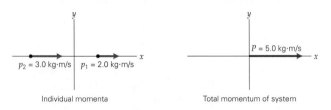

Individual momenta   Total momentum of system

■ **Newton's second law in terms of momentum (for a particle):**

$$\vec{\mathbf{F}}_{net} = \frac{\Delta \vec{\mathbf{p}}}{\Delta t} \tag{6.3}$$

■ The **impulse–momentum theorem** relates the impulse acting on an object to its change in momentum:

$$\text{Impulse} = \vec{\mathbf{F}}_{avg}\Delta t = \Delta \vec{\mathbf{p}}_o = m\vec{\mathbf{v}} - m\vec{\mathbf{v}}_o \tag{6.5}$$

■ **Conservation of linear momentum:** In the absence of a net external force, the total linear momentum of a system is conserved:

$$\vec{\mathbf{P}} = \vec{\mathbf{P}}_o \tag{6.7}$$

■ **In an elastic collision, the total kinetic energy of the system is conserved.**

■ **Momentum is conserved in both elastic and inelastic collisions**. In a completely inelastic collision, objects stick together after impact.

■ **Conditions for an elastic collision:**

$$\vec{\mathbf{P}}_f = \vec{\mathbf{P}}_i \\ K_f = K_i \tag{6.8}$$

■ **Conditions for an inelastic collision:**

$$\vec{\mathbf{P}}_f = \vec{\mathbf{P}}_i \\ K_f < K_i \tag{6.9}$$

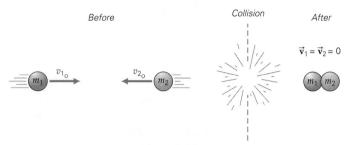

■ **Final velocity in a head-on, two-body completely inelastic collision ($v_{2_o} = 0$):**

$$v = \left(\frac{m_1}{m_1 + m_2}\right)v_{1_o} \tag{6.10}$$

■ **Ratio of kinetic energies in a head-on, two-body completely inelastic collision ($v_{2_o} = 0$):**

$$\frac{K_f}{K_i} = \frac{m_1}{m_1 + m_2} \tag{6.11}$$

■ **Final velocities in a head-on, two-body elastic collision**

$$v_1 = \left(\frac{m_1 - m_2}{m_1 + m_2}\right)v_{1_o} + \left(\frac{2m_2}{m_1 + m_2}\right)v_{2_o} \tag{6.14}$$

$$v_2 = \left(\frac{2m_1}{m_1 + m_2}\right)v_{1_o} - \left(\frac{m_1 - m_2}{m_1 + m_2}\right)v_{2_o} \tag{6.15}$$

■ The **center of mass** is the point at which all of the mass of an object or system may be considered to be concentrated. The center of mass does not necessarily lie within an object. (The **center of gravity** is the point at which all the weight may be considered to be concentrated.)

■ **Coordinates of the center of mass (using signs for directions):**

$$X_{CM} = \frac{\Sigma m_i x_i}{M} \tag{6.19}$$

Learning Path Questions and Exercises

For instructor-assigned homework, go to www.masteringphysics.com

6.1 LINEAR MOMENTUM

1. Linear momentum has units of (a) N/m, (b) kg · m/s, (c) N/s, (d) all of the preceding.

2. Linear momentum is (a) always conserved, (b) a scalar quantity, (c) a vector quantity, (d) unrelated to force.

3. A net force on an object can cause (a) an acceleration, (b) a change in momentum, (c) a change in velocity, (d) all of the preceding.

4. A change in momentum requires which of the following: (a) an unbalanced force, (b) a change in velocity, (c) an acceleration, or (d) any of these?

6.2 IMPULSE

5. Impulse has units (a) of kg · m/s, (b) of N · s, (c) the same as momentum, (d) all of the preceding.

6. Impulse is equal to (a) $F \, \Delta x$, (b) the change in kinetic energy, (c) the change in momentum, (d) $\Delta p / \Delta t$.

7. Impulse (a) is the time rate of change of momentum, (b) is the force per unit time, (c) has the same units as momentum, (d) none of these.

6.3 CONSERVATION OF LINEAR MOMENTUM

8. The conservation of linear momentum is described by (a) the momentum–impulse theorem, (b) the work–energy theorem, (c) Newton's first law, (d) conservation of energy.

9. The linear momentum of an object is conserved if (a) the force acting on the object is conservative, (b) a single, unbalanced internal force is acting on the object, (c) the mechanical energy is conserved, (d) none of the preceding.

10. Internal forces do not affect the conservation of momentum because (a) they cancel each other, (b) their effects are canceled by external forces, (c) they can never produce a change in velocity, (d) Newton's second law is not applicable to them.

6.4 ELASTIC AND INELASTIC COLLISIONS

11. Which of the following is *not* conserved in an inelastic collision: (a) momentum, (b) mass, (c) kinetic energy, or (d) total energy?

12. A rubber ball of mass m traveling horizontally with a speed v hits a wall and bounces back with the same speed. The change in momentum is (a) mv, (b) $-mv$, (c) $-mv/2$, (d) $+2mv$.

13. In a head-on elastic collision, a mass m_1 strikes a stationary mass m_2. There is a complete transfer of energy if (a) $m_1 = m_2$, (b) $m_1 \gg m_2$, (c) $m_1 \ll m_2$, (d) the masses stick together.

14. The condition for a two-object inelastic collision is (a) $K_f < K_i$, (b) $p_i \neq p_f$, (c) $m_1 > m_2$, (d) $v_1 < v_2$.

6.5 CENTER OF MASS

15. The center of mass of an object (a) always lies at the center of the object, (b) is at the location of the most massive particle in the object, (c) always lies within the object, (d) none of the preceding.

16. The center of mass and center of gravity coincide (a) if the acceleration due to gravity is constant, (b) if momentum is conserved, (c) if momentum is not conserved, (d) only for irregularly shaped objects.

6.1 LINEAR MOMENTUM

1. In a football game, does a fast-running running back always have more linear momentum than a slow-moving, more massive lineman? Explain.

2. Two objects have the same momentum. Do they necessarily have the same kinetic energy? Explain.

3. Two objects have the same kinetic energy. Do they necessarily have the same momentum? Explain.

6.2 IMPULSE

4. "Follow-through" is very important in many sports, such as in serving a tennis ball. Explain how follow-through can increase the speed of the tennis ball when it is served.

5. A karate student tries *not* to follow through in order to break a board, as shown in ▼Fig. 6.27. How can the abrupt stop of the hand (with no follow-through) generate so much force?

◀ FIGURE 6.27
A karate chop See Conceptual Question 5 and Exercise 22.

6. Explain the difference for each of the following pairs of actions in terms of impulse: (a) a golfer's long drive and a short chip shot; (b) a boxer's jab and a knockout punch; (c) a baseball player's bunting action and a home-run swing.

7. When jumping from a height to the ground, it is advised to land with the legs bent rather than stiff-legged. Why is this?

8. In the Revolutionary War, the Americans had an advantage in using long rifles, instead of the smooth bore muskets used by the British. The long rifle had a barrel of 48 in. or more, whereas the musket barrel length was on the order of 30 in. (▼Fig. 6.28). The long rifle had a much greater range than the musket. Explain this greater range in terms of work and impulse. (The rifling grooves in the barrel of a rifle cause the bullet to spin, which gyroscopically improves stability and accuracy. See Section 8.5.)

▲ FIGURE 6.28 **Long rifle** See Conceptual Question 8.

6.3 CONSERVATION OF LINEAR MOMENTUM

9. An airboat of the type used in swampy and marshy areas is shown in ▼Fig. 6.29. Explain the principle of its propulsion. Using the concept of conservation of linear momentum, determine what would happen to the boat if a sail were installed behind the fan.

▲ FIGURE 6.29 **Fan propulsion** See Conceptual Question 9.

10. Imagine yourself standing in the middle of a frozen lake. The ice is so smooth that it is frictionless. How could you get to shore? (You couldn't walk. Why?)

11. A stationary object receives a direct hit by another object moving toward it. Is it possible for both objects to be at rest after the collision? Explain.

12. Does the conservation of momentum follow from Newton's third law?

6.4 ELASTIC AND INELASTIC COLLISIONS

13. Since $K = p^2/2m$, how can kinetic energy be lost in an inelastic collision while the total momentum is still conserved? Explain.

14. Can all of the kinetic energy be lost in the collision of two objects? Explain.

15. Automobiles used to have firm steel bumpers for safety. Today auto bumpers are made out of materials that crumple or collapse on sufficient impact. Why is this?

16. Two balls of equal mass collide head on in a completely inelastic collision and come to rest. (a) Is the kinetic energy conserved? (b) Is the momentum conserved? Explain.

6.5 CENTER OF MASS

17. ▼Figure 6.30 shows a flamingo standing on one of its two legs, with its other leg lifted. What can you say about the location of the flamingo's center of mass?

◀ FIGURE 6.30
Delicate balance See Conceptual Question 17.

18. Two identical objects are located a distance d apart. If one of the objects remains at rest and the other moves away with a constant velocity, what is the effect on the CM of the system?

6.6 JET PROPULSION AND ROCKETS

19. Rockets used in the space program are generally multistage—that is, they have stacked stages, each with its own engine and propellant. Starting with the bottom, the stages are jettisoned when they have run out of fuel. The first stage is usually the largest, the second stage above it is the next largest, and so on. What are the advantages of multistage rockets and stage sizes?

Integrated Exercises (IEs) are two-part exercises. The first part typically requires a conceptual answer choice based on physical thinking and basic principles. The following part is quantitative calculations associated with the conceptual choice made in the first part of the exercise.

Throughout the text, many exercise sections will include "paired" exercises. These exercise pairs, identified with red numbers, *are intended to assist you in problem solving and learning. In a pair, the first exercise (even numbered) is worked out in the Study Guide so that you can consult it should you need assistance in solving it. The second exercise (odd numbered) is similar in nature, and its answer is given in Appendix VII at the back of the book.*

1. • If a 60-kg woman is riding in a car traveling at 90 km/h, what is her linear momentum relative to (a) the ground and (b) the car?

2. • The linear momentum of a runner in a 100-m dash is 7.5×10^2 kg·m/s. If the runner's speed is 10 m/s, what is his mass?

3. • Find the magnitude of the linear momentum of (a) a 7.1-kg bowling ball traveling at 12 m/s and (b) a 1200-kg automobile traveling at 90 km/h.

4. • In a football game, a lineman usually has more mass than a running back. (a) Will a lineman always have greater linear momentum than a running back? Why? (b) Who has greater linear momentum, a 75-kg running back running at 8.5 m/s or a 120-kg lineman moving at 5.0 m/s?

5. •• A 0.150-kg baseball traveling with a horizontal speed of 4.50 m/s is hit by a bat and then moves with a speed of 34.7 m/s in the opposite direction. What is the change in the ball's momentum?

6. •• A 15.0-g rubber bullet hits a wall with a speed of 150 m/s. If the bullet bounces straight back with a speed of 120 m/s, what is the change in momentum of the bullet?

7. **IE** •• Two protons approach each other with different speeds. (a) Will the magnitude of the total momentum of the two-proton system be (1) greater than the magnitude of the momentum of either proton, (2) equal to the difference between the magnitudes of momenta of the two protons, or (3) equal to the sum of the magnitudes of momenta of the two protons? Why? (b) If the speeds of the two protons are 340 m/s and 450 m/s, respectively, what is the total momentum of the two-proton system? [*Hint*: Find the mass of a proton in one of the tables inside the backcover.]

8. •• How much momentum is acquired by a 75-kg skydiver in free fall in 2.0 minutes after jumping from the plane?

9. •• A 5.0-g bullet with a speed of 200 m/s is fired horizontally into a 0.75-kg wooden block at rest on a table. If the block containing the bullet slides a distance of 0.20 m before coming to rest, (a) what is the coefficient of kinetic friction between the block and the table? (b) What fraction of the bullet's energy is dissipated in the collision?

10. •• Two runners of mass 70 kg and 60 kg, respectively, have a total linear momentum of 350 kg·m/s. The heavier runner is running at 2.0 m/s. Determine the possible velocities of the lighter runner.

11. •• A 0.20-kg billiard ball traveling at a speed of 15 m/s strikes the side rail of a pool table at an angle of 60° (▼Fig. 6.31). If the ball rebounds at the same speed and angle, what is the change in its momentum?

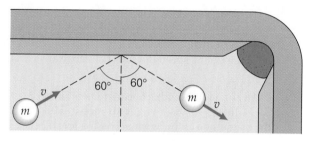

▲ **FIGURE 6.31** **Glancing collision** See Exercises 11, 12, and 33.

12. •• Suppose the billiard ball in Fig. 6.31 approaches the rail at a speed of 15 m/s and an angle of 60°, as shown, but rebounds at a speed of 10 m/s and an angle of 50°. What is the change in momentum in this case? [*Hint*: Use components.]

13. •• A loaded tractor-trailer with a total mass of 5000 kg traveling at 3.0 km/h hits a loading dock and comes to a stop in 0.64 s. What is the magnitude of the average force exerted on the truck by the dock?

14. •• A 2.0-kg mud ball drops from rest at a height of 15 m. If the impact between the ball and the ground lasts 0.50 s, what is the average net force exerted by the ball on the ground?

15. **IE** •• In football practice, two wide receivers run different pass receiving patterns. One with a mass of 80.0 kg runs at 45° northeast at a speed of 5.00 m/s. The second receiver (mass of 90.0 kg) runs straight down the field (due east) at 6.00 m/s. (a) What is the direction of their total momentum: (1) exactly northeast, (2) to the north of northeast, (3) exactly east, or (4) to the east of northeast? (b) Justify your answer in part (a) by actually computing their total momentum.

16. •• A major league catcher catches a fastball moving at 95.0 mi/h and his hand and glove recoil 10.0 cm in bringing the ball to rest. If it took 0.00470 s to bring the ball (with a mass of 250 g) to rest in the glove, (a) what are the magnitude and direction of the change in momentum of the ball? (b) Find the average force the ball exerts on the hand and glove.

17. ••• At a basketball game, a 120-lb cheerleader is tossed vertically upward with a speed of 4.50 m/s by a male

cheerleader. (a) What is the cheerleader's change in momentum from the time she is released to just before being caught if she is caught at the height at which she was released? (b) Would there be any difference if she were caught 0.30 m below the point of release? If so, what is the change then?

18. ••• A ball of mass 200 g is released from rest at a height of 2.00 m above the floor and it rebounds straight up to a height of 0.900 m. (a) Determine the ball's change in momentum due to its contact with the floor. (b) If the contact time with the floor was 0.0950 s, what was the average force the floor exerted on the ball, and in what direction?

6.2 IMPULSE

19. • When tossed upward and hit horizontally by a batter, a 0.20-kg softball receives an impulse of 3.0 N·s. With what horizontal speed does the ball move away from the bat?

20. • An automobile with a linear momentum of 3.0×10^4 kg · m/s is brought to a stop in 5.0 s. What is the magnitude of the average braking force?

21. • A pool player imparts an impulse of 3.2 N·s to a stationary 0.25-kg cue ball with a cue stick. What is the speed of the ball just after impact?

22. •• For the karate chop in Fig. 6.27, assume that the hand has a mass of 0.35 kg and that the speeds of the hand just before and just after hitting the board are 10 m/s and 0, respectively. What is the average force exerted by the fist on the board if (a) the fist follows through, so the contact time is 3.0 ms, and (b) the fist stops abruptly, so the contact time is only 0.30 ms?

23. IE •• When bunting, a baseball player uses the bat to change both the speed and direction of the baseball. (a) Will the magnitude of the change in momentum of the baseball before and after the bunt be (1) greater than the magnitude of the momentum of the baseball either before or after the bunt, (2) equal to the difference between the magnitudes of momenta of the baseball before and after the bunt, or (3) equal to the sum of the magnitudes of momenta of the baseball before and after the bunt? Why? (b) The baseball has a mass of 0.16 kg; its speeds before and after the bunt are 15 m/s and 10 m/s, respectively; the bunt lasts 0.025 s. What is the change in momentum of the baseball? (c) What is the average force on the ball by the bat?

24. IE •• A car with a mass of 1500 kg is rolling on a level road at 30.0 m/s. It receives an impulse with a magnitude of 2000 N · s and its speed is reduced as much as possible by an impulse of this size. (a) Was this impulse caused by (1) the driver hitting the accelerator, (2) the driver putting on the brakes, or (3) the driver turning the steering wheel? (b) What was the car's speed after the impulse was applied?

25. •• An astronaut (mass of 100 kg, with equipment) is headed back to her space station at a speed of 0.750 m/s but at the wrong angle. To correct her direction, she fires rockets from her backpack at right angles to her motion

for a brief time. These directional rockets exert a constant force of 100.0 N for only 0.200 s. [Neglect the small loss of mass due to burning fuel and assume the impulse is at right angles to her initial momentum.] (a) What is the magnitude of the impulse delivered to the astronaut? (b) What is her new direction (relative to the initial direction)? (c) What is her new speed?

26. •• A volleyball is traveling toward you. (a) Which action will require a greater force on the volleyball, your catching the ball or your hitting the ball back? Why? (b) A 0.45-kg volleyball travels with a horizontal velocity of 4.0 m/s over the net. You jump up and hit the ball back with a horizontal velocity of 7.0 m/s. If the contact time is 0.040 s, what was the average force on the ball?

27. •• A boy catches—with bare hands and his arms rigidly extended—a 0.16-kg baseball coming directly toward him at a speed of 25 m/s. He emits an audible "Ouch!" because the ball stings his hands. He learns quickly to move his hands with the ball as he catches it. If the contact time of the collision is increased from 3.5 ms to 8.5 ms in this way, how do the magnitudes of the average impulse forces compare?

28. •• A one-dimensional impulse force acts on a 3.0-kg object as diagrammed in ▾Fig. 6.32. Find (a) the magnitude of the impulse given to the object, (b) the magnitude of the average force, and (c) the final speed if the object had an initial speed of 6.0 m/s.

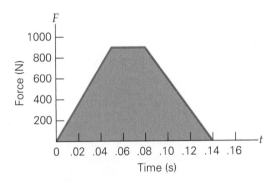

▲ FIGURE 6.32 **Force-versus-time graph** See Exercise 28.

29. •• A 0.45-kg piece of putty is dropped from a height of 2.5 m above a flat surface. When it hits the surface, the putty comes to rest in 0.30 s. What is the average force exerted on the putty by the surface?

30. •• A 50-kg driver sits in her car waiting for the traffic light to change. Another car hits her from behind in a head-on, rear-end collision and her car suddenly receives an acceleration of 16 m/s². If all of this takes place in 0.25 s, (a) what is the impulse on the driver? (b) What is the average force exerted on the driver, and what exerts this force?

31. •• An incoming 0.14-kg baseball has a speed of 45 m/s. The batter hits the ball, giving it a speed of 60 m/s. If the contact time is 0.040 s, what is the average force of the bat on the ball?

32. •• At a shooting competition, a contestant fires and a 12.0-g bullet leaves the rifle with a muzzle speed of 130 m/s. The bullet hits the thick target backing and

stops after traveling 4.00 cm. Assuming a uniform acceleration, (a) what is the impulse on the target? (b) What is the average force on the target?

33. •• If the billiard ball in Fig. 6.31 is in contact with the rail for 0.010 s, what is the magnitude of the average force exerted on the ball? (See Exercise 11.)

34. •• A 15000-N automobile travels at a speed of 45 km/h northward along a street, and a 7500-N sports car travels at a speed of 60 km/h eastward along an intersecting street. (a) If neither driver brakes and the cars collide at the intersection and lock bumpers, what will the velocity of the cars be immediately after the collision? (b) What percentage of the initial kinetic energy will be lost in the collision?

35. ••• In a simulated head-on crash test, a car impacts a wall at 25 mi/h (40 km/h) and comes abruptly to rest. A 120-lb passenger dummy (with a mass of 55 kg), without a seatbelt, is stopped by an air bag, which exerts a force on the dummy of 2400 lb. How long was the dummy in contact with the air bag while coming to a stop?

36. ••• A baseball player pops a pitch straight up. The ball (mass 200 g) was traveling horizontally at 35.0 m/s just before contact with the bat, and 20.0 m/s just after contact. Determine the direction and magnitude of the impulse delivered to the ball by the bat.

6.3 CONSERVATION OF LINEAR MOMENTUM

37. • A 60-kg astronaut floating at rest in space outside a space capsule throws his 0.50-kg hammer such that it moves with a speed of 10 m/s relative to the capsule. What happens to the astronaut?

38. • In a pairs figure-skating competition, a 65-kg man and his 45-kg female partner stand facing each other on skates on the ice. If they push apart and the woman has a velocity of 1.5 m/s eastward, what is the velocity of her partner? (Neglect friction.)

39. •• To get off a frozen, frictionless lake, a 65.0-kg person takes off a 0.150-kg shoe and throws it horizontally, directly away from the shore with a speed of 2.00 m/s. If the person is 5.00 m from the shore, how long does he take to reach it?

40. IE •• An object initially at rest explodes and splits into three fragments. The first fragment flies off to the west, and the second fragment flies off to the south. The third fragment will fly off toward a general direction of (1) southwest, (2) north of east, (3) either due north or due east. Why? (b) If the object has a mass of 3.0 kg, the first fragment has a mass of 0.50 kg and a speed of 2.8 m/s, and the second fragment has a mass of 1.3 kg and a speed of 1.5 m/s, what are the speed and direction of the third fragment?

41. •• Consider two string-suspended balls, both with a mass of 0.15 kg. (Similar to the arrangement in Fig. 6.15, but with only two balls.) One ball is pulled back in line with the other so it has a vertical height of 10 cm, and is then released. (a) What is the speed of the ball just before hitting the stationary one? (b) If the collision is completely inelastic, to what height do the balls swing?

42. •• A cherry bomb explodes into three pieces of equal mass. One piece has an initial velocity of 10 m/s $\hat{x}$. Another piece has an initial velocity of 6.0 m/s $\hat{x} - 3.0$ m/s $\hat{y}$. What is the velocity of the third piece?

43. •• Two ice skaters not paying attention collide in a completely inelastic collision. Prior to the collision, skater 1, with a mass of 60 kg, has a velocity of 5.0 km/h eastward, and moves at a right angle to skater 2, who has a mass of 75 kg and a velocity of 7.5 km/h southward. What is the velocity of the skaters after collision?

44. •• Two balls of equal mass (0.50 kg) approach the origin along the positive x- and y-axes at the same speed (3.3 m/s). (a) What is the total momentum of the system? (b) Will the balls necessarily collide at the origin? What is the total momentum of the system after both balls have passed through the origin?

45. •• A 1200-kg car moving to the right with a speed of 25 m/s collides with a 1500-kg truck and locks bumpers with the truck. Calculate the velocity of the combination after the collision if the truck is initially (a) at rest, (b) moving to the right with a speed of 20 m/s, and (c) moving to the left with a speed of 20 m/s.

46. •• A 10-g bullet moving horizontally at 400 m/s penetrates a 3.0-kg wood block resting on a horizontal surface. If the bullet slows down to 300 m/s after emerging from the block, what is the speed of the block immediately after the bullet emerges (▼Fig. 6.33)?

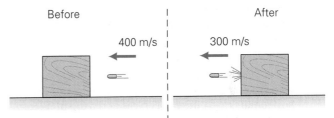

Before   After

400 m/s   300 m/s

▲ FIGURE 6.33 **Momentum transfer?** See Exercise 46.

47. •• An explosion of a 10.0-kg bomb releases only two separate pieces. The bomb was initially at rest and a 4.00-kg piece travels westward at 100 m/s immediately after the explosion. (a) What are the speed and direction of the other piece immediately after the explosion? (b) How much kinetic energy was released in this explosion?

48. •• A 1600-kg (empty) truck rolls with a speed of 2.5 m/s under a loading bin, and a mass of 3500 kg is deposited into the truck. What is the truck's speed immediately after loading?

49. IE •• A new crowd control method utilizes "rubber" bullets instead of real ones. Suppose that, in a test, one of these "bullets" with a mass of 500 g is traveling at 250 m/s to the right. It hits a stationary target head-on. The target's mass is 25.0 kg and it rests on a smooth surface. The bullet bounces backward (to the left) off the target at 100 m/s. (a) Which way must the target move after the collision: (1) right, (2) left, (3) it could be stationary, or (4) you can't tell from the data given? (b) Determine the recoil speed of the target after the collision.

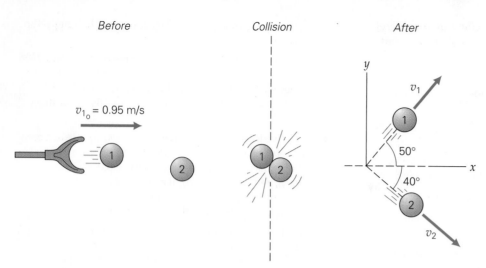

▲ FIGURE 6.34 **Another glancing collision** See Exercise 53.

50. •• For a movie scene, a 75-kg stuntman drops from a tree onto a 50-kg sled that is moving on a frozen lake with a velocity of 10 m/s toward the shore. (a) What is the speed of the sled after the stuntman is on board? (b) If the sled hits the bank and stops, but the stuntman keeps on going, with what speed does he leave the sled? (Neglect friction.)

51. •• A 90-kg astronaut is stranded in space at a point 6.0 m from his spaceship, and he needs to get back in 4.0 min to control the spaceship. To get back, he throws a 0.50-kg piece of equipment so that it moves at a speed of 4.0 m/s directly away from the spaceship. (a) Does he get back in time? (b) How fast must he throw the piece of equipment so he gets back in time?

52. ••• A projectile that is fired from a gun has an initial velocity of 90.0 km/h at an angle of 60.0° above the horizontal. When the projectile is at the top of its trajectory, an internal explosion causes it to separate into two fragments of equal mass. One of the fragments falls straight downward as though it had been released from rest. How far from the gun does the other fragment land?

53. ••• A moving shuffleboard puck has a glancing collision with a stationary puck of the same mass, as shown in ▲ Fig. 6.34. If friction is negligible, what are the speeds of the pucks after the collision?

54. ••• A small asteroid (mass of 10 g) strikes a glancing blow at a satellite in empty space. The satellite was initially at rest and the asteroid was traveling at 2000 m/s. The satellite's mass is 100 kg. The asteroid is deflected 10° from its original direction and its speed decreases to 1000 m/s, but neither object loses mass. Determine the (a) direction and (b) speed of the satellite after the collision.

55. ••• A *ballistic pendulum* is a device used to measure the velocity of a projectile—for example, the muzzle velocity of a rifle bullet. The projectile is shot horizontally into, and becomes embedded in, the bob of a pendulum, as illustrated in ▶ Fig. 6.35. The pendulum swings upward to some height h, which is measured. The masses of the

block and the bullet are known. Using the laws of momentum and energy, show that the initial velocity of the projectile is given by $v_\mathrm{o} = [(m + M)/m]\sqrt{2gh}$.

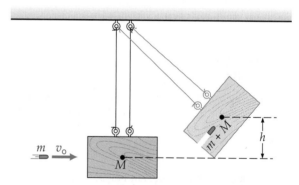

▲ FIGURE 6.35 **A ballistic pendulum** See Exercises 55 and 73.

6.4 ELASTIC AND INELASTIC COLLISIONS

56. •• For the apparatus in Fig. 6.15, one ball swinging in at a speed of $2v_\mathrm{o}$ will not cause two balls to swing out with speeds v_o. (a) Which law of physics precludes this situation from happening: the law of conservation of momentum or the law of conservation of mechanical energy? (b) Prove this law mathematically.

57. •• A proton of mass m moving with a speed of 3.0×10^6 m/s undergoes a head-on elastic collision with an alpha particle of mass $4m$, which is initially at rest. What are the velocities of the two particles after the collision?

58. •• A 4.0-kg ball with a velocity of 4.0 m/s in the $+x$-direction collides head-on elastically with a stationary 2.0-kg ball. What are the velocities of the balls after the collision?

59. •• A dropped rubber ball hits the floor with a speed of 8.0 m/s and rebounds to a height of 0.25 m. What fraction of the initial kinetic energy was lost in the collision?

60. •• At a county fair, two children ram each other head-on while riding on the bumper cars. Jill and her car, traveling left at 3.50 m/s, have a total mass of 325 kg. Jack and his car, traveling to the right at 2.00 m/s, have a total mass of 290 kg. Assuming the collision to be elastic, determine their velocities after the collision.

61. •• In a high-speed chase, a policeman's car bumps a criminal's car directly from behind to get his attention. The policeman's car is moving at 40.0 m/s to the right and has a total mass of 1800 kg. The criminal's car is initially moving in the same direction at 38.0 m/s. His car has a total mass of 1500 kg. Assuming an elastic collision, determine their two velocities immediately after the bump.

62. **IE** •• ▼ Fig. 6.36 shows a bird catching a fish. Assume that initially the fish jumps up and that the bird coasts horizontally and does not touch the water with its feet or flap its wings. (a) Is this kind of collision (1) elastic, (2) inelastic, or (3) completely inelastic? Why? (b) If the mass of the bird is 5.0 kg, the mass of the fish is 0.80 kg, and the bird coasts with a speed of 6.5 m/s before grabbing, what is the speed of the bird after grabbing the fish?

◀ **FIGURE 6.36**
Elastic or inelastic?
See Exercise 62.

63. •• A 1.0-kg object moving at 10 m/s collides with a stationary 2.0-kg object as shown in ▼ Fig. 6.37. If the collision is perfectly inelastic, how far along the inclined plane will the combined system travel? (Neglect friction.)

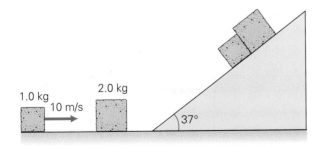

1.0 kg
10 m/s
2.0 kg
37°

▲ **FIGURE 6.37** **How far is up?** See Exercises 63 and 67.

64. •• In a pool game, a cue ball traveling at 0.75 m/s hits the stationary eight ball. The eight ball moves off with a velocity of 0.25 m/s at an angle of 37° relative to the cue ball's initial direction. Assuming that the collision is inelastic, at what angle will the cue ball be deflected, and what will be its speed?

65. •• Two balls approach each other as shown in ▼ Fig. 6.38, where $m = 2.0$ kg , $v = 3.0$ m/s , $M = 4.0$ kg , and $V = 5.0$ m/s. If the balls collide and stick together at the origin, (a) what are the components of the velocity v of the balls after collision, and (b) what is the angle θ?

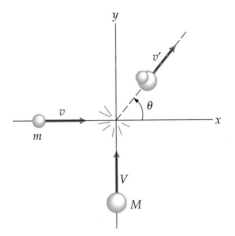

◀ **FIGURE 6.38**
A completely inelastic collision
See Exercise 65.

66. **IE** •• A car traveling east and a minivan traveling south collide in a completely inelastic collision at a perpendicular intersection. (a) Right after the collision, will the car and minivan move toward a general direction (1) south of east, (2) north of west, or (3) either due south or due east? Why? (b) If the initial speed of the 1500-kg car was 90.0 km/h and the initial speed of the 3000-kg minivan was 60.0 km/h, what is the velocity of the vehicles immediately after collision?

67. •• A 1.0-kg object moving at 2.0 m/s collides elastically with a stationary 1.0-kg object, similar to the situation shown in Fig. 6.37. How far will the initially stationary object travel along a 37° inclined plane? (Neglect friction.)

68. •• A fellow student states that the total momentum of a three-particle system ($m_1 = 0.25$ kg , $m_2 = 0.20$ kg , and $m_3 = 0.33$ kg) is initially zero. He calculates that after an inelastic triple collision the particles have velocities of 4.0 m/s at 0°, 6.0 m at 120°, and 2.5 m/s at 230°, respectively, with angles measured from the +x-axis. Do you agree with his calculations? If not, assuming the first two answers to be correct, what should be the momentum of the third particle so the total momentum is zero?

69. •• A freight car with a mass of 25 000 kg rolls down an inclined track through a vertical distance of 1.5 m. At the bottom of the incline, on a level track, the car collides and couples with an identical freight car that was at rest. What percentage of the initial kinetic energy is lost in the collision?

70. ••• In nuclear reactors, subatomic particles called *neutrons* are slowed down by allowing them to collide with the atoms of a moderator material, such as carbon atoms, which are 12 times as massive as neutrons. (a) In a head-on elastic collision with a carbon atom, what percentage of a neutron's energy is lost? (b) If the neutron has an initial speed of 1.5×10^7 m/s, what will be its speed after collision?

71. ••• In a noninjury chain-reaction accident on a foggy freeway, car 1 (mass of 2000 kg) moving at 15.0 m/s to the right elastically collides with car 2, initially at rest. The mass of car 2 is 1500 kg. In turn, car 2 then goes on to lock bumpers (that is, it is a completely inelastic collision) with car 3, which has a mass of 2500 kg and was also at rest. Determine the speed of all cars immediately after this unfortunate accident.

72. ••• Pendulum 1 is made of a 1.50-m string with a small Super Ball attached as a bob. It is pulled aside 30° and released. At the bottom of its arc, it collides with another pendulum bob of the same length, but the second pendulum has a bob made from a Super Ball whose mass is twice that of the bob of pendulum 1. Determine the angles to which both pendulums rebound (when they come to rest) after they collide and bounce back.

73. ••• Show that the fraction of kinetic energy lost in a ballistic-pendulum collision (as in Fig. 6.35) is equal to $M/(m + M)$.

6.5   CENTER OF MASS

74. • (a) The center of mass of a system consisting of two 0.10-kg particles is located at the origin. If one of the particles is at (0, 0.45 m), where is the other? (b) If the masses are moved so their center of mass is located at (0.25 m, 0.15 m), can you tell where the particles are located?

75. • (a) Find the center of mass of the Earth–Moon system. [*Hint*: Use data from the tables on the inside cover of the book, and consider the distance between the Earth and Moon to be measured from their centers.] (b) Where is that center of mass relative to the surface of the Earth?

76. •• Find the center of mass of a system composed of three spherical objects with masses of 3.0 kg, 2.0 kg, and 4.0 kg and centers located at $(-6.0 \text{ m}, 0)$, $(1.0 \text{ m}, 0)$, and $(3.0 \text{ m}, 0)$, respectively.

77. •• Rework Exercise 52, using the concept of the center of mass, and compute the distance the other fragment landed from the gun.

78. IE •• A 3.0-kg rod of length 5.0 m has at opposite ends point masses of 4.0 kg and 6.0 kg. (a) Will the center of mass of this system be (1) nearer to the 4.0-kg mass, (2) nearer to the 6.0-kg mass, or (3) at the center of the rod? Why? (b) Where is the center of mass of the system?

79. •• A piece of uniform sheet metal measures 25 cm by 25 cm. If a circular piece with a radius of 5.0 cm is cut from the center of the sheet, where is the sheet's center of mass now?

80. •• Locate the center of mass of the system shown in ▼Fig. 6.39 (a) if all of the masses are equal; (b) if $m_2 = m_4 = 2m_1 = 2m_3$; (c) if $m_1 = 1.0 \text{ kg}$, $m_2 = 2.0 \text{ kg}$, $m_3 = 3.0 \text{ kg}$, and $m_4 = 4.0 \text{ kg}$.

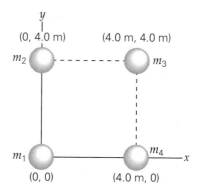

▲ FIGURE 6.39   **Where's the center of mass?** See Exercise 80.

81. •• Two cups are placed on a uniform board that is balanced on a cylinder (▼Fig. 6.40). The board has a mass of 2.00 kg and is 2.00 m long. The mass of cup 1 is 200 g and it is placed 1.05 m to the left of the balance point. The mass of cup 2 is 400 g. Where should cup 2 be placed for balance (relative to the right end of the board)?

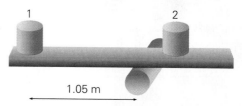

▲ FIGURE 6.40   **Don't let it roll** See Exercise 81.

82. •• Two skaters with masses of 65 kg and 45 kg, respectively, stand 8.0 m apart, each holding one end of a piece of rope. (a) If they pull themselves along the rope until they meet, how far does each skater travel? (Neglect friction.) (b) If only the 45-kg skater pulls along the rope until she meets her friend (who just holds onto the rope), how far does each skater travel?

83. ••• Three particles, each with a mass of 0.25 kg, are located at $(-4.0 \text{ m}, 0)$, $(2.0 \text{ m}, 0)$, and $(0, 3.0 \text{ m})$ and are acted on by forces $\vec{F}_1 = (-3.0 \text{ N})\hat{y}$, $\vec{F}_2 = (5.0 \text{ N})\hat{y}$, and $\vec{F}_3 = (4.0 \text{ N})\hat{x}$, respectively. Find the acceleration (magnitude and direction) of the center of mass of the system. [*Hint*: Consider the components of the acceleration.]

PULLING IT TOGETHER: MULTICONCEPT EXERCISES

The Multiconcept Exercises require the use of more than one fundamental concept for their understanding and solution. The concepts may be from this chapter, but may include those from previous chapters.

84. A 170-g hockey puck sliding on ice perpendicularly impacts a flat piece of sideboard. Its incoming momentum is 6.10 kg · m/s. It rebounds along its incoming path after having suffered a momentum change (magnitude) of 8.80 kg · m/s. (a) If the impact with the board took 35.0 ms, determine the average force (including direction) exerted by the puck on the board. (b) Determine the final momentum of the puck. (c) Was this collision elastic or inelastic? Prove your answer mathematically.

85. You are traveling north and make a 90° right-hand turn east on a flat road while driving a car that has a total weight of 3600 lb. Before the turn, the car was traveling at 40 mi/h, and after the turn is completed you have slowed to 30 mi/h. If the turn took 4.25 s to complete, determine the following: (a) the car's change in kinetic energy, (b) the car's change in momentum (including direction), and (c) the average net force exerted on the car during the turn (including direction).

86. **IE** In the radioactive decay of a nucleus of an atom called americium-241 (symbol ^{241}Am, mass of 4.03×10^{-25} kg), it emits an alpha particle (designated as α) with a mass of 6.68×10^{-27} kg to the right with a kinetic energy of 8.64×10^{-13} J. (This is typical of nuclear energies, small on the everyday scale.) The remaining nucleus is neptunium-237 (^{237}Np) and has a mass of 3.96×10^{-25} kg. Assume the initial nucleus was at rest. (a) Will the neptunium nucleus have (1) more, (2) less, or (3) the same amount of kinetic energy compared to the alpha particle? (b) Determine the kinetic energy of the ^{237}Np nucleus afterward.

87. A youth hockey player with a mass of 30.0 kg is initially moving at 2.00 m/s to the east. He intercepts and catches on the stick a puck initially moving at 35.0 m/s at an angle of $\theta = 60°$ (▼Fig. 6.41). Assume that the puck's mass is 0.18 kg and the player and puck form a single object for a few seconds. (a) Determine the direction angle and speed of the puck and skater after the collision. (b) Was this collision elastic or inelastic? Prove your answer with numbers.

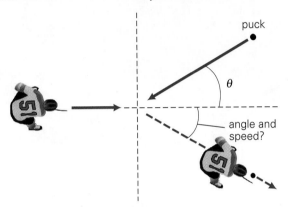

▲ **FIGURE 6.41** **A player–puck collision** See Exercise 87.

88. **IE** In a laboratory setup, two frictionless carts are placed on a horizontal surface. Cart A has a mass of 500 g and cart B's mass is 1000 g. Between them is placed an ideal (very light) spring and they are squeezed together carefully, thereby compressing the spring by 5.50 cm. Both carts are then released and B's recoil speed is measured to be 0.55 m/s. (a) Will cart A's speed be (1) greater than, (2) less than, or (3) the same as B's speed? Explain. (b) Determine B's recoil speed to see if your conjecture in (a) was correct. (c) Determine the spring constant of the spring.

7 Circular Motion and Gravitation

PHYSICS FACTS

✦ An ultracentrifuge can spin samples with a force of 15 000 g. Such force is needed for harvesting protein precipitates, bacteria, and other cells.

✦ Newton coined the word *gravity* from *gravitas*, the Latin word for "weight" or "heaviness."

✦ If you want to "lose" weight, go to the Earth's equator. Because the Earth bulges slightly, the acceleration due to gravity is slightly less there and you would weigh less (but your mass would still be the same).

✦ The centripetal force that keeps planets in orbit is supplied by gravity. The centripetal force that keeps atomic electrons in orbit about the nuclear proton(s) is supplied by the electrical force. The electrical force between an electron and a proton is on the order of 10^{40} times greater than the gravitational force between them (Section 15.3).

Ｐeople often say that rides like the spirling circular one in the chapter-opening photograph "defy gravity." Of course, you know that in reality, gravity cannot be defied; it commands respect. There is nothing that will shield you from it and no place in the universe where you can go to be entirely free of gravity.

Circular motion is everywhere, from atoms to galaxies, from flagella of bacteria to Ferris wheels. Two terms are frequently used to describe such motion. In general, we say that an object *rotates* when the axis of rotation lies within the body and that it *revolves* when the axis is outside the body. Thus, the Earth rotates on its axis and revolves about the Sun.

Such motion is in two dimensions, and so can be described by rectangular components as used in Chapter 3. However, it is usually more convenient to describe circular motion in terms of angular quantities that will be introduced in this chapter. Being familiar with the description of circular motion will make the study of rotating rigid bodies in Chapter 8 much easier.

Gravity plays a major role in determining the motions of the planets, since it supplies the force necessary to maintain their orbits. Newton's law of gravitation will be considered in this chapter. This law describes the fundamental force of gravity, and will be used to analyze planetary motion. The same considerations will help you understand the motions of Earth satellites, which include one natural satellite (the Moon) and many artificial ones.

7.1 Angular Measure*

LEARNING PATH QUESTIONS

➡ The radian is defined by what two quantities?
➡ How many radians are there in a full circle?

Motion is described as a change of position with time. (Section 2.1). As you might guess, *angular speed* and *angular velocity* also involve a time rate of change of position, which is expressed by an *angular change*. Consider a particle traveling in a circular path, as shown in ▶ Fig. 7.1. At a particular instant, the particle's position (P) may be designated by the Cartesian coordinates x and y. However, the position may also be designated by the *polar coordinates* r and θ. The distance r extends from the origin, and the angle θ is commonly measured counterclockwise from the positive x-axis. The transformation equations that relate one set of coordinates to the other are

$$x = r \cos \theta \qquad (7.1a)$$

$$y = r \sin \theta \qquad (7.1b)$$

as can be seen from the x- and y-coordinates of point P in Fig. 7.1.

Note that r is the same for any point on a given circle. As a particle travels in a circle, the value of r is constant, and only θ changes with time. Thus, circular motion can be described by using one polar coordinate (θ) that changes with time, instead of two Cartesian coordinates (x) and (y), both of which change with time.

Analogous to linear displacement is **angular displacement**, the magnitude of which is given by

$$\Delta\theta = \theta - \theta_0 \qquad (7.2)$$

or simply $\Delta\theta = \theta$ when $\theta_0 = 0°$. (The direction of the angular displacement will be explained in the next section on angular velocity.) A unit commonly used to express angular displacement is the degree (°); there are 360° in one complete circle.[†]

It is important to be able to relate the angular description of circular motion to the orbital or tangential description—that is, to relate the angular displacement to the arc length s. The *arc length* is the distance traveled along the circular path, and the angle θ is said to *subtend* (define) the arc length. A quantity that is very convenient for relating

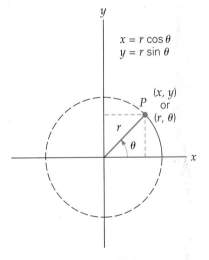

▲ **FIGURE 7.1 Polar coordinates**
A point (P) may be described by polar coordinates instead of Cartesian coordinates—that is, by (r, θ) instead of (x, y). For a circle, θ is the angular distance and r is the radial distance. The two types of coordinates are related by the transformation equations $x = r \cos \theta$ and $y = r \sin \theta$.

*Here and throughout the text, angles will be considered exact, that is, they do not determine the number of significant figures.

[†]A degree may be divided into the smaller units of minutes (1 degree = 60 minutes) and seconds (1 minute = 60 seconds). These divisions have nothing to do with time units.

▶ **FIGURE 7.2 Radian measure**
Angular displacement may be measured either in degrees or in radians (rad). An angle θ is subtended by an arc length s. When $s = r$, the angle subtending s is defined to be 1 rad. More generally, $\theta = s/r$, where θ is in radians. One radian is equal to 57.3°.

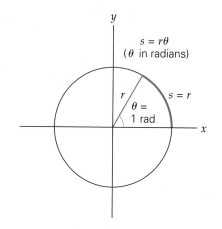

angle to arc length is the **radian (rad)**. The angle in radians is given by the ratio of the arc length (s) and the radius (r)—that is, θ (in radians) equals s/r. When $s = r$, the angle is equal to one radian, $\theta = s/r = r/r = 1$ rad (▲ Fig. 7.2).

Thus, (with the angle in radians),

$$s = r\theta \tag{7.3}$$

which is an important relationship between the circular arc length s and the radius of the circle r. (Notice that since $\theta = s/r$, the angle in radians is the ratio of two lengths. This means that a radian measure is a pure number—that is, it is dimensionless and has no units.)

To get a general relationship between radians and degrees, consider the distance around a complete circle (360°). For one full circle, with $s = 2\pi r$ (the circumference), there are a total of $\theta = s/r = 2\pi r/r = 2\pi$ rad in 360°, that is,

$$2\pi \text{ rad} = 360°$$

This relationship can be used to obtain convenient conversions of common angles (◀ Table 7.1). Also, dividing both sides of this relationship by 2π, the degree value of 1 rad is obtained:

$$1 \text{ rad} = 360°/2\pi = 57.3°$$

Notice in Table 7.1 that the angles in radians are expressed in terms of π explicitly, for convenience.

TABLE 7.1
Equivalent Degree and Radian Measures

Degrees	Radians
360°	2π
180°	π
90°	$\pi/2$
60°	$\pi/3$
57.3°	1
45°	$\pi/4$
30°	$\pi/6$

EXAMPLE 7.1 | ## Finding Arc Length: Using Radian Measure

A spectator standing at the center of a circular running track observes a runner start a practice race 256 m due east of her own position (▶ Fig. 7.3). The runner runs on the track to the finish line, which is located due north of the observer's position. What is the distance of the run?

THINKING IT THROUGH.   Note that the subtending angle of the section of circular track is $\theta = 90°$. The arc length (s) can be found, since the radius r of the circle is known.

SOLUTION.   Listing what is given and what is to be found,

Given: $r = 256$ m   *Find:* s (arc length)
           $\theta = 90° = \pi/2$ rad

Simply using Eq. 7.3 to find the arc length,

$$s = r\theta = (256 \text{ m})\left(\frac{\pi}{2}\right) = 402 \text{ m}$$

Note that the unitless rad is omitted, and the equation is dimensionally correct.

FOLLOW-UP EXERCISE.   What path length would the runner have traveled when going an angular distance of 210° around the track? (*Answers to all Follow-Up Exercises are given in Appendix VI at the back of the book.*)

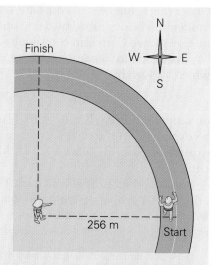

▲ **FIGURE 7.3 Arc length—found by means of radians** See Example text for description.

EXAMPLE 7.2 | How Far Away? A Useful Approximation

A sailor sights a distant tanker ship and finds that it subtends an angle of 1.15° as illustrated in ▸ Fig. 7.4a. He knows from the shipping charts that the tanker is 150 m in length. Approximately how far away is the tanker?

THINKING IT THROUGH. Note that in the accompanying Learn by Drawing 7.1, The Small Angle Approximation, for small angles, the arc length approximates the y-length of the triangle (the opposite side from θ), or $s \approx y$. Hence, if the length and the angle are known, the radial distance can be found, which is approximately equal to the tanker's distance from the sailor.

THINKING IT THROUGH. To approximate the distance, the ship's length is taken to be nearly equal to the arc length subtended by the measured angle. This approximation is good for small angles.

SOLUTION. The data are as follows:

Given: $\theta = 1.15°(1 \text{ rad}/57.3°)$ *Find:* r (radial distance)
$= 0.0201 \text{ rad}$
$L = 150 \text{ m}$

Knowing the arc length and angle, Eq. 7.3 can be used to find r.

$$r = \frac{s}{\theta} \approx \frac{150 \text{ m}}{0.0201} = 7.46 \times 10^3 \text{ m} = 7.46 \text{ km}$$

(Note that the unitless rad is omitted.)

The distance r is an approximation, obtained by assuming that for small angles, the arc length s and the straight-line chord length L are very nearly the same length (Fig. 7.4b). How good is this approximation? To check, let's compute the actual distance d to the middle of the ship. From the geometry, $\tan(\theta/2) = (L/2)/d$, so

$$d = \frac{L}{2\tan(\theta/2)} = \frac{150 \text{ m}}{2\tan(1.15°/2)} = 7.47 \times 10^3 \text{ m} = 7.47 \text{ km}$$

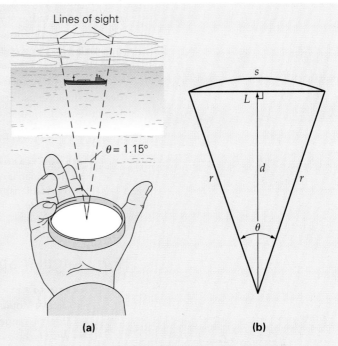

(a) **(b)**

▲ **FIGURE 7.4 Angular distance** For small angles, the arc length is approximately a straight line, or the chord length. Knowing the length of the tanker, how far away it can be found by measuring its angular size. See Example text for description. (Drawing not to scale for clarity.)

The first calculation is a pretty good approximation—the values derived by the two methods are nearly equal.

FOLLOW-UP EXERCISE. As pointed out, the approximation used in this Example is for *small* angles. You might wonder what is small. To investigate this question, what would be the percentage error of the approximated distance to the tanker for angles of 10° and 20°?

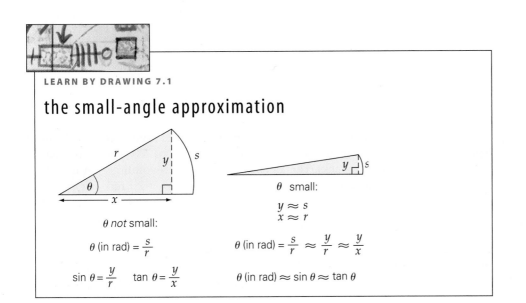

LEARN BY DRAWING 7.1

the small-angle approximation

θ *not* small:

$\theta \text{ (in rad)} = \frac{s}{r}$

$\sin\theta = \frac{y}{r}$ $\tan\theta = \frac{y}{x}$

θ small:

$y \approx s$
$x \approx r$

$\theta \text{ (in rad)} = \frac{s}{r} \approx \frac{y}{r} \approx \frac{y}{x}$

$\theta \text{ (in rad)} \approx \sin\theta \approx \tan\theta$

PROBLEM-SOLVING HINT

In computing trigonometric functions such as tan θ or sin θ, the angle may be expressed in degrees or radians; for example, sin 30° = sin[($\pi/6$) rad] = sin(0.524 rad) = 0.500. When finding trig functions with a calculator, note that there is usually a way to change the angle entry between *deg* and *rad* modes. Hand calculators commonly are set in *deg* (degree) mode, so if you want to find the value of, say, sin(1.22 rad), first change to *rad* mode and enter sin 1.22, and sin(1.22 rad) = 0.939. (Or you could convert rads to degrees first and use *deg* mode.) Some calculators may have a third mode, *grad*. The grad is a little-used angular unit. A grad is 1/100 of a right (90°) angle; that is, there are 100 grads in a right angle.

DID YOU LEARN?

➥ The radian is given by the ratio of the arc length (s) and the radius (r). That is, θ (in radians) = s/r.

➥ There are 2π radians (rad) in a circle. That is, 2π rad = 360°, and 1 rad = 57.3°.

7.2 Angular Speed and Velocity

LEARNING PATH QUESTIONS

In Circular Motion:

➥ How are tangential speed (v) and angular speed (ω) related?

➥ What is the relationship between period (T) and frequency (f)?

➥ How is the angular speed related to the period (T) and frequency (f)?

The description of circular motion in angular form is analogous to the description of linear motion. In fact, you'll notice that the equations are almost identical mathematically, with different symbols being used to indicate that the quantities have different meanings. The lowercase Greek letter omega with a bar over it is used to represent **average angular speed ($\overline{\omega}$)**, the magnitude of the angular displacement divided by the total time to travel the angular distance:

$$\overline{\omega} = \frac{\Delta\theta}{\Delta t} = \frac{\theta - \theta_0}{t - t_0} \quad \textit{(average angular speed)} \tag{7.4}$$

It is commonly said that the units of angular speed are radians per second. Technically, the unit is $1/s$ or s^{-1} since the radian is unitless. But it is useful to keep the rad to indicate that the quantity is angular speed. The **instantaneous angular speed (ω)** is given by considering a very small time interval—that is, as Δt approaches zero.

As in the linear case, if the angular speed is *constant*, then $\overline{\omega} = \omega$. Taking θ_0 and t_0 to be zero in Eq. 7.4,

$$\omega = \frac{\theta}{t} \quad \text{or} \quad \theta = \omega t \quad \textit{(instantaneous angular speed)} \tag{7.5}$$

SI unit of angular speed: radians per second (rad/s or s^{-1})

Another common descriptive unit for angular speed is revolutions per minute (rpm); for example, a CD (compact disc) rotates at a speed of 200–500 rpm (depending on the location of the track). This nonstandard unit of revolutions per minute is readily converted to radians per second, since 1 revolution = 2π rad. For example, (150 rev/min)(2π rad/rev)(1 min/ 60 s) = 5.0π rad/s (= 16 rad/s).*

The **average angular velocity** and the **instantaneous angular velocity** are analogous to their linear counterparts. Angular velocity is associated with angular displacement. Both are vectors and thus have direction; however, this directionality is, by convention, specified in a special way. In one-dimensional, or linear, motion, a

---

*It is often convenient to leave the angular speed with π in symbol form, in this case, 5.0π rad/s.

particle can go only in one direction or the other (+ or −), so the displacement and velocity vectors can have only these two directions. In the angular case, a particle moves one way or the other, but the motion is along its *circular path*.

Thus, the angular displacement and angular velocity vectors of a particle in circular motion can have only two directions, which correspond to going around the circular path with either increasing or decreasing angular displacement from θ_0—that is, counterclockwise or clockwise. Let's focus on the angular velocity vector $\vec{\omega}$. (The direction of the angular displacement will be the same as that of the angular velocity. Why?)

The *direction* of the angular velocity vector is given by a *right-hand rule* as illustrated in ▶ Fig. 7.5a. When the fingers of your right hand are curled in the direction of the circular motion, your extended thumb points in the direction of $\vec{\omega}$. Note that since circular motion can be in only one of two circular *senses*, clockwise or counterclockwise, then plus and minus signs can be used to distinguish circular rotation directions. It is customary to take a counterclockwise rotation as positive (+), since positive angular distance (and displacement) is conventionally measured counterclockwise from the positive x-axis.

Why not just designate the direction of the angular velocity vector to be either clockwise or counterclockwise? This designation is not used because clockwise (cw) and counterclockwise (ccw) are directional senses or indications rather than actual directions. These rotational senses are like right and left. If you faced another person and each of you were asked whether something was on the right or left, your answers would disagree. Similarly, if you held this book up toward a person facing you and rotated it, would it be rotating cw or ccw for both of you? Check it out. We can use cw and ccw to indicate rotational "directions" when they are specified relative to a reference—for example, the positive x-axis.

Referring to Fig. 7.5, imagine yourself being first on one side of one of the rotating disks and then on the other. Then apply the right-hand rule on both sides. You should find that the direction of the angular velocity vector is the same for both locations (because it is referenced to the right hand). Relative to this vector—for example, looking at the tip—there is no ambiguity in using + and − to indicate rotational senses or directions.

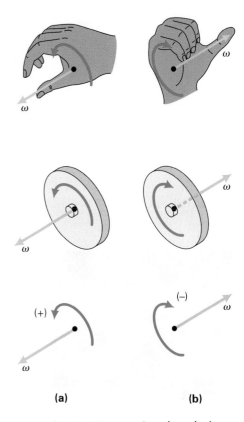

(a)   **(b)**

▲ **FIGURE 7.5 Angular velocity**
The direction of the angular velocity vector for an object in rotational motion is given by the right-hand rule: When the fingers of the right hand are curled in the direction of the rotation, the extended thumb points in the direction of the angular velocity vector. Circular senses or directions are commonly indicated by **(a)** plus and **(b)** minus signs.

RELATIONSHIP BETWEEN TANGENTIAL AND ANGULAR SPEEDS

A particle moving in a circle has an instantaneous velocity tangential to its circular path. For a constant angular velocity, the particle's *orbital speed*, or **tangential speed (v_t)**, the magnitude of the tangential velocity, is also constant. How the angular and tangential speeds are related is revealed by starting with Eq. 7.3 ($s = r\theta$) and Eq. 7.5 ($\theta = \omega t$):

$$s = r\theta = r(\omega t)$$

The arc length, or distance, is also given by

$$s = v_t t$$

Combining the equations for s gives the relationship between the tangential speed (v) and the angular speed (ω),

$$v_t = r\omega \qquad \begin{array}{l}\textit{(tangential speed relation} \\ \textit{to angular speed for circular motion)}\end{array} \qquad (7.6)$$

where ω is in radians per second. Equation 7.6 holds in general for instantaneous tangential and angular speeds for solid- or rigid-body rotation about a fixed axis, even when ω might vary with time.

Note that all the particles of a solid object rotating with constant angular velocity have the same angular speed, but the tangential speeds are different at different distances from the axis of rotation (▼Fig. 7.6a).

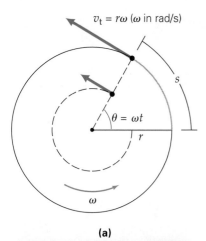

$v_t = r\omega$ (ω in rad/s)

$\theta = \omega t$

r

s

ω

(a)

(b)

▲ **FIGURE 7.6 Tangential and angular speeds** **(a)** Tangential and angular speeds are related by $v_t = r\omega$, where ω is in radians per second. Note that all of the particles of an object rotating about a fixed axis travel in circles. All the particles have the same angular speed ω, but particles at different distances from the axis of rotation have different tangential speeds. **(b)** Sparks from a grinding wheel provide a graphic illustration of instantaneous tangential velocity. (Why do the paths curve slightly?)

EXAMPLE 7.3 | **Merry-Go-Rounds: Do Some Go Faster Than Others?**

An amusement park merry-go-round at its constant operational speed makes one complete rotation in 45 s. Two children are on horses, one at 3.0 m from the center of the ride and the other farther out, 6.0 m from the center. What are (a) the angular speed and (b) the tangential speed of each child?

THINKING IT THROUGH. The angular speed of each child is the same, since both children make a complete rotation in the same time. However, the tangential speeds will be different, because the radii are different. That is, the child at the greater radius travels in a larger circle during the rotation time and thus must travel faster.

SOLUTION.

Given: $\theta = 2\pi$ rad (one rotation)
$t = 45$ s
$r_1 = 3.0$ m
$r_2 = 6.0$ m

Find: (a) ω_1 and ω_2 (angular speeds)
(b) v_1 and v_2 (tangential speeds)

(a) As noted, $\omega_1 = \omega_2$, that is, both riders rotate at the same angular speed. All points on the merry-go-round travel through 2π rad in the time it takes to make one rotation. The angular speed can be found from Eq. 7.5 (constant ω) as

$$\omega = \frac{\theta}{t} = \frac{2\pi \text{ rad}}{45 \text{ s}} = 0.14 \text{ rad/s}$$

Hence, $\omega = \omega_1 = \omega_2 = 0.14$ rad/s.

(b) The tangential speed is different at different radial locations on the merry-go-round. All of the "particles" making up the merry-go-round go through one rotation in the same amount of time. Therefore, the farther a particle is from the center, the longer its circular path, and the greater its tangential speed, as Eq. 7.6 indicates. (See also Fig. 7.6a.) Thus,

$$v_{t_1} = r_1\omega = (3.0 \text{ m})(0.14 \text{ rad/s}) = 0.42 \text{ m/s}$$

and

$$v_{t_2} = r_2\omega = (6.0 \text{ m})(0.14 \text{ rad/s}) = 0.84 \text{ m/s}$$

(Note that the rad has been dropped from the answer. Why?)
Then, a rider on the outer part of the ride has a greater tangential speed than a rider closer to the center. Here, rider 2 has a radius twice that of rider 1 and therefore goes twice as fast.

FOLLOW-UP EXERCISE. (a) On an old 45-rpm record, the beginning track is 8.0 cm from the center, and the end track is 5.0 cm from the center. What are the angular speeds and the tangential speeds at these distances when the record is spinning at 45 rpm? (b) For races on oval tracks, why do inside and outside runners have different starting points (called a "staggered" start), such that some runners start "ahead" of others?

PERIOD AND FREQUENCY

Some other quantities commonly used to describe circular motion are period and frequency. The time it takes an object in circular motion to make one complete revolution (or rotation), or *cycle*, is called the **period** (*T*). For example, the period of revolution of the Earth about the Sun is one year, and the period of the Earth's axial rotation is 24 h.* The standard unit of period is the second (s). The period is sometimes given in seconds per cycle (s/cycle).

Closely related to the period is the **frequency** (*f*), which is the number of revolutions, or cycles, made in a given time, generally a second. For example, if a particle traveling uniformly in a circular orbit makes 5.0 revolutions in 2.0 s, the frequency (of revolution) is $f = 5.0$ rev/2.0 s = 2.5 rev/s, or 2.5 cycles/s (cps, or cycles per second).

Revolution and *cycle* are merely descriptive terms used for convenience and are *not* units. Without these descriptive terms, it can be seen that the unit of frequency is inverse seconds (1/s, or s⁻¹), which is called the **hertz (Hz)** in the SI.†

---

*The discussion applies to rotations as well as revolutions. Revolutions will be used as a general term, as is commonly done, for example a CD rotates at so many revolution per minute (rpm).

†Named for Heinrich Hertz (1857–1894), a German physicist and pioneering investigator of electromagnetic waves, which also are characterized by frequency.

The two quantities are inversely related by

$$f = \frac{1}{T} \quad \begin{array}{l}\text{(relationship of}\\ \text{frequency and period)}\end{array} \qquad (7.7)$$

SI unit of frequency: hertz (Hz, $1/s$ or s^{-1})

where the period is in seconds and the frequency is in hertz, or inverse seconds.

For uniform circular motion, the tangential orbital speed is related to the period T by $v = 2\pi r/T$—that is, the distance traveled in one revolution divided by the time for one revolution (one period). The frequency can also be related to the angular speed.

Since an angular distance of 2π rad is traveled in one period (by definition of the period), then

$$\omega = \frac{2\pi}{T} = 2\pi f \quad \begin{array}{l}\text{(angular speed}\\ \text{in terms of period and frequency)}\end{array} \qquad (7.8)$$

Notice that ω and f have the same units, $\omega = 1/s$ and $f = 1/s$. This notation can easily cause confusion, which is why the unitless radian (rad) term is often used in angular speed (rad/s) and cycles in frequency (cycles/s).

EXAMPLE 7.4 | Frequency and Period: An Inverse Relationship

A compact disc (CD) rotates in a player at a constant speed of 200 rpm. What are the CD's (a) frequency and (b) period of revolution?

THINKING IT THROUGH. The relationships for the frequency (f), the period (T), and the angular frequency (ω), are expressed in Eqs. 7.7 and 7.8, so these equations can be used.

SOLUTION. The angular speed is not in standard units and so must be converted. Revolutions per minute (rpm) is converted to radians per second (rad/s).

Given:
$$\omega = \left(\frac{200\ \text{rev}}{\text{min}}\right)\left(\frac{1\ \text{min}}{60\ \text{s}}\right)\left(\frac{2\pi\ \text{rad}}{\text{rev}}\right) = 20.9\ \text{rad/s}$$

Find: (a) f (frequency)
(b) T (period)

[Note that the above unit conversion could be done using one convenient factor: $1(\text{rev/min}) = (\pi/30)\ \text{rad/s}]^*$

(a) Rearranging Eq. 7.8 and solving for f,

$$f = \frac{\omega}{2\pi} = \frac{20.9\ \text{rad/s}}{2\pi\ \text{rad/cycle}} = 3.33\ \text{Hz}$$

The units of 2π are rad/cycle or revolution, so the result is in cycles/second or inverse seconds, which is the hertz.

(b) Equation 7.8 could be used to find T, but Eq. 7.7 is a bit simpler:

$$T = \frac{1}{f} = \frac{1}{3.33\ \text{Hz}} = 0.300\ \text{s}$$

Thus, the CD takes 0.300 s to make one revolution. (Notice that since Hz = $1/s$, the equation is dimensionally correct.)

$^*(1\ \text{rev/min})[(1\ \text{min/60 s})(2\pi\ \text{rad/rev})] = (\pi/30)\ \text{rad/s}$. That is, $(\pi/30\ \text{rad/s})/\text{rpm}$.

FOLLOW-UP EXERCISE. If the period of a particular CD is 0.500 s, what is the CD's angular speed in revolutions per minute?

DID YOU LEARN?
➡ The tangential and angular speeds are directly proportional: $v_t = r\omega$.
➡ Period and frequency are inversely proportional: $T = 1/f$, or $f = 1/T$.
➡ The angular speed (ω) is inversely proportional to the period (T) and directly proportional to the frequency (f), that is, $\omega = 2\pi/T = 2\pi f$.

7.3 Uniform Circular Motion and Centripetal Acceleration

LEARNING PATH QUESTIONS
➡ What is necessary for uniform circular motion?
➡ How is it known that there is an acceleration for uniform circular motion?

A simple, but important, type of circular motion is **uniform circular motion**, which occurs when an object moves at *a constant speed in a circular path*. An example of this motion may be a car going around a circular track at a constant speed.

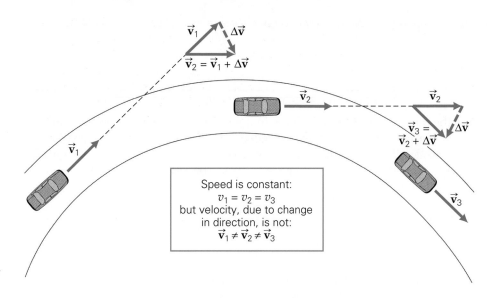

► **FIGURE 7.7 Uniform circular motion** The speed of an object in uniform circular motion is constant, but the object's velocity changes in the direction of motion. Thus, there is an acceleration.

Speed is constant:
$v_1 = v_2 = v_3$
but velocity, due to change in direction, is not:
$\vec{\mathbf{v}}_1 \neq \vec{\mathbf{v}}_2 \neq \vec{\mathbf{v}}_3$

The motion of the Moon around the Earth is approximated by uniform circular motion. Such motion is curvilinear, and as discussed in Section 3.1 there must be an acceleration. But what are its magnitude and direction?

CENTRIPETAL ACCELERATION

The acceleration of uniform circular motion is not in the same direction as the instantaneous velocity (which is tangent to the circular path at any point). If it were, the object would speed up, and the circular motion wouldn't be uniform. Recall that acceleration is the time rate of change of velocity and that velocity has both *magnitude* and *direction*. In uniform circular motion, the direction of the velocity is continuously changing, which is a clue to the direction of the acceleration. (▲Fig. 7.7).

The velocity vectors at the beginning and end of a time interval give the change in velocity, or $\Delta\vec{\mathbf{v}}$, via vector subtraction. All of the instantaneous velocity vectors have the same magnitude or length (constant speed), but they differ in direction. Note that because $\Delta\vec{\mathbf{v}}$ is not zero, there must be an acceleration ($\vec{\mathbf{a}} = \Delta\vec{\mathbf{v}}/\Delta t$).

As illustrated in ◄Fig. 7.8, as Δt (or $\Delta\theta$) becomes smaller, $\Delta\vec{\mathbf{v}}$ points more toward the center of the circular path. As Δt approaches zero, the instantaneous change in the velocity, and therefore the acceleration, points directly toward the center of the circle. As a result, the acceleration in uniform circular motion is called **centripetal acceleration (a_c)**, which means "center-seeking" acceleration (from the Latin *centri*, "center," and *petere*, "to fall toward" or "to seek").

The centripetal acceleration must be directed radially inward, that is, with no component in the direction of the perpendicular (tangential) velocity, or else the magnitude of that velocity would change (▼Fig. 7.9). Note that for an object in

▼ **FIGURE 7.8 Analysis of centripetal acceleration (a)** The velocity vector of an object in uniform circular motion is constantly changing direction. **(b)** As Δt, the time interval for $\Delta\theta$, is taken to be smaller and smaller and approaches zero, $\Delta\vec{\mathbf{v}}$ (the change in the velocity, and therefore an acceleration) is directed toward the center of the circle. The result is a centripetal, or center-seeking, acceleration that has a magnitude of $a_c = v^2/r$.

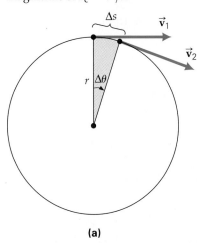

(a)

$\vec{\mathbf{v}}_2 = \vec{\mathbf{v}}_1 + \Delta\vec{\mathbf{v}}$ or
$\vec{\mathbf{v}}_2 - \vec{\mathbf{v}}_1 = \Delta\vec{\mathbf{v}}$

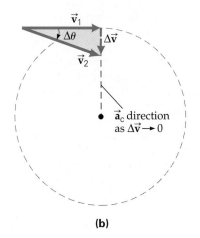

(b)

► **FIGURE 7.9 Centripetal acceleration** For an object in uniform circular motion, the centripetal acceleration is directed radially inward. There is no acceleration component in the tangential direction; if there were, the magnitude of the velocity (tangential *speed*) would change.

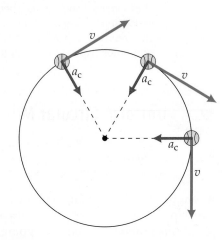

uniform circular motion, the direction of the centripetal acceleration is continuously changing. In terms of x- and y-components, a_x and a_y are not constant. (Can you describe how this differs from the acceleration in projectile motion?)

The magnitude of the centripetal acceleration can be deduced from the small shaded triangles in Fig. 7.8. (For very short time intervals, the arc length Δs is almost a straight line—the chord.) These two triangles are similar triangles, because each has a pair of equal sides surrounding the same angle $\Delta \theta$. (Note that the velocity vectors have the same magnitude.) Thus, Δv is to v as Δs is to r, which can be written as*

$$\frac{\Delta v}{v} \approx \frac{\Delta s}{r}$$

The arc length Δs is the distance traveled in time Δt, thus, $\Delta s = v\Delta t$ so

$$\frac{\Delta v}{v} \approx \frac{\Delta s}{r} = \frac{v\Delta t}{r}$$

and

$$\frac{\Delta v}{\Delta t} \approx \frac{v^2}{r}$$

Then, as Δt approaches zero, this approximation becomes exact. The instantaneous centripetal acceleration, $a_c = \Delta v/\Delta t$, thus has a magnitude of

$$a_c = \frac{v^2}{r} \qquad \text{\textit{(magnitude of centripetal acceleration in terms of tangential speed)}} \qquad (7.9)$$

Using Eq. 7.6 ($v = r\omega$), the equation for centripetal acceleration can also be written in terms of the angular speed:

$$a_c = \frac{v^2}{r} = \frac{(r\omega)^2}{r} = r\omega^2 \qquad \text{\textit{(magnitude of centripetal acceleration in terms of angular speed)}} \qquad (7.10)$$

Orbiting satellites have centripetal accelerations, and a down-to-Earth medical application of centripetal acceleration is discussed in Insight 7.1, The Centrifuge: Separating Blood Components.

*The subscript t will be dropped with the understanding that v is tangential speed in ufiorm circular motion.

INSIGHT 7.1 | ## The Centrifuge: Separating Blood Components

The centrifuge is a machine with rotating parts that is used to separate particles of different sizes and densities suspended in a liquid (or a gas). For example, cream is separated from milk by centrifuging, and blood components are separated in centrifuges in medical laboratories (see Fig. 7.10).

There is a much slower process to separate blood components. They will eventually settle in layers in a vertical test tube—a process called *sedimentation*—under the influence of normal gravity alone. The viscous drag of the plasma on the particles is analogous to (but much greater than) the air resistance that determines the terminal velocity of falling objects (Section 4.6). Red blood cells settle in the bottom layer of the tube, because they have a greater terminal velocity than do the white blood cells and platelets and reach the bottom sooner. The white cells settle in the next layer and the platelets settle on top. However, gravitational sedimentation is generally a very slow process.

The erythrocite sedimentation rate (ESR) has some diagnostic value, but clinicians usually do not want to wait a long time to see the fractional volume of red cells (erythrocites) in the blood or separate them from the plasma. And so a centrifuge is used to speed up the process. Centrifuge tubes are pivoted and spin horizontally. The resistance of the fluid medium on the particles supplies the centripetal acceleration that keeps them moving in slowly widening circles as they travel toward the bottom of the tube. The bottom of the tube itself must exert a strong force on the contents as a whole and must be strong enough not to break.

Laboratory centrifuges commonly operate at speeds sufficient to produce centripetal accelerations thousands of times larger than g. (See Example 7.5.) Since the principle of the centrifuge involves centripetal acceleration, perhaps "centripuge" would be a more descriptive name.

EXAMPLE 7.5 | A Centrifuge: Centripetal Acceleration

A laboratory centrifuge like that shown in ▸ Fig. 7.10 operates at a rotational speed of 12 000 rpm. (a) What is the magnitude of the centripetal acceleration of a red blood cell at a radial distance of 8.00 cm from the centrifuge's axis of rotation? (b) How does this acceleration compare with g?

THINKING IT THROUGH. Here, the angular speed and the radius are given, so the magnitude of the centripetal acceleration can be computed directly from Eq. 7.10. The result can be compared with g by using $g = 9.80$ m/s^2.

SOLUTION. The data are as follows:

Given:
$$\omega = (1.20 \times 10^4 \text{ rpm})\left[\frac{(\pi/30)\text{rad/s}}{\text{rpm}}\right]$$
$$= 1.26 \times 10^3 \text{ rad/s}$$
(using a single conversion factor)
$$r = 8.00 \text{ cm} = 0.0800 \text{ m}$$

Find: (a) a_c
(b) how a_c compares with g

(a) The centripetal acceleration is found from Eq. 7.10:

$$a_c = r\omega^2 = (0.0800 \text{ m})(1.26 \times 10^3 \text{ rad/s})^2 = 1.27 \times 10^5 \text{ m/s}^2$$

(b) Using the relationship $1\,g = 9.80$ m/s^2 to express a_c in terms of g,

$$a_c = (1.27 \times 10^5 \text{ m/s}^2)\left(\frac{1\,g}{9.80 \text{ m/s}^2}\right) = 1.30 \times 10^4\,g\ (= 13\,000\,g!)$$

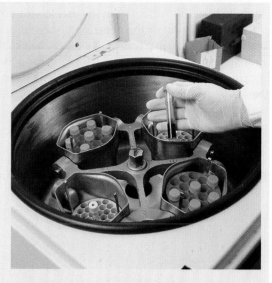

▲ **FIGURE 7.10 Centrifuge** Centrifuges are used to separate particles of different sizes and densities suspended in liquids. For example, red and white blood cells can be separated from each other and from the plasma that makes up the liquid portion of the blood in the centrifuge tube. When spinning, the tubes are horizontal.

FOLLOW-UP EXERCISE. What angular speed in revolutions per minute would give a centripetal acceleration of $1\,g$ at the radial distance in this Example, and, taking gravity into account, what would be the resultant acceleration?

CENTRIPETAL FORCE

For an acceleration to exist, there must be a net force. Thus, for a centripetal (inward) acceleration to exist, there must be a **centripetal force** (net inward force). Expressing the magnitude of this force in terms of Newton's second law ($\vec{F}_{net} = m\vec{a}$) and inserting the expression for centripetal acceleration from Eq. 7.9 for magnitude,

$$F_c = ma_c = \frac{mv^2}{r} \quad \textit{(magnitude of centripetal force)} \tag{7.11}$$

The centripetal force, like the centripetal acceleration, is directed radially toward the center of the circular path.

CONCEPTUAL EXAMPLE 7.6 | Breaking Away

A ball attached to a string is swung with uniform motion in a horizontal circle above a person's head (▸ Fig. 7.11a). If the string breaks, which of the trajectories shown in Fig. 7.11b (viewed from above) would the ball follow?

REASONING AND ANSWER. When the string breaks, the centripetal force goes to zero. There is no force in the outward direction, so the ball could not follow trajectory *a*. Newton's

first law states that if no force acts on an object in motion, the object will continue to move in a straight line. This factor rules out trajectories *b*, *d*, and *e*.

It should be evident from the previous discussion that at any instant (including the instant when the string breaks), the isolated ball has a horizontal, tangential velocity. The downward force of gravity acts on it, but this force affects only its vertical motion, which is not visible in Fig. 7.11b.

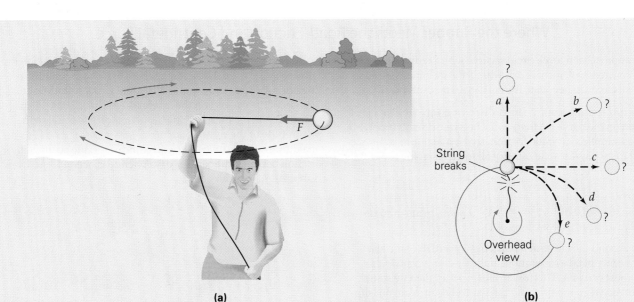

▲ **FIGURE 7.11 Centripetal force (a)** A ball is swung in a horizontal circle. **(b)** If the string breaks and the centripetal force goes to zero, what happens to the ball? See Example text for description.

The ball thus flies off tangentially and is essentially a horizontal projectile (with $v_{x_o} = v$, $v_{y_o} = 0$, and $a_y = -g$).

Viewed from above, the ball would appear to follow the path labeled c.

FOLLOW-UP EXERCISE.   If you swing a ball in a horizontal circle about your head, can the string be exactly horizontal? (See Fig. 7.11a.) Explain your answer. [*Hint*: Analyze the forces acting on the ball.]

Keep in mind that, in general, a net force applied at an angle to the direction of motion of an object produces changes in the magnitude *and* direction of the velocity. However, when a net force of constant magnitude is continuously applied at an angle of 90° to the direction of motion (as is centripetal force), only the direction of the velocity changes. This is because there is no force component parallel to the velocity. Also notice that because the centripetal force is always perpendicular to the direction of motion, this force does no work. (Why?) Therefore, *a centripetal force does not change the kinetic energy or speed of the object*.

Note that the centripetal force in the form $F_c = mv^2/r$ is *not* a new individual force, but rather the cause of the centripetal acceleration, and is supplied by either a real force or the vector sum of several forces.

The force supplying the centripetal acceleration for satellites is gravity. In Conceptual Example 7.6, it was the tension in the string. Another force that often supplies centripetal acceleration is friction. Suppose that an automobile moves into a level, circular curve. To negotiate the curve, the car must have a centripetal acceleration, which is supplied by the force of friction between the tires and the road.

However, this static friction (why static?) has a maximum limiting value. If the speed of the car is high enough or the curve is sharp enough, the friction will not be sufficient to supply the necessary centripetal acceleration, and the car will skid outward from the center of the curve. If the car moves onto a wet or icy spot, the friction between the tires and the road may be reduced, allowing the car to skid at an even lower speed. (Banking a curve helps vehicles negotiate the curve without slipping.)

EXAMPLE 7.7 | # Where the Rubber Meets the Road: Friction and Centripetal Force

A car approaches a level, circular curve with a radius of 45.0 m. If the concrete pavement is dry, what is the maximum speed at which the car can negotiate the curve at a constant speed?

THINKING IT THROUGH.   The car is in uniform circular motion on the curve, so there must be a centripetal force. This force is supplied by static friction, so the maximum static frictional force provides the centripetal force when the car is at its maximum tangential speed.

SOLUTION.

Given: $r = 45.0$ m   *Find:* v (maximum speed)
    $\mu_s = 1.20$ (from Table 4.1)

To go around the curve at a particular speed, the car must have a centripetal acceleration, and therefore a centripetal force must act on it. This inward force is supplied by static friction between the tires and the road. (The tires are not slipping or skidding relative to the road.)

Recall from Section 4.6 that the maximum frictional force is given by $f_{s_{max}} = \mu_s N$ (Eq. 4.7), where N is the magnitude of the normal force on the car and is equal in magnitude to the weight of the car, mg, on the level road (why?). Thus the magnitude of the maximum static frictional force is equal to the magnitude of the centripetal force ($F_c = mv^2/r$). From this the maximum speed can be found. To find $f_{s_{max}}$, the coefficient of friction between rubber and dry concrete is needed, and from Table 4.1, $\mu_s = 1.20$. Then,

$$f_{s_{max}} = F_c$$

$$\mu_s N = \mu_s mg = \frac{mv^2}{r}$$

So

$$v = \sqrt{\mu_s rg} = \sqrt{(1.20)(45.0 \text{ m})(9.80 \text{ m/s}^2)} = 23.0 \text{ m/s}$$

(about 83 km/h, or 52 mi/h).

FOLLOW-UP EXERCISE.   Would the centripetal force be the same for all types of vehicles as in this Example? Explain.

The proper safe speed for driving on a highway curve is an important consideration. The coefficient of friction between tires and the road may vary, depending on weather, road conditions, the design of the tires, the amount of tread wear, and so on. When a curved road is designed, safety may be promoted by banking, or inclining, the roadway. This design reduces the chances of skidding because the normal force exerted on the car by the road then has a component toward the center of the curve that reduces the need for friction. In fact, for a circular curve with a given banking angle and radius, there is one speed for which no friction is required at all. This condition is used in banking design. (See Conceptual Question 12 at the end of the chapter.)

Let's look at one more example of centripetal force, this time with two objects in uniform circular motion. Example 7.8 will help give a better understanding of the motions of satellites in circular orbits, discussed in a later section.

EXAMPLE 7.8 | # Strung Out: Centripetal Force and Newton's Second Law

Suppose that two masses, $m_1 = 2.5$ kg and $m_2 = 3.5$ kg, are connected by light strings and are in uniform circular motion on a horizontal frictionless surface as illustrated in ▼ Fig. 7.12, where $r_1 = 1.0$ m and $r_2 = 1.3$ m. The tension forces acting on the masses are $T_1 = 4.5$ N and $T_2 = 2.9$ N, which are the respective tensions in the strings. Find the magnitude of the centripetal acceleration and the tangential speed of (a) mass m_2 and (b) mass m_1.

▶ **FIGURE 7.12 Centripetal force and Newton's second law** See Example text for description.

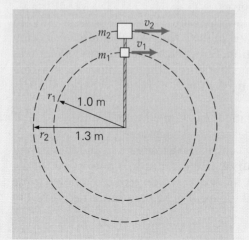

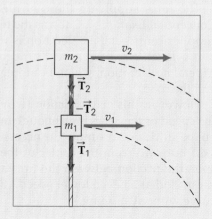

THINKING IT THROUGH. The centripetal forces on the masses are supplied by the tensions (T_1 and T_2) in the strings. By isolating the masses, a_c for each mass can be found, because the net force on a mass is equal to the mass's centripetal force ($F_c = ma_c$). The tangential speeds can then be found, since the radii are known ($a_c = v^2/r$).

SOLUTION.

Given: $r_1 = 1.0$ m and $r_2 = 1.3$ m   *Find:* a_c (centripetal acceleration) and v (tangential speed)
$m_1 = 2.5$ kg and $m_2 = 3.5$ kg   (a) m_2
$T_1 = 4.5$ N   (b) m_1
$T_2 = 2.9$ N

(a) By isolating m_2 in the figure, it can seen that the centripetal force is provided by the tension in the string. (T_2 is the only force acting on m_2 toward the center of its circular path.) Thus,

$$T_2 = m_2 a_{c_2}$$

and

$$a_{c_2} = \frac{T_2}{m_2} = \frac{2.9 \text{ N}}{3.5 \text{ kg}} = 0.83 \text{ m/s}^2$$

where the acceleration is toward the center of the circle.
The tangential speed of m_2 can be found from $a_c = v^2/r$:

$$v_2 = \sqrt{a_{c_2} r_2} = \sqrt{(0.83 \text{ m/s}^2)(1.3 \text{ m})} = 1.0 \text{ m/s}$$

(b) The situation is a bit different for m_1. In this case, two radial forces are acting on m_1: the string tensions T_1 (inward) and $-T_2$ (outward). By Newton's second law, in order to have a centripetal acceleration, there must be a net force, which is given by the difference in the two tensions, so we expect $T_1 > T_2$, and

$$F_{\text{net}_1} = +T_1 + (-T_2) = m_1 a_{c_1} = \frac{m_1 v_1^2}{r_1}$$

where the radial direction (toward the center of the circular path) is taken to be positive. Then

$$a_{c_1} = \frac{T_1 - T_2}{m_1} = \frac{4.5 \text{ N} - 2.9 \text{ N}}{2.5 \text{ kg}} = 0.64 \text{ m/s}^2$$

and

$$v_1 = \sqrt{a_{c_1} r_1} = \sqrt{(0.64 \text{ m/s}^2)(1.0 \text{ m})} = 0.80 \text{ m/s}$$

FOLLOW-UP EXERCISE. Notice in this Example that the centripetal acceleration of m_2 is greater than that of m_1 yet $r_2 > r_1$, and $a_c \propto 1/r$. Is something wrong here? Explain.

INTEGRATED EXAMPLE 7.9 | Center-Seeking Force: One More Time

A 1.0-m cord is used to suspend a 0.50-kg tetherball from the top of the pole. After being hit several times, the ball goes around the pole in uniform circular motion with a tangential speed of 1.1 m/s at an angle of 20° relative to the pole. (a) The force that supplies the centripetal acceleration is (1) the weight of the ball, (2) a component of the tension force in the string, (3) the total tension in the string. (b) What is the magnitude of the centripetal force?

(A) CONCEPTUAL REASONING. The centripetal force, being a "center-seeking" force, is directed perpendicularly toward the pole, about which the ball is in circular motion. As suggested in the problem-solving procedures provided in Section 1.7, it is almost always helpful to sketch a diagram, such as that in ▶ Fig. 7.13.
Immediately, it can be seen that (1) and (3) are not correct, as these forces

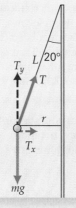

▶ **FIGURE 7.13 Ball on a string** See Example text for description.

are not directly toward the circle's center located on the pole. (mg and T_y are equal and opposite, because there is no acceleration in the y-direction.) The answer is obviously (2), with a component of the tension force, T_x, supplying the centripetal force.

(B) QUANTITATIVE REASONING AND SOLUTION. T_x supplies the centripetal force, and the given data are for the dynamical form of the centripetal force, that is, $T_x = F_c = mv^2/r$ (Eq. 7.11).

Given: $L = 1.0$ m   *Find:* F_c (magnitude of the
$v_t = 1.1$ m/s   centripetal force)
$m = 0.50$ kg
$\theta = 20°$

As pointed out previously, the magnitude of the centripetal force may be found using Eq. 7.11:

$$F_c = T_x = \frac{mv^2}{r}$$

But the radial distance r is needed. From the figure, this quantity can be seen to be $r = L \sin 20°$, so

$$F_c = \frac{mv^2}{L \sin 20°} = \frac{(0.50 \text{ kg})(1.1 \text{ m/s})^2}{(1.0 \text{ m})(0.342)} = 1.8 \text{ N}$$

FOLLOW-UP EXERCISE. (a) What is the magnitude of the tension T in the string? (b) What is the period of the ball's rotation?

7.4   Angular Acceleration

LEARNING PATH QUESTIONS
➟ How is the tangential acceleration (a_t) related to the angular acceleration (α)?
➟ The linear and angular kinematic equations for constant accelerations have similar form. What are the analogous quantities?

As you might have guessed, there is another type of acceleration besides linear, and that is *angular acceleration*. This quantity is the time rate of change of angular velocity. In circular motion, if there is an angular acceleration, the motion is not uniform, because the speed and/or direction would be changing. Analogous to the linear case, the magnitude of the **average angular acceleration ($\bar{\alpha}$)** is given by

$$\bar{\alpha} = \frac{\Delta\omega}{\Delta t}$$

where the bar over the alpha indicates that it is an average value, as usual. Taking $t_o = 0$, and if the angular acceleration is constant, so that $\bar{\alpha} = \alpha$, then

$$\alpha = \frac{\omega - \omega_o}{t} \quad \textit{(constant angular acceleration)}$$

SI unit of angular acceleration: radians per second squared (rad/s^2)

and rearranging,

$$\omega = \omega_o + \alpha t \quad \begin{array}{l}\textit{(constant angular}\\\textit{acceleration only)}\end{array} \tag{7.12}$$

No boldface vector symbols with overarrows are used in Eq. 7.12, because, plus and minus signs will be used to indicate angular directions, as described earlier. As in the case of linear motion, if the angular acceleration increases the angular velocity, both quantities have the same sign, meaning that their vector directions are the same (that is, α is in the same direction as ω as given by the right-hand rule). If the angular acceleration decreases the angular velocity, then the two quantities have opposite signs, meaning that their vectors are opposed (that is, α is in the direction opposite to ω as given by the right-hand rule, or is an angular deceleration, so to speak).

EXAMPLE 7.10 | A Rotating CD: Angular Acceleration

A CD accelerates uniformly from rest to its operational speed of 500 rpm in 3.50 s. (a) What is the angular acceleration of the CD during this time? (b) What is the angular acceleration of the CD after this time? (c) If the CD comes uniformly to a stop in 4.50 s, what is its angular acceleration during this part of the motion?

THINKING IT THROUGH.   (a) Once given the initial and final angular velocities, the constant (uniform) angular acceleration can be calculated (Eq. 7.12), since the amount of time during which the CD accelerates is known. (b) Keep in mind that the operational angular speed is constant. (c) Everything is given for Eq. 7.12, but a negative result should be expected. Why?

SOLUTION.
Given: $\omega_o = 0$

$\omega = (500 \text{ rpm})\left[\dfrac{(\pi/30) \text{ rad/s}}{\text{rpm}}\right] = 52.4 \text{ rad/s}$

$t_1 = 3.50 \text{ s (starting up)}$
$t_2 = 4.50 \text{ s (coming to a stop)}$

Find: (a) α (during startup)
(b) α (in operation)
(c) α (in coming to a stop)

(a) Using Eq. 7.12, the acceleration during startup is

$$\alpha = \frac{\omega - \omega_o}{t_1} = \frac{52.4 \text{ rad/s} - 0}{3.50 \text{ s}} = 15.0 \text{ rad/s}^2$$

in the direction of the angular velocity.

(b) After the CD reaches its operational speed, the angular velocity remains constant, so $\alpha = 0$.

(c) Again using Eq. 7.12, but this time with $\omega_o = 500$ rpm and $\omega = 0$.

$$\alpha = \frac{\omega - \omega_o}{t_2} = \frac{0 - 52.4 \text{ rad/s}}{4.50 \text{ s}} = -11.6 \text{ rad/s}^2$$

where the minus sign indicates that the angular acceleration is in the direction opposite that of the angular velocity (which is taken as +).

FOLLOW-UP EXERCISE. (a) What are the directions of the $\vec{\omega}$ and $\vec{\alpha}$ vectors in part (a) of this Example if the CD rotates clockwise when viewed from above? (b) Do the directions of these vectors change in part (c)? Explain.

As with arc length and angle ($s = r\theta$) and tangential and angular speeds ($v = r\omega$), there is a relationship between the magnitudes of the tangential acceleration and the angular acceleration. The **tangential acceleration** (a_t) is associated with changes in tangential speed and hence continuously changes direction. The magnitudes of the tangential and angular accelerations are related by a factor of r. For circular motion with a constant radius r,

$$a_t = \frac{\Delta v}{\Delta t} = \frac{\Delta(r\omega)}{\Delta t} = \frac{r\Delta\omega}{\Delta t} = r\alpha$$

so

$$a_t = r\alpha \quad \textit{(magnitude of tangential acceleration)} \quad (7.13)$$

The tangential acceleration (a_t) is written with a subscript t to distinguish it from the radial, or centripetal, acceleration (a_c). Centripetal acceleration is necessary for circular motion, but tangential acceleration is not. For uniform circular motion, there is no angular acceleration ($\alpha = 0$) or tangential acceleration, as can be seen from Eq. 7.13. There is only centripetal acceleration (▶Fig. 7.14a).

However, when there is an angular acceleration α (and therefore a tangential acceleration of magnitude $a_t = r\alpha$), there is a change in *both* the angular *and* tangential velocities. As a result, the centripetal acceleration $a_c = v^2/r = r\omega^2$ must increase or decrease if the object is to maintain the same circular orbit (that is, if r is to stay the same). When there are both tangential and centripetal accelerations, the instantaneous acceleration is their vector sum (Fig. 7.14b).

The tangential acceleration vector and the centripetal acceleration vector are perpendicular to each other at any instant, and the acceleration is $\vec{a} = a_t\hat{t} + a_c\hat{r}$, where $\hat{t}$ and $\hat{r}$ are unit vectors directed tangentially and radially inward, respectively. You should be able to find the magnitude of $\vec{a}$ and the angle it makes relative to $\vec{a}_t$ by using trigonometry (Fig. 7.14b).

Other equations for angular kinematics can be derived, as was done for the linear equations in Section 2.4. That development will not be shown here; the set of angular equations with their linear counterparts for constant accelerations is listed in ▼Table 7.2. A quick review of Section 2.4 (with a change of symbols) will show you how the angular equations are derived.

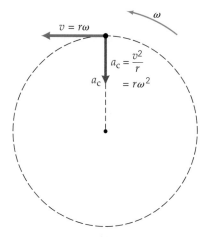

(a) Uniform circular motion
$(\alpha = 0)$

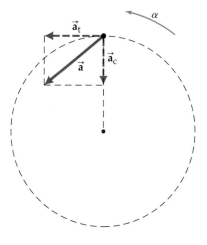

(b) Nonuniform circular motion
$(\vec{a} = \vec{a}_t + \vec{a}_c)$

▲ **FIGURE 7.14 Acceleration and circular motion (a)** In uniform circular motion, there is centripetal acceleration, but no angular acceleration $(\alpha = 0)$ or tangential acceleration $(a_t = r\alpha = 0)$. **(b)** In nonuniform circular motion, there are angular and tangential accelerations, and the total acceleration is the vector sum of the tangential and centripetal accelerations.

TABLE 7.2 **Equations for Linear and Angular Motion with Constant Acceleration***

Linear	Angular	
$x = \bar{v}t$	$\theta = \bar{\omega}t$	(1)
$\bar{v} = \dfrac{v + v_o}{2}$	$\bar{\omega} = \dfrac{\omega + \omega_o}{2}$	(2)
$v = v_o + at$	$\omega = \omega_o + \alpha t$	(3)
$x = x_o + v_o t + \dfrac{1}{2}at^2$	$\theta = \theta_o + \omega_o t + \dfrac{1}{2}\alpha t^2$	(4)
$v^2 = v_o^2 + 2a(x - x_o)$	$\omega^2 = \omega_o^2 + 2\alpha(\theta - \theta_o)$	(5)

*The first equation in each column is general, that is, not limited to situations where the acceleration is constant.

EXAMPLE 7.11 | **Even Cooking: Rotational Kinematics**

A microwave oven has a 30-cm-diameter rotating plate for even cooking. The plate accelerates from rest at a uniform rate of 0.87 rad/s^2 for 0.50 s before reaching its constant operational speed. (a) How many revolutions does the plate make before reaching its operational speed? (b) What are the operational angular speed of the plate and the operational tangential speed at its rim?

THINKING IT THROUGH. This Example involves the use of the angular kinematic equations (Table 7.2). In (a), the angular distance θ will give the number of revolutions. For (b), first find ω and then $v = r\omega$.

SOLUTION. Listing the given data and what is to be found:

Given: $d = 30$ cm,   **Find:** (a) θ (in revolutions)
$r = 15$ cm $= 0.15$ m   (b) ω and v (angular
(radius)   and tangential
$\omega_\mathrm{o} = 0$ (at rest)   speeds,
$\alpha = 0.87$ rad/s^2 respectively)
$t = 0.50$ s

(a) To find the angular distance θ *in radians*, use Eq. 4 from Table 7.2 with $\theta_\mathrm{o} = 0$:

$$\theta = \omega_\mathrm{o}t + \tfrac{1}{2}\alpha t^2 = 0 + \tfrac{1}{2}(0.87 \text{ rad/s}^2)(0.50 \text{ s})^2 = 0.11 \text{ rad}$$

Since 2π rad $= 1$ rev,

$$\theta = (0.11 \text{ rad})\left(\frac{1 \text{ rev}}{2\pi \text{ rad}}\right) = 0.018 \text{ rev}$$

so the plate reaches its operational speed in only a small fraction of a revolution.

(b) From Table 7.2, it can be seen that Eq. 3 gives the angular speed, and

$$\omega = \omega_\mathrm{o} + \alpha t = 0 + (0.87 \text{ rad/s}^2)(0.50 \text{ s}) = 0.44 \text{ rad/s}$$

Then, Eq. 7.6 gives the tangential speed at the rim radius:

$$v = r\omega = (0.15 \text{ m})(0.44 \text{ rad/s}) = 0.066 \text{ m/s}$$

FOLLOW-UP EXERCISE. (a) When the oven is turned off, the plate makes half a revolution before stopping. What is the plate's angular acceleration during this period? (b) How long does it take to stop?

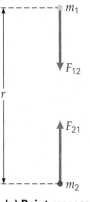

(a) Point masses

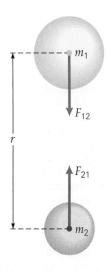

(b) Homogeneous spheres

$$F_{12} = F_{21} = \frac{Gm_1 m_2}{r^2}$$

DID YOU LEARN?
➥ The tangential acceleration (a_t) and angular acceleration (α) are directly proportional, $a_\mathrm{t} = r\alpha$ (similar to speeds: $v_\mathrm{t} = r\omega$).
➥ The linear quantities x, v, and a are respectively analogous to the angular quantities θ, ω and α.

7.5 Newton's Law of Gravitation

LEARNING PATH QUESTIONS
➥ What is meant by an inverse-square relationship?
➥ How does the acceleration due to gravity vary with altitude above the Earth's surface?

Another of Isaac Newton's many accomplishments was the formulation of the **universal law of gravitation**. This law is very powerful and fundamental. Without it, for example, we would not understand the cause of tides or know how to put satellites into particular orbits around the Earth. This law allows us to analyze the motions of planets, comets, stars, and even galaxies. The word *universal* in the name indicates that it is believed to apply everywhere in the universe. (This term highlights the importance of the law, but for brevity, it is common to refer simply to *Newton's law of gravitation* or the *law of gravitation*.)

Newton's law of gravitation in mathematical form gives a simple relationship for the gravitational interaction between two particles, or point masses, m_1 and m_2 separated by a distance r (◀Fig. 7.15a). Basically, every particle in the universe has an attractive gravitational interaction with every other particle because of their masses. The forces of mutual interaction are equal and opposite, forming a force pair as described by Newton's third law (Section 4.4), that is, $\vec{\mathbf{F}}_{12} = -\vec{\mathbf{F}}_{21}$ in Fig. 7.15a.

◀**FIGURE 7.15 Universal law of gravitation** **(a)** Any two particles, or point masses, are gravitationally attracted to each other with a force that has a magnitude given by Newton's universal law of gravitation. **(b)** For homogeneous spheres, the masses may be considered to be concentrated at their centers.

The gravitational attraction, or force (F), decreases as the square of the distance (r^2) between two point masses increases; that is, the magnitude of the gravitational force and the distance separating the two particles are related as follows:

$$F \propto \frac{1}{r^2}$$

(This type of relationship is called an *inverse-square*, that is, F is inversely proportional to r^2.)

Newton's law also correctly postulates that the gravitational force, or attraction of a body, depends on the body's mass—the greater the mass, the greater the attraction. However, because gravity is a mutual interaction between masses, it should be directly proportional to both masses, that is, to their product ($F \propto m_1 m_2$).

Hence, **Newton's law of gravitation** has the form $F \propto m_1 m_2 / r^2$. Expressed as an equation with a constant of proportionality, the magnitude of the mutually attractive gravitational force (F_g) between two masses is given by

$$F_g = \frac{Gm_1m_2}{r^2} \qquad \textit{(Newton's law of gravitation)} \qquad (7.14)$$

where G is a constant called the **universal gravitational constant** and has a value of

$$G = 6.67 \times 10^{-11}\,\text{N} \cdot \text{m}^2/\text{kg}^2$$

This constant is often referred to as "big G" to distinguish it from "little g," the acceleration due to gravity. Note from Eq. 7.14 that F_g approaches zero only when r becomes infinitely large. That is, the gravitational force has, or acts over, an *infinite range*.

How did Newton come to his conclusions about the force of gravity? Legend has it that his insight came after he observed an apple fall from a tree to the ground. Newton had been wondering what supplied the centripetal force to keep the Moon in orbit and might have had this thought: "If gravity attracts an apple toward the Earth, perhaps it also attracts the Moon, and the Moon is 'falling', or accelerating toward the Earth, under the influence of gravity" (▶ Fig. 7.16).

Whether or not the legendary apple did the trick, Newton assumed that the Moon and the Earth were attracted to each other and could be treated as point masses, with their total masses concentrated at their centers (Fig 7.15b). The inverse-square relationship had been speculated on by some of his contemporaries. Newton's achievement was demonstrating that the relationship could be deduced from one of Johannes Kepler's laws of planetary motion (Section 7.6).

Newton expressed Eq. 7.14 as a proportion ($F_g \propto m_1 m_2 / r^2$) because he did not know the value of G. It was not until 1798 (seventy-one years after Newton's death) that the value of the universal gravitational constant was experimentally determined by an English physicist, Henry Cavendish. Cavendish used a sensitive balance to measure the gravitational force between separated spherical masses (as illustrated in Fig. 7.15b). If F, r, and the m's are known, G can be computed from Eq. 7.14.

As mentioned earlier, Newton considered the nearly spherical Earth and Moon to be point masses located at their respective centers. It took him some years, using mathematical methods he developed, to prove that this is the case only for spherical, *homogeneous* objects.* The concept is illustrated in ▼Fig. 7.17.

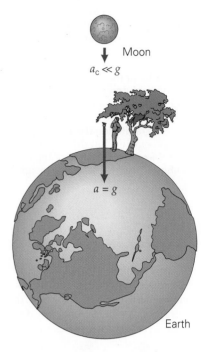

▲ **FIGURE 7.16 Gravitational insight?** Newton developed his law of gravitation while studying the orbital motion of the Moon. According to legend, his thinking was spurred when he observed an apple falling from a tree. He supposedly wondered whether the force causing the apple to accelerate toward the ground could extend to the Moon and cause it to "fall" or accelerate toward Earth, that is, supply its orbital centripetal acceleration.

---

*For a homogeneous sphere, the equivalent point mass is located at the center of mass. However, this is a special case. The center of gravitational force and the center of mass of a configuration of particles or an object do not generally coincide.

▶ **FIGURE 7.17 Uniform spherical masses**
(a) Gravity acts between any two particles. The resultant gravitational force exerted on an object outside a homogeneous sphere by two particles at symmetric locations within the sphere is directed toward the center of the sphere.
(b) Because of the sphere's symmetry and uniform distribution of mass, the net effect is as though all the mass of the sphere were concentrated as a particle at its center. For this special case, the gravitational center of force and center of mass coincide, but this is generally not true for other objects. (Only a few of the red force arrows are shown because of space considerations.)

(a)   **(b)**

EXAMPLE 7.12 | ## Greater Gravitational Attraction?

The gravitational attractions of the Sun and the Moon give rise to ocean tides. It is sometimes said that since the Moon is closer to the Earth than the Sun, the Moon's gravitational attraction is much stronger, and therefore has a greater influence on ocean tides. Is this true?

THINKING IT THROUGH.   To see if this is true, the gravitational attractions of the Moon and the Sun on the Earth can be easily calculated using Newton's law of gravitation. The masses and distances are given on the inside back cover of the book. (Assume that the bodies are solid, homogenous spheres.)

SOLUTION.   No data are given, so this must be available from references:

Given: From tables on the inside back cover:

$m_E = 6.0 \times 10^{24}$ kg (mass of the Earth)

$m_M = 7.4 \times 10^{22}$ kg (mass of the Moon)

$m_S = 2.0 \times 10^{30}$ kg (mass of the Sun)

$r_{EM} = 3.8 \times 10^8$ m (average distance between)

$r_{ES} = 1.5 \times 10^8$ km (average distance between)

Find: F_{EM} (gravitational force, Earth–Moon)
   F_{ES} (gravitational force, Earth–Sun)

The average distances are taken to be the distance from the center of one to the center of the other. Using Eq. 7.14, remembering to change kilometers to meters:

$$F_{EM} = \frac{Gm_1 m_2}{r^2} = \frac{GM_E m_M}{r_{EM}^2}$$

$$= \frac{(6.67 \times 10^{-11}\,\text{N} \cdot \text{m}^2/\text{kg}^2)(6.0 \times 10^{24}\,\text{kg})(7.4 \times 10^{22}\,\text{kg})}{(3.8 \times 10^8\,\text{m})^2}$$

$$= 2.1 \times 10^{20}\,\text{N} \quad (\text{Earth–Moon})$$

$$F_{ES} = \frac{Gm_E m_s}{r_{ES}^2} = \frac{(6.67 \times 10^{-11}\,\text{N} \cdot \text{m}^2/\text{kg}^2)(6.0 \times 10^{24}\,\text{kg})(2.0 \times 10^{30}\,\text{kg})}{(1.5 \times 10^{11}\,\text{m})^2}$$

$$= 3.6 \times 10^{22}\,\text{N} \quad (\text{Earth–Sun})$$

So, the gravitational attraction of the Sun on the Earth is much greater than that of the Moon on the Earth, on the order of 100 times greater. But it is well known that the Moon has the major influence on tides. How is this with less gravitational attraction? Basically, it is because the gravitational *differential* of the Moon with less gravitational attraction is greater. That is, the ocean water on the side of the Earth toward the Moon is closer and the gravitational attraction forms a tidal bulge. The Earth is less attracted toward the Moon, but it is some-

what displaced, leaving the least attracted water on the opposite side of the Moon where another tidal bulge is formed. As the Moon revolves about the Earth, the tidal bulges tag along, and there are two high tides (bulges) and two low tides daily. (Actually, the two high tides are 12 h, 25 min apart.)

Even though the Sun has greater gravitational attraction, the differential distances from the Sun to the water-Earth-water are miniscule, and so the Sun has little effect on the daily tides.

FOLLOW-UP EXERCISE. The gravitational attraction of the Earth on the Moon provides the centripetal force that keeps the Moon revolving in its orbit. It is sometimes said the Moon is "falling" (accelerating) toward the Earth. What is the magnitude of the Moon's acceleration in "falling" toward the Earth? And with this acceleration, why doesn't the Moon get closer to the Earth?

The acceleration due to gravity at a particular distance from a planet can also be investigated by using Newton's second law of motion and the law of gravitation. The magnitude of the acceleration due to gravity, which will generally be written as a_g at a distance r from the center of a spherical mass M, is found by setting the force of gravitational attraction due to that spherical mass equal to ma_g. This is the net force on an object of mass m at a distance r:

$$ma_g = \frac{GmM}{r^2}$$

Then, the acceleration due to gravity at any distance r from the planet's center is

$$a_g = \frac{GM}{r^2} \qquad (7.15)$$

Notice that a_g is proportional to $1/r^2$, so the farther away an object is from the planet, the smaller its acceleration due to gravity and the smaller the attractive force (ma_g) on the object. The force is directed toward the center of the planet.

Equation 7.15 can be applied to the Moon or any planet. For example, taking the Earth to be a point mass M_E located at its center and R_E as its radius, we obtain the acceleration due to gravity at the Earth's surface ($a_{g_E} = g$) by setting the distance r to be equal to R_E:

$$a_{g_E} = g = \frac{GM_E}{R_E^2} \qquad (7.16)$$

This equation has several interesting implications. First, it reveals that taking g to be constant everywhere on the surface of the Earth involves the assumption that the Earth has a homogeneous distribution of mass and that the distance from the center of the Earth to any location on its surface is the same. These two assumptions are not exactly true. Therefore, taking g to be a constant is an approximation, but one that works pretty well for most situations.

Also, you can see why the acceleration due to gravity is the same for all free-falling objects—that is, independent of the mass of the object. The mass of the object doesn't appear in Eq. 7.16, so all objects in free fall accelerate at the same rate.

Finally, if you're observant, you'll notice that Eq. 7.16 can be used to compute the mass of the Earth. All of the other quantities in the equation are measurable and their values are known, so M_E can readily be calculated. This is what Cavendish did after he determined the value of G experimentally.

The acceleration due to gravity does vary with altitude. At a distance h above the Earth's surface, $r = R_E + h$. The acceleration is then given by

$$a_g = \frac{GM_E}{(R_E + h)^2} \qquad (7.17)$$

PROBLEM-SOLVING HINT

When comparing accelerations due to gravity or gravitational forces, you will often find it convenient to work with ratios. For example, comparing a_g with g (Eqs. 7.15 and 7.16) for the Earth gives

$$\frac{a_g}{g} = \frac{GM_E/r^2}{GM_E/R_E^2} = \frac{R_E}{r^2} = \left(\frac{R_E}{r}\right)^2 \quad \text{or} \quad \frac{a_g}{g} = \left(\frac{R_E}{r}\right)^2$$

(continued on next page)

Note how the constants cancel out. Taking $r = R_E + h$ you can easily compute a_g/g, or the acceleration due to gravity at some altitude h above the Earth compared with g on the Earth's surface (9.80 m/s^2).

Because R_E is very large compared with everyday altitudes above the Earth's surface, the acceleration due to gravity does not decrease very rapidly with height. At an altitude of 16 km (10 mi, about twice as high as modern jet airliners fly), $a_g/g = 0.99$, and thus a_g is still 99% of the value of g at the Earth's surface. At an altitude of 320 km (200 mi), a_g is 91% of g. This is the approximate altitude of an orbiting space shuttle. (So "floating" astronauts in an Earth-orbiting space station do have weight. The so-called "weightless" condition is discussed in Section 7.6.)

EXAMPLE 7.13 | ## Geosynchronous Satellite Orbit

Some communication and weather satellites are launched into circular orbits above the Earth's equator so they are *synchronous* (from the Greek *syn-*, same, and *chronos*, time) with the Earth's rotation. That is, they remain "fixed" or "hover" over one point on the equator. At what altitude are these geosynchronous satellites?

THINKING IT THROUGH. To remain above one location at the equator, the period of the satellite's revolution must be the same as the Earth's period of rotation—24 h. Also, the centripetal force keeping the satellite in orbit is supplied by the gravitational force of the Earth, $F_g = F_c$. The distance between the center of the Earth and the satellite is $r = R_E + h$. (R_E is the radius of the Earth and h is the height or altitude of the satellite above the Earth's surface.)

SOLUTION. Listing the known data,

Given: T (period) $= 24$ h $= 8.64 \times 10^4$ s
$\quad\quad\quad r = R_E + h$

Find: h (altitude)

From solar system data inside back cover:
$R_E = 6.4 \times 10^3$ km $= 6.4 \times 10^6$ m
$M_E = 6.0 \times 10^{24}$ kg

Setting the magnitudes of gravitational force and the motional centripetal force equal ($F_g = F_c$), where m is the mass of the satellite, and putting the values in terms of angular speed,

$$F_g = F_c$$

$$\frac{GmM_E}{r^2} = \frac{mv^2}{r} = \frac{m(r\omega)^2}{r} = mr\omega^2$$

and

$$r^3 = \frac{GM_E}{\omega^2} = GM_E\left(\frac{T}{2\pi}\right)^2 = \left(\frac{GM_E}{4\pi^2}\right)T^2$$

using the relationship $\omega = 2\pi/T$. Then substituting values:

$$r^3 = \frac{(6.67 \times 10^{-11}\,\text{N}\cdot\text{m}^2/\text{kg}^2)(6.0 \times 10^{24}\,\text{kg})(8.64 \times 10^4\,\text{s})^2}{4\pi^2}$$

$$= 76 \times 10^{21}\,\text{m}^3$$

And taking the cube root:

$$r = \sqrt[3]{76 \times 10^{21}\,\text{m}^3}$$

$$= 4.2 \times 10^7\,\text{m}$$

So,

$$h = r - R_E = 4.2 \times 10^7\,\text{m} - 0.64 \times 10^7 = 3.6 \times 10^7\,\text{m}$$

$$= 3.6 \times 10^4\,\text{km}\ (= 22\,000\,\text{mi})$$

FOLLOW-UP EXERCISE. Show that the period of a satellite in orbit close to the Earth's surface ($h \ll R_E$) may be approximated by $T^2 \approx 4R_E$ and compute T. (Neglect air resistance.)

Another aspect of the decrease of g with altitude concerns potential energy. In Section 5.5, it was learned that $U = mgh$ for an object at a height h above some zero reference point, since g is essentially constant near the Earth's surface. This potential energy is equal to the work done in raising the object a distance h above the Earth's surface in a *uniform* gravitational field.

But what if the change in altitude is so large that g cannot be considered constant while work is done in moving an object, such as a satellite? In this case, the equation $U = mgh$ doesn't apply. In general, it can be shown (using mathematical methods beyond the scope of this book) that the **gravitational potential energy (U)** of two point masses separated by a distance r is given by

$$U = -\frac{Gm_1m_2}{r} \qquad (7.18)$$

The minus sign in Eq. 7.18 arises from the choice of the zero reference point (the point where $U = 0$), which is $r = \infty$ (infinity).

In terms of the Earth and a mass m at an altitude h above the Earth's surface,

$$U = -\frac{Gm_1m_2}{r} = -\frac{GmM_E}{R_E + h} \qquad (7.19)$$

where r is the distance separating the Earth's center and the mass. This means that on the Earth we can visualize ourselves as being in a negative gravitational potential energy well (▼Fig. 7.18) that extends to infinity, because the force of gravity has an infinite range. As h increases, so does U. That is, U becomes *less negative*, or gets closer to zero (that is, more positive), corresponding to a higher position in the potential energy well.

Thus, when gravity does negative work (an object moves higher in the well) or gravity does positive work (an object falls lower in the well), there is a *change* in potential energy. As with finite potential energy wells, this *change* in energy is one of the most important things in analyzing situations such as these.

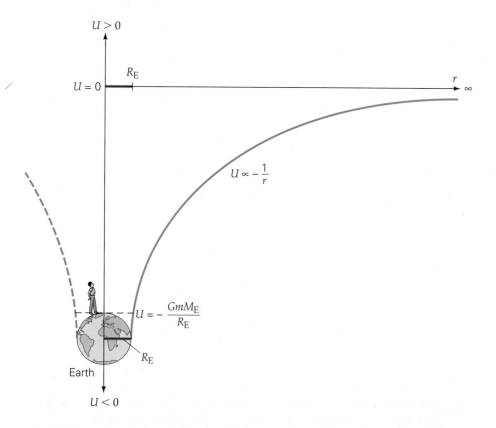

▲ **FIGURE 7.18 Gravitational potential energy well** On the Earth, we can visualize ourselves as being in a negative gravitational potential energy well. As with an actual well or hole in the ground, work must be done against gravity to get higher in the well. The potential energy of an object increases as the object moves higher in the well. This means that the value of U becomes less negative. The top of the Earth's gravitational well is at infinity, where the gravitational potential energy is, by choice, zero.

EXAMPLE 7.14 | ## Different Orbits: Change in Gravitational Potential Energy

Two 50-kg satellites move in circular orbits about the Earth at altitudes of 1000 km (about 620 mi) and 37 000 km (about 23 000 mi), respectively. The lower one monitors particles about to enter the atmosphere, and the higher, geosynchronous one takes weather pictures from its stationary position with respect to the Earth's surface over the equator (see Example 7.13). What is the difference in the gravitational potential energies of the two satellites in their respective orbits?

THINKING IT THROUGH.   The potential energies of the satellites are given by Eq. 7.19. Since an increase in altitude (h) results in a less negative value of U, the satellite with the greater h is higher in the gravitational-potential energy well and has more gravitational potential energy.

SOLUTION.   Listing the data so we can better see what's given (with two significant figures),

Given: $m = 50$ kg
$h_1 = 1000$ km $= 1.0 \times 10^6$ m
$h_2 = 37\,000$ km $= 37 \times 10^6$ m
$M_E = 6.0 \times 10^{24}$ kg (from the inside the back cover of the book)
$R_E = 6.4 \times 10^6$ m

Find: ΔU (difference in potential energy)

The difference in the gravitational potential energy can be computed directly from Eq. 7.19. Keep in mind that the potential energy is the energy of position, so we compute the potential energies for each position or altitude and subtract one from the other. Thus,

$$\Delta U = U_2 - U_1 = -\frac{GmM_E}{R_E + h_2} - \left(-\frac{GmM_E}{R_E + h_1}\right) = GmM_E\left(\frac{1}{R_E + h_1} - \frac{1}{R_E + h_2}\right)$$

$$= (6.67 \times 10^{-11}\,\text{N}\cdot\text{m}^2/\text{kg}^2)(50\ \text{kg})(6.0 \times 10^{24}\ \text{kg})$$

$$\times \left[\frac{1}{6.4 \times 10^6\,\text{m} + 1.0 \times 10^6\,\text{m}} - \frac{1}{6.4 \times 10^6\,\text{m} + 37 \times 10^6\,\text{m}}\right]$$

$$= +2.2 \times 10^9\,\text{J}$$

Because ΔU is positive, m_2 is higher in the gravitational potential energy well than m_1. Note that even though both U_1 and U_2 are negative, U_2 is "more positive," or "less negative," and closer to zero. Thus, it takes more energy to get a satellite farther from the Earth.

FOLLOW-UP EXERCISE.   Suppose that the altitude of the higher satellite in this Example were doubled, to 72 000 km. Would the difference in the gravitational potential energies of the two satellites then be twice as great? Justify your answer.

Substituting the gravitational potential energy (Eq. 7.18) into the equation for the total mechanical energy gives the equation a different form than it had in Chapter 5. For example, the total mechanical energy of a mass m_1 moving at a distance r from mass m_2 is

$$E = K + U = \tfrac{1}{2}m_1v^2 - \frac{Gm_1m_2}{r} \tag{7.20}$$

This equation and the principle of the conservation of energy can be applied to the Earth's motion about the Sun by neglecting other gravitational forces. The Earth's orbit is not quite circular, but slightly elliptical. At *perihelion* (the point of the Earth's closest approach to the Sun), the mutual gravitational potential energy is less (a larger negative value) than it is at *aphelion* (the point farthest from the Sun). Therefore, as can be seen from Eq. 7.20 in the form $\tfrac{1}{2}m_1v^2 = E + Gm_1m_2/r$, where E is constant, the Earth's kinetic energy and orbital speed are greatest at perihelion (the smallest value of r) and least at aphelion (the greatest value of r). Or, in general, the Earth's orbital speed is greater when it is nearer the Sun than when it is farther away.

Mutual gravitational potential energy also applies to a group, or *configuration*, of more than two masses. That is, there is gravitational potential energy due to the several masses in a configuration, because work was needed to be done in bringing the masses together. Suppose that there is a single fixed mass m_1, and another mass m_2 is brought close to m_1 from an infinite distance (where $U = 0$). The work done against the attractive force of gravity is negative (why?) and equal to the change in the mutual potential energy of the masses, which are now separated by a distance r_{12}; that is, $U_{12} = -Gm_1m_2/r_{12}$.

If a third mass m_3 is brought close to the other two fixed masses, there are then two forces of gravity acting on m_3, so $U_{13} = -Gm_1m_3/r_{13}$ and $U_{23} = -Gm_2m_3/r_{23}$. The total gravitational potential energy of the configuration is therefore

$$U = U_{12} + U_{13} + U_{23}$$

$$= -\frac{Gm_1m_2}{r_{12}} - \frac{Gm_1m_3}{r_{13}} - \frac{Gm_2m_3}{r_{23}} \qquad (7.21)$$

A fourth mass could be brought in to further prove the point, but this development should be sufficient to suggest that the total gravitational potential energy of a configuration of particles is equal to the sum of the individual potential energies for all pairs of particles.

EXAMPLE 7.15 | Total Gravitational Potential Energy: Energy of Configuration

Three masses are in a configuration as shown in ▶ Fig. 7.19. What is their total gravitational potential energy?

THINKING IT THROUGH. Equation 7.21 applies, but be sure to keep your masses and their distances distinct.

SOLUTION. From the figure, the data are:

Given: $m_1 = 1.0\text{ kg}$ *Find:* U (total gravitational
$m_2 = 2.0\text{ kg}$ potential energy)
$m_3 = 2.0\text{ kg}$
$r_{12} = 3.0\text{ m}; r_{13} = 4.0\text{ m}; r_{23} = 5.0\text{ m}$
(3–4–5 right triangle)

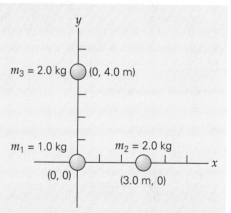

▲ **FIGURE 7.19 Total gravitational potential energy** See Example text for description.

Eq. 7.21 can be used directly, since only three masses are used in this Example. (Note that Eq. 7.21 can be extended to any number of masses.) Then,

$$U = U_{12} + U_{13} + U_{23}$$

$$= -\frac{Gm_1m_2}{r_{12}} - \frac{Gm_1m_3}{r_{13}} - \frac{Gm_2m_3}{r_{23}}$$

$$= (6.67 \times 10^{-11}\text{ N}\cdot\text{m}^2/\text{kg}^2)$$

$$\times \left[-\frac{(1.0\text{ kg})(2.0\text{ kg})}{3.0\text{ m}} - \frac{(1.0\text{ kg})(2.0\text{ kg})}{4.0\text{ m}} - \frac{(2.0\text{ kg})(2.0\text{ kg})}{5.0\text{ m}} \right]$$

$$= -1.3 \times 10^{-10}\text{ J}$$

FOLLOW-UP EXERCISE. Explain what the *negative* potential energy in this Example means in physical terms.

Many of the effects of gravity are familiar to us. When lifting an object, it may be thought of as being heavy, but work is being done against gravity. Gravity causes rocks to tumble down and causes mudslides. But gravity is often put to use. For example, fluids from bottles used for intravenous infusions flow because of gravity. An extraterrestrial application of gravity is given in Insight 7.2, Space Exploration: Gravity Assists.

INSIGHT 7.2 | Space Exploration: Gravity Assists

After a seven-year, 3.5-billion-km (2.2-billion-mi) journey, the *Cassini-Huygens* spacecraft arrived at Saturn in July 2004, having made two Venus flybys, a Jupiter flyby, and one Earth flyby (Fig. 1).* Why was the spacecraft launched toward Venus, an inner planet, in order to go to Saturn, an outer planet?

Although space probes can be launched from the Earth with current rocket technology, there are limitations—in particular, fuel versus payload: the more fuel, the smaller the payload. Using rockets alone, planetary spacecraft are realistically limited to visiting Venus, Mars, and Jupiter. The other planets could not be reached by a spacecraft of reasonable size without taking decades to get there.

So how did *Cassini* get to Saturn in 2004, almost seven years after its 1997 launch? This was accomplished by using gravity in a clever scheme called *gravity assist.* By using gravity assists, missions to all of the planets in our solar system are possible. Rocket energy is needed to get a spacecraft to the first planet, and after that, the energy is more or less "free." Basically, during a planetary flyby (or swing-by), there is an exchange of energy between the planet and the spacecraft, which enables the spacecraft to increase its speed relative to the Sun. (This phenomenon is sometimes called a *slingshot effect.*)

Let's take a brief look at the physics of this ingenious use of gravity. Imagine the *Cassini* spacecraft making a swing-by of Jupiter. Recall from Section 6.2 that a collision is an interaction of objects in which there is an exchange of momentum and energy. Technically, in a swing-by, a spacecraft is having a "collision" with a planet.

When the spacecraft approaches from "behind" the planet and leaves in "front" (relative to the planet's motional direction), the gravitational interaction gives rise to a change in momentum—that is, there is a greater magnitude afterward and the direction is different. Then there is a $\Delta \vec{p}$ in the general "forward" direction of the spacecraft. Since $\Delta \vec{p} \propto \vec{F}$, a force acting on the craft gives it a "kick" of energy in that direction. So positive net work is done and there is an increase in kinetic energy ($W_{net} = \Delta K > 0$ by the work–energy theorem). The spacecraft leaves with more energy, a greater speed, and a new direction. (If the swing-by occurred in the opposite direction, the spacecraft would slow down.)

Momentum and energy are conserved in this elastic collision, and the planet gets an equal and opposite change in momentum, giving a retarding effect. But because the planet's mass is so much larger than that of the spacecraft, the effect on the planet is negligible.

To help you grasp the idea of a gravity assist, consider the analogous roller derby "slingshot maneuver" illustrated in Fig. 2. The skaters interact, and skater S comes out of the "flyby" with increased speed. Here, the change in momentum of the "slinger," skater J, would probably be noticeable, but that would not be so for Jupiter or any other planet.

More recently, Jupiter was called in again for a gravity assist in 2007 for the *New Horizons* spacecraft on its way toward the first close-up observation of the dwarf planet Pluto and one of its moons. Launched in 2006, *New Horizons* is expected to complete its 5-billion-km (3-billion-mi) journey in 2015.

*Giovanni Cassini (1625–1712) was a French–Italian astronomer who studied Saturn, discovering four of its moons and that Saturn's rings are separated into two parts by a narrow gap, now called the *Cassini Division.* The *Cassini-Huygens* spacecraft released a Huygens probe to Saturn's moon Titan, which was discovered by the Dutch scientist Christiaan Huygens (1629–1695).

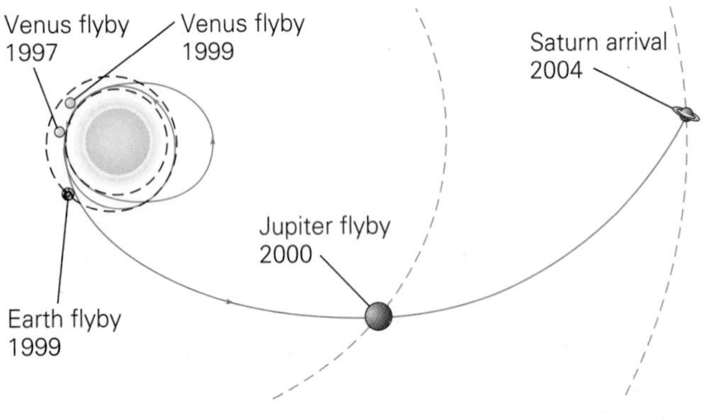

Venus flyby 1997   Venus flyby 1999

Saturn arrival 2004

Jupiter flyby 2000

Earth flyby 1999

▲ **FIGURE 1** *Cassini-Huygens* **spacecraft trajectory** See text for description.

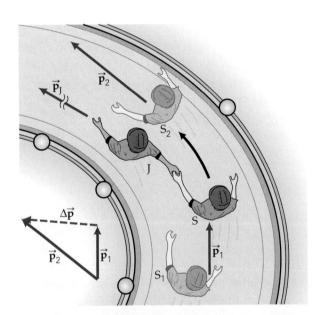

▲ **FIGURE 2** **Skating swing-by** Analgous to a planetary swing-by is a roller derby "slingshot maneuver." Skater J slings skater S, who comes out of the "flyby" with greater speed than she had before (S_1 S, and S_2 sequence). In this case, the change in momentum on skater J, the slinger, would probably be noticeable, but it is not for planets. (Why?)

7.6   Kepler's Laws and Earth Satellites

LEARNING PATH QUESTIONS

➡ Which of Kepler's laws tells that a planet's speed varies in different parts of its orbit?
➡ What is meant by the Earth's escape speed?

The force of gravity determines the motions of the planets and satellites and holds the solar system (and galaxy) together. A general description of planetary motion had been set forth shortly before Newton's time by the German astronomer and mathematician Johannes Kepler (1571–1630). Kepler formulated three *empirical* laws from observational data gathered during a twenty-year period by the Danish astronomer Tycho Brahe (1546–1601).

Kepler went to Prague to assist Brahe, who was the official mathematician at the court of the Holy Roman Emperor. Brahe died the next year, and Kepler succeeded him, inheriting his records of the positions of the planets. Analyzing these data, Kepler announced the first two of his three laws in 1609 (the year Galileo built his first telescope). These laws were applied initially only to Mars. Kepler's third law came ten years later.

Interestingly enough, Kepler's laws of planetary motion, which took him about fifteen years to deduce from observed data, can now be derived theoretically with a page or two of calculations. These three laws apply not only to planets, but also to any system composed of a body revolving about a much more massive body to which the inverse-square law of gravitation applies (such as the Moon, artificial Earth satellites, and solar-bound comets).

Kepler's first law (the law of orbits):

Planets move in elliptical orbits, with the Sun at one of the focal points.

An ellipse, shown in ▼Fig. 7.20a, has, in general, an oval shape, resembling a flattened circle. In fact, a circle is a special case of an ellipse in which the focal points, or *foci* (plural of *focus*), are at the same point (the center of the circle). Although the orbits of the planets are elliptical, most do not deviate very much from circles (Mercury and the dwarf planet Pluto are notable exceptions; see "Eccentricity", Appendix III.) For example, the difference between the perihelion and aphelion of the Earth (its closest and farthest distances from the Sun, respectively) is about 5 million km. This distance may sound like a lot, but it is only a little more than 3% of 150 million km, which is the average distance between the Earth and the Sun.

Kepler's second law (the law of areas):

A line from the Sun to a planet sweeps out equal areas in equal lengths of time.

This law is illustrated in Fig. 7.20b. Since the time to travel the different orbital distances (s_1 and s_2) is the same such that the areas swept out (A_1 and A_2) are equal, this law tells you that the orbital speed of a planet varies in different parts of its orbit. Because a planet's orbit is elliptical, its orbital speed is greater when it is closer to the Sun than when it is farther away. The conservation of energy was used in Section 7.5 (Eq. 7.20) to deduce this relationship for the Earth.

Kepler's third law (the law of periods):

The square of the orbital period of a planet is directly proportional to the cube of the average distance of the planet from the Sun; that is, $T^2 \propto r^3$.

▶ FIGURE 7.20 **Kepler's first and second laws of planetary motion** (a) In general, an ellipse has an oval shape. The sum of the distances from the focal points F to any point on the ellipse is constant: $r_1 + r_2 = 2a$. Here, $2a$ is the length of the line joining the two points on the ellipse at the greatest distance from its center, called the *major axis*. (The line joining the two points closest to the center is b, the *minor axis*.) Planets revolve about the Sun in elliptical orbits for which the Sun is at one of the focal points and nothing is at the other. **(b)** A line joining the Sun and a planet sweeps out equal areas in equal times. Since $A_1 = A_2$, a planet travels faster along s_1 than along s_2.

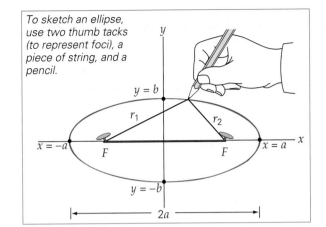

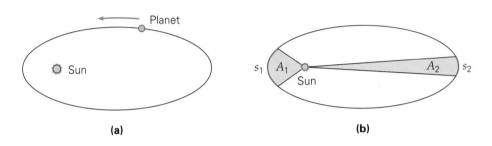

(a)   (b)

Kepler's third law is easily derived for the special case of a planet with a circular orbit, using Newton's law of gravitation. Since the centripetal force is supplied by the force of gravity, the expressions for these forces can be set equal:

$$\underset{\substack{\text{centripetal}\\\text{force}}}{\frac{m_p v^2}{r}} = \underset{\substack{\text{gravitational}\\\text{force}}}{\frac{G m_p M_S}{r^2}}$$

and

$$v = \sqrt{\frac{GM_S}{r}}$$

In these equations, m_p and M_S are the masses of the planet and the Sun, respectively, and v is the planet's orbital speed. But $v = 2\pi r/T$ (circumference/period = distance/time), so

$$\frac{2\pi r}{T} = \sqrt{\frac{GM_S}{r}}$$

Squaring both sides and solving for T^2 gives

$$T^2 = \left(\frac{4\pi^2}{GM_S}\right) r^3$$

or

$$T^2 = K r^3 \tag{7.22}$$

The constant K for solar-system planetary orbits is easily evaluated from orbital data (for T and r) for the Earth: $K = 2.97 \times 10^{-19}\ \text{s}^2/\text{m}^3$. As an exercise, you might wish to convert K to the more useful units of y^2/km^3. (*Note:* This value of K applies to all the planets in our solar system, but does *not* apply to planet satellites as Example 7.16 will show.)

If you look inside the back cover and in Appendix III, you will find the masses of the Sun and the planets of the solar system. How were these masses determined? The following Example shows how Kepler's third law can be used to do this.

EXAMPLE 7.16 | **By Jove!**

The planet Jupiter (Roman name Jove) is the largest in the solar system, both in volume and mass. Jupiter has 63 known moons, the four largest having been discovered by Galileo in 1610. Two of these moons, Io and Europa, are shown in ▶Fig. 7.21. Given that Io is an average distance of 4.22×10^5 km from Jupiter and has an orbital period of 1.77 days, compute the mass of Jupiter.

THINKING IT THROUGH. Given the values for Io's distance from the planet (r) and period (T), this would appear to be an application of Kepler's third law, and it is. However, keep in mind that the M_S in Eq. 7.22 is the mass of the Sun, which the planets orbit. The third law can be applied to any satellite, as long as the M is that of the body being orbited by the satellite. In this case, it will be M_J, the mass of Jupiter.

▲ **FIGURE 7.21** **Jupiter and moons** Two of Jupiter's moons discovered by Galileo, Europa and Io, are shown here. Europa is on the left, and Io on the right over the Great Red Spot. Io and Europa are comparable in size to our Moon. The Great Red Spot, roughly twice the size of the Earth, is believed to be a huge storm, similar to a hurricane on the Earth.

SOLUTION.

Given: $r = 4.22 \times 10^5 \text{ km} = 4.22 \times 10^8 \text{ m}$ *Find:* M_J (mass of Jupiter)
$T = 1.77 \text{ days } (8.64 \times 10^4 \text{ s/day}) = 1.53 \times 10^5 \text{ s}$

With r and T known, K can be found in Eq. 7.22 (written K_I, indicating it is for Io-Jupiter)*

$$K_I = \frac{T^2}{r^3} = \frac{(1.53 \times 10^5 \text{ s})^2}{(4.22 \times 10^8 \text{ m})^3} = 3.11 \times 10^{-16} \text{ s}^2/\text{m}^3$$

Then, writing K_I explicitly, $K_I = \dfrac{4\pi^2}{GM_J}$, and

$$M_J = \frac{4\pi^2}{GK_I} = \frac{4\pi^2}{(6.67 \times 10^{-11} \text{ N} \cdot \text{m}^2/\text{kg}^2)(3.11 \times 10^{-16} \text{ s}^2/\text{m}^3)} = 1.90 \times 10^{27} \text{ kg}$$

*Note that this is different from the K for planets orbiting about the Sun.

FOLLOW-UP EXERCISE. Compute the mass of the Sun from Earth's orbital data.

EARTH'S SATELLITES

We are only a little more than half a century into the space age. Since the 1950s, numerous uncrewed satellites have been put into orbit about the Earth, and now astronauts regularly spend weeks or months in orbiting space laboratories.

Putting a spacecraft into orbit about the Earth (or any planet) is an extremely complex task. However, a basic understanding of the method may be obtained from fundamental principles of physics. First, suppose that a projectile could be given the initial speed required to take it just to the top of the Earth's potential energy well. At the exact top of the well, which is an infinite distance away ($r = \infty$), the potential energy is zero. By the conservation of energy and Eq. 7.18,

initial   *final*
$$K_o + U_o = K + U$$

or

$$\text{initial} \qquad\qquad \text{final}$$

$$\tfrac{1}{2}mv_{esc}^2 - \frac{GmM_E}{R_E} = 0 + 0$$

where v_{esc} is the **escape speed**—that is, the initial speed needed to escape from the surface of the Earth. The final energy is zero, since the projectile stops at the top of the well (at very large distances, and it is barely moving, $K \approx 0$), and $U = 0$ there. Solving for v_{esc} gives

$$v_{esc} = \sqrt{\frac{2GM_E}{R_E}} \qquad\qquad (7.23)$$

Since $g = GM_E/R_E^2$ (Eq. 7.17), it is convenient to write

$$v_{esc} = \sqrt{2gR_E} \qquad\qquad (7.24)$$

Although derived here for the Earth, this equation may be used generally to find the escape speeds for other planets and our Moon (using their accelerations due to gravity and radii). The escape speed for Earth turns out to be 11 km/s, or about 7 mi/s.

A tangential speed less than the escape speed is required for a satellite to orbit. Consider the centripetal force on a satellite in circular orbit about the Earth. Since the centripetal force on the satellite is supplied by the gravitational attraction between the satellite and the Earth, the quantities are equal and:

$$F_c = \frac{mv^2}{r} = \frac{GmM_E}{r^2}$$

Then

$$v = \sqrt{\frac{GM_E}{r}} \qquad\qquad (7.25)$$

where $r = R_E + h$. For example, suppose that a satellite is in a circular orbit at an altitude of 500 km (about 300 mi); its tangential speed must be

$$v = \sqrt{\frac{GM_E}{r}} = \sqrt{\frac{GM_E}{R_E + h}} = \sqrt{\frac{(6.67 \times 10^{-11}\ \text{N}\cdot\text{m}^2/\text{kg}^2)(6.0 \times 10^{24}\ \text{kg})}{(6.4 \times 10^6\ \text{m} + 5.0 \times 10^5\ \text{m})}}$$

$$= 7.6 \times 10^3\ \text{m/s} = 7.6\ \text{km/s} \ (\text{about } 4.7\ \text{mi/s})$$

This speed is about 27 000 km/h, or 17 000 mi/h. As can be seen from Eq. 7.25, the required circular orbital speed *decreases* with altitude (greater r).

In practice, a satellite is given a tangential speed by a component of the thrust from a rocket stage (▶Fig. 7.22a). The inverse-square relationship of Newton's law of gravitation means that the satellite orbits that are possible about a massive planet or star are ellipses, of which a circular orbit is a special case. This condition is illustrated in Fig. 7.22b for the Earth, using the previously calculated values. If a satellite is not given a sufficient tangential speed, it will fall back to the Earth (and possibly be burned up while falling through the atmosphere). If the tangential speed reaches the escape speed, the satellite will leave its orbit and go off into space.

Finally, the total energy of an orbiting satellite in circular orbit is

$$E = K + U = \tfrac{1}{2}mv^2 - \frac{GmM_E}{r} \qquad\qquad (7.26)$$

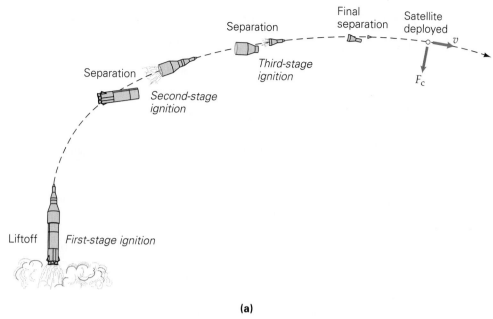

(a)

◀ **FIGURE 7.22** **Satellite orbits**
(a) A satellite is put into orbit by
giving it a tangential speed sufficient for maintaining an orbit at a
particular altitude. The higher the
orbit, the smaller the tangential
speed. **(b)** At an altitude of 500 km,
a tangential speed of 7.6 km/s is
required for a circular orbit. With a
tangential speed between 7.6 km/s
and 11 km/s (the escape speed), the
satellite would move out of the circular orbit. Since it would not have
the escape speed, it would "fall"
around the Earth in an elliptical
orbit, with the Earth's center at one
focal point. A tangential speed less
than 7.6 km/s would also give an
elliptical path about the center of
the Earth, but because the Earth is
not a point mass, a certain minimum speed would be needed to
keep the satellite from striking the
Earth's surface.

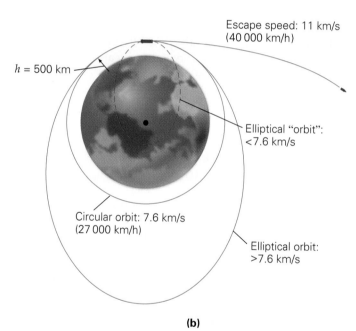

(b)

Substituting the expression for v from Eq. 7.25 into the kinetic energy term in Eq.
7.26 gives

$$E = \frac{GmM_E}{2r} - \frac{GmM_E}{r}$$

Thus,

$$E = -\frac{GmM_E}{2r} \qquad \text{(total energy of an Earth-orbiting satellite)} \qquad (7.27)$$

Note that the total energy of the satellite is negative: *more work is required to put a
satellite into a higher orbit, where it has more potential and total energy.* The total

TABLE 7.3   Relationship of Radius, Speed, and Energy for Circular Orbital Motion

	Increasing r (larger orbit)	Decreasing r (smaller orbit)
ω	decreases	increases
v	decreases	increases
K	decreases	increases
U	increases (smaller negative value)	decreases (larger negative value)
$E \, (= K + U)$	increases (smaller negative value)	decreases (larger negative value)

energy E increases as its *numerical value* becomes smaller—that is, less negative—as the satellite goes to a higher orbit toward the zero potential at the top of the well. That is, the farther a satellite is from Earth, the greater its total energy. The relationship of speed and energy to orbital radius is summarized in ▲ Table 7.3.

To help understand why the total energy increases when its value becomes less negative, think of a change in energy from, say, 5.0 J to 10 J. This change would be an increase in energy. Similarly, a change from -10 J to -5.0 J would also be an increase in energy, even though the *absolute* value has decreased:

$$\Delta U = U - U_0 = -5.0\,\text{J} - (-10\,\text{J}) = +5.0\,\text{J}$$

Also note from the development of Eq. 7.27 that the kinetic energy of an orbiting satellite, $K = \frac{1}{2}mv^2 = GmM_E/2r$, is equal to the absolute value of the satellite's total energy:

$$K = \frac{GmM_E}{2r} = |E| \tag{7.28}$$

The absolute value is taken because the kinetic energy is always positive.

Adjustments in satellite altitude (r) are made by applying forward or reverse thrusts. For example, a reverse thrust, provided by the engines of docked cargo ships, was used to put the Russian space station *Mir* into lower orbits and ultimately led to its final destruction in March 2001. A final thrust sent the station into a decaying orbit and into our atmosphere. Most of the 120-ton *Mir* burned up in the atmosphere; however, some pieces did fall into the Pacific Ocean.

The advent of the space age and the use of orbiting satellites have brought us the terms *weightlessness* and *zero gravity*, because astronauts appear to "float" about in orbiting spacecraft (▶ Fig. 7.23a). However, these terms are misnomers. As mentioned earlier in the chapter, gravity is an infinite-range force, and the Earth's gravity acts on a spacecraft and astronauts, supplying the centripetal force necessary to keep them in orbit. Gravity there is not zero, so there must be weight.*

A better term to describe the floating effect of astronauts in orbiting spacecraft would be *apparent weightlessness*. The astronauts "float" because both the astronauts and the spacecraft are centripetally accelerating (or "falling") toward the Earth at the same rate. To help you understand this effect, consider the analogous situation of a person standing on a scale in an elevator (Fig. 7.23b). The "weight" measurement that the scale registers is actually the normal force N of the scale on the person. In a nonaccelerating elevator ($a = 0$), $N = mg = w$, and N is equal to the true weight of the individual. However, suppose the elevator is descending with an acceleration a, where $a < g$. As the vector diagram in the figure shows,

$$mg - N = ma$$

*Another term used to describe astronaut "floating" is *microgravity*, implying that it is caused by an apparent large reduction in gravity. This too is a misnomer. Using Eq. 7.18, at a typical satellite altitude of 300 km, it can be shown that the reduction in the acceleration due to gravity is about 10%.

▶ **FIGURE 7.23 Apparent weightlessness** (a) An astronaut "floats" in a spacecraft, seemingly in a weightless condition. (He is not being held up.) (b) In a stationary elevator (top), a scale reads the passenger's true weight. The weight reading is the reaction force N of the scale on the person. If the elevator is descending with an acceleration $a < g$ (middle), the reaction force and apparent weight are less than the true weight. If the elevator were in free fall ($a = g$; bottom), the reaction force and indicated weight would be zero, since the scale would be falling as fast as the person.

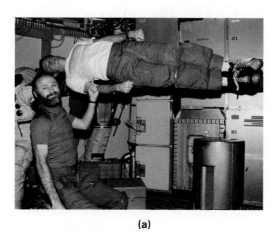

(a)

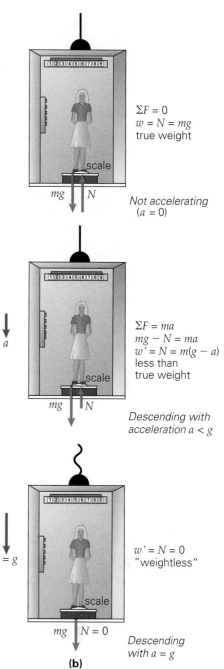

$\Sigma F = 0$
$w = N = mg$
true weight

Not accelerating
($a = 0$)

$\Sigma F = ma$
$mg - N = ma$
$w' = N = m(g - a)$
less than
true weight

Descending with
acceleration $a < g$

$w' = N = 0$
"weightless"

$N = 0$

Descending
with $a = g$

(b)

and the *apparent* weight w' is

$$w' = N = m(g - a) < mg$$

where the downward direction is taken as positive in this instance. With a downward acceleration a, we see that N is less than mg, hence the scale indicates that the person weighs less than his or her true weight. Note that the *apparent acceleration* due to gravity is $g' = g - a$.

Now suppose the elevator were in free fall, with $a = g$. As you can see, N (and thus the apparent weight w') would be zero. Essentially, the scale is accelerating, or falling, at the same rate as the person. The scale may indicate a "weightless" condition ($N = 0$), but gravity still acts, as would be noted by the sudden stop at the bottom of the shaft. (See Insight 7.3, "Weightlessness": Effects on the Human Body.)

Space has been called the *final frontier*. Someday, instead of brief stays in Earth-orbiting spacecraft, there may be permanent space colonies with "artificial" gravity in the future. One proposal is to have a huge, rotating space colony in the form of a wheel—somewhat like an automobile tire, with the inhabitants living inside the tire. As you know, centripetal force is necessary to keep an object in rotational circular motion. On the rotating Earth, that force is supplied by gravity, and we refer to it as *weight*. We exert a force on the ground, and the normal force (by Newton's third law) exerted upward on our feet is what is actually sensed and gives the feeling of "having our feet on solid ground."

In a rotating space colony, the situation is somewhat reversed. The rotating colony would supply the centripetal force on the inhabitants, and the centripetal force would be perceived as a normal force acting on the soles of the feet, providing artificial gravity. Rotation at the proper speed would produce a simulation of "normal" gravity ($a_c \approx g = 9.80 \text{ m/s}^2$) within the colony wheel. Note that in the colonists' world, "down" would be outward, toward the periphery of the space station, and "up" would always be inward, toward the axis of rotation (▼Fig. 7.24).

DID YOU LEARN

➥ Kepler's second law (law of areas) states that a planet sweeps out equal orbital areas equal times. Since planets have elliptical orbits, this means the speed varies in different parts of the orbit.

➥ The initial speed of a projectile needed to raise it to the top of the Earth's gravitational potential energy well is termed the escape speed. (This is an infinite distance, so such an initial speed is not practical.)

INSIGHT 7.3 | "Weightlessness": Effects on the Human Body

Astronauts spend weeks and months in orbiting spacecraft and space stations. Although gravity acts on them, the astronauts experience long durations of "zero gravity" (zero-g),* due to the centripetal motion. On Earth, gravity provides the force that causes our muscles and bones to develop to the proper strength so we may function in our environment. That is, our muscles and bones must be strong enough for us to be able to walk, lift objects, and so on. And we exercise and eat properly to maintain our ability to function against the pull of gravity.

However, in a zero-g environment, muscle atrophy occurs quickly, because the body perceives no need for muscles. That is, muscles lose mass if there is no need for them to respond to gravity. In zero-g, muscle mass may deplete as much as 5% per week. Bone loss also occurs at a rate of about 1% per month. Models show that the total bone loss could reach 40−60%. The bone loss raises the calcium level in the blood, which may lead to kidney stones.

*This term will be used here for description, with the understanding that it is *apparent* zero-g.

The circulatory system is affected, too. On Earth, gravity causes blood to pool in our feet. When we stand, the blood pressure in our feet (about 200 mm Hg) is much greater than that in our heads (60–80 mm Hg), because of the downward force of gravity. (See Section 9.2 for a discussion of the measurement of blood pressure.) In the zero-g experienced by astronauts, this force is not present, and the blood pressure equalizes throughout the body at about 100 mm Hg. This condition causes fluid to flow from the legs to the head, and gives rise to the so-called puffy face and bird leg syndromes. The veins in the neck and face stand out more than usual and the eyes become red and swollen. An astronaut's legs become thinner, because the blood flow to them is no longer gravity-assisted, and it is difficult for the heart to pump blood to them (Fig. 1).

Even more serious, the condition of above-normal blood pressure in the head is interpreted by the brain as indicating that there is too much blood in the body, and blood production is signaled to slow down. Astronauts can lose up to 22% of their blood as a result of the equalized body pressure in zero-g. Also, with equalized blood pressure, the heart doesn't work as hard, and the heart muscles may atrophy.

▶ **FIGURE 7.24 Space colony and artificial gravity (top)** It has been suggested that a space colony could be housed in a huge, rotating wheel as in this artist's conception. The rotation would supply the "artificial gravity" for the colonists. **(bottom) (a)** In the frame of reference of someone in a rotating space colony, centripetal force, coming from the normal force N of the floor, would be perceived as weight sensation or artificial gravity. We are used to feeling N upward on our feet to balance gravity. Rotation at the proper speed would simulate normal gravity. To an outside observer, a dropped ball would follow a tangential straight-line path, as shown. **(b)** A colonist on board the space colony would observe the ball to fall downward as in a normal gravitational situation.

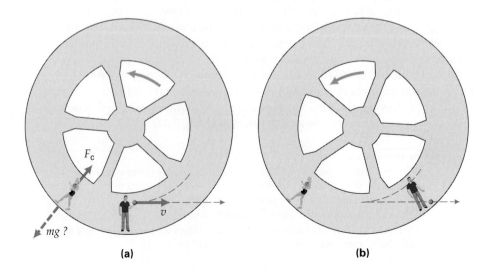

(a)   (b)

All of these phenomena explain why astronauts undergo rigorous physical fitness programs before going into space and exercise in space using elastic restraints. On returning to Earth, their bodies have to readjust to a normal "9.8 m/s² g" environment. Each of the bodily losses requires a different recovery time. Blood volume is typically restored in a few days with astronauts drinking lots of liquids. Most muscles are regenerated in a month or so, depending on the length of stay in zero-g. Bone recovery takes much longer. Astronauts spending three to six months in space may require two or three years to regain the lost bone, if it is regained at all. Exercise and nutrition are very important in all the recovery processes.

There is much to learn about the effects of zero-g—or even reduced-g. Uncrewed spacecraft have visited Mars, with the aim of one day sending astronauts to the Red Planet. This task would involve perhaps a six-month trip in zero-g and, on arrival, a Martian surface gravity that is only 38% of the Earth's gravity. No one yet understands completely the effects that such a space journey might have on an astronaut's body.

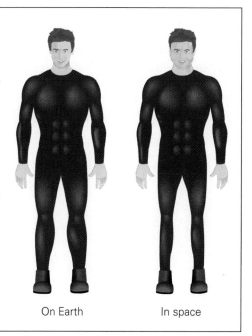

▶ **FIGURE 1** **Puffy face syndrome** In zero-g, without a gravity gradient the blood pressure equalizes throughout the body and fluid flows from the legs to the head, giving rise to the so-called puffy face syndrome. An astronaut's legs become thinner (bird leg syndrome) because the blood flow to them is no longer gravity-assisted and it is difficult for the heart to pump blood to them.

On Earth   In space

PULLING IT TOGETHER | ## Swinging and Releasing a Ball

A boy swings a 0.500-kg ball in a horizontal circle at a constant speed as illustrated in ▶ Fig. 7.25. The cord is 1.20 m long and makes an angle of 5.00° below the horizontal of the hand. Assume the boy's hand remains still and neglect air resistance. (a) What is the string tension? (b) What are the ball's tangential speed, angular speed, frequency, and period? (c) What are the ball's tangential, angular, and centripetal accelerations? (d) If the boy's hand is 2.20 m above the ground and he releases the ball, how long will it take to hit the ground, and how far horizontally from the release point will it travel?

THINKING IT THROUGH.   This example involves angular and tangential kinematics, Newton's second law, centripetal force, and projectile motion. (a) The tension is the force acting on the ball via the cord. Since the ball's weight is known, Newton's laws and a free-body diagram should enable the tension to be found. (b) The horizontal component of the tension is the centripetal force, which is related to the ball's speed. From this, the angular speed, frequency, and period can be determined. (c) Since tangential speed is constant, no tangential or angular

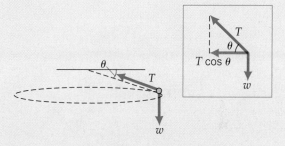

▲ **FIGURE 7.25** **Around it goes** The ball is swung in a horizontal circle at a constant speed. The angle θ is greatly exaggerated for clarity. A free-body diagram is shown on the right. See Example text for description.

acceleration would be expected. However, there will be a centripetal acceleration due to the ball's directional change. This can be determined from the tangential speed. (d) Once it is released, the ball becomes a two-dimensional projectile, so a quick look at Section 3.4 may be in order.

SOLUTION.

Given: $m = 0.500$ kg   **Find:**   (a) T (string tension)
   $L = 1.20$ m   (b) v (tangential speed), ω (angular speed), f (frequency), T (period)
   $\theta = 5.00°$ (c) a_t (tangential acceleration), α (angular acceleration), a_c (centripetal acceleration)
   $h = 2.20$ m   (d) t (time), x (distance)

(a) The free-body diagram in Fig. 7.25 shows two forces acting on the ball: the string tension (T) and the downward pull of gravity (weight, w). You should be able to show that the horizontal and vertical components of the tension force are $T \cos \theta$ and $T \sin \theta$, respectively.

Since there is no vertical acceleration, summing the vertical forces enables the tension to be found:

$$\sum F_y = T \sin \theta - w = ma_y = 0$$

(continued on next page)

and

$$T = \frac{mg}{\sin\theta} = \frac{(0.500\text{ kg})(9.80\text{ m/s}^2)}{\sin 5.00°} = 56.2\text{ N}$$

(b) Summing the forces towards the center of the dotted circle (and designating this as the positive direction) gives the centripetal force on the ball, which can then be used to find its tangential speed. Noting that the circle radius is $r = L\cos\theta$, we have

$$F_c = T\cos\theta = ma_c$$

$$= m\frac{v^2}{r} = m\frac{v^2}{L\cos\theta}$$

Solving for v,

$$v = \sqrt{\frac{TL}{m}}\cos\theta$$

$$= \sqrt{\frac{(56.2\text{ N})(1.20\text{ m})}{0.500\text{ kg}}}\cos 5.00° = 11.6\text{ m/s}$$

From this, the angular speed can be found since it is inversely related to the radius:

$$\omega = \frac{v}{r} = \frac{v}{L\cos\theta} = \frac{11.6\text{ m/s}}{(1.20\text{ m})\cos 5.00°} = 9.70\text{ rad/s}$$

The period (T) is the time for one complete orbit, or for an angular displacement of 2π radians. Recall that the angular speed is $\omega = \Delta\theta/\Delta t$, so

$$\omega = \frac{\Delta\theta}{\Delta t} = \frac{2\pi\text{ rad}}{T}$$

and

$$T = \frac{2\pi\text{ rad}}{9.70\text{ rad/s}} = 0.648\text{ s}$$

and the frequency is $f = \dfrac{1}{T} = \dfrac{1\text{ cycle}}{0.648\text{ s}} = 1.54\text{ Hz}$.

(c) The ball's angular and tangential accelerations are both zero because the ball has no change in angular or tangential speeds. However, its centripetal acceleration is not zero, because the tangential *velocity* is changing due to a continual directional change. Thus

$$a_c = \frac{v^2}{r} = \frac{v^2}{L\cos\theta}$$

$$= \frac{(11.6\text{ m/s})^2}{(1.20\text{ m})\cos 5.00°} = 113\text{ m/s}^2$$

This is more than 11g!

(d) When the ball is released, it becomes a projectile with an initial horizontal velocity with components of $v_{x_o} = 11.6$ m/s and $v_{y_o} = 0$ m/s. The initial height of the ball is slightly less than the 2.20 m since $h = 2.20\text{ m} - L\sin 5.00° = 2.20\text{ m} - 0.105\text{ m} = 2.095\text{ m}$. Choosing the coordinate origin on the ground directly below the release point, the vertical motion is described by

$$y = y_o + v_{y_o}t - \tfrac{1}{2}gt^2 = h + 0 - \tfrac{1}{2}gt^2$$

Since the ground is at $y = 0$, the time in the air after release is

$$t = \sqrt{\frac{2h}{g}} = \sqrt{\frac{2(2.095\text{ m})}{9.80\text{ m/s}^2}} = 0.654\text{ s}$$

Once the ball is in projectile motion there is no horizontal acceleration. Hence the horizontal travel distance is

$$x = v_{x_o}t$$

$$= (11.6\text{ m/s})(0.654\text{ s}) = 7.59\text{ m}$$

Learning Path Review

- The **radian (rad)** is a measure of angle; 1 rad is the angle of a circle subtended by an arc length (s) equal to the radius (r):

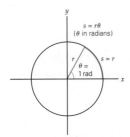

Arc Length (angle in radians):

$$s = r\theta \qquad (7.3)$$

Angular Kinematic Equations for $\theta_o = 0$ and $t_o = 0$ (see Table 7.2 for linear analogues):

$$\theta = \overline{\omega}t \quad \text{(in general, not limited to constant acceleration)} \quad (7.5)$$

$$\left.\begin{array}{l} \overline{\omega} = \dfrac{\omega + \omega_o}{2} \\[2mm] \omega = \omega_o + \alpha t \\[2mm] \theta = \theta_o + \omega_o t + \tfrac{1}{2}\alpha t^2 \\[2mm] \omega^2 = \omega_o^2 + 2\alpha(\theta - \theta_o) \end{array}\right\} \text{constant acceleration only}$$

$$\begin{array}{r} (2,\text{ Table 7.2}) \\ (7.12) \\ (4,\text{ Table 7.2}) \\ (5,\text{ Table 7.2}) \end{array}$$

- **Tangential speed** (v_t) and **angular speed** (ω) for circular motion are directly proportional, with the radius r being the constant of proportionality:

$$v_t = r\omega \qquad (7.6)$$

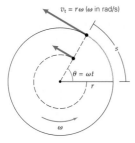

- The **frequency** (f) and **period** (T) are inversely related:

$$f = \frac{1}{T} \qquad (7.7)$$

- **Angular speed** (*with uniform circular motion*) in terms of period (T) and frequency (f):

$$\omega = \frac{2\pi}{T} = 2\pi f \qquad (7.8)$$

- In uniform circular motion, a **centripetal acceleration** (a_c) is required and is always directed toward the center of the circular path, and its magnitude is given by:

$$a_c = \frac{v^2}{r} = r\omega^2 \qquad (7.10)$$

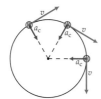

- A **centripetal force**, F_c, (the net force directed toward the center of a circle) is a requirement for circular motion, the magnitude of which is

$$F_c = ma_c = \frac{mv^2}{r} \qquad (7.11)$$

- **Angular acceleration** (α) is the time rate of change of angular velocity and is related to the **tangential acceleration** (a_t) in magnitude by

$$a_t = r\alpha \qquad (7.13)$$

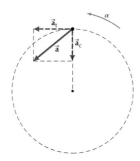

Nonuniform circular motion
($\vec{a} = \vec{a}_t + \vec{a}_c$)

- According to **Newton's law of gravitation**, every particle attracts every other particle in the universe with a force that is proportional to the masses of both particles and inversely proportional to the square of the distance between them:

$$F_g = \frac{Gm_1 m_2}{r^2}$$

$(G = 6.67 \times 10^{-11}\,\mathrm{N \cdot m^2/kg^2}) \qquad (7.14)$

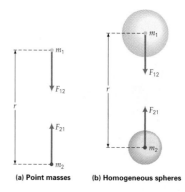

(a) Point masses   **(b) Homogeneous spheres**

- **Acceleration due to gravity at an altitude h:**

$$a_g = \frac{GM_E}{(R_E + h)^2} \qquad (7.17)$$

- **Gravitational potential energy of two particles:**

$$U = -\frac{Gm_1 m_2}{r} \qquad (7.18)$$

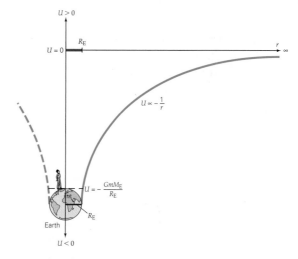

- **Kepler's first law (law of orbits):** Planets move in elliptical orbits, with the Sun at one of the focal points.

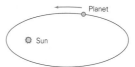

- **Kepler's second law (law of areas):** A line from the Sun to a planet sweeps out equal areas in equal lengths of time.

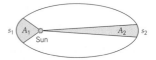

- **Kepler's third law (law of periods):**

$$T^2 = Kr^3 \qquad (7.22)$$

(K depends on the mass of the object orbited; for objects orbiting the Sun, $K = 2.97 \times 10^{-19}\,\mathrm{s^2/m^3}$.)

- **Escape speed** (*from the Earth*) is

$$v_{\mathrm{esc}} = \sqrt{\frac{2GM_E}{R_E}} = \sqrt{2gR_E} \qquad (7.23, 7.24)$$

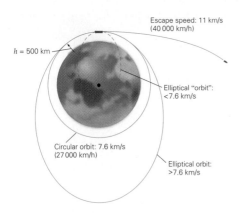

Escape speed: 11 km/s
(40 000 km/h)

$h = 500$ km

Elliptical "orbit":
<7.6 km/s

Circular orbit: 7.6 km/s
(27 000 km/h)

Elliptical orbit:
>7.6 km/s

■ Earth's satellites are in a negative potential energy well; the higher the object in the well, the greater the object's potential energy and the less its kinetic energy.

■ **Energy of a satellite orbiting Earth:**

$$E = -\frac{GmM_E}{2r} \tag{7.27}$$

$$K = |E| \tag{7.28}$$

Learning Path Questions and Exercises   For instructor-assigned homework, go to www.masteringphysics.com

MULTIPLE CHOICE QUESTIONS

7.1 ANGULAR MEASURE

1. The radian unit is a ratio of (a) degree/time, (b) length, (c) length/length (d) length/time.

2. For the polar coordinates of a particle traveling in a circle, the variables are (a) both r and θ, (b) only r, (c) only θ, (d) none of the preceding.

3. Which of the following is the greatest angle: (a) $3\pi/2$ rad, (b) $5\pi/8$ rad, or (c) 220°?

7.2 ANGULAR SPEED AND VELOCITY

4. Viewed from above, a turntable rotates counterclockwise. The angular velocity vector is then (a) tangential to the turntable's rim, (b) out of the plane of the turntable, (c) counterclockwise, (d) none of the preceding.

5. The frequency unit of hertz is equivalent to (a) that of the period, (b) that of the cycle, (c) radian/s, (d) s^{-1}.

6. The unit of angular speed is (a) rad, (b) s^{-1} (c) s, (d) rad/rpm.

7. The particles in a uniformly rotating object all have the same (a) angular acceleration, (b) angular speed, (c) tangential velocity, (d) both (a) and (b).

7.3 UNIFORM CIRCULAR MOTION AND CENTRIPETAL ACCELERATION

8. Uniform circular motion requires (a) centripetal acceleration, (b) angular speed, (c) tangential velocity, (d) all of the preceding.

9. In uniform circular motion, there is a (a) constant velocity, (b) constant angular velocity, (c) zero acceleration, (d) nonzero tangential acceleration.

10. If the centripetal force on a particle in uniform circular motion is increased, (a) the tangential speed will remain constant, (b) the tangential speed will decrease, (c) the radius of the circular path will increase, (d) the tangential speed will increase and/or the radius will decrease.

7.4 ANGULAR ACCELERATION

11. The unit of angular acceleration is (a) s^{-2}, (b) rpm, (c) rad^2/s, (d) s^2.

12. The angular acceleration in circular motion (a) is equal in magnitude to the tangential acceleration divided by the radius, (b) increases the angular velocity if both angular velocity and angular acceleration are in the same direction, (c) has units of s^{-2}, (d) all of the preceding.

13. In circular motion, the tangential acceleration (a) does not depend on the angular acceleration, (b) is constant, (c) has units of s^{-2}, (d) none of these.

14. For uniform circular motion, (a) $\alpha = 0$, (b) $\omega = 0$, (c) $r = 0$, (d) none of the preceding.

7.5 NEWTON'S LAW OF GRAVITATION

15. The gravitational force is (a) a linear function of distance, (b) an inverse function of distance, (c) an inverse function of distance squared, (d) sometimes repulsive.

16. The acceleration due to gravity of an object on the Earth's surface (a) is a universal constant, like G, (b) does not depend on the Earth's mass, (c) is directly proportional to the Earth's radius, (d) does not depend on the object's mass.

17. Compared with its value on the Earth's surface, the value of the acceleration due to gravity at an altitude of one Earth radius is (a) the same, (b) two times as great, (c) one-half as great, (d) one-fourth as great.

7.6 KEPLER'S LAWS AND EARTH SATELLITES

18. A new planet is discovered and its period determined. The new planet's distance from the Sun could then be found by using Kepler's (a) first law, (b) second law, (c) third law.

19. As a planet moves in its elliptical orbit, (a) its speed is constant. (b) its distance from the Sun is constant, (c) it moves faster when it is closer to the Sun, (d) it moves slower when it is closer to the Sun.

20. When a satellite is put into a higher circular orbit, its kinetic energy (a) increases, (b) decreases, (c) remains the same.

CONCEPTUAL QUESTIONS

7.1 ANGULAR MEASURE

1. Why does 1 rad equal 57.3°? Wouldn't it be more convenient to have an even number of degrees?

2. A wheel rotates about a rigid axis through its center. Do all points on the wheel travel the same distance? How about the same *angular* distance?

7.2 ANGULAR SPEED AND VELOCITY

3. Do all points on a wheel rotating about a fixed axis through its center have the same angular velocity? The same tangential speed? Explain.

4. When "clockwise" or "counterclockwise" is used to describe rotational motion, why is a phrase such as "viewed from above" added?

5. Imagine yourself standing on the edge of an operating merry-go-round. How would your tangential speed be affected if you walked toward the center? (Watch out for the horses going up and down.)

6. A car's speedometer is set to read in relationship to the angular speed of the rear wheels. If for winter the tires are changed to larger diameter all-weather tires, would this affect the speedometer reading? Explain. How about the odometer?

7.3 UNIFORM CIRCULAR MOTION AND CENTRIPETAL ACCELERATION

7. The spin cycle of a washing machine is used to extract water from recently washed clothes. Explain the physical principle(s) involved.

8. Can a car be moving with a constant speed of 100 km/h and still be accelerating? Explain.

9. The apparatus illustrated in ▼Fig. 7.26 is used to demonstrate forces in a rotating system. The floats are in jars of water. When the arm is rotated, which way will the floats move? Does it make a difference which way the arm is rotated?

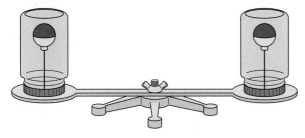

▲ FIGURE 7.26 **When set in motion, a rotating system** See Conceptual Question 9.

10. On the rotating Earth, at what location(s) would a person have (a) the greatest and (b) the least centripetal acceleration? How does the centripetal acceleration for a person at 40° N latitude compare to that of a person at 40° S latitude? (What supplies the centripetal acceleration?)

11. When rounding a curve in a fast-moving car, we experience a feeling of being thrown outward (▶Fig. 7.27). It is sometimes said that this effect occurs because of an outward centrifugal (center-fleeing) force. However, in

▲ FIGURE 7.27 **A center-fleeing force?** See Conceptual Question 11.

terms of Newton's laws for a ground-based observer, this pseudo, or false, force doesn't really exist. Analyze the situation in the figure to show that this is the case (that is, that the force does not exist). [*Hint*: Start with Newton's first law.]

12. Many curves have banked turns, which allow the cars to travel faster around the curves than if the road were flat. Actually, cars could also make turns on these banked curves if there were no friction at all. Explain this statement using the free-body diagram shown in ▼Fig. 7.28.

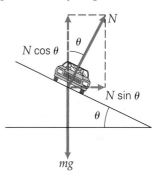

▲ FIGURE 7.28 **Banking safety** See Conceptual Question 12.

7.4 ANGULAR ACCELERATION

13. A car increases its speed when it is on a circular track. Does the car have centripetal acceleration? How about angular acceleration? Explain.

14. Is it possible for a car traveling on a circular track to have angular acceleration, but not centripetal acceleration? Explain.

15. Is it possible for a car traveling on a circular track to have a change in tangential acceleration and no change centripetal acceleration?

7.5 NEWTON'S LAW OF GRAVITATION

16. Astronauts in a spacecraft orbiting the Earth or out for a "space walk" (▼Fig. 7.29) are seen to "float" in midair. This phenomenon is sometimes referred to as *weightlessness* or *zero gravity* (zero-*g*). Are these terms correct? Explain why an astronaut appears to float in or near an orbiting spacecraft.

▲ **FIGURE 7.29** **Out for a walk** Why does this astronaut seem to "float"? See Conceptual Question 16.

17. If the mass of the Moon were doubled, how would this affect its orbit?

18. Weighing yourself at a park in Ecuador through which the equator runs, you would find that you weigh slightly less than normal. Why is this?

19. If the cup in ▼Fig. 7.30 were dropped, no water would run out. Explain.

▲ **FIGURE 7.30** **Let it go** See Conceptual Question 19.

20. Can you determine the mass of the Earth simply by measuring the gravitational acceleration near the Earth's surface? If yes, give the details.

EXERCISES

*Integrated Exercises (**IE**s) are two-part exercises. The first part typically requires a conceptual answer choice based on physical thinking and basic principles. The following part is quantitative calculations associated with the conceptual choice made in the first part of the exercise.*

Throughout the text, many exercise sections will include "paired" exercises. These exercise pairs, identified with **red numbers,** *are intended to assist you in problem solving and learning. In a pair, the first exercise (even numbered) is worked out in the Study Guide so that you can consult it should you need assistance in solving it. The second exercise (odd numbered) is similar in nature, and its answer is given in Appendix VII at the back of the book.*

1. • The Cartesian coordinates of a point on a circle are (1.5 m, 2.0 m). What are the polar coordinates (r, θ) of this point?

2. • The polar coordinates of a point are (5.3 m, 32°). What are the point's Cartesian coordinates?

3. • Convert the following angles from degrees to radians, to two significant figures: (a) 15°, (b) 45°, (c) 90°, and (d) 120°.

4. • Convert the following angles from radians to degrees: (a) $\pi/6$ rad, (b) $5\pi/12$ rad, (c) $3\pi/4$ rad, and (d) π rad.

5. • Express the following angles in degrees, radians, and/or revolutions (rev) as appropriate: (a) 105°, (b) 1.8 rad, and (c) 5/7 rev.

6. • You measure the length of a distant car to be subtended by an angular distance of 1.5°. If the car is actually 5.0 m long, approximately how far away is the car?

7. • How large an angle in radians and degrees does the diameter of the Moon subtend to a person on the Earth?

8. •• The hour, minute, and second hands on a clock are 0.25 m, 0.30 m, and 0.35 m long, respectively. What are the distances traveled by the tips of the hands in a 30-min interval?

9. •• A car with a 65-cm-diameter wheel travels 3.0 km. How many revolutions does the wheel make in this distance?

10. •• Two gear wheels with radii of 25 cm and 60 cm have interlocking teeth. How many radians does the smaller wheel turn when the larger wheel turns 4.0 rev?

11. •• You ordered a 12-in. pizza for a party of five. For the pizza to be distributed evenly, how should it be cut in triangular pieces? (▶Fig. 7.31)?

▶ **FIGURE 7.31** **Tough pizza to cut** See Exercise 11.

12. **IE** •• To attend the 2000 Summer Olympics, a fan flew from Mosselbaai, South Africa (34°S, 22°E) to Sydney, Australia (34°S, 151°E). (a) What is the smallest angular distance the fan has to travel: (1) 34°, (2) 12°, (3) 117°, or (4) 129°? Why? (b) Determine the approximate shortest flight distance, in kilometers.

13. **IE** •• A bicycle wheel has a small pebble embedded in its tread. The rider sets the bike upside down, and accidentally bumps the wheel, causing the pebble to move through an arc length of 25.0 cm before coming to rest. In that time, the wheel spins 35°. (a) The radius of the wheel is therefore (1) more than 25.0 cm, (2) less than 25.0 cm, (3) equal to 25.0 cm. (b) Determine the radius of the wheel.

14. •• At the end of her routine, an ice skater spins through 7.50 revolutions with her arms always fully outstretched at right angles to her body. If her arms are 60.0 cm long, through what arc length distance do the tips of her fingers move during her finish?

15. ••• (a) Could a circular pie be cut such that all of the wedge-shaped pieces have an arc length along the outer crust equal to the pie's radius? (b) If not, how many such pieces could you cut, and what would be the angular dimension of the final piece?

16. ••• Electrical wire with a diameter of 0.50 cm is wound on a spool with a radius of 30 cm and a height of 24 cm. (a) Through how many radians must the spool be turned to wrap one even layer of wire? (b) What is the length of this wound wire?

17. ••• A yo-yo with an axle diameter of 1.00 cm has a 90.0-cm length of string wrapped around it many times in such a way that the string completely covers the surface of its axle, but there are no double layers of string. The outermost portion of the yo-yo is 5.00 cm from the center of the axle. (a) If the yo-yo is dropped with the string fully wound, through what angle does it rotate by the time it reaches the bottom of its fall? (b) How much arc length has a piece of the yo-yo on its outer edge traveled by the time it bottoms out?

7.2 ANGULAR SPEED AND VELOCITY

18. • A computer DVD-ROM has a variable angular speed from 200 rpm to 450 rpm. Express this range of angular speed in radians per second.

19. • A race car makes two and a half laps around a circular track in 3.0 min. What is the car's average angular speed?

20. • What are the angular speeds of the (a) second hand, (b) minute hand, and (c) hour hand of a clock? Are the speeds constant?

21. • What is the period of revolution for (a) a 9500-rpm centrifuge and (b) a 9500-rpm computer hard disk drive?

22. •• Determine which has the greater angular speed: particle A, which travels 160° in 2.00 s, or particle B, which travels 4π rad in 8.00 s.

23. •• The tangential speed of a particle on a rotating wheel is 3.0 m/s. If the particle is 0.20 m from the axis of rotation, how long will the particle take to make one revolution?

24. •• A merry-go-round makes 24 revolutions in a 3.0-min ride. (a) What is its average angular speed in rad/s? (b) What are the tangential speeds of two people 4.0 m and 5.0 m from the center, or axis of rotation?

25. •• In Exercise 13, suppose the wheel took 1.20 s to stop after it was bumped. Assume as you face the plane of the wheel, it was rotating counterclockwise. During this time, determine (a) the average angular speed and tangential speed of the pebble, (b) the average angular speed and tangential speed of a piece of grease on the wheel's axle (radius 1.50 cm), and (c) the direction of their respective angular velocities.

26. IE •• The Earth rotates on its axis once a day and revolves around the Sun once a year. (a) Which is greater, the rotating angular speed or the revolving angular speed? Why? (b) Calculate both angular speeds in rad/s.

27. •• A little boy jumps onto a small merry-go-round (radius of 2.00 m) in a park and rotates for 2.30 s through an arc length distance of 2.55 m before coming to rest. If he landed (and stayed) at a distance of 1.75 m from the central axis of rotation of the merry-go-round, what was his average angular speed and average tangential speed?

28. ••• The driver of a car sets the cruise control and ties the steering wheel so that the car travels at a uniform speed of 15 m/s in a circle with a diameter of 120 m. (a) Through what angular distance does the car move in 4.00 min? (b) What arc length does it travel in this time?

29. ••• In a noninjury, noncontact skid on icy pavement on an empty road, a car spins 1.75 revolutions while it skids to a halt. It was initially moving at 15.0 m/s, and because of the ice it was able to decelerate at a rate of only 1.50 m/s². Viewed from above, the car spun clockwise. Determine its average angular velocity as it spun and slid to a halt.

7.3 UNIFORM CIRCULAR MOTION AND CENTRIPETAL ACCELERATION

30. • An Indy car with a speed of 120 km/h goes around a level, circular track with a radius of 1.00 km. What is the centripetal acceleration of the car?

31. • A wheel of radius 1.5 m rotates at a uniform speed. If a point on the rim of the wheel has a centripetal acceleration of 1.2 m/s², what is the point's tangential speed?

32. • A rotating cylinder about 16 km long and 7.0 km in diameter is designed to be used as a space colony. With what angular speed must it rotate so that the residents on it will experience the same acceleration due to gravity as on Earth?

33. •• An airplane pilot is going to demonstrate flying in a tight vertical circle. To ensure that she doesn't black out at the bottom of the circle, the acceleration must not exceed 4.0*g*. If the speed of the plane is 50 m/s at the bottom of the circle, what is the minimum radius of the circle so that the 4.0*g* limit is not exceeded?

34. •• Imagine that you swing about your head a ball attached to the end of a string. The ball moves at a constant speed in a horizontal circle. (a) Can the string be exactly horizontal? Why or why not? (b) If the mass of the ball is 0.250 kg, the radius of the circle is 1.50 m, and it takes 1.20 s for the ball to make one revolution, what is the ball's tangential speed? (c) What centripetal force are you imparting to the ball via the string?

35. •• In Exercise 34, if you supplied a tension force of 12.5 N to the string, what angle would the string make relative to the horizontal?

36. •• A car with a constant speed of 83.0 km/h enters a circular flat curve with a radius of curvature of 0.400 km. If the friction between the road and the car's tires can supply a centripetal acceleration of 1.25 m/s², does the car negotiate the curve safely? Justify your answer.

37. IE ●● A student is to swing a bucket of water in a vertical circle without spilling any (▼ Fig. 7.32). (a) Explain how this task is possible. (b) If the distance from his shoulder to the center of mass of the bucket of water is 1.0 m, what is the minimum speed required to keep the water from coming out of the bucket at the top of the swing?

▲ **FIGURE 7.32 Weightless water?** See Exercise 37.

38. ●● In performing a "figure 8" maneuver, a figure skater wants to make the top part of the 8 approximately a circle of radius 2.20 m. He needs to glide through this part of the figure at approximately a constant speed, taking 4.50 s. His skates digging into the ice are capable of providing a maximum centripetal acceleration of 3.25 m/s². Will he be able to do this as planned? If not, what adjustment can he make if he wants this part of the figure to remain the same size (assume the ice conditions and skates don't change)?

39. ●● A light string of length of 56.0 cm connects two small square blocks, each with a mass of 1.50 kg. The system is placed on a slippery (frictionless) sheet of horizontal ice and spun so that the two blocks rotate uniformly about their common center of mass, which itself does not move. They are supposed to rotate with a period of 0.750 s. If the string can exert a force of only 100 N before it breaks, determine whether this string will work.

40. IE ●● A jet pilot puts an aircraft with a constant speed into a vertical circular loop. (a) Which is greater, the normal force exerted on the seat by the pilot at the bottom of the loop or that at the top of the loop? Why? (b) If the speed of the aircraft is 700 km/h and the radius of the circle is 2.0 km, calculate the normal forces exerted on the seat by the pilot at the bottom and top of the loop. Express your answer in terms of the pilot's weight.

41. ●●● A block of mass m slides down an inclined plane into a loop-the-loop of radius r (▼ Fig. 7.33). (a) Neglecting friction, what is the minimum speed the block must have

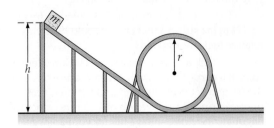

▲ **FIGURE 7.33 Loop-the-loop** See Exercise 41.

at the highest point of the loop in order to stay in the loop? [*Hint*: What force must act on the block at the top of the loop to keep the block on a circular path?] (b) At what vertical height on the inclined plane (in terms of the radius of the loop) must the block be released if it is to have the required minimum speed at the top of the loop?

42. ●●● For a scene in a movie, a stunt driver drives a 1.50×10^3 kg SUV with a length of 4.25 m around a circular curve with a radius of curvature of 0.333 km (▼ Fig. 7.34). The vehicle is to be driven off the edge of a gully 10.0 m wide, and land on the other side 2.96 m below the initial side. What is the minimum centripetal acceleration the SUV must have in going around the circular curve to clear the gully and land on the other side?

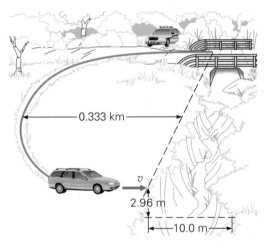

▲ **FIGURE 7.34 Over the gully** See Exercise 42.

43. ●●● Consider a simple pendulum of length L that has a small mass (the bob) of mass m attached to the end of its string. If the pendulum starts out horizontally and is released from rest, show that (a) the speed at the bottom of the swing is $v_{max} = \sqrt{2gL}$ and (b) the tension in the string at that point is three times the weight of the bob, or $T_{max} = 3mg$. [*Hint*: Use conservation of energy to determine the speed at the bottom and centripetal force ideas and a free-body diagram to determine the tension at the bottom.]

7.4 ANGULAR ACCELERATION

44. ● A CD originally at rest reaches an angular speed of 40 rad/s in 5.0 s. (a) What is the magnitude of its angular acceleration? (b) How many revolutions does the CD make in the 5.0 s?

45. ● A merry-go-round accelerating uniformly from rest achieves its operating speed of 2.5 rpm in 5 revolutions. What is the magnitude of its angular acceleration?

46. ●● A flywheel rotates with an angular speed of 25 rev/s. As it is brought to rest with a constant acceleration, it turns 50 rev. (a) What is the magnitude of the angular acceleration? (b) How much time does it take to stop?

47. IE ●● A car on a circular track accelerates from rest. (a) The car experiences (1) only angular acceleration, (2) only centripetal acceleration, (3) both angular and centripetal accelerations. Why? (b) If the radius of the track is 0.30 km and the magnitude of the constant angular acceleration is 4.5×10^{-3} rad/s^2, how long does the car take to make one lap around the track? (c) What is the total (vector) acceleration of the car when it has completed half of a lap?

48. •• Show that for a constant acceleration,

$$\theta = \theta_o + \frac{(\omega^2 - \omega_o^2)}{2\alpha}$$

49. •• The blades of a fan running at low speed turn at 250 rpm. When the fan is switched to high speed, the rotation rate increases uniformly to 350 rpm in 5.75 s. (a) What is the magnitude of the angular acceleration of the blades? (b) How many revolutions do the blades go through while the fan is accelerating?

50. •• In the spin-dry cycle of a modern washing machine, a wet towel with a mass of 1.50 kg is "stuck to" the inside surface of the perforated (to allow the water out) washing cylinder. To have decent removal of water, damp/wet clothes need to experience a centripetal acceleration of at least 10g. Assuming this value, and that the cylinder has a radius of 35.0 cm, determine the constant angular acceleration of the towel required if the washing machine takes 2.50 s to achieve its final angular speed.

51. ••• A pendulum swinging in a circular arc under the influence of gravity, as shown in ▼Fig. 7.35, has both centripetal and tangential components of acceleration. (a) If the pendulum bob has a speed of 2.7 m/s when the cord makes an angle of $\theta = 15°$ with the vertical, what are the magnitudes of the components at this time? (b) Where is the centripetal acceleration a maximum? What is the value of the tangential acceleration at that location?

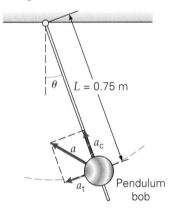

▲ FIGURE 7.35 A swinging pendulum See Exercise 51.

52. ••• A simple pendulum of length 2.00 m is released from a horizontal position. When it makes an angle of 30° from the vertical, determine (a) its angular acceleration, (b) its centripetal acceleration, and (c) the tension in the string. Assume the bob's mass is 1.50 kg.

7.5 NEWTON'S LAW OF GRAVITATION

53. • From the known mass and radius of the Moon (see the tables inside the back cover of the book), compute the value of the acceleration due to gravity, g_M, at the surface of the Moon.

54. • The gravitational forces of the Earth and the Moon are attractive, so there must be a point on a line joining their centers where the gravitational forces on an object cancel. How far is this distance from the Earth's center?

55. •• Four identical masses of 2.5 kg each are located at the corners of a square with 1.0-m sides. What is the net force on any one of the masses?

56. •• The average density of the Earth is 5.52 g/cm^3. Assuming this is a uniform density, compute the value of G.

57. •• A 100-kg object is taken to a height of 300 km above the Earth's surface. (a) What is the object's mass at this height? (b) What is the object's weight at this height?

58. •• A man has a mass of 75 kg on the Earth's surface. How far above the surface of the Earth would he have to go to "lose" 10% of his body weight?

59. •• It takes 27 days for the Moon to orbit the Earth in a nearly circular orbit of radius 3.80×10^5 km. (a) Show in symbol notation that the mass of the Earth can be found using these data. (b) Compute the Earth's mass and compare with the value given inside the back cover of the book.

60. IE •• Two objects are attracting each other with a certain gravitational force. (a) If the distance between the objects is halved, the new gravitational force will (1) increase by a factor of 2, (2) increase by a factor of 4, (3) decrease by a factor of 2, (4) decrease by a factor of 4. Why? (b) If the original force between the two objects is 0.90 N, and the distance is tripled, what is the new gravitational force between the objects?

61. •• During the Apollo lunar explorations of the late 1960s and early 1970s, the main section of the spaceship remained in orbit about the Moon with one astronaut in it while the other two astronauts descended to the surface in the landing module. If the main section orbited about 50 mi above the lunar surface, determine that section's centripetal acceleration.

62. •• Referring to Exercise 61, determine the (a) the gravitational potential energy, (b) the total energy, and (c) the energy needed to "escape" the Moon for the main section of the lunar exploration mission in orbit. Assume the mass of this section is 5000 kg.

63. ••• The diameter of the Moon's (nearly circular) orbit about the Earth is 3.6×10^5 km and it takes 27 days for one orbit. What is (a) the Moon's tangential speed, (b) its kinetic energy, (c) the system potential energy and system total energy?

64. ••• (a) What is the mutual gravitational potential energy of the configuration shown in ▼Fig. 7.36 if all the masses are 1.0 kg? (b) What is the gravitational force per unit mass at the center of the configuration?

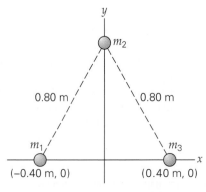

▲ FIGURE 7.36 Gravitational potential, gravitational force, and center of mass See Exercise 64.

65. **IE ●●●** A deep space probe mission is planned to explore the composition of interstellar space. Assuming the three most important objects in the solar system for this project are the Sun, the Earth, and Jupiter, (a) what would be the distance of the Earth relative to Jupiter that would result in the lowest escape speed needed if the probe is to be launched from the Earth: (1) the Earth should be as close as possible to Jupiter, (2) the Earth should be as far as possible from Jupiter, or (3) the distance of the Earth relative to Jupiter doesn't matter? (b) Estimate the least escape speed for this probe, assuming planetary circular orbits, and only the Earth, Sun, and Jupiter are important. (See data in Appendix III.) Comment on which of the three objects, if any, determines most of the escape speed.

7.6 KEPLER'S LAWS AND EARTH SATELLITES

66. ● An instrument package is projected vertically upward to collect data near the top of the Earth's atmosphere (at an altitude of about 900 km). (a) What initial speed is required at the Earth's surface for the package to reach this height? (b) What percentage of the escape speed is this initial speed?

67. ● What is the orbital speed of a geosynchronous satellite? (See Example 7.13.)

68. ●● In the year 2056, Martian Colony I wants to put a Mars-synchronous communication satellite in orbit about Mars to facilitate communications with the new bases being planned on the Red Planet. At what distance above the Martian equator would this satellite be placed? (To a good approximation, the Martian day is the same length as that of the Earth's.)

69. ●● The asteroid belt that lies between Mars and Jupiter may be the debris of a planet that broke apart or that was not able to form as a result of Jupiter's strong gravitation. An average asteroid has a period of about 5.0 y. Approximately how far from the Sun would this "fifth" planet have been?

70. ●● Using a development similar to Kepler's law of periods for planets orbiting the Sun, find the required altitude of geosynchronous satellites above the Earth. [*Hint*: The period of such satellites is the same as that of the Earth.]

71. ●● Venus has a rotational period of 243 days. What would be the altitude of a synchronous satellite for this planet (similar to geosynchronous satellite on the Earth)?

72. ●●● A small space probe is put into circular orbit about a newly discovered moon of Saturn. The moon's radius is known to be 550 km. If the probe orbits at a height of 1500 km above the moon's surface and takes 2.00 Earth days to make one orbit, determine the moon's mass.

PULLING IT TOGETHER: MULTICONCEPT EXERCISES

The Multiconcept Exercises require the use of more than one fundamental concept for their understanding and solution. The concepts may be from this chapter, but may include those from previous chapters.

73. Just an instant before reaching the very bottom of a semicircular section of a roller coaster ride, the automatic emergency brake inadvertently goes on. Assume the car has a total mass of 750 kg, the radius of that section of the track is 55.0 m, and the car entered the bottom after descending vertically (from rest) 25.0 m on a frictionless straight incline. If the braking force is a steady 1700 N, determine (a) the car's centripetal acceleration (including direction), (b) the normal force of the track on the car, (c) the tangential acceleration of the car (including direction), and (d) the total acceleration of the car.

74. A car accelerates uniformly from rest, and is initially pointed north. It then travels in a quarter circle taking 10.0 s and reaching a final speed of 30.0 m/s traveling due east. (a) What is the radius of its path? At 5.0 s from the start, determine (b) the car's tangential acceleration (including direction), (c) the car's centripetal acceleration (including direction), and (d) the car's velocity.

75. A simple pendulum consists of a light string 1.50 m long with a small 0.500-kg mass attached. The pendulum starts out at 45° below the horizontal and is given an initial downward speed of 1.50 m/s. At the bottom of the arc, determine (a) the centripetal acceleration of the bob and (b) the tension in the string.

76. To see in principle how astronomers determine stellar masses, consider the following. Unlike our solar system, many systems have two or more stars. If there are two, it is a *binary* star system. The simplest possible case is that of two identical stars in a circular orbit about their common center of mass midway between them (small black dot in ▼Fig. 7.37). Using telescopic measurements, it is sometimes possible to measure the distance, D, between the star's centers and the time (orbital period), T, for one orbit. Assume uniform circular motion and the following data. The stars have the same mass, the distance between them is one billion km $\left(1.0 \times 10^9 \text{ km}\right)$, and the time each takes for one orbit is 10.0 Earth years. Determine the mass of each star.

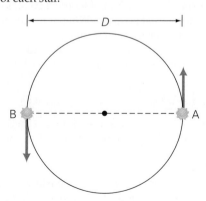

▲ **FIGURE 7.37** **Binary stars** See Exercise 76.

77. As an example of the effect of a meteor impact on a space probe, consider the following idealized situation. Assume the probe is a uniform sphere of iron with a radius of 0.550 m. It is initially at rest when it is struck head on by a 100-g meteoroid traveling at 1.2 km/s. (a) If the meteor embeds itself into the probe, calculate the center of mass speed of the combined system afterward. (b) Determine the percentage of kinetic energy left after the collision. (c) How would the energy analysis differ if the meteor struck the probe off center? *Hint:* The density of iron is given in Table 9.2.

78. The acceleration due to gravity near a planet's surface is known to be 3.00 m/s^2. If the escape speed from the planet is 8.42 km/s, (a) determine its radius. (b) Find the mass of the planet. (c) If a probe is launched from its surface with a speed twice the escape speed and then coasts outward, neglecting other nearby astronomical bodies, what will be its speed when it is very far from the planet? (Neglect any atmospheric effects also.) (d) Under these launch conditions, at what distance will its speed be equal to the escape speed?

79. A newly discovered asteroid is being tracked by astronomers for possible crossing points with Earth's orbit. They have determined that its elliptical orbit brings it as close as 16.0 million km to the Sun and its farthest distance from the Sun is 317 million km. Ignore any gravitational affects from objects other than the Sun. Its speed at closest approach to the Sun is 126 km/s (a) How many times larger is the system's (asteroid plus Sun) gravitational potential energy (magnitude) when the asteroid is at its maximum distance, compared to its minimum distance? (b) What is the system's total energy if the asteroid's mass is 3.35×10^{13} kg? (c) What is the asteroid's minimum speed? (d) How long does it take to orbit the Sun? *Hint:* You will need the Sun's mass. [Note: the R in Kepler's third law refers to the average distance from the Sun if the orbit is not circular, that is, $(R_{max} + R_{min})/2$.]

8 Rotational Motion and Equilibrium

PHYSICS FACTS

- If it were not for torques supplied by our muscles, we would be without body mobility.

- Antilock brakes are used on cars because the rolling stopping distance is less than that of a locked-brake stopping distance.

- The Earth's rotational axis, which is tilted $23\frac{1}{2}^{\circ}$, precesses (rotates about the vertical) with a period of 26 000 years. As a result, Polaris, toward which the axis currently points, has not always been, nor will it always be, the North Star.

- Only one side of the Moon is seen from the Earth because the Moon's period of rotation is the same as its period of revolution.

- The planet Uranus' spin axis is almost in the plane of its orbit. As a result, Uranus rotates on its side while revolving around the Sun.

- Some figure skaters in jumps reach rotational speeds on the order of 7 rev/s, or 420 rpm (revolutions per minute). Some automobiles engines have idle speeds of 600–800 rpm.

t's always a good idea to keep your equilibrium—but it's more important in some situations than in others. When looking at the chapter-opening photo, your first reaction is probably to wonder how does this tightrope walker traversing the Niagara River at Horseshoe Falls keep from falling. Presumably, the pole must help— but in what way? You'll find out in this chapter.

It might be said that the tightrope walker is in equilibrium. Translational equilibrium ($\sum \vec{\mathbf{F}}_i = 0$) was discussed in Section 4.5, but here there is another consideration, namely, rotation. Should the walker start to fall (and it is hoped he doesn't), there would be a sideways rotation about the wire (ropes are rarely used anymore). To avoid this calamity, another condition must be met: rotational equilibrium, which will be considered in this chapter.

The tightrope walker is striving to avoid rotational motion. But rotational motion is very important in physics, because rotating objects are all around us: wheels on vehicles, gears and pulleys in machinery, planets in our solar system, and even many bones in the human body. (Can you think of bones that rotate in sockets?)

Fortunately, the equations describing rotational motion can be written as almost direct analogues of those for translational (linear) motion. In Section 7.4, this similarity was pointed out with respect to the linear and angular kinematic equations. With the addition of equations describing rotational dynamics, you will be able to analyze the general motions of real objects that can rotate, as well as translate.

8.1 Rigid Bodies, Translations, and Rotations

LEARNING PATH QUESTIONS

➡ How are rigid body translational motion and rotational motion characterized in terms of object particles?

➡ What is the instantaneous axis of a rolling object?

➡ What are the conditions for rolling without slipping?

In previous chapters, it was convenient to consider motion with the understanding that an object can be represented by a particle located at the center of mass of the object. Rotation, or spinning, was not a consideration, because a particle, or point mass, has no physical dimensions. Rotational motion becomes relevant when analyzing the motion of a solid, extended object or *rigid body*, which is the focus of this chapter.

> A **rigid body** is an object or a system of particles in which the distances between particles are fixed (remain constant).

A quantity of liquid water is not a rigid body, but the ice that would form if the water were frozen would be. The discussion of rigid body rotation is therefore restricted to solids. Actually, the concept of a rigid body is an idealization. In reality, the particles (atoms and molecules) of a solid vibrate constantly. Also, solids can undergo elastic (and inelastic) deformations in collisions (Section 6.4). Even so, most solids can be considered rigid bodies for purposes of analyzing rotational motion.

A rigid body may be subject to either or both of two types of motions: *translational* and *rotational*. Translational motion is basically the linear motion studied in previous chapters. If an object has only (pure) **translational motion**, *every particle in it has the same instantaneous velocity*, which means that the object is not rotating (▼ Fig. 8.1a).

An object may have only (pure) **rotational motion** (motion about a fixed axis), and *all of the particles of the object have the same instantaneous angular velocity* and travel in circles about the axis of rotation (Fig. 8.1b).*

---

*The words *rotation* and *revolution* are commonly used synonymously. In general, this book uses *rotation* when the axis of rotation goes through the body (for example, the Earth's rotation on its axis, in a period of 24 h) and *revolution* when the axis is outside the body (for example, the revolution of the Earth about the Sun, in a period of 365 days).

▶ **FIGURE 8.1 Rolling—a combination of translational and rotational motions** (a) In pure translational motion, all the particles of an object have the same instantaneous velocity. (b) In pure rotational motion, all the particles of an object have the same instantaneous angular velocity. (c) Rolling is a combination of translational and rotational motions. Summing the velocity vectors for these two motions shows that the point of contact (for a sphere) or the line of contact (for a cylinder) is instantaneously at rest. (d) The line of contact for a cylinder (or, for a sphere, a line through the point of contact) is called the *instantaneous axis of rotation*. Note that the center of mass of a rolling object on a level surface moves linearly and remains over the point or line of contact. (Is the ω vector in the right direction?)

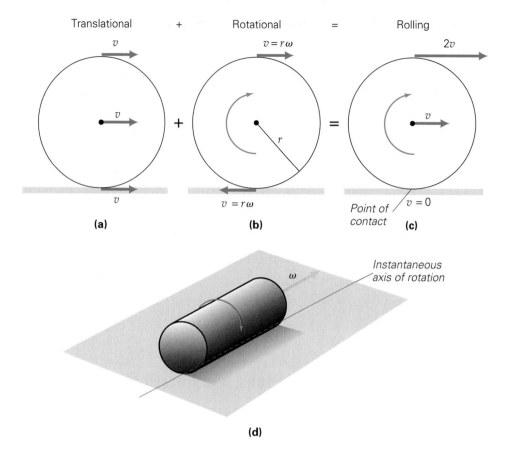

(d)

General rigid body motion is a combination of both translational and rotational motions. When you throw a ball, the translational motion is described by the motion of its center of mass (as in projectile motion). But the ball may also spin, or rotate, and it usually does. A common example of rigid body motion involving both translation and rotation is rolling, as illustrated in Fig. 8.1c. The combined motion of any point or particle is given by the vector sum of the particle's instantaneous velocity vectors. (Three points or particles are shown in the figure—one at the top, one in the middle, and one at the bottom of the object.)

At each instant, a rolling object rotates about an **instantaneous axis of rotation** through the point of contact of the object with the surface it is rolling on (for a sphere) or along the line of contact of the object with the surface (for a cylinder; Fig. 8.1d). The location of this axis changes with time. However, note in Fig. 8.1c that the point or line of contact of the body with the surface is instantaneously at rest (and thus has zero velocity), as can be seen from the vector addition of the combined motions at that point. Also, the point on the top has twice the tangential speed (2v) of the middle (center-of-mass) point (v), because the top point is twice as far away from the instantaneous axis of rotation as the middle point. (With a radius r, for the middle point, $r\omega = v$, and for the top point, $2r\omega = 2v$).

When an object rolls without slipping, for example, when a ball (or cylinder) rolls in a straight line on a flat surface, it turns through an angle θ, and a point (or line) on the object that was initially in contact with the surface moves through an arc distance s (◀Fig. 8.2). And from Section 7.1, $s = r\theta$ (Eq. 7.3). The center of mass of the ball is directly over the point of contact and moves a linear distance s. Then

$$v_{CM} = \frac{s}{t} = \frac{r\theta}{t} = r\omega$$

where $\omega = \theta/t$. In terms of the speed of the center of mass and the angular speed ω the **condition for rolling without slipping** is

$$v_{CM} = r\omega \quad \textit{(rolling, no slipping)} \qquad (8.1)$$

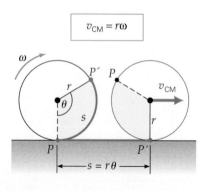

▲ **FIGURE 8.2 Rolling without slipping** As an object rolls without slipping, the length of the arc between two points of contact on the circumference is equal to the linear distance traveled. (Think of paint coming off a roller.) This distance is $s = r\theta$. The speed of the center of mass is $v_{CM} = r\omega$.

The condition for rolling without slipping is also expressed by

$$s = r\theta \quad \textit{(rolling, no slipping)} \tag{8.1a}$$

where s is the distance the object rolls (the distance the center of mass moves).

By carrying Eq. 8.1 one step further, an expression for the time rate of change of the velocity can be obtained. Assuming the object started from rest ($\omega_o = 0$), then $\Delta v_{CM}/\Delta t = v_{CM}/t = (r\omega)/t$, which yields an equation for *accelerated rolling without slipping*:

$$a_{CM} = \frac{v_{CM}}{t} = \frac{r\omega}{t} = r\alpha$$

or

$$a_{CM} = r\alpha \quad \textit{(accelerated rolling without slipping)} \tag{8.1b}$$

where $\alpha = \omega/t$ (for a constant α with ω_o assumed to be zero).

Essentially, an object will roll without slipping if the coefficient of static friction between the object and surface is great enough to prevent slippage. There may be a combination of rolling and slipping motions—for example, the slipping of a car's wheels when traveling through mud or ice. If there is rolling and slipping, then there is no clear relationship between the translational and rotational motions, and $v_{CM} = r\omega$ does *not* hold.

INTEGRATED EXAMPLE 8.1	Rolling without Slipping

A cylinder rolls on a horizontal surface without slipping. (a) At any point in time, the tangential speed of the top of the cylinder is (1) v, (2) $r\omega$. (3) $v+r\omega$, or (4) zero. (b) The cylinder has a radius of 12 cm and a center-of-mass speed of 0.10 m/s as it rolls without slipping. If it continues to travel at this speed for 2.0 s, through what angle does the cylinder rotate during this time?

(A) CONCEPTUAL REASONING. Since the cylinder rolls without slipping, the relationship $v_{CM} = r\omega$ applies. As was shown in Fig. 8.1, the speed at the point of contact is zero, v for the center (of mass), and $2v$ at the top. With $v = 0$ at point of contact, and $v = r\omega$, the answer is (3), $v + r\omega = v + v = 2v$.

(B) QUANTITATIVE REASONING AND SOLUTION. Since the radius and the translational speed are known, the angular speed can be calculated from the nonslipping condition, $v_{CM} = r\omega$. With this relationship and the time, the angle of rotation may be calculated. Listing the data:

Given: $r = 12$ cm $= 0.12$ m   **Find:** θ (angle of rotation)
$\qquad v_{CM} = 0.10$ m/s
$\qquad t = 2.0$ s

Using $v_{CM} = r\omega$ to find the angular speed,

$$\omega = \frac{v_{CM}}{r} = \frac{0.10 \text{ m/s}}{0.12 \text{ m}} = 0.83 \text{ rad/s}$$

Then,

$$\theta = \omega t = (0.83 \text{ m/s})(2.0 \text{ s}) = 1.7 \text{ rad}$$

The cylinder makes a little over one quarter of a rotation. (Right? Check it yourself.)

FOLLOW-UP EXERCISE. How far does the CM of the cylinder travel linearly in part (b) of this Example? Find the distance by using two different methods: translational and rotational. (*Answers to all Follow-Up Exercises are given in Appendix VI at the back of the book.*)

DID YOU LEARN

➥ In pure translational motion, every particle of an object has the same instantaneous velocity; in pure rotational motion, every particle has the same instantaneous angular velocity.

➥ The instantaneous axis of a rolling object is the point or line of contact about which the object rotates at each instant.

➥ The conditions for rolling without slipping are $v_{CM} = r\omega$ or $s = r\theta$. For accelerated rotational motion without slipping, $a_{CM} = r\alpha$.

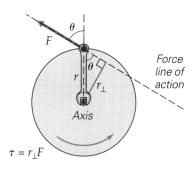

$$\tau = r_\perp F$$

(a) Counterclockwise torque

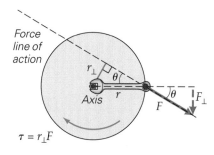

$$\tau = r_\perp F$$

(b) Smaller clockwise torque

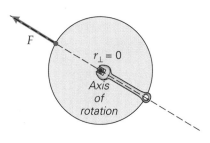

(c) Zero torque

▲ **FIGURE 8.3 Torque and moment arm (a)** The perpendicular distance $r_\perp$ from the axis of rotation to the line of action of a force is called the *moment arm* (or *lever arm*) and is equal to $r \sin \theta$ (where θ is the angle between the line of r, or radial vector $\vec{r}$, and the force vector $\vec{F}$). The magnitude of the torque (τ), or twisting force, that produces rotational motion is $\tau = r_\perp F$. **(b)** The same force in the opposite direction with a smaller moment arm produces a smaller torque in the opposite direction. Note that $r_\perp F = rF_\perp$ or $(r \sin \theta)F = r(F \sin \theta)$. **(c)** When a force acts through the axis of rotation, $r_\perp = 0$ and $\tau = 0$.

8.2 Torque, Equilibrium, and Stability

LEARNING PATH QUESTIONS

➡ What is necessary for rational motion?
➡ What is necessary for mechanical equilibrium?
➡ What is necessary for stable equilibrium?

TORQUE

As with translational motion, a force is necessary to produce a change in rotational motion. However, the rate of change of rotational motion depends not only on the magnitude of the force, but also on the perpendicular distance of its line of action from the axis of rotation, $r_\perp$ (◀Fig. 8.3a, b). The line of action of a force is an imaginary line extending through the force vector arrow—that is, an extended line along which the force acts. (Note that if force is applied at the axis of rotation, $r_\perp$ is zero and there is no rotation about that axis.)

Figure 8.3 shows that $r_\perp = r \sin \theta$, where r is the straight-line distance between the axis of rotation and the force line of action and θ is the angle between the line of r (or radial vector $\vec{r}$) and the force vector $\vec{F}$. The perpendicular distance $r_\perp$ is called the **moment arm** or **lever arm**.

The product of the force and the lever arm is called **torque** ($\vec{\tau}$), from the Latin *torquere*, meaning "to twist." The magnitude of the torque provided by the force is

$$\tau = r_\perp F = rF \sin \theta \qquad (8.2)$$

SI unit of torque: meter × newton (m · N)

(The symbolism $r_\perp F$ is commonly used to denote torque, but also note from Fig. 8.3b that $r_\perp F = rF_\perp$.) The SI units of torque are meter × newton (m · N) the same as the units of work, $W = Fd$ (N · m or J). However, the units of torque are usually written in reverse order as m · N to avoid confusion. But keep in mind that torque is *not* work, and its unit is *not* the joule.

Rotational acceleration is not *always* produced when a force acts on a stationary rigid body. From Eq. 8.2, it can be seen that when the force acts through the axis of rotation such that $\theta = 0$, then $\tau = 0$ (Fig. 8.3c). Also, when $\theta = 90°$, the torque is at a maximum and the force acts perpendicularly to r. The angular acceleration depends on *where* a perpendicular force is applied (and therefore on the length of the lever arm). As a practical example, think of applying a force to a heavy glass door that swings in and out. Where you apply the force makes a great difference in how easily the door opens or rotates (through the hinge axis). Have you ever tried to open such a door and inadvertently pushed on the side near the hinges? This force produces a small torque and thus little or no rotational acceleration.

Torque in rotational motion can be thought of as the analogue of force in translational motion. An unbalanced or net force changes translational motion, whereas an unbalanced or net torque changes rotational motion. Torque is a vector. Its direction is always perpendicular to the plane of the force and moment arm and is given by a *right-hand rule* similar to that for angular velocity given in Section 7.2. If the fingers of the right hand are curled around the axis of rotation in the direction that the torque would produce a rotational (angular) acceleration, the extended thumb points in the direction of the torque. A sign convention, as in the case of linear motion, can be used to represent torque directions, as will be discussed shortly.

EXAMPLE 8.2 | Lifting and Holding: Muscle Torque at Work

In the human body torques produced by the contraction of muscles cause some bones to rotate at joints. For example, when you lift something with your forearm, a torque is applied on the lower arm by the biceps muscle (▶Fig. 8.4).

With the axis of rotation through the elbow joint and the muscle attached 4.0 cm from the joint, what are the magnitudes of the muscle torques for cases (a) and (b) in Fig. 8.4 if the muscle exerts a force of 600 N?

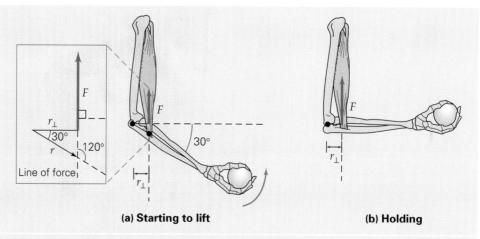

▶ FIGURE 8.4 **Human torque**
In our bodies, torques produced by
the contraction of muscles cause
bones to rotate at joints. Here the
bicep muscles supply the force. See
Example text for description.

(a) Starting to lift

(b) Holding

THINKING IT THROUGH. As in many rotational situations, it is important to know the orientations of the $\vec{r}$ and $\vec{F}$ vectors so that the angle *between* them can be found to determine the lever arm. Note in the inset in Fig. 8.4a that if the tails of the $\vec{r}$ and $\vec{F}$ vectors were put together, the angle between them would be greater than 90°, that is, 30° + 90°=120°. In Fig. 8.4b, the angle is 90°. This Example demonstrates an important point, namely that θ is the angle *between* the radial vector $\vec{r}$ and the force vector $\vec{F}$.

SOLUTION. First listing the data given here and in the figure:

Given: $r = 4.0$ cm $= 0.040$ m

$F = 600$ N

$\theta_a = 30° + 90° = 120°$ (note the 90° right angle in the boxed figure.)

$\theta_b = 90°$

Find: (a) τ_a (muscle torque magnitude) for Fig. 8.4a
(b) τ_b (muscle torque magnitude) for Fig. 8.4b

(a) In this case, $\vec{r}$ is directed along the forearm, so the angle between the $\vec{r}$ and $\vec{F}$ vectors is $\theta_a = 120°$. Using Eq. 8.2,

$$\tau_a = rF \sin(120°) = (0.040 \text{ m})(600 \text{ N})(0.866) = 21 \text{ m} \cdot \text{N}$$

at the instant in question.

(b) Here, the distance r and the line of action of the force are perpendicular ($\theta_b = 90°$), and $r_\perp = r \sin 90° = r$. Then,

$$\tau_b = r_\perp F = rF = (0.040 \text{ m})(600 \text{ N}) = 24 \text{ m} \cdot \text{N}$$

The torque is greater in (b). This is to be expected because the maximum value of the torque (τ_{max}) occurs when $\theta = 90°$.

FOLLOW-UP EXERCISE. In part (a) of this Example, there must have been a net torque, since the ball was accelerated upward by a rotation of the forearm. In part (b), the ball is just being held and there is no rotational acceleration, so there is no net torque on the system. Identify the other torque(s) in each case.

CONCEPTUAL EXAMPLE 8.3 | My Aching Back

A person bends over as shown in ▼Fig. 8.5a. For most of us, the center of gravity of the human body is in or near the chest region. When bending over, the force of gravity on the person's upper torso, acting through its center of gravity, gives rise to a torque that tends to produce rotation about an axis at the base of the spine that could cause us to fall over—but this doesn't usually happen. So why don't we fall when bending over like this? (Consider only the upper torso.)

REASONING AND ANSWER. If this were the only torque acting, we would indeed fall when bending forward. But since there is no rotation and fall, another force must be producing a torque such that the net torque is zero. Where does this other torque come from? Obviously from inside the body through a complicated combination of back muscles.

Representing the vector sum of all the back muscle forces as the net force F_b (as shown in Fig. 8.5b), it can be seen that the back muscles exert a force that counterbalances the torque on the torso's center of gravity.

FOLLOW-UP EXERCISE. Suppose the person was bent over holding a heavy object he had just picked up. How would this affect the back muscle force?

(continued on next page)

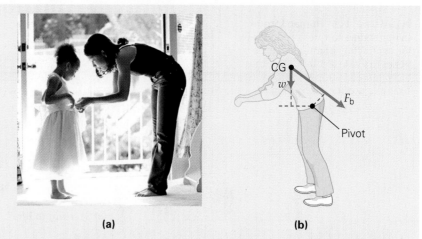

(a)   **(b)**

▲ **FIGURE 8.5 Torque but no rotation** **(a)** When bending over, a person's weight, acting through the upper torso's center of gravity, gives rise to a counterclockwise torque that tends to produce rotation about an axis at the base of the spine. **(b)** However, the back muscles attached between the shoulders combine to produce a force, F_b, and the resulting clockwise torque counterbalances that of gravity, such that the net torque is zero.

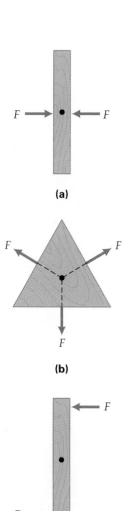

(a)

(b)

(c)

▲ **FIGURE 8.6 Equilibrium and forces** Forces with lines of action through the same point are said to be *concurrent*. The resultants of the concurrent forces acting on the objects in **(a)** and **(b)** are zero ($\vec{F}_{net} = \Sigma\vec{F}_i = 0$), and the objects are in static equilibrium, because the net torque *and* net force are zero. In **(c)**, the object is in *translational* equilibrium, but it will undergo angular acceleration; thus, the object is *not* in rotational equilibrium.

Before considering rotational dynamics with net torques and rotational motions, let's look at a situation in which the forces and torques acting on an object are balanced, and the object is in equilibrium.

EQUILIBRIUM

In general, equilibrium means that forces and torques are in balance. Unbalanced forces produce translational accelerations, but *balanced* forces produce the condition called *translational equilibrium*. Similarly, unbalanced torques produce rotational accelerations, but *balanced* torques produce *rotational equilibrium*.

According to Newton's first law of motion, when the sum of the forces acting on a body is zero, the body remains either at rest (static) or in motion with a constant velocity. In either case, the body is said to be in **translational equilibrium** (Section 4.5). Stated another way, the *condition for translational equilibrium* is that the net force on a body is zero; that is, $\vec{F}_{net} = \Sigma\vec{F}_i = 0$. It should be apparent that this condition is satisfied for the situations illustrated in ◄Fig. 8.6a and b. Forces with lines of action through the same point are called **concurrent forces**. When these forces vectorially add to zero, as in Fig. 8.6a and b, the body is in translational equilibrium.

But what about the situation pictured in Fig. 8.6c? Here, $\Sigma\vec{F}_i = 0$, but the opposing forces will cause the object to rotate, and it will clearly not be in a state of static equilibrium. (Such a pair of equal and opposite forces that do not have the same line of action is called a *couple*.) Thus, the condition $\Sigma\vec{F}_i = 0$ is a necessary, but *not sufficient*, condition for static equilibrium.

Since $\vec{F}_{net} = \Sigma\vec{F}_i = 0$ is the condition for translational equilibrium, you might predict (and correctly so) that $\vec{\tau}_{net} = \Sigma\vec{\tau}_i = 0$ is the *condition for rotational equilibrium*. That is, if the sum of the *torques* acting on an object is zero, then the object is in **rotational equilibrium**—it remains rotationally at rest or rotates with a constant angular velocity.

Thus, there are actually *two* equilibrium conditions. Taken together, they define **mechanical equilibrium**. *A body is said to be in mechanical equilibrium when the conditions for both translational and rotational equilibrium are satisfied:*

$$\vec{F}_{net} = \Sigma\vec{F}_i = 0 \quad \textit{(for translational equilibrium)} \tag{8.3}$$

$$\vec{\tau}_{net} = \Sigma\vec{\tau}_i = 0 \quad \textit{(for rotational equilibrium)}$$

A rigid body in mechanical equilibrium may be either at rest or moving with a constant linear and/or angular velocity. An example of the latter is an object rolling (rotating) without slipping on a level surface, with the center of mass of the object having a constant velocity. Of greater practical interest is **static equilibrium**, the condition that exists when a rigid body remains at rest—that is, a body for which $v = 0$ and $\omega = 0$. There are many instances in which we do not want things to move, and this absence of motion can occur only if the equilibrium conditions are satisfied. It is particularly comforting to know, for example, that a bridge over which cars are crossing is in static equilibrium and not subject to translational or rotational motions.

Let's consider examples of static translational equilibrium and static rotational equilibrium separately and then an example in which both apply.

EXAMPLE 8.4 | **Translational Static Equilibrium: No Translational Acceleration or Motion**

A picture hangs motionless on a wall as shown in ▼ Fig. 8.7a. If the picture has a mass of 3.0 kg, what are the magnitudes of the tension forces in the wires?

▶ **FIGURE 8.7 Translational static equilibrium (a)** Since the picture hangs motionless on the wall, the sum of the forces acting on it must be zero. The forces are concurrent, with their lines of action passing through a common point at the nail. **(b)** In the free-body diagram, all the forces are represented as acting at the common point. T_1 and T_2 have been moved to this point for convenience. Note, however, that the forces shown are acting on the *picture*. See Example text for description.

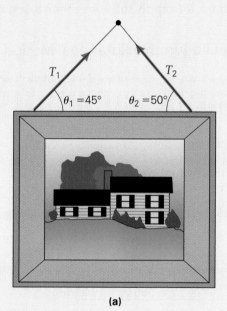
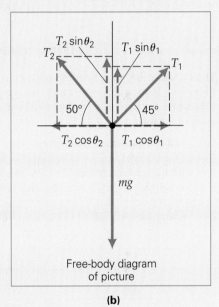

(a)

(b)

THINKING IT THROUGH. Since the picture remains motionless, it must be in static equilibrium, so applying the conditions for mechanical equilibrium should give equations that yield the tensions. Note that all the forces (tension and weight forces) are concurrent; that is, their lines of action pass through a common point, the nail. Because of this, the condition for rotational equilibrium ($\sum \vec{\tau}_i = 0$) is automatically satisfied. With respect to the axis of rotation, the moment arms ($r_\perp$) of the forces are zero, and therefore the torques are zero. Thus, only translational equilibrium needs to be considered.

SOLUTION.

Given: $\theta_1 = 45°, \theta_2 = 50°$ *Find:* T_1 and T_2
         $m = 3.00$ kg

It is helpful to isolate the forces acting on the picture in a free-body diagram, as was done in Section 4.5 for force problems (Fig. 8.7b). The diagram shows the concurrent forces acting through their common point. Note that all the force vectors have been moved to that point, which is taken as the origin of the coordinate axes. The weight force mg acts downward.

With the system in static equilibrium, the net force on the picture is zero; that is, $\sum \vec{F}_i = 0$. Thus, the sums of the rectangular components are also zero: $\sum \vec{F}_x = 0$ and $\sum \vec{F}_y = 0$. Then (using $\pm$ for directions),

$$\sum F_x: \ +T_1 \cos \theta_1 - T_2 \cos \theta_2 = 0 \qquad (1)$$

$$\sum F_y: \ +T_1 \sin \theta_1 + T_2 \sin \theta_2 - mg = 0 \qquad (2)$$

Then, solving for T_2 in Eq. 1 (or T_1 if you like),

$$T_2 = T_1 \left(\frac{\cos \theta_1}{\cos \theta_2} \right) \qquad (3)$$

Then substituting Eq. 3 into Eq. 2 so as to eliminate T_2 and solving for T_1 with a little algebra,

$$T_1 \left[\sin 45° + \left(\frac{\cos 45°}{\cos 50°} \right) \sin 50° \right] - mg =$$

$$T_1 \left[0.707 + \left(\frac{0.707}{0.643} \right) (0.766) \right] - (3.00 \text{ kg})(9.80 \text{ m/s}^2) = 0$$

(continued on next page)

and

$$T_1 = \frac{29.4 \text{ N}}{1.55} = 19.0 \text{ N}$$

Then, using Eq. 2 to find T_2,

$$T_2 = T_1\left(\frac{\cos \theta_1}{\cos \theta_2}\right) = 19.0 \text{ N}\left(\frac{0.707}{0.643}\right) = 20.9 \text{ N}$$

FOLLOW-UP EXERCISE.   Analyze the situation in Fig. 8.7 that would result if the wires were at equal angles and were shortened such that the angles were decreased, but kept equal. Carry your analysis to the limit where the angles approach zero. Is the answer realistic?

As pointed out earlier, torque is a vector and therefore has direction. Similar to linear motion (Section 2.2), in which plus and minus signs were used to express opposite directions (for example, $+x$ and $-x$), torque directions can be designated as being plus or minus, depending on the rotational acceleration they tend to produce. The rotational "directions" are taken as clockwise or counterclockwise around the axis of rotation. A torque that tends to produce a counterclockwise rotation will be taken as positive $(+)$, and a torque that tends to produce a clockwise rotation will be taken as negative $(-)$. (See the right-hand rule in Section 7.2.) To illustrate, let's apply this convention to the situation in Example 8.5.

EXAMPLE 8.5 | Rotational Static Equilibrium: No Rotational Motion

Three masses are suspended from a meterstick as shown in ▼Fig. 8.8a. How much mass must be suspended on the right side for the system to be in static equilibrium? (Neglect the mass of the meterstick.)

THINKING IT THROUGH.   As the free-body diagram (Fig. 8.8b) shows, the translational equilibrium condition will be satisfied with the upward normal force $\vec{\mathbf{N}}$ balancing the downward weight forces, so long as the stick remains horizontal. But $\vec{\mathbf{N}}$ is not known if m_3 is unknown, so applying the condition for rotational equilibrium should give the required value

of m_3. (Note that the lever arms are measured from the pivot point, the center of the meterstick.)

SOLUTION.   From the figure the data are (using cgs units for convenience):

Given: $m_1 = 25 \text{ g}$ *Find:* m_3 (unknown mass)
     $r_1 = 50 \text{ cm}$
     $m_2 = 75 \text{ g}$
     $r_2 = 30 \text{ cm}$
     $r_3 = 35 \text{ cm}$

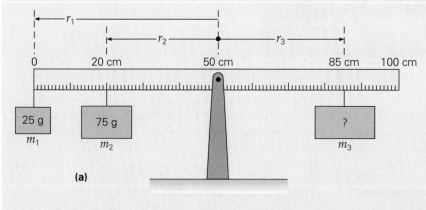

(a)

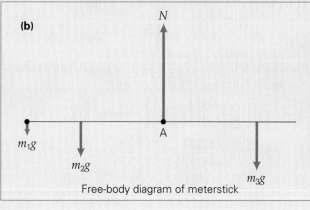

(b)

Free-body diagram of meterstick

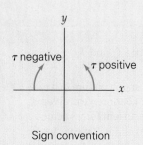

Sign convention

◀**FIGURE 8.8 Rotational static equilibrium** For the meterstick to be in rotational equilibrium, the sum of the torques acting about any selected axis must be zero. **(b)** Here the axis is taken to be through point A. (The mass of the meterstick is considered negligible.)

Because the condition for translational equilibrium ($\sum \vec{F}_i = 0$) is satisfied (there is no $\vec{F}_{net}$ in the y-direction), $N - Mg = 0$, or $N = Mg$ where M, is the total mass. This is true no matter what the total mass may be—that is, regardless of how much mass is added for m_3. However, unless the proper mass for m_3 is placed on the right side, the stick will experience a net torque and begin to rotate.

Notice that the masses on the left side produce torques that would tend to rotate the stick counterclockwise, and a mass on the right side produces a torque that would tend to rotate the stick clockwise. The condition for rotational equilibrium by is applied by summing the torques about an axis. Here, this axis is conveniently taken to be through the center of the stick at the 50-cm position, or point A in Fig. 8.8b. Then, noting that N passes through the axis of rotation ($r_\perp = 0$) and produces no torque,

$$\sum \tau_i: \quad \tau_1 + \tau_2 + \tau_3 = +r_1 F_1 + r_2 F_2 - r_3 F_3 \quad \text{(using sign convention for torque vectors)}$$

$$= r_1(m_1 g) + r_2(m_2 g) - r_3(m_3 g) = 0$$

Noting that the g's cancel and solving for m_3,

$$m_3 = \frac{m_1 r_1 + m_2 r_2}{r_3} = \frac{(25 \text{ g})(50 \text{ cm}) + (75 \text{ g})(30 \text{ cm})}{35 \text{ cm}} = 100 \text{ g}$$

(The mass of the stick was neglected. If the stick is uniform, however, its mass will not affect the equilibrium, as long as the pivot point is at the 50-cm mark. Why?)

FOLLOW-UP EXERCISE. The axis of rotation could have been taken through any point along the stick. That is, if a system is in static rotational equilibrium, the condition $\sum \vec{\tau}_i = 0$ holds for *any* axis of rotation. Show that the preceding statement is true for the system in this Example by taking the axis of rotation through the left end of the stick ($x = 0$).

In general, the conditions for both translational and rotational equilibrium need to be written explicitly to solve a statics problem. Example 8.6 is one such case.

EXAMPLE 8.6 | **Static Equilibrium: No Translation, No Rotation**

A ladder with a mass of 15 kg rests against a smooth wall (▶ Fig. 8.9a). A painter who has a mass of 78 kg stands on the ladder as shown in the figure. What is the magnitude of frictional force that must act on the bottom of the ladder to keep it from slipping?

THINKING IT THROUGH. Here there are a variety of forces and torques. However, the ladder will not slip as long as the conditions for static equilibrium are satisfied. Summing both the forces and torques to zero should enable us to solve for the necessary frictional force. Also, as will be seen, choosing a convenient axis of rotation, such that one or more τ's are zero in the summation of the torques, can simplify the torque equation.

SOLUTION.

Given: $m_\ell = 15$ kg   *Find:* f_s (force of static
        $m_m = 78$ kg   friction)
        Distances given
        in figure

Because the wall is smooth, there is negligible friction between it and the ladder, and only the normal reaction force of the wall (N_w) acts on the ladder at this point (Fig. 8.9b).

In applying the conditions for static equilibrium, any axis of rotation may be chosen. (The conditions must hold for all parts of a system that is in static equilibrium; that is, there can't be motion in any part of the system.) Note that choosing an axis at the end of the ladder where it touches the ground

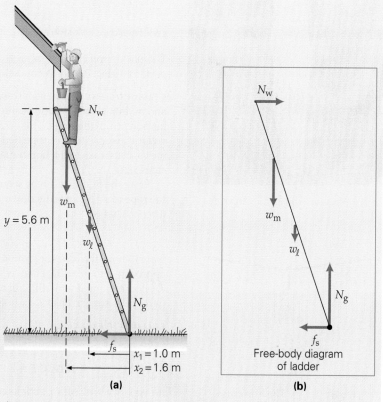

▲ **FIGURE 8.9 Static equilibrium** For the painter's sake, the ladder has to be in static equilibrium; that is, both the sum of the forces and the sum of the torques must be zero. See Example text for description.

(continued on next page)

eliminates the torques due to f_s and N_g, since the moment arms are zero. Then writing the equations for force components and torque (using mg for w):

$$\Sigma F_x: \quad N_w - f_s = 0$$

$$\Sigma F_y: \quad N_g - m_m g - m_\ell g = 0$$

and

$$\Sigma \tau_i: \quad (m_\ell g)x_1 + (m_m g)x_2 + (-N_w y) = 0$$

The weight of the ladder is considered to be concentrated at its center of gravity. Solving the third equation for N_w and

substituting the given values for the masses and distances yields,

$$N_w = \frac{(m_\ell g)x_1 + (m_m g)x_2}{y}$$

$$= \frac{(15 \text{ kg})(9.8 \text{ m/s}^2)(1.0 \text{ m}) + (78 \text{ kg})(9.8 \text{ m/s}^2)(1.6 \text{ m})}{5.6 \text{ m}}$$

$$= 2.4 \times 10^2 \text{ N}$$

Then from the ΣF_x equation,

$$f_s = N_w = 2.4 \times 10^2 \text{ N}$$

FOLLOW-UP EXERCISE. In this Example, would the frictional force between the ladder and the ground (call it f_{s_1}) remain the same if there were friction between the wall and the ladder (call it f_{s_2})? Justify your answer.

PROBLEM-SOLVING HINT

As the preceding Examples have shown, a good procedure to follow in working problems involving static equilibrium is as follows:

1. Sketch a space diagram of the problem.
2. Draw a free-body diagram, showing and labeling all external forces and, if necessary, resolving the forces into x- and y-components.
3. Apply the equilibrium conditions. Sum the forces: $\Sigma \vec{F_i} = 0$, usually in component form; $\Sigma \vec{F_x} = 0$ and $\Sigma \vec{F_y} = 0$. Sum the torques: $\Sigma \vec{\tau_i} = 0$. Remember to select an appropriate axis of rotation to reduce the number of terms as much as possible. Use $\pm$ sign conventions for both $\vec{F}$ and $\vec{\tau}$.
4. Solve for the unknown quantities.

(a)

(b)

mg

▲ **FIGURE 8.10 The iron cross**
(a) The static iron cross gymnastic position is one of the most strenuous and difficult to perform. **(b)** An analogous situation of a weight suspended on a rope tied at both ends. See Conceptual Example 8.7.

CONCEPTUAL EXAMPLE 8.7 | **No Net Torque: The Iron Cross**

The static "iron cross" gymnastic position is one of the most strenuous and difficult to perform (◄Fig. 8.10a). What makes it so difficult?

REASONING AND ANSWER. The gymnast must be extremely strong in order to achieve and maintain such a static position because it requires a huge muscle force to suspend his body from the rings. This can be shown by considering an analogous situation of a weight suspended on a rope tied at each end (Fig. 8.10b). The closer the rope (or gymnast's arms) gets to the horizontal position, the more force is needed to keep the weight suspended.

From the figure, it can be seen that the vertical components of the tension force (T) in the rope must balance the downward weight force. (T is analogous to the arm muscle forces.) That is,

$$2T \sin \theta = mg$$

Note that for the rope (or gymnast's arms) to become horizontal, the angle must approach zero ($\theta \to 0$). Then, what happens to the tension force T? As $\theta \to 0$, then $T \to \infty$, making it very difficult to achieve a small θ, and impossible for the rope (and gymnast's arms to be exactly horizontal).

FOLLOW-UP EXERCISE. What is the least stressful position for a gymnast on the rings?

STABILITY AND CENTER OF GRAVITY

The equilibrium of a particle or a rigid body can be either stable or unstable in a gravitational field. For rigid bodies, these categories of equilibria are conveniently analyzed in terms of the center of gravity of the body. Recall from Section 6.5 that the **center of gravity** is the point at which all the weight of an object may be considered to be acting as if the object were a particle. When the acceleration due to gravity is constant, the center of gravity and the center of mass coincide.

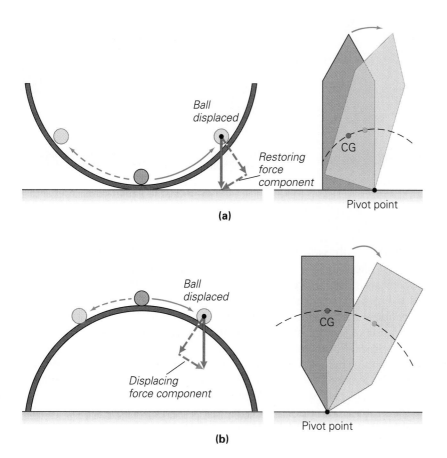

Ball displaced

Restoring force component

(a)

CG

Pivot point

Ball displaced

Displacing force component

(b)

CG

Pivot point

◀ **FIGURE 8.11 Stable and unstable equilibria (a)** When an object is in stable equilibrium, any small displacement from an equilibrium position results in a force or torque that tends to return the object to that position. A ball in a bowl (left) returns to the bottom after being displaced. Analogously, the center of gravity (CG) of an extended object (right) can be thought of as being on an inverted potential energy "bowl": A small displacement raises the CG, increasing the object's potential energy. When released, the object will rotate about the pivot point back to its stable equilibrium position. **(b)** For an object in unstable equilibrium, any small displacement from its equilibrium position results in a force or torque that tends to take the object farther away from that position. The ball on top of an overturned bowl (left) is in unstable equilibrium. For an extended object (right), the CG can be thought of as being on an inverted potential energy bowl. A small displacement lowers the CG, decreasing the object's potential energy.

If an object is in **stable equilibrium**, *any small displacement results in a restoring force or torque, which tends to return the object to its original equilibrium position.* As illustrated in ▲Fig. 8.11a, a ball in a bowl is in stable equilibrium. Analogously, the center of gravity of an extended body on the right is in stable equilibrium. Any slight displacement raises its center of gravity (CG), and a restoring gravitational force tends to return it to the position of minimum potential energy. This force actually produces a restoring torque that is due to a component of the weight force and that tends to rotate the object about a pivot point back to its original position.

For an object in **unstable equilibrium**, *any small displacement from equilibrium results in a torque that tends to rotate the object farther away from its equilibrium position.* This situation is illustrated in Fig. 8.11b. Note that the center of gravity of the object is at the top of an overturned, or inverted, potential energy bowl; that is, the potential energy is at a maximum in this case. Small displacements or slight disturbances have profound effects on objects that are in unstable equilibrium. It doesn't take much to cause such an object to change its position.

Yet even if the angular displacement of an object in stable equilibrium is quite substantial, the object will still be restored to its equilibrium position. As you might have surmised, the **condition for stable equilibrium** is:

> An object is in stable equilibrium as long as its center of gravity after a small displacement still lies above and inside the object's original base of support. That is, the line of action of the weight force through the center of gravity intersects the original base of support.

When this is the case, there will always be a restoring gravitational torque (▼Fig. 8.12a). However, when the center of gravity or center of mass falls outside the base of support, over goes the object—because of a gravitational torque that rotates it away from its equilibrium position (Fig. 8.12b).

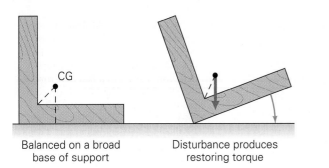

Balanced on a broad base of support

Disturbance produces restoring torque

(a) Stable Equilibrium

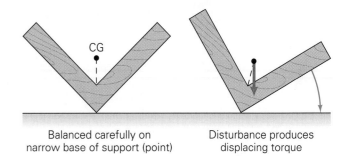

Balanced carefully on narrow base of support (point)

Disturbance produces displacing torque

(b) Unstable Equilibrium

▲ FIGURE 8.12 **Examples of stable and unstable equilibria (a)** When the center of gravity is above and inside an object's base of support, the object is in stable equilibrium. There is a restoring torque when the object is displaced. Note how the line of action of the weight intersects the original base of support after the displacement. **(b)** When the center of gravity lies outside the base of support, the object is unstable. (There is a displacing torque.)

Rigid bodies with wide bases and low centers of gravity are therefore most stable and least likely to tip over. This relationship is evident in the design of high-speed race cars, which have wide wheel bases and centers of gravity close to the ground (▼Fig. 8.13a). SUVs, on the other hand, can roll over more easily. Why? And how about the acrobat in Fig. 8.13b?

The location of the center of gravity of the human body has an effect on certain physical abilities. For example, women can generally bend over and touch their toes or touch their palms to the floor more easily than can men, who often fall over trying. On the average, men have higher centers of gravity (larger shoulders) than do women (larger pelvises), so it is more likely that a man's center of gravity will be outside his base of support when he bends over. Conceptual Example 8.8 gives another real-life example of equilibrium and stability.

▶ FIGURE 8.13 **Stable and unstable (a)** Race cars are very stable because of their wide wheel bases and low center of gravity. **(b)** The acrobat's base of support is very narrow: the small area of head-to-head contact. As long as his center of gravity remains above the head area, he is in equilibrium, but a displacement of only a few centimeters would probably be enough to topple him. (Why he is in a spread-eagle position will become clearer in Section 8.3.)

(a)

(b)

CONCEPTUAL EXAMPLE 8.8 | ## The Center-of-Gravity Challenge

A female student issues a challenge to a male student. She states that she can perform a simple physical feat that he can't. She places a straight-back chair (like most kitchen chairs) with its back against a wall. He is to face the wall and stand next to the chair with his toes touching the wall, then step two foot-lengths backward. (That is, he is to bring the toe of one foot behind the heel of the other foot twice and end up with his feet together, away from the wall.) Next, he is to lean forward and

place the top of his head against the wall, reach over to bring the chair directly in front of him, and place one hand on each side of the chair (▶ Fig. 8.14a). Finally, without moving his feet, he is to stand up while lifting the chair. The female student demonstrates this and easily stands up.

Most males can't perform this feat, but most females can. Why?

REASONING AND ANSWER. When the male student bends over and tries to lift the chair, he is in unstable equilibrium (but fortunately, using his head he doesn't fall over). That is, the center of gravity of the male student/chair system falls outside (in front of) the system's base of support—his feet. Males tend to have a higher center of gravity (larger shoulders and narrower pelvis) than do females (narrow shoulders and larger pelvis).

When the female student bends over and lifts the chair, the center of gravity of the female student/chair system does not fall outside the system's base of support (her feet). She is in stable equilibrium and so is able to stand up from the bent position while she lifts the chair.

But wait! The male student applies physics and swings the chair back (Fig. 8.14b). The combined center of gravity is now over his base of support, and he can stand while holding the chair.

FOLLOW-UP EXERCISE. Why might some males be able to stand while lifting the chair, and some females not be able to do this?

(a)   **(b)**

▲ FIGURE 8.14 **The challenge** **(a)** The male student leans forward with his head on the wall. He is to lift the chair and stand up—but he can't. Yet the female student can easily perform this simple feat. **(b)** But wait. He applies principles of physics and swings the chair back, and he can stand. Why?

Another classic example of equilibrium is the Leaning Tower of Pisa (▶ Fig. 8.15a), from which Galileo allegedly performed his "free-fall" experiments. (See Chapter 2, Insight 2.1, Galileo Galilei and the Leaning Tower of Pisa.) The tower started leaning before its completion in 1350 CE because of the soft subsoil beneath it. In 1990, the lean was about 5.5° from the vertical (about 5 m, or 17 ft, at the top) with an average increase in the lean of about 1.2 mm a year.

Attempts have been made to stop the lean increase. In 1930, cement was injected under the base, but the lean continued to increase. In the 1990s, major actions were taken. The tower was cabled back and counterweights were added to the high side (Fig. 8.15b). Drilling was done diagonally below the foundation on the high side so as to create cavities from which soil could be removed. The tower settled back to about a 5° lean, or a shift of about 40 cm at the top. Moral of the story: Keep that center of gravity above the base of support.

(a)

EXAMPLE 8.9 | Stack Them Up: Center of Gravity

Uniform, identical bricks 20 cm long are stacked so that 4.0 cm of each brick extends beyond the brick beneath, as shown in ▼ Fig. 8.16a. How many bricks can be stacked in this way before the stack falls over?

THINKING IT THROUGH. As each brick is added, the center of mass (or center of gravity) of the stack moves to the right. The stack will be stable as long as the combined center of mass (CM) is over the base of support—the bottom brick. All of the bricks have the same mass, and the center of mass of each is located at its midpoint. So the horizontal location of the stack's CM must be computed as bricks are added, until the CM extends beyond the base. The location of the CM was discussed in Section 6.5 (see Eq. 6.19).

SOLUTION.

Given: brick length = 20 cm   *Find:* maximum number of bricks that yields stability displacement of each brick = 4.0 cm

(continued on next page)

(b)

▲ FIGURE 8.15 **Hold it stable!** **(a)** The Leaning Tower of Pisa. Although leaning, it is still in stable equilibrium. Why? **(b)** Tons of lead counterweight were used in an effort to help correct the tower's lean and keep it in stable equilibrium.

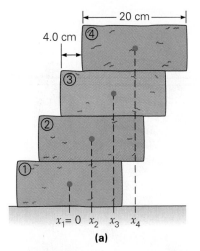

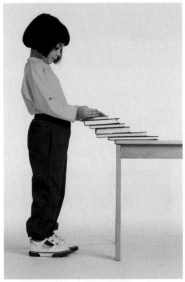

(b)

▲ **FIGURE 8.16 Stack them up!**
(a) How many bricks can be
stacked like this before the stack
falls? See Example 8.9. **(b)** Try a
similar experiment with books.

Taking the origin to be at the center of the bottom brick, the horizontal coordinate of the center of mass (or center of gravity) for the first two bricks in the stack is given by Eq. 6.19, where $m_1 = m_2 = m$ and x_2 is the displacement of the second brick:

$$X_{CM_2} = \frac{mx_1 + mx_2}{m + m} = \frac{m(x_1 + x_2)}{2m} = \frac{x_1 + x_2}{2} = \frac{0 + 4.0 \text{ cm}}{2} = 2.0 \text{ cm}$$

The masses of the bricks cancel out (since they are all the same). For three bricks,

$$X_{CM_3} = \frac{m(x_1 + x_2 + x_3)}{3m} = \frac{0 + 4.0 \text{ cm} + 8.0 \text{ cm}}{3} = 4.0 \text{ cm}$$

For four bricks,

$$X_{CM_4} = \frac{m(x_1 + x_2 + x_3 + x_4)}{4m} = \frac{0 + 4.0 \text{ cm} + 8.0 \text{ cm} + 12.0 \text{ cm}}{4} = 6.0 \text{ cm}$$

and so on.

This series of results shows that the center of mass of the stack moves horizontally 2.0 cm for each brick added to the bottom one. For a stack of six bricks, the center of mass is 10 cm from the origin and directly over the edge of the bottom brick (2.0 cm × 5 *added* bricks = 10 cm, which is half the length of the bottom brick), so the stack is just at unstable equilibrium. The stack may not topple if the sixth brick is positioned very carefully, but it is doubtful that this could be done in practice. A seventh brick would definitely cause the stack to fall off the bottom brick. Why? (As shown in Fig. 8.16b, you can try it yourself and stack them up using books. Don't let the librarian catch you.)

FOLLOW-UP EXERCISE.   If the bricks in this Example were stacked so that, alternately, 4.0 cm and 6.0 cm extended beyond the brick beneath, how many bricks could be stacked before the stack toppled?

For another case of stability, see Insight 8.1, Stability in Action.

DID YOU LEARN?
➥ A net torque is necessary for rotational motion, and torque is the product of the applied force and the lever arm.
➥ The conditions for mechanical equilibrium are $\vec{F}_{net} = \Sigma\vec{F}_i = 0$ and $\vec{\tau}_{net} = \Sigma\vec{\tau}_i = 0$.
➥ As long as the center of gravity of an object lies above and inside the original base of support, the object is in stable equilibrium.

8.3 Rotational Dynamics

LEARNING PATH QUESTIONS
➥ What does the moment of inertia measure?
➥ What is the rotational form of Newton's second law?

MOMENT OF INERTIA

Torque is the rotational analogue of force in linear motion, and a net torque produces rotational motion. To analyze this relationship, consider a constant net force acting on a particle of mass m about the given axis (◀Fig. 8.17). The magnitude of the torque on the particle is

$$\tau_{net} = r_\perp F = rF_\perp = rma_\perp = mr^2\alpha \quad \text{(torque on a particle)} \quad (8.4)$$

where $a_\perp = a_t = r\alpha$ is the tangential acceleration (a_t, Eq. 7.13). For the rotation of a rigid body about a fixed axis, this equation can be applied to each particle in the object and the results summed over the entire body (n particles) to find the total

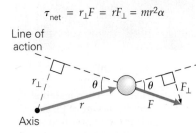

▲ **FIGURE 8.17 Torque on a
particle** The magnitude of the
torque on a particle of mass m is
$\tau = mr^2\alpha$. See text for desscription.

torque. Since all the particles of a rotating rigid body have the same angular acceleration, we can simply add the individual torque magnitudes:

$$\tau_{net} = \Sigma\tau_i = \tau_1 + \tau_2 + \tau_3 + \cdots + \tau_n$$
$$= m_1 r_1^2 \alpha + m_2^2 r_2 \alpha + m_3 r_3^2 \alpha + \cdots + m_n r_n^2 \alpha$$
$$= (m_1 r_1^2 + m_2 r_2^2 + m_3 r_3^2 + \cdots + m_n r_n^2)\alpha$$
$$= (\Sigma m_i r_i^2)\alpha \tag{8.5}$$

But for a rigid body, the masses ($m_i's$) and the distances from the axis of rotation ($r_i's$) do not change. Therefore, the quantity in the parentheses in Eq. 8.5 is constant, and it is called the **moment of inertia, I** (for a given axis):

$$I = \Sigma m_i r_i^2 \quad \textit{(moment of inertia)} \tag{8.6}$$

SI unit of moment of inertia: kilogram-meters squared $(kg \cdot m^2)$

The magnitude of the net torque can be conveniently written as

$$\tau_{net} = I\alpha \quad \textit{(net torque on a rigid body)} \tag{8.7}$$

This is the *rotational form of Newton's second law* ($\vec{\tau}_{net} = I\vec{\alpha}$, in vector form). Keep in mind that, as a *net* force is necessary to produce a translational acceleration, a *net* torque (τ_{net}) is necessary to produce an angular acceleration.

By comparing the rotational form of Newton's second law ($\vec{\tau}_{net} = I\vec{\alpha}$) with the translational form ($\vec{F}_{net} = m\vec{a}$), where m is a measure of translational inertia, it can be seen that the moment of inertia I is a measure of **rotational inertia,** *or a body's tendency to resist change in its rotational motion.* Although I is constant for a rigid body and is the rotational analogue of mass, unlike the mass of a particle, the moment of inertia of a body is referenced to a particular axis and can have different values for different axes.

The moment of inertia also depends on the mass distribution of the body *relative* to its axis of rotation. It is easier (that is, it takes less torque) to give an object an angular acceleration about some axes than about others. The following Example illustrates this point.

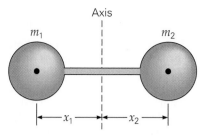

(a) $m_1 = m_2 = 30$ kg
$x_1 = x_2 = 0.50$ m

(b) $m_1 = 40$ kg, $m_2 = 10$ kg
$x_1 = x_2 = 0.50$ m

(c) $m_1 = m_2 = 30$ kg
$x_1 = x_2 = 1.5$ m

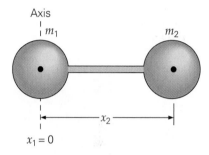

(d) $m_1 = m_2 = 30$ kg
$x_1 = 0, x_2 = 3.0$ m

(e) $m_1 = 40$ kg, $m_2 = 10$ kg
$x_1 = 0, x_2 = 3.0$ m

▲ **FIGURE 8.18 Moment of inertia** The moment of inertia depends on the distribution of mass relative to a particular axis of rotation and, in general, has a different value for each axis. This difference reflects the fact that objects are easier or more difficult to rotate about certain axes. See Example 8.10.

EXAMPLE 8.10 | Rotational Inertia: Mass Distribution and Axis of Rotation

Find the moment of inertia about the axis indicated for each of the one-dimensional dumbbell configurations in ▶ Fig. 8.18. (Neglect the mass of the connecting bar, and give your answers to three significant figures for comparison.)

THINKING IT THROUGH. This is a direct application of Eq. 8.6 for cases with different masses and distances. It will show that the moment of inertia of an object depends on the axis of rotation and on the mass distribution relative to the axis of rotation. The sum for I will include only two terms (two masses).

SOLUTION.

Given: Values of m and r in the figure.

Find: I (moment of inertia)

With $I = m_1 r_1^2 + m_2 r_2^2$:

(a) $I = (30 \text{ kg})(0.50 \text{ m})^2 + (30 \text{ kg})(0.50 \text{ m})^2 = 15.0 \text{ kg} \cdot m^2$

(b) $I = (40 \text{ kg})(0.50 \text{ m})^2 + (10 \text{ kg})(0.50 \text{ m})^2 = 12.5 \text{ kg} \cdot m^2$

(c) $I = (30 \text{ kg})(1.5 \text{ m})^2 + (30 \text{ kg})(1.5 \text{ m})^2 = 135 \text{ kg} \cdot m^2$

(d) $I = (30 \text{ kg})(0 \text{ m})^2 + (30 \text{ kg})(3.0 \text{ m})^2 = 270 \text{ kg} \cdot m^2$

(e) $I = (40 \text{ kg})(0 \text{ m})^2 + (10 \text{ kg})(3.0 \text{ m})^2 = 90.0 \text{ kg} \cdot m^2$

This Example clearly shows how the moment of inertia depends on mass *and* its distribution relative to a particular axis of rotation. In general, the moment of inertia is larger the farther the mass is from the axis of rotation. This principle is important in the design of flywheels, which are used in automobiles to keep the engine running smoothly between cylinder firings. The mass of a flywheel is concentrated near the rim, giving a large moment of inertia, which resists changes in motion.

FOLLOW-UP EXERCISE. In parts (d) and (e) of this Example, would the moments of inertia be different if the axis of rotation went through m_2? Explain.

| ## Stability in Action

When riding a bicycle and going around a curve or making a turn on a level surface, the rider instinctively leans into the curve (Fig. 1). Why? You might think that leaning over, rather than remaining upright, is more likely to cause a spill. However, leaning really does increase stability—it's all a matter of torques.

When a vehicle goes around a level circular curve, a centripetal force is needed to keep the vehicle on the road, as was learned in Section 7.3. This force is generally supplied by the force of static friction between the tires and the road. As illustrated in Fig. 2a, the force $\vec{R}$ of the ground on the bicycle provides the required centripetal force ($\vec{R}_x = \vec{F}_c = \vec{f}_s$) to round the curve, and the normal force ($\vec{R}_y = \vec{N}$).

Suppose the rider tried to remain upright while going around the curve with these forces operative, as shown in Fig. 2a. Note that the line of action of $\vec{R}$ does not go through the system's center of gravity (indicated by a dot). With an axis of rotation through the center of gravity, a counterclockwise

torque would tend to rotate the bicycle in such a way that the wheels would slide inward underneath the rider. However, if the rider leans inward at the proper angle (Fig. 2b), both the line of action of $\vec{R}$ and the weight force act through the center of gravity, and there is no rotational instability (as the gentleman on the bicycle well knew).

There is still a torque on the leaning rider, however. Indeed, when the rider leans into the curve, the weight force gives rise to a torque about an axis through the point of contact with the ground. This torque, along with the turning of the handlebars, causes the bicycle to turn. If the bicycle were not moving, there would be a rotation about this axis, and the bicycle and rider would fall over.

The need to lean into a curve is readily apparent in bicycle and motorcycle races on level tracks. Things can be made easier for the riders if tracks or roadways are banked to provide a natural lean (Section 7.3).

▲ **FIGURE 1   Leaning into a curve** When rounding a curve or making a turn, a bicycle rider must lean into the curve. (This rider could have told you why.)

(a)   (b)

▲ FIGURE 2   **Make that turn** See text for description.

| ## Balancing Act: Locating the Center of Gravity

(a) A rod with a movable ball, like that shown in ▶ Fig. 8.19, is more easily balanced if the ball is in a higher position. Is this because, when the ball in a higher position, (1) the system has a higher center of gravity and more stability; (2) the center of gravity is off the vertical, and there is less torque and a smaller angular acceleration; (3) the center of gravity is closer to the axis of rotation; or (4) the moment of inertia about the axis of rotation is larger? (b) Suppose the distance of the ball from the finger for the farthest position in Fig. 8.19 is $r_2 = 60$ cm and the distance to the closest position is $r_1 = 20$ cm. When the rod rotates, how many times greater is the angular acceleration of the rod with the ball at the closest position than that with the ball at the farthest position? (Neglect the mass of the rod.)

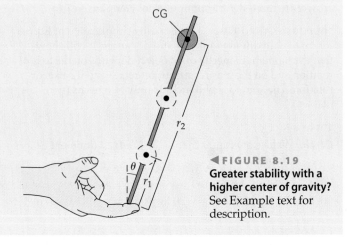

◀ **FIGURE 8.19**
Greater stability with a higher center of gravity? See Example text for description.

(A) CONCEPTUAL REASONING. With the ball at any position and the rod vertical, the system is in unstable equilibrium. As learned in Section 8.2, rigid bodies with wide bases and *low* centers of gravity are more stable, so answer (1) isn't correct. Any slight movement would cause the rod to rotate about an axis through its point of contact. With the center of gravity (CG) at a higher position and off the vertical, there would be a greater lever arm (and thus a *greater* torque), so (2) is also incorrect. With the ball in a higher position, the center of gravity is *farther* from the axis of rotation, which makes (3) incorrect. This leaves (4) by a process of elimination, but let's justify it as the correct answer.

Moving the CG farther from the axis of rotation has an interesting consequence: a greater moment of inertia, or resistance to change in rotational motion, and hence a smaller angular acceleration. However, with the ball in a higher position, as the rod starts to rotate there is a greater torque. The net result is that the increased moment of inertia produces an even greater resistance to rotational motion and hence a smaller angular acceleration. [Note that the torque ($\tau = rF \sin \theta$) varies as r, whereas the moment of inertia ($I = mr^2$) varies as r^2, and so has a larger increase with increasing r. What effect does $\sin \theta$ have?]

Then, the smaller the angular acceleration, the more time there is to adjust your hand under the rod to balance it by bringing the finger and the axis of rotation under the center of gravity. The torque is then zero and the rod is again in equilibrium, albeit unstable. And so, the answer is (4).

(B) QUANTITATIVE REASONING AND SOLUTION. Being asked how many times greater or less something is compared to something else usually implies the use of a ratio in which

some quantity (or quantities) that is not given cancels. Note that the mass of the ball is not given, which would be needed to compute the gravitational torque (τ). Also, the angle θ is not given. So it is best to start with basic equations and see what happens.

Given: $r_1 = 20$ cm *Find:* How many times greater the
$r_2 = 60$ cm rod's angular acceleration is with the ball at r_1 compared to when it is at r_2

The magnitude of the angular acceleration is given by Eq. 8.7, $\alpha = \tau_{net}/I$. So attention turns to the torque τ_{net} and the moment of inertia I. From the basic equations of the chapter, $\tau_{net} = r_\perp F = rF \sin \theta$ (Eq. 8.2), or $\tau_{net} = rmg \sin \theta$ where $F = mg$ in this case, with m being the mass of the ball. Similarly, $I = mr^2$ (Eq. 8.6). Thus,

$$\alpha = \frac{\tau_{net}}{I} = \frac{rmg \sin \theta}{mr^2} = \frac{g \sin \theta}{r}$$

(Note that the angular acceleration α is inversely proportional to the lever arm r; that is, the longer the lever arm, the smaller the angular acceleration.) The $\sin \theta$ is still there, but note what happens when the ratio of the angular accelerations is formed:

$$\frac{\alpha_1}{\alpha_2} = \frac{g \sin \theta / r_1}{g \sin \theta / r_2} = \frac{r_2}{r_1} = \frac{60 \text{ cm}}{20 \text{ cm}} = 3 \quad \text{or} \quad \alpha_2 = \frac{\alpha_1}{3}$$

Hence, the angular acceleration of the rod with the ball at the upper position is one-third that with the ball at the lower position.

FOLLOW-UP EXERCISE. When walking on a thin bar or rail, such as a railroad rail, you have probably found that it helps to hold your arms outstretched. Similarly, tightrope walkers often carry long poles, as in the chapter-opening photo. How does this posture and the pole help maintain balance?

As Integrated Example 8.11 shows, the moment of inertia is an important consideration in rotational motion. By changing the axis of rotation and the relative mass distribution, the value of I can be changed and the motion affected. You were probably told to do this when playing softball or baseball as a child. When at bat, children are often instructed to "choke up" on the bat—to move their hands farther up on the handle.

Now you know why. In doing so, the child moves the axis of rotation of the bat closer to the more massive end of the bat (or its center of mass). Hence, the moment of inertia of the bat is decreased (smaller r in the mr^2 term). Then, when a swing is taken, the angular acceleration is greater. The bat gets around quicker, and the chance of hitting the ball before it goes past is greater. A batter has only a fraction of a second to swing, and with $\theta = \frac{1}{2}\alpha t^2$, a larger α allows the bat to rotate more quickly (swing faster).

PARALLEL AXIS THEOREM

Calculations of the moments of inertia of most extended rigid bodies require math that is beyond the scope of this book. The results for some common shapes are given in ▼Fig. 8.20. The rotational axes are generally taken along axes of symmetry—that is, axes running through the center of mass that give a symmetrical mass distribution. One exception is the rod with an axis of rotation through one end (Fig. 8.20c). This axis is parallel to an axis of rotation through the center of mass of

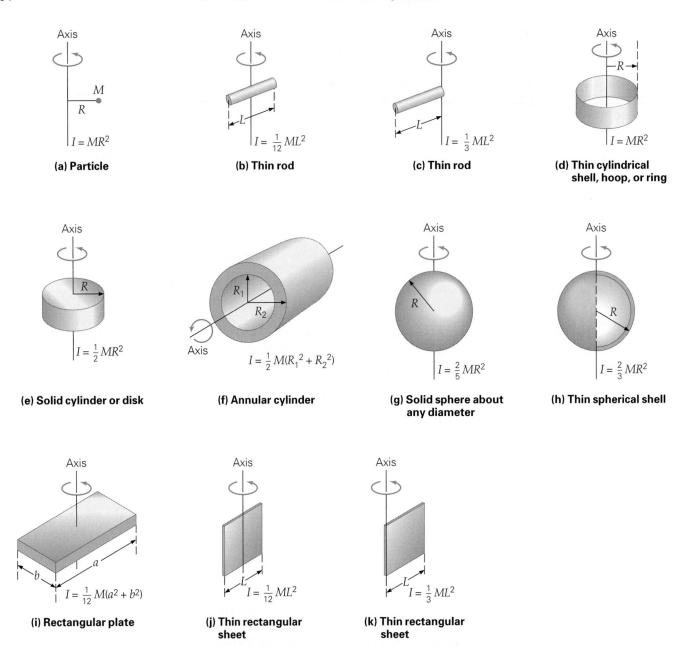

(a) Particle

$$I = MR^2$$

(b) Thin rod

$$I = \frac{1}{12} ML^2$$

(c) Thin rod

$$I = \frac{1}{3} ML^2$$

(d) Thin cylindrical shell, hoop, or ring

$$I = MR^2$$

(e) Solid cylinder or disk

$$I = \frac{1}{2} MR^2$$

(f) Annular cylinder

$$I = \frac{1}{2} M(R_1{}^2 + R_2{}^2)$$

(g) Solid sphere about any diameter

$$I = \frac{2}{5} MR^2$$

(h) Thin spherical shell

$$I = \frac{2}{3} MR^2$$

(i) Rectangular plate

$$I = \frac{1}{12} M(a^2 + b^2)$$

(j) Thin rectangular sheet

$$I = \frac{1}{12} ML^2$$

(k) Thin rectangular sheet

$$I = \frac{1}{3} ML^2$$

▲ **FIGURE 8.20** Moments of inertia of some uniform density objects with common shapes

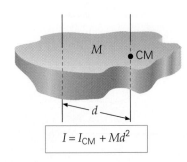

$$I = I_{CM} + Md^2$$

▲ **FIGURE 8.21 Parallel axis theorem** The moment of inertia about an axis parallel to another through the center of mass of a body is $I = I_{CM} + Md^2$, where M is the total mass of the body and d is the distance between the two axes.

the rod (Fig. 8.20b). The moment of inertia about such a parallel axis is given by a useful theorem called the **parallel axis theorem**, namely,

$$I = I_{CM} + Md^2 \tag{8.8}$$

where I is the moment of inertia about an axis that is parallel to one through the center of mass and at a distance d from it, I_{CM} is the moment of inertia about an axis through the center of mass, and M is the total mass of the body (◄Fig. 8.21). For the axis through the end of the rod (Fig. 8.20c), the moment of inertia is obtained by applying the parallel axis theorem to the thin rod in Fig. 8.20b:

$$I = I_{CM} + Md^2 = \frac{1}{12} ML^2 + M\left(\frac{L}{2}\right)^2 = \frac{1}{12} ML^2 + \frac{1}{4} ML^2 = \frac{1}{3} ML^2$$

APPLICATIONS OF ROTATIONAL DYNAMICS

The rotational form of Newton's second law allows us to analyze dynamic rotational situations. Examples 8.12 and 8.13 illustrate how this is done. In such situations, it is very important to make certain that all the data are properly listed to help with the increasing number of variables.

EXAMPLE 8.12 | Opening the Door: Torque in Action

A student opens a uniform 12-kg door by applying a constant force of 40 N at a perpendicular distance of 0.90 m from the hinges (▶ Fig. 8.22). If the door is 2.0 m in height and 1.0 m wide, what is the magnitude of its angular acceleration? (Assume that the door rotates freely on its hinges.)

THINKING IT THROUGH. From the given information, the applied net torque can be calculated. To find the angular acceleration of the door, the moment of inertia is needed. This can be calculated, since the door's mass and dimensions are known.

SOLUTION. From the information given in the problem,

Given: $M = 12$ kg   *Find:* α (magnitude of angular
      $F = 40$ N   acceleration)
      $r_\perp = r = 0.90$ m
      $h = 2.0$ m (door height)
      $w = 1.0$m (door width)

▲ **FIGURE 8.22 Torque in action**
See Example text for description.

The rotational form of Newton's second law can be applied, $\tau_{net} = I\alpha$, where I is about the hinge axis. τ_{net} can be found from the given data, so the problem boils down to determining the moment of inertia of the door.

   Looking at Fig. 8.20, it can be seen that case (k) applies to a door (treated as a uniform rectangle) rotating on hinges, so $I = \frac{1}{3}ML^2$, where $L = w$, the width of the door. Then,

$$\tau_{net} = I\alpha$$

or

$$\alpha = \frac{\tau_{net}}{I} = \frac{r_\perp F}{\frac{1}{3}ML^2} = \frac{3rF}{Mw^2} = \frac{3(0.90\text{ m})(40\text{ N})}{(12\text{ kg})(1.0\text{ m})^2} = 9.0 \text{ rad/s}^2$$

FOLLOW-UP EXERCISE. In this Example, if the constant torque were applied through an angular distance of 45° and then removed, how long would the door take to swing completely open (90°)? Neglect friction.

In problems involving pulleys in Section 4.5, the mass (and hence the inertia) of the pulley was neglected in order to simplify things. Now you know how to include those quantities and can treat pulleys more realistically, as seen in the next Example.

EXAMPLE 8.13 | Pulleys Have Mass, Too—Taking Pulley Inertia into Account

A block of mass m hangs from a string wrapped around a frictionless, disk-shaped pulley of mass M and radius R, as shown in ▼ Fig. 8.23. If the block descends from rest under the influence of gravity, what is the magnitude of its linear acceleration? (Neglect the mass of the string.)

THINKING IT THROUGH. Real pulleys have mass and rotational inertia, which affect their motion. The suspended mass (via the string) applies a torque to the pulley. Here the rotational form of Newton's second law can be used to find the angular acceleration of the pulley and then its tangential acceleration, which is the same in magnitude as the linear acceleration of the block. (Why?) No numerical values are given so the answer will be in symbol form.

(continued on next page)

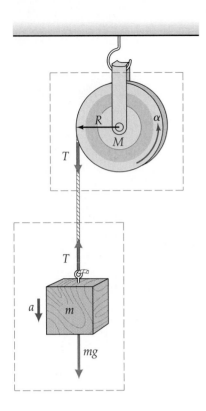

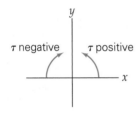

▲ **FIGURE 8.23 Pulley with inertia** Taking the mass, or rotational inertia, of a pulley into account allows a more realistic description of the motion. The directional sign convention for torque is shown. See Example 8.13.

SOLUTION. The linear acceleration of the block depends on the angular acceleration of the pulley, so we look at the pulley system first. The pulley is treated as a disk and thus has a moment of inertia $I = \frac{1}{2}MR^2$ (Fig. 8.20e). A torque due to the tension force in the string (T) acts on the pulley. With $\tau = I\alpha$ (considering only the upper dashed box in Fig. 8.23),

$$\tau_{\text{net}} = r_\perp F = RT = I\alpha = \left(\tfrac{1}{2}MR^2\right)\alpha$$

such that

$$\alpha = \frac{2T}{MR}$$

The linear acceleration of the block and the angular acceleration of the pulley are related by $a = R\alpha$, where a is the tangential acceleration, and

$$a = R\alpha = \frac{2T}{M} \tag{1}$$

But T is unknown. Looking at the descending mass (the lower dashed box) and summing the forces in the vertical direction (choosing down as positive) gives

$$mg - T = ma$$

or

$$T = mg - ma \tag{2}$$

Using Eq. 2 to eliminate T from Eq. 1 yields

$$a = \frac{2T}{M} = \frac{2(mg - ma)}{M}$$

And solving for a,

$$a = \frac{2mg}{(2m + M)} \tag{3}$$

Note that if $M \to 0$ (as in the case of ideal, massless pulleys in previous chapters), then $I \to 0$ and $a = g$ (from Eq. 3). Here, however, $M \neq 0$, so $a < g$. (Why?)

FOLLOW-UP EXERCISE. Pulleys can be analyzed even more realistically. In this Example, friction was neglected, but practically, a frictional torque (τ_f) exists and should be included. What would be the form, as in Eq. 3, of the angular acceleration in this case? Show that your result is dimensionally correct.

In pulley problems, as before, the mass of the string will be neglected—an approach that still gives a good approximation if the string is relatively light. Taking the mass of the string into account would give a continuously varying mass hanging on the pulley, thus producing a variable torque. Such problems are beyond the scope of this book.

Suppose you had masses suspended from each side of a pulley. Here, you would have to compute the net torque. If the values of the masses are unknown, so which way the pulley would rotate cannot be determined, then simply assume a direction. As in the linear case, if the result came out with the opposite sign, it would indicate that you had assumed the wrong direction.

PROBLEM-SOLVING HINT

For problems such as those of Examples 8.13 and 8.14, dealing with coupled rotational and translational motions, keep in mind that with no string slippage, the magnitudes of the accelerations are usually related by $a = r\alpha$, while $v = r\omega$ relates the magnitudes of the velocities at any instant of time. Applying Newton's second law (in rotational or linear form) to different parts of the system gives equations that can be combined by using such relationships. Also, for rolling without slipping, $a = r\alpha$ and $v = r\omega$, relate the angular quantities to the linear motion of the center of mass.

Another application of rotational dynamics is the analysis of motion of objects that can roll.

CONCEPTUAL EXAMPLE 8.14 | Applying a Torque One More Time: Which Way Does the Yo-Yo Roll?

The string of a yo-yo sitting on a level surface is pulled as shown in ▼ Fig. 8.24. Will the yo-yo roll (a) toward the person or (b) away from the person?

▶ **FIGURE 8.24**
Pulling the yo-yo's string See Conceptual Example text for description.

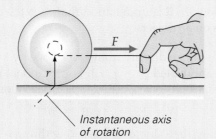

Instantaneous axis of rotation

REASONING AND ANSWER. Apply the physics just studied to the situation. Note that the instantaneous axis of rotation is along the line of contact of the yo-yo with the surface. If you had a stick standing vertically in place of the $\vec{r}$ vector and pulled on a string attached to the top of the stick in the direction of $\vec{F}$, which way would the stick rotate? Of course, it would rotate clockwise (about its instantaneous axis of rotation). The yo-yo reacts similarly; that is, it rolls in the direction

of the pull, so the answer is (a). (Get a yo-yo and try it if you're a nonbeliever.)

There is more interesting physics in our yo-yo situation. The pull force is not the only force acting on the yo-yo; there are three others. Do they contribute torques? Let's identify these forces. There's the weight of the yo-yo and the normal force from the surface. Also, there is a horizontal force of static friction between the yo-yo and the surface. (Otherwise the yo-yo would slide rather than roll.) But these three forces act through the line of contact or through the instantaneous axis of rotation, so they produce no torques here. (Why?)

What would happen if the angle of the string or pull force were increased (relative to the horizontal), as illustrated in ▼ Fig. 8.25a? The yo-yo would still roll to the right. As can be seen in Fig. 8.25b, at some critical angle θ_c the line of force goes through the axis of rotation, and the net torque on the yo-yo becomes zero, so the yo-yo does not roll.

If this critical angle is exceeded (Fig. 8.25c), the yo-yo will begin to roll counterclockwise, or to the left. Note that the line of action of the force is on the other side of the axis of rotation from that in Fig. 8.26a and that the lever arm ($r_\perp$) has changed directions, resulting in a reversed net torque direction.

▶ **FIGURE 8.25 The angle makes a difference** (a) With the line of action to the left of the instantaneous axis, the yo-yo rolls to the right. (b) At a critical angle θ_c the line of action passes through the axis, and the yo-yo is in equilibrium. (c) When the line of action is to the right of the axis, the yo-yo rolls to the left.

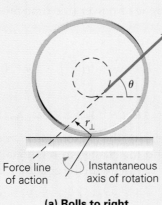

Force line of action   Instantaneous axis of rotation

(a) Rolls to right

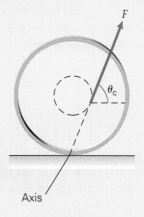

Axis

(b) $\theta = \theta_c$, in rotational equilibrium does not roll

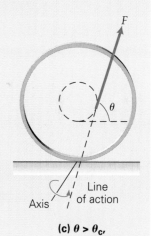

Axis   Line of action

(c) $\theta > \theta_c$, rolls to left ($r_\perp$ to the right)

FOLLOW-UP EXERCISE. Suppose you set the yo-yo string at the critical angle, with the string over a round, horizontal bar at the appropriate height, and you suspend a weight on the end of the string to supply the force for the equilibrium condition. What will happen if you then pull the yo-yo toward you, away from its equilibrium position, and release it?

DID YOU LEARN?
➡ The moment of inertia is a measure of rotational inertia—a body's tendency to resist rotational motion.
➡ The rotational form of Newton's second law is $\vec{\tau}_{net} = I\vec{\alpha}$ (analogous to $\vec{F}_{net} = m\vec{a}$).

8.4 Rotational Work and Kinetic Energy

LEARNING PATH QUESTIONS

➡ How do the equations for rotational work and power compare to their translational analogues?

➡ What is the total kinetic energy of a rolling object (without slipping)?

This section gives the rotational analogues of various equations of linear motion associated with work and kinetic energy for constant torques. Because their development is similar to that given for their linear counterparts, detailed discussion is not needed. As in Section 5.1, it is understood that W is the net work if more than one force or torque acts on an object.

Rotational Work   We can go directly from work done by a force to work done by a torque, since the two are related ($\tau = r_\perp F$). For rotational motion, the **rotational work**, $W = Fs$, done by a single force F acting tangentially along an arc length s is

$$W = Fs = F(r_\perp \theta) = \tau\theta$$

where θ is in radians. Thus, for a single torque acting through an angle of rotation θ,

$$W = \tau\theta \quad \textit{(rotational work for a single force)} \tag{8.9}$$

In this book, both the torque (τ) and angular displacement (θ) vectors are almost always along the fixed axis of rotation, so you will not need to be concerned about parallel components, as you were for translational work. The torque and angular displacement may be in opposite directions, in which case the torque does negative work and slows the rotation of the body. Negative rotational work is analogous to F and d being in opposite directions for translational motion.

Rotational Power   An expression for the instantaneous **rotational power** (P), the rotational analogue of power (the time rate of doing work), is easily obtained from Eq. 8.9:

$$P = \frac{W}{t} = \tau\left(\frac{\theta}{t}\right) = \tau\omega \quad \textit{(rotational power)} \tag{8.10}$$

THE WORK—ENERGY THEOREM AND KINETIC ENERGY

The relationship between the net rotational work done on a rigid body (more than one force acting) and the change in rotational kinetic energy of the body can be derived as follows, starting with the equation for rotational work:

$$W_{\text{net}} = \tau\theta = I\alpha\theta$$

Since we assume the torques are due only to constant forces, α is constant. But from rotational kinematics in Chapter 7, it is known that for a constant angular acceleration, $\omega^2 = \omega_0^2 + 2\alpha\theta$, and

$$W_{\text{net}} = I\left(\frac{\omega^2 - \omega_0^2}{2}\right) = \tfrac{1}{2}I\omega^2 - \tfrac{1}{2}I\omega_0^2$$

From Eq. 5.6 (work–energy), $W_{\text{net}} = \Delta K$. Therefore,

$$W_{\text{net}} = \tfrac{1}{2}I\omega^2 - \tfrac{1}{2}I\omega_0^2 = K - K_\text{o} = \Delta K \tag{8.11}$$

Then the expression for **rotational kinetic energy** is

$$K = \tfrac{1}{2}I\omega^2 \quad \textit{(rotational kinetic energy)} \tag{8.12}$$

Thus, *the net rotational work done on an object is equal to the change in the rotational kinetic energy of the object* (with zero linear kinetic energy). Consequently, to change the rotational kinetic energy of an object, a net torque must be applied.

TABLE 8.1   Translational and Rotational Quantities and Equations

Translational		Rotational	
Force:	$\vec{\mathbf{F}}$	Torque (magnitude):	$\tau = rF \sin \theta$
Mass (inertia):	m	Moment of inertia:	$I = \sum m_i r_i^2$
Newton's second law:	$\vec{\mathbf{F}}_{net} = m\vec{\mathbf{a}}$	Newton's second law:	$\vec{\tau}_{net} = I\vec{\alpha}$
Work:	$W = Fd$	Work:	$W = \tau\theta$
Power:	$P = Fv$	Power:	$P = \tau\omega$
Kinetic energy:	$K = \frac{1}{2}mv^2$	Kinetic energy:	$K = \frac{1}{2}I\omega^2$
Work–energy theorem:	$W_{net} = \frac{1}{2}mv^2 - \frac{1}{2}mv_0^2 = \Delta K$	Work–energy theorem:	$W_{net} = \frac{1}{2}I\omega^2 - \frac{1}{2}I\omega_0^2 = \Delta K$
Linear momentum:	$\vec{\mathbf{p}} = m\vec{\mathbf{v}}$	Angular momentum:	$\vec{\mathbf{L}} = I\vec{\omega}$

It is possible to derive the expression for the kinetic energy of a rotating rigid body (about a fixed axis) directly. Summing the instantaneous kinetic energies of the body's individual particles relative to the fixed axis gives

$$K = \tfrac{1}{2}\sum m_i v_i^2 = \tfrac{1}{2}\left(\sum m_i r_i^2\right)\omega^2 = \tfrac{1}{2}I\omega^2$$

where, for each particle of the body, $v_i = r_i\omega$. So, Eq. 8.12 doesn't represent a new form of energy; rather, it is simply another expression for kinetic energy, in a form that is more convenient for rigid body rotation.

A summary of translational and rotational analogues is given in ▲Table 8.1. (The table also contains angular momentum, which will be discussed in Section 8.5.)

When an object has both translational and rotational motion, its total kinetic energy may be divided into parts to reflect the two kinds of motion. For example, for a cylinder rolling without slipping on a level surface, the motion is purely rotational relative to the instantaneous axis of rotation (the point or line of contact), which is instantaneously at rest. The total kinetic energy of the rolling cylinder is

$$K = \tfrac{1}{2}I_i\omega^2$$

where I_i is the moment of inertia about the instantaneous axis. This moment of inertia about the point of contact (the axis) is given by the parallel axis theorem (Eq. 8.8), $I_i = I_{CM} + MR^2$, where R is the radius of the cylinder. Then

$$K = \tfrac{1}{2}I_i\omega^2 = \tfrac{1}{2}\left(I_{CM} + MR^2\right)\omega^2 = \tfrac{1}{2}I_{CM}\omega^2 + \tfrac{1}{2}MR^2\omega^2$$

But since there is no slipping, $v_{CM} = R\omega$, and the total K is

$$K = \tfrac{1}{2}I_{CM}\omega^2 + \tfrac{1}{2}Mv_{CM}^2 \quad \text{(rolling, no slipping)} \tag{8.13}$$

$$\underset{K}{\text{total}} = \underset{K_r}{\text{rotational}} + \underset{K_t}{\text{translational}}$$

Note that although a cylinder was used as an example here, this is a general result and applies to any object that is rolling without slipping.

Thus, *the total kinetic energy of such an object is the sum of two contributions: the translational kinetic energy of the object's center of mass and the rotational kinetic energy of the object relative to a horizontal axis through its center of mass.*

EXAMPLE 8.15 | Division of Energy: Rotational and Translational

A uniform, solid 1.0-kg cylinder rolls without slipping at a speed of 1.8 m/s on a flat surface. (a) What is the total kinetic energy of the cylinder? (b) What percentage of this total is rotational kinetic energy?

THINKING IT THROUGH. The cylinder has both rotational and translational kinetic energies, so Eq. 8.13 applies, and its terms are related by the condition of rolling without slipping.

(continued on next page)

SOLUTION.

Given: $M = 1.0$ kg

$v_{CM} = 1.8$ m/s

$I_{CM} = \frac{1}{2}MR^2$ (from Fig. 8.20e)

Find: (a) K (total kinetic energy)

(b) $\frac{K_r}{K}$ ($\times 100\%$) (percentage of rotational energy)

(a) The cylinder rolls without slipping, so the condition $v_{CM} = R\omega$ applies. Then the total kinetic energy (K) is the sum of the rotational kinetic energy K_r and the translational kinetic energy K_t of the center of mass, K_{CM} (Eq. 8.13):

$$K = \tfrac{1}{2}I_{CM}\omega^2 + \tfrac{1}{2}Mv_{CM}^2 = \tfrac{1}{2}\left(\tfrac{1}{2}MR^2\right)\left(\frac{v_{CM}}{R}\right)^2 + \tfrac{1}{2}Mv_{CM}^2 = \tfrac{1}{4}Mv_{CM}^2 + \tfrac{1}{2}Mv_{CM}^2$$

$$= \tfrac{3}{4}Mv_{CM}^2 = \tfrac{3}{4}(1.0\text{ kg})(1.8\text{ m/s})^2 = 2.4\text{ J}$$

(b) The rotational kinetic energy K_r of the cylinder is the first term of the preceding equation, so, forming a ratio in symbol form,

$$\frac{K_r}{K} = \frac{\tfrac{1}{4}Mv_{CM}^2}{\tfrac{3}{4}Mv_{CM}^2} = \tfrac{1}{3}(\times 100\%) = 33\%$$

Thus, the total kinetic energy of the cylinder is made up of rotational and translational parts, with one-third being rotational.

Note that in part (b) the radius of the cylinder was not needed, nor was the mass. Because a ratio was used, these quantities canceled. However, *don't* think that this exact division of energy is a general result. It is easy to show that the percentage is different for objects with different moments of inertia. For example, you should expect a rolling sphere to have a smaller percentage of rotational kinetic energy than a cylinder has, because the sphere has a smaller moment of inertia ($I = \tfrac{2}{5}MR^2$).

FOLLOW-UP EXERCISE. Potential energy can be brought into the act by applying the conservation of energy to an object rolling up or down an inclined plane. In this Example, suppose that the cylinder rolled up a 20° inclined plane without slipping. (a) At what vertical height (measured by the vertical distance of its CM) on the plane does the cylinder stop? (b) To find the height in part (a), you probably equated the initial total kinetic energy to the final gravitational potential energy. That is, the total kinetic energy was reduced by the work done by gravity. However, a frictional force also acts (to prevent slipping). Is there not work done here, too?

EXAMPLE 8.16 | **Rolling Down**

A uniform cylindrical hoop is released from rest at a height of 0.25 m near the top of an inclined plane (▼ Fig. 8.26). If the hoop rolls down the plane without slipping and no energy is lost due to friction, what is the linear speed of the cylinder's center of mass at the bottom of the incline?

THINKING IT THROUGH. Here, gravitational potential energy is converted into kinetic energy—both rotational and translational. The conservation of (mechanical) energy applies, since W_f (frictional work) is zero.

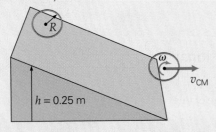

▲ **FIGURE 8.26 Rolling motion and energy** When an object rolls down an inclined plane, potential energy is converted to translational *and* rotational kinetic energy. This makes the rolling slower than frictionless sliding.

SOLUTION.

Given: $h = 0.25$ m

$I_{CM} = MR^2$ (from Fig. 8.20d)

Find: v_{CM} (speed of CM)

The total mechanical energy of the cylinder is conserved:

$$E_o = E$$

Since $v_o = 0$ at the top of the incline and assuming that $U = 0$ at the bottom, this becomes

$$U_o = K$$

$$\underset{\substack{\text{initially} \\ \text{at rest}}}{Mgh} = \underset{\text{at bottom of incline}}{\tfrac{1}{2}I_{CM}\omega^2 + \tfrac{1}{2}Mv_{CM}^2}$$

Using the rolling condition $v_{CM} = R\omega$ gives

$$Mgh = \tfrac{1}{2}(MR^2)\left(\frac{v_{CM}}{R}\right)^2 + \tfrac{1}{2}Mv_{CM}^2 = Mv_{CM}^2$$

Solving for v_{CM},

$$v_{CM} = \sqrt{gh} = \sqrt{(9.8\text{m/s}^2)(0.25\text{m})} = 1.6\text{ m/s}$$

A FIXED RACE

As Example 8.16 shows for an object rolling down an incline without slipping, v_{CM} is independent of M and R. The masses and radii cancel, so all objects of a particular shape (with the same equation for the moment of inertia) roll with the same speed, regardless of their size or density. But the rolling speed does vary with the moment of inertia, which varies with an object's shape. Therefore, rigid bodies with different shapes roll with different speeds. For example, if you released a cylindrical hoop, a solid cylinder, and a uniform sphere at the same time from the top of an inclined plane, the sphere would win the race to the bottom, followed by the cylinder, with the hoop coming in last—every time!

You can try this as an experiment with a couple of cans of food or other cylindrical containers—one full of some solid material (in effect, a rigid body) and one empty and with the ends cut out—and a smooth, solid ball. Remember that the masses and the radii make no difference. You might think that an annular cylinder (a hollow cylinder with inner and outer radii that are appreciably different—Fig. 8.20f) would be a possible front-runner, or "front-roller," in such a race, but it wouldn't be. The rolling race down an incline is fixed even when you vary the masses and the radii.

Another aspect of rolling is discussed in Insight 8.2, Slide or Roll to a Stop? Antilock Brakes.

DID YOU LEARN?
→ Rotational work is $W = \tau\theta$ (compared to translational $W = Fd$), and rotational power is $P = \tau\omega$ (compared to translational $W = Fv$).
→ The total kinetic energy of a rolling object is the sum of two contributions: the translational kinetic energy of the object's center of mass and the rotational kinetic energy relative to a horizontal axis through the CM that is, $K = \frac{1}{2}I_{CM}\omega^2 + \frac{1}{2}Mv_{CM}^2$.

8.5 Angular Momentum

LEARNING PATH QUESTIONS
→ How is angular momentum related to torque, and what is the linear analogy?
→ When is angular momentum conserved?

Another important quantity in rotational motion is angular momentum. Recall from Section 6.1 how the linear momentum of an object is changed by a force. Analogously, changes in angular momentum are associated with torques. As has been learned, torque is the product of a moment arm and a force. In a similar manner, **angular momentum** (L) is the product of a moment arm (r) and a linear momentum (p). For a particle of mass m, the magnitude of the linear momentum is $p = mv$, where $v = r\omega$. The magnitude of the angular momentum is

$$L = r_\perp p = mr_\perp v = mr_\perp^2 \omega \quad \text{(single-particle angular momentum)} \quad (8.14)$$

SI unit of angular momentum: kilogram-meters squared per second $(\text{kg} \cdot \text{m}^2/\text{s})$, where v is the speed of the particle $r_\perp$ is the moment arm, and ω is the angular speed.

INSIGHT 8.2 | Slide or Roll to a Stop? Antilock Brakes

While driving, in an emergency you may instinctively jam on the brakes, trying to come to a quick stop—that is, to stop in the shortest distance. But with the wheels locked, the car skids or slides, often out of control. In this case, the force of sliding friction is acting on the wheels.

To prevent skidding, you may have learned to pump the brakes in order to roll rather than slide to a stop, particularly on wet or icy roads. Most newer automobiles have a computerized antilock braking system (ABS) that pumps the brakes automatically. When the brakes are applied firmly and the car begins to slide, sensors in the wheels note the sliding motion, and a computer takes control of the braking system. It momentarily releases the brakes and then varies the brake fluid pressure with a pumping action (up to thirteen times per second) so that the wheels will continue to roll without slipping.

In the absence of sliding, both rolling friction and static friction act. In many cases, however, the force of rolling friction is small, and only static friction need be taken into account. The ABS works to keep static friction near the maximum, $f_s \approx f_{s_{max}}$ which you can't do easily by foot.

Does sliding instead of rolling make a big difference in an automobile's stopping distance? We can calculate the difference by assuming that rolling friction is negligible. Although the external force of static friction does no work to dissipate energy in slowing a car (this is done internally by friction on the brake pads), it does determine whether the wheels roll or slide.

In Example 2.8, a vehicle's stopping distance was given by

$$x = \frac{v_0^2}{2a}$$

By Newton's second law, the net force in the horizontal direction is $F = f = \mu N = \mu m g = ma$, and the stopping acceleration is then $a = \mu g$. Thus,

$$x = \frac{v_0^2}{2\mu g} \tag{1}$$

But, as was noted in Section 4.6, the coefficient of sliding (kinetic) friction is generally less than that of static friction; that is, $\mu_k < \mu_s$. The general difference between rolling stops and sliding stops can be seen by using the same initial velocity v_0 for both cases. Then, using Eq. 1 to form a ratio,

$$\frac{x_{roll}}{x_{slide}} = \frac{\mu_k}{\mu_s} \quad \text{or} \quad x_{roll} = \left(\frac{\mu_k}{\mu_s}\right) x_{slide}$$

From Table 4.1, the value of μ_k for rubber on wet concrete is 0.60, and the value of μ_s for these surfaces is 0.80. Using these values for a comparison of the stopping distances gives

$$x_{roll} = \left(\frac{0.60}{0.80}\right) x_{slide} = (0.75) x_{slide}$$

Thus, the car comes to a rolling stop in 75% of the distance required for a sliding stop—for example, 15 m instead of 20 m. Although this distance may vary for different conditions, it could be an important, perhaps lifesaving, difference.

For circular motion, $r_\perp = r$, since $\vec{v}$ is perpendicular to $\vec{r}$. For a system of particles making up a rigid body, all the particles travel in circles, and the magnitude of the total angular momentum is

$$L = (\Sigma m_i r_i^2)\omega = I\omega \quad \textit{(rigid body angular momentum)} \tag{8.15}$$

which, for rotation about a fixed axis, is (in vector notation)

$$\vec{L} = I\vec{\omega} \tag{8.16}$$

Thus, $\vec{L}$ is in the direction of the angular velocity vector ($\vec{\omega}$). This direction is given by the right-hand rule (Section 8.2).

For linear motion, the change in the total linear momentum of a system is related to the net external force by $\vec{F}_{net} = \Delta\vec{P}/\Delta t$. Angular momentum is analogously related to net torque (in magnitude form):

$$\tau_{net} = I\alpha = \frac{I\Delta\omega}{\Delta t} = \frac{\Delta(I\omega)}{\Delta t} = \frac{\Delta L}{\Delta t}$$

That is,

$$\tau_{net} = \frac{\Delta L}{\Delta t} \tag{8.17}$$

Thus, the net torque is equal to *the time rate of change of angular momentum*. In other words, a net torque results in a *change* in angular momentum.

CONSERVATION OF ANGULAR MOMENTUM

Equation 8.17 was derived using $\tau_{net} = I\alpha$, which applies to a rigid system of particles or a rigid body having a constant moment of inertia. However, Eq. 8.17 is a general equation that also applies to even nonrigid systems of particles. In such a system, there may be a change in the internal mass distribution and a change in the moment of inertia.

If the net torque on a system is zero, then, by Eq. 8.17, $\vec{\tau}_{net} = \Delta\vec{L}/\Delta t = 0$, and

$$\Delta\vec{L} = \vec{L} - \vec{L}_o = I\vec{\omega} - I_o\vec{\omega}_o = 0$$

or in magnitude,

$$I\omega = I_o\omega_o \qquad (8.18)$$

Thus, the condition for the **conservation of angular momentum** is as follows:

> In the absence of an external, unbalanced torque, the total (vector) angular momentum of a system is conserved (remains constant).

Just as the internal forces cannot change a systems's linear momentum, neither can internal torques change a system's angular momentum.

For a rigid body with a constant moment of inertia (that is, $I = I_o$), the angular speed remains constant ($\omega = \omega_o$) in the absence of a net torque. But it is possible for the moment of inertia to change in some systems, giving rise to a change in the angular speed, as the following Example illustrates.

EXAMPLE 8.17 | Pull It Down: Conservation of Angular Momentum

A small ball at the end of a string that passes through a tube is swung in a circle, as illustrated in ▶ Fig. 8.27. When the string is pulled downward through the tube, the angular speed of the ball increases. (a) Is the increase in angular speed caused by a torque due to the pulling force? (b) If the ball is initially moving at 2.8 m/s in a circle with a radius of 0.30 m, what will be its tangential speed if the string is pulled down to reduce the radius of the circle to 0.15 m? (Neglect the mass of the string.)

THINKING IT THROUGH. (a) A force is applied to the ball via the string, but consider the axis of rotation. (b) In the absence of a net torque, the angular momentum is conserved (Eq. 8.18), and the tangential speed is related to the angular speed by $v = r\omega$.

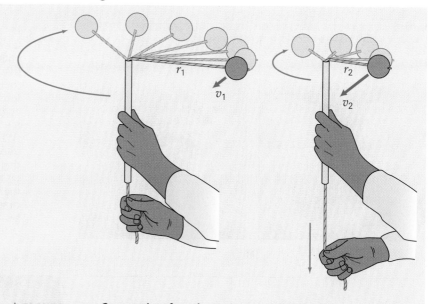

▲ **FIGURE 8.27 Conservation of angular momentum** When the string is pulled downward through the tube, the revolving ball speeds up. See Example text for description.

SOLUTION.

Given: $r_1 = 0.30$ m *Find:* (a) Cause of the increase in angular speed
$r_2 = 0.15$ m (b) v_2 (final tangential speed)
$v_1 = 2.8$ m/s

(a) The change in the angular velocity, or an angular acceleration, is not caused by a torque due to the pulling force. The force on the ball, as transmitted by the string (tension), acts through the axis of rotation, and therefore the torque is zero. Because the rotating portion of the string is shortened, the moment of inertia of the ball ($I = mr^2$, from Fig. 8.20a), decreases. Because of the absence of an external torque, the angular momentum ($I\omega$) of the ball is conserved, and if I is reduced, ω must increase.

(continued on next page)

(b) Because the angular momentum is conserved, we can equate the magnitudes of the angular momenta:

$$I_0\omega_0 = I\omega$$

Then, using $I = mr^2$ and $\omega = v/r$ gives

$$mr_1 v_1 = mr_2 v_2$$

and

$$v_2 = \left(\frac{r_1}{r_2}\right)v_1 = \left(\frac{0.30 \text{ m}}{0.15 \text{ m}}\right)2.8 \text{ m/s} = 5.6 \text{ m/s}$$

When the radial distance is shortened, the ball speeds up (accelerates).

FOLLOW-UP EXERCISE. Let's look at the situation in this Example in terms of work and energy. If the initial speed is the same and the vertical pulling force is 7.8 N, what is the final speed of the 0.10-kg ball?

Example 8.17 should help you understand Kepler's law of equal areas (Section 7.6) from another viewpoint. A planet's angular momentum is conserved to a good approximation by neglecting the weak gravitational torques from other planets. (The Sun's gravitational force on a planet produces little or no torque. Why?) When a planet is closer to the Sun in its elliptical orbit and so has a shorter moment arm, its speed is greater, by the conservation of angular momentum. [This is the basis of Kepler's second law (the law of areas).] Similarly, when an orbiting satellite's altitude varies during the course of an elliptical orbit about a planet, the satellite speeds up or slows down in accordance with the same principle.

REAL-LIFE ANGULAR MOMENTUM

A popular demonstration of the conservation of angular momentum is shown in ▼Fig. 8.28a. A person sitting on a stool that rotates holds weights with his arms outstretched and is started slowly rotating. An external torque to start this rotation

(a)

◄ **FIGURE 8.28 Change in moment of inertia**
(a) When the student spins slowly with masses in outstretched arms, his moment of inertia is relatively large. (The masses are farther from the axis of rotation.) Note that he is isolated, with no external torques (neglecting friction) acting on him, so his angular momentum, $L = I\omega$ is conserved. Pulling his arms inward decreases his moment of inertia. (Why?) Consequently, ω must increase, and he goes into a dizzying spin. **(b)** Ice skaters similarly change their moment of inertia to increase ω in doing spins. **(c)** The same principle helps explain the violence of the winds that spiral around the center of a hurricane. As air rushes in toward the low pressure center of the storm, the air's rotational speed must increase for angular momentum to be conserved.

(b)   **(c)**

must be supplied by someone else, because the person on the stool cannot initiate the motion by himself. (Why not?) Once rotating, if the person brings his arms inward, the angular speed increases and he spins much faster. Extending his arms again slows him down. Can you explain this phenomenon?

If L is constant, what happens to ω when I is made smaller by reducing r? The angular speed must increase to compensate and keep L constant. Ice skaters and ballerinas perform dizzying spins by pulling in and raising their arms to reduce their moment of inertia (Fig. 8.28b). Similarly, a diver spins during a high dive by tucking in the body and limbs, greatly decreasing his or her moment of inertia. The enormous wind speeds of tornadoes and hurricanes represent another example of the same effect (Fig. 8.28c).

Angular momentum also plays a role in ice-skating jumps in which the skater spins in the air, such as a triple axel or triple lutz. A torque applied on the jump gives the skater angular momentum, and the arms and legs are drawn into the body, which, as in spinning on one's toes, decreases the moment of inertia and increases the angular speed so that multiple spins can be made during the jump. To land with a smaller rate of spin, the skater opens the arms and projects the non-landing leg. You may have noticed that most jump landings proceed in a curved arc, which allows the skater to gain control.

EXAMPLE 8.18 | A Skater Model

Real-life situations are generally complicated, but some can be approximately analyzed by using simple models. Such a model for a skater's spin is shown in ▶ Fig. 8.29, with a cylinder and rods representing the skater. In (a), the skater goes into the spin with the "arms" out, and in (b) the "arms" are over the head to achieve a faster spin by the conservation of angular momentum. If the initial spin rate is 1 revolution per 1.5 s, what is the angular speed when the arms are tucked in?

THINKING IT THROUGH. The body and arms of a skater are approximated by the cylinder and rods, for which the moments of inertia are known (Fig. 8.20). Special attention must be given to finding the moment of inertia of the arms around the axis of rotation (through the cylinder). This can be done by applying the parallel axis theorem (Eq. 8.8).

With the angular momentum conserved, $L = L_0$ or $I\omega = I_0\omega_0$. Knowing the initial angular speed, and given quantities to evaluate the moments of inertia (Fig. 8.29), the final angular speed can be found.

SOLUTION. Listing the given data (see Fig. 8.29):

Given: $\omega_0 = (1 \text{ rev}/1.5 \text{ s})(2\pi \text{ rad/rev}) = 4.2 \text{ rad/s}$ *Find:* ω (final angular speed)
$M_c = 75 \text{ kg}$ (cylinder or body)
$M_r = 5.0 \text{ kg}$ (one rod or arm)
$R = 20 \text{ cm} = 0.20 \text{ m}$
$L = 80 \text{ cm} = 0.80 \text{ m}$
Momenta of inertia (from Fig. 8.20).
cylinder: $I_c = \frac{1}{2}M_cR^2$ rod: $I_r = \frac{1}{12}M_rL^2$

Let's first compute the moments of inertia of the system using the parallel axis theorem, $I = I_{CM} + Md^2$ (Eq. 8.8).
  Before: The I_c of the cylinder is straightforward (Fig. 8.20e):

$$I_c = \frac{1}{2}M_cR^2 = \frac{1}{2}(75 \text{ kg})(0.20 \text{ m})^2 = 1.5 \text{ kg} \cdot \text{m}^2$$

Referencing the moment of inertia of a horizontal rod (Fig. 8.29a) to the cylinder's axis of rotation using the parallel axis theorem:

$I_r = I_{CM(rod)} + Md^2$
   $= \frac{1}{12}M_rL^2 + M_r(R + L/2)^2$ (where the parallel axis through the CM of the rod is a distance of $R + L/2$ from the axis of rotation)
   $= \frac{1}{12}(5.0 \text{ kg})(0.80 \text{ m})^2 + (5.0 \text{ kg})(0.20 \text{ m} + 0.40 \text{ m})^2 = 2.1 \text{ kg} \cdot \text{m}^2$

And, $I_0 = I_c + 2I_r = 1.5 \text{ kg} \cdot \text{m}^2 + 2(2.1 \text{ kg} \cdot \text{m}^2) = 5.7 \text{ kg} \cdot \text{m}^2$

(continued on next page)

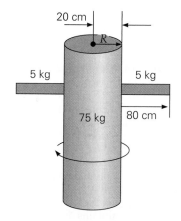

20 cm

R

5 kg   5 kg

75 kg   80 cm

(a) Arms extended (not to scale)

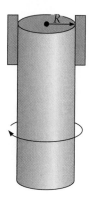

R

(b) Arms overhead

▲ **FIGURE 8.29 Skater model.** Change in moment of inertia and spin. See Example 8.18.

After: In Fig. 8.29b, treating an arm mass as if its center of mass is now only about 20 cm from the axis of rotation, the moment of inertia of each arm is $I = M_r R^2$ (Fig. 8.20b), and,

$$I = I_c + 2(M_r R^2) = 1.5 \text{ kg} \cdot \text{m}^2 + 2(5.0 \text{ kg} \cdot \text{m}^2)(0.20 \text{ m})^2 = 1.9 \text{ kg} \cdot \text{m}^2$$

Then with the conservation of angular momentum, $L = L_o$ or $I\omega = I_o\omega_o$ and

$$\omega = \left(\frac{I_a}{I_b}\right)\omega_o = \left(\frac{5.7 \text{ kg} \cdot \text{m}^2}{1.9 \text{ kg} \cdot \text{m}^2}\right)(4.2 \text{ rad/s}) = 13 \text{ rad/s}$$

So the angular speed increases by a factor of 3.

FOLLOW-UP EXERCISE.   Suppose a skater with 75% of the mass of the skater in the Exercise did a spin. What would be the spin rate ω in this case? (Consider all masses to be reduced by 75%.)

Angular momentum, $\vec{L}$, is a vector, and when it is conserved or constant, its magnitude *and* direction must remain unchanged. Thus, when no external torques act, the direction of $\vec{L}$ is fixed in space. This is the principle behind passing a football accurately, as well as that behind the movement of a gyrocompass (▼Fig. 8.30). A football is normally passed with a spiraling rotation. This spin, or gyroscopic action, stabilizes the ball's spin axis in the direction of motion. Similarly, rifle bullets are set spinning by the rifling in the barrel for directional stability.

The $\vec{L}$ vector of a spinning gyroscope in the compass is set in a particular direction (usually north). In the absence of external torques, the compass direction remains fixed, even though its carrier (an airplane or ship, for example) changes

▶ **FIGURE 8.30 Constant direction of angular momentum** When angular momentum is conserved, its direction is constant in space. **(a)** This principle can be demonstrated by a passed football. **(b)** Gyroscopic action also occurs in a gyroscope, a rotating wheel that is universally mounted on gimbals (rings) so that it is free to turn about any axis. When the frame moves, the wheel maintains its direction. This is the principle of the gyrocompass.

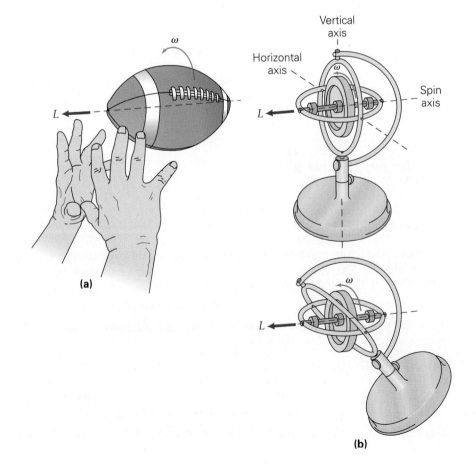

directions. You may have played with a toy gyroscope that is set spinning and placed on a pedestal. In a "sleeping" condition, the gyro stands straight up with its angular momentum vector fixed in space for some time. The gyro's center of gravity is on the axis of rotation, so there is no net torque due to its weight.

However, the gyroscope eventually slows down because of friction, causing $\vec{L}$ to tilt. In watching this motion, you may have noticed that the spin axis revolves, or *precesses*, about the vertical axis. It revolves tilted over, so to speak (Fig. 8.30b). Since the gyroscope precesses, the angular momentum vector $\vec{L}$ is no longer constant in direction, indicating that a torque must be acting to produce a change ($\Delta\vec{L}$) with time.

As can be seen from the figure, the torque arises from the vertical component of the weight force, since the center of gravity no longer lies directly above the point of support or on the vertical axis of rotation. The instantaneous torque is such that the gyroscope's axis moves or precesses about the vertical axis.

In a similar manner, the Earth's rotational axis precesses. The Earth's spin axis is tilted $23\frac{1}{2}°$ with respect to a line perpendicular to the plane of its revolution about the Sun; the axis precesses about this line (▶Fig. 8.31). The precession is due to slight gravitational torques exerted on the Earth by the Sun and the Moon.

The period of the precession of the Earth's axis is about 26 000 years, so the precession has little day-to-day effect. However, it does have an interesting long-term effect. Polaris will not always be (nor has it always been) the North Star—that is, the star toward which the Earth's axis of rotation points. About 5000 years ago, Alpha Draconis was the North Star, and 5000 years from now it will be Alpha Cephei, which is at an angular distance of about 68° away from Polaris on the circle described by the precession of the Earth's axis.

There are some other long-term torque effects on the Earth and the Moon. Did you know that the Earth's daily spin rate is slowing down and hence the days are getting longer? Also, that the Moon is receding, or getting farther away, from the Earth? This is due primarily to ocean tidal friction, which gives rise to a slowing torque. As a result, the Earth's spin angular momentum, and therefore its rate of rotation, is changing. The slowing rate of rotation causes the average day to be longer. This century will be about 25 s longer than the previous century.

But this is an average rate. At times, the Earth's rotation speeds up for relatively short periods. This increase is thought to be associated with the rotational inertia of the liquid layer of the Earth's core. (See the Chapter 13 Insight 13.1, Earthquakes, Seismic Waves, and Seismology.)

The tidal torque on the Earth results chiefly from the Moon's gravitational attraction, which is the main cause of ocean tides. This torque is *internal* to the Earth–Moon system, so the total angular momentum of that system is conserved. Since the Earth is losing angular momentum, the Moon must be gaining angular momentum to keep the total angular momentum of the system constant. The Earth loses rotational (spin) angular momentum, and the Moon gains orbital angular momentum. As a result, the Moon drifts slightly farther from Earth and its orbital speed decreases. The Moon moves away from the Earth at about 4 cm per year. Thus, the Moon moves in a slowly widening spiral.

Finally, a common example in which angular momentum is an important consideration is the helicopter. What would happen if a helicopter had a single rotor? Since the motor supplying the torque is internal, the angular momentum would be conserved. Initially, $\vec{L} = 0$; hence, to conserve the total angular momentum of the system (rotor plus body), the separate angular momenta of the rotor and body would have to be in opposite directions to cancel. Thus, on takeoff, the rotor would rotate one way and the helicopter body the other, which is not a desirable situation.

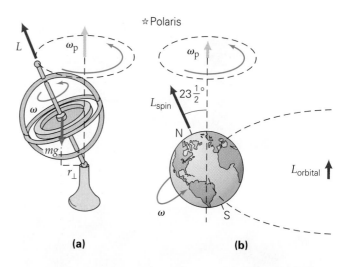

(a)

(b)

▲ **FIGURE 8.31 Precession** An external torque causes a change in angular momentum. **(a)** For a spinning gyroscope, this change is directional, and the axis of rotation precesses at angular acceleration ω_p about a vertical line. (The torque due to the weight force would point out of the page as drawn here, as would $\Delta\vec{L}$.) Note that although there is a torque that would topple a nonspinning gyroscope, a spinning gyroscope doesn't fall. **(b)** Similarly, the Earth's axis precesses because of gravitational torques caused by the Sun and the Moon. We don't notice this motion because the period of precession is about 26 000 years.

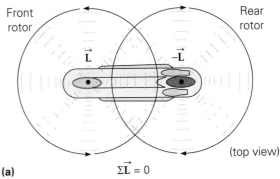

Front rotor

Rear rotor

$\vec{L}$ $-\vec{L}$

(top view)

$\Sigma\vec{L} = 0$

(a)

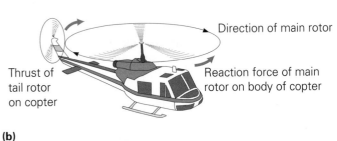

Direction of main rotor

Thrust of
tail rotor
on copter

Reaction force of main
rotor on body of copter

(b)

▲ **FIGURE 8.32 Different rotors**
See text for description.

To prevent this situation, helicopters have two rotors. Large helicopters have two overlapping rotors (▲ Fig. 8.32a). The oppositely rotating rotors cancel each other's angular momenta, so the helicopter body does not have to rotate to provide canceling angular momentum. The rotors are offset at different heights so that they do not collide.

Small helicopters with a single overhead rotor have small "antitorque" tail rotors (Fig. 8.32b). The tail rotor produces a thrust like a propeller and supplies the torque to counterbalance the torque produced by the overhead rotor. The tail rotor also helps in steering the craft. By increasing or decreasing the tail rotor's thrust, the helicopter turns (rotates) one way or the other.

DID YOU LEARN?

➡ The net torque is equal to the time rate of change of angular momentum, $\vec{\tau}_{net} = \Delta\vec{L}/\Delta t$. The net force is equal to the time rate of change of linear momentum, $\vec{F}_{net} = \Delta\vec{P}/\Delta t$.

➡ The angular momentum is conserved in the absence of a net torque, $\vec{\tau}_{net} = \Delta\vec{L}/\Delta t = 0$, and $\Delta\vec{L} = 0$. There is no change in the angular momentum, so it is conserved.

PULLING IT TOGETHER | Making a "Lazy Susan" Lazier?

"Lazy Susan" is the name for a small rotatable disk placed in the center of a table for the easy delivery of appetizers or condiments to those sitting around the table. Assume such a lazy Susan has a frictionless axis though its center on which to rotate, and consists of a circular piece of wood (density of 700 kg/m^3) that is 60.0 cm in diameter and 1.00 cm thick. It is set spinning initially so that it makes one complete revolution in 5.00 s. A small 100-g mass (piece of food) is dropped from just above the perimeter of the rotating disk and it sticks to the surface. (Neglect the speed of the mass as it hits the disk.)

Before the mass lands, (a) what are the frequency and angular speed of the lazy Susan? (b) What is the moment of inertia of the disk about its central axis? (c) What is its kinetic energy and angular momentum? (d) After the small mass lands, where is the new center of mass located? (e) What is the system's final kinetic energy and angular momentum? Is either conserved? Explain.

THINKING IT THROUGH.   This example demonstrates the concepts of angular momentum, moment of inertia, rotational kinetic energy, and center of mass. (a) The frequency and angular speed are inversely related to the period, which is given, and thus can be directly determined. (b) The moment of inertia depends on the mass and radius of the disk. The mass is determined by its volume and density. (c) Once the moment of inertia and angular speed are known, the rotational kinetic energy and angular momentum follow directly. (d) The sticky mass will move the center of mass from the center of the disk toward the perimeter. The exact location can be determined by recalling the definition of center of mass from Section 8.4. (e) Angular momentum is conserved because the net torque on the system is zero (why?). This enables the determination of the final (slower) angular speed and the final kinetic energy. It is expected that the final kinetic energy will be less than the initial kinetic energy due to the inelastic collision that takes place.

SOLUTION.

Given: $m = 100 \text{ g} = 0.100 \text{ kg}$
$d = 60.0 \text{ cm} = 0.600 \text{ m (diameter)}$
$h = 1.00 \text{ cm} = 0.0100 \text{ m (thickness)}$
$\rho = 700 \text{ kg/m}^3$
$T = 5.00 \text{ s (period)}$

Find: (a) f (frequency) and ω (angular speed)
(b) I (moment of inertia)
(c) K_o (initial kinetic energy) and L_o (angular momentum)
(d) location of center of mass
(e) K_f (final kinetic energy) and L_f (final angular momentum)
Are they conserved?

(a) Both the initial frequency and angular speed can be determined from the period (T), which is the time for one complete rotation of the disk. Thus the initial frequency is

$$f = \frac{1}{T}$$

$$= \frac{1 \text{ rev}}{5.00 \text{ s}} = 0.200 \text{ Hz}$$

and the initial angular speed is

$$\omega_o = 2\pi f$$
$$= 2\pi(0.200 \text{ Hz}) = 1.26 \text{ rad/s}$$

(b) The moment of inertia (I) of a circular disk is $\frac{1}{2}MR^2$ (see Fig. 8.20). To find the mass from the density, the disk volume is needed:

$$V = \pi R^2 h$$
$$= \pi(0.300 \text{ m})^2(0.0100 \text{ m}) = 2.83 \times 10^{-3} \text{ m}^3$$

Then the mass is,

$$M = \rho V$$
$$= (700 \text{ kg/m}^3)(2.83 \times 10^{-3} \text{ m}^3)$$
$$= 1.98 \text{ kg}$$

Finally, the initial moment of inertia is

$$I_o = \frac{1}{2}MR^2$$
$$= \frac{1}{2}(1.98 \text{ kg})(0.300 \text{ m})^2$$
$$= 8.91 \times 10^{-2} \text{ kg} \cdot \text{m}^2$$

(c) The initial rotational kinetic energy and angular momentum are

$$K_o = \frac{1}{2}I_o\omega_o^2$$
$$= \frac{1}{2}(8.91 \times 10^{-2} \text{ kg} \cdot \text{m}^2)(1.26 \text{ rad/s})^2$$
$$= 7.07 \times 10^{-2} \text{ J}$$

and

$$L_o = I_o\omega_o$$
$$= (8.91 \times 10^{-2} \text{ kg} \cdot \text{m}^2)(1.26 \text{ rad/s})$$
$$= 0.112 \text{ kg} \cdot \text{m}^2/\text{s}$$

(d) The center of mass is initially at the geometric center of the disk. With the small mass on the perimeter, the center of mass moves radially outward from the center toward the perimeter a distance x_{cm}. This is determined by treating the uniform disk as if all of its mass (M) were at its center of mass ($x = 0$) in combination with a point mass (m) located at ($x = R = 0.300 \text{ m}$):

$$x_{CM} = \frac{mR + M(0)}{m + M} = \frac{(0.100 \text{ kg})(0.300 \text{ m})}{0.100 \text{ kg} + 1.98 \text{ kg}} = 1.44 \times 10^{-2} \text{ m}$$

(e) Because the system of the disk and small mass has a net torque of zero on it (each has an equal but opposite torque on it at the collision), the net torque on the system is zero. Thus the angular momentum of the system stays constant, and $L_o = L_f$ or $I_o\omega_o = I_f\omega_f$. This can be used to find the final angular speed if the final moment of inertia is known. This is just the original disk moment of inertia plus a term for a point mass located on the perimeter. Therefore,

$$I_f = I_o + mR^2$$
$$= 8.91 \times 10^{-2} \text{ kg} \cdot \text{m}^2 + (0.100 \text{ kg})(0.300 \text{ m})^2$$
$$= 9.81 \times 10^{-2} \text{ kg} \cdot \text{m}^2$$

Now the final angular speed can be found using angular momentum conservation:

$$\omega_f = \frac{I_o}{I_f}\omega_o$$
$$= \frac{(8.91 \times 10^{-2} \text{ kg} \cdot \text{m}^2)}{(9.81 \times 10^{-2} \text{ kg} \cdot \text{m}^2)}(1.26 \text{ rad/s})$$
$$= 1.14 \text{ rad/s}$$

and the final rotational kinetic energy is

$$K_f = \frac{1}{2}I_f\omega_f^2$$
$$= \frac{1}{2}(9.81 \times 10^{-2} \text{ kg} \cdot \text{m}^2)(1.14 \text{ rad/s})^2$$
$$= 6.37 \times 10^{-2} \text{ J}$$

This amounts to a system loss of about 10%, so the kinetic energy is not conserved. A loss of kinetic energy is expected, since this was an inelastic collision. Some of the energy was converted into the heating of the small ball and also perhaps sound. However, since there are no external torques, the angular momentum is conserved (Section 8.5).

Learning Path Review

- In **pure translational motion**, all of the particles that make up a rigid object have the same instantaneous velocity.

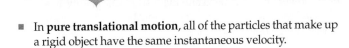
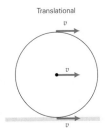

Translational

- In **pure rotational motion (about a fixed axis)**, all of the particles that make up a rigid object have the same instantaneous angular velocity.

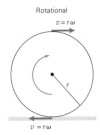

Rotational

Condition for rolling without slipping:

$$v_{CM} = r\omega \tag{8.1}$$

$$(\text{or } s = r\theta \quad \text{or} \quad a_{CM} = r\alpha)$$

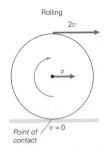

Rolling

- **Torque** ($\vec{\tau}$), the rotational analogue of force, is the product of a force and a moment arm, or lever arm.

Torque (magnitude):

$$\tau = r_{\perp} F = rF \sin\theta \tag{8.2}$$

(Direction given by right-hand rule.)

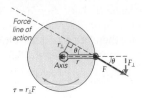

- **Mechanical equilibrium** requires that the net force, or summation of the forces, be zero (translational equilibrium) and that the net torque, or summation of the torques, be zero (rotational equilibrium).

Conditions for translational and rotational mechanical equilibrium, respectively:

$$\vec{F}_{net} = \Sigma\vec{F}_i = 0 \quad \text{and} \quad \vec{\tau}_{net} = \Sigma\vec{\tau}_i = 0 \tag{8.3}$$

- An object is in **stable equilibrium** as long as its center of gravity, upon small displacement, lies above and inside the object's original base of support.

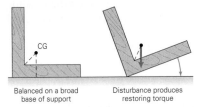

Balanced on a broad   Disturbance produces
base of support   restoring torque

Stable Equilibrium

- **Moment of inertia (I)** is the rotational analogue of mass and is given by

$$I = \Sigma m_i r_i^2 \tag{8.6}$$

Rotational form of Newton's second law:

$$\vec{\tau}_{net} = I\vec{\alpha} \tag{8.7}$$

Parallel axis theorem:

$$I = I_{CM} + Md^2 \tag{8.8}$$

$$I = I_{CM} + Md^2$$

Rotational work:

$$W = \tau\theta \tag{8.9}$$

Rotational power:

$$P = \tau\omega \tag{8.10}$$

Work–energy theorem (rotational):

$$W_{net} = \tfrac{1}{2}I\omega^2 - \tfrac{1}{2}I\omega_0^2 = \Delta K_r \tag{8.11}$$

Rotational kinetic energy:

$$K = \tfrac{1}{2}I\omega^2 \tag{8.12}$$

Kinetic energy of a rolling object (no slipping):

$$K = \tfrac{1}{2}I_{CM}\omega^2 + \tfrac{1}{2}Mv_{CM}^2 \tag{8.13}$$

■ **Angular momentum:** The product of a moment arm and linear momentum or the product of a moment of inertia and angular velocity.

Angular momentum of a particle in circular motion (magnitude):

$$L = r_\perp p = mr_\perp v = mr_\perp^2 \omega \qquad (8.14)$$

Angular momentum of a rigid body:

$$\vec{\mathbf{L}} = I\vec{\boldsymbol{\omega}} \qquad (8.16)$$

Torque as change in angular momentum (vector form):

$$\vec{\boldsymbol{\tau}}_{\text{net}} = \frac{\Delta \vec{\mathbf{L}}}{\Delta t} \qquad (8.17)$$

Conservation of angular momentum (with $\vec{\boldsymbol{\tau}}_{\text{net}} = 0$):

$$L = L_\text{o} \quad \text{or} \quad I\omega = I_\text{o}\omega_\text{o} \qquad (8.18)$$

Learning Path Questions and Exercises

For instructor-assigned homework, go to www.masteringphysics.com

MULTIPLE CHOICE QUESTIONS

8.1 RIGID BODIES, TRANSLATIONS, AND ROTATIONS

1. In pure rotational motion of a rigid body, (a) all the particles of the body have the same angular velocity, (b) all the particles of the body have the same tangential velocity, (c) acceleration is always zero, (d) there are always two simultaneous axes of rotation.

2. For an object with only rotational motion, all particles of the object have the same (a) instantaneous velocity, (b) average velocity, (c) distance from the axis of rotation, (d) instantaneous angular velocity.

3. The condition for rolling without slipping is (a) $a_c = r\omega^2$, (b) $v_{\text{CM}} = r\omega$, (c) $F = ma$, (d) $a_c = v^2/r$.

4. A rolling object (a) has an axis of rotation through the axis of symmetry, (b) has a zero velocity at the point or line of contact, (c) will slip if $s = r\theta$, (d) all of the preceding.

5. For the tires on your rolling, but skidding car, (a) $v_{\text{CM}} = r\omega$, (b) $v_{\text{CM}} > r\omega$, (c) $v_{\text{CM}} < r\omega$, (d) none of the preceding.

8.2 TORQUE, EQUILIBRIUM, AND STABILITY

6. It is possible to have a net torque when (a) all forces act through the axis of rotation, (b) $\Sigma \vec{\mathbf{F}}_i = 0$, (c) an object is in rotational equilibrium, (d) an object remains in unstable equilibrium.

7. If an object in unstable equilibrium is displaced slightly, (a) its potential energy will decrease, (b) the center of gravity is directly above the axis of rotation, (c) no gravitational work is done, (d) stable equilibrium follows.

8. Torque has the same units as (a) work, (b) force, (c) angular velocity, (d) angular acceleration.

8.2 ROTATIONAL DYNAMICS

9. The moment of inertia of a rigid body (a) depends on the axis of rotation, (b) cannot be zero, (c) depends on mass distribution, (d) all of the preceding.

10. Which of the following best describes the physical quantity called torque: (a) rotational analogue of force, (b) energy due to rotation, (c) rate of change of linear momentum, or (d) force that is tangent to a circle?

11. In general, the moment of inertia is greater when (a) more mass is farther from the axis of rotation, (b) more mass is closer to the axis of rotation, (c) it makes no difference.

12. A solid sphere (radius R) and an annular cylinder (radius $2R$) with equal masses are released simultaneously from the top of a frictionless inclined plane. Then, (a) the sphere reaches the bottom first, (b) the cylinder reaches the bottom first, (c) they reach the bottom together.

13. The moment of inertia about an axis parallel to the axis through the center of mass depends on (a) the mass of the rigid body, (b) the distance between the axes, (c) the moment of inertia about the axis through the center of mass, (d) all of the preceding.

8.4 ROTATIONAL WORK AND KINETIC ENERGY

14. From $W = \tau\theta$, the unit of rotational work is the (a) watt, (b) N·m, (c) kg·rad/s², (d) N·rad.

15. A bowling ball rolls without slipping on a flat surface. The ball has (a) rotational kinetic energy, (b) translational kinetic energy, (c) both translational and rotational kinetic energy, (d) neither translational nor rotational kinetic energy.

16. A rolling cylinder on a level surface has (a) rotational kinetic energy, (b) translational kinetic energy, (c) both translational and rotational kinetic energies.

8.5 ANGULAR MOMENTUM

17. The units of angular momentum are (a) N·m, (b) kg·m/s², (c) kg·m²/s, (d) J·m.

18. The Earth's orbital speed is greatest about (a) March 21, (b) June 21, (c) Sept. 21, (d) Dec. 21.

19. The angular momentum may be increased by (a) decreasing the moment of inertia, (b) decreasing the angular velocity, (c) increasing the product of the angular momentum and moment of inertia, (d) none of these.

CONCEPTUAL QUESTIONS

8.1 RIGID BODIES, TRANSLATIONS, AND ROTATIONS

1. Suppose someone in your physics class says that it is possible for a rigid body to have translational motion and rotational motion at the same time. Would you agree? If so, give an example.

2. For a rolling cylinder, what would happen if the tangential speed v were less than $r\omega$? Is it possible for v to be greater than $r\omega$? Explain.

3. If the top of your automobile tire is moving with a speed of v, what is the reading of your speedometer?

8.2 TORQUE, EQUILIBRIUM, AND STABILITY

4. A small force and a large force produce torques. Can you tell which one will have the larger torque? Explain.

5. In cutting large trees, loggers first notch or make a V-cut on the side of the tree in the desired direction of fall and then cut from the other side. Why is this? Is there any danger for the logger to be on the opposite side of the tree of the direction of fall?

6. Explain the balancing acts in ▼Fig. 8.33. Where are the centers of gravity?

▲ **FIGURE 8.33 Balancing acts** *Left*: A toothpick on the rim of the glass supports a fork and spoon. *Right*: Toy birds balance on their beaks. See Conceptual Question 6.

7. "Popping a wheelie" is a motorcycle stunt in which the front end of the cycle rises up from the ground on a fast start, and can remain there for some distance. Explain the physics involved in this stunt.

8. A yo-yo is thrown downward with a rotational spin. Reaching the bottom of the string, it climbs back upward. Is the rotational direction reversed at the bottom? Explain.

9. In the cases of both stable and unstable equilibrium, a small displacement of the center of gravity causes gravitational work to be done. (See the balls and bowls in Fig. 8.11.) However, there is another type of equilibrium in which the displacement of the center of mass involves no gravitational work and the displaced center of gravity essentially moves in a straight line. This is called *neutral equilibrium*. Give an example of an object in neutral equilibrium.

8.3 ROTATIONAL DYNAMICS

10. (a) Does the moment of inertia of a rigid body depend in any way on the center of mass of the body? Explain. (b) Can a moment of inertia have a negative value? If so, explain what this would mean.

11. Why does the moment of inertia of a rigid body have different values for different axes of rotation? What does this mean physically?

12. Two cylinders of equal mass sitting on a horizontal surface are made from materials with different densities. (a) Which cylinder will have the greater moment of inertia about an axis passing horizontally through the center? (b) Which cylinder will have the greater moment of inertia about an axis along the surface of contact?

13. Here is an interesting experiment you can try for yourself at home. Prepare a hard-boiled egg and have a raw egg available. Set them both spinning on the kitchen table. Stop both eggs quickly, and the release both. You will notice the hard-boiled one remains at rest, whereas the raw one starts spinning again. Explain.

14. Why does jerking a paper towel from a roll cause the paper to tear more easily than pulling it smoothly? Will the amount of paper on the roll affect the results?

15. Tightrope walkers are continually in danger of falling (unstable equilibrium). Commonly, a performer carries a long pole while walking the tight rope, as shown in the chapter-opening photo. What is the purpose of the pole? (In walking along a narrow board or rail, you probably extend your arms for the same reason.)

16. A solid cylinder and an annular cylinder of equal mass are rolling on the floor with the same speed. (a) If the solid cylinder's radius is equal to the annular cylinder's inner radius, which cylinder would be harder to stop? Explain. (b) Would it make any stopping difference if the solid cylinder's radius were equal to the annular cylinder's outer radius? Justify your answer explicitly.

8.4 ROTATIONAL WORK AND KINETIC ENERGY

17. Can you increase the rotational kinetic energy of a wheel without changing its translational kinetic energy? Explain.

18. In order to produce fuel-efficient vehicles, automobile manufacturers want to minimize rotational kinetic energy and maximize translational kinetic energy when a car is traveling. If you were the designer of wheels of a certain diameter, how would you design them?

19. What is required to produce a change in rotational kinetic energy?

8.5 ANGULAR MOMENTUM

20. A child stands on the edge of a rotating playground merry-go-round (the hand-driven type). He then starts to walk toward the center of the merry-go-round. This can result in a dangerous situation. Why?

21. The release of vast amounts of carbon dioxide may result in an increase in the Earth's average temperature through the so-called greenhouse effect and cause melting of the polar ice caps. If this occurred and the ocean level rose substantially, what effect would it have on the Earth's rotation?

22. In the classroom demonstration illustrated in ▶Fig. 8.34, a person on a rotating stool holds a rotating bicycle

wheel by handles attached to the wheel. When the wheel is held horizontally, she rotates one way (clockwise as viewed from above). When the wheel is turned over, she rotates in the opposite direction. Explain why this occurs. [*Hint*: Consider angular momentum vectors.]

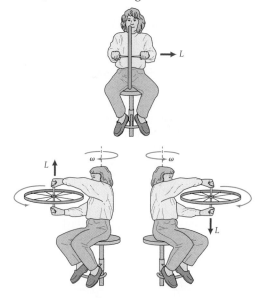

▲ **FIGURE 8.34 Faster rotation** See Conceptual Question 22.

23. Cats usually land on their feet when they fall, even if held upside down when dropped (▶Fig. 8.35). While a cat is falling, there is no external torque and its center of mass falls as a particle. How can cats turn themselves over while falling?

24. Two ice skaters that weigh the same skate toward each other with the same mass and same speed on parallel paths. As they pass each other, they link arms. (a) What is the velocity of their center of mass after they link arms? (b) What happens to their initial, translational kinetic energies?

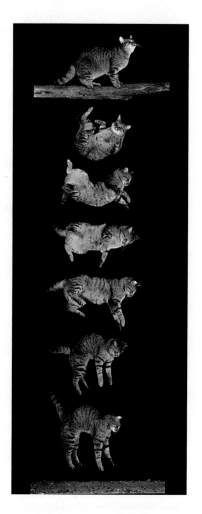

▲ **FIGURE 8.35 A double rotation** See Conceptual Question 23.

EXERCISES

*Integrated Exercises (**IEs**) are two-part exercises. The first part typically requires a conceptual answer choice based on physical thinking and basic principles. The following part is quantitative calculations associated with the conceptual choice made in the first part of the exercise.*

*Throughout the text, many exercise sections will include "paired" exercises. These exercise pairs, identified with **red numbers**, are intended to assist you in problem solving and learning. In a pair, the first exercise (even numbered) is worked out in the Study Guide so that you can consult it should you need assistance in solving it. The second exercise (odd numbered) is similar in nature, and its answer is given in Appendix VII at the back of the book.*

8.1 RIGID BODIES, TRANSLATIONS, AND ROTATIONS

1. • A wheel rolls uniformly on level ground without slipping. A piece of mud on the wheel flies off when it is at the 9 o'clock position (rear of wheel). Describe the subsequent motion of the mud.

2. • A rope goes over a circular pulley with a radius of 6.5 cm. If the pulley makes 4 revolutions without the rope slipping, what length of rope passes over the pulley?

3. • A wheel rolls 5 revolutions on a horizontal surface without slipping. If the center of the wheel moves 3.2 m, what is the radius of the wheel?

4. •• A bowling ball with a radius of 15.0 cm travels down the lane so that its center of mass is moving at 3.60 m/s. The bowler estimates that it makes about 7.50 complete revolutions in 2.00 seconds. Is it rolling without slipping? Prove your answer, assuming that the bowler's quick observation limits answers to two significant figures.

5. •• A ball with a radius of 15 cm rolls on a level surface, and the translational speed of the center of mass is 0.25 m/s. What is the angular speed about the center of mass if the ball rolls without slipping?

6. **IE** •• (a) When a disk rolls without slipping, should the product $r\omega$ be (1) greater than, (2) equal to, or (3) less than v_{CM}? (b) A disk with a radius of 0.15 m rotates through 270° as it travels 0.71 m. Does the disk roll without slipping? Prove your answer.

7. ••• A bocce ball with a diameter of 6.00 cm rolls without slipping on a level lawn. It has an initial angular speed of 2.35 rad/s and comes to rest after 2.50 m. Assuming constant deceleration, determine (a) the magnitude of its angular deceleration and (b) the magnitude of the maximum tangential acceleration of the ball's surface (tell where that part is located).

8. ••• A cylinder with a diameter of 20 cm rolls with an angular speed of 0.050 rad/s on a level surface. If the cylinder experiences a uniform tangential acceleration of 0.018 m/s^2 without slipping until its angular speed is 1.2 rad/s, through how many complete revolutions does the cylinder rotate during the time it accelerates?

8.2 TORQUE, EQUILIBRIUM, AND STABILITY

9. • In Fig. 8.4a, if the arm makes a 37° angle with the horizontal and a torque of 18 m · N is to be produced, what force must the biceps muscle supply?

10. • The drain plug on a car's engine has been tightened to a torque of 25 m · N. If a 0.15-m-long wrench is used to change the oil, what is the minimum force needed to loosen the plug?

11. • In Exercise 10, due to limited work space, you must crawl under the car. The force thus cannot be applied perpendicularly to the length of the wrench. If the applied force makes a 30° angle with the length of the wrench, what is the force required to loosen the drain plug?

12. • How many different positions of stable equilibrium and unstable equilibrium are there for a cube? Consider each surface, edge, and corner to be a different position.

13. IE •• Two children are sitting on opposite ends of a uniform seesaw of negligible mass. (a) Can the seesaw be balanced if the masses of the children are different? How? (b) If a 35-kg child is 2.0 m from the pivot point (or fulcrum), how far from the pivot point will her 30-kg playmate have to sit on the other side for the seesaw to be in equilibrium?

14. • A uniform meterstick pivoted at its center, as in Example 8.5, has a 100-g mass suspended at the 25.0-cm position. (a) At what position should a 75.0-g mass be suspended to put the system in equilibrium? (b) What mass would have to be suspended at the 90.0-cm position for the system to be in equilibrium?

15. •• A worker applies a horizontal force to the top edge of a crate to get it to tip forward (▼ Fig. 8.36). If the create

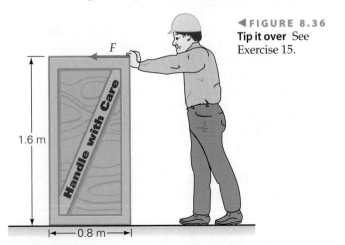

◄ FIGURE 8.36
Tip it over See Exercise 15.

has a mass of 100 kg and is 1.6 m tall and 0.80 m in depth and width, what is the minimum force needed to make the crate start tipping? (Assume the center of mass of the crate is at its center and static friction great enough to prevent slipping.)

16. •• Show that the balanced meterstick in Example 8.5 is in static rotational equilibrium about a horizontal axis through the 100-cm end of the stick.

17. IE •• Telephone and electrical lines are allowed to sag between poles so that the tension will not be too great when something hits or sits on the line. (a) Is it possible to have the lines perfectly horizontal? Why or why not? (b) Suppose that a line were stretched almost perfectly horizontally between two poles that are 30 m apart. If a 0.25-kg bird perches on the wire midway between the poles and the wire sags 1.0 cm, what would be the tension in the wire? (Neglect the mass of the wire.)

18. •• In ▼ Fig. 8.37, what is the force F_m supplied by the deltoid muscle so as to hold up the outstretched arm if the mass of the arm is 3.0 kg? (F_j is the joint force on the bone of the upper arm—the humerus.)

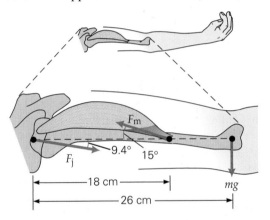

▲ FIGURE 8.37 **Arm in static equilibrium** See Exercise 18.

19. •• In Figure 8.4b, determine the force exerted by the bicep muscle, assuming that the hand is holding a ball with a mass of 5.00 kg. Assume that the mass of the forearm is 8.50 kg with its center of mass located 20.0 cm away from the elbow joint (the black dot in the figure). Assume also that the center of mass of the ball in the hand is 30.0 cm away from the elbow joint. (The muscle contact is 4.00 cm from the elbow joint; Example 8.2.)

20. •• A bowling ball (mass 7.00 kg and radius 17.0 cm) is released so fast that it skids without rotating down the lane (at least for a while). Assume the ball skids to the right and the coefficient of sliding friction between the ball and the lane surface is 0.400. (a) What is the direction of the torque exerted by the friction on the ball about the center of mass of the ball? (b) Determine the magnitude of this torque (again about the ball's center of mass).

21. •• A variation of Russell traction (► Fig. 8.38) supports the lower leg in a cast. Suppose that the patient's leg and cast have a combined mass of 15.0 kg and m_1 is 4.50 kg. (a) What is the reaction force of the leg muscles to the traction? (b) What must m_2 be to keep the leg horizontal?

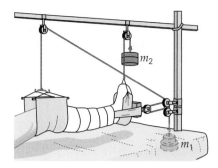

▲ FIGURE 8.38 **Static traction** See Exercise 21.

22. •• In doing physical therapy for an injured knee joint, a person raises a 5.0-kg weighted boot as shown in ▼Fig. 8.39. Compute the torque due to the boot for each position shown.

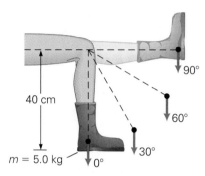

▲ FIGURE 8.39 **Torque in physical therapy** See Exercise 22.

23. •• An artist wishes to construct a birds and bees mobile, as shown in ▼Fig. 8.40. If the mass of the bee on the lower left is 0.10 kg and each vertical support string has a length of 30 cm, what are the masses of the other birds and bees? (Neglect the masses of the bars and strings.)

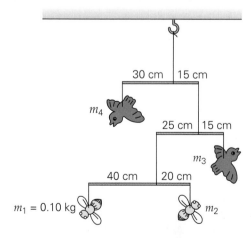

▲ FIGURE 8.40 **Birds and bees** See Exercise 23.

24. IE •• The location of a person's center of gravity relative to his or her height can be found by using the arrangement shown in ▶Fig. 8.41. The scales are initially adjusted to zero with the board alone. (a) Would you expect the location of the center of gravity to be (1) midway between the scales, (2) toward the scale at the person's head, or (3) toward the scale at the person's feet?

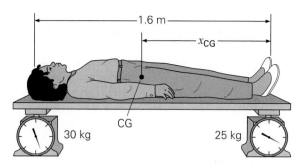

▲ FIGURE 8.41 **Locating the center of gravity** See Exercise 24.

Why? (b) Locate the center of gravity of the person relative to the horizontal dimension.

25. •• (a) How many uniform, identical textbooks of width 25.0 cm can be stacked on top of each other on a level surface without the stack falling over if each successive book is displaced 3.00 cm in width relative to the book below it? (b) If the books are 5.00 cm thick, what will be the height of the center of mass of the stack above the level surface?.

26. •• If four metersticks were stacked on a table with 10 cm, 15 cm, 30 cm, and 50 cm, respectively, hanging over the edge, as shown in ▼Fig. 8.42, would the top meterstick remain on the table?

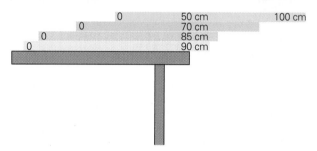

▲ FIGURE 8.42 **Will they fall off?** See Exercise 26.

27. •• A 10.0-kg solid uniform cube with 0.500-m sides rests on a level surface. What is the minimum amount of work necessary to put the cube into an unstable equilibrium position?

28. •• While standing on a long board resting on a scaffold, a 70-kg painter paints the side of a house, as shown in ▼Fig. 8.43. If the mass of the board is 15 kg, how close to the end can the painter stand without tipping the board over?

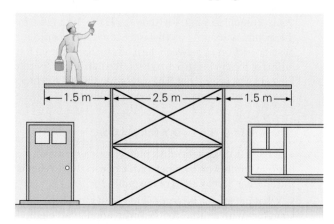

▲ FIGURE 8.43 **Not too far!** See Exercise 28.

29. •• A mass is suspended by two cords as shown in ▼Fig. 8.44. What are the tensions in the cords?

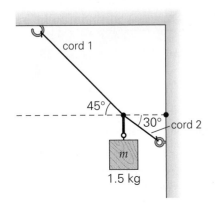

◀ FIGURE 8.44
A lot of tension See Exercises 29 and 30.

30. •• If the cord attached to the vertical wall in Fig. 8.44 were horizontal (instead of at a 30° angle), what would the tensions in the cords be?

31. •• A force is applied to a cord wrapped around a solid 2.0-kg cylinder as shown in ▼Fig. 8.45. Assuming the cylinder rolls without slipping, what is the force of friction acting on the cylinder?

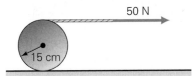

◀ FIGURE 8.45 **No slipping** See Exercise 31.

32. IE ••• In a circus act, a uniform board (length 3.00 m, mass 35.0 kg) is suspended from a bungie-type rope at one end, and the other end rests on a concrete pillar. When a clown (mass 75.0 kg) steps out halfway onto the board, the board tilts so the rope end is 30° from the horizontal and the rope stays vertical. (a) In which situation will the rope tension be larger: (1) the board without the clown on it, (2) the board with the clown on it, or (3) you can't tell from the data given? (b) Calculate the force exerted by the rope in both situations.

33. IE ••• The forces acting on Einstein and the bicycle (Fig. 2 of the Insight 8.1, Stability in Action) are the total weight of Einstein and the bicycle (mg) at the center of gravity of the system, the normal force (N) exerted by the road, and the force of static friction (f_s) acting on the tires due to the road. (a) If Einstein is to maintain balance, should the tangent of the lean angle $\theta(\tan \theta)$ be (1) greater than, (2) equal to, or (3) less than f_s/N? (b) The angle θ in the picture is about 11°. What is the minimum coefficient of static friction μ_s between the road and the tires? (c) If the radius of the circle is 6.5 m, what is the maximum speed of Einstein's bicycle? [*Hint*: The net torque about the center of gravity must be zero for rotational equilibrium.]

8.3 ROTATIONAL DYNAMICS

34. • A fixed 0.15-kg solid-disk pulley with a radius of 0.075 m is acted on by a net torque of 6.4 m · N. What is the angular acceleration of the pulley?

35. • What net torque is required to give a uniform 20-kg solid ball with a radius of 0.20 m an angular acceleration of 20 rad/s²?

36. • For the system of masses shown in ▼Fig. 8.46, find the moment of inertia about (a) the x-axis, (b) the y-axis, and (c) an axis through the origin and perpendicular to the page (z-axis). Neglect the masses of the connecting rods.

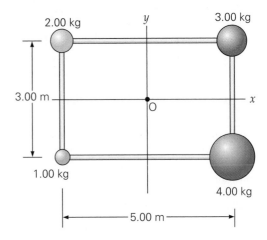

▲ FIGURE 8.46 **Moments of inertia about different axes** See Exercise 36.

37. •• A 2000-kg Ferris wheel accelerates from rest to an angular speed of 20 rad/s in 12 s. Approximate the Ferris wheel as a circular disk with a radius of 30 m. What is the net torque on the wheel?

38. IE •• Two objects of different masses are joined by a light rod. (a) Is the moment of inertia about the center of mass the minimum or the maximum? Why? (b) If the two masses are 3.0 kg and 5.0 kg and the length of the rod is 2.0 m, find the moments of inertia of the system about an axis perpendicular to the rod, through the center of the rod and the center of mass.

39. •• Two masses are suspended from a pulley as shown in ▼Fig. 8.47 (the Atwood machine revisited; see Chapter 4, Exercise 55). The pulley itself has a mass of

◀ FIGURE 8.47 **The Atwood machine revisited** See Exercise 39.

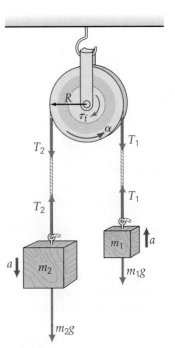

0.20 kg, a radius of 0.15 m, and a constant torque of 0.35 m · N due to the friction between the rotating pulley and its axle. What is the magnitude of the acceleration of the suspended masses if $m_1 = 0.40$ kg and $m_2 = 0.80$ kg? (Neglect the mass of the string.)

40. ●● To start her lawn mower, Julie pulls on a cord that is wrapped around a pulley. The pulley has a moment of inertia about its central axis of $I = 0.550$ kg · m² and a radius of 5.00 cm. There is an equivalent frictional torque impeding her pull of $\tau_f = 0.430$ m · N. To accelerate the pulley at $\alpha = 4.55$ rad/s², (a) how much torque does Julie need to apply to the pulley? (b) How much tension must the rope exert?

41. ●● For the system shown in ▼Fig. 8.48, $m_1 = 8.0$ kg, $m_2 = 3.0$ kg, $\theta = 30°$, and the radius and mass of the pulley are 0.10 m and 0.10 kg, respectively. (a) What is the acceleration of the masses? (Neglect friction and the string's mass.) (b) If the pulley has a constant frictional torque of 0.050 m · N when the system is in motion, what is the acceleration of the masses? [Hint: Isolate the forces. The tensions in the strings are different. Why?]

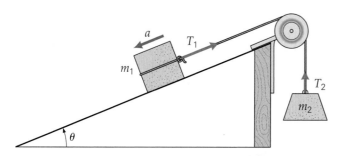

▲ FIGURE 8.48 **Inclined plane and pulley** See Exercise 41.

42. ●● A meterstick pivoted about a horizontal axis through the 0-cm end is held in a horizontal position and let go. (a) What is the initial tangential acceleration of the 100-cm position? Are you surprised by this result? (b) Which position has a tangential acceleration equal to the acceleration due to gravity?

43. ●● Pennies are placed every 10 cm on a meterstick. One end of the stick is put on a table and the other end is held horizontally with a finger, as shown in ▼Fig. 8.49. If the finger is pulled away, what happens to the pennies?

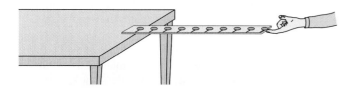

▲ FIGURE 8.49 **Money left behind?** See Exercise 43.

44. ●●● A uniform 2.0-kg cylinder of radius 0.15 m is suspended by two strings wrapped around it (▶Fig. 8.50). As the cylinder descends, the strings unwind from it. What is the acceleration of the center of mass of the cylinder? (Neglect the mass of the string.)

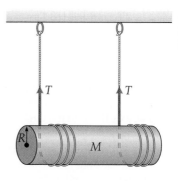

◀ FIGURE 8.50 **Unwinding with gravity** See Exercise 44.

45. ●●● A planetary space probe is in the shape of a cylinder. To protect it from heat on one side (from the Sun's rays), operators on the Earth put it into a "barbecue mode," that is, they set it rotating about its long axis. To do this, they fire four small rockets mounted tangentially as shown in ▼Fig. 8.51 (the probe is shown coming toward you). The object is to get the probe to rotate completely once every 30 s, starting from no rotation at all. They wish to do this by firing all four rockets for a certain length of time. Each rocket can exert a thrust of 50.0 N. Assume the probe is a uniform solid cylinder with a radius of 2.50 m and a mass of 1000 kg and neglect the mass of each rocket engine. Determine the amount of time the rockets need to be fired.

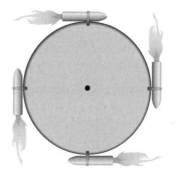

◀ FIGURE 8.51 **Space probe in the "barbecue mode"** See Exercise 45.

46. IE ●●● A ball of radius R and mass M rolls down an incline of angle θ. (a) For the ball to roll without slipping, should the tangent of the maximum angle of incline ($\tan \theta$) be equal to (1) $3 \mu_s/2$, (2) $5 \mu_s/2$, (3) $7 \mu_s/2$, or (4) $9 \mu_s/2$? Here, μ_s is the coefficient of static friction. (b) If the ball is made of wood and the surface is also wood, what is the maximum angle of incline? [Hint: See Table 4.1.]

8.4 ROTATIONAL WORK AND KINETIC ENERGY

47. ● A constant retarding torque of 12 m · N stops a rolling wheel of diameter 0.80 m in a distance of 15 m. How much work is done by the torque?

48. ● A person opens a door by applying a 15-N force perpendicular to it at a distance 0.90 m from the hinges. The door is pushed wide open (to 120°) in 2.0 s. (a) How much work was done? (b) What was the average power delivered?

49. IE ● In Fig. 8.23, a mass m descends a vertical distance from rest. (Neglect friction and the mass of the string.) (a) From the conservation of mechanical energy, will the

linear speed of the descending mass be (1) greater than, (2) equal to, or (3) less than $\sqrt{2gh}$? Why? (b) If $m = 1.0$ kg, $M = 0.30$ kg, and $R = 0.15$ m, what is the linear speed of the mass after it has descended a vertical distance of 2.0 m from rest?

50. • A constant torque of $10 \text{ m} \cdot \text{N}$ is applied to the rim of a 10-kg uniform disk of radius 0.20 m. What is the angular speed of the disk about an axis through its center after it rotates 2.0 revolutions from rest?

51. • A 2.5-kg pulley of radius 0.15 m is pivoted about an axis through its center. What constant torque is required for the pulley to reach an angular speed of 25 rad/s after rotating 3.0 revolutions, starting from rest?

52. •• A solid ball of mass m rolls along a horizontal surface with a translational speed of v. What percent of its total kinetic energy is translational?

53. •• Estimate the ratio of the translational kinetic energy of the Earth as it orbits the Sun to the rotational kinetic energy it has about its N–S axis.

54. •• You wish to accelerate a small merry-go-round from rest to a rotational speed of one-third of a revolution per second by pushing tangentially on it. Assume the merry-go-round is a disk with a mass of 250 kg and a radius of 1.50 m. Ignoring friction, how hard do you have to push tangentially to accomplish this in 5.00 s? (Use energy methods and assume a constant push on your part.)

55. •• A pencil 18 cm long stands vertically on its point end on a horizontal table. If it falls over without slipping, with what tangential speed does the eraser end strike the table?

56. •• A uniform sphere and a uniform cylinder with the same mass and radius roll at the same velocity side by side on a level surface without slipping. If the sphere and the cylinder approach an inclined plane and roll up it without slipping, will they be at the same height on the plane when they come to a stop? If not, what will be the percentage difference of the heights?

57. •• A hoop starts from rest at a height 1.2 m above the base of an inclined plane and rolls down under the influence of gravity. What is the linear speed of the hoop's center of mass just as the hoop leaves the incline and rolls onto a horizontal surface? (Neglect friction.)

58. •• A cylindrical hoop, a cylinder, and a sphere of equal radius and mass are released at the same time from the top of an inclined plane. Using the conservation of mechanical energy, show that the sphere always gets to the bottom of the incline first with the fastest speed and that the hoop always arrives last with the slowest speed.

59. •• For the following objects, which all roll without slipping, determine the rotational kinetic energy about the center of mass as a percentage of the total kinetic energy: (a) a solid sphere, (b) a thin spherical shell, and (c) a thin cylindrical shell.

60. •• An industrial flywheel with a moment of inertia of $4.25 \times 10^2 \text{ kg} \cdot \text{m}^2$ rotates with a speed of 7500 rpm. (a) How much work is required to bring the flywheel to rest? (b) If this work is done uniformly in 1.5 min, how much power is required?

61. ••• A hollow, thin-shelled ball and a solid ball of equal mass are rolled up an inclined plane (without slipping) with both balls having the same initial velocity at the bottom of the plane. (a) Which ball rolls higher on the incline before coming to rest? (b) Do the radii of the balls make a difference? (c) After stopping, the balls roll back down the incline. By the conservation of energy, both balls should have the same speed when reaching the bottom of the incline. Show this explicitly.

62. ••• In a tumbling clothes dryer, the cylindrical drum (radius 50.0 cm and mass 35.0 kg) rotates once every second. (a) Determine the rotational kinetic energy about its central axis. (b) If it started from rest and reached that speed in 2.50 s, determine the average net torque on the dryer drum.

63. ••• A steel ball rolls down an incline into a loop-the-loop of radius R (▼Fig. 8.52a). (a) What minimum speed must the ball have at the top of the loop in order to stay on the track? (b) At what vertical height (h) on the incline, in terms of the radius of the loop, must the ball be released in order for it to have the required minimum speed at the top of the loop? (Neglect frictional losses.) (c) Figure 8.52b shows the loop-the-loop of a roller coaster. What are the sensations of the riders if the roller coaster has the minimum speed or a greater speed at the top of the loop? [*Hint:* In case the speed is below the minimum, seat and shoulder straps hold the riders in.]

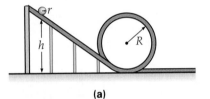

(a)

◀ **FIGURE 8.52**
Loop-the-loop and rotational speed See Exercise 63.

(b)

8.5 ANGULAR MOMENTUM

64. • What is the angular momentum of a 2.0-g particle moving counterclockwise (as viewed from above) with an angular speed of 5π rad/s in a horizontal circle of radius 15 cm? (Give the magnitude and direction.)

65. • A 10-kg rotating disk of radius 0.25 m has an angular momentum of $0.45 \text{ kg} \cdot \text{m}^2/\text{s}$ What is the angular speed of the disk?

66. •• Compute the ratio of the magnitudes of the Earth's orbital angular momentum and its rotational angular momentum. Are these momenta in the same direction?

67. •• The Earth revolves about the Sun and spins on its axis, which is tilted $23 \frac{1}{2}°$ to its orbital plane. (a) Assuming a circular orbit, what is the magnitude of the angular momentum associated with the Earth's orbital motion about the Sun? (b) What is the magnitude of the angular momentum associated with the Earth's rotation on its axis?

68. •• The period of the Moon's rotation is the same as the period of its revolution: 27.3 days (sidereal). What is the

angular momentum for each rotation and revolution? (Because the periods are equal, we see only one side of the Moon from Earth.)

69. **IE** •• Circular disks are used in automobile clutches and transmissions. When a rotating disk couples to a stationary one through frictional force, the energy from the rotating disk can transfer to the stationary one. (a) Is the angular speed of the coupled disks (1) greater than, (2) less than, or (3) the same as the angular speed of the original rotating disk? Why? (b) If a disk rotating at 800 rpm couples to a stationary disk with three times the moment of inertia, what is the angular speed of the combination?

70. •• An ice skater has a moment of inertia of $100 \text{ kg} \cdot \text{m}^2$ when his arms are outstretched and a moment of inertia of $75 \text{ kg} \cdot \text{m}^2$ when his arms are tucked in close to his chest. If he starts to spin at an angular speed of 2.0 rps (revolutions per second) with his arms outstretched, what will his angular speed be when they are tucked in?

71. •• An ice skater spinning with outstretched arms has an angular speed of 4.0 rad/s. She tucks in her arms, decreasing her moment of inertia by 7.5%. (a) What is the resulting angular speed? (b) By what factor does the skater's kinetic energy change? (Neglect any frictional effects.) (c) Where does the extra kinetic energy come from?

72. •• A billiard ball at rest is struck (bold arrow in ▼Fig. 8.53) by a cue with an average force of 5.50 N lasting for 0.050 s. The cue contacts the ball's surface so that the lever arm is half the radius of the ball, as shown. If the cue ball has a mass of 200 g and a radius of 2.50 cm, determine the angular speed of the ball immediately after the blow. (Neglect friction.)

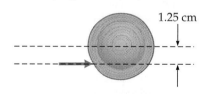

1.25 cm

▲ **FIGURE 8.53 Cueing low** See Exercise 72.

73. ••• A comet approaches the Sun as illustrated in ▶Fig. 8.54 and is deflected by the Sun's gravitational attraction. This event is considered a collision, and *b* is called the *impact parameter*. Find the distance of closest

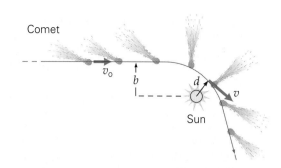

▲ **FIGURE 8.54 A comet "collision"** See Exercise 73.

approach (*d*) in terms of the impact parameter and the velocities (v_0 at large distances and v at closest approach). Assume that the radius of the Sun is negligible compared to *d*. (As the figure shows, the tail of a comet always "points" away from the Sun.)

74. ••• While repairing his bicycle, a student turns it upside down and sets the front wheel spinning at 2.00 rev/s. Assume the wheel has a mass of 3.25 kg and all of the mass is located on the rim, which has a radius of 41.0 cm. To slow the wheel, he places his hand on the tire, thereby exerting a tangential force of friction on the wheel. It takes 3.50 s to come to rest. Use the change in angular momentum to determine the force he exerts on the wheel. Assume the frictional force of the axle is negligible.

75. **IE** ••• A kitten stands on the edge of a lazy Susan (a turntable). Assume that the lazy Susan has frictionless bearings and is initially at rest. (a) If the kitten starts to walk around the edge of the lazy Susan, the lazy Susan will (1) remain lazy and stationary, (2) rotate in the direction opposite that in which the kitten is walking, or (3) rotate in the direction the kitten is walking. Explain. (b) The mass of the kitten is 0.50 kg, and the lazy Susan has a mass of 1.5 kg and a radius of 0.30 m. If the kitten walks at a speed of 0.25 m/s, relative to the ground, what will be the angular speed of the lazy Susan? (c) When the kitten has walked completely around the edge and is back at its starting point, will that point be above the same point on the ground as it was at the start? If not, where is the kitten relative to the starting point? (Speculate on what might happen if everyone on the Earth suddenly started to run eastward. What effect might this have on the length of a day?)

PULLING IT TOGETHER: MULTICONCEPT EXERCISES

The Multiconcept Exercises require the use of more than one fundamental concept for their understanding and solution. The concepts may be from this chapter, but may include those from previous chapters.

76. **IE** A small heavy object of mass *m* is attached to a thin string to make a simple pendulum whose length is *L*. When the object is pulled aside by a horizontal force *F* it is in static equilibrium and the string makes a constant angle θ from the vertical. (a) The tension in the string should be (1) the same as, (2) greater than, or (3) less than the object's weight, *mg*. (b) Use the force condition for static equilibrium (along with a free-body diagram of the object) to prove that the string tension is $T = \dfrac{mg}{\cos \theta} > mg$.

Use the same procedure to show that $F = mg \tan \theta$. (c) Prove the same result for *F* as in part (b) using the torque condition, summing the torques about the string's tied end. Explain why you *cannot* use this method to determine the string tension.

77. A bowling ball with a diameter of 21.6 cm is rolling down a level alley surface at 12.7 m/s without slipping. Assume the ball is uniform and made of plastic with a density of 800 kg/m^3. (a) What is the angular speed of the ball?

(b) Calculate the speed (relative to the alley surface) of a point on top of the ball directly above the contact point on the floor. (c) What is the ball's linear kinetic energy? (d) If it now starts to roll up a 30° incline, how far up the incline will it travel before it stops?

78. A solid cylindrical 10-kg roll of roofing paper with a radius of 15 cm, starting from rest rolls down a roof with a 20° incline (▼Fig. 8.55). (a) If the cylinder rolls 4.0 m without slipping, what is the angular speed about its center when leaving the roof? (b) If the roof edge of the house is 6.0 m above level ground, how far from the edge of the roof does the cylindrical roll land? (Figure not to scale.)

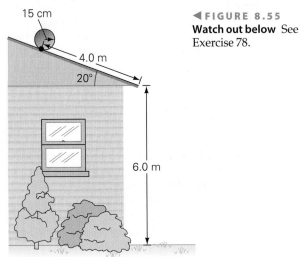

◀ FIGURE 8.55
Watch out below See Exercise 78.

79. A flat cylindrical grinding wheel is spinning at 2000 rpm (clockwise when viewed head-on) when its power is suddenly turned off. Normally, if left alone, it takes 45.0 s to coast to rest. Assume the grinder has a moment of inertia of 2.43 kg · m². (a) Determine its angular acceleration during this process. (b) Determine the tangential acceleration of a point on the grinding wheel if the wheel is 7.5 cm in diameter. (c) The slowing down is caused by a frictional torque on the axle of the wheel. The axle is 1.00 cm in diameter. Determine the frictional force on the axle. (d) How much work was done by friction on the system?

80. Modern bowling alleys have automatic ball returns. The ball is lifted to a height of 2.00 m at the end of the alley and, starting from rest, rolls down a ramp. It continues to roll horizontally and eventually rolls up a ramp at the other end that is 0.500 m off the ground. Assuming the mass of the bowling ball is 7.00 kg and its radius is 16.0 cm, determine (a) the rotation rate of the ball during the middle horizontal travel, (b) its linear speed during the middle horizontal travel, and (c) the final rotation rate and linear speed.

81. In a "modern art" exhibit, a multicolored empty industrial wire spool is suspended from two light wires as shown in ▼Fig. 8.56. The spool has a mass of 50.0 kg, with an outer diameter of 75.0 cm and an inner axle diameter of 18.0 cm. One wire (#1) is attached tangentially to the axle and makes a 10° angle with the vertical. The other wire (#2) is attached tangentially to the outer edge and makes an unknown angle θ with the vertical. Determine the tension in each wire and the angle θ.

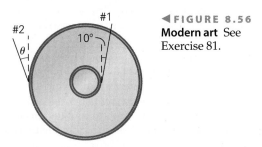

◀ FIGURE 8.56
Modern art See Exercise 81.

82. A flat, solid cylindrical grinding wheel with a diameter of 20.2 cm is spinning at 3000 rpm when its power is suddenly turned off. A workman continues to press his tool bit toward the wheel's center at the wheel's circumference so as to continue to grind as the wheel coasts to a stop. If the wheel has a moment of inertia of 4.73 kg · m², (a) determine the necessary torque that must be exerted by the workman to bring it to rest in 10.5 s. Ignore any friction at the axle. (b) If the coefficient of kinetic friction between the tool bit and the wheel surface is 0.85, how hard must the workman push on the bit?

83. A uniform sphere of mass 2.50 kg and radius 15.0 cm is released from rest at the top of an incline that is 5.25 m long and makes an angle of 35° with the horizontal. Assuming it rolls without slipping, (a) determine its total kinetic energy at the bottom of the incline. (b) Determine its rotational kinetic energy at the bottom of the incline. (c) What type of friction, static or kinetic, is acting on the surface of the sphere? Explain. (d) Determine the force of friction in part (d).

84. A stationary ice skater with a mass of 80.0 kg and a moment of inertia (about her central vertical axis) of 3.00 kg · m² catches a baseball with her outstretched arm. The catch is made at a distance of 1.00 m from the central axis. The ball has a mass of 145 g and is traveling at 20.0 m/s before the catch. (a) What linear speed does the system (skater + ball) have after the catch? (b) What is the angular speed of the system (skater + ball) after the catch? (c) What percentage of the ball's initial kinetic energy is lost during the catch? Neglect friction with the ice.

9 Solids and Fluids

PHYSICS FACTS

✦ The Mariana Trench in the Pacific Ocean is the deepest known point on the Earth. It is about 11 km (6.8 mi) below sea level. At this depth, the ocean water exerts a pressure of about 108 MPa (15 900 lb/in^2), or more than 1000 atmospheres of pressure.

✦ Legend has it that Archimedes, who is credited with the principle of buoyancy, was given the problem of determining whether the king's gold crown was pure gold or contained some silver. According to a Roman account, the solution came to him when he got into a full bath. On immersing, he noticed that water overflowed the tub. Quantities of pure gold and silver equal in weight to the king's crown were each put into bowls filled with water, and the silver caused more water to overflow. When the crown was tested, more water overflowed than for the pure gold, which implied some silver content.

Shown in the chapter-opening photo are solid cliffs, water, and unseen air that makes gliding possible. We walk on the solid surface of the Earth and in our daily lives use solid objects of all sorts, from scissors to computers. But we are surrounded by fluids—liquids and gases—some of which are indispensable. Without water, survival would be for only a few days at most. By far the most abundant substance in our bodies is water, and it is in the watery environment of our cells that all chemical processes on which life depends take place. Also, the gaseous oxygen in the air is essential for life processes.

On the basis of general physical distinctions, matter is commonly divided into three phases: solid, liquid, and gas. A *solid* has a definite shape and volume. A *liquid* has a fairly definite volume, but assumes the shape of its container. A *gas* takes on the shape and volume of its container. Solids and liquids are sometimes called *condensed matter*. In this chapter, a different classification scheme will be used and matter will be considered in terms of solids and fluids. Liquids and gases are referred to collectively as fluids. A **fluid** is a substance that can flow; liquids and gases qualify, but solids do not.

A simplistic description of solids is that they are made up of particles called atoms that are held rigidly together by interatomic forces. In Section 8.1, the concept of an ideal rigid body was used to describe rotational motion. Real solid bodies are not absolutely rigid and can be elastically deformed by external forces. Elasticity usually brings to mind a rubber band or spring that will resume its original dimensions even after being greatly deformed. In fact, all materials—even very hard steel—are elastic to some degree. But, as will be learned, such deformation has an *elastic limit*.

Fluids, however, have little or no elastic response to a force. Instead, the force merely causes an unconfined fluid to flow. This chapter pays particular attention to the behavior of fluids, shedding light on such questions as how hydraulic lifts work, why icebergs and ocean liners float, and what "10W-30" on a can of motor oil means. You'll also discover why the person in the chapter-opening photo can soar, with the aid of a suitably shaped piece of plastic.

Because of their fluidity, liquids and gases have many properties in common, and it is convenient to study them together. But there are important differences as well. For example, liquids are not very compressible, whereas gases are easily compressed.

9.1   Solids and Elastic Moduli

LEARNING PATH QUESTIONS

➡ What do stress and strain measure?
➡ What is an elastic modulus?
➡ What are the types of elastic moduli associated with materials?

As stated previously, all solid materials are elastic to some degree. That is, a body that is slightly deformed by an applied force will return to its original dimensions or shape when the force is removed. The deformation may not be noticeable for many materials, but it's there.

You may be able to visualize why materials are elastic if you think in terms of the simplistic model of a solid in ◄Fig. 9.1. The atoms of the solid substance are imagined to be held together by springs. The elasticity of the springs represents the resilient nature of the interatomic forces. The springs resist permanent deformation, as do the forces between atoms. The elastic properties of solids are commonly discussed in terms of stress and strain. **Stress** is a measure of the force causing a deformation. **Strain** is a relative measure of the deformation a stress causes. Quantitatively, *stress is the applied force per unit cross-sectional area*:

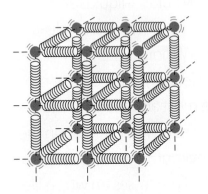

▲ FIGURE 9.1   A springy solid
The elastic nature of interatomic forces is indicated by simplistically representing them as springs, which, like the forces, resist deformation.

$$\text{stress} = \frac{F}{A} \qquad (9.1)$$

SI unit of stress : newton per square meter (N/m^2)

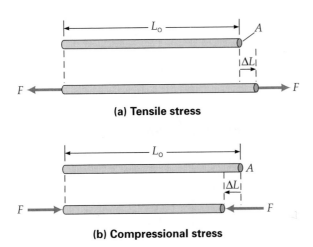

(a) Tensile stress

(b) Compressional stress

◀ **FIGURE 9.2 Tensile and compressional stresses** Tensile and compressional stresses are due to forces applied normally to the surface area of the ends of bodies. **(a)** A tension, or tensile stress, tends to increase the length of an object. **(b)** A compressional stress tends to shorten the length. ($\Delta L = L - L_o$) can be positive, as in (a), or negative, as in (b). The sign is not needed in Eq. 9.2, so the absolute value, $|\Delta L|$, is used.

Here, F is the magnitude of the applied force normal (perpendicular) to the cross-sectional area. Equation 9.1 shows that the SI units for stress are newtons per square meter (N/m^2).

As illustrated in ▲Fig. 9.2, a force applied to the ends of a rod gives rise to either a *tensile stress* (an elongating tension, $\Delta L > 0$) or a *compressional stress* (a shortening tension, $\Delta L < 0$), depending on the direction of the force. In both these cases, the *tensile strain* is the ratio of the change in length ($\Delta L = L - L_o$) to the original length (L_o), without regard to the sign, so the absolute value, $|\Delta L|$, is used. Then,

$$\text{strain} = \frac{|\text{change in length}|}{\text{original length}} = \frac{|\Delta L|}{L_o} = \frac{|L - L_o|}{L_o} \qquad (9.2)$$

Strain is a positive unitless quantity

Thus the strain is the *fractional change* in length. For example, if the strain is 0.05, the length of the material has changed by 5% of the original length.

As might be expected, the resulting strain depends on the applied stress. For relatively small stresses, this is a direct proportion, that is, stress $\propto$ strain. For relatively small stresses, this is a direct (or linear) proportion. The constant of proportionality, which depends on the nature of the material, is called the **elastic modulus**, that is,

$$\text{stress} = \text{elastic modulus} \times \text{strain}$$

or

$$\text{elastic modulus} = \frac{\text{stress}}{\text{strain}} \qquad (9.3)$$

SI unit of elastic modulus: newton per square meter (N/m^2)

The elastic modulus is the stress divided by the strain, and the elastic modulus has the same units as stress. (Why?)

Three general types of elastic moduli (plural of *modulus*) are associated with stresses that produce changes in length, shape, and volume. These are called *Young's modulus*, the *shear modulus*, and the *bulk modulus*, respectively.

CHANGE IN LENGTH: YOUNG'S MODULUS

▼Figure 9.3 is a typical graph of the tensile stress versus the strain for a metal rod. The curve is a straight line up to a point called the *proportional limit*. Beyond this point, the strain begins to increase more rapidly to another critical point called the **elastic limit**. If the tension is removed at this point, the material will return to its original length. If the tension is applied beyond the elastic limit and then removed, the material will recover somewhat, but will retain some permanent deformation.

▶ FIGURE 9.3   **Stress versus strain**
A plot of stress versus strain for a typical metal rod is a straight line up to the proportional limit. Then elastic deformation continues until the elastic limit is reached. Beyond that, the rod will be permanently deformed and will eventually fracture or break.

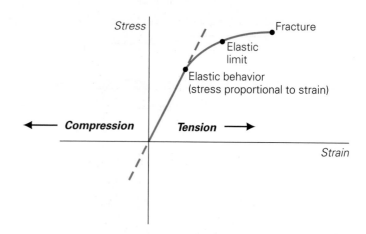

The straight-line part of the graph shows a direct proportionality between stress and strain. This relationship, first formalized by the English physicist Robert Hooke in 1678, is known as *Hooke's law*. (It is the same general relationship as that given for a spring in Section 5.2—see Fig. 5.5.) The elastic modulus for a tension or a compression is called **Young's modulus (Y):** *

$$\underset{\text{stress}}{\frac{F}{A}} = Y\left(\underset{\text{strain}}{\frac{\Delta L}{L_{\mathrm{o}}}}\right) \quad \text{or} \quad Y = \frac{F/A}{\Delta L/L_{\mathrm{o}}} \tag{9.4}$$

SI unit of Young's modulus: newton per square meter $(\mathrm{N/m^2})$

The units of Young's modulus are the same as those of stress, newtons per square meter $(\mathrm{N/m^2})$, since the strain is unitless. Some typical values of Young's modulus are given in ▾ Table 9.1.

TABLE 9.1   Elastic Moduli for Various Materials (in $\mathrm{N/m^2}$)

Substance	Young's modulus (*Y*)	Shear modulus (*S*)	Bulk modulus (*B*)
Solids			
Aluminum	7.0×10^{10}	2.5×10^{10}	7.0×10^{10}
Bone (limb)	Tension: 1.5×10^{10}	1.2×10^{10}	
	Compression: 9.3×10^{9}		
Brass	9.0×10^{10}	3.5×10^{10}	7.5×10^{10}
Copper	11×10^{10}	3.8×10^{10}	12×10^{10}
Glass	5.7×10^{10}	2.4×10^{10}	4.0×10^{10}
Iron	15×10^{10}	6.0×10^{10}	12×10^{10}
Nylon	5.0×10^{9}	8.0×10^{8}	
Steel	20×10^{10}	8.2×10^{10}	15×10^{10}
Liquids			
Alcohol, ethyl			1.0×10^{9}
Glycerin			4.5×10^{9}
Mercury			26×10^{9}
Water			2.2×10^{9}

*Thomas Young (1773–1829) was an English physician and physicist who also demonstrated the wave nature of light. See Young's double-slit experiment, Section 24.1.

To obtain a conceptual or physical understanding of Young's modulus, let's solve Eq. 9.4 for ΔL:

$$\Delta L = \left(\frac{FL_o}{A}\right)\frac{1}{Y} \quad \text{or} \quad \Delta L \propto \frac{1}{Y}$$

Hence, the larger the Young's modulus of a material, the smaller its change in length (with other parameters being equal).

EXAMPLE 9.1 | **Pulling My Leg: Under a Lot of Stress**

The femur (upper leg bone) is the longest and strongest bone in the body. Taking a typical femur to be approximately circular in cross-section with a radius of 2.0 cm, how much force would be required to extend a patient's femur by 0.010% while in horizontal traction?

THINKING IT THROUGH. Equation 9.4 should apply, but where does the percentage increase fit in? This question can be answered as soon as it is recognized that the $\Delta L/L_o$ term is the *fractional* increase in length. For example, if you had a spring with a length of 10 cm (L_o) and you stretched it 1.0 cm (ΔL), then $\Delta L/L_o = 1.0 \text{ cm}/10 \text{ cm} = 0.10$. This ratio can readily be changed to a percentage, and the spring's length was increased by 10%. So the percentage increase is really just the value of the $\Delta L/L_o$ term (multiplied by 100%).

SOLUTION. Listing the data,

Given: $r = 2.0 \text{ cm} = 0.020 \text{ m}$ *Find:* F (tensile force)
$\Delta L/L_o = 0.010\% = 1.0 \times 10^{-4}$
$Y = 1.5 \times 10^{10} \text{ N/m}^2$ (for bone, from Table 9.1)

Using Eq. 9.4,

$$F = Y(\Delta L/L_o)A = Y(\Delta L/L_o)\pi r^2$$
$$= (1.5 \times 10^{10} \text{ N/m}^2)(1.0 \times 10^{-4})\pi(0.020 \text{ m})^2 = 1.9 \times 10^3 \text{ N}$$

How much force is this? Quite a bit—in fact, more than 400 lb. The femur is a pretty strong bone.

FOLLOW-UP EXERCISE. A total mass of 16 kg is suspended from a 0.10-cm-diameter steel wire. (a) By what percentage does the length of the wire increase? (b) The tensile or ultimate strength of a material is the maximum stress the material can support before breaking or fracturing. If the tensile strength of the steel wire in (a) is $4.9 \times 10^8 \text{ N/m}^2$, how much mass could be suspended before the wire would break? *(Answers to all Follow-Up Exercises are given in Appendix VI at the back of the book.)*

Most types of bone are composed of protein collagen fibers that are tightly bound together and overlapping. Collagen has great tensile strength, and the calcium salts within the collagen give bone great compressional strength. Collagen also makes up cartilage, tendons, and skin, which have good tensile strength.

CHANGE IN SHAPE: SHEAR MODULUS

Another way an elastic body can be deformed is by a *shear stress*. In this case, the deformation is due to an applied force that is *tangential* to the surface area (▸ Fig. 9.4a). A change in shape results without a change in volume. The *shear strain* is given by x/h, where x is the relative displacement of the faces and h is the distance between them.

The shear strain may be defined in terms of the *shear angle* ϕ. As Fig. 9.4b shows, $\tan \phi = x/h$. But the shear angle is usually quite small, so a good approximation is $\tan \phi \approx \phi \approx x/h$, where ϕ is in radians.* (If $\phi = 10°$,

*See the Chapter 7 Learn by Drawing 7.1, The Small-Angle Approximation.

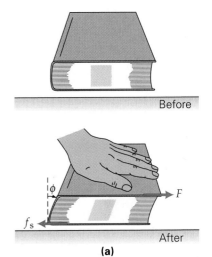

Before

After

(a)

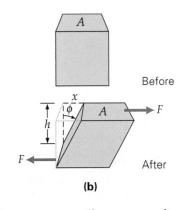

Before

After

(b)

▲ **FIGURE 9.4 Shear stress and strain** (a) A shear stress is produced when a force is applied tangentially to a surface area. (b) The strain is measured in terms of the relative displacement of the object's faces, or the shear angle ϕ.

for example, there is only 1.0% difference between ϕ and $\tan \phi$). The **shear modulus (S)**, sometimes called the *modulus of rigidity*, is then

$$S = \frac{F/A}{x/h} \approx \frac{F/A}{\phi} \tag{9.5}$$

SI unit of shear modulus: newton per square meter (N/m^2)

Note in Table 9.1 that the shear modulus is generally less than Young's modulus. In fact, S is approximately $Y/3$ for many materials, which indicates a greater response to a shear stress than to a tensile stress. Note also that the inverse relationship $\phi \approx 1/S$ is similar to that pointed out previously for Young's modulus.

A shear stress may be of the torsional type, resulting from the twisting action of a torque. For example, a torsional shear stress may shear off the head of a bolt that is being tightened.

Liquids do not have shear moduli (or Young's moduli)—hence the gaps in Table 9.1. A shear stress cannot be effectively applied to a liquid or a gas because fluids deform continuously in response. That is, *fluids cannot support a shear.*

CHANGE IN VOLUME: BULK MODULUS

Suppose that a force directed inward acts over the entire surface of a body (▾Fig. 9.5). Such a *volume stress* is often applied by pressure transmitted by a fluid. An elastic material will be compressed by a volume stress; that is, the material will show a change in volume, but not in general shape, in response to a pressure change Δp. [Pressure (p) is force per unit area, Section 9.2.] The change in pressure is equal to the volume stress, or $\Delta p = F/A$. The *volume strain* is the ratio of the volume change (ΔV) to the original volume (V_0). The **bulk modulus (B)** is then

$$B = \frac{F/A}{-\Delta V/V_0} = -\frac{\Delta p}{\Delta V/V_0} \tag{9.6}$$

SI unit of bulk modulus: newton per square meter (N/m^2)

The minus sign is introduced to make B a positive quantity, since $\Delta V = V - V_0$ is negative for an increase in external pressure (when Δp is positive). Similarly to the previous moduli relationships, $\Delta V \propto 1/B$.

Bulk moduli of selected solids and liquids are listed in Table 9.1. Gases also have bulk moduli, since they can be compressed. For a gas, it is common to talk about the reciprocal of the bulk modulus, which is called the **compressibility (k)**:

$$k = \frac{1}{B} \quad \text{(compressibility for gases)} \tag{9.7}$$

The change in volume ΔV is thus directly proportional to the compressibility k.

Solids and liquids are relatively incompressible and thus have small values of compressibility. Conversely, gases are easily compressed and have large compressibilities, which vary with pressure and temperature.

▶ **FIGURE 9.5 Volume stress and strain (a)** A volume stress is applied when a normal force acts over an entire surface area, as shown here for a cube. This type of stress most commonly occurs in gases. **(b)** The resulting strain is a change in volume.

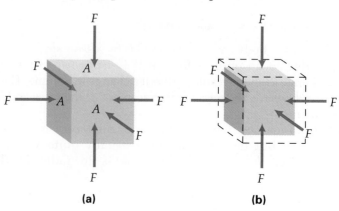

(a)   (b)

| EXAMPLE 9.2 | Compressing a Liquid: Volume Stress and Bulk Modulus |

By how much should the pressure on a liter of water be changed to compress it by 0.10%?

THINKING IT THROUGH. Similarly to the fractional change in length, $\Delta L / L_o$, the fractional change in volume is given by $-\Delta V / V_o$, which may be expressed as a percentage. The pressure change can then be found from Eq. 9.6. Compression implies a negative ΔV.

SOLUTION.

Given: $-\Delta V / V_o = 0.0010$ (or 0.10%)   *Find:* Δp
$V_o = 1.0 \text{ L} = 1000 \text{ cm}^3$
$B_{H_2O} = 2.2 \times 10^9 \text{ N/m}^2$
(from Table 9.1)

Note that $-\Delta V / V_o$ is the *fractional* change in the volume. With $V_o = 1000 \text{ cm}^3$, the change (reduction) in volume is

$$-\Delta V = 0.0010 \, V_o = 0.0010(1000 \text{ cm}^3) = 1.0 \text{ cm}^3$$

However, the change in volume is not needed. The fractional change, as listed in the given data, can be used directly in Eq. 9.6 to find the increase in pressure:

$$\Delta p = B\left(\frac{-\Delta V}{V_o}\right) = (2.2 \times 10^9 \text{ N/m}^2)(0.0010) = 2.2 \times 10^6 \text{ N/m}^2$$

(This increase is about twenty-two times normal atmospheric pressure. Not too compressible.)

FOLLOW-UP EXERCISE. If an extra $1.0 \times 10^6 \text{ N/m}^2$ of pressure above normal atmospheric pressure is applied to a half liter of water, what is the change in the water's volume?

DID YOU LEARN?

⇒ Stress is a measure of the force causing a deformation. Strain is a relative measure of the deformation a stress causes.
⇒ An elastic modulus is stress divided by strain, with units of N/m^2.
⇒ The three major moduli are Young's modulus (change in length), shear modulus (change in area), and bulk modulus (change in volume).

9.2 Fluids: Pressure and Pascal's Principle

LEARNING PATH QUESTIONS

⇒ How is applied pressure transmitted in an enclosed fluid?
⇒ What is the difference between absolute pressure and gauge pressure?

A force can be applied to a solid at a point of contact, but this won't work with a fluid, since a fluid cannot support a shear. With fluids, a force must be applied over an area. Such an application of force is expressed in terms of **pressure,** or the *force per unit area*:

$$p = \frac{F}{A} \tag{9.8a}$$

SI unit of pressure: newton per square meter (N/m^2), or pascal (Pa)

Pressure has SI units of newton per square meter (N/m^2), or **pascal (Pa)**, in honor of the French scientist and philosopher Blaise Pascal (1623–1662), who studied fluids and pressure. By definition,*

$$1 \text{ Pa} = 1 \text{ N/m}^2$$

The force in Eq. 9.8a is understood to be acting normally (perpendicularly) to the surface area. F may be the perpendicular component of a force that acts at an angle to a surface (▸Fig. 9.6).

As Figure 9.6 shows, in the more general case,

$$p = \frac{F_\perp}{A} = \frac{F \cos \theta}{A} \tag{9.8b}$$

Pressure is a scalar quantity (with magnitude only), even though the force producing it is a vector.

*Notice that the unit of pressure is equivalent to energy per volume, $N/m^2 = N \cdot m/m^3 = J/m^3$, an energy density.

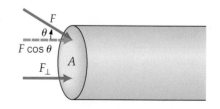

$$p = \frac{F_\perp}{A} = \frac{F \cos \theta}{A}$$

▲ **FIGURE 9.6 Pressure** Pressure is usually written $p = F/A$, where it is understood that F is the force or component of force normal to the surface. In general, then, $p = (F \cos \theta)/A$.

In the British system, a common unit of pressure is pound per square inch (lb/in², or psi). Other units, some of which will be introduced later, are used in special applications. Before going on, here's a "solid" example of the relationship between force and pressure.

CONCEPTUAL EXAMPLE 9.3 | ## Force and Pressure: Taking a Nap on a Bed of Nails

Suppose you are getting ready to take a nap, and you have a choice of lying stretched out on your back on (a) a bed of nails, (b) a hardwood floor, or (c) a couch. Which one would you choose for the most comfort, and *why*?

REASONING AND ANSWER. The comfortable choice is quite apparent—the couch. But here, the conceptual question is *why*.

First let's look at the prospect of lying on a bed of nails, an old trick that originated in India and used to be demonstrated in carnival sideshows (See Fig. 9.28). There is really no trick here, just physics—namely, force and pressure. It is the force per unit area, or pressure ($p = F/A$), that determines whether a nail will pierce the skin. The force is determined by the weight of the person lying on the nails. The area is determined by the *effective* area of the nails in contact with the skin (neglecting one's clothes).

If there were only one nail, the person's weight would not be supported by the nail, and with such a small area, the pressure would be very great—a situation in which the lone nail would pierce the skin. However, when a bed of nails is used, the same force (weight) is distributed over hundreds of nails, which gives a relatively large effective area of contact. The pressure is then reduced to a level at which the nails do not pierce the skin.

When you are lying on a hardwood floor, the area in contact with your body is appreciable and the pressure is reduced, but it still may be a bit uncomfortable. Parts of your body, such as your neck and the small of your back, are *not* in contact with a surface, but they would be on a couch. On a soft couch, the body sinks into it and the contact surface is greater, therefore reduced pressure and more comfort. So (c) is the answer.

FOLLOW-UP EXERCISE. What are a couple of important considerations in constructing a bed of nails to lie on?

Now, let's take a quick review of density, which is an important consideration in the study of fluids. Recall from Section 1.4 that the density (ρ) of a substance is defined as mass per *unit* volume (Eq. 1.1):

$$\text{density} = \frac{\text{mass}}{\text{volume}}$$

$$\rho = \frac{m}{V}$$

SI unit of density: kilogram per cubic meter (kg/m^3)

(common cgs unit: gram per cubic centimeter, or g/cm^3)

The densities of some common substances are given in ▼Table 9.2.

Water has a density of $1.00 \times 10^3 \, kg/m^3$ (or $1.00 \, g/cm^3$) from the original definition of the kilogram (Section 1.2). Mercury has a density of $13.6 \times 10^3 \, kg/m^3$ (or $13.6 \, g/cm^3$). Hence, mercury is 13.6 times as dense as water. Gasoline, however, is less dense than water. See Table 9.2. (*Note*: Be careful not to confuse the symbol for density, ρ (Greek rho), with that for pressure, p.)

TABLE 9.2   ## Densities of Some Common Substances (in kg/m³)

Solids	Density (ρ)	Liquids	Density (ρ)	Gases*	Density (ρ)
Aluminum	2.7×10^3	Alcohol, ethyl	0.79×10^3	Air	1.29
Brass	8.7×10^3	Alcohol, methyl	0.82×10^3	Helium	0.18
Copper	8.9×10^3	Blood, whole	1.05×10^3	Hydrogen	0.090
Glass	2.6×10^3	Blood plasma	1.03×10^3	Oxygen	1.43
Gold	19.3×10^3	Gasoline	0.68×10^3	Water vapor (100 °C)	0.63
Ice	0.92×10^3	Kerosene	0.82×10^3		
Iron (and steel)	7.8×10^3 (general value)	Mercury	13.6×10^3		
Lead	11.4×10^3	Seawater (4° C)	1.03×10^3		
Silver	10.5×10^3	Water, fresh (4° C)	1.00×10^3		
Wood, oak	0.81×10^3				

*At 0 °C and 1 atm, unless otherwise specified.

INSIGHT 9.1 | Osteoporosis and Bone Mineral Density (BMD)

Bone is a living, growing tissue. Your body is continuously taking up old bone (resorption) and making new bone tissue. In the early years of life, bone growth is greater than bone loss. This continues until a peak bone mass is reached as a young adult. After this, bone growth is slowly outpaced by bone loss. Bones naturally become less dense and weaker with age. Osteoporosis ("porous bone") occurs when bones deteriorate to the point where they are easily fractured (Fig. 1).

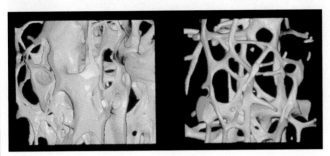

▲ FIGURE 1 **Bone mass loss** An X-ray micrograph of the bone structure of the vertebrae of a 50-year-old (left) and a 70-year-old (right). Osteoporosis, a condition characterized by bone weakening caused by loss of bone mass, is evident for the vertebrae on the right.

Osteoporosis and low bone mass affect an estimated 24 million Americans, most of whom are women. Osteoporosis results in an increased risk of bone fractures, particularly of the hip and the spine. Many women take calcium supplements to help prevent this.

To understand how bone density is measured, let's first distinguish between *bone* and *bone tissue*. Bone is the solid material composed of a protein matrix, most of which has calcified. Bone tissue includes the marrow spaces within the matrix. (Marrow is the soft, fatty, vascular tissue in the interior cavities of bones and is a major site of blood cell production.) The marrow volume varies with the bone type.

If the volume of an intact bone is measured (for example, by water displacement), then the *bone tissue density* can be computed, commonly in grams per cubic centimeter, after the bone is weighed to determine mass. If you burn the bone, weigh the remaining ash, and divide by the volume of the overall bone (bone tissue), you get the *bone tissue mineral density*, which is commonly called the **bone mineral density (BMD)**.

To measure the BMD of bones *in vivo*, types of radiation transmission through the bone are measured, which is related to the amount of bone mineral present. Also, a "projected" area of the bone is measured. Using these measurements, a projected BMD is computed in units of mg/cm^2. Figure 2 illustrates the magnitude of the effect of bone density loss with aging.

The diagnosis of osteoporosis relies primarily on the measurement of BMD. The mass of a bone, measured by a BMD test (also called a *bone densitometry test*), generally correlates to the bone strength. It is possible to predict fracture risk, much as blood pressure measurements can help predict stroke risk. Bone density testing is recommended for all women age 65 and older, and for younger women at an increased risk of osteoporosis. This testing also applies to men. Osteoporosis is often thought to be a woman's disease, but 20% of osteoporosis cases occur in men. A BMD test cannot predict the certainty of developing a fracture, but only predicts the degree of risk.

So how is BMD measured? This is where the physics comes in. Various instruments, divided into *central devices* and *peripheral devices*, are used. Central devices are used primarily to measure the bone density of the hip and spine. Peripheral devices are smaller, portable machines that are used to measure the bone density in such places as the heel or finger.

The most widely used central device relies on *dual energy X-ray absorptiometry* (DXA), which uses X-ray imaging to measure bone density. (See Section 20.4 for a discussion of X-rays.) The DXA scanner produces two X-ray beams of different energy levels. The amount of X-rays that pass through a bone is measured for each beam; the amounts vary with the density of bone. The calculated bone density is based on the difference between the two beams. The procedure is nonintrusive and takes 10–20 min, and the X-ray exposure is usually about one-tenth of that of a chest X-ray (Fig. 3).

▶ FIGURE 2 **Bone density loss with aging** An illustration of how normal bone density loss for a female hip bone increases with age (scale on right). Osteopenia refers to decreased calcification or bone density. A person with osteopenia is at risk for developing osteoporosis, a condition that causes bones to become brittle and prone to fracture.

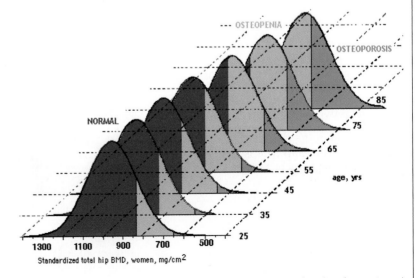

(continued on next page)

A common peripheral device uses *quantitative ultrasound* (QUS). Instead of using X-rays, a bone density screening is made using high-frequency sound waves (ultrasound). (See Section 14.1 for a discussion of ultrasound.) QUS measurements are usually done on the heel. The test takes only a minute or two, and the devices are now found in some pharmacies or drugstores. Its purpose is to tell you if you are "at risk" and may need a more thorough DXA test.

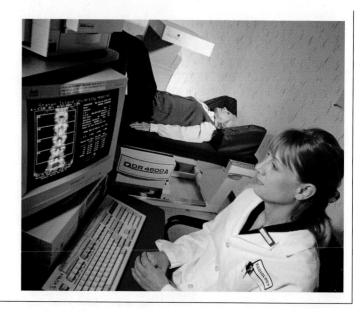

▶ **FIGURE 3** **Osteoporosis bone scanning**
A technician runs an X-ray bone scan on an elderly patient to check for osteoporosis. X-ray images are displayed on the monitor; these images can confirm the presence of osteoporosis. Such bone densitometry tests can also be used to diagnoses rickets, a children's disease characterized by softening of the bones.

Density is a measure of the compactness of the matter of a substance—the greater the density, the more matter or mass in a given volume. Notice that density quantifies the amount or mass per unit volume. For an important density consideration, see Insight 9.1, Osteoporosis and Bone Mineral Density (BMD).

PRESSURE AND DEPTH

If you have gone scuba diving, you well know that pressure increases with depth, having felt the increased pressure on your eardrums. An opposite effect is commonly felt when you fly in a plane or ride in a car going up a mountain. With increasing altitude, your ears may "pop" because of *reduced* external air pressure.

How the pressure in a fluid varies with depth can be demonstrated by considering a container of liquid at rest. Imagine that you can isolate a rectangular column of water, as shown in ▼Fig. 9.7. Then the force on the bottom of the container below the column (or the hand) is equal to the weight of the liquid making up the column: $F = w = mg$. Since density is $\rho = m/V$, the mass in the column is equal to the density times the volume; that is, $m = \rho V$. (The liquid is assumed incompressible, so ρ is constant.)

▶ **FIGURE 9.7** **Pressure and depth** The extra pressure at a depth h in a liquid is due to the weight of the liquid above: $p = \rho gh$, where ρ is the density of the liquid (assumed to be constant). This is shown for an imaginary rectangular column of liquid.

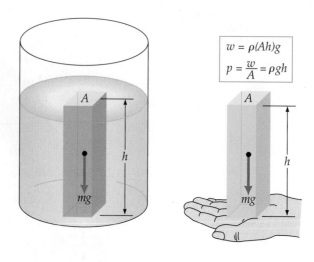

$$w = \rho(Ah)g$$
$$p = \frac{w}{A} = \rho gh$$

The volume of the isolated liquid column is equal to the height of the column times the area of its base, or $V = hA$. Thus,

$$F = w = mg = \rho Vg = \rho ghA$$

With $p = F/A$, the pressure at a depth h due to the weight of the column is

$$p = \rho gh \qquad (9.9)$$

This is a general result for incompressible liquids. The pressure is the same everywhere on a horizontal plane at a depth h (with ρ and g constant). Note that Eq. 9.9 is independent of the base area of the rectangular column. The whole cylindrical column of the liquid in the container in Fig. 9.7 could have been taken with the same result.

The derivation of Eq. 9.9 did not take into account pressure being applied to the open surface of the liquid. This factor adds to the pressure at a depth h to give a *total* pressure of

$$p = p_o + \rho gh \qquad \begin{array}{c}\text{(incompressible liquid}\\\text{at constant density)}\end{array} \qquad (9.10)$$

where p_o is the pressure applied to the liquid surface (that is, the pressure applied at $h = 0$). For an open container, $p_o = p_a$, atmospheric pressure, or the weight (force) per unit area due to the gases in the atmosphere above the liquid's surface. The average atmospheric pressure at sea level is sometimes used as a unit, called an **atmosphere (atm)**:

$$1 \text{ atm} = 101.325 \text{ kPa} = 1.01325 \times 10^5 \text{ N/m}^2 = 14.7 \text{ lb/in}^2$$

The measurement of atmospheric pressure will be described shortly.

EXAMPLE 9.4 | A Scuba Diver: Pressure and Force

(a) What is the total pressure on the back of a scuba diver in a lake at a depth of 8.00 m? (b) What is the force on the diver's back due to the water alone? (Take the surface of the back to be a rectangle 60.0 cm by 50.0 cm.)

THINKING IT THROUGH. (a) This is a direct application of Eq. 9.10 in which p_o is taken as the atmospheric pressure p_a. (b) Knowing the area and the pressure due to the water, the force can be found from the definition of pressure, $p = F/A$.

SOLUTION.

Given: $h = 8.00$ m
$A = 60.0 \text{ cm} \times 50.0 \text{ cm}$
$= 0.600 \text{ m} \times 0.500 \text{ m} = 0.300 \text{ m}^2$
$\rho_{H_2O} = 1.00 \times 10^3 \text{ kg/m}^3$ (from Table 9.2)
$p_a = 1.01 \times 10^5 \text{ N/m}^2$

Find: (a) p (total pressure)
(b) F (force due to water)

(a) The total pressure is the sum of the pressure due to the water and the atmospheric pressure (p_a). By Eq. 9.10, this is

$$p = p_a + \rho gh$$
$$= (1.01 \times 10^5 \text{ N/m}^2) + (1.00 \times 10^3 \text{ kg/m}^3)(9.80 \text{ m/s}^2)(8.00 \text{ m})$$
$$= (1.01 \times 10^5 \text{ N/m}^2) + (0.784 \times 10^5 \text{ N/m}^2) = 1.79 \times 10^5 \text{ N/m}^2 \text{ (or Pa)}$$
$$\text{(expressed in atmospheres)} \approx 1.8 \text{ atm}$$

This is also the inward pressure on the diver's eardrums.

(b) The pressure p_{H_2O} due to the water alone is the ρgh portion of the preceding equation, so $p_{H_2O} = 0.784 \times 10^5 \text{ N/m}^2$. Then, $p_{H_2O} = F/A$, and

$$F = p_{H_2O} A = (0.784 \times 10^5 \text{ N/m}^2)(0.300 \text{ m}^2)$$
$$= 2.35 \times 10^4 \text{ N} \quad \text{(or } 5.29 \times 10^3 \text{ lb—about 2.6 tons!)}$$

FOLLOW-UP EXERCISE. You might question the answer to part (b) of this Example—how could the diver support such a force? To get a better idea of the forces our bodies can support, what would be the force on the diver's back at the water surface from atmospheric pressure alone? How do you suppose our bodies can support such forces or pressures?

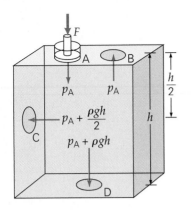

▲ FIGURE 9.8 Pascal's principle
The pressure applied at point A is fully transmitted to all parts of the fluid and to the walls of the container. There is also pressure due to the weight of the fluid above at different depths (for instance, $\rho gh/2$ at C and ρgh at D).

PASCAL'S PRINCIPLE

When the pressure (for example, air pressure) is increased on the entire open surface of an incompressible liquid at rest, the pressure at any point in the liquid or on the boundary surfaces increases by the same amount. The effect is the same if pressure is applied to any surface of an enclosed fluid by means of a piston (◄Fig. 9.8). The transmission of pressure in fluids was studied by Pascal, and the observed effect is called **Pascal's principle**:

> Pressure applied to an enclosed fluid is transmitted undiminished to every point in the fluid and to the walls of the container.

For an incompressible liquid, the change in pressure is transmitted essentially instantaneously. For a gas, a change in pressure will generally be accompanied by a change in volume or temperature (or both), but after equilibrium has been reestablished, Pascal's principle remains valid.

Common practical applications of Pascal's principle include the hydraulic braking systems used on automobiles. Through tubes filled with brake fluid, a force on the brake pedal transmits a force to the wheel brake cylinder. Similarly, hydraulic lifts and jacks are used to raise automobiles and other heavy objects (▼Fig. 9.9).

Using Pascal's principle, it can be shown how such systems allow us not only to transmit force from one place to another, but also to multiply that force. The input pressure p_i supplied by compressed air for a garage lift, for example, gives an input force F_i on a small piston area A_i (Fig. 9.9). The full magnitude of the pressure is transmitted to the output piston, which has an area A_o. Since $p_i = p_o$, it follows that

$$\frac{F_i}{A_i} = \frac{F_o}{A_o}$$

and

$$F_o = \left(\frac{A_o}{A_i}\right)F_i \quad \text{(hydraulic force multiplication)} \tag{9.11}$$

With A_o larger than A_i, then F_o will be larger than F_i. The input force is greatly multiplied if the input piston has a relatively small area.

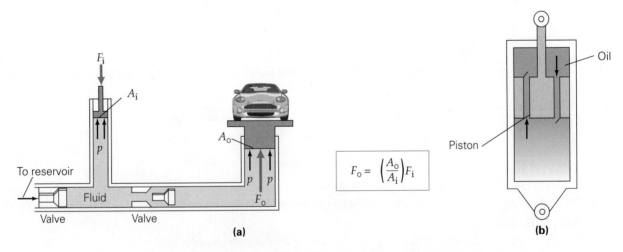

▲ FIGURE 9.9 The hydraulic lift and shock absorbers (a) Because the input and output pressures are equal (Pascal's principle), a small input force gives a large output force proportional to the ratio of the piston areas. **(b)** A simplified exposed view of one type of shock absorber. (See Example 9.5 for description.)

EXAMPLE 9.5 | The Hydraulic Lift: Pascal's Principle

A garage lift has input and lift (output) pistons with diameters of 10 cm and 30 cm, respectively. The lift is used to hold up a car with a weight of 1.4×10^4 N. (a) What is the magnitude of the force on the input piston? (b) What pressure is applied to the input piston?

THINKING IT THROUGH. (a) Pascal's principle, as expressed in the hydraulic Eq. 9.11, has four variables, and three are given (areas via diameters). (b) The pressure is simply $p = F/A$.

SOLUTION.

Given: $d_i = 10$ cm $= 0.10$ m *Find:* (a) F_i (input force)
  $d_o = 30$ cm $= 0.30$ m   (b) p_i (input pressure)
  $F_o = 1.4 \times 10^4$ N

(a) Rearranging Eq. 9.11 and using $A = \pi r^2 = \pi d^2/4$ for the circular piston $(r = d/2)$ gives

$$F_i = \left(\frac{A_i}{A_o}\right)F_o = \left(\frac{\pi d_i^2/4}{\pi d_o^2/4}\right)F_o = \left(\frac{d_i}{d_o}\right)^2 F_o$$

or

$$F_i = \left(\frac{0.10 \text{ m}}{0.30 \text{ m}}\right)^2 F_o = \frac{F_o}{9} = \frac{1.4 \times 10^4 \text{ N}}{9} = 1.6 \times 10^3 \text{ N}$$

The input force is one-ninth of the output force; in other words, the force was multiplied by 9 (that is, $F_o = 9F_i$).

(Note that we didn't really need to write the complete expressions for the areas. The area of a circle is proportional to the square of the diameter of the circle. If the ratio of the piston diameters is 3 to 1, the ratio of their areas must therefore be 9 to 1, and this ratio is used directly in Eq. 9.11.)

(b) Then applying Eq. 9.8a:

$$p_i = \frac{F_i}{A_i} = \frac{F_i}{\pi r_i^2} = \frac{F_i}{\pi (d_i/2)^2} = \frac{1.6 \times 10^3 \text{ N}}{\pi (0.10 \text{ m})^2/4}$$
$$= 2.0 \times 10^5 \text{ N/m}^2 \quad (= 200 \text{ kPa})$$

This pressure is about 30 lb/in², a common pressure used in automobile tires and about twice atmospheric pressure (which is approximately 100 kPa, or 15 lb/in².)

FOLLOW-UP EXERCISE. Pascal's principle is used in shock absorbers on automobiles and on the landing gear of airplanes. (The polished steel piston rods can be seen above the wheels on aircraft.) In these devices, a large force (the shock produced on hitting a bump in the road or on an airport runway at high speed) must be reduced to a safe level by removing energy. Basically, fluid is forced by the motion of a large-diameter piston through small channels in the piston on each stroke cycle (Fig. 9.9b).

Note that the valves allow for fluid through the channel, which creates resistance to the motion of the piston (effectively the reverse of the situation in Fig. 9.9a). The piston goes up and down, dissipating the energy of the shock. This is called *damping* (Section 13.2). Suppose that the input piston of a shock absorber on a jet plane has a diameter of 8.0 cm. What would be the diameter of an output channel that would reduce the force by a factor of 10?

As Example 9.5 shows, forces produced by pistons relate directly to their diameters: $F_i = (d_i/d_o)^2 F_o$ or $F_o = (d_o/d_i)^2 F_i$. By making $d_o \gg d_i$, huge factors of force multiplication can be obtained, as is typical for hydraulic presses, jacks, and earth-moving equipment. (The shiny input piston rods are often visible on front loaders and backhoes.) Inversely, force reductions may be obtained by making $d_i > d_o$, as in Follow-Up Exercise 9.5.

However, don't think that you are getting something for nothing with large force multiplications. Energy is still a factor, and it can never be multiplied by a machine. (Why not?) Looking at the work involved and assuming that the work output is equal to the work input, $W_o = W_i$ (an ideal condition—why?). Then, with $W = Fx$. (Eq. 5.1),

$$F_o x_o = F_i x_i$$

or

$$F_o = \left(\frac{x_i}{x_o}\right)F_i$$

where x_o and x_i are the output and input distances moved by the respective pistons.

Thus, the output force can be much greater than the input force only if the input distance is much greater than the output distance. For example, if $F_o = 10F_i$, then $x_i = 10x_o$, and the input piston must travel 10 times the distance of the output piston. *Force is multiplied at the expense of distance.*

PRESSURE MEASUREMENT

Pressure can be measured by mechanical devices that are often spring loaded (such as a tire gauge). Another type of instrument, called a manometer, uses a liquid—usually mercury—to measure pressure. An *open-tube manometer* is illustrated in ▼Fig. 9.10a. One end of the U-shaped tube is open to the atmosphere, and the other is connected to the container of gas whose pressure is to be measured. The liquid in the U-tube acts as a reservoir through which pressure is transmitted according to Pascal's principle.

The pressure of the gas (p) is balanced by the weight of the column of liquid (of height h, the difference in the heights of the columns) and the atmospheric pressure (p_a) on the open liquid surface:

$$p = p_a + \rho g h \qquad\qquad (9.12)$$

The pressure p is called the **absolute pressure**.

You may have measured pressure using pressure gauges; a tire gauge used to measure air pressure in automobile tires is a common example (Fig. 9.10b). Such gauges, quite appropriately, measure **gauge pressure**. A pressure gauge registers only the pressure *above* (or *below*) atmospheric pressure. Hence, to get the absolute pressure (p), you have to add the atmospheric pressure (p_a) to the gauge pressure (p_g):

$$p = p_a + p_g$$

For example, suppose your tire gauge reads a pressure of 200 kPa ($\approx$ 30 lb/in^2). The absolute pressure within the tire is then $p = p_a + p_g = 101$ kPa $+ 200$ kPa $= 301$ kPa, where normal atmospheric pressure is about 101 kPa (14.7 lb/in^2), as will be shown shortly.

The gauge pressure of a tire keeps the tire rigid or operational. In terms of the more familiar pounds per square inch (psi, or lb/in^2), a tire with a gauge pressure of 30 psi has an absolute pressure of about 45 psi (30 + 15, with atmospheric pressure = 15 psi). Hence, the pressure on the inside of the tire is 45 psi, and that

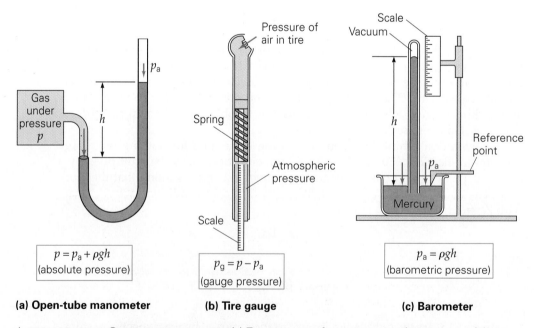

(a) Open-tube manometer   **(b) Tire gauge**   **(c) Barometer**

▲ **FIGURE 9.10 Pressure measurement (a)** For an open-tube manometer, the pressure of the gas in the container is balanced by the pressure of the liquid column and atmospheric pressure acting on the open surface of the liquid. The absolute pressure of the gas equals the sum of the atmospheric pressure (p_a) and $\rho g h$ the gauge pressure. **(b)** A tire gauge measures gauge pressure, the difference between the pressure in the tire and atmospheric pressure: $p_{gauge} = p - p_a$. Thus, if a tire gauge reads 200 kPa (30 lb/in^2), the actual pressure within the tire is 1 atm higher, or 300 kPa. **(c)** A barometer is a closed-tube manometer that is exposed to the atmosphere and thus reads only atmospheric pressure.

on the outside is 15 psi. The Δp of 30 psi keeps the tire inflated. If you open the valve or get a puncture, the internal and external pressures equalize and you have a flat!

Atmospheric pressure can be measured with a *barometer*. The principle of a mercury barometer is illustrated in Fig. 9.10c. The device was invented by Evangelista Torricelli (1608–1647), Galileo's successor as professor of mathematics at an academy in Florence. A simple barometer consists of a tube filled with mercury that is inverted into a reservoir. Some mercury runs from the tube into the reservoir, but a column supported by the air pressure on the surface of the reservoir remains in the tube. This device can be considered a *closed-tube manometer*, and the pressure it measures is just the atmospheric pressure, since the gauge pressure (the pressure *above* atmospheric pressure) is zero.

The atmospheric pressure is then equal to the pressure due to the weight of the column of mercury, or

$$p = \rho g h \qquad (9.13)$$

A *standard atmosphere* is defined as the pressure supporting a column of mercury exactly 76 cm in height at sea level and at 0 °C. (For a common biological atmospheric effect because of pressure changes, see Insight 9.2, An Atmospheric Effect: Possible Earaches.)

Changes in atmospheric pressure can be observed as changes in the height of a column of mercury. These changes are due primarily to high- and low-pressure air masses that travel across the country. Atmospheric pressure is commonly reported in terms of the height of the barometer column, and weather forecasters say that the barometer is rising or falling. That is,

$$1 \text{ atm (about 101 kPa)} = 76 \text{ cm Hg} = 760 \text{ mm Hg}$$

$$= 29.92 \text{ in. Hg (about 30 in. Hg)}$$

In honor of Torricelli, a pressure supporting 1 mm of mercury is given the name *torr*:

$$1 \text{ mm Hg} \equiv 1 \text{ torr}$$

and

$$1 \text{ atm} = 760 \text{ torr*}$$

*In the SI, one atmosphere has a pressure of $1.013 \times 10^5 \text{ N/m}^2$, or about 10^5 N/m^2. Meteorologists use yet another nonstandard unit of pressure called the *millibar* (mb). A *bar* is defined to be 10^5 N/m^2, and because 1 bar = 1000 mb, 1 atm = 1 bar = 1000 mb. Small changes in atmospheric pressure are more easily reported using the millibar.

INSIGHT 9.2 | **An Atmospheric Effect: Possible Earaches**

Variations in atmospheric pressure can have a common physiological effect: changes in pressure in the ears with changes in altitude. This "plugging up" and "popping" of the ears is frequently experienced in ascents and descents on mountain roads or on airplanes. The eardrum, so important to your hearing, is a membrane that separates the middle ear from the outer ear. [See Fig. 1 in the Chapter 14 (Sound) Insight 14.2, The Physiology and Physics of the Ear and Hearing, to view the anatomy of the ear.] The middle ear is connected to the throat by the Eustachian tube, the end of which is normally closed. The tube opens during swallowing or yawning to permit air to escape, so the internal and external pressures are equalized.

However, when climbing relatively quickly in an airplane or in a car in a mountainous region, the air pressure outside the ear may be less than that in the middle ear. This difference in pressure forces the eardrum outward. If the outward pressure is not relieved, you may soon have an earache. The pressure is relieved by a "pushing" of air through the Eustachian tube into the throat, which produces a popping sound. We often swallow or yawn to assist this process. Similarly, when descending, the higher outside pressure at lower altitudes needs to be equalized with the lower pressure in the middle ear. Swallowing allows air to flow into the middle ear in this case.

Nature takes care of us, but it is important to understand what is going on. If you have a throat infection, the opening of the Eustachian tube to the throat might be swollen, partially blocking the tube. You may be tempted to hold your nose and blow with your mouth closed in order to clear your ears. Don't do it! You could blow mucus into the inner ear and cause a painful inner-ear infection. Instead, swallow hard several times and give some big yawns to help open the Eustachian tube and equalize the pressure.

▶ **FIGURE 9.11 Aneroid barometer** Changes in atmospheric pressure on a sensitive metal diaphragm are reflected on the dial face of the barometer. A fair weather prediction is generally associated with high barometric pressures, and rainy weather with low barometric pressures.

Because mercury is highly toxic, it is sealed inside a barometer. A safer and less expensive device that is widely used to measure atmospheric pressure is the *aneroid* ("without fluid") *barometer*. In an aneroid barometer, a sensitive metal diaphragm on an evacuated container (something like a drumhead) responds to pressure changes, which are indicated on a dial. This is the kind of barometer you frequently find in homes in decorative wall mountings (◀Fig. 9.11).

Since air is compressible, the atmospheric density and pressure are greatest at the Earth's surface and decrease with altitude. We live at the bottom of the atmosphere, but don't notice its pressure very much in our daily activities. Remember that our bodies are composed largely of fluids, which exert a matching outward pressure. Indeed, the external pressure of the atmosphere is so important to our normal functioning that it is taken with us wherever we can. For example, the pressurized suits worn by astronauts in space or on the Moon are needed to provide an external pressure similar to that on the Earth's surface.

A very important gauge pressure reading is discussed in Insight 9.3, Blood Pressure, Intraocular Pressure, and Measurement. Read this before going on to Example 9.6.

INSIGHT 9.3 | ## Blood Pressure and Intraocular Pressure

BLOOD PRESSURE

Basically, a pump is a machine that transfers mechanical energy to a fluid, thereby increasing the pressure and causing the fluid to flow. One pump that is of interest to everyone is the heart, a muscular pump that drives blood throughout the body's circulatory network of arteries, capillaries, and veins. With each pumping cycle, the human heart's interior chambers enlarge and fill with freshly oxygenated blood from the lungs (Fig. 1).

The human heart contains two pairs of chambers: two ventricles and two atria. When the ventricles contract, blood is forced out through the arteries. Smaller and smaller arteries branch off from the main ones, until the very small capillaries are reached. There, oxygen and nutrients being carried by the blood are exchanged with the surrounding tissues, and carbon dioxide (a waste gas) is picked up. The blood then flows into the veins to the lungs to expel carbon dioxide, and then back to the heart to complete the circuit.

When the ventricles contract, forcing blood into the arterial system, the pressure in the arteries increases sharply. The maximum pressure achieved during ventricular contraction is called the *systolic pressure*. When the ventricles relax, the arterial pressure drops, and the lowest pressure before the next contraction, called the *diastolic pressure*, is reached. (These pressures are named after two parts of the pumping cycle, *systole* and *diastole*.)

▶ **FIGURE 1 The heart as a pump** The human heart is analogous to a mechanical force pump. Its pumping action, consisting of **(a)** intake and **(b)** output, gives rise to variations in blood pressure.

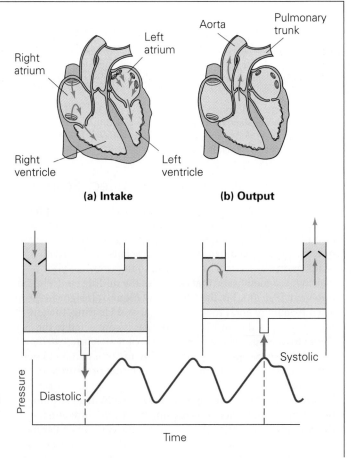

The walls of the arteries have considerable elasticity and expand and contract with each pumping cycle. This alternating expansion and contraction can be felt as a *pulse* in an artery near the surface of the body. For example, the radial artery near the surface of the wrist is commonly used to measure a person's pulse. The pulse rate is equal to the ventricular contraction rate, and hence the pulse rate indicates the heart rate.

Taking a person's blood pressure involves measuring the pressure of the blood on the arterial walls, usually in the arm. This is done with a *sphygmomanometer*. (The Greek word *sphygmo* means "pulse.") An inflatable cuff is wrapped around the arm and inflated to shut off the blood flow temporarily. The cuff pressure is slowly released, and the artery is monitored with a stethoscope (Fig. 2). Soon blood is just forced through the constricted artery. This flow is turbulent and gives rise to a specific sound with each heartbeat. When the sound is first heard, the systolic pressure is noted on the gauge. When the turbulent beats disappear because blood begins to flow smoothly, the diastolic pressure is taken.

Blood pressure is commonly reported by giving the systolic and diastolic pressures, separated by a slash—for example, 120/80 (mm Hg, read as "120 over 80"). (The gauge in Fig. 2 is an aneroid type; older types of sphygmomanometers used a mercury column to measure blood pressure.) Normal blood pressure ranges are 120–139 for systolic and 80–89 for diastolic. (Blood pressure is a gauge pressure. Why?)

High blood pressure is a common health problem. The elastic walls of the arteries expand under the hydraulic force of the blood pumped from the heart. Their elasticity may diminish with age, however. Cholesterol deposits can narrow and roughen the arterial passageways, impeding the blood flow and giving rise to a form of arteriosclerosis, or hardening of the arteries. Because of these defects, the driving pressure must increase to maintain a normal blood flow. The heart must work harder, which places a greater demand on the heart muscles. A relatively slight decrease in the effective cross-sectional area of a blood vessel has a rather large effect (an increase) on the flow rate, as will be shown in Section 9.4.

INTRAOCULAR PRESSURE

Another commonly measured pressure is intraocular pressure (IOP), or pressure of the eye. Elevated intraocular pressure can cause the damage of the optic nerve. Glaucoma is associated with elevated eye pressure and can cause the loss of vision.

The process of measuring intraocular pressure is referred to as tonometry. There are various types of devices, called tonometers, used to make this measurement. One of the most common instruments is a hand-held device called the Tono-Pen AVIA®.

After the eye has been numbed with anesthetic drops, the tonometer's tip is gently placed against the front surface (cornea) of the eye (Fig. 3). The cornea bends under the force applied by the tip of the Tono-Pen. Once the cornea passes a flattened stage, it becomes slightly indented and a pressure transducer in the Pen measures the force required to reach this state. The result is displayed on a digital readout in mm Hg. The procedure is painless and takes only a few seconds. Normal eye pressures range from 10 to 20 mm Hg.

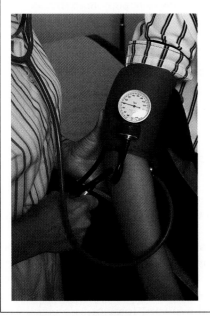

◄ **FIGURE 2**
Measuring blood pressure The pressure is indicated on the gauge in millimeters Hg.

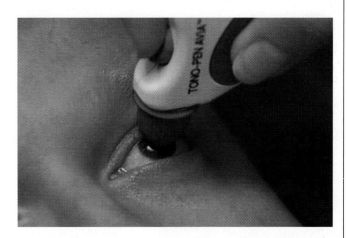

▲ **FIGURE 3** **Intraocular pressure.** Intraocular (eye) pressure being measured with a Tono-Pen AVIA®. See text for description.

EXAMPLE 9.6 | An IV: A Gravity Assist

An IV (*intravenous injection*) is a type of gravity assist quite different from that discussed for space probes in Section 7.5. Consider a hospital patient who receives an IV under gravity flow, as shown in ▼ Fig. 9.12. If the blood gauge pressure in the vein is 20.0 mm Hg, above what height should the bottle be placed for the IV blood transfusion to function properly?

THINKING IT THROUGH. The fluid gauge pressure at the bottom of the IV tube must be greater than the pressure in the vein and can be computed from Eq. 9.9. (The liquid is assumed to be incompressible.)

(continued on next page)

SOLUTION.
Given: p_v = 20.0 mm Hg (vein gauge pressure)   *Find:* h (height for $p_v > 20$ mm Hg)
$\rho = 1.05 \times 10^3$ kg/m^3
(whole blood density from Table 9.2)

First, the common medical unit of mm Hg (or torr) needs to be changed to the SI unit of pascal (Pa, or N/m^2):

$$p_v = (20.0 \text{ mm Hg})[133 \text{ Pa}/(\text{mm Hg})] = 2.66 \times 10^3 \text{ Pa}$$

Then, for $p > p_v$,

$$p = \rho g h > p_v$$

or

$$h > \frac{p_v}{\rho g} = \frac{2.66 \times 10^3 \text{ Pa}}{(1.05 \times 10^3 \text{ kg/m}^3)(9.80 \text{ m/s}^2)} = 0.259 \text{ m} \ (\approx 26 \text{ cm})$$

The IV bottle needs to be at least 26 cm above the injection site.

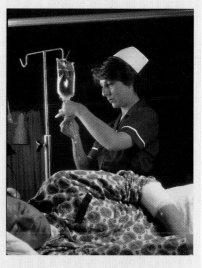

▲ **FIGURE 9.12 What height is needed?** See Example text for description.

FOLLOW-UP EXERCISE. The normal (gauge) blood pressure range is commonly reported as 120/80 (in millimeters Hg). Why is the blood pressure of 20 mm Hg in this Example so low?

DID YOU LEARN
➡ By Pascal's principle, pressure applied to an enclosed fluid is transmitted to every point in the fluid and the walls of the container.
➡ Absolute pressure is the measured pressure plus atmospheric pressure. Gauge pressure is that measured above (or below) atmospheric pressure, so absolute pressure is gauge pressure plus atmospheric pressure.

9.3 Buoyancy and Archimedes' Principle

LEARNING PATH QUESTIONS
➡ What is meant by buoyant force?
➡ What does Archimedes' principle tell us?
➡ Under what conditions will an object float in a fluid?

When placed in a fluid, an object will either sink or float. This is most commonly observed with liquids; for example, objects float or sink in water. But the same effect occurs in gases: A falling object sinks in the atmosphere, while other objects float (◄Fig. 9.13).

Things float because they are buoyant, or are buoyed up. For example, if you immerse a cork in water and release it, the cork will be buoyed up to the surface and float there. From your knowledge of forces, you know that such motion requires an upward net force on an object. That is, there must be an upward force acting on the object that is greater than the downward force of its weight. The forces are equal when the object floats in equilibrium. The upward force resulting from an object being wholly or partially immersed in a fluid is called the **buoyant force**.

How the buoyant force comes about can be seen by considering a buoyant object being held under the surface of a fluid (►Fig. 9.14a). The pressures on the upper and lower surfaces of the block are $p_1 = \rho_f g h_1$ and $p_2 = \rho_f g h_2$, respectively, where ρ_f is the density of the fluid. Thus, there is a pressure difference $\Delta p = p_2 - p_1 = \rho_f g(h_2 - h_1)$ between the top and bottom of the block, which gives an upward force (the buoyant force) F_b. This force is balanced by the applied force and the weight of the block.

▲ **FIGURE 9.13 Fluid buoyancy** The air is a fluid in which objects such as this dirigible float. The helium inside the blimp is less dense than the surrounding air, and displacing its volume of air, the blimp is supported by the resulting buoyant force.

It is not difficult to derive an expression for the magnitude of the buoyant force. Pressure is force per unit area. Thus, if both the top and bottom areas of the block are A, the magnitude of the net buoyant force in terms of the pressure difference is

$$F_b = p_2 A - p_1 A = (\Delta p)A = \rho_f g(h_2 - h_1)A$$

Since $(h_2 - h_1)A$ is the volume of the block and hence the volume of fluid displaced by the block, V_f, the expression for F_b may be written as

$$F_b = \rho_f g V_f$$

But $\rho_f V_f$ is simply the mass of the fluid displaced by the block, m_f. Thus, the expression for the buoyant force becomes $F_b = m_f g$: The magnitude of the buoyant force is equal to the weight of the fluid displaced by the block (Fig. 9.14b). This general result is known as **Archimedes' principle**:

> A body immersed wholly or partially in a fluid experiences a buoyant force equal in magnitude to the weight of the *volume of fluid* that is displaced:

$$F_b = m_f g = \rho_f g V_f \tag{9.14}$$

Archimedes (287–212 BCE), a Greek scientist, was given the task of determining whether a gold crown made for the king was pure gold or contained a quantity of silver. Legend has it that the solution came to him upon immersing himself in a full bath. (See the Physics Facts at the beginning of the chapter.) It is said that he was so excited he jumped out and ran home through the streets of the city (unclothed) shouting "Eureka! Eureka!" (Greek for "I have found it"). Archimedes' solution to the problem involved density and volume and it may have gotten him thinking about buoyancy.

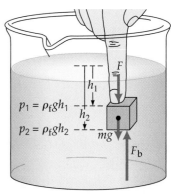

$$\Delta p = \rho_f g(h_2 - h_1)$$

(a)

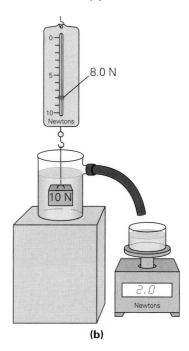

(b)

▲ **FIGURE 9.14 Buoyancy and Archimedes' principle (a)** A buoyant force arises from the difference in pressure at different depths. The pressure on the bottom of the submerged block (p_2) is greater than that on the top (p_1), so there is a (buoyant) force directed upward. (Shifted for clarity.)
(b) Archimedes' principle: The buoyant force on the object is equal to the weight of the volume of fluid displaced. (The scale is set to read zero when the container is empty.)

INTEGRATED EXAMPLE 9.7 | ## Lighter Than Air: Buoyant Force

A spherical helium-filled weather balloon has a radius of 1.10 m. (a) Does the buoyant force on the balloon depend on the density of (1) helium, (2) density of air, or (3) the weight of the rubber "skin"? [$\rho_{air} = 1.29$ kg/m^3 and $\rho_{He} = 0.180$ kg/m^3.] (b) Compute the magnitude of the buoyant force on the balloon. (c) The balloon's rubber skin has a mass of 1.20 kg. When released, what is the magnitude of the balloon's initial acceleration if it carries a payload with a mass of 3.52 kg?

(A) CONCEPTUAL REASONING. The upward buoyant force has nothing to do with the helium or rubber skin and is equal to the weight of the displaced air, which can be found from the balloon's volume and the density of air. So the answer is (2).

(B, C) QUANTITATIVE REASONING AND SOLUTION.

Given: $\rho_{air} = 1.29$ kg/m^3 **Find:** (b) F_b (buoyant force)
$\rho_{He} = 0.180$ kg/m^3 (c) a (initial acceleration)
$m_s = 1.20$ kg
$m_p = 3.52$ kg
$r = 1.10$ m

(b) The volume of the balloon is

$$V = (4/3)\pi r^3 = (4/3)\pi(1.10 \text{ m})^3 = 5.58 \text{ m}^3$$

Then the buoyant force is equal to the weight of the air displaced:

$$F_b = m_{air}g = (\rho_{air}V)g = (1.29 \text{ kg/m}^3)(5.58 \text{ m}^3)(9.80 \text{ m/s}^2) = 70.5 \text{ N}$$

(c) Draw a free-body diagram. There are three weight forces downward—those of the helium, the rubber skin, and the payload—and the upward buoyant force. Sum these forces to find the net force, and then use Newton's second law to find the acceleration.

(continued on next page)

The weights of the helium, rubber skin, and payload are as follows:

$$w_{He} = m_{He}g = (\rho_{He}V)g = (0.180 \text{ kg/m}^3)(5.58 \text{ m}^3)(9.80 \text{ m/s}^2) = 9.84 \text{ N}$$

$$w_s = m_sg = (1.20 \text{ kg})(9.80 \text{ m/s}^2) = 11.8 \text{ N}$$

$$w_p = m_pg = (3.52 \text{ kg})(9.80 \text{ m/s}^2) = 35.5 \text{ N}$$

Summing the forces (taking upward as positive),

$$F_{net} = F_b - w_{He} - w_s - w_p = 70.5 \text{ N} - 9.84 \text{ N} - 11.8 \text{ N} - 35.5 \text{ N} = 13.4 \text{ N}$$

and with the masses found from the weights:

$$a = \frac{F_{net}}{m_{total}} = \frac{F_{net}}{m_{He} + m_s + m_p} = \frac{13.4 \text{ N}}{1.00 \text{ kg} + 1.20 \text{ kg} + 3.52 \text{ kg}} = 2.34 \text{ m/s}^2$$

FOLLOW-UP EXERCISE. As the balloon rises, it eventually stops accelerating and rises at a constant velocity for a short time, then starts sinking toward the ground. Explain this behavior in terms of atmospheric density and temperature. [*Hint*: Temperature and air density decrease with altitude. The pressure of a quantity of gas is directly proportional to temperature.]

EXAMPLE 9.8 | ## Your Buoyancy in Air

Air is a fluid and our bodies displace air. And so, a buoyant force is acting on each of us. Estimate the magnitude of the buoyant force on a 75-kg person due to the air displaced.

THINKING IT THROUGH. The key word here is *estimate*, because not much data are given. We know that the buoyant force is $F_b = \rho_a gV$, where ρ_a is the density of air (which can be found in Table 9.2), and V is the volume of the air dis-

placed, which is the same as the volume of the person. The question is, how do we find the volume of the person?

The mass is given, and if the density of the person were known, the volume could be found ($\rho = m/V$) or $V = m/\rho$. Here is where the estimate comes in. Most people can barely float in water, so the density of the human body is about that of water, $\rho = 1000 \text{ kg/m}^3$. Using this estimate, the buoyant force can also be estimated.

SOLUTION.

Given: $m = 75 \text{ kg}$ *Find:* F_b (buoyant force)

$\rho_a = 1.29 \text{ kg/m}^3$ (Table 9.2)

$\rho_p = 1000 \text{ kg/m}^3$ (estimated density of person)

First, let's find the volume of the person:

$$V_p = \frac{m}{\rho_p} = \frac{75 \text{ kg}}{1000 \text{ kg/m}^3} = 0.075 \text{ m}^3$$

Then,

$$F_b = \rho_a gV_p = (1.29 \text{ kg/m}^3)(9.8 \text{ m/s}^2)(0.075 \text{ m}^3)$$

$$= 0.95 \text{ N} \; (\approx 1.0 \text{ N or } 0.225 \text{ lb})$$

This amount is not much when you weigh yourself. But it does mean that your weight is ≈ 0.2 lb more than the scale reading.

FOLLOW-UP EXERCISE. Estimate the buoyant force on a helium-filled weather balloon that has a diameter on the order of a meteorologist's arm span (arms held horizontally), and compare with the result in the Example.

INTEGRATED EXAMPLE 9.9 | ## Weight and Buoyant Force: Archimedes' Principle

A container of water with an overflow tube, similar to that shown in Fig. 9.14b, sits on a scale that reads 40 N. The water level is just below the exit tube in the side of the container. (a) An 8.0-N cube of wood is placed in the container. The water displaced by the floating cube runs out the exit tube into another container that is not on the scale. Will the scale

reading then be (1) exactly 48 N, (2) between 40 N and 48 N, (3) exactly 40 N, or (4) less than 40 N? (b) Suppose you pushed down on the wooden cube with your finger such that the top surface of the cube was even with the water level. How much force would have to be applied if the wooden cube measured 10 cm on a side?

(A) CONCEPTUAL REASONING. By Archimedes' principle, the block is buoyed upward with a force equal in magnitude to the weight of the water displaced. Since the block floats, the upward buoyant force must balance the weight of the cube and so has a magnitude of 8.0 N. Thus, a volume of water weighing 8.0 N is displaced from the container as 8.0 N of weight is added to the container. The scale still reads 40 N, so the answer is (3).

Note that the upward buoyant force and the block's weight act *on the block*. The reaction force (pressure) of the block *on the water* is transmitted to the bottom of the container (Pascal's principle) and is registered on the scale. (Make a sketch showing the forces on the cube.)

(B) QUANTITATIVE REASONING AND SOLUTION. Here three forces are acting on the stationary cube: the buoyant force upward and the weight and the force applied by the finger downward. The weight of the cube is known, so to find the applied finger force, we need to determine the buoyant force on the cube.

Given: $\ell = 10$ cm $= 0.10$ m (side length of cube)
        $w = 8.0$ N (weight of cube)

Find: F_f (downward applied force necessary to put cube even with water level)

The summation of the forces acting on the cube is $\Sigma F_y = +F_b - w - F_f = 0$, where F_b is the upward buoyant force and F_f is the downward force applied by the finger. Hence, $F_f = F_b - w$. As we know, the magnitude of the buoyant force is equal to the weight of the water the cube displaces, which is given by $F_b = \rho_f g V_f$ (Eq. 9.14). The density of the fluid is that of water, which is known (1.0×10^3 kg/m^3, Table 9.2), so

$$F_b = \rho_f g V_f = (1.0 \times 10^3 \text{ kg/m}^3)(9.8 \text{ m/s}^2)(0.10 \text{ m})^3 = 9.8 \text{ N}$$

Thus,

$$F_f = F_b - w = 9.8 \text{ N} - 8.0 \text{ N} = 1.8 \text{ N}$$

FOLLOW-UP EXERCISE. In part (a), would the scale still read 40 N if the object had a density greater than that of water? In part (b), what would the scale read?

BUOYANCY AND DENSITY

It is commonly said that helium and hot-air balloons float because they are lighter than air. To be technically correct, it should be said that the balloons are *less dense than air*. An object's density will tell you whether it will sink or float in a fluid, as long as you also know the density of the fluid. Consider a solid uniform object that is totally immersed in a fluid. The weight of the object is

$$w_o = m_o g = \rho_o V_o g$$

The weight of the volume of fluid displaced, or the magnitude of the buoyant force, is

$$F_b = w_f = m_f g = \rho_f V_f g$$

If the object is *completely submerged*, $V_f = V_o$. Dividing the second equation by the first gives

$$\frac{F_b}{w_o} = \frac{\rho_f}{\rho_o} \quad \text{or} \quad F_b = \left(\frac{\rho_f}{\rho_o}\right) w_o \quad \textit{(object completely submerged)} \tag{9.15}$$

Thus, if ρ_o is less than ρ_f, then F_b will be greater than w_o, and the object will be buoyed to the surface and float. If ρ_o is greater than ρ_f, then F_b will be less than w_o, and the object will sink. If ρ_o equals ρ_f, then F_b will be equal to w_o, and the object will remain in equilibrium at any submerged depth (as long as the density of the fluid is constant). If the object is not uniform, so that its density varies over its volume, then the density of the object in Eq. 9.15 is the average density.

Expressed in words, these three conditions are as follows:

An object will float in a fluid if the average density of the object is less than the density of the fluid ($\rho_o < \rho_f$)

An object will sink in a fluid if the average density of the object is greater than the density of the fluid ($\rho_o > \rho_f$)

An object will be in equilibrium at any submerged depth in a fluid if the average density of the object and the density of the fluid are equal ($\rho_o = \rho_f$)

See ▶ Fig. 9.15 for an example of the last condition.

A quick look at Table 9.2 will tell you whether an object will float in a fluid, regardless of the shape or volume of the object. The three conditions just stated also apply to a fluid in a fluid, provided that the two are immiscible (do not mix). For example, you might think that cream is "heavier" than skim milk, but that's not so: Since cream floats on milk, it is less dense than milk.

▲ **FIGURE 9.15 Equal densities and buoyancy** This soft drink contains colored gelatin beads that remain suspended for months with virtually no change. What is the density of the beads compared to the density of the drink?

DEMONSTRATION 2 | **Buoyancy and Density**

This demonstration of buoyancy shows that the overall density of a can of Diet Coke is less than that of water while the density of a can of a Classic Coke is greater. Consider the following questions: Does one can have a greater volume of metal? higher gas pressure inside? more fluid volume? Do calories make a difference? Investigate the possibilities to determine the reason(s) for the different densities.

The can of Classic Coke sinks and the can of Diet Coke floats.

In general, the densities of objects or fluids will be assumed to be uniform and constant in this book. (The density of the atmosphere does vary with altitude, but is relatively constant near the surface of the Earth.) In any event, in practical applications it is the *average* density of an object that often matters with regard to floating and sinking. For example, an ocean liner is, on average, less dense than water, even though it is made of steel. Most of its volume is occupied by air, so the liner's average density is less than that of water. Similarly, the human body has air-filled spaces, so most of us float in water. The surface depth at which a person floats depends on his or her density. (Why?)

In some instances, the overall density of an object is purposefully varied. For example, a submarine submerges by flooding its tanks with seawater (called "taking on ballast"), which increases its average density. When the sub is ready to surface, the water is pumped out of the tanks, so the average density of the sub becomes less than that of the surrounding seawater.

Similarly, many fish control their depths by using their *swim bladders* or *gas bladders*. A fish changes or maintains buoyancy by regulating the volume of gas in the gas bladder. Maintaining neutral buoyancy (neither rising nor sinking) is important because it allows the fish to stay at a particular depth for feeding. Some fish may move up and down in the water in search of food. Instead of using up energy to swim up and down, the fish alters its buoyancy to rise and sink.

This is accomplished by adjusting the quantities of gas in the gas bladder. Gas is transferred from the gas bladder to the adjoining blood vessels and back again. Deflating the bladder decreases the volume and increases the average density, and the fish sinks. Gas is forced into the surrounding blood vessels and carried away.

Conversely, to inflate the bladder, gases are forced into the bladder from the blood vessels, thereby increasing the volume and decreasing the average density, and the fish rises. These processes are complex, but Archimedes' principle is being applied in a biological setting.

EXAMPLE 9.10 | **Float or Sink? Comparison of Densities**

A uniform solid cube of material 10.0 cm on each side has a mass of 700 g. (a) Will the cube float in water? (b) If so, how much of its volume would be submerged?

THINKING IT THROUGH. (a) The question is whether the density of the material the cube is made of is greater or less than that of water, so we compute the cube's density. (b) If the cube floats, then the buoyant force and the cube's weight are equal.

Both of these forces are related to the cube's volume, so we can write them in terms of that volume and equate them.

SOLUTION. It is sometimes convenient to work in cgs units in comparing small quantities. For densities in grams per cubic centimeter, divide the values in Table 9.2 by 10^3, or drop the "$\times 10^3$" from the values given for solids and liquids, and replace with "$\times 10^{-3}$" for gases.

Given: $m = 700$ g   **Find:** (a) Whether the cube will float in water
$L = 10.0$ cm   (b) The percentage of the volume submerged
$\rho_{H_2O} = 1.00 \times 10^3$ kg/m^3 if the cube does float
$= 1.00$ g/cm^3 (Table 9.2)

(a) The density of the cube is

$$\rho_c = \frac{m}{V_c} = \frac{m}{L^3} = \frac{700 \text{ g}}{(10.0 \text{ cm})^3} = 0.700 \text{ g/cm}^3 < \rho_{H_2O} = 1.00 \text{ g/cm}^3$$

Since ρ_c is less than ρ_{H_2O} the cube will float.

(b) The weight of the cube is $w_c = \rho_c g V_c$. When the cube is floating, it is in equilibrium, which means that its weight is balanced by the buoyant force. That is, $F_b = \rho_{H_2O} g V_{H_2O}$, where V_{H_2O} is the volume of water the submerged part of the

cube displaces. Equating the expressions for weight and buoyant force gives

$$\rho_{H_2O} g V_{H_2O} = \rho_c g V_c$$

or

$$\frac{V_{H_2O}}{V_c} = \frac{\rho_c}{\rho_{H_2O}} = \frac{0.700 \text{ g/cm}^3}{1.00 \text{ g/cm}^2} = 0.700$$

Thus, $V_{H_2O} = 0.70 V_c$, and 70% of the cube is submerged.

FOLLOW-UP EXERCISE.   Most of an iceberg floating in the ocean is submerged (▼ Fig. 9.16). The visible portion is the proverbial "tip of the iceberg." What percentage of an iceberg's volume is seen above the surface? (*Note:* Icebergs are frozen *fresh* water floating in salty sea water.)

A quantity called specific gravity is related to density. It is commonly used for liquids, but also applies to solids. The **specific gravity (sp. gr.)** of a substance is equal to the ratio of the density of the substance (ρ_s) to the density of water (ρ_{H_2O}) at 4 °C, the temperature for maximum density:

$$\text{sp. gr.} = \frac{\rho_s}{\rho_{H_2O}}$$

Because it is a ratio of densities, specific gravity has no units. In cgs units, $\rho_{H_2O} = 1.00$ g/cm^3, so

$$\text{sp. gr.} = \frac{\rho_s}{1.00} = \rho_s \quad (\rho_s \text{ in g/cm}^3 \text{ only})$$

That is, the specific gravity of a substance is equal to the numerical value of its density *in cgs units*. For example, if a liquid has a density of 1.5 g/cm^3, its specific gravity is 1.5, which tells you that it is 1.5 times as dense as water. (As pointed out earlier, to get density values for solids and liquids in grams per cubic centimeter, divide the value in Table 9.2 by 10^3.)

DID YOU LEARN?
➥ Buoyant force is the upward force resulting from an object being wholly or partially submerged in a fluid.
➥ Archimedes' principle allows the buoyant force to be measured: The magnitude of the buoyant force on an object is equal to the weight of the volume of fluid displaced.
➥ An object will float if its average density is less than that of the fluid.

▲ **FIGURE 9.16 The tip of the iceberg** The vast majority of an iceberg's bulk is underneath the water, as illustrated here in a false photo. See Example 9.10 Follow-Up Exercise.

9.4 Fluid Dynamics and Bernoulli's Equation

LEARNING PATH QUESTIONS
➥ What are the characteristics of ideal fluid flow?
➥ What does the equation of continuity tell about incompressible fluid flow?
➥ On what is Bernoulli's equation based?

In general, fluid motion is difficult to analyze. For example, think of trying to describe the motion of a particle (a molecule, as an approximation) of water in a rushing stream. The overall motion of the stream may be apparent, but a mathematical description of the motion of any one particle of it may be virtually impossible

Streamlines

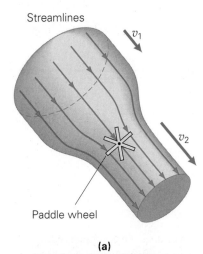

Paddle wheel

(a)

(b)

▲ FIGURE 9.17 **Streamline flow**
(a) Streamlines never cross and are closer together in regions of greater fluid velocity. The stationary paddle wheel indicates that the flow is irrotational, or without whirlpools and eddy currents. **(b)** The smoke from an extinguished candle begins to rise in nearly streamline flow, but quickly becomes rotational and turbulent.

because of eddy currents (small whirlpool motions), the gushing of water over rocks, frictional drag on the stream bottom, and so on. A basic description of fluid flow is conveniently obtained by ignoring such complications and considering an ideal fluid. Actual fluid flow can then be approximated with reference to this simpler theoretical model.

In this simplified approach to fluid dynamics, it is customary to consider four characteristics of an **ideal fluid**. In such a fluid, flow is (1) *steady*, (2) *irrotational*, (3) *nonviscous*, and (4) *incompressible*.

Condition 1: *Steady flow* means that all the particles of a fluid have the same velocity as they pass a given point.

Steady flow might be called smooth or regular flow. The path of steady flow can be depicted in the form of **streamlines** (◀Fig. 9.17a). Every particle that passes a particular point moves along a streamline. That is, every particle moves along the same path (streamline) as particles that passed by earlier. Streamlines never cross; if they did, a particle would have alternative paths and abrupt changes in its velocity, in which case the flow would not be steady.

Steady flow requires low velocities. For example, steady flow is approximated by the flow relative to a canoe that is gliding slowly through still water. When the flow velocity is high, eddies tend to appear, especially near boundaries, and the flow becomes turbulent, as in Fig. 9.17b.

Streamlines also indicate the relative magnitude of the velocity of a fluid. The velocity is greater where the streamlines are closer together. Notice this effect in Fig. 9.17a. The reason for it will be explained shortly.

Condition 2: *Irrotational flow* means that a fluid element (a small volume of the fluid) has no net angular velocity. This condition eliminates the possibility of whirlpools and eddy currents. (Nonturbulent flow.)

Consider the small paddle wheel in Fig. 9.17a. With a zero net torque, the wheel does not rotate. Thus, the flow is irrotational.

Condition 3: *Nonviscous flow* means that viscosity is negligible.

Viscosity refers to a fluid's internal friction, or resistance to flow. (For example, honey has a much greater viscosity than water.) A truly nonviscous fluid would flow freely with no internal energy loss. Also, there would be no frictional drag between the fluid and the walls containing it. In reality, when a liquid flows through a pipe, the speed is lower near the walls because of frictional drag and is higher toward the center of the pipe. (Viscosity is discussed in more detail in Section 9.5.)

Condition 4: *Incompressible flow* means that the fluid's density is constant.

Liquids can usually be considered incompressible. Gases, by contrast, are quite compressible. Sometimes, however, gases approximate incompressible flow—for example, air flowing relative to the wings of an airplane traveling at low speeds. Theoretical or ideal fluid flow is not characteristic of most real situations, but the analysis of ideal flow provides results that approximate, or generally describe, a variety of applications. Usually, this analysis is derived, not from Newton's laws, but instead from two basic principles: conservation of mass and conservation of energy.

EQUATION OF CONTINUITY

If there are no losses of fluid within a uniform tube, the mass of fluid flowing into the tube in a given time must be equal to the mass flowing out of the tube in the same time (by the conservation of mass). For example, in ▶Fig. 9.18a, the mass (Δm_1) entering the tube during a short time (Δt) is

$$\Delta m_1 = \rho_1 \Delta V_1 = \rho_1 (A_1 \Delta x_1) = \rho_1 (A_1 v_1 \Delta t)$$

where A_1 is the cross-sectional area of the tube at the entrance and, in a time Δt, a fluid particle moves a distance equal to $v_1 \Delta t$. Similarly, the mass leaving the tube in the same interval is (Fig. 9.18b)

$$\Delta m_2 = \rho_2 \Delta V_2 = \rho_2 (A_2 \Delta x_2) = \rho_2 (A_2 v_2 \Delta t)$$

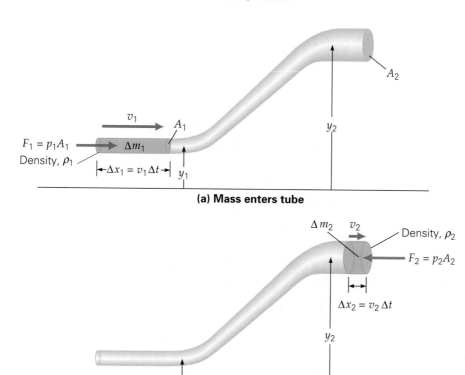

◀ **FIGURE 9.18 Flow continuity**
Ideal fluid flow can be described in
terms of the conservation of mass
by the equation of continuity. See
text for description.

(a) Mass enters tube

(b) Mass exits tube

Since the mass is conserved, $\Delta m_1 = \Delta m_2$, and it follows that

$$\rho_1 A_1 v_1 = \rho_2 A_2 v_2 \quad \text{or} \quad \rho A v = \text{constant} \qquad (9.16)$$

This general result is called the **equation of continuity**.

For an incompressible fluid, the density ρ is constant, so

$$A_1 v_1 = A_2 v_2 \quad \text{or} \quad A v = \text{constant} \quad \textit{(for an incompressible fluid)} \qquad (9.17)$$

This is sometimes called the **flow rate equation**. Av is called the *volume rate of flow*, and is the volume of fluid that passes by a point in the tube per unit time. (The units of Av are $\text{m}^2 \cdot \text{m/s} = \text{m}^3/\text{s}$, volume per time.)

Note that the flow rate equation shows that the fluid speed is greater where the cross-sectional area of the tube is smaller. That is,

$$v_2 = \left(\frac{A_1}{A_2}\right) v_1$$

and v_2 is greater than v_1 if A_2 is less than A_1. This effect is evident in the common experience that the speed of water is greater from a hose fitted with a nozzle than that from the same hose without a nozzle (▶ Fig. 9.19).

The flow rate equation can be applied to the flow of blood in your body. Blood flows from the heart into the aorta. It then makes a circuit through the circulatory system, passing through arteries, arterioles (small arteries), capillaries, and venules (small veins) and back to the heart through veins. The speed is lowest in the capillaries. Is this a contradiction? No: The *total* area of the capillaries is much larger than that of the arteries or veins, so the flow rate equation is still valid.

▲ **FIGURE 9.19 Flow rate** By the
flow rate equation, the speed of a
fluid is greater when the cross-
sectional area of the tube through
which the fluid is flowing is smaller.
Think of a hose that is equipped
with a nozzle such that the cross-
sectional area of the hose is made
smaller.

EXAMPLE 9.11 | Blood Flow: Cholesterol and Plaque

High cholesterol in the blood can cause fatty deposits called plaques to form on the walls of blood vessels. Suppose a plaque reduces the effective radius of an artery by 25%. How does this partial blockage affect the speed of blood through the artery?

THINKING IT THROUGH. The flow rate equation (Eq. 9.17) applies, but note that no values of area or speed are given. This indicates that we should use ratios.

(continued on next page)

SOLUTION. Taking the unclogged artery to have a radius r_1, that the plaque then reduces the effective radius to r_2.

Given: $r_2 = 0.75r_1$ (for a 25% reduction)   **Find:** v_2

Writing the flow rate equation in terms of the radii,

$$A_1v_1 = A_2v_2$$
$$(\pi r_1^2)v_1 = (\pi r_2^2)v_2$$

Rearranging and canceling,

$$v_2 = \left(\frac{r_1}{r_2}\right)^2 v_1$$

From the given information, $r_1/r_2 = 1/0.75$, so

$$v_2 = (1/0.75)^2 v_1 = 1.8v_1$$

Hence, the speed through the clogged artery increases by 80%.

FOLLOW-UP EXERCISE. By how much would the effective radius of an artery have to be reduced to have a 50% increase in the speed of the blood flowing through it?

EXAMPLE 9.12 | **Speed of Blood in the Aorta**

Blood flows at a rate of 5.00 L/min through an aorta with a radius of 1.00 cm. What is the speed of blood flow in the aorta?

THINKING IT THROUGH. It is noted that the flow rate is a volume flow rate, which implies the use of the flow rate equation (Eq. 9.17), Av = constant. Since the constant is in terms of volume/time, the given flow rate is the constant.

SOLUTION. Listing the data:

Given: Flow rate = 5.00 L/min   **Find:** v (blood
$r = 1.00$ cm $= 1.00 \times 10^{-2}$ m   speed)

Let's first find the cross-sectional area of the circular aorta.

$$A = \pi r^2 = (3.14)(1.00 \times 10^{-2}\,\text{m})^2 = 3.14 \times 10^{-4}\,\text{m}^2$$

Then the (volume) flow rate needs to be put into standard units.

$$5.00\,\text{L/min} = (5.00\,\text{L/min})(10^{-3}\,\text{m}^3/\text{L})(1\,\text{min}/60\,\text{s})$$
$$= 8.33 \times 10^{-5}\,\text{m}^3/\text{s}$$

Using the flow rate equation,

$$v = \frac{\text{constant}}{A} = \frac{8.33 \times 10^{-5}\,\text{m}^3/\text{s}}{3.14 \times 10^{-4}\,\text{m}^2} = 0.265\,\text{m/s}$$

FOLLOW-UP EXERCISE. Constrictions of the arteries occur with hardening of the arteries. If the radius of the aorta in this Example were constricted to 0.900 cm, what would be the percentage change in blood flow?

BERNOULLI'S EQUATION

The conservation of energy or the general work–energy theorem leads to another relationship that has great generality for fluid flow. This relationship was first derived in 1738 by the Swiss mathematician Daniel Bernoulli (1700–1782) and is named for him. Bernoulli's result was

$$W_{\text{net}} = \Delta K + \Delta U$$
$$\frac{\Delta m}{\rho}(p_1 - p_2) = \tfrac{1}{2}\Delta m(v_2^2 - v_1^2) + \Delta mg(y_2 - y_1)$$

where Δm is a mass increment as in the derivation of the continuity equation.

Note that in working with a fluid, the terms in Bernoulli's equation are work or energy per unit volume (J/m^3). That is, $W = F\Delta x = p(A\Delta x) = p\Delta V$ and therefore $p = W/\Delta V$ (work/volume). Similarly, with $\rho = m/V$, we have $\tfrac{1}{2}\rho v^2 = \tfrac{1}{2}mv^2/V$ (energy/volume) and $\rho gy = mgy/V$ (energy/volume).

Canceling each Δm and rearranging gives the common form of **Bernoulli's equation**:

$$p_1 + \tfrac{1}{2}\rho v_1^2 + \rho gy_1 = p_2 + \tfrac{1}{2}\rho v_2^2 + \rho gy_2 \tag{9.18}$$

or

$$p + \tfrac{1}{2}\rho v^2 + \rho gy = \text{constant}$$

Bernoulli's equation, or principle, can be applied to many situations. For example, for a fluid at rest ($v_2 = v_1 = 0$). Bernoulli's equation becomes

$$p_2 - p_1 = \rho g(y_1 - y_2)$$

This is the pressure–depth relationship derived earlier (Eq. 9.10). Also, if there is horizontal flow ($y_1 = y_2$), then $p + \tfrac{1}{2}\rho v^2$ = constant, which indicates that

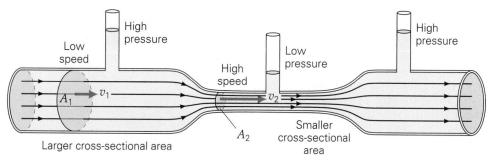

◀ FIGURE 9.20 Flow rate and pressure Taking the horizontal difference in flow heights to be negligible in a constricted pipe, we obtain, for Bernoulli's equation, $p + \frac{1}{2}\rho v^2$ = constant. In a region of smaller cross-sectional area, the flow speed is greater (see flow rate equation); from Bernoulli's equation, the pressure in that region is lower than in other regions.

◀ FIGURE 9.20 Flow rate and pressure Taking the horizontal difference in flow heights to be negligible in a constricted pipe, we obtain, for Bernoulli's equation, $p + \frac{1}{2}\rho v^2$ = constant. In a region of smaller cross-sectional area, the flow speed is greater (see flow rate equation); from Bernoulli's equation, the pressure in that region is lower than in other regions.

the pressure decreases if the speed of the fluid increases (and vice versa). This effect is illustrated in ▲Fig. 9.20, where the difference in flow heights through the pipe is considered negligible (so the $\rho g y$ term drops out).

Chimneys and smokestacks are tall in order to take advantage of the more consistent and higher wind speeds at greater heights. The faster the wind blows over the top of a chimney, the lower the pressure, and the greater the pressure difference between the bottom and top of the chimney. Thus, the chimney "draws" exhaust out more efficiently. Bernoulli's equation and the continuity equation (Av = constant) also tell you that if the cross-sectional area of a pipe is reduced so that the speed of the fluid passing through it is increased, then the pressure is reduced.

The Bernoulli effect (as it is sometimes called) gives a *simplistic* explanation for the lift of an airplane. Ideal airflow over an airfoil or wing is shown in ▶Fig. 9.21. (Turbulence is neglected.) The wing is curved on the top side and is angled relative to the incident streamlines. As a result, the streamlines above the wing are closer together than those below, which causes a higher air speed and lower pressure above the wing. With a higher pressure on the bottom of the wing, there is a net upward force, or *lift*.

This rather common explanation of lift is termed simplistic because Bernoulli's effect does not apply to the situation. Bernoulli's principle requires the conditions of both ideal fluid flow and energy conservation within the system, neither of which is satisfied in aircraft flying conditions. It is perhaps better to rely on Newton's laws, which always must be satisfied. Basically, the wing deflects the airflow downward, giving rise to a downward change in the airflow momentum and a downward force (Newton's second law). This results in an upward reaction force on the wing (Newton's third law). When this upward force exceeds the weight of the plane, there is enough lift for takeoff and flight.

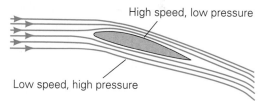

▲ FIGURE 9.21 Airplane lift—Bernoulli's principle in action Because of the shape and orientation of an airfoil or airplane wing, the air streamlines are closer together, and the air speed is greater above the wing than below it. By Bernoulli's principle, the resulting pressure difference supplies part of the upward force called the lift. (But, Bernoulli's principle is not applicable, see text.)

EXAMPLE 9.13 | Flow Rate from a Tank: Bernoulli's Equation

A cylindrical tank containing water has a small hole punched in its side below the water level, and water runs out (▶Fig. 9.22). What is the approximate initial flow rate of water out of the tank in terms of the heights shown?

THINKING IT THROUGH. Equation 9.17 ($A_1v_1 = A_2v_2$) is the flow rate equation, where Av has units of m^3/s, or volume/time. The v terms can be related by Bernoulli's equation, which also contains y, and can be used to find differences in height. The areas are not given, so relating the v terms might require some sort of approximation, as will be seen. (Note that the *approximate* initial flow rate is wanted.)

SOLUTION.

Given: No specific values are given, so symbols will be used.

Find: An expression for the approximate initial water flow rate from the hole

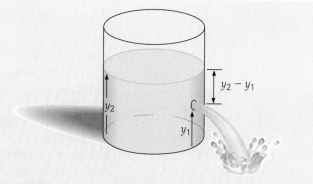

▲ FIGURE 9.22 Fluid flow from a tank The flow rate is given by Bernoulli's equation. See Example text for description.

(continued on next page)

Bernoulli's equation,

$$p_1 + \tfrac{1}{2}\rho v_1^2 + \rho g y_1 = p_2 + \tfrac{1}{2}\rho v_2^2 + \rho g y_2$$

can be used. Note that $y_2 - y_1$ is just the height of the surface of the liquid above the hole. The atmospheric pressures acting on the open surface and at the hole, p_1 and p_2, respectively, are essentially equal and cancel from the equation, as does the density, so

$$v_1^2 - v_2^2 = 2g(y_2 - y_1)$$

By the equation of continuity (the flow rate equation, Eq. 9.17), $A_1v_1 = A_2v_2$, where A_2 is the cross-sectional area of the

tank and A_1 is that of the hole. Since A_2 is much greater than A_1, then v_1 is much greater than v_2 (initially, $v_2 \approx 0$). So, to a good approximation,

$$v_1^2 = 2g(y_2 - y_1) \quad \text{or} \quad v_1 = \sqrt{2g(y_2 - y_1)}$$

The flow rate (volume/time) is then

$$\text{flow rate} = A_1v_1 = A_1\sqrt{2g(y_2 - y_1)}$$

Given the area of the hole and the height of the liquid above it, the initial speed of the water coming from the hole and the flow rate can be found. (What happens as the water level falls?)

FOLLOW-UP EXERCISE.   What would be the percentage change in the initial flow rate from the tank in this Example if the diameter of the small circular hole were increased by 30.0%?

CONCEPTUAL EXAMPLE 9.14 | A Stream of Water: Smaller and Smaller

You have probably observed that a steady stream of water flowing out of a kitchen faucet gets smaller the farther the water falls from the faucet. Why does that happen?

REASONING AND ANSWER.   This effect can be explained by Bernoulli's principle. As the water falls, it accelerates and its speed

increases. Then, by Bernoulli's principle, the liquid pressure inside the stream decreases. (See Fig. 9.20.) A pressure difference between that inside stream and the atmospheric pressure on the outside is thus created. As a result, there is an increasing inward force as the stream falls, so it becomes smaller. Eventually, the stream may get so thin that it breaks up into individual droplets.

FOLLOW-UP EXERCISE.   The equation of continuity can also be used to explain this stream effect. Give this explanation.

DID YOU LEARN

➥ Ideal fluid flow is steady, irrrotational, nonviscous, and incompressible.

➥ For an incompressible fluid, the volume flow rate (Av) is constant.

➥ Bernoulli's equation is based on the conservation of energy or the general work–energy theorem.

---

*9.5   Surface Tension, Viscosity, and Poiseuille's Law

LEARNING PATH QUESTIONS

➥ What is surface tension?

➥ What is viscosity and how does it arise?

➥ In Poiseuille's law, what is the effect of the radius of the flow tube?

SURFACE TENSION

The molecules of a liquid exert small attractive forces on each other. Even though molecules are electrically neutral overall, there is often some slight asymmetry of charge that gives rise to attractive forces between them (called *van der Waals forces*).* Within a liquid, any molecule is completely surrounded by other molecules, and the net force is zero (▸Fig. 9.23a). However, for molecules at the surface of the liquid, there is no attractive force acting from above the surface. (The effect of air molecules is small and considered negligible.) As a result, net forces act upon the molecules of the surface layer, due to the attraction of neighboring molecules just below the surface. This inward pull on the surface molecules causes the surface of the liquid to contract and to resist being stretched or broken, a property called **surface tension**.

If a sewing needle is carefully placed on the surface of a bowl of water, the surface acts like an elastic membrane under tension. There is a slight depression in

---

*After Johannes van der Waals (1837–1923), a Dutch scientist who first postulated an intermolecular force.

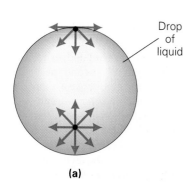

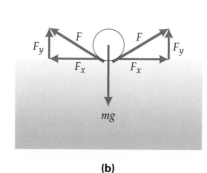

(a)   (b)   (c)

▲ **FIGURE 9.23 Surface tension (a)** The net force on a molecule in the interior of a liquid is zero, because the molecule is surrounded by other molecules. However, a nonzero fluid force acts on a molecule at the surface, due to the attractive forces of the neighboring molecules just below the surface. **(b)** For an object such as a needle to form a depression on the surface, work must be done, since more interior molecules must be brought to the surface to increase its area. As a result, the surface area acts like a stretched elastic membrane, and the weight of the object is supported by the upward components of the surface tension. **(c)** Insects such as this water strider can walk on water because of the upward components of the surface tension, much as you might walk on a large trampoline. Note the depressions in the surface of the liquid where the legs touch it.

the surface, and molecular forces along the depression act at an angle to the surface (Fig. 9.23b). The vertical components of these forces balance the weight (mg) of the needle, and the needle "floats" on the surface. Similarly, surface tension supports the weight of a water strider (Fig. 9.23c).

The net effect of surface tension is to make the surface area of a liquid as small as possible. That is, a given volume of liquid tends to assume the shape that has the least surface area. As a result, drops of water and soap bubbles have spherical shapes, because a sphere has the smallest surface area for a given volume (▼Fig. 9.24). In forming a drop or bubble, surface tension pulls the molecules together to minimize the surface area. (See Insight 9.4, The Lungs and Baby's First Breath for an example of surface tension in respiration.)

VISCOSITY

All real fluids have an internal resistance to flow, or **viscosity**, which can be considered to be friction between the molecules of a fluid. In liquids, viscosity is caused by short-range cohesive forces, and in gases, it is caused by collisions between molecules. (See the discussion of air resistance in Section 4.6.) The viscous drag for both liquids and gases depends on their speeds and may be directly proportional to it in some cases. However, the relationship varies with the conditions; for example, the drag is approximately proportional to either v^2 or v^3 in turbulent flow.

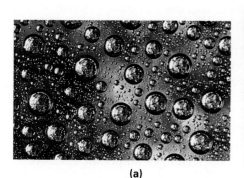

(a)   (b)

◄ **FIGURE 9.24 Surface tension at work** Because of surface tension, **(a)** water droplets and **(b)** soap bubbles tend to assume the shape that minimizes their surface area—that of a sphere.

INSIGHT 9.4 | **The Lungs and Baby's First Breath**

Respiration, or breathing, is vital to life. It is a fascinating procedure that supplies oxygen to the blood and carries away carbon dioxide—and a lot of physics is involved.

The process of respiration involves the lowering of the diaphragm to increase the volume of the thoracic cavity. Figure 1 shows a bell jar model of respiration. By the ideal gas law (Section 10.3), the lowering of the diaphragm and the increasing of the volume of the thoracic cavity lower the pressure ($p \propto 1/V$), and air is inhaled. The inhalation process inflates the alveoli—small balloonlike structures in the lungs, as illustrated in Fig. 2a. (Figure 2b shows an illustration of a damaged lung, the cause and effects of which will be discussed shortly.)

The oxygen exchange with the blood takes place across the membrane surfaces of the alveoli. The total membrane surface in the lungs is on the order of 100 m^2, with a thickness of less than a millionth of a meter ($<$1 μm, micrometer), making the gas exchange very efficient. The behavior of the alveoli may be described by Laplace's law and surface tension.*

Laplace's law states that the larger a spherical membrane, the greater the wall tension required to withstand the pres-

*Pierre-Simon de Laplace (1749–1827) was a French astronomer and mathematician.

sure of an internal fluid. That is, the wall tension is directly proportional to the spherical radius. So when the alveoli inflate, there is greater tension. Once they are inflated, exhalation is accomplished when the diaphragm relaxes and the wall tension of the alveoli acts to force the air out. Also, there is a fluid coating on the alveoli, which is a *surfactant*—a substance that lowers surface tension. A reduction in surface tension makes it easier to inflate the alveoli on inhalation.

The pulmonary disease *emphysema*, most common in long-term smokers, results from an enlargement of the alveoli as some are destroyed and others either enlarge or combine (Fig. 2b). Normally it would take twice the pressure to inflate a membrane with twice the radius. The enlarged alveoli provide less recoil on exhalation, and a person with emphysema has difficulty breathing as well as reduced oxygen exchange.

Now, about a baby's first breath. Most everyone knows that it is much more difficult to blow up a balloon for the first time than to blow it up again. This is because the applied pressure does not create much tension in the balloon to start the stretching process. It takes a greater tension increase to expand a small balloon than to expand a large balloon. Consider the tension ratios for a 3-cm expansion in radius—say, from 1 cm to 4 cm $\left(\frac{4}{1} = 4\right)$ and 10 cm to 13 cm $\left(\frac{13}{10} = 1.3\right)$.

In a newborn baby, the alveoli are small and collapsed and must be inflated with an initial inhalation. The traditional practice to accomplish this are slaps on the baby's bottom to make the newborn cry and inhale.

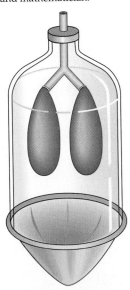 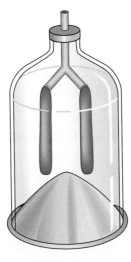

(a) Inhalation   **(b) Exhalation**

▲ FIGURE 1 **Bell jar model of respiration** **(a)** Lowering the diaphragm (rubber sheet) and increasing the volume of the thoracic cavity lowers the pressure and air is inhaled into the lungs (balloons). **(b)** When the diaphragm moves upward, the process is reversed and air is exhaled.

 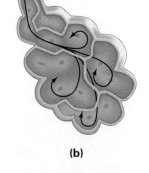

(a)   **(b)**

▲ FIGURE 2 **Alveoli** **(a)** Inhalation inflates the alveoli, balloonlike structures of the lungs. There are between 300 million and 400 million alveoli in each lung. **(b)** Pulmonary disease can cause the enlargement of the alveoli as some are destroyed and others enlarge or combine. As a result, there is less oxygen exchange and a shortness of breath.

Internal friction causes the layers of a fluid to move relative to each other in response to a shear stress. This layered motion, called *laminar flow*, is characteristic of steady flow for viscous liquids at low velocities (▶Fig. 9.25a). At higher velocities, the flow becomes rotational, or *turbulent*, and difficult to analyze.

Since there are shear stresses and shear strains (deformation) in laminar flow, the viscous property of a fluid can be described by a coefficient, like the elastic moduli discussed in Section 9.1. Viscosity is characterized by a *coefficient of viscosity*, η (the Greek letter eta), commonly referred to as simply the viscosity.

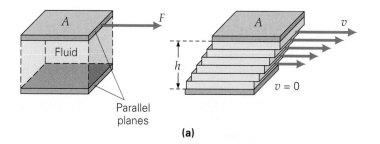

(a)

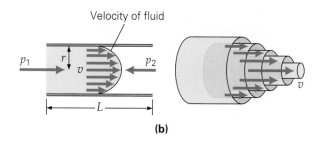

Velocity of fluid

(b)

◀ **FIGURE 9.25 Laminar flow**
(a) A shear stress causes layers of a fluid to move over each other in laminar flow. The shear force and the flow rate depend on the viscosity of the fluid. **(b)** For laminar flow through a pipe, the speed of the fluid is less near the walls of the pipe than near the center because of frictional drag between the walls and the fluid.

The coefficient of viscosity is, in effect, the ratio of the shear stress to the rate of change of the shear strain (since motion is involved). Unit analysis shows that the SI unit of viscosity is the pascal-second (Pa·s) This combined unit is called the *poiseuille* (Pl), in honor of the French scientist Jean Poiseuille (1797–1869), who studied the flow of liquids, particularly blood. (Poiseuille's law on flow rate will be presented shortly.) The cgs unit of viscosity is the *poise* (P). A smaller multiple, the *centipoise* (cP), is widely used because of its convenient size; $1 \text{ P} = 10^2 \text{ cP}$.

The viscosities of some fluids are listed in ▶ Table 9.3. The greater the viscosity of a liquid, which is easier to visualize than that of a gas, the greater the shear stress required to get the layers of the liquid to slide along each other. Note, for example, the large viscosity of glycerin compared to that of water.*

As you might expect, viscosity, and thus fluid flow, varies with temperature, which is evident from the old saying, "slow as molasses in January." A familiar application is the viscosity grading of motor oil used in automobiles. In winter, a low-viscosity, or relatively thin, oil should be used (such as SAE grade 10W or 20W), because it will flow more readily, particularly when the engine is cold at startup. In summer, a higher viscosity, or thicker, oil is used (SAE 30, 40, or even 50).†

Seasonal changes in the grade of motor oil are not necessary if you use the multigrade, year-round oils. These oils contain additives called viscosity improvers, which are polymers whose molecules are long, coiled chains. An increase in temperature causes the molecules to uncoil and intertwine. Thus, the normal decrease in viscosity is counteracted. The action is reversed on cooling, and the oil maintains a relatively small viscosity range over a large temperature range. Such motor oils are graded, for example, as SAE 10W-30 ("ten-W-thirty").

TABLE 9.3   Viscosities of Various Fluids*

Fluid	Viscosity (η) [Poiseuille (Pl)]
Liquids	
Alcohol, ethyl	1.2×10^{-3}
Blood, whole (37 °C)	1.7×10^{-3}
Blood plasma (37 °C)	2.5×10^{-3}
Glycerin	1.5×10^{-3}
Mercury	1.55×10^{-3}
Oil, light machine	1.1
Water	1.00×10^{-3}
Gases	
Air	1.9×10^{-5}
Oxygen	2.2×10^{-5}

*At 20 °C unless otherwise indicated.

*If you want to think about a substance with a very large viscosity, consider glass. It has been said that the glass in the stained glass windows of medieval churches has "flowed" over time, such that the panes are now thicker at the bottom than at the top. However a more recent analysis indicates that window glass may flow over incredibly long periods that exceed the limits of human history. On human time scales, such a flow would not be evident. [See E. D. Zanotto, *American Journal of Physics*, 66 (May 1998), 392–395.]

†*SAE* stands for *Society of Automotive Engineers*, an organization that designates the grades of motor oils based on their viscosity.

POISEUILLE'S LAW

Viscosity makes analyzing fluid flow difficult. For example, when a fluid flows through a pipe, there is frictional drag between the liquid and the walls, and the fluid speed is greater toward the center of the pipe (Fig. 9.25b). In practice, this effect makes a difference in a fluid's *average flow rate* $Q = A\bar{v} = \Delta V/\Delta t$ (see Eq. 9.17), which describes the volume (ΔV) of fluid flowing past a given point during a time Δt. The SI unit of flow rate is cubic meters per second (m^3/s). The flow rate depends on the properties of the fluid and the dimensions of the pipe, as well as on the pressure difference (Δp) between the ends of the pipe.

Jean Poiseuille studied flow in pipes and tubes, assuming a constant viscosity and steady or laminar flow. He derived the following relationship, known as **Poiseuille's law**, for the flow rate:

$$Q = \frac{\Delta V}{\Delta t} = \frac{\pi r^4 \Delta p}{8\eta L} \tag{9.19}$$

Here, r is the radius of the pipe and L is its length.

As expected, the flow rate is inversely proportional to the viscosity (η) and the length of the pipe. Also as expected, the flow rate is directly proportional to the pressure difference Δp between the ends of the pipe. Somewhat surprisingly, however, the flow rate is proportional to r^4, which makes it more highly dependent on the radius of the tube than might have been thought.

An application of fluid flow in a medical IV was examined in Example 9.6. However, Poiseuille's law, which incorporates the flow rate, affords more reality to this application, as the next Example shows.

EXAMPLE 9.15 | ### Poiseuille's Law: A Blood Transfusion

A hospital patient needs a blood IV (intravenous) transfusion, which will be administered through a vein in the arm via a gravity IV. The physician wishes to have 500 cc of whole blood delivered over a period of 10 min by an 18-gauge needle with a length of 50 mm and an inner diameter of 1.0 mm. At what height above the arm should the bag of blood be hung? (Assume a venous blood pressure of 15 mm Hg.)

THINKING IT THROUGH. This is an application of Poiseuille's law (Eq. 9.19) to find the pressure needed at the inlet of the needle that will provide the required flow rate (Q). Note that $\Delta p = p_{in} - p_{out}$ (inlet pressure minus outlet pressure). Knowing the inlet pressure, the required height of the bag can be found, as in Example 9.6. (*Caution*: There are a lot of nonstandard units here, and some quantities are assumed to be known from tables.)

SOLUTION.   First writing the given (and known) quantities and converting to standard SI units:

Given: $\Delta V = 500 \text{ cc} = 500 \text{ cm}^3 \, (1 \text{ m}^3/10^6 \text{cm}^3)$ *Find:* h (height of bag)

$\qquad\qquad = 5.00 \times 10^{-4} \text{ m}^3$

$\qquad \Delta t = 10 \text{ min} = 600 \text{ s} = 6.00 \times 10^2 \text{ s}$

$\qquad L = 50 \text{ mm} = 5.0 \times 10^{-2} \text{ m}$

$\qquad d = 1.0 \text{ mm, or } r = 0.50 \text{ mm} = 5.0 \times 10^{-4} \text{ m}$

$\qquad p_{out} = 15 \text{ mm Hg} = 15 \text{ torr } (133 \text{ Pa/torr}) = 2.0 \times 10^3 \text{ Pa}$

$\qquad \eta = 1.7 \times 10^{-3} \text{ Pl (whole blood, from Table 9.3)}$

The flow rate is

$$Q = \frac{\Delta V}{\Delta t} = \frac{5.00 \times 10^{-4} \text{ m}^3}{6.00 \times 10^2 \text{ s}} = 8.33 \times 10^{-7} \text{ m}^3/s$$

Inserting this number into Eq. 9.19 and solving for Δp:

$$\Delta p = \frac{8\eta L Q}{\pi r^4} = \frac{8(1.7 \times 10^{-3} \text{ Pl})(5.0 \times 10^{-2} \text{ m})(8.33 \times 10^{-7} \text{ m}^3/s)}{\pi(5.0 \times 10^{-4} \text{ m})^4} = 2.9 \times 10^3 \text{ Pa}$$

With $\Delta p = p_{in} - p_{out}$,

$$p_{in} = \Delta p + p_{out} = (2.9 \times 10^3 \text{ Pa}) + (2.0 \times 10^3 \text{ Pa}) = 4.9 \times 10^3 \text{ Pa}$$

Then, to find the height of the bag that will deliver this amount of pressure, we use $p_{in} = \rho g h$ (where $\rho_{whole\ blood} = 1.05 \times 10^3 \text{ kg/m}^3$ from Table 9.2). Thus,

$$h = \frac{p_{in}}{\rho g} = \frac{4.9 \times 10^3 \text{ Pa}}{(1.05 \times 10^3 \text{ kg/m}^3)(9.80 \text{ m/s}^2)} = 0.48 \text{ m}$$

Hence, for the prescribed flow rate, the bag of blood should be hung about 48 cm above the needle in the arm.

FOLLOW-UP EXERCISE.   Suppose the physician wants to follow up the blood transfusion with 500 cc of saline solution at the same rate of flow. At what height should the saline bag be placed? (The *isotonic* saline solution administered by IV is a 0.85% aqueous salt solution, which has the same salt concentration as do body cells. To a good approximation, saline has the same density as water.)

Gravity flow IVs are still used, but with modern technology, the flow rates of IVs are now often controlled and monitored by machines (►Fig. 9.26).

DID YOU LEARN?

➥ Surface tension arises from the inward pull on the surface molecules of a liquid, which causes the surface to contract and resist being stretched or broken.

➥ Viscosity, the internal resistance to fluid flow, is caused by short-range cohesive forces between molecules in a liquid and by molecular collisions in a gas.

➥ Poiseuille's law indicates the flow rate in a pipe or tube depends highly on the radius, r^4.

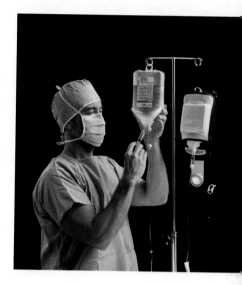

►**FIGURE 9.26 IV technology**
The mechanism of intravenous injection is still a gravity assist, but IV flow rates are now commonly controlled and monitored by machines.

PULLING IT TOGETHER | ## Sunken Treasure

A Spanish galleon is about to be boarded by bloodthirsty pirates in the shallows of a Caribbean island. To save a box of treasure on board, the captain orders his crew to secretly toss the box overboard, planning to come back for it later. The rectangular box is waterproof and measures 40.0 cm by 25.0 cm by 30.0 cm. It is made of wood and has mostly gold pieces inside, resulting in an average box density three times that of seawater.

Sinking below the surface, the box moves at a constant vertical velocity of 1.15 m/s for 12.0 m (that's 2 fathoms for pirates) before hitting the bottom. (a) Draw the free-body diagram for the box, (b) determine the magnitudes of the forces on the box, and (c) calculate the work done by each force and the net work done on the box. (d) Calculate the change in the box's gravitational potential energy. (e) What is the change in the box's total energy and what happens to it?

THINKING IT THROUGH.   This example uses the concepts of Newton's first law, buoyant force, and density, along with work and energy. (a) Constant velocity means that the box has no net force on it. The free-body diagram should show this. There is an upward buoyant force and a downward pull of gravity (weight). Since the box sinks with a constant velocity, there must be a third upward force to make the net force zero. (b) The box's weight can be found from the volume and density, the buoyant force from Archimedes' principle, and the water drag force is the difference. (c) All the forces are constant; thus, work can be determined by using the definition in Section 5.1. (d) Change in gravitational potential energy is discussed in Section 5.4. (e) The box's kinetic energy is constant; thus, by the work–energy theorem, the net work will be zero.

SOLUTION.

Given: $l \times w \times h = 40.0 \text{ cm} \times 25.0 \text{ cm} \times 30.0 \text{ cm}$
$\qquad\qquad = 0.400 \text{ m} \times 0.250 \text{ m} \times 0.300 \text{ m}$ (box dimensions)
$\quad \rho = 3\rho_{sw} = 3.09 \times 10^3 \text{ kg/m}^3$ (density; ρ_{sw} from Table 9.2)
$\quad v = 1.15 \text{ m/s}$ downward (box's velocity)
$\quad \Delta y = -12.0 \text{ m}, d = 12.0$ (box's vertical movement)

Find: (a) free-body diagram
(b) w_{box}, F_b, F_{drag} (forces on box)
(c) W_{grav}, W_b, W_{drag} (work done by each force)
(d) ΔU_g (change in potential energy)
(e) ΔE (change in total energy and what happened to it)

(continued on next page)

(a) Since the box's acceleration is zero, the net force on it must be zero. The upward buoyant force (F_b) is less than the downward pull of gravity (weight). Thus there must be a water (fluid) drag force (F_{drag}) upward to help cancel the weight force. See ▶ Fig. 9.27.

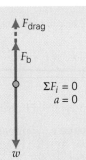

(b) The weight of the box depends on its volume and density. Its volume is

$$V_{box} = l \times w \times h = (0.400 \text{ m})(0.250 \text{ m})(0.300 \text{ m}) = 3.00 \times 10^{-2} \text{ m}^3$$

Thus its mass is

$$m_{box} = \bar{\rho}_{box} V_{box} = (3.09 \times 10^3 \text{ kg/m}^3)(3.00 \times 10^{-2} \text{ m}^3) = 92.7 \text{ kg}$$

and its weight is

$$w_{box} = m_{box} g = (92.7 \text{ kg})(9.80 \text{ m/s}^2) = 908 \text{ N}$$

The buoyant force, by Archimedes's principle, is equal in magnitude to the weight of the seawater displaced. Since the box is fully submerged, the volume of displaced seawater is the same as the volume of the box. Therefore,

$$F_b = w_{sw} = \rho_{sw} V_{sw} g = (1.03 \times 10^3 \text{ kg/m}^3)(3.00 \times 10^{-2} \text{ m}^3)(9.80 \text{ m/s}^2) = 303 \text{ N}$$

From the free-body diagram, the upward fluid drag force is the difference between these two forces, or

$$F_{drag} = w_{box} - w_{sw} = 908 \text{ N} - 303 \text{ N} = 605 \text{ N}$$

▲ **FIGURE 9.27 Free-body diagram for the sinking chest** The chest sinks with a constant velocity so the sum of the forces on it is zero. See Example text for description.

(c) The work done by a constant force is given by $W = Fd \cos \theta$, where θ is the angle between the displacement and the force direction (Section 5.1). The weight is in the same direction as that of the box's displacement, so $\theta = 0°$ and

$$W_{grav} = Fd \cos \theta = (908 \text{ N})(12.0 \text{ m}) \cos 0° = +1.09 \times 10^4 \text{ J}$$

Both the buoyant force and the fluid drag force will do negative work because they are exactly opposite the direction of the displacement. Hence the work done by the buoyancy force is

$$W_{buoy} = Fd \cos \theta = (303 \text{ N})(12.0 \text{ m}) \cos 180° = -3.64 \times 10^3 \text{ J}$$

and the work done by the fluid drag force is

$$W_{drag} = Fd \cos \theta = (605 \text{ N})(12.0 \text{ m}) \cos 180° = -7.26 \times 10^3 \text{ J}$$

The net work done on the box is zero, consistent with the work–energy theorem, since its kinetic energy does not change.

$$W_{net} = \sum_i W_i = W_{grav} + W_{buoy} + W_{drag} = 1.09 \times 10^4 \text{ J} - 3.64 \times 10^3 \text{ J} - 7.27 \times 10^3 \text{ J} = 0$$

(For the operation, the 1.09×10^4 J is converted to 10.9×10^3 J. Why?)

(d) The change in the box's gravitational potential energy is

$$\Delta U_g = m_{box} g \Delta y = (92.7 \text{ kg})(9.80 \text{ m/s}^2)(-12.0 \text{ m}) = -1.09 \times 10^4 \text{ J}$$

(e) The box's kinetic energy does not change, but its potential energy decreases, thus its total energy decreases, since

$$\Delta E = \Delta K + \Delta U = 0 + (-1.09 \times 10^4 \text{ J}) = -1.09 \times 10^4 \text{ J}$$

This energy is gained by the seawater in the form of increased thermal energy (it is slightly warmed) and turbulence (kinetic energy of the water).

Learning Path Review

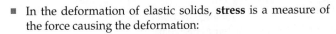

■ In the deformation of elastic solids, **stress** is a measure of the force causing the deformation:

$$\text{stress} = \frac{F}{A} \quad (9.1)$$

Strain is a relative measure of the deformation a stress causes:

$$\text{strain} = \frac{\text{change in length}}{\text{original length}} = \frac{\Delta L}{L_o} = \frac{|L - L_o|}{L_o} \quad (9.2)$$

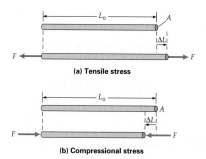

(a) Tensile stress

(b) Compressional stress

■ An **elastic modulus** is the ratio of stress to strain.

Young's modulus:

$$Y = \frac{F/A}{\Delta L/L_o} \qquad (9.4)$$

Shear modulus:

$$S = \frac{F/A}{x/h} \approx \frac{F/A}{\phi} \qquad (9.5)$$

Before

Before

After

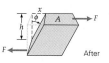

After

Bulk modulus:

$$B = \frac{F/A}{-\Delta V/V_o} = -\frac{\Delta p}{\Delta V/V_o} \qquad (9.6)$$

■ Pressure is the force per unit area:

$$p = \frac{F}{A} \qquad (9.8a)$$

Pressure–depth relationship (for an incompressible fluid at constant density):

$$p = p_o + \rho g h \qquad (9.10)$$

■ **Pascal's principle.** Pressure applied to an enclosed fluid is transmitted undiminished to every point in the fluid and to the walls of the container.

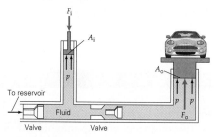

■ **Archimedes' principle.** A body immersed wholly or partially in a fluid is buoyed up by a force equal in magnitude to the weight of the volume of fluid displaced.

Buoyant force:

$$F_b = m_f g = \rho_f g V_f \qquad (9.14)$$

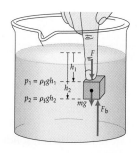

$p_1 = \rho_f g h_1$

$p_2 = \rho_f g h_2$

■ An object will float in a fluid if the average density of the object is less than the density of the fluid. If the average density of the object is greater than the density of the fluid, the object will sink.

■ For an ideal fluid, the flow is (1) steady, (2) irrotational, (3) nonviscous, and (4) incompressible. The following equations describe such a flow:

Equation of continuity:

$$\rho_1 A_1 v_1 = \rho_2 A_2 v_2 \quad \text{or} \quad \rho A v = \text{constant} \qquad (9.16)$$

Flow rate equation (for an incompressible fluid):

$$A_1 v_1 = A_2 v_2 \quad \text{or} \quad A v = \text{constant} \qquad (9.17)$$

Bernoulli's equation (for an incompressible fluid):

$$p_1 + \tfrac{1}{2}\rho v_1^2 + \rho g y_1 = p_2 + \tfrac{1}{2}\rho v_2^2 + \rho g y_2$$

or

$$p + \tfrac{1}{2}\rho v^2 + \rho g y = \text{constant} \qquad (9.18)$$

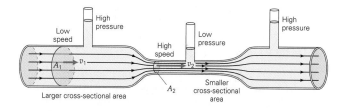

■ Bernoulli's equation is a statement of the conservation of energy for a fluid.

■ **Surface tension:** The inward pull on the surface molecules of a liquid that causes the surface to contract and resist being stretched or broken.

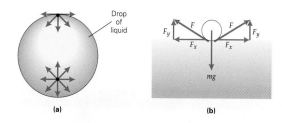

(a)

(b)

■ **Viscosity:** A fluid's internal resistance to flow. All real fluids have a nonzero viscosity.

■ **Poiseuille's law** (flow rate in pipes and tubes for fluids with constant viscosity and steady or laminar flow):

$$Q = \frac{\Delta V}{\Delta t} = \frac{\pi r^4 \Delta p}{8 \eta L} \qquad (9.19)$$

Learning Path Questions and Exercises

MULTIPLE CHOICE QUESTIONS

9.1 SOLIDS AND ELASTIC MODULI

1. The pressure on an elastic body is described by (a) a modulus, (b) work, (c) stress, (d) strain.
2. Shear moduli are not zero for (a) solids, (b) liquids, (c) gases, (d) all of these.
3. A relative measure of deformation is (a) a modulus, (b) work, (c) stress, (d) strain.
4. The volume stress for the bulk modulus is (a) Δp, (b) ΔV, (c) V_o, (d) $\Delta V/V_o$.

9.2 FLUIDS: PRESSURE AND PASCAL'S PRINCIPLE

5. For a liquid in an open container, the total pressure at any depth depends on (a) atmospheric pressure, (b) liquid density, (c) acceleration due to gravity, (d) all of the preceding.
6. For the pressure–depth relationship for a fluid ($p = \rho g h$), it is assumed that (a) the pressure decreases with depth, (b) a pressure difference depends on the reference point, (c) the fluid density is constant, (d) the relationship applies only to liquids.
7. When measuring automobile tire pressure, what type of pressure is this: (a) gauge, (b) absolute, (c) relative, or (d) all of the preceding?

9.3 BUOYANCY AND ARCHIMEDES' PRINCIPLE

8. A wood block floats in a swimming pool. The buoyant force exerted on the block by water depends on (a) the volume of water in the pool, (b) the volume of the wood block, (c) the volume of the wood block under water, (d) all of the preceding
9. If a submerged object displaces an amount of liquid of greater weight than its own and is then released, the object will (a) rise to the surface and float, (b) sink, (c) remain in equilibrium at its submerged position.
10. A rock is thrown into a lake. While sinking, the buoyant force (a) is zero, (b) decreases, (c) increases, (d) remains constant.

11. A glass containing an ice cube is filled to the brim and the cube floats on the surface. When the ice cube melts, (a) water will spill over the sides of the glass, (b) the water level decreases, (c) the water level is at the top of the glass without any spill.
12. Comparing an object's average density (ρ_o) to that of a fluid (ρ_f). what is the condition for the object to float: (a) $\rho_o < \rho_f$, or (b) $\rho_f < \rho_o$?
13. A block of material of known density (ρ_b) floats two-thirds submerged in a liquid of unknown density (ρ_o). Using Archimedes' principle, the unknown liquid density is (a) $\rho_u = \frac{2}{3}\rho_b$, (b) $\rho_u = \frac{3}{2}\rho_b$, (c) $\rho_u = \frac{1}{3}\rho_b$, (d) $\rho_u = 3\rho_b$.

9.4 FLUID DYNAMICS AND BERNOULLI'S EQUATION

14. If the speed at some point in a fluid changes with time, the fluid flow is *not* (a) steady, (b) irrotational, (c) incompressible, (d) nonviscous.
15. An ideal fluid is not (a) steady, (b) compressible, (c) irrotational, (d) nonviscous.
16. Bernoulli's equation is based primarily on (a) Newton's laws, (b) conservation of momentum, (c) a nonideal fluid, (d) conservation of energy.
17. According to Bernoulli's equation, if the pressure on the liquid in Fig. 9.20 is increased, (a) the flow speed always increases, (b) the height of the liquid always increases, (c) both the flow speed and the height of the liquid may increase, (d) none of the preceding.

*9.5 SURFACE TENSION, VISCOSITY, AND POISEUILLE'S LAW

18. Water droplets and soap bubbles tend to assume the shape of a sphere. This effect is due to (a) viscosity, (b) surface tension, (c) laminar flow, (d) none of the preceding.
19. Some insects can walk on water because (a) the density of water is greater than that of the insect, (b) water is viscous, (c) water has surface tension, (d) none of the preceding.
20. The viscosity of a fluid is due to (a) forces causing friction between the molecules, (b) surface tension, (c) density, (d) none of the preceding.

CONCEPTUAL QUESTIONS

9.1 SOLIDS AND ELASTIC MODULI

1. Which has a greater Young's modulus, a steel wire or a rubber band? Explain.
2. Why are scissors sometimes called shears? Is this a descriptive name in the physical sense?

3. Ancient stonemasons sometimes split huge blocks of rock by inserting wooden pegs into holes drilled in the rock and then pouring water on the pegs. Can you explain the physics that underlies this technique? [*Hint*: Think about sponges and paper towels.]

9.2 FLUIDS: PRESSURE AND PASCAL'S PRINCIPLE

4. ▼Figure 9.28 shows a famous "bed of nails" trick. The woman lies on a bed of nails with a cinder block on her chest. A person hits the anvil with a sledgehammer. The nails do not pierce the woman's skin. Explain why.

▲ FIGURE 9.28 **A bed of nails** See Conceptual Question 4.

5. Automobile tires are inflated to about 30 lb/in², whereas thin bicycle tires are inflated to 90 to 115 lb/in²—at least three times as much pressure! Why?

6. (a) Why is blood pressure usually measured at the arm? (b) Suppose the pressure reading were taken on the calf of the leg of a standing person. Would there be a difference, in principle? Explain.

7. What kind of pressure does a sphygmomanometer measure?

8. What is the principle of drinking through a straw? (Liquids aren't "sucked" up.)

9. What is the absolute pressure inside a flat tire?

10. (a) Two dams form artificial lakes of equal depth. However, one lake backs up 15 km behind the dam, and the other backs up 50 km behind. What effect does the difference in length have on the pressures on the dams? (b) Dams are usually thicker at the bottom. Why?

11. Water towers (storage tanks) are generally bulb shaped, as shown in ▼Fig. 9.29. Wouldn't it be better to have a cylindrical storage tank of the same height? Explain.

◄FIGURE 9.29 **Why a bulb-shaped water tower?** See Conceptual Question 11.

12. A water dispenser for pets contains an inverted plastic bottle, as shown in ▼Fig. 9.30. (The water is dyed blue for contrast.) When a certain amount of water is drunk from the bowl, more water flows automatically from the bottle into the bowl. The bowl never overflows. Explain the operation of the dispenser. Does the height of the water in the bottle depend on the surface area of the water in the bowl?

◄FIGURE 9.30 **Pet barometer** See Conceptual Question 12.

9.3 BUOYANCY AND ARCHIMEDES' PRINCIPLE

13. (a) What is the most important factor in constructing a life jacket that will keep a person afloat? (b) Why is it so easy to float in Utah's Great Salt Lake?

14. An ice cube floats in a glass of water. As the ice melts, how does the level of the water in the glass change? Would it make any difference if the ice cube were hollow? Explain.

15. Ocean-going ships in port are loaded to the so-called *Plimsoll mark*, which is a line indicating the maximum safe loading depth. However, in New Orleans, located at the mouth of the Mississippi River, where the water is brackish (partly salty and partly fresh), ships are loaded until the Plimsoll mark is somewhat below the water line. Why?

16. A heavy object is dropped into a lake. As it descends below the surface, does the pressure on it increase? Does the buoyant force on the object increase?

17. Ocean liners weigh thousands of tons. How are they made to float?

18. Two blocks of equal volume, one iron and one aluminum, are dropped into a body of water. Which block will experience the greater buoyant force? Why?

19. An inventor comes up with an idea for a perpetual motion machine, as illustrated in ▼Fig. 9.31. It contains a

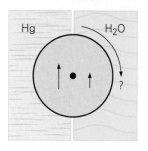

◄FIGURE 9.31 **Perpetual motion?** See Conceptual Question 19.

sealed chamber with mercury (Hg) in one half and water (H_2O) in the other. A cylinder is mounted in the center and is free to rotate. The inventor reasons that since mercury is much denser than water (13.6 g/cm^3 to 1.00 g/cm^3), the weight of the mercury displaced by half the cylinder is much greater than the water displaced by the other half.

Therefore, the buoyant force on the mercury side is greater than that on the water side—more than thirteen times greater. The difference in forces and torques should cause the cylinder to rotate—perpetually. Would you invest any money in this invention? Why or why not?

9.4 FLUID DYNAMICS AND BERNOULLI'S EQUATION

20. The speed of blood flow is greater in arteries than in capillaries. However, the flow rate equation (Av = constant) seems to predict that the speed should be greater in the smaller capillaries. Can you resolve this apparent inconsistency?

21. When driving your car on an interstate at the posted speed limit (of course) and an 18-wheeler quickly passes you going in the opposite direction, you feel an force toward the truck. Why is this?

22. A pump spray bottle or "atomizer" operates by the Bernoulli principle. Explain how this works.

23. Whea a large on-coming truck passes you on a highway, you may feel your car sway toward the truck. Why is this?

24. (a) If an Indy racer had a flat bottom, it would be highly unstable (like an airplane wing) due to the lift it gets when it moves at a high speed. To increase friction and stability, the bottom has a concave section called the *Venturi tunnel* (▼Fig. 9.32). (a) In terms of Bernoulli's

equation, explain how this concavity supplies extra downward force to the car in addition to that supplied by the front and rear wings. (b) What is the purpose of the "spoiler" on the back of the racer?

25. Here are two common demonstrations of Bernoulli effects: (a) If you hold a narrow strip of paper in front of your mouth and blow over the top surface, the strip will rise (▼Fig. 9.33a). (Try it.) Why? (b) A plastic egg is supported vertically by a stream of air from a tube (Fig. 9.33b). The egg will not move away from the midstream position. Why not?

(a)   **(b)**

▲ FIGURE 9.33 **Bernoulli effects** See Conceptual Question 25.

*9.5 SURFACE TENSION, VISCOSITY, AND POISEUILLE'S LAW

26. A motor oil is labeled 10W-40. What do the numbers 10 and 40 measure? How about the W?

27. Why are clothes generally washed in hot water and with detergent?

▲ FIGURE 9.32 **Venturi tunnel and spoiler** See Conceptual Question 24.

Integrated Exercises (IEs) are two-part exercises. The first part typically requires a conceptual answer choice based on physical thinking and basic principles. The following part is quantitative calculations associated with the conceptual choice made in the first part of the exercise.

Throughout the text, many exercise sections will include "paired" exercises. These exercise pairs, identified with **red numbers,** *are intended to assist you in problem solving and learning. In a pair, the first exercise (even numbered) is worked out in the Study Guide so that you can consult it should you need assistance in solving it. The second exercise (odd numbered) is similar in nature, and its answer is given in Appendix VII at the back of the book. Use as many significant figures as you need to show small changes.*

9.1 SOLIDS AND ELASTIC MODULI

1. • A tennis racket has nylon strings. If one of the strings with a diameter of 1.0 mm is under a tension of 15 N, how much is it lengthened from its original length of 40 cm?

2. • Suppose you use the tip of one finger to support a 1.0-kg object. If your finger has a diameter of 2.0 cm, what is the stress on your finger?

3. • A 2.5-m nylon fishing line used to hold up a 8.0-kg fish has a diameter of 1.6 mm. How much is the line elongated?

4. • A 5.0-m-long rod is stretched 0.10 m by a force. What is the strain in the rod?

5. • A 250-N force is applied at a 37° angle to the surface of the end of a square bar. The surface is 4.00 cm on a side. What are (a) the compressional stress and (b) the shear stress on the bar?

6. •• A 4.0-kg object is supported by an aluminum wire of length 2.0 m and diameter 2.0 mm. How much will the wire stretch?

7. •• A copper wire has a length of 5.0 m and a diameter of 3.0 mm. Under what load will its length increase by 0.30 mm?

8. •• A metal wire 1.0 mm in diameter and 2.0 m long hangs vertically with a 6.0-kg object suspended from it. If the wire stretches 1.4 mm under the tension, what is the value of Young's modulus for the metal?

9. **IE** •• When railroad tracks are installed, gaps are left between the rails. (a) Should a greater gap be used if the rails are installed on (1) a cold day or (2) a hot day? Or (3) does the temperature not make any difference? Why? (b) Each steel rail is 8.0 m long and has a cross-sectional area of 0.0025 m². On a hot day, each rail thermally expands as much as 3.0×10^{-3} m If there were no gaps between the rails, what would be the force on the ends of each rail?

10. •• A rectangular steel column (20.0 cm × 15.0 cm) supports a load of 12.0 metric tons. If the column is 2.00 m in length before being stressed, what is the decrease in length?

11. **IE** •• A bimetallic rod as illustrated in ▶ Fig. 9.34 is composed of brass and copper. (a) If the rod is subjected to a compressive force, will the rod bend toward the brass or the copper? Why? (b) Justify your answer mathematically if the compressive force is 5.00×10^4 N.

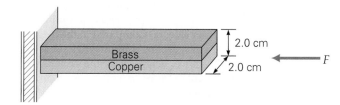

▲ **FIGURE 9.34 Bimetallic rod and mechanical stress** See Exercise 11.

12. **IE** •• Two same-size metal posts, one aluminum and one copper, are subjected to equal shear stresses. (a) Which post will show the larger deformation angle, (1) the copper post or (2) the aluminum post? Or (3) Is the angle the same for both? Why? (b) By what factor is the deformation angle of one post greater than the other?

13. •• A 85.0-kg person stands on one leg and 90% of the weight is supported by the upper leg connecting the knee and hip joint—the femur. Assuming the femur is 0.650 m long and has a radius of 2.00 cm, by how much is the bone compressed?

14. •• Two metal plates are held together by two steel rivets, each of diameter 0.20 cm and length 1.0 cm. How much force must be applied parallel to the plates to shear off both rivets?

15. **IE** •• (a) Which of the liquids in Table 9.1 has the greatest compressibility? Why? (b) For equal volumes of ethyl alcohol and water, which would require more pressure to be compressed by 0.10%, and how many times more?

16. •• How much pressure would be required to compress a quantity of mercury by 0.010%?

17. ••• A brass cube 6.0 cm on each side is placed in a pressure chamber and subjected to a pressure of 1.2×10^7 N/m² on all of its surfaces. By how much will each side be compressed under this pressure?

18. ••• A cylindrical eraser of negligible mass is dragged across a paper at a constant velocity to the right by its pencil. The coefficient of kinetic friction between eraser and paper is 0.650. The pencil pushes down with 4.20 N. The height of the eraser is 1.10 cm and its diameter is 0.760 cm. Its top surface is displaced horizontally 0.910 mm relative to the bottom. Determine the shear modulus of the eraser material.

19. ••• A 45-kg traffic light is suspended from two steel cables of equal length and radii 0.50 cm. If each cable makes a 15° angle with the horizontal, what is the fractional increase in their length due to the weight of the light?

9.2 FLUIDS: PRESSURE AND PASCAL'S PRINCIPLE

20. **IE •** In his original barometer, Pascal used water instead of mercury. (a) Water is less dense than mercury, so the water barometer would have (1) a higher height than, (2) a lower height than, or (3) the same height as the mercury barometer. Why? (b) How high would the water column have been?

21. **•** If you dive to a depth of 10 m below the surface of a lake, (a) what is the pressure due to the water alone? (b) What is the absolute pressure at that depth?

22. **IE •** In an open U-tube, the pressure of a water column on one side is balanced by the pressure of a column of gasoline on the other side. (a) Compared to the height of the water column, the gasoline column will have (1) a higher height, (2) a lower height, or (3) the same height. Why? (b) If the height of the water column is 15 cm, what is the height of the gasoline column?

23. **•** A 75.0-kg athlete performs a single-hand handstand. If the area of the hand in contact with the floor is 125 cm², what pressure is exerted on the floor?

24. **•** A rectangular fish tank measuring 0.75 m × 0.50 m is filled with water to a height of 65 cm. What is the gauge pressure on the bottom of the tank?

25. **•** (a) What is the absolute pressure at a depth of 10 m in a lake? (b) What is the gauge pressure?

26. **••** The gauge pressure in both tires of a bicycle is 690 kPa. If the bicycle and the rider have a combined mass of 90.0 kg, what is the area of contact of *each* tire with the ground? (Assume that each tire supports half the total weight of the bicycle.)

27. **••** In a sample of seawater taken from an oil spill, an oil layer 4.0 cm thick floats on 55 cm of water. If the density of the oil is 0.75×10^3 kg/m³, what is the absolute pressure on the bottom of the container?

28. **IE ••** In a lecture demonstration, an empty can is used to demonstrate the force exerted by air pressure (▼Fig. 9.35). A small quantity of water is poured into the can, and the water is brought to a boil. Then the can is sealed with a *rubber stopper*. As you watch, the can is slowly crushed with sounds of metal bending. (Why is a rubber stopper used as a safety precaution?) (a) This is because of (1) thermal expansion and contraction, (2) a higher steam pressure inside the can, (3) a lower pressure inside the can as steam condenses. Why? (b) Assuming the dimensions of the can are 0.24 m × 0.16 m × 0.10 m and the inside of the can is in a perfect vacuum, what is the total force exerted on the can by the air pressure?

▲ FIGURE 9.35 **Atmospheric pressure** See Exercise 28.

29. **••** What is the fractional decrease in pressure when a barometer is raised 40.0 m to the top of a building? (Assume that the density of air is constant over that distance.)

30. **••** To drink a soda (assume same density as water) through a straw requires that you lower the pressure at the top of the straw. What does the pressure need to be at the top of a straw that is 15.0 cm above the surface of the soda in order for the soda to reach your lips?

31. **••** During a plane flight, a passenger experiences ear pain due to a head cold that has clogged his Eustachian tubes. Assuming the pressure in his tubes remained at 1.00 atm (from sea level) and the cabin pressure is maintained at 0.900 atm, determine the air pressure force (including its direction) on one eardrum, assuming it has a diameter of 0.800 cm.

32. **••** Here is a demonstration Pascal used to show the importance of a fluid's pressure on the fluid's depth (▼Fig. 9.36): An oak barrel with a lid of area 0.20 m² is filled with water. A long, thin tube of cross-sectional area 5.0×10^{-5} m² is inserted into a hole at the center of the lid, and water is poured into the tube. When the water reaches 12 m high, the barrel bursts. (a) What was the weight of the water in the tube? (b) What was the pressure of the water on the lid of the barrel? (c) What was the net force on the lid due to the water pressure?

▲ FIGURE 9.36 **Pascal and the bursting barrel** See Exercise 32.

33. **••** The door and the seals on an aircraft are subject to a tremendous amount of force during flight. At an altitude of 10 000 m (about 33 000 ft), the air pressure outside the airplane is only 2.7×10^4 N/m² while the inside is still at normal atmospheric pressure, due to pressurization of the cabin. Calculate the force due to the air pressure on a door of area 3.0 m².

34. **••** The pressure exerted by a person's lungs can be measured by having the person blow as hard as possible into one side of a manometer. If a person blowing into one side of an open tube manometer produces an 80-cm difference between the heights of the columns of water in the manometer arms, what is the gauge pressure of the lungs?

35. •• In a head-on auto collision, the driver, who had his air bags disconnected, hits his head on the windshield, fracturing his skull. Assuming the driver's head has a mass of 4.0 kg, the area of the head to hit the windshield to be 25 cm^2, and an impact time of 3.0 ms, with what speed does his head hit the windshield? (Take the compressive fracture strength of the cranial bone to be 1.0×10^8 Pa.)

36. •• A cylinder has a diameter of 15 cm (▼Fig. 9.37). The water level in the cylinder is maintained at a constant height of 0.45 m. If the diameter of the spout pipe is 0.50 cm, how high is h, the vertical stream of water? (Assume the water to be an ideal fluid.)

◀ **FIGURE 9.37**
How high a fountain?
See Exercise 36.

37. •• In 1960, the U.S. Navy's bathyscaphe *Trieste* (a submersible) descended to a depth of 10 912 m (about 35 000 ft) into the Mariana Trench in the Pacific Ocean. (a) What was the pressure at that depth? (Assume that seawater is incompressible.) (b) What was the force on a circular observation window with a diameter of 15 cm?

38. •• The output piston of a hydraulic press has a cross-sectional area of 0.25 m^2. (a) How much pressure on the input piston is required for the press to generate a force of 1.5×10^6 N? (b) What force is applied to the input piston if it has a diameter of 5.0 cm?

39. •• A hydraulic lift in a garage has two pistons: a small one of cross-sectional area 4.00 cm^2 and a large one of cross-sectional area 250 cm^2. (a) If this lift is designed to raise a 3500-kg car, what minimum force must be applied to the small piston? (b) If the force is applied through compressed air, what must be the minimum air pressure applied to the small piston?

40. •• The Magdeburg water bridge is a channel bridge over the River Elbe in Germany (▼Fig. 9.38). Its dimen-

▲ **FIGURE 9.38 Water bridge** See Exercise 40.

sions are length 918 m, width 43.0 m, and depth 4.25 m. (a) When filled with water, what is the weight of the water? (b) What is the pressure on the bridge floor?

41. ••• A hypodermic syringe has a plunger of area 2.5 cm^2 and a 5.0×10^{-3}-cm^2 needle. (a) If a 1.0-N force is applied to the plunger, what is the gauge pressure in the syringe's chamber? (b) If a small obstruction is present at the end of the needle, what force does the fluid exert on it? (c) If the blood pressure in a vein is 50 mm Hg, what force must be applied on the plunger so that fluid can be injected into the vein?

42. ••• A funnel has a cork blocking its drain tube. The cork has a diameter of 1.50 cm and is held in place by static friction with the sides of the drain tube. When water is added to a height of 10.0 cm above the cork, it comes flying out of the tube. Determine the maximum force of static friction between the cork and drain tube. Neglect the weight of the cork.

9.3 BUOYANCY AND ARCHIMEDES' PRINCIPLE

43. IE • (a) If the density of an object is exactly equal to the density of a fluid, the object will (1) float, (2) sink, (3) stay at any height in the fluid, as long as it is totally immersed. (b) A cube 8.5 cm on each side has a mass of 0.65 kg. Will the cube float or sink in water? Prove your answer.

44. • A rectangular boat, as illustrated in ▼Fig. 9.39, is overloaded such that the water level is just 1.0 cm below the top of the boat. What is the combined mass of the people and the boat?

0.30 m

2.0 m

4.5 m

▲ **FIGURE 9.39 An overloaded boat** See Exercise 44.

45. •• An object has a weight of 8.0 N in air. However, it apparently weighs only 4.0 N when it is completely submerged in water. What is the density of the object?

46. •• When a 0.80-kg crown is submerged in water, its apparent weight is measured to be 7.3 N. Is the crown pure gold?

47. •• A steel cube 0.30 m on each side is suspended from a scale and immersed in water. What will the scale read?

48. •• A solid ball has a weight of 3.0 N. When it is submerged in water, it has an apparent weight of 2.7 N. What is the density of the ball?

49. •• A wood cube 0.30 m on each side has a density of 700 kg/m^3 and floats levelly in water. (a) What is the distance from the top of the wood to the water surface? (b) What mass has to be placed on top of the wood so that its top is just at the water level?

50. •• (a) Given a piece of metal with a light string attached, a scale, and a container of water in which the piece of metal can be submersed, how could you find the volume of the piece without using the variation in the water level? (b) An object has a weight of 0.882 N. It is suspended from a scale, which reads 0.735 N when the piece is submerged in water. What are the volume and density of the piece of metal?

51. •• An aquarium is filled with a liquid. A cork cube, 10.0 cm on a side, is pushed and held at rest completely submerged in the liquid. It takes a force of 7.84 N to hold it under the liquid. If the density of cork is 200 kg/m³, find the density of the liquid.

52. •• A block of iron quickly sinks in water, but ships constructed of iron float. A solid cube of iron 1.0 m on each side is made into sheets. To make these sheets into a hollow cube that will not sink, what should be the minimum length of the sides of the sheets?

53. •• Plans are being made to bring back the zeppelin, a lighter-than-air airship like the Goodyear blimp that carries passengers and cargo, but is filled with helium, not flammable hydrogen as was used in the ill-fated *Hindenburg*. One design calls for the ship to be 110 m long and to have a total mass (without helium) of 30.0 metric tons. Assuming the ship's "envelope" to be cylindrical, what would its diameter have to be so as to lift the total weight of the ship and the helium?

54. •• A girl floats in a lake with 97% of her body beneath the water. What are (a) her mass density and (b) her weight density?

55. ••• A spherical navigation buoy is tethered to the lake floor by a vertical cable (▼ Fig. 9.40). The outside diameter of the buoy is 1.00 m. The interior of the buoy consists of an aluminum shell 1.0 cm thick, and the rest is solid plastic. The density of aluminum is 2700 kg/m³ and the density of the plastic is 200 kg/m³. The buoy is set to float exactly halfway out of the water. Determine the tension in the cable.

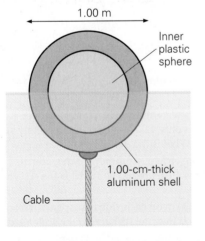

1.00 m

Inner plastic sphere

◄ **FIGURE 9.40**
It's a buoy See Exercise 55.

1.00-cm-thick aluminum shell

Cable

56. ••• ▶ Figure 9.41 shows a simple laboratory experiment. Calculate (a) the volume and (b) the density of the suspended sphere. (Assume that the density of the sphere is uniform and that the liquid in the beaker is water.) (c) Would you be able to make the same determinations if the liquid in the beaker were mercury? (See Table 9.2.) Explain.

▲ **FIGURE 9.41** **Dunking a sphere** See Exercise 56.

9.4 FLUID DYNAMICS AND BERNOULLI'S EQUATION

57. • An ideal fluid is moving at 3.0 m/s in a section of a pipe of radius 0.20 m. If the radius in another section is 0.35 m, what is the flow speed there?

58. IE • (a) If the radius of a pipe narrows to half of its original size, will the flow speed in the narrow section (1) increase by a factor of 2, (2) increase by a factor of 4, (3) decrease by a factor of 2, or (4) decrease by a factor of 4? Why? (b) If the radius widens to three times its original size, what is the ratio of the flow speed in the wider section to that in the narrow section?

59. •• Water flows through a horizontal tube similar to that in Fig. 9.20. However in this case, the constricted part of the tube is half the diameter of the larger part. If the water speed is 1.5 m/s in the larger parts of the tube, by how much does the pressure drop in the constricted part? Express the final answer in atmospheres.

60. •• The speed of blood in a major artery of diameter 1.0 cm is 4.5 cm/s. (a) What is the flow rate in the artery? (b) If the capillary system has a total cross-sectional area of 2500 cm², the average speed of blood through the capillaries is what percentage of that through the major artery? (c) Why must blood flow at low speed through the capillaries?

61. •• The blood flow speed through an aorta with a radius of 1.00 cm is 0.265 m/s. If hardening of the arteries causes the aorta to be constricted to a radius of 0.800 cm, by how much would the blood flow speed increase?

62. •• Using the data and result of Exercise 61, calculate the pressure difference between the two areas of the aorta. (Blood density: $\rho = 1.05 \times 10^3$ kg/m³.)

63. •• In a dramatic lecture demonstration, a physics professor blows hard across the top of a copper penny that is at rest on a level desk. By doing this at the right speed, he can get the penny to accelerate vertically, into the airstream, and then deflect it into a tray, as shown in ▶ Fig. 9.42. Assuming the diameter of a penny is 1.90 cm and its mass is 2.50 g, what is the minimum airspeed needed to lift the penny off the tabletop? Assume the air under the penny remains at rest.

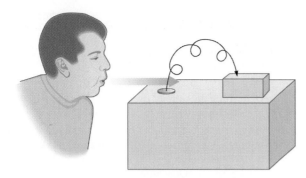

▲ **FIGURE 9.42 A big blow** See Exercise 63.

64. ●● The spout heights in the container in ▼Fig. 9.43 are 10 cm, 20 cm, 30 cm, and 40 cm. The water level is maintained at a 45-cm height by an outside supply. (a) What is the speed of the water out of each hole? (b) Which water stream has the greatest range relative to the base of the container? Justify your answer.

▲ **FIGURE 9.43 Streams as projectiles** See Exercise 64.

65. ●● In Conceptual Example 9.14, it was explained why a stream of water from a faucet necks down into a smaller cross-sectional area as it descends. Suppose at the top of the stream it has a cross-sectional area of 2.0 cm², and a vertical distance 5.0 cm below the cross-sectional area of the stream is 0.80 cm². What is (a) the speed of the water and (b) the flow rate?

66. ●● Water flows at a rate of 25 L/min through a horizontal 7.0-cm-diameter pipe under a pressure of 6.0 Pa. At one point, calcium deposits reduce the cross-sectional area of the pipe to 30 cm². What is the pressure at this point? (Consider the water to be an ideal fluid.)

67. ●●● As a fire-fighting method, a homeowner in the deep woods rigs up a water pump to bring water from a lake that is 10.0 m below the level of the house. If the pump is capable of producing a gauge pressure of 140 kPa, at what rate (in L/s) can water be pumped to the house assuming the hose has a radius of 5.00 cm?

68. ●●● A Venturi meter can be used to measure the flow speed of a liquid. A simple such device is shown in ▼Fig. 9.44. Show that the flow speed of an ideal fluid is given by

$$v_1 = \sqrt{\frac{2g\Delta h}{(A_1^2/A_2^2) - 1}}$$

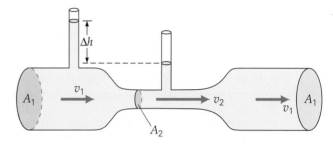

▲ **FIGURE 9.44 A flow speed meter** See Exercise 68.

*9.5 SURFACE TENSION, VISCOSITY, AND POISEUILLE'S LAW

69. ●● The pulmonary artery, which connects the heart to the lungs, is about 8.0 cm long and has an inside diameter of 5.0 mm. If the flow rate in it is to be 25 mL/s, what is the required pressure difference over its length?

70. ●● A hospital patient receives a quick 500-cc blood transfusion through a needle with a length of 5.0 cm and an inner diameter of 1.0 mm. If the blood bag is suspended 0.85 m above the needle, how long does the transfusion take? (Neglect the viscosity of the blood flowing in the plastic tube between the bag and the needle.)

71. ●● A nurse needs to draw 20.0 cc of blood from a patient and deposit it into a small plastic container whose interior is at atmospheric pressure. He inserts the needle end of a long tube into a vein where the average gauge pressure is 30.0 mm Hg. This allows the internal pressure in the vein to push the blood into the collection container. The needle is 0.900 mm in diameter and 2.54 cm long. The long tube is wide and smooth enough that we can assume its resistance is negligible, and that all the resistance to blood flow occurs in the narrow needle. How long does it take him to collect the sample?

72. ●● What is the difference in volume (due only to pressure changes, and not temperature or other factors) between 1000 kg of water at the surface (assume 4 °C) of the ocean and the same mass at the deepest known depth, 8.00 km? (Mariana Trench, assume 4 °C also.)

The Multiconcept Exercises require the use of more than one fundamental concept for their understanding and solution. The concepts may be from this chapter, but may include those from previous chapters.

73. A rock is suspended from a string in air. The tension in the string is 2.94 N. When the rock is then dunked into a liquid and the string is allowed to go slack, it sinks and comes to rest on a spring with a spring constant of 200 N/m. The spring's final compression is 1.00 cm. If the density of the rock is 2500 kg/m^3, what is the density of the liquid?

74. An unevenly weighted baton (cylindrical in shape) consists of two sections: a denser (lower) section and a less dense (upper) section. When placed in water, it is upright and barely floats. The baton has a diameter of 2.00 cm; its lower part is made of steel with a density of 7800 kg/m^3, and the upper part is made of wood with a density of 810 kg/m^3. The steel part has a length of 5.00 cm. Find the length of the wooden section.

75. (a) Referring to the metal rod in Figure 9.2a (under tensile stress), show that Eq. 9.4 can be rewritten to resemble a Hooke's law type of spring relationship for the rod. That is, show that it can be written as $F = k\,\Delta L$, where k is the "effective" spring constant for the rod. Express k symbolically in terms of the rod's cross-sectional area A, its Young's modulus Y, and its unstressed length L_o and show that it has the proper SI units. (b) Now consider a thin rod of iron that is subjected to a tensile force of 2.00×10^3 N. If it has a cross-section of radius 1.00 cm and an unstressed length of 25.0 cm, determine its effective spring constant. (c) By how much does this rod stretch when this force is applied? (d) How much work is done by this stretching force? [*Hint:* Remember the expression for work done on a spring.]

76. The ocean can be as deep as 10 km. (a) Assuming the density for seawater is constant at the value given in Table 9.2, what is the absolute pressure at such depths? (b) What would be the percentage change in volume of a cube of aluminum that measured 1.00 m on a side when at the ocean surface? (c) By how much did the aluminum cube's volume change?

77. In preparation for its tire rotation, a car weighing 2.25 tons is placed on a hydraulic garage lift. The mechanic then raises the car 30.0 cm. (a) Calculate the work done on the car when it is lifted. (b) Assuming no frictional losses in the hydraulic fluid, how much work was done by the lift on the *input* side? (c) What was the force on the input side if its piston moved 52.5 cm? (d)

Determine the ratio of the input side area to that of the lifting side (output) area?

78. A spherical object has an outside diameter of 48.0 cm. Its outer shell is composed of aluminum and is 2.00 cm thick. The remainder is uniform plastic with a density of 800 kg/m^3. (a) Determine the object's average density. (b) Will this object float by itself in fresh water? Explain your reasoning. (c) If it does float, how much of it is above the water surface? If it doesn't float, determine the force required to keep it from sinking if it is entirely submerged.

79. As a medical technologist, you are attending to a worker who has been wounded by an accidental industrial explosion. After measuring her arterial blood pressure to be 132/86, you determine her major wound to be a small circular puncture of an artery, with an estimated diameter of 0.25 mm. Determine (a) the maximum speed at which blood is flowing out of the puncture and (b) the maximum rate (in cc/min) at which she is losing blood through it.

80. An engineer is designing a water filter that works by forcing water through a circular plate that has many identical holes in it. The plate is to be welded into a pipe so the water stream and the plate have the same 2.54-cm diameter. (See ▼Figure 9.45.) Before it hits the filter, the water is to be traveling at 75.0 cm/s. The holes are planned to be circular and 0.100 mm in diameter. (a) If the holes are to cover 65% of the total plate area, how many of them will you need? (b) Assuming that initially none of the holes are plugged, what is the flow speed just after the water leaves the holes? (c) Later, if 25% of the holes are completely plugged with gunk and the water speed before the filter has not changed, what will be the flow speed of the water upon leaving the filter area? (d) Compare the flow rate in this system (in liters per minute) before and after some of the holes plug up.

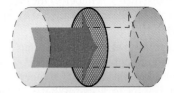

▲ **FIGURE 9.45** **Filter that water** See Exercise 80.

10 Temperature and Kinetic Theory

PHYSICS FACTS

- The Celsius and Fahrenheit temperature scales have equal readings at −40 degrees, that is, −40 °C = −40 °F.
- The lowest possible temperature is absolute zero (−273.15 °C). There is no known upper limit on temperature.
- The Golden Gate Bridge over San Francisco Bay varies in length by almost 1 m between summer and winter (thermal expansion).
- While the normal average human body temperature is 37 °C (98.6 °F), the normal skin temperature is only 33 °C (91 °F). The skin temperature depends on air temperature and time spent in that environment.
- Almost all substances have positive coefficients of thermal expansion (expanding on heating). A few have negative coefficients (contraction on heating). Water contracts on heating from 0 °C to 4 °C.

Global warning has becoming a very popular topic lately due to the increasing evidence that human activities have accelerated the increase of the atmospheric temperature of the Earth. Increasing global temperatures will cause polar ice caps to melt and sea levels to rise. The NASA satellite photographs (chapter-opener photographs) taken since 1979 clearly show the shrinking of the Arctic ice caps. The photograph on the top was taken in 1979 and the one on the bottom in 2005.

Temperature and heat are frequent subjects of conversation, but if you had to explain what the words really mean, you might find yourself at a loss. We use various types of thermometers to measure temperatures, which provide an objective equivalent for our sensory experience of hot and cold. A temperature change generally results from the addition or removal of heat. Temperature, therefore, is related to heat. But how? And what is heat? In this chapter, you'll find that the answers to such questions lead to an understanding of some far-reaching physical principles.

An early theory of heat considered it to be a fluid-like substance called *caloric* (from the Latin word *calor*, meaning "heat") that could be made to flow into and out of a body. Even though this theory has been abandoned, we still speak of heat as "flowing" from one object to another. Heat is now known to be energy in transit, and temperature and thermal properties are explained by considering the atomic and molecular behavior of substances. This and the next two chapters examine the nature of temperature and heat in terms of microscopic (molecular) theory and macroscopic observations. Here, you'll explore the nature of heat and the ways temperature is measured. You'll also encounter the gas laws, which explain not only the pressure increase of a hot automobile tire, but also more important phenomena, such as how our lungs supply us with the oxygen we need to live.

10.1   Temperature and Heat

LEARNING PATH QUESTIONS

➡ What is the difference between temperature and heat?
➡ What types of energy make up the internal energy of a diatomic gas?
➡ Does higher temperature mean a system having more internal energy?

A good way to begin studying thermal physics is with definitions of temperature and heat. **Temperature** is a relative measure, or indication, of hotness or coldness. A hot stove is said to have a high temperature and an ice cube to have a low temperature. An object that has a higher temperature than another object is said to be hotter, or the other object is said to be colder. Note that *hot* and *cold* are relative terms, like *tall* and *short*. We can perceive temperature by touch. However, this temperature sense is somewhat unreliable, and its range is too limited to be useful for scientific purposes.

Heat is related to temperature and describes the process of energy transfer from one object to another. That is, **heat** is *the net energy transferred from one object to another because of a temperature difference.* Heat is energy in transit, so to speak. Once transferred, the energy becomes part of the total energy of the molecules of the object or system, that is, its **internal energy**. So heat (energy) transfers between objects can result in internal energy changes.*

On a microscopic level, temperature is associated with molecular motion. In kinetic theory (Section 10.5), which treats gas molecules as point particles, temperature is shown to be a measure of the average *translational* kinetic energy of the molecules. However, diatomic gas, besides having such translational "temperature" kinetic energy, also may have kinetic energy due to rotations

---

*Note: Some of the energy may go into doing work and not into internal energy (Section 12.2).

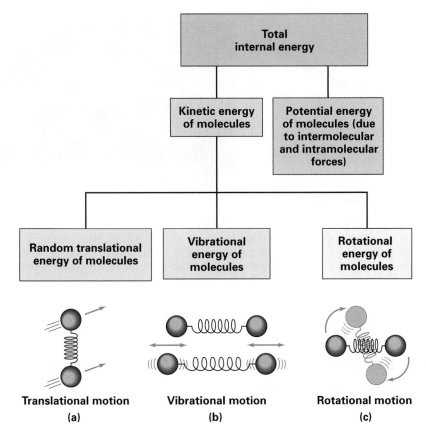

▲ **FIGURE 10.1** **Diatomic molecular motions** The total internal energy is made up of kinetic and intermolecular and intramolecular potential energy. The kinetic energy has the following forms: **(a)** translational kinetic energy. **(b)** linear vibrational kinetic energy and **(c)** rotational kinetic energy.

and vibrations, as well as potential energy due to intermolecular and intramolecular interactions. The total internal energy is the sum of all such energies (▲Fig. 10.1).

Note that a higher temperature does not necessarily mean that one system has a greater internal energy than another. For example, in a classroom on a cold day, the air temperature is relatively high compared to that of the outdoor air. But all that cold air outside the classroom has far more internal energy than does the warm air inside, simply because there is so much *more* of it. If this were not the case, heat pumps would not be practical (Section 12.5). In other words, the internal energy of a system also depends on its mass, or the number of molecules in the system.

When heat is transferred between two objects, regardless of whether they are touching, the objects are said to be in *thermal contact*. When there is no longer a net heat transfer between objects in thermal contact, they have come to the same temperature and are said to be in *thermal equilibrium*.

DID YOU LEARN?

➥ Heat is the net energy transferred between objects due to temperature differences, and temperature is an indication of the average translational kinetic energy of the molecules.

➥ The total internal energy of a diatomic gas may consist of translational kinetic energy, vibrational kinetic energy, rotational kinetic energy, and potential energy due to attractive forces between the atoms.

➥ The mass or number of molecules in a system is a factor in determining the total internal energy.

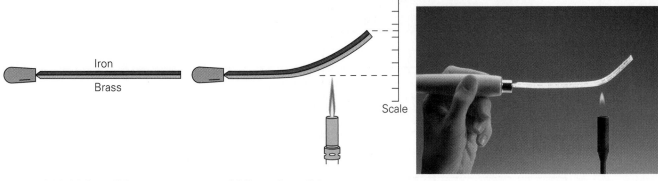

(a) Initial condition   **(b) Heated condition**

▲ FIGURE 10.2 **Thermal expansion (a)** A bimetallic strip is made of two strips of different metals bonded together. **(b)** When such a strip is heated, it bends because of unequal expansions of the two metals. Here, brass expands more than iron, so the deflection is toward the iron. The deflection of the end of a strip could be used to measure temperature.

(a)

(b)

▲ FIGURE 10.3 **Bimetallic coil** Bimetallic coils are used in **(a)** dial thermometers (the coil is in the center) and **(b)** household thermostats (the coil is to the right). Thermostats are used to regulate a heating or cooling system, turning off and on as the temperature of the room changes. The expansion and contraction of the coil causes the tilting of a glass vial containing mercury, which makes and breaks electrical contact.

10.2 The Celsius and Fahrenheit Temperature Scales

LEARNING PATH QUESTIONS

➥ How is a temperature scale constructed?

➥ How do you convert a temperature on the Fahrenheit scale to a temperature on the Celsius scale?

➥ How do you convert a temperature on the Celsius scale to a temperature on the Fahrenheit scale?

A measure of temperature is obtained by using a **thermometer**, a device constructed to make use of some property of a substance that changes with temperature. Fortunately, many physical properties of materials change sufficiently with temperature to be used as the bases for thermometers. By far the most obvious and commonly used property is **thermal expansion** (Section 10.4), a change in the dimensions or volume of a substance that occurs when the temperature changes.

Almost all substances expand with increasing temperature, and do so to different extents. They also contract with decreasing temperature. (Thermal expansion refers to both expansion and contraction; contraction is considered a negative expansion.) Because some metals expand more than others, a bimetallic strip (a strip made of two different metals bonded together) can be used to measure temperature changes. As heat is added, the composite strip will bend away from the side made of the metal that expands more (▲Fig. 10.2). Coils formed from such strips are used in dial thermometers and in common household thermostats (◄Fig. 10.3).

A common thermometer is the liquid-in-glass type, which is based on the thermal expansion of a liquid. A liquid in a glass bulb expands into a glass stem, rising in a capillary bore (a thin tube). Mercury and alcohol (usually dyed red to make it more visible) are the liquids used in most liquid-in-glass thermometers. These substances are chosen because of their relatively large thermal expansion and because they remain liquids over normal temperature ranges.

Thermometers are calibrated so that a numerical value can be assigned to a given temperature. For the definition of any standard scale or unit, two fixed reference points are needed. The ice point and the steam point of water at standard atmospheric pressure are two convenient fixed points. More commonly known as the freezing and boiling points, these are the temperatures at which pure water freezes and boils, respectively, under a pressure of 1 atm (standard pressure).

The two most familiar temperature scales are the **Fahrenheit temperature scale*** (used in the United States) and the **Celsius temperature scale**† (used in the rest of the world). As shown in ▸Fig. 10.4, the ice and steam points have values of 32 °F and 212 °F, respectively, on the Fahrenheit scale and 0 °C and 100 °C, respectively, on the Celsius scale. On the Fahrenheit scale, there are 180 equal intervals, or degrees (°F), between the two reference points; on the Celsius scale, there are 100 degrees (°C). Therefore, since $180/100 = 9/5 = 1.8$, one Celsius degree is almost twice as large as one Fahrenheit degree.

A relationship for converting between the two scales can be obtained from a graph of Fahrenheit temperature (T_F) versus Celsius temperature (T_C), such as the one in ▾Fig. 10.5. The equation of the straight line (in slope-intercept form, $y = mx + b$) is $T_F = (180/100)T_C + 32$, and

$$T_F = \tfrac{9}{5}T_C + 32$$

or

$$T_F = 1.8T_C + 32$$

(Celsius-to-Fahrenheit conversion)   (10.1)

where $\tfrac{9}{5} = 1.8$ is the slope of the line and 32 is the intercept on the vertical axis. Thus, to change from a Celsius temperature (T_C) to its equivalent Fahrenheit temperature (T_F), you simply first multiply the Celsius reading by $\tfrac{9}{5}$ and then add 32.

The equation can be solved for T_C to convert from Fahrenheit to Celsius:

$$T_C = \tfrac{5}{9}(T_F - 32) \quad \textit{(Fahrenheit-to-Celsius conversion)} \quad (10.2)$$

Therefore, to change from a Fahrenheit temperature (T_F) to its equivalent Celsius temperature (T_C), you first subtract 32 from the Fahrenheit reading and then multiply by $\tfrac{5}{9}$.

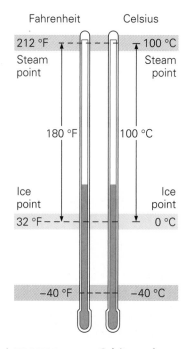

▲ **FIGURE 10.4 Celsius and Fahrenheit temperature scales** Between the ice and steam fixed points, there are 100 degrees on the Celsius scale and 180 degrees on the Fahrenheit scale. Thus, a Celsius degree is 1.8 times as large as a Fahrenheit degree.

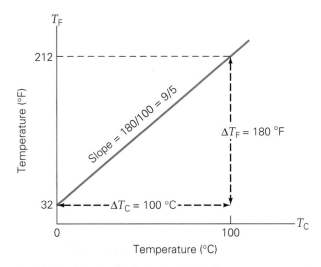

▲ **FIGURE 10.5 Fahrenheit versus Celsius** A plot of Fahrenheit temperature versus Celsius temperature gives a straight line of the general form $y = mx + b$, where $T_F = \tfrac{9}{5}T_C + 32$.

*Daniel Gabriel Fahrenheit (1686–1736), a German instrument maker, constructed the first alcohol thermometer (1709) and mercury thermometer (1714). The freezing and boiling points of water were measured to be 32 °F and 212 °F.

†Anders Celsius (1701–1744), a Swedish astronomer, invented the Celsius temperature scale with a 100-degree interval between the freezing and boiling point of water (0 °C and 100 °C).

EXAMPLE 10.1 | Converting Temperature Scale Readings: Fahrenheit and Celsius

What are (a) the typical room temperature of 20 °C and a cold temperature of −15 °C on the Fahrenheit scale, and (b) another cold temperature of −10 °F and normal body temperature, 98.6 °F, on the Celsius scale?

THINKING IT THROUGH. These are direct applications of Eqs. 10.1 and 10.2.

SOLUTION.

Given: (a) T_C = 20 °C and   *Find:*   for each temperature,
         T_C = −15 °C   (a) T_F
    (b) T_F = −10 °F and   (b) T_C
        T_F = 98.6 °F

(a) Equation 10.1 is for changing Celsius readings to Fahrenheit:

$$20 \text{ °C:} \quad T_F = \tfrac{9}{5}T_C + 32 = \left[\tfrac{9}{5}(20) + 32\right] \text{ °F} = 68 \text{ °F}$$

(This typical room temperature of 20 °C is a good one to remember.)

$$-15 \text{ °C:} \quad T_F = \tfrac{9}{5}T_C + 32 = \left[\tfrac{9}{5}(-15) + 32\right] \text{ °F} = 5.0 \text{ °F}$$

(b) Equation 10.2 changes Fahrenheit to Celsius:

$$-10 \text{ °F:} \quad T_C = \tfrac{5}{9}(T_F - 32) = \left[\tfrac{5}{9}(-10 - 32)\right] \text{ °C} = -23 \text{ °C}$$

$$98.6 \text{ °F:} \quad T_C = \tfrac{5}{9}(T_F - 32) = \left[\tfrac{5}{9}(98.6 - 32)\right] \text{ °C} = 37.0 \text{ °C}$$

Note that one Celsius degree is 1.8 times (almost twice) as large as one Fahrenheit degree. For example, a body temperature of 40.0 °C represents an elevation of 3.0 °C over normal body temperature. However, on the Fahrenheit scale, this is an increase of 3.0 × 1.8 °F = 5.4 °F, or a temperature of (98.6 + 5.4) °F = 104.0 °F.

FOLLOW-UP EXERCISE. Convert the following temperatures: (a) −40 °F to Celsius and (b) −40 °C to Fahrenheit. (*Answers to all Follow-Up Exercises are given in Appendix VI at the back of the book.*)

PROBLEM-SOLVING HINT

Because Eqs. 10.1 and 10.2 are so similar, it is easy to miswrite them. Since they are equivalent, you need to know only one of them—say, Celsius to Fahrenheit, Eq. 10.1, $T_F = \tfrac{9}{5}T_C + 32$. Solving this equation for T_C algebraically gives Eq. 10.2. A good way to make sure that you have written the conversion equation correctly is to test it with a known temperature, such as the boiling point of water. For example, T_F = 212 °F, so

$$T_C = \tfrac{5}{9}(T_F - 32) = \left[\tfrac{5}{9}(212 - 32)\right] \text{ °C} = \tfrac{5}{9}(180) \text{ °C} = 100 \text{ °C}$$

Thus, we know the equation is correct.

Liquid-in-glass thermometers are adequate for many temperature measurements, but problems arise when highly accurate determinations are needed. A material may not expand uniformly over a wide temperature range. When calibrated to the ice and steam points, an alcohol thermometer and a mercury thermometer have the same readings at those points, but because alcohol and mercury have different expansion properties, the thermometers will not have exactly the same reading at an intermediate temperature, such as room temperature. For very sensitive temperature measurements and to define intermediate temperatures precisely, some other type of thermometer must be used. One such thermometer, a *gas thermometer*, is discussed in the next section.

INSIGHT 10.1 | **Human Body Temperature**

We commonly take "normal" human body temperature to be 98.6 °F or 37.0 °C. The source of this value is a study of human temperature readings done in 1868—more than 135 years ago. A more recent study, conducted in 1992, notes that the 1868 study used thermometers that were not as accurate as modern electronic (digital) thermometers. The new study has some interesting results.

The normal human body temperature from oral measurements varies among individuals over a range of about 96 °F to 101 °F, with an average temperature of 98.2 °F. After strenuous exercise, the oral temperature can rise as high as 103 °F. When the body is exposed to cold, oral temperatures can fall below 96 °F. A rapid drop in temperature of (2 to 3) °F produces uncontrollable shivering. The skeletal muscles contract and so do the tiny muscles attached to the hair follicles. The result is "goose bumps."

Your body temperature is typically lowest in the morning, after you have slept and your digestive processes are at a low point. Normal body temperature generally rises during the day to a peak and then recedes. The 1992 study also indicated that women have a slightly higher average body temperature than do men (98.4 °F versus 98.1 °F).

What about the extremes? A fever temperature is typically between 102 °F and 104 °F. A body temperature above 106 °F is extremely dangerous. At such temperatures, the enzymes that take part in certain chemical reactions in the body begin to be inactivated, and a total breakdown of body chemistry can result. On the cold side, decreased body temperature results in memory lapses and slurred speech, muscular rigidity, erratic heartbeats, and loss of consciousness. Below 78 °F, death occurs due to heart failure. However, mild hypothermia

(lower-than-normal body temperature) can be beneficial. A decrease in body temperature slows down the body's chemical reactions, and cells use less oxygen than they normally do. This effect is applied in some surgeries (Fig. 1). A patient's body temperature may be lowered significantly to avoid damage to the brain and to the heart, which must be stopped during some procedures.

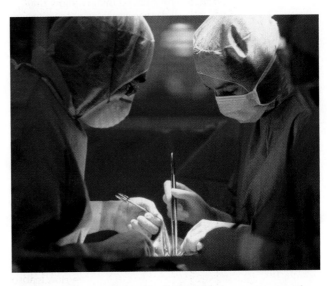

FIGURE 1 **Lower than normal** During some surgeries, the patient's body temperature is lowered to slow down the body's chemical reactions and to reduce the need for blood to supply oxygen to the tissues.

INSIGHT 10.2 | **Warm-Blooded Versus Cold-Blooded**

With few exceptions, all mammals and birds are warm-blooded and all fish, reptiles, amphibians, and insects are cold-blooded. The difference is that warm-blooded creatures try to maintain their bodies at a relatively constant temperature, while cold-blooded creatures take on the temperature of their surroundings (Fig. 1).

Warm-blooded creatures maintain a relatively constant body temperature by generating their own heat when in a cold environment and by cooling themselves when in a hot environment. To generate heat, warm-blooded animals convert food into energy. To stay cool on hot days, they sweat, pant, or get wet and thereby remove heat by water evaporation. Primates (humans, apes, monkeys, and so on) have sweat glands all over their bodies. Dogs and cats have sweat

glands only in their feet. Pigs and whales have no sweat glands. Pigs generally rely on wallowing in mud for cooling, and whales can change water depths for temperature changes or seasonally migrate.

Also, some animals have fur coats for warmth in the winter and shed them to cool off in the summer. Warm-blooded animals can shiver to activate certain muscles to increase metabolism and thereby generate heat. Some birds (and some people) migrate between colder and warmer regions.

The body temperature of cold-blooded creatures changes with the temperature of their environment. They are very active in warm environments and are sluggish when it is cold.

(continued on next page)

This is because their muscle activity depends on chemical reactions that vary with temperature. Cold-blooded creatures often bask in the sun to warm up to increase their metabolism. Fish can change water depths or seasonally migrate. Frogs, toads, and lizards hibernate during winter. To stay warm, honeybees crowd together and rapidly flap their wings to generate heat.

Some animals do not fall into the strict definitions of being warm-blooded or cold-blooded. Bats, for example, are mammals that cannot maintain a constant body temperature, and they cool off when not active. Some warm-blooded animals, such as bears, groundhogs, and gophers, hibernate in winter. During the hibernation period, they live off stored body fat; their body temperatures may drop as much as 10 °C (18 °F).

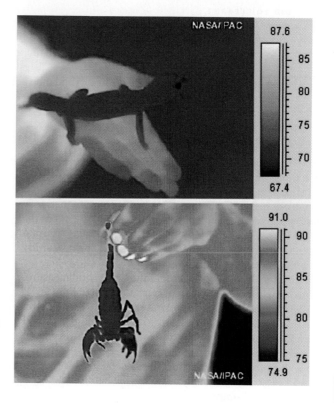

FIGURE 1 **Warm-blooded and cold-blooded** The infrared images show that cold-blooded creatures take on the temperature of their surroundings. Both the gecko and the scorpion are at the same temperature (color) as the air surrounding them. Notice the difference between these cold-blooded creatures and the warm-blooded humans holding them.

10.3 Gas Laws, Absolute Temperature, and the Kelvin Temperature Scale

LEARNING PATH QUESTIONS

⇒ What are the three common forms of the ideal gas law?
⇒ How is absolute zero determined?
⇒ How do you convert a temperature on the Celsius scale to a temperature on the Kelvin scale?

Whereas different liquid-in-glass thermometers show slightly different readings for temperatures other than fixed points because of the liquids' different expansion properties, a thermometer that uses a gas gives the same readings regardless of the gas used. The reason is that at very low densities all gases exhibit the same expansion behavior.

The variables that describe the behavior of a given quantity (mass) of gas are pressure, volume, and temperature (p, V, and T). When temperature is held constant, the pressure and volume of a quantity of gas are related as follows:

$$pV = \text{constant} \quad \text{or} \quad p_1V_1 = p_2V_2 \quad \textit{(at constant temperature)} \qquad (10.3)$$

That is, the product of pressure and volume is a constant. This relationship is known as *Boyle's law*, after Robert Boyle (1627–1691), the English chemist who discovered it.

When the pressure is held constant, the volume of a quantity of gas is related to the *absolute* temperature (to be defined shortly):

$$\frac{V}{T} = \text{constant} \quad \text{or} \quad \frac{V_1}{T_1} = \frac{V_2}{T_2} \quad \textit{(at constant pressure)} \qquad (10.4)$$

That is, the ratio of the volume to the temperature is a constant. This relationship is known as *Charles's law*, named for the French scientist Jacques Charles (1746–1823), who took early hot-air balloon flights and was therefore quite interested in the relationship between the volume and temperature of a gas. A popular demonstration of Charles's law is shown in ▸Fig. 10.6.

Low-density gases obey these laws, which may be combined into a single relationship. Since pV = constant and V/T = constant for a given quantity of gas, pV/T must also equal a constant. This relationship is the **ideal gas law**:

$$\frac{pV}{T} = \text{constant} \quad \text{or} \quad \frac{p_1 V_1}{T_1} = \frac{p_2 V_2}{T_2} \qquad \text{\textit{(ideal gas law, ratio form)}} \qquad (10.5)$$

That is, the ratio pV/T at one time (t_1) is the same as at another time (t_2), or at any other time, as long as the quantity (number of molecules or mass) of gas does not change.

This relationship can be written in a more general form that applies not just to a given quantity of a single gas, but to any quantity of any low-pressure, dilute gas. With a quantity of gas determined by the number of molecules (N) in the gas (that is, $pV/T \propto N$), it follows that

$$\frac{pV}{T} = Nk_B \quad \text{or} \quad pV = Nk_B T \qquad \text{\textit{(ideal gas law)}} \qquad (10.6)$$

where k_B is a constant of proportionality known as the *Boltzmann's constant*: $k_B = 1.38 \times 10^{-23}$ J/K.*

The K stands for temperature on the Kelvin scale, discussed shortly. Note that the mass of the sample does not appear explicitly in Eq. 10.6. However, the number of molecules N in a sample of a gas is proportional to the total mass of the gas. The ideal gas law, sometimes called the *perfect gas law*, applies to real gases with low pressures and densities, and describes the behavior of most gases fairly accurately at normal densities.

(a)   (b)

▲ **FIGURE 10.6 Charles's law in action** Demonstrations of the relationship between the volume and the temperature of a quantity of gas. A weighted balloon, initially at room temperature, is placed in a beaker of water. **(a)** When ice is placed in the beaker and the temperature falls, the balloon's volume decreases. **(b)** When the water is heated and the temperature rises, the balloon's volume increases.

DEMONSTRATION 3 | **Boyle's Shaving Cream**

A demonstration that shows Boyle's law, $p \propto 1/V$, the inverse relationship of pressure (p) and volume (V).

As the pressure is reduced, the swirl of cream grows in volume from the expansion of the air bubbles trapped in the cream.

In a dramatic volume reduction, the shaving cream cannot stand up to the sudden increase in pressure when the chamber is vented to the atmosphere.

*Named after the Austrian physicist Ludwig Boltzmann (1844–1906), who made important contributions in determining this constant.

MACROSCOPIC FORM OF THE IDEAL GAS LAW

Equation 10.6 is a *microscopic* (*micro* means extremely small) form of the ideal gas law in that it refers specifically to the number of molecules, N. However, the law can be rewritten in a *macroscopic* (*macro* means large) form, which involves quantities that can be measured with everyday laboratory equipment. The ideal gas law in this form is

$$pV = nRT \quad \textit{(ideal gas law)} \tag{10.7}$$

using nR rather than Nk_B for convenience since $n \propto N$. Here, n is the number of moles (mol) of the gas, a quantity defined next, and R is called the *universal gas constant*:

$$R = 8.31 \text{ J/(mol} \cdot \text{K)}$$

A **mole** (abbreviated mol) of a substance is defined as the quantity that contains **Avogadro's number** N_A of molecules:

$$N_A = 6.02 \times 10^{23} \text{ molecules/mol}$$

Thus, n and N in the two forms of the ideal gas law are related by $N = nN_A$. From Eq. 10.7, it can be shown that 1 mol of *any* gas occupies 22.4 L at 0 °C and 1 atm. These conditions, 0 °C and 1 atm, are known as *standard temperature and pressure (STP)*.

It is important to note what these equations for the macroscopic (Eq. 10.7) and microscopic (Eq. 10.6) forms of the ideal gas law represent. For the macroscopic form of the ideal gas law, the constant $R = pV/(nT)$ has units of J/(mol·K). For the microscopic form of the law, $k_B = pV/(NT)$, with units of J/(molecule·K). Note that the difference between the macroscopic and microscopic forms of the ideal gas law is moles versus the number of molecules, and gas quantities are usually measured in moles in the laboratory.

To use Eq. 10.7, we need to know the number of moles of a quantity of gas. This is done by finding the *molar mass*, M, of a compound or element. Molar mass is the mass of one mole of substance, so $M = mN_A$, where m is the *molecular mass* or the mass of one molecule. Because molecular masses are so small in relation to the SI standard kilogram, another unit, the *atomic mass unit* (u), is used:

$$1 \text{ atomic mass unit (u)} = 1.66054 \times 10^{-27} \text{ kg*}$$

The molecular mass is determined from the chemical formula and the atomic masses of the atoms. (The latter are listed in Appendix IV and are commonly rounded to the nearest one half.) For example, water, H_2O, with two hydrogen atoms and one oxygen atom, has a molecular mass of $2(m_H) + 1(m_O) = 2(1.0 \text{ u}) + 1(16.0 \text{ u}) = 18.0 \text{ u}$, because the atomic mass of each hydrogen atom is 1.0 u and that of an oxygen atom is 16.0 u. Then, one mole of water has a molar mass of $(18 \text{ u})(1.66054 \times 10^{-27} \text{ kg/u})(6.02 \times 10^{23}/\text{mol}) = 0.0180 \text{ kg/mol} = 18.0 \text{ g/mol}$. Similarly, the oxygen we breathe, O_2, has a molecular mass of $2 \times 16.0 \text{ u} = 32.0 \text{ u}$. Hence, one mole of oxygen has a mass of 32.0 g.

The reverse calculation can also be made. For example, suppose you want to know the mass of a water molecule (H_2O). As was just seen, the molar mass of water is 18.0 g, or 18.0 g/mol. The molecular mass (m) is then given by

$$m_{H_2O} = \frac{M \text{ (molar mass)}}{N_A} = \frac{(18.0 \text{ g/mol})}{6.02 \times 10^{23} \text{ molecules/mol}}$$

$$= 2.99 \times 10^{-23} \text{ g/molecule} = 2.99 \times 10^{-26} \text{ kg/molecule}$$

ABSOLUTE ZERO AND THE KELVIN TEMPERATURE SCALE

The product of the pressure and the volume of a sample of ideal gas is directly proportional to the temperature of the gas: $pV \propto T$. This relationship allows a gas to be used to measure temperature in a *constant volume gas thermometer*. Holding the volume of the gas constant, which can be done easily in a rigid container, means that $p \propto T$ (▶Fig. 10.7). Then using a constant volume gas thermometer,

---

*The atomic mass unit is based on assigning a carbon-12 atom the value of exactly 12 u.

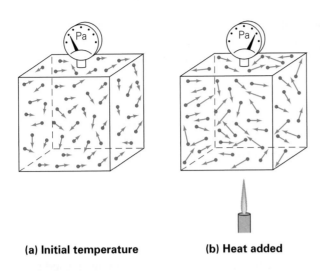

(a) Initial temperature   **(b) Heat added**

◄ **FIGURE 10.7 Constant volume gas thermometer** Such a thermometer indicates temperature as a function of pressure, since, for a low-density gas, $p \propto T$. **(a)** At some initial temperature, the pressure reading has a certain value. **(b)** When the gas thermometer is heated, the pressure (and temperature) reading is higher, because, on average, the gas molecules are moving faster.

one reads the temperature in terms of pressure. A plot of pressure versus temperature gives a straight line in this case (▼Fig. 10.8a).

As can be seen in Fig. 10.8b, measurements of real gases (plotted data points) deviate from the values predicted by the ideal gas law at very low temperatures. This is because the gases liquefy at such temperatures. However, the relationship is linear over a large temperature range, and it looks as though the pressure might reach zero with decreasing temperature if the gas were to ramain in its gaseous state.

The absolute minimum temperature for an ideal gas is therefore inferred by extrapolating, or extending the straight line to the axis, as in Fig. 10.8b. This temperature is found to be $-273.15\,°C$ and is designated as **absolute zero**. Absolute zero is believed to be the lower limit of temperature, but it has never been attained. In fact, there is a law of thermodynamics that says it never can be achieved (Section 12.5).* There is no known upper limit to temperature. For example, the temperatures at the centers of some stars are estimated to be greater than 100 million degrees Celsius.

Absolute zero is the foundation of the **Kelvin temperature scale**, named after the British scientist Lord Kelvin who proposed it in 1848.[†] On this scale, $-273.15\,°C$

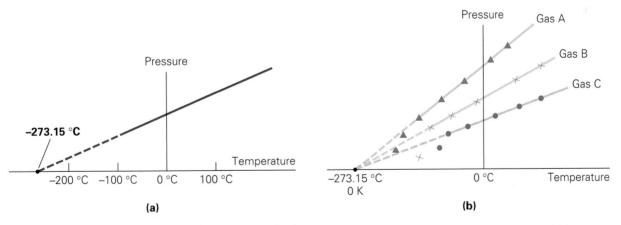

(a)   **(b)**

▲ **FIGURE 10.8 Pressure versus temperature** **(a)** A low-density gas kept at a constant volume gives a straight line on a graph of p versus T, that is, $p = (Nk_B/V)T$. When the line is extended to the zero pressure value, a temperature of $-273.15\,°C$ is obtained, which is taken to be absolute zero. **(b)** Extrapolation of lines for all low-density gases indicates the same absolute zero temperature. The actual behavior of gases deviates from this straight-line relationship at low temperatures because the gases start to liquefy.

*At the time of this writing, the lowest overall average thermodynamic temperature that scientists have been able to attain is $450 \times 10^{-12}\,K$, that is, 450 pK (picokelvins) above absolute zero.

[†]Lord Kelvin, born William Thomson (1824–1907), developed devices to improve telegraphy and the compass and was involved in the laying of the first transatlantic cable. When he received his title, it is said that he considered choosing Lord Cable or Lord Compass, but decided on Lord Kelvin, after a river that runs near the University of Glasgow in Scotland, where he was a professor of physics for fifty years.

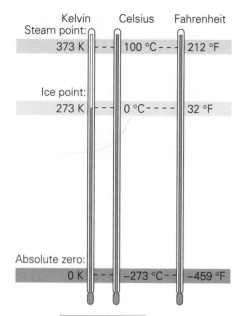

$$T_K = T_C + 273$$

▲ **FIGURE 10.9 The Kelvin temperature scale** The lowest temperature on the Kelvin scale (corresponding to $-273.15\,°C$) is absolute zero. A unit interval on the Kelvin scale, called a kelvin and abbreviated K, is equivalent to a temperature change of $1\,°C$, thus, $T_K = T_C + 273.15$. (The constant is usually rounded to 273 for convenience.) For example, a temperature of $0\,°C$ is equal to 273 kelvins.

is taken as the zero point—that is, as 0 K (◄Fig. 10.9). The size of a single unit of Kelvin temperature is the same as that of the degree Celsius, so temperatures on these scales are related by

$$T = T_C + 273.15 \quad \textit{(Celsius-to-Kelvin conversion)} \qquad (10.8)$$

where T is the temperature in **kelvins** (for example, a temperature of 300 kelvins). The kelvin is abbreviated as K (*not* degrees Kelvin, °K). For general calculations, it is common to round the 273.15 in Eq. 10.8 to 273, that is,

$$T = T_C + 273 \quad \textit{(for general calculations)} \qquad (10.8a)$$

The absolute Kelvin scale is the official SI temperature scale; however, the Celsius scale is used in most parts of the world for everyday temperature readings. The absolute temperature in kelvins is used primarily in scientific applications.

PROBLEM-SOLVING HINT

Keep in mind that Kelvin temperatures *must* be used with any form of the ideal gas law. It is a common mistake to use Celsius or Fahrenheit temperatures. Suppose you used a Celsius temperature of $T = 0\,°C$ in the gas law. You would have $pV = 0$, which makes no sense, since neither p nor V is zero at the ice point of water.

Note that there can be no negative temperatures on the Kelvin scale if absolute zero is the lowest possible temperature. That is, the Kelvin scale doesn't have an arbitrary zero temperature somewhere within the scale as on the Fahrenheit and Celsius scales—zero K is absolute zero, period.

EXAMPLE 10.2 | Deepest Freeze: Absolute Zero on the Fahrenheit Scale

What is absolute zero on the Fahrenheit scale?

THINKING IT THROUGH.   This requires the conversion of 0 K to the Fahrenheit scale. But first a conversion to the Celsius scale is in order. (Why?)

SOLUTION.

Given: $T = 0\,K$ **Find:** T_F

Temperatures on the Kelvin scale are related directly to Celsius temperatures by $T = T_C + 273.15$ (Eq. 10.8 for accuracy), so first we convert 0 K to a Celsius value:

$$T_C = T - 273.15 = (0 - 273.15)\,°C = -273.15\,°C$$

Then, converting to Fahrenheit (Eq. 10.1) gives

$$T_F = \tfrac{9}{5}T_C + 32 = \left[\tfrac{9}{5}(-273.15) + 32\right]\,°F = -459.67\,°F$$

Thus, absolute zero is about $-460\,°F$.

FOLLOW-UP EXERCISE.   There is an absolute temperature scale associated with the Fahrenheit temperature scale called the Rankine scale. A Rankine degree is the same size as a Fahrenheit degree, and absolute zero is taken as 0 °R (zero degree Rankine). Write the conversion equations between (a) the Rankine and the Fahrenheit scales, (b) the Rankine and the Celsius scales, and (c) the Rankine and the Kelvin scales.

Initially, gas thermometers were calibrated by using the ice and steam points. The Kelvin scale uses absolute zero and a second fixed point adopted in 1954 by the International Committee on Weights and Measures. This second fixed point is the **triple point of water**, at which water coexists simultaneously in equilibrium as a solid (ice), liquid (water), and gas (water vapor). The triple point occurs at a unique set of values for temperature and pressure—a temperature of 0.01 °C and a pressure of 4.58 mm Hg (611.73 Pa)—and provides a reproducible reference temperature for the Kelvin scale. The temperature of the triple point on the Kelvin scale was assigned a value of 273.16 K. The SI kelvin unit is then defined in terms of the temperature at the triple point of water.*

The Kelvin temperature scale has special significance. As will be seen in Section 10.5, the absolute temperature is directly proportional to the internal energy of an ideal gas and so can be used as an indication of that energy. There are no negative values on the absolute scale. Negative absolute temperatures would imply negative internal energy for the gas, a meaningless concept.

Now let's use various forms of the ideal gas law, which requires absolute temperatures.

EXAMPLE 10.3 | ## The Ideal Gas Law: Using Absolute Temperatures

A quantity of ideal gas in a rigid container is initially at room temperature (20 °C) and a particular pressure (p_1). If the gas is heated to a temperature of 60 °C, by what factor does the pressure change?

THINKING IT THROUGH. The law requires *absolute* temperatures, so we need to change the Celsius temperatures to kelvins. A "factor" of change implies a ratio (p_2/p_1), so the ideal gas law in ratio form (Eq. 10.5) should apply. Note that the container is rigid, which means that $V_1 = V_2$.

SOLUTION.

Given: $T_1 = 20 \, °C = (20 + 273) \, K = 293 \, K$ *Find:* p_2/p_1 (pressure ratio or factor)
            $T_2 = 60 \, °C = (60 + 273) \, K = 333 \, K$
            $V_1 = V_2$

Since the factor by which the pressure changes is wanted, p_2/p_1 is written as a ratio. For example, if $p_2/p_1 = 2$, then $p_2 = 2p_1$, or the pressure would change (increase) by a factor of 2. Using the ideal gas law in ratio form (Eq. 10.5) $p_2V_2/T_2 = p_1V_1/T_1$, we have, with $V_1 = V_2$,

$$p_2 = \left(\frac{T_2}{T_1}\right)p_1 = \left(\frac{333 \, K}{293 \, K}\right)p_1 = 1.14p_1$$

So, p_2 is 1.14 times p_1; that is, the pressure increases by a factor of 1.14, or 14%. (What would the factor be if the Celsius temperatures were *incorrectly* used? It would be much larger: $(60 \, °C)/(20 \, °C) = 3$, or $p_2 = 3p_1$. Wrong.)

FOLLOW-UP EXERCISE. If the gas in this Example is heated from an initial temperature of 20 °C (room temperature) so that the pressure increases by a factor of 1.26, what is the final Celsius temperature?

EXAMPLE 10.4 | ## The Ideal Gas Law: How Much Oxygen?

A patient receiving breathing therapy purchased a filled M9 (medical size 9) oxygen (O_2) tank. The tank has a volume of 2.5 L and is filled with pure oxygen to an absolute pressure of 100 atm at 20 °C. What is the mass of the oxygen in the tank?

THINKING IT THROUGH. Since the mass of oxygen (O_2) is to be determined, the number of moles of the gas in the tank

needs to be calculated first. That implies that the macroscopic form of the ideal gas law (Eq. 10.7) should be used. Then the molar mass of oxygen (O_2) can be used to calculate the mass. In addition, the units of pressure and volume should be in standard SI units, and temperature should be in kelvin.

SOLUTION.

Given: $T = 20 \, °C = (20 + 273) \, K = 293 \, K$ *Find:* m (mass of oxygen)
           $p = 100 \, atm = (100)(1.01 \times 10^5 \, Pa) = 1.01 \times 10^7 \, Pa$
           $V = 2.5 \, L = 0.0025 \, m^3$ (because $1 \, m^3 = 1000 \, L$)

Using the macroscopic form of the ideal gas law (Eq. 10.7) and solving for the number of moles, n,

$$n = \frac{pV}{RT} = \frac{(1.01 \times 10^7 \, Pa)(0.0025 \, m^3)}{[8.31 \, J/(mol \cdot K)](293 \, K)} = 10.4 \, mol$$

Oxygen (O_2) has a molecular mass of $2 \times 16.0 \, u = 32.0 \, u$, so its molar mass is 32.0 g/mol. Therefore, the mass of oxygen in the tank is

$$m = (10.4 \, mol)(32.0 \, g/mol) = 333 \, g = 0.333 \, kg$$

FOLLOW-UP EXERCISE. When the patient breathes the oxygen, it is depressurized to 1 atm. What is the volume of the oxygen at this pressure?

---

*The 273.16 value given here for the triple point temperature and the −273.15 value, as determined in Fig. 10.8, indicate different things. The −273.15 °C is taken as 0 K. The 273.16 K (or 0.01 °C) is a different reading on a different temperature scale.

10.4 Thermal Expansion

LEARNING PATH QUESTIONS

➥ What is the fundamental cause of thermal expansion?

➥ If water is heated from 0 °C to 4 °C, will it expand or contract?

➥ Which temperature scale(s) can be used in calculating the change in temperature (ΔT) in thermal expansion when the unit of the coefficient of thermalexpansion is 1/°C?

Changes in the dimensions and volumes of materials are common thermal effects. As you learned earlier, thermal expansion provides a means of measuring temperature. The thermal expansion of gases is generally described by the ideal gas law and is very obvious. Less dramatic, but by no means less important, is the thermal expansion of liquids and solids (discussed in Section 9.1).

Thermal expansion results from a change in the average distance separating the atoms of a substance as it is heated. The atoms are held together by bonding forces, which can be simplistically represented as springs in a simple model of a solid. (See Fig. 9.1.) With increased temperature, the atoms vibrate back and forth over greater distances. With wider vibrations in all dimensions, the solid expands as a whole.

The change in one dimension of a solid (length, width, or thickness) is called *linear* expansion. For small temperature changes, linear expansion (or contraction) is approximately proportional to ΔT, or $T - T_o$ (▾Fig. 10.10a). The *fractional* change in length is $(L - L_o)/L_o$ or $\Delta L/L_o$, where L_o is the original length of the solid at the initial temperature.* This ratio is related to the change in temperature by

$$\frac{\Delta L}{L_o} = \alpha\Delta T \quad \text{or} \quad \Delta L = \alpha L_o\Delta T \quad \textit{(linear expansion)} \tag{10.9}$$

where α is the **thermal coefficient of linear expansion**. Note that the unit of α is inverse temperature: inverse degree Celsius (1/°C, or °C^{-1}). Values of α for some materials are given in ▸Table 10.1.

▾**FIGURE 10.10 Thermal expansion (a)** Linear expansion is proportional to the temperature change; that is, the change in length ΔL is proportional to ΔT, and $\Delta L/L_o = \alpha\Delta T$, where α is the thermal coefficient of linear expansion. **(b)** For isotropic expansion, the thermal coefficient of area expansion is approximately 2α. **(c)** The thermal coefficient of volume expansion for solids is about 3α.

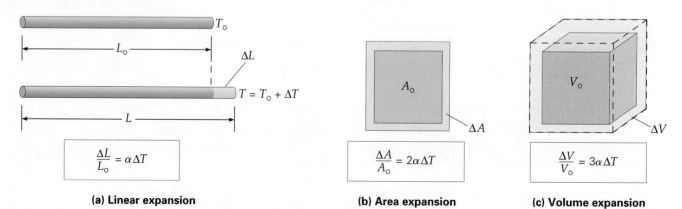

(a) Linear expansion

$$\frac{\Delta L}{L_o} = \alpha\Delta T$$

(b) Area expansion

$$\frac{\Delta A}{A_o} = 2\alpha\Delta T$$

(c) Volume expansion

$$\frac{\Delta V}{V_o} = 3\alpha\Delta T$$

*A fractional change may also be expressed as a percent change. For example, by analogy, if you invested \$100 (\$$_o$) and made \$10 (Δ\$), then the fractional change would be $\Delta\$/\$_o = 10/100 = 0.10$, or an increase (percent change) of 10%

TABLE 10.1 Values of Thermal Expansion Coefficients (in $°C^{-1}$) for Some Materials at 20 °C

Material	Coefficient of Linear Expansion (α)	Material	Coefficient of Volume Expansion (β)
Aluminum	24×10^{-6}	Alcohol, ethyl	1.1×10^{-4}
Brass	19×10^{-6}	Gasoline	9.5×10^{-4}
Brick or concrete	12×10^{-6}	Glycerin	4.9×10^{-4}
Copper	17×10^{-6}	Mercury	1.8×10^{-4}
Glass, window	9.0×10^{-6}	Water	2.1×10^{-4}
Glass, Pyrex	3.3×10^{-6}		
Gold	14×10^{-6}	Air (and most other gases at 1 atm)	3.5×10^{-3}
Ice	52×10^{-6}		
Iron and steel	12×10^{-6}		

A solid may have different coefficients of linear expansion for different directions, but for simplicity it will be assumed that the same coefficient applies to all directions (in other words, that solids show *isotropic* expansion). Also, the coefficient of expansion may vary slightly for different temperature ranges. Since this variation is negligible for most common applications, α will be considered to be constant and independent of temperature.

Equation 10.9 can be rewritten to give the final length (L) after a change in temperature:

$$L - L_o = \alpha L_o \Delta T \quad \text{so} \quad L = L_o + \alpha L_o \Delta T$$

or

$$L = L_o(1 + \alpha \Delta T) \qquad (10.10)$$

Equation 10.10 can be used to compute the thermal expansion of *areas* of flat objects. Since area (A) is length squared (L^2) for a square,

$$A = L^2 = L_o^2(1 + \alpha \Delta T)^2 = A_o(1 + 2\alpha \Delta T + \alpha^2 \Delta T^2)$$

where A_o is the original area. Because the values of α for solids are much less than 1 ($\sim 10^{-5}$), as shown in Table 10.1, the second-order term (containing $\alpha^2 \approx (10^{-5})^2 = 10^{-10} \ll 10^{-5}$) can be dropped with negligible error. As a first-order approximation, then, and with the understanding that the change in area $\Delta A = A - A_o$, we have

$$A = A_o(1 + 2\alpha \Delta T) \quad \text{or} \quad \frac{\Delta A}{A_o} = 2\alpha \Delta T \quad \textit{(area expansion)} \qquad (10.11)$$

Thus, the **thermal coefficient of area expansion** (Fig. 10.10b) is twice as large as the coefficient of linear expansion. (That is, it is equal to 2α). This relationship is valid for all flat shapes. (See Learn by Drawing 10.1, Thermal Area Expansion.)

Similarly, a first-order expression for thermal *volume* expansion is

$$V = V_o(1 + 3\alpha \Delta T) \quad \text{or} \quad \frac{\Delta V}{V_o} = 3\alpha \Delta T \quad \textit{(volume expansion)} \qquad (10.12)$$

The **thermal coefficient of volume expansion** (Fig. 10.10c) is equal to 3α (for isotropic solids).

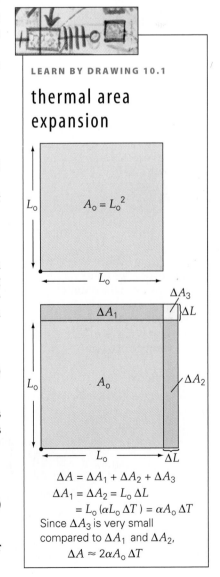

LEARN BY DRAWING 10.1

thermal area expansion

$$A_o = L_o^2$$

$$\Delta A = \Delta A_1 + \Delta A_2 + \Delta A_3$$
$$\Delta A_1 = \Delta A_2 = L_o \Delta L$$
$$= L_o(\alpha L_o \Delta T) = \alpha A_o \Delta T$$

Since ΔA_3 is very small compared to ΔA_1 and ΔA_2,

$$\Delta A \approx 2\alpha A_o \Delta T$$

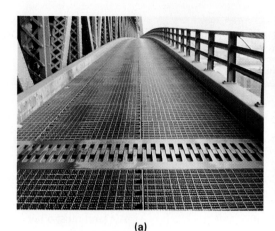

(a)   **(b)**

▲ **FIGURE 10.11 Expansion gaps** **(a)** Expansion gaps are built into bridge roadways to prevent contact stresses produced by thermal expansion. **(b)** These loops in oil pipelines serve a similar purpose. As hot oil passes through them, the pipes expand, and the loops take up the extra length. The loops also accommodate expansions resulting from day–night and seasonal temperature variations.

Keep in mind that the equations for thermal expansions are approximations (why?), so they may apply only in certain situations.

The thermal expansion of materials is an important consideration in construction. Seams are put in concrete highways and sidewalks to allow room for expansion and to prevent cracking. Expansion gaps in large bridges and between railroad rails are necessary to prevent damage (▲Fig. 10.11a). The Golden Gate Bridge across San Francisco Bay varies in length by about 1 m between summer and winter. Similarly, expansion loops are found in oil pipelines (Fig. 10.11b). The height of the Eiffel Tower in Paris varies 0.36 cm for each degree Celsius change.

The thermal expansion of steel beams and girders can cause tremendous pressures, as the following Example shows.

EXAMPLE 10.5 | ## Temperature Rising: Thermal Expansion and Stress

A steel beam is 5.0 m long at a temperature of 20 °C (68 °F). On a hot day, the temperature rises to 40 °C (104 °F). (a) What is the change in the beam's length due to thermal expansion? (b) Suppose that the ends of the beam are initially in contact with rigid vertical supports. How much force will the expanded beam exert on the supports if the beam has a cross-sectional area of 60 cm²?

THINKING IT THROUGH. (a) This is a direct application of Eq. 10.9. (b) As the constricted beam expands, it applies a stress, and hence a force, to the supports. For linear expansion, Young's modulus (Section 9.1) should come into play.

SOLUTION.

Given: $L_0 = 5.0$ m
$T_0 = 20$ °C
$T = 40$ °C
$\alpha = 12 \times 10^{-6}$ °C^{-1} (from Table 10.1)
$A = (60 \text{ cm}^2)\left(\dfrac{1 \text{ m}}{100 \text{ cm}}\right)^2 = 6.0 \times 10^{-3} \text{ m}^2$

Find: (a) ΔL (change in length)
(b) F (force)

(a) Using Eq. 10.9 to find the change in length with $\Delta T = T - T_0 = 40$ °C $- 20$ °C $= 20$ °C, we have

$$\Delta L = \alpha L_0 \Delta T = (12 \times 10^{-6} \text{ °C}^{-1})(5.0 \text{ m})(20 \text{ °C}) = 1.2 \times 10^{-3} \text{ m} = 1.2 \text{ mm}$$

This may not seem like much of an expansion, but it can give rise to a great deal of force if the beam is constrained and kept from expanding, as part (b) will show.

(b) By Newton's third law, if the beam is kept from expanding, the force the beam exerts on its constraint supports is equal to the force exerted by the supports to prevent the beam from expanding by a length ΔL. This is the same as the force that would be required to compress the beam by that length. Using Young's modulus and Eq. 9.4 with $Y = 20 \times 10^{10}$ N/m² (Table 9.1), the stress on the beam is

$$\frac{F}{A} = \frac{Y\Delta L}{L_0} = \frac{(20 \times 10^{10}\,\text{N/m}^2)(1.2 \times 10^{-3}\,\text{m})}{5.0\,\text{m}} = 4.8 \times 10^7\,\text{N/m}^2$$

The force is then

$$F = (4.8 \times 10^7\,\text{N/m}^2)A = (4.8 \times 10^7\,\text{N/m}^2)(6.0 \times 10^{-3}\,\text{m}^2)$$

$$= 2.9 \times 10^5\,\text{N (about 65 000 lb, or 32.5 tons!)}$$

FOLLOW-UP EXERCISE. Expansion gaps between identical steel beams laid end to end are specified to be 0.060% of the length of a beam at the installation temperature. With this specification, what is the temperature range for noncontact expansion?

CONCEPTUAL EXAMPLE 10.6 | **Larger or Smaller? Area Expansion**

A circular piece is cut from a flat metal sheet (▶ Fig. 10.12a). If the sheet is then heated in an oven, the size of the hole will (a) become larger, (b) become smaller, (c) remain unchanged.

REASONING AND ANSWER. It is a common misconception to think that the area of the hole will shrink because the metal expands inwardly around it. To counter this misconception, think of the piece of metal removed from the hole rather than of the hole itself. This piece would expand with increasing temperature. The metal in the heated sheet reacts as if the piece that was removed were still part of it. (Think of putting the piece of metal back into the hole after heating, as in Fig. 10.12b, or consider drawing a circle on an uncut metal sheet and heating it.) So the answer is (a).

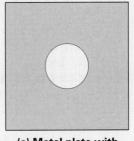

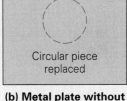

(a) Metal plate with hole

(b) Metal plate without hole

Circular piece replaced

▲ **FIGURE 10.12 A larger or smaller hole?** See Example text for description.

FOLLOW-UP EXERCISE. A student is trying to fit a bearing onto a shaft. The inside diameter of the bearing is just slightly smaller than the outside diameter of the shaft. Should the student heat the bearing or the shaft in order to fit the shaft inside the bearing?

Fluids (liquids and gases), like solids, normally expand with increasing temperature. Because fluids have no definite shape, only volume expansion (and not linear or area expansion) is meaningful. The expression is

$$\frac{\Delta V}{V_0} = \beta\Delta T \quad \textit{(fluid volume expansion)} \qquad (10.13)$$

where β is the coefficient of volume expansion for fluids. Note in Table 10.1 that the values of β for fluids are typically larger than the values of 3α for solids.

Unlike most liquids, water exhibits an anomalous expansion in volume near its ice point. The volume of a given amount of water decreases as it is cooled from room temperature, until its temperature reaches 4 °C (▼ Fig. 10.13a). Below 4 °C, the volume increases, and therefore the density decreases (Fig. 10.13b). This means that water has its maximum density ($\rho = m/V$) at 4 °C (actually, 3.98 °C).

When water freezes, its molecules form a hexagonal (six-sided) lattice pattern. (This is why snowflakes have hexagonal shapes.) It is the open structure of this lattice that gives water its unusual property of expanding on freezing and being less dense as a solid than as a liquid. (This is why ice floats in water and frozen water pipes burst—water expands by about 9% on freezing.)

This property has an important environmental effect: Bodies of water such as lakes and ponds freeze at the top first, and the ice that forms floats. As a lake cools toward 4 °C, water near the surface contracts and becomes denser, and sinks. The warmer, less dense water near the bottom rises. However, once the colder water

▶ **FIGURE 10.13 Thermal expansion of water** Water exhibits nonlinear expansion behavior near its ice point. **(a)** Above 4 °C (actually, 3.98 °C), water expands with increasing temperature, but from 4 °C down to 0 °C, it expands with decreasing temperature. **(b)** As a result, water has its maximum density near 4 °C.

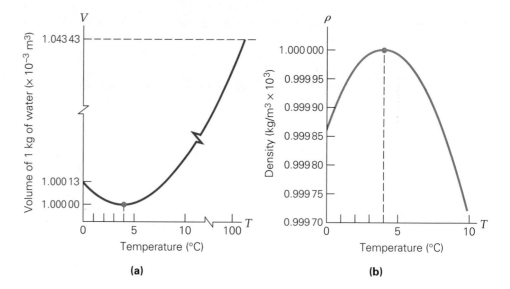

(a)   (b)

on top reaches temperatures below 4 °C, it becomes less dense and remains at the surface, where it freezes. If water did not have this property, lakes and ponds would freeze from the bottom up, which would destroy much of their animal and plant life (and would make ice skating a lot less popular). There would also be no oceanic ice caps at the polar regions. Instead, there would be a thick layer of ice at the bottom of the ocean, covered by a layer of water.

DID YOU LEARN?
➥ Thermal expansion is caused by the change in the average distance between the atoms in a substance when the temperature is changed.
➥ Water exhibits an anomalous expansion in volume between 0 °C and 4 °C; that is, water will actually contract if heated from 0 °C to 4 °C.
➥ Either the Kelvin or Celsius temperature scales can be used for the ΔT in thermal expansion. The Fahrenheit temperature scale cannot be used.

10.5 The Kinetic Theory of Gases

LEARNING PATH QUESTIONS
➥ What fundamental physical quantity does the absolute temperature of a gas determine?
➥ If a sample of gas with more massive molecules and another with less massive molecules are at the same temperature, which gas molecules will have a higher rms speed?
➥ For a given monatomic gas, what is the relationship between its absolute temperature and its total internal energy?

If the molecules of a sample of gas are viewed as colliding particles, the laws of mechanics can be applied to each molecule of the gas. Then the gas's microscopic characteristics, such as velocity and kinetic energy, can be described in terms of molecular motion. Because of the large number of particles involved, however, a statistical approach is employed for such a microscopic description.

One of the major accomplishments of theoretical physics was to do exactly that—derive the ideal gas law from mechanical principles. This derivation led to a new interpretation of temperature in terms of the translational kinetic energy of the gas molecules. As a theoretical starting point, the molecules of an ideal gas are viewed as point masses in random motion with relatively large distances separating them so molecular collisions can be neglected.

In this section, we consider primarily the kinetic theory of *monatomic* (single-atom) gases, such as He, and learn about the internal energy of such a gas. In the next section, the internal energy of *diatomic* (two-atom molecules) gases, such as O_2, will be considered.

According to the **kinetic theory of gases**, the molecules of an ideal gas undergo perfectly elastic collisions (discussed in Section 6.4) with the walls of its container. From Newton's laws of motion, the force on the walls of the container can be calculated from the change in momentum of the gas molecules when they collide with the walls (▶Fig. 10.14). If this force is expressed in terms of pressure (force/area), the following equation is obtained (see Appendix II for derivation):

$$pV = \tfrac{1}{3}Nmv_{rms}^2 \qquad (10.14)$$

Here, V is the volume of the container or gas, N is the number of gas molecules in the closed container, m is the mass of a gas molecule, and v_{rms} is the average speed of the molecules, but a special kind of average. It is obtained by averaging the squares of the speeds and then taking the square root of the average—that is, $\sqrt{\overline{v^{-2}}} = v_{rms}$. As a result, v_{rms} is called the *root-mean-square (rms)* speed.

Solving Eq. 10.6 for pV and equating the resulting expression with Eq. 10.14 shows how temperature came to be interpreted as a measure of translational kinetic energy: $pV = Nk_BT = \tfrac{1}{3}Nmv_{rms}^2$ or

$$\tfrac{1}{2}mv_{rms}^2 = \tfrac{3}{2}k_BT \qquad \text{(for ideal gases)} \qquad (10.15)$$

Thus, the absolute temperature of a gas is directly proportional to its average random kinetic energy (per molecule), since $\overline{K} = \tfrac{1}{2}mv_{rms}^2 = \tfrac{3}{2}k_BT$.

INTEGRATED EXAMPLE 10.7	Molecular Speed: Relation to Absolute Temperature

A helium molecule (He) in a helium balloon is at 20 °C. (a) If it is heated to 40 °C, its rms speed will (1) double, (2) increase by less than a factor of 2, (3) be half as much, (4) decrease by less than a factor of 2. Explain. (b) Calculate the rms speeds at these two temperatures. (Take the mass of the helium molecule to be 6.65×10^{-27} kg).

(A) CONCEPTUAL REASONING According to Eq. 10.15, the absolute temperature is proportional to the square of the rms speed, $T \propto v_{rms}^2$, or the rms speed is proportional to the square root of the absolute temperature, $v_{rms} \propto \sqrt{T}$. Therefore, a higher temperature will increase the rms speed, thus (3) and (4) are not possible.

When the temperature increases from 20 °C to 40 °C, the absolute temperature increases only from $(273 + 20)$ K = 293 K to $(273 + 40)$ K = 313 K, not even close to doubling. Furthermore, even if the absolute temperature were to double, the square root of it would not double either (but it would still increase). Thus, the answer is (2) increase by less than a factor of 2.

(B) QUANTITATIVE REASONING AND SOLUTION All the data needed to solve for the average speed in Eq. 10.15 are given. The Celsius temperatures must be changed to kelvins.

Given: $m = 6.65 \times 10^{-27}$ kg   *Find:* v_{rms} (rms speed)
  $T_1 = 20$ °C $= (273 + 20)$ K $= 293$ K
  $T_2 = 40$ °C $= (273 + 40)$ K $= 313$ K

Rearranging Eq. 10.15,

For 20 °C: $v_{rms} = \sqrt{\dfrac{3k_BT}{m}} = \sqrt{\dfrac{3(1.38 \times 10^{-22}\ \text{J/K})(293\ \text{K})}{6.65 \times 10^{-27}\ \text{kg}}}$

  $= 1.35 \times 10^3$ m/s $= 1.35$ km/s (3020 mph)

For 40 °C: $v_{rms} = \sqrt{\dfrac{3(1.38 \times 10^{-23}\ \text{J/K})(313\ \text{K})}{6.65 \times 10^{-27}\ \text{kg}}} = 1.40 \times 10^3$ m/s

  $= 1.40$ km/s (3130 mph)

FOLLOW-UP EXERCISE. In this Example, if the rms speed is to double its value at 20 °C, what would be the new Celsius temperature?

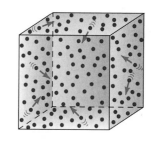

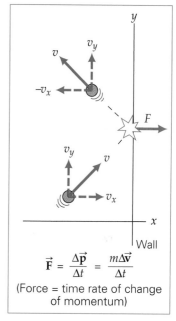

$$\vec{F} = \dfrac{\Delta\vec{p}}{\Delta t} = \dfrac{m\Delta\vec{v}}{\Delta t}$$

(Force = time rate of change of momentum)

▲ **FIGURE 10.14 Kinetic theory of gases** The pressure a gas exerts on the walls of a container is due to the force resulting from the change in momentum of the gas molecules that collide with the wall. The wall exerts a force (action) on the molecule to change its momentum. The molecule exerts a reaction force on the wall. The force exerted by an individual molecule is equal to the time rate of change of momentum; that is, $\vec{F} = \Delta\vec{p}/\Delta t = m\Delta\vec{v}/\Delta t$, where $\vec{p} = m\vec{v}$. The sum of the instantaneous normal components of the collision forces gives rise to the average pressure on the wall.

Interestingly, Eq. 10.15 predicts that at absolute zero ($T = 0$ K), all translational molecular motion of a gas would cease. According to classical theory, this would correspond to absolute zero energy. However, modern quantum theory says that there would still be some zero-point motion and a corresponding minimum *zero-point energy*. Basically, absolute zero is the temperature at which all the energy that *can* be removed from an object has been removed.

INTERNAL ENERGY OF MONATOMIC GASES

Because the "particles" in an ideal monatomic gas do not vibrate or rotate, as explained previously, the total translational kinetic energy of all the molecules is equal to the total internal energy of the gas. That is, the gas's internal energy is all "temperature" energy (Section 10.1). With N molecules in a system, Eq. 10.15 can be used, expressing the energy per molecule, to write an equation for the total internal energy U:

$$U = N\left(\tfrac{1}{2}mv_{\text{rms}}^2\right) = \tfrac{3}{2}Nk_BT = \tfrac{3}{2}nRT \quad \textit{(for monatomic gases)} \quad (10.16)$$

Thus, it can be seen that the internal energy of an ideal monatomic gas is directly proportional to its absolute temperature. (In Section 10.6, it will be learned that this is true regardless of the molecular structure of the gas. However, the expression for U will be a bit different for gases that are not monatomic.) This means that if the absolute temperature of a gas is doubled, for example, from 200 K to 400 K, then the internal energy of the gas is also doubled.

DIFFUSION

We depend on our sense of smell to detect odors, such as the smell of smoke from something burning. That you can smell something from a distance implies that molecules get from one place to another in the air—from the source to your nose. This process of random molecular mixing in which particular molecules move from a region where they are present in higher concentration to one where they are in lower concentration is called **diffusion**. Diffusion also occurs readily in liquids; think about what happens to a drop of ink in a glass of water (▾Fig. 10.15). It even occurs to some degree in solids.

  The rate of diffusion for a particular gas depends on the rms speed of its molecules. Even though gas molecules have large average speeds (Example 10.8), their average positions change slowly, and the molecules do not fly from one side of a

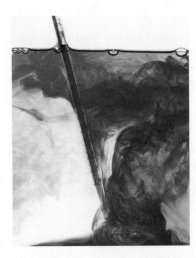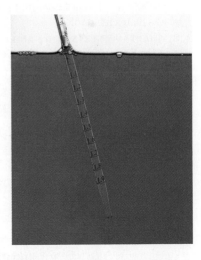

▲ **FIGURE 10.15 Diffusion in liquids** Random molecular motion would eventually distribute the dye throughout the water. Here there is some distribution due to mixing, and the ink colors the water after a few minutes. The distribution would take more time by diffusion only.

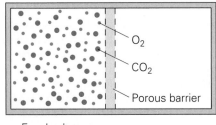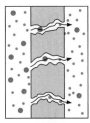

Equal volumes
of O_2 and CO_2

Diffusion
through barrier

▲ FIGURE 10.16 **Separation by gaseous diffusion** The molecules of both gases diffuse through the porous barrier, but because oxygen molecules have the greater average speed, more of them pass through. Thus, over time, there is a greater concentration of oxygen molecules on the other side of the barrier.

room to the other. Instead, there are frequent collisions, and as a result, the molecules "drift" rather slowly. For example, suppose someone opened a bottle of ammonia on the other side of a closed room. It would take some time for the ammonia to diffuse across the room until you could smell it.

The kinetic theory of gases says that the average translational kinetic energy (per molecule) of a gas is proportional to the absolute temperature of the gas: $\frac{1}{2}mv_{rms}^2 = \frac{3}{2}k_BT$. So on the average, the molecules of different gases (having different masses) move at different speeds at a given temperature. As you might expect, because they move faster, less massive gas molecules diffuse faster than do more massive gas molecules.

For instance, at a particular temperature, molecules of oxygen (O_2) move faster on the average than do the more massive molecules of carbon dioxide (CO_2), so oxygen can diffuse through a barrier faster than carbon dioxide can. Suppose that a mixture of equal volumes of oxygen and carbon dioxide is contained on one side of a porous barrier (▲ Fig. 10.16). After a while, some O_2 molecules and some CO_2 molecules will have diffused through the barrier, but more oxygen than carbon dioxide. Purer oxygen can be obtained by repeating the separation process many times. Separation by gaseous diffusion is a key process in obtaining enriched uranium, which was used in the first atomic bomb and in early nuclear reactors that generate electricity (Section 30.2).

Fluid diffusion is very important to organisms. In plant photosynthesis, carbon dioxide from the air diffuses into leaves, and oxygen and water vapor diffuse out. The diffusion of a liquid across a permeable membrane with a concentration gradient (a concentration difference) is called **osmosis**, a process that is vital in living cells. Osmotic diffusion is also important to kidney functioning: Tubules in the kidneys concentrate waste matter from the blood in much the same way that oxygen is removed from mixtures. (See the accompanying Insight 10.3, Physiological Diffusion in Life Processes, for other examples of diffusion.)

Osmosis is the tendency for the solvent of a solution, such as water, to diffuse across a semipermeable membrane from the side where the solvent is at higher concentration to the side where it is at lower concentration. When pressure is applied to the side with the lower concentration, the diffusion is reversed—a process called *reverse osmosis*. Reverse osmosis is used in desalination plants to provide freshwater from seawater in dry coastal regions and in drinking water purification.

DID YOU LEARN?

➥ The absolute temperature of a gas is directly proportional to the average translational kinetic energy of the gas molecules. That is, if the absolute temperature is doubled, the average kinetic energy of the molecules will also double.

➥ The two samples of gas molecules have the same average translational kinetic energy because they are at the same temperature. Therefore, the less massive molecules will have a higher rms speed because $\bar{K} = \frac{1}{2}mv_{rms}^2$.

➥ The total internal energy of a monatomic gas is directly proportional to the absolute temperature, $U = \frac{3}{2}nRT$. When the absolute temperature doubles, the gas will have twice as much total internal energy.

INSIGHT 10.3 | Physiological Diffusion in Life Processes

Diffusion plays a central role in many life processes. For example, consider a cell membrane in the lung. Such a membrane is permeable to a number of substances, any of which will diffuse through the membrane from a region where its concentration is high to a region where its concentration is low. Most important, the lung membrane is permeable to oxygen (O_2), and the transfer of O_2 across the membrane occurs because of a concentration gradient.

The blood carried to the lungs is low in O_2, having given up the oxygen during its circulation through the body to tissues requiring O_2 for metabolism. Conversely, the air in the lungs is high in O_2, because there is a continuous exchange of fresh air in the breathing process. As a result of this concentration difference, or gradient, O_2 diffuses from the space within the lungs into the blood that flows through the lung tissue, and the blood leaving the lungs is high in O_2.

Exchanges between the blood and the tissues occur across capillary walls, and diffusion again is a major factor. The chemical composition of arterial blood is regulated to maintain the proper concentrations of particular solutes (substances dissolved in the blood solution), so diffusion takes place in the appropriate directions across capillary walls. For example, as cells take up O_2 and nutrients, including glucose (blood

sugar), the blood continuously brings in fresh supplies of the substances to maintain the concentration gradient needed for diffusion to the cells. The continuous production of carbon dioxide (CO_2) and metabolic wastes in the cells produces concentration gradients in the opposite direction for these substances. They therefore diffuse out of the cells into the blood, to be carried away from the tissues by the circulatory system.

During periods of physical exertion, cellular activity increases. More O_2 is used up and more CO_2 is produced, thereby increasing the concentration gradients and the diffusion rates. How do the lungs respond to an increased demand for O_2 to the blood? As you might expect, the rate of diffusion depends on the surface area and thickness of the lung membrane. Deeper breathing during exercise causes the alveoli (small air sacs in the lungs) to increase in volume. The alveolar surface area increases accordingly, and the thickness of the membrane wall decreases, allowing more rapid diffusion.

Also, the heart works harder during exercise, and the blood pressure is raised. The increased pressure forces open capillaries that are normally closed during rest or mild activity. As a result, the total exchange area between the blood and cells is increased. Each of these changes helps expedite the exchange of gases during exercise.

*10.6 Kinetic Theory, Diatomic Gases, and the Equipartition Theorem

LEARNING PATH QUESTIONS

➥ What is a diatomic molecule, and what are some examples?

➥ What is the essence of the equipartition theorem?

➥ How does the equipartition theorem apply to monatomic and diatomic gases?

In the real world, most gases are *not* monatomic gases. Monatomic gases are elements known as *noble* or *inert* gases, because they do not readily combine with other atoms. These elements are found on the far right side of the periodic table: helium, neon, argon, krypton, xenon, and radon.

However, the mixture of gases we breathe (collectively known as "air") consists mainly of diatomic molecules of nitrogen (N_2, 78% by volume) and oxygen (O_2, 21% by volume). Each of these gases has two identical atoms chemically bonded together to form a single molecule. How do we deal with these more complicated molecules in terms of the kinetic theory of gases? [There are even more complicated gas molecules consisting of more than two atoms, such as carbon dioxide (CO_2). However, because of the complexity of such gas molecules, our discussion will be limited to diatomic molecules.]

THE EQUIPARTITION THEOREM

As was learned in Section 10.5, the translational kinetic energy of a gas is determined by the gas's temperature. Thus, for any type of gas, regardless of how many atoms make up its molecules, it is *always true* that the average *translational* kinetic energy per molecule is still proportional to the temperature of the gas (Eq. 10.15): $\frac{1}{2}mv_{rms}^2 = \frac{3}{2}k_BT$ (for all gases).

Recall that for monatomic gases, the total internal energy U consists solely of translational kinetic energy. For diatomic molecules, this is not true, because a diatomic molecule is free to rotate and vibrate in addition to moving linearly. Therefore, these extra forms of energy must be taken into account. The expression given in Eq. 10.16 ($U = \frac{3}{2}Nk_BT$) for monatomic gases, which assumes that the total energy is due only to translational kinetic energy, therefore does *not* hold for diatomic gases.

Scientists wondered exactly how the expression for the internal energy of a diatomic gas might differ from that for a monatomic gas. In looking at the derivation of Eq. 10.16 from the kinetic theory, they realized that the factor of 3 in that equation was due to the fact that the gas molecules had three independent linear ways (dimensions) of moving. Thus, for each molecule, there were three independent ways of possessing kinetic energy: with x, y, and z linear motion. Each independent way a molecule has for possessing energy is called a **degree of freedom**.

According to this scheme, a monatomic gas has only three degrees of freedom, since its molecules can move only linearly and can possess kinetic energy in three dimensions.

On the basis of the understanding of monatomic gases and their three degrees of freedom, the **equipartition theorem** was proposed. (As the name implies, the total energy of a gas or molecule is "partitioned," or divided, equally for each degree of freedom.) That is,

> On average, the total internal energy U of an ideal gas is divided equally among each degree of freedom its molecules possess. Furthermore, each degree of freedom contributes $\frac{1}{2}Nk_BT$ (or $\frac{1}{2}nRT$) to the total internal energy of the gas.

THE INTERNAL ENERGY OF A DIATOMIC GAS

To use the equipartition theorem to calculate the internal energy of a diatomic gas such as oxygen, it must be realized that U now includes all the available degrees of freedom. A diatomic molecule could rotate (see Fig. 10.1), thus having rotational kinetic energies about three independent axes of rotations (three more degrees of freedom). A diatomic gas might also vibrate, thus having vibrational kinetic and potential energies (two additional degrees of freedom). Altogether, a diatomic molecule should have seven degrees of freedom.

Consider a symmetric diatomic molecule—for example, O_2. A classical model describes such a diatomic molecule as though the molecules were particles connected by a rigid rod (▶Fig. 10.17). The rotational moment of inertia, I, has the same value about each of the axes (x and y) that pass perpendicularly through the center of the rod. The moment of inertia about the z-axis is essentially zero. (Why?) Thus, only two degrees of freedom are associated with the rotational kinetic energies of diatomic molecules.

Furthermore, quantum theory predicts (and experiment verifies) that for normal (room) temperatures, the vibrational kinetic energy and potential energy are much smaller than the translational and rotational kinetic energies and therefore can be ignored. Thus, the total internal energy of a diatomic gas is composed of the internal energies due to the three linear degrees of freedom and the two rotational degrees of freedom, for a total of five degrees of freedom. Hence,

$$U = K_{\text{trans}} + K_{\text{rot}} = 3\left(\tfrac{1}{2}nRT\right) + 2\left(\tfrac{1}{2}nRT\right)$$
$$= \tfrac{5}{2}nRT = \tfrac{5}{2}Nk_BT$$

(for diatomic gases) (10.17)

Thus, a given monatomic sample of gas at normal room temperature has 40% less internal energy than a similar diatomic sample at the same temperature. Or, equivalently, the monatomic sample possesses only 60% of the internal energy of the diatomic sample.

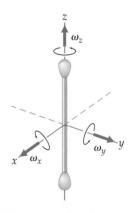

▲ **FIGURE 10.17 Model of a diatomic gas molecule** A dumbbell-like molecule can rotate about three axes. The moment of inertia, I, about the x- and y-axes is the same. The masses (molecules) on the ends of the rod are point-like particles, so the moment of inertia about the z-axis I_z is negligible compared to I_x and I_y.

EXAMPLE 10.8 | Monatomic versus Diatomic: Are Two Atoms Better Than One?

More than 99% of the air we breathe consists of diatomic gases, mainly nitrogen (N_2, 78%) and oxygen (O_2, 21%). There are traces of other gases, one of which is radon (Rn), a monatomic gas arising from radioactive decay of uranium in the ground. (a) Calculate the total internal energy of 1.00-mol samples each of oxygen and radon at room temperature (20 °C). (b) For each sample, calculate the amount of internal energy associated with molecular *translational* kinetic energy.

THINKING IT THROUGH.   (a) We have to consider the number of degrees of freedom in a monatomic gas and a diatomic gas in computing the internal energy U. (b) Only three linear degrees of freedom contribute to the translational kinetic energy portion (U_{trans}) of the internal energy.

SOLUTION.   Listing the data and converting to kelvins because internal energy is expressed in terms of absolute temperature:

Given: $n = 1.00$ mol   *Find:*   (a) U (for O_2 and Rn samples)
       $T = (20 + 273)$ K $= 293$ K   (b) U_{trans} (for O_2 and Rn at 20 °C)

(a) Let's compute the total internal energy of the (monatomic) radon sample first, using Eq. 10.16:

$$U_{Rn} = \tfrac{3}{2}nRT = \tfrac{3}{2}(1.00 \text{ mol})[8.31 \text{ J/(mol} \cdot \text{K)}](293 \text{ K}) = 3.65 \times 10^3 \text{ J}$$

The (diatomic) oxygen will also include internal energy stored as two extra degrees of freedom, due to rotation. Thus, we have

$$U_{O_2} = \tfrac{5}{2}nRT = \tfrac{5}{2}(1.00 \text{ mol})[8.31 \text{ J/(mol} \cdot \text{K)}](293 \text{ K}) = 6.09 \times 10^3 \text{ J}$$

As we have seen, even though each sample has the same number of molecules and the same temperature, the oxygen sample has about 67% more total internal energy.

(b) For (monatomic) radon, all the internal energy is in the form of translational kinetic energy; hence, the answer is the same as in part (a):

$$U_{trans} = U_{Rn} = 3.65 \times 10^3 \text{ J}$$

For (diatomic) oxygen, only $\tfrac{3}{2}nRT$ of the total internal energy $\left(\tfrac{5}{2}nRT\right)$ is in the form of translational kinetic energy, so the answer is the same as for radon; that is, $U_{trans} = 3.65 \times 10^3$ J for both gas samples.

FOLLOW-UP EXERCISE.   (a) In this Example, how much energy is associated with the rotational motion of the oxygen molecules? (b) Which sample has the higher rms speed? (*Note*: The mass of one radon atom is about seven times the mass of an oxygen molecule.) Explain your reasoning.

DID YOU LEARN?

➡ A diatomic molecule consists of two atoms in a molecule. For example, oxygen gas (O_2) and nitrogen gas (N_2) both have two atoms (O and N, respectively) in the gas molecule.

➡ The equipartition theorem states that each degree of freedom of gases contributes equally to the total internal energy by an amount equal to $\tfrac{1}{2}nRT$.

➡ A monatomic gas has three (3) degrees of translational freedom, so its total internal energy is $U = 3 \times \tfrac{1}{2}nRT = \tfrac{3}{2}nRT$. A diatomic gas has three (3) degrees of translational freedom plus two (2) degrees of rotational freedom, so its total internal energy is $U = 5 \times \tfrac{1}{2}nRT = \tfrac{5}{2}nRT$.

PULLING IT TOGETHER | Temperatures, Ideal Gases, and Internal Energies

A quantity of 2.0 moles of an ideal monatomic gas at a pressure of 1.5 atm is confined in a volume of 0.040 m^3. It is then heated so its internal energy doubles. What is its final Fahrenheit temperature?

THINKING IT THROUGH.   This example involves the macroscopic ideal gas law (Eq. 10.7), the internal energy of

monatomic gas relationship (Eq. 10.16), and temperature conversion between the Kelvin and Fahrenheit scales. In order to find the final temperature, the initial absolute temperature is first found using the ideal gas law. Then the internal energy of monatomic gas formula can be used to find the final absolute temperature. Finally, the absolute temperature is converted to the Fahrenheit scale.

SOLUTION. Eq. 10.7 and Eq. 10.16 can be used. Also, pressure needs to be converted from atm to Pa.

Given: $n = 2.0$ mol
      $V_1 = V_2 = 0.040$ m^3 (confined)
      $p_1 = 1.5$ atm $= (1.5)(1.01 \times 10^5$ Pa$) = 1.515 \times 10^5$ Pa
      $U_2 = 2U_1$

Find: $T_{2(F)}$ (final Fahrenheit temperature)

Rearranging Eq. 10.7 to find T_1:

$$T_1 = \frac{pV}{nR} = \frac{(1.515 \times 10^5 \text{ Pa})(0.040 \text{ m}^3)}{(2.0 \text{ mol})[8.31 \text{ J/(mol} \cdot \text{K})]} = 364.6 \text{ K}$$

According to Eq. 10.16, $U = \frac{3}{2}nRT$, doubling U will require T to double as well.

$$U_2 = 2U_1, \quad \text{so} \quad T_2 = 2T_1 = 2(364.6 \text{ K}) = 729.2 \text{ K}$$

Converting T_2 to Celsius scale: $T_{2(C)} = (729.2 - 273)$ °C $= 456.2$ °C.

Finally, converting $T_{2(C)}$ to the Fahrenheit scale:

$$T_{2(F)} = \frac{9}{5}T_{2(C)} + 32 = \left[\frac{9}{5}(456.2) + 32\right] \text{ °F} = 853 \text{ °F}$$

Learning Path Review

- **Celsius–Fahrenheit conversions:**

$$T_F = \frac{9}{5}T_C + 32 \quad \text{or} \quad T_F = 1.8T_C + 32 \qquad (10.1)$$

$$T_C = \frac{5}{9}(T_F - 32) \qquad (10.2)$$

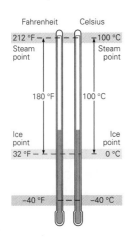

- **Heat** is the net energy transferred from one object to another because of temperature differences. Once transferred, the energy becomes part of the internal energy of the object (or system).

- The **ideal (or perfect) gas law** relates the pressure, volume, and absolute temperature of an ideal gas.

  Ideal (or perfect) gas law (always use absolute temperatures):

$$\frac{p_1 V_1}{T_1} = \frac{p_2 V_2}{T_2} \quad \text{or} \quad pV = Nk_B T \qquad (10.5, 10.6)$$

  or

$$pV = nRT \qquad (10.7)$$

  where $k_B = 1.38 \times 10^{-23}$ J/K and $R = 8.31$ J/(mol$\cdot$K)

- **Absolute zero (0 K)** corresponds to -273.15 °C.

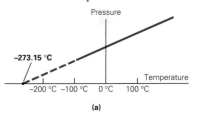

(a)

Celsius–Kelvin conversion:

$$T = T_C + 273.15 \qquad (10.8)$$

$$T = T_C + 273 \quad (\textit{for general calculations}) \qquad (10.8a)$$

- **Thermal coefficients of expansion** relate the fractional change in dimension(s) to a change in temperature.

  Thermal expansion of solids:

$$\text{linear:} \quad \frac{\Delta L}{L_o} = \alpha \Delta T \quad \text{or} \quad L = L_o(1 + \alpha \Delta T) \qquad (10.9, 10.10)$$

$$\text{area:} \quad \frac{\Delta A}{A_o} = 2\alpha \Delta T \quad \text{or} \quad A = A_o(1 + 2\alpha \Delta T) \qquad (10.11)$$

$$\text{volume:} \quad \frac{\Delta V}{V_o} = 3\alpha \Delta T \quad \text{or} \quad V = V_o(1 + 3\alpha \Delta T) \qquad (10.12)$$

Thermal volume expansion of fluids:

$$\frac{\Delta V}{V_0} = \beta \Delta T \qquad (10.13)$$

- According to the **kinetic theory of gases**, the absolute temperature of a gas is directly proportional to the average translational kinetic energy per molecule.

Results of kinetic theory of gases:

$$pV = \frac{1}{3} N m v_{\text{rms}}^2 \qquad (10.14)$$

$$\frac{1}{2} m v_{\text{rms}}^2 = \frac{3}{2} k_B T \quad \text{(for ideal gases)} \qquad (10.15)$$

$$U = \frac{3}{2} N k_B T = \frac{3}{2} n R T \quad \text{(for monatomic gases)} \qquad (10.16)$$

$$U = \frac{5}{2} N k_B T = \frac{5}{2} n R T \quad \text{(for diatomic gases)} \qquad (10.17)$$

Learning Path Questions and Exercises

For instructor-assigned homework, go to www.masteringphysics.com

MULTIPLE CHOICE QUESTIONS

10.1 TEMPERATURE AND HEAT AND
10.2 THE CELSIUS AND FAHRENHEIT TEMPERATURE SCALES

1. Temperature is associated with molecular (a) kinetic energy, (b) potential energy, (c) momentum, (d) all of the preceding.

2. What types of energies can make up the internal energy of a diatomic gas: (a) rotational kinetic energy, (b) translational kinetic energy, (c) vibrational kinetic energy, or (d) all of the preceding?

3. An object at a higher temperature (a) must, (b) may, or (c) must not have more internal energy than another object at a lower temperature.

10.3 GAS LAWS, ABSOLUTE TEMPERATURE, AND THE KELVIN TEMPERATURE SCALE

4. The temperature used in the ideal gas law must be expressed on which scale: (a) Celsius, (b) Fahrenheit, (c) Kelvin, or (d) any of the preceding?

5. If a low-pressure gas at constant volume were to reach absolute zero, (a) its pressure would reach zero, (b) its pressure would reach infinity, (c) its mass would disappear, or (d) its mass would be infinite.

6. When the temperature of a quantity of gas is increased, (a) the pressure must increase, (b) the volume must increase, (c) both the pressure and volume must increase, (d) none of the preceding.

10.4 THERMAL EXPANSION

7. What is the predominant cause of thermal expansion: (a) atom sizes change, (b) atom shapes change, or (c) the distances between atoms change?

8. Are the units of the thermal coefficient of linear expansion (a) m/°C, (b) m²/°C, (c) m · °C, or (d) 1/°C?

9. Which of the following describes the behavior of water density in the temperature range of 0 °C to 4 °C: (a) increases with increasing temperature, (b) remains constant, (c) decreases with increasing temperature, or (d) none of the preceding?

10.5 THE KINETIC THEORY OF GASES

10. If the average kinetic energy of the molecules in an ideal gas initially at 20 °C doubles, what is the final temperature of the gas: (a) 10 °C, (b) 40 °C, (c) 313 °C, or (d) 586 °C?

11. If the temperature of a quantity of ideal gas is raised from 100 K to 200 K, is the internal energy of the gas (a) doubled, (b) halved, (c) unchanged, or (d) none of the preceding?

12. Two different gas samples are at the same temperature. The more massive gas molecules will have (a) a higher, (b) a lower, or (c) the same rms speed as that of the less massive gas molecules.

*10.6 KINETIC THEORY, DIATOMIC GASES, AND THE EQUIPARTITION THEOREM

13. Which of the following is a diatomic molecule: (a) He, (b) N_2, (c) CO_2, or (d) Ne?

14. A diatomic gas such as O_2 near room temperature has an internal energy of (a) $\frac{3}{2} nRT$, (b) $\frac{5}{2} nRT$, (c) $\frac{7}{2} nRT$, or (d) none of the preceding.

15. On average, is the total internal energy of a gas divided equally among (a) each molecule, (b) each degree of freedom, (c) translational motion, rotational motion, and vibrational motion, or (d) none of the preceding?

CONCEPTUAL QUESTIONS

10.1 TEMPERATURE AND HEAT AND
10.2 THE CELSIUS AND FAHRENHEIT TEMPERATURE SCALES

1. Heat flows spontaneously from a body at a higher temperature to one at a lower temperature that is in thermal

contact with it. Does heat always flow from a body with more internal energy to one with less internal energy? Explain.

2. What is the hottest (highest temperature) item in a home? [*Hint:* Think about this one, and maybe a light will come on.]

3. The tires of commercial jumbo jets are inflated with pure nitrogen, not air. Why? [*Hint*: air contains moisture.]

4. When temperature changes during the day, which scale, Celsius or Fahrenheit, will read a smaller change? Explain.

5. What types of energy make up the internal energy of monatomic gases? How about diatomic gases?

10.3 GAS LAWS, ABSOLUTE TEMPERATURE, AND THE KELVIN TEMPERATURE SCALE

6. A type of constant volume gas thermometer is shown in ▼Fig. 10.18. Describe how it operates.

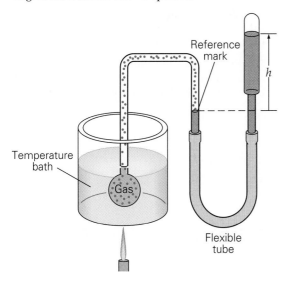

▲ FIGURE 10.18 **A type of constant volume gas thermometer** See Conceptual Question 6.

7. Describe how a constant pressure gas thermometer might be constructed.

8. In terms of the ideal gas law, what would a temperature of absolute zero imply? How about a negative absolute temperature?

9. Excited about a New Year's Eve party in Times Square, you pump up ten balloons in your warm apartment and take them to the cold square. However, you are very disappointed with your decorations. Why?

10. Which has more molecules, 1 mole of oxygen or 1 mole of nitrogen? Explain.

10.4 THERMAL EXPANSION

11. A cube of ice sits on a bimetallic strip at room temperature (▼Fig. 10.19). What will happen if (a) the upper strip is aluminum and the lower strip is brass, and (b) the upper strip is iron and the lower strip is copper? (c) If the cube is made of a hot metal rather than ice and the two strips are brass and copper, which metal should be on top to keep the cube from falling off?

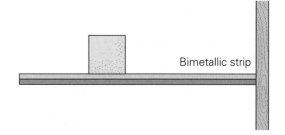

▲ FIGURE 10.19 **Which way will the cube go?** See Conceptual Question 11.

12. A solid metal disk rotates freely, so the conservation of angular momentum applies (Chapter 8). If the disk is heated while it is rotating, will there be any effect on the rate of rotation (the angular speed)?

13. A demonstration of thermal expansion is shown in ▼Fig. 10.20. (a) Initially, the ball fits through the ring made of the same metal. When the ball is heated (b), it does not fit through the ring (c). If both the ball and the ring are heated, the ball again fits through the ring. Explain what is being demonstrated.

14. A circular ring of iron has a tight-fitting iron bar across its diameter, as illustrated in ▼Fig. 10.21. If the arrangement is heated in an oven to a high temperature, will the circular ring be distorted? What if the bar is made of aluminum?

◄FIGURE 10.21 **Stress out of shape?** See Conceptual Question 14.

▶ FIGURE 10.20 **Ball-and-ring expansion** See Conceptual Question 13 and Exercise 45.

(a)   (b)   (c)

15. We often use hot water to loosen tightly sealed metal lids on glass jars. Explain why this works.

10.5 THE KINETIC THEORY OF GASES

16. Gas sample has twice as much average translational kinetic energy as gas sample B. What can be said about the absolute temperatures of the gas samples?

17. Equal volumes of helium gas (He) and neon gas (Ne) at the same temperature (and pressure) are on opposite sides of a porous membrane (▼Fig. 10.22). Describe what happens after a period of time, and why.

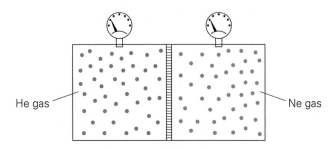

He gas   Ne gas

▲ FIGURE 10.22 **What happens as time passes?** See Conceptual Question 17.

18. Natural gas is odorless; to alert people to gas leaks, the gas company inserts an additive that has a distinctive scent. When there is a gas leak, the additive reaches your nose before the gas does. What can you conclude about the masses of the additive molecules and gas molecules?

*10.6 KINETIC THEORY, DIATOMIC GASES, AND THE EQUIPARTITION THEOREM

19. If a monatomic gas and a diatomic gas have the same average kinetic energy per molecule will they have the same temperature? Explain.

20. Why does a diatomic gas have more internal energy than a monatomic gas with the same number of molecules at the same temperature?

21. A monatomic gas and a diatomic gas both have n moles and are at temperature T. What is the difference in their internal energies? Express your answer in n, R, and T.

EXERCISES*

Integrated Exercises (IEs) are two-part exercises. The first part typically requires a conceptual answer choice based on physical thinking and basic principles. The following part is quantitative calculations associated with the conceptual choice made in the first part of the exercise.

Many exercise sections include "paired" exercises. These exercise pairs, identified with red numbers, are intended to assist you in problem solving and learning. In a pair, the first exercise (even numbered) is worked out in the Study Guide so that you can consult it should you need assistance in solving it. The second exercise (odd numbered) is similar in nature, and its answer is given in Appendix VII at the back of the book.

10.1 TEMPERATURE AND HEAT AND
10.2 THE CELSIUS AND FAHRENHEIT TEMPERATURE SCALES

1. • A person running a fever has a body temperature of 40 °C. What is this temperature on the Fahrenheit scale?

2. • Convert the following to Celsius readings: (a) 80 °F, (b) 0 °F, and (c) −10 °F.

3. • Convert the following to Fahrenheit readings: (a) 120 °C (b) 12 °C and (c) −5 °C

4. • Which is the lower temperature: (a) 245 °C or 245 °F? (b) 200 °C or 375 °F?

5. • The coldest inhabited village in the world is Oymyakon, a town located in eastern Siberia, where it gets as cold as −94 °F. What is this temperature on the Celsius scale?

6. • The highest and lowest recorded air temperatures in the world are, respectively, 58 °C (Libya, 1922) and −89 °C (Antarctica, 1983). What are these temperatures on the Fahrenheit scale?

7. • The highest and lowest recorded air temperatures in the United States are, respectively, 134 °F (Death Valley, California, 1913) and −80 °F (Prospect Creek, Alaska, 1971). What are these temperatures on the Celsius scale?

8. •• During open-heart surgery it is common to cool the patient's body down to slow body processes and gain an extra margin of safety. A drop of 8.5 °C is typical in these types of operations. If a patient's normal body temperature is 98.2 °F, what is her final temperature in both Celsius and Fahrenheit?

9. •• In the troposphere (the lowest part of the atmosphere), the temperature decreases rather uniformly with altitude at a so-called "lapse" rate of about 6.5 °C/km. What are the temperatures (a) near the top of the troposphere (which has an average thickness of 11 km) and (b) outside a commercial aircraft flying at a cruising altitude of 34 000 ft? (Assume that the ground temperature is normal room temperature.)

10. IE •• The temperature drops from 60 °F during the day to 35 °F during the night. (a) The corresponding temperature drop on the Celsius scale is (1) greater than, (2) the

*Assume all temperatures to be exact, and neglect significant figures for small changes in dimension.

same as, or (3) less than . Explain. (b) Compute the temperature drop on the Celsius scale.

11. **IE ••** There is one temperature at which the Celsius and Fahrenheit scales have the same reading. (a) To find that temperature, would you set (1) $5T_F = 9T_C$ (2) $9T_F = 5T_C$ or (3) $T_F = T_C$? Why? (b) Find the temperature.

12. **••** (a) The largest temperature drop recorded in the United States in one day occurred in Browning, Montana, in 1916, when the temperature went from 7 °C to −49 °C. What is the corresponding change on the Fahrenheit scale? (b) On the Moon, the average surface temperature is 127 °C during the day and −183 °C during the night. What is the corresponding change on the Fahrenheit scale?

13. **•••** Astronomers know that the temperatures of stellar interiors are "extremely high." By this they mean they can convert from Fahrenheit to Celsius temperature using a rough rule of thumb:

$$T(\text{in °C}) \approx \tfrac{1}{2}T(\text{in °F})$$

(a) Determine the exact fraction (it isn't $\tfrac{1}{2}$) and (b) the percentage error astronomers make by using $\tfrac{1}{2}$ at high temperatures.

14. **IE •••** Fig. 10.5 is a plot of Fahrenheit temperature versus Celsius temperature. (a) Is the value of the y-intercept found by setting (1) $T_F = T_C$, (2) $T_C = 0$, or (3) $T_F = 0$? Why? (b) Compute the value of the y-intercept. (c) What would be the slope and y-intercept if the graph were plotted the opposite way (Celsius versus Fahrenheit)?

10.3 GAS LAWS, ABSOLUTE TEMPERATURE, AND THE KELVIN TEMPERATURE SCALE

15. **•** Convert the following temperatures to absolute temperatures in kelvins: (a) 0 °C, (b) 100 °C, (c) 20 °C, and (d) −35 °C.

16. **•** Convert the following temperatures to Celsius: (a) 0 K, (b) 250 K, (c) 273 K, and (d) 325 K.

17. **•** (a) Derive an equation for converting Fahrenheit temperatures directly to absolute temperatures in kelvins. (b) Which is the lower temperature, 300 °F or 300 K?

18. **•** When lightning strikes, it can heat the air around it to more than 30 000 K, five times the surface temperature of the Sun. (a) What is this temperature on the Fahrenheit and Celsius scales? (b) The temperature is sometimes reported to be 30 000 °C. Assuming that 30 000 K is correct, what is the percentage error of this Celsius value?

19. **•** How many moles are in (a) 40 g of water, (b) 245 g of CO_2 (carbon dioxide), (c) 138 g of N_2 (nitrogen), and (d) 56 g of O_2 (oxygen) at STP?

20. **IE •** (a) In a constant volume gas thermometer, if the pressure of the gas decreases, has the temperature of the gas (1) increased, (2) decreased, or (3) remained the same? Why? (b) The initial absolute pressure of a gas is 1000 Pa at room temperature (20 °C). If the pressure increases to 1500 Pa, what is the new Celsius temperature?

21. **•** If the pressure of an ideal gas is doubled while its absolute temperature is halved, what is the ratio of the final volume to the initial volume?

22. **••** Show that 1.00 mol of ideal gas under STP occupies a volume of 0.0224 m³ = 22.4 L.

23. **••** What volume is occupied by 160 g of oxygen under a pressure of 2.00 atm and a temperature of 300 K?

24. **••** An athlete has a large lung capacity, 7.0 L. Assuming air to be an ideal gas, how many molecules of air are in the athlete's lungs when the air temperature in the lungs is 37 °C under normal atmospheric pressure?

25. **••** Is there a temperature that has the same numerical value on the Kelvin and the Fahrenheit scales? Justify your answer.

26. **••** A husband buys a helium-filled anniversary balloon for his wife. The balloon has a volume of 3.5 L in the warm store at 74 °F. When he takes it outside, where the temperature is 48 °F, he finds it has shrunk. By how much has the volume decreased?

27. **••** An automobile tire is filled to an absolute pressure of 3.0 atm at a temperature of 30 °C. Later it is driven to a place where the temperature is only −20 °C. What is the absolute pressure of the tire at the cold place? (Assume that the air in the tire behaves as an ideal gas and the volume is constant.)

28. **••** On a warm day (92 °F), an air-filled balloon occupies a volume of 0.200 m³ and has a pressure of 20.0 lb/in². If the balloon is cooled to 32 °F in a refrigerator while its pressure is reduced to 14.7 lb/in², what is the volume of the air in the container? (Assume that the air behaves as an ideal gas.)

29. **••** A steel-belted radial automobile tire is inflated to a gauge pressure of 30.0 lb/in² when the temperature is 61 °F. Later in the day, the temperature rises to 100 °F. Assuming the volume of the tire remains constant, what is the tire's pressure at the elevated temperature? [*Hint*: Remember that the ideal gas law uses absolute pressure.]

30. **••** A scuba diver takes a tank of air on a deep dive. The tank's volume is 10 L and it is completely filled with air at an absolute pressure of 232 atm at the start of the dive. The air temperature at the surface is 94 °F and the diver ends up in deep water at 60 °F. Assuming thermal equilibrium and neglecting air loss, determine the absolute internal pressure of the air when it is cold.

31. **IE ••** (a) If the temperature of an ideal gas increases and its volume decreases, will the pressure of the gas (1) increase, (2) remain the same, or (3) decrease? Why? (b) The Kelvin temperature of an ideal gas is doubled and its volume is halved. How is the pressure affected?

32. **••** If 2.4 m³ of a gas initially at STP is compressed to 1.6 m³ and its temperature is raised to 30 °C, what is its final pressure?

33. **IE ••** The pressure on a low-density gas in a cylinder is kept constant as its temperature is increased. (a) Does the volume of the gas (1) increase, (2) decrease, or (3) remain the same? Why? (b) If the temperature is increased from 10 °C to 40 °C, what is the percentage change in the volume of the gas?

34. **•••** A diver releases an air bubble of volume 2.0 cm³ from a depth of 15 m below the surface of a lake, where the temperature is 7.0 °C. What is the volume of the bubble when it reaches just below the surface of the lake, where the temperature is 20 °C?

35. ••• (a) Show that for the Kelvin temperature range

$$T \gg 273 \text{ K}, \quad T \approx T_C \approx \tfrac{5}{9}T_F$$

(b) For room temperature, what percentage error would result from using this estimation to determine the Kelvin temperature? (c) For a typical stellar interior temperature of 10 million °F, what is the percentage error in the Kelvin temperature? (Carry as many significant figures as needed.)

10.4 THERMAL EXPANSION

36. • A steel beam 10 m long is installed in a structure at 20 °C. What is the beam's change in length when the temperature reaches (a) −25 °C and (b) 45 °C?

37. IE • An aluminum tape measure is accurate at 20 °C. (a) If the tape measure is placed in a freezer, would it read (1) high, (2) low, or (3) the same? Why? (b) If the temperature of the freezer is −5.0 °C, what would be the stick's percentage error because of thermal contraction?

38. • Concrete highway slabs are poured in lengths of 5.00 m. How wide should the expansion gaps between the slabs be at a temperature of 20 °C to ensure that there will be no contact between adjacent slabs over a temperature range of −25 °C to 45 °C?

39. • A man's gold wedding ring has an inner diameter of 2.4 cm at 20 °C. If the ring is dropped into boiling water, what will be the change in the inner diameter of the ring?

40. •• A circular steel plate of radius 15 cm is cooled from 350 °C to 20 °C. By what percentage does the plate's area decrease?

41. •• What temperature change would cause a 0.20% increase in the volume of a quantity of water that was initially at 20 °C?

42. •• A piece of copper tubing used in plumbing has a length of 60.0 cm and an inner diameter of 1.50 cm at 20 °C. When hot water at 85 °C flows through the tube, what are (a) the tube's new length and (b) the change in its cross-sectional area? Does the latter affect the flow speed?

43. •• A pie plate is filled up to the brim with pumpkin pie filling. The pie plate is made of Pyrex and its expansion can be neglected. It is a cylinder with an inside depth of 2.10 cm and an inside diameter of 30.0 cm. It is prepared at a room temperature of 68 °F and placed in an oven at 400 °F. When it taken out, 151 cc of the pie filling has flowed out and over the rim. Determine the coefficient of volume expansion of the pie filling, assuming it is a fluid.

44. IE •• A circular piece is cut from an aluminum sheet at room temperature. (a) When the sheet is then placed in an oven, will the hole (1) get larger, (2) get smaller, or (3) remain the same? Why? (b) If the diameter of the hole is 8.00 cm at 20 °C and the temperature of the oven is 150 °C, what will be the new area of the hole?

45. IE •• In Fig. 10.20, the steel ring of diameter 2.5 cm is 0.10 mm smaller in diameter than the steel ball at 20 °C. (a) For the ball to go through the ring, should you heat (1) the ring, (2) the ball, or (3) both? Why? (b) What is the minimum required temperature?

46. •• When exposed to sunlight, a hole in a sheet of copper expands in diameter by 0.153% compared to its diameter at 68 °F. What is the Celsius temperature of the copper sheet in the sun?

47. •• One morning, an employee at a rental car company fills a car's steel gas tank to the top and then parks the car a short distance away. (a) That afternoon, when the temperature increases, will any gas overflow? Why? (b) If the temperatures in the morning and afternoon are, respectively, 10 °C and 30 °C and the gas tank can hold 25 gal in the morning, how much gas will be lost? (Neglect the expansion of the tank.)

48. •• A copper block has an internal spherical cavity with a 10-cm diameter (▼ Fig. 10.23). The block is heated in an oven from 20 °C to 500 K. (a) Does the cavity get larger or smaller? (b) What is the change in the cavity's volume?

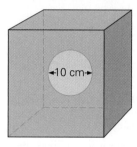

◀ **FIGURE 10.23**
A hole in a block See Exercise 48.

◀10 cm▶

49. ••• A brass rod has a circular cross-section of radius 5.00 cm. The rod fits into a circular hole in a copper sheet with a clearance of 0.010 mm completely around it when both the rod and the sheet are at 20 °C. (a) At what temperature will the clearance be zero? (b) Would such a tight fit be possible if the sheet were brass and the rod were copper?

50. ••• An aluminum rod is measured with a steel tape at 20 °C, and the length of the rod is found to be 75 cm. What length will the tape indicate when both the rod and the tape are at (a) −10 °C? (b) 50 °C? [*Hint:* Both the rod and tape will either expand or shrink as temperature changes. Keep as many significant figures as needed to express the answer.]

51. ••• Table 10.1 states that the (experimental) coefficient of volume expansion β for air (and most other ideal gases at 1 atm and 20 °C) is $3.5 \times 10^{-3}/°C$. Use the definition of the volume expansion coefficient to show that this value can, to a very good approximation, be predicted from the ideal gas law and that the result holds for all ideal gases, not just air.

52. ••• A Pyrex beaker that has a capacity of 1000 cm^3 at 20 °C contains 990 cm^3 of mercury at that temperature. Is there some temperature at which the mercury will completely fill the beaker? Justify your answer. (Assume that no mass is lost by vaporization and include the expansion of the beaker.)

10.5 THE KINETIC THEORY OF GASES

53. • If the average kinetic energy per molecule of a monatomic gas is 7.0×10^{-21} J, what is the Celsius temperature of the gas?

54. • What is the average kinetic energy per molecule in a monatomic gas at (a) 10 °C and (b) 90 °C?

55. IE • If the Celsius temperature of a monatomic gas is doubled, (a) will the internal energy of the gas (1) double,

(2) increase by less than a factor of 2, (3) be half as much, or (4) decrease by less than a factor of 2? Why? (b) If the temperature is raised from 20 °C to 40 °C, what is the ratio of the final internal energy to initial internal energy?

56. ● What is the rms speed of the molecules in low-density oxygen gas at 0 °C? (The mass of an oxygen molecule, O_2, is 5.31×10^{-26} kg).

57. ● (a) What is the average kinetic energy per molecule of a monatomic gas at a temperature of 25 °C? (b) What is the rms speed of the molecules if the gas is helium? (A helium molecule consists of a single atom of mass 6.65×10^{-27} kg).

58. ●● (a) Estimate the total amount of translational kinetic energy in a small classroom at normal room temperature. Assume the room measures 4.00 m by 10.0 m by 3.00 m. (b) If this energy were all harnessed, how high would it be able to lift an elephant with a mass of 1200 kg?

59. ●● A quantity of an ideal gas is at 0 °C. An equal quantity of another ideal gas is at twice the absolute temperature. What is its Celsius temperature?

60. IE ●● A sample of oxygen (O_2) and another sample of nitrogen (N_2) are at the same temperature. (a) The rms speed of the nitrogen sample is (1) greater than, (2) the same as, or (3) less than the rms speed of the oxygen sample. Explain. (b) Calculate the ratio of the rms speed in the nitrogen sample to in the oxygen sample.

61. ●● If 2.0 mol of oxygen gas is confined in a 10-L bottle under a pressure of 6.0 atm, what is the average kinetic energy of an oxygen molecule?

62. ●● If the temperature of an ideal gas increases from 300 K to 600 K, what happens to the rms speed of the gas molecules?

63. ●● If the temperature of an ideal gas is raised from 25 °C to 100 °C, how much faster is the new rms speed of the gas molecules?

64. ●● If the rms speed of the molecules in an ideal gas at 20 °C increases by a factor of 2, what is the new Celsius temperature?

65. IE ●●● During the race to develop the atomic bomb in World War II, it was necessary to separate a lighter isotope of uranium (U-235 was the fissionable one needed for bomb material) from a heavier variety (U-238). The uranium was converted into a gas, uranium hexafluoride (UF_6), and the two uranium isotopes were separated by *gaseous diffusion* using the difference in their rms speeds. As a two-component molecular mixture at room temperature, which of the two types of molecules would be moving faster, on average: (1) $^{235}UF_6$ or (2) $^{238}UF_6$. Or (3) would they move equally fast? Explain. (b) Determine the ratio of their rms speeds, light molecule to heavy molecule. Treat the molecules as ideal gases and neglect rotations and/or vibrations of the molecules. The masses of the three atoms in atomic mass units are 238 and 235 for the two uranium isotopes and 19 for fluorine.

*10.6 KINETIC THEORY, DIATOMIC GASES, AND THE EQUIPARTITION THEOREM

66. ● What is the total internal energy of 1.00 mol of 30 °C He gas and O_2 gas, respectively?

67. ● If 1.0 mol of a monatomic gas has a total internal energy of 5.0×10^3 J at a certain temperature, what is the total internal energy of 1.0 mol of a diatomic gas at the same temperature?

68. ●● For an average molecule of N_2 gas at 10 °C, what are its (a) translational kinetic energy, (b) rotational kinetic energy, and (c) total energy? Repeat for He gas at the same temperature.

69. ●● A diatomic gas has a certain total kinetic energy at 25 °C. If a monatomic gas of the same number of molecules as the diatomic gas has the same total kinetic energy, what is the Celsius temperature of the monatomic gas?

The Multiconcept Exercises require the use of more than one fundamental concept for their understanding and solution. The concepts may be from this chapter, but may include those from previous chapters.

70. IE (a) When cooled, the densities of most objects (1) increase, (2) decrease, (3) stay the same. (b) By what percentage does the density of a bowling ball change (assuming it is a uniform sphere) when it is taken from room temperature (68 °F) into the cold night air in Nome, Alaska (−40 °F). Assume the ball is made out of a material that has a linear coefficient of expansion α of 75.2×10^{-6}/°C.

71. When a full copper kettle is tipped vertically at room temperature (68 °F), water initially pours out of its spout at 100 cm³/s (cubic centimeters per second). By what percentage will this change if the kettle instead contains boiling water at 212 °F? Assume that the only significant change is due to the change in size of the spout.

72. An ideal gas sample occupies a container of volume 0.75 L at STP. Find (a) the number of moles and (b) the number of moleculesin in the sample. (c) If the gas is carbon monoxide (CO), what is the sample's mass?

73. 2.00 mol of a monatomic gas at atmospheric pressure has a total internal energy of 7.48×10^3 J. What is the volume occupied a rigid cylinder by the gas?

74. An ideal gas in a cylinder is at 20 °C and 2.0 atm. If it is heated so its rms speed increases by 20%, what is its new pressure?

75. IE The escape speed from the Earth is about 11 000 m/s (Section 7.5). Assume that for a given type of gas to eventually escape the Earth's atmosphere, its average molecular speed must be about 10% of the escape speed. (a) Which gas would be more likely to escape the Earth: (1) oxygen, (2) nitrogen, or (3) helium? (b) Assuming a temperature of −40 °F in the upper atmosphere, determine the rms speed of a molecule of oxygen. Is it enough to escape the Earth? (Data: The mass of an oxygen molecule is 5.34×10^{-26} kg, that of a nitrogen molecule is 4.68×10^{-26} kg, and that of a helium molecule is 6.68×10^{-27} kg.

11 Heat

PHYSICS FACTS

- With a skin temperature of 34 °C (93.2 °F), a person sitting in a room at 23 °C (73.4 °F) will lose about 100 J of heat per second, which is the power output approximately equal to that of a 100-W lightbulb. This is why a closed room full of people tends to get very warm.

- A couple of inches of fiberglass in the attic can cut heat loss by as much as 90% (see Example 11.7).

- If the Earth did not have an atmosphere (hence no greenhouse effect), its average surface temperature would be 30 °C (86 °F) lower than it is now. That would freeze liquid water and basically eliminate life as we know it.

- Most metals are excellent thermal conductors. However, stainless steel is a relatively poor conductor; it conducts only about 5% as much as copper.

- During a race on a hot day, a professional cyclist can lose as much as 7 L of water in 3 h to evaporation in getting rid of the heat generated by this vigorous activity.

Heat is crucial to our existence. Our bodies must balance heat loss and gain to stay within the narrow temperature range necessary for life. This thermal balance is delicate and any disturbance can have serious consequences. Sickness can disrupt the balance, and as a result, our bodies produce a chill or fever.

To maintain our health, we exercise by doing mechanical work such as lifting weights and riding bicycles. Our bodies convert food energy (chemical potential) to mechanical work; however, this process is not perfect. That is, the body cannot convert all the food energy into mechanical work—in fact, it converts less than 20% depending on which muscle groups are doing the work. The rest becomes heat transferred to the environment. The leg muscles are the largest and most efficient in performing mechanical work; for example, cycling and running are relatively efficient processes. The arm and shoulder muscles are less efficient; hence, snow shoveling is a low-efficiency exercise. The body must have special cooling mechanisms to get rid of excess thermal energy generated during intense exercise. The most efficient mechanism is through perspiring, or the evaporation of water. The Olympic marathon champion Gezahgne Abera tries to promote cooling and evaporation by pouring water over his head, as shown in the chapter-opening photograph.

On a larger scale, heat exchanges are important to our planet's ecosystem. The average temperature of the Earth, so critical to our environment and to the survival of the organisms that inhabit it, is maintained through a heat exchange balance. Each day, a vast quantity of solar energy reaches our planet's atmosphere and surface. Scientists are concerned that a buildup of atmospheric "greenhouse" gases, a product of our industrial society, could significantly raise the Earth's average temperature. This change would undoubtedly have a negative effect on life on the Earth.

On a more practical level, most of us know to be very careful while handling anything that has recently been in contact with a flame or other source of heat. Yet while the copper bottom of a steel pot on the stove can be very hot, the steel pot handle is only warm to the touch. Sometimes direct contact isn't necessary for heat to be transmitted, but how was heat transferred? And why was the steel handle not nearly as hot as the pot? The answer has to do with thermal conduction, as you will learn.

In this chapter, what heat is and how it is measured will be discussed. Also studied are the various mechanisms by which heat is transferred from one object to another. This knowledge will allow you to explain many everyday phenomena, as well as provide a basis for understanding the conversion of thermal energy into useful mechanical work.

11.1 Definition and Units of Heat

LEARNING PATH QUESTIONS
➡ What is heat?
➡ What are four common units of heat?
➡ What is the mechanical equivalent of heat?

Like work, heat is related to a transfer of energy. In the 1800s, it was thought that heat described the amount of energy an object possessed, but this is not true. Rather, **heat** is the name used to describe a type of energy *transfer*. "Heat," or "heat energy," is the energy added to, or removed from, the total internal energy of an object due to temperature differences.

Heat then is energy *in transit*, and is measured in the standard SI unit, the joule (J). However, other nonstandard, commonly used units of heat are also defined. An important one is the **kilocalorie (kcal)** (▼Fig. 11.1a):

One kilocalorie (kcal) is defined as the amount of heat needed to raise the temperature of 1 kg of water by 1 °C (from 14.5 °C to 15.5 °C).

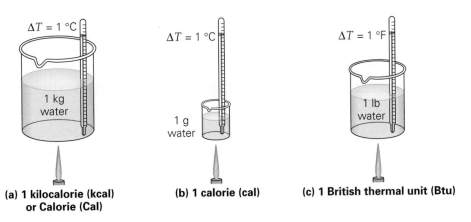

▶ FIGURE 11.1 **Units of heat**
(a) A kilocalorie raises the temperature of 1 kg of water by 1 °C.
(b) A calorie raises the temperature of 1 g of water by 1 °C. **(c)** A Btu raises the temperature of 1 lb of water by 1 °F. (Not drawn to scale.)

(a) 1 kilocalorie (kcal) or Calorie (Cal)

(b) 1 calorie (cal)

(c) 1 British thermal unit (Btu)

(This kilocalorie is technically known as the "15° kilocalorie.") The temperature range is specified because the energy needed varies slightly with temperature—a variation so small that it can be ignored for our purposes.

For smaller quantities, the **calorie (cal)** is sometimes used (1 kcal = 1000 cal). One calorie is the amount of heat needed to raise the temperature of 1 g of water by 1 °C (from 14.5 °C to 15.5 °C) (Fig. 11.1b).

A familiar use of the larger unit, the kilocalorie, is for specifying the energy values of foods. In this context, the word is usually shortened to *Calorie* (Cal). That is, people on diets really count kilocalories. This quantity refers to the food energy that is available for conversion to heat to be used for mechanical movement, to maintain body temperature, or to increase body mass. The capital C distinguishes the larger kilocalorie, from the smaller calorie. They are sometimes referred to as "big Calorie" and "little calorie." (In some countries, the joule is used for food values—see ◀ Fig. 11.2.)

A unit of heat sometimes used in industry is the **British thermal unit (Btu)**. One Btu is the amount of heat needed to raise the temperature of 1 lb of water by 1 °F (from 63 °F to 64 °F; Fig. 11.1c), and 1 Btu = 252 cal = 0.252 kcal. If you buy an air conditioner or an electric heater, you will find that it is rated in Btu, which is really Btu per hour—in other words, a power rating. For example, window air conditioners range from 4000 to 25 000 Btu/h. This specifies the rate at which the appliance can transfer heat.

▲ FIGURE 11.2 **It's a joule!** In Australia, diet drinks are labeled as being "low joule." In Germany, the labeling is a bit more specific: "Less than 4 kilojoules (1 kcal) in 0.3 Liter." How does this labeling compare to that for diet drinks in the United States?

THE MECHANICAL EQUIVALENT OF HEAT

The idea that heat is actually a transfer of energy is the result of work by many scientists. Some early observations were made by the American Benjamin Thompson (Count Rumford), 1753–1814, while he was supervising the boring of cannon barrels in Germany. Rumford noticed that water put into the bore of the cannon (to prevent overheating during drilling) boiled away and had to be replenished. The theory of heat at that time pictured it as a "caloric fluid," which flowed from hot objects to colder ones. Rumford did several experiments to detect "caloric fluid" by measuring changes in the weights of heated substances. Since no weight change was detected, he concluded that the mechanical work done by friction was actually responsible for the heating of the water.

This conclusion was later proven quantitatively by the English scientist James Joule (after whom the unit of energy is named; see Section 5.6). Using the apparatus illustrated in ◀Fig. 11.3, Joule demonstrated that when a given amount of mechanical work was done, the water was heated, as indicated by an increase in its temperature. He found that for every 4186 J of work done, the temperature of the water rose 1 °C per kg, or that 4186 J was equivalent to 1 kcal:

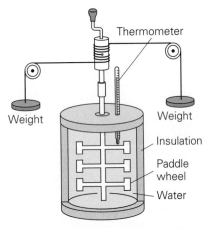

▲ FIGURE 11.3 **Joule's apparatus for determining the mechanical equivalent of heat** As the weights descend, the paddle wheels churn the water, and the mechanical energy, or work, is converted into heat energy, raising the temperature of the water. For every 4186 J of work done, the temperature of the water rises 1 °C per kilogram. Thus, 4186 J is equivalent to 1 kcal.

$$1 \text{ kcal} = 4186 \text{ J} = 4.186 \text{ kJ} \quad \text{or} \quad 1 \text{ cal} = 4.186 \text{ J}$$

This relationship is called the **mechanical equivalent of heat**. Example 11.1 illustrates an everyday use of these conversion factors.

EXAMPLE 11.1 | Working Off That Birthday Cake: Mechanical Equivalent of Heat to the Rescue

At a birthday party, a student eats a piece of cake (food energy value of 200 Cal). To prevent this energy from being stored as fat, she takes a stationary bicycle workout class right after the party. This exercise requires the body to do work at an average rate of 200 watts. How long must the student bicycle to achieve her goal of "working off" the cake's energy?

THINKING IT THROUGH. Power is the rate at which the student does work, and the watt (W) is its SI unit (1 W = 1 J/s; Section 5.6). To find the time it will take to do this work, the food energy content is expressed in joules and the definition of average power, $\overline{P} = W/t$ (work/time) is used.

SOLUTION. The work required to "burn up" the energy content of the cake is at least 200 Cal. Listing the data given and converting to SI units (remember that Cal means kcal):

Given: $W = (200 \text{ kcal})\left(\dfrac{4186 \text{ J}}{\text{kcal}}\right) = 8.37 \times 10^5 \text{ J}$ **Find:** t (time to "burn up" 200 Cal)

$\overline{P} = 200 \text{ W} = 200 \text{ J/s}$

Rearranging the equation for average power,

$$t = \frac{W}{\overline{P}} = \frac{8.37 \times 10^5 \text{ J}}{200 \text{ J/s}} = 4.19 \times 10^3 \text{ s} = 69.8 \text{ min} = 1.16 \text{ h}$$

FOLLOW-UP EXERCISE. If the 200 Cal in this Example were used to increase the student's gravitational potential energy, how high would she rise? (Assume her mass is 60 kg.) *(Answers to all Follow-Up Exercises are given in Appendix VI at the back of the book.)*

DID YOU LEARN?

➥ Heat is a form of energy. It is the energy added to or removed from an object due to temperature differences.

➥ Four common units of heat are the joule, kilocalorie (Cal), calorie, and Btu (British thermal unit).

➥ The mechanical equivalent of heat relates mechanical energy (work) and heat energy. The relationship is 1 cal = 4.186 J or 1 kcal = 4186 J.

11.2 Specific Heat and Calorimetry

LEARNING PATH QUESTIONS

➥ How is specific heat of a substance defined?

➥ If two different substances (with different specific heats) have equal mass and receive equal amounts of heat, which substance will experience a larger temperature change?

➥ What is the fundamental physical principle that supports the calorimetry technique?

SPECIFIC HEATS OF SOLIDS AND LIQUIDS

Recall from Chapter 10 that when heat is added to a solid or liquid, the energy may go toward increasing the average molecular kinetic energy (temperature change) and also toward increasing the potential energy associated with the molecular bonds (phase change). Different substances have different molecular configurations and bonding patterns. Thus, if equal amounts of heat are added to equal masses of different substances, the resulting temperature changes will *not* generally be the same.

Specific heat capacity, or simply **specific heat (c)** is defined as the heat (transfer) required to raise (or lower) the temperature of 1 kg of a substance by 1 °C. The SI units of specific heat are J/(kg·K) or J/(kg·°C), because 1 K = 1 °C. Specific heat is a characteristic of the substance type. The specific heats of some common substances are given in ▼Table 11.1. Specific heats vary slightly with temperature, but they can be considered constant for our purpose.

The amount of heat (Q) required to change the temperature of a substance of mass m by a temperature difference of $\Delta T (T_f - T_i)$ is then

$$Q = cm\Delta T \quad \text{or} \quad c = \frac{Q}{m\Delta T} \qquad \textit{(specific heat)} \qquad (11.1)$$

TABLE 11.1 Specific Heats of Various Substances (Solids and Liquids) at 20 °C and 1 atm

Substance	Specific Heat (c)	
	J/(kg · °C)	kcal/(kg · °C) or cal/(g · °C)
Solids		
Aluminum	920	0.220
Copper	390	0.0932
Glass	840	0.201
Ice (−10 °C)	2100	0.500
Iron or steel	460	0.110
Lead	130	0.0311
Soil (average)	1050	0.251
Wood (average)	1680	0.401
Human body (average)	3500	0.84
Liquids		
Ethyl alcohol	2450	0.585
Glycerin	2410	0.576
Mercury	139	0.0332
Water (15 °C)	4186	1.000
Gas		
Steam (Water vapor, 100 °C)	2000	0.48

The larger the specific heat of a substance, the more heat must be transferred to or taken from it (per kilogram of mass) to change its temperature by a given amount. That is, a substance with a higher specific heat requires more heat for a given temperature change and mass than one with a lower specific heat. Table 11.1 shows that metals have specific heats considerably lower than that of water. Thus it takes only a small amount of heat to produce a relatively large temperature increase in a metal object, compared to the same mass of water.

Compared to most common materials, water has a very large specific heat of 4186 J/(kg · °C), or 1.00 kcal/(kg · °C). You have been the victim of the high specific heat of water if you have ever burned your mouth on a baked potato or the hot cheese on a pizza. These foods have high water content, and due to water's high specific heat, they don't cool off as quickly as some other drier foods do. The large specific heat of water is also responsible for the mild climate of places near large bodies of water. (See Section 11.4 for more details.)

Note from Eq. 11.1 that when there is a temperature increase, ΔT is positive $(T_f > T_i)$, then Q is positive. This condition corresponds to energy being *added* to a system or object. Conversely, ΔT and Q are negative when energy is *removed from* a system or object. This sign convention will be used throughout this book.

EXAMPLE 11.2 | Birthday Cake Revisited: Specific Heat for a Warm Bath?

At the birthday party in Example 11.1, a student ate a piece of cake (200 Cal). To get an idea of the magnitude of this amount of energy there is in that piece of cake, the student would like to know how much water at 20 °C can be brought to 45 °C (enough for a bath?). Can you help her out?

THINKING IT THROUGH.   The heat energy from the cake is used to heat water from 20 °C to 45 °C. Using the mechanical equivalent of heat and Eq. 11.1, the mass of water can be found.

SOLUTION. Listing the data given and converting to SI units (remember that Cal means kcal):

Given: $Q = (200 \text{ kcal})\left(\dfrac{4186 \text{ J}}{\text{kcal}}\right) = 8.37 \times 10^5 \text{ J}$ **Find:** m (mass of water)

$T_i = 20 \text{ °C}$
$T_f = 45 \text{ °C}$
$c = 4186 \text{ J}/(\text{kg} \cdot \text{°C})$ (from Table 11.1)

From Eq. 11.1, $Q = cm\Delta T = cm(T_f - T_i)$. Solving for m gives

$$m = \frac{Q}{c\Delta T} = \frac{8.37 \times 10^5 \text{ J}}{[4186 \text{ J}/(\text{kg} \cdot \text{°C})](45 \text{ °C} - 20 \text{ °C})} = 8.00 \text{ kg}$$

This mass of water occupies more than 2 gallons—quite a bit, but it may not be enough for a bath. It does show that "a piece of cake" contains a fair amount of energy.

FOLLOW-UP EXERCISE. In this Example, how would the answer change if the water was initially at a temperature of 5 °C rather than 20 °C?

INTEGRATED EXAMPLE 11.3 | ## Cooking Class 101: Studying Specific Heats While Learning How to Boil Water

To prepare pasta, you bring a pot of water from room temperature (20 °C) to its boiling point (100 °C). The pot itself has a mass of 0.900 kg, is made of steel, and holds 3.00 kg of water. (a) Which of the following is true: (1) the pot requires more heat than the water, (2) the water requires more heat than the pot, or (3) they require the same amount of heat? (b) Determine the required heat for both the water and the pot, and the ratio Q_w / Q_{pot}.

(A) CONCEPTUAL REASONING. The temperature increase is the same for the water and the pot. Thus, the required heat is affected by the product of mass and specific heat. There is 3.00 kg of water to heat. This is more than three times the mass of the pot. From Table 11.1, the specific heat of water is about nine times larger than that of steel. Both factors together indicate that the water will require significantly more heat than the pot, so the answer is (2).

(B) QUANTITATIVE REASONING AND SOLUTION. The heat needed can be found using Eq. 11.1, after looking up the specific heats in Table 11.1. The temperature change is easily determined from the initial and final values.

Listing the data given:

Given: $m_{pot} = 0.900 \text{ kg}$ **Find:** Q_w, Q_{pot} and Q_w/Q_{pot} (the required heat for
$m_w = 3.00 \text{ kg}$ the water and the pot, and the heat ratio)
$c_{pot} = 460 \text{ J/kg} \cdot \text{°C}$ (from Table 11.1)
$c_w = 4186 \text{ J/kg} \cdot \text{°C}$ (from Table 11.1)
$\Delta T = T_f - T_i = 100 \text{ °C} - 20 \text{ °C} = 80 \text{ °C}$

In general, the amount of heat is given by $Q = cm\Delta T$. The temperature increase (ΔT) for both objects is 80 °C. Thus, the heat for the water is

$$Q_w = c_w m_w \Delta T_w$$
$$= [4186 \text{ J}/(\text{kg} \cdot \text{°C})](3.00 \text{ kg})(80 \text{ °C}) = 1.00 \times 10^6 \text{ J}$$

and the heat required for the pot is

$$Q_{pot} = c_{pot} m_{pot} \Delta T_{pot}$$
$$= [460 \text{ J}/(\text{kg} \cdot \text{°C})](0.900 \text{ kg})(80 \text{ °C}) = 3.31 \times 10^4 \text{ J}$$

Therefore,

$$\frac{Q_w}{Q_{pot}} = \frac{1.00 \times 10^6 \text{ J}}{3.31 \times 10^4 \text{ J}} = 30.2$$

Hence the water requires more than thirty times the heat required for the pot, because it has more mass and a greater specific heat.

FOLLOW-UP EXERCISE. (a) In this Example, if the pot were the same mass but instead made out of aluminum, would the heat ratio (water to pot) be smaller or larger than the answer for the steel pot? Explain. (b) Verify your answer.

▲ FIGURE 11.4 Calorimetry apparatus The calorimetry cup (center, with black insulating ring) goes into the larger container. The cover with the thermometer and stirrer is seen at the right. Metal shot or pieces of metal are heated in the small cup (with the handle) in the steam generator on the tripod.

CALORIMETRY

Calorimetry is a technique that quantitatively measures heat exchanges. Such measurements are made by using an instrument called a *calorimeter* (cal-oh-RIM-i-ter), usually an insulated container that allows little heat exchange with the environment (ideally none). A simple laboratory calorimeter is shown in ◄Fig. 11.4.

The specific heat of a substance can be determined by measuring the masses and temperature changes of the objects involved and using Eq. 11.1.* Usually the unknown is the unknown specific heat, *c*. Typically, a substance of known mass and temperature is put into a quantity of water in a calorimeter. The water is at a different temperature from that of the substance, usually a lower one. The principle of the conservation of energy is then applied to determine the substance's specific heat, *c*. This procedure is called the *method of mixtures*. Example 11.4 illustrates the use of this procedure. Such heat exchanges are simply the applications of the conservation of energy. The total of all the heat losses ($Q < 0$) must have the same absolute value as all the heat gains ($Q > 0$). This means the algebraic sum of all the heat transfers must equal zero, or $\Sigma Q_i = 0$, assuming negligible heat exchange with the environment.

| EXAMPLE 11.4 | Calorimetry Using the Method of Mixtures |

Students in a physics lab are to determine the specific heat of copper experimentally. They place 0.150 kg of copper shot into boiling water and let it stay for a while, so as to reach a temperature of 100 °C. Then they carefully pour the hot shot into a calorimeter cup (Fig. 11.4) containing 0.200 kg of water at 20.0 °C. The final temperature of the mixture in the cup is measured to be 25.0 °C. If the aluminum cup has a mass of 0.0450 kg, what is the specific heat of copper? (Assume that there is no heat exchange with the surroundings.)

THINKING IT THROUGH. The conservation of heat energy is involved: $\Sigma Q_i = 0$, taking into account the correct positive and negative signs. In calorimetry problems, it is important to identify and label all of the quantities with proper signs. Identification of the heat gains and losses is crucial. You will probably use this method in the laboratory.

SOLUTION. The subscripts Cu, w, and Al will be used to refer to the copper, water, and aluminum calorimeter cup, respectively. The subscripts *h*, *i*, and *f* will refer to the temperature of the *h*ot metal shot, the water (and cup) *i*nitially at room temperature, and the *f*inal temperature of the system, respectively. With this notation,

Given: $m_{Cu} = 0.150$ kg
$m_w = 0.200$ kg
$c_w = 4186$ J/(kg · °C) (from Table 11.1)
$m_{Al} = 0.0450$ kg
$c_{Al} = 920$ J/(kg · °C) (from Table 11.1)
$T_h = 100$ °C (initial temperature of Cu shot),
$T_i = 20.0$ °C, and
$T_f = 25.0$ °C

Find: c_{Cu} (specific heat)

If there is no heat exchange with the surroundings, the system's total energy is conserved, $\Sigma Q_i = 0$, and

$$\Sigma Q_i = Q_w + Q_{Al} + Q_{Cu} = 0$$

Substituting the relationship in Eq. 11.1 for these heats,

$$c_w m_w \Delta T_w + c_{Al} m_{Al} \Delta T_{Al} + c_{Cu} m_{Cu} \Delta T_{Cu} = 0$$

or

$$c_w m_w (T_f - T_i) + c_{Al} m_{Al} (T_f - T_i) + c_{Cu} m_{Cu} (T_f - T_h) = 0$$

Here, the water and aluminum cup, initially at T_i, are heated to T_f, so $\Delta T_w = \Delta T_{Al} = (T_f - T_i)$. The copper initially at T_h is cooled to T_f, so $\Delta T_{Cu} = (T_f - T_h)$ and this is a negative quantity, indicating a temperature drop for the copper. Solving for c_{Cu},

$$c_{Cu} = -\frac{(c_w m_w + c_{Al} m_{Al})(T_f - T_i)}{m_{Cu}(T_f - T_h)}$$

$$= -\frac{\{[4186 \text{ J}/(\text{kg} \cdot °\text{C})](0.200 \text{ kg}) + [920 \text{ J}/(\text{kg} \cdot °\text{C})](0.0450 \text{ kg})\}(25.0 °\text{C} - 20.0 °\text{C})}{(0.150 \text{ kg})(25.0 °\text{C} - 100 °\text{C})}$$

$$= 390 \text{ J}/(\text{kg} \cdot °\text{C})$$

---

*In this section, calorimetry will *not* involve phase changes, such as ice melting or water boiling. These effects are discussed in Section 11.3.

Notice that the proper use of signs resulted in a positive answer for c_{Cu}, as required. If, for example, the Q_{Cu} term had not had the correct sign, the answer would have been negative—a big clue that you had an initial sign error.

FOLLOW-UP EXERCISE. In this example, what would the final equilibrium temperature be if the calorimeter (water and cup) initially had been at a warmer 30 °C?

SPECIFIC HEAT OF GASES

When heat is added to or removed from most materials, they expand or contract. During expansion, for example, the materials would then do work on the environment. For most solids and liquids, this work is negligible, because the volume changes are very small (Section 10.4). This is why this effect wasn't included in our discussion of specific heat of solids and liquids.

However, for gases, expansion and contraction *can* be significant. It is therefore important to specify the *conditions* under which heat is transferred when referring to a gas. If heat is added to a gas at constant volume (a *rigid* container), the gas does no work. (Why?) In this case, all of the heat goes into increasing the gas's internal energy and, therefore, to increasing its temperature. However, if the same amount of heat is added at constant pressure (a *nonrigid* container allowing a volume change), a portion of the heat is converted to work as the gas expands. Thus, not all of the heat will go into the gas's internal energy. This process results in a *smaller* temperature change than occurred during the constant volume process.

To designate the physical quantities that are held constant while heat is added to or removed from a gas, a subscript notation will be used: c_p means specific heat under conditions of constant pressure (p), and c_v means specific heat under conditions of constant volume (v). The specific heat for water vapor (H_2O) given in Table 11.1 is the specific heat under constant pressure (c_p).

An important result is that for a particular gas, c_p is *always* greater than c_v. This is true because for a specific mass of gas, $c \propto Q/\Delta T$. Since for a given Q, ΔT_v is as large as it can be, c_v will be less than c_p. In other words, $\Delta T_v > \Delta T_p$. Specific heats of gases play an important role in adiabatic thermodynamic processes. (See Section 12.3.)

DID YOU LEARN?

➥ The specific heat of a substance is defined as the amount of heat required to change the temperature of 1 kg substance by 1 °C or 1 K.

➥ For equal amounts of heat and mass, a substance with a smaller specific heat will experience a larger temperature change.

➥ The fundamental physical principle behind calorimetry is the conservation of energy. The total heat lost by components in an isolated system must equal the heat gained by other commponents within the system so $\Sigma Q_i = 0$ or there is no net heat exchange within the system.

(a) Solid

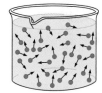

(b) Liquid

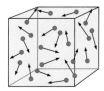

(c) Gas

▲ **FIGURE 11.5 Three phases of matter** **(a)** The molecules of a solid are held together by bonds; consequently, a solid has a definite shape and volume. **(b)** The molecules of a liquid can move more freely, so a liquid has a definite volume and assumes the shape of its container. **(c)** The molecules of a gas interact weakly and are separated by relatively large distances; thus, a gas has no definite shape or volume, unless it is confined in a container.

11.3 Phase Changes and Latent Heat

LEARNING PATH QUESTIONS

➥ What are the three common phases of matter?

➥ When a substance undergoes a phase change, what happens to its temperature?

➥ How is the latent heat of fusion (vaporization) of a substance defined?

Matter normally exists in one of three traditional *phases*: solid, liquid, or gas (▶Fig. 11.5). However, this division into three common phases is only approximate since there are other phases, such as a plasma phase and a superconducting phase. The phase that a substance is in depends on the substance's internal energy (as indicated by its temperature) and the pressure on it. However, it is likely that you think of adding or removing heat as the way to change the phase of a substance.

In the **solid phase**, molecules are held together by attractive forces, or bonds (Fig. 11.5a). Adding heat causes increased motion about the molecular equilibrium positions. If enough heat is added to provide sufficient energy to break the intermolecular bonds, most solids undergo a phase change and become liquids. The temperature at which this phase change occurs is called the **melting point**. The temperature at which a liquid becomes a solid is called the **freezing point**. In general, these temperatures are the same for a given substance, but they can differ slightly.

In the **liquid phase**, molecules of a substance are relatively free to move and a liquid assumes the shape of its container (Fig. 11.5b). In certain liquids, there may be some locally ordered structure, giving rise to liquid crystals, such as those used in LCDs (liquid crystal displays) of calculators and computer displays (Section 24.4).

Adding even more heat increases the motion of the molecules of a liquid. When they have enough energy to become separated, the liquid changes to the **gaseous (vapor) phase**. This change may occur slowly, by *evaporation*, or rapidly, at a particular temperature called the **boiling point**. The temperature at which a gas condenses into a liquid is the **condensation point**.

Some solids, such as dry ice (solid carbon dioxide), mothballs, and certain air fresheners, change directly from the solid to the gaseous phase at standard pressure. This process is called **sublimation**. Like the rate of evaporation, the rate of sublimation increases with the temperature of the surrounding medium. A phase change from a gas to a solid is called *deposition*. Frost, for example, is solidified water vapor (gas) deposited on grass, car windows, and other objects. Frost is *not* frozen dew (liquid water), as is sometimes mistakenly assumed.

LATENT HEAT

In general, when heat is transferred to a substance, its temperature increases as the average kinetic energy per molecule increases. However, when heat is added (or removed) during a phase change, the temperature of the substance does *not* change. For example, if heat is added to a quantity of ice at -10 °C, the temperature of the ice increases until it reaches its melting point of 0 °C. At this point, the addition of more heat does not increase the ice's temperature, but causes it to melt, or change phase. (The heat must be added slowly so that the ice and melted water remain in thermal equilibrium, otherwise, the ice water can warm above 0 °C even though the ice remains at 0 °C.) Only after the ice is completely melted does adding more heat cause the temperature of the water to rise.

A similar situation occurs during the liquid–gas phase change at the boiling point. Adding more heat to boiling water only causes more vaporization. A temperature increase occurs only *after* the water is completely boiled, resulting in *superheated steam*. Keep in mind that ice can be colder than 0 °C and steam can be hotter than 100 °C.

During a phase change, the heat goes into breaking the attractive bonds and separating molecules rather than into increasing the temperature (increasing the potential, rather than kinetic, energies). The heat required for a phase change is called the **latent heat (L)**, which is defined as the magnitude of the heat needed per unit mass to induce a phase change:

$$L = \frac{|Q|}{m} \quad \text{(latent heat)} \tag{11.2}$$

where m is the mass of the substance. Latent heat has the SI unit of joule per kilogram (J/kg), or kilocalorie per kilogram kcal/kg.

The latent heat for a solid–liquid phase change is called the **latent heat of fusion** (L_f), and that for a liquid–gas phase change is called the **latent heat of vaporization** (L_v.) These quantities are often referred to as simply the *heat of fusion*

TABLE 11.2 Temperatures of Phase Changes and Latent Heats for Various Substances (at 1 atm)

Substance	Melting Point	L_f J/kg	L_f kcal/kg	Boiling Point	L_v J/kg	L_v kcal/kg
Alcohol, ethyl	−114 °C	1.0×10^5	25	78 °C	8.5×10^5	204
Gold	1063 °C	0.645×10^5	15.4	2660 °C	15.8×10^5	377
Helium*	—	—	—	−269 °C	0.21×10^5	5
Lead	328 °C	0.25×10^5	5.9	1744 °C	8.67×10^5	207
Mercury	−39 °C	0.12×10^5	2.8	357 °C	2.7×10^5	65
Nitrogen	−210 °C	0.26×10^5	6.1	−196 °C	2.0×10^5	48
Oxygen	−219 °C	0.14×10^5	3.3	−183 °C	2.1×10^5	51
Tungsten	3410 °C	1.8×10^5	44	5900 °C	48.2×10^5	1150
Water	0 °C	3.33×10^5	80	100 °C	22.6×10^5	540

*Not a solid at a pressure of 1 atm; melting point is −272 °C at 26 atm.

and the *heat of vaporization*. The latent heats of some substances, along with their melting and boiling points, are given in ▲Table 11.2. (The latent heat for the less common solid–gas phase change is called the *latent heat of sublimation* and is symbolized by L_s.) As you might expect, the latent heat (in joules per kilogram) is the amount of energy per kilogram *given up* when the phase change is in the opposite direction, that is, from liquid to solid or gas to liquid.

A more useful form of Eq. 11.3 is given by solving for Q and including a positive/negative sign for the two possible directions of heat flow:

$$Q = \pm mL \quad \text{(signs with latent heat)} \tag{11.3}$$

This equation is more practical for problem solving because in calorimetry problems, you are typically interested in applying conservation of energy in the form of $\Sigma Q_i = 0$. The positive/negative sign ($\pm$) must be explicitly expressed because heat can flow either into (+) or out of (−) the object or system of interest.

When solving calorimetry problems involving phase changes, you must be careful to use the correct sign for those terms, in agreement with our sign conventions (▼Fig. 11.6). For example, if water is condensing from steam into liquid droplets, *removal* of heat is involved, necessitating the choice of the *negative* sign.

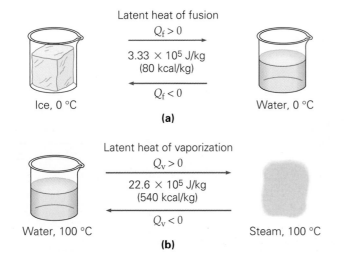

Latent heat of fusion
$Q_f > 0$
3.33×10^5 J/kg
(80 kcal/kg)
$Q_f < 0$

Ice, 0 °C Water, 0 °C

(a)

Latent heat of vaporization
$Q_v > 0$
22.6×10^5 J/kg
(540 kcal/kg)
$Q_v < 0$

Water, 100 °C Steam, 100 °C

(b)

◀**FIGURE 11.6 Phase changes and latent heats (a)** At 0 °C, 3.33×10^5 J must be added to 1 kg of ice or removed from 1 kg of liquid water to change its phase. **(b)** At 100 °C, 22.6×10^5 J must be added to 1 kg of liquid water or removed from 1 kg of steam to change its phase.

Recall in Section 11.2 that there were no phase changes, the expression for heat $Q = cm\Delta T$ automatically gave the correct sign for Q from the sign of ΔT. But there is no ΔT during a phase change. *Choosing the correct sign is up to you.*

For water, the latent heats of fusion and vaporization are

$$L_f = 3.33 \times 10^5 \, J/kg$$
$$L_v = 22.6 \times 10^5 \, J/kg$$

The accompanying Learn by Drawing 11.1, From Cold Ice to Hot Steam is numerically expressed in Example 11.5 and shows explicitly the two types of heat terms (specific heat and latent heat) that must be employed in the general situation when any of the objects undergo a temperature change *and* a phase change.

LEARN BY DRAWING 11.1

from cold ice to hot steam

It can be helpful to focus on the fusion and vaporization of water graphically. To heat a piece of cold ice at −10 °C all the way to hot steam at 110 °C, five separate specific heat and latent heat calculations are necessary. (Most freezers are at a temperature of about −10 °C.) At the phase change (0 °C and 100 °C) heat is added without a temperature change. Once each phase change is complete, adding more heat causes the temperature to increase. The slopes of the lines in the drawings are not all the same, which indicates that the specific heats of the various phases are not the same. (Why do different slopes mean different specific heats?) The numbers come from Example 11.5.

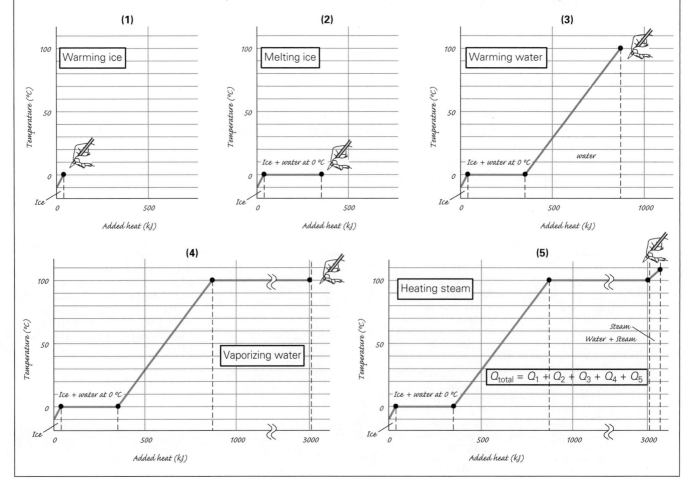

EXAMPLE 11.5 | From Cold Ice to Hot Steam

Heat is added to 1.00 kg of cold ice at $-10\,°C$. How much heat is required to change the cold ice to hot steam at $110\,°C$?

THINKING IT THROUGH. Five steps are involved: (1) heating ice to its melting point (specific heat of ice), (2) melting ice to water at $0\,°C$ (latent heat, a phase change), (3) heating water to its boiling point (specific heat of water), (4) vaporizing water to steam (water vapor) at $100\,°C$ (latent heat, a phase change), and (5) heating steam (specific heat of steam). The key idea here is that the temperature does not change during a phase change. [Refer to Learn by Drawing 11.1, From Cold Ice to Hot Steam.]

SOLUTION.

Given: $m = 1.00$ kg
$T_i = -10\,°C$
$T_f = 110\,°C$
$L_f = 3.33 \times 10^5$ J/kg (from Table 11.2)
$L_v = 22.6 \times 10^5$ J/kg (from Table 11.2)
$c_{ice} = 2100$ J/(kg·°C) (from Table 11.1)
$c_{water} = 4186$ J/(kg·°C) (from Table 11.1)
$c_{steam} = 2000$ J/(kg·°C) (from Table 11.1)

Find: Q_{total} (total heat required)

1. $Q_1 = c_{ice}m\Delta T_1 = [2100\ \text{J/(kg·°C)}](1.00\ \text{kg})[0\,°C - (-10\,°C)]$ *(heating ice)*

   $= +2.10 \times 10^4$ J

2. $Q_2 = +mL_v = (1.00\ \text{kg})(3.33 \times 10^5\ \text{J/kg}) = +3.33 \times 10^5$ J *(melting ice)*

3. $Q_3 = c_{water}m\Delta T_2 = [4186\ \text{J/(kg·°C)}](1.00\ \text{kg})(100\,°C - 0\,°C)$ *(heating water)*

   $= +4.19 \times 10^5$ J

4. $Q_4 = +mL_v = (1.00\ \text{kg})(22.6 \times 10^5\ \text{J/kg}) = +2.26 \times 10^6$ J *(vaporizing water)*

5. $Q_5 = c_{steam}m\Delta T_3 = [2000\ \text{J/(kg·°C)}](1.00\ \text{kg})(110\,°C - 100\,°C)$ *(heating steam)*

   $= +2.00 \times 10^4$ J

The total heat required is

$$Q_{total} = \Sigma Q_i = 2.10 \times 10^4\ \text{J} + 3.33 \times 10^5\ \text{J} + 4.19 \times 10^5\ \text{J} + 2.26 \times 10^6\ \text{J} + 2.00 \times 10^4\ \text{J}$$

$$= 3.05 \times 10^6\ \text{J}$$

The latent heat of vaporization is, by far, the largest. It is actually greater than the sum of the other four terms.

FOLLOW-UP EXERCISE. How much heat must a freezer remove from liquid water (initially at $20\,°C$) to create 0.250 kg of ice at $-10\,°C$?

PROBLEM-SOLVING HINT

Note that the latent heat must be computed at each phase change. It is a common error to use the specific heat equation with a temperature interval *that includes* a phase change. Also, a complete phase change cannot be assumed until you have checked for it numerically. (See Example 11.6.)

Technically, the freezing and boiling points of water ($0\,°C$ and $100\,°C$, respectively) apply only at 1 atm of pressure. Phase change temperatures generally vary with pressure. For example, the boiling point of water decreases with decreasing pressure. At high altitudes, where there is lower atmospheric pressure, the boiling point of water is lowered. For example, at Pikes Peak, Colorado, at an elevation of about 4300 m, the atmospheric pressure is about 0.79 atm and water boils at about $94\,°C$ rather than at $100\,°C$. The lower temperature lengthens the cooking time of food. Conversely, some cooks use a pressure cooker to *reduce* cooking time—by increasing the pressure, a pressure cooker raises the boiling point.

The freezing point of water actually *decreases* with increasing pressure. This inverse relationship is characteristic of only a very few substances, including water (Section 10.4), that expand when they freeze.

EXAMPLE 11.6 | ## Practical Calorimetry: Using Phase Changes to Save a Life

Organ transplants are becoming commonplace. Many times, the procedure involves removing a healthy organ from a deceased person and flying it to the recipient. During that time, to prevent its deterioration, the organ is packed in ice in an insulated container. Assume that a human liver has a mass of 0.500 kg and is initially at 29 °C. The specific heat of the liver is 3500 J/(kg · °C). The liver is surrounded by 2.00 kg of ice initially at −10 °C. Calculate the final equilibrium temperature.

THINKING IT THROUGH.   Clearly, the liver will cool, and the ice will warm. However, it is not clear what temperature the

ice will reach. If it gets to the freezing point, it will begin to melt, and a phase change must be considered. If all of it melts, then additional heat required to warm that water to a temperature above 0 °C must be considered. Thus, care must be taken, since it *cannot* be assumed that all the ice melts, or even that the ice reaches its melting point. Hence, the calorimetry equation (conservation of energy) cannot be written down until the terms in it are determined. First a review of the *possible* heat transfers is needed. Only then can the final temperature be determined.

SOLUTION.   Listing the data given and the information obtained from tables,

Given: $m_l = 0.500$ kg   *Find:* T_f (the final temperature of the system)
$m_{ice} = 2.00$ kg
$c_l = 3500$ J/(kg · °C)
$c_{ice} = 2100$ J/(kg · °C) (from Table 11.1)
$L_f = 3.33 \times 10^5$ J/(kg · °C) (from Table 11.2)

The amount of heat required to bring the ice from −10 °C to 0 °C is

$$Q_{ice} = c_{ice}m_{ice}\Delta T_{ice} = [2100 \text{ J}/(\text{kg} \cdot °\text{C})](2.00 \text{ kg})[0 °\text{C} - (-10 °\text{C})] = +4.20 \times 10^4 \text{ J}$$

Since this heat must come from the liver, the *maximum* heat available from the liver needs to be calculated; that is, if its temperature drops all the way from 29 °C to 0 °C:

$$Q_{l,max} = c_l m_l \Delta T_{l,max} = [3500 \text{ J}/(\text{kg} \cdot °\text{C})](0.500 \text{ kg})(0 °\text{C} - 29 °\text{C}) = -5.08 \times 10^4 \text{ J}$$

This *is* enough heat to bring the ice to 0 °C. If 4.20×10^4 J of heat flows into the ice (bringing it to 0 °C), the liver is still not at 0 °C Then how much ice melts? This depends on how much more heat can be transferred from the liver.

   How much more heat Q' would be transferred from the liver if its temperature were to drop to 0 °C? This value is just the maximum amount minus the heat that went into warming the ice, or

$$Q' = |Q_{l,max}| - 4.20 \times 10^4 \text{ J}$$
$$= 5.08 \times 10^4 \text{ J} - 4.20 \times 10^4 \text{ J} = 8.8 \times 10^3 \text{ J}$$

Compare this with the magnitude of the heat needed to melt the ice completely ($|Q_{melt}|$) to decide whether this can, in fact, happen. The heat required to melt *all* the ice is

$$|Q_{melt}| = +m_{ice}L_{ice} = +(2.00 \text{ kg})(3.33 \times 10^5 \text{ J/kg}) = +6.66 \times 10^5 \text{ J}$$

Since this amount of heat is much larger than the amount available from the liver, only part of the ice melts. In the process, the temperature of the liver has dropped to 0 °C, and the remainder of the ice is at 0 °C. Since everything in the "calorimeter" is at the same temperature, heat flow stops, and the final system temperature, T_f, is 0 °C. Thus the final result is that the liver is in a container with ice and some liquid water, all at 0 °C. Since the container is a very good insulator, it will prevent any inward heat flow that might raise the liver's temperature. It is therefore expected that the liver will arrive at its destination in good shape.

FOLLOW-UP EXERCISE.   (a) In this Example, how much of the ice melts? (b) If the ice originally had been at its melting point (0 °C), what would the equilibrium temperature have been?

PROBLEM-SOLVING HINT

Notice that in Example 11.6 numbers were *not* plugged directly into the $\Sigma Q_i = 0$ equation, which is equivalent to assuming that all the ice melts. In fact, if this step had been done, we would have been on the wrong track. For calorimetry problems *involving phase changes*, a careful step-by-step numerical "accounting" procedure should be followed until all of the pieces of the system are at the same temperature. At that point, the problem is over, because no more heat exchanges can happen.

EVAPORATION

The **evaporation** of water from an open container becomes evident only after a relatively long period of time. This phenomenon can be explained in terms of the kinetic theory (Section 10.5). The molecules in a liquid are in motion at different speeds. A faster-moving molecule near the surface may momentarily leave the liquid. If its speed is not too large, the molecule will return to the liquid, because of the attractive forces exerted by the other molecules. Occasionally, however, a molecule has a large enough speed to leave the liquid entirely. The higher the temperature of the liquid, the more likely this phenomenon is to occur.

The escaping molecules take their energy with them. Since those molecules with greater-than-average energy are the ones most likely to escape, the average molecular energy, and thus the temperature of the remaining liquid, will be reduced. That is, *evaporation is a cooling process* for the object from which the molecules escape. You have probably noticed this phenomenon when drying off after a bath or shower. You can read more about this in Insight 11.1, Physiological Regulation of Body Temperature.

DID YOU LEARN?

➥ The three common phases of matter are the solid phase, the liquid phase, and the gaseous phase.

➥ The temperature of matter during a phase change remains constant. If the phase change is between solid and liquid or vice versa, the temperature will remain at the melting or freezing point; if the phase change is between liquid and gas or vice versa, the temperature will remain at the boiling or condensation point.

➥ The latent heat of fusion (vaporization) of a substance is the heat required to change 1 kg of the substance from solid to liquid (liquid to gas) or vice versa while the temperature remains constant at the freezing (boiling) point.

INSIGHT 11.1 | ## Physiological Regulation of Body Temperature

Being warm-blooded, humans must maintain a narrow range of body temperature. (See Chapter 10 Insight 10.1, Human Body Temperature.) The generally accepted value for the average normal body temperature is 37.0 °C (98.6 °F). However, it can be as low as 35.5 °C (95.9 °F) in the early morning hours on a cold day and as high as 39.5 °C (103 °F) during intense exercise on a hot day. For females, the body temperature at rest rises very slightly after ovulation as a result of a rise in the hormone progesterone. This information can be used to predict on which day ovulation will occur in the next cycle.

When the ambient temperature is lower than the body temperature, the body loses heat. If the body loses too much heat, a circulatory mechanism causes a reduction of blood flow to the skin in order to reduce heat loss. A physiological response to this mechanism is to increase heat generation (and thus warm the body) through shivering by "burning" the body's reserves of carbohydrate or fat. If the body temperature drops below 33 °C (91.4 °F), *hypothermia* may result, which can cause severe thermal injuries to organs and even death.

At the other extreme, if the body is subjected to ambient temperatures higher than the body temperature, along with intense exercise, the body can overheat. *Heat stroke* is a prolonged elevation of body temperature above 40 °C (104 °F). If the body becomes overheated, the blood vessels to the skin dilate, carrying more warm blood to the skin, enabling the interior of the body and organs to remain cooler. (The person's face may turn red.)

Usually, radiation, conduction, natural convection (discussed later in Section 11.4), and possibly slight evaporation of perspiration on the skin are sufficient to maintain a heat loss at a rate that keeps our body temperature in the safe range. However, when the ambient temperature becomes too high, these mechanisms cannot do the job completely. To avoid heat stroke, as a last resort, the body produces heavy perspiration (the most efficient mechanism). The evaporation of water from the skin removes a lot of heat, thanks to the large value of the latent heat of vaporization of water.

Evaporation draws heat from our skin and thus cools our bodies. The removal of a minimum of 2.26×10^6 J of heat from the body is required to evaporate each kilogram (liter) of water.* For the body of a 75-kg person (which is mainly composed of water), the heat loss due to evaporation of 1 kg of water could lower the body temperature as much as

$$\Delta T = \frac{Q}{cm} = \frac{2.26 \times 10^6 \text{ J}}{[4186 \text{ J}/(\text{kg} \cdot °\text{C})](75 \text{ kg})} = 7.2 °\text{C}$$

During a race on a hot day, a professional cyclist can lose as much as 7.0 kg of water in 3.5 h via evaporation. This heat transfer through perspiration is the mechanism that enables the body to keep its temperature in the safe range.

On a summer day, a person may stand in front of a fan and remark how "cool" the blowing air feels. But the fan is merely blowing hot air from one place to another. The air *feels* cool because it is relatively drier (has low humidity compared to the body's perspiration), and therefore its flow promotes evaporation, which removes heat.

*The actual latent heat of vaporization of perspiration is about 2.42×10^6 J/(kg·°C)—greater than the 2.26×10^6 J/(kg·°C) value used here for temperature at 100 °C. Why?

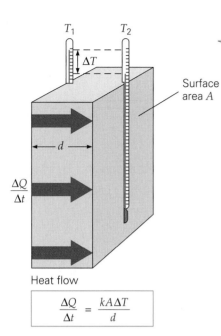

$$\frac{\Delta Q}{\Delta t} = \frac{kA\Delta T}{d}$$

▲ FIGURE 11.7 **Thermal conduction** Heat conduction is characterized by the time rate of heat flow ($\Delta Q/\Delta t$) in a material with a temperature difference across it of ΔT. For a slab of material, $\Delta Q/\Delta t$ is directly proportional to the cross-sectional area (A) and the thermal conductivity (k) of the material; it is inversely proportional to the thickness of the slab (d).

11.4 Heat Transfer

LEARNING PATH QUESTIONS

➥ What are the three mechanisms of heat transfer?

➥ Should house insulation materials have high or low thermal conductivity?

➥ Which mechanism of heat transfer does not require a transport medium?

CONDUCTION

You can keep a pot of coffee hot on an electric stove because heat is conducted through the bottom of the coffeepot from the hot metal burner. The process of **conduction** results from molecular interactions. Molecules at a higher-temperature region on an object move relatively rapidly. They collide with, and transfer some of their energy to, the less energetic molecules in a nearby cooler part of the object. In this way, energy is conductively transferred from a higher-temperature region to a lower-temperature region—transfer as a result of a temperature difference.

Solids can be divided into two general categories: metals and nonmetals. Metals are generally good conductors of heat, or **thermal conductors**. Metals have a large number of electrons that are free to move around (not permanently bound to a particular molecule or atom). These free electrons (rather than the interaction between adjacent atoms) are primarily responsible for the good heat conduction in metals. Nonmetals, such as wood and cloth, have relatively few free electrons. The absence of this transfer mechanism makes them poor heat conductors relative to metals. A poor heat conductor is called a **thermal insulator**.

In general, the ability of a substance to conduct heat depends on the substance's phase. Gases are poor thermal conductors; their molecules are relatively far apart, and collisions are therefore infrequent. Liquids and solids are better thermal conductors than gases, because their molecules are closer together and can interact more readily.

Heat conduction is usually described using the *time rate* of heat flow ($\Delta Q/\Delta t$) in a material for a given temperature difference (ΔT), as illustrated in ◂Fig. 11.7. Experiment has established that the rate of heat flow through a substance depends on the temperature difference between its boundaries. Heat conduction also depends on the size and shape of the object as well as its composition.

Experimentally, it was found that the heat flow rate ($\Delta Q/\Delta t$ in J/s or W) through a slab of material is directly proportional to the material's surface area (A) and the temperature difference across its ends (ΔT), and is inversely proportional to its thickness (d). That is,

$$\frac{\Delta Q}{\Delta t} \propto \frac{A\Delta T}{d}$$

Using a constant of proportionality k allows us to write the relation as an equation:

$$\frac{\Delta Q}{\Delta t} = \frac{kA\Delta T}{d} \quad \textit{(conduction only)} \qquad (11.4)$$

The constant k, called the **thermal conductivity**, characterizes the heat-conducting ability of a material and depends only on the type of material. The greater the value of k for a material, the better it will conduct heat, all other factors being equal. The units of k are $J/(m \cdot s \cdot °C) = W/(m \cdot °C)$. The thermal conductivities of various substances are listed in ▸Table 11.3. These values vary slightly with temperature, but can be considered constant over normal temperature ranges.

Compare the relatively large thermal conductivities of the good thermal conductors, the metals, with the relatively small thermal conductivities of some good thermal insulators, such as Styrofoam and wood. Some stainless steel cooking pots have copper bottoms (◂Fig. 11.8). Being a good conductor of heat, the copper conducts heat faster to the food being cooked and also promotes the distribution of heat over the bottom of a pot for even cooking. Conversely, Styrofoam is a good insulator, mainly because it contains small, trapped pockets of air, thus reducing conduction and convection losses (discussed in the next section). When you step on a tile floor with one

▲ FIGURE 11.8 **Copper-bottomed pots** Copper is used on the bottoms of some stainless steel pots and saucepans. The high thermal conductivity of copper ensures the rapid and even spread of heat from the burner; the low thermal conductivity of stainless steel retains the heat in the pot and keeps the handle not too hot to touch. (The thermal conductivity of stainless steel is only 12% of that of copper.)

TABLE 11.3 Thermal Conductivities of Some Substances

Substance	Thermal Conductivity, k	
	J/(m·s·°C) or W/(m·°C)	kcal/(m·s·°C)
Metals		
Aluminum	240	5.73×10^{-2}
Copper	390	9.32×10^{-2}
Iron	80	1.9×10^{-2}
Stainless steel	16	3.8×10^{-3}
Silver	420	10×10^{-2}
Liquids		
Transformer oil	0.18	4.3×10^{-5}
Water	0.57	14×10^{-5}
Gases		
Air	0.024	0.57×10^{-5}
Hydrogen	0.17	4.1×10^{-5}
Oxygen	0.024	0.57×10^{-5}
Other Materials		
Brick	0.71	17×10^{-5}
Concrete	1.3	31×10^{-5}
Cotton	0.075	1.8×10^{-5}
Fiberboard	0.059	1.4×10^{-5}
Floor tile	0.67	16×10^{-5}
Glass (typical)	0.84	20×10^{-5}
Glass wool	0.042	1.0×10^{-5}
Goose down	0.025	0.59×10^{-5}
Human tissue (average)	0.20	4.8×10^{-5}
Ice	2.2	53×10^{-5}
Styrofoam	0.042	1.0×10^{-5}
Wood, oak	0.15	3.6×10^{-5}
Wood, pine	0.12	2.9×10^{-5}
Vacuum	0	0

bare foot and on an adjacent rug with the other bare foot, you feel that the tile is "colder" than the rug. However, both the tile and rug are actually at the same temperature. But the tile is a much better thermal conductor, so it transfers heat from your foot more efficiently than the rug, making your foot on tile feel colder.

EXAMPLE 11.7 | Thermal Insulation: Helping Prevent Heat Loss

A room with a pine ceiling that measures 3.0 m by 5.0 m and is 2.0 cm thick has a layer of glass wool insulation above it that is 6.0 cm thick (▼Fig. 11.9a). On a cold day, the temperature inside the room at ceiling height is 20 °C, and the temperature in the attic above the insulation layer is 8.0 °C. Assuming that the temperatures remain constant and heat loss is due to conduction only, how much energy does the layer of insulation save in 1.0 h?

THINKING IT THROUGH. Here there are two materials, so Eq. 11.4 is applied for two different thermal conductivities (k). We want to find $\Delta Q/\Delta t$ for the combination so that ΔQ can be found for $\Delta t = 1.0$ h. The situation is a bit complicated, because the heat flows through two materials. But at a steady rate, *the heat flows must be the same through both*. (Why?) To find the energy saved in 1.0 h, the heat conducted in this time both without and with the layer of insulation needs to be calculated.

(continued on next page)

▶ **FIGURE 11.9 Insulation and thermal conductivity (a), (b)** Attics should be insulated to prevent loss of heat by the mechanism of conduction. See Example 11.7 and Insight 11.2, Physics, the Construction Industry, and Energy Conservation. **(c)** This thermogram of a house allows us to visualize the house's heat loss. Blue represents the areas that have the lowest rate of heat leaking; white, pink, and red indicate areas with increasingly larger heat losses. (Red areas have the most loss.) What recommendations would you make to the owner of this house to save both money and energy? (Compare this figure with Fig. 11.15.)

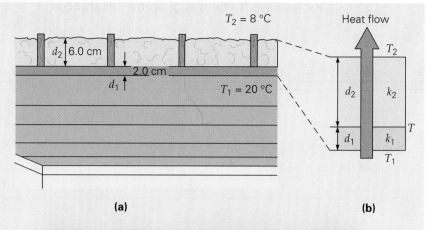

(a)   **(b)**

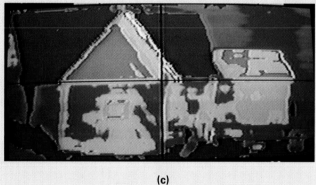

(c)

SOLUTION.   Listing the data, computing some of the quantities in Eq. 11.4, and making conversions:

Given: $A = 3.0\,\text{m} \times 5.0\,\text{m} = 15\,\text{m}^2$ **Find:**   Energy saved in 1.0 h
  $d_1 = 2.0\,\text{cm} = 0.020\,\text{m}$
  $d_2 = 6.0\,\text{cm} = 0.060\,\text{m}$
  $\Delta T = T_1 - T_2 = 20\,°\text{C} - 8.0\,°\text{C} = 12\,°\text{C}$
  $\Delta t = 1.0\,\text{h} = 3.6 \times 10^3\,\text{s}$
  $k_1 = 0.12\,\text{J}/(\text{m} \cdot \text{s} \cdot °\text{C})\,(\text{wood, pine})$ $\Big\}$(from Table 11.3)
  $k_2 = 0.042\,\text{J}/(\text{m} \cdot \text{s} \cdot °\text{C})\,(\text{glass wool})$

(In working such problems with several given quantities, it is especially important to label all the data correctly.)

First, let's consider how much heat would be conducted in 1.0 h through the wooden ceiling without insulation. Since Δt is known, Eq. 11.4* can be rearranged to find ΔQ_c (heat conducted through the wooden ceiling alone, assuming the same ΔT):

$$\Delta Q_c = \left(\frac{k_1 A \Delta T}{d_1}\right)\Delta t = \left\{\frac{[0.12\,\text{J}/(\text{m} \cdot \text{s} \cdot °\text{C})](15\,\text{m}^2)(12\,°\text{C})}{0.020\,\text{m}}\right\}(3.6 \times 10^3\,\text{s}) = 3.9 \times 10^6\,\text{J}$$

Now we need to find the heat conducted through the ceiling *and* the insulation layer together. Let T be the temperature at the interface of the materials and T_1 and T_2 be the warmer and cooler temperatures, respectively (Fig. 11.9b). Then

$$\frac{\Delta Q_1}{\Delta t} = \frac{k_1 A(T_1 - T)}{d_1} \quad \text{and} \quad \frac{\Delta Q_2}{\Delta t} = \frac{k_2 A(T - T_2)}{d_2}$$

T is not known, but when the conduction is steady, the flow rates are the same for both materials; that is, $\Delta Q_1/\Delta t = \Delta Q_2/\Delta t$, or

$$\frac{k_1 A(T_1 - T)}{d_1} = \frac{k_2 A(T - T_2)}{d_2}$$

The A's cancel, and solving for T gives

$$T = \frac{k_1 d_2 T_1 + k_2 d_1 T_2}{k_1 d_2 + k_2 d_1} = \frac{[0.12\,\text{J}/(\text{m} \cdot \text{s} \cdot °\text{C})](0.060\,\text{m})(20\,°\text{C}) + [0.042\,\text{J}/(\text{m} \cdot \text{s} \cdot °\text{C})](0.020\,\text{m})(8.0\,°\text{C})}{[0.12\,\text{J}/(\text{m} \cdot \text{s} \cdot °\text{C})](0.060\,\text{m}) + [0.042\,\text{J}/(\text{m} \cdot \text{s} \cdot °\text{C})](0.020\,\text{m})} = 18.7\,°\text{C}$$

*Equation 11.4 can be extended to any number of layers or slabs of materials: $\Delta Q/\Delta t = A(T_2 - T_1)/\Sigma(d_i/k_i)$. (See Insight 11.2, Physics, the Construction Industry, and Energy Conservation, involving insulation in building construction.)

Since the flow rate through the wood and the insulation is the same, we can use the expression for either material to calculate it. Let's use the expression for the wood ceiling. Here, care must be taken to use the correct ΔT. The temperature at the wood–insulation interface is 18.7 °C; thus,

$$\Delta T_{\text{wood}} = |T_1 - T| = |20\ °C - 18.7\ °C| = 1.3\ °C$$

Therefore, the heat flow rate is

$$\frac{\Delta Q_1}{\Delta t} = \frac{k_1 A |\Delta T_{\text{wood}}|}{d_1} = \frac{[0.12\ \text{J}/(\text{m} \cdot \text{s} \cdot °C)](15\ \text{m}^2)(1.3\ °C)}{0.020\ \text{m}} = 1.2 \times 10^2\ \text{J/s (or W)}$$

In 1.0 h, the heat loss with insulation in place is

$$\Delta Q_1 = \frac{\Delta Q_1}{\Delta t} \times \Delta t = (1.2 \times 10^2\ \text{J/s})(3600\ \text{s}) = 4.3 \times 10^5\ \text{J}$$

This value represents a decreased heat loss of

$$\Delta Q_c - \Delta Q_1 = 3.9 \times 10^6\ \text{J} - 4.3 \times 10^5\ \text{J} = 3.5 \times 10^6\ \text{J}$$

This amount represents an energy savings of $\dfrac{3.5 \times 10^6\ \text{J}}{3.9 \times 10^6\ \text{J}} \times (100\%) = 90\%$.

FOLLOW-UP EXERCISE. Verify that the heat flow rate through the insulation is the same as that through the wood $(1.2 \times 10^2\ \text{J/s})$ in this Example.

INSIGHT 11.2 | # Physics, the Construction Industry, and Energy Conservation

Many homeowners have found it cost-effective to provide their homes with better insulation, especially in the last few years due to the higher energy costs. To quantify the insulating properties of various materials, the insulation and construction industries do not use thermal conductivity, k. Rather, they use a quantity called *thermal resistance*, which is related to the *inverse* of k.

To see how these two quantities are related, consider Eq. 11.4 rewritten as

$$\frac{\Delta Q}{\Delta t} = \left(\frac{k}{d}\right) A \Delta T = \left(\frac{1}{R_t}\right) A \Delta T$$

where the *thermal resistance* is $R_t = d/k$. Note that R_t depends not only on the material's properties (expressed in the thermal conductivity, k), but also on its thickness, d. R_t is a measure of how "resistant" to heat flow a certain thickness of material is. The heat flow rate, $\Delta Q/\Delta t$, is inversely related to the thermal resistance: More thermal resistance results in less heat flow. More resistance is attained using *thicker* material with a *low* conductivity.

For homeowners, the lesson is clear. To reduce heat flow (and thus minimize heat loss in the winter and heat gain in the summer), they should reduce areas of low thermal resistance, such as windows, or at least increase the windows' resistance by switching to double or triple panes. Similarly, increasing the thermal resistance by adding or upgrading insulation to walls is the way to go. Lastly, changing interior temperature requirements (changing $\Delta T = |T_{\text{exterior}} - T_{\text{interior}}|$) can make a big difference. In the summer, homeowners should raise the thermostat setting on their air conditioning (lowering ΔT by increasing T_{interior}), and in winter, they should lower the ther-

FIGURE 1 Differences in R-values For insulation blankets made of identical materials, the R-values are proportional to the materials' thickness.

mostat setting on their heating system (also lowering ΔT but by decreasing T_{interior}).

Insulation and building materials are classified according to their *R-values*, that is, their thermal resistance values. In the United States, the units of R_t are $\text{ft}^2 \cdot \text{h} \cdot °\text{F/Btu}$. While these units may seem awkward, the important point is that they are proportional to the thermal resistance of the material. Thus, wall insulation with a value of R-30 (meaning $R_t = 30\ \text{ft}^2 \cdot \text{h} \cdot °\text{F/Btu}$) is about 2.3 times (or 30/13) as resistive as insulation with a value of R-13. A photo showing various types of insulation is shown in Fig. 1.

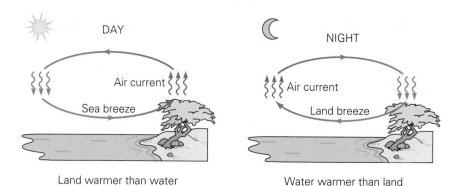

▲ **FIGURE 11.10 Convection cycles** During the day, natural convections give rise to sea breezes near large bodies of water. At night, the pattern of circulation is reversed, and the land breezes blow. The temperature differences between land and water are the result of their specific heat differences. Water has a much larger specific heat, so the land warms up more quickly during the day. At night, the land cools more quickly, while the water remains warmer, because of its larger specific heat.

CONVECTION

In general, compared with solids, liquids and gases are not good thermal conductors. However, the mobility of molecules in fluids permits heat transfer by another process—convection. (A fluid is a substance that can flow, and hence includes both liquids and gases.) **Convection** is heat transfer as a result of mass transfer, which can be natural or forced.

Natural convection occurs in liquids and gases. For example, when cold water is in contact with a hot object, such as the bottom of a pot on a stove, heat is transferred to the water adjacent to the pot by conduction. Since the water at the bottom is warmer, its density is lower, causing it to rise to the top. The top water, being cooler, has a higher density, so it sinks to the bottom. This sets up a natural convection. Such convections are also important in atmospheric processes, as illustrated in ▲ Fig. 11.10. During the day, the ground heats up more quickly than do large bodies of water, as you may have noticed if you have been to the beach. This phenomenon occurs mainly because the water has a higher specific heat than land. The air in contact with the warm ground is heated and expands, becoming less dense. As a result, the warm air rises (air currents) and, to fill the space, other air moves horizontally (winds)—creating a sea breeze near a large body of water. Cooler air descends, and a thermal convection cycle is set up, which transfers heat away from the land. At night, the ground loses its heat more quickly than the water, and the surface of the water is warmer than the land. As a result, the current is reversed. Since the prevailing jet streams over the Northern Hemisphere flow mostly from west to east, west coasts usually have a milder climate than east coasts. The winds move the Pacific ocean air with more constant temperature toward the west coasts.

In *forced convection*, the fluid is moved mechanically. Common examples of forced convection systems are forced-air heating systems in homes (◄Fig. 11.11), the human circulatory system, and the cooling system of an automobile engine. The human body loses a great deal of heat when the surroundings are colder than the body. The internally generated heat is transferred close to the surface of the skin by blood circulation. From the skin, the heat is conducted to the air or lost by radiation (the other heat transfer mechanism, to be discussed shortly). This circulatory system is highly adjustable; blood flow can be increased or decreased to specific areas depending on needs.

Coolant is circulated (pumped) through most automobile cooling systems. (Some smaller engines are air-cooled.) The coolant carries engine heat to the radiator (a form of *heat exchanger*), where forced-air flow produced by the fan and car movement carries it away. The *radiator* of an automobile is actually misnamed—most of the heat is transferred from it by forced convection rather than by radiation.

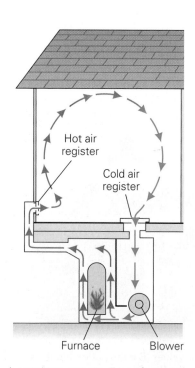

▲ **FIGURE 11.11 Forced convection** Houses are commonly heated by forced convection. Registers or gratings in the floors or walls allow heated air to enter and cooler air to return to the heat source. (Can you explain why the registers are located near the floor?)

CONCEPTUAL EXAMPLE 11.8 | Foam Insulation: Better Than Air?

Foam insulation is sometimes blown into the space between the inner and outer walls of a house. Since air is a better thermal insulator than foam (Table 11.3), why is the foam insulation needed: (a) to prevent loss of heat by conduction, (b) to prevent loss of heat by convection, or (c) for fireproofing?

REASONING AND ANSWER. Foams will generally burn, so (c) isn't likely to be the answer. Air is a poor thermal conductor, even poorer than foam (Styrofoam—see Table 11.3), so the answer can't be (a). However, as a gas, the air is subject to convection *within the wall space*. In the winter, the air near the warm inner wall is heated and rises, thus setting up a convection cycle in the space and transferring heat to the cold outer wall. In the summer, with air conditioning, the heat loss cycle is reversed. Foam blocks the movement of air and thus stops such convection cycles. Hence, the answer is (b).

FOLLOW-UP EXERCISE. Thermal underwear and thermal blankets are loosely knit with lots of small holes. Wouldn't they be more effective if the material were closely knit?

RADIATION

Conduction and convection require some material as a transport medium. The third mechanism of heat transfer needs no medium; it is called **radiation**, which refers to energy transfer by electromagnetic waves (Section 20.4). Heat is transferred to the Earth from the Sun through empty space by radiation. Visible light and other forms of electromagnetic radiation are commonly referred to as *radiant energy*.

You have experienced heat transfer by radiation if you've ever stood near an open fire (▶Fig. 11.12). You can feel the heat on your exposed hands and face. This heat transfer is not due to convection or conduction, since heated air rises and air is a poor conductor. Visible radiation is emitted from the burning material, but most of the heating effect comes from the invisible **infrared radiation**. You feel this radiation because it is absorbed by water molecules in your skin. The water molecule has an internal vibration whose frequency coincides with that of infrared radiation, which is therefore readily absorbed. (This effect is called *resonance absorption*. The electromagnetic wave drives the molecular vibration, and energy is transferred to the molecule, somewhat like pushing a swing. See Chapter 13 on oscillations for more details.) Heat transfer by radiation can play a practical role in daily living (▶Fig. 11.13).

Infrared radiation is sometimes referred to as "heat radiation" or thermal radiation. You may have noticed the reddish infrared lamps used to keep food warm in cafeterias. Heat transfer by infrared radiation is also important in maintaining our planet's warmth by a mechanism known as the *greenhouse effect*. This important environmental topic is discussed in Insight 11.3, The Greenhouse Effect.

Although infrared radiation is invisible to the human eye, it can be detected by other means. Infrared detectors can measure temperature remotely (▼Fig. 11.14).

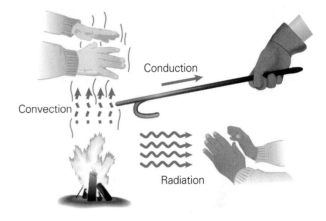

▲ **FIGURE 11.12 Heating by conduction, convection, and radiation** The hands on top of the flame are warmed by the convection of rising hot air (and some radiation). The gloved hand is warmed by conduction. The hands to the right of the flame are warmed by radiation.

▲ **FIGURE 11.13 A practical application of heat transfer by radiation** A Tibetan teakettle is heated by focusing sunlight, using a metal reflector.

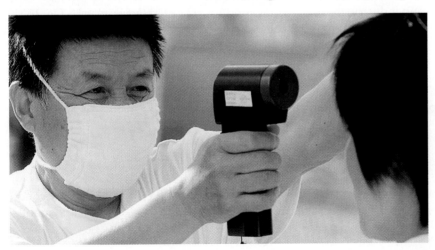

◀ **FIGURE 11.14 Detecting SARS** Infrared thermometers were used to measure body temperature during the severe acute respiratory syndrome (SARS) outbreak in 2003.

INSIGHT 11.3 | The Greenhouse Effect

The *greenhouse effect* helps regulate the Earth's long-term average temperature, which has been fairly constant for some centuries. When a portion of the solar radiation (mostly visible light) reaches and warms the Earth's surface, the Earth, in turn, reradiates energy in the form of infrared radiation (IR). As the reradiated IR radiation passes back through the atmosphere, some of the radiation is absorbed by the *greenhouse gases* there—primarily water vapor, carbon dioxide (CO_2), and methane. These gases are selective absorbers: They absorb radiation at certain IR wavelengths but not at others (Fig. 1a). Without this absorption, the IR radiation would go back into space and life on the Earth would probably not exist, because the average surface temperature would be a cold −18 °C, rather than the present 15 °C.

Why is this phenomenon called the *greenhouse effect?* The reason is that the atmosphere functions *somewhat* like the gases in a greenhouse. In general, visible radiation is transmitted but infrared radiation is selectively absorbed by the gas in a greenhouse (Fig. 1b) so the greenhouse heats up by trapping the reradiated infrared radiation inside. The glass enclosure also keeps warm air from escaping upward, resulting in the elimination of heat loss by convection. It is quite warm in a greenhouse on a sunny day, even in winter. We have all observed this warming effect—for example, in a closed car on a sunny but cold day.

The problem on Earth is that human activities since the beginning of the industrial age have been accelerating greenhouse warming. With the combustion of hydrocarbon fuels (gas, oil, coal, and so on), vast amounts of CO_2 and other greenhouse gases are vented into the atmosphere, where they trap increasingly more IR radiation. There is grave concern that the result of this trend is increasing the Earth's average surface temperature—*global warming.* Such an increase could dramatically affect agricultural production and world food supplies. It can also cause partial melting of the polar ice caps. Sea levels would then rise, flooding low-lying regions and endangering coastal ports and population centers.

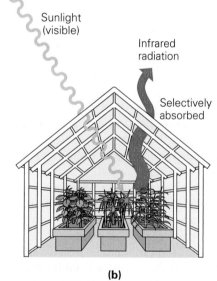

FIGURE 1 The greenhouse effect
(a) The greenhouse gases of the atmosphere, particularly water vapor, methane, and carbon dioxide, are selective absorbers with absorption properties similar to those of the glass used in greenhouses. Visible light is transmitted and heats the Earth's surface, while some of the infrared radiation that is re-emitted is absorbed and trapped in the Earth's atmosphere. **(b)** A greenhouse operates in a similar way.

Sunlight (visible)

Infrared radiation

Selectively absorbed

Atmospheric gases

Sunlight (visible)

Infrared radiation

Selectively absorbed

(a)   **(b)**

Also, cameras using special infrared films take pictures consisting of contrasting bright and dark areas that correspond to regions of higher and lower temperatures, respectively. Special instruments that apply such *thermography* are used in medicine and industry; the images they produce are called *thermograms* (▶Fig. 11.15).

A new application of thermograms is for security. An infrared camera takes a picture of an individual using the unique heat pattern emitted by the facial blood vessels. A computer then compares the picture with an earlier stored image.

The rate at which an object radiates energy has been found to be proportional to the fourth power of the object's absolute temperature (T^4). This relationship is expressed in an equation known as **Stefan's law,***

$$P = \frac{\Delta Q}{\Delta t} = \sigma A e T^4 \quad \text{(radiation only)} \tag{11.5}$$

where $P(\Delta Q/\Delta t)$ is the power radiated in watts (W), or joules per second (J/s). A is the object's surface area and T is its temperature in Kelvin. The symbol σ (the

*Developed by the Austrian physicist Joseph Stefan (1835–1893).

Greek letter sigma) is the *Stefan–Boltzmann constant*: $\sigma = 5.67 \times 10^{-8}\,\text{W}/(\text{m}^2 \cdot \text{K}^4)$. The **emissivity** (*e*) is a unitless number between 0 and 1 that is a characteristic of the material. Dark surfaces have emissivities close to 1, and shiny surfaces have emissivities close to 0. The emissivity of human skin is about 0.70.

Dark surfaces not only are better emitters of radiation, but also are good absorbers. This must be the case because to maintain a constant temperature, the incident energy absorbed must equal the emitted energy. *Thus, a good absorber is also a good emitter.* An ideal, or perfect, absorber (and emitter) is referred to as a **black body** (*e* = 1.0). Shiny surfaces are poor absorbers, since most of the incident radiation is reflected. This fact can be demonstrated easily, as shown in ▶Fig. 11.16. (Can you see why it is better to wear light-colored clothes in the summer and dark-colored clothes in the winter?)

When an object is in thermal equilibrium with its surroundings, its temperature is constant; thus, it must be emitting and absorbing radiation at the same rate. However, if the temperatures of the object and its surroundings are different, there will be a net flow of radiant energy. If an object is at a temperature T and its surroundings are at a temperature T_s, the net rate of energy loss or gain per unit time (power) is given by

$$P_{\text{net}} = \sigma Ae(T_s^4 - T^4) \qquad (11.6)$$

Note that if T_s is less than T, then P will be negative, indicating a net heat energy loss, in keeping with our heat flow sign convention. Keep in mind that the temperatures used in calculating radiated power are the absolute temperatures in kelvins.

You may have noticed in Section 10.1 that heat was defined as the net energy transfer due to temperature differences. The word *net* here is important. It is possible to have energy transfer between an object and its surroundings, or between objects, at the same temperature. Note that if $T_s = T$ (that is, there is no temperature difference), there is a continuous exchange of radiant energy, but there is no *net* change of the internal energy of the object.

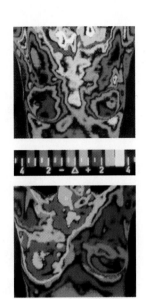

▲ **FIGURE 11.15 Applied thermography** Thermograms can be used to detect breast cancer by showing tumor regions that are higher in temperature than normal. The upper photo shows a thermogram scan of a woman without breast cancer. The lower photo shows the result for a woman with breast cancer. The "hot spots" in this scan tell the physician where the cancer resides.

EXAMPLE 11.9	**Body Heat: Radiant Heat Transfer**

Suppose that your skin has an emissivity of 0.70, a temperature of 34 °C, and a total area of 1.5 m². How much net energy per second will be radiated from your skin if the ambient room temperature is 20 °C?

THINKING IT THROUGH. Everything is given for us to find P_{net} from Eq. 11.6. The net radiant energy transfer is between the skin and the surroundings. We must remember to work with temperatures in kelvins.

SOLUTION.

Given: $T_s = (20 + 273)\,\text{K} = 293\,\text{K}$ *Find:* P_{net} (net power)

$T = (34 + 273)\,\text{K} = 307\,\text{K}$

$e = 0.70$

$A = 1.5\,\text{m}^2$

$\sigma = 5.67 \times 10^{-8}\,\text{W}/(\text{m}^2 \cdot \text{K}^4)$ (known)

Using Eq. 11.6 directly,

$$P_{\text{net}} = \sigma Ae(T_s^4 - T^4) = [5.67 \times 10^{-8}\,\text{W}/(\text{m}^2 \cdot \text{K}^4)](1.5\,\text{m}^2)(0.70)[(293\,\text{K})^4 - (307\,\text{K})^4]$$
$$= -90\,\text{W (or } -90\,\text{J/s})$$

Thus, 90 J of energy is radiated, or *lost* (as indicated by the minus sign), each second. That is, the human body loses heat at a rate that is close to that of a 100-W lightbulb. No wonder a room full of people can get warm.

FOLLOW-UP EXERCISE. (a) In this Example, suppose the skin had been exposed to an ambient room temperature of only 10 °C. What would the rate of heat loss be? (b) Elephants have huge body masses and large daily caloric food intakes. Can you explain how their huge ear flaps (large surface area) might help stabilize their body temperature?

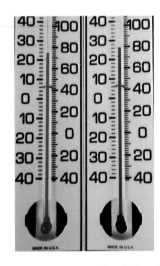

▲ **FIGURE 11.16 Good absorber** Black objects are generally good absorbers of radiation. The bulb of the thermometer on the right has been painted black. Note the difference in temperature readings.

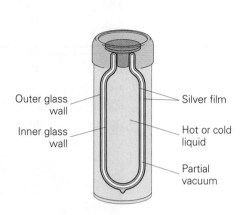

▲ FIGURE 11.17 **Thermal insulation**
The Thermos bottle minimizes all three mechanisms of heat transfer. See text for description.

Outer glass wall
Inner glass wall
Silver film
Hot or cold liquid
Partial vacuum

▲ FIGURE 11.18 **A dark robe in the desert?** Dark objects absorb more radiation than do lighter ones, and they become hotter. What's going on here? See the book for an explanation.

PROBLEM-SOLVING HINT

Note that in Example 11.9, the fourth powers of the temperatures were found first, and then their difference was found. It is *not* correct to find the temperature difference and then raise it to the fourth power: $T_s^4 - T^4 \neq (T_s - T)^4$.

Let's look at a few more real-life examples of heat transfer. In the spring, a late frost could kill the buds on fruit trees. To save the buds, some growers spray water on the trees to form ice before a hard frost occurs. Using ice to save buds? Ice is a relatively poor (and inexpensive) thermal conductor, so it has an insulating effect. It will maintain the buds' temperature at 0 °C, not going below that value, and therefore protects the buds.

Another method to protect orchards from freezing is the use of smudge pots, containers in which material is burned to create a dense cloud of smoke. At night, when the Sun-warmed ground cools off by radiation, the cloud absorbs this heat and reradiates it back to the ground. Thus, the ground takes longer to cool, hopefully without reaching freezing temperatures before the Sun comes up.

A Thermos bottle (▲Fig. 11.17) keeps cold beverages cold and hot ones hot. It consists of a double-walled, partially evacuated container with silvered walls (mirrored interior). The bottle is constructed to minimize all three mechanisms of heat transfer. The double-walled and partially evacuated container counteracts conduction and convection because both processes depend on a medium to transfer the heat (the double walls are more for holding the partially evacuated region than for reducing conduction and convection). The mirrored interior minimizes loss by radiation. The stopper on top of the thermos stops convection off the top of the liquid as well.

Look at ▲Fig. 11.18. Why would anyone wear a dark robe in the desert? It was previously learned that dark objects absorb radiation (Fig. 11.16). Wouldn't a white robe be better? A dark robe definitely absorbs more radiant energy and warms the air inside near the body. But note that the robe is open at the bottom. The warm air rises (since it is less dense) and exits at the neck area, and outside cooler air enters the robe at the bottom—natural convection air circulation.

Finally, consider some of the thermal factors involved in "passive" solar house design used as far back as in ancient China (▶Fig. 11.19). The term *passive* means that the design elements require no active use of energy. In Beijing, China, for

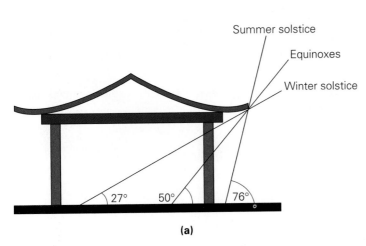

(a)   **(b)**

▲ **FIGURE 11.19 Aspects of passive solar design in ancient China** **(a)** In summer, with the sun angle high, the overhangs provide shade to the building. The brick and mud walls are thick to reduce conductive heat flow to the interior. In winter, the sun angle is low, so the sunlight streams into the building, especially with the help of the upward curved overhangs. The leaves of nearby deciduous trees provide additional shade in the summer but allow sunlight in when they have dropped their leaves in the winter. **(b)** A photo of such a building in Beijing, China, in December.

example, the angles of the sunlight are 76°, 50°, and 27° above the horizon at the summer solstice, the spring and fall equinoxes, and the winter solstice, respectively. With a proper combination of column height and roof overhang length, a maximum amount of sunlight is allowed *into* the building in the winter, but most of the sunlight will *not* reach the inside of the building in the summer. The overhangs of the roofs are also curved upward, not just for good looks, but also for letting the maximum amount of light into the building in the winter. Trees planted on the south side of the building can also play important roles in both summer and winter. In the summer, the leaves block and filter the sunlight; in the winter, the dropped leaves will let sunlight through.

DID YOU LEARN?
➥ The three mechanisms of heat transfer are conduction, convection, and radiation.
➥ Insulation materials should have low thermal conductivities so as to conduct less heat.
➥ Radiation does not require a transport medium, that is, radiation heat transfer can occur in a vacuum.

PULLING IT TOGETHER | ## Ice Skating and Latent Heat

The world record in 500-m speed skating is 34.03 s, set by Canadian skater Jeremy Wotherspoon, who has a mass of 82.0 kg. Assume his final speed as he crossed the finish line is his average speed of the race. In coasting to a smooth stop, 40% of the frictional heat generated by the skate blades goes into melting the ice (assumed to be at 0 °C). How much ice is melted after he crossed the finish line? Where does the other 60% of the energy go?

THINKING IT THROUGH. This example involves kinetic energy, energy conversion, and latent heat. In order to find the mass of ice melted, the energy lost when Jeremy coasted to a stop needs to be determined. Then 40% of this energy loss became heat to melt the ice using latent heat.

SOLUTION. Eq. 11.3 and the kinetic energy formula, $K = \frac{1}{2}mv^2$, are used. The latent heat of fusion of ice can be looked up in Table 11.3.

Given: $\quad m = 82.0$ kg $\qquad\qquad$ *Find:* $\quad m_{ice}$
$$v_o = \frac{500 \text{ m}}{34.03 \text{ s}} = 14.69 \text{ m/s}$$ (mass of ice melted)
$\quad v = 0$
$\quad L_f = 3.33 \times 10^5 \text{ J/kg}$
(from Table 11.2)

(continued on next page)

The change in kinetic energy (energy loss) is equal to

$$\Delta K = K - K_o = \tfrac{1}{2}mv^2 - \tfrac{1}{2}m_o v^2$$
$$= \tfrac{1}{2}(82.0 \text{ kg})(0)^2 - \tfrac{1}{2}(82.0 \text{ kg})(14.69 \text{ m/s})^2$$
$$= -8.85 \times 10^3 \text{ J}$$

The negative sign means energy is being lost. Since 40% of this energy loss became heat used to melt ice, so

$$Q = (0.40)|\Delta K| = (0.40)(8.85 \times 10^3 \text{ J}) = 3.54 \times 10^3 \text{ J}$$

Using Eq. 11.3,

$$m_{ice} = \frac{Q}{L_f} = \frac{3.54 \times 10^3 \text{ J}}{3.33 \times 10^5 \text{ J/kg}} = 0.0106 \text{ kg} = 10.6 \text{ g}$$

The rest of the energy loss (60%) went into heating the skates, generating noise, etc.

Learning Path Review

- **Heat** (Q) is the energy exchanged between objects, commonly because they are at different temperatures.

- The **specific heat** (c) tells how much heat is needed to raise the temperature of 1 kg of a particular material by 1 °C It is a characteristic of the type of material and is defined by

$$c = \frac{Q}{m\Delta T} \qquad (11.1)$$

- **Calorimetry** is a technique that uses heat transfer between objects, most commonly to measure specific heats of materials. It is based on conservation of energy, written as $\Sigma Q_i = 0$, assuming no heat losses or gains to the environment.

- **Latent heat** (L) is the heat required to change the phase of an object *per kilogram* of mass. During the phase change, the temperature of the system does not change. Its general definition is

$$L = \frac{|Q|}{|m|} \quad \text{or} \quad Q = \pm mL \qquad (11.2, 11.3)$$

- Heat transfer due to direct contact of objects that have different temperatures is called **conduction**. The rate of heat flow by conduction through a slab of material is given by

$$\frac{\Delta Q}{\Delta t} = \frac{kA\Delta T}{d} \qquad (11.4)$$

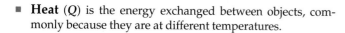

- **Convection** refers to heat transfer due to mass movement of gas or liquid molecules. *Natural convection* is driven by density differences caused by temperature differences. In *forced convection*, the movement is driven by mechanical means.

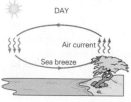

- **Radiation** refers to heat transferred by electromagnetic radiation between objects that have different temperatures, usually an object and its surroundings. The rate of transfer is given by

$$P_{net} = \sigma Ae(T_s^4 - T^4) \qquad (11.6)$$

where σ is the Stefan–Boltzmann constant, $5.67 \times 10^{-8} \text{ W/(m}^2 \cdot \text{K}^4)$.

Learning Path Questions and Exercises* For instructor-assigned homework, go to www.masteringphysics.com

11.1 DEFINITION AND UNITS OF HEAT

1. The SI unit of heat energy is the (a) calorie, (b) kilocalorie, (c) Btu, (d) joule.

2. Which of the following is the largest unit of heat energy: (a) calorie, (b) Btu, (c) joule, or (d) kilojoule?

3. The mechanical equivalent of heat is (a) 1 kcal = 4.186 J, (b) 1 J = 4.186 cal, (c) 1 cal = 4.186 J, (d) 1 Cal = 4.186 J.

11.2 SPECIFIC HEAT AND CALORIMETRY

4. The amount of heat necessary to change the temperature of 1 kg of a substance by 1 °C is called the substance's (a) specific heat, (b) latent heat, (c) heat of combustion, (d) mechanical equivalent of heat.

5. The same amount of heat Q is added to two objects of the same mass. If object 1 experienced a greater temperature change than object 2, that is, $\Delta T_1 > \Delta T_2$, then (a) $c_1 > c_2$, (b) $c_1 < c_2$, (c) $c_1 = c_2$.

6. The fundamental physical principle for calorimetry is (a) Newton's second law, (b) conservation of momentum, (c) conservation of energy, (d) equilibrium.

7. For gases, which of the following is true about the specific heat under constant pressure, c_p, and specific heat under constant volume, c_v: (a) $c_p > c_v$, (b) $c_p = c_v$, or (c) $c_p < c_v$?

11.3 PHASE CHANGES AND LATENT HEAT

8. The units of latent heat are (a) $1/°C$, (b) $J/(kg \cdot °C)$, (c) $J/°C$, (d) J/kg.

9. Latent heat is always (a) part of the specific heat, (b) related to the specific heat, (c) the same as the mechanical equivalent of heat, (d) none of the preceding.

10. When a substance undergoes a phase change, the added heat changes (a) the temperature, (b) the kinetic energy, (c) the potential energy, (d) the mass of the substance.

11.4 HEAT TRANSFER

11. House insulation materials should have (a) high thermal conductivity, (b) low thermal conductivity, (c) high emissivity, (d) low emissivity.

12. Which of the following is the dominant heat transfer mechanism by which the Earth receives energy from the Sun: (a) conduction, (b) convection, (c) radiation, or (d) all of the preceding?

13. Water is a poor heat conductor, but a pot of water can be heated more quickly than you might think. This fast heating time is mainly due to heat (a) conduction, (b) convection, (c) radiation, (d) all of the preceding.

11.1 DEFINITION AND UNITS OF HEAT

1. Discuss the difference between a calorie and a Calorie.

2. What is the main difference between internal energy and heat?

3. If someone says that a hot object contains more heat than a cold one, would you agree? Why?

11.2 SPECIFIC HEAT AND CALORIMETRY

4. At a lake, does the lake water or the lake beach get hotter during a summer day? Which gets colder during a winter night? Explain.

5. Equal amounts of heat are added to two different objects at the same initial temperature. What factors can cause the final temperature of the two objects to be different?

6. Many people have performed firewalking, in which a bed of red-hot coals (temperature over 2000 °F) is walked on with bare feet. (You should not try this at home!) How is this possible? [*Hint:* Human tissues largely consist of water.]

7. A hot steel ball is dropped into a cold aluminum cup containing some water. (Assume the system is an isolated.) If the ball loses 400 J of heat, what can be said according to calorimetry?

11.3 PHASE CHANGES AND LATENT HEAT

8. You are monitoring the temperature of some cold ice cubes (−5.0 °C) in a cup as the ice and cup are heated. Initially, the temperature rises, but it stops at 0 °C. After a while, it begins rising again. Is anything wrong with the thermometer? Explain.

9. Discuss the energy conversion in the process of adding heat to an object that is undergoing a phase change.

10. In general, you would get a more severe burn from steam at 100 °C than from the same mass of hot water at 100 °C. Why?

11. When you breathe out in the winter, you can see your breath, like fog. Explain.

*Neglect heat losses to the external environment in the questions and exercises unless instructed otherwise, and consider all temperatures to be exact.

11.4 HEAT TRANSFER

12. A plastic ice cube tray and a metal ice cube tray are removed from the same freezer, at the same initial temperature. However, the metal one feels cooler to the touch. Why?

13. Why is the warning shown on the highway road sign in ▶Fig. 11.20 necessary?

14. Polar bears have an excellent heat insulation system. (Sometimes even infrared cameras cannot detect them.) Polar bear hairs are actually hollow inside. Explain how this helps the bears maintain their body temperature in the cold winter.

15. Explain how the Thermos bottle shown in Fig. 11.17 can minimize all mechanisms of heat transfer.

▲ FIGURE 11.20 **A cold warning** See Conceptual Question 13.

11.1 DEFINITION AND UNITS OF HEAT

1. • A window air conditioner has a rating of 20 000 Btu/h. What is this rating in watts?

2. • A person goes on a 1500-Cal-per-day diet to lose weight. What is his daily energy allowance expressed in joules?

3. • A typical NBA basketball player will do about 3.00×10^6 J of work per hour. Express this work in Calories.

4. •• A typical person's normal metabolic rate (the rate at which food/stored energy is consumed) is about 4×10^5 J/h, and the average food energy in a Big Mac is 600 Calories. If a person lived on nothing but Big Macs, how many per day would he or she have to eat to maintain a constant body weight?

5. •• A student ate a Thanksgiving dinner that totaled 2800 Cal. He wants to use up all that energy by lifting a 20-kg mass a distance of 1.0 m. Assume that he lifts the mass with constant velocity and no work is required in lowering the mass. (a) How many times must he lift the mass? (b) If he can lift and lower the mass once every 5.0 s, how long does this exercise take?

11.2 SPECIFIC HEAT AND CALORIMETRY

6. • It takes 2.0×10^6 J of heat to bring a quantity of water from 20 °C to a boil. What is the mass of water?

7. IE • The temperature of a lead block and a copper block, both 1.0 kg and at 20 °C, is to be raised to 100 °C. (a) The copper will require (1) more heat, (2) the same heat, (3) less heat than the lead. Why? (b) Calculate the difference between the heat required for the two blocks to prove your answer to part (a).

*Assume all temperatures to be exact.

8. • A 5.00-g pellet of aluminum reaches a final temperature of 63 °C when gaining 200 J of heat. What is its initial temperature?

9. • Blood can carry excess heat from the interior to the surface of the body, where the heat is transferred to the outside environment. If 0.250 kg of blood at a temperature of 37.0 °C flows to the surface and loses 1500 J of heat, what is the temperature of the blood when it flows back into the interior? Assume blood has the same specific heat as water.

10. IE •• Equal amounts of heat are added to an aluminum block and a copper block of different masses to achieve the same temperature increase. (a) The mass of the aluminum block is (1) more, (2) the same, (3) less than the mass of the copper block. Why? (b) If the mass of the copper block is 3.00 kg, what is the mass of the aluminum block?

11. •• A modern engine of alloy construction consists of 25 kg of aluminum and 80 kg of iron. How much heat does the engine absorb as its temperature increases from 20 °C to 100 °C as it warms up to operating temperature?

12. IE •• Equal amounts of heat are added to different quantities of copper and lead. The temperature of the copper increases by 5.0 °C and the temperature of the lead by 10 °C. (a) The lead has (1) a greater mass than the copper, (2) the same amount of mass as the copper, (3) less mass than the copper. (b) Calculate the mass ratio of the lead to the copper to prove your answer to part (a).

13. IE •• Initially at 20 °C, 0.50 kg of aluminum and 0.50 kg of iron are heated to 100 °C. (a) The aluminum gains (1) more heat than the iron, (2) the same amount of heat as the iron, (3) less heat than the iron. Why? (b) Calculate the difference in heat required to prove your answer to part (a).

14. •• A 0.20-kg glass cup at 20 °C is filled with 0.40 kg of hot water at 90 °C. Neglecting any heat losses to the environment, what is the equilibrium temperature of the water?

15. •• A 0.250-kg coffee cup at 20 °C is filled with 0.250 kg of brewed coffee at 100 °C. The cup and the coffee come to thermal equilibrium at 80 °C. If no heat is lost to the environment, what is the specific heat of the cup material? [Hint: Consider the coffee essentially to be water.]

16. •• An aluminum spoon at 100 °C is placed in a Styrofoam cup containing 0.200 kg of water at 20 °C. If the final equilibrium temperature is 30 °C and no heat is lost to the cup itself or the environment, what is the mass of the aluminum spoon?

17. •• A student doing an experiment pours 0.150 kg of heated copper shot into a 0.375-kg aluminum calorimeter cup containing 0.200 kg of water. The cup and water are both initially at 25 °C. The mixture (and the cup) comes to thermal equilibrium at 28 °C. What was the initial temperature of the shot?

18. •• At what average rate would heat have to be removed from 1.5 L of (a) water and (b) mercury to reduce the liquid's temperature from 20 °C to its freezing point in 3.0 min?

19. •• When resting, a person gives off heat at a rate of about 100 W. If the person is submerged in a tub containing 150 kg of water at 27 °C and the heat from the person goes only into the water, how many hours will it take for the water temperature to rise to 28 °C?

20. •• To determine the specific heat of a new metal alloy, 0.150 kg of the substance is heated to 400 °C and then placed in a 0.200-kg aluminum calorimeter cup containing 0.400 kg of water at 10.0 °C. If the final temperature of the mixture is 30.5 °C, what is the specific heat of the alloy? (Ignore the calorimeter stirrer and thermometer.)

21. IE •• In a calorimetry experiment, 0.50 kg of a metal at 100 °C is added to 0.50 kg of water at 20 °C in an aluminum calorimeter cup. The cup has a mass of 0.250 kg. (a) If some water splashed out of the cup when the metal was added, the measured specific heat will appear to be (1) higher, (2) the same, (3) lower than the value calculated for the case in which the water does not splash out. Why? (b) If the final temperature of the mixture is 25 °C, and no water splashed out, what is the specific heat of the metal?

22. ••• Lead pellets of total mass 0.60 kg are heated to 100 °C and then placed in a well-insulated aluminum cup of mass 0.20 kg that contains 0.50 kg of water initially at 17.3 °C. What is the equilibrium temperature of the mixture?

23. ••• A student mixes 1.0 L of water at 40 °C with 1.0 L of ethyl alcohol at 20 °C. Assuming that no heat is lost to the container or the surroundings, what is the final temperature of the mixture? [Hint: See Table 11.1.]

24. ••• We all have had the experience that a room full of people always feels warmer than when the room is empty. Ten people are in a 4.0 m × 6.0 m × 3.0 m room at 20 °C. If each person gives off heat at a rate of about 100 W and there is no heat loss to the outside of the room, what is the temperature of the room after 10 min? At 20 °C, the density of air is 1.2 kg/m³ and its specific heat at constant pressure is 1005 J/(kg·°C).

11.3 PHASE CHANGES AND LATENT HEAT

25. • How much heat is required to melt a 2.5-kg block of ice at 0 °C?

26. • How much heat is required to boil away 1.50 kg of water that is initially at 100 °C?

27. IE • (a) Converting 1.0 kg of water at 100 °C to steam at 100 °C requires (1) more heat, (2) the same amount of heat, (3) less heat than converting 1.0 kg of ice at 0 °C to water at 0 °C. Explain. (b) Calculate the difference in heat required to prove your answer to part (a).

28. • Water is boiled to add moisture to the air in the winter to help a congested person breathe better. Calculate the heat required to boil away 1.0 L of water that is initially at 50 °C.

29. • An artist wants to melt some lead to make a statue. How much heat must be added to 0.75 kg of lead at 20 °C to cause it to melt completely?

30. • First calculate the heat that needs to be removed to convert 1.0 kg of steam at 100 °C to water at 40 °C and then compute the heat that needs to be removed to lower the temperature of water at 100 °C to water at 40 °C. Compare the two results. Are you surprised?

31. • How much heat is required to completely boil away 0.50 L of liquid nitrogen at −196 °C? (Take the density of liquid nitrogen to be 0.80×10^3 kg/m³.)

32. IE •• An alcohol rub can rapidly decrease body (skin) temperature. (a) This is because of (1) the cooler temperature of the alcohol, (2) the evaporation of alcohol, (3) the high specific heat of the human body. (b) To decrease the body temperature of a 65-kg person by 1.0 °C, what mass of alcohol must be evaporated from the person's skin? Ignore the heat involved in raising the temperature of alcohol to its boiling point (why?) and approximate the human body as water.

33. IE •• Heat has to be removed to condense mercury vapor at a temperature of 630 K into liquid mercury. (a) This heat involves (1) only specific heat, (2) only latent heat, or (3) both specific and latent heats. Explain. (b) If the mass of the mercury vapor is 15 g, how much heat would have to be removed?

34. •• If 0.050 kg of ice at 0 °C is added to 0.300 kg of water at 25 °C in a 0.100-kg aluminum calorimeter cup, what is the final temperature of the water?

35. •• How much ice (at 0 °C) must be added to 0.500 kg of water at 100 °C in a 0.200-kg aluminum calorimeter cup to end up with all liquid at 20 °C?

36. •• Ice (initially at 0 °C) is added to 0.75 L of tea at 20 °C to make the coldest possible iced tea. If enough ice is added so the final mixture is all liquid, how much liquid is in the pitcher when this condition occurs?

37. •• To cool a very hot piece of 4.00-kg steel at 900 °C, the steel is put into a 5.00-kg water bath at 20 °C . What is the final temperature of the steel-water mixture?

38. •• Steam at 100 °C is bubbled into 0.250 kg of water at 20 °C in a calorimeter cup, where it condenses into liquid form. How much steam will have been added when the water in the cup reaches 60 °C? (Ignore the effect of the cup.)

39. IE •• Evaporation of water from our skin is a very important mechanism for controlling body temperature. (a) This is because (1) water has a high specific heat, (2) water has a high latent heat of vaporization, (3) water contains more heat when hot, (4) water is a good heat conductor. (b) In a 3.5-h intense cycling race, a cyclist can loses 7.0 kg of water through perspiration. Estimate how much heat the cyclist loses in the process.

40. IE ••• A 0.400-kg piece of ice at −10 °C is placed in an equal mass of water at 30 °C. (a) When thermal equilibrium is reached between the two, (1) all the ice will melt, (2) some of the ice will melt, (3) none of the ice will melt. (b) How much ice melts?

41. ••• One kilogram of a substance experimentally shows the *T*-versus-*Q* graph in ▼Fig. 11.21. (a) What are its melting and boiling points? In SI units, what are (b) the specific heats of the substance during its various phases and (c) the latent heats of the substance at the various phase changes?

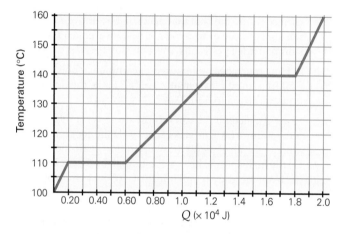

▲ **FIGURE 11.21 Temperature versus heat input** See Exercise 41.

42. ••• In an experiment, a 0.150-kg piece of a ceramic material at 20 °C is placed in liquid nitrogen at its boiling point to cool in a perfectly insulated flask, which allows the gaseous N_2 to immediately escape. How many liters of liquid nitrogen will be boiled away during this operation? (Take the specific heat of the ceramic material to be that of glass and the density of liquid nitrogen to be $0.80 \times 10^3 \, \text{kg/m}^3$.)

11.4 HEAT TRANSFER

43. • The single glass pane in a window has dimensions of 2.00 m by 1.50 m and is 4.00 mm thick. How much heat will flow through the glass in 1.00 h if there is a temperature difference of 2 °C between the inner and outer surfaces? (Consider conduction only.)

44. IE • Assume that a tile floor and an oak floor each have the same temperature and thickness. (a) Compared with the oak floor, the tile floor will conduct heat away from your bare feet (1) faster, (2) at the same rate, (3) slower. Why? (b) Calculate the ratio of the rate of heat flow of the tile floor to that of the oak floor.

45. IE • A house can have a brick wall or a concrete wall with the same thickness. (a) Compared with the concrete wall, the brick wall will conduct heat away from the house (1) faster, (2) at the same rate, (3) slower. Why? (b) Calculate the ratio of the rate of heat flow of the brick wall to that of the concrete wall.

46. • Assume a goose has a 2.0-cm-thick layer of feather down (on average) and a body surface area of 0.15 m². What is the rate of heat loss (conduction only) if the goose, with a body temperature of 41 °C, is outside on a winter day when the air temperature is 11 °C?

47. • Assume that your skin has an emissivity of 0.70, a normal temperature of 34 °C, and a total exposed area of 0.25 m². How much heat energy per second do you lose due to radiation if the outside temperature is 22 °C?

48. • The U.S. five-cent coin, the nickel, has a mass of 5.1 g, a volume of 0.719 cm³, and a total surface area of 8.54 cm². Assuming that a nickel is an ideal radiator, how much radiant energy per second comes from the nickel, if it is at 20 °C?

49. IE •• An aluminum bar and a copper bar of identical cross-sectional area have the same temperature difference between their ends and conduct heat at the same rate. (a) The copper bar is (1) longer, (2) of the same length, (3) shorter than the aluminum bar. Why? (b) Calculate the ratio of the length of the copper bar to that of the aluminum bar.

50. •• A copper teakettle has a circular bottom 30.0 cm in diameter that has a uniform thickness of 2.50 mm. It sits on a burner whose temperature is 150 °C. (a) If the teakettle is full of boiling water, what is the rate of heat conduction through its bottom? (b) Assuming that the heat from the burner is the only heat input, how much water is boiled away in 5.0 min? Is your answer unreasonably large? If yes, explain why.

51. •• Assuming that the human body has a 1.0-cm-thick layer of skin tissue and a surface area of 1.5 m², estimate the rate at which heat is conducted from inside the body to the surface if the skin temperature is 34 °C. (Assume a normal body temperature of 37 °C for the temperature of the interior.)

52. IE •• The emissivity of an object is 0.50. (a) Compared with a perfect blackbody at the same temperature, this object would radiate (1) more power, (2) the same amount of power, (3) less power. Why? (b) Calculate the ratio of the power radiated by the blackbody to that radiated by the object.

53. •• A lamp filament radiates energy at a rate of 100 W when the temperature of the surroundings is 20 °C, and only 99.5 W when the surroundings are at 30 °C. If the temperature of the filament is the same in each case, what is its temperature in Celsius?

54. IE •• (a) If the Kelvin temperature of an object is doubled, its radiated power increases by (1) 2, (2) 4, (3) 8, (4) 16 times. Explain. (b) If its temperature is increased from 20 °C to 40 °C, by how much does the radiated power change?

55. •• A certain object with a surface temperature of 100 °C is radiating heat at a rate of 200 J/s. To double the object's rate of radiation energy, what should be its surface temperature in Celsius?

56. **IE ••** The thermal insulation used in building is commonly rated in terms of its *R-value*, defined as d/k, where d is the thickness of the insulation in inches and k is its thermal conductivity. (See Insight 11.2 on p. 403.) In the United States, R-values are expressed in British units. For example, 3.0 in. of foam plastic would have an R-value of $3.0/0.30 = 10$, where $k = 0.30$ Btu·in./(ft²·h·°F). This value is expressed as R-10. (a) Better insulation has a (1) high, (2) low, or (3) zero R-value. Explain. (b) What thicknesses of (1) styrofoam and (2) brick would give an R-value of R-10?

57. **IE ••** A piece of pine 14 in. thick has an R-value of 19. (a) For glass wool to have the same R-value, its thickness should be (a) thicker than, (2) the same as, (3) thinner than 14 in. Why? (b) Calculate the required thickness of such a piece of glass wool. (See Exercise 56 and Insight 11.2.)

58. **••** Solar heating takes advantage of solar collectors such as the type shown in ▼Fig. 11.22. During daylight hours, the average intensity of solar radiation at the top of the atmosphere is about 1400 W/m². About 50% of this radiation reaches the Earth during daylight hours. (The rest is reflected, scattered, absorbed, and so on.) How much heat energy would be received, on average, by the cylindrical collector shown in the figure during 10 h of daylight?

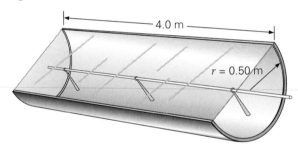

▲ **FIGURE 11.22 Solar collector and solar heating**
See Exercise 58.

For Exercises 59–64, read Example 11.7 and the footnote on p. 402.

59. **•••** A large window measures 2.0 m by 3.0 m. At what rate will heat be conducted through the window when the room temperature is 20 °C and the outside temperature is 0 °C if (a) the window consists of a single pane of glass 4.0 mm thick and (b) the window instead has a double pane of glass (a "thermopane"), in which each pane is 2.0 mm thick, with an intervening air space of 1.0 mm? (Assume that there is a constant temperature difference and consider conduction only.)

60. **•••** The lowest natural temperature ever recorded on the Earth was at Vostok, a Russian Antarctic station, when a temperature of −89.4 °C (−129 °F) was recorded on July 21, 1983. A typical person has a body temperature of 37.0 °C, skin tissue 0.0250 m thick, and a total skin surface area of 1.50 m². (a) What would be the rate of heat loss of a naked human? (b) What would be the rate of heat loss of a human wearing a 0.100-m-thick goose down jacket and pants capable of covering the whole body?

61. **•••** The wall of a house is composed of a solid concrete block with an outside brick veneer and is faced on the inside with fiberboard, as illustrated in ▼Fig. 11.23. If the outside temperature on a cold day is −10 °C and the inside temperature is 20 °C, how much energy is conducted through the wall in 1.0 h if it measures 3.5 m by 5.0 m?

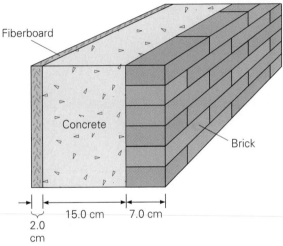

▲ **FIGURE 11.23 Thermal conductivity and heat loss**
See Exercise 61.

62. **•••** Suppose you wished to cut the heat loss through the wall in Exercise 61 in half by installing insulation. What thickness of Styrofoam should be placed between the fiberboard and concrete block to accomplish this goal?

63. **•••** A steel cylinder of radius 5.0 cm and length 4.0 cm is placed in end-to-end thermal contact with a copper cylinder of the same dimensions. If the free ends of the two cylinders are maintained at constant temperatures of 95 °C (steel) and 15 °C (copper), how much heat will flow through the cylinders in 20 min?

64. **•••** In Exercise 63, what is the temperature at the interface of the cylinders?

PULLING IT TOGETHER: MULTICONCEPT EXERCISES

The Multiconcept Exercises require the use of more than one fundamental concept for their understanding and solution. The concepts may be from this chapter, but may include those from previous chapters.

65. A 0.60 kg piece of ice at 14 °F is placed in 0.30 kg of water at 323 K. How much liquid is left when the system reaches thermal equilibrium?

66. A large Styrofoam cooler has a surface area of 1.0 m² and a thickness of 2.5 cm. If 5.0 kg of ice at 0 °C is stored inside and the outside temperature is a constant 35 °C, how long does it take for all the ice to melt? (Consider conduction only.)

67. A 1600-kg automobile traveling at 55 mph brakes smoothly to a stop. Assume 40% of the heat generated in stopping the car is dissipated in the front steel brake disks. Each front disk has a mass of 3.0 kg. What is the temperature rise of the front brake disks during the stop?

68. A waterfall is 75 m high. If 20% of the gravitational potential energy of the water went into heating the water, by how much would the temperature of the water, increase in going from the top of the falls to the bottom? [*Hint:* Consider a kilogram of water going over the falls.]

69. A 0.030-kg lead bullet hits a steel plate, both initially at 20 °C. The bullet melts and splatters on impact. (This action has been photographed.) Assuming that 80% of the bullet's kinetic energy goes into increasing its temperature and then melting it, what is the *minimum* speed it must have to melt on impact?

70. A cyclist with a total skin area of 1.5 m² is riding a bicycle on a day when the air temperature is 20 °C and her skin temperature is 34 °C. The cyclist does work at about 200 W (moving the pedals) but her efficiency is only about 20% in terms of converting energy into mechanical work. Estimate the amount of water this cyclist must evaporate per hour (through perspiration) to get rid of the excess body heat she produces. Assume a skin emissivity of 0.70.

71. A 200-kg cast iron machine part at 500 °C is left to cool at room temperature. Assume the machine part is a cube and has an emissivity of 0.780. At what rate is the machine part initially losing heat due to radiation? [*Hint:* The density of iron can be found in Table 9.2.]

12 Thermodynamics†

PHYSICS FACTS

✦ An automobile with a typical thermodynamic efficiency of one-fifth will lose about one-third of its energy through the exhaust, another one-third to the coolant, and about one-tenth to the surroundings.

✦ In Europe, more than 52% of cars sold in the first half of 2007 were diesel-powered. In the United States, fewer than 3% of cars sold are diesel-powered, but diesel sales are projected to triple to 9% by 2013 due to diesel engine's higher efficiency.

✦ The efficiency of the human body can be as high as 20% when large muscle groups, such as leg muscles are used, but as low as 3% to 5% when only the small muscle groups, such as arm muscles, are used.

✦ The brain makes up 2% of a person's weight, but consumes 20% of the body's energy. The average power consumption of a typical adult is 100 W, with the brain consuming 20 W.

As the word implies, **thermodynamics** deals with the transfer (dynamics) of heat (the Greek word for "heat" is *therme*). The development of thermodynamics started about 200 years ago out of efforts to develop heat engines. The steam engine was one of the first such devices, designed to convert heat to mechanical work. Steam engines in factories and locomotives powered the Industrial Revolution, which changed the world.

Automobiles are very useful tools for civilized society. However, with decreasing resources, soaring oil prices, and concern about

†The mathematics needed in this chapter involves natural logarithms (ln) and common logarithms (log). You may want to review these in Appendix I.

increasing greenhouse gas emissions, automobile manufacturers have been striving to produce cars using the highest possible thermodynamic efficiencies. The 2010 Honda Accord Diesel is powered by a clean 2.2-liter diesel engine, shown in the chapter-opening photograph. Its real-world fuel economy exceeds 50 mpg (miles per gallon) highway. Honda has said that the car has achieved 62.8 mpg in tests.

In this chapter, you'll learn under what conditions, and with what efficiency, heat can be exploited to perform work in the human body and in machines as different as automobile engines and home freezers. The laws governing such energy conversions include some of the most general and far-reaching laws in all of physics. Although our study is primarily concerned with heat and work, thermodynamics is a broad and comprehensive science that includes a great deal more than heat engine theory. In this chapter, the laws on which thermodynamics is based, as well as the concept of entropy, will be presented.

12.1 Thermodynamic Systems, States, and Processes

LEARNING PATH QUESTIONS

- What is a thermodynamic system?
- What is meant by the state of a thermodynamic system?
- What is a thermodynamic process?

Thermodynamics is a field that describes systems with so many particles—think of the number of molecules in a gas sample—that using ordinary dynamics (Newton's laws) to keep track of them is impossible. Therefore, even though the underlying physics is the same as for other systems, we generally use alternative (macroscopic) variables, such as pressure and temperature, to describe thermodynamic systems as a whole. Because of this difference in language, it is important to become familiar with the terms and definitions at the outset.

The term **system**, as used in thermodynamics, refers to a definite quantity of matter enclosed by boundaries or surfaces, either real or imaginary. For example, a quantity of gas in the piston cylinder of an engine has real boundaries, and imaginary boundaries enclose a cubic meter of air in a room.

The interchange of energy between a system and its surroundings is very important. This exchange may occur through a transfer of heat and/or the performance of mechanical work. For example, if a gasoline and air mixture is ignited inside a piston of an engine, it can expand and do work by exerting a force through a distance on the piston.

If no heat is transferred into or out of a system, it is said to be a **thermally isolated system**. However, work may be done on a thermally isolated system, thus transferring energy to it. For example, a thermally isolated syringe (perhaps surrounded by heavy insulation) filled with gas can be compressed by an external force exerted on a plunger. Work is thus done on the system, and as we know, work is a way of transferring energy.

When heat does enter or leave a system, it is usually taken in from or given up to the surroundings, or to what is called a heat reservoir. A **heat reservoir** is a system assumed to have unlimited heat capacity. Any amount of heat can be withdrawn from or added to a heat reservoir without appreciably changing its temperature. For example, pouring a bottle of warm water into a cold lake does not noticeably raise the lake's temperature. This cold lake is an example of a low-temperature heat reservoir.

STATE OF A SYSTEM

Just as there are kinematic equations to describe the motion of an object, there are **equations of state** to describe the conditions of thermodynamic systems. Such an equation expresses a mathematical relationship between the thermodynamic variables that describe the system. The ideal gas law, $pV = nRT$ (Section 10.3), is an example of an equation of state. This expression establishes a relationship among the pressure (p), volume (V), absolute temperature (T), and number of moles (n, or equivalently, N, the number of molecules, since from Section 10.3, $N = nN_A$) of a gas. These ideal gas quantities are examples of *state variables*. Clearly, different states have different sets of values for these variables.

For a quantity of ideal gas, a set of these three variables (p, V, and T) that satisfies the ideal gas law specifies its state completely as long as the system is in thermal equilibrium. Such a system is said to be in a definite *state*. It is convenient to plot the states according to the thermodynamic coordinates (p, V, T), much as graphs using Cartesian coordinates (x, y, z) are plotted. A general two-dimensional illustration of such a plot is shown in ▸ Fig. 12.1.

Just as the coordinates (x, y) specify individual points on a Cartesian graph, the coordinates (V, p) specify individual *states* on the p–V graph or diagram. This is because the ideal gas law, $pV = nRT$, can be solved for the unique temperature of a gas if the gas's pressure, volume, and number of molecules or moles in the sample are known. In other words, on a p–V diagram, each "coordinate" or point gives the pressure and volume of a gas directly, and the temperature of the gas indirectly. Thus, to describe a gas completely, only a p–V plot is necessary. In some cases, however, it can be instructive to refer to other plots, such as p–T or T–V plots. (Notice that Fig. 12.1b could illustrate a phenomenon that you might be familiar with—reduction of the pressure of a gas, resulting in its expansion.)

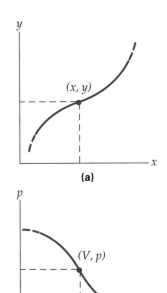

▲ **FIGURE 12.1 Graphing states** **(a)** On a Cartesian graph, the coordinates (x, y) represent an individual point. **(b)** Similarly, on a p–V graph or diagram, the coordinates (V, p) represent a particular state of a system. (It is common to say p–V, rather than V–p, because the plot is a p vs. V graph.)

PROCESSES

A **process** is any *change* in the state, or the thermodynamic coordinates, of a system. For instance, when an ideal gas undergoes a process, its state variables p, V, and T will, in general, all change. Suppose a gas initially in state 1, described by state variables (p_1, V_1, T_1), changes to a second state, state 2. Then state 2 will, in general, be described by a different set of state variables (p_2, V_2, T_2). A system that has undergone a change of state has been subjected to a *thermodynamic process.*

Processes are classified as either reversible or irreversible. Suppose that a system of gas in equilibrium (with known p, V, and T values) is allowed to expand quickly when the pressure on it is reduced. The state of the system will change rapidly and unpredictably, but eventually the system will reach a different state of equilibrium, with another set of thermodynamic coordinates. On a p–V diagram (▸ Fig. 12.2), the initial and final states (labeled 1 and 2, respectively) are known, but what happened in between them is not. This type of process is called an **irreversible process**—a process for which the intermediate steps are nonequilibrium states. "Irreversible" does not mean that the system can't be taken back to the initial state; it means only that the process path can't be retraced, because of the nonequilibrium conditions that existed. An explosion is an example of an irreversible process.

If, however, the gas changes state very, very slowly, passing from one equilibrium state to a neighboring one and eventually arriving at the final state (see Fig. 12.2, initial and final states 3 and 4, respectively), then the process path is known. In such a situation, the system could be brought back to its initial conditions by "traveling" the path in the opposite direction, re-creating every intermediate state (again, in many small steps) along the way. Such a process is called a **reversible process**. In practice, a perfectly reversible process cannot be achieved.

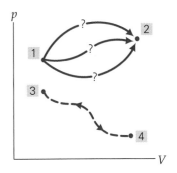

▲ **FIGURE 12.2 Paths of reversible and irreversible processes** If a gas quickly goes from state 1 to state 2, the process is irreversible, since we do not know the "path." If, however, the gas is taken through many closely spaced equilibrium states (as in going from state 3 to state 4), the process is reversible in principle. Reversible means "exactly retraceable."

All real thermodynamic processes are irreversible to some degree, because they follow complicated paths with many intermediate nonequilibrium states. However, the concept of an ideal reversible process is useful and will be the primary tool in discussing the thermodynamics of an ideal gas.

DID YOU LEARN?

➥ A thermodynamic system is simply a quantity of matter enclosed in real or imaginary boundaries or surfaces.

➥ The state of a thermodynamic system describes the conditions of the system. Some common variables used to specify the state of an ideal gas are pressure (p), volume (V), and temperature (T).

➥ A thermodynamic process changes the system from one state to another. For an ideal gas, a process changes one set of values for pressure, volume, and temperature (p, V, T) to another set.

12.2 The First Law of Thermodynamics

LEARNING PATH QUESTIONS

➥ What is the first law of thermodynamics?

➥ How can the internal energy of a system be changed?

➥ How can the work done by an ideal gas based on a process curve on a p–V diagram be calculated?

Recall from Section 5.1 that work describes the transfer of energy from one object to another by application of a force. For example, when you push on a chair initially at rest and set it into motion, some of the work done on the chair (exerting a force through a distance) goes into increasing its kinetic energy. At the same time, you lose stored (chemical) energy in your body in doing so. For example, when a gas (enclosed in a cylinder and fitted with a piston) is allowed to expand, the gas does work on the piston at the expense of some of its internal energy. From Chapters 10 and 11, we know there is a second way to change the energy of a system—by adding or removing heat energy. Thus, internal energy is lost by a hot object when the heat is transferred to a cold object, which then gains internal energy. This process changes both objects' internal energies, but in opposite ways.

Although the actual process cannot be seen, heat transfer is really the same concept as mechanical work, but on a microscopic (atomic) level. During a conduction process, for example, energy is transferred from a hot object to a cold object, because the faster-vibrating atoms of the hot object do work on the slower atoms of the cold object (▼Fig. 12.3). This energy is then transferred farther into the volume of the cold object as more work is done on the neighboring (slower-vibrating) atoms. This ongoing process is the "flow" or "transfer" of energy observed macroscopically as heat transfer.

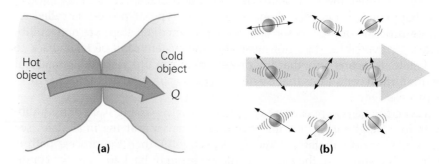

(a)   **(b)**

▲ **FIGURE 12.3 Heat flow (via conduction) on the atomic scale (a)** Macroscopically, heat is transferred by conduction from the hot object to the cold one. **(b)** On the atomic scale, heat conduction is explained as the energy transfer from the more energetic atoms (in the hot object) to the less energetic atoms (in the cold object). This transfer of energy from an atom to its neighbor results in the heat transfer observed in part (a).

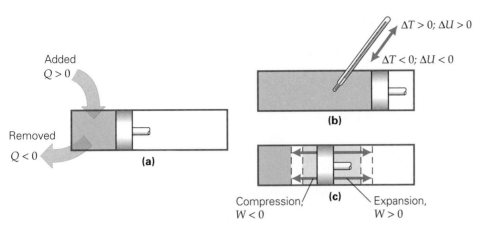

▲ **FIGURE 12.4 Sign conventions for Q, W, and ΔU (a)** If heat flows into a system, Q is positive. For heat flowing out, Q is designated as negative. **(b)** The experimental way to tell if a gas's internal energy changes is to take its temperature, assuming there is no phase change. Since internal energy is determined by temperature, a rise or fall in one of these quantities implies a similar rise or fall in the other. **(c)** If a gas expands, the work W it does is positive. If the gas is compressed, the work done by the gas is negative.

The **first law of thermodynamics** describes how work and heat are related to a system's internal energy. This law is a statement of *energy conservation* in terms of thermodynamic variables. It relates the change in internal energy (ΔU) *of a system* to the work (W) done *by or on that system* and the heat energy transferred (Q) *to or from that system*. Depending on the conditions, heat transfer Q can result in a change in that system's internal energy, ΔU. However, because of the heat transfer, the system might do work on the environment. Thus, heat transferred to a system can end up in change in the internal energy of the system and/or work done by the system. This is just a statement of energy conservation. Therefore, the first law of thermodynamics can be written as

$$Q = \Delta U + W \quad \text{(the first law of thermodynamics)} \quad (12.1)$$

As always, it is important to remember what the symbols mean and what their sign conventions denote (shown in ▲Fig. 12.4). Q is the net heat *added to or removed from the system*, ΔU is the change in internal energy *of the system* and W is the work done *by the system* (on the environment).* For example, a sample of gas may absorb 1000 J of heat and do 400 J of work on the environment, thus leaving 600 J as the increase in the gas's internal energy. If the gas were to do more than 400 J of work, less energy would go to the internal energy of the gas. The first law does *not* tell you the values of ΔU or W in processes. These amounts depend, as will be seen, on the system's conditions or the specific process involved (constant pressure, constant volume, and so on) as the heat energy is transferred (Section 12.3).

It is important to note that heat flow is *not* necessary for temperature to change. When a soda bottle is opened, as shown in ▶Fig. 12.5, the gas inside the bottle expands because it is at a higher pressure than the atmosphere. In doing so it does (positive) work on the surroundings (the atmospheric gases) and its internal energy decreases. This is because the net heat flow is zero in this process. Since $\Delta U = Q - W$, then ΔU is negative (U decreases) if $Q = 0$ and W is positive. This reduction in internal energy will result in a temperature drop which in turn will cause the water vapor in the bottled gas to condense into a cloud of tiny liquid

*In some chemistry and engineering books, the first law of thermodynamics is written as $Q = \Delta U - W'$. The two equations are the same, but each has a different emphasis. In this expression, W' means the work done *by the environment on the system* and is thus the negative of our work W (why?), or $W = -W'$. The first law was discovered by researchers interested in building heat engines (Sections 12.5 and 12.6). Their emphasis was on finding the work done *by* the system, W, not W'. Since our main concern is to understand heat engines, the historical definition is adopted: W *means the work done by the system*.

▲ **FIGURE 12.5 Temperature decrease without removing heat** The gas does positive work on the outside air upon the opening of the bottle. This results in a decrease of both its internal energy and temperature.

water droplets. You can demonstrate this high-pressure cooling by putting your palm near your mouth and blowing air with your mouth opened wide. You feel a gush of warm air (roughly at body temperature). However, if you repeat this with your lips puckered so as to increase the pressure, the air will feel cooler.

Although the above discussion of the first law of thermodynamics was primarily about gas systems, the law holds for any systems. Consider the application of the first law of thermodynamics to exercise and weight loss in Example 12.1.

EXAMPLE 12.1 | Energy Balancing: Exercising Using Physics

A 65-kg worker shovels coal for 3.0 h. During the shoveling, the worker did work at an average rate of 20 W and lost heat to the environment at an average rate of 480 W. Ignoring the loss of water by the evaporation of perspiration from his skin, how much fat will the worker lose? The energy value of fat (E_f) is 9.3 kcal/g.

THINKING IT THROUGH.   Since the time duration of the shoveling, the rate of work being done (power), and the rate of heat loss are known, the total work done and the heat can be calculated. Then the change in internal energy can be found using the first law of thermodynamics. This change in internal energy (a decrease) results in a loss of fat.

SOLUTION.   Listing the given values, and converting power to work and heat units:

Given: $W = Pt = (20 \text{ J/s})(3.0 \text{ h})(3600 \text{ s/h}) = 2.16 \times 10^5 \text{ J}$ **Find:** m (mass of fat burned)
(W is positive because work is done *by* the worker)
$Q = -(480 \text{ J/s})(3.0 \text{ h})(3600 \text{ s/h}) = -5.18 \times 10^6 \text{ J}$
(Q is negative because heat is lost)
$E_f = 9.3 \text{ kcal/g} = 9.3 \times 10^3 \text{ kcal/kg} = (9.3 \times 10^3 \text{ kcal/kg})(4186 \text{ J/kcal})$
$= 3.89 \times 10^7 \text{ J/kg}$

From the first law of thermodynamics, $Q = \Delta U + W$.

$$\Delta U = Q - W = -5.18 \times 10^6 \text{ J} - 2.16 \times 10^5 \text{ J} = -5.40 \times 10^6 \text{ J}$$

Thus the mass of fat loss is

$$m = \frac{|\Delta U|}{E_f} = \frac{5.40 \times 10^6 \text{ J}}{3.89 \times 10^7 \text{ J/kg}} = 0.14 \text{ kg}$$

That is about a third of a pound, or about 5 ounces (140 g).

FOLLOW-UP EXERCISE.   How much fat would be lost if the worker were playing basketball for 3.0 h, doing work at a rate of 120 W and generating heat at a rate of 600 W? *(Answers to all Follow-Up Exercises are given in Appendix VI at the back of the book.)*

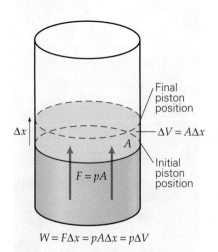

$$W = F\Delta x = pA\Delta x = p\Delta V$$

▲ **FIGURE 12.6 Work in thermodynamic terms** If a gas expands by a very small amount and does so slowly, its pressure remains constant. The small amount of work done by the gas is $p\Delta V$.

When applying the first law of thermodynamics, the proper use of signs (shown in Fig. 12.4) cannot be overemphasized. The signs for work are easy to remember if you keep in mind that positive work is done by a force that acts generally in the direction of the displacement, such as when a gas expands. Similarly, negative work means that the force acts generally opposite to the direction of the displacement, as when a gas contracts.

But how do you compute the work done by the gas? To answer this question, consider a cylindrical piston with end area A, containing a known sample of gas (◄Fig. 12.6). Let us imagine that the gas is allowed to expand over a very small distance Δx. If the volume of the gas does not change appreciably, then the pressure remains constant. In moving the piston slowly and steadily outward, the gas does positive work on the piston. Thus from the definition of work,

$$W = F\Delta x \cos \theta = F\Delta x \cos 0° = F\Delta x$$

In terms of pressure, $P = F/A$ or $F = pA$. Substituting for F, we have

$$W = pA\Delta x$$

But $A\Delta x$ is the volume of a cylinder with end area A and height Δx. Here, that volume represents the change in volume of the gas, or $\Delta V = A\Delta x$, and

$$W = p\Delta V$$

Note that the work done in Fig. 12.6 is positive because ΔV is positive. If the gas contracts, the work is negative because the volume change is negative ($\Delta V < 0$).

Of course, gases don't always change their volumes by small amounts and aren't usually subject to constant pressure. In fact, changes in volume and pressure can be significant. How is the calculation of work handled under these circumstances? The answer is seen in ▸Fig. 12.7. Here, we have a reversible path on a p–V diagram. Notice that during each small step, the pressure remains approximately constant. Therefore, for each step, we approximate the work done by $p\Delta V$. Graphically, this quantity is just the area of a small narrow rectangle, extending from the process curve to the V-axis. To approximate the *total* work, we add up these small amounts of work or $W \approx \Sigma(p\Delta V)$. To get an exact value, think of the area as made up of a very large number of very thin rectangles. As the number of rectangles becomes infinitely large, each rectangle's thickness approaches zero. This process involves calculus and is beyond the scope of this book. However, it should be clear that the following is true:

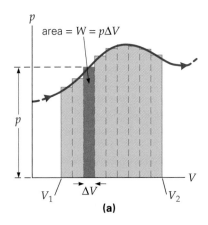
(a)

> The work done by a system is equal to the area under the process curve on a p–V diagram.

Before discussing specific types of processes, note that there is a fundamental difference between U and both Q and W. Any system "contains" a certain amount of internal energy U. However, it is wrong to say that a system "possesses" certain "amounts" of heat or work, as these quantities represent energy *transfers*, not total energies. A further distinction is that both Q and W depend on the path the gas takes from its initial to its final state, whereas ΔU does not. The heat added to, or removed from, a system *depends on the conditions* under which this transfer is done (Section 11.2). Similarly, from the area-under-the-curve representation, the work *depends on the path* (▾Fig. 12.8). For example, more work is done if the process takes place at higher pressures. This situation is represented as a larger area, with more work done by a larger force with the same volume change.

Contrast these properties with those of ΔU for an ideal gas, for example. To find ΔU, we need know only the internal energies at the ends of the path. This is because for an ideal gas (with a fixed number of moles), the internal energy depends only on the absolute temperature of the gas. For that case (see Section 10.5), $U \propto nRT$; thus, $\Delta U = U_2 - U_1$ depends only on ΔT. To summarize, the change in the internal energy, ΔU, is *independent* of the process path for an ideal gas, whereas Q and W both *depend* on the path.

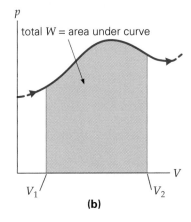
(b)

▲**FIGURE 12.7 Thermodynamic work as the area under the process curve** (a) If a gas expands by a significant amount, the work done can be computed by treating the expansion in little steps, each one yielding a small amount of work. The total work is determined (approximately) by adding up the many rectangular strips. **(b)** If the number of rectangular strips becomes large, and each one becomes very thin, the calculation of the area becomes exact. The work done is equal to the area between the process curve and the V-axis.

DID YOU LEARN?
➡ The first law of thermodynamics is the law of conservation of energy applied to thermodynamic systems. It relates the change in internal energy ΔU, work W, and heat Q.
➡ The internal energy of a system can be changed by either work done or heat exchange.
➡ The work done by an ideal gas is equal to the area under the process curve on a p–V diagram.

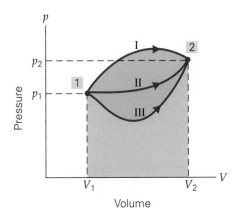

◀**FIGURE 12.8 Thermodynamic work depends on the process path** This graph shows the work done by a gas as it expands the same amount, but by three different processes. The work done during process I is larger than the work during process II, which in turn is larger than the work during process III. Fundamentally, applying a larger force (pressure) through the same distance (volume change) requires more work. Process I includes the blue, green, and pink areas; process II includes just the green and pink areas; and process III includes just the pink area.

12.3 Thermodynamic Processes for an Ideal Gas

LEARNING PATH QUESTIONS

➥ What are the four important thermodynamic processes?
➥ What is the change in internal energy of an ideal gas after an isothermal process?
➥ What is the work done by a gas in an isometric process?

The first law of thermodynamics can be applied to several processes for a system consisting of an ideal gas. Note that in three of the processes, one thermodynamic variable is kept constant. Such processes have names that begin with *iso-* (from the Greek *isos*, meaning "equal").

ISOTHERMAL PROCESS

An **isothermal process** is a constant-temperature process (*iso* for equal, *thermal* for temperature). In this case, the process path is called an *isotherm*, or a curve of constant temperature. (See ▼Fig. 12.9.) The ideal gas law may be rewritten as $p = nRT/V$. Since the gas remains at constant temperature, nRT is a constant. Therefore, p is inversely proportional to V—that is, $p \propto 1/V$, which is a hyperbola. (Recall that a hyperbola is written as $y = a/x$ or $y \propto 1/x$, and it plots as a downward curve.)

In the expansion from state 1 (initial) to state 2 (final) in Fig. 12.9, heat is added to the system, while both the pressure and volume vary in such a way as to keep the temperature constant. Positive work is done by the expanding gas. On an isotherm, $\Delta T = 0$; therefore, $\Delta U = 0$. The heat added to the gas is exactly equal to the amount of work done by the gas, and none of the heat goes into increasing the gas's internal energy. See the Learn by Drawing 12.1, Leaning on Isotherms.

In terms of the first law of thermodynamics,

$$Q = \Delta U + W = 0 + W$$

or

$$Q = W \quad \text{(ideal gas isothermal process)} \tag{12.2}$$

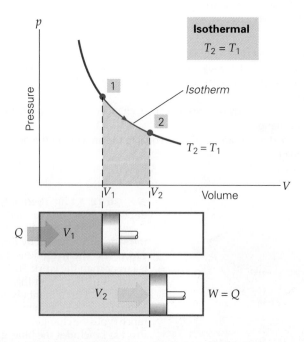

▲ **FIGURE 12.9   Isothermal (constant temperature) process** All of the heat added to the gas goes into doing work (the expanding gas moves the piston): Because $\Delta T = 0$, then $\Delta U = 0$, and from the first law of thermodynamics, $Q = W$. As always, the work is equal to the area (shaded) under the isotherm on the p–V diagram.

The magnitude of the work done by the gas is equal to the area under the curve (requiring calculus to compute), which can be written as follows.

$$W_{\text{isothermal}} = nRT \ln\left(\frac{V_2}{V_1}\right) \quad \text{(ideal gas isothermal process)} \quad (12.3)$$

Since the product nRT is a constant along a given isotherm, the work done depends on the ratio of the endpoint volumes.

PROBLEM-SOLVING HINT

In Eq. 12.3, the function "ln" stands for *natural logarithm.* Recall that *common logarithms* ("log") are referenced to the base 10 (see Appendix I). For this type, the exponent of the base 10 is the logarithm of the number in question. For example, $100 = 10^2$, so the logarithm of 100 is 2, or, in equation terms, $\log 100 = 2$. In general, if $y = 10^x$, then x is the logarithm of y, or $x = \log y$. The natural logarithm is similar, except it uses a different base, e, which is an irrational number ($e \approx 2.7183$). As a check, find the natural logarithm of 100 on your calculator. (The answer is $\ln 100 = 4.605$).

ISOBARIC PROCESS

A constant-pressure process is called an **isobaric process** (*iso* for equal, and *bar* for pressure).* An isobaric process for an ideal gas is illustrated in ▾Fig. 12.10. On a *p–V* diagram, an isobaric process is represented by a horizontal line called an *isobar.* When heat is added to or removed from an ideal gas at constant pressure, the ratio V/T remains constant ($V/T = nR/p = \text{constant}$). As the heated gas expands, its temperature must increase, and the gas crosses to higher temperature isotherms. This temperature increase means that the internal energy of the gas increases, since $\Delta U \propto \Delta T$.

As can be seen from the isobar in Fig. 12.10, the area representing the work is rectangular. Thus, the work is relatively easy to compute (length times width):

$$W_{\text{isobaric}} = p(V_2 - V_1) = p\Delta V \quad \text{(ideal gas isobaric process)} \quad (12.4)$$

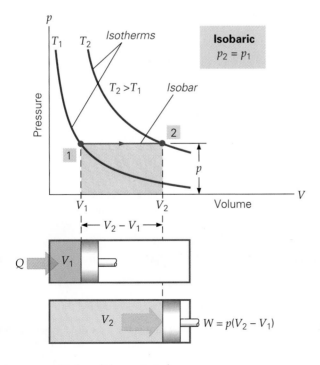

◀**FIGURE 12.10 Isobaric (constant pressure) process** The heat added to the gas in the frictionless piston goes into work done by the gas and into changing the internal energy of the gas: $Q = \Delta U + W$. The work is equal to the area under the isobar (from state 1 to state 2 here, shown as shaded) on the *p–V* diagram. Note the two isotherms. They are not part of the isobaric process, but they show us that the temperature rises during the isobaric expansion.

*Pressure can be measured in bars (1 bar = 1 atm).

For example, when heat is added to or removed from a gas under isobaric conditions, the gas's internal energy changes *and* the gas expands or contracts, doing positive or negative work, respectively. (See Integrated Example 12.2 for the signs.) This relationship can be written, using the first law of thermodynamics, with the work expression appropriate for isobaric conditions (Eq. 12.4):

$$Q = \Delta U + W = \Delta U + p\Delta V \quad \textit{(ideal gas isobaric process)} \quad (12.5)$$

To see a detailed comparison of an isobaric process and an isothermal process, consider the following Integrated Example.

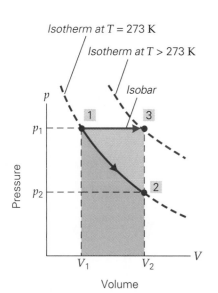

Isotherm at T = 273 K

Isotherm at T > 273 K

Isobar

▲ **FIGURE 12.11 Comparing work** In Integrated Example 12.2, the gas does positive work while expanding. It does more work under isobaric conditions (from state 1 to state 3) than under isothermal conditions (from state 1 to state 2) because the pressure remains constant on the isobar but decreases along the isotherm. (Compare areas under the curves.)

INTEGRATED EXAMPLE 12.2 | **Isotherms versus Isobars: Which Area?**

Two moles of a monatomic ideal gas, initially at 0 °C and 1.00 atm, are expanded to twice their original volume, using two different processes. They are expanded isothermally or isobarically, both starting in the same initial state. (a) Does the gas (1) do more work during the isothermal process, (2) do more work during the isobaric process, or (3) do the same work during both processes? Explain. (b) To prove your answer, determine the work done by the gas in each process.

(A) CONCEPTUAL REASONING. As shown in ◄Fig. 12.11, both processes involve expansion. The isobar is horizontal, and the isotherm is a decreasing hyperbola. Thus, the gas does more work during the isobaric expansion (more area under the curve). Fundamentally, this is because the isobaric process is done at higher (constant) pressure than the isothermal process (where the pressure drops as the gas expands according to gas law). In both cases, the work is positive. (How do we know this?) Thus, the correct answer to part (a) is (2), that the isobaric process does more work.

(B) QUANTITATIVE REASONING AND SOLUTION. If the volumes are known, Eqs. 12.3 and 12.4 can be used. These quantities can be calculated from the ideal gas law. Listing the data,

Given: $p_1 = 1.00 \text{ atm} = 1.01 \times 10^5 \text{ N/m}^2$ *Find:* $W_{\text{isothermal}}$ and W_{isobar}
$T_1 = 0 °C = 273 \text{ K}$ (the work done during the
$n = 2.00 \text{ mol}$ (see Section 10.3)   isothermal and isobaric
$V_2 = 2V_1$ processes)

For the isotherm, use Eq. 12.3 (the natural logarithm of the volume ratio is ln 2 = 0.693):

$$W_{\text{isothermal}} = nRT \ln\left(\frac{V_2}{V_1}\right) = (2.00 \text{ mol})[8.31 \text{ J/(mol·K)}](273 \text{ K})(\ln 2)$$
$$= +3.14 \times 10^3 \text{ J}$$

For the isobar, we need to know the two volumes. Using the ideal gas law,

$$V_1 = \frac{nRT_1}{p_1} = \frac{(2.00 \text{ mol})[8.31 \text{ J/(mol·K)}](273 \text{ K})}{1.01 \times 10^5 \text{ N/m}^2} = 4.49 \times 10^{-2} \text{ m}^3$$

and therefore

$$V_2 = 2V_1 = 8.98 \times 10^{-2} \text{ m}^3$$

The work is given by Eq. 12.4 as

$$W_{\text{isobar}} = p(V_2 - V_1) = (1.01 \times 10^5 \text{ N/m}^2)(8.98 \times 10^{-2} \text{ m}^3 - 4.49 \times 10^{-2} \text{ m}^3)$$
$$= +4.53 \times 10^3 \text{ J}$$

This amount is larger than the isothermal work, as expected from part (a).

FOLLOW-UP EXERCISE. In this Example, what is the heat flow in each process?

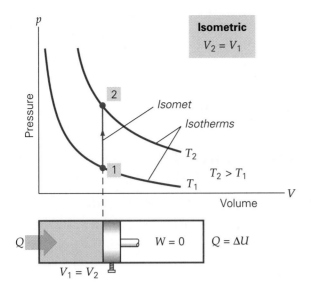

◀**FIGURE 12.12 Isometric (constant volume) process** All of the heat added to the gas goes into increasing the gas's internal energy, because there is no work done ($W = 0$); thus, $Q = \Delta U$. (Notice the locking nut on the piston, which prevents any movement.) Again, the isotherms, although not part of the isometric process, tell us visually that the temperature of the gas rises.

ISOMETRIC PROCESS

An **isometric process** (short for *isovolumetric*, or constant-volume, process), sometimes called an *isochoric process*, is a constant-volume process. As illustrated in ▲Fig. 12.12, the process path on a *p–V* diagram is a vertical line, called an *isomet*. No work is done, since the area under such a curve is zero. (There is no displacement, as there is no change in volume.) Because the gas does not do any work, if heat is added it must go completely into increasing the gas's internal energy and, therefore, its temperature. In terms of the first law of thermodynamics,

$$Q = \Delta U + W = \Delta U + 0 = \Delta U$$

and thus

$$Q = \Delta U \quad \text{(ideal gas isometric process)} \quad (12.6)$$

Consider the following example of an isometric process in action.

EXAMPLE 12.3 | ## A Practical Isometric Exercise: How *Not* to Recycle a Spray Can

Many "empty" aerosol cans contain remnant propellant gases under approximately 1 atm of pressure (assume 1.00 atm) at 20 °C. They display the warning "Do not dispose of this can in an incinerator or open fire." (a) Explain why it is dangerous to throw such a can into a fire. (b) If there are 0.0100 moles of monatomic gas in the can and its temperature rises to 2000 °F, how much heat was added to the gas? (c) What is the final pressure of the gas?

THINKING IT THROUGH. This is an isovolumetric process; hence, all the heat goes into increasing the gas's internal energy. A pressure rise is expected, which is where the danger lies. The change in internal energy can be calculated with Eq. 10.16. The final pressure can be obtained using the ideal gas law.

SOLUTION. Listing the data and converting given temperatures into kelvins (again, for qualitative reasoning, refer to the Learn by Drawing 12.1, Learning on Isotherms);

Given: $p_1 = 1.00 \text{ atm} = 1.01 \times 10^5 \text{ N/m}^2$
$V_1 = V_2$
$T_1 = 20\,°C = 293 \text{ K}$
$T_2 = 2000\,°F = 1.09 \times 10^3\,°C$
$\quad = 1.37 \times 10^3 \text{ K}$
$n = 0.0100 \text{ mol}$

Find: (a) Explain the danger in heating the can.
(b) Q (heat added to gas)
(c) p_2 (final pressure of gas)

(continued on next page)

(a) When heat is added, it all goes into increasing the gas's internal energy. Because at constant volume, pressure is proportional to temperature, the final pressure will be greater than 1.00 atm. The danger is that the container could explode into metallic fragments like a grenade if the maximum design pressure of the container is exceeded.

(b) To calculate the heat, we use the first law of thermodynamics. Recall that the work done in an isometric process is zero

(why?). $\Delta U = Q - W = Q - 0 = Q$ or $Q = \Delta U$. From Eq. 10.16, $U = \frac{3}{2}nRT$, then

$$\begin{aligned} \Delta U &= \tfrac{3}{2}nR\Delta T \\ &= \tfrac{3}{2}(0.0100 \text{ mol})[8.31 \text{ J}/(\text{mol}\cdot\text{K})](1.37 \times 10^3 \text{ K} - 293 \text{ K}) \\ &= 134 \text{ J}. \end{aligned}$$

(c) The final pressure of the gas is determined directly from the ideal gas law:

$$\frac{p_2 V_2}{T_2} = \frac{p_1 V_1}{T_1} \quad \text{or} \quad p_2 = p_1\left(\frac{V_1}{V_2}\right)\left(\frac{T_2}{T_1}\right) = (1.00 \text{ atm})\left(\frac{V_1}{V_1}\right)\left(\frac{1.37 \times 10^3 \text{ K}}{293 \text{ K}}\right) = 4.68 \text{ atm}$$

FOLLOW-UP EXERCISE.   Suppose the can were designed to withstand pressures up to 3.50 atm. What would be the highest Celsius temperature it could reach without exploding?

ADIABATIC PROCESS

In an **adiabatic process**, no heat is transferred into or out of the system. That is, $Q = 0$ (▼Fig. 12.13). (The Greek word *adiabatos* means "impassable.") This condition is satisfied for a thermally isolated system, one completely surrounded by "perfect" insulation. This is an ideal situation; for real-life conditions, adiabatic processes can only be approximated. For example, nearly adiabatic processes can take place if the changes occur rapidly enough so that there isn't time for significant heat to flow into or out of the system. In other words, *quick* processes can approximate adiabatic conditions.

The curve for this process is called an *adiabat*. During an adiabatic process, all three thermodynamic coordinates (p, V, T) change. For example, if the pressure on a gas is reduced, the gas expands. However, no heat flows into the gas. Without a compensating input of heat, work is done at the expense of the gas's internal energy. Therefore, ΔU must be negative. Since the internal energy, and thus the temperature, both decrease, such an expansion is a cooling process. Similarly, an adiabatic compression is a warming process (temperature increase).

From the first law of thermodynamics, an adiabatic process can be described by

$$Q = 0 = \Delta U + W$$

▶ **FIGURE 12.13 Adiabatic (no heat transfer) process** In an adiabatic process (shown here for a cylinder with heavy insulation), no heat is added to or removed from the system; thus, $Q = 0$. During expansion (shown here), positive work is done by the gas at the expense of its internal energy: $W = -\Delta U$. The pressure, volume, and temperature all change in the process. The work done by the gas is the shaded area between the adiabat and the V-axis.

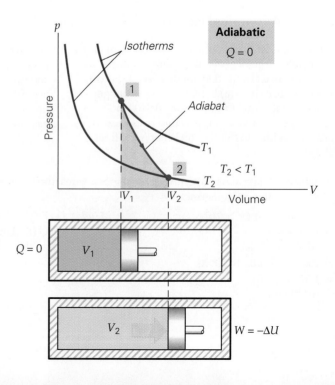

or

$$\Delta U = -W \quad \text{(adiabatic process)} \tag{12.7}$$

For completeness, we state some other relationships in an adiabatic process. An important factor is the ratio of the gas's molar specific heats, defined by a dimensionless quantity $\gamma = c_p/c_v$, where c_p and c_v are the specific heats at constant pressure and volume, respectively. For the two common types of gas molecules, monatomic and diatomic, the values of γ are about 1.67 and 1.40, respectively. The volume and pressure at any two points on an adiabat are related by

$$p_1 V_1^\gamma = p_2 V_2^\gamma \quad \text{(ideal gas adiabatic process)} \tag{12.8}$$

The work done by an ideal gas during an adiabatic process can be shown to be

$$W_{\text{adiabatic}} = \frac{p_1 V_1 - p_2 V_2}{\gamma - 1} \quad \text{(ideal gas adiabatic process)} \tag{12.9}$$

To clear up confusion that often occurs between isotherms and abiabats, see Integrated Example 12.4.

INTEGRATED EXAMPLE 12.4 | **Adiabats versus Isotherms:**
Two Different Processes That Are Often Confused

A sample of helium gas expands to triple its initial volume adiabatically in one case and isothermally in another. In both cases, it starts from the same initial state. The sample contains 2.00 mol of helium ($\gamma = 1.67$), initially at 20 °C and 1.00 atm. (a) Does the gas, (1) do more work during the adiabatic process, (2) do more work during the isothermal process, or (3) do the same work during both processes? (b) Calculate the work done during each process to verify your reasoning in part (a).

(A) CONCEPTUAL REASONING. To determine graphically which process involves more work, note the areas under the process curves (see Fig. 12.13). The area under the isothermal process

curve is larger; thus, the gas does more work during its isothermal expansion, and the correct answer is (2). Physically, the isothermal expansion involves more work because the pressures are always higher during the isothermal expansion than during the adiabatic one.

(B) QUANTITATIVE REASONING AND SOLUTION. To determine the isothermal work, the ratio of the final to initial volumes is needed and is given. For the adiabatic work, the ratio of specific heats, γ, is important, as are the final pressure and volume. The final pressure can be found using Eq. 12.8, and the ideal gas law enables the determination of the final volume.

Listing the given values and converting the temperature into kelvins:

Given: $p_1 = 1.00 \text{ atm} = 1.01 \times 10^5 \text{ N/m}^2$
$n = 2.00 \text{ mol}$
$T_1 = (20 + 273) \text{ K} = 293 \text{ K}$
$V_2 = 3V_1$
$\gamma = 1.67$

Find: $W_{\text{isothermal}}$ and $W_{\text{adiabatic}}$
(work done during each process)

The data necessary to calculate the isothermal work from Eq. 12.3 are given. The volume ratio is 3 and $\ln 3 = 1.10$, so

$$W_{\text{isothermal}} = nRT \ln\left(\frac{V_2}{V_1}\right)$$

$$= (2.00 \text{ mol})[8.31 \text{ J/(mol} \cdot \text{K)}](293 \text{ K})(\ln 3) = +5.35 \times 10^5 \text{ J}$$

For the adiabatic process, the work can be determined from Eq. 12.9, but first the final pressure and volume are needed. The final pressure can be determined from a ratio form of Eq. 12.8:

$$p_2 = p_1\left(\frac{V_1}{V_2}\right)^\gamma = p_1\left(\frac{V_1}{3V_1}\right)^\gamma = p_1\left(\frac{1}{3}\right)^{1.67} = 0.160p_1$$

$$= (0.160)(1.01 \times 10^5 \text{ N/m}^2) = 1.62 \times 10^4 \text{ N/m}^2$$

The initial volume is determined from the ideal gas law:

$$V_1 = \frac{nRT_1}{p_1} = \frac{(2.00 \text{ mol})[8.31 \text{ J/(mol} \cdot \text{K)}](293 \text{ K})}{1.01 \times 10^5 \text{ N/m}^2}$$

$$= 4.82 \times 10^{-2} \text{ m}^3$$

(continued on next page)

Therefore, $V_2 = 3V_1 = 0.145 \, \text{m}^3$. Then, applying Eq. 12.9,

$$W_{\text{adiabatic}} = \frac{p_1 V_1 - p_2 V_2}{\gamma - 1}$$

$$= \frac{(1.01 \times 10^5 \, \text{N/m}^2)(4.82 \times 10^{-2} \, \text{m}^3) - (1.62 \times 10^4 \, \text{N/m}^2)(0.145 \, \text{m}^3)}{1.67 - 1}$$

$$= +3.76 \times 10^3 \, \text{J}$$

As expected, this result is less than the isothermal work.

FOLLOW-UP EXERCISE. In this Example, (a) calculate the final temperature of the gas in the adiabatic expansion. (b) During the adiabatic expansion, determine the gas's change in internal energy using its temperature change (helium is a monatomic gas). Does it equal the negative of the work done (as calculated in the Example)? Explain.

LEARN BY DRAWING 12.1

leaning on isotherms

When you are analyzing thermodynamic processes, it is sometimes hard to keep track of the signs of heat flow (Q), work (W), and internal energy change (ΔU). One method that can help with this bookkeeping is to superimpose a series of isotherms on the p–V diagram you are working with (as in Figs. 12.9 through 12.13). This method is useful even if the situation you are studying does not involve isothermal processes.

Before starting, recall that an isothermal process is one in which the temperature remains constant:

1. In an isothermal process for an ideal gas, ΔU is zero. (Why?)
2. Since T is constant, pV must also be constant, since, from the ideal gas law (Eq. 10.3), $pV = nRT = $ constant. You may recall from algebra that $p = $ constant$/V$ is the equation of a hyperbola. Thus, on a p–V diagram, an isothermal process is described by a hyperbola. The farther from the axes the hyperbola is, the higher the temperature it represents (Fig. 1).
   To take advantage of these properties, follow these steps:

■ Sketch a set of isotherms for a series of increasing temperatures on the p–V diagram (Fig. 1).
■ Then sketch the process you are analyzing—for example, the isobar shown in Fig. 2. Isomet = constant volume (vertical line), isobar = constant pressure

(horizontal line), and adiabat = no heat flow (downward sloping curve, steeper than an isotherm).

■ Next, use the graphs to determine the signs of W and ΔU. W is represented by the area under the p–V curve for the process represented, and its sign is determined by whether the gas expanded (positive) or was compressed (negative). The sign of ΔT will be clear from the isotherms, since they serve as a temperature scale. For example, a rise in T implies an increase in U.
■ Last, determine the sign of Q from the first law of thermodynamics, $Q = \Delta U + W$. From the sign of Q, it should be clear whether heat was transferred into or out of the system.

The example in Fig. 2 shows the power of this visual approach. Here, we are to decide whether heat flows into or out of the gas during an isobaric expansion. Expansion implies positive work done by the gas. But what is the direction of the heat flow (or is it zero)? After sketching the isobar, it can be seen that it crosses from lower-temperature isotherms to higher-temperature ones. Hence, there is a temperature increase, and ΔU is positive. From $Q = \Delta U + W$, we see that Q is the sum of two positive quantities, ΔU and W. Therefore, Q must be positive, which means that heat enters the gas.

As an exercise, try analyzing an isometric process using this graphical approach. See also Integrated Examples 12.2 and 12.4.

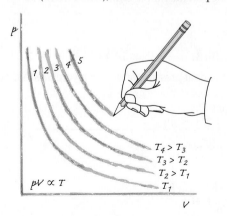

FIGURE 1 **Isotherms on a p–V diagram**

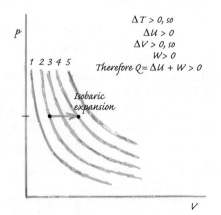

FIGURE 2 **An isobaric expansion**

TABLE 12.1   Important Thermodynamic Processes

Process	Definition	Characteristic	Result of the First Law
Isothermal	T = constant	$\Delta U = 0$	$Q = W$
Isobaric	p = constant	$W = p\Delta V$	$Q = \Delta U + p\Delta V$
Isometric	V = constant	$W = 0$	$Q = \Delta U$
Adiabatic	$Q = 0$		$\Delta U = -W$

As a final summary, the characteristics and consequences of these thermodynamic processes are listed in ▲ Table 12.1.

DID YOU LEARN?

➡ The four important thermodynamic processes are isobaric, isothermal, isometric, and adiabatic.

➡ The change in internal energy of an ideal gas after an isothermal process is equal to zero because its internal energy depends only on its temperature, which is a constant in an isothermal process.

➡ The work done by a gas in an isometric process is zero because there is no volume change (no movement, no work).

12.4 The Second Law of Thermodynamics and Entropy

LEARNING PATH QUESTIONS

➡ If the entropy of a system increases, is the system becoming more ordered or more disordered?

➡ Under what condition, if any, can heat be transferred from a cooler object to a warmer object?

➡ Can heat energy be completely converted to useful mechanical work in a thermodynamic cycle?

Suppose that a piece of hot metal is placed in an insulated container of cold water. Heat will be transferred from the metal to the water, and the two will eventually reach thermal equilibrium at some intermediate temperature. For a thermally isolated system, the system's total energy remains constant. Could heat have been transferred from the cold water to the hot metal instead? This process would not happen naturally. But if it did, the total energy of the system would still remain constant, and this "impossible" inverse process would *not* violate energy conservation or the first law of thermodynamics.

There must be another principle that specifies the *direction* in which a process can take place. This principle is embodied in the **second law of thermodynamics**, which states that certain processes do not take place, or have never been observed to take place, even though they may be consistent with the first law.

There are several equivalent statements of the second law, which are worded according to their application. One applicable to the aforementioned situation is as follows:

Heat will not flow spontaneously from a cooler body to a warmer body.

An equivalent alternative statement of the second law involves thermal cycles. A *thermal cycle* typically consists of several separate thermal processes after which the system ends up back at its starting conditions. If the system is a gas, this means the same p–V–T state from which it started. The second law, stated in terms of a thermal cycle (operating as a heat engine; see Section 12.5), is as follows:

In a thermal cycle, heat energy cannot be completely transformed into mechanical work.

In general, the second law of thermodynamics applies to all forms of energy. It is considered true because no one has ever found an exception to it. If it were not true, a perpetual motion machine could have been built. Such a machine could first transform heat completely into work and motion (mechanical energy), with no energy loss. The mechanical energy could then be transformed back into heat and be used to reheat the reservoir from where the heat came originally (again with no loss). Since the processes could be repeated indefinitely, the machine would run perpetually, just shifting energy back and forth. All of the energy is accounted for, so this situation does *not* violate the first law. However, it is obvious that real machines are always less than 100% efficient (even if there were no friction)—that is, the work output is always less than the energy input. Another statement of the second law is therefore as follows:

> It is impossible to construct an operational perpetual motion machine.

Attempts have been made to construct such perpetual machines, with no success.*

It would be convenient to have some way of expressing the *direction* of a process in terms of the thermodynamic properties of a system. One such property is temperature. In analyzing a conductive heat transfer process, you need to know the temperatures of the system and its surroundings. Knowing the temperature difference between the two processes allows you to state the direction in which the heat transfer will spontaneously take place. Another useful quantity, particularly during the discussion of heat engines, is entropy.

ENTROPY

A quantity that indicates the *natural direction* of a process was first described by Rudolf Clausius (1822–1888), a German physicist. This quantity is called **entropy**. Entropy is a multifaceted concept, with various different physical interpretations:

- Entropy is a measure of a system's ability to do useful work. As a system loses the ability to do work, its entropy increases.

- Entropy determines the direction of time. It is "time's arrow" that points out the forward flow of events, distinguishing past events from future ones.

- Entropy is a measure of disorder. A system naturally moves toward greater disorder, or disarray. The more order there is, the lower the system's entropy.

- The entropy of the universe is increasing.

All of these statements (and others) turn out to be equally valid interpretations of entropy and are physically equivalent, as will be seen in the upcoming discussions. First, however, the definition of the change in entropy is introduced. The change in a system's entropy (ΔS) when an amount of heat (Q) is added or removed by a reversible process at a constant temperature is

$$\Delta S = \frac{Q}{T} \quad \begin{array}{l} \textit{(change in entropy} \\ \textit{at constant temperature)} \end{array} \tag{12.10}$$

SI unit of entropy: joule per kelvin (J/K)

The temperature T must be in kelvins. ΔS is positive if a system absorbs heat ($Q > 0$) and negative if a system loses heat ($Q < 0$). If the temperature changes during the process, calculating the change in entropy requires advanced mathematics. Our discussions will be limited to isothermal processes or those involving small temperature changes. For the latter, entropy changes can be approximated by average temperatures, as in Example 12.6. But first, let's look at an example of a change in entropy and how it is interpreted.

---

*Although perpetual motion *machines* cannot exist, (very nearly) perpetual motion is known to exist—for example, the planets have been in motion around the Sun for about 5 billion years.

EXAMPLE 12.5 | Change in Entropy: An Isothermal Process

While doing physical exercise at a temperature of 34 °C, an athlete loses 0.400 kg of water per hour by the evaporation of perspiration from his skin. Estimate the change in entropy of the water as it vaporizes. The latent heat of vaporization of perspiration is about 24.2×10^5 J/kg.

THINKING IT THROUGH. A phase change occurs at constant temperature; hence, Eq. 12.10 ($\Delta S = Q/T$) applies after converting temperature to kelvins. From Eq. 11.2 ($Q = mL_v$) the amount of heat added can be computed.

SOLUTION. From the statement of the problem,

Given: $m = 0.400$ kg
$T = (34 + 273)$ K $= 307$ K
$L_v = 24.2 \times 10^5$ J/kg

Find: ΔS (change in entropy)

Since a phase change occurs, latent heat is absorbed by water:

$$Q = mL_v = (0.400 \text{ kg})(24.2 \times 10^5 \text{ J/kg}) = 9.68 \times 10^5 \text{ J}$$

Then

$$\Delta S = \frac{Q}{T} = \frac{+9.68 \times 10^5 \text{ J}}{307 \text{ K}} = +3.15 \times 10^3 \text{ J/K}$$

Q is positive, because heat is added to water. The change in entropy, then, is also positive, and the entropy of the water increases. This outcome is reasonable, because a gaseous state is more random (disordered) than a liquid state.

FOLLOW-UP EXERCISE. What is the change in entropy of a 1.00-kg water sample when it freezes to form ice at 0 °C?

EXAMPLE 12.6 | A Warm Spoon into Cool Water: System Entropy Increase or Decrease?

A metal spoon at 24 °C is immersed in 1.00 kg of water at 18 °C. The system (spoon and water) is thermally isolated and comes to equilibrium at a temperature of 20 °C. (a) Find the approximate change in the entropy of the system. (b) Repeat the calculation, assuming, although this can't happen, that the water temperature dropped to 16 °C and the spoon's temperature increased to 28 °C. Comment on how entropy shows that the situation in part (b) cannot happen.

THINKING IT THROUGH. The system is thermally isolated, so there is heat exchange only between the spoon and the water,

that is, $Q_s + Q_w = 0$, where the subscripts s and w stand for spoon and water, respectively. Q_w can be determined from the known water mass, specific heat, and temperature change. Therefore, both Q values (equal but opposite signs) can be determined. Strictly speaking, Eq. 12.10 cannot be used because it is applicable only for constant temperature processes. However, here the temperature changes are small, so a good approximation for ΔS can be obtained by using each object's *average* temperature $\overline{T}$.

SOLUTION. Using i and f to stand for *initial* and *final*, respectively,

Given: $T_{s,i} = 24$ °C
$T_{w,i} = 18$ °C
$m_w = 1.00$ kg
$c_w = 4186$ J/(kg · °C) (from Table 11.1)
(a) $T_f = 20$ °C
(b) $T_{s,f} = 28$ °C; $T_{w,f} = 16$ °C

Find: (a) ΔS (change in entropy of the system in a realistic situation)
(b) ΔS (change in entropy of the system in an unrealistic situation)

(a) The amount of heat transferred (Q) needs to be determined in order to solve for ΔS. With $\Delta T_w = T_f - T_{w,i} = 20$ °C $- 18$ °C $= +2.0$ °C, the heat gained by the water is from Eq. 11.1.

$$Q_w = c_w m_w \Delta T = [4186 \text{ J/(kg·C°)}](1.00 \text{ kg})(2.0 \text{ °C})$$
$$= +8.37 \times 10^3 \text{ J}$$

This quantity is also the magnitude of the heat *lost* by the metal. Therefore,

$$Q_s = -8.37 \times 10^3 \text{ J}$$

The average temperatures are

$$\overline{T}_w = \frac{T_{w,i} + T_f}{2} = \frac{18 \text{ °C} + 20 \text{ °C}}{2} = 19 \text{ °C} = 292 \text{ K}$$

$$\overline{T}_s = \frac{T_{s,i} + T_f}{2} = \frac{24 \text{ °C} + 20 \text{ °C}}{2} = 22 \text{ °C} = 295 \text{ K}$$

Then using these average temperatures and Eq. 12.10 to compute the approximate entropy changes for the water and the metal:

$$\Delta S_w \approx \frac{Q_w}{\overline{T}_w} = \frac{+8.37 \times 10^3 \text{ J}}{292 \text{ K}} = +28.7 \text{ J/K}$$

$$\Delta S_s \approx \frac{Q_s}{\overline{T}_s} = \frac{-8.37 \times 10^3 \text{ J}}{295 \text{ K}} = -28.4 \text{ J/K}$$

The change in the entropy of the *system* is the sum of these, or

$$\Delta S = \Delta S_w + \Delta S_s \approx +28.7 \text{ J/K} - 28.4 \text{ J/K} = +0.3 \text{ J/K}$$

The entropy of the spoon decreased, because heat was lost. The entropy of the water increased by a greater amount, so overall, the system's entropy increased.

(continued on next page)

(b) Although this situation conserves energy, it violates the second law of thermodynamics. To see this violation in terms of entropy, let's repeat the foregoing calculation, using the second set of numbers. With $\Delta T_w = T_f - T_{w,i} = 16\,°C - 18\,°C = -2.0\,°C$, the heat lost by the water is

$$Q_w = c_w m_w \Delta T = [4186\,J/(kg\cdot°C)(1.00\,kg)](-2.0\,°C)$$
$$= -8.37 \times 10^3\,J$$

Again using the average temperatures, $\overline{T}_w = 17\,°C = 290\,K$ and $\overline{T}_s = 26\,°C = 299\,K$, to compute the approximate entropy changes for the water and the metal spoon:

$$\Delta S_w \approx \frac{Q_w}{\overline{T}_w} = \frac{-8.37 \times 10^3\,J}{290\,K} = -28.9\,J/K$$

$$\Delta S_s \approx \frac{Q_s}{\overline{T}_s} = \frac{+8.37 \times 10^3\,J}{299\,K} = +28.0\,J/K$$

The change in the entropy of the *system* is:

$$\Delta S = \Delta S_w + \Delta S_s \approx -28.9\,J/K + 28.0\,J/K = -0.9\,J/K$$

In this unrealistic scenario, the entropy of the metal increased, but the entropy of the water decreased by a greater amount, and the total system entropy decreased.

FOLLOW-UP EXERCISE. What should the initial temperatures in this Example be to make the overall system entropy change zero? Explain in terms of heat transfers.

Note that the entropy change of the system in Example 12.6a is positive, because the process is a *natural* one. That is, it is a process that is always observed to occur. In general, the direction of any process is toward an increase in total system entropy. That is, *the entropy of an isolated system never decreases.* Another way to state this observation is to say that *the entropy of an isolated system increases for every natural process* ($\Delta S > 0$). In coming to an intermediate temperature, the water and spoon in Example 12.6a are undergoing a natural process. The process in 12.6b would never be observed, and the decrease in system entropy indicates this. Similarly, water at room temperature in an isolated ice cube tray will not naturally (spontaneously) turn into ice.

However, if a system is *not* isolated, it may undergo a decrease in entropy. For example, if the ice cube tray filled with water is instead put into a freezer compartment, the water will freeze, undergoing a decrease in entropy. But there will be a *larger increase in entropy somewhere else* in the universe. In this case, the freezer warms the kitchen as it freezes the ice, and the total entropy of the system (ice plus kitchen) actually increases.

Thus, a statement of the second law of thermodynamics in terms of entropy (for natural processes) is:

The total entropy of the universe increases in every *natural* process.

Processes exist for which the entropy is constant. One obvious such process is any adiabatic process, since $Q = 0$. In this case, $\Delta S = Q/T = 0$. Similarly, any reversible isothermal expansion that is followed immediately by an isothermal compression along the same path has zero net entropy change. This last example is true because the two heat flows are the same, but opposite in sign, and the temperatures are also the same; thus, $\Delta S = Q/T + (-Q/T) = 0$. With the realization that, under some circumstances, it *is* possible to have $\Delta S = 0$, the previous statement of the second law of thermodynamics can be generalized to include all possible processes. This is as follows:

During any process, the entropy of the universe can only increase or remain constant ($\Delta S \geq 0$).

To appreciate one of the many alternative (and equivalent) interpretations of entropy, consider the foregoing statements rewritten in terms of order and disorder. Here, entropy is interpreted as a measure of the disorder of a system. Thus, a larger value for entropy means more disorder (or, equivalently, less order):

All naturally occurring processes move toward a state of greater disorder or disarray.

A working definition of order and disorder may be extracted from everyday observations. Suppose you are making a pasta salad and have chopped tomatoes ready to toss into the cooked pasta. Before you mix the pasta and tomatoes, there is a relative amount of order; that is, the ingredients are separate and unmixed. Upon mixing, the separate ingredients become one dish, and there is less order (or more disorder, if you prefer). The pasta salad, once mixed, will never separate into the individual ingredients on its own (that is, by a natural process—one that happens on its own). Of course, you could go in and pick out the individual tomato pieces, but that would not be a natural process. Similarly, broken eyeglasses do not, on their own, go back together. See Global Warming: Some Inconvenient Facts.

DID YOU LEARN?

➡ An increase in entropy is an indication of the increasing disorder of a system. Therefore, a system is becoming more disordered if its entropy increases.

➡ Heat can be transferred from a cooler object to a warmer object, but this is not a spontaneous process. In other words, work has to be done in order to move heat from a cooler object to a warmer object.

➡ According to the second law of thermodynamics, heat energy cannot be completely transformed into mechanical work.

INSIGHT 12.1 | # Global Warming: Some Inconvenient Facts

Human activities such as driving cars and running refrigerators release heat and greenhouse gases into the Earth atmosphere. (See Insight 11.3, The Greenhouse Effect.) There is no doubt that these activities will cause the global temperature to rise. There are, however, hot debates about whether these activities are causing *significant and noticeable* global warming. In this Insight, some facts about global warming are given.

Average temperatures have climbed 0.6 °C ± 0.2 °C around the world since 1880, with much of this increase (about 0.2 °C to 0.3 °C) occurring over the past 25 years (the period with the most credible data), according to NASA's Goddard Institute for Space Studies. The rate of warming is increasing. The past 25 years were the hottest in 400 years (Figure 1).

The average temperatures in the Arctic have risen at twice the global average, according to the multinational Arctic Climate Impact Assessment report compiled between 2000 and 2004. This temperature increase is causing the Arctic ice to rapidly disappear. (See the Chapter 10 opening satellite photographs that show the decreasing Arctic ice.)

The global average sea level has been rising at an average rate of (1 to 2) mm/year over the past 100 years, which is significantly faster than the rate averaged over the last several thousand years. Coral reefs, which are highly sensitive to small changes in water temperature, suffered the worst die-off ever recorded in 1998.

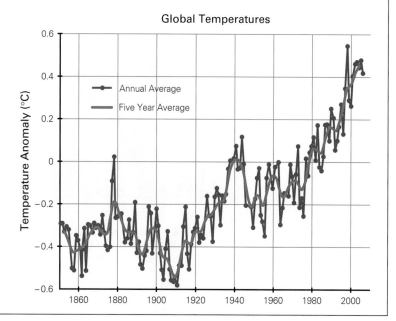

FIGURE 1 Global temperature This graph shows global average temperatures as compiled by the Hadley Centre for Climate Prediction and Research of the UK Meteorological Office. Temperature anomaly is the difference from long-term average temperatures defined by NCDC (National Climate Data Center).

12.5   Heat Engines and Thermal Pumps

LEARNING PATH QUESTIONS

➥ What is the working principle of a heat engine?

➥ How is the thermal efficiency of a heat engine calculated in terms of the heat absorbed from the hot reservoir and the heat expelled to the cold reservoir?

➥ How does a refrigerator work?

A **heat engine** is any device that converts heat energy into work. Since the second law of thermodynamics prohibits complete conversion of heat energy into work in a heat engine, some of the heat input will unfortunately be lost and not go into work. For our purposes, a heat engine is any device that takes heat from a high-temperature source (a hot, or high-temperature, reservoir), converts some of it to useful work, and expels the rest to its surroundings (a cold, or low-temperature, reservoir). For example, most turbines that generate electricity (Section 20.2) are heat engines, using heat from various sources such as oil, gas, coal, or energy released in nuclear reactions (Section 30.2). They might be cooled by river water, for example, thus losing heat to this low-temperature reservoir. A generalized heat engine is represented in ▼Fig. 12.14a. (We will not be concerned with the mechanical details of an engine, such as pistons and cylinders, only thermodynamic processes.)

A few reminders about our sign convention are in order before starting to analyze heat engines. For engines, we are interested primarily in the work W done *by the gas*. During expansion, the gas does positive work. Similarly, during a compression, the work done *by the gas* is negative. Also, it will be assumed that the "working substance" (the material that absorbs the heat and does the work) behaves like an ideal gas. The fundamental physics on which heat engines are based is the same regardless of the working substance. However, using ideal gases makes the mathematics easier.

Adding heat to a gas can produce work. But since a *continuous* output is usually wanted, practical heat engines operate in a **thermal cycle**, or a series of processes that brings the system back to its original condition. Cyclic heat engines include steam engines and internal combustion engines, such as automobile engines.

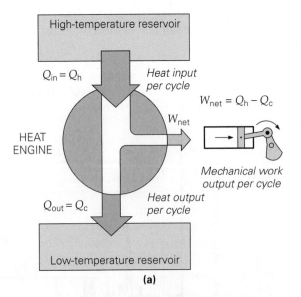
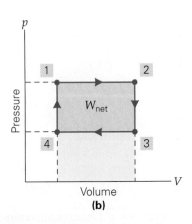

(a)   **(b)**

▲ **FIGURE 12.14   Heat engine   (a)** Energy flow for a generalized cyclic heat engine. Note that the width of the arrow representing Q_h (heat flow out of hot reservoir) is equal to the combined widths of the arrows representing W_{net} and Q_c (heat flow into cold reservoir), reflecting the conservation of energy: $Q_h = Q_c + W_{net}$. **(b)** This specific cyclic process consists of two isobars and two isomets. The net work output per cycle is the area of the rectangle formed by the process paths. (See Example 12.11 for the analysis of this particular cycle.)

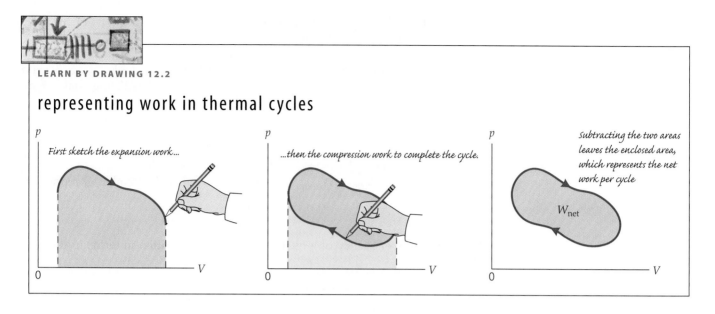

LEARN BY DRAWING 12.2

representing work in thermal cycles

First sketch the expansion work…

…then the compression work to complete the cycle.

Subtracting the two areas leaves the enclosed area, which represents the net work per cycle

W_{net}

An idealized, rectangular thermodynamic cycle is shown in Fig. 12.14b. It consists of two isobars and two isomets. When these processes occur in the sequence indicated, the system goes through a cycle (1–2–3–4–1), returning to its original condition. When the gas expands (during 1 to 2), it does (positive) work equal to the area under the isobar. Doing positive work is exactly the desired output of an engine. (Think of a car engine piston moving the crankshaft.) However, there must be a compression of the gas (during 3 to 4) to bring it back to its initial conditions. During this phase, the work done by the gas is negative, which is *not* the purpose of an engine. In a sense, a portion of the positive work done by the gas is "cancelled" by the negative work done during the compression.

From this discussion, it can be seen that the important quantity in engine design is the *net work* (W_{net}) per cycle. This quantity is represented graphically as the area enclosed by the process curves that make up the cycle in the accompanying Learn By Drawing 12.2, Representing Work in Thermal Cycles. (In Fig 12.14b, the area is rectangular.) When the paths are not straight lines, numerical calculations of the areas may be difficult, but the concept is the same.

THERMAL EFFICIENCY

The **thermal efficiency (ε)** of a heat engine is defined as

$$\varepsilon = \frac{\text{net work output}}{\text{heat input}} = \frac{W_{net}}{Q_{in}} \quad \begin{array}{l}\textit{(thermal efficiency}\\ \textit{of a heat engine)}\end{array} \quad (12.11)$$

Efficiency tells us how much useful work (W_{net}) the engine does in comparison with the input heat it receives (Q_{in}). For example, modern automobile engines have an efficiency of about 20% to 25%. This means that only about one-fourth of the heat generated by igniting the air–gasoline mixture is actually converted into mechanical work, which turns the car wheels and so on. Alternatively, you could say that the engine wastes about three-fourths of the heat, eventually transferring it to the atmosphere through the hot exhaust system, radiator system, and metal engine.

For one cycle of an ideal gas heat engine, W_{net} is determined by applying the first law of thermodynamics to the complete cycle. Recall that our heat sign convention designates Q_{out} as negative. For our discussion of heat engines and pumps, *all heat symbols (all Q's) will represent magnitude only.* Therefore, Q_{out} is written as $-Q_c$ (the negative of a positive quantity Q_c to indicate flow *out* of the engine into a cold reservoir). Q_{in} is positive by our sign convention and is shown as $+Q_h$ (to indicate flow *to* the engine from the hot gas ignition).

Applying the first law of thermodynamics to the expansion part of the cycle and showing the work done by the gas as $W = +W_{\text{expansion}}$, then $\Delta U_{\text{h}} = +Q_{\text{h}} - W_{\text{expansion}}$. For the compression part of the cycle, the work done by the gas is shown explicitly as being negative ($W = -W_{\text{compression}}$ and $\Delta U_{\text{c}} = -Q_{\text{c}} + W_{\text{compression}}$). Adding these equations and realizing that for an ideal gas, $\Delta U_{\text{cycle}} = \Delta U_{\text{h}} + \Delta U_{\text{c}} = 0$ (why?),

$$0 = (Q_{\text{h}} - Q_{\text{c}}) + (W_{\text{compression}} - W_{\text{expansion}})$$

or

$$W_{\text{expansion}} - W_{\text{compression}} = Q_{\text{h}} - Q_{\text{c}}$$

However, $W_{\text{net}} = W_{\text{exp}} - W_{\text{comp}}$, and the final result is (remember Q represents magnitude here)

$$W_{\text{net}} = Q_{\text{h}} - Q_{\text{c}}$$

So the thermal efficiency of a heat engine can be rewritten in terms of the heat flows as

$$\varepsilon = \frac{W_{\text{net}}}{Q_{\text{h}}} = \frac{Q_{\text{h}} - Q_{\text{c}}}{Q_{\text{h}}} = 1 - \frac{Q_{\text{c}}}{Q_{\text{h}}} \qquad \begin{array}{l}\textit{(efficiency of} \\ \textit{an ideal gas heat engine)}\end{array} \qquad (12.12)$$

Like mechanical efficiency, thermal efficiency is a dimensionless fraction and is commonly expressed as a percentage. Equation 12.12 indicates that a heat engine could have 100% efficiency if Q_{c} were zero. This condition would mean that no heat energy would be lost and all the input (hot reservoir) heat would be converted to useful work. However, this situation is impossible according to the second law of thermodynamics. In 1851, this observation led Lord Kelvin (who developed the Kelvin temperature scale; Section 10.3) to state the second law in yet another physically equivalent manner:

No cyclic heat engine can convert its heat input completely to work.

From Eq. 12.12 it can be seen that to maximize the work output per cycle of a heat engine, $Q_{\text{c}}/Q_{\text{h}}$ must be minimized, which increases the efficiency.

Almost all automobile gasoline engines use a *four-stroke cycle*. An approximation of this important cycle involves the steps shown in ▼Fig. 12.15, along with a *p–V* diagram of the thermodynamic processes that make up the cycle. This theoretical

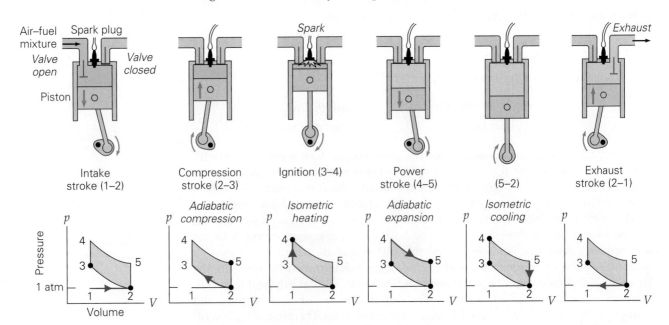

▲ **FIGURE 12.15 The four-stroke cycle of a heat engine** The steps of the four-stroke Otto cycle. The piston moves up and down twice each cycle, for a total of four strokes per cycle. See text for description.

cycle is called the *Otto cycle*, named for the German engineer Nikolaus Otto (1832–1891), who built one of the first successful gasoline engines.

During the intake stroke (1–2), an isobaric expansion, the air–fuel mixture is admitted at atmospheric pressure through the open intake valve as the piston drops. This mixture is compressed adiabatically (quickly) on the compression stroke (2–3). This step is followed by fuel ignition (3–4, when the spark plug fires, giving an isometric pressure rise). Next, an adiabatic expansion occurs during the power stroke (4–5). Following this step is an isometric cooling of the system when the piston is at its lowest position (5–2). The final, exhaust stroke is along the isobaric leg of the Otto cycle (2–1). Notice that it takes two up and down motions of the piston to produce one power stroke.

EXAMPLE 12.7 | ## Thermal Efficiency: What You Get Out of What You Put In

The small, gasoline-powered engine of a leaf blower absorbs 800 J of heat energy from a high-temperature reservoir (the ignited gas–air mixture) and exhausts 700 J to a low-temperature reservoir (the outside air, through its cooling fins). What is the engine's thermal efficiency?

THINKING IT THROUGH. The definition of thermal efficiency of a heat engine (Eq. 12.12) can be used if W_{net} is determined. (Keep in mind that the Q's mean heat magnitudes.)

SOLUTION.

Given: $Q_h = 800$ J   *Find:* ε (thermal efficiency)
$\quad\quad\quad\ Q_c = 700$ J

The net work done by the engine per cycle is

$$W_{net} = Q_h - Q_c = 800\,\text{J} - 700\,\text{J} = 100\,\text{J}$$

Therefore, the thermal efficiency is

$$\varepsilon = \frac{W_{net}}{Q_h} = \frac{100\,\text{J}}{800\,\text{J}} = 0.125 \text{ (or 12.5\%)}$$

FOLLOW-UP EXERCISE. (a) What would be the net work per cycle of the engine in this Example if the efficiency were raised to 15% and the input heat per cycle were raised to 1000 J? (b) How much heat would be exhausted in this case?

Here is a more practical application of a small heat engine.

EXAMPLE 12.8 | ## Thermal Efficiency: Pumping Water

A gasoline-powered water pump can pump 7.6×10^3 kg (about 2000 gal) of water from a basement floor to the ground outside the house in each hour while consuming 1.0 gal of gasoline. Assume the energy content of gasoline is 1.3×10^6 J/gal and the basement floor is 3.0 m below the ground. (a) What is the thermal efficiency of the water pump? (b) How much heat is wasted to the environment in 1.0 h? Ignore frictional losses and assume no change in the kinetic energy of the water.

THINKING IT THROUGH. The definition of thermal efficiency of a heat engine applies (Eq. 12.12). However, we need to calculate the heat input from the energy content of gasoline as well as net work output from raising the water to increase its potential energy (Eq. 5.8).

SOLUTION. The heat input is from the energy in 1.0 gal of gasoline.

Given: $Q_h = (1.0\,\text{gal})(1.3 \times 10^6\,\text{J/gal}) = 1.3 \times 10^6$ J   *Find:* (a) ε (thermal efficiency)
$\quad\quad\quad\ m = 7.6 \times 10^3$ kg   (b) Q_c (heat to environment)
$\quad\quad\quad\ \Delta y = 3.0$ m

(a) The work output is equal to the increase in potential energy of the water:

$$W_{net} = mg\Delta y = (7.6 \times 10^3\,\text{kg})(9.80\,\text{m/s}^2)(3.0\,\text{m}) = 2.2 \times 10^5\,\text{J}$$

The thermal efficiency is then

$$\varepsilon = \frac{W_{net}}{Q_h} = \frac{2.2 \times 10^5\,\text{J}}{1.3 \times 10^6\,\text{J}} = 0.17 \text{ (or 17\%)}$$

(b) The heat exhausted to the environment in 1 h is

$$Q_c = Q_h - W_{net} = 1.3 \times 10^6\,\text{J} - 2.2 \times 10^5\,\text{J} = 1.1 \times 10^6\,\text{J}$$

FOLLOW-UP EXERCISE. If the heat exhausted to the environment were completely absorbed by the pumped water, what would be the temperature change of the water?

INSIGHT 12.2 | # Thermodynamics and the Human Body

Like the bodies of all other organisms, the human body is not a closed system. We must consume food and oxygen to survive. Both the first and second laws of thermodynamics have interesting implications for our bodies.

The human body metabolizes the chemical energy stored in food and/or the fatty tissues of the body. This is quite an efficient process; typically 95% of the energy content in food is eventually metabolized. Some of this metabolized energy is converted into work, W, to circulate blood, perform daily tasks, and so on. The rest is lost to the environment in the form of heat, Q. For a typical 65-kg person, about 80 J of work per second is needed just to keep the body parts, such as the liver, brain, and skeletal muscles, performing their functions.

The first law of thermodynamics, or the law of conservation of energy, can be written as $\Delta U = Q - W$.

Here ΔU is the change in internal energy of the body, which could come from two sources: the already consumed food or the body's stored fat. Thus we can write $\Delta U = \Delta U_{food} + \Delta U_{fat}$. Hence ΔU is a negative quantity, because as the energy stored in food and fat is converted to heat and work, our bodies have less energy stored (until more food is consumed again). Since Q is heat lost to the environment, it is also a negative quantity.

The human body is an example of a biological heat engine. The energy source is the energy metabolized from food and fatty tissues. Some of this energy is converted into work, and the rest is expelled to the environment in the form of heat. This situation is directly analogous to a heat engine taking in heat from the hot reservoir, doing mechanical work, and exhausting waste heat into the environment. Thus the efficiency of the human body is

$$\varepsilon = \frac{\text{work output}}{|\text{internal energy loss}|} = \frac{W}{|\Delta U|}$$

Since W, Q, and ΔU vary widely from one activity to another, efficiency is often determined by using the time rate of these quantities—that is, the work per unit time (power P), $W/\Delta t$, and the energy consumed per unit time (metabolic rate), $|\Delta U|/\Delta t$:

$$\varepsilon = \frac{W}{|\Delta U|} = \frac{W/\Delta t}{(|\Delta U|/\Delta t)} = \frac{P}{(|\Delta U|/\Delta t)}$$

The power exerted during a particular activity such as running or cycling can be measured by a device called a *dynamometer*. The metabolic rate has been found to be directly

FIGURE 1 Measuring energy consumed and work performed A cyclist is tested with a breathing device and a dynamometer so that both her power and metabolic rate can be measured.

proportional to the rate of oxygen consumption, so this rate $(|\Delta U|/\Delta t)$ can be measured by using breathing devices, as in Fig. 1. Thus, the efficiency of the body performing different activities can be determined by measuring the rate of oxygen consumption associated with each separate activity.

The efficiency of the human body, for the most part, depends on muscle activity and which muscles are used. The largest muscles in the body are leg muscles, so if an activity uses these muscles, the efficiency associated with that activity is relatively high. For example, some professional bicycle racers can achieve efficiencies as high as 20%, generating more than 2 hp of power in short bursts. (A typical table saw delivers about 2 hp.) Arm muscles, conversely, are relatively small, so activities such as bench pressing have efficiencies of less than 5%. Like any other heat engine, the human body can never achieve 100% efficiency. When people exercise, a lot of waste heat is generated; they must get rid of it through processes such as perspiring to avoid overheating. Read more about this in Chapter 11 Insight 11.1, Physiological Regulation of Body Temperature.

THERMAL PUMPS: REFRIGERATORS, AIR CONDITIONERS, AND HEAT PUMPS

The function performed by a thermal pump is basically the reverse of that of a heat engine. The name **thermal pump** is a generic term for *any* device that transfers heat from a low-temperature reservoir to a high-temperature reservoir (►Fig. 12.16a), including refrigerators, air conditioners, and heat pumps. For such a transfer to occur, there must be work input. Since the second law of thermodynamics says that heat will not *spontaneously* flow from a cold body to a hot body, the means for this process to happen must be provided; that is, work must be done on the system.

A familiar example of a thermal pump is an air conditioner. By using input work from electrical energy, heat is transferred from the inside of the house (low-temperature reservoir) to the outside of the house (high-temperature reservoir), as

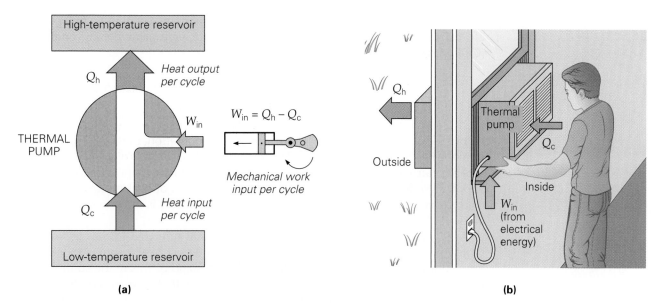

▲ FIGURE 12.16 **Thermal pumps** (a) An energy flow diagram for a generalized cyclic thermal pump. The width of the arrow representing Q_h, the heat transferred into the high-temperature reservoir, is equal to the combined widths of the arrows representing W_{in} and Q_c, reflecting the conservation of energy: $Q_h = W_{in} + Q_c$. (b) An air conditioner is an example of a thermal pump. Using the input work, it transfers heat (Q_c) from a low-temperature reservoir (inside the house) to a high-temperature reservoir (outside).

shown in Fig. 12.16b. A refrigerator (▼Fig. 12.17) uses the exact same principles and processes. With the work performed by the compressor (W_{in}), heat (Q_c) is transferred to the evaporator coils inside of the refrigerator. The combination of this heat and work (Q_h) is then expelled to the outside of the refrigerator through the condenser.

In essence, a refrigerator or air conditioner pumps heat *up* a temperature gradient, or "hill." (Think of pumping water up an actual hill against the force of gravity.) The cooling efficiency of this operation is based on the amount of heat *extracted* from the low-temperature reservoir (the refrigerator, the freezer, or the inside of a house), Q_c, compared with the work W_{in} needed to do so. Since a practical refrigerator operates in a cycle to provide continuous removal of heat, $\Delta U = 0$ for the cycle. Then, by the conservation of energy (the first law of thermodynamics), $Q_c + W_{in} = Q_h$, where Q_h is the heat ejected to the high-temperature reservoir, or the outside.

The measure of an air conditioner's or a refrigerator's performance is defined differently from that of a heat engine, because of the difference in their functions. For the cooling appliances, the efficiency is expressed as a **coefficient of performance (COP)**.

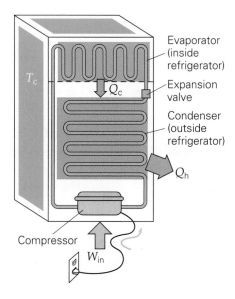

◄FIGURE 12.17 **Refrigerator** Heat (Q_c) is carried away from the interior by the refrigerant as latent heat. This heat energy and that of the work input (W_{in}) are discharged from the condenser to the surroundings (Q_h). A refrigerator can be thought of as a remover of heat (Q_c) from an already cold region (its interior) *or* as a heat pump that adds heat (Q_h) to an already warm area (the kitchen).

Since the purpose is to extract the most heat (Q_c to make or keep things cold) per unit of work input (W_{in}), the coefficient of performance for a refrigerator or air conditioner (COP_{ref}) is the ratio of these two quantities:

$$COP_{ref} = \frac{Q_c}{W_{in}} = \frac{Q_c}{Q_h - Q_c} \quad \text{(refrigerator or air conditioner)} \quad (12.13)$$

Thus, the greater the COP, the better the performance—that is, more heat is extracted for each unit of work done. For normal operation, the work input is less than the heat removed, so the COP is greater than 1. The COPs of typical refrigerators and air conditioners range from 3 to 5, depending on operating conditions and design details. This range means that the amount of heat removed from the cold reservoir (the refrigerator, freezer, or house interior) is three to five times the amount of work needed to remove it.

Any machine that transfers heat in the opposite direction to that in which it would naturally flow is called a *thermal pump*. The term **heat pump** is specifically applied to commercial devices used to cool homes and offices in the summer and to heat them in the winter. The summer operation is that of an air conditioner. In this mode, it cools the interior of the house and heats the outdoors. Operating in its winter heating mode, a heat pump heats the interior and cools the outdoors, usually by taking heat energy from the cold air or ground.

For a heat pump in its heating mode, the heat *output* (to warm something up or keep something warm) is the item of interest, so the COP for heat pump (hp) is defined differently than that of a refrigerator or air conditioner. As you might guess, it is the ratio of Q_h to W_{in} (the heating you get for the work put in), or

$$COP_{hp} = \frac{Q_h}{W_{in}} = \frac{Q_h}{Q_h - Q_c} \quad \text{(heat pump in heating mode)} \quad (12.14)$$

where, again, $Q_c + W_{in} = Q_h$ is used. Typical COPs for heat pumps range between 2 and 4, again depending on the operating conditions and design.

Compared with electrical heating, heat pumps are very efficient. For each unit of electric energy consumed, a heat pump typically pumps in from two to four times as much heat as direct electric heating systems provide. Some heat pumps use water from underground reservoirs, wells, or buried loops of pipe as a low-temperature reservoir. These heat pumps are more efficient than the ones that use the outside air, because water has a larger specific heat than air, and the average temperature difference between the water and the inside air is usually smaller.

EXAMPLE 12.9 | **Air Conditioner/Heat Pump: Thermal Switch Hitting**

A thermal pump operating as an air conditioner in summer extracts 1000 J of heat from the interior of a house for every 400 J of electric energy required to operate it. Determine (a) the air conditioner's COP and (b) its COP if it runs as a heat pump in the winter. Assume it is capable of moving the same amount of heat for the same amount of electric energy, regardless of the direction in which it runs.

THINKING IT THROUGH.   The input work and input heat in part (a) is known, so the definition of COP for a refrigerator (Eq. 12.13) can be applied. For the reverse operation, it is the output heat that is important, so we use the definition of COP for a heat pump (Eq. 12.14).

SOLUTION.

Given: $Q_c = 1000\,J$ *Find:*   (a) COP_{ref} (COP of air conditioner)
  $W_{in} = 400\,J$ (b) COP_{hp} (COP of heat pump)

(a) From Eq. 12.13, the COP for this thermal pump operating as an air conditioner is

$$COP_{ref} = \frac{Q_c}{W_{in}} = \frac{1000\,J}{400\,J} = 2.5$$

(b) When the thermal pump operates as a heat pump, the relevant heat is the output heat, which can be calculated from the conservation of energy:

$$Q_h = Q_c + W_{in} = 1000\,J + 400\,J = 1400\,J$$

Thus, the COP for this engine operating as a heat pump in winter is, from Eq. 12.14,

$$COP_{hp} = \frac{Q_h}{W_{in}} = \frac{1400\,J}{400\,J} = 3.5$$

FOLLOW-UP EXERCISE.   (a) Suppose you redesigned the thermal pump in this Example to perform the same operation, but with 25% less work input. What would be the new values of the two COPs? (b) Which COP would have the larger percentage increase?

12.6 The Carnot Cycle and Ideal Heat Engines

LEARNING PATH QUESTIONS

➥ Since a Carnot engine cannot be built, what then is the significance of a Carnot engine?

➥ If two Carnot engines are operating at the same hot reservoir temperature but different cold reservoir temperatures, which Carnot engine is more thermodynamically efficient?

➥ Is it possible to reach absolute zero?

Lord Kelvin's statement of the second law of thermodynamics says that any *cyclic* heat engine, regardless of its design, must always exhaust some heat energy (Section 12.5). But how much heat must be lost in the process? In other words, what is the *maximum* possible efficiency of a heat engine? In designing heat engines, engineers strive to make them as efficient as possible, but there must be some theoretical limit, and, according to the second law, it must be less than 100%.

Sadi Carnot (1796–1832), a French engineer, studied this limit. The first thing he sought was the thermodynamic cycle an *ideal* heat engine would use, that is, the most efficient cycle. Carnot found that the ideal heat engine absorbs heat from a *constant* high-temperature reservoir (T_h) and exhausts it to a *constant* low-temperature reservoir (T_c). These processes are ideally reversible isothermal processes and may be represented as two isotherms on a p–V diagram. But what are the processes that complete the cycle? Carnot showed that these processes are reversible adiabatic processes. As we saw in Section 12.3, the curves on a p–V diagram are called adiabats and are steeper than isotherms (▼Fig. 12.18a). An irreversible heat engine operating between two heat reservoirs at constant temperatures cannot have an efficiency greater than that of a reversible heat engine operating between the same two temperatures.

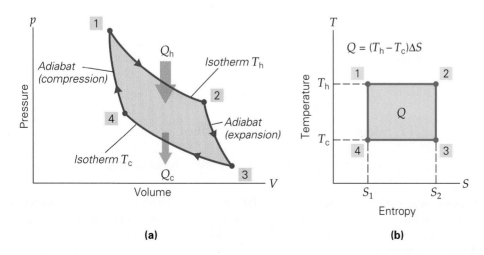

▲ **FIGURE 12.18 The Carnot cycle** **(a)** The Carnot cycle consists of two isotherms and two adiabats. Heat is absorbed during the isothermal expansion and exhausted during the isothermal compression. **(b)** On a T–S diagram, the Carnot cycle forms a rectangle, the area of which is equal to Q.

Thus, the ideal **Carnot cycle** consists of two isotherms and two adiabats and is conveniently represented on a T–S diagram, where it forms a rectangle (Fig. 12.18b). The area under the upper isotherm (1–2) is the heat added to the system from the high-temperature reservoir: $Q_h = T_h \Delta S$. Similarly, the area under the lower isotherm (3–4) is the heat exhausted: $Q_c = T_c \Delta S$. Here, Q_h and Q_c are the heat transfers at *constant* temperatures (T_h and T_c, respectively). There is no heat transfer ($Q = 0$) during the adiabatic legs of the cycle. (Why?)

The difference between these heat transfers is the work output, which is equal to the area enclosed by the process paths (the shaded areas on the diagrams):

$$W_{net} = Q_h - Q_c = (T_h - T_c)\Delta S$$

Since ΔS is the same for both isotherms (see Fig. 12.18b, processes 1–2 and 3–4), the ΔS expressions can be used to relate the temperatures and heats. That is, since

$$\Delta S = \frac{Q_h}{T_h} \quad \text{and} \quad \Delta S = \frac{Q_c}{T_c}$$

then

$$\frac{Q_h}{T_h} = \frac{Q_c}{T_c} \quad \text{or} \quad \frac{Q_c}{Q_h} = \frac{T_c}{T_h}$$

This equation can be used to express the efficiency of an ideal heat engine in terms of temperature. From Eq. 12.12, this ideal **Carnot efficiency (ε_c)** is

$$\varepsilon_C = 1 - \frac{Q_c}{Q_h} = 1 - \frac{T_c}{T_h}$$

or

$$\varepsilon_C = 1 - \frac{T_c}{T_h} \qquad \textit{(Carnot efficiency, ideal heat engine)} \qquad (12.15)$$

where, as usual, the fractional efficiency is often expressed as a percentage. Note that T_c and T_h must be expressed in kelvins.

The Carnot efficiency expresses the theoretical upper limit on the thermodynamic efficiency of a cyclic heat engine operating between two known temperatures. In practice, this limit can never be achieved because no real engine processes are reversible. A true Carnot engine cannot be built, because the necessary reversible processes can only be approximated.

However, the Carnot efficiency does illustrate a general idea: The greater the difference in the temperatures of the heat reservoirs, the greater the Carnot efficiency. For example, if T_h is twice T_c, or $T_c/T_h = 0.50$, the Carnot efficiency is

$$\varepsilon_C = 1 - \frac{T_c}{T_h} = 1 - 0.50 = 0.50(\times 100\%) = 50\%$$

However, if T_h is four times T_c, or $T_c/T_h = 0.25$, then

$$\varepsilon_C = 1 - \frac{T_c}{T_h} = 1 - 0.25 = 0.75(\times 100\%) = 75\%$$

Since a heat engine can never attain 100% thermal efficiency, it is useful to compare its actual efficiency ε with its theoretical maximum efficiency, that of a Carnot cycle, ε_C. To see the importance of this concept in more detail, study the next Example carefully.

There are also Carnot COPs for refrigerators and heat pumps. (See Exercise 66.)

EXAMPLE 12.10 | Carnot Efficiency: The Dream Measure of Efficiency for Any Real Engine

An engineer is designing a cyclic heat engine to operate between the temperatures of 150 °C and 27 °C, (a) What is the maximum theoretical efficiency that can be achieved? (b) Suppose the engine, when built, does 1000 J of work per cycle for every 5000 J of input heat per cycle. What is its efficiency, and how close is it to the Carnot efficiency?

THINKING IT THROUGH. The maximum efficiency for specific high and low temperatures is given by Eq. 12.15. Remember, we must convert to absolute temperatures. In part (b), we calculate the actual efficiency and compare it with our answer in part (a).

SOLUTION.

Given: $T_h = (150 + 273)\,\text{K} = 423\,\text{K}$ *Find:* (a) ε_C (Carnot efficiency)
$T_c = (27 + 273)\,\text{K} = 300\,\text{K}$ (b) ε (actual efficiency) and compare it with ε_c
$W_{net} = 1000\,\text{J}$
$Q_h = 5000\,\text{J}$

(a) Using Eq. 12.15 to find the maximum theoretical efficiency,

$$\varepsilon_C = 1 - \frac{T_c}{T_h} = 1 - \frac{300\,\text{K}}{423\,\text{K}} = 0.291(\times 100\%) = 29.1\%$$

(b) The actual efficiency is, from Eq. 12.12,

$$\varepsilon = \frac{W_{net}}{Q_h} = \frac{1000\,\text{J}}{5000\,\text{J}} = 0.200 \ (\text{or } 20.0\%)$$

Thus,

$$\frac{\varepsilon}{\varepsilon_C} = \frac{0.200}{0.291} = 0.687 \ (\text{or } 68.7\%)$$

In other words, the heat engine is operating at 68.7% of its theoretical maximum. That's pretty good.

FOLLOW-UP EXERCISE. If the operating high temperature of the engine in this Example were increased to 200 °C, what would be the change in the theoretical efficiency?

THE THIRD LAW OF THERMODYNAMICS

Another inference might be drawn from the expression for the Carnot efficiency (Eq. 12.15). It would seem possible to have ε_C equal to 100%, if only T_c could be absolute zero. (See Section 10.3.) However, absolute zero has never been achieved, although ultralow-temperature (cryogenic) experiments have come within 450 pK (4.5×10^{-10} K) of it. Apparently, reducing the temperature of a system already close to absolute zero in a finite number of steps is impossible. This is embodied in the **third law of thermodynamics** which, simply stated, reads:

It is impossible to reach absolute zero in a finite number of thermal processes.

DID YOU LEARN?

➥ A Carnot engine sets the theoretical upper limit of the thermodynamic efficiency of a heat engine operating between two temperatures. In other words, one should never expect a heat engine to have an efficiency equal to or higher than the Carnot efficiency.

➥ Since the Carnot efficiency is equal to $\varepsilon_C = 1 - T_c/T_h$, the engine with a lower cold reservoir temperature will have a smaller T_c/T_h or greater $1 - T_c/T_h$, and therefore a higher efficiency.

➥ The third law of thermodynamics states that it is impossible to reach absolute zero. If it were possible, then the Carnot efficiency would be 100%, or heat could be completely converted to work, which is a violation of the second law of thermodynamics.

PULLING IT TOGETHER | Ideal Gas Law, Thermodynamics, and Thermal Efficiency

Assume you have 0.100 mol of an ideal monatomic gas that follows the cycle given in Fig. 12.14b and that the pressure and temperature at the lower left-hand corner of that figure are 1.00 atm and 20 °C, respectively. Further assume that the pressure doubles during the isometric process and the volume also doubles during the isobaric expansion. What would be the thermal efficiency of this cycle?

THINKING IT THROUGH. This example combines thermal efficiency (Eq. 12.12), thermal dynamic processes, work, internal energy, heat, and the ideal gas law. Care, however, needs to be taken because heat exchanges can occur during more than one of the processes in the cycle. To determine heat input during the isobaric expansion, the change in internal energy and thus the change in temperature are needed. So it seems likely that the temperatures at all four corners of the cycle will be needed. These can be calculated using the ideal gas law. The four thermodynamic processes involved are two isobaric and two isometric processes.

SOLUTION.   The four corners are labeled with numbers as shown in Figure 12.14b. Listing the data given and converting to SI units,

Given: $p_4 = p_3 = 1.00 \text{ atm} = 1.01 \times 10^5 \text{ N/m}^2$ **Find:** ε (thermal efficiency)
$n = 0.100 \text{ mol}$
$T_4 = 20 \text{ °C} = 293 \text{ K}$
$p_1 = p_2 = 2.00 \text{ atm} = 2.02 \times 10^5 \text{ N/m}^2$
$V_2 = V_3 = 2V_4 = 2V_1$

First, the volumes and temperatures at the corners are computed, using the ideal gas law:

$$V_4 = V_1 = \frac{nRT_1}{p_1} = \frac{(0.100 \text{ mol})[8.31 \text{ J/(mol} \cdot \text{K)}](293 \text{ K})}{1.01 \times 10^5 \text{ N/m}^2} = 2.41 \times 10^{-3} \text{ m}^3$$

Therefore,

$$V_2 = V_3 = 2V_1 = 4.82 \times 10^{-3} \text{ m}^3$$

During isometric processes, temperature (absolute in kelvins) is directly proportional to pressure ($p/T = \text{constant}$), and during isobaric processes, temperature is directly proportional to volume ($V/T = \text{constant}$). Therefore,

$$T_1 = 2T_4 = 586 \text{ K}$$
$$T_2 = 2T_1 = 1172 \text{ K}$$
$$T_3 = \tfrac{1}{2}T_2 = 586 \text{ K}$$

Now the heat transfers can be calculated. $W = 0$ during the 4–1 process, and for a monatomic gas, $\Delta U = \tfrac{3}{2}nR\Delta T$. Therefore,

$$Q_{41} = \Delta U_{41} = \tfrac{3}{2}nR\Delta T_{41} = \tfrac{3}{2}(0.100 \text{ mol})[8.31 \text{ J/(mol} \cdot \text{K)}](586 \text{ K} - 293 \text{ K}) = +365 \text{ J}$$

During the 1–2 process, the gas expands and its internal energy increases. The work done by the gas is

$$W_{12} = p_1 \Delta V_{12} = (2.02 \times 10^5 \text{ N/m}^2)(4.82 \times 10^{-3} \text{ m}^3 - 2.41 \times 10^{-3} \text{ m}^3) = +487 \text{ J}$$

Since work was done *and* the internal energy increased,

$$Q_{12} = \Delta U_{12} + W_{12} = \tfrac{3}{2}nR\Delta T_{12} + 487 \text{ J}$$
$$= \tfrac{3}{2}(0.100 \text{ mol})[8.31 \text{ J/(mol} \cdot \text{K)}](1172 \text{ K} - 586 \text{ K}) + 487 \text{ J}$$
$$= +730 \text{ J} + 487 \text{ J} = +1.22 \times 10^3 \text{ J}$$

Thus the total heat input per cycle, Q_h, is

$$Q_h = Q_{41} + Q_{12} = 1.59 \times 10^3 \text{ J}$$

To find the net work, we need the area enclosed by the cycle. Therefore,

$$W_{\text{net}} = (\Delta p_{23})(\Delta V_{12}) = (1.01 \times 10^5 \text{ N/m}^2)(2.41 \times 10^{-3} \text{ m}^3) = +243 \text{ J}$$

and the efficiency is

$$\varepsilon = \frac{W_{\text{net}}}{Q_h} = \frac{243 \text{ J}}{1.59 \times 10^3 \text{ J}} = 0.153 \text{ or } 15.3\%$$

Learning Path Review

■ The **first law of thermodynamics** is a statement of the conservation of energy for a thermodynamic system. Expressed in equation form, it relates the change in a system's internal energy to the heat flow and the work done by it and is written as

$$Q = \Delta U + W \qquad (12.1)$$

■ Some **thermodynamic processes** (for gases) are

isothermal: a process at constant temperature (T = constant)

isobaric: a process at constant pressure (p = constant)

isometric: a process at constant volume (V = constant)

adiabatic: a process involving no heat flow ($Q = 0$)

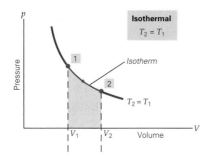

■ The expressions for **thermodynamic work** done by an ideal gas during various processes are

$$W_{\text{isothermal}} = nRT \ln\left(\frac{V_2}{V_1}\right) \quad \textit{(ideal gas isothermal process)} \quad (12.3)$$

$$W_{\text{isobaric}} = p(V_2 - V_1) = p\Delta V \quad \textit{(ideal gas isobaric process)} \quad (12.4)$$

$$W_{\text{adiabatic}} = \frac{p_1 V_1 - p_2 V_2}{\gamma - 1} \quad \textit{(ideal gas adiabatic process)} \quad (12.9)$$

(In the adiabatic process, $\gamma = c_p/c_v$ is the ratio of specific heats at constant pressure and volume, respectively.)

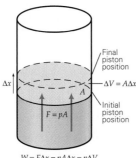

■ The **second law of thermodynamics** determines whether a process can take place naturally or, alternatively, specifies the direction a process can take.

■ **Entropy (S)** is a measure of the disorder of a system. The **change in entropy** of an object at constant temperature is given by

$$\Delta S = \frac{Q}{T} \qquad (12.10)$$

The total entropy of the universe increases in every natural process.

■ A **heat engine** is a device that converts heat into work. Its **thermal efficiency** ε is the ratio of work output to heat input, or

$$\varepsilon = \frac{W_{\text{net}}}{Q_h} = \frac{Q_h - Q_c}{Q_h} = 1 - \frac{Q_c}{Q_h} \qquad (12.12)$$

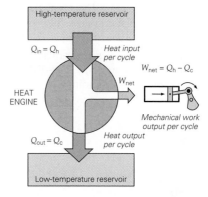

■ A **thermal pump** is a device that transfers heat energy from a low-temperature reservoir to a high-temperature reservoir. The coefficient of performance (COP) is the ratio of heat transferred to the input work. The COP differs depending on whether the thermal pump is used as a heat pump or as an air conditioner/refrigerator.

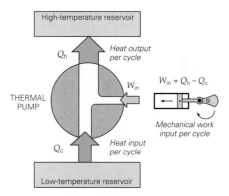

■ A **Carnot cycle** is a theoretical heat engine cycle consisting of two isotherms and two adiabats. Its efficiency is the highest possible efficiency that any heat engine could have, operating between two temperature extremes. The efficiency of a Carnot cycle is

$$\varepsilon_C = 1 - \frac{T_c}{T_h} \qquad (12.15)$$

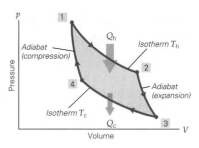

Learning Path Questions and Exercises

For instructor-assigned homework, go to www.masteringphysics.com

MULTIPLE CHOICE QUESTIONS

12.1 THERMODYNAMIC SYSTEMS, STATES, AND PROCESSES

1. On a p–V diagram, a reversible process is a process (a) whose path is known, (b) whose path is unknown, (c) for which the intermediate steps are nonequilibrium states, (d) none of the preceding.

2. There may be an exchange of heat with the surroundings for (a) a thermally isolated system, (b) a completely isolated system, (c) a heat reservoir, (d) none of the preceding.

3. Only initial and final states are known for irreversible processes on (a) p–V diagrams, (b) p–T diagrams, (c) V–T diagrams, (d) all of the preceding.

12.2 THE FIRST LAW OF THERMODYNAMICS AND 12.3 THERMODYNAMIC PROCESSES FOR AN IDEAL GAS

4. There is no heat flow into or out of the system in an (a) isothermal process, (b) adiabatic process, (c) isobaric process, (d) isometric process.

5. If the work done by a system is equal to zero, the process is (a) isothermal, (b) adiabatic, (c) isobaric, (d) isometric.

6. According to the first law of thermodynamics, if work is done on a system, then (a) the internal energy of the system must change, (b) heat must be transferred from the system, (c) the internal energy of the system may change and/or heat may be transferred from the system, (d) heat must be transferred to the system.

7. When heat is added to a system of an ideal gas during the process of an isothermal expansion, (a) work is done by the system, (b) the internal energy increases, (c) work is done on the system, (d) the internal energy decreases.

12.4 THE SECOND LAW OF THERMODYNAMICS AND ENTROPY

8. In any natural process, the overall change in the entropy of the universe could not be (a) negative, (b) zero, (c) positive.

9. For which type of thermodynamic process is the change in entropy equal to zero: (a) isothermal, (b) isobaric, (c) isometric, or (d) none of the preceding?

10. Which one of the following statements is a violation of the second law of thermodynamics: (a) heat flows naturally from hot to cold, (b) heat can be completely converted to mechanical work, (c) the entropy of the universe can never decrease, or (d) it is not possible to construct a perpetual motion engine?

11. An ideal gas is compressed isothermally. The change in entropy of the gas for this process is (a) positive, (b) negative, (c) zero, (d) none of the preceding.

12.5 HEAT ENGINES AND THERMAL PUMPS

12. If the first law of thermodynamics is applied to a heat engine, the result is (a) $W_{net} = Q_h + Q_c$, (b) $W_{net} = Q_h - Q_c$, (c) $W_{net} = Q_c - Q_h$, (d) $Q_c = 0$.

13. For a cyclic heat engine, (a) $\varepsilon = 1$, (b) $Q_h = W_{net}$, (c) $\Delta U = W_{net}$, (d) $Q_h > Q_c$.

14. A thermal pump (a) is rated by thermal efficiency, (b) requires work input, (c) has $Q_h = Q_c$, (d) has COP = 1.

15. Which of the following determines the thermal efficiency of a heat engine: (a) $Q_c \times Q_h$, (b) Q_c/Q_h, (c) $Q_h - Q_c$, or (d) $Q_h + Q_c$?

12.6 THE CARNOT CYCLE AND IDEAL HEAT ENGINES

16. The Carnot cycle consists of (a) two isobaric and two isothermal processes, (b) two isometric and two adiabatic processes, (c) two adiabatic and two isothermal processes, (d) four arbitrary processes that return the system to its initial state.

17. Which of the following temperature reservoir relationships would yield the lowest efficiency for a Carnot engine: (a) $T_c = 0.15T_h$, (b) $T_c = 0.25T_h$, (c) $T_c = 0.50T_h$, or (d) $T_c = 0.90T_h$?

18. For a heat engine that operates between two reservoirs of temperatures T_c and T_h, the Carnot efficiency is the (a) highest possible value, (b) lowest possible value, (c) average value, (d) none of the preceding.

19. If absolute zero were reached, then the Carnot efficiency could be (a) 0%, (b) 50%, (c) 75%, (d) 100%.

CONCEPTUAL QUESTIONS

12.1 THERMODYNAMIC SYSTEMS, STATES, AND PROCESSES

1. Explain why the process shown in Fig. 12.1b is *not* that for an ideal gas at constant temperature.

2. Does an irreversible process mean the system cannot return to its original state? Explain.

3. What are the four state variables, used in this chapter, for ideal gases?

12.2 THE FIRST LAW OF THERMODYNAMICS AND 12.3 THERMODYNAMIC PROCESSES FOR AN IDEAL GAS

4. On a p–V diagram, sketch a cyclic process that consists of an isothermal expansion followed by an isobaric compression, and lastly followed by an isometric process.

5. In ▼Fig. 12.19, the plunger of a syringe is pushed in quickly, and the small pieces of paper in the syringe catch fire. Explain this phenomenon using the first law of thermodynamics. (Similarly, in a diesel engine, there are no spark plugs. How can the air–fuel mixture ignite?)

▲ FIGURE 12.19 **Syringe fire** See Conceptual Question 5.

6. Discuss heat, work, and the change in internal energy of your body when you shovel snow.

7. In an adiabatic process, there is no heat exchange between the system and the environment, but the temperature of the ideal gas changes. How can this be? Explain.

8. In an isobaric process, an ideal gas sample can do work on the environment but its temperature also increases. How can this be?

9. An ideal gas initially at temperature T_o, pressure p_o, and volume V_o is compressed to one-half its initial volume. As shown in ▼Fig. 12.20, process 1 is adiabatic, 2 is isothermal, and 3 is isobaric. Rank the work done on the gas and the final temperatures of the gas, from highest to lowest, for all three processes, and explain how you decided upon your rankings.

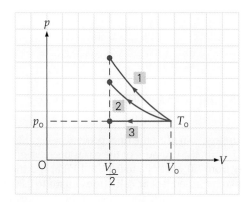

▲ FIGURE 12.20 **Thermodynamic processes** See Conceptual Question 9.

10. If ideal gas sample A receives more heat than ideal gas sample B, will sample A experience a higher increase in internal energy? Explain.

12.4 THE SECOND LAW OF THERMODYNAMICS AND ENTROPY

11. Heat is converted to mechanical energy in many applications, such as cars. Is this a violation of the second law of thermodynamics? Explain.

12. Does the entropy of each of the following objects increase or decrease? (a) *Ice* as it melts; (b) *water vapor* as it condenses; (c) *water* as it is heated on a stove; (d) *food* as it is cooled in a refrigerator.

13. When a quantity of hot water is mixed with a quantity of cold water, the combined system comes to thermal equilibrium at some intermediate temperature. How does the entropy of the system (both liquids) change?

14. A student challenges the second law of thermodynamics by saying that entropy does not have to increase in all situations, such as when water freezes to ice. Is this challenge valid? Why or why not?

15. Is a living organism an open system or an isolated system? Explain.

16. A student tries to cool his dormitory room by opening the refrigerator door. Will that work? Explain.

12.5 HEAT ENGINES AND THERMAL PUMPS

17. What happens to the pressure and internal energy of a cyclic heat engine after a complete cycle?

18. Lord Kelvin's statement of the second law of thermodynamics as applied to heat engines ("No heat engine operating in a cycle can convert its heat input completely to work") refers to their operation *in a cycle.* Why is the phrase "in a cycle" included?

19. If heat engine A absorbs more heat than heat engine B from a hot reservoir, will engine A necessarily do more net work than engine B? Explain your reasoning.

20. The heat output of a thermal pump is greater than the energy used to operate the pump. Does this device violate the first law of thermodynamics?

21. The maximum efficiency of a heat engine is 1 (or 100%). Can the COP of a thermal pump be greater than 1? Explain.

12.6 THE CARNOT CYCLE AND IDEAL HEAT ENGINES

22. Diesel engines are more efficient than gasoline engines. Which type of engine wold you expect to run hotter? Why?

23. If you have the choice of running your heat engine between either of the following two sets of temperatures for the cold and hot reservoirs, which would you choose, and why: between 100 °C and 300 °C, or between 50 °C and 250 °C?

24. Carnot engine A operates at a higher hot reservoir temperature than Carnot engine B. Will engine A necessarily have a higher Carnot efficiency? Explain.

EXERCISES*

Integrated Exercises (**IE**s) *are two-part exercises. The first part typically requires a conceptual answer choice based on physical thinking and basic principles. The following part is quantitative calculations associated with the conceptual choice made in the first part of the exercise.*

Many exercise sections include "paired" exercises. These exercise pairs, identified with red numbers, *are intended to assist you in problem solving and learning. In a pair, the first exercise (even numbered) is worked out in the Study Guide so that you can consult it should you need assistance in solving it. The second exercise (odd numbered) is similar in nature, and its answer is given in Appendix VII at the back of the book.*

12.2 THE FIRST LAW OF THERMODYNAMICS
AND
12.3 THERMODYNAMIC PROCESSES FOR AN IDEAL GAS

1. • While playing in a tennis match, you lost 6.5×10^5 J of heat, and your internal energy also decreased by 1.2×10^6 J. How much work did you do in the match?

2. **IE** • A rigid container contains 1.0 mol of an ideal gas that slowly receives 2.0×10^4 J of heat. (a) The work done by the gas is (1) positive, (2) zero, (3) negative. Why? (b) What is the change in the internal energy of the gas?

3. **IE** • A quantity of ideal gas goes through an isothermal process and does 400 J of net work. (a) The internal energy of the gas is (1) higher than, (2) the same as, (3) less than when it started. Why? (b) Is a net amount of heat added to or removed from the system, and how much is involved?

4. • An ideal gas goes through a thermodynamic process in which 500 J of work is done on the gas and the gas loses 300 J of heat. What is the change in internal energy of the gas?

5. **IE** • While doing 500 J of work, an ideal gas expands adiabatically to 1.5 times its initial volume. (a) The temperature of the gas (1) increases, (2) remains the same, (3) decreases. Why? (b) What is the change in the internal energy of the gas?

6. **IE** • An ideal gas expands from 1.0 m^3 to 3.0 m^3 at atmospheric pressure while absorbing 5.0×10^5 J of heat in the process. (a) The temperature of the system (1) increases, (2) stays the same, (3) decreases. Explain. (b) What is the change in internal energy of the system?

7. •• An ideal gas is under an initial pressure of 2.45×10^4 Pa and occupies a volume of 0.20 m^3. The slow addition of 8.4×10^3 J of heat to this gas causes it to expand isobarically to a volume of 0.40 m^3. (a) How much work is done by the gas in the process? (b) Does the internal energy of the gas change? If so, by how much?

8. •• An Olympic weight lifter lifts 145 kg a vertical distance of 2.1 m. When he does so, 6.0×10^4 J of heat is transferred to air through perspiration. Does he gain or lose internal energy and how much?

9. **IE** •• An ideal gas is taken through the reversible processes shown in ▶ Fig. 12.21. (a) Is the overall change in the internal energy of the gas (1) positive, (2) zero, or (3) negative? Explain. (b) In terms of state variables p and V, how much work is done by or on the gas, and (c) what is the net heat transfer in the overall process?

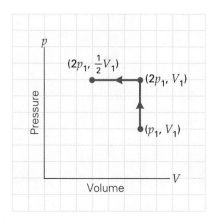

▲ **FIGURE 12.21** A p–V **diagram for an ideal gas** See Exercise 9.

10. •• A fixed quantity of gas undergoes the reversible changes illustrated in the p–V diagram in ▼Fig. 12.22. How much work is done in each process?

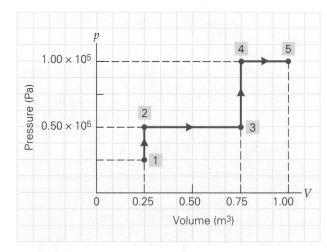

▲ **FIGURE 12.22** A p–V **diagram and work** See Exercises 10 and 11.

11. •• Suppose that after the final process in Fig. 12.22 (see Exercise 10), the pressure of the gas is decreased isometrically from 1.0×10^5 Pa to 0.70×10^5 Pa, and then the gas is compressed isobarically from 1.0 m^3 to 0.80 m^3. What is the total work done in all of these processes, including 1 through 5?

12. **IE** •• A gas is enclosed in a cylindrical piston with a 12.0-cm radius. Heat is slowly added to the gas while the pressure is maintained at 1.00 atm. During the process,

*Take temperatures and efficiencies to be exact.

the piston moves 6.00 cm. (a) This is an (1) isothermal, (2) isobaric, (3) adiabatic process. Explain. (b) If the heat transferred to the gas during the expansion is 420 J, what is the change in the internal energy of the gas?

13. **IE ●●** 2.0 mol of an ideal gas expands isothermally from a volume of 20 L to 40 L at 20 °C. (a) The work done by the gas is (1) positive, (2) negative, (3) zero. Explain. (b) What is the magnitude of the work?

14. **●●** A monatomic ideal gas ($\gamma = 1.67$) is compressed adiabatically from a pressure of 1.00×10^5 Pa and volume of 240 L to a volume of 40.0 L. (a) What is the final pressure of the gas? (b) How much work is done on the gas?

15. **●●** An ideal gas sample expands isothermally by tripling its volume and doing 5.0×10^4 J of work at 40 °C. (a) How many moles of gas are there in the sample? (b) Was heat added to or removed from the sample, and how much?

16. **IE ●●●** The temperature of 2.0 mol of ideal gas is increased from 150 °C to 250 °C by two different processes. In process A, 2500 J of heat is added to the gas; in process B, 3000 J of heat is added. (a) In which case is more work done: (1) process A, (2) process B, or (3) the same amount of work is done? Explain. [*Hint:* See Eq. 10.16.] (b) Calculate the change in internal energy and work done for each process.

17. **IE ●●●** One handred moles of a monatomic gas is compressed as shown on the *p–V* diagram in ▼Fig. 12.23. (a) Is the work done by the gas (1) positive, (2) zero, or (3) negative? Why? (b) What is the work done by the gas? (c) What is the change in temperature of the gas? (d) What is the change in internal energy of the gas? (e) How much heat is involved in the process?

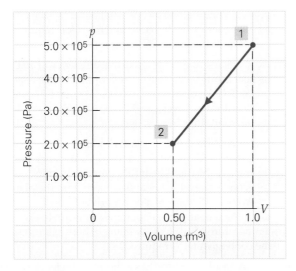

▲ **FIGURE 12.23 A variable *p–V* process and work** See Exercise 17.

18. **●●●** One mole of an ideal gas is taken through the cyclic process shown in ▶Fig. 12.24. (a) Compute the work involved for each of the four processes. (b) Find ΔU, W, and Q for the complete cycle. (c) What is T_3?

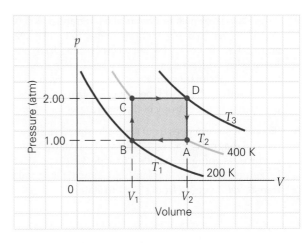

▲ **FIGURE 12.24 A cyclic process** See Exercise 18.

12.4 THE SECOND LAW OF THERMODYNAMICS AND ENTROPY

19. **●** What is the change in entropy of mercury vapor ($L_v = 2.7 \times 10^5$ J/kg) when 0.50 kg of it condenses to a liquid at its boiling point of 357 °C?

20. **IE ●** 2.0 kg of ice melts completely into liquid water at 0 °C. (a) The change in entropy of the ice (water) in this process is (1) positive, (2) zero, (3) negative. Explain. (b) What *is* the change in entropy of the ice (water)?

21. **IE ●** A process involves 1.0 kg of steam condensing to water at 100 °C. (a) The change in entropy of the steam (water) is (1) positive, (2) zero, (3) negative. Why? (b) What *is* the change in entropy of the steam (water)?

22. **●** During a liquid-to-solid phase change of a substance, its change in entropy is -4.19×10^3 J/K. If 1.67×10^6 J of heat is removed in the process, what is the freezing point of the substance in degrees Celsius?

23. **●●** In an isothermal expansion at 27 °C, an ideal gas does 60 J of work. What is the change in entropy of the gas?

24. **IE ●●** One mole of an ideal gas undergoes an isothermal compression at 0 °C, and 7.5×10^3 J of work is done in compressing the gas. (a) Will the entropy of the gas (1) increase, (2) remain the same, or (3) decrease? Why? (b) What is the change in entropy of the gas?

25. **IE ●●** A quantity of an ideal gas undergoes an isothermal expansion at 20 °C and does 3.0×10^3 J of work on its surroundings in the process. (a) Will the entropy of the gas (1) increase, (2) remain the same, or (3) decrease? Explain. (b) What is the change in the entropy of the gas?

26. **●●** In the winter, heat from a house with an inside temperature of 18 °C leaks out at a rate of 2.0×10^4 J/s. The outside temperature is 0 °C. (a) What is the change in entropy per second of the house? (b) What is the total change in entropy per second of the house–outside system?

27. **IE ●●** An isolated system consists of two very large thermal reservoirs at constant temperatures of 100 °C and 0 °C. Assume the reservoirs made contact and 1000 J of heat flew from the cold reservoir to the hot reservoir spontaneously. (a) The total change in entropy of the isolated system (both reservoirs) would be (1) positive, (2) zero, (3) negative. Explain. (b) Calculate the total change in entropy of this isolated system.

28. **IE ••** Two large heat reservoirs at temperatures 200 °C and 60 °C, respectively, are brought into thermal contact, and 1.50×10^3 J of heat spontaneously flows from one to the other with no significant temperature change. (a) The change in the entropy of the two-reservoir system is (1) positive, (2) zero, (3) negative. Explain. (b) Calculate the change in the entropy of the two-reservoir system.

29. **IE •••** A system goes from state 1 to state 3 as shown on the *T–S* diagram in ▼Fig. 12.25. (a) The heat transfer for the process going from state 2 to state 3 is (1) positive, (2) zero, (3) negative. Explain. (b) Calculate the total heat transferred in going from state 1 to state 3.

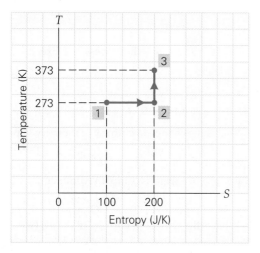

▲ **FIGURE 12.25 Entropy and heat** See Exercises 29 and 30.

30. **IE •••** Suppose that the system described by the *T–S* diagram in Fig. 12.25 is returned to its original state, state 1, by a reversible process depicted by a straight line from state 3 to state 1. (a) The change in entropy of the system for this overall cyclic process is (1) positive, (2) zero, (3) negative. Explain. (b) How much heat is transferred in the cyclic process? [*Hint*: See Example 12.6.]

31. **•••** A 50.0-g ice cube at 0 °C is placed in 500 mL of water at 20 °C. *Estimate* the change in entropy (after all the ice has melted) (a) for the ice, (b) for the water, and (c) for the ice–water system.

12.5 HEAT ENGINES AND THERMAL PUMPS

32. **•** If an engine does 200 J of net work and exhausts 800 J of heat per cycle, what is its thermal efficiency?

33. **•** A gasoline engine has a thermal efficiency of 28%. If the engine absorbs 2000 J of heat per cycle, (a) what is the net work output per cycle? (b) How much heat is exhausted per cycle?

34. **•** A heat engine with a thermal efficiency of 20% does 500 J of net work each cycle. How much heat per cycle is lost to the low-temperature reservoir?

35. **•** An internal combustion engine with a thermal efficiency of 15.0% absorbs 1.75×10^5 J of heat from the hot reservoir. How much heat is lost by the engine in each cycle?

36. **IE •** The heat output of a particular engine is 7.5×10^3 J per cycle, and the net work out is 4.0×10^3 J per cycle. (a) The heat input is (1) less than 4.0×10^3 J, (2) between 4.0×10^3 J and 7.5×10^3 J, (3) greater than 7.5×10^3 J. Explain. (b) What is the heat input and thermal efficiency of the engine?

37. **••** A gasoline engine burns fuel that releases 3.3×10^8 J of heat per hour. (a) What is the energy input during a 2.0-h period? (b) If the engine delivers 25 kW of power during this time, what is its thermal efficiency?

38. **IE ••** A steam engine is to have its thermal efficiency improved from 8.00% to 10.0% while continuing to produce 4500 J of useful work each cycle. (a) Does the ratio of the heat output to heat input (1) increase, (2) remain the same, or (3) decrease? Why? (b) What is the *change* in Q_c/Q_h in this example?

39. **IE ••** An engineer redesigns a heat engine and improves its thermal efficiency from 20% to 25%. (a) Does the ratio of the heat input to heat output (1) increase, (2) remain the same, or (3) decrease? Explain. (b) What is the engine's change in Q_h/Q_c?

40. **••** When running, a refrigerator exhausts heat to the kitchen at a rate of 10 kW when the required input work is done at a rate of 3.0 kW. (a) At what rate is heat removed from its cold interior? (b) What is the COP of the refrigerator?

41. **••** A refrigerator with a COP of 2.2 removes 4.2×10^5 J of heat from its interior each cycle. (a) How much heat is exhausted each cycle? (b) What is the total work input in joules for 10 cycles?

42. **••** An air conditioner has a COP of 2.75. What is the power rating of the unit if it is to remove 1.00×10^7 J of heat from a house interior in 20 min?

43. **••** A heat pump removes 2.2×10^3 J of heat from the outdoors and delivers 4.3×10^3 J of heat to the inside of a house each cycle. (a) How much work is required per cycle? (b) What is the COP of this pump?

44. **••** A steam engine has a thermal efficiency of 15.0%. If its heat input for each cycle is supplied by the condensation of 8.00 kg of steam at 100 °C. (a) what is the net work output per cycle, and (b) how much heat is lost to the surroundings in each cycle?

45. **•••** A coal-fired power plant produces 900 MW of electric power and operates at a thermal efficiency of 25% (a) What is the input heat rate from the burning coal? (b) What is the rate of heat discharge from the plant? (c) Water at 15 °C from a nearby river is used to cool the discharged heat. If the cooling water is not to exceed a temperature of 40 °C, how many gallons per minute of the cooling water is required?

46. **•••** A gasoline engine has a thermal efficiency of 25.0%. If heat is expelled from the engine at a rate of 1.50×10^6 J/h, how long does the engine take to perform a task that requires an amount of work of 1.5×10^6 J?

47. **•••** A four-stroke engine runs on the Otto cycle. It delivers 150 hp at 3600 rpm. (a) How many cycles are in 1 min? (b) If the thermal efficiency of the engine is 20%, what is the heat input per minute? (c) How much heat is wasted (per minute) to the environment?

12.6 THE CARNOT CYCLE AND IDEAL HEAT ENGINES

48. • A Carnot engine has an efficiency of 35% and takes in heat from a high-temperature reservoir at 178 °C. What is the Celsius temperature of the engine's low-temperature reservoir?

49. • A steam engine operates between 100 °C and 20 °C. What is the Carnot efficiency of the ideal engine that operates between these temperatures?

50. • It has been proposed that temperature differences in the ocean could be used to run a heat engine to generate electricity. In tropical regions, the water temperature is about 25 °C at the surface and about 5 °C at very deep depths. (a) What would be the maximum theoretical efficiency of such an engine? (b) Would a heat engine with such a low efficiency be practical? Explain.

51. • What is the Celsius temperature of the hot reservoir of a Carnot engine that is 32% efficient and has a 20 °C cold reservoir?

52. • An engineer wants to run a heat engine with an efficiency of 40% between a high-temperature reservoir at 300 °C and a low-temperature reservoir. What is the maximum Celsius temperature of the low-temperature reservoir?

53. •• A Carnot engine with an efficiency of 40% operates with a low-temperature reservoir at 40 °C and exhausts 1200 J of heat each cycle. What are (a) the heat input per cycle and (b) the Celsius temperature of the high-temperature reservoir?

54. •• A Carnot engine takes 2.7×10^4 J of heat per cycle from a high-temperature reservoir at 320 °C and exhausts some of it to a low-temperature reservoir at 120 °C How much net work is done by the engine per cycle?

55. **IE** •• A Carnot engine takes in heat from a reservoir at 350 °C and has an efficiency of 35%. The exhaust temperature is not changed and the efficiency is increased to 40%, (a) The temperature of the hot reservoir is (1) lower than (2) equal to, (3) higher than 350 °C. Explain. (b) What is the new Celsius temperature of the hot reservoir?

56. •• An inventor claims to have created a heat engine that produces 10.0 kW of power for a 15.0-kW heat input while operating between reservoirs at 27 °C and 427 °C. (a) Is this claim valid? (b) To produce 10.0 kW of power, what is the minimum heat input required?

57. •• An inventor claims to have developed a heat engine that, each cycle, takes in 5.0×10^5 J of heat from a high-temperature reservoir at 400 °C and exhausts 2.0×10^5 J to the surroundings at 125 °C. Would you invest your money in the production of this engine? Explain.

58. •• A heat engine operates at a thermal efficiency that is 45% of the Carnot efficiency. If the temperatures of the high-temperature and low-temperature reservoirs are 400 °C and 50 °C, respectively, what are the Carnot efficiency and the thermal efficiency of the engine?

59. •• A heat engine's thermal efficiency is 70.0% of the Carnot efficiency of an engine operating between temperatures of 80 °C and 375 °C.(a) What is the Carnot efficiency of the heat engine? (b) If the heat engine absorbs heat at a rate of 50 kW, at what rate is heat exhausted?

60. •• In each cycle, a Carnot engine takes 800 J of heat from a high-temperature reservoir and discharges 600 J to a low-temperature reservoir. What is the ratio of the temperature of the high-temperature reservoir to that of the low-temperature reservoir?

61. **IE** •• A Carnot engine operating between reservoirs at 27 °C and 227 °C does 1500 J of work in each cycle. (a) The change in entropy for the engine for each cycle is (1) negative, (2) zero, (3) positive. Why? (b) What is the heat input of the engine?

62. •• The *autoignition temperature* of a fuel is defined as the temperature at which a fuel–air mixture would self-explode and ignite. Thus, it sets an upper limit on the temperature of the hot reservoir in an automobile engine. The autoignition temperatures for commonly available gasoline and diesel fuel are about 495 °F and 600 °F, respectively. What are the maximum Carnot efficiencies of a gasoline engine and a diesel engine if the cold reservoir temperature is 40 °C?

63. •• Because of limitations on materials, the maximum temperature of the superheated steam used in a turbine for the generation of electricity is about 540 °C. (a) If the steam condenser operates at 20 °C, what is the maximum Carnot efficiency of a steam turbine generator? (b) The actual efficiency of such generators is about 35% to 40%. What does this range tell you?

64. •• The working substance of a cyclic heat engine is 0.75 kg of an ideal gas. The cycle consists of two isobaric processes and two isometric processes, as shown in ▼Fig. 12.26. What would be the efficiency of a Carnot engine operating with the same high-temperature and low-temperature reservoirs?

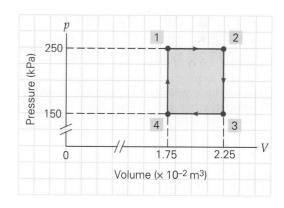

▲ FIGURE 12.26 **Thermal efficiency** See Exercise 64.

65. **IE** •• Equation 12.15 shows that the greater the temperature difference between the reservoirs of a heat engine, the greater the engine's Carnot efficiency. Suppose you had the choice of raising the temperature of the high-temperature reservoir by a certain number of kelvins or lowering the temperature of the low-temperature reservoir by the same number of kelvins. (a) To produce the largest increase in efficiency, you should choose (1) to raise the high-temperature reservoir, (2) to lower the low-temperature reservoir, (3) both 1 and 2 produce the same change in efficiency, so it does not matter which you choose. Explain. (b) Prove your answer to part (a) mathematically.

66. ••• There is a Carnot coefficient of performance (COP_C) for an ideal, or Carnot, refrigerator. (a) Show that this quantity is given by

$$COP_C = \frac{T_c}{T_h - T_c}$$

(b) What does this tell you about adjusting the temperatures for the maximum COP of a refrigerator? (Can you guess the equation for the COP_C for a heat pump?)

67. ••• A salesperson tells you that a new refrigerator with a high COP removes 2.6×10^3 J (each cycle) from the inside of the refrigerator at 5.0 °C and expels 2.8×10^3 J into the 30 °C kitchen. (a) What is the refrigerator's COP? (b) Is this scenario possible? Justify your answer. (See Exercise 66.)

68. ••• An ideal heat pump is equivalent to a Carnot engine running in reverse. (a) Show that the Carnot COP of the heat pump is

$$COP_C = \frac{1}{\varepsilon_C}$$

where ε_C is the Carnot efficiency of the heat engine. (b) If a Carnot engine has an efficiency of 40%, what would be the COP_C when it runs in reverse as a heat pump? (See Exercise 66.)

PULLING IT TOGETHER: MULTICONCEPT EXERCISES

The Multiconcept Exercises require the use of more than one fundamental concept for their understanding and solution. The concepts may be from this chapter, but may include those from previous chapters.

69. A heat engine with a thermal efficiency of 25% is used to hoist 2.5-kg bricks to an elevation of 3.0 m. If the engine expels heat to the environment at a rate of 1.2×10^6 J/h, how many bricks can the engine hoist in 2.0 h?

70. When cruising at 75 mi/h on a highway, a car's engine develops 45 hp. If this engine has a thermodynamic efficiency of 25% and 1 gal of gasoline has an energy content of 1.3×10^8 J, what is the fuel efficiency (in miles per gallon) of this car?

71. A gram of water (volume of 1.00 cm³) at 100 °C is converted to 1.67×10^3 cm³ of steam at atmospheric pressure. What is the change in the internal energy of the water (steam)?

72. In a highly competitive game, a basketball player can produce 300 W of power. Assuming the efficiency of the player's "engine" is 15% and heat dissipates primarily through the evaporation of perspiration, what mass of perspiration is evaporated per hour?

73. A Carnot engine is to produce 100 J of work per cycle. If 300 J of heat is exhausted to a 27 °C cold reservoir per cycle, what is the change in entropy of the hot reservoir per cycle?

74. A quantity of an ideal gas at an initial pressure of 2.00 atm undergoes an adiabatic expansion to atmospheric pressure. What is the ratio of the final temperature to the initial temperature of the gas?

75. A 100-MW power generating plant has an efficiency of 40%. If water is used to carry off the wasted heat and the temperature of the water is not to increase by more than 10 °C, what mass of water must flow through the plant each second?

76. An ice machine is to convert 10 °C water to 0 °C ice. If the machine has a COP of 2.0 and consumes electrical power at a rate of 1.0 kW, how much ice can it make in 1.0 h?

13 Vibrations and Waves

PHYSICS FACTS

✦ Disturbances set up waves. Soldiers marching across older wooden bridges are told to break step and not march in a periodic cadence. Such a cadence might correspond to a natural frequency of the bridge, resulting in resonance and large oscillations that could damage the bridge and even cause it to collapse. This happened in 1850 in France. About 500 soldiers marching across a suspension bridge over a river caused a resonant vibration that rose to such a level that the bridge collapsed. Over 200 soldiers drowned.

✦ *Tidal waves* are not related to tides. A more appropriate name for them is the Japanese name *tsunami*, which means "harbor wave." The waves are generated by subterranean earthquakes and can race across the ocean at speeds up to 960 km/h, with little surface evidence. When a tsunami reaches the shallow coast, friction slows the wave down, at the same time causing it to roll up into a 5- to 30-m-high wall of water that crashes down on the shoreline.

The chapter-opening photograph depicts what a lot of people probably first think of when hearing the word *wave*. We're all familiar with ocean waves and their smaller relatives, the ripples that form on the surface of a lake or pond when something disturbs the surface. However, the waves that are most important to us, as well as most interesting to physicists, either are invisible or don't look like water waves. Sound, for example, is a wave. Perhaps most surprisingly, light is a wave. In fact, all electromagnetic radiations are waves— radio waves, microwaves, X-rays, and so on. Whenever you peer

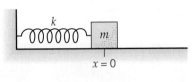

(a) Equilibrium

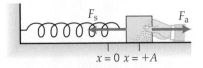

(b) $t = 0$ Just before release

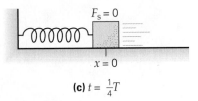

(c) $t = \frac{1}{4}T$

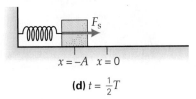

(d) $t = \frac{1}{2}T$

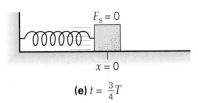

(e) $t = \frac{3}{4}T$

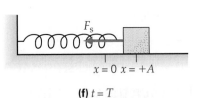

(f) $t = T$

▲ **FIGURE 13.1 Simple harmonic motion (SHM)** When an object on a spring **(a)** is at its equilibrium position ($x = 0$) and then **(b)** is displaced and released, the object undergoes SHM (assuming no frictional losses). The time it takes to complete one cycle is the period of oscillation (T). (Here, F_s is the spring force and F_a is the applied force.) **(c)** At $t = T/4$, the object is back at its equilibrium position; **(d)** at $t = T/2$, it is at $x = -A$. **(e)** During the next half cycle, the motion is to the right; **(f)** at $t = T$, the object is back at its initial ($t = 0$) starting position as in **(b)**.

through a microscope, put on a pair of glasses, or look at a rainbow, you are experiencing wave energy in the form of light. In Section 28.1, you'll learn how even moving particles have wavelike properties. But first we need to look at the basic description of waves.

In general, waves are related to vibrations or oscillations—back-and-forth motion—such as that of a mass on a spring or a swinging pendulum, and fundamental to such motions are restoring forces or torques. In a material medium, the restoring force is provided by intermolecular forces. If a molecule is disturbed, restoring forces from interactions with its neighbors tend to return the molecule to its original position, and it begins to oscillate. In so doing, it affects adjacent molecules, which are in turn set into oscillation, and so on. This is referred to as *propagation*. One might ask, "What is propagated by the molecules in a material?" The answer is *energy*. A single disturbance, which happens when you give the end of a stretched rope a quick shake, gives rise to a *wave pulse*. A continuous, repetitive disturbance gives rise to a continuous propagation of energy called a *wave motion*. But before looking at waves in media, it is helpful to analyze the oscillations of a single mass.

13.1 Simple Harmonic Motion

LEARNING PATH QUESTIONS

➡ What is the type of force necessary for an object to be in simple harmonic motion?
➡ At what position of mass is the total energy of a mass–spring system a maximum?
➡ At what position pf mass is the speed of a mass–spring system a maximum?

The motion of an oscillating object depends on the restoring force that makes the object go back and forth. It is convenient to begin to study such motion by considering the simplest type of force acting along the *x*-axis: a force that is directly proportional to the object's displacement from equilibrium. A common example is the (ideal) spring force, described by **Hooke's law** (Section 5.2),

$$F_s = -kx \quad \text{(Hooke's law)} \tag{13.1}$$

where k is the spring constant representing the stiffness of the spring. The negative sign indicates that the spring force is always opposite to its displacement. That is, the force always tends to *restore* the object to the spring's equilibrium position.

Suppose that an object on a horizontal frictionless surface is connected to a spring as shown in ◄Fig. 13.1. When the object is displaced to one side of its equilibrium position and released, it will move back and forth—that is, it will vibrate, or oscillate. Here, an oscillation or a vibration is clearly a *periodic motion*—a motion that repeats itself again and again along the same path. For linear oscillations, like those of an object attached to a spring, the path may be back and forth or up and down. For the angular oscillation of a pendulum, the path is back and forth along a circular arc.

The motion under the influence of the type of force described by Hooke's law is called **simple harmonic motion (SHM)**. This is because the force is the simplest restoring force and the motion can be described by harmonic functions (sine and cosine functions), as will be seen later in the chapter. The change in position of an object in SHM from its equilibrium position is the object's **displacement**, a vector quantity (Section 2.2). Often, the equilibrium position is chosen to be at the origin, so $x_o = 0$; then the displacement $\Delta x = x - x_o = x$. Note in Fig. 13.1 that the displacement can be either positive or negative, which indicates direction. The maxi-

TABLE 13.1	Terms Used to Describe Simple Harmonic Motion

displacement—the change in position of an object measured from its equilibrium position ($x - x_o = x$ with $x_o = 0$).

amplitude (A)—the magnitude of the maximum displacement, or the maximum distance, of an object from its equilibrium position.

period (T)—the time for one complete cycle of motion.

frequency (f)—the number of cycles per second (in hertz or inverse seconds, where $f = 1/T$).

mum displacements are $+A$ and $-A$ (Fig. 13.1b, d). The magnitude of the maximum displacement, or the maximum distance of an object from its equilibrium position, is called the object's **amplitude (A)**, a scalar quantity that expresses the distance of both extreme displacements from the equilibrium position.

Besides the amplitude, two other important quantities used in describing an oscillation are its period and frequency. The **period (T)** is the time it takes the object to complete one cycle of motion. A cycle is a *complete* round trip, or motion through a complete oscillation. For example, if an object starts at $x = A$ (Fig. 13.1b), then when it returns to $x = A$ (as in Fig. 13.1f), it will have completed one cycle in one period. If an object were initially at $x = 0$ when disturbed, then its second return to this point would mark a cycle. (Why a *second* return?) In either case, the object would travel a distance of $4A$ during one cycle. Can you show this?

The **frequency (f)** is the number of cycles per second. The frequency and the period are inversely proportional, that is,

$$f = \frac{1}{T} \quad \text{(frequency and period)} \tag{13.2}$$

SI unit of frequency: hertz (Hz), or cycles per second (cycles/s or 1/s or s^{-1}).

The inverse relationship is reflected in the units. The period is the number of seconds per cycle, and the frequency is the number of cycles per second. For example, if $T = \frac{1}{2}$ s/cycle, then it completes 2 cycles each second, or $f = 2$ cycles/s.

The standard unit of frequency is the **hertz (Hz)**, which is 1 cycle per second.* From Eq. 13.2, frequency has the unit inverse seconds (1/s, or s^{-1}), since the period is a measure of time. Although a cycle is not really a unit, it is convenient at times to express frequency in cycles per second to help with unit analysis. This is similar to the way the radian (rad) is used in the description of circular motion in Sections 7.1 and 7.2.

The terms used to describe SHM are summarized in ▲Table 13.1.

ENERGY AND SPEED OF A MASS—SPRING SYSTEM IN SHM

Recall from Section 5.4 that the potential energy stored in a spring that is stretched or compressed a distance x from equilibrium (chosen to be $x_o = 0$) is

$$U = \frac{1}{2}kx^2 \quad \text{(potential energy of a deformed spring)} \tag{13.3}$$

The *change* in potential energy of an object oscillating on a spring is related to the work done by the spring force. An object with mass m oscillating on a spring also has kinetic energy. The kinetic and potential energies together give the total mechanical energy E of the system:

$$E = K + U = \frac{1}{2}mv^2 + \frac{1}{2}kx^2 \quad \text{(total energy)} \tag{13.4}$$

*The unit is named for Heinrich Hertz (1857–1894), a German physicist and early investigator of electromagnetic waves.

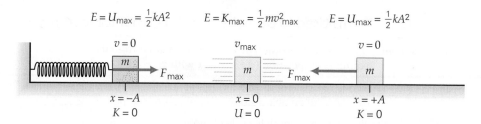

▲ **FIGURE 13.2 Oscillations and energy** For a mass oscillating in SHM on a spring (on a frictionless surface), the total energy at the amplitude positions ($\pm A$) is all potential energy (U_{max}), and $E = \frac{1}{2}kA^2$, which is the total energy of the system. At the center position ($x = 0$), the total energy is all kinetic energy ($E = \frac{1}{2}mv_{max}^2$, where m is the mass of the block). How is the total energy divided at locations somewhere between $x = 0$ and $x = \pm A$?

When the object is at one of its maximum displacements, $x = +A$ or $-A$, it is instantaneously at rest, $v = 0$ (▲Fig. 13.2). Thus, all the energy is in the form of potential energy (U_{max}) at this location; that is,

$$E = \tfrac{1}{2}m(0)^2 + \tfrac{1}{2}k(\pm A)^2 = \tfrac{1}{2}kA^2$$

or

$$E = \tfrac{1}{2}kA^2 \quad \begin{array}{l}\textit{(total energy of an object}\\ \textit{in SHM on a spring)}\end{array} \qquad (13.5)$$

This outcome is a general result for SHM:

> The total energy of an object in simple harmonic motion is directly proportional to the square of the amplitude.

Equation 13.5 allows us to express the velocity of an object oscillating on a spring as a function of position:

$$E = K + U \quad \text{or} \quad \tfrac{1}{2}kA^2 = \tfrac{1}{2}mv^2 + \tfrac{1}{2}kx^2$$

Solving for v^2 and taking the square root:

$$v = \pm\sqrt{\frac{k}{m}(A^2 - x^2)} \quad \textit{(velocity of an object in SHM)} \qquad (13.6)$$

where the positive and negative signs indicate the direction of the velocity. Note that at $x = \pm A$, the velocity is zero, since the object is instantaneously at rest at its maximum displacement from equilibrium.

Note also that when the oscillating object passes through its equilibrium position ($x = 0$), its potential energy is zero. At that instant, the energy is all kinetic, and the object is traveling at its maximum speed v_{max}. The expression for the energy in this case is

$$E = \tfrac{1}{2}kA^2 = \tfrac{1}{2}mv_{max}^2$$

and

$$v_{max} = \sqrt{\frac{k}{m}}\,A \quad \begin{array}{l}\textit{(maximum speed}\\ \textit{of mass on a spring)}\end{array} \qquad (13.7)$$

In the next Example, as well as in the accompanying Learn by Drawing 13.1, Oscillating in a Parabolic Potential Well, you can visualize the continuous trade-off between kinetic and potential energy.

EXAMPLE 13.1 | A Block and a Spring: Simple Harmonic Motion

A block with a mass of 0.50 kg sitting on a frictionless surface is connected to a light spring that has a spring constant of 180 N/m (see Fig. 13.1). If the block is displaced 15 cm from its equilibrium position and released, what are (a) the total energy of the system, (b) the speed of the block when it is 10 cm from its equilibrium position, and (c) the maximum speed of the block?

THINKING IT THROUGH. The total energy depends on the spring constant (k) and the amplitude (A), which are given. At $x = 10$ cm, the speed should be less than the maximum speed. (Why?)

SOLUTION. First the given data and what is to be found are listed, as usual. The initial position corresponds to the amplitude. (Why?) The speeds can be calculated with Eq. 13.6.

Given: $m = 0.50$ kg
$k = 180$ N/m
$A = 15$ cm $= 0.15$ m
$x = \pm 10$ cm $= \pm 0.10$ m

Find: (a) E (total energy)
(b) v (speed)
(c) v_{max}

(a) The total energy is given by Eq. 13.5:

$$E = \tfrac{1}{2}kA^2 = \tfrac{1}{2}(180 \text{ N/m})(0.15 \text{ m})^2 = 2.0 \text{ J}$$

(b) The instantaneous speed of the block at a distance 10 cm from the equilibrium position is given by Eq. 13.6. As you can see, whether $x = +0.10$ m or $x = -0.10$ m makes no difference because of x^2 in Eq. 13.6.

$$v = \sqrt{\frac{k}{m}(A^2 - x^2)} = \sqrt{\frac{180 \text{ N/m}}{0.50 \text{ kg}}[(0.15 \text{ m})^2 - (\pm 0.10 \text{ m})^2]} = \sqrt{4.5 \text{ m}^2/\text{s}^2} = 2.1 \text{ m/s}$$

(c) The maximum speed occurs at $x = 0$, so Eq. 13.2 becomes

$$v_{max} = \sqrt{\frac{180 \text{ N/m}}{0.50 \text{ kg}}[(0.15 \text{ m})^2 - (0)^2]} = 2.8 \text{ m/s}$$

You could also use Eq. 13.7 directly to calculate v_{max}.

FOLLOW-UP EXERCISE. What is the magnitude of the acceleration when $x = \pm 0.10$ m? What is the maximum acceleration? *(Answers to all Follow-Up Exercises are given in Appendix VI at the back of the book.)*

DID YOU LEARN?
➥ The type of force in SHM has to be a restoring force, such as that described by Hooke's law. The direction of the force has to be toward the equilibrium position at all times.
➥ The total energy of a mass–spring system is constant, so it is the same at any position. However, the kinetic energy and potential energy that make up the total energy do vary at different positions.
➥ The speed of a mass–spring system is at a maximum at the equilibrium position, where the potential energy is zero. The energy is all kinetic; therefore, the speed is at a maximum.

13.2 Equations of Motion

LEARNING PATH QUESTIONS
➥ What are the two possible trigonometric functions of the equations of motion that could be used for simple harmonic motion?
➥ Does the oscillation period or frequency of a mass–spring system or a simple pendulum depend on the amplitude of vibration?
➥ How do the motions of two objects in SHM compare if they have a phase difference of 180°?

The equation that gives an object's position as a function of time is referred to as the **equation of motion**. For example, the equation of motion with a constant linear acceleration is $x = x_0 + v_0 t + \tfrac{1}{2}at^2$, where v_0 is the initial velocity (Section 2.4). However, the acceleration is not constant in simple harmonic motion, so the kinematic equations of Section 2.4 do not apply to this case.

The equation of motion for an object in SHM can be found from a relationship between simple harmonic and uniform circular motions. SHM can be simulated

oscillating in a parabolic potential well

A way to visualize the conservation of energy in simple harmonic motion is shown in Fig. 1. The potential energy of a mass–spring system can be sketched on a plot of energy (E) versus position (x). Since $U = \frac{1}{2}kx^2 \propto x^2$, the graph is a *parabola*.

In the absence of nonconservative forces, the total energy of the system, E, is constant. But E is the sum of the kinetic and potential energies. During the oscillations, there is a continuous trade-off between the two types of energies, but their sum remains constant. Mathematically, this relationship is written as $E = K + U$. In Fig. 2, the potential energy U (indicated by a blue arrow) is represented by the vertical distance from the x-axis.

Since E is constant and independent of x, it is plotted as a horizontal line (shown in green). The kinetic energy is the part of the total energy that is *not* potential energy; that is, $K = E - U$. It can be graphically interpreted (purple arrow) as the vertical distance between the potential energy parabola and the horizontal green total energy line. As the object oscillates on the x-axis, the energy trade-offs can be visualized as the lengths of the two arrows change.

A general location, x_1, is shown in Fig. 2. Neither the kinetic energy nor the potential energy is at its maximum value of E there. These maximum values occur instead at $x = 0$ and $x = \pm A$, respectively. The motion cannot exceed $x = \pm A$, because that would imply a negative kinetic energy, which is physically impossible. (Why?) The amplitude positions are sometimes called the *endpoints* of the motion, because they are the locations where the speed is instantaneously zero and the object reverses direction.

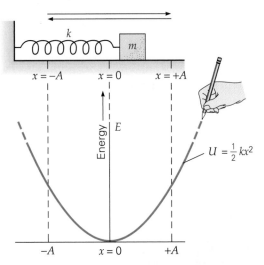

FIGURE 1   The potential energy "well" of a spring–mass system The potential energy of a spring that is stretched or compressed from its equilibrium position ($x = 0$) is a parabola, since $U \propto x^2$. At $x = \pm A$, all of the system's energy is potential.

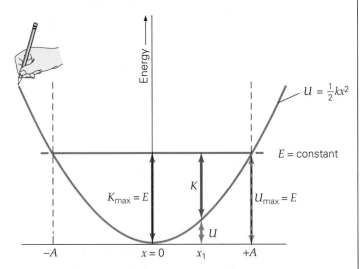

FIGURE 2   Energy transfers as the spring–mass system oscillates The vertical distance from the x-axis to the parabola is the system's potential energy. The remainder—the vertical distance between the parabola and the horizontal line representing the system's constant total energy E—is the system's kinetic energy (K).

by a component of uniform circular motion, as illustrated in ▶Fig. 13.3. As the object moves in uniform circular motion (with constant angular speed ω) in a vertical plane, its shadow moves back and forth vertically, following the same path as the object on the spring, which is in simple harmonic motion. Since the shadow and the object have the same position at any time, it follows that the equation of motion for the shadow of the object in circular motion is the same as the equation of motion for the oscillating object on the spring.

From the reference circle in Fig. 13.3b, the y-coordinate (position) of the object is given by

$$y = A \sin \theta$$

But the object moves with a constant angular velocity of magnitude ω. In terms of the angular distance θ, assuming that $\theta_0 = 0$ at $t = 0$, then $\theta = \omega t$, so

$$y = A \sin \omega t \qquad \begin{array}{l}\textit{(SHM for } y_0 = 0.\\ \textit{initial upward motion)}\end{array} \qquad (13.8)$$

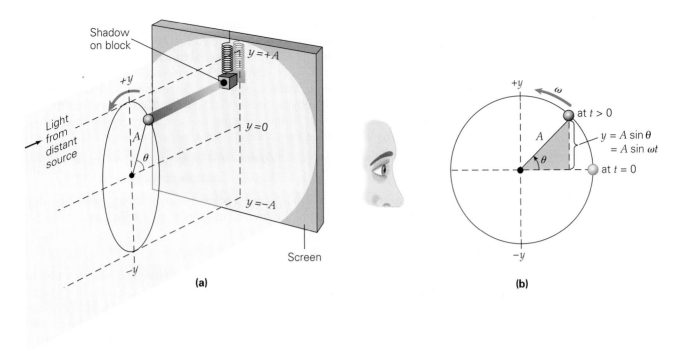

▲ **FIGURE 13.3 Reference circle for vertical motion** **(a)** The shadow of an object in uniform circular motion has the same vertical motion as an object oscillating on a spring in simple harmonic motion. **(b)** The motion can be described by $y = A \sin \theta = A \sin \omega t$ (assuming that $y = 0$ at $t = 0$).

Note that as t increases from zero, y increases in the positive direction, so the equation describes initial upward motion.

With Eq. 13.8 as the equation of motion, the mass *must* be initially at $y_\mathrm{o} = 0$. But what if the mass on the spring were initially at the amplitude position $+A$? In that case, the sine equation would not describe the motion, because it does *not* describe the *initial condition*—that is, $y_\mathrm{o} = +A$ at $t_\mathrm{o} = 0$. So another equation of motion is needed, and $y = A \cos \omega t$ applies. By this equation, at $t_\mathrm{o} = 0$, the mass is at $y_\mathrm{o} = A \cos \omega t = A \cos \omega(0) = +A$, and the cosine equation correctly describes the initial conditions (▼ Fig. 13.4):

$$y = A \cos \omega t \quad \begin{array}{l} \text{(initial downward motion} \\ \text{with } y_\mathrm{o} = +A) \end{array} \tag{13.9}$$

Here, the initial motion is downward, because, for times shortly after $t_\mathrm{o} = 0$, the value of y decreases. If the amplitude were $-A$, the mass would initially be at the bottom and the initial motion would be upward.

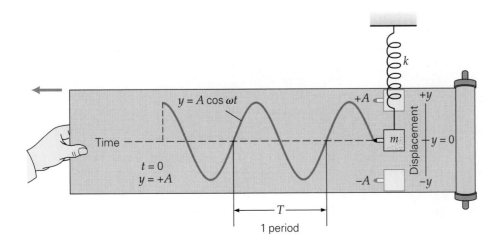

◀ **FIGURE 13.4 Sinusoidal equation of motion** As time passes, the oscillating object traces out a sinusoidal curve on the moving paper. In this case, $y = A \cos \omega t$, because the object's initial displacement is $y_\mathrm{o} = +A$.

Thus, the equation of motion for an oscillating object may be either a sine or a cosine function. Both of these functions are referred to as being *sinusoidal*. That is, simple harmonic motion is described by a sinusoidal function of time.

The angular speed ω (in rad/s) of the *reference circle object* (Fig. 13.3) is called the *angular frequency* of the oscillating object, since $\omega = 2\pi f$, where f is the frequency of revolution or rotation of the object (Section 7.2). Figure 13.3 shows that the frequency of the "orbiting" object is the same as the frequency of the oscillating object on the spring. Thus, using $f = 1/T$, Eq. 13.8 may be written as

$$y = A \sin(2\pi ft) = A \sin\left(\frac{2\pi t}{T}\right) \quad \begin{array}{l}\textit{(SHM for } y_\mathrm{o} = 0 \\ \textit{initial upward motion)}\end{array} \tag{13.10}$$

Note that this equation is for initial upward motion, because after $t_\mathrm{o} = 0$, the value of y increases positively. For initial downward motion, $y = -A \sin(2\pi ft)$.

INITIAL CONDITIONS AND PHASE

You may be wondering how to decide whether to use a sine or cosine function to describe a particular case of simple harmonic motion. In general, the form of the function is determined by the initial displacement and velocity of the object: the *initial conditions* of the system. These initial conditions are the values of the displacement and velocity at $t = 0$; taken together, they tell how the system is initially set into motion.

Let's look at four special cases. If an object in vertical SHM has an initial displacement of $y = 0$ at $t = 0$ and moves initially upward, the equation of motion is $y = A \sin \omega t$ (▾Fig. 13.5a). Note that $y = A \cos \omega t$ does not satisfy the initial condition, because $y_\mathrm{o} = A \cos \omega t = A \cos \omega(0) = A$, since $\cos 0 = 1$.

▶ **FIGURE 13.5 Initial conditions and equations of motion** The initial conditions (y_o and t_o) determine the form of the equation of motion—for the cases shown here, either a sine or a cosine. For $t_\mathrm{o} = 0$, the initial displacements are **(a)** $y_\mathrm{o} = 0$, **(b)** $y_\mathrm{o} = +A$, **(c)** $y_\mathrm{o} = 0$, and **(d)** $y_\mathrm{o} = -A$. The equations of motion must match the initial conditions. (See book for description.)

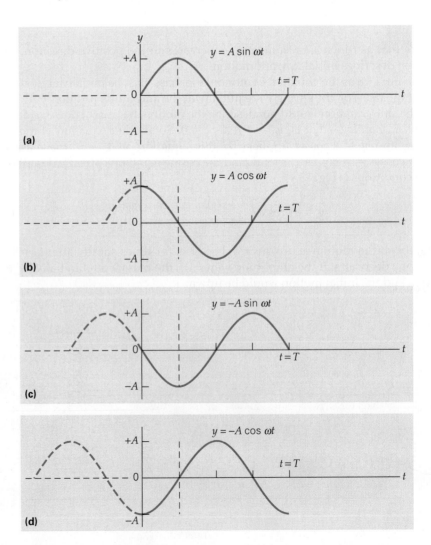

Suppose that the object is initially released ($t = 0$) from its positive amplitude position ($+A$), as in the case of the object on a spring shown in Fig. 13.4. Here, the equation of motion is $y = A \cos \omega t$ (Fig. 13.5b). This expression satisfies the initial condition: $y_0 = A \cos \omega(0) = A$.

The other two cases are (1) $y = 0$ at $t = 0$, with motion initially downward (for an object on a spring) or in the negative direction (for horizontal SHM), and (2) $y = -A$ at $t = 0$, meaning that the object is initially at its negative amplitude position. These motions are described by $y = -A \sin \omega t$ and $y = -A \cos \omega t$, respectively, as illustrated in Figs. 13.5c and d.

Only these four initial conditions will be considered in our study. Should y_0 have a value other than 0 or $\pm A$, the equation of motion is somewhat complicated. Note in Fig. 13.5 that if the curves are extended in the negative direction of the horizontal axis (dashed purple lines), they all have the same shape, but have been "shifted," so to speak. In (a) and (b), one curve is ahead of the other by 90°, or $\frac{1}{4}$ cycle. That is, the two curves are shifted by a quarter cycle with respect to one another. The oscillations are then said to have a *phase difference* of 90°. In (a) and (c), the curves are shifted 180° and are 180° out of phase. (Note in this case that the oscillations are opposite: When one mass is going up, the other is going down.) What about the oscillations in (a) and (d)?

DEMONSTRATION 4 | ## Simple Harmonic Motion (SHM) and Sinusoidal Oscillation

A salt-filled funnel oscillates, suspended from two strings.

A demonstration to show that SHM can be represented by a sinusoidal function. A "graph" of the function is generated with an analogue of a strip chart recorder.

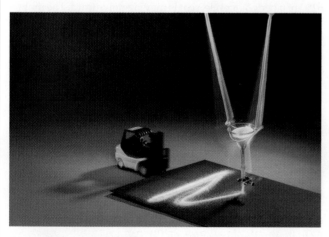

Away we go. The salt falls on a black-painted poster board that will be pulled in a direction perpendicular to the plane of the funnel's oscillation.

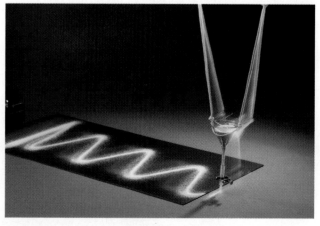

The salt trail traces out a plot of displacement versus time, or $y = A \sin(\omega t + \delta)$. Note that in this case the phase constant is about $\delta = 90°$ and $y = A \cos \omega t$. (Why?)

A figure with a 360° (or 0°) phase shift is not shown, because this would be the same as that in (a). When two objects in SHM have the same equation of motion, they are said to be oscillating *in phase*, which means that they are oscillating together with identical motions. Objects with a 180° phase shift or difference are said to be *completely out of phase* and will always be going in opposite directions and be at opposite amplitudes at the same time.

Hence, we may write in general,

$$y = \pm A \sin \omega t = \pm A \sin(2\pi ft) = \pm A \sin\left(\frac{2\pi t}{T}\right)$$

(+ for initial motion upward with $y_o = 0$; − for initial motion downward with $y_o = 0$) (13.8a)

By a similar development, Eq. 13.9 has the general form

$$y = \pm A \cos \omega t = \pm A \cos(2\pi ft) = \pm A \cos\left(\frac{2\pi t}{T}\right)$$

(+ for initial motion downward with $y_o = +A$; − for initial motion upward with $y_o = -A$) (13.9b)

The next Example demonstrates the usage of the equation of motion for SHM.

EXAMPLE 13.2 | **An Oscillating Mass: Applying the Equation of Motion**

A mass on a spring oscillates vertically with an amplitude of 15 cm, and a frequency of 0.20 Hz. At $t = 0$, it is at $y_o = 0$ and it moves initially in the upward direction. (a) Write the equation of motion for this oscillation. (b) What are the position and direction of motion of the mass at $t = 3.1$ s? (c) How many oscillations (cycles) does the mass make in a time of 12 s?

THINKING IT THROUGH. For part (a), the equation of motion has to be a sine function because $y_o = 0$. (Why not cosine?)

Since the mass is initially moving upward, the equation of motion should be of the form of Eq. 13.8 or Eq. 13.10. Part (b) is then the application of the equation of motion. In part (c), the number of oscillations means the number of cycles, and recall that frequency is sometimes expressed in cycles per second. Hence, multiplying the frequency by the time will give the number of cycles or oscillations.

SOLUTION.

Given: $A = 15$ cm $= 0.15$ m   **Find:** (a) equation of motion
  $f = 0.20$ Hz   (b) y (position and direction of motion)
  (b) $t = 3.1$ s   (c) n (number of oscillations or cycles)
  (c) $t = 12$ s

(a) First, since the frequency f is given, it is convenient to use the equation of motion in the form $y = A \sin 2\pi ft$ (Eq. 13.10). As can be seen from the equation, at $t_o = 0$, $y_o = 0$, so initially the mass is at the zero (equilibrium) position. At $t > 0$, $y > 0$, so it is moving upward. Using the known quantities, the equation of motion is then

$$y = (0.15 \text{ m}) \sin[2\pi(0.20 \text{ Hz})t] = (0.15 \text{ m}) \sin[(0.4\pi \text{ rad/s})t]$$

(b) At $t = 3.1$ s,

$$y = (0.15 \text{ m}) \sin[(0.4\pi \text{ rad/s})(3.1 \text{ s})] = (0.15 \text{ m}) \sin(3.9 \text{ rad})$$
$$= -0.10 \text{ m}$$

So the mass is at $y = -0.10$ m at $t = 3.1$ s. But what is its direction of motion? Let's look at the period (T) and see what part of its cycle the mass is in. By Eq. 13.2,

$$T = \frac{1}{f} = \frac{1}{0.20 \text{ Hz}} = 5.0 \text{ s}$$

In $t = 3.1$ s, the mass has gone through 3.1 s/5.0 s = 0.62, or 62%, of a period or cycle, so it is moving downward. The motion is up ($\frac{1}{4}$ cycle) and back ($\frac{1}{4}$ cycle) to $y_o = 0$ in $\frac{1}{2}$, or 50%, of the cycle, and therefore downward during the next $\frac{1}{4}$ cycle.]

(c) The number of oscillations (cycles) is equal to the product of the frequency (cycles/s) and the elapsed time (s), both of which are given:

$$n = ft = (0.20 \text{ cycles/s})(12 \text{ s}) = 2.4 \text{ cycles}$$

or with $f = 1/T$,

$$n = \frac{t}{T} = \frac{12 \text{ s}}{5.0 \text{ s}} = 2.4 \text{ cycles}$$

(Note that *cycle* is not a unit and is used only for convenience.)

Thus, the mass has gone through two complete cycles and 0.4 of another, which means that it is on its way back to $y_o = 0$ from its amplitude position of $+A$. (Why?)

FOLLOW-UP EXERCISE. Find what is asked for in this Example at times (1) $t = 4.5$ s and (2) $t = 7.5$ s.

On what does the period of oscillation depend? Let us compute the period of the spring–mass system by comparing it to the reference circle in uniform circular motion. (See Fig. 13.3.) Note that the time for the object in the reference circle to make one complete "orbit" is exactly the time it takes for the oscillating object to make one complete cycle. Thus, all we need is the time for one orbit around the reference circle, and we have the period of oscillation. Because the object "orbiting" the reference circle is in uniform circular motion at a constant speed equal to the maximum speed of oscillation v_{max}, the object travels a distance of one circumference in one period. Then $t = d/v$, where $t = T$, the circumference is d, and v is v_{max} given by Eq. 13.7; that is,

$$T = \frac{d}{v_{max}} = \frac{2\pi A}{A\sqrt{k/m}}$$

or

$$T = 2\pi\sqrt{\frac{m}{k}} \quad \text{\textit{(period of object oscillating on a spring)}} \tag{13.11}$$

Because the amplitudes canceled out in Eq. 13.11, *the period and frequency are independent of the amplitude of the simple harmonic motion.*

From Eq. 13.11 it can be seen that the greater the mass, the longer the period and the greater the spring constant (or the stiffer the spring), the shorter the period. It is the *ratio* of mass to stiffness that determines the period. Thus, an increase in mass can be offset by using a stiffer spring.

Since $f = 1/T$,

$$f = \frac{1}{2\pi}\sqrt{\frac{k}{m}} \quad \text{\textit{(frequency of mass oscillating on a spring)}} \tag{13.12}$$

Thus, the greater the spring constant (the stiffer the spring), the more frequently the system vibrates, as expected.

Also, note that since $\omega = 2\pi f$, we may write

$$\omega = \sqrt{\frac{k}{m}} \quad \text{\textit{(angular frequency of mass oscillating on a spring)}} \tag{13.13}$$

As another example, a simple pendulum (a small, heavy object on a string) will undergo simple harmonic motion for small angles of oscillation. The period of a simple pendulum oscillating through a small angle $\theta < 10°$ is given, to a good approximation, by

$$T = 2\pi\sqrt{\frac{L}{g}} \quad \text{\textit{(period of a simple pendulum)}} \tag{13.14}$$

where L is the length of the pendulum and g is the acceleration due to gravity. A pendulum-driven clock that is not properly rewound and is running down would still keep correct time, because the period would remain unchanged as the amplitude decreased. As shown by Eq. 13.14, the period is independent of amplitude.

An important difference between the period of the mass–spring system and that of the pendulum is that the latter is independent of the mass of the pendulum bob. (See Eq. 13.11 and Eq. 13.14.) Can you explain why? Think about what supplies the restoring force for the pendulum's oscillations. It is the gravitational force. Hence, the acceleration (along with the velocity and period) is expected to be independent of mass. That is, the gravitational force automatically provides the same acceleration to different bob masses on pendulums with the same length. Similar effects occur in free fall (Section 2.5) and with blocks sliding and cylinders rolling down inclines (Sections 4.5 and 8.4, respectively).

Let's take a look at two Examples related to the frequency and period of SHM.

EXAMPLE 13.3 | Fun with a Pothole: Frequency and Spring Constant

A typical family automobile has a mass of 1500 kg. Assume that the car has one spring on each wheel, that the springs are identical, and that the mass is equally distributed over the four springs. (a) What is the spring constant of each spring if the empty car bounces up and down 1.2 times each second when hit a pothole? (b) What will be the car's oscillation frequency when four 75-kg people are in the car?

THINKING IT THROUGH. (a) The frequency is given by Eq. 13.12 and the spring constant can then be found. Keep in mind that each spring will carry 1/4 of the total mass of the car. (b) Once the spring constant is found, Equation 13.1 can be used again to find the new frequency. The spring constant is the same with or without the people in the car.

SOLUTION. The data are listed below, where m represents the mass on each individual spring.

Given: (a) $m = \dfrac{1500 \text{ kg}}{4} = 375 \text{ kg}$

$f_1 = 1.2 \text{ Hz}$

(b) $m = \dfrac{1500 \text{ kg} + 4(75 \text{ kg})}{4} = 450 \text{ kg}$

Find: (a) k (spring constant)

(b) f (new frequency)

(a) Using Eq. 13.12 to solve for the spring constant, k.

$$f = \frac{1}{2\pi}\sqrt{\frac{k}{m}} \quad \text{so} \quad k = 4\pi^2 f^2 m = 4\pi^2 (1.2 \text{ Hz})^2 (375 \text{ kg}) = 2.13 \times 10^4 \text{ N/m}$$

(b) The new frequency is found again from Eq. 13.12.

$$f = \frac{1}{2\pi}\sqrt{\frac{k}{m}} = \frac{1}{2\pi}\sqrt{\frac{2.13 \times 10^4 \text{ N/m}}{450 \text{ kg}}} = 1.1 \text{ Hz}$$

FOLLOW-UP EXERCISE. Research has shown that the human body feels most comfortable if the oscillation frequency of a car is 1.0 Hz. What spring constant would you use for a half-loaded car (two 75-kg people)?

EXAMPLE 13.4 | Fun with a Pendulum: Frequency and Period

A helpful older brother takes his sister to play on the swings in the park. He pushes her from behind on each return. Assuming that the swing behaves as a simple pendulum with a length of 2.50 m, (a) what would be the frequency of the oscillations, and (b) what would be the interval between the brother's pushes?

THINKING IT THROUGH. (a) The period is given by Eq. 13.14, and the frequency and period are inversely related: $f = 1/T$. (b) Since the brother pushes from one side on each return, he must push once every cycle that is completed, so the time between his pushes is equal to the swing's period.

SOLUTION.

Given: $L = 2.50 \text{ m}$

Find: (a) f (frequency)
(b) T (period)

(a) We can take the reciprocal of Eq. 13.14 to solve directly for the frequency:

$$f = \frac{1}{T} = \frac{1}{2\pi}\sqrt{\frac{g}{L}} = \frac{1}{2\pi}\sqrt{\frac{9.80 \text{ m/s}^2}{2.50 \text{ m}}} = 0.315 \text{ Hz}$$

(b) The period is then found from the frequency:

$$T = \frac{1}{f} = \frac{1}{0.315 \text{ Hz}} = 3.17 \text{ s}$$

The brother must push every 3.17 s to maintain a steady swing (and to keep his sister from complaining).

VELOCITY AND ACCELERATION IN SHM

Expressions for the velocity and acceleration of an object in SHM can also be obtained. Using advanced mathematics, one can show that $v = \Delta y/\Delta t = \Delta(A \sin \omega t)/\Delta t$ in the limit as Δt goes to zero gives the following expression for the instantaneous velocity:

$$v = \omega A \cos \omega t \quad \begin{array}{l}\text{(vertical velocity if } v_o \text{ is upward at} \\ t_o = 0, y_o = 0) \end{array} \quad (13.15)$$

The acceleration can be found by using Newton's second law with the spring force $F_s = -ky$:

$$a = \frac{F_s}{m} = \frac{-ky}{m} = -\frac{k}{m}A \sin \omega t$$

Since $\omega = \sqrt{k/m}$,

$$a = -\omega^2 A \sin \omega t = -\omega^2 y \quad \begin{array}{l}\text{(vertical acceleration if } v_o \text{ is} \\ \text{upward at } t_o = 0, y_o = 0) \end{array} \quad (13.16)$$

Note that the functions for the velocity and acceleration are out of phase with that for the displacement. Since the velocity is 90° out of phase with the displacement, the speed is greatest when $\cos \omega t = \pm 1$ at $y = 0$, that is, when the oscillating object is passing through its equilibrium position. The acceleration is 180° out of phase with the displacement (as indicated by the negative sign on the right-hand side of Equation 13.16). Therefore, the magnitude of the acceleration is a maximum when $\sin \omega t = \pm 1$ at $y = \pm A$, that is, when the displacement is a maximum, or when the object is at an amplitude position. At any position except the equilibrium position, the directional sign of the acceleration is opposite that of the displacement, as it should be for an acceleration resulting from a restoring force. At the equilibrium position, both the displacement and acceleration are zero. (Can you see why?)

Note also that the acceleration in SHM is not constant with time. Hence, the kinematic equations for acceleration (Chapter 2) *cannot* be used, since they describe constant acceleration.

DAMPED HARMONIC MOTION

Simple harmonic motion with constant amplitude implies that there are no losses of energy, but in practical applications there are always some frictional losses. Therefore, to maintain a constant amplitude motion, energy must be added to a system by some external driving force, such as someone pushing a swing. Without a driving force, the amplitude and the energy of an oscillator decrease with time, giving rise to **damped harmonic motion** (▼Fig. 13.6a). The time required for the oscillations to cease, or damp out, depends on the magnitude and type of the damping force (such as air resistance).

In many applications involving continuous periodic motion, damping is unwanted and necessitates an energy input. However, in some instances, damping is desirable. For example, the dial in a spring-operated bathroom scale oscillates briefly before stopping at a weight reading. If not properly damped, these oscillations would continue for some time, and you would have to wait before you could read your weight. Shock absorbers provide damping in the suspension systems of

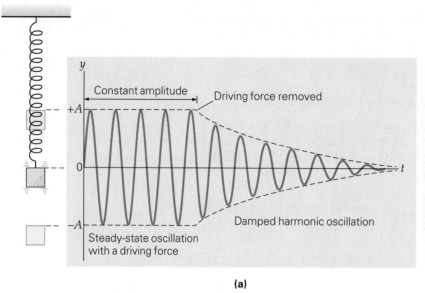

(a)   (b)

▲ **FIGURE 13.6** **Damped harmonic motion** **(a)** When a driving force adds energy to a system in an amount equal to the energy losses of the system, the oscillation is steady with a constant amplitude. When the driving force is removed, the oscillations decay (that is, they are damped), and the amplitude decreases nonlinearly with time. **(b)** In some applications, damping is desirable and even promoted, as with shock absorbers in automobile suspension systems. Otherwise, the passengers would be in for a bouncy ride.

automobiles (Fig. 13.6b; also see Fig. 9.9b). Without "shock absorbers" to dissipate energy after hitting a bump, the ride would be bouncy. In California, many new buildings incorporate damping mechanisms (giant shock absorbers) to dampen their oscillatory motion after they are set in motion by earthquake waves.

DID YOU LEARN?
➡ The functions of the equations of motion for simple harmonic motion are sinusoidal (harmonic), that is, they are either sine or cosine functions.
➡ The period or frequency of a mass–spring system or a simple pendulum does not depend on the amplitude of vibration.
➡ With a phase difference of 180°, the oscillations of two objects in SHM are out of phase. That is, when one object is at a maximum, the other is at a minimum and vice versa.

13.3 Wave Motion

LEARNING PATH QUESTIONS
➡ What are the most common four parameters that describe a wave?
➡ What is meant by a transverse wave?
➡ What is meant by a longitudinal wave?

The world is full of waves of various types; some examples are water waves, sound waves, waves generated by earthquakes, and light waves. All waves result from a disturbance, the source of the wave. In this chapter, the focus will be mechanical waves—those that are propagated in some medium. (Light waves, which do not require a propagating medium, will be considered in more detail in later chapters.)

When a medium is disturbed, energy is imparted to it. The addition of the energy sets some of the particles in the medium vibrating. Because the particles are linked by intermolecular forces, the oscillation of each particle affects that of its neighbors. The added energy propagates, or spreads, by means of interactions among the particles of the medium. An analogy to this process is shown in ◀Fig. 13.7, where the "particles" are dominoes. As each domino falls, it topples the one next to it. Thus, energy is transferred from domino to domino, and the disturbance propagates through the medium—the energy travels, not the medium.

▲ **FIGURE 13.7** **Energy transfer** The propagation of a disturbance, or a transfer of energy through space, is seen in a row of falling dominoes.

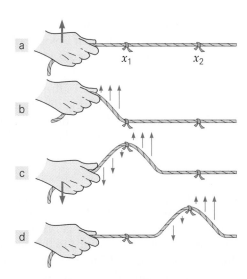

In this case, there is no restoring force between the dominoes, so they do not oscillate. Therefore, the disturbance moves in space, but it does not repeat itself in time at any one location.

Similarly, if the end of a stretched rope is given a quick shake, the disturbance transfers energy from the hand to the rope, as illustrated in ▲Fig. 13.8. The forces acting between the "particles" in the rope cause them to move in response to the motion of the hand, and a *wave pulse* travels down the rope. Each "particle" goes up and then back down as the pulse passes by. This motion of individual particles and the propagation of the wave pulse as a whole can be observed by tying pieces of ribbon onto the rope (at x_1 and x_2 in the figure). As the disturbance passes point x_1, the ribbon rises and falls, as do the rope's "particles." Later, the same thing happens to the ribbon at x_2, which indicates that the energy disturbance is propagating or traveling along the rope.

In a continuous material medium, particles interact with their neighbors, and restoring forces cause them to oscillate when they are disturbed. Thus, a disturbance not only propagates through space, but also may be repeated over and over in time at each position. Such a regular, rhythmic disturbance in both time and space is called a **wave**, and the transfer of energy is said to take place by means of **wave motion**.

A continuous wave motion, or *periodic wave*, requires a disturbance from an oscillating source (▼Fig. 13.9). In this case, the particles move up and down continuously. If the driving source is such that a constant amplitude is maintained (the source oscillates in simple harmonic motion), the resulting particle motion is also simple harmonic.

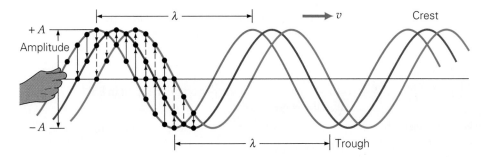

▲ FIGURE 13.9 **Periodic wave** A continuous harmonic disturbance can set up a sinusoidal wave in a stretched rope, and the wave travels down the rope with wave speed v. Note that the "particles" in the rope oscillate vertically in simple harmonic motion. The distance between two successive points that are in phase (for example, at two crests) on the waveform is the wavelength λ of the wave. Can you tell how much time has elapsed, as a fraction of the period T, between the first (red) and last (blue) waves?

Such periodic wave motion will have sinusoidal forms (sine or cosine) in both time and space. Being *sinusoidal in space* means that if you took a photograph of the wave at any instant ("freezing" it in time), you would see a sinusoidal waveform (such as one of the curves in Fig. 13.9). However, if you looked at a single point in space as a wave passed by, you would see a particle of the medium oscillating up and down *sinusoidally with time,* like the mass on a spring discussed in Section 13.2. (For example, imagine looking through a thin slit at a fixed location on the moving paper in Fig. 13.4. The wave trace would be seen rising and falling like a particle in SHM.)

WAVE CHARACTERISTICS

Specific quantities are used to describe sinusoidal waves. As with a particle in simple harmonic motion, the **amplitude (*A*)** of a wave is the magnitude of the maximum displacement, or the maximum distance, from the particle's equilibrium position (Fig. 13.9). This quantity corresponds to the height of a wave crest or the depth of a trough. Recall from Section 13.2 that, in SHM, the total energy of the oscillator is proportional to the square of the amplitude. Similarly, the energy *transported* by a wave is proportional to the square of its amplitude ($E \propto A^2$). Note the difference, though: A wave is one way of *transmitting* energy through space, whereas an oscillator's energy is localized in space.

For a periodic wave, the distance between two successive crests (or troughs) is called the **wavelength (*λ*)** (see Fig. 13.9). Actually, it is the distance between any two successive parts of the wave that are in phase (that is, that are at identical points on the waveform). The crest and trough positions are usually used for convenience.

The **frequency (*f*)** of a periodic wave is the number of waves per second—that is, the number of complete waveforms, or wavelengths, that pass by a given point during each second. The frequency of the wave is the same as the frequency of the SHM source that created it.

A periodic wave is said to possess a **period (*T*)**. The period $T = 1/f$ is the time for one complete waveform (a wavelength) to pass by a given point. Since a wave moves, it also has a **wave speed (*v*)** (or velocity if the wave's direction is specified). Any particular point on the wave (such as a crest) travels a distance of one wavelength λ in a time of one period *T*. Then, since $v = d/t$ and $f = 1/T$,

$$v = \frac{\lambda}{T} = \lambda f \quad \text{(wave speed)} \tag{13.17}$$

Note that the dimensions of *v* are correct (length/time). In general, the wave speed depends on the nature of the medium, in addition to the source frequency *f*.

EXAMPLE 13.5 | ## Dock of the Bay: Finding Wave Speed

A person on a pier observes a set of incoming water waves that have a sinusoidal form with a distance of 5.6 m between the crests. If a wave laps against the pier every 2.0 s, what are (a) the frequency and (b) the speed of the waves?

THINKING IT THROUGH. Since the period and wavelength are known, the definition of frequency and Eq. 13.17 for wave speed can be used.

SOLUTION. The distance between crests is the wavelength, so we have the following information:

Given: $\lambda = 5.6$ m   *Find:* (a) *f* (frequency)
         $T = 2.0$ s   (b) *v* (wave speed)

(a) The lapping indicates the arrival of a wave crest; hence, 2.0 s is the wave period—the time it takes to travel one wavelength (the crest-to-crest distance). Then

(b) The frequency or the period can be used in Eq. 13.17 to find the wave speed:

$$v = \lambda f = (5.6 \text{ m})(0.50 \text{ s}^{-1}) = 2.8 \text{ m/s}$$

Alternatively,

$$v = \frac{\lambda}{T} = \frac{5.6 \text{ m}}{2.0 \text{ s}} = 2.8 \text{ m/s}$$

FOLLOW-UP EXERCISE. On another day, the person measures the speed of sinusoidal water waves at 2.5 m/s. (a) How far does a wave crest travel in 4.0 s? (b) If the distance between successive crests is 6.3 m, what is the frequency of these waves?

TYPES OF WAVES

In general, waves may be divided into two types, based on the direction of the particles' oscillations relative to that of the wave velocity. In a **transverse wave**, the particle motion is perpendicular to the direction of the wave velocity. The wave produced in a stretched string (Fig. 13.9) is an example of a transverse wave, as is the wave shown in ▸ Fig. 13.10a. A transverse wave is sometimes called a *shear wave*, because the disturbance supplies a force that tends to shear the medium—to separate layers of that medium at a right angle to the direction of the wave velocity. Shear waves can propagate only in solids, since a liquid or a gas cannot support a shear. That is, a liquid or a gas does not have sufficient restoring forces between its particles to propagate a transverse wave.

In a **longitudinal wave**, the particle oscillation is parallel to the direction of the wave velocity. A longitudinal wave can be produced in a stretched spring by moving the coils back and forth along the spring axis (Fig. 13.10b). Alternating pulses of compression and relaxation travel along the spring. A longitudinal wave is sometimes called a *compressional wave*, because the force tends to compress the medium.

Sound waves in air are another example of longitudinal waves. A periodic disturbance produces compressions in the air. Between the compressions are *rarefactions*—regions where the density of the air is reduced, or rarefied. A loudspeaker oscillating back and forth, for example, can create these compressions and rarefactions, which travel out into the air as sound waves. (Sound is discussed in detail in Chapter 14.)

Longitudinal waves can propagate in solids, liquids, and gases, because all phases of matter can be compressed to some extent. The propagations of transverse and longitudinal waves in different media give information about the Earth's interior structure, as discussed in Insight 13.1, Earthquakes, Seismic Waves, and Seismology.

The sinusoidal profile of water waves might make you think that they are transverse waves. Actually, they reflect a combination of longitudinal and transverse motions (▾Fig. 13.11). The particle motion may be nearly circular at the surface and becomes more elliptical with depth, eventually becoming longitudinal. A hundred meters or so below the surface of a large body of water, the wave disturbances have little effect. For example, a submarine at these depths is undisturbed by large waves on the ocean's surface. As a wave approaches shallower water near shore, the water particles have difficulty completing their elliptical paths. When the water becomes too shallow, the particles can no longer move through the bottom parts of their paths, and the wave breaks. Its crest falls forward to form breaking surf as the waves' kinetic energy is transformed into potential energy—a water "hill" that eventually topples over.

DID YOU LEARN?

➥ The four parameters that describe a wave are amplitude *A*, period *T*, frequency *f*, and speed *v*, where $v = \lambda f$.

➥ In a transverse wave, the particle motion is perpendicular to the wave velocity. An example is a wave on a stretched string.

➥ In a longitudinal wave, the particle oscillation is parallel to the wave velocity. An example is a sound wave.

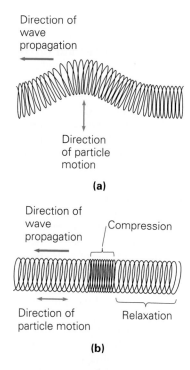

▲ **FIGURE 13.10 Transverse and longitudinal waves** **(a)** In a transverse wave, the motion of the particles is perpendicular to the direction of the wave velocity, as shown here in a spring for a wave moving to the left. **(b)** In a longitudinal wave, the particle motion is parallel to (or *along*) the direction of the wave velocity. Here, a wave pulse also moves to the left. Can you explain the motion of the wave *source* for both types of waves?

▾ **FIGURE 13.11 Water waves** Water waves are a combination of longitudinal and transverse motions. **(a)** At the surface, the water particles move in circles, but their motions become more longitudinal with depth. **(b)** When a wave approaches the shore, the lower particles are forced into steeper paths until, finally, the wave breaks or falls over to form surf.

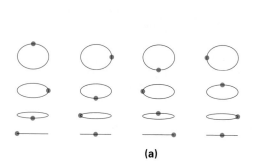

(a)

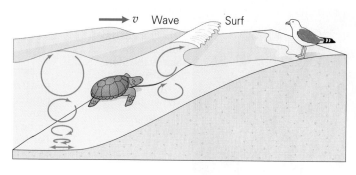

(b)

| # Earthquakes, Seismic Waves, and Seismology

The structure of the Earth's interior is something of a mystery. The deepest mine shafts and drillings extend only a few kilometers into the Earth, compared with a depth of about 6400 km to the Earth's center. Using waves to probe the Earth's structure is one way to investigate it further. Waves generated by earthquakes have proved to be especially useful for this purpose. Seismology is the study of these waves, called *seismic waves.*

Earthquakes are caused by the sudden release of built-up stress along cracks and faults, such as the famous San Andreas Fault in California (Fig. 1). According to the geological theory of plate tectonics, the outer layer of the Earth consists of rigid plates—huge slabs of rock that move very slowly relative to one another. Stresses continuously build up, particularly along boundaries between plates.

When slippage of plates finally occurs, the energy from this stress-relieving event propagates outward as (seismic) waves from a site below the surface called the *focus*. The point on the Earth's surface directly above the focus is called the *epicenter*, and receives the greatest impact of a quake. Seismic waves are of two general types: surface waves and body waves. *Surface waves*, which move along the Earth's surface, account for most earthquake damage (Fig. 2). *Body waves* travel through the Earth and are both longitudinal and transverse waves. The compressional (longitudinal) waves are called *P waves*, and the shear (transverse) waves are called *S waves* (Fig. 3).

P and S stand for *primary* and *secondary* and indicate the waves' relative speeds (actually, their arrival times at monitoring stations). In general, primary waves travel through materials faster than do secondary waves, so are detected first. An earthquake's rating on the Richter scale is related to the energy released in the form of seismic waves.

Seismic stations around the world monitor P and S waves with sensitive detecting instruments called *seismographs.* From the data gathered, the paths of the waves through the Earth can be mapped, and thereby learn about the interior structure of our planet. The Earth's interior seems to be divided into three general regions: the crust, the mantle, and the core, which itself has a solid inner region and a liquid outer region.*

The locations of these regions' boundaries are determined in part by *shadow zones*, regions where no waves of a particular type are detected. These zones appear because, although longi-

tudinal waves can travel through solids *or* liquids, transverse waves can travel only through solids. When an earthquake occurs, P waves are detected on the side of the Earth opposite the focus, but S waves are not. (See Fig. 3.) The absence of S waves in a shadow zone leads to the conclusion that the Earth must have a region near its center that is in the liquid phase.

When the transmitted P waves enter and leave the liquid region, they are refracted (bent). This refraction gives rise to a P-wave shadow zone, which indicates that only the outer part of the core is liquid. As you will learn in Chapter 19, the combination of a liquid outer core and the Earth's rotation may be responsible for the Earth's magnetic field.

FIGURE 2 Bad vibrations Earthquake damage caused by the surface waves of a major shock that struck Kobe, Japan, in January 1995.

*In most places, the crust is about 24–30 km (15–20 mi) thick; the mantle is 2900 km (1800 mi) thick; and the core has a radius of 3450 km (2150 mi). The solid inner core has a radius of about 1200 km (750 mi).

FIGURE 1 The San Andreas Fault A small section of the fault, which runs through the San Francisco Bay area as well as across the more rural regions of California, is shown here.

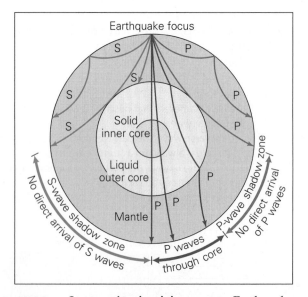

FIGURE 3 Compressional and shear waves Earthquakes produce waves that travel through the Earth. Because transverse (S) waves are not detected on the opposite side of the Earth, scientists believe that at least part of the Earth's core is a viscous liquid under high pressures and temperatures. The waves bend continuously, or refract, because their speed varies with depth.

13.4 Wave Properties

⟶ What is the principle that wave action obeys that result in constructive or destructive interference?

⟶ What is meant by a total destructive interference?

⟶ What phenomenon explains why you can hear "around corners"—for example, why you can hear people around the corner of a building talking but not see them?

Among the properties exhibited by all waves are superposition, interference, reflection, refraction, dispersion, and diffraction.

SUPERPOSITION AND INTERFERENCE

When two or more waves meet or pass through the same region of a medium, they pass through each other and each wave proceeds without being altered. While they are in the same region, the waves are said to be interfering.

What happens during interference? That is, what does the combined waveform look like? The relatively simple answer is given by the **principle of superposition**:

> At any time, the combined waveform of two or more interfering waves is given by the sum of the displacements of the individual waves at each point in the medium.

Using the principle of superposition, **interference** is illustrated in ▼Fig. 13.12. The displacement of the combined waveform at any point is $y = y_1 + y_2$, where y_1 and y_2 are the displacements of the individual pulses at that point. (Directions are indicated by positive and negative values.) Interference, then, is the physical addition of waves. In adding waves, we must take into account the possibility that

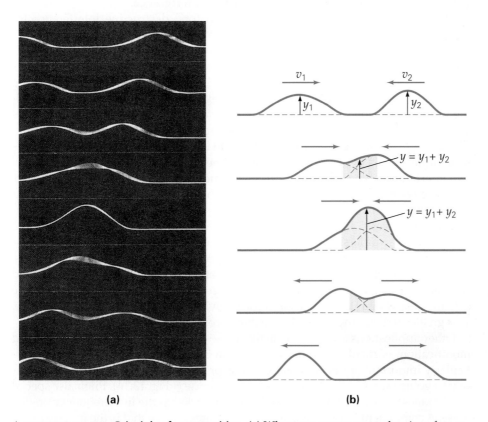

(a)   (b)

▲ FIGURE 13.12 **Principle of superposition** **(a)** When two waves meet, they interfere as shown in the photographs. **(b)** The beige tint marks the area where the two waves, moving in opposite directions, overlap and combine. The displacement at any point on the combined wave is equal to the sum of the displacements on the individual waves: $y = y_1 + y_2$.

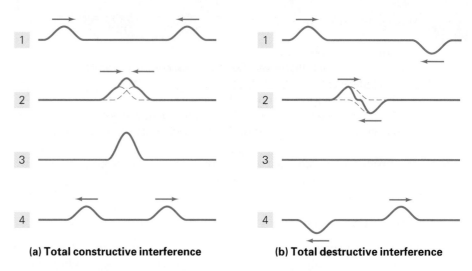

(a) Total constructive interference   (b) Total destructive interference

▲ **FIGURE 13.13** **Interference** **(a)** When two wave pulses of the same amplitude meet and are in phase, they interfere constructively. When the pulses are exactly superimposed (3), total constructive interference occurs. **(b)** When the interfering pulses are 180° out of phase and exactly superimposed (3), total destructive interference occurs.

they are producing disturbances in opposite directions. In other words, we must treat the disturbances in terms of vector addition.

In the figure, the vertical displacements of the two pulses are in the same direction, and the amplitude of the combined waveform is greater than that of either pulse. This situation is called **constructive interference**. Conversely, if one pulse has a negative displacement, the two pulses tend to cancel each other when they overlap, and the amplitude of the combined waveform is smaller than that of either pulse. This situation is called **destructive interference**.

The special cases of total constructive and total destructive interference for traveling wave pulses of the same width and amplitude are shown in ▲Fig. 13.13. At the instant these interfering waves exactly overlap (crest coinciding with crest), the amplitude of the combined waveform is twice that of either individual wave. This case is referred to as **total constructive interference**. When the interfering pulses have opposite displacements and are exactly superimposed (crest coinciding with trough), the waveforms momentarily disappear; that is, the amplitude of the combined wave is zero. This case is called **total destructive interference**.

The word *destructive* unfortunately tends to imply that the energy, as well as the form of the waves, is destroyed. This is not the case. At the point of total destructive interference, when the net wave shape and, hence, potential energy are zero, the wave energy is stored in the medium completely in the form of kinetic energy. That is, the straight string has instantaneous velocity.

There are several practical applications of destructive interference. One of these is automobile mufflers. Exhaust gases from the engine passing from a high pressure in the cylinders to normal atmospheric pressure would produce loud noises. Typically, a muffler consists of a metal jacket containing perforated pipes and chambers. The pipes and chambers are arranged so that the pressure waves of the exhaust gases are reflected back and forth, giving rise to destructive interference. This greatly reduces the noise of the exhaust coming from the tail pipe.

Other applications are termed "active noise cancellation." This involves sound modification, particularly sound cancellation by electro-acoustical means. A particularly important application is for airline or helicopter pilots who need to hear what's going on around them over the engine noise (▶Fig. 13.14). Pilots use special headphones mounted with a microphone that picks up the low-frequency engine noise. A component in the headphone then creates a wave that is the inverse of the engine noise. This is played back through the headphones, and destructive interference produces a quieter background. A pilot can then better hear the mid- and high-frequency sounds, such as conversation and instrument warning sounds.

(a)

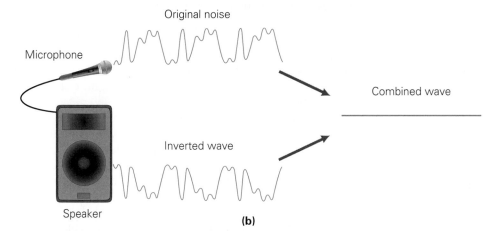

Original noise

Microphone

Combined wave

Inverted wave

Speaker

(b)

▲ FIGURE 13.14 **Destructive interference in action** **(a)** Pilots use headphones mounted with a microphone that picks up low-frequency engine noise. **(b)** A wave is generated that is inverse that of the engine noise. When played back through the headphones, destructive interference produces less engine noise. This process is called "active noise cancellation."

REFLECTION, REFRACTION, DISPERSION, AND DIFFRACTION

Besides meeting other waves, waves can (and do) meet objects or a boundary with another medium. In such cases, several things may occur. One of these is **reflection**, which occurs when a wave strikes an object or comes to a boundary with another medium and is at least partly diverted back into the original medium. An echo is the reflection of sound waves, and mirrors reflect light waves.

Two cases of reflection are illustrated in ▸ Fig. 13.15. If the end of the string is fixed, the reflected pulse is inverted (Fig. 13.15a). This is because the pulse causes the string to exert an upward force on the wall, and the wall exerts an equal and opposite downward force on the string (by Newton's third law). The downward force creates the downward, or inverted, reflected pulse. If the end of the string is free to move, then the reflected pulse is not inverted. (There is no phase shift.) This is illustrated in Fig. 13.15b, which shows the string attached to a light ring that can move freely on a smooth pole. The ring is accelerated upward by the front portion of the incoming pulse and then comes back down, thus creating a noninverted reflected pulse.

More generally, when a wave strikes a boundary, the wave is not completely reflected. Instead, some of the wave's energy is reflected and some is transmitted or absorbed. When a wave crosses a boundary into another medium, its speed generally changes because the new material has different characteristics. When

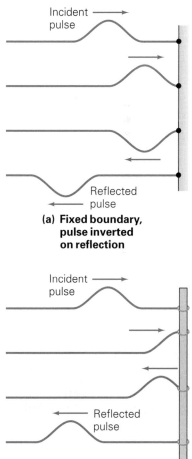

▼ FIGURE 13.15 **Reflection** **(a)** When a wave (pulse) on a string is reflected from a fixed boundary, the reflected wave is inverted. **(b)** If the string is free to move at the boundary, the phase of the reflected wave is not shifted from that of the incident wave.

Incident pulse

Reflected pulse

(a) Fixed boundary, pulse inverted on reflection

Incident pulse

Reflected pulse

(b) Free (movable) boundary, pulse not inverted on reflection

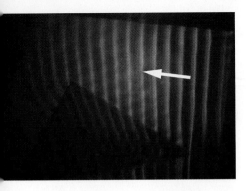

▲ **FIGURE 13.16 Refraction** The refraction of water waves is shown from overhead. As the crests approach the triangle from the right, the wave speed slows because it is entering shallow water. Thus, the wave changes its direction.

entering the medium obliquely (at an angle), the transmitted wave moves in a direction different from that of the incident wave. This phenomenon is called **refraction** (◄Fig. 13.16).

Since refraction depends on changes in the speed of the wave, you might be wondering which physical parameters determine the wave speed. Generally, there are two types of situations. The simplest kind of wave is one whose speed does *not* depend on its wavelength (or frequency). All such waves travel at the same speed, determined solely by the properties of the medium. These waves are called *nondispersive waves*, because they do not disperse, or spread apart from one another. An example of a nondispersive transverse wave is a wave on a string, whose speed, as we shall see, is determined only by the tension and mass density of the string (Section 13.5). Sound is a nondispersive longitudinal wave; the speed of sound (in air) is determined only by the compressibility and density of the air. Indeed, if the speed of sound did depend on the frequency, at the back of the symphony hall you might hear the violins well before the cellos, even though the two sound waves were in perfect synchronization when they left the orchestra.

When the wave speed *does* depend on wavelength (or frequency), the waves are said to exhibit **dispersion**: waves of different frequencies spread apart from one another. An example of dispersion is light waves. When they enter some media, they are spread out or dispersed. This is the basis for prisms separating sunlight into a color spectrum and for the formation of a rainbow, as will be seen in Section 22.5. Dispersion will be most important in the study of light, but waves other than light can also be dispersive under the right conditions.

Diffraction refers to the bending of waves around an edge of an object and is not related to refraction. For example, if you stand along an outside wall of a building near the corner of the street, you can hear people talking around the corner. Assuming that there are no reflections or air motion (wind), this would not be possible if the sound waves traveled in a straight line. As the sound waves pass the corner, instead of being sharply cut off, they "wrap around" the edge, and you can hear the sound.

In general, the effects of diffraction are evident only when the size of the diffracting object or opening is about the same as or smaller than the wavelength of the waves. The dependence of diffraction on the wavelength and size of the object or opening is illustrated in ▼Fig. 13.17. For many waves, diffraction is negligible under normal circumstances. For instance, visible light has wavelengths on the order of 10^{-6} m. Such wavelengths are much too small to exhibit diffraction when they pass through common-sized openings, such as an eyeglass lens.

Reflection, refraction, dispersion, and diffraction will be considered in more detail when we study light waves in Chapters 22 and 24.

DID YOU LEARN?
➡ The principle of superposition explains constructive and destructive wave interference.
➡ When two waves of the same wavelength and amplitude have opposite displacements (a 180° phase difference), the wave resulting from superposition has zero amplitude.
➡ Wave diffraction causes sound to be "bent" around corners.

▶ **FIGURE 13.17 Diffraction** Diffraction effects are greatest when the opening (or object) is about the same size as or smaller than the wavelength of the waves. **(a)** With an opening much larger than the wavelength of these plane water waves, diffraction is noticeable only near the edges. **(b)** With an opening about the same size as the wavelength of the waves, diffraction produces nearly semicircular waves.

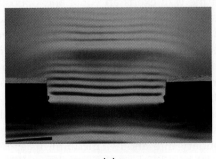

(a)

(b)

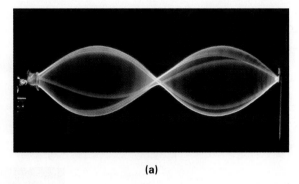

(a)

▲ **FIGURE 13.18 Standing waves (a)** Standing waves are formed by interfering waves traveling in opposite directions. **(b)** Conditions of destructive and constructive interference recur as each wave travels a distance of $\lambda/4$ in a time $t = T/4$. The velocities of the rope's particles are indicated by the arrows. This motion gives rise to standing waves with stationary nodes and maximum amplitude antinodes.

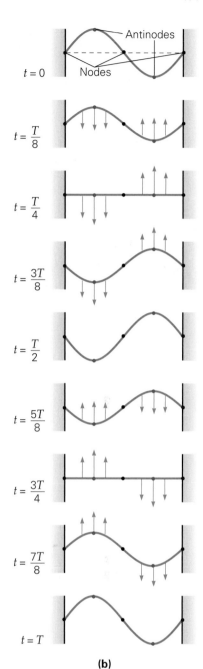

(b)

13.5 Standing Waves and Resonance

LEARNING PATH QUESTIONS

➡ How is a standing wave formed?

➡ What are the nodes and antinodes of a standing wave?

➡ Under what condition does resonance occur for a particular oscillatory system?

If you shake one end of a stretched rope that is fixed at the other end, waves travel along the rope to the fixed end and are reflected back. The waves going to and from the fixed end interfere with each other. In most cases, the combined waveforms have a changing, jumbled appearance. But if the rope is shaken at just the right frequency, a steady waveform, or series of uniform loops, appears to stand in place along the rope. Appropriately, this phenomenon is called a **standing wave** (▲ Fig. 13.18). It arises because of interference with the reflected waves, which have the same wavelength, amplitude, and speed as the incident waves. Since the two identical waves travel in opposite directions, the net energy flow in the rope is zero. In effect, the energy is standing in the loops.

Some points on the rope remain stationary at all times and are called **nodes**. At these points, the displacements of the interfering waves are *always* equal and opposite. Thus, by the principle of superposition, the interfering waves cancel each other completely at these points, and the rope does not undergo displacement there. At all other points, the rope oscillates back and forth at the same frequency. The points of maximum amplitude, where constructive interference is greatest, are called **antinodes**. As you can see in Fig. 13.18a, adjacent antinodes are separated by a half-wavelength ($\lambda/2$), or one loop; adjacent nodes are also separated by a half-wavelength.

Standing waves can be generated in a rope by more than one driving frequency; the higher the frequency, the more oscillating half-wavelength loops there are in the rope. The only requirement is that an integer number of half-wavelengths "fit" the length of the rope. The frequencies at which standing waves are produced are called **natural frequencies**, or **resonant frequencies**. The resulting standing wave patterns are called *normal*, or *resonant*, *modes of vibration*. In general, all systems that oscillate have one or more natural frequencies, which depend on such factors as mass, elasticity or restoring force, and geometry (boundary conditions). The natural frequencies of a system are sometimes called its *characteristic* frequencies.

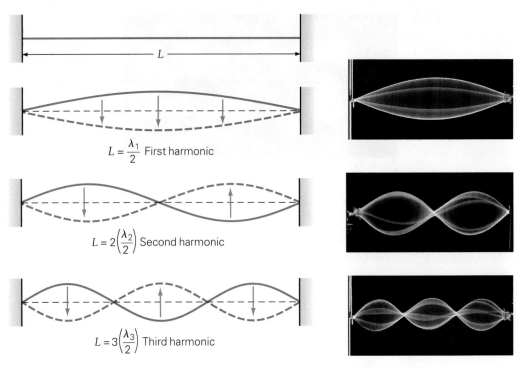

$$L = \frac{\lambda_1}{2} \ \text{First harmonic}$$

$$L = 2\left(\frac{\lambda_2}{2}\right) \ \text{Second harmonic}$$

$$L = 3\left(\frac{\lambda_3}{2}\right) \ \text{Third harmonic}$$

▲ FIGURE 13.19 **Natural frequencies** A stretched string can have standing waves only at certain frequencies. These correspond to the numbers of half-wavelength loops that will fit along the length of string between the nodes at the fixed ends.

A stretched string or rope can be analyzed to determine its natural frequencies. The boundary condition is that the ends are fixed; thus, there must be a node at each end. The number of closed segments or loops of a standing wave that will fit between the nodes at the ends (along the length of the string) is equal to an integral number of *half*-wavelengths (▲Fig. 13.19). Note that $L = (\lambda_1/2)$, $L = 2(\lambda_2/2)$, $L = 3(\lambda_3/2)$, $L = 4(\lambda_4/2)$, and so on. In general,

$$L = n\left(\frac{\lambda_n}{2}\right) \quad \text{or} \quad \lambda_n = \frac{2L}{n} \quad (\text{for } n = 1, 2, 3, \dots)$$

The natural frequencies of oscillation are therefore

$$f_n = \frac{v}{\lambda_n} = n\left(\frac{v}{2L}\right) = nf_1 \quad \text{for } n = 1, 2, 3, \dots \qquad \begin{array}{l}\textit{(natural frequencies} \\ \textit{for a stretched string)}\end{array} \quad (13.18)$$

where v is the speed of waves on a string. The lowest natural frequency ($f_1 = v/2L$ for $n = 1$) is called the **fundamental frequency.** All of the other natural frequencies are integral multiples of the fundamental frequency f_1: $f_n = nf_1$ (for $n = 1, 2, 3, \dots$). The set of frequencies $f_1, f_2 = 2f_1, f_3 = 3f_1, \dots$ is called a **harmonic series**: f_1 (the fundamental frequency) is the *first harmonic*, f_2 the *second harmonic*, and so on.

Strings that are fixed at each end are found in stringed musical instruments such as violins, pianos, and guitars. When such a string is disturbed—that is, plucked, struck, or bowed, the resulting vibration generally includes several higher harmonics in addition to the first harmonic. The number of harmonics depends on how and where the string is disturbed. It is the combination of harmonic frequencies that gives a particular instrument its characteristic sound quality (more on this in Section 14.6). As Eq. 13.18 shows, all harmonic frequencies depend on the length of the string. Different notes are obtained on a particular string of a violin by touching the string at a particular location so as to change its vibrating length (◄Fig. 13.20).

▲ FIGURE 13.20 **Fundamental frequencies** Performers of stringed instruments such as the violin use their fingers to stop or fret the strings. By pressing a string against the fingerboard, the player reduces the amount of its length that is free to vibrate. This reduction increases the harmonic frequencies of the string and the pitch of the tones it produces.

Natural frequencies also depend on other parameters, such as mass and force, which affect the wave speed in the string. For a stretched string, the wave speed (v) can be shown to be

$$v = \sqrt{\frac{F_T}{\mu}} \quad \text{(wave speed in a stretched string)} \qquad (13.19)$$

where F_T is the tension in the string and μ is the linear mass density (mass per unit length, $\mu = m/L$). (We use F_T, rather than the T in previous chapters, so as not to confuse the tension with the period T.) Thus, Eq. 13.18 can be written as

$$f_n = n\left(\frac{v}{2L}\right) = \frac{n}{2L}\sqrt{\frac{F_T}{\mu}} = nf_1 \quad (\text{for } n = 1, 2, 3\ldots) \qquad (13.20)$$

Note that the greater the linear mass density of a string, the lower its natural frequencies. As you may know, the low-note strings on a violin or guitar are thicker, or more massive, than the high-note strings. By tightening a string, all the frequencies of that string are increased. Changing the tension in the string is how violinists, for example, tune their instruments before a performance.

EXAMPLE 13.6 | **A Piano String: Fundamental Frequency and Harmonics**

A piano string with a length of 1.15 m and a mass of 20.0 g is under a tension of 6.30×10^3 N. (a) What is the fundamental frequency of the string? (b) What are the frequencies of the next two harmonics?

THINKING IT THROUGH. The linear mass density can be determined from the data. This, along with the given string tension can be used to find the fundamental frequency. With this, the harmonics can be calculated.

SOLUTION.

Given: $L = 1.15$ m
$m = 20.0$ g $= 0.0200$ kg
$F_T = 6.30 \times 10^3$ N

Find: (a) f_1 (fundamental frequency)
(b) f_2 and f_3 (frequencies of next two harmonics)

(a) The linear mass density of the string is

$$\mu = \frac{m}{L} = \frac{0.0200 \text{ kg}}{1.15 \text{ m}} = 0.0174 \text{ kg/m}$$

Then, using Eq. 13.20, we have

$$f_1 = \frac{1}{2L}\sqrt{\frac{F_T}{\mu}} = \frac{1}{2(1.15 \text{ m})}\sqrt{\frac{6.30 \times 10^3 \text{ N}}{0.0174 \text{ kg/m}}} = 262 \text{ Hz}$$

This is approximately the frequency of middle C (C_4) on a piano.

(b) Since $f_2 = 2f_1$ and $f_3 = 3f_1$ it follows that

$$f_2 = 2f_1 = 2(262 \text{ Hz}) = 524 \text{ Hz}$$

and

$$f_3 = 3f_1 = 3(262 \text{ Hz}) = 786 \text{ Hz}$$

The second harmonic corresponds approximately to C_5 on a piano, since, by definition, the frequency doubles with each octave (every eighth white key).

FOLLOW-UP EXERCISE. A musical note is referenced to the fundamental frequency, or first harmonic. In musical terms, the second harmonic is called the first overtone, the third harmonic is the second overtone, and so on. If an instrument has a third overtone with a frequency of 880 Hz, what is the frequency of the first overtone?

RESONANCE

When an oscillating system is driven at one of its natural, or resonant, frequencies, the maximum amount of energy is transferred to the system. The natural frequencies of a system are the frequencies at which the system "prefers" to vibrate, so to speak. The condition of driving a system at a natural frequency is referred to as **resonance**.

A common example of a system in mechanical resonance is someone being pushed on a swing. Basically, a swing is a simple pendulum and has only one resonant frequency for a given length $[f = 1/T = \sqrt{g/L}/(2\pi)]$. If you push the swing with this frequency and in phase with its motion, its amplitude and energy

▲ **FIGURE 13.21** **Resonance in the playground** The swing behaves like a pendulum in SHM. To transfer energy efficiently, the man must time his pushes to the natural frequency of the swing.

increase (◄ Fig. 13.21). If you push at a slightly different frequency, the energy transfer is no longer a maximum. (What do you think happens if you push with the resonant frequency, but 180° out of phase with the swing's motion?)

Unlike a simple pendulum, a stretched string has many natural frequencies. Almost any driving frequency will cause a disturbance in the string. However, if the frequency of the driving force is not equal to one of the natural frequencies, the resulting wave will be relatively small and jumbled. By contrast, when the frequency of the driving force matches one of the natural frequencies, the maximum amount of energy is transferred to the string. A steady standing wave pattern results, with the amplitude at the antinodes becoming relatively large.

When a large number of soldiers march over a small bridge, they are generally ordered to break step. The reason is that the marching (stepping) frequency may correspond to one of the natural frequencies of the bridge and set it into resonant vibration, which could cause it to collapse. This actually occurred on a suspension bridge in England in 1831. The bridge was weak and in need of repair, but it collapsed as a direct result of the resonance vibrations induced by the marching soldiers—with some injuries.

Another incident of a bridge vibrating was the collapse of the "Galloping Gertie," the Tacoma Narrows Bridge (in Washington State). On November 7, 1940, winds with speeds of 65–72 km/h (40–45 mi/h) started the main span vibrating. The bridge, 855 m (2800 ft) long and 12 m (39 ft) wide, had been opened to traffic only four months earlier. The main span vibrated in two different modes: a transverse mode of frequency 0.6 Hz and a torsional (twisting) mode of frequency 0.2 Hz. The main span finally collapsed (▼ Fig. 13.22). The cause of the collapse is quite complicated, but the energy provided by the wind was a major factor.

Mechanical resonance is not the only type of resonance. When you tune a radio, you are changing the resonant frequency of an electrical circuit (Section 21.5) so that it will be driven by, or will pick up, a signal at the frequency of the station you want.

DID YOU LEARN?

➡ A standing wave is the result of the superposition or interference of two waves of the same wavelength, amplitude, and speed, but traveling in opposite directions.

➡ Nodes are points that remain stationary at all times, and antinodes correspond to points having maximum amplitudes.

➡ Resonance occurs when the external driving frequency is equal to the natural frequency or resonant frequency of a system. Maximum energy is transferred to the system, so the amplitude of the system is at a maximum.

▶ **FIGURE 13.22** **Galloping Gertie** The collapse of the Tacoma Narrows Bridge on November 7, 1940, is captured in this frame from a movie camera.

PULLING IT TOGETHER | Frequency, Density, and Diameter of a Guitar String

Suppose you want to increase the fundamental frequency of a guitar string. (a) Would you (1) loosen the string to decrease its tension, (2) tighten the string to increase its tension, (3) use another string of the same material with a smaller diameter at the same tension, or (4) use another string of the same material with a larger diameter at the same tension? (b) In actuality, you want to go from the A note (220 Hz) below middle C to the A note (440 Hz) above middle C. Show that a string with half the diameter of the same material will have double the fundamental frequency.

(A) CONCEPTUAL REASONING. The fundamental frequency of a stretched string is given by Eq. 13.20:

$$f = \frac{1}{2L}\sqrt{\frac{F_T}{\mu}} \quad \text{(for } n = 1\text{)}$$

The frequency of the string is proportional to the *square root* of the tension force F_T, so loosening the string—that is, decreasing F_T—would not increase the frequency. Increasing the tension does increases the frequency, so (2) is correct. Please note that doubling the tension will not double the frequency because $\sqrt{2F_T} \neq 2\sqrt{F_T}$.

Since the strings are of the same material (same density ρ), the greater the diameter of a string, the greater its volume, the greater its mass, and therefore the greater its mass per unit length (greater μ). Hence, a thinner string, with a smaller μ, will vibrate at a higher frequency, and answer (3) is also correct.

(B) QUANTITATIVE REASONING AND SOLUTION. At first, one might think of computing the frequencies directly, using Eq. 13.20. But this can't be done, because not enough data are given, which usually implies the use of a ratio. Since the linear mass density, $\mu = m/L$, depends on mass, and mass in turn depends on diameter, the frequency equation needs to be expressed in terms of the diameter of a string.

Recall that the mass of a string depends on its density ρ and volume V; that is, $\rho = m/V$, or $m = \rho V$. Then, the volume V of a length (L) of wire, which may be thought of as a long cylinder with circular cross-section A, can be determined by $V = AL$. The circular area is proportional to the square of the diameter of the wire, $A = \left(\frac{\pi d^2}{4}\right)$. This is the key to our proof.

Given: $f = \frac{1}{2L}\sqrt{\frac{F_T}{\mu}}$ (for $n = 1$, Eq. 13.20)   *Find:* Prove that if $d_2 = d_1/2$, then $f_2 = 2f_1$.

$d_2 = d_1/2$

The linear mass density of the wire string can be expressed in terms of its density and volume, the latter of which is proportional to the square of the string's diameter:

$$\mu = \frac{m}{L} = \frac{\rho V}{L} = \frac{\rho AL}{L} = \rho\left(\frac{\pi d^2}{4}\right)$$

Putting this into Eq. 13.20 yields

$$f = \frac{1}{2L}\sqrt{\frac{F_T}{\mu}} = \frac{1}{2L}\sqrt{\frac{4F_T}{\rho\pi d^2}} = \left(\frac{1}{L}\sqrt{\frac{F_T}{\rho\pi}}\right)\frac{1}{d}$$

The quantities in the brackets are constant, and $f \propto 1/d$, so, in ratio form,

$$\frac{f_2}{f_1} = \frac{d_1}{d_2} = \frac{0.30\text{ cm}}{0.15\text{ cm}} = 2 \quad \text{and} \quad f_2 = 2f_1$$

Learning Path Review

- **Simple harmonic motion (SHM)** requires a restoring force directly proportional to the displacement, such as an ideal spring force, which is given by Hooke's law.

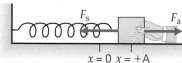

Hooke's law:

$$F_s = -kx \tag{13.1}$$

- The **frequency (f)** and **period (T)** for SMH are inversely related.

Frequency and period for SHM:

$$f = \frac{1}{T} \tag{13.2}$$

- In general, the total energy of an object in SHM is directly proportional to the square of the amplitude.

Total energy of a spring and mass in SHM:

$$E = \tfrac{1}{2}kA^2 = \tfrac{1}{2}mv^2 + \tfrac{1}{2}kx^2 \tag{13.4–5}$$

■ The form of an equation of motion for an object in SHM depends on the object's initial (y_o) displacement.

Equations of motion for SHM:

$$y = \pm A \sin \omega t = \pm A \sin(2\pi f t) = \pm A \sin\left(\frac{2\pi t}{T}\right) \quad (13.8a)$$

+ *for initial motion upward with $y_o = 0$*
− *for initial motion downward with $y_o = 0$*

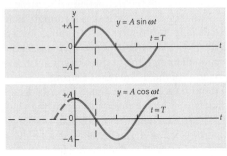

$$y = \pm A \cos \omega t = \pm A \cos(2\pi f t) = \pm A \cos\left(\frac{2\pi t}{T}\right) \quad (13.9a)$$

+ *for initial motion downward with $y_o = +A$*
− *for initial motion upward with $y_o = -A$*

Velocity of a mass oscillating on a spring:

$$v = \pm\sqrt{\frac{k}{m}(A^2 - x^2)} \quad (13.6)$$

Period of a mass oscillating on a spring:

$$T = 2\pi\sqrt{\frac{m}{k}} \quad (13.11)$$

Angular frequency of a mass oscillating on a spring:

$$\omega = 2\pi f = \sqrt{\frac{k}{m}} \quad (13.13)$$

Period of a simple pendulum (small-angle approximation):

$$T = 2\pi\sqrt{\frac{L}{g}} \quad (13.14)$$

Velocity of a mass in SHM:

$$v = \omega A \cos \omega t \quad \begin{array}{l}\text{(vertical velocity if } v_o \\ \text{is upward at } t_o = 0, y_o = 0)\end{array} \quad (13.15)$$

Acceleration of a mass in SHM:

$$a = -\omega^2 A \sin \omega t = -\omega^2 y \quad \begin{array}{l}\text{(vertical acceleration if } v_o \\ \text{is upward at } t_o = 0, y_o = 0)\end{array} \quad (13.16)$$

■ A wave is a disturbance in time and space; energy is transferred or propagated by wave motion.

Wave speed:

$$v = \frac{\lambda}{T} = \lambda f \quad (13.17)$$

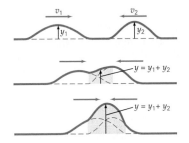

■ At any time, the combined waveform of two or more interfering waves is given by the sum of the displacements of the individual waves at each point in the medium.

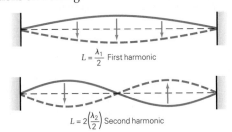

■ At natural frequencies, **standing waves** can form on a string as a result of the **interference** of two waves of identical wavelength, amplitude, and speed traveling in opposite directions on a string.

$$L = \frac{\lambda_1}{2} \text{ First harmonic}$$

$$L = 2\left(\frac{\lambda_2}{2}\right) \text{ Second harmonic}$$

Natural frequencies for a stretched string:

$$f_n = n\left(\frac{v}{2L}\right) = \frac{n}{2L}\sqrt{\frac{F_T}{\mu}} = nf_1 \quad (\text{for } n = 1, 2, 3, \ldots) \quad (13.20)$$

Learning Path Questions and Exercises   For instructor-assigned homework, go to www.masteringphysics.com

MULTIPLE CHOICE QUESTIONS

13.1 SIMPLE HARMONIC MOTION

1. For a particle in SHM, the force on it (F) and its displacement from its equilibrium position (x) are (a) in the same direction, (b) opposite in direction, (c) perpendicular to each other, (d) none of the preceding.

2. The maximum kinetic energy of a mass–spring system in SHM is equal to (a) A, (b) A^2, (c) kA, (d) $kA^2/2$.

3. If the frequency of a system in SHM is doubled, the period of the system is (a) doubled, (b) halved, (c) four times as large, (d) one-quarter as large.

4. When a particle in a horizontal SHM is at the equilibrium position, the kinetic energy of the system is (a) zero, (b) at a maximum, (c) half the maximum value, (d) none of the preceding.

13.2 EQUATIONS OF MOTION

5. The equation of motion for a particle in SHM (a) is a sine or cosine function, (b) is a tangent or cotangent function, (c) could be any mathematical function, (d) gives the velocity of the particle as a function of time.

6. For the SHM equation $y = A \sin[(200\pi \text{ rad/s})t]$, the frequency of oscillation, f, is (a) 50 Hz, (b) 100 Hz, (c) 200 Hz, (d) 200π Hz.

7. For the SHM equation $y = A \sin(2\pi t/T)$, the y-position of the object three-quarters of the period after the motion starts is (a) $+A$, (b) $-A$, (c) $A/2$, (d) 0.

13.3 WAVE MOTION

8. Wave motion in a material medium involves (a) the propagation of a disturbance, (b) interparticle interactions, (c) the transfer of energy, (d) all of the preceding.

9. For a longitudinal wave, the direction between the wave velocity and particle oscillation is (a) perpendicular, (b) parallel, (c) 45°, or (d) none of the preceding.

10. A water wave is (a) transverse, (b) longitudinal, (c) a combination of transverse and longitudinal, (d) none of the preceding.

13.4 WAVE PROPERTIES

11. When two waves meet each other and interfere, the resultant waveform is determined by (a) reflection, (b) refraction, (c) diffraction, (d) superposition.

12. When two identical waves of the same wavelength (λ) and amplitude (A) interfere in phase, the amplitude of the resulting wave is (a) A, (b) $2A$, (c) $3A$, (d) $4A$.

13. You can often hear people talking from around a corner of a building. This is due primarily to (a) reflection, (b) refraction, (c) interference, (d) diffraction.

13.5 STANDING WAVES AND RESONANCE

14. For two traveling waves to form standing waves, the waves must have the same (a) wavelength, (b) amplitude, (c) speed, (d) all of the preceding.

15. The points of zero amplitude on a rope that is supporting a standing wave waveform are called (a) nodes, (b) antinodes, (c) fundamentals, (d) resonance points.

16. For a standing wave on a rope, the distance between two adjacent antinodes is (a) 1/4 wavelength, (b) 1/2 wavelength, (c) one wavelength, (d) two wavelengths.

17. When a stretched violin string oscillates in its second harmonic mode, the standing wave in the string will exhibit (a) 1/4 wavelength, (b) 1/2 wavelength, (c) one wavelength, (d) two wavelengths.

CONCEPTUAL QUESTIONS

13.1 SIMPLE HARMONIC MOTION

1. If the amplitude of a particle in SHM is doubled, how are (a) the total energy and (b) the maximum speed affected?

2. How does the speed of a mass in SHM change as the mass leaves its equilibrium position? Explain.

3. A mass–spring system in SHM has an amplitude A and period T. How long does the mass take to travel a distance A? How about $2A$?

4. A tennis player uses a racket to bounce a ball up and down with a constant period. Is this a simple harmonic motion? Explain.

13.2 EQUATIONS OF MOTION

5. If a mass–spring system were taken to the Moon, would the period of the system change? How about the period of a pendulum taken to the Moon? Explain.

6. If you want to increase the frequency of vibration of a mass–spring system, would you increase or decrease the mass? Explain.

7. If the length of a pendulum is doubled, what is the ratio of the new period to the old one?

8. Would the period of a pendulum in an upward-accelerating elevator be increased or decreased compared with its period in a nonaccelerating elevator? Explain.

9. One simple harmonic motion is described by a sine function, $y = A \sin(\omega t)$, and another is described by a cosine function, $y = A \cos(\omega t)$. Discuss the differences in their initial position, velocity, and acceleration.

13.3 WAVE MOTION

10. When a wave pulse travels along a rope, what travels with the wave motion and what does not travel with the wave?

11. ▼ Figure 13.23 shows pictures of two mechanical waves. Identify each as being transverse or longitudinal.

12. What type(s) of wave(s), transverse or longitudinal, will propagate through (a) solids, (b) liquids, and (c) gases?

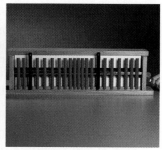

▲ FIGURE 13.23 **Transverse or longitudinal?** See Conceptual Exercise 11.

13. Standing on a hill and looking at a tall wheat field, you see a beautiful wave traveling across the field whenever a breeze blows. What type of wave is this? Explain.

13.4 WAVE PROPERTIES

14. What is cancelled out when destructive interference occurs? What happens to the wave energy in such a situation? Explain.

15. Dolphins and bats determine the location of their prey by emitting ultrasonic sound waves. Which wave phenomenon is involved?

16. If sound waves were dispersive (that is, if the speed of sound depended on its frequency), what would be the consequences of someone listening to an orchestra in a concert hall?

13.5 STANDING WAVES AND RESONANCE

17. Can harmonic sound of any frequency be generated and heard from a violin string with a fixed tension? Explain.

18. If they have the same tension and length, will a thicker or a thinner guitar string sound higher in frequency? Why?

19. A child's swing (a pendulum) has only one natural frequency, f_1, yet it can be driven or pushed smoothly at frequencies of $f_1/2$, $f_1/3$, and $2f_1$. How is this possible?

20. By rubbing the circular lip of a wide, thin wine glass with a moist finger, you can make the glass "sing." (Try it.) (a) What causes this? (b) What would happen to the frequency of the sound if you added water to the glass?

EXERCISES

*Integrated Exercises (**IE**s) are two-part exercises. The first part typically requires a conceptual answer choice based on physical thinking and basic principles. The following part is quantitative calculations associated with the conceptual choice made in the first part of the exercise.*

*Many exercise sections include "paired" exercises. These exercise pairs, identified with **red numbers**, are intended to assist you in problem solving and learning. In a pair, the first exercise (even numbered) is worked out in the Study Guide so that you can consult it should you need assistance in solving it. The second exercise (odd numbered) is similar in nature, and its answer is given in Appendix VII at the back of the book.*

13.1 SIMPLE HARMONIC MOTION

1. • A particle oscillates in SHM with an amplitude A. What is the total *distance* (in terms of A) the particle travels in three periods?

2. • If it takes a particle in SHM 0.50 s to travel from the equilibrium position to the maximum displacement (amplitude), what is the period of oscillation?

3. • A 0.75-kg object oscillating on a spring completes a cycle every 0.50 s. What is the frequency of this oscillation?

4. • A particle in simple harmonic motion has a frequency of 40 Hz. What is the period of this oscillation?

5. • The frequency of a simple harmonic oscillator is doubled from 0.25 Hz to 0.50 Hz. What is the change in its period?

6. • An object of mass 0.50 kg is attached to a spring with spring constant 10 N/m. If the object is pulled down 0.050 m from the equilibrium position and released, what is its maximum speed?

7. • An object of mass 1.0 kg is attached to a spring with spring constant 15 N/m. If the object has a maximum speed of 0.50 m/s, what is the amplitude of oscillation?

8. •• Atoms in a solid are in continuous vibrational motion due to thermal energy. At room temperature, the amplitude of these atomic vibrations is typically about 10^{-9}cm, and their frequency is on the order of 10^{12} Hz. (a) What is the approximate period of oscillation of a typical atom? (b) What is the maximum speed of such an atom?

9. •• A particle of mass 0.10 kg is attached to a spring of spring constant 10 N/m. If the maximum acceleration of the particle is 5.0 m/s², what is the maximum speed of the particle?

10. **IE** •• (a) At what position is the magnitude of the force on a mass in a mass–spring system minimum: (1) $x = 0$, (2) $x = -A$, or (3) $x = +A$? Why? (b) If $m = 0.500$ kg, $k = 150$ N/m, and $A = 0.150$ m, what are the magnitude of the force on the mass and the acceleration of the mass at $x = 0, 0.050$ m, and 0.150 m?

11. **IE** •• (a) At what position is the speed of a mass in a mass–spring system maximum: (1) $x = 0$, (2) $x = -A$, or (3) $x = +A$? Why? (b) If $m = 0.250$ kg, $k = 100$ N/m, and $A = 0.10$ m for such a system, what is the mass's maximum speed?

12. •• A mass–spring system is in SHM in the horizontal direction. If the mass is 0.25 kg, the spring constant is 12 N/m, and the amplitude is 15 cm, (a) what is the maximum speed of the mass, and (b) where does this occur? (c) What is the speed at a half-amplitude position?

13. •• A horizontal spring on a frictionless level air track has a 0.150-kg object attached to it and it is stretched 6.50 cm. Then the object is given an outward initial velocity of 2.20 m/s. If the spring constant is 35.2 N/m, determine how much farther the spring stretches.

14. •• A 0.25-kg object is suspended on a light spring of spring constant 49 N/m. The spring is then compressed to a position 15 cm above the stretched equilibrium position. How much more energy does the system have at the compressed position than at the stretched equilibrium position?

15. •• A 0.25-kg object is suspended on a light spring of spring constant 49 N/m and the system is allowed to come to rest at its equilibrium position. The object is then pulled down 0.10 m from the equilibrium position and released. What is the speed of the object when it goes through the equilibrium position?

16. •• A 0.350-kg block moving vertically upward collides with a light vertical spring and compresses it 4.50 cm before coming to rest. If the spring constant is 50.0 N/m, what was the initial speed of the block? (Ignore energy losses to sound and other factors during the collision.)

17. ••• A 75-kg circus performer jumps from a 5.0-m height onto a trampoline and stretches it downward 0.30 m. Assuming that the trampoline obeys Hooke's law, (a) how far will it stretch if the performer jumps from a height of 8.0 m? (b) How far will the trampoline stretch if the performer stands still on it while taking a bow?

18. ••• A vertical spring has a 0.200-kg mass attached to it. The mass is released from rest and falls 22.3 cm before stopping. (a) Determine the spring constant. (b) Determine the speed of the mass when it has fallen only 10.0 cm.

19. ••• A 0.250-kg ball is dropped from a height of 10.0 cm onto a spring, as illustrated in ▼Fig. 13.24. If the spring has a spring constant of 60.0 N/m, (a) what distance will the spring be compressed? (Neglect energy loss during collision.) (b) On recoiling upward, how high will the ball go?

◄**FIGURE 13.24**
How far down?
See Exercise 19.

13.2 EQUATIONS OF MOTION

20. • A 0.50-kg mass oscillates in simple harmonic motion on a spring with a spring constant of 200 N/m. What are (a) the period and (b) the frequency of the oscillation?

21. • The simple pendulum in a tall clock is 0.75 m long. What are (a) the period and (b) the frequency of this pendulum?

22. • How much mass should be at the end of a spring ($k = 100$ N/m) in order to have a period of 2.0 s?

23. • If the frequency of a mass–spring system is 1.50 Hz and the mass on the spring is 5.00 kg, what is the spring constant?

24. • A breeze sets a suspended lamp into oscillation. If the period is 1.0 s, what is the distance from the ceiling to the lamp at the lowest point? Assume that the lamp is a point mass and acts as a simple pendulum.

25. • Write the general equation of motion for a mass that is on a horizontal frictionless surface and is connected to a spring at equilibrium (a) if the mass is initially pulled in the $+x$ axis from the spring (stretched) and released, and (b) if the mass is pushed in the $-x$ axis toward the spring (compressed) and released.

26. • The equation of motion for an oscillator in vertical SHM is given by $y = (0.10 \text{ m}) \sin[(100 \text{ rad/s})t]$. What are the (a) amplitude, (b) frequency, and (c) period of this motion?

27. • The displacement of an object is given by $y = (5.0 \text{ cm}) \cos[(20\pi \text{ rad/s})t]$. What are the object's (a) amplitude, (b) frequency, and (c) period of oscillation?

28. • If the displacement of an oscillator in SHM is described by the equation $y = (0.25 \text{ m}) \cos[(314 \text{ rad/s})t]$, where y is in meters and t is in seconds, what is the position of the oscillator at (a) $t = 0$, (b) $t = 5.0$ s, and (c) $t = 15$ s?

29. •• The equation of motion of a SHM oscillator is $x = (0.50 \text{ m}) \sin(2\pi f)t$, where x is in meters and t is in seconds. If the position of the oscillator is at $x = 0.25$ m at $t = 0.25$ s, what is the frequency of the oscillator?

30. IE •• The oscillations of two oscillating mass–spring systems are graphed in ▼Fig. 13.25. The mass in System A is four times that in System B. (a) Compared with System B, System A has (1) more, (2) the same, or (3) less energy. Why? (b) Calculate the ratio of energy between System B and System A.

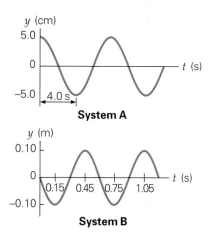

▲**FIGURE 13.25 Wave energy and equation of motion** See Exercises 30, 42, and 43.

31. •• Show that the total energy of a mass–spring system in simple harmonic motion is given by $\frac{1}{2}m\omega^2 A^2$.

32. •• Show that for a pendulum to oscillate at the same frequency as a mass on a spring, the pendulum's length must be given by $L = mg/k$.

33. •• The velocity of a vertically oscillating mass–spring system is given by $v = (0.650 \text{ m/s}) \sin[(4 \text{ rad/s})t]$. Determine (a) the amplitude and (b) the maximum acceleration of this oscillator.

34. IE •• (a) If the mass in a mass–spring system is halved, the new period is (1) 2, (2) $\sqrt{2}$, (3) $1/\sqrt{2}$, (4) $1/2$ times the old period. Why? (b) If the initial period is 3.0 s and the mass is reduced to 1/3 of its initial value, what is the new period?

35. IE •• (a) If the spring constant in a mass–spring system is halved, the new period is (1) 2, (2) $\sqrt{2}$, (3) $1/\sqrt{2}$, (4) 1/2 times the old period. Why? (b) If the initial period is 2.0 s and the spring constant is reduced to 1/3 of its initial value, what is the new period?

36. •• Students use a simple pendulum with a length of 36.90 cm to measure the acceleration of gravity at the location of their school. If it takes 12.20 s for the pendulum to complete ten oscillations, what is the experimental value of g at the school?

37. •• The equation of motion of a particle in vertical SHM is given by $y = (10 \text{ cm}) \sin[(0.50 \text{ rad/s})t]$. What are the particle's (a) displacement, (b) velocity, and (c) acceleration at $t = 1.0$ s?

38. •• What is the maximum elastic potential energy of a simple horizontal mass–spring oscillator whose equation of motion is given by $x = (0.350 \text{ m}) \sin[(7 \text{ rad/s})t]$? The mass on the end of the spring is 0.900 kg.

39. •• Two masses oscillate on light springs. The second mass is half of the first and its spring constant is twice that of the first. Which system will have the greater frequency, and what is the ratio of the frequency of the second mass to that of the first mass?

40. •• During an earthquake, the floor of an apartment building is measured to oscillate in approximately simple harmonic motion with a period of 1.95 seconds and an amplitude of 8.65 cm. Determine the maximum speed and acceleration of the floor during this motion.

41. IE •• (a) If a pendulum clock were taken to the Moon, where the acceleration due to gravity is only one-sixth (assume the figure to be exact) that on the Earth, will the period of vibration (1) increase, (2) remain the same, or (3) decrease? Why? (b) If the period on the Earth is 2.0 s, what is the period on the Moon?

42. •• The motion of a particle is described by the curve for System A in Fig. 13.25. (a) Write the equation of motion in terms of a sine or cosine function. (b) If the spring constant is 20 N/m, what is the mass of the object?

43. •• The motion of a 0.25-kg mass oscillating on a light spring is described by the curve for System B in Fig. 13.25. (a) Write the equation for the displacement of the mass as a function of time. (b) What is the spring constant of the spring?

44. ••• The forces acting on a simple pendulum are shown in ▼Fig. 13.26. (a) Show that, for the small angle approximation ($\sin \theta \approx \theta$), the force producing the motion has the same form as Hooke's law. (b) Show by analogy with a mass on a spring that the period of a simple pendulum is given by $T = 2\pi \sqrt{L/g}$. [*Hint:* Think of the effective spring constant.]

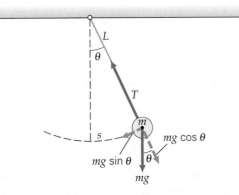

▲ **FIGURE 13.26 SHM of a pendulum** See Exercise 44.

45. ••• The acceleration as a function of time of a mass–spring system is given by $a = (0.60 \text{ m/s}^2) \sin[(2 \text{ rad/s})t]$. If the spring constant is 10 N/m, what are (a) the amplitude, (b) the initial velocity and (c) the mass of the object?

46. ••• A clock uses a pendulum that is 75 cm long. The clock is accidentally broken, and when it is repaired, the length of the pendulum is shortened by 2.0 mm. Consider the pendulum to be a simple pendulum. (a) Will the repaired clock gain or lose time? (b) By how much will the time indicated by the repaired clock differ from the correct time (taken to be the time determined by the original pendulum in 24 h)? (c) If the pendulum rod were metal, would the surrounding temperature make a difference in the timekeeping of the clock? Explain.

47. ••• The velocity of a vertically oscillating 5.00-kg mass on a spring is given by $v = (-0.600 \text{ m/s}) \sin[(6 \text{ rad/s})t]$. (a) Determine the equation of motion (y). (b) Where does the motion start and in what direction does the object move initially and with what speed? (c) Determine the period of the motion. (d) Determine the maximum force on the mass.

13.3 WAVE MOTION

48. • A sound wave has a speed of 340 m/s in air. If this wave produces a tone with a frequency of 1000 Hz, what is its wavelength?

49. • A wave on a rope that measures 10 m long takes 2.0 s to travel the whole rope. If the wavelength of the wave is 2.5 m, what is the frequency of oscillation of any piece of the rope?

50. • A student reading his physics book on a lake dock notices that the distance between two incoming wave crests is about 0.75 m, and he then measures the time of arrival between the crests to be 1.6 s. What is the approximate speed of the waves?

51. • Dolphins and bats determine the location of their prey using echolocation (see Conceptual Question 15). If it takes 15 ms for a bat to receive the ultrasonic sound wave reflected off a mosquito, how far is the mosquito from the bat? Take the speed of sound as 345 m/s.

52. • Light waves travel in a vacuum at a speed of 3.00×10^8 m/s. The frequency of blue light is about 6×10^{14} Hz. What is the approximate wavelength of the light?

53. •• A sonar generator on a submarine produces periodic ultrasonic waves at a frequency of 2.50 MHz. The wavelength of the waves in seawater is 4.80×10^{-4} m. When the generator is directed downward, an echo reflected from the ocean floor is received 10.0 s later. How deep is the ocean at that point?

54. •• The range of sound frequencies audible to the human ear extends from about 20 Hz to 20 kHz. If the speed of sound in air is 345 m/s, what are the wavelength limits of this audible range?

55. IE •• The AM frequencies on a radio dial range from 550 kHz to 1600 kHz, and the FM frequencies range from 88.0 MHz to 108 MHz. All of these radio waves travel at a speed of 3.00×10^8 m/s (speed of light). (a) Compared with the FM frequencies, the AM frequencies have (1) longer, (2) the same, or (3) shorter wavelengths. Why? (b) What are the wavelength ranges of the AM band and the FM band?

56. •• ▾Fig. 13.27a shows a snapshot of a wave traveling on a rope, and Fig. 13.27b describes the position as a function of time of a point on the rope. (a) What is the amplitude of the traveling wave? (b) What is the wavelength of the wave? (c) What is the period of the wave? (d) What is the wave speed?

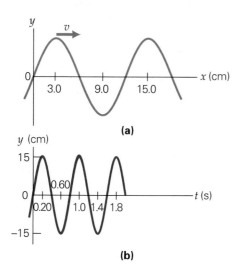

▲ FIGURE 13.27 **How high and how fast?** See Exercise 56.

57. •• Assume that P and S (primary and secondary) waves from an earthquake with a focus near the Earth's surface travel through the Earth at nearly constant but different average speeds. A monitoring station that is 1000 km from the epicenter detected the S wave to arrive at 42 s after the arrival of the P wave. If the P wave has an average speed of 8.0 km/s, what is the average speed of the S wave?

58. ••• The speed of longitudinal waves traveling in a long, solid rod is given by $v = \sqrt{Y/\rho}$, where Y is Young's modulus and ρ is the density of the solid. If a disturbance has a frequency of 40 Hz, what is the wavelength of the waves it produces in (a) an aluminum rod and (b) a copper rod? [*Hint*: See Tables 9.1 and 9.2.]

59. ••• Fred strikes a steel train rail with a hammer at a frequency of 2.50 Hz, and Wilma puts her ear to the rail 1.0 km away. (a) How long after the first strike does Wilma hear the sound? (b) What is the time interval between the successive sound pulses she hears? [*Hint*: See Tables 9.1 and 9.2 and Exercise 58.]

60. ••• Refer to the wave shown in Fig. 13.27 (Exercise 56). (a) Locate the points on the rope that have a maximum speed. Determine (b) the maximum speed, and (c) the distance between successive high and low spots on the string.

13.5 STANDING WAVES AND RESONANCE

61. • If the frequency of the third harmonic of a vibrating string is 600 Hz, what is the frequency of the first harmonic?

62. • The fundamental frequency of a stretched string is 150 Hz. What are the frequencies of (a) the second harmonic and (b) the third harmonic?

63. • If the frequency of the fifth harmonic of a vibrating string is 425 Hz, what is the frequency of the second harmonic?

64. • A standing wave is formed in a stretched string that is 3.0 m long. What are the wavelengths of (a) the first harmonic and (b) the second harmonic?

65. • If the wavelength of the third harmonic on a string is 5.0 m, what is the length of the string?

66. IE •• A piece of steel string is under tension. (a) If the tension doubles, the transverse wave speed (1) doubles, (2) halves, (3) increases by $\sqrt{2}$, (4) decreases by $\sqrt{2}$. Why? (b) If the linear mass density of a 10.0-m length of string is 0.125 kg/m and it is under a tension of 9.00 N, what is the transverse wave speed in the string? (c) What are its waves' natural frequencies?

67. •• On a violin, a correctly tuned A string has a frequency of 440 Hz. If an A string produces sound at 450 Hz under a tension of 500 N, what should the tension be to produce the correct frequency?

68. •• Will a standing wave be formed in a 4.0-m length of stretched string that transmits waves at a speed of 12 m/s if it is driven at a frequency of (a) 15 Hz or (b) 20 Hz?

69. •• Two waves of equal amplitude and frequency of 250 Hz travel in opposite directions at a speed of 150 m/s in a string. If the string is 0.90 m long, for which harmonic mode is the standing wave set up in the string?

70. •• A university physics professor buys 100 m of string and determines its total mass to be 0.150 kg. This string is used to set up a standing wave laboratory demonstration between two posts 3.0 m apart. If the desired second harmonic frequency is 35 Hz, what should be the required string tension?

71. IE •• String A has twice the tension but half the linear mass density as string B, and both strings have the same length. (a) The frequency of the first harmonic on string A is (1) four times, (2) twice, (3) half, (4) 1/4 times that of string B. Explain. (b) If the lengths of the strings are 2.5 m and the wave speed on string A is 500 m/s, what are the frequencies of the first harmonic on both strings?

72. •• You are setting up two standing string waves. You have a length of uniform piano wire that is 3.0 m long and has a mass of 0.150 kg. You cut this into two lengths, one of 1.0 m and the other of 2.0 m, and place each length under tension. What should be the ratio of tensions (expressed as short to long) so that their fundamental frequencies are the same?

73. IE •• A violin string is tuned to a certain frequency (first harmonic or the fundamental frequency). (a) If a violinist wants a higher frequency, should the string be (1) lengthened, (2) kept the same length, or (3) shortened? Why? (b) If the string is tuned to 520 Hz and the violinist puts a finger down on the string one-eighth of the string length from the neck end, what is the frequency of the string when the instrument is played this way?

74. ••• A tight uniform string with a length of 1.80 m is tied down at both ends and placed under a tension of 100 N. When it vibrates in its third harmonic (draw a sketch), the sound given off has a frequency of 75.0 Hz. What is the mass of the string?

75. ••• In a common laboratory experiment on standing waves, the waves are produced in a stretched string by an electrical vibrator that oscillates at 60 Hz (▼Fig. 13.28). The string runs over a pulley, and a hanger is suspended from the end. The tension in the string is varied by adding weights to the hanger. If the active length of the string (the part that vibrates) is 1.5 m and this length of the string has a mass of 0.10 g, what masses must be suspended to produce the first four harmonics in that length?

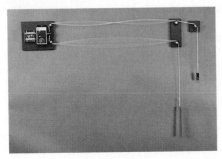

▲ **FIGURE 13.28 Standing waves on strings** Twin vibrating strings with standing waves. This demonstration model allows you to vary the string's tension, length, and type (linear mass density). Also, the vibration frequency can be adjusted. See Exercise 75.

76. ••• A student uses a 2.00-m-long steel string with a diameter of 0.90 mm for a standing wave experiment. The tension on the string is tweaked so that the second harmonic of this string vibrates at 25.0 Hz. (a) Calculate the tension the string is under. (b) Calculate the first harmonic frequency for this string. (c) If you wanted to increase the first harmonic frequency by 50%, what would be the tension in the string? [*Hint*: See Table 9.2]

PULLING IT TOGETHER: MULTICONCEPT EXERCISES

The Multiconcept Exercises require the use of more than one fundamental concept for their understanding and solution. The concepts may be from this chapter, but may include those from previous chapters.

77. To study the effects of acceleration on the period of oscillation, a student puts a grandfather clock with a 0.9929-m-long pendulum inside an elevator. Find the period of the grandfather clock (a) when the elevator is stationary, (b) when the elevator is accelerating upward at 1.50 m/s^2, (c) when the elevator is accelerating downward at 1.50 m/s^2, (d) when the cable on the elevator breaks and the elevator simply falls, and (e) when the elevator is moving upward at a constant speed of 5.00 m/s .

78. A 0.500-kg mass is attached to a vertical spring and the system is allowed to come to equilibrium. The mass is then given an initial downward speed of 1.50 m/s. The mass travels downward 25.3 cm before stopping and returning. (a) Determine the spring constant. (b) What is its speed after it falls 15.0 cm? (c) What is the acceleration of the mass at the very bottom of the motion?

79. During an earthquake, a house plant of mass 15.0 kg in a tall building oscillates with a horizontal amplitude of 10.0 cm at 0.50 Hz. What are the magnitudes of (a) the maximum velocity, (b) the maximum acceleration, and (c) the maximum force on the plant? (Assume SHM.)

80. A 2.0-kg mass resting on a horizontal frictionless surface is connected to a fixed spring. The mass is displaced 16 cm from its equilibrium position and released. At $t = 0.50$ s, the mass is 8.0 cm from its equilibrium position (and has not passed through it yet). (a) What is the period of oscillation of the mass? (b) What are the speed of the mass and the force on the mass at $t = 0.50$ s?

81. A simple pendulum is set into small-angle motion, making a maximum angle with the vertical of 5°. Its period is 2.21 s. (a) Determine its length. (b) Determine its maximum speed. (c) What is the acceleration of the pendulum bob when it is at the lowest position?

82. Spring A (50.0 N/m) is attached to the ceiling. The top of spring B (30.0 N/m) is hooked onto the bottom of spring A. Then a 0.250-kg mass is then attached to the bottom of Spring B. (a) How far will the object fall until it reaches equilibrium? (b) What is the period of the resulting oscillation?

14 Sound†

PHYSICS FACTS

◆ Sound is (a) the physical propagation of a disturbance (energy) in a medium, and (b) the physiological and psychological response generally to pressure waves.

◆ Humans cannot hear sounds with frequencies below 20 Hz—infrasound. Both elephants and rhinoceroses communicate by infrasound. Infrasound is produced by avalanches, meteors, tornadoes, earthquakes, and ocean waves.

◆ The normal audible frequency range of human hearing is between 20 Hz and 20 kHz.

◆ The visible part of the outer ear is called the *pinna*, or ear flap. Many animals move the ear flap in order to focus their hearing in a certain direction. Humans cannot do so—but some people can wiggle their ears.

◆ Ultrasound (frequency > 20 kHz) is used to make fetal images—"baby's first picture."

◆ Loud noise exposure—for example, from rock bands—is a common cause of tinnitus, or ringing in the ears.

The band shown in the chapter-opening photo is clearly giving good vibrations! We owe a lot to sound waves. Not only do they provide us with one of our main sources of enjoyment in the form of music, but they also bring us a wealth of vital information about our environment, from the chime of a doorbell to the warning shrill of a police siren to the song of a bird. Indeed, sound waves are the basis for our major form of communication—speech. These waves can also constitute highly irritating distractions (noise). But sound waves become music, speech, or noise only when our ears perceive

†The mathematics needed in this chapter involves common logarithms (base 10). You may want to review these functions in Appendix I.

them. Physically, sound is simply waves that propagate in solids, liquids, and gases. Without a medium, there can be no sound; in a vacuum, as in outer space, there is utter silence.

This distinction between the sensory and physical meanings of sound provides an answer to the old philosophical question: If a tree falls in the forest where there is no one to hear it, is there sound? The answer depends on how sound is defined—the answer is no if thinking in terms of sensory hearing, but yes if considering physical waves.

Since sound waves are all around us most of the time, we are exposed to many interesting sound phenomena. Some of the most important of these will be considered in this chapter.

14.1 Sound Waves

LEARNING PATH QUESTIONS

➥ How are sound waves generated, and what type of wave is sound?
➥ What are the regions and divisions of the sound frequency spectrum?

For sound waves to exist there must be a disturbance or vibrations in some medium. This disturbance may be the clapping of hands or the skidding of tires as a car comes to a sudden stop. Under water, you can hear the click of rocks against one another. If you put your ear to a thin wall, you can hear sounds from the other side of the wall. **Sound waves** in gases and liquids (both are fluids, Chapter 9) are primarily longitudinal waves. However, sound disturbances moving through solids can have both longitudinal and transverse components. The intermolecular interactions in solids are much stronger than in fluids and allow transverse components to propagate.

The characteristics of sound waves can be visualized by considering those produced by a tuning fork, which is essentially a metal bar bent into a U shape (▾Fig. 14.1). The prongs, or tines, vibrate when struck. The fork vibrates at its fundamental frequency, so a single tone is heard. (A *tone* is sound with a definite frequency.) The vibrations disturb the air, producing alternating high-pressure regions called *condensations* and low-pressure regions called *rarefactions*. Assuming the fork vibrates continually, the disturbances propagate outward, and a series of them can be described by a sinusoidal wave (Fig. 14.1b).

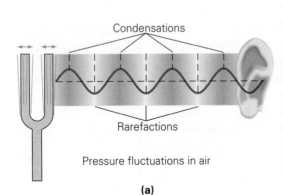

Condensations

Rarefactions

Pressure fluctuations in air

(a)

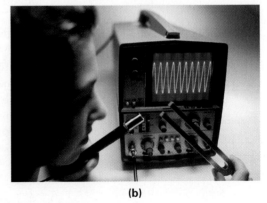

(b)

▲ **FIGURE 14.1 Vibrations make waves (a)** A vibrating tuning fork disturbs the air, producing alternating high-pressure regions (condensations) and low-pressure regions (rarefactions), which form sound waves. **(b)** After being picked up by a microphone, the pressure variations are converted to electrical signals. When these signals are displayed on an oscilloscope, the sinusoidal waveform is evident.

When the disturbances traveling through the air reach the ear, the eardrum (a thin membrane) is set into vibration by the pressure variations. On the other side of the eardrum, tiny bones (the hammer, anvil, and stirrup) carry the vibrations to the inner ear, where they are picked up by the auditory nerve.

Characteristics of the ear limit the perception of sound. Only sound waves with frequencies between about 20 Hz and 20 kHz (kilohertz) initiate nerve impulses that are interpreted by the human brain as sound. This frequency range is called the **audible region** of the **sound frequency spectrum** (▸Fig. 14.2). Hearing is most acute in the 1000 Hz–10 000 Hz range, with speech mainly in the frequency range between 300 Hz–3400 Hz (that used for the telephone).

INFRASOUND

Sound wave frequencies lower than 20 Hz are in the **infrasonic region** (infrasound). Waves in this region, which humans are unable to hear, are found in nature. Longitudinal waves generated by earthquakes have infrasonic frequencies, and these waves are used to study the Earth's interior (see Chapter 13 Insight 13.1, Earthquakes, Seismic Waves, and Seismology). Infrasonic waves are also generated by wind and weather patterns. Elephants and cattle have hearing responses in the infrasonic region and may give early warnings of earthquakes and weather disturbances, such as tornadoes. (Elephants can detect sounds with frequencies as low as 1 Hz, but the pigeon takes the infrasound hearing prize, being able to detect sound frequencies as low as 0.1 Hz.) It has been found that the vortex of a tornado produces infrasound, and the frequency changes—low frequencies when the vortex is small and higher frequencies when the vortex is large. Infrasound can be detected miles away from a tornado, and so may be a method for gaining increased warning times for tornado approaches.

Nuclear explosions produce infrasound, and after the Nuclear Test Ban Treaty of 1963, infrasound listening stations were set up to detect possible violations. Now these stations can be used to detect other sources such as earthquakes and tornadoes.

ULTRASOUND

Above 20 kHz in the sound frequency spectrum is the **ultrasonic region** (ultrasound). Ultrasonic waves can be generated by high-frequency vibrations in crystals. Ultrasonic waves cannot be detected by humans, but can be by other animals. The audible region for dogs extends to about 40 kHz, so ultrasonic or "silent" whistles can be used to call dogs without disturbing people. Cats and bats have even higher audible ranges, up to about 70 kHz and 100 kHz, respectively.

There are many practical applications of ultrasound. Because ultrasound can travel for kilometers in water, it is used in sonar to detect underwater objects and their ranges (distances), much like radar uses radio waves. Sound pulses generated by the sonar apparatus are reflected by underwater objects, and the resulting echoes are picked up by a detector. The time required for a sound pulse to make one round trip, together with the speed of sound in water, gives the distance or range of the object. Sonar is also widely used by fishermen to detect schools of fish, and in a similar manner, ultrasound is used in autofocus cameras. Distance measurements allow focal adjustments to be made.

There are applications of ultrasonic sonar in nature. Sonar appeared in the animal kingdom long before it was developed by human engineers. On their nocturnal hunting flights, bats use a kind of natural sonar to navigate in and out of their caves and to locate and catch flying insects (▾Fig. 14.3a). The bats emit pulses of ultrasound and track their prey by means of the reflected echoes. The technique is known as *echolocation*. The auditory system and data-processing capabilities of bats are truly amazing. (Note the size of the bat's ears in Fig. 14.3b.)

On the basis of the intensity of the echo, a bat can tell how big an insect is—the smaller the insect, the less intense the echo. The direction of motion of an insect is sensed by the frequency of the echo. If an insect is moving away from the bat, the returning echo will have a lower frequency. If the insect is moving toward the bat,

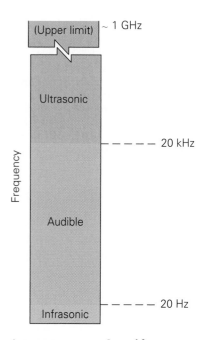

▲ **FIGURE 14.2 Sound frequency spectrum** The audible region of sound for humans lies between about 20 Hz and 20 kHz. Below this is the infrasonic region, and above it is the ultrasonic region. The upper limit is about 1 GHz, because of the elastic limitations of materials.

▶ FIGURE 14.3 **Echolocation**
(a) With the aid of their own natural sonar systems, bats hunt flying insects. The bats emit pulses of ultrasonic waves, which lie within their audible region, and use the echoes reflected from their prey to guide their attack. **(b)** Note the size of the bat's ears—good for ultrasonic hearing. Do you know why bats roost hanging upside down? See text for the answer.

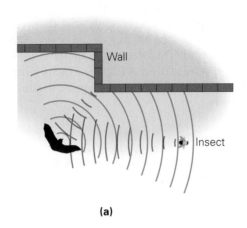

(a)

(b)

the echo will have a higher frequency. The change in frequency is known as the *Doppler effect*, which is presented in more detail in Section 14.5. Dolphins also use ultrasonic sonar to locate objects. This is very efficient since sound travels almost five times as fast in water as in air.

INSIGHT 14.1 | # Ultrasound in Medicine

Probably the best known applications of ultrasound are in medicine. For instance, ultrasound is used to obtain an image of a fetus, avoiding potentially dangerous X-rays. Ultrasonic generators (transducers) made of piezoelectric materials produce high-frequency pulses that are used to scan the designated region of the body.* When the pulses encounter a boundary between two tissues that have different densities, the pulses are reflected (Fig. 1a). These reflections are monitored by a receiving transducer, and a computer constructs an image from the reflected signals. Images of the fetus are recorded several times each second

*When an electric field is applied to a piezoelectric material, it undergoes mechanical distortion. Periodic applications allow the generation of ultrasonic waves. Conversely, when the material experiences wave pressure, an electric voltage develops. This allows the detection of ultrasonic waves.

as the transducer is scanned across the mother's abdomen. A still shot, or "echogram," of a fetus is shown in Fig. 1b. A developing fetus, which is surrounded by a sac containing the amniotic fluid, can be distinguished from other anatomical features, and the position, size, sex, and possible abnormalities may be detected.

Ultrasound can be used to assess stroke risk. Plaque deposits may accumulate on the inner walls of blood vessels and restrict blood flow. One of the major causes of stroke is the obstruction of the carotid artery in the neck, which directly affects the blood supply to the brain. The presence and severity of such obstructions may be detected by using ultrasound (Fig. 2). An ultrasonic generator is placed on the neck, and the reflections from blood cells moving through the artery are monitored to determine the rate of blood flow, thereby providing an indication of the severity of any blockage. This procedure involves shifting the frequency of the reflected waves, as described by

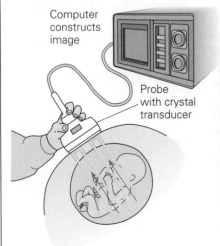

Computer constructs image

Probe with crystal transducer

(a)

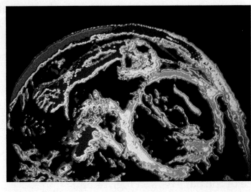

(b)

FIGURE 1 **Ultrasound in use**
(a) Ultrasound generated by transducers, which convert electrical oscillations into mechanical vibrations and vice versa, is transmitted through tissue and is reflected from internal structures. The reflected waves are detected by the transducers, and the signals are used to construct an image, or echogram. **(b)** An echogram of a well-developed fetus.

The bat, the only mammal to have evolved true flight, is a much maligned and feared creature. However, because they feed on tons of insects yearly, bats save the environment from a lot of insecticides. "Blind as a bat" is a common expression, yet bats have fairly good vision, which complements their use of echolocation. Finally, do you know why bats roost and hang upside down (Fig. 14.3b)? That is their takeoff position. Unlike birds, bats can't launch themselves from the ground. Their wings don't produce enough lift to allow takeoff directly from the ground, and their legs are so small and underdeveloped that they can't run to build up takeoff speed. So they use their claws to hang, and fall into flight when they are ready to fly.

Ultrasound is used to clean teeth with ultrasonic toothbrushes. In industrial and home applications, ultrasonic baths are used to clean metal machine parts, dentures, and jewelry. The high-frequency (short-wavelength) ultrasound vibrations loosen particles in otherwise inaccessible places. Perhaps the best known medical application of ultrasound is to view a fetus without exposing it to harmful X-rays. (See Insight 14.1, Ultrasound in Medicine.) Also, ultrasound is used to diagnose gallstones and kidney stones, and can be used to break these up by a technique called *lithotripsy* (Greek, "stone breaking").

Ultrasonic frequencies extend into the megahertz (MHz) range, but the sound frequency spectrum does not continue indefinitely. There is an upper limit of about 10^9 Hz, or 1 GHz (gigahertz), which is determined by the upper limit of the elasticity of the materials through which the sound propagates.

---

the *Doppler effect*. (More on this effect in Section 14.5, along with more on Doppler "flow meters.")

Another widely used ultrasonic device is the ultrasonic scalpel, which uses ultrasonic energy for both precise cutting and coagulation. Vibrating at about 55 kHz, the scalpel makes small incisions, at the same time causing a protein clot to form that seals blood vessels—"bloodless" surgery, so to speak. The ultrasonic scalpel has been used in gynecological procedures such as the removal of fibroid tumors, in tonsillectomies, and in many other types of surgical procedures.[†]

[†]One of the most remarkable and complicated inventions of nature is the blood clot. It can be life-saving, as when it magically forms and stops a site of bleeding, or it can be life-threatening, as when it blocks an artery in the heart or the brain.

In cases of uncontrolled bleeding, such as blunt trauma resulting from a car accident or severe wounds received in combat, rapid hemostasis (termination of bleeding) is essential. Solutions being investigated and developed include the use of diagnostic ultrasound to detect the site of bleeding and high-intensity focused ultrasound (HIFU) to induce hemostasis by ultrasonic cauterization. In China, ultrasound-guided HIFU has been used successfully for several years and is becoming the treatment of choice for many forms of cancer.

Note: Adapted from the plenary lecture given by Dr. Lawrence A. Crum at the 18th International Congress on Acoustics in Kyoto, Japan, in the summer of 2004. Professor Crum is at the Applied Physics Laboratory at the University of Washington in Seattle, Washington.

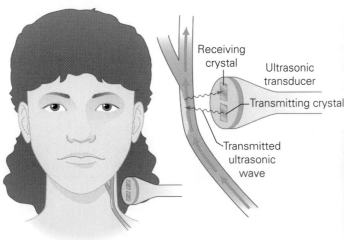
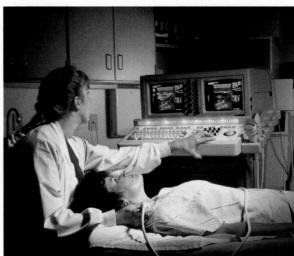

FIGURE 2 Carotid artery blockage? Ultrasound is used to measure blood flow through the carotid artery to see if there is a blockage. See text for description.

14.2 The Speed of Sound

LEARNING PATH QUESTIONS

➥ In what type of media is the speed of sound greatest and why?

➥ On what does the speed of sound generally depend?

In general, the speed at which a disturbance moves through a medium depends on physical quantities, elasticity and density of the medium. For example, as learned in Section 13.5, the wave speed in a stretched string is given by $v = \sqrt{F_T/\mu}$, where F_T is the tension in the string and μ is the linear mass density of the string.

Similar expressions describe wave speeds in solids and liquids, for which the elasticity is expressed in terms of moduli (Section 9.1). In general, the speed of sound in a solid and in a liquid is given by $v = \sqrt{Y/\rho}$ and $v = \sqrt{B/\rho}$, respectively, where Y is Young's modulus, B is the bulk modulus, and ρ is the density. The speed of sound in a gas is inversely proportional to the square root of the molecular mass, but the complex equation will not be presented here.

Solids are generally more elastic than liquids, which in turn are more elastic than gases. In a highly elastic material, the restoring forces between the atoms or molecules cause a disturbance to propagate faster. Thus, the speed of sound is generally about two to four times as fast in solids as in liquids and about ten to fifteen times as fast in solids as in gases such as air (▾ Table 14.1).

The speed of sound generally depends on the temperature in a gaseous medium. In dry air, for example, the speed of sound is 331 m/s (about 740 mi/h) at

TABLE 14.1   Speed of Sound in Various Media (Typical Values)

Medium	Speed (m/s)
Solids	
Aluminum	5100
Copper	3500
Iron	4500
Glass	5200
Polystyrene	1850
Zinc	3200
Liquids	
Alcohol, ethyl	1125
Mercury	1400
Water	1500
Gases	
Air (0 °C)	331
Air (100 °C)	387
Helium (0 °C)	965
Hydrogen (0 °C)	1284
Oxygen (0 °C)	316

0 °C. As the temperature increases, so does the speed of sound. For *normal environmental temperatures*, the speed of sound in air increases by about 0.6 m/s for each degree Celsius above 0 °C. Thus, a good approximation of the speed of sound in air for a particular (environmental) temperature is given by

$$v = (331 + 0.6T_C) \text{ m/s} \quad \textit{(speed of sound in dry air)} \quad (14.1)$$

where T_C is the air temperature in degrees Celsius.* Although not written explicitly, the units associated with the factor 0.6 are meters per second per Celsius degree $[\text{m/(s} \cdot {}^\circ\text{C)}]$.

Let's take a comparative look at the speed of sound in different media.

EXAMPLE 14.1 | **Solid, Liquid, Gas: Speed of Sound in Different Media**

From their material properties, find the speed of sound in (a) a solid copper rod, (b) liquid water, and (c) air at room temperature (20 °C).

THINKING IT THROUGH. We know that the speed of sound in a solid or a liquid depends on the elastic modulus and the density of the solid or liquid. These values are available in Tables 9.1 and 9.2. The speed of sound in air is given by Eq. 14.1.

SOLUTION.

Given: $Y_{Cu} = 11 \times 10^{10} \text{ N/m}^2$
$B_{H_2O} = 2.2 \times 10^9 \text{ N/m}^2$
$\rho_{Cu} = 8.9 \times 10^3 \text{ kg/m}^3$
$\rho_{H_2O} = 1.0 \times 10^3 \text{ kg/m}^3$
(values from Tables 9.1 and 9.2)
$T_C = 20 \text{ °C (for air)}$

Find: (a) v_{Cu} (speed in copper)
(b) v_{H_2O} (speed in water)
(c) v_{air} (speed in air)

(a) To find the speed of sound in a copper rod, the expression $v = \sqrt{Y/\rho}$ is used.

$$v_{Cu} = \sqrt{\frac{Y}{\rho}} = \sqrt{\frac{11 \times 10^{10} \text{ N/m}^2}{8.9 \times 10^3 \text{ kg/m}^3}} = 3.5 \times 10^3 \text{ m/s}$$

(b) For water, $v = \sqrt{B/\rho}$:

$$v_{H_2O} = \sqrt{\frac{B}{\rho}} = \sqrt{\frac{2.2 \times 10^9 \text{ N/m}^2}{1.0 \times 10^3 \text{ kg/m}^3}} = 1.5 \times 10^3 \text{ m/s}$$

(c) For air at 20 °C, by Eq. 14.1,

$$v_{air} = (331 + 0.6T_C) \text{ m/s} = [331 + 0.6(20)] \text{ m/s} = 343 \text{ m/s} = 3.43 \times 10^2 \text{ m/s}$$

FOLLOW-UP EXERCISE. In this Example, how many times faster is the speed of sound in copper (a) than in water and (b) than in air (at room temperature)? Compare your results with the values given at the beginning of the section. (*Answers to all Follow-Up Exercises are given in Appendix VI at the back of the book.*)

A generally useful approximate value for the speed of sound in air is $\frac{1}{3}$ km/s (or $\frac{1}{5}$ mi/s). Using this value, you can, for example, estimate how far away lightning is by counting the number of seconds between the time you observe the flash and the time you hear the associated thunder. Because the speed of light is so fast, you see the lightning flash almost instantaneously. The sound waves of the thunder travel relatively slowly, at about $\frac{1}{3}$ km/s. For example, if the interval between the two events is measured to be 6 s (often by counting "one thousand one, one thousand two, …"), the lightning stroke was about 2 km away ($\frac{1}{3}$ km/s $\times$ 6 s = 2 km, or $\frac{1}{5}$ mi/s $\times$ 6 s = 1.2 mi).

You may also have noticed the delay in the arrival of sound relative to that of light at a baseball game. If sitting in the outfield stands, you see the batter hit the ball before you hear the crack of the bat.

---

*A better approximation of these and higher temperatures is given by the expression

$$v = \left(331\sqrt{1 + \frac{T_C}{273}}\right) \text{ m/s}$$

In Table 14.1, see v for air at 100 °C, which is outside the normal environmental temperature range.

EXAMPLE 14.2 | **Good Approximations?**

(a) Show how good the approximations of $\frac{1}{3}$ km/s and $\frac{1}{5}$ mi/s are for the speed of sound. Use room temperature and dry air conditions. (b) Find the percent error of each compared to the exact value.

THINKING IT THROUGH. Taking the actual speed of sound to be given by Eq. 14.1, and converting $\frac{1}{3}$ km/s and $\frac{1}{5}$ mi/s to m/s, comparisons can be made.

SOLUTION. Listing what is given, along with the calculation of the speed of sound:

Given: $T_C = 20\,°C$ (room temperature)
$v = (331 + 0.6T_C)$ m/s
$\quad = [331 + 0.6(20)]$ m/s $= 343$ m/s
$v_{km} = \frac{1}{3}$ km/s
$v_{mi} = \frac{1}{5}$ mi/s

Find: (a) How approximations compare to actual value
(b) Percent errors

(a) Doing the conversions to standard units:

$$v_{km} = \frac{1}{3}\text{ km/s }(10^3\text{ m/km}) = 333\text{ m/s}$$

$$v_{mi} = \frac{1}{5}\text{ mi/s }(1609\text{ m/mi}) = 322\text{ m/s}$$

The approximations are somewhat reasonable, with v_{km} being the better of the two.

(b) The percent error is given by the absolute difference of the values, divided by the accepted value times 100%. So (where units cancel),

$$v_{km} = \frac{1}{3}\text{ km/s:}\quad \%\text{ error} = \frac{|343 - 333|}{343} \times 100\% = \frac{10}{343} \times 100\% = 2.9\%$$

$$v_{mi} = \frac{1}{5}\text{ mi/s:}\quad \%\text{ error} = \frac{|343 - 322|}{343} \times 100\% = \frac{21}{343} \times 100\% = 6.1\%$$

The kilometers per second approximation is considerably better.

FOLLOW-UP EXERCISE. Suppose the thunderstorm and lightning occurs on a very hot day with a dry air temperature of 38 °C. Would the percent errors in the Example increase or decrease? Justify your answer.

The speed of sound in air depends on various factors. Temperature is the most important, but there are other considerations, such as the homogeneity and composition of the air. For example, the air composition may not be "normal" in a polluted area. These effects will not be considered here. However, the dependence of the speed of sound on humidity is considered conceptually in the following Example.

CONCEPTUAL EXAMPLE 14.3 | **Speed of Sound: Sound Traveling Far and Wide**

Note that the speed of sound in *dry* air for a given temperature is given to a good approximation by Eq. 14.1. However, the moisture content of the air (humidity) varies, and this variation affects the speed of sound. At the same temperature, would sound travel faster in (a) dry air or (b) moist air?

REASONING AND ANSWER. According to an old folklore saying, "Sound traveling far and wide, a stormy day will betide." This saying implies that sound travels faster on a highly humid day, when a storm or precipitation is likely. But is the saying true?

Near the beginning of this section, it was pointed out that the speed of sound in a gas is inversely proportional to the square root of the molecular mass of the gas. So at constant pressure, is moist air more or less dense than dry air?

In a volume of moist air, a large number of water (H_2O) molecules occupy the space normally occupied by either nitrogen (N_2) or oxygen (O_2) molecules, which make up 98% of the air. Water molecules are less massive than both nitrogen and oxygen molecules. [From Section 10.3, the molecular (formula) masses are H_2O, 18 g/mol; N_2, 28 g/mol; and O_2, 32 g/mol.] Thus, the average molecular mass of a volume of moist air is less than that of dry air, and the speed of sound is greater in moist air.

This situation can be looked at in another way. Since water molecules are less massive, they have less inertia and respond to the sound wave faster than nitrogen or oxygen molecules do. The water molecules therefore propagate the disturbance faster.*

FOLLOW-UP EXERCISE. Considering only molecular masses, would you expect the speed of sound to be greatest in nitrogen, oxygen, or helium (at the same temperature and pressure)? Explain.

*Humidity was included here as an interesting consideration for the speed of sound in air. However, henceforth, in computing the speed of sound in air at a certain temperature, only dry air will be considered (Eq. 14.1) unless otherwise stated.

Always keep in mind that our discussion generally assumes ideal conditions for the propagation of sound. Actually, the speed of sound depends on many things, one of which is humidity, as the preceding Conceptual Example shows. A variety of other properties affect the propagation of sound. As an example, let's ask the question, "Why do ships' foghorns have such a low pitch or frequency?" The answer is that low-frequency sound waves travel farther than high-frequency ones under identical conditions.

This effect is explained by a couple of characteristics of sound waves. First, sound waves are attenuated (that is, lose energy) because of the viscosity of the air (Section 9.5). Second, sound waves tend to interact with oxygen and water molecules in the air. The combined result of these two properties is that the total attenuation of sound in air depends on the frequency of the sound: the higher the frequency, the more the attenuation and the shorter the distance traveled. It turns out that the attenuation increases as the *square* of the frequency. For example, a 200-Hz sound will travel 16 times as far as an 800-Hz sound to obtain the same attenuation. So, low-frequency foghorns are used. Because of this wave dependence on frequency, you might notice that when a storm's lightning is farther away, the thunder you generally hear is a low-frequency rumble. (See Insight 14.2, The Physiology and Physics of the Ear and Hearing.)

INSIGHT 14.2 | The Physiology and Physics of the Ear and Hearing

The ear consists of three basic parts: the outer ear, the middle ear, and the inner ear (Fig. 1). The visible part of the ear is the *pinna* (or ear flap), and it collects and focuses sound waves. Many animals can move the ear flap in order to focus their hearing in a particular direction; humans have generally lost this ability and must turn the head. The sound enters the ear and travels through the *ear canal* to the *eardrum* of the middle ear.

The eardrum is a membrane that vibrates in response to the pressure variations of impinging sound waves. The vibrations are transmitted through the middle ear by an intricate set of three bones called the *malleus*, or hammer; the *incus*, or anvil; and the *stapes*, or stirrup. These bones form a linkage to the *oval window*, the opening to the inner ear. The eardrum transmits sound vibrations to the bones of the middle ear, which in turn transmits the vibrations through the oval window to the fluid of the inner ear.

The inner ear consists of the *semicircular canals*, the *cochlea*, and the *auditory nerve*. The semicircular canals and the cochlea are filled with a water-like liquid. The liquid and the nerve cells in the semicircular canal play no role in the process of hearing but serve to detect rapid movements and assist in maintaining balance.

The inner surface of the cochlea, a snail-shaped organ, is lined with more than 25 000 hairlike nerve cells. These nerve cells differ from each other slightly in length and have different degrees of resiliency to the fluid waves passing through the cochlea. Different hair cells are sensitive to particular frequencies of waves. When the frequency of a compressional wave matches the natural frequency of hair cells, the cells resonate (Section 13.5) with a larger amplitude of vibration. This causes the release of electrical impulses from the nerve cells, which are transmitted to the auditory nerve. The auditory nerve carries the signals to the brain, where they are interpreted as sound.

The hair cells of the cochlea are very critical to hearing. Damage to those cells can give rise to *tinnitus*, or "ringing in the ears." Exposure to loud noises is a common cause of tinnitus and often leads to hearing loss as well. After a loud rock concert in an enclosed room, people often experience a temporary ringing in the ears and slight loss of hearing. Hair cells can be dam-

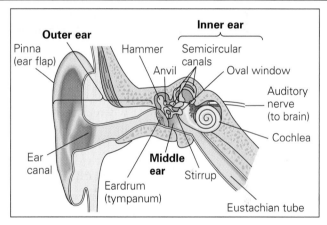

FIGURE 1 **Anatomy of the human ear** The ear converts pressure waves in the air into electrical nerve impulses that are interpreted as sounds by the brain. See text for description.

aged by loud noises temporarily or permanently. Over time, loud sounds can cause permanent injury because hair cells are lost. Because the hair cells are (resonance) frequency specific, a person may be unable to hear sounds at particular frequencies.

In a quiet room, put both thumbs in your ears firmly and listen. Do you hear a low pulsating sound? You are hearing the sound, at about 25 Hz, made by the contracting and relaxing of the muscle fibers in your hands and arms. Although in the audible range, these sounds are not normally heard, because the human ear is relatively insensitive to low-frequency sounds.

The middle ear is connected to the throat by the Eustachian tube, the end of which is normally closed. It opens during swallowing and yawning to permit air to enter and leave, so that internal and external pressures are equalized. You have probably experienced a "stopping up" of your ears with a sudden change in atmospheric pressure (for example, during rapid ascents or descents in elevators or airplanes). Swallowing opens the Eustachian tubes and relieves the excess pressure difference on the middle ear. (See the Chapter 9 Insight 9.2, An Atmospheric Effect: Possible Earaches.)

14.3 Sound Intensity and Sound Intensity Level

LEARNING PATH QUESTIONS

➥ What does sound intensity mean physically?
➥ Does doubling the sound intensity double the intensity level?
➥ How does the difference in intensity levels affect the sound intensity?

Wave motion involves the propagation of energy. The rate of energy transfer is expressed in terms of **intensity**, which is the energy transported per unit time across a unit area. Since energy divided by time is power, intensity is power divided by area:

$$\text{intensity} = \frac{\text{energy/time}}{\text{area}} = \frac{\text{power}}{\text{area}} \quad \left[I = \frac{E/t}{A} = \frac{P}{A} \right]$$

The standard units of intensity (power/area) are watts per square meter (W/m^2).

Consider a point source that sends out spherical sound waves, as shown in ▼Fig. 14.4. If there are no losses, the sound intensity at a distance R from the source is

$$I = \frac{P}{A} = \frac{P}{4\pi R^2} \quad \textit{(point source only)} \tag{14.2}$$

where P is the power of the source and $4\pi R^2$ is the area of a sphere of radius R, through which the sound energy passes perpendicularly.

The intensity of a point source of sound is therefore *inversely proportional to the square of the distance from the source* (an inverse-square relationship). Two intensities at different distances from a point source of constant power can be compared as a ratio:

$$\frac{I_2}{I_1} = \frac{P/(4\pi R_2^2)}{P/(4\pi R_1^2)} = \frac{R_1^2}{R_2^2}$$

or

$$\frac{I_2}{I_1} = \left(\frac{R_1}{R_2} \right)^2 \quad \textit{(point source only)} \tag{14.3}$$

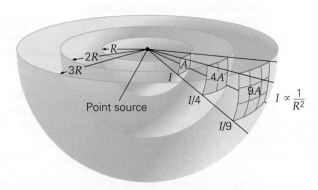

▲ **FIGURE 14.4 Intensity of a point source** The energy emitted from a point source spreads out equally in all directions. Since intensity is power divided by area, $I = P/A = P/(4\pi R^2)$, where the area is that of a spherical surface. The intensity then decreases with the distance from the source as $1/R^2$ (figure not to scale).

Suppose that the distance from a point source is doubled; that is, $R_2 = 2R_1$ or $R_1/R_2 = \frac{1}{2}$. Then

$$\frac{I_2}{I_1} = \left(\frac{R_1}{R_2}\right)^2 = \left(\frac{1}{2}\right)^2 = \frac{1}{4}$$

and

$$I_2 = \frac{I_1}{4}$$

Since the intensity decreases by a factor of $1/R^2$, doubling the distance decreases the intensity to a quarter of its original value.

A good way to understand this inverse-square relationship intuitively is to look at the geometry of the situation. As Fig. 14.4 shows, the greater the distance from the source, the larger the area over which a given amount of sound energy is spread, and thus the lower its intensity. (Imagine having to paint two walls of different areas. If you had the same amount of paint to use on each, you'd have to spread it more thinly over the larger wall.) Since this area increases as the square of the radius R, the intensity decreases accordingly—that is, as $1/R^2$.

Sound intensity is perceived by the ear as **loudness**. On the average, the human ear can detect sound waves (at 1 kHz) with an intensity as low as 10^{-12} W/m^2. This intensity (I_o) is referred to as the *threshold of hearing*. Thus, for us to hear a sound, it must not only have a frequency in the audible range, but also be of sufficient intensity. As the intensity is increased, the perceived sound becomes louder. At an intensity of 1.0 W/m^2, the sound is uncomfortably loud and may be painful to the ear. This intensity (I_p) is called the *threshold of pain*.

Note that the thresholds of pain and hearing differ by a factor of 10^{12}:

$$\frac{I_p}{I_o} = \frac{1.0 \text{ W/m}^2}{10^{-12} \text{ W/m}^2} = 10^{12}$$

That is, the intensity at the threshold of pain is a *trillion* times that at the threshold of hearing. Within this enormous range, the perceived loudness is not directly proportional to the intensity. That is, if the intensity is doubled, the perceived loudness does not double. In fact, a doubling of perceived loudness corresponds approximately to a tenfold increase in intensity. For example, a sound with an intensity of 10^{-5} W/m^2 would be perceived to be twice as loud as one with an intensity of 10^{-6} W/m^2. (The smaller the negative exponent, the larger the intensity.)

SOUND INTENSITY LEVEL: THE BEL AND THE DECIBEL

It is convenient to compress the large range of sound intensities by using a logarithmic scale (base 10) to express *intensity levels* (not to be confused with sound intensity in W/m^2). The **intensity level** of a sound must be referenced to a standard intensity, which is taken to be that of the threshold of hearing, $I_o = 10^{-12}$ W/m^2. Then, for any intensity I, the intensity level is the logarithm (or log) of the ratio of I to I_o, that is, $\log I/I_o$. For example, if a sound has an intensity of $I = 10^{-6}$ W/m^2,

$$\log \frac{I}{I_o} = \log \left(\frac{10^{-6} \text{ W/m}^2}{10^{-12} \text{ W/m}^2}\right) = \log 10^6 = 6 \text{ B}$$

(Recall that $\log_{10} 10^x = x$.) The exponent of the power of 10 in the final log term is taken to have a unit called the **bel (B)**.* Thus, a sound with an intensity of 10^{-6} W/m^2 has an intensity level of 6 B on this scale. That way, the intensity range from 10^{-12} W/m^2 to 1.0 W/m^2 is compressed into a scale of intensity levels ranging from 0 B to 12 B.

*The bel was named in honor of Alexander Graham Bell, who got the first patent on the telephone.

▶ FIGURE 14.5 **Sound intensity levels and the decibel scale** The intensity levels of some common sounds on the decibel (dB) scale.

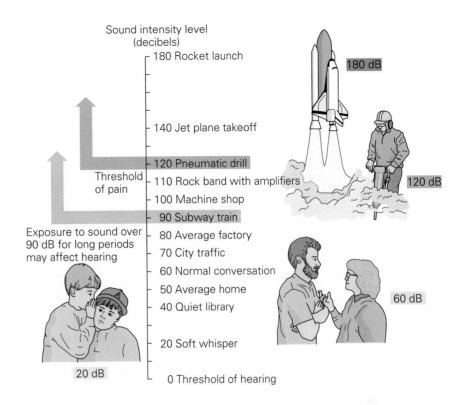

Sound intensity level
(decibels)

- 180 Rocket launch   180 dB

- 140 Jet plane takeoff

- 120 Pneumatic drill   120 dB

Threshold of pain
- 110 Rock band with amplifiers
- 100 Machine shop
- 90 Subway train

Exposure to sound over 90 dB for long periods may affect hearing
- 80 Average factory
- 70 City traffic
- 60 Normal conversation   60 dB
- 50 Average home
- 40 Quiet library

- 20 Soft whisper

20 dB
- 0 Threshold of hearing

A finer intensity scale is obtained by using a smaller unit, the **decibel (dB)**, which is a tenth of a bel. The range from 0 to 12 B corresponds to 0 to 120 dB. In this case, the equation for the relative **sound intensity level**, or **decibel level (β)**, is

$$\beta = 10 \log \frac{I}{I_o} \qquad (14.4)$$

where $I_o = 10^{-12}$ W/m². Note that the sound intensity level (in decibels, which are dimensionless) is *not* the same as the sound intensity (in watts per square meter).

The decibel intensity scale and familiar sounds at some intensity levels are shown in ▲ Fig. 14.5. Taking the decibel prize is the blue whale, which can produce sounds up to 188 dB in a frequency range of 10 Hz to 40 Hz. The sounds are transmitted hundreds of miles underwater. The blue whale also takes the size prize, being the largest creature ever known to have existed on the Earth—reaching up to a length of 33 m (108 ft) and a weight of 145 tons. By comparison, the largest dinosaur had a length of about 22 m (72 ft) and a weight of about 36 tons.

EXAMPLE 14.4 | Sound Intensity Levels: Using Logarithms

What are the intensity levels of sounds with intensities of (a) 10^{-12} W/m² and (b) 5.0×10^{-6} W/m²?

THINKING IT THROUGH. The sound intensity levels can be found by using Eq. 14.4.

SOLUTION.

Given: (a) $I = 10^{-12}$ W/m²
(b) $I = 5.0 \times 10^{-6}$ W/m²

Find: (a) β (sound intensity level)
(b) β

(a) Using Eq. 14.4,

$$\beta = 10 \log \frac{I}{I_o} = 10 \log\left(\frac{10^{-12} \text{ W/m}^2}{10^{-12} \text{ W/m}^2}\right) = 10 \log 1 = 0 \text{ dB}$$

The intensity 10^{-12} W/m² is the same as that at the threshold of hearing. (Recall that log 1 = 0, since 1 = 10^0 and log 10^0 = 0.) Note that an intensity level of 0 dB does *not* mean that there is no sound.

(b) $\beta = 10 \log \dfrac{I}{I_o} = 10 \log \left(\dfrac{5.0 \times 10^{-6} \, \text{W/m}^2}{10^{-12} \, \text{W/m}^2} \right)$

$\qquad = 10 \log(5.0 \times 10^6) = 10(\log 5.0 + \log 10^6) = 10(0.70 + 6.0) = 67 \, \text{dB}$

FOLLOW-UP EXERCISE. Note in this Example that the intensity of $5.0 \times 10^{-6} \, \text{W/m}^2$ is halfway between 10^{-6} and 10^{-5} (or 60 and 70 dB), yet this intensity does not correspond to a midway value of 65 dB. (a) Why? (b) What intensity *does* correspond to 65 dB? (Compute it to three significant figures.)

EXAMPLE 14.5 | **Intensity Level Differences: Using Ratios**

(a) What is the difference in the intensity levels if the intensity of a sound is doubled? (b) By what factors does the intensity increase for intensity level *differences* ($\Delta\beta$) of 10 dB and 20 dB?

THINKING IT THROUGH. (a) If the sound intensity is doubled, $I_2 = 2I_1$, or $I_2/I_1 = 2$, then Eq. 14.4 can be used to find the intensity level difference, $\beta_2 - \beta_1$. Recall that $\log a - \log b = \log a/b$. (b) Here it is important to note that these values are intensity level *differences*, $\Delta\beta = \beta_2 - \beta_1$, *not* intensity *levels*. The equation developed in part (a) will work. (Why?)

SOLUTION. Listing the data,

Given: (a) $I_2 = 2I_1$ *Find:* (a) $\Delta\beta$ (intensity level difference)
      (b) $\Delta\beta = 10$ dB   (b) I_2/I_1 (factors of intensity increase)
          $\Delta\beta = 20$ dB

(a) Using Eq. 14.4 and the relationship $\log a - \log b = \log a/b$, for the intensity level difference

$$\Delta\beta = \beta_2 - \beta_1 = 10[\log(I_2/I_o) - \log(I_1/I_o)] = 10 \log[(I_2/I_o)/(I_1/I_o)] = 10 \log I_2/I_1$$

Then,

$$\Delta\beta = 10 \log \dfrac{I_2}{I_1} = 10 \log 2 = 3 \, \text{dB}$$

Thus, doubling the intensity increases the intensity level by 3 dB (such as an increase from 55 dB to 58 dB).

(b) For a 10-dB difference,

$$\Delta\beta = 10 \, \text{dB} = 10 \log \dfrac{I_2}{I_1} \quad \text{and} \quad \log \dfrac{I_2}{I_1} = 1.0$$

Since $\log 10^1 = 1$, the intensity ratio is 10:1 because

$$\dfrac{I_2}{I_1} = 10^1 \quad \text{and} \quad I_2 = 10 I_1$$

Similarly, for a 20-dB difference,

$$\Delta\beta = 20 \, \text{dB} = 10 \log \dfrac{I_2}{I_1} \quad \text{and} \quad \log \dfrac{I_2}{I_1} = 2.0$$

Since $\log 10^2 = 2$,

$$\dfrac{I_2}{I_1} = 10^2 \quad \text{and} \quad I_2 = 100 I_1$$

Thus, an intensity level difference of 10 dB corresponds to changing (increasing or decreasing) the intensity by a factor of 10. An intensity level difference of 20 dB corresponds to changing the intensity by a factor of 100.

You should be able to guess the factor that corresponds to an intensity level difference of 30 dB. In general, the factor of the intensity change is $10^{\Delta\beta}$, where $\Delta\beta$ is the difference in levels of bels. Since 30 dB = 3 B and $10^3 = 1000$, the intensity changes by a factor of 1000 for an intensity level difference of 30 dB.

FOLLOW-UP EXERCISE. A $\Delta\beta$ of 20 dB and a $\Delta\beta$ of 30 dB correspond to factors of 100 and 1000, respectively, in intensity changes. Does a $\Delta\beta$ of 25 dB correspond to an intensity change factor of 500? Justify your answer.

EXAMPLE 14.6 | **Combined Sound Levels: Adding Intensities**

Sitting at a sidewalk restaurant table, a friend talks to you in normal conversation (60 dB). At the same time, the intensity level of the street traffic reaching you is also 60 dB. What is the total intensity level of the combined sounds?

THINKING IT THROUGH. It is tempting simply to add the two sound intensity levels together and say that the total is 120 dB. But intensity levels in decibels are logarithmic, so you can't add them in the normal way. However, intensities (I) can

be added arithmetically, since energy and power are scalar quantities. Then the combined intensity level can be found from the sum of the intensities.

SOLUTION. Listing the data and what is to be found:

Given: $\beta_1 = 60$ dB   *Find:* Total β
       $\beta_2 = 60$ dB

(continued on next page)

First let's find the intensities associated with the intensity levels:

$$\beta_1 = 60 \text{ dB} = 10 \log \frac{I_1}{I_o} = 10 \log\left(\frac{I_1}{10^{-12} \text{ W/m}^2}\right)$$

By inspection,

$$I_1 = 10^{-6} \text{ W/m}^2$$

That is, $I_1 = 10^{-6}$ W/m² for the 10 log term to be equal to 60 dB.

Similarly, $I_2 = 10^{-6}$ W/m², since both intensity levels are 60 dB. So the total intensity is

$$I_{\text{total}} = I_1 + I_2 = 1.0 \times 10^{-6} \text{ W/m}^2 + 1.0 \times 10^{-6} \text{ W/m}^2 = 2.0 \times 10^{-6} \text{ W/m}^2$$

Then, converting back to intensity level,

$$\beta = 10 \log \frac{I_{\text{total}}}{I_o} = 10 \log\left(\frac{2.0 \times 10^{-6} \text{ W/m}^2}{10^{-12} \text{ W/m}^2}\right) = 10 \log(2.0 \times 10^6)$$

$$= 10(\log 2.0 + \log 10^6) = 10(0.30 + 6.0) = 63 \text{ dB}$$

This value is a long way from 120 dB. Notice that the combined intensities doubled the intensity value, and the intensity level increased by 3 dB, in agreement with our finding in part (a) of Example 14.5.

FOLLOW-UP EXERCISE.   In this Example, suppose the added noise gave a total that *tripled* the sound intensity level of the conversation. What would be the total combined intensity level in this case?

PROTECT YOUR HEARING

Hearing may be damaged by excessive noise, so our ears sometimes need protection from continuous loud sounds (▼ Fig. 14.6). Hearing damage depends on the sound intensity level (decibel level) and the exposure time. The exact combinations vary for different people, but a general guide to noise levels is given in ▶ Table 14.2.

Studies have shown that sound levels of 90 dB and above will damage receptor nerves in the ear, resulting in a loss of hearing. At 90 dB, it takes 8 h or less for damage to occur. In general, if the sound level is increased by 5 dB, the safe exposure time is cut in half. For example, if a sound level of 95 dB (that of a very loud lawn mower or motorcycle) takes 4 h to damage your hearing, then a sound level of 105 dB takes only 1 h to do damage. Because of the detrimental effects on hearing, the U.S. government has set occupational noise exposure limits.

▲ **FIGURE 14.6 Protect your hearing** Continuous loud sounds can damage hearing, so our ears may need protection, as shown here. Note in Table 14.2 that the intensity level of lawn mowers is on the order of 90 dB.

TABLE 14.2   Sound Intensity Levels and Ear Damage Exposure Times

Sound	Decibels (dB)	Examples	Time of Nonstop Exposure That Can Cause Damage
Faint	30	Quiet library, whispering	
Moderate	60	Normal conversation, sewing machine	
Very loud	80	Heavy traffic, noisy restaurant, screaming child	10 hours
	90	Lawn mower, motorcycle, loud party	Less than 8 hours
	100	Chainsaw, subway train, snowmobile	Less than 2 hours
Extremely loud	110	Stereo headset at full blast, rock concert	30 minutes
	120	Dance clubs, car stereos, action movies, some musical toys	15 minutes
	130	Jackhammer, loud computer games, loud sporting events	Less than 15 minutes
Painful	140	Boom stereos, gunshot blast, firecrackers	Only seconds (for example, hearing loss can occur from a few shots of a high-powered gun if protection is not worn)

Courtesy of The EAR Foundation.

DID YOU LEARN?

➡ Sound intensity expresses the rate of energy transfer (energy/time/area or power/area).

➡ Doubling the sound intensity increases the intensity level by 3 dB. The decibel scale is a logarithmic scale, not a linear scale.

➡ An intensity level difference of 10 dB corresponds to an intensity increase (or decrease) of a factor of 10; a 20-dB difference to a factor of 100, and so on. In general, the factor of intensity change is $10^{\Delta\beta}$, where β is in bels.

14.4 Sound Phenomena

LEARNING PATH QUESTIONS

➡ What is the difference between sound reflection, refraction, and diffraction?

➡ When does total constructive interference occur?

➡ When does total destructive interference occur?

REFLECTION, REFRACTION, AND DIFFRACTION

An echo is a familiar example of the *reflection* of sound—sound "bouncing" off a surface. Sound *refraction* is less obvious than reflection, but you may have experienced it on a calm summer evening, when it is possible to hear distant voices or other sounds that ordinarily would not be audible. This effect is due to the refraction, or bending (change in direction), of the sound waves as they pass from one region into a region where the air density is different. The effect is similar to what would happen if the sound passed into another medium.

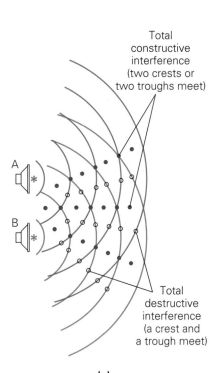

Total
constructive
interference
(two crests or
two troughs meet)

Total
destructive
interference
(a crest and
a trough meet)

(a)

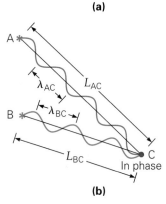

(b)

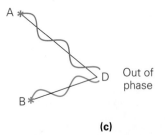

Out of
phase

(c)

▲ **FIGURE 14.8 Interference**
(a) Sound waves from two point sources spread out and interfere. **(b)** At points where the waves arrive in phase (with a zero phase difference), such as point C, constructive interference occurs. **(c)** At points where the waves arrive completely out of phase (with a phase difference of 180°), such as point D, destructive interference occurs. The phase difference at a particular point depends on the path lengths the waves travel to reach that point.

▲ **FIGURE 14.7 Sound refraction** Sound travels more slowly in the cool air near the water surface than in the upper, warmer air. As a result, the waves are refracted, or bent downward. This bending increases the intensity of the sound at a distance where it otherwise might not be heard.

The required conditions for sound to be refracted downward are a layer of cooler air near the ground or water and a layer of warmer air above it, which provides a wave speed change. These conditions occur frequently over bodies of water, which cool after sunset (▲Fig. 14.7). As a result of the cooling, the waves are refracted in an arc that may allow a distant person to receive an increased intensity of sound.

Another phenomenon is *diffraction*, described in Section 13.4. Sound may be diffracted, or spread out, around corners or around an object. We usually think of waves as traveling in straight lines. However, you can hear someone you cannot see standing around a corner. This direction change is different from that of refraction, in which no obstacle causes the bending.

Reflection, refraction, and diffraction are described in a general sense here for sound. These phenomena are important considerations for light waves as well, and will be discussed more fully in Chapters 22 and 24.

INTERFERENCE

Like waves of any kind, sound waves *interfere* when they meet. Suppose that two loudspeakers separated by some distance emit sound waves in phase at the same frequency. If we consider the speakers to be point sources, then the waves will spread out spherically and interfere (◀Fig. 14.8a). The lines from a particular speaker represent wave crests (or condensations), and the troughs (or rarefactions) lie in the intervening white areas.

In particular regions of space, there will be constructive or destructive interference. But, if two waves meet in a region where they are exactly in phase (two crests or two troughs coincide), there will be **total constructive interference** (Fig. 14.8b). Notice that the waves have the same motion at point C in the figure. If, instead, the waves meet such that the crest of one coincides with the trough of the other (at point D), the two waves will cancel each other out (Fig. 14.8c). The result will be **total destructive interference**. (See superposition in Section 13.4.)

It is convenient to describe the path lengths traveled by the waves in terms of wavelength (λ) to determine whether they arrive in phase. Consider the waves arriving at point C in Fig. 14.8b. The path lengths in this case are $L_{AC} = 4\lambda$ and $L_{BC} = 3\lambda$. The **phase difference** ($\Delta\theta$) is related to the **path length difference** (ΔL) by the simple relationship

$$\Delta\theta = \frac{2\pi}{\lambda}(\Delta L) \quad \begin{array}{l}\textit{(phase difference} \\ \textit{and path length difference)}\end{array} \quad (14.5)$$

Since 2π rad is equivalent, in angular terms, to a full wave cycle or wavelength, multiplying the path length difference by $2\pi/\lambda$ gives the phase difference in radians. For the example illustrated in Fig. 14.8b,

$$\Delta\theta = \frac{2\pi}{\lambda}(L_{AC} - L_{BC}) = \frac{2\pi}{\lambda}(4\lambda - 3\lambda) = 2\pi \text{ rad}$$

When $\Delta\theta = 2\pi$ rad, the waves are shifted by one wavelength. This is the same as $\Delta\theta = 0°$, so the waves are in phase. Thus, the waves interfere constructively in the region of point C, increasing the intensity, or loudness, of the sound detected there.

From Eq. 14.5, it can be seen that the sound waves are in phase at any point where the path length difference is zero or an integral multiple of the wavelength. That is,

$$\Delta L = n\lambda \quad (n = 0, 1, 2, 3, \ldots) \quad \begin{array}{l}\textit{(condition for}\\ \textit{constructive interference)}\end{array} \quad (14.6)$$

A similar analysis of the situation in Fig. 14.8c, where $L_{AD} = 2\frac{3}{4}\lambda$ and $L_{BD} = 2\frac{1}{4}\lambda$, gives

$$\Delta\theta = \frac{2\pi}{\lambda}\left(2\frac{3}{4}\lambda - 2\frac{1}{4}\lambda\right) = \pi \text{ rad}$$

or $\Delta\theta = 180°$. At point D, the waves are completely out of phase, and destructive interference occurs in this region.

Sound waves will be out of phase at any point where the path length difference is an odd number of half-wavelengths ($\lambda/2$), or

$$\Delta L = m\left(\frac{\lambda}{2}\right) \quad (m = 1, 3, 5, \ldots) \quad \begin{array}{l}\textit{(condition for}\\ \textit{destructive interference)}\end{array} \quad (14.7)$$

At these points, a softer, or less intense, sound will be heard or detected. If the amplitudes of the waves are exactly equal, the destructive interference is total and no sound is heard.

Destructive interference of sound waves provides a way to reduce loud noises, which can be distracting and cause hearing discomfort. The procedure is to have a reflected wave or an introduced wave with a phase difference that cancels out the original sound as much as possible. Ideally, this would be 180° out of phase with the undesirable noise. A couple of such applications, automobile mufflers and pilot headphones, were discussed in Section 13.4.

EXAMPLE 14.7 | **Pump Up the Volume: Sound Interference**

At an open-air concert on a hot day (with an air temperature of 25 °C), you sit 7.00 m and 9.10 m, respectively, from a pair of speakers, one at each side of the stage. A musician, warming up, plays a single 494-Hz tone. What do you hear in terms of intensity? (Consider the speakers to be point sources.)

THINKING IT THROUGH. The sound waves from the speakers will interfere. Is the interference, on which the intensity depends, constructive, destructive, or something in between? This depends on the path length difference, which can be computed from the given distances.

SOLUTION.

Given: $d_1 = 7.00$ m and $d_2 = 9.10$ m
$f = 494$ Hz
$T = 25\,°C$

Find: ΔL (path length difference in wavelength units to determine interference)

The path length difference (2.10 m) between the waves arriving at your location must be expressed in terms of the wavelength of the sound. To do this, we first need to know the wavelength. Given the frequency, the wavelength can be found from the relationship $\lambda = v/f$, provided that the speed

of sound, v, at the given temperature is known. The speed v can be found by using Eq. 14.1:

$$v = (331 + 0.6T_C) \text{ m/s} = [331 + 0.6(25)] \text{ m/s} = 346 \text{ m/s}$$

(continued on next page)

The wavelength of the sound waves is then

$$\lambda = \frac{v}{f} = \frac{346 \text{ m/s}}{494 \text{ Hz}} = 0.700 \text{ m}$$

Thus, the distances in terms of wavelength are

$$d_1 = (7.00 \text{ m})\left(\frac{\lambda}{0.700 \text{ m}}\right) = 10.0\lambda \quad \text{and} \quad d_2 = (9.10 \text{ m})\left(\frac{\lambda}{0.700 \text{ m}}\right) = 13.0\lambda$$

The path length difference in terms of wavelengths is

$$\Delta L = d_2 - d_1 = 13.0\lambda - 10.0\lambda = 3.0\lambda$$

This is an integral number of wavelengths ($n = 3$), so constructive interference occurs. The sounds of the two speakers reinforce each other, and you hear an intense (loud) tone at 494 Hz.

FOLLOW-UP EXERCISE.   Suppose that in this Example the tone traveled to a person sitting 7.00 m and 8.75 m, respectively, from the two speakers. What would be the situation in that case?

Another interesting interference effect occurs when two tones of nearly the same frequency ($f_1 \approx f_2$) are sounded simultaneously. The ear senses pulsations in loudness known as **beats**. The human ear can detect as many as seven beats per second. A greater number of beats per second sounds "smooth" (continuous, without any pulsations).

Suppose that two sinusoidal waves with the same amplitude, but slightly different frequencies, interfere (▼Fig. 14.9a). Figure 14.9b represents the resulting sound wave. The amplitude of the combined wave varies sinusoidally, as shown by the black curves (known as *envelopes*) that outline the wave.

What does this variation in amplitude mean in terms of what the listener perceives? A listener will hear a pulsating sound (beats), as determined by the envelopes. The maximum amplitude is 2*A* (at the point where the maxima of the two original waves interfere constructively). Detailed mathematics shows that a listener will hear the beats at a frequency called the **beat frequency (f_b)**, given by

$$f_b = |f_1 - f_2| \tag{14.8}$$

The absolute value is taken because the frequency f_b cannot be negative, even if $f_2 > f_1$. A negative beat frequency would be meaningless.

Beats can be produced when tuning forks of nearly the same frequency are vibrating at the same time. For example, using forks with frequencies of 516 Hz and 513 Hz, one can generate a beat frequency of $f_b = 516 \text{ Hz} - 513 \text{ Hz} = 3 \text{ Hz}$, and three beats are heard each second. Musicians tune two stringed instruments to the same note by adjusting the tensions in the strings until the beats disappear ($f_1 = f_2$).

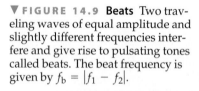

▼ **FIGURE 14.9 Beats** Two traveling waves of equal amplitude and slightly different frequencies interfere and give rise to pulsating tones called beats. The beat frequency is given by $f_b = |f_1 - f_2|$.

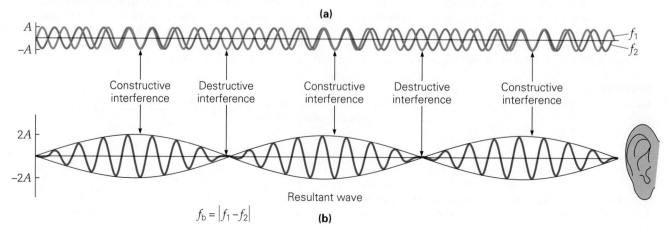

(a)

Constructive interference　Destructive interference　Constructive interference　Destructive interference　Constructive interference

Resultant wave

$f_b = |f_1 - f_2|$

(b)

➡ Reflection is the "bouncing off" of sound from a surface; refraction is the "bending" (change in direction) of sound when passing into a region of different wave speeds; diffraction is the spreading out of sound around corners or objects.
➡ Total constructive interference occurs when two waves meet and are exactly in phase (two crests or two troughs coincide).
➡ Total destructive interference occurs when two waves meet and are exactly out of phase (a crest of one wave coincides with a trough of the other wave).

14.5 The Doppler Effect

LEARNING PATH QUESTIONS

➡ What gives rise to the Doppler effect?
➡ What does an aircraft sonic boom indicate?
➡ What meant by Mach number?

Standing along a highway, the **pitch** (the perceived frequency) of the sound of the horn of a moving car or truck is heard to be higher as the vehicle approaches and lower as it recedes. Variations in the frequency of the motor noise can also be heard when a race car passes by in going around a track. A variation in the perceived sound frequency due to the motion of the source is an example of the **Doppler effect**. (The Austrian physicist Christian Doppler (1803–1853) first described this effect.)

As ▼Fig. 14.10 shows, the sound waves emitted by a moving source tend to bunch up in front of the source and spread out in back. The Doppler shift in frequency can be found by assuming that the air is at rest in a reference frame such as that depicted in ▼Fig. 14.11. The speed of sound in air is v, and the speed of the moving source is v_s. The frequency of the sound produced by the source is f_s. In one period, $T = 1/f_s$, a wave crest moves a distance $d = vT = \lambda$. (The sound wave would travel this distance in still air in any case, regardless of whether the source is moving.) But in one period, the source travels a distance $d_s = v_sT$ before emitting another wave crest. The distance between the successive wave crests is thus shortened to a wavelength λ':

$$\lambda' = d - d_s = vT - v_sT = (v - v_s)T = \frac{v - v_s}{f_s}$$

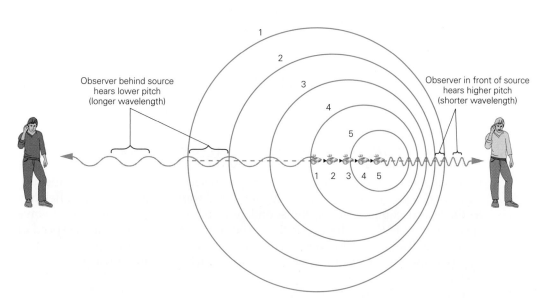

▲ **FIGURE 14.10 The Doppler effect for a moving source** The sound waves bunch up in front of a moving source—the whistle—giving a higher frequency there. The waves trail out behind the source, giving a lower frequency there.

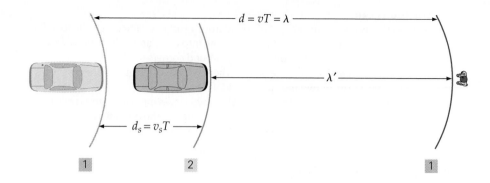

▶ FIGURE 14.11 **The Doppler effect and wavelength** Sound from a moving car's horn travels a distance d in a time T. During this time, the car (the source) travels a distance d_s before putting out a second pulse, thereby shortening the observed wavelength of the sound in the approaching direction.

The frequency heard by the observer (f_o) is related to the shortened wavelength by $f_o = v/\lambda'$, and substituting λ' gives

$$f_o = \frac{v}{\lambda'} = \left(\frac{v}{v - v_s}\right)f_s$$

or

$$f_o = \left(\frac{1}{1 - \dfrac{v_s}{v}}\right)f_s \qquad \begin{array}{l}\textit{(source moving toward}\\ \textit{a stationary observer}\\ \textit{where } v_s = \textit{speed of source}\\ \textit{and } v = \textit{speed of sound)}\end{array} \qquad (14.9)$$

Since $1 - (v_s/v)$ is less than 1, f_o is greater than f_s in this situation. For example, suppose that the speed of the source is a tenth of the speed of sound; that is, $v_s = v/10$, or $v_s/v = \frac{1}{10}$. Then, by Eq. 14.9, $f_o = \frac{10}{9}f_s$.

Similarly, when the source is moving away from the observer ($\lambda' = d + d_s$), the observed frequency is given by

$$f_o = \left(\frac{v}{v + v_s}\right)f_s = \left(\frac{1}{1 + \dfrac{v_s}{v}}\right)f_s \qquad \begin{array}{l}\textit{(source moving away}\\ \textit{from a stationary}\\ \textit{observer)}\end{array} \qquad (14.10)$$

Here, f_o is less than f_s. (Why?)

Combining Eqs. 14.9 and 14.10 yields a general equation for the observed frequency with a moving source and a stationary observer:

$$f_o = \left(\frac{v}{v \pm v_s}\right)f_s = \left(\frac{1}{1 \pm \dfrac{v_s}{v}}\right)f_s \qquad \left\{\begin{array}{l}- \textit{for source moving toward}\\ \quad \textit{stationary observer}\\ + \textit{for source moving away}\\ \quad \textit{from stationary observer}\end{array}\right. \qquad (14.11)$$

As you might expect, the Doppler effect also occurs with a moving observer and a stationary source, although this situation is a bit different. As the observer moves toward the source, the distance between successive wave crests is the normal wavelength (or $\lambda = v/f_s$), but the measured wave speed is different. Relative to the approaching observer, the sound from the stationary source has a wave speed of $v' = v + v_o$, where v_o is the speed of the observer and v is the speed of sound in still air. (The observer moving toward the source is moving in a direction opposite that of the propagating waves and thus meets more wave crests in a given time.)

With $\lambda = v/f_s$ the observed frequency is then

$$f_o = \frac{v'}{\lambda} = \left(\frac{v + v_o}{v}\right)f_s$$

or

$$f_o = \left(1 + \frac{v_o}{v}\right)f_s \qquad \begin{array}{l}\textit{(observer moving toward a}\\ \textit{stationary source where}\\ v_o = \textit{speed of observer and}\\ v = \textit{speed of sound)}\end{array} \qquad (14.12)$$

Similarly, for an observer moving away from a stationary source, the perceived wave speed is $v' = v - v_o$ and

$$f_o = \frac{v'}{\lambda} = \left(\frac{v - v_o}{v}\right)f_s$$

or

$$f_o = \left(1 - \frac{v_o}{v}\right)f_s \qquad \begin{array}{l}\textit{(observer moving away from a}\\ \textit{stationary source)}\end{array} \qquad (14.13)$$

Equations 14.12 and 14.13 can be combined into a general equation for a moving observer and a stationary source:

$$f_o = \left(\frac{v \pm v_o}{v}\right)f_s = \left(1 \pm \frac{v_o}{v}\right)f_s \qquad \left\{\begin{array}{l}- \textit{for observer moving}\\ \quad \textit{toward stationary source}\\ + \textit{for observer moving away}\\ \quad \textit{from stationary source}\end{array}\right. \qquad (14.14)$$

PROBLEM-SOLVING HINT

You may find it difficult to remember whether a plus or minus sign is used in the general equations for the Doppler effect. Let your experience help you. For the common case of a stationary observer, the frequency of the sound increases when the source approaches, so the denominator in Eq. 14.11 must be smaller than the numerator. Accordingly, in this case you use the minus sign. When the source is receding, the frequency is lower. The denominator in Eq. 14.11 must then be larger than the numerator, and you use the plus sign. Similar reasoning will help you choose a plus or minus sign for the numerator in Eq. 14.14. See Eq. 14.14a in footnote on the next page.

EXAMPLE 14.8 | On the Road Again: The Doppler Effect

As a truck traveling at 96 km/h approaches and passes a person standing along the highway, the driver sounds the horn. If the horn has a frequency of 400 Hz, what are the frequencies of the sound waves heard by the person (a) as the truck approaches and (b) after it has passed? (Assume that the speed of sound is 346 m/s.)

THINKING IT THROUGH. This situation is an application of the Doppler effect, Eq. 14.11, with a moving source and a stationary observer. In such problems, it is important to identify the data correctly.

SOLUTION. In Doppler situations, it is important to clearly list what is to be found.

Given: $v_s = 96$ km/h $= 27$ m/s
$f_s = 400$ Hz
$v = 346$ m/s

Find: (a) f_o (observed frequency while truck is approaching)
(b) f_o (observed frequency while truck is moving away)

(a) From Eq. 14.11 with a minus sign (source approaching stationary observer),

$$f_o = \left(\frac{v}{v - v_s}\right)f_s = \left(\frac{346 \text{ m/s}}{346 \text{ m/s} - 27 \text{ m/s}}\right)(400 \text{ Hz}) = 434 \text{ Hz}$$

(b) A plus sign is used in Eq. 14.11 when the source is moving away:

$$f_o = \left(\frac{v}{v + v_s}\right)f_s = \left(\frac{346 \text{ m/s}}{346 \text{ m/s} + 27 \text{ m/s}}\right)(400 \text{ Hz}) = 371 \text{ Hz}$$

FOLLOW-UP EXERCISE. Suppose that the observer in this Example were initially moving toward and then past a stationary 400-Hz source at a speed of 96 km/h. What would be the observed frequencies? (Would they differ from those for the moving source?)

There are also cases in which both the source and the observer are moving, either toward or away from one another. These will not be considered mathematically here, but will be conceptually in the next Conceptual Example.*

CONCEPTUAL EXAMPLE 14.9 | ## It's All Relative: Moving Source and Moving Observer

Suppose a sound source and an observer are moving away from each other in opposite directions, each at half the speed of sound in air. In this case, the observer would (a) receive sound with a frequency higher than the source frequency, (b) receive sound with a frequency lower than the source frequency, (c) receive sound with the same frequency as the source frequency, or (d) receive no sound from the source.

REASONING AND ANSWER.   As we know, when a source moves away from a stationary observer, the observed frequency is lower (Eq. 14.10). Similarly, when an observer moves away from a stationary source, the observed frequency is also lower (Eq. 14.13). With both source and observer moving away from each other in opposite directions, the combined effect would make the observed frequency even less, so neither (a) nor (c) is the answer.

It would appear that (b) is the correct answer, but (d) must logically eliminated for completeness. Remember that the speed of sound relative to the air is constant. Therefore, (d) would be correct *only if the observer is moving faster than the speed of sound* relative to the air. Since the observer is moving at only half the speed of sound, (b) is the correct answer.

FOLLOW-UP EXERCISE.   In this Example, what would be the result if both the source and the observer were traveling in the same direction with the same subsonic speed? (*Subsonic*, as opposed to *supersonic*, refers to a speed that is less than the speed of sound in air.)

The Doppler effect also applies to light waves, although the equations describing the effect are different from those just given. When a distant light source such as a star moves away from us, the frequency of the light we receive from it is lowered. That is, the light is shifted toward the red (long-wavelength) end of the spectrum, an effect known as a *Doppler red shift*. Similarly, the frequency of light from an object approaching us is increased—the light is shifted toward the blue (short-wavelength) end of the spectrum, producing a *Doppler blue shift*. The magnitude of the shift is related to the speed of the source.

The Doppler shift of light from astronomical objects is very useful to astronomers. The rotation of a planet, a star, or some other body can be established by looking at the Doppler shifts of light from opposite sides of the object: because of the rotation, one side is receding (and hence is red-shifted) and the other is approaching (and thus is blue-shifted). Similarly, the Doppler shifts of light from stars in different regions of our galaxy, the Milky Way, indicate that the galaxy is rotating.

You have been subjected to a practical application of the Doppler effect if you have ever been caught speeding in your car by police radar, which uses reflected radio waves. (*Radar* stands for *r*adio *d*etecting *a*nd *r*anging and is similar to underwater sonar, which uses ultrasound.) If radio waves are reflected from a parked car, the reflected waves return to the source with the same frequency. But for a car that is moving toward a patrol car, the reflected waves have a higher frequency, or are Doppler-shifted.

Actually, there is a double Doppler shift: In receiving the wave, the moving car acts like a moving observer (the first Doppler shift), and in reflecting the wave, the car acts like a moving source emitting a wave (the second Doppler shift). The magnitudes of the shifts depend on the speed of the car. A computer quickly calculates this speed and displays it for the police officer.

For other important medical and weather applications of the Doppler effect, see Insight 14.3, Doppler Applications: Blood Cells and Raindrops.

*In the case of both the observer and source moving,

$$f_o = \left(\frac{v \pm v_o}{v \mp v_s} \right) f_s \qquad (14.14a)$$

From experience, you should know there would be a frequency increase when the observer and source approach each other, and vice versa. That is, the upper signs in the numerator and denominator apply if the observer and source move toward each other, and the lower signs apply if they are moving away. (See Exercise 54.)

SONIC BOOMS

Consider a jet plane that can travel at supersonic speeds. As the speed of a moving source of sound approaches the speed of sound, the waves ahead of the source come close together (▼Fig. 14.12a). When a plane is traveling at the speed of sound, the waves can't outrun it, and they pile up in front. At supersonic speeds, the waves overlap. This overlapping of a large number of waves produces many points of constructive interference, forming a large pressure ridge, or *shock wave*. This kind of wave is sometimes called a *bow wave* because it is analogous to the wave produced by the bow of a boat moving through water at a speed greater than the speed of the water waves. Figure 14.12b shows the shock wave of a bullet traveling at 500 m/s.

From aircraft traveling at supersonic speed, the shock wave trails out to the sides and downward. When this pressure ridge passes over an observer on the ground, the large concentration of energy produces what is known as a **sonic boom**. There is really a double boom, because shock waves are formed at both ends of the aircraft. Under certain conditions, the shock waves can break windows and cause other damage to structures on the ground. (Sonic booms are no longer heard as frequently as in the past. Pilots are now instructed to fly supersonically only at high altitudes and away from populated areas.)

On a smaller scale, you have probably heard a "mini" sonic boom—the "crack" of a whip. This must mean that the whip's tip has somehow attained supersonic speed. How does this happen? Whips generally taper down from the handle to

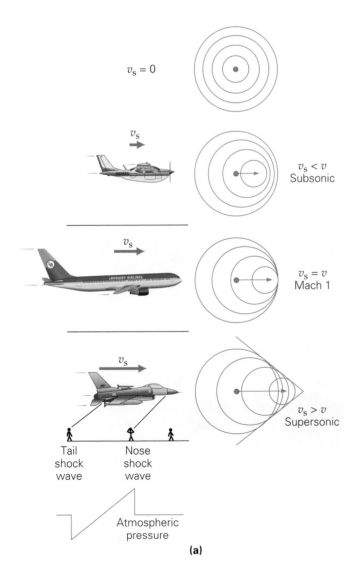

(b)

▲ **FIGURE 14.12 Bow waves and sonic booms (a)** When an aircraft exceeds the speed of sound in air, v_s, the sound waves form a pressure ridge, or shock wave. As the trailing shock wave passes over the ground, observers hear a sonic boom (actually, two booms, because shock waves are formed at the front and tail of the plane). **(b)** A bullet traveling at a speed of 500 m/s. Note the shock waves produced (and the turbulence behind the bullet). The image was made by using interferometry with polarized light and a pulsed laser, with an exposure time of 20 ns.

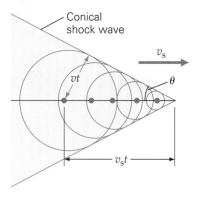

▲ FIGURE 14.13 Shock wave cone and Mach number When the speed of the source (v_s) is greater than the speed of sound in air (v), the interfering spherical sound waves form a conical shock wave that appears as a V-shaped pressure ridge when viewed in two dimensions. The angle θ is given by $\sin \theta = v/v_s$, and the inverse ratio v_s/v is called the Mach number.

the tip, which may have several frayed strands. When the whip is given a flick of the wrist, a wave pulse is sent down the length of the whip. Treating the whip pulse as a string wave pulse, recall that the speed of the pulse depends inversely on the linear mass density, which decreases toward the whip's tip. Thus the speed of the pulse increases to the point that at the tip, it is greater than the speed of sound. The "crack" is made by the air rushing back into the region of reduced pressure created by the final flip of the whip's tip, much as the sonic boom from a supersonic jet trails behind the jet.

A common misconception is that a sonic boom is heard only when a plane breaks the sound barrier. As an aircraft approaches the speed of sound, the pressure ridge in front of it is essentially a barrier that must be overcome with extra power. However, once supersonic speed is reached, the barrier is no longer there, and the shock waves, continuously created, trail behind the plane, producing booms for everyone along its ground path.

Ideally, the sound waves produced by a supersonic aircraft form a cone-shaped shock wave (◄Fig. 14.13). The waves travel outward with a speed v, and the speed of the source (plane) is v_s. Note from the figure that the angle between a line tangent to the spherical waves and the line along which the plane is moving is given by

$$\sin \theta = \frac{vt}{v_s t} = \frac{v}{v_s} = \frac{1}{M} \qquad (14.15)$$

INSIGHT 14.3 | Doppler Applications: Blood Cells and Raindrops

BLOOD CELLS

As learned in Insight 14.1, Ultrasound in Medicine, ultrasound provides a variety of uses in the medical field. Since the Doppler effect can be used to detect and provide information on moving objects, it is used to examine blood flow in the major arteries and veins of the arms and legs (Fig. 1a). In this application, the Doppler effect is used to measure the blood flow speed. Ultrasound reflects from red blood cells with a change in frequency according to the speed of the cells. The overall flow speed helps physicians diagnose such things as blood clots, arterial occlusion (closing), and venous insufficiency. Ultrasound procedures offer a less invasive alternative to other diagnostic procedures, such as arteriography (X-ray pictures of an artery after the injection of a dye).

Another medical use of ultrasound is the echocardiogram, which is an examination of the heart. On a monitor, this ultrasonic technique can display the beating movements of the heart, and the physician can see the heart's chambers, valves, and blood flow into and out of the organ (Fig. 1b).

RAINDROPS

Radar has been used since the early 1940s to provide information about rainstorms and other forms of precipitation.

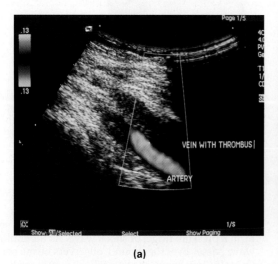

(a)

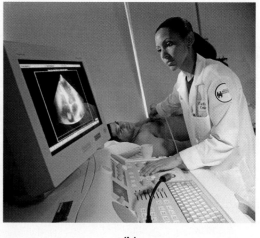

(b)

FIGURE 1 (a) Blood flow and blockage This Doppler ultrasound scan shows a deep vein thrombosis in a patient's leg. The thrombus (clot) blocking the vein is the dark area right of center. Blood flow in an adjacent artery (orange) is slowed due to the clot. In extreme cases, a clot can break away and be carried to the lungs, where it can block an artery and cause a potentially fatal pulmonary embolism (blockage of a blood vessel). **(b) Echocardiogram** This ultrasonic procedure can display the beating movements of the heart, the heart chambers, valves, and blood flow as it makes it way in and out of the organ.

The inverse ratio of the speeds is called the **Mach number (M)**, named after Ernst Mach (1838–1916), an Austrian physicist who used it in studying supersonics, and is given by

$$M = \frac{v_s}{v} = \frac{1}{\sin \theta} \qquad (14.16)$$

If v equals v_s, the plane is flying at the speed of sound, and the Mach number is 1 (that is, $v_s/v = 1$). Therefore, a Mach number less than 1 indicates a subsonic speed, and a Mach number greater than 1 indicates a supersonic speed. In the latter case, the Mach number tells the speed of the aircraft in terms of a multiple of the speed of sound. A Mach number of 2, for instance, indicates a speed twice the speed of sound. Note that since $\sin \theta \leq 1$, no shock wave can exist unless $M \geq 1$.

DID YOU LEARN?

➥ The Doppler effect is caused by the relative motion of a sound source and observer, giving rise to a variation in perceived frequency.

➥ Sonic booms are caused by a pressure ridge generated by supersonic aircraft—one flying faster than the speed of sound.

➥ The Mach number (M) is given by the ratio of the speed of the sound source (v_s) and the speed of sound (v), so, for example, if $M = v_s/v = 2$, then $v_s = 2v$ and the source is traveling twice the speed of sound.

This information is obtained from the intensity of the reflected signal. Such conventional radars can also detect the hooked (rotational) "signature" of a tornado, but only after the storm is well developed.

A major improvement in weather forecasting came about with the development of a radar system that could measure the Doppler frequency shift in addition to the magnitude of the echo signal reflected from precipitation (usually raindrops). The Doppler shift is related to the velocity of the precipitation blown by the wind.

A Doppler-based radar system (Fig. 2a) can penetrate a storm and monitor its wind speeds. The direction of a storm's wind-driven rain gives a wind "field" map of the affected region. Such maps provide strong clues of developing tornadoes, so meteorologists can detect tornadoes much earlier than was ever before possible (Fig. 2b). With Doppler radar, forecasters have been able to predict tornadoes as much as 20 min before they touch down, compared with just over 2 min with conventional radar. Doppler radar has saved many lives with this increased warning time. The National Weather Service has a network of Doppler radars around the United States, and Doppler radar scans are now common on both TV weather forecasts and the Internet.

Doppler radars installed at major airports have another use: to detect wind shears. Several airplane crashes and near-crashes have been attributed to downward wind bursts (also known as microbursts or downbursts). Such strong downdrafts cause wind shears capable of forcing landing aircraft to crash. Wind bursts generally result from high-speed downdrafts in the turbulence of thunderstorms, but they can also occur in clear air when rain evaporates high above ground. Since Doppler radar can detect the wind speed and the direction of raindrops in clouds, as well as dust and other objects floating in the air, it can provide an early warning against dangerous wind shear conditions. Two or three radar sites are needed to detect motions in two or three directions (dimensions), respectively.

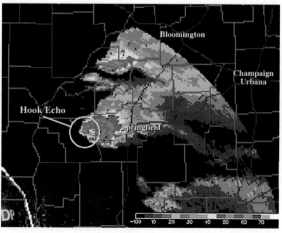

FIGURE 2 Doppler radar (a) A Doppler radar installation. **(b)** Doppler radar depicts the precipitation inside a thunderstorm. A hook echo is a signature of a possible tornado.

(a)

(b)

14.6 Musical Instruments and Sound Characteristics

LEARNING PATH QUESTIONS

➡ In what sense are standing waves associated with musical instruments?

➡ What are the sensory effects of sound intensity, frequency, and waveform?

Musical instruments provide good examples of standing waves and boundary conditions. On some stringed instruments, different notes are produced by using finger pressure to vary the lengths of the strings (◀Fig. 14.14). As was learned in Section 13.5, the natural frequencies of a stretched string (fixed at each end, as is the case for the strings on an instrument) are $f_n = n(v/2L)$, from Eq. 13.20. The speed of the wave in the string is given by $v = \sqrt{F_T/\mu}$. Initially adjusting the tension in a string tunes it to a particular (fundamental) frequency. Then the effective length of the string is varied by finger location and pressure.

Standing waves can be set up in air columns. You have probably done this in blowing across the open top of a soda bottle, producing an audible tone (▼Fig. 14.15a)*. The blowing across the bottle excites the fundamental mode of the column of air in the bottle. The frequency of the tone depends on the length of the

* A more complicated phenomenon, called Helmholtz resonance, occurs here, but for simplicity and standing waves, assume the bottle to be a circular cylinder. (See Conceptual Question 20 and answer to the Follow-up Exercise.)

▲ **FIGURE 14.14 A shorter vibrating string, a higher frequency** Different notes are produced on stringed instruments such as guitars, violins, and cellos by placing a finger on a string to change its effective, or vibrating, length.

(a)

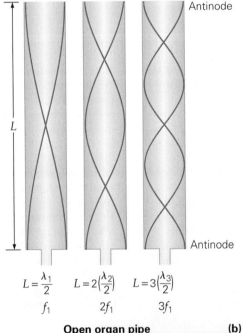

$$L = \frac{\lambda_1}{2} \qquad L = 2\left(\frac{\lambda_2}{2}\right) \qquad L = 3\left(\frac{\lambda_3}{2}\right)$$
$$f_1 \qquad\qquad 2f_1 \qquad\qquad 3f_1$$

Open organ pipe

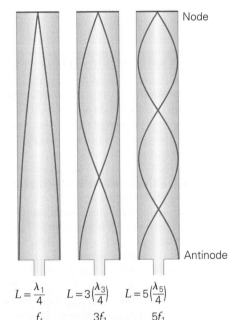

$$L = \frac{\lambda_1}{4} \qquad L = 3\left(\frac{\lambda_3}{4}\right) \qquad L = 5\left(\frac{\lambda_5}{4}\right)$$
$$f_1 \qquad\qquad 3f_1 \qquad\qquad 5f_1$$

(b)   Closed organ pipe

(c)

▲ **FIGURE 14.15 Standing waves (a)** When air is blown across the open top of a bottle, the air flow can cause an audible tone. **(b)** Longitudinal standing waves (illustrated here as sinusoidal curves) are formed in vibrating air columns in pipes. An open pipe has antinodes at both open ends. A closed pipe has node at the closed end and an antinode at the open end. **(c)** A modern pipe organ. The pipes can be open or closed.

air column. With more liquid in the bottle, the air column is shorter and the frequency higher. There will be an antinode at the open end of the bottle and a node at the liquid surface or the bottom of an empty bottle.

Standing waves are the basis of wind musical instruments. For example, consider a pipe organ with fixed lengths of pipe, which may be open or closed (Fig. 14.15b). An open pipe is open at both ends, and a closed pipe is closed at one end and open at the other (the end with the antinode). Analysis similar to that done in Section 13.5 for a stretched string with the proper boundary conditions shows that the natural frequencies of the pipes are

$$ f_n = \frac{v}{\lambda_n} = n\left(\frac{v}{2L}\right) = nf_1 \quad (n = 1, 2, 3, \dots) \qquad \begin{array}{l}\textit{(natural frequencies}\\ \textit{for an open pipe–open}\\ \textit{on both ends)}\end{array} \quad (14.17) $$

and

$$ f_m = \frac{v}{\lambda_m} = m\left(\frac{v}{4L}\right) = mf_1 \quad (m = 1, 3, 5, \dots) \qquad \begin{array}{l}\textit{(natural frequencies}\\ \textit{for a closed pipe–}\\ \textit{closed on one end)}\end{array} \quad (14.18) $$

where v is the speed of sound in air. Note that the natural frequencies depend on the length of the pipe. This is an important consideration in a pipe organ (Fig. 14.15c), particularly in selecting the dominant or fundamental frequency. (The diameter of the pipe is also a factor, but is not considered in this simple analysis.)

The same physical principles apply to wind and brass instruments. In all of these, human breath is used to create standing waves in an open tube. Most such instruments allow the player to vary the effective length of the tube and thus the pitch produced—either with the help of slides or valves that vary the actual length of tubing in which the air can resonate, as in most brasses, or by opening and closing holes in the tube, as in woodwinds (▼Fig. 14.16).

Recall from Section 13.5 that a musical note or tone is referenced to the fundamental vibrational frequency of an instrument. In musical terms, the first overtone is the second harmonic, the second overtone is the third harmonic, and so on. Note that for a closed organ pipe (Eq. 14.18), the even harmonics are missing.

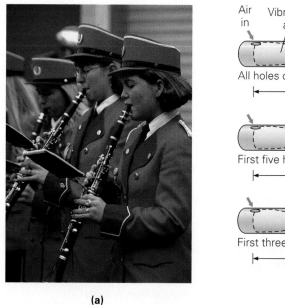

(a)

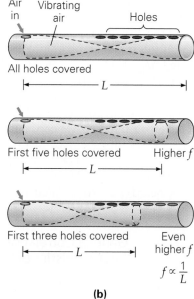

(b)

▲ FIGURE 14.16 **Wind instruments** (a) Wind instruments, such as clarinets, are essentially open tubes. (b) The effective length of the air column, and hence the pitch of the sound, is varied by opening and closing holes along the tube. The frequency f is inversely proportional to the effective length L of the air column.

EXAMPLE 14.10 | Pipe Dreams: Fundamental Frequency

A particular open organ pipe has a length of 0.653 m. Taking the speed of sound in air to be 345 m/s, what is the fundamental frequency of this pipe?

THINKING IT THROUGH.   The fundamental frequency ($n = 1$) of an open pipe is given directly by Eq. 14.17. Physically, there is a half-wavelength ($\lambda/2$) in the length of the pipe, so $\lambda = 2L$.

SOLUTION.

Given: $L = 0.653$ m
          $v = 345$ m/s (speed of sound)

Find: f_1 (fundamental frequency)

With $n = 1$ in Eq. 14.17,

$$f_1 = \frac{v}{2L} = \frac{345 \text{ m/s}}{2(0.653 \text{ m})} = 264 \text{ Hz}$$

This frequency is middle C (C_4).

FOLLOW-UP EXERCISE.   A closed organ pipe has a fundamental frequency of 256 Hz. What would be the frequency of its first overtone? Is this frequency audible?

Perceived sounds are described by terms whose meanings are similar to those used to describe the physical properties of sound waves. Physically, a wave is generally characterized by *intensity*, *frequency*, and *waveform* (harmonics). The corresponding terms used to describe the sensations of the ear are *loudness*, *pitch*, and *quality* (or timbre). These general correlations are shown in ▼Table 14.3. However, the correspondence is not perfect. The physical properties are objective and can be measured directly. The sensory effects are subjective and vary from person to person. (Think of temperature as measured by a thermometer and by the sense of touch.)

Sound intensity and its measurement on the decibel scale were covered in Section 14.3. Loudness is related to intensity, but the human ear responds differently to sounds of different frequencies. For example, two tones with the same intensity (in watts per square meter) but different frequencies might be judged by the ear to be different in loudness.

Frequency and *pitch* are often used synonymously, but again there is an objective–subjective difference: If the same low-frequency tone is sounded at two intensity levels, most people say that the more intense sound has a lower pitch, or perceived frequency.

The curves in the graph of intensity level versus frequency shown in ▶Fig. 14.17 are called *equal-loudness contours* (or Fletcher–Munson curves, after the researchers who generated them). These contours join points representing intensity–frequency combinations that a person with average hearing judges to be equally loud. The top curve shows that the decibel level of the threshold of pain (120 dB) does not vary a great deal over the normal hearing range, regardless of the frequency of the sound. In contrast, the threshold of hearing, represented by the lowest contour, varies widely with frequency. For a tone with a frequency of 2000 Hz, the threshold of hearing is 0 dB, but a 20-Hz tone would have to have an intensity level of over 70 dB just to be heard (the extrapolated *y*-intercept of the lowest curve).

TABLE 14.3   General Correlation between Perceptual and Physical Characteristics of Sound

Sensory Effect	Physical Wave Property
Loudness	Intensity
Pitch	Frequency
Quality (timbre)	Waveform (harmonics)

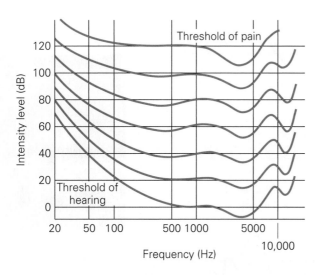

▲ FIGURE 14.17 **Equal-loudness contours** The curves indicate tones that are judged to be equally loud, although they have different frequencies and intensity levels. For example, on the lowest contour, a 1000-Hz tone at 0 dB sounds as loud as a 50-Hz tone at 40 dB. Note that the frequency scale is logarithmic to compress the large frequency range.

It is interesting to note the dips (or minima) in the curves. The hearing curves show a significant dip in the 2000–5000 Hz range, the ear being most sensitive around 4000 Hz. A tone with a frequency of 4000 Hz can be heard at intensity levels *below* 0 dB. The high sensitivity in the 2000–5000 Hz region is very important for the understanding of speech. (Why?) Another dip in the curves, or region of sensitivity, occurs at about 12 000 Hz.

The minima occur as a result of resonance in a closed cavity in the auditory canal (similar to a closed pipe). The length of the cavity is such that it has a fundamental resonance frequency of about 4000 Hz, resulting in extra sensitivity. As in a closed cavity, the next natural frequency is the third harmonic (see Eq. 14.18), which is three times the fundamental frequency, or about 12 000 Hz.

EXAMPLE 14.11 | The Human Ear Canal: Standing Waves

Consider the human ear canal to be a cylindrical tube of length 2.54 cm (1.0 in.; see Fig. 1 in Insight 14.2). What would be the lowest sound resonance frequency? Take the air temperature in the canal to be 37 °C.

THINKING IT THROUGH. The auditory ear canal as described is essentially a closed pipe—open at one end (outer ear canal) and closed at the other (eardrum). The lowest-frequency resonance standing wave that will fit in the pipe is $L = \lambda/4$ (Fig. 14.15), and $\lambda = 4L$. Then, the frequency is given by $f_1 = v/\lambda_1 = v/4L$ (Eq. 14.18), where v is the speed of sound in air.

SOLUTION.

Given: $L = 2.54$ cm $= 0.0254$ m $\qquad$ *Find:* f_1 (lowest resonance frequency)
$ T = 37$ °C, (normal body temperature)

First finding the speed of sound at 37 °C,

$$v = (331 + 0.6T_C) \text{ m/s} = [331 + 0.6(37)] \text{ m/s} = 353 \text{ m/s}$$

and

$$f_1 = \frac{v}{4L} = \frac{353 \text{ m/s}}{4(0.0254 \text{ m})} = 3.47 \times 10^3 \text{ Hz} = 3.47 \text{ kHz}$$

Compare with the curves in Fig. 14.17, and note the dip in the curves at about this frequency. How about the other dip just above 10 kHz? Check out the next natural frequency for the ear canal, f_3.

FOLLOW-UP EXERCISE. Children have smaller ear canals than adults, on the order of 1.30 cm in length. What is the lowest fundamental frequency for a child's ear canal? Use the same air temperature as in the Example. (Note: With growth the ear canal lengthens, and it has been experimentally determined that the "adult" ear canal length and lowest fundamental frequency is reached at about age 7.)

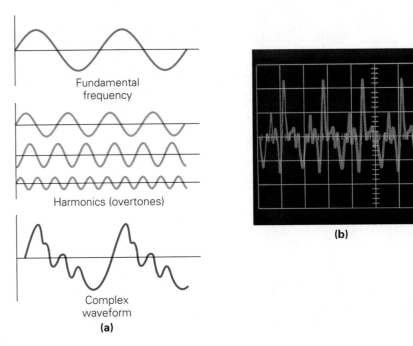

▲ FIGURE 14.18 Waveform and quality (a) The superposition of sounds of different frequencies and amplitudes gives a complex waveform. The harmonics, or overtones, determine the quality of the sound. **(b)** The waveform of a violin tone is displayed on an oscilloscope.

The **quality** of a tone is the characteristic that enables it to be distinguished from another tone of basically the same intensity and frequency. Tone quality depends on the waveform—specifically, the number of harmonics (overtones) present and their relative intensities (▲Fig. 14.18). The tone of a voice depends in large part on the vocal resonance cavities. One person can sing a tone with the same basic frequency and intensity as another, but different combinations of overtones give the two voices different qualities.

The notes of a musical scale correspond to certain frequencies; as we saw in Example 14.10, middle C (C_4) has a frequency of 264 Hz. When a note is played on an instrument, its assigned frequency is that of the first harmonic, which is the fundamental frequency. (The second harmonic is the first overtone, the third harmonic is the second overtone, and so on.) The fundamental frequency is dominant over the accompanying overtones that determine the sound quality of the instrument. Recall from Section 13.5 that the overtones that are produced depend on how an instrument is played. Whether a violin string is plucked or bowed, for example, can be discerned from the quality of identical notes.

DID YOU LEARN?
➥ In musical instruments, different standing waves produce different notes.
➥ The perception or sensory characteristics of sound are loudness (intensity), pitch (frequency), and quality (waveform–harmonics).

PULLING IT TOGETHER | A "Sound" Thermometer?

As you are standing in a doorway with the outside cold air on one side and the inside warm air on the other, you simultaneously hear sound from two different organ pipes. Pipe A is outside in the cold air, and pipe B is inside in the room temperature air (20 °C). Pipe A is closed at one end, while pipe B is open at both ends. When both are excited in their funda-

mental modes, a beat frequency of 2.00 Hz is produced. The length of pipe A is 1.00 m and that of pipe B is 2.10 m. See ▶Fig. 14.19. (a) What is the frequency of the sound produced by pipe B? (b) What are the two possible beat frequencies of the sound produced by pipe A? (c) What are the two possible temperatures of the air in the region of pipe A?

THINKING IT THROUGH. The concepts in this example include resonant frequencies in open and closed pipes, beats, and the speed of sound as a function of temperature. (a) Fig 14.19 shows the standing wave in pipe B and gives the wavelength of the wave. Since the speed of sound is known from the room temperature, the frequency of pipe B can be found. (b) Beats tell only the amount by which pipe A's frequency differs from pipe B's frequency, so there are two possible answers. (c) Fig. 14.19 shows the standing wave in pipe A and gives its wavelength. From that and the two possible frequencies in part (b), two possible speeds for sound can be found. Those speeds can then be translated into temperatures because the speed of sound depends on the temperature.

SOLUTION.

Given: pipe A (closed), $L = 1.00$ m
pipe B (open), $L = 2.10$ m
beat frequency, $f_b = 2.00$ Hz
pipe B is in room temperature air (20 °C)

Find: (a) f_B (frequency of pipe B)
(b) f_A (two possible frequencies of pipe A)
(c) T_C (two possible temperatures for pipe A)

(a) Fig. 14.19 shows pipe B's standing wave to be a half-wavelength. This is because it is in its fundamental (lowest frequency, longest wavelength) mode. Therefore, $\lambda_B = 2L_B = 4.20$ m. The speed of sound in B's locale is:

$$v_B = (331 + 0.6T_C) \text{ m/s} = 331 + 0.6(20 \text{ °C}) = 343 \text{ m/s}$$

From this, the frequency of B's sound can be found:

$$f_B = \frac{v_B}{\lambda_B} = \frac{343 \text{ m/s}}{4.20 \text{ m}} = 81.7 \text{ Hz}$$

(b) The beat frequency tells only that pipe A's frequency differs from that of pipe B by 2.00 Hz; therefore, there are two choices:

$$f_A = 81.7 \text{ Hz} \pm 2.00 \text{ Hz}$$
$$= 83.7 \text{ Hz} \quad \text{or} \quad 79.7 \text{ Hz}$$

(c) Fig 14.19 shows pipe A's standing wave to be a quarter-wavelength. This is because it is in its fundamental (lowest frequency, longest wavelength) mode. Therefore, $\lambda_A = 4L_A = 4.00$ m. Since there are two possible frequencies for pipe A, the speed of sound in A's locale also has two possibilities. These are,

$$v_A = f_A\lambda_A = (83.7 \text{ Hz})(4.00 \text{ m}) = 334.8 \text{ m/s}$$

and

$$v_A = f_A\lambda_A = (79.7 \text{ Hz})(4.00 \text{ m}) = 318.8 \text{ m/s}$$

The dependence of the speed of sound on air temperature is given by $v = (331 + 0.6T_C)$ m/s. Solving for T_C for the two possible air temperatures for pipe A gives:

$$T_C = \frac{v_A - 331 \text{ m/s}}{0.6} = \frac{334.8 \text{ m/s} - 331 \text{ m/s}}{0.6} = 6.3 \text{ °C}$$

and

$$T_C = \frac{v_A - 331 \text{ m/s}}{0.6} = \frac{318.8 \text{ m/s} - 331 \text{ m/s}}{0.6} = -20.3 \text{ °C}$$

Without any further information, that is the best that can be done. The latter temperature is a bit cold for playing an organ outside, so the first temperature is probably the correct one. (You should be able to show that 6.3 °C = 43.3 °F and −20.3 °C = −4.54 °F

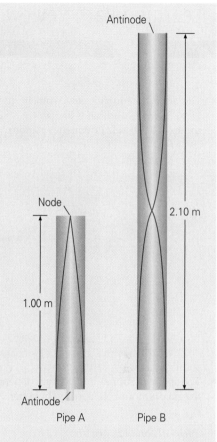

▲ **FIGURE 14.19 Closed and open pipes; nodes and antinodes** The fundamental modes in pipe A and pipe B.

Learning Path Review

- The sound frequency spectrum is divided into infrasonic ($f < 20$ Hz), audible (20 Hz $< f < 20$ kHz), and ultrasonic ($f > 20$ kHz) frequency regions.

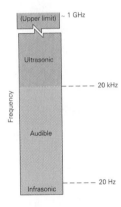

- The speed of sound in a medium depends on the elasticity of the medium and its density. In general, $v_{solids} > v_{liquids} > v_{gases}$.

Speed of sound in dry air:

$$v = (331 + 0.6T_C) \text{ m/s} \qquad (14.1)$$

- The intensity of a point source is inversely proportional to the square of the distance from the source.

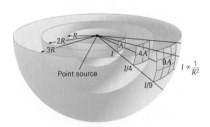

Intensity of a point source:

$$I = \frac{P}{4\pi R^2} \quad \text{and} \quad \frac{I_2}{I_1} = \left(\frac{R_1}{R_2}\right)^2 \qquad (14.2, 14.3)$$

- The sound intensity level is a logarithmic function of the sound intensity and is expressed in decibels (dB).

Intensity level (in decibels, dB):

$$\beta = 10 \log \frac{I}{I_o} \quad \text{where} \quad I_o = 10^{-12} \text{ W/m}^2 \qquad (14.4)$$

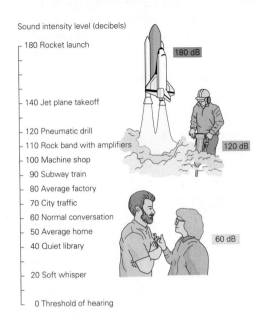

- Sound wave interference of two point sources depends on phase difference as related to path length difference. Sound waves that arrive at a point in phase reinforce each other (constructive interference); sound waves that arrive at a point out of phase cancel each other (destructive interference).

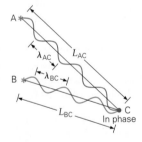

Phase difference (where ΔL is the path length difference):

$$\Delta \theta = \frac{2\pi}{\lambda}(\Delta L) \qquad (14.5)$$

Condition for constructive interference:

$$\Delta L = n\lambda \qquad (n = 0, 1, 2, 3, \dots) \qquad (14.6)$$

Condition for destructive interference:

$$\Delta L = m\left(\frac{\lambda}{2}\right) \quad (m = 1, 3, 5, \dots) \qquad (14.7)$$

Beat frequency:

$$f_b = |f_1 - f_2| \qquad (14.8)$$

- The Doppler effect depends on the velocities of the sound source and observer relative to still air. When the relative motion of the source and observer is toward each other, the observed pitch increases; when the relative motion of the source and observer is away from each other, the observed pitch decreases.

Doppler effect:

Moving source, stationary observer

$$f_o = \left(\frac{v}{v \pm v_s}\right)f_s = \left(\frac{1}{1 \pm \frac{v_s}{v}}\right)f_s \quad (14.11)$$

$$\begin{cases} - \text{ for source moving toward stationary observer} \\ + \text{ for source moving away from stationary obsever} \end{cases}$$

where v_s = speed of source

and v = speed of sound

Moving observer, stationary source

$$f_o = \left(\frac{v \pm v_o}{v}\right)f_s = \left(1 \pm \frac{v_o}{v}\right)f_s \quad (14.14)$$

$$\begin{cases} + \text{ for observer moving toward stationary source} \\ - \text{ for observer moving away from stationary source} \end{cases}$$

where v_o = speed of observer

and v = speed of sound

Moving observer and moving source

$$f_o = \left(\frac{v \pm v_o}{v \mp v_s}\right)f_s \quad (14.14a)$$

Upper signs if moving toward each other

Lower signs if moving away from each other

Angle for conical shock wave:

$$\sin \theta = \frac{vt}{v_s t} = \frac{v}{v_s} = \frac{1}{M} \quad (14.15)$$

Mach number:

$$M = \frac{v_s}{v} = \frac{1}{\sin \theta} \quad (14.16)$$

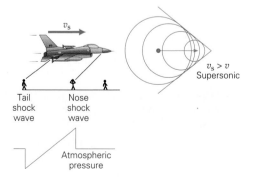

Natural frequencies of an open organ pipe—open on both ends:

$$f_n = n\left(\frac{v}{2L}\right) = nf_1 \quad (n = 1, 2, 3, \dots) \quad (14.17)$$

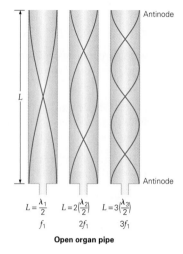

$$L = \frac{\lambda_1}{2} \qquad L = 2\left(\frac{\lambda_2}{2}\right) \qquad L = 3\left(\frac{\lambda_3}{2}\right)$$
$$f_1 \qquad 2f_1 \qquad 3f_1$$

Open organ pipe

Natural frequencies of a closed organ pipe—closed on one end:

$$f_m = m\left(\frac{v}{4L}\right) = mf_1 \quad (m = 1, 3, 5, \dots) \quad (14.18)$$

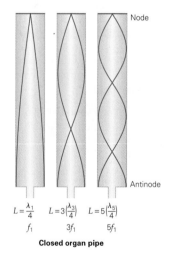

$$L = \frac{\lambda_1}{4} \qquad L = 3\left(\frac{\lambda_3}{4}\right) \qquad L = 5\left(\frac{\lambda_5}{4}\right)$$
$$f_1 \qquad 3f_1 \qquad 5f_1$$

Closed organ pipe

Learning Path Questions and Exercises

For instructor-assigned homework, go to www.masteringphysics.com

14.1 SOUND WAVES AND
14.2 THE SPEED OF SOUND

1. A sound wave with a frequency of 15 Hz is in what region of the sound spectrum: (a) audible, (b) infrasonic, (c) ultrasonic, or (d) supersonic?

2. A sound wave in air (a) is longitudinal, (b) is transverse, (c) has longitudinal and transverse components, (d) travels faster than a sound wave through a liquid.

3. The speed of sound is generally greatest in (a) solids, (b) liquids, (c) gases, (d) a vacuum.

4. The speed of sound in air (a) is about 1/3 km/s, (b) is about 1/5 mi/s, (c) depends on temperature, (d) all of the preceding.

5. The speed of sound in water is about 4.5 times that in air. A single-frequency sound source in air is f_o. On penetrating water, the frequency of the sound will be (a) $4f_o$, (b) $f_o/4$, (c) f_o.

14.3 SOUND INTENSITY AND SOUND INTENSITY LEVEL

6. If the air temperature increases, would the sound intensity from a constant-output point source (a) increase, (b) decrease, or (c) remain unchanged?

7. The decibel scale is referenced to a standard intensity of (a) 1.0 W/m^2, (b) 10^{-12} W/m^2, (c) normal conversation, (d) the threshold of pain.

8. If the intensity level of a sound at 20 dB is increased to 40 dB, the intensity would increase by a factor of (a) 10, (b) 20, (c) 40, (d) 100.

9. The intensity of a sound wave is directly proportional to the (a) amplitude, (b) frequency, (c) square of the amplitude, (d) square of the frequency.

10. A sound with an intensity level of 30 dB is how many times more intense than the threshold of hearing: (a)10, (b) 100, (c) 1000, or (d) 3000?

14.4 SOUND PHENOMENA AND
14.5 THE DOPPLER EFFECT

11. Constructive and destructive interference of sound waves depends on (a) the speed of sound, (b) diffraction, (c) phase difference, (d) all of the preceding.

12. Beats are the direct result of (a) interference, (b) refraction, (c) diffraction, (d) the Doppler effect.

13. Police radar makes use of (a) refraction, (b) the Doppler effect, (c) interference, (d) sonic boom.

14.6 MUSICAL INSTRUMENTS AND SOUND CHARACTERISTICS

14. Given open and closed pipes of the same length, which would have the lowest natural frequency: (a) the open pipe, (b) the closed pipe, or (c) they both would have the same low frequency?

15. The human ear can hear tones best at (a) 1000 Hz, (b) 4000 Hz, (c) 6000 Hz, (d) all frequencies.

16. Equal loudness curves vary with sound (a) quality, (b) harmonics, (c) waveform, (d) pitch.

17. The quality of sound depends on its (a) waveform, (b) frequency, (c) speed, (d) intensity.

14.1 SOUND WAVES AND
14.2 THE SPEED OF SOUND

1. Suggest a possible explanation of why some flying insects produce buzzing sounds and some do not.

2. Explain why sound travels faster in warmer air than in colder air.

3. Why does the speed of sound vary with temperature?

4. The speed of sound in air depends on temperature. What effect, if any, should humidity have?

5. What is the difference between ultrasonic and supersonic?

14.3 SOUND INTENSITY AND SOUND INTENSITY LEVEL

6. What is the difference between sound intensity and sound intensity level?

7. Where is the intensity greater and by what factor: (1) at a point a distance R from a power source P, or (2) at a point a distance $2R$ from a power source of $2P$? Explain.

8. The Richter scale, used to measure the intensity level of earthquakes, is a logarithmic scale, as is the decibel scale. Why are such logarithmic scales used?

9. Can there be negative decibel levels, such as −10 dB? If so, what would these mean?

14.4 SOUND PHENOMENA
AND
14.5 THE DOPPLER EFFECT

10. Do interference beats have anything to do with the "beat" of music? Explain.

11. (a) Is there a Doppler effect if a sound source and an observer are moving with the same velocity? (b) What would be the effect if a moving source accelerated toward a stationary observer?

12. As a person walks between a pair of loudspeakers that produce tones of the same amplitude and frequency, he hears a varying sound intensity. Explain.

13. How can Doppler radar used in weather forecasting measure both the location and internal motions of a storm?

14. A stationary sound source and a stationary observer are a fixed distance apart. However, the air between them is moving toward the observer with a constant speed. How is the frequency received by the observer affected? Explain.

15. Can a jet pilot flying faster than the speed of sound hear sound? Explain.

14.6 MUSICAL INSTRUMENTS AND
SOUND CHARACTERISTICS

16. (a) Why does it seem particularly quiet after a snowfall? (b) Why do empty rooms sound hollow? (c) Why do people's voices sound fuller or richer when they sing in the shower?

17. The frets on a guitar finger board are spaced closer together the farther they are from the neck. Why is this? What would be the result if they were evenly spaced?

18. Is it possible for an open organ pipe and a closed organ pipe, each of the same length, to produce notes of the same frequency? Justify your answer.

19. How would an increase in air temperature affect the frequencies of an organ pipe?

20. When you blow across the mouth of an empty soda bottle, a particular tone is produced. If the bottle is filled to one-third its height with water, how would the tone be affected? Explain. How about if it were filled to one-half? (Consider only standing waves in the bottle.)*

21. A crystal wine glass is partially filled with water. A person wets her finger and rubs it around the rim of the glass, which produces sound. Why is this?

22. Why are there no even harmonics in a closed organ pipe?

*A more complicated phenomenon, called Helmholtz resonance, occurs here, but for simplicity and standing waves, assume the bottle to be a circular cylinder. (See the answer to the Follow-up Exercise.)

EXERCISES

Integrated Exercises (IEs) are two-part exercises. The first part typically requires a conceptual answer choice based on physical thinking and basic principles. The following part is quantitative calculations associated with the conceptual choice made in the first part of the exercise.

Throughout the text, many exercise sections will include "paired" exercises. These exercise pairs, identified with red numbers*, are intended to assist you in problem solving and learning. In a pair, the first exercise (even numbered) is worked out in the Study Guide so that you can consult it should you need assistance in solving it. The second exercise (odd numbered) is similar in nature, and its answer is given in Appendix VII at the back of the book.*

14.1 SOUND WAVES
AND
14.2 THE SPEED OF SOUND

1. • What is the speed of sound in air at (a) 10 °C and (b) 20 °C?

2. • The speed of sound in air on a summer day is 350 m/s. What is the air temperature?

3. • Sonar is used to map the ocean floor. If an ultrasonic signal is received 2.0 s after it is emitted, how deep is the ocean floor at that location?

4. • What temperature change from 0 °C would increase the speed of sound by 1.0%?

5. • The wave speed in a liquid is given by $v = \sqrt{B/\rho}$, where B is the bulk modulus of the liquid and ρ is its density. Show that this equation is dimensionally correct. What about $v = \sqrt{Y/\rho}$ for a solid? (Y is Young's modulus.)

6. •• A 0.75-m-long metal rod is dropped vertically onto a marble floor. When the rod strikes the floor, it is determined electronically that the impact produces a 4-kHz tone. What is the speed of sound in the rod?

7. IE •• A tuning fork vibrates at a frequency of 256 Hz. (a) When the air temperature increases, the wavelength of the sound from the tuning fork (1) increases, (2) remains the same, (3) decreases. Why? (b) If the temperature rises from 0 °C to 20 °C, what is the change in the wavelength?

8. •• Particles approximately 3.0×10^{-2} cm in diameter are to be scrubbed loose from machine parts in an aqueous ultrasonic cleaning bath. Above what frequency should the bath be operated to produce wavelengths of this size and smaller?

9. •• Medical ultrasound uses a frequency of around 20 MHz to diagnose human conditions and ailments. (a) If the speed of sound in tissue is 1500 m/s, what is the smallest detectable object? (b) If the penetration depth is about 200 wavelengths, how deep can this instrument penetrate?

10. •• Brass is an alloy of copper and zinc. Does the addition of zinc to copper cause an increase or decrease in the speed of sound in brass rods compared to copper? Explain.

11. •• The speed of sound in steel is about 4.50 km/s. A steel rail is struck with a hammer, and an observer 0.400 km away has one ear to the rail. (a) How much time will elapse from the time the sound is heard through the rail until the time it is heard through the air? Assume that the air temperature is 20 °C and that no wind is blowing. (b) How much time would elapse if the wind were blowing toward the observer at 36.0 km/h from where the rail was struck?

12. •• A person holds a rifle horizontally and fires at a target. The bullet has a muzzle speed of 200 m/s, and the person hears the bullet strike the target 1.00 s after firing it. The air temperature is 72 °F. What is the distance to the target?

13. •• A freshwater dolphin sends an ultrasonic sound to locate a prey. If the echo off the prey is received by the dolphin 0.12 s after being sent, how far is the prey from the dolphin?

14. •• A submarine on the ocean surface receives a sonar echo indicating an underwater object. The echo comes back at an angle of 20° above the horizontal and the echo took 2.32 s to get back to the submarine. What is the object's depth?

15. •• The speed of sound in human tissue is on the order of 1500 m/s. A 3.50-MHz probe is used for an ultrasonic procedure. (a) If the effective physical depth of the ultrasound is 250 wavelengths, what is the physical depth in meters? (b) What is the time lapse for the ultrasound to make a round trip if reflected from an object at the effective depth? (c) The smallest detail capable of being detected is on the order of one wavelength of the ultrasound. What would this be?

16. •• The size of your eardrum (the tympanum; see Fig. 1 in Insight 14.2, The Physiology and Physics of the Ear and Hearing) partially determines the upper frequency limit of your audible region, usually between 16 000 Hz and 20 000 Hz. If the wavelength is on the order of twice the diameter of the eardrum and the air temperature is 20 °C, how wide is your eardrum? Is your answer reasonable?

17. IE ••• On hiking up a mountain that has several overhanging cliffs, a climber drops a stone at the first cliff to determine its height by measuring the time it takes to hear the stone hit the ground. (a) At a second cliff that is twice the height of the first, the measured time of the sound from the dropped stone is (1) less than double, (2) double, or (3) more than double that of the first. Why? (b) If the measured time is 4.8 s for the stone dropping from the first cliff, and the air temperature is 20 °C, how high is the cliff? (c) If the height of a third cliff is three times that of the first one, what would be the measured time for a stone dropped from that cliff to reach the ground?

18. ••• A bat moving at 15.0 m/s emits a high-frequency sound as it approaches a wall that is 25.0 m away. Assuming that the bat continues straight toward the wall, how far away is it when it receives the echo? (Assume the air temperature in the cave to be 0 °C.)

19. ••• Sound propagating through air at 30 °C passes through a vertical cold front into air that is 4.0 °C. If the sound has a frequency of 2500 Hz, by what percentage does its wavelength change in crossing the boundary?

14.3 SOUND INTENSITY AND SOUND INTENSITY LEVEL

20. • Calculate the intensity generated by a 1.0-W point source of sound at a location (a) 3.0 m and (b) 6.0 m from it.

21. IE • (a) If the distance from a point sound source triples, the sound intensity will be (1) 3, (2) 1/3, (3) 9, (4) 1/9 times the original value. Why? (b) By how much must the distance from a point source be increased to reduce the sound intensity by half?

22. • Assuming that the diameter of your eardrum is 1 cm (see Exercise 16), what is the sound power received by the eardrum at the threshold of (a) hearing and (b) pain?

23. • A middle C note (262 Hz) is sounded on a piano to help tune a violin string. When the string is sounded, nine beats are heard in 3.0 s. (a) How much is the violin string off tune? (b) Should the string be tightened or loosened to sound middle C?

24. • Calculate the intensity level for (a) the threshold of hearing and (b) the threshold of pain.

25. • Find the intensity levels in decibels for sounds with intensities of (a) 10^{-2} W/m^2, (b) 10^{-6} W/m^2, and (c) 10^{-15} W/m^2.

26. •• At Cape Canaveral, on blastoff a rocket produces an intensity level of 160 dB as measured 10 m from the rocket. What would be the intensity level at 100 m away? (Assume no energy is lost due to reflections, etc.)

27. IE •• (a) If the power of a sound source doubles, the intensity level at a certain distance from the source (1) increases, (2) exactly doubles, or (3) decreases. Why? (b) What are the intensity levels at a distance of 10 m from a 5.0-W and a 10-W source, respectively?

28. •• The intensity levels of two people holding a conversation are 60 dB and 70 dB, respectively. What is the intensity of the combined sounds?

29. •• A point source emits radiation in all directions at a rate of 7.5 kW. What is the intensity of the radiation 5.0 m from the source?

30. •• Two sound sources have intensities of 10^{-9} W/m^2 and 10^{-6} W/m^2, respectively. Which source is more intense and by how many times more?

31. •• Average speech has an intensity level of about 60 dB. Assuming that 20 people all speak at 60 dB, what is the total sound intensity?

32. •• A rock band (with loud speakers) has an average intensity level of 110 dB at a distance of 15 m from the band. Assuming the sound is radiated equally over a hemisphere in front of the band, what is the total power output?

33. •• A person has a hearing loss of 30 dB for a particular frequency. What is the sound intensity that is heard at this frequency that has an intensity of the threshold of pain?

TABLE 14.4 Takeoff and Landing Noise Levels for Some Common Commercial Jet Aircraft* (See Exercise 34)

Aircraft	Takeoff Noise (dB)	Landing Noise (dB)
737	85.7–97.7	99.8–105.3
747	89.5–110.0	103.8–107.8
DC-10	98.4–103.0	103.8–106.6
L-1011	95.9–99.3	101.4–102.8

*Noise level readings are taken from 198 m (650 ft). The range depends on the aircraft model and the type of engine used.

34. •• Noise levels for some common aircraft are given in ▲ Table 14.4. What are the lowest and highest intensities for (a) takeoff and (b) landing for these planes?

35. IE •• If the distance to a sound source is halved, (a) will the sound intensity level change by a factor of (1) 2, (2) 1/2, (3) 4, (4) 1/4, or (5) none of the preceding? Why? (b) What is the change in the sound intensity level?

36. •• A compact speaker puts out 100 W of sound power. (a) Neglecting losses to the air, at what distance would the sound intensity be at the pain threshold? (b) Neglecting losses to the air, at what distance would the sound intensity be that of normal speech? Does your answer seem reasonable? Explain.

37. •• What is the intensity level of a 23-dB sound after being amplified (a) ten thousand times, (b) a million times, (c) a billion times?

38. •• In a neighborhood challenge to see who can climb a tree the fastest, you are ready to climb. Your friends have surrounded you in a circle as a cheering section; each individual alone would cause a sound intensity level of 80 dB at your location. If the actual sound level at your location is 87 dB, how many people are rooting for you?

39. IE •• A dog's bark has a sound intensity level of 40 dB. (a) If two of the same dogs were barking, the intensity level is (1) less than 40 dB, (2) between 40 dB and 80 dB, (3) 80 dB. (b) What would be the intensity level?

40. •• At a rock concert, the average sound intensity level for a person in a front-row seat is 110 dB for a single band. If all the bands scheduled to play produce sound of that same intensity, how many of them would have to play simultaneously for the sound level to be at or above the threshold of pain?

41. •• At a distance of 12.0 m from a point source, the intensity level is measured to be 70 dB. At what distance from the source will the intensity level be 40 dB?

42. •• At a Fourth of July celebration, a firecracker explodes (▶ Fig. 14.20). Considering the firecracker to be a point source, what are the intensities heard by observers at points B, C, and D, relative to that heard by the observer at A?

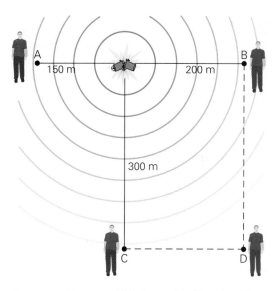

▲ **FIGURE 14.20 A big bang** See Exercise 42.

43. •• An office in an e-commerce company has fifty computers, which generate a sound intensity level of 40 dB (from the keyboards). The office manager tries to cut the noise to half as loud by removing twenty-five computers. Does he achieve his goal? What is the intensity level generated by twenty-five computers?

44. ••• A 1000-Hz tone from a loudspeaker has an intensity level of 100 dB at a distance of 2.5 m. If the speaker is assumed to be a point source, how far from the speaker will the sound have intensity levels (a) of 60 dB and (b) barely high enough to be heard?

45. ••• During practice in a huddle, a quarterback shouts the play in anticipation of crowd noise during the actual game. To a receiver 0.750 m away from the quarterback in the huddle, it seems as loud as the noise from a screaming child. When they get into practice formation, the quarterback yells at twice the output power, yet the instructions seem only about as loud as normal conversation. Use typical values in Table 14.2 to estimate how far from the quarterback the receiver is in the formation.

46. ••• A bee produces a buzzing sound that is barely audible to a person 3.0 m away. How many bees would have to be buzzing at that distance to produce a sound with an intensity level of 50 dB?

14.4 SOUND PHENOMENA
AND
14.5 THE DOPPLER EFFECT

47. • A violinist and a pianist simultaneously sound notes with frequencies of 436 Hz and 440 Hz, respectively. What beat frequency will the musicians hear?

48. IE • A violinist tuning her instrument to a piano note of 264 Hz detects three beats per second. (a) The frequency of the violin could be (1) less than 264 Hz, (2) equal to 264 Hz, (3) greater than 264 Hz, (4) both (1) and (3). Why? (b) What are the possible frequencies of the violin tone?

49. • What is the frequency heard by a person driving 60 km/h directly toward a factory whistle ($f = 800$ Hz) if the air temperature is 0 °C?

50. IE • On a day with a temperature of 20 °C and no wind blowing, the frequency heard by a moving person from a 500-Hz stationary siren is 520 Hz. (a) The person is (1) moving toward, (2) moving away from, or (3) stationary relative to the siren. Explain. (b) What is the person's speed?

51. •• While standing near a railroad crossing, you hear a train horn. The frequency emitted by the horn is 400 Hz. If the train is traveling at 90.0 km/h and the air temperature is 25 °C, what is the frequency you hear (a) when the train is approaching and (b) after it has passed?

52. •• Two identical strings on different cellos are tuned to the 440-Hz A note. The peg holding one of the strings slips, so its tension is decreased by 1.5%. What is the beat frequency heard when the strings are then played together?

53. •• How fast, in kilometers per hour, must a sound source be moving toward you to make the observed frequency 5.0% greater than the true frequency? (Assume that the speed of sound is 340 m/s.)

54. IE •• You are driving east at 25.0 m/s as you notice an ambulance traveling west toward you at 35.0 m/s. The sound you detect from the sirens has a frequency of 300 Hz. (a) Is the true frequency of the sirens (1) greater than 300 Hz, (2) less than 300 Hz, or (3) exactly 300 Hz? (b) Determine the true frequency of the sirens. Assume normal room temperature.

55. •• The frequency of an ambulance siren is 700 Hz. What are the frequencies heard by a stationary pedestrian as the ambulance approaches and moves away from her at a speed of 90.0 km/h? (Assume that the air temperature is 20 °C.)

56. •• A jet flies at a speed of Mach 2.0. What is the half-angle of the conical shock wave formed by the aircraft? Can you tell the speed of the shock wave?

57. IE •• A fighter jet flies at a speed of Mach 1.5. (a) If the jet were to fly faster than Mach 1.5, the half-angle of the conical shock wave would (1) increase, (2) remain the same, (3) decrease. Why? (b) What is the half-angle of the conical shock wave formed by the jet plane at Mach 1.5?

58. •• The half-angle of the conical shock wave formed by a supersonic jet is 30°. What are (a) the Mach number of the aircraft and (b) the actual speed of the aircraft if the air temperature is −20 °C?

59. •• An observer is traveling between two identical sources of sound (frequency 100 Hz). His speed is 10.0 m/s as he approaches one and recedes from the other. (a) What frequency tone does he hear from each source? (b) How many beats per second does he hear? Assume normal room temperature.

60. ••• A bystander hears a siren vary in frequency from 476 Hz to 404 Hz as a fire truck approaches, passes by, and moves away on a straight street (▼Fig. 14.21). What is the speed of the truck? (Take the speed of sound in air to be 343 m/s.)

▲ **FIGURE 14.21 The siren's wail** See Exercise 60.

61. ••• Bats emit sounds of frequencies around 35.0 kHz and use echolocation to find their prey. If a bat is moving with a speed of 12.0 m/s toward a hovering, stationary insect, (a) what is the frequency received by the insect if the air temperature is 20 °C? (b) What frequency of the reflected sound is heard by the bat? (c) If the insect were initially moving directly away from the bat, would this affect the frequencies? Explain.

62. ••• A supersonic jet flies directly overhead relative to an observer, at an altitude of 2.0 km (▼Fig. 14.22). When the observer hears the first sonic boom, the plane has flown a horizontal distance of 2.5 km at a constant speed. (a) What is the angle of the shock wave cone? (b) At what Mach number is the plane flying? (Assume that the speed of sound is at an average constant temperature of 15 °C.)

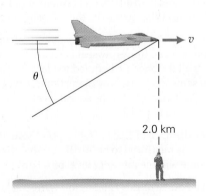

▲ **FIGURE 14.22 Faster than a speeding bullet** See Exercise 62.

14.6 MUSICAL INSTRUMENTS AND SOUND CHARACTERISTICS

63. • The first three natural frequencies of an organ pipe are 126 Hz, 378 Hz, and 630 Hz. (a) Is the pipe an open or a closed pipe? (b) Taking the speed of sound in air to be 340 m/s, find the length of the pipe.

64. • A closed organ pipe has a fundamental frequency of 528 Hz (a C note) at 20 °C. What is the fundamental frequency of the pipe when the temperature is 0 °C?

65. • The human ear canal is about 2.5 cm long. It is open at one end and closed at the other. (See Fig. 1 in Insight 14.2) (a) What is the fundamental frequency of the ear canal at 20 °C? (b) To what frequency is the ear most sensitive? (c) If a person's ear canal is longer than 2.5 cm, is the fundamental frequency higher or lower than that in part (a)? Explain.

66. •• An organ pipe that is closed at one end has a length of 0.80 m. At 20 °C, what is the distance between a node and an adjacent antinode for (a) the second harmonic and (b) the third harmonic?

67. •• An open organ pipe and an organ pipe that is closed at one end both have lengths of 0.52 m at 20 °C. What is the fundamental frequency of each pipe?

68. •• An open organ pipe is 0.50 m long. If the speed of sound is 340 m/s, what are the pipe's fundamental frequency and the frequencies of the first two overtones?

69. •• An organ pipe that is closed at one end is 1.10 m long. It is oriented vertically and filled with carbon dioxide gas (which is denser than air and thus will stay in the pipe). A tuning fork with a frequency of 60.0 Hz can be used to set up a standing wave in the fundamental mode. What is the speed of sound in carbon dioxide?

70. •• An open organ pipe 0.750 m long has its first overtone at a frequency of 441 Hz. What is the temperature of the air in the pipe?

71. IE •• When all of its holes are closed, a flute is essentially a tube that is open at both ends, with the length measured from the mouthpiece to the far end (as in Fig. 14.16b). If a hole is open, then the length of the tube is effectively measured from the mouthpiece to the hole. (a) Is the position at the mouthpiece (1) a node, (2) an antinode, or (3) neither a node nor an antinode? Why? (b) If the lowest fundamental frequency on a flute is 262 Hz, what is the minimum length of the flute at 20 °C? (c) If a note of frequency 440 Hz is to be played, which hole should be open? Express your answer as a distance from the hole to the mouthpiece.

72. ••• An organ pipe that is closed at one end is filled with helium. The pipe has a fundamental frequency of 660 Hz in air at 20 °C. What is the pipe's fundamental frequency with the helium in it?

73. ••• An open organ pipe, in its fundamental mode, has a length of 50.0 cm. A second pipe, closed at one end, is also in its fundamental mode. A beat frequency of 2.00 Hz is heard. Determine the possible lengths of the closed pipe. Assume normal room temperature.

74. ••• Bats typically give off an ultrahigh-frequency sound at about 50 000 Hz. If a bat is approaching a stationary object at 18.0 m/s, what will be the reflected frequency it detects? Assume the air in the cave is at 5 °C. [*Hint:* You will need to apply the Doppler equations twice. Why?]

PULLING IT TOGETHER: MULTICONCEPT EXERCISES

The Multiconcept Exercises require the use of more than one fundamental concept for their understanding and solution. The concepts may be from this chapter, but may include those from previous chapters.

75. At a rock concert there are two main speakers, each putting out 500 W of sound power. You are 5.00 m from one and 10.0 m from the other. (a) What are the sound intensities at your location due to each speaker and what is the total sound intensity? (b) What are the sound intensity levels at your location due to each speaker and what is the total sound intensity level? (c) About how long can you sit there without suffering permanent hearing damage?

76. You hear sound from two organ pipes that are equidistant from you. Pipe A is open at one end and closed at the other, while pipe B is open at both ends. When both are oscillating in their first-overtone mode, you hear a beat frequency of 5.0 Hz. Assume normal room temperature. (a) If the length of pipe A is 1.00 m, calculate the possible lengths of pipe B. (b) Assuming your shortest length for pipe B, what would the beat frequency be (assuming both are still in their first-overtone modes) on a hot desert summer day with a temperature of 40 °C?

77. IE An open organ pipe with a length of 50.0 cm is oscillating in its second-overtone or third-harmonic mode. Assume the air to be at room temperature and the pipe to be at rest in still air. A person moves toward this pipe at 2.00 m/s and, at the same time, away from a highly reflective wall. (a) Will the observer hear beats: (1) yes, (2) no, or (3) can't tell from the data given? (b) Calculate the frequency of sound emitted. (c) Calculate the beat frequency the observer would hear. [*Hint:* There are two frequencies, one directly from the pipe and one from the wall.]

78. IE A whale is swimming at a steady speed either directly at or directly away from an underwater cliff (you don't know which). When the whale is 300 m from the cliff, it emits a sound and it hears the echo 0.399 s later. (a) Which way is the whale traveling: (i) toward the cliff or (ii) away from the cliff? Explain your reasoning. (b) How fast is the whale traveling? (c) If the emitted sound has a frequency of 12.1 kHz, by how much has the frequency *changed* by the time the whale hears the echo?

79. **IE** A pair of speakers are separated by 4.00 m. Speaker A puts out a constant volume of sound at a total power of 36.0 W. Speaker B operates at 100 W. You are located at 3.00 m directly in front of speaker A with the line connecting you to A being perpendicular to the line joining the speakers.

Neglecting sound energy absorption by the air, (a) how do the sound intensities at your location compare: (i) $I_B > I_A$, (ii) $I_B < I_A$, (iii) $I_B = I_A$, (iv) you can't tell from the data given? (*Hint:* Draw a sketch of the arrangement. You do NOT need to calculate each intensity; use ratio reasoning.) (b) Compute each speaker's intensity at your location. Do your results confirm your answer to part (a)? (c) Compute the intensity level of each speaker and the total intensity level at your location.

80. An unstretchable steel string is used to replace a broken violin string. A length of 5.00 m of this string has a mass of 25.0 g. When in place, the new string will be 30.0 cm long and oscillate at 256 Hz in its fundamental mode. (a) After it is in place, what tension must the new string be placed under? (b) Assuming normal room temperature, what is the wavelength of the sound emitted by the new string in its fundamental mode? (c) If you wanted to decrease this sound wavelength by 5.00%, what would you do to the tension: increase or decrease it? Explain. (d) Determine the required tension in part (c).

Appendices

APPENDIX I*   Mathematical Review (with Examples and Exercises) for College Physics

Note: Answers to the odd exercises for Appendix 1 can be found in Appendix VII.

A Symbols, Arithmetic Operations, Exponents, and Scientific Notation

COMMONLY USED SYMBOLS IN RELATIONSHIPS

$=$ means two quantities are equal, such as $2x = y$.

$\equiv$ means "defined as," such as the definition of pi:

$$\pi \equiv \frac{\text{circumference of a circle}}{\text{the diameter of that circle}}.$$

$\approx$ means approximately equal, as in $30 \text{ m/s} \approx 60 \text{ m/h}$.

$\neq$ means inequality, such as $\pi \neq 22/7$.

$\geq$ means that one quantity is greater than or equal to another. For example, if the age of the universe ≥ 10 billion years its minimum age is 10 billion years.

$\leq$ means that one quantity is less than or equal to another. For example, if a lecture room holds ≤ 45 students, the maximum number of students is 45.

$>$ means that one quantity is greater than another, such as 14 eggs > 1 dozen eggs.

$\gg$ means that one quantity is *much* greater than another. For example, the number of people on Earth $\gg 1$ million.

$<$ means that one quantity is less than another, such as $3 \times 10^{22} < 10^{24}$.

$\ll$ means that one quantity is *much* less than another, such as $10 \ll 10^{11}$.

$\propto$ means proportional to. That is, if $y = 2x$ then $y \propto x$. This means that if x is increased by a certain multiplicative factor, y is also increased the same way. For example, if $y = 3x$, if x is changed by a factor of n (that is, if x becomes nx), then so is y, because $y' = 3x' = (3nx) = n(3x) = ny$.

ΔQ means "change in the quantity Q." This means "final minus initial." For example, if the value V of an investor's stock portfolio in the morning is $V_i = \$10\,100$ and at the close of trading it is $V_f = \$10\,050$, then

$$\Delta V = V_f - V_i = \$10\,050 - \$10\,100 = -\$50$$

The Greek letter capital sigma (Σ) indicates the sum of a series of values for the quantity Q_i where $i = 1, 2, 3, \ldots, N$, that is,

$$\sum_{i=1}^{N} Q_i = Q_1 + Q_2 + Q_3 + \cdots Q_N.$$

$|Q|$ denotes the absolute value of a quantity Q without a sign. If Q is positive then $|Q| = Q$; if Q is negative then $|Q| = -Q$. Thus $|-3| = 3$.

APPENDIX I-A EXERCISES ON SYMBOL USAGE

1. What values of x satisfy $3 \leq |x| \leq 8$?
2. What integer is closest to $\sqrt{10}$?
3. If at the end of the weekend you count your widgets and find $\Delta w = -10$ and the number of widgets on Friday was 500, how many do you have on Monday morning?
4. Give a reasonable range of values for the number z that satisfies $1 < z \ll 100$.
5. If $y \propto x^2$ and the value of x doubles, what happens to the value of y?
6. What is the value of $\dfrac{\sum_{i=1}^{3} 3^i}{10}$?

ARITHMETIC OPERATIONS AND THEIR ORDER OF USAGE

Basic arithmetic operations are addition ($+$) subtraction ($-$), multiplication ($\times$ or $\cdot$), and division ($/$ or $\div$). Another common operation, exponentiation (x^n), involves raising a quantity (x) to a power (n). If several of these operations are included in one equation, they are performed in this order: (a) parentheses, (b) exponentiation, (c) multiplication, (d) division, (e) addition and (f) subtraction.

A handy mnemonic used to remember this order is: "**P**lease **E**xcuse **M**y **D**ear **A**unt **S**ally," where the capital letters

---

*This appendix does not include a discussion of significant figures, since a thorough discussion is presented in Section 1.6.

stand for the various operations in order: **P**arentheses, **E**xponents, **M**ultiplication, **D**ivision, **A**ddition, **S**ubtraction. Note that operations within parentheses are always first, so to be on the safe side, appropriate use of parentheses is encouraged. For example, $24^2/8 \cdot 4 + 12$ could be evaluated several ways. However, according to the agreed-on order, it has a unique value: $24^2/8 \cdot 4 + 12 = 576/8 \cdot 4 + 12 = 576/32 + 12 = 18 + 12 = 30$. To avoid possible confusion, the quantity could be written using two sets of parentheses as follows: $(24^2/(8 \cdot 4)) + 12 = [576/(32)] + 12 = 18 + 12 = 30$.

APPENDIX I-A EXERCISES ON ARITHMETIC OPERATIONS

1. Insert parentheses so $3^2 + 4^2/5^3 - \sqrt{4} + 6$ yields $+15$ without any questions.
2. Evaluate $2^3 \cdot 3/4 + 5/2 \times 4 - 1$.
3. Evaluate $2 \times 4 + 7 - 6^2/3 \times 2$.
4. How would you write $3^2 + 4^2 \cdot 1^3 - \sqrt{4} + 7$ to guarantee that anyone evaluating the expression would obtain zero even if he or she didn't know the ordering rules?

EXPONENTS AND EXPONENTIAL NOTATION

Exponents and exponential notation are very important when employing scientific notation (see the next section). You should be familiar with power and exponential notation (both positive and negative, fractional and integral) such as the following:

$$x^0 = 1$$

$$x^1 = x \qquad x^{-1} = \frac{1}{x}$$

$$x^2 = x \cdot x \qquad x^{-2} = \frac{1}{x^2} \quad x^{1/2} = \sqrt{x}$$

$$x^3 = x \cdot x \cdot x \qquad x^{-3} = \frac{1}{x^3} \quad x^{1/3} = \sqrt[3]{x} \quad \text{etc.}$$

Exponents combine according to the following rules:

$$x^a \cdot x^b = x^{(a+b)} \qquad x^a/x^b = x^{(a-b)} \qquad (x^a)^b = x^{ab}$$

APPENDIX I-A EXERCISES ON EXPONENTS AND EXPONENTIAL NOTATION

1. What is the value of $\dfrac{2^3}{2^4}$?
2. Evaluate $3^3 \times 9^{-1/2}$.
3. Find the value(s) of $3^4 \times \sqrt{4^6}$.
4. What is $(\sqrt{10})^4$?

SCIENTIFIC NOTATION (POWERS-OF-10 NOTATION)

In physics, many quantities have values that are very large or very small. To express them, **scientific notation** is frequently used. This notation is sometimes referred to as powers-of-10 notation for obvious reasons. (See the previous section for a discussion of exponents.) When the number 10 is squared or cubed, we can write it as $10^2 = 10 \times 10 = 100$ or $10^3 = 10 \times 10 \times 10 = 1000$. You can see that the number of zeros is equal to the power of 10. Thus 10^{23} is a compact way of expressing the number 1 followed by 23 zeros.

A number can be represented in many different ways—all of which are correct. For example, the distance from the Earth to the Sun is 93 million miles. This value can be written as 93 000 000 miles. Expressed in a more compact scientific

notation, there are many correct forms, such as 93×10^6 miles, 9.3×10^7 miles, or 0.93×10^8 miles. Any of these is correct, although 9.3×10^7 is preferred, because when using powers-of-10 notation, it is customary to leave only one digit to the left of the decimal point, in this case 9. (This is called customary or standard form.) Note that the exponent, or power of 10, changes when the decimal point of the prefix number is shifted.

Negative powers of 10 also can be used. For example, $10^{-2} = \dfrac{1}{10^2} = \dfrac{1}{100} = 0.01$. So, if a power of 10 has a negative exponent, the decimal point may be shifted to the left once for each power of 10. For example, 5.0×10^{-2} is equal to 0.050 (two shifts to the left).

The decimal point of a quantity expressed in powers-of-10 notation may be shifted to the right or left irrespective of whether the power of 10 is positive or negative. General rules for shifting the decimal point are as follows:

1. The exponent, or power of 10, is *increased* by 1 for every place the decimal point is shifted to the *left*.
2. The exponent, or power of 10, is *decreased* by 1 for every place the decimal point is shifted to the *right*.

This is simply a way of saying that as the coefficient (prefix number) gets smaller, the exponent gets correspondingly larger, and vice versa. Overall, the number is the same.

When multiplying using this notation, the exponents are added. Thus $10^2 \times 10^4 = (100)(10\,000) = 1\,000\,000 = 10^6 = 10^{2+4}$. Division follows similar rules using negative exponents, for example, $\dfrac{10^5}{10^2} = \dfrac{100\,000}{100} = 1\,000 = 10^3 = 10^{5+(-2)}$.

Care should be taken when adding and subtracting numbers written in scientific notation. Before doing so, all numbers must be converted to the same power of 10. For example,

$$1.75 \times 10^3 - 5.0 \times 10^2 = 1.75 \times 10^3 - 0.50 \times 10^3$$
$$= (1.75 - 0.50) \times 10^3 = 1.25 \times 10^3.$$

APPENDIX I-A EXERCISES ON SCIENTIFIC NOTATION (POWERS-OF-10 NOTATION)

1. Express your weight (in pounds) in scientific notation.
2. The circumference of the Earth is about 40 000 km. Express this in scientific notation.
3. Evaluate and express the answers to the following in scientific notation:
   (a) $\dfrac{12.1}{1.10 \times 10^{-1}}$, (b) $\dfrac{14 \times 10^{15}}{0.70 \times 10^{19}}$, (c) 14.1×23.2, (d) $\dfrac{975}{0.00541}$.
4. Find the value of $(1.44 \times 10^2)^{1/2}$ in scientific notation.
5. What is the value of $(3.0 \times 10^8)^2$ in scientific notation?
6. Evaluate and express the answers to the following in scientific notation:
   (a) $0.12 + 1.1 \times 10^{-1}$, (b) $14.0 \times 10^{15} - 700 \times 10^{13}$, (c) $-20 + 1.4 \times 10^2$, (d) $-1.70 \times 10^{-1} - 1.40 \times 10^{-2}$.

B Algebra and Common Algebraic Relationships

GENERAL

The basic rule of algebra, used for solving equations, is that if you perform any legitimate operation on both sides of an equation, it remains an equation, or equality. (An example of an illegal operation is dividing by zero; why?) Thus such operations as adding a number to both sides, cubing both sides, and dividing both sides by the same number all maintain the equality.

For example, suppose you want to solve $\dfrac{x^2 + 6}{2} = 11$ for x.

To do this, first multiply both sides by 2, giving $\left(\dfrac{x^2 + 6}{\cancel{2}}\right) \times \cancel{2} =$

$11 \times 2 = 22$ or $x^2 + 6 = 22$. Then subtract 6 from both sides to obtain $x^2 + 6 - 6 = 22 - 6 = 16$ or $x^2 = 16$. Finally, taking the square root of both sides, the solutions are $x = \pm 4$.

SOME USEFUL RESULTS

Many times *the square of the sum and/or difference of two numbers* is required. For any numbers a and b:

$$(a \pm b)^2 = a^2 \pm 2ab + b^2$$

Similarly, *the difference of two squares* can be factored:

$$(a^2 - b^2) = (a + b)(a - b)$$

A quadratic equation is one that can be expressed in the form $ax^2 + bx + c = 0$. In this form it can always be solved (usually for two different roots) using the *quadratic formula*:

$x = \dfrac{-b \pm \sqrt{b^2 - 4ac}}{2a}$. In kinematics this result can be especially useful as it is common to have equations of this form to solve for example: $4.9t^2 - 10t - 20 = 0$. Just insert the coefficients (making sure to include the sign) and solve for t (here t represents the time for a ball to reach the ground when thrown straight upward from the edge of a cliff; see Chapter 2). The result is

$$t = \dfrac{10 \pm \sqrt{10^2 - 4(4.9)(-20)}}{2(4.9)} = \dfrac{10 \pm 22.2}{9.8}$$

$$= +3.3\,\text{s} \quad \text{or} \quad -1.2\,\text{s}$$

In all such problems, time is "stopwatch" time and starts at zero; hence the negative answer can be ignored as physically unreasonable although it is a solution to the equation.

SOLVING SIMULTANEOUS EQUATIONS

Occasionally solving a problem might require solving two or more equations simultaneously. In general if there are N unknowns in a problem, exactly N independent equations will be needed. If there are less than N equations, there are not enough for a complete solution. If there are more than N equations, then some are redundant, and a solution is usually still possible, although more complicated. In general in this textbook, such concerns will usually be with two simultaneous equations, and both will be linear. Linear equations are of the form $y = mx + b$. Recall that when plotted on an x–y Cartesian coordinate system, the result is a straight line with a slope of $m\ (=\Delta y/\Delta x)$ and a y-intercept of b, as shown for the red line here.

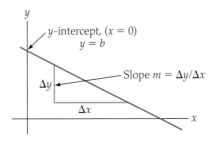

To solve two linear equations simultaneously *graphically*, simply plot them on the axes and evaluate the coordinates at their intersection point. While this can always be done in principle, it is only an approximate answer and usually takes quite a bit of time.

The most common (and exact) method of solving simultaneous equations involves the use of algebra. Essentially, you solve one equation for an unknown and substitute the result into the other equation, ending up with one equation and one unknown. Suppose you have two equations and two unknown quantities (x and y), but in general, any two unknown quantities:

$$3y + 4x = 4 \quad \text{and} \quad 2x - y = 2$$

Solving the second equation for y yields $y = 2x - 2$. Substituting this value for y into the first equation, $3(2x - 2) + 4x = 4$. Thus, $10x = 10$ and $x = 1$. Putting this value into the second of the original two equations, the result is $2(1) - y = 2$ and therefore $y = 0$. (Of course, at this point a good double-check is to substitute the answers and see if they solve both equations.)

APPENDIX I-B EXERCISES ON ALGEBRA

1. Expand $(y - 2x)^2$.
2. Express $x^2 - 4x + 4$ as a product of two factors.
3. Solve the following equation for the time t (starting at zero and measured in seconds) for an upward thrown ball to reach a certain height: $4.9t^2 + 10 - 30t = 0$. How many physically reasonable roots are there?
4. (a) Show that a quadratic equation has real roots only if $b^2 \geq 4ac$. (b) Under what conditions (for a, b, and c) are the two roots identical?
5. Solve these equations simultaneously using algebra: $2x - 3y = 7$ and $3y + 5x = 7$.
6. Solve the two equations in Exercise 5 approximately using graphing methods.

C Geometric Relationships

In physics and many other areas of science, it is important to know how to find circumferences, areas, and volumes of some common shapes. Here are some equations for such shapes.

CIRCUMFERENCE (c), AREA (A), AND VOLUME (V)

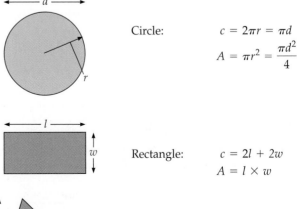

Circle: $\quad c = 2\pi r = \pi d$

$\quad A = \pi r^2 = \dfrac{\pi d^2}{4}$

Rectangle: $\quad c = 2l + 2w$

$\quad A = l \times w$

Triangle: $\quad A = \dfrac{1}{2}ab$

Sphere: $A = 4\pi r^2$

$V = \dfrac{4}{3}\pi r^3$

Cylinder: $A = \pi r^2$ (end)

$A = 2\pi rh$ (body)

$V = \pi r^2 h$

For practice, try the following exercises.

APPENDIX I-C EXERCISES ON GEOMETRIC RELATIONSHIPS

1. Estimate the volume of a bowling ball in (a) cubic centimeters and (b) cubic inches.

2. A square hole has a side measuring 5.0 cm. What is the area of the end of a cylindrical rod that will barely fit into this hole?

3. A glass of water has an interior diameter of 4.5 cm and contains a column of water 4.0 in. high. What volume of water does it contain in liters?

4. What is the *total* surface area of a pancake that has a 16 cm diameter and is 8.0 mm thick?

5. Compute the volume of the pancake in Exercise 4 in cubic centimeters.

D Trigonometric Relationships

Understanding elementary trigonometry is crucial in physics, especially since many of the quantities are vectors. Here is a summary of definitions of the three most common trig functions, which you should commit to memory.

DEFINITIONS OF TRIGONOMETRIC FUNCTIONS

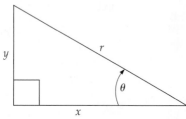

$\sin \theta = \dfrac{y}{r} \quad \cos \theta = \dfrac{x}{r} \quad \tan \theta = \dfrac{\sin \theta}{\cos \theta} = \dfrac{y}{x}$

$\theta°$ (rad)	$\sin \theta$	$\cos \theta$	$\tan \theta$
0° (0)	0	1	0
30° ($\pi/6$)	0.500	0.866	0.577
45° ($\pi/4$)	0.707	0.707	1.00
60° ($\pi/3$)	0.866	0.500	1.73
90° ($\pi/2$)	1	0	$\to \infty$

For very small angles,

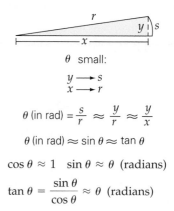

θ small:

$y \longrightarrow s$
$x \longrightarrow r$

θ (in rad) $= \dfrac{s}{r} \approx \dfrac{y}{r} \approx \dfrac{y}{x}$

θ (in rad) $\approx \sin \theta \approx \tan \theta$

$\cos \theta \approx 1 \quad \sin \theta \approx \theta$ (radians)

$\tan \theta = \dfrac{\sin \theta}{\cos \theta} \approx \theta$ (radians)

The sign of a trigonometric function depends on the quadrant, or the signs of x and y. For example, in the second quadrant x is negative and y is positive; therefore, $\cos \theta = x/r$ is negative and $\sin \theta = y/r$ is positive. [Note that r (shown as the dashed lines) is always taken as positive.] In the figure, the red lines are positive and the blue lines negative.

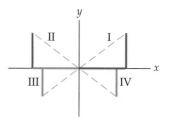

SOME USEFUL TRIGONOMETRIC IDENTITIES

$$1 = \sin^2 \theta + \cos^2 \theta$$

$$\sin 2\theta = 2 \sin \theta \cos \theta$$

$$\cos 2\theta = \cos^2 \theta - \sin^2 \theta = 2 \cos^2 \theta - 1 = 1 - 2 \sin^2 \theta$$

$$\sin^2 \theta = \frac{1}{2}(1 - \cos 2\theta)$$

$$\cos^2 \theta = \frac{1}{2}(1 + \cos 2\theta)$$

For half-angle ($\theta/2$) identities, simply replace θ with $\theta/2$; for example,

$$\sin^2 (\theta/2) = \frac{1}{2}(1 - \cos \theta)$$

$$\cos^2 (\theta/2) = \frac{1}{2}(1 + \cos \theta)$$

Trigonometric values of sums and differences of angles are sometimes of interest. Here are several basic relationships.

$$\sin(\alpha \pm \beta) = \sin \alpha \cos \beta \pm \cos \alpha \sin \beta$$

$$\cos(\alpha \pm \beta) = \cos \alpha \cos \beta \mp \sin \alpha \sin \beta$$

$$\tan(\alpha \pm \beta) = \frac{\tan \alpha \pm \tan \beta}{1 \mp \tan \alpha \tan \beta}$$

LAW OF COSINES

For a triangle with angles A, B, and C with opposite sides a, b, and c, respectively:

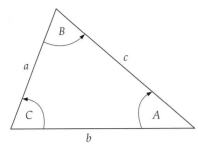

$$a^2 = b^2 + c^2 - 2bc \cos A \quad \text{(with similar results for}$$
$$b^2 = \cdots \text{ and } c^2 = \cdots)$$

If $A = 90°$ then $\cos A = 0$ and this reduces to the Pythagorean theorem as it should: $a^2 = b^2 + c^2$.

LAW OF SINES

For a triangle with angles A, B, and C with opposite sides a, b, and c, respectively:

$$\frac{a}{\sin A} = \frac{b}{\sin B} = \frac{c}{\sin C}$$

APPENDIX I-D EXERCISES ON TRIGONOMETRY

1. From ground level you find you must look at an upward angle of 60° to see the very top of a building that is 50 m away from you. (a) How high is the building? (b) How far is the top of the building from you?

2. On an x–y Cartesian set of axes, a point is at $x = -2.5$ and $y = -4.2$. (a) What quadrant is it in? (b) What is the angle between the −x-axis and the line drawn from the origin to the given point? (Express the answer in degrees and radians.)

3. Use the sine equation for the sum of two angles, one angle being 30°, the other 60°, to show that the sine of a 90° angle is 1.00.

4. Assume that the Earth's orbit about the Sun is a circle with a radius of 150 million km. (a) Calculate the *arc length* distance traveled by the Earth about the Sun in four months. (b) Using the laws of sines and/or cosines, determine the *straight-line* distance between the beginning and end points of this arc.

5. A right triangle has a hypotenuse of length 11.0 cm and one angle of 25°. Determine (a) the lengths of the other two sides of the triangle, (b) its area, and (c) its perimeter.

E Logarithms

Presented here are some of the fundamental definitions and relationships for logarithms. Logarithms are commonly used in science, so it is important that you know what they are and how to use them. Logarithms are useful because, among other things, they allow you to more easily multiply and divide very large and very small numbers.

DEFINITION OF A LOGARITHM

If a number x is written as another number a to some power n, as $x = a^n$, then n is defined to be the *logarithm of the number x to the base a*. This is written compactly as

$$n \equiv \log_a x.$$

COMMON LOGARITHMS

If the base a is 10, the logarithms are called *common logarithms*. When the abbreviation *log* is used, without a base specified, base 10 is assumed. If another base is being used, it will be specifically shown. For example, $1000 = 10^3$; therefore, $3 = \log_{10} 1000$, or simply $3 = \log 1000$. This is read "3 is the log of 1000."

IDENTITIES FOR COMMON LOGARITHMS

For any two numbers x and y:

$$\log(10^x) = x$$
$$\log(xy) = \log x + \log y$$
$$\log\left(\frac{x}{y}\right) = \log x - \log y$$
$$\log(x^y) = y \log x$$

NATURAL LOGARITHMS

The natural logarithm uses as its base the irrational number e. To six significant figures, its value is $e \approx 2.71828\ldots$. Fortunately, most calculators have this number (along with other irrational numbers, such as pi) in their memories. (You should be able to find both e and π on yours.) The natural logarithm received its name because it occurs naturally when describing a quantity that grows or decays at a constant percentage (rate). The natural logarithm is abbreviated *ln* to distinguish it from the common logarithm, *log*. That is, $\log_e x \equiv \ln x$, and if $n = \ln x$, then $x = e^n$. Similarly to the common logarithm, we have the following relationships for any two numbers x and y:

$$\ln(e^x) = x$$
$$\ln(xy) = \ln x + \ln y$$
$$\ln\left(\frac{x}{y}\right) = \ln x - \ln y$$
$$\ln(x^y) = y \ln x$$

Occasionally you must convert between the two types of logarithms. For that, the following relationships can be handy:

$$\log x = 0.43429 \ln x$$
$$\ln x = 2.3026 \log x$$

For practice with logarithms of both types, try the following exercises:

APPENDIX I-E EXERCISES ON LOGARITHMS

1. Use your calculator to find the following: (a) log 20, (b) log 50, (c) log 2500, and (d) log 3.

2. (a) Explain why numbers less than 1 have a negative logarithm. (b) Does it make sense to talk about $\log(-100)$? Explain.

3. Use your calculator to find the following: (a) ln 20, (b) ln 2, (c) ln 100, and (d) ln 3.

4. Double check your answers for ln 20 and log 20 from Exercises 1 and 3 using the relationships $\log x = 0.43429 \ln x$ and $\ln x = 2.3026 \log x$.

5. Show that the rules for combining logarithms work for the following by evaluating each side and showing an equivalence: (a) $\log 1500 = \log(15 \times 100)$, (b) $\log 6400 = \log(64/0.010)$, and (c) $\log 8 = \log(2^3)$.

6. Show that the rules for combining logarithms work for the following by evaluating each side and showing an equivalence: (a) $\ln 4 = \ln(2 \times 2)$, (b) $\ln 20 = 2.3026 \log(2 \times 10)$, and (c) $\log 49 = 0.43429 \ln(7^2)$.

7. In describing the growth of a bacteria colony, the number of bacteria N at any given time t (from the start of observation) can be written in terms of the number at the start, N_o, as follows: $N = N_o e^{0.020t}$, where t is in minutes. How many minutes does it take the colony to double in population?

8. In describing the decay of a radioactive sample of atomic nuclei, the number of undecayed nuclei N at any given time t (from the start of observation) can be written in terms of the number at the start, N_o, as follows: $N = N_o e^{-0.050t}$, where t is in years. How many years does it take until only one-tenth of the original number of nuclei remain?

APPENDIX II   Kinetic Theory of Gases

The basic assumptions are as follows:

1. All the molecules of a pure gas have the same mass (m) and are in continuous and completely random motion. (The mass of each molecule is so small that the effect of gravity on it is negligible.)

2. The gas molecules are separated by large distances and occupy a volume that is negligible compared with these distances.

3. The molecules exert no forces on each other except when they collide.

4. Collisions of the molecules with one another and with the walls of the container are perfectly elastic.

The magnitude of the force exerted on the wall of the container by a gas molecule colliding with it is $F = \Delta p/\Delta t$. Assuming that the direction of the velocity (v_x) is normal to the wall, the magnitude of the average force is

$$F = \frac{\Delta(mv)}{\Delta t} = \frac{mv_x - (-mv_x)}{\Delta t} = \frac{2mv_x}{\Delta t} \tag{1}$$

After striking one wall of the container, which, for convenience, is assumed to be a cube with sides of dimensions L, the molecule recoils in a straight line. Suppose that the molecule reaches the opposite wall without colliding with any other molecules along the way. The molecule then travels the distance L in a time equal to L/v_x. After the collision with that wall, again assuming no collisions on the return trip, the round trip will take $\Delta t = 2L/v_x$. Thus, the number of collisions per unit time a molecule makes with a particular wall is $v_x/(2L)$, and the average force of the wall from successive collisions is

$$F = \frac{2mv_x}{\Delta t} = \frac{2mv_x}{2L/v_x} = \frac{mv_x^2}{L} \tag{2}$$

The random motions of the many molecules produce a relatively constant force on the walls, and the pressure (p) is the total force on a wall divided by the wall's area:

$$p = \frac{\sum F_i}{L^2} = \frac{m\left(v_{x_1}^2 + v_{x_2}^2 + v_{x_3}^2 + \cdots\right)}{L^3} \tag{3}$$

The subscripts refer to individual molecules.

The average of the squares of the speeds is given by

$$\overline{v_x^2} = \frac{v_{x_1}^2 + v_{x_2}^2 + v_{x_3}^2 + \cdots}{N}$$

where N is the number of molecules in the container. In terms of this average, Eq. 3 can be written as

$$p = \frac{Nm\overline{v_x^2}}{L^3} \tag{4}$$

However, the molecules' motions occur with equal frequency along any one of the three axes, so $\overline{v_x^2} = \overline{v_y^2} = \overline{v_z^2}$ and $\overline{v^2} = \overline{v_x^2} + \overline{v_y^2} + \overline{v_z^2} = 3\overline{v_x^2}$. Then

$$\sqrt{\overline{v^2}} = v_{rms}$$

where v_{rms} is called the root-mean-square (rms) speed. Substituting this result into Eq. 4 and replacing L^3 with V (since L^3 is the volume of the cubical container) gives

$$pV = \tfrac{1}{3}Nmv_{rms}^2 \tag{5}$$

This result is correct even though collisions between molecules were ignored. Statistically, these collisions average out, so the number of collisions with each wall is as described. This result is also independent of the shape of the container. A cube merely simplifies the derivation.

We now combine this result with the empirical perfect gas law:

$$pV = Nk_B T = \tfrac{1}{3}Nmv_{rms}^2$$

The average kinetic energy per gas molecule is thus proportional to the absolute temperature of the gas:

$$\overline{K} = \tfrac{1}{2}mv_{rms}^2 = \tfrac{3}{2}k_B T \tag{6}$$

The collision time is negligible compared with the time between collisions. Some kinetic energy will be momentarily converted to potential energy during a collision; however, this potential energy can be ignored, because each molecule spends a negligible amount of time in collisions. Therefore, by this approximation, the total kinetic energy is the internal energy of the gas, and the internal energy of a perfect gas is directly proportional to its absolute temperature.

APPENDIX III   Planetary Data

Name	Equatorial Radius (km)	Mass (Compared with Earth's)*	Mean Density (× 10³ kg/m³)	Surface Gravity (Compared with Earth's)	Semimajor Axis × 10⁶ km	Semimajor Axis AU†	Orbital Period Years	Orbital Period Days	Eccentricity	Inclination to Ecliptic
Mercury	2439	0.0553	5.43	0.378	57.9	0.3871	0.24084	87.96	0.2056	7°00′26″
Venus	6052	0.8150	5.24	0.894	108.2	0.7233	0.615 15	224.68	0.0068	3°23′40″
Earth	6378.140	1	5.515	1	149.6	1	1.000 04	365.25	0.0167	0°00′14″
Mars	3397.2	0.1074	3.93	0.379	227.9	1.5237	1.8808	686.95	0.0934	1°51′09″
Jupiter	71 398	317.89	1.36	2.54	778.3	5.2028	11.862	4337	0.0483	1°18′29″
Saturn	60 000	95.17	0.71	1.07	1427.0	9.5388	29.456	10 760	0.0560	2°29′17″
Uranus	26 145	14.56	1.30	0.8	2871.0	19.1914	84.07	30 700	0.0461	0°48′26″
Neptune	24 300	17.24	1.8	1.2	4497.1	30.0611	164.81	60 200	0.0100	1°46′27″
Pluto††	1500–1800	0.02	0.5–0.8	~0.03	5913.5	39.5294	248.53	90 780	0.2484	17°09′03″

*Planet's mass/Earth's mass, where $M_E = 6.0 \times 10^{24}$ kg.
†Astronomical unit: 1 AU = 1.5×10^8 km, the average distance between the Earth and the Sun.
††Pluto is now classified as a "dwarf" planet.

APPENDIX IV   Alphabetical Listing of the Chemical Elements (The periodic table is provided inside the back cover.)

Element	Symbol	Atomic Number (Proton Number)	Atomic Mass	Element	Symbol	Atomic Number (Proton Number)	Atomic Mass	Element	Symbol	Atomic Number (Proton Number)	Atomic Mass
Actinium	Ac	89	227.0278	Hafnium	Hf	72	178.49	Praseodymium	Pr	159	140.9077
Aluminum	Al	13	26.981 54	Hahnium	Ha	105	(262)	Promethium	Pm	61	(145)
Americium	Am	95	(243)	Hassium	Hs	108	(265)	Protactinium	Pa	91	231.0359
Antimony	Sb	51	121.757	Helium	He	2	4.002 60	Radium	Ra	88	226.0254
Argon	Ar	18	39.948	Holmium	Ho	67	164.9304	Radon	Rn	86	(222)
Arsenic	As	33	74.9216	Hydrogen	H	1	1.007 94	Rhenium	Re	75	186.207
Astatine	At	85	(210)	Indium	In	49	114.82	Rhodium	Rh	45	102.9055
Barium	Ba	56	137.33	Iodine	I	53	126.9045	Rubidium	Rb	37	85.4678
Berkelium	Bk	97	(247)	Iridium	Ir	77	192.22	Ruthenium	Ru	44	101.07
Beryllium	Be	4	9.01218	Iron	Fe	26	55.847	Rutherfordium	Rf	104	(261)
Bismuth	Bi	83	208.9804	Krypton	Kr	36	83.80	Samarium	Sm	62	150.36
Bohrium	Bh	107	(264)	Lanthanum	La	57	138.9055	Scandium	Sc	21	44.9559
Boron	B	5	10.81	Lawrencium	Lr	103	(260)	Seaborgium	Sg	106	(263)
Bromine	Br	35	79.904	Lead	Pb	82	207.2	Selenium	Se	34	78.96
Cadmium	Cd	48	112.41	Lithium	Li	3	6.941	Silicon	Si	14	28.0855
Calcium	Ca	20	40.078	Lutetium	Lu	71	174.967	Silver	Ag	47	107.8682
Californium	Cf	98	(251)	Magnesium	Mg	12	24.305	Sodium	Na	11	22.989 77
Carbon	C	6	12.011	Manganese	Mn	25	54.9380	Strontium	Sr	38	87.62
Cerium	Ce	58	140.12	Meitnerium	Mt	109	(268)	Sulfur	S	16	32.066
Cesium	Cs	55	132.9054	Mendelevium	Md	101	(258)	Tantalum	Ta	73	180.9479
Chlorine	Cl	17	35.453	Mercury	Hg	80	200.59	Technetium	Tc	43	(98)
Chromium	Cr	24	51.996	Molybdenum	Mo	42	95.94	Tellurium	Te	52	127.60
Cobalt	Co	27	58.9332	Neodymium	Nd	60	144.24	Terbium	Tb	65	158.9254
Copper	Cu	29	63.546	Neon	Ne	10	20.1797	Thallium	Tl	81	204.383
Curium	Cm	96	(247)	Neptunium	Np	93	237.048	Thorium	Th	90	232.0381
Dubnium	Db	105	(262)	Nickel	Ni	28	58.69	Thulium	Tm	69	168.9342
Dysprosium	Dy	66	162.50	Niobium	Nb	41	92.9064	Tin	Sn	50	118.710
Einsteinium	Es	99	(252)	Nitrogen	N	7	14.0067	Titanium	Ti	22	47.88
Erbium	Er	68	167.26	Nobelium	No	102	(259)	Tungsten	W	74	183.85
Europium	Eu	63	151.96	Osmium	Os	76	190.2	Uranium	U	92	238.0289
Fermium	Fm	100	(257)	Oxygen	O	8	15.9994	Vanadium	V	23	50.9415
Fluorine	F	9	18.998 403	Palladium	Pd	46	106.42	Xenon	Xe	54	131.29
Francium	Fr	87	(223)	Phosphorus	P	15	30.973 76	Ytterbium	Yb	70	173.04
Gadolinium	Gd	64	157.25	Platinum	Pt	78	195.08	Yttrium	Y	39	88.9059
Gallium	Ga	31	69.72	Plutonium	Pu	94	(244)	Zinc	Zn	30	65.39
Germanium	Ge	32	72.561	Polonium	Po	84	(209)	Zirconium	Zr	40	91.22
Gold	Au	79	196.9665	Potassium	K	19	39.0983				

APPENDIX V   Properties of Selected Isotopes

Atomic Number (Z)	Element	Symbol	Mass Number (A)	Atomic Mass*	Abundance (%) or Decay Mode† (if Radioactive)	Half-Life (if Radioactive)
0	(Neutron)	n	1	1.008 665	β^-	10.6 min
1	Hydrogen	H	1	1.007 825	99.985	
	Deuterium	D	2	2.014 102	0.015	
	Tritium	T	3	3.016 049	β^-	12.33 y
2	Helium	He	3	3.016 029	0.00014	
			4	4.002 603	≈ 100	
3	Lithium	Li	6	6.015 123	7.5	
			7	7.016 005	92.5	
4	Beryllium	Be	7	7.016 930	EC, γ	53.3 d
			8	8.005 305	2α	6.7×10^{-17} s
			9	9.012 183	100	
5	Boron	B	10	10.012 938	19.8	
			11	11.009 305	80.2	
			12	12.014 353	β^-	20.4 ms
6	Carbon	C	11	11.011 433	β^+, EC	20.4 ms
			12	12.000 000	98.89	
			13	13.003 355	1.11	
			14	14.003 242	β^-	5730 y
7	Nitrogen	N	13	13.005 739	β^-	9.96 min
			14	14.003 074	99.63	
			15	15.000 109	0.37	
8	Oxygen	O	15	15.003 065	β^+, EC	122 s
			16	15.994 915	99.76	
			18	17.999 159	0.204	
9	Fluorine	F	19	18.998 403	100	
10	Neon	Ne	20	19.992 439	90.51	
			22	21.991 384	9.22	
11	Sodium	Na	22	21.994 435	β^+, EC, γ	2.602 y
			23	22.989 770	100	
			24	23.990 964	β^-, γ	15.0 h
12	Magnesium	Mg	24	23.985 045	78.99	
13	Aluminum	Al	27	26.981 541	100	
14	Silicon	Si	28	27.976 928	92.23	
			31	30.975 364	β^-, γ	2.62 h
15	Phosphorus	P	31	30.973 763	100	
			32	31.973 908	β^-	14.28 d
16	Sulfur	S	32	31.972 072	95.0	
			35	34.969 033	β^-	87.4 d
17	Chlorine	Cl	35	34.968 853	75.77	
			37	36.965 903	24.23	
18	Argon	Ar	40	39.962 383	99.60	
19	Potassium	K	39	38.963 708	93.26	
			40	39.964 000	β^-, EC, γ, β^+	1.28×10^9 y
20	Calcium	Ca	40	39.962 591	96.94	
24	Chromium	Cr	52	51.940 510	83.79	
25	Manganese	Mn	55	54.938 046	100	
26	Iron	Fe	56	55.934 939	91.8	
27	Cobalt	Co	59	58.933 198	100	
			60	59.933 820	β^-, γ	5.271 y
28	Nickel	Ni	58	57.935 347	68.3	
			60	59.930 789	26.1	
			64	63.927 968	0.91	
29	Copper	Cu	63	62.929 599	69.2	
			64	63.929 766	β^-, β^+	12.7 h
			65	64.927 792	30.8	
30	Zinc	Zn	64	63.929 145	48.6	
			66	65.926 035	27.9	
33	Arsenic	As	75	74.921 596	100	
35	Bromine	Br	79	78.918 336	50.69	
36	Krypton	Kr	84	83.911 506	57.0	
			89	88.917 563	β^-	3.2 min

Atomic Number (Z)	Element	Symbol	Mass Number (A)	Atomic Mass*	Abundance (%) or Decay Mode[†] (if Radioactive)	Half-Life (if Radioactive)
38	Strontium	Sr	86	85.909 273	9.8	
			88	87.905 625	82.6	
			90	89.907 746	β^-	28.8 y
39	Yttrium	Y	89	89.905 856	100	
43	Technetium	Tc	98	97.907 210	β^-, γ	4.2×10^6 y
47	Silver	Ag	107	106.905 095	51.83	
			109	108.904 754	48.17	
48	Cadmium	Cd	114	113.903 361	28.7	
49	Indium	In	115	114.903 88	95.7; β^-	5.1×10^{14} y
50	Tin	Sn	120	119.902 199	32.4	
53	Iodine	I	127	126.904 477	100	
			131	130.906 118	β^-, γ	8.04 d
54	Xenon	Xe	132	131.904 15	26.9	
			136	135.907 22	8.9	
55	Cesium	Cs	133	132.905 43	100	
56	Barium	Ba	137	136.905 82	11.2	
			138	137.905 24	71.7	
			144	143.922 73	β^-	11.9 s
61	Promethium	Pm	145	144.912 75	EC, α, γ	17.7 y
74	Tungsten	W	184	183.950 95	30.7	
76	Osmium	Os	191	190.960 94	β^-, γ	15.4 d
			192	191.961 49	41.0	
78	Platinum	Pt	195	194.964 79	33.8	
79	Gold	Au	197	196.966 56	100	
80	Mercury	Hg	202	201.970 63	29.8	
81	Thallium	Tl	205	204.974 41	70.5	
			210	209.990 069	β^-	1.3 min
82	Lead	Pb	204	203.973 044	β^-, 1.48	1.4×10^{17} y
			206	205.974 46	24.1	
			207	206.975 89	22.1	
			208	207.976 64	52.3	
			210	209.984 18	α, β^-, γ	22.3 y
			211	210.988 74	β^-, γ	36.1 min
			212	211.991 88	β^-, γ	10.64 h
			214	213.999 80	β^-, γ	26.8 min
83	Bismuth	Bi	209	208.980 39	100	
			211	210.987 26	α, β^-, γ	2.15 min
84	Polonium	Po	210	209.982 86	α, γ	138.38 d
			214	213.995 19	α, γ	164 μs
86	Radon	Rn	222	222.017 574	α, β	3.8235 d
87	Francium	Fr	223	223.019 734	α, β^-, γ	21.8 min
88	Radium	Ra	226	226.025 406	α, γ	1.60×10^3 y
			228	228.031 069	β^-	5.76 y
89	Actinium	Ac	227	227.027 751	α, β^-, γ	21.773 y
90	Thorium	Th	228	228.028 73	α, γ	1.9131 y
			232	232.038 054	100; α, γ	1.41×10^{10} y
92	Uranium	U	232	232.037 14	α, γ	72 y
			233	233.039 629	α, γ	1.592×10^5 y
			235	235.043 925	0.72; α, γ	7.038×10^8 y
			236	236.045 563	α, γ	2.342×10^7 y
			238	238.050 786	99.275; α, γ	4.468×10^9 y
			239	239.054 291	β^-, γ	23.5 min
93	Neptunium	Np	239	239.052 932	β^-, γ	2.35 d
94	Plutonium	Pu	239	239.052 158	α, γ	2.41×10^4 y
95	Americium	Am	243	243.061 374	α, γ	7.37×10^3 y
96	Curium	Cm	245	245.065 487	α, γ	8.5×10^3 y
97	Berkelium	Bk	247	247.070 03	α, γ	1.4×10^3 y
98	Californium	Cf	249	249.074 849	α, γ	351 y
99	Einsteinium	Es	254	254.088 02	α, γ, β^-	276 d
100	Fermium	Fm	253	253.085 18	EC, α, γ	3.0 d

*The masses given throughout this table are those for the neutral atom, including the Z electrons.
[†]"EC" stands for electron capture.

APPENDIX VI   Answers to Follow-up Exercises

CHAPTER 1

1.1 $L = 10$ m

1.2 No. $m = (m/s)^2$ or $m = s^2$

1.3 30 days (24 h/day)(60 min/h)(60 s/min) $= 2.6 \times 10^6$ s

1.4 13.3 times

1.5 1 m$^3 = 10^6$ cm^3

1.6 2-L cost: $1.35/2 L = $0.68/L; 1/2 gal (4 qt/gal)(1 L/1.894 qt) $= 1.894$ L, cost: $1.32/1.894 L = $0.70/L. Costs are different because of rounding errors, but still there is a 2 cent difference.

1.7 (a) 7.0×10^5 kg^2 (b) 3.02×10^2 (no units)

1.8 (a) 23.70 (b) 22.09

1.9 $V = \pi r^2 h = \pi(0.490$ m$)^2 (1.28$ m$) = 0.965$ m^3

1.10 11.6 m

1.11 750 cm$^3 = 7.50 \times 10^{-4}$ m$^3 \approx 10^{-3}$ m^3, $m = \rho V \approx (10^3$ kg/m$^3)(10^{-3}$ m$^3) = 1$ kg (By direct calculation, $m = 0.79$ kg.)

1.12 $V \approx 10^{-2}$ m^3, cells/vol $\approx 10^4$ cells/mm^3 $(10^9$ mm^3/m$^3) = 10^{13}$ cells/m^3, and (cells/vol) (vol) $\approx 10^{11}$ white cells

CHAPTER 2

2.1 $\Delta t = (8 \times 5.0$ s$) + (7 \times 10$ s$) = 110$ s

2.2 (a) $s_1 = 2.00$ m/s; $s_2 = 1.52$ m/s; $s_3 = 1.72$ m/s $\neq 0$, although the velocity is zero

2.3 No. If the velocity is also in the negative direction, the object will speed up.

2.4 9.0 m/s in the direction of the original motion

2.5 Yes, 96 m. (A lot quicker, isn't it?)

2.6 No, always more than one unknown variable

2.7 No, changes x_o positions, but separation distance is the same

2.8 $x = v^2/2a$, $x_B = 48.6$ m, $x_C = 39.6$ m; the Blazer should not tailgate within at least 9.0 m.

2.9 1.16 s longer

2.10 The time for the bill to fall its length is 0.179 s. This time is less than the average reaction time (0.192 s) computed in the Example, so most people cannot catch the bill.

2.11 $y_u = y_d = 5.12$ m, as measured from reference $y = 0$ at the release point

2.12 $t = \dfrac{v - v_o}{-g_M} = \dfrac{-4.6 \text{ m/s} - (-1.5 \text{ m/s})}{-1.6 \text{ m/s}^2} = 1.9$ s

CHAPTER 3

3.1 $v_x = -0.40$ m/s, $v_y = +0.30$ m/s; the distance is unchanged.

3.2 $x = 9.00$ m, $y = 12.6$ m (same)

3.3 $\vec{v} = (0)\hat{x} + (3.7$ m/s$)\hat{y}$

3.4 $\vec{C} = (-7.7$ m$)\hat{x} + (-4.3$ m$)\hat{y}$

3.5 (a) $y_o = +25$ m and $y = 0$; the equation is the same. (b) $\vec{v} = (8.25$ m/s$)\hat{x} + (-22.1$ m/s$)\hat{y}$

3.6 Both increase sixfold.

3.7 (a) If not, the stone would hit the side of the block. (b) Eq. 3.11 does not apply; the initial and final heights are not the same. $R = 15$ m, which is way off the 27-m answer.

3.8 The ball thrown at 45°. It would have a greater initial velocity.

3.9 At the top of the parabolic arc, the player's vertical motion is zero and is very small on either side of this maximum height. Here, the player's horizontal velocity component dominates, and he moves horizontally, with little motion in the vertical direction. This gives the illusion of "hanging" in the air.

3.10 4.15 m from the net

3.11 $v_{bs}t = (2.33$ m/s$)(225$ m$) = 524$ m

3.12 14.5° W of N

CHAPTER 4

4.1 6.0 m/s in the direction of the net force

4.2 (a) 11 lb (b) Weight in pounds ≈ 2.2 lb/kg.

4.3 8.3 N

4.4 (a) 50° above the $+x$-axis (b) x- and y-components reversed: $\vec{v} = (9.8$ m/s$)\hat{x} + (4.5$ m/s$)\hat{y}$

4.5 Yes, mutual gravitational attractions between the briefcase and the Earth.

4.6 (a) $m_2 > 1.7$ kg (b) $\theta < 17.5°$

4.7 (a) 7.35 N (b) Neglecting air resistance, 7.35 N, downward

4.8 Increase; $\tan\theta = \dfrac{T}{mg} = \dfrac{55 \text{ N}}{(5.0 \text{ kg})(9.8 \text{ m/s}^2)} = 1.1, \theta = 48°$

4.9 (a) $F_1 = 3.5w$; even greater than F_2 (b) $\Sigma F_y = ma$, and F_1 and F_2 would both increase.

4.10 $\mu_s = 1.41\mu_k$ (for three cases in Table 4.1)

4.11 No. F varies with angle, with the angle for minimum applied force being around 33° in this case. (Greater forces are required for 20° and 50°.) In general, the optimum angle depends on the coefficient of friction.

4.12 Friction is kinetic, and f_k is in the $+x$ direction. Acceleration in the $-x$ direction.

4.13 Air resistance depends not only on speed, but also on size and shape. If the heavier ball were larger, it would have more exposed area to collide with air molecules, and the retarding force would increase faster. Depending on the size difference, the heavier ball might reach terminal velocity first, and the lighter ball would strike the ground first. Alternatively, the balls might reach terminal velocity together.

CHAPTER 5

5.1 -2.0 J

5.2 $d = \dfrac{W}{F \cos\theta} = \dfrac{3.80 \times 10^4 \text{ J}}{(189 \text{ N})(0.866)} = 232$ m

5.3 No, speed would decrease and it would stop moving.

5.4 $W_{x_1} = 0.034$ J, $W_x = 0.64$ J (measured from x_o)

5.5 No, $W_2/W_1 = 4$, or four times as much

5.6 Here $m_s = m_g/2$ as before. However, $v_s/v_g = (6.0$ m/s$)/(4.0$ m/s$) = \frac{3}{2}$. Using a ratio, $K_s/K_g = \frac{9}{8}$, and the safety still has more kinetic energy than the guard. (Answer could also be obtained from direct calculations of kinetic energies, but for a relative comparison, a ratio is usually quicker.)

5.7 $W_3/W_2 = 1.4$, or 40% larger; more work, but a smaller percentage increase

5.8 $\Delta U = mgh = (60$ kg$)(9.8$ m/s$^2)(1000$ m$) \sin 10° = 10.2 \times 10^4$ J; yes, doubled

5.9 $\Delta K_{total} = 0$, $\Delta U_{total} = 0$

5.10 Without friction, the liquid would oscillate back and forth between the containers.

5.11 9.9 m/s

5.12 No. $E_o = E$ or $\frac{1}{2}mv_o^2 + mgh = \frac{1}{2}mv^2$. The mass cancels and the speed is independent of mass. (Recall that in free fall, all objects or projectiles fall with the same vertical acceleration g—see Section 2.5.)

5.13 0.025 m

5.14 (a) 59% (b) $E_{loss}/t = mg(y/t) = mgv = (60\ mg)$ J/s

5.15 Block would stop in rough area.

5.16 52%

5.17 (a) Same work in twice the time (b) Same work in half the time

5.18 (a) No. (b) Creation of energy

CHAPTER 6

6.1 5.0 m/s. Yes, this is 18 km/h or 11 mi/h, a speed at which humans can run.

6.2 (1) Ship the greatest KE (2) Bullet the least KE

6.3 $(-3.0 \text{ kg} \cdot \text{m/s})\hat{\mathbf{x}} + (4.0 \text{ kg} \cdot \text{m/s})\hat{\mathbf{y}}$

6.4 It would increase to 60 m/s, and a greater speed means a longer drive, ideally. (There is also a directional consideration.)

6.5 $F_{\text{avg}} = \dfrac{\Delta p}{\Delta t} = \dfrac{-310 \text{ kg} \cdot \text{m/s}}{0.600 \text{ s}} = -517 \text{ N}$

6.6 (a) No, for the m_1/m_2 system, external force on block. Yes, for the m_1/m_2 Earth system. But with m_2 attached to the Earth, the mass of this part of the system would be vastly greater than that of m_2, so its change in velocity would be negligible. (b) Assuming the ball is tossed in the +direction: for the tosser, $v_t = -0.50$ m/s; for the catcher, $v_c = 0.48$ m/s. For the ball: $p = 0, +25 \text{ kg} \cdot \text{m/s}, +1.2 \text{ kg} \cdot \text{m/s}$.

6.7 No. Energy went into work of breaking the brick, and some was lost as heat and sound.

6.8 No.

6.9 No. All of the kinetic energy cannot be lost to make the dent. The momentum after the collision cannot be zero, since it was not zero initially. Thus, the balls must be moving and have kinetic energy. This can also be seen from Eq. 6.11: $K_f/K_i = m_1/(m_1+m_2)$, and K_f cannot be zero (unless m_1 is zero, which is not possible).

6.10 $x_1 = v_1 t = (-0.80 \text{ m/s})(2.5 \text{ s}) = -2.0 \text{ m}, x_2 = v_2 t = (1.2 \text{ m/s})(2.5 \text{ s}) = 3.0 \text{ m}$
$\Delta x = x_2 - x_1 = 3.0 \text{ m} - (-2.0 \text{ m}) = 5.0 \text{ m}$. The objects are 5.0 m apart.

6.11 (a) $\Delta p_1 = p_{1_f} - p_{1_o} = 32 \text{ kg} \cdot \text{m/s} - 40 \text{ kg} \cdot \text{m/s} = -8.0 \text{ kg} \cdot \text{m/s}$
$\Delta p_2 = p_{2_f} - p_{2_o} = 13 \text{ kg} \cdot \text{m/s} - 5.0 \text{ kg} \cdot \text{m/s} = +8.0 \text{ kg} \cdot \text{m/s}$
(b) $\Delta p_1 = p_{1_f} - p_{1_o} = (-20 \text{ kg} \cdot \text{m/s}) - (12 \text{ kg} \cdot \text{m/s}) = -32 \text{ kg} \cdot \text{m/s}$
$\Delta p_2 = p_{2_f} - p_{2_o} = (8.0 \text{ kg} \cdot \text{m/s}) - (-24 \text{ kg} \cdot \text{m/s}) = +32 \text{ kg} \cdot \text{m/s}$

6.12 $p_{1_o} = mv_{1_o}, p_{2_o} = -mv_{2_o}$ and $p_1 = mv_1 = -mv_{2_o}$,
$p_2 = mv_2 = mv_{1_o}$, so conserved. $K_i = \dfrac{m}{2}\left(v_{1_o}^2 + v_{2_o}^2\right)$ and
$K_f = \dfrac{m}{2}(v_1^2 + v_2^2) = \dfrac{m}{2}\left[\left(-v_{2_o}\right)^2 + \left(v_{1_o}\right)^2\right]$, so conserved.

6.13 All of the balls swing out, but to different degrees. With $m_1 > m_2$, the stationary ball (m_2) moves off with a greater speed after collision than the incoming, heavier ball (m_1), and the heavier ball's speed is reduced after collision, in accordance with Eq. 6.16 (see Fig. 6.14b). Hence, a "shot" of momentum is passed along the row of balls with equal mass (see Fig. 6.14a), and the end ball swings out with the same speed as was imparted to m_2. Then, the process is repeated: m_1, *now moving more slowly*, collides again with the initial ball in the row (m_2), and another, but smaller, shot of momentum is passed down the row. The new end ball in the row receives less kinetic energy than the one that swung out just a moment previously, and so doesn't swing as high. This process repeats itself instantaneously for each ball, with the observed result that all of the balls swing out to different degrees.

6.14 $X_{\text{CM}} = \dfrac{(\text{same as in example}) + (8.0 \text{ kg})x_4}{(\text{same as in example}) + (8.0 \text{ kg})} =$
$\dfrac{0 + (8.0 \text{ kg})x_4}{19 \text{ kg}} = +1.0 \text{ m}$
$x_4 = \left(\dfrac{19}{8}\right) \text{ m} = 2.4 \text{ m}$

6.15 $(X_{\text{CM}}, Y_{\text{CM}}) = (0.47 \text{ m}, 0.10 \text{ m})$; same location as in Example, two-thirds of the length of the bar from m_1. *Note:* The location of the CM does not depend on the frame of reference.

6.16 Yes, the CM does not move.

CHAPTER 7

7.1 $210° (2\pi \text{ rad}/360°)(256 \text{ m}) = 938 \text{ m}$

7.2 (a) 0.35% for 10° (b) 1.2% for 20°

7.3 (a) 4.7 rad/s, 0.38 m/s; 4.7 rad/s, 0.24 m/s (b) To equalize the running distances, because the curved sections of the track have different radii and thus different lengths

7.4 120 rpm

7.5 106 rpm

7.6 The string cannot be exactly horizontal; it must make some small angle to the horizontal so that there will be an upward component of the tension force to balance the ball's weight.

7.7 No; it depends on mass: $F_c = \mu_s mg$.

7.8 No. Both masses have the same angular frequency or speed ω, and $a_c = r\omega^2$, so actually $a_c \propto r$. Remember, $v = 2\pi r/T$, and note that $v_2 > v_1$, with $a_c = v^2/r$.

7.9 (a) $T = 5.2$ N (b) 1.9 s, $v = r\left(\dfrac{2\pi}{T}\right)$, and $r = L \sin 20°$

7.10 (a) The directions of ω and α would be downward, perpendicular to the plane of the CD. (b) Negative α, which means it is the opposite direction of ω.

7.11 -0.031 rad/s^2

7.12 $2.8 \times 10^{-3} \text{ m/s}^2$. Centripetal acceleration keeps Moon in circular orbit.

7.13 $T^2 = \left(\dfrac{4\pi^2}{GM_E}\right)r^3 = \left(\dfrac{4\pi^2}{GM_E}\right)(R_E + h)^3 \approx \left(\dfrac{4\pi^2}{g}\right)R_E \approx 4R_E$
$T = 2\sqrt{R_E} = 2(6.4 \times 10^6 \text{ m})^{\frac{1}{2}} = 5.1 \times 10^3 \text{ s}$ (Why are the units not consistent?)

7.14 No, they do not vary linearly. $\Delta U = 2.4 \times 10^9$ J, only a 9.1% increase

7.15 This is the amount of *negative* work done by an external force or agent when the masses are brought together. To separate the masses by infinite distances, an equal amount of positive work (against gravity) would have to be done.

7.16 $T^2 = \left(\dfrac{4\pi^2}{GM_S}\right)r^3$ and $M_S = \dfrac{4\pi^2 r^3}{GT^2} =$
$\dfrac{4\pi^2(1.50 \times 10^{11} \text{ m})^3}{(6.67 \times 10^{-11} \text{ N} \cdot \text{m}^2/\text{kg}^2)(3.16 \times 10^7 \text{ s})^2} = 2.00 \times 10^{30} \text{ kg}$

CHAPTER 8

8.1 $s = r\omega = 5(0.12 \text{ m})(1.7) = 0.20 \text{ m}$;
$s = v_{\text{CM}}t = (0.10 \text{ m/s})(2.00 \text{ s}) = 0.20 \text{ m}$

8.2 The weights of the balls and the forearm produce torques that tend to cause rotation in the direction opposite that of the applied torque.

8.3 More strain

8.4 $T \propto 1/\sin\theta$, and as θ gets smaller, so does $\sin\theta$, and T increases. In the limit $\sin\theta \to 0$ and $T \to$ infinity (unrealistic).

8.5 $\Sigma\tau$: $Nx - m_1gx_1m_2gx_2 - m_3gx_3 = (200\text{ g})g(50\text{ cm}) - (25\text{ g})g(0\text{ cm}) - (75\text{ g})g(20\text{ cm}) - (100\text{ g})g(85\text{ cm}) = 0$, where $N = Mg$

8.6 No. With f_{s_1}, the reaction force N would not generally be the same (f_{s_2} and N are perpendicular components of the force exerted on the ladder by the wall). In this case, we still have $N = f_{s_1}$, but $Ny - (m_1g)x_1 - (m_mg)x_m - f_{s_2}$, and $x_3 = 0$.

8.7 Hanging vertically

8.8 Male: lighter upper torso; female: heavier lower torso

8.9 Five bricks

8.10 (d) No (equal masses) (e) Yes; with larger mass farther from axis of rotation, $I = 360\text{ kg}\cdot\text{m}^2$

8.11 The long pole (or your extended arms) increases the moment of inertia by placing more mass farther from the axis of rotation (the tightrope or rail). When the walker leans to the side, a gravitational torque tends to produce a rotation about the axis of rotation, causing a fall. However, with a greater rotational inertia (greater I), the walker has time to shift his or her body so that the center of gravity is again over the rope or rail and thus again in (unstable) equilibrium. With very flexible poles, the CG may be below the wire, thus ensuring stability.

8.12 $t = 0.63$ s

8.13 $\alpha = \dfrac{2\,mg - (2\tau_f R)}{(2\,m + M)R}$; $\dfrac{N}{\text{kg}\cdot\text{m}}$; $\dfrac{N}{\text{kg}\cdot\text{m}} = \dfrac{\text{kg}\cdot\text{m/s}^2}{\text{kg}\cdot\text{m}} = \dfrac{1}{\text{s}^2}$

8.14 The yo-yo would roll back and forth, oscillating about the critical angle.

8.15 (a) 0.24 m (b) The force of *static* friction, f_s, acts at the point of contact, which is always instantaneously at rest, and so does no work. Some frictional work may be done due to rolling friction, but this is considered negligible for hard objects and surfaces.

8.16 $v_{CM} = 2.2$ m/s; using a ratio, 1.4 times greater; no rotational energy

8.17 You already know the answer: 5.6 m/s. (It doesn't depend on the mass of the ball.)

8.18 $M_a = 0.75(75\text{ kg}) = 56$ kg; $L_1 = 13$ kg·m²/s and $L_2 = (1.3\text{ kg}\cdot\text{m}^2)\omega$ [math not shown]; $L_2 = L_1$ or $(1.3\text{ kg}\cdot\text{m}^2)\omega = 13$ kg·m²/s and $\omega = 10$ rad/s

CHAPTER 9

9.1 (a) +0.10% (b) 39 kg

9.2 2.3×10^{-4} L, or 2.3×10^{-7} m³

9.3 (1) Having enough nails and (2) having them all of equal height and with not so sharp a point. This could be achieved by filing off the tips of the nails so as to have a "uniform" surface. Also, this would increase the effective area.

9.4 3.03×10^4 N (or 6.82×10^3 lb—about 3.4 tons!) This is roughly the force on your back right now. Our bodies don't collapse under atmospheric pressure because cells are filled with incompressible fluids (mostly water), bone, and muscle, which react with an equal outward pressure (equal and opposite forces). As with forces, it is a pressure *difference* that gives rise to dynamic effects.

9.5 $d_o = \sqrt{\dfrac{F_o}{F_i}}d_i = \sqrt{\dfrac{1}{10}}(8.0\text{ cm}) = 2.5$ cm

9.6 Pressure in veins is lower than that in arteries (120/80).

9.7 As the balloon rises, the buoyant force decreases as a result of the temperature decrease (less helium pressure, less volume) and the less dense air ($F_b = m_f g = \rho_f g V_f$). When the net force is zero, the velocity is constant. The cooling effect continues with altitude and the balloon will start to sink when the net force is negative.

9.8 $r \approx 1.0$ m; $F_b = \rho g V = \rho g\left(\dfrac{4}{3}\pi r^3\right) =$

$(1.29\text{ kg/m}^3)\left(\dfrac{4g\pi}{3}\right)(1.0\text{ m})^3 = 53$ N, much more

9.9 (a) The object would sink, so the buoyant force is less than the object's weight. Hence, the scale would have a reading greater than 40 N. Note that with a greater density, the object would not be as large and less water would be displaced. (b) 41.8 N.

9.10 11%

9.11 -18%

9.12 $r = 9.00 \times 10^{-3}$ m, $v = \dfrac{\text{constant}}{A} = \dfrac{8.33 \times 10^{-5}\text{ m}^3/\text{s}}{\pi(9.00 \times 10^{-3}\text{ m})^2}$
$= 0.327$ m/s; 23%

9.13 69%

9.14 As the water falls, speed (v) increases and area (A) must decrease to have Av equal a constant.

9.15 0.38 m

CHAPTER 10

10.1 (a) -40 °C (b) -40 °F (You should immediately know the answer—this is the temperature at which the Fahrenheit and Celsius temperatures are numerically equal.)

10.2 (a) $T_R = T_F + 460$ (b) $T_R = \frac{9}{5}T_C + 492$ (c) $T_R = \frac{9}{5}T$

10.3 96 °C

10.4 0.250 m³ $= 250$ L

10.5 50 °C

10.6 The student should heat the bearing so its inside diameter will be larger.

10.7 899 °C

10.8 (a) The rotational kinetic energy of the oxygen molecules is the difference between the total energies of oxygen and radon molecules, 2.44×10^3 J. (b) The oxygen molecule is less massive and has the higher v_{rms}.

CHAPTER 11

11.1 1.42×10^3 m

11.2 5.00 kg

11.3 (a) The ratio will be smaller because the specific heat of aluminum is greater than that of copper. (b) $Q_w/Q_{pot} = 15.2$

11.4 The final temperature T_f is expected to be higher because the water was at a higher initial temperature. $T_f = 34.4$ °C

11.5 -1.09×10^5 J (negative because heat is lost or removed)

11.6 (a) 2.64×10^{-2} kg or 26.4 g of ice melts. (b) The final temperature is still 0 °C because the liver cannot lose enough heat to melt all the ice, even if the ice started at 0 °C. The final result is an ice/water/liver system at 0 °C, but with more water than in the Example.

11.7 1.1×10^5 J/s (difference due to rounding)

11.8 No, the air spaces provide good insulation because air is a poor thermal conductor. The many small pockets of air between the body and the outer garment form an insulating layer that minimizes conduction and so decreases the loss of body heat. (There is little convection in the small spaces.)

11.9 (a) -1.5×10^2 J/s or -1.5×10^2 W (b) The huge ear flaps have large surface area so more heat can be radiated out.

CHAPTER 12

12.1 0.20 kg

12.2 In both cases, heat flows into the gas. During the isothermal expansion, $Q = W = +3.14 \times 10^3$ J. During the isobaric expansion, $W = +4.53 \times 10^3$ J and $T_2 = 2T_1 = 546$ K.

$\Delta U = U_2 - U_1 = \frac{3}{2}nR(T_2 - T_1) = +6.80 \times 10^3 \text{ J}$;   therefore, $Q = \Delta U + W = +1.13 \times 10^4 \text{ J}$.

12.3 753 °C

12.4 (a) 142 K or −131 °C (b) For monatomic gas, $\Delta U = (3/2)nR\Delta T = -3.76 \times 10^3 \text{ J}$. This should be the same as $-W$ since, for an adiabatic process, $Q = 0 = \Delta U + W$; therefore, $\Delta U = -W$. The slight difference is due to rounding.

12.5 $-1.22 \times 10^3 \text{ J/K}$

12.6 Overall zero entropy change requires $|\Delta S_w| = |\Delta S_s|$ or $|Q_w/T_w| = |Q_s/T_s|$. Because the system is isolated, the magnitudes of the two heat flows *must* be the same, $|Q_w| = |Q_s|$. Thus no overall entropy change requires the water and the spoon to have the same average temperature, $\overline{T}_w = \overline{T}_s$. This is not possible, unless they are initially at the *same* temperature. Thus, this can only happen if there is no net heat flow.

12.7 (a) 150 J/cycle (b) 850 J/cycle

12.8 0.035 °C

12.9 (a) The new values are $COP_{ref} = 3.3$ and $COP_{hp} = 4.3$. (b) The COP of the air conditioner has the largest percentage increase.

12.10 It would show an increase of 7.5%.

CHAPTER 13

13.1 36 m/s²; 54 m/s²

13.2 (1) $y = -0.088$ m, going up; $n = 0.90$ (2) $y = 0$, going up; $n = 1.5$

13.3 $1.6 \times 10^4 \text{ N/m}$

13.4 9.75 m/s²; no. Since this is less than the accepted value at sea level, the park is probably at an altitude above sea level.

13.5 (a) 10 m (b) 0.40 Hz

13.6 440 Hz

CHAPTER 14

14.1 (a) 2.3 (b) 10.2

14.2 $v = (331 + 0.6 T_C)$ m/s $= [331 + 0.6(38°)] = 354$ m/s; increase

14.3 It would be greatest in He, because it has the smallest molecular mass. (It would be lowest in oxygen, which has the largest molecular mass.)

14.4 (a) The dB scale is logarithmic, not linear.

14.4 (b) $3.16 \times 10^{-6} \text{ W/m}^2$

14.5 No, $I_2 = (316)I_1$

14.6 65 dB

14.7 Destructive interference: $\Delta L = 2.5\lambda = 5(\lambda/2)$, and $m = 5$. No sound would be heard if the waves from the speakers had equal amplitudes. Of course, during a concert the sound would not be single-frequency tones but would have a variety of frequencies and amplitudes. Listeners at certain locations might not hear certain parts of the audible spectrum, but this probably wouldn't be noticed.

14.8 Toward, 431 Hz; past, 369 Hz

14.9 With the source and the observer traveling in the same direction at the same speed, their relative velocity would be zero. That is, the observer would consider the source to be stationary. Since the speed of the source and observer is subsonic, the sound from the source would overtake the observer without a shift in frequency. Generally, for motions involved in a Doppler shift, the word *toward* is associated with an *increase* in frequency and *away* with a *decrease* in frequency. Here, the source and observer remain a constant distance apart. (What would be the case if the speeds were supersonic?)

14.10 768 Hz; yes

14.11 $f_1 = \dfrac{v}{4L} = \dfrac{353 \text{ m/s}}{4(0.0130 \text{ m})} = 6790$ Hz

CHAPTER 15

15.1 $1.52 \times 10^{-20}\%$

15.2 No. If the comb were positive, it would polarize the paper in the reverse way and still attract it.

15.3 $\vec{F}_1$ has a magnitude of 3.8×10^{-7} N at 57° above the $+x$-axis or $\vec{F}_1 = (-0.22\ \mu N)\hat{x} + (0.32\ \mu N)\hat{y}$.

15.4 0.12 m or 12 cm

15.5 $F_e/F_g = 4.2 \times 10^{42}$ or $F_e = 4.2 \times 10^{42}F_g$. The magnitude of the electrical force is the same as that between a proton and electron because they all have the same (magnitude) charge. However, the gravitational force is reduced because the masses are now two low-mass electrons rather than an electron and a much more massive proton.

15.6 The field is zero at 0.60 m (or 60 cm) to the left of q_1.

15.7 $\vec{E} = (-797 \text{ N/C})\hat{x} + (359 \text{ N/C})\hat{y}$ or $E = 874$ N/C at an angle of 24.2° above the $-x$-axis

15.8 (a) The larger of the two fields is due to the closer positive end and points upward. The smaller field due to the negative end points downward, thus the field line is straight and vertically upward away from the positive end. (b) The larger of the two fields is due to the closer negative end and points upward. The smaller field due to the positive end points downward, thus the field line is straight and vertically upward toward the negative end. (c) Both fields point downward, thus the field line is straight and downward, away from the positive end and toward the negative end.

15.9 (a) The electric field is upward from ground to cloud. (b) 2.3×10^3 C

15.10 Positive charge would reside completely on the outside surface, thus only the electroscope attached to the outside surface would show deflection.

15.11 Negative—since the field lines end at negative charges, they are all inward relative to the Gaussian surface

CHAPTER 16

16.1 (a) ΔU_e would double to $+7.20 \times 10^{-18}$ J because the particle's charge is doubled. (b) ΔV is unchanged because it is not related to the particle. (c) $v = 4.65 \times 10^4$ m/s.

16.2 6.63×10^7 m/s

16.3 (a) It has moved further from a positive charge (the proton) and thus has moved to a region of lower electric potential. (b) $\Delta U_e = +3.27 \times 10^{-18}$ J

16.4 6.60×10^{-20} C

16.5 (a) 2.22 m (b) The one closest to the Earth's surface is at a higher potential. (c) No. You can only tell the separation distance between the two surfaces, not their absolute location.

16.6 (a) Surface 1 is at a higher electric potential than surface 2 because it is closer to the positively charged surface. (b) At large distances, the charged object would "look like" a point charge, thus the equipotential surfaces gradually become spherical as the distance from the object gets larger.

16.7 $d = 8.9 \times 10^{-16}$ m, which is much smaller than the size of an atom (or a nucleus for that matter), and thus this design is completely unfeasible.

16.8 7.90×10^3 V

16.9 The capacitance decreases as the spacing d increases. Since the voltage across the capacitor remains constant, this means that the charge on the capacitor would have to decrease, thus charge would flow off of the capacitor. $\Delta Q = -3.30 \times 10^{-12}$ C.

16.10 $U_{\text{parallel}} = 1.20 \times 10^{-4}$ J and $U_{\text{series}} = 5.40 \times 10^{-4}$ J, so the parallel arrangement stores more energy.

16.11 (a) 0.50 μF (b) The energy stored in capacitor 3 would be six times that stored in capacitor 1.

CHAPTER 17

17.1 The result is the same: $V_{\text{AB}} = V$.

17.2 About 32 years

17.3 100 V

17.4 From $R = \rho L/A$, if resistivity is doubled and length halved, the numerator stays the same. If the diameter is halved, the area *decreases* by a factor of 4. Thus the resistance increases by a factor of 4, to 3.0×10^3 Ω. The current in the second fish is $I = V/R = 0.133$ A or 133 mA.

17.5 0.67 Ω. The material with the largest temperature coefficient of resistivity makes a more sensitive thermometer because it produces a larger (and thus more accurate to measure) change in resistance for a given temperature change.

17.6 (a) $R_1 = V^2/P_1 = 11.0$ Ω and $R_2 = 0.900R_1 = 9.92$ Ω (b) $I_1 = V/R_1 = 10.5$ A and $I_2 = 1.11I_1 = 11.6$ A

17.7 (a) The heat needed is $Q = mc\Delta T = 1.67 \times 10^5$ J. Thus $P = Q/t = 1.67 \times 10^5$ J$/180$ s $= 930$ W $= 0.930$ kW. So $R = V^2/P = (120 \text{ V})^2/1.67 \times 10^5$ J $= 15.5$ Ω. (b) Two cups a day for 30 days (60 cups) means a total operation time for the heater of 3.00 min $\times$ 60 $= 180$ min $= 3.00$ h. Hence the energy usage is (0.930 kW)(3.00 h) $= 2.79$ kWh, which gives a monthly cost of 2.79 kWh $\times$ (0.15/kWh) $\approx$ 42 cents.

17.8 8.3 hours

17.9 At best, power plants produce electric energy with efficiencies of 35% (ignoring transmission losses). Thus in terms of primary fuels, the maximum efficiency of any electrical appliance is 35%. However, natural gas is delivered at essentially no energy cost. At the point of delivery, it is burned and can deliver, at least theoretically, up to 100% of its heat content to the task at hand. For example, a well-insulated water heater will be able to absorb about 95% of the energy heat delivered to it. Thus the overall electrical efficiency would be 0.95 (35%) or about 34%. For the gas version, it would be 95% efficient.

CHAPTER 18

18.1 (a) Series: $P_1 = 4.0$ W, $P_2 = 8.0$ W, $P_3 = 12$ W; parallel: $P_1 = 1.4 \times 10^2$ W, $P_2 = 72$ W, $P_3 = 48$ W (b) In series, the most power is dissipated in the largest resistance. In parallel, the most power is dissipated in the least resistance. (c) Series: total resistor power is 24 W, and $P_b = I_b V_b = (2.0 \text{ A})(12 \text{ V}) = 24$ W, so yes, as required by energy conservation. Parallel: total resistor power is $P_{\text{tot}} = 2.6 \times 10^2$ W, and $P_b = I_b V_b = (22 \text{ A})(12 \text{ V}) = 2.6 \times 10^2$ W (to two significant figures), so yes, as required by energy conservation

18.2 (a) The voltage across the open socket will be 120 V. (b) The voltage across the remaining bulbs will be zero.

18.3 $P_1 = I_1^2 R_1 = 54.0$ W, $P_2 = I_2^2 R_2 = 9.0$ W, $P_3 = I_3^2 R_3 = 0.87$ W, $P_4 = I_4^2 R_4 = 2.55$ W, and $P_5 = I_5^2 R_5 = 5.63$ W. The sum is 72.1 W (three significant figures). The power output of the battery is $P_b = I_b V_b = (3.00 \text{ A})(24.0 \text{ V}) = 72.0$ W. This equality (slight difference is due to rounding) is just conservation of energy on a per unit time (power) basis.

18.4 (a) If R_2 is increased, then the equivalent parallel resistance of R_2 and R_1 will increase. Thus the total circuit resistance

will increase, resulting in a reduction in the total circuit current. Since the current in R_3 is the same as the total current, I_3 will decrease. From this, V_3 should decrease. Therefore, V_1 and V_2 should both increase (they are equal) and $V = V_2 + V_3 = a$ constant. Since R_1 has not changed, due to the voltage increase, I_1 should increase. Since I_3 decreases and I_1 increases, it must be (since $I_3 = I_1 + I_2$) that I_2 decreases. (b) Recalculation confirms these predictions: $I_1 = 0.51$ A (increase), $I_2 = 0.38$ A (decrease), and $I_3 = 0.89$ A (decrease).

18.5 At the junction, we still have $I_1 = I_2 + I_3$ (Eq. 1). Using the loop theorem around loop 3 in the clockwise direction (all numbers are volts, deleted for convenience): $6 - 6I_1 - 9I_2 = 0$ (Eq. 2). For loop 1, the result is $6 - 6I_1 - 12 - 2I_3 = 0$ (Eq. 3). Solve Eq. 1 for I_2 and substitute into Eq. 2. Then solve Eq. 2 and Eq. 3 simultaneously for I_1 and I_3. All answers are the same as in the Example, as they should be.

18.6 (a) The maximum energy storage at 9.00 V is 4.05 J. At 7.20 V, the capacitor stores only 2.59 J or 64% of the maximum. This is because the energy storage varies as the *square* of the voltage across the capacitor and $0.8^2 = 0.64$. (b) 8.64 V, because the voltage does not rise linearly, but levels off in an exponential fashion

18.7 10 A

18.8 $I_G = 3.99 \times 10^{-4}$ A (small, as expected), $I_R = 2.9996$ A ≈ 3 A (the current in the external resistor is barely affected, as expected), and $V_R = 5.9992$ V ≈ 6 V (as expected, the voltage across R is barely affected by connecting the voltmeter, which is by design)

CHAPTER 19

19.1 East, since reversing both the velocity direction and the sign of the charge leaves the direction the same

19.2 (a) Using the force right-hand rule, the proton would initially deflect in the negative x-direction. (b) 0.21 T

19.3 0.500 V

19.4 (a) At the poles the magnetic field is perpendicular to the ground. Since the current is parallel to the ground, according to the force right-hand rule the force on the wire would be in a plane parallel to the ground. Thus it would not be able to cancel the downward force of gravity. (b) The wire's mass is 0.041 g, which is unrealistically low for a wire that is 1 meter long.

19.5 (a) At 45°, the torque is 0.27 m·N, or about 71% of the maximum torque. (b) 30°

19.6 (a) South (b) 38 A

19.7 1500 turns

19.8 (a) The force becomes repulsive. You should be able to show this by using the right-hand source and force rules. (b) The force is proportional to the product of the currents, which will increase by a factor of 9. To offset this, the distance between the wires must increase by a factor of 9 to 27 mm. (c) The fields are proportional to the currents (which triple) and inversely proportional to the distance (which increases by a factor of 9), so the field from each wire will be 3/9 or only 1/3 as large as before. (Why does the force remain the same?)

19.9 The permeability would only have to be 40% of the value in the Example, or $\mu \geq 480\mu_o = 6.0 \times 10^{-4}$ T·m/A.

CHAPTER 20

20.1 (a) Clockwise (b) 0.335 mA

20.2 Any way that will increase the flux, such as increasing the loop area or the number of loops. Changing to a lower resistance would also help.

20.3 7.36×10^{-4} T

20.4 1.5 m/s

20.5 0.28 m

20.6 (a) 6.1×10^3 J (b) 5.0×10^2 J, so about 12 times more energy is used during startup

20.7 (a) She would use a step-up transformer because European appliances are designed to work at 240 V, which is twice the U.S. voltage of 120 V. (b) The output current would be 1500 W/240 V or 6.25 A. Thus the input current would be 12.5 A. (Voltage would be stepped up by a factor of 2, and thus the input current is twice as large as the output current.)

20.8 (a) Higher voltages allow for lower current usage. This in turn reduces joule heat losses in the delivery wires and in the motor windings, making more energy available for doing mechanical work and therefore a higher efficiency. (b) Since the voltage is doubled, the current is halved. The heat loss in the wire is proportional to the *square* of the current. Thus, losses will be cut by a factor of 4 or be reduced to 25% of their value at 120 V.

20.9 0.38 cm/s

20.10 (a) With increasing distance, the Sun's light intensity (energy per second per unit area) drops, thus so would the force due to the light pressure on the sail. In turn, the ship's acceleration would be reduced. (b) You would need to somehow enlarge the sail area to catch more light.

CHAPTER 21

21.1 (a) 0.25A (b) 0.35 A (c) 9.6×10^2 Ω, larger than the 240 Ω required for a bulb of the same power in the United States. The voltage in Great Britain is larger than that in the United States. Thus to keep the power constant, the current must be reduced by using a larger resistance.

21.2 If the resistance of the appliance is constant, the power will quadruple since $P \propto V^2$. Even if the resistance increased, the power would probably be much more than the appliance was designed for and it would likely burn out, or at least blow a fuse.

21.3 (a) $\sqrt{2}(120 \text{ V}) = 170$ V (b) 120 Hz

21.4 (a) $\sqrt{2}(2.55 \text{ A}) = 3.61$ A (b) 180 Hz

21.5 (a) The current would increase to 0.896 A. (b) With a frequency increase, the capacitive reactance X_C decreases. Resistance is independent of frequency, thus it remains constant. Thus the total circuit impedance decreases because of the capacitor.

21.6 (a) In an RLC circuit, the phase angle ϕ depends on the difference $X_L - X_C$. If you increase the frequency, X_L would increase and X_C would decrease, thus their difference would increase and so would ϕ. (b) $\phi = 84.0°$, an increase as expected

21.7 6.98 W

21.8 (a) If you have a receiver tuned to a frequency *between* the two station frequencies, you would not receive the maximum strength signal from either station, but there might be enough power from each to hear them simultaneously. (b) 651 kHz

CHAPTER 22

22.1 Light travels in straight lines and is reversible. If you can see someone in a mirror, that person can see you. Conversely, if you can't see the trucker's mirror, then he or she can't see your image in that mirror and won't know that your car is behind the truck.

22.2 $n = 1.25$ and $\lambda_m = 400$ nm

22.3 By Snell's law, $n_2 = 1.24$ so $v = c/n_2 = 2.42 \times 10^8$ m/s.

22.4 With a greater n, θ_2 is smaller so the refracted light inside the glass is toward the lower left. Therefore, the lateral displacement is larger. 0.72 cm.

22.5 (a) The frequency of the light is unchanged in different media, so the emerging light has the same frequency as that of the source. (b) In general, the index of refraction of glass is greater than that of water and air has the lowest index of refraction. Therefore, the wavelength of light decreases from water to glass, and then increases from glass to air, or $\lambda_{air} > \lambda_{water} > \lambda_{glass}$.

22.6 Because of total internal reflections, the diver could not see anything above water. Instead, he would see the reflection of something on the sides and/or bottom of the pool. (Use reverse ray tracing.)

22.7 $n = 1.4574$. Green light will be refracted more than red light as green has a shorter wavelength, thus greater n than red light. By Snell's law, green will have a smaller angle of refraction so it is refracted more.

CHAPTER 23

23.1 No effect. Note that the solution to the Example does not depend on the distance from the mirror. The geometry of the situation is the same regardless of the distance.

23.2 $d_i \approx 60$ cm; real, inverted, and magnified

23.3 $d_i = d_o$ and $M = -1$; real, inverted, and same size

23.4 The virtual image is also always upright and reduced.

23.5 $d_i = -20$ cm (in front of the lens); virtual, upright, and magnified

23.6 $d_o = 2f = 24$ cm

23.7 Blocking off half of the lens would result in half the *amount* of light focused at the image plane, so the resulting image would be less bright but still full size.

23.8 The virtual image is also always upright and reduced.

23.9 3 cm behind L_2; real, inverted, and reduced ($M_{total} = -0.75$)

23.10 (a) If the lens is immersed in water, Eq. 23.8 should be modified to $\frac{1}{f} = (n/n_m - 1)\left(\frac{1}{R_1} + \frac{1}{R_2}\right)$, where $n_m = 1.33$ (water). Since $n = 1.52 > n_m = 1.33$, the lens is still converging. (b) $P = 0.238$ D; $f = 4.20$ m

CHAPTER 24

24.1 $\Delta y = y_r - y_b = 1.2 \times 10^{-2}$ m = 1.2 cm

24.2 Twice as thick, $t = 199$ nm

24.3 In brass instruments, the sound comes from a relatively large, flared opening. Thus there is little diffraction, so most of the energy is radiated in the forward direction. In woodwind instruments, much of the sound comes from tone holes along the column of the instrument. These holes are small compared to the wavelength of the sound, so there is appreciable diffraction. As a result, the sound is radiated in nearly all directions, even backward.

24.4 The width would increase by a factor of $700/550 = 1.27$

24.5 $\Delta\theta_2 = \theta_2(700 \text{ nm}) - \theta_2(400 \text{ nm}) = 47.8° - 25.0° = 22.8°$

24.6 45°

24.7 $\theta_2 = 41.2°$

24.8 589 nm; yellow

CHAPTER 25

25.1 It wouldn't work; a real image would form on person's side of lens ($d_i = +0.76$ m).

25.2 For an object at $d_o = 25$ cm, the image for the left eye would be formed at 1.0 m; this is beyond the near point for that eye, so the object could be seen clearly. The image for the right eye would be formed at 0.77 m; this is inside the near point for that eye, so the object would not be seen clearly.

25.3 Lens for near-point viewing, 2.0 cm longer

25.4 Length doubles to 40 cm

25.5 $f_i = 8.0$ cm

25.6 The erecting lens (of focal length f_i) should go between the objective and the eyepiece, positioned a distance of $2f_i$ from the image formed by the objective, which acts as an object. The erecting lens then produces an inverted image of the same size at $2f_i$ on the opposite side of the lens, which acts as an object for the eyepiece. The use of the erecting lens lengthens the telescope by $4f_i = 16$ cm, so the total length is slightly less than 25 cm + 16 cm = 41 cm.

25.7 3.4 cm

25.8 2.8×10^{-7} rad, so the GMT will have 10 times the resolution of the HST

CHAPTER 26

26.1 Light waves from two simultaneous events on the y-axis meet at some midpoint receptor on the y-axis. Since there is no relative motion along that axis, a simultaneous recording of the two events will also be recorded along the y'-axis. Hence the two observers agree on simultaneity for this situation.

26.2 $v = 0.9995c$; no, not twice as fast. Travel is limited to less than c, so this is only about a 0.15% increase.

26.3 (a) 0.667 μs (b) 0.580 μs The observer watching the ship measures the proper time interval. To the person on the ship, that time interval is dilated or lengthened.

26.4 $v = 0.991c$

26.5 The traveler measures the proper time interval of 20.0 y, but Earth inhabitants measure the dilated version of this (why?). The gamma factor is based on a recalculated value for traveler speed $v = 0.90504c$. Keeping five places after the decimal and rounding to three significant figures, $\gamma = 1/\sqrt{1 - (0.99504c/c)^2} = 10.04988$ and $\Delta t = \gamma \Delta t_o = (10.04988)(20.0 \text{ y}) = 200.99 \text{ y} \approx 201$ y.

26.6 (a) 1.17 MeV (b) 0.207 MeV

26.7 (a) $0.319mc^2$ (b) $3.33mc^2$

26.8 1.95×10^7 more massive

26.9 $u = +0.69c$, which, as expected, is lower in magnitude than the nonrelativistic (and wrong) result of $0.80c$.

CHAPTER 27

27.1 (a) 3000 K (b) 10 000 K

27.2 496 nm, which is shorter than the 550 nm used in the Example, indicating a higher photon energy. Thus the maximum kinetic energy of the photoelectrons is higher, requiring an increased stopping voltage.

27.3 2.50 V

27.4 (a) The ratio is $\Delta\lambda/\lambda_o = 324$, meaning a wavelength increase of 3.24×10^4%. (b) Percentagewise, this is a much larger increase than in the Example because the wavelength of the incoming light (λ_o) is much smaller for gamma rays.

27.5 (a) 7.29×10^5 m/s (b) 1.51 eV = 2.42×10^{-19} J

27.6 365 nm (UV)

27.7 The least energetic photon in the Lyman series results in a transition from $n = 2$ (first excited state) to the ground state $n = 1$. This results in a photon of energy 10.2 eV. The wavelength of this light is 122 nm, which is UV.

27.8 (a) There are six possible transitions from the $n = 4$ state to the ground state, and thus the emitted light has six different possible wavelengths. (b) If the atom is excited from the ground state to the first excited state ($n = 1$ to $n = 2$), then it has no choice but to emit a single photon during the de-excitation state ($n = 2$ to $n = 1$) because there are no intermediate states.

CHAPTER 28

28.1 (a) 8.8×10^{-33} m/s (b) 2.3×10^{33} s, or 7.2×10^{25} y. This is about 4.8×10^{15} times longer than the age of the universe. This movement would definitely not be noticeable.

28.2 The proton's de Broglie wavelength is 4.1×10^{-12} m, or about twenty times smaller than atomic spacing distances. With a wavelength much smaller than the atomic spacing, these protons would *not* be expected to exhibit significant diffraction effects.

28.3 Only five electrons could be accommodated in the $3d$ subshell if there were no spin (with spin there can be ten).

28.4 (a) $1s^2 2s^2 2p^6$ (b) $-2e$ or -3.2×10^{-19} C

28.5 1.2×10^6 m/s

CHAPTER 29

29.1 $^{12}_{6}\text{C} + ^{4}_{2}\text{He} \longrightarrow ^{16}_{8}\text{O}$, thus the resulting nucleus is oxygen-16.

29.2 (a) Since $^{23}_{11}\text{Na}$ is the stable isotope with 11 protons and 12 neutrons, $^{22}_{11}\text{Na}$ is 1 neutron shy of being stable. In other words, it is proton-rich or neutron-poor. Thus, the expected decay mode is β^+ or positron decay. (b) Neglecting the emitted neutrino, the decay is $^{22}_{11}\text{Na} \longrightarrow ^{22}_{10}\text{Ne} + ^{0}_{+1}\text{e}$. The daughter nucleus is neon-22.

29.3 (a) 48 d because reducing the activity by a factor of 64 requires six half-lives; $1/2^6 = 1/64$ (b) The process of excretion from the body can also remove ^{131}I.

29.4 The closest integer is 20, since $2^{20} \approx 1.05 \times 10^6$. Thus it takes about 20 half-lives, or about 560 y.

29.5 The measurement can be made as far back as four ^{14}C half-lives, or 2.3×10^4 y or 23 000 y.

29.6 (a) ^{40}K is an odd-odd nucleus (19 protons, 21 neutrons) and thus is unstable. ^{41}K, an odd-even potassium isotope (19 protons, 22 neutrons), is the likely candidate for the remainder of the stable potassium. ^{43}K would have too many neutrons (24) compared to protons (19) for this region of the periodic chart. (b) Using $N = N_0 e^{-\lambda t}$, it follows that $N_0/N = e^{\lambda t} = 12.8$. Thus there would have been about 13 times more ^{40}K (than exists now) at the formation of the Earth.

29.7 (a) Starting with 29 protons and 29 neutrons (why?), we have the following candidates: $^{58}_{29}\text{Cu}$, $^{59}_{29}\text{Cu}$, $^{60}_{29}\text{Cu}$, $^{61}_{29}\text{Cu}$, $^{62}_{29}\text{Cu}$, $^{63}_{29}\text{Cu}$, $^{64}_{29}\text{Cu}$, etc. Now delete the odd–odd isotopes (why?) to get the most likely (stable) isotopes: $^{59}_{29}\text{Cu}$, $^{61}_{29}\text{Cu}$, $^{63}_{29}\text{Cu}$, $^{65}_{29}\text{Cu}$, etc. (b) Further trimming of the list can be done by deleting those with $N \approx Z$ (why?) and those with N significantly larger than Z (why?). Since Z should be just a bit smaller than N in this mass region, we expect neutron numbers in the mid-30s. Hence a good guess would be just $^{63}_{29}\text{Cu}$ and $^{65}_{29}\text{Cu}$. According to Appendix V, these are, in fact, the only two stable isotopes of copper.

29.8 (a) The result for ^3He is 2.573 MeV/nucleon, which is considerably smaller than the 7.075 MeV/nucleon value for ^4He. (b) ^4He is the more tightly bound of the two. Unlike ^3He, all protons and neutrons in ^4He *are* paired, resulting in a more tightly bound nucleus.

29.9 The absorbed dose is 0.0215 Gy or 2.15 rad. Since the RBE for gamma rays is 1, the effective dose is 0.0215 Sv or 2.15 rem or only about one-seventh of the dose from the beta radiation.

CHAPTER 30

30.1 $Q = -15.63$ MeV, so it is endoergic (takes energy to happen).

30.2 The increase in mass has an energy equivalent of 1.193 MeV. The rest of the incident kinetic energy (1.534 MeV − 1.193 MeV, or 0.341 MeV) must be distributed between the kinetic energies of the proton and the oxygen-17.

30.3 There are about 1.00×10^{38} proton cycles per second.

30.4 Because the beta particle has little energy and therefore little momentum, the neutrino and the daughter nucleus would have to recoil in (almost) opposite directions to conserve linear momentum. This assumes the original nucleus had zero linear momentum.

30.5 Since the negative pion (π^-) has a charge of $-e$, its quark structure must be $\bar{u}d$. Using similar reasoning, the quark structure of the positive pion (π^+) is $u\bar{d}$. Thus the antiparticle of the π^+ would have the composition of $\bar{u}\bar{\bar{d}}$. However, the antiquark of an antiquark is just the original quark—for example $(\bar{\bar{u}}) = u$. Hence the antiparticle of the π^+ has the quark structure $\bar{u}\bar{\bar{d}} = \bar{u}d$, which is the π^- quark structure, as expected.

APPENDIX VII   Answers to Odd-Numbered Questions and Exercises

CHAPTER 1 MULTIPLE CHOICE

1. (c)
3. (c)
5. (b)
7. (a)
9. (c)
11. (d)
13. (a)
15. (a)
17. (b)
19. (c)

CHAPTER 1 CONCEPTUAL QUESTIONS

1. Because there are no more fundamental units. The units of all quantities can be expressed in terms of the fundamental, or base, units.

3. The mean solar day replaced the original definition. No, because this has been replaced by atomic clocks.

5. No, because 3 cm is over an inch. Ladybugs are on the order of several millimeters long. Yes, a 10-kg salmon would weigh on the order of 22 pounds, which is typical for a medium-sized fish like that.

7. The metric ton is actually a misnomer; it is not a weight unit but a mass unit, defined as the mass of one cubic meter of water. But $1 m^3 = 1000 L$ and 1 L of water has a mass of 1 kg. So one metric ton is equal to 1000 kg.

9. No, unit analysis can only tell if it is dimensionally correct. There may be missing dimensionless factors such as π.

11. π is dimensionless and therefore also unitless because it is defined as the ratio of two lengths, circumference to diameter.

13. Yes, whether you multiply or divide should be consistent with unit analysis for the final answer.

15. To provide an estimate of the accuracy of a quantity.

17. For (a) and (b), the result should have the least number of significant figures. For (c) and (d), the result should have the least number of decimal places.

19. See the six steps as listed in Chapter 1.

21. The accuracy of the answer is expected to be within a factor of 10 of the correct answer.

23. Since a liter is close to a quart and there are four quarts in a gallon, this volume is about 75 gallons, which is *not* reasonable for a car (but might be for a large truck or other large vehicle).

CHAPTER 1 EXERCISES

1. In the decimal system (base 10), a dime is valued at 10¢ and a dollar is valued at 10 dimes, or 100¢. By analogy, a duodecimal system would have a dime worth 12¢ and a dollar worth 12 "dimes," or $1.44 in decimal dollars. Then a penny would be $\frac{1}{144}$ of a decimal dollar.

3. (a) 40 Mb (b) 5.722×10^{-4} L (c) 268.4 cm (d) 5.5 kilobucks

5. 48 kg

7. (a) 8.0 L (b) 8.0 kg

9. (d)

11. *a*: 1/m; *b*: dimensionless; *c*: m

13. No, $V = \pi d^3/6$

15. Yes, $m^2 = \frac{1}{2}m(m + m) = m^2 + m^2$

17. (a) $kg \cdot m^2/s$ (b) The unit of $L^2/(2mr^2)$ is $(kg \cdot m^2/s)^2/(kg \cdot m^2) = kg \cdot m^2/s^2$, which is the unit of kinetic energy, K. (c) $kg \cdot m^2$

19. 39.6 m

21. 37 000 000 times

23. (a) 91.5 m by 48.8 m (b) 27.9 cm to 28.6 cm

25. 474 g

27. 320 m

29. (a) (1) 1 m/s (b) 33.6 mi/h

31. (a) 77.3 kg (b) 0.0773 m^3 or about 77.3 L

33. 6.5×10^3 L/day

35. 1.9×10^{10}/s

37. 15.0 min of arc

39. (a) $1.5 \times 10^5 m^3$ (b) 1.5×10^8 kg (c) 3.3×10^8 lb

41. 5.05 cm; 5.05×10^{-1} dm; 5.05×10^{-2} m

43. (a) 4 (b) 3 (c) 5 (d) 2

45. (a) 96 (b) 0.0021 (c) 9400 (d) 0.00034

47. $6.08 \times 10^{-2} m^2$

49. (a) The smallest division is (2) cm, as the last digit is estimated. (b) $0.946 m^2$

51. (a) 14.7 (b) 11.4 (c) $0.20 m^2$ (d) 0.82

53. (a) $2.0 kg \cdot m/s$ (b) $2.1 kg \cdot m/s$ (c) No, the results are not the same. The difference comes from rounding differently.

55. 100 kg

57. (a) 52% (b) Total fat = 64 g; saturated fat = 20 g

59. $5.4 \times 10^3 kg/m^3$

61. 0.87 m

63. The area of a 12" (6" radius) pizza is $\pi(6 in.)^2 \approx 113 in.^2$ The total area of two 8" pizzas is $2 \times \pi(4 in.)^2 = 101 in.^2$ The offer is *not* a good deal!

65. 25 min

67. (a) 1950 hairs (b) 2.0×10^4 hairs

69. (a) (3) Less than the 190 mi/h because more time is spent at lower speeds (b) 187 mi/h

71. (a) The answer is (2) between 5° and 7°. (b) 6.2°

73. 31.8 m

75. (a) 19.3 cm (b) 1.07, or the outer area is 7% larger than the inner area

77. (a) 94.2 s (b) 7.00×10^4 gal

CHAPTER 2 MULTIPLE CHOICE

1. (a)
3. (c)
5. (c)
7. (d)
9. (d)
11. (c)
13. (c)
15. (a)
17. (d)
19. (a)

CHAPTER 2 CONCEPTUAL QUESTIONS

1. Yes, for a round trip. No; distance is always greater than or equal to the magnitude of displacement.

3. The distance traveled is greater than or equal to 300 m. The object could travel in a variety of ways as long as it ends up at 300 m north. If the object travels straight north, then the minimum distance is 300 m.

5. Yes, this is possible. The jogger can jog in the opposite direction during part of the jog (negative instantaneous velocity) as long as the overall jog is in the forward direction (positive average velocity).

7. Not necessarily. The change in velocity is the key. If a fast-moving object does not change its velocity, its acceleration is zero. However, if a slow-moving object changes its velocity, it will have some non-zero acceleration.

9. In part (a), the object accelerates uniformly first, maintains constant velocity (zero acceleration) for a while, and then accelerates uniformly at the same rate as in the first segment. In part (b), the object accelerates uniformly.

11. Assuming uniform accelerations, both cars have the same *average* speed but travel unequal distances. Car A will take less time to reach the line because it has less distance to travel. Since the *change* in velocities are the same, car A will have a higher rate of change of velocity, thus A's acceleration is greater than B's. Since car A travels half the distance as B at the same average speed as A, it will take half as long to finish as B. Thus A will have twice the acceleration as B.

13. Not necessarily, because even if the acceleration is negative; the object can still have positive velocity (meaning it is slowing) and the result could be a positive value for *x*.

15. Yes, if the displacement is negative, meaning the object accelerates to the left.

17. No, since one value of the instantaneous velocity does *not* tell you if the velocity is changing. It could be zero just for an instant and not zero either before or after that instant, thus it could be changing and the object could be accelerating. You need two values of instantaneous velocity to determine if an object is accelerating.

19. Since the first stone has been accelerating downward for a longer time, it will always have a higher speed and thus as time goes on, it will have fallen further, and thus the gap between them (Δy) will increase.

CHAPTER 2 EXERCISES

1. Half lap: 300 m; full lap: 0 m

3. 9.8 s

5. 125 min

7. Average speed is 75 km/h; average velocity is zero.

9. (a) 1.4 h, (b) 0.27 h, or about 16 min

11. (a) The correct choice is (3) between 40 m and 60 m. (b) 45 m at 27° west of north

13. (a) 2.7 cm/s (b) 1.9 cm/s
15. (a) $\bar{s}_{0-2.0\,s} = 1.0$ m/s; $\bar{s}_{2.0\,s-3.0\,s} = 0$;
$\bar{s}_{3.0\,s-4.5\,s} = 1.3$ m/s; $\bar{s}_{4.5\,s-6.5\,s} = 2.8$ m/s;
$\bar{s}_{6.5\,s-7.5\,s} = 0$; $\bar{s}_{7.5\,s-9.0\,s} = 1.0$ m/s
(b) $\bar{v}_{0-2.0\,s} = 1.0$ m/s; $\bar{v}_{2.0\,s-3.0\,s} = 0$;
$\bar{v}_{3.0\,s-4.5\,s} = 1.3$ m/s; $\bar{v}_{4.5\,s-6.5\,s} = -2.8$ m/s;
$\bar{v}_{6.5\,s-7.5\,s} = 0$; $\bar{v}_{7.5\,s-9.0\,s} = 1.0$ m/s
(c) $v_{1.0\,s} = \bar{s}_{0-2.0\,s} = 1.0$ m/s;
$v_{2.5\,s} = \bar{s}_{2.0\,s-3.0\,s} = 0$; $v_{4.5\,s} = 0$;
$v_{6.0\,s} = \bar{s}_{4.5\,s-6.5\,s} = -2.8$ m/s
(d) $v_{4.5\,s-9.0\,s} = -0.89$ m/s
17. (a) $x_o = 10$ m (b) $\Delta x = -6.0$ m (c) 4.5 s
after the start
19. 1 month
21. (a) 500 km at 37° east of north
(b) 400 km/h at 37° east of north (c) 560 km/h
(d) Since speed involves total distance, which
is greater than the magnitude of the displacement, the average speed does not equal the
magnitude of the average velocity.
23. 2.32 m/s^2
25. 3.7 s
27. -2.1 m/s^2
29. (a) 190 m (b) 11 s (c) 33 m/s
31. (a) 26.3 m/s (b) 11.1 s
33. $a_{0-4.0} = 2.0$ m/s^2; $a_{4.0-10.0} = 0$ m/s^2;
$a_{10.0-18.0} = -1.0$ m/s^2. The object accelerates at
2.0 m/s^2 first, next moves at constant velocity,
then decelerates at 1.0 m/s^2.
35. (a)

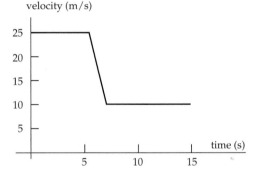

velocity (m/s)

(b) 10 m/s (c) 2.4×10^2 m (d) 17 m/s
37. No, the acceleration must be at least
9.9 m/s^2.
39. (a) 1.8 m/s^2 (b) 6.3 s
41. (a) 81.4 km/h (b) 0.794 s
43. 3.09 s and 13.7 s. The 13.7 s answer is
physically possible but not likely in reality.
After 3.09 s, it is 175 m from where the reverse
thrust was applied, but the rocket keeps traveling forward while slowing down. Finally it
stops. However, if the reverse thrust is continuously applied (which is possible, but not
likely), it will reverse its direction and be back
to 175 m from the point where the initial
reverse thrust was applied, a process that
would take 13.7 s.
45. No, since $a = 3.33$ m/s$^2 < 4.90$ m/s^2
47. 2.2×10^5 m/s^2
49. No, since 13.3 m > 13 m
51. (a) Total area = Triangle + Rectangle =
$\frac{1}{2}ab + = \frac{1}{2}(v - v_o)t + v_ot = \frac{1}{2}(v + v_o)t =$
$v_{avg}t = \Delta x$ (b) 96 m
53. (a) (3) $v_1 > \frac{1}{2}v_2$ (b) 9.22 m/s, 13.0 m/s
55. (a) -12 m/s; -4.0 m/s (b) -18 m (c) 50 m

57. (a) 12.2 m/s, 16.4 m/s (b) 24.8 m (c) 4.07 s
59. (a) -27 m/s (b) -38 m
61. (a) A straight line (linear), with a downward slope of $-g$ and starting at the origin
(b) A downward curving parabola (starting at
the origin, assuming the initial location was
chosen as $y = 0$) with an initial slope of zero
(since the initial velocity was zero)
63. Twice as fast and thus four times higher
65. 67 m
67. 39.4 m
69. (a) The correct answer is (1) less than
95%, as the height depends on the initial speed
squared. (b) 3.61 m
71. (a) 5.00 s (b) 36.5 m/s
73. 1.49 m above the top of the window
75. (a) 155 m/s (b) 2.22×10^3 m (c) 28.7 s
77. (a) 8.45 s (b) $x_M = 157$ m; $x_C = 132$ m
(c) 13 m
79. (a) 38.7 m/s (b) 15.5 s (c) 19.2 m/s
81. (a) 119 m (b) 4.92 s (c) Lois: 48.2 m/s;
Superman: 73.8 m/s
83. (a) -297 m/s (b) 3.66 m/s^2 (c) 108 s
85. -1.43 m/s^2
87. (a) 4.06×10^3 m (b) 33.2 s (c) 862 s

CHAPTER 3 MULTIPLE CHOICE

1. (a)
3. (c)
5. (c)
7. (d)
9. (c)
11. (b)
13. (d)

CHAPTER 3 CONCEPTUAL QUESTIONS

1. The answer is no to both. The component
of a vector can never be greater than the magnitude of a vector since the magnitude is the
hypotenuse of the triangle representing the
vector and its component.
3. (a) Its velocity either increases (it speeds
up) or decreases (it slows down) in magnitude
only. (b) It follows in a parabolic path. (c) It
moves in a along a circular path.
5. (a) No, a vector cannot be less than one of
its components in magnitude. (b) Yes, a vector
can have the same magnitude as one of its
components if all the other components are
zero.
7. No, a vector quantity cannot be added to
a scalar quantity.
9. Yes, they are all equal. Only two things
determine a vector—the magnitude (the length
of the arrow) and the direction (the direction of
the arrow).
11. The horizontal motion does not affect the
vertical motion. The vertical motion of the ball
projected horizontally is identical to that of the
ball dropped.
13. In both cases, aim at the target. When the
gun is sighted-in, it corrects for the distance the
bullet falls on its way to the target. It still falls
the same distance whether it is traveling
upward or downward.
15. (a) Zero (b) 4.0 m/s
17. When the player is driving to the basket
for a lay-up, she already has an upward

motion. Since the ball is with the player, the
ball already has a velocity relative to the
ground as the player jumps.

CHAPTER 3 EXERCISES

1. (a) 210 km/h (b) 54 km/h
3. (a) The magnitude of the acceleration is
(3) between 4.0 m/s^2 and 7.0 m/s^2. The
hypotenuse of a right triangle can never be
smaller than either of its sides (therefore,
$a > 4.0$ m/s^2) and also can't be greater than the
sum of its two sides (therefore, $a < 7.0$ m/s^2).
(b) 5.0 m/s at an angle of 53° above the $+x$-axis
5. (a) 6.0 m/s (b) 3.6 m/s
7. (a) 70 m (b) The time on the long side is
0.57 min, and the short side takes 0.43 min.
9. 2.5 m at an angle of 53° above the $+x$-axis
11. (a) The angle is specified from the vertical, but also means 70° from the horizontal,
thus the vertical component is the largest or (1)
$v_y > v_x$. (b) $v_x = 306$ m/s, $v_y = 840$ m/s, and
$D_y = 50.4$ km.
13. (12 m, -6.0 m)
15. (a) $\bar{v}_x = 32.9$ m/s, $\bar{v}_y = 6.10$ m/s
(b) $\bar{v} = 33.5$ m/s, $\theta = 10.5°$ above the horizontal (c) The total path length is not known and
the average speed is defined in terms of the
total path length.
17. (a) Yes, vector addition is associative.
(b) See the diagrams below.

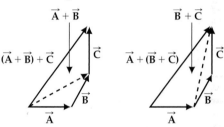

19. (a) $4.0\,\hat{x} + 2.0\,\hat{y}$ (b) $|A + B| = 4.5$ at an
angle of 27° above the $+x$-axis
21. 6.4
23. (a) $(-3.4$ cm$)\,\hat{x} + (-2.9$ cm$)\,\hat{y}$ (b) 4.5 cm,
63° above $-x$-axis (c) $(4.0$ cm$)\,\hat{x} + (-6.9$ cm$)\,\hat{y}$
25. (a) $5.0\hat{x} + 3.0\hat{y}$ (b) $-3.0\hat{x} + 7.0\hat{y}$
(c) $-5.0\hat{x} - 3.0\hat{y}$
27. (a) See the diagram below.
(b) $\vec{A} + \vec{B} = (-5.1)\,\hat{x} + (-3.1)\,\hat{y}$ or the sum's
magnitude is $|\vec{A} + \vec{B}| = 5.9$ at an angle $\theta = 31°$
below the $-x$-axis.

29. 16 m/s at an angle of 79° above the
$-x$-axis
31. $\vec{A}$ and $\vec{B}$ must be oppositely directed.
33. $\vec{C} = \vec{B} - \vec{A}$; $\vec{C} - \vec{B} = -\vec{A}$;
$\vec{E} - \vec{D} + \vec{C} = 3\vec{B} - 2\vec{A}$
35. (a) They are in the same direction.
(b) They are in opposite directions. (c) They are
perpendicular.

37. (a) The two vectors are in the same direction. In this case, the sum's magnitude is 35.0 m. **(b)** The two vectors are oppositely directed. The sum's magnitude is 5.0 m. **(c)** In general, when any two vectors are in the same direction, the magnitude of the resultant (sum vector) is the sum of the magnitudes of those vectors. When any two vectors are in opposite directions, the magnitude of the resultant (sum vector) is the absolute value of the difference of the magnitudes of those vectors.

39. (a) From the sketch below, the general direction of the thunderstorm's velocity is (2) north of west. **(b)** 26.7 mi/h, at an angle θ of 37.6° north of west

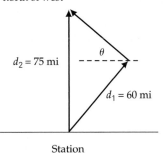

$d_2 = 75$ mi

θ

$d_1 = 60$ mi

Station

41. (a) The answer is (2), their magnitudes are equal. **(b)** 41 N down
43. $F_3 = 242$ N, at an angle of 48° below the $-x$-axis.
45. It falls 2.7×10^{-13} m. This is a very small distance. Therefore, the answer is no, the designer need not worry about gravitational effects.
47. (a) 31 m vertically below its original position; 13 m horizontally displaced from its original position **(b)** (5.0 m/s) $\hat{x}$ + (−25 m/s) $\hat{y}$
49. (a) 0.123 s **(b)** 1.10 m
51. (a) 46° above the horizontal **(b)** 1.5 km above the soldiers
53. 1.3 m from where the car was when the ball was thrown
55. 3.8×10^2 m
57. (a) 0.43 s **(b)** 3.5 m/s
59. 63°
61. $v_0 = 40.9$ m/s at 11.9° above the horizontal
63. (a) The shot would be in the air (1) a longer time. This is because when the shot returns to ear level, it would have spent the same time as a projectile launched at the same angle from ground level. **(b)** The range of the shot is 13.3 m. Its velocity just before impact is (11.3 m/s) $\hat{x}$ + (−7.46 m/s) $\hat{y}$.
65. (a) 0.967 s **(b)** 4.18 m **(c)** 12.8 m/s
67. 10.9 m/s
69. 6.7 s
71. (a) +55 km/h (same direction as truck velocity) **(b)** −35 km/h (opposite the truck velocity)
73. 4.0 min
75. (a) From the sketch, the general direction of the swimmer's velocity is (1) north of east. **(b)** 0.25 m/s at $\theta = 37°$ north of east

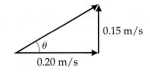

0.15 m/s

θ

0.20 m/s

77. 14° from vertical
79. Use the subscripts b = boat, w = water, and g = ground. For the boat to go straight across, v_{bw} must be the hypotenuse of a right triangle. So it must be greater in magnitude than v_{wg}. Hence if $v_{wg} > v_{bw}$, the boat cannot make the trip directly across the river.

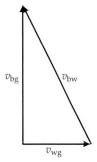

v_{bg} v_{bw}

v_{wg}

81. (a) −125 km/h (negative sign indicates that A is approaching B in a southerly direction) **(b)** 3.5 m/s² to the north **(c)** 3.5 m/s² to the north
83. 0.56 m/s
85. (a) See sketch below for coordinate choice. $\vec{v}_o = (16.4$ m/s$) \hat{x} + (−11.5$ m/s$) \hat{y}$ and $\vec{v}_f = (9.06$ m/s$) \hat{x} + (4.23$ m/s$) \hat{y}$

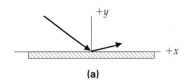

$+y$

$+x$

(a)

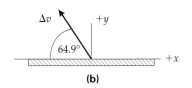

Δv $+y$

64.9°

$+x$

(b)

(b) $\Delta\vec{v} = (−7.34$ m/s$) \hat{x} + (15.7$ m/s$) \hat{y}$
(c) $|\Delta\vec{v}| = 17.3$ m/s at an angle of 64.9° from the wall. See sketch above for orientation.
87. (a) The flatbed observer sees the ball go straight up and come back down, in the absence of air resistance. The parabolic ground observer sees the ball move in a paraboic arc because to that observer it has not only an initial vertical velocity but also a horizontal component.
(b) 5.10 s **(c)** For an observer in the car, $v = 0$ at the highest point because that person is moving along with the ball horizontally. For an observer on the ground, the ball has a horizontal velocity of 12.0 m/s at the highest point.
(d) $\theta = 64.4°$ above the horizontal **(e)** The car moves horizontally 61.2 m. The distance moved by the ball relative to the car is zero because it lands at the same spot *on the car*.
89. (a) (−5.5 m/s) $\hat{x}$ + (5.2 m/s) $\hat{y}$ **(b)** Time for car A to get to point ⊗: $t_A = 10$ s. During that time, car B will travel a distance of $d_B = 300$ m. Since the angled ramp is longer than the straight ramp, car B will not be at point ⊗ when car A is there and they will *not* collide at point ⊗. **(c)** The length of the 10° ramp is $d_B = 355$ m. So the time for car B to reach point ⊗ is $t_B = 11.8$ s. During this time,

A will travel a distance of 413 m. Therefore, when B reaches ⊗, A will have traveled an extra distance of 63 m. There will be no collision and car A is 63 m ahead of B.

CHAPTER 4 MULTIPLE CHOICE

1. **(d)**
3. **(c)**
5. **(d)**
7. **(c)**
9. **(c)**
11. **(c)**
13. **(d)**
15. **(c)**
17. **(a)**

CHAPTER 4 CONCEPTUAL QUESTIONS

1. (a) No, just no *net* force **(b)** No, just constant velocity, no acceleration
3. No, same mass, same inertia
5. (a) Balloon moves forward. The air has more inertia and tends to stay at the rear of the car. **(b)** Balloon moves backward. The air has more inertia and tends to stay at the front of the car.
7. (a) Gradually increase the downward pull of the lower string. For balance, the tension in the upper string must equal this pull plus the weight of the object, so it will have more tension in it and break first. **(b)** Pull the lower string with a sudden jerk. By Newton's third law, the object will tend to remain at rest, so the tension in the upper string will not increase as much as the tension in the lower string, breaking the lower one.
9. Zero; 70 kg
11. No, since both its mass and the force of gravity on it decrease with time.
13. The forces act on different objects (one on horse, one on cart) and therefore cannot cancel.
15.

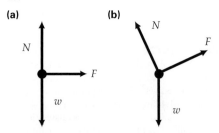

(a)   **(b)**

N N F

F

w w

The force F is the combination of the pushing by the seat back and the friction force by the seat surface.
17. When the arms are quickly raised (accelerated upward), an upward force must be exerted on them. This force, ultimately, is provided by the normal force of the scale upward on the feet. From Newton's third law, the person must push down on the scale (more than usual with arms at rest), so the scale reading increases. Conversely, a downward force is required to lower the arms, and that results in a decrease in the scale reading. The scale reads the push down on it, which is not necessarily the same magnitude as the person's weight.
19. This is because kinetic friction (sliding) is less than the maximum static friction (barely not sliding, antilock keeps it rolling). A greater

friction force can decrease the stopping distance.

21. (a) No, there is no inconsistency. Here the friction force *opposes* slipping. (b) Wind can increase or decrease air friction depending on wind directions. If wind is in the direction of motion, air resistance decreases, and the opposite is true if the wind is opposite the object's velocity.

23. Run at a speed you can estimate (say 2 m/s, then slide and measure the distance to stop. Then setting the final speed equal to zero in $v^2 = v_0^2 + 2a(x - x_o)$ enables you to solve for your average acceleration. Now use Newton's second law based on the fact that the normal fore is the same magnitude as your weight: $F_{net,x} = f_k = \mu_k mg = ma$ gives the value for $\mu_k = a/g$.

CHAPTER 4 EXERCISES

1. $m_{Al} = 1.4 m_w$

3. 0.64 m/s^2

5. (a) 2.0 kg: 20 N; 6.0 kg: 59 N (b) Same for both: 9.80 m/s^2

7. $(-7.6 \text{ N})\,\hat{\mathbf{x}} + (0.64 \text{ N})\,\hat{\mathbf{y}}$

9. (a) The correct choice is (3), the tension is the same in both situations. (b) $T = w = 25$ lb

11. (a) The correct choice is (2), the second quadrant. (b) 184 N, 12.5° above the negative x-axis

13. (a) 1.8 m/s^2 (b) 4.5 N

15. (a) The correct answer is (3) 6.0 kg; because mass is a measure of inertia it does not change. (b) 9.8 N

17. (a) The correct answer is (1) on the Earth. 1 lb is equivalent to 454 g, or 454 g has a weight of 1 lb. (b) 5.4 kg (2.0 lb)

19. (a) The correct answer is (4) one-fourth as great. (b) 4.0 m/s^2

21. 1.23 m/s^2

23. 2.40 m/s^2

25. (a) 30 N (b) -4.60 m/s^2 (downward)

27. (a) The ball tends to remain at rest as the RV accelerates forward, so the ball hangs backward. (b) 0.51 m/s^2

29. $(-2.6 \text{ N})\,\hat{\mathbf{x}} + (1.5 \text{ N})\,\hat{\mathbf{y}}$

31. (a) The correct choice is (2), two forces act on the book, the downward gravitational force (weight, w) due to the Earth and the normal force upward exerted by the surface, N. (b) The reaction to the weight (force) is an upward force on the Earth by the book. The reaction force to the normal force is a downward force on the surface by the book.

33. (a) The correct choice is (3), the force the blocks exert forward on him. (b) 3.08 m/s^2

35. The correct answer is (a), (4) the pull of the rope on the girl. (b) 264 N

37. 585 N

39. (a) N is the measured weight and can take any value, depending on acceleration. Thus the answer is (4) all of the preceding. (b) The elevator must be accelerating downward at 1.8 m/s^2.

41. (a) 0.96 m/s^2 (b) 2.6×10^2 N

43. 123 N up the incline

45. 64 m

47. (a) The correct answer is (3), both the tree separation and sag. (b) 6.1×10^2 N

49. 2.63 m/s^2

51. The tension in both upper cords is 69 N. The tension in the cord between A and B is 49 N. The tension in both lower cords is 57 N. The tension in the cord attached to the mass is 98 N.

53. 1.50×10^3 N

55. (a) The two objects have the same acceleration of 1.8 m/s^2. (b) 6.4 N

57. (a) 1.62 s (b) m_1 will ascend 1.19 m above the floor before stopping.

59. (a) 97.5 N (b) 82.5 N

61. (a) 1.2 m/s^2, m_1 up the incline and m_2 vertically down (b) 21 N

63. (a) 1.1 m/s^2. (b) Since the force of friction is below the maximum value of static friction, the object will not accelerate.

65. 2.7×10^2 N

67. (a) 38 m (b) 53 m

69. 33 m

71. 0.77 m

73. The coefficient of kinetic friction can be found by applying Newton's second law with zero acceleration both parallel to the incline and perpendicular to it. The result is $\mu_k = \sin\theta/\cos\theta = \tan\theta$.

75. (a) μ_s is independent of area, mass, etc., so it is still 30°. (b) 0.58

77. (a) The incline is not frictionless. (b) 33 N

79. (a) m_2 can be anywhere between 0.72 kg and 1.7 kg. (b) m_2 can be anywhere between 0.88 kg and 1.5 kg.

81. (a) 0.179 kg (b) 0.862 m/s^2

83. (a) 5.5 m/s^2 (b) 173 N

85. (a) -8.24 m/s^2 (deceleration) (b) 0.841 (c) 68.8 mi/h

87. (a) 513 N in the direction of opposite the puck's initial velocity (b) 0.151 m/s

CHAPTER 5 MULTIPLE CHOICE

1. (d)

3. (b)

5. (d)

7. (d)

9. (c)

11. (d)

13. (d)

15. (d)

17. (a)

19. (c)

21. (b)

CHAPTER 5 CONCEPTUAL QUESTIONS

1. (a) Yes, the lifter does positive work in raising the barbell from the floor to the position shown. (b) Yes, positive work is done by the force exerted by the weightlifter. (c) No work is done, so it is less. (d) Yes, but the (positive work) is done by gravity, not the weightlifter because he is not exerting a force on it.

3. Positive on the way down and negative on the way up. No, it is not constant because the angle between the plane's displacement and the force of gravity continually changes. The maximum (positive) value is halfway up and the minimum (negative) value is halfway down. At the very top and bottom, the work done by the force of gravity is zero.

5. Work isn't "possessed" by an object. However, work is the term used to describe one way of *transferring* energy, and thus work can lead to *changes* in its energy, but work is not energy.

7. From equilibrium the work is proportional to the square of the stretch. Thus the 4.0-cm stretch *from the relaxed position* requires four times the work as the 2.0-cm stretch *from the relaxed postion*. Hence the stretch from 2.0 cm to 4.0 cm requires three times as much work as a 2.0-cm stretch from equilibrium.

9. The work done yields an increase in kinetic energy (from zero). Since kinetic energy depends on the square of the speed, the (net) work to get to half speed is one-fourth of the "full speed" work, or $W/4$.

11. They have the same kinetic energy since it depends linearly on mass and on the square of the speed. Mathematically,
$K_A = \frac{1}{2}(4m_B)v_A^2 = 2m_B v_A^2$ and
$K_B = \frac{1}{2}m_B(2v_A)^2 = 2m_B v_A^2$.

13. Both are correct. Potential energy is defined with respect to a reference point. Depending on where the reference point is located, the potential energy of the notebook on a table can be positive (with respect to floor), zero (with respect to table), or even negative (with respect to ceiling).

15. The initial potential energy is equal to the final potential energy, so the final height is equal to the initial height.

17. Yes. When the upward-thrown ball is at its maximum height, its velocity is zero, so actually they have the same (total) mechanical energy everywhere.

19. (a) No, efficiency is the ratio of work output to energy input and is not related to time. (b) No. If same efficiency, work depends on time of operation, and the less powerful one could do more work over a longer time. If operating for the same time, a less powerful one could do more work if more efficient.

CHAPTER 5 EXERCISES

1. 5.0 N

3. 1.8×10^3 J

5. 3.7 J

7. 2.3×10^3 J

9. -2.2 J

11. (a) Negative work because the force acts opposite the motion (b) 1.47×10^5 J

13. (a) The correct choice is (1), positive because the force of friction acts in the direction of motion. (b) 0.0147 J

15. 1.35×10^4 J

17. 80 N/m

19. 1.25×10^5 N/m

21. (a) The correct choice is (1) $\sqrt{2}$, because when W doubles, x becomes $\sqrt{2}$ times larger. (b) 900 J

23. (a) The correct choice is (1), more than because the force is greater and the displacement is the same. (b) 0–10 cm: 0.25 J; 10–20 cm: 0.75 J

25. (a) 4.5 J (b) 3.5 J

27. 6.0 J

29. (a) The correct answer is (3), 75%. $K_o = \frac{1}{2}mv_o^2$, thus reducing v to $v_o/2$ reduces the

kinetic energy to $0.25K_o$, or 25% of the original. So 75% of the initial kinetic energy is lost.
(b) $K_o = (1/2)(0.20 \text{ kg})(10 \text{ m/s})^2 = 10 \text{ J}$ and $K = (1/2)(0.20 \text{ kg})(5.0 \text{ m/s})^2 = 2.5 \text{ J}$. So $\Delta K = -7.5 \text{ J (loss of 7.5 J)}$.

31. (a) 45 J **(b)** 21 m/s
33. 200 m
35. 2.0×10^3 m
37. 2.9 J
39. 4.7 J
41. (a) 0.16 J **(b)** -0.32 J
43. 18 J
45. (a) $K_o = 15.0 \text{ J}$, $U_o = 0$, $E_o = K_o + U_o = 15.0 \text{ J}$
(b) $E = E_o = 15.0 \text{ J}$. Since $U = (0.300 \text{ kg})(9.80 \text{ m/s}^2)(2.50 \text{ m}) = 7.35 \text{ J}$, then $K = E - U = 7.65 \text{ J}$. **(c)** At maximum height, $K = 0$; therefore, $U = 15.0 \text{ J}$ and $E = 15.0 \text{ J}$.
47. Conservation of energy gives $K_i + U_i = K_f + U_f$, $\therefore \frac{1}{2}mv_i^2 + mgy_i = \frac{1}{2}mv_f^2 + mgy_f$. Mass is not needed; the final speed is $v_f = \sqrt{v_i^2 + 2g(y_i - y_f)} = \sqrt{(12 \text{ m/s})^2 + 2(9.8 \text{ m/s}^2)(20 \text{ m})} = 23 \text{ m/s}$.
49. 0.176 m
51. (a) 4.4 m/s **(b)** 3.7 m/s **(c)** $\Delta E = -0.59 \text{ J}$ to heat and sound
53. (a) 11 m/s **(b)** No **(c)** 7.7 m/s
55. (a) 2.7 m/s **(b)** 0.38 m **(c)** 29°
57. (a) 160 N/m **(b)** 0.33 kg **(c)** $+0.13$ J
59. 1.9 m/s
61. 12 m/s
63. 97 W
65. 5.7×10^{-5} W
67. 2.70×10^3 N
69. 137 kg
71. 48.7%
73. 5.0×10^2 W
75. (a) Use the initial kinetic energy and launch angle to find the initial speed component in the horizontal direction. At the top, that value *is* the total speed. Use that to determine the kinetic energy at the peak. The change in gravitational potential energy can then be found. From that the object height of 19.0 m above the launch point and 39.0 m above the lake at its peak can be calculated. **(b)** Find the vertical velocity component at launch, set it equal to zero, and use vertical kinematic equations to solve for the same maximum height. **(c)** The loss in gravitational potential energy between the launch point and the lake surface can be found from the 20.0-m elevation drop. Added to the initial kinetic energy, this gives the final kinetic energy, and from that the speed can be found. It is 35.9 m/s. **(d)** The horizontal component is the same at impact as at launch. The vertical component at impact is given by the vertical kinematics equation after a drop in elevation of 20.0 m. Finding the speed by the Pythagorean theorem gives 35.9 m/s, as expected.
77. (a) At equilibrium the net force on the mass is zero, thus the spring force cancels the pull of gravity or $F_{sp} = kd = w = mg$. Solving, $d = mg/k$ **(b)** The spring pulls up as the mass moves down to the new equilibrium position, thus the work done by the spring is equal to the negative of its change in potential energy,

or $W_{sp} = -\Delta U_{sp} = -\frac{1}{2}kd^2 = \frac{1}{2}k\left(\frac{mg}{k}\right)^2 = -\frac{m^2g^2}{2k}$. **(c)** $W_g = mgd = mg\left(\frac{mg}{k}\right) = \frac{m^2g^2}{k}$.
The sum of the two works is not the net work, which must be zero from the work-energy theorem since the kinetic energy of the mass did not change overall. The explanation is that there is a missing *third work done by the hand* that supports the mass (pushing upward) on its way down to equilibrium where it is released and stays. Without the hand, the mass would speed up and continue below equilibrium and oscillate. **(d)** Since $W_{net} = \Delta K = 0$, then $W_{sp} + W_g + W_{hand} = 0$ or $W_{hand} = -(W_{sp} + W_g) = -\left(-\frac{m^2g^2}{2k} + \frac{m^2g^2}{k}\right) = -\frac{m^2g^2}{2k}$.
The negative work is because the hand force is upward while the object moved downward. To find the average hand force, use $W_{hand} = -\overline{F}_{hand}\, d$ and solve $\overline{F}_{hand} = -\frac{(-m^2g^2/2k)}{mg/k} = \frac{mg}{2} = \frac{w}{2}$.
79. (a) Total mechanical energy is conserved and it gains gravitational potential energy; its kinetic energy must be less, so (1) is correct. **(b)** The initial kinetic energy is 205 J. The kinetic energy at the top is 120 J, based on its speed at the top, which is the same as the initial velocity's horizontal component.

CHAPTER 6 MULTIPLE CHOICE

1. (b)
3. (d)
5. (d)
7. (c)
9. (d)
11. (c)
13. (a)
15. (d)

CHAPTER 6 CONCEPTUAL QUESTIONS

1. No, mass is also a factor in momentum and can offset speed.
3. No, both depend on mass and kinetic energy depends on speed squared; momentum is linearly related to speed.
5. By stopping, the contact time is short. From the impulse momentum theorem, a shorter contact time results in a greater force if all other factors (m, v_o, v) remain the same.
7. With a stiff-legged landing, the force of the floor on the falling jumper acts only for a short time, thus requiring a large force to change the jumper's momentum. When landing with bent legs, the force acts over a longer time, and hence a smaller force is needed.
9. Air is forced backward and the boat moves forward to conserve momentum. If a sail were installed behind the fan on the boat, the boat would not accelerate forward because the backward force on the sail (and hence the boat) from the air that hits it would cancel the forward force on the fan (and hence the boat).
11. No, it is not possible. Before the hit, the two-object system has (non-zero) momentum because one is moving. To conserve momen-

tum, the system must have momentum after the hit. Therefore, it is not possible for both to be at rest because that would mean zero total momentum afterward.
13. This is because momentum is a vector and kinetic energy is a scalar and the conservation requirements for each are different. For example, two objects with the same but oppositely directed momentum have non-zero total kinetic energy but zero total momentum. After they collide and stick, both stop, resulting in zero total kinetic energy and zero total momentum. Therefore, kinetic energy is not conserved but momentum is.
15. The modern auto bumper crumples upon impact, thus increasing the time of impact. This results in a reduction of the force on the car in a given situation, compared to old rigid bumpers. Force reduction means that all objects in the car require less force to accelerate with the car, and thus the passengers and car contents are less likely to be hurt, killed, or damaged.
17. The flamingo's center of mass is directly above the foot on the ground since it is in equilibrium.
19. The larger lower stages carry more fuel to produce greater impulses on the successively smaller upper stages. Once the fuel in each succesive lower stage is gone, that stage can be jettisoned and does not need to be accelerated after that.

CHAPTER 6 EXERCISES

1. (a) 1.5×10^3 kg $\cdot$ m/s **(b)** Zero
3. (a) 85 kg $\cdot$ m/s **(b)** 3.0×10^4 kg $\cdot$ m/s
5. 5.88 kg $\cdot$ m/s in the direction opposite $\vec{v}_o$
7. (a) The magnitude of the total momentum of the two-proton system will be (2) equal to the difference between the magnitudes of their momenta. Momentum is a vector quantity. When two momenta are opposite, the magnitude of the addition of the two momenta is equal to the difference of the magnitudes of the two momenta. Because $\vec{p}_1$ and $\vec{p}_2$ are opposite, $P = p_1 + (-p_2) = p_1 - p_2$.
(b) 1.84×10^{-25} kg $\cdot$ m/s in the direction of the faster proton
9. (a) 0.45 **(b)** 99%
11. $\Delta \vec{p} = (-3.0 \text{ kg} \cdot \text{m/s})\,\hat{\mathbf{y}}$
13. 6.5×10^3 N
15. (a) The direction will be (4) to the east of northeast from a vector sketch.
(b) $(823 \text{ kg} \cdot \text{m/s})\,\hat{\mathbf{x}} + (283 \text{ kg} \cdot \text{m/s})\,\hat{\mathbf{y}}$ and since the x-component is larger than the y-component, the total momentum is to the east of northeast.
17. (a) 491 kg $\cdot$ m/s downward **(b)** Yes, there would be a difference. $\Delta p = 524$ kg $\cdot$ m/s downward
19. 15 m/s
21. 13 m/s
23. (a) The magnitude of the change in momentum of the baseball is (3) equal to the sum of the momenta of the baseball before and after the bunt. Momentum is a vector quantity. When two momenta are opposite, the magnitude of the difference of the two momenta is equal to the sum of the magnitudes of the two

momenta. Because $\vec{p}_1$ and $\vec{p}_2$ are opposite, $\Delta p = p_1 - (-p_2) = p_1 + p_2$. **(b)** 4.0 kg·m/s in the direction opposite $\vec{v}_o$ **(c)** 1.6×10^2 N in the direction opposite $\vec{v}_o$

25. (a) 20.0 N·s **(b)** 14.9° **(c)** 0.776 m/s
27. 1.1×10^3 N, 4.7×10^2 N
29. 15 N upward
31. 3.7×10^2 N
33. 3.0×10^2 N
35. 0.057 s
37. He moves at a speed of 0.083 m/s in the opposite direction.
39. 1.08×10^3 s = 18.1 min
41. (a) 1.4 m/s **(b)** 2.5 cm
43. 4.7 km/h at an angle of 62° south of east
45. (a) 11 m/s to the right **(b)** 22 m/s to the right **(c)** $v = 0$, or it is at rest
47. (a) 66.7 m/s east **(b)** 3.33×10^4 J
49. (a) The target must move to the (1) right to conserve the total momentum. **(b)** 7.00 m/s
51. (a) It takes him 4.5 min. So the answer is no, he does not get back in time. **(b)** 4.5 m/s
53. $v_1 = 0.61$ m/s and $v_2 = 0.73$ m/s
55. First use energy conservation to find the speed of the bullet and the bob just after collision from the swing motion. The speed becomes the speed at the start of the swing. So $\frac{1}{2}(m+M)v^2 + (m+M)g(0) = \frac{1}{2}(m+M)(0)^2 + (m+M)g(h)$; therefore, $v = \sqrt{2gh}$. Now apply momentum conservation: $Mv_o + M(0) = (m+M)\,v = (m+M)\sqrt{2gh}$, so $v_o = \frac{m+M}{m}\sqrt{2gh}$.
57. $v_p = -1.8 \times 10^6$ m/s (opposite its initial velocity); $v_\alpha = 1.2 \times 10^6$ m/s (in direction of proton's original velocity)
59. 0.92, or 92%
61. $v_p = 38.2$ m/s, $v_c = 40.2$ m/s, both in the same direction as initially moving
63. 0.94 m
65. (a) $v'_x = 1.0$ m/s; $v'_y = 3.3$ m/s **(b)** $\theta = 73°$
67. 0.34 m
69. 50%
71. $v_1 = 2.14$ m/s; $v_2 = 17.1$ m/s; $v_{23} = 6.41$ m/s
73. From momentum conservation, $mv_o + M(0) = (m+M)v$, $\therefore v = \frac{mv_o}{m+M}$. $K_o = \frac{1}{2}m_1 v_o^2$ and $K = \frac{1}{2}(m+M)v^2 = \frac{1}{2}(m+M)\left(\frac{m_1 v_o}{m+M}\right)^2 = \frac{1}{2}\frac{(mv_o)^2}{m+M}$. The fraction of kinetic energy lost is $\frac{|\Delta K|}{K_o} =$

$\frac{K_o - K}{K_o} = 1 - \frac{K}{K_o} = 1 - \frac{\frac{1}{2}\frac{(mv_o)^2}{m+M}}{\frac{1}{2}mv_o^2} = 1 - \frac{m}{m+M} = \frac{M}{m+M}$.

75. (a) 4.6×10^6 m from the center of the Earth **(b)** 1.8×10^6 m below the surface of the Earth
77. 82.8 m
79. The CM of both the square sheet and the circle is at the center of the square. So from symmetry, the CM of the remaining portion is still at the center of the sheet.
81. 0.175 m

83. $(5.3 \text{ m/s}^2)\,\hat{x} + (2.7 \text{ m/s}^2)\,\hat{y}$, or $A_{cm} = \sqrt{(5.3 \text{ m/s}^2)^2 + (2.7 \text{ m/s}^2)^2} = 5.9 \text{ m/s}^2$ at $\theta = \tan\left[\frac{2.7}{5.3}\right] = 27°$ above the positive x-axis
85. (a) -1.14×10^5 J **(b)** $(2.19 \times 10^4 \text{ kg·m/s})\,\hat{x} - (2.92 \times 10^4 \text{ kg·m/s})\,\hat{y}$ or 3.65×10^4 kg·m/s at an angle of 53.1° south of east **(c)** $(5.15 \times 10^3 \text{ N})\,\hat{x} - (6.87 \times 10^3 \text{ N})\,\hat{y}$ or 8.59×10^3 N at an angle of 53.1° south of east
87. (a) 1.89 m/s, 5.48° **(b)** $K < K_o$, inelastic

CHAPTER 7 MULTIPLE CHOICE

1. (c)
3. (a)
5. (d)
7. (d)
9. (b)
11. (a)
13. (d)
15. (c)
17. (d)
19. (c)

CHAPTER 7 CONCEPTUAL QUESTIONS

1. Since 2π rad $= 360°$, hence 1 rad $= 360°/2\pi = 57.3°$
3. Yes, since a wheel is rigid they all sweep through the same angle. No, they do not have the same tangential speed, because their distances to the center of the wheel may be different.
5. Your tangential speed would decrease, because it depends linearly on the distance from the center, if the angular speed is constant.
7. There is insufficient centripetal force (provided by friction and adhesive forces) on the water drops, so the water drops fly out along a tangent away from the accelerating clothes, leaving the clothes drier.
9. The floats will point in direction of acceleration, in this case centripetally or inward. The principle is the same as that of the accelerometer in Fig. 4.26. No, the rotation direction does not make a difference since the centripetal acceleration is always inward.
11. The body's inertia means that it has a tendency to keep moving along a straight line, and the car makes a turn (centripetally accelerates) because of the centripetal force (friction) between the tires and the road. So passengers will feel as if they are "thrown outward" when actually it is the car that is accelerating away from their straight-line tendency.
13. Yes, a car in circular motion always has centripetal acceleration. Yes, it also has angular acceleration as its speed is increasing
15. No. When the tangential acceleration changes, the angular acceleration changes, resulting in a change in tangential speed and therefore a change in centripetal acceleration.
17. It would not appreciably affect its orbit, which is determined by the mass of the Earth. This assumes the Moon's mass is negligible compared to that of the Earth. In actuality, there would be a small effect because the center of mass of the Earth–Moon system would

be slightly closer to the Moon, but this effect would be very slight.
19. If the cup is released, water will not run out since both the cup and the water are in free fall.
21. (a) Zero, because the gravitational force and the satellite displacement are perpendicular ($\theta = 90°$), thus $W_g = F_g d \cos\theta = 0$. **(b)** No. The person is still in free fall and will eventually hit the (stopped) elevator floor.
23. (a) No, you cannot speed up and stay in the same orbit with one rocket burst. Once you speed up, you will be in a different orbit. **(b)** You must decrease the orbital radius to increase speed and then increase the orbital radius to get back into the original orbit to catch the equipment.

CHAPTER 7 EXERCISES

1. (2.5 m, 53°)
3. (a) 0.26 rad **(b)** 0.79 rad **(c)** 1.6 rad **(d)** 2.1 rad
5. (a) 1.83 rad, 0.292 rev **(b)** 103°, 0.286 rev **(c)** 4.49 rad, 257°
7. 9.2×10^{-3} rad or 0.53°
9. 1.5×10^3 rev
11. 0.19-m arc length, 72°
13. (a) The correct answer is (1), more than 25.0 cm **(b)** 40.9 cm
15. (a) No, since 57.3° does not yield an integer when divided into 360° **(b)** Six such pieces and one 0.28-radian leftover piece.
17. (a) 180 rad **(b)** 9.00 m
19. 0.087 rad/s
21. (a) 4.80×10^{-3} s **(b)** 6.32×10^{-3} s
23. 0.42 s
25. (a) $\overline{\omega} = 0.509$ rad/s and $\overline{v} = 20.8$ cm/s **(b)** $\overline{\omega}$ is still 0.509 rad/s, but $\overline{v} = 0.764$ cm/s **(c)** The direction is toward the observer.
27. $\overline{v} = 1.11$ m/s and $\overline{\omega} = 0.634$ rad/s
29. 1.10 rad/s, down
31. 1.3 m/s
33. 64 m
35. 11.3°
37. (a) The force of gravity (weight) is supplying the centripetal force. **(b)** 3.1 m/s
39. The required tension is only 29.5 N, so the string will work (not break).
41. (a) $v = \sqrt{rg}$ **(b)** $h = (5/2)r$
43. Expected derivation results are given in the exercise statement.
45. 1.1×10^{-3} rad/s^2
47. (a) The correct choice is (3), both angular and centripetal accelerations. There is always centripetal acceleration in circular motion. When the car increases its speed on a circular track, there is also angular acceleration. **(b)** 53 s **(c)** $\vec{a} = -(8.5 \text{ m/s}^2)\hat{r} - (1.4 \text{ m/s}^2)\hat{t}$
49. (a) $\alpha = 1.82$ rad/s^2 **(b)** 28.7 revolutions
51. (a) $a_t = 2.5$ m/s^2 and $a_c = 9.7$ m/s^2 **(b)** At the lowest point of the swing, since v is maximum there and $a_t = 0$ because the velocity is not changing
53. $g_M = 1.6$ m/s^2
55. 8.0×10^{-10} N, toward opposite corner
57. (a) 100 kg **(b)** 894 N
59. (a) Equate the gravitational force on the Moon to its centripetal acceleration: $GmM/r^2 = mv^2/r$, where m and M are the masses of the Moon and Earth, respectively.

Assuming circular orbits, $v = 2\pi r/T$; now substitute this into the first equation and solve for M: $M = 4\pi^2 R^3/(GT^2)$. **(b)** 6.0×10^{24} kg

61. 1.5 m/s^2

63. (a) 1.0×10^3 m/s **(b)** 3.9×10^{28} J **(c)** -7.8×10^{28} J **(d)** -3.9×10^{28} J

65. (a) Since the escape speed depends on gravitational potential energy, Jupiter should be as far as way as possible. So the answer is (2) Earth should be as far as possible from Jupiter. Earth should be directly opposite Jupiter when the launch occurs.
(b) 4.3×10^4 m/s; the Sun determines most of the escape speed.

67. 3.13 km/s

69. 4.4×10^{11} m

71. 1.53×10^9 m

73. (a) 8.91 m/s^2, toward the center **(b)** 1.40×10^4 N, toward the car **(c)** 2.27 m/s^2, opposite the velocity **(d)** 9.19 m/s^2, $\theta = 75.7°$

75. (a) $a_c = 7.24$ m/s^2 **(b)** $T = 8.52$ N

77. (a) 0.069 m/s **(b)** 0.0057% **(c)** If the meteoroid hit off-center, the probe would feel a torque and begin to spin, leaving it with some rotational kinetic energy in addition to its translational kinetic energy. Thus a larger percentage of the initial kinetic energy would remain.

79. (a) 19.8 times larger **(b)** -1.34×10^{22} J **(c)** 6.53 km/s **(d)** 428 d or 1.17 y

CHAPTER 8 MULTIPLE CHOICE

1. (a)
3. (b)
5. (b)
7. (a)
9. (d)
11. (a)
13. (d)
15. (c)
17. (c)
19. (c)

CHAPTER 8 CONCEPTUAL QUESTIONS

1. Yes, it is possible and the rolling motion of a tire is a good example.

3. The speedometer reading is $v/2$ because the distance from the ground for the top of the tire is twice that for the center of the tire (the point at which the tire makes contact with the ground is the instantaneous axis of rotation for the tire), and the speedometer reads the speed of the center of the tire (v_{CM}), not the top.

5. The uncut portion of the trunk, just inside the notch, will act like a hinge so the falling tree will rotate about that portion and will not bind the saw. There is still danger in standing too close behind the tree because the "hinge" will eventually break and the base of the tree may kick back and hit the logger.

7. As the rear wheels exert a backward frictional force on the roadbed, the reaction force (roadbed forward on the tire) creates a frictional torque that causes the motorcycle to rotate upward until it reaches a point where the gravitational torque balances the frictional torque.

9. An example is a cylinder rolling on a level surface.

11. The moment of inertia of an object depends not only on its mass, but also on how it is distributed. Physically, this means that in a given situation, an object's angular acceleration depends on where its axis of rotation is located.

13. The hardboiled egg is essentially a solid object, and when it is brought to rest by an external torque, it will have no angular momentum. When released, its angular momentum will remain at zero. For the raw egg, the shell loses all its angular momentum, but the liquid center (which is only loosely connected to the solid shell) does not experience enough torque to stop it completely. Thus the liquid part of the raw egg maintains some angular momentum. When the egg as a whole is released, connection to the shell transfers some angular momentum to the latter and the egg as a whole will continue to rotate, although slower than initially (less angular momentum).

15. The pole serves to increase the moment of inertia. If the walker starts to rotate (fall), the angular acceleration will be decreased, hopefully giving more time to recover.

17. Yes. For example, a bicycle wheel mounted on a fixed axis can be spun faster, thus increasing its rotational kinetic energy while maintaining its translational kinetic energy at zero.

19. A net *rotational* work is required to produce a change in rotational kinetic energy. Rotational work is done by a torque acting through an angular displacement.

21. The polar ice caps (with almost zero moment of inertia because they are close to the Earth's rotation axis) will spread out as liquid into the oceans and thus increase the moment of inertia of the Earth. To conserve angular momentum, the Earth's rotation rate would slow. The result would be a longer day.

23. The cat manipulates its body and tail to change its rotational inertia as it falls. Because the force of gravity exerts no net torque on the cat, its angular momentum stays constant. Thus by adjusting its moment of inertia, the cat can control its angular speed and hopefully reach the ground with its feet (more or less) pointed downward.

CHAPTER 8 EXERCISES

1. At the nine-o'clock position, the velocity is straight upward. So it is a "free fall" with an initial upward velocity. It will rise, reach a maximum height, and then fall back down.

3. 0.10 m

5. 1.7 rad/s

7. (a) 0.0331 rad/s^2 **(b)** 1.99×10^{-3} m/s^2

9. 5.6×10^2 N

11. 3.3×10^2 N

13. (a) Yes, the seesaw can be balanced if the lever arms are appropriate for the weights of the children because torque is equal to force times the lever arm. **(b)** 2.3 m

15. 245 N

17. (a) No, it is not possible to have the lines perfectly horizontal, because the weight has to be supported by upward components of these tensions in the lines. **(b)** 1.8×10^3 N > 400 lb

19. 784 N

21. (a) 88.2 N **(b)** 10.5 kg

23. $m_2 = 0.20$ kg, $m_3 = 0.50$ kg, $m_4 = 0.40$ kg

25. (a) Nine books **(b)** 22.5 cm

27. 10.2 J

29. $T_2 = 40$ N, $T_1 = 49$ N

31. 16.7 N

33. (a) $\tan \theta$ should be (2) equal to f_s/N. **(b)** 0.19 **(c)** 3.5 m/s

35. 0.64 m $\cdot$ N

37. 1.5×10^5 m $\cdot$ N

39. 1.2 m/s^2

41. (a) 0.89 m/s^2 **(b)** 0.84 m/s^2

43. According to Exercise 42, the 66.7-cm mark will fall with acceleration equal to g. Below that mark, the acceleration is less than g and above that mark, the acceleration is greater than g. Therefore, the pennies at the 10-, 20-, 30-, 40-, 50-, and 60-cm marks will be slowed (acceleration-wise) by the meterstick. The pennies at the 70-, 80-, 90-, and 100-cm marks will separate from the meterstick and thus fall with acceleration equal to g.

45. 1.31 s

47. 4.5×10^2 J

49. (a) The answer is (3) less than $\sqrt{2gh}$. If it were a point mass, then its speed would be $\sqrt{2gh}$. However, when the object rotates, it has rotational kinetic energy. Therefore, its translational energy (and therefore linear speed) is less than if it did not rotate. **(b)** 5.8 m/s

51. 0.47 m $\cdot$ N

53. $K_t/K_r = 1.0 \times 10^4$

55. 2.3 m/s

57. 3.4 m/s

59. (a) 29% **(b)** 40% **(c)** 50%

61. (a) The hollow thin-shelled and the solid ball have the same translational kinetic energy since they start with the same initial linear speed. However, their moments of inertia are different: $I_{solid} = 2mR^2/5$ is less than $I_{hollow} = 2mR^2/3$. Thus, the hollow ball has more *rotational* kinetic energy and thus more *total* kinetic energy. Therefore, this is converted into *more* gravitational potential energy at the top; hence, $h_{hollow} > h_{solid}$. **(b)** No, the radii do not make any difference. The ball's radius affects its moment of inertia (which is proportional to r^2) and *inversely* affects its angular speed. Since the ball's rotational energy is proportional to the moment of inertia and the *square* of the angular speed, the radius cancels out. **(c)** When both balls return they will have the same translational and rotational kinetic energies they had at the start, and thus the same speed as the initial speed. (Use energy conservation.)

63. (a) $\sqrt{gR}$ **(b)** $2.7R$ **(c)** At the minimum speed, the centripetal force is supplied entirely by gravity, hence the rider feels no (downward) normal force and might incorrectly say he or she feels "weightless."

65. 1.4 rad/s

67. (a) 2.67×10^{40} kg $\cdot$ m^2/s **(b)** 7.06×10^{33} kg $\cdot$ m^2/s

69. (a) The angular speed of the coupled disks is (2) less than the angular speed of the original rotating disk. This is because the moment of inertia of two coupled disks is greater than that of one of them, and to conserve angular momentum, the coupled disks must rotate at a slower angular speed. **(b)** 200 rev/min

71. (a) 4.3 rad/s **(b)** $K = 1.1K_o$ **(c)** The work done by the skater

73. $d = b(v_o/v)$

75. (a) The correct answer is (2), rotate in the direction opposite that of the cat, to conserve angular momentum (initially zero). **(b)** 0.56 rad/s **(c)** No, it will be an angular distance of 2.1 radians from where it was initially.

77. (a) 118 rad/s **(b)** 25.4 m/s **(c)** 340 J **(d)** 23.0 m

79. (a) -4.65 rad/s^2 (deceleration) **(b)** -0.175 m/s^2 (deceleration) **(c)** 2.26×10^3 N **(d)** -5.33×10^4 J

81. $\theta = 46.3°$, $T_1 = 426$ N, $T_2 = 102$ N

83. (a) 73.8 J **(b)** 21.1 J **(c)** The friction is static since the ball rolls without slipping. **(d)** 4.02 N

CHAPTER 9 MULTIPLE CHOICE

1. (c)
3. (d)
5. (d)
7. (a)
9. (a)
11. (c)
13. (b)
15. (b)
17. (c)
19. (c)

CHAPTER 9 CONCEPTUAL QUESTIONS

1. Steel wire has a greater Young's modulus. Young's modulus is a measure of the ratio of stress over strain. For a given stress, a material with a greater Young's modulus will have a smaller strain. Steel will have a smaller strain here.

3. Through capillary action, the wooden peg absorbs water and swells and splits the rock.

5. Bicycle tires have a much smaller contact area with the ground, so they need a higher pressure to balance the weight of the bicycle and the rider.

7. It measures gauge pressure. Blood pressure is gauge pressure since it is the pressure *above* atmospheric pressure in the closed circulatory system that is relevant.

9. The absolute pressure is 1 atmosphere but the gauge pressure is zero. The pressure in the flat tire is the same as that of the outside.

11. Pressure depends only on depth. To get the same water pressure, spherical tanks do not need as much water as cylindrical tanks. Also, spherical shapes can distribute pressure more evenly so will reduce the risk of tank damage by water pressure.

13. **(a)** A life jacket must have lower density than water, such that the average density of a person and a jacket is less than the density of water. **(b)** Salt water has a higher density, so it can exert a larger buoyant force.

15. The water in the Mississippi River at New Orleans is partly fresh (or less salty). Sea-

water has a higher density; therefore, less seawater than harbor water needs to be displaced to exert the same buoyancy force. Thus when the ship is in seawater (open ocean), the Plimsoll mark will be *above* the water level.

17. They are designed to displace a large volume of water with empty space inside, thus producing a large buoyant force. In essence, the large percentage of air makes the average density of the liners less than that of the water.

19. No, because the forces are perpendicular to all of the surfaces, so they won't rotate the cylinder.

21. The air between you and the truck is now flowing through a narrower opening and therefore flowing faster (compared to the air on the outside surfaces). Thus there is a reduced air pressure in the space between you and the truck, compared to the pressure on the outside surfaces. This results in pressure differential forcing both you and the truck toward one another. Since the truck is usually much more massive than your car, it is usually you who will feel the effect.

23. The Bernoulli effect is responsible. As the wind moves across a roof, the pressure is lowered there compared to the still air inside the house. Thus the pressure difference will tend to force the roof upward.

25. **(a)** The air flow above the paper decreases the pressure there. This creates a pressure difference, and a lift force results. **(b)** The egg is kept aloft by the pressure of the air coming out of the end of the tube. If the egg moves to one side (partially out of the stream), there is then a slower flow speed around the outside of the egg and thus a larger pressure on that side. This creates an an inward pressure difference that forces the egg back to midstream.

27. To reduce surface tension

CHAPTER 9 EXERCISES

1. 1.5 mm
3. 1.9 cm
5. **(a)** 9.40×10^4 N/m^2 **(b)** 1.25×10^5 N/m^2
7. 47 N
9. **(a)** The correct choice is (1), a cold day, because tracks expand when temperature increases **(b)** 1.9×10^5 N
11. **(a)** Bends toward brass, because the stresses are the same for both, and brass has a smaller Young's modulus. Brass will have a greater strain $\Delta L/L_o$, so it will be compressed more. Therefore, the brass will be shorter than the copper. **(b)** Brass: $\Delta L/L_o = 2.8 \times 10^{-3}$; copper: $\Delta L/L_o = 2.3 \times 10^{-3}$
13. 4.2×10^{-7} m
15. **(a)** Ethyl alcohol has the greatest compressibility, because it has the smallest bulk modulus B. The smaller the B, the greater the compressibility. **(b)** $\Delta p_w/\Delta p_{ea} = 2.2$
17. 3.2×10^{-6} m
19. 5.4×10^{-5}
21. **(a)** 9.8×10^4 Pa **(b)** 2.0×10^5 Pa
23. 5.88×10^4 Pa
25. **(a)** 1.99×10^5 N/m^2 **(b)** 9.80×10^4 N/m^2
27. 1.07×10^5 Pa
29. 0.50%

31. 0.51 N
33. 2.2×10^5 N (about 50,000 lb!)
35. 1.9×10^2 m/s
37. **(a)** 1.1×10^8 Pa **(b)** 1.9×10^6 N
39. 549 N, 1.37×10^6 Pa
41. **(a)** 4.0×10^3 N/m^2 **(b)** 2.0×10^{-3} N **(c)** 1.7 N
43. **(a)** The correct answer is (3), stay at any height in the fluid, because the weight is exactly balanced by the buoyant force. **(b)** It will sink, since $W > F_b$.
45. 2.0×10^3 kg/m^3
47. 1.8×10^3 N
49. **(a)** 0.09 m **(b)** 8.1 kg
51. 1.00×10^3 kg/m^3 (probably water)
53. 17.7 m
55. 8.1×10^2 N
57. 0.98 m/s
59. -0.167 atm
61. 0.149 m/s
63. 14.5 m/s
65. **(a)** 0.43 m/s **(b)** 8.6×10^{-5} m^3/s
67. 71.9 L/s
69. 3.5×10^2 Pa
71. 13.5 s
73. 8.0×10^2 kg/m^3
75. **(a)** The Young's modulus definition may be rewritten and solved for F: $F = (AY/L_0)\Delta L = k\Delta L$, where $k = AY/L_0$ is the effective spring constant, which is constant as long as the area is constant. **(b)** 1.88×10^8 N/m **(c)** 1.06×10^{-5} m **(d)** 1.06×10^{-2} J
77. **(a)** 6.00×10^3 J **(b)** 6.00×10^3 J **(c)** 1.29 tons **(d)** 0.571
79. **(a)** 5.78 m/s **(b)** 17.0 cm^3 per minute

CHAPTER 10 MULTIPLE CHOICE

1. (a)
3. (b)
5. (a)
7. (c)
9. (a)
11. (a)
13. (b)
15. (b)

CHAPTER 10 CONCEPTUAL QUESTIONS

1. Not necessarily, because internal energy does not depend solely on temperature. It also depends on mass.

3. Air contains water vapor and it may freeze at high altitudes where the temperature is low.

5. Monatomic molecules behave like point masses so they can have only translational kinetic energy. In addition to translational kinetic energy, diatomic molecules can have rotational and vibrational kinetic energy because this type of molecule can rotate and the atoms can vibrate.

7. When the pressure of the gas is held constant, if the temperature increases or decreases, so does the volume. Therefore, a gas's temperature can be measured by monitoring its volume.

9. The balloons collapsed. Due to the decrease in temperature, the volume decreases.

11. (a) Ice moves upward. **(b)** Ice moves downward. **(c)** Copper

13. When the ball alone is heated, it expands and cannot go through the ring. When the ring is heated, it expands and the hole gets larger so the ball can go through again.

15. Most metals have a higher coefficient of thermal expansion than glass. The lid expands more than glass so it becomes easier to loosen the lid.

17. The gases diffuse through the porous membrane, but the helium gas diffuses faster because its atoms have a smaller mass than neon and thus, on average, are traveling at a faster (rms) speed. Eventually, there will be equal concentrations of gases on both sides of the container.

19. Yes, because for all gases, their average *translational* kinetic energy per molecule is determined by their absolute temperature.

21. $\Delta U = U_{\text{diatomic}} - U_{\text{monatomic}} = (5/2)nRT - (3/2)nRT = nRT$

CHAPTER 10 EXERCISES

1. 104 °F
3. (a) 248 °F **(b)** 53.6 °F **(c)** 23.0 °F
5. −70 °C
7. 56.7 °C, −62 °C
9. (a) −51.5 °C **(b)** −47.3 °C
11. (a) The correct choice is (3) $T_F = T_C$, because we want to find the one temperature at which the Celsius and Fahrenheit scales have the same reading. **(b)** −40 °C = −40 °F
13. $0.555T_F$; the percentage difference is 11%.
15. (a) 273 K **(b)** 373 K **(c)** 293 K **(d)** 238 K
17. (a) $T_K = \frac{5}{9}T_F + 255.37$ **(b)** 300 K is lower.
19. (a) 2.22 mol **(b)** 5.57 mol **(c)** 4.93 mol **(d)** 1.75 mol
21. 1/4
23. 0.0618 m³
25. 574.58 K
27. 2.5 atm
29. 33.4 lb/in.²
31. (a) The temperature will (1) increase. **(b)** $p_2 = 4p_1$
33. (a) The correct answer is (1), increase, because with $p = p_o$, $p_oV_o/T_o = pV/T$ becomes $V/V_o = Tp_o/(T_op) = T/T_o$, or volume is proportional to temperature. **(b)** 10.6%
35. (a) If $T \gg 273$ K, ignore the 273 when converting from T_C to T. Similarly, if $T_C \gg 0$ °C and $T_F \gg 32$ °F, ignore the 32 when converting from T_F to T_C. Thus $T = T_C + 273 \approx T_C = \frac{5}{9}(T_F - 32) \approx \frac{5}{9}T_F$.
(b) 87% **(c)** 4.6×10^{-3}%
37. (a) The correct choice is (1), high, because the tape shrinks. One division on the tape (it is now less than one true division due to shrinkage) still reads one division. **(b)** 0.060%
39. 0.0027 cm
41. 9.5 °C
43. 5.52×10^{-4}/°C
45. (a) The correct answer is (1), the ring, so it expands, then the ball can go through. **(b)** 353 °C
47. (a) There will be a gas spill, because the coefficient of volume expansion is greater for gasoline than for steel. **(b)** 0.48 gal

49. (a) 116 °C **(b)** No
51. Start with the definition: $\beta = \dfrac{\Delta V}{V_0\Delta T}$.
For an ideal gas: $p_oV_o = nRT_o$ and $p_oV = nRT$ (constant pressure).
Therefore, $\Delta V = V - V_o = \dfrac{nRT}{p_o} - \dfrac{nRT_o}{p_o}$
$= \dfrac{nR\Delta T}{p_o} = \dfrac{V_o\Delta T}{T_o}$ (since $p_o = \dfrac{nRT_o}{V_o}$).
So $\dfrac{\Delta V}{V_o} = \dfrac{\Delta T}{T_o}$. Therefore, $\beta = \dfrac{\Delta V}{V_o\Delta T} =$
$\dfrac{\Delta T}{T_o} \times \dfrac{1}{\Delta T} = \dfrac{1}{T_o} = \dfrac{1}{293 \text{ K}} = 3.41 \times 10^{-3}$/K = 3.41×10^{-3}/°C. This is in good agreement with the 3.5×10^{-3}/°C value.
53. 65 °C
55. (a) The internal energy will (2) increase by less than a factor of 2. This is because the internal energy is proportional to the Kelvin temperature, and doubling the Celsius temperature will increase, but not double, the Kelvin temperature. **(b)** 1.07
57. (a) 6.17×10^{-21} J **(b)** 1.36×10^3 m/s
59. 273 °C
61. 7.5×10^{-21} J
63. 1.12 times as fast
65. (a) The answer is (1) $^{235}UF_6$. Since both samples have the same average (per molecule) kinetic energy, the one with the lowest molecular mass ($^{235}UF_6$) must be moving faster on average. **(b)** 1.0043
67. 8.3×10^3 J
69. 224 °C
71. 0.27%
73. 0.0494 m³
75. (a) The answer is (3), helium, because at the same temperature, helium with the smaller mass will have the higher rms speed.
(b) 425 m/s < 1100 m/s

CHAPTER 11 MULTIPLE CHOICE

1. (d)
3. (c)
5. (b)
7. (a)
9. (d)
11. (b)
13. (b)

CHAPTER 11 CONCEPTUAL QUESTIONS

1. 1 Cal = 1000 cal
3. No, heat is the energy added to or removed from an object (thus *changing* its internal thermal energy). But a cold object *could* have more internal energy simply because it has more mass than one at a higher temperature. In other words, besides temperature, mass is also a factor in determining total internal energy.
5. Temperature change is also determined by mass and specific heat ($\Delta T = Q/mc$). Thus the final temperature of two objects *can* be different, even if the heat transfers (Q) and initial temperatures are the same.
7. Only that together, the cold water and cup, gain 400 J of heat. Since both the cup and water will experience the same temperature

change, most of the 400 J will end up in the water because of its high specific heat compared to aluminum.
9. The (heat) energy added is called the latent heat. It is the energy required to change the phase of a substance. The energy goes into breaking attractive bonds between molecules and separating them rather than into increasing temperature (kinetic energy of the molecules).
11. The water molecules in your breath condense, which looks like steam or fog.
13. The bridge is exposed to the cold air above and below, while the road is exposed only above. Also, the road can receive heat energy from the ground, whereas the lower surface of the bridge is in contact with cold air and transfer of thermal energy (conduction) from gas to solid is less than from solid to solid. Thus the water on the bridge could be frozen while that on the roadway is still liquid.
15. The low-pressure gas trapped between the walls of the bottle is a poor conductor of heat, so conduction is very low. This gas is a partial vacuum, so it cannot transfer much heat by convection. The interior has a silver film coating that minimizes radiation losses or gains.

CHAPTER 11 EXERCISES

1. 5.86×10^3 W
3. 720 Cal
5. (a) 60 000 times **(b)** 83 h
7. (a) The answer is (1) because the specific heat of copper is 3 times that of lead, so it will require 3 times as much heat, everything else being equal. **(b)** $Q_{Cu} - Q_{Pb} = +2.1 \times 10^4$ J
9. 35.6 °C
11. 4.8×10^6 J
13. (a) The answer is (1) since the mass and temperature change are the same for both, the aluminum gains twice as much heat as the iron because its specific heat is twice as great. **(b)** $Q_{Al} - Q_{Fe} = +1.8 \times 10^4$ J
15. 1.4×10^2 J/(kg · °C)
17. 88.6 °C
19. 1.7 h
21. (a) The answer is (1), higher, because if some water splashed out, there will be less water to absorb the heat. The final temperature will be higher, thus the metal's temperature drop will be smaller and its calculated specific heat value will be higher than the correct value. **(b)** 3.1×10^2 J/(kg · °C)
23. 34 °C
25. 8.3×10^5 J
27. (a) Converting 1.0 kg of water at 100 °C to steam at 100 °C requires (1) more heat, because the heat of vaporization L_v is greater than the heat of fusion L_f. **(b)** Vaporization by 1.93×10^6 J more
29. 4.9×10^4 J
31. 8.0×10^4 J
33. (a) The correct choice is (2), only latent heat, because the boiling point of mercury is 357 °C = 630 K so it is already at its boiling temperature. **(b)** 4.1×10^3 J
35. 0.437 kg
37. 195 °C

39. (a) This is because (2) water has a high latent heat of vaporization. Thus when water evaporates, a lot of heat is removed from the skin. **(b)** 1.6×10^7 J
41. (a) 110 °C and 140 °C **(b)** For solid: 1.0×10^2 J/(kg·°C); for liquid: 2.0×10^2 J/(kg·°C); for gas: 1.0×10^2 J/(kg·°C) **(c)** For fusion: 4.0×10^3 J/kg; for vaporization: 6.0×10^3 J/kg
43. 4.54×10^6 J
45. (a) The answer is (3) since, everything else being equal, the rate of heat conduction is directly proportional to the thermal conductivity of the material. Since the brick has a lower thermal conductivity than the concrete, it will conduct heat at a *slower* rate than the concrete. **(b)** 0.55
47. 13 J
49. (a) The answer is (1) longer, because copper has a higher thermal conductivity **(b)** 1.63
51. 90 J/s
53. 411 °C
55. 171 °C
57. (a) (3) Thinner than, because glass wool has a lower thermal conductivity **(b)** 4.9 in.
59. (a) 2.5×10^4 J/s **(b)** 2.6×10^3 J/s
61. 3.4×10^6 J
63. 7.8×10^5 J
65. 0.45 kg
67. 69 °C
69. 4.0×10^2 m/s
71. 8.07×10^3 W

CHAPTER 12 MULTIPLE CHOICE

1. (a)
3. (d)
5. (d)
7. (a)
9. (d)
11. (b)
13. (d)
15. (b)
17. (d)
19. (d)

CHAPTER 12 CONCEPTUAL QUESTIONS

1. For an ideal gas, if T is a constant, $p \propto 1/V$. (That is, pressure varies inversely with volume.) The figure in question does *not* display this property.
3. Pressure p, volume V, absolute temperature T, and the quantity of gas (number of moles n or number of molecules N) are the state variables.
5. This is an adiabatic compression. When the plunger is pushed in, the work done goes into increasing the internal energy of the air. The increase in internal energy increases its temperature and eventually that of the paper above its "flash point," and it catches fire.
7. This is possible through work. Since $Q = \Delta U + W$, $\Delta U = -W$ when $Q = 0$.
9. Work: 1, 2, 3. Work is equal to the area under the curve in the p-V diagram. The area under 1 is the greatest and the area under 3 is the smallest. Final temperature: 1, 2, 3. According to the ideal gas law, the temperature of a

gas is proportional to the product of pressure and volume, $pV = nRT$. Since the final volume is the same for all three processes, the higher the pressure, the higher the final temperature.
11. The conversion of heat to mechanical energy does not violate the second law of thermodynamics. The law only implies that the conversion can never be 100% efficient.
13. From the second law of thermodynamics, the entropy increases. The cold water gains more entropy than that lost by the hot water.
15. A living organism is an open system because it has to obtain energy from outside.
17. Both pressure and internal energy return to the original values they had at the start of the cycle.
19. Not necessarily. The amount of work also depends on the efficiency of the engine.
21. Yes, it is actually *always* greater than 1. In practical terms, the COP of a heat pump is the ratio of the heat of interest (that moved into the already hotter reservoir, to keep a room warm, say, when it is cold outside) compared to the work required to accomplish this, or $\text{COP}_{hp} = \dfrac{Q_h}{W_{in}}$. However, this work is *always* less than the heat moved into the hot reservoir (since $Q_h = W_{in} + Q_c > W_{in}$); thus this ratio is *always* > 1.
23. For 100 °C and 300 °C, $\varepsilon_C = 35\%$. For 50 °C and 250 °C, $\varepsilon_C = 38\%$. Choose 50 °C and 250 °C for higher efficiency.

CHAPTER 12 EXERCISES

1. 5.5×10^5 J
3. (a) The answer is (2). During an isothermal process, the temperature, and hence the internal energy, of the gas remains constant. **(b)** $Q = \Delta U + W_{net} = 0 + W_{net} = W_{net} = +400$ J, so 400 J of heat are added.
5. (a) The answer is (3), decreases, as $Q = 0$ and W is positive. **(b)** -500 J
7. (a) 4.9×10^3 J **(b)** $\Delta U = 3.5 \times 10^3$ J
9. (a) The answer is (2), zero, because $\Delta T = 0$ **(b)** $-p_1 V_1$ (on the gas) **(c)** $-p_1 V_1$ (out of the gas)
11. 3.6×10^4 J
13. (a) The answer is (1). Since the volume of the gas has increased, this gas must have done work *on* its environment, making the work positive. **(b)** 3.4×10^3 J
15. (a) 17 mol **(b)** 5.0×10^4 J of heat are added to the gas.
17. (a) The correct answer is (3). The work is negative because the volume is decreasing, indicating that the gas is being compressed. Thus work is done *on* the gas. **(b)** -1.8×10^5 J. The minus sign means that work is done *on* the gas. **(c)** -480 K **(d)** -6.0×10^5 J **(e)** -7.8×10^5 J
21. (a) The correct choice is (3). The molecules of the condensed liquid water are more ordered than the randomly moving molecules of steam, so the order has increased, meaning that the *disorder* has *decreased*. Therefore, the entropy has decreased. Alternatively, heat flows out of the steam, so Q is negative, making the entropy change negative, which means

that the entropy has decreased. **(b)** -6.1×10^3 J/K
23. 0.20 J/K
25. (a) The correct choice is (1). Without changing its temperature, the expanded gas occupies a larger volume than before, making it more *disordered*. Therefore, its entropy has increased. **(b)** 10 J/K
27. (a) The correct choice is (3), the change in entropy is *negative*. This is because the entropy decrease of the cold reservoir exceeds the increase of entropy of the hot reservoir. **(b)** -0.98 J/K
29. (a) The correct answer is (2), zero, $\Delta S = 0$. **(b)** 2.73×10^4 J
31. (a) 61.0 J/K **(b)** -57.8 J/K **(c)** 3.2 J/K
33. (a) 5.6×10^2 J **(b)** 1.4×10^3 J
35. 1.49×10^5 J
37. (a) 6.6×10^8 J **(b)** 27%
39. (a) The correct answer is (1). To increase the efficiency, we must decrease the ratio Q_c/Q_h, or *increase* the ratio Q_h/Q_c. **(b)** $+0.0833$
41. (a) 6.1×10^5 J **(b)** 1.9×10^6 J
43. (a) 2.1×10^3 J **(b)** 2.0
45. (a) 3.6×10^3 MW **(b)** 2.7×10^3 MW **(c)** 4.1×10^5 gal/min
47. (a) 1800 **(b)** 3.4×10^7 J **(c)** 2.7×10^7 J
49. 21.4%
51. 158 °C
53. (a) 2000 J **(b)** 249 °C
55. (a) The correct choice is (3). Decreasing the ratio T_c/T_h increases the efficiency and *increasing* T_h will decrease this ratio. **(b)** 402 °C
57. $\varepsilon_C = 59.1\%$ and $\varepsilon = 60\%$. This engine is not possible because its claimed efficiency is higher than the Carnot efficiency (upper limit).
59. (a) 45.5% **(b)** 34.1 kW
61. (a) The correct choice is (2), zero, because many quantities such as temperature, pressure, volume, internal energy, and entropy return to original value after each cycle. **(b)** 3750 J
63. (a) 64% **(b)** ε_C is the upper limit of efficiency. In reality, a lot more energy is lost than in the ideal situation.
65. (a) You should choose (2) to lower the low-temperature reservoir. **(b)** If T_h is raised by ΔT, then $\varepsilon_{C1} = 1 - \dfrac{T_c}{T_h + \Delta T}$; if instead T_c is lowered by the same ΔT, then $\varepsilon_{C2} = 1 - \dfrac{T_c - \Delta T}{T_h}$. Their difference is $\varepsilon_{C1} - \varepsilon_{C2} =$
$$\left(1 - \frac{T_c}{T_h + \Delta T}\right) - \left(1 - \frac{T_c - \Delta T}{T_h}\right)$$
$$= \frac{T_c - \Delta T}{T_h} - \frac{T_c}{T_h + \Delta T}$$
$$= \frac{(T_c - \Delta T)(T_h + \Delta T) - T_h T_c}{T_h(T_h + \Delta T)}$$
$$= \frac{(T_c - T_h)\Delta T + (\Delta T)^2}{T_h(T_h + \Delta T)} < 0. \text{ So } \varepsilon_{C1} < \varepsilon_{C2}.$$
67. (a) 13 **(b)** No, $\text{COP}_C = 11$
69. 1.1×10^4 bricks
71. 2.09×10^3 J
73. -1.00 J/K
75. 1.43×10^3 kg/s

CHAPTER 13 MULTIPLE CHOICE

1. (b)
3. (b)
5. (a)
7. (b)
9. (b)
11. (d)
13. (d)
15. (a)
17. (c)

CHAPTER 13 CONCEPTUAL QUESTIONS

1. (a) Four times as large (b) twice as large
3. $T/4, T/2$
5. For a spring-mass system, the period is independent of gravitational acceleration. So the answer is no. For a pendulum, the period *does* depend on gravitational acceleration. The period actually increases on the Moon due to the lower value of gravitational acceleration. Thus the answer for a pendulum is yes.
7. $T = 2\pi\sqrt{L/g} \propto \sqrt{L}$, so the period is $\sqrt{2}$ times as large.
9. For the sine function, the initial position is zero because $\sin 0 = 0$. The initial velocity is a maximum because $\cos 0 = 1$ (its maximum value). The acceleration is zero because it is proportional to $\sin \omega t$, which is zero. For the cosine function, things are reversed. The initial position and acceleration are at their maximum values, but the initial velocity is zero.
11. The one on the top is transverse and the one on the bottom is longitudinal.
13. This is a longitudinal wave, because the direction of the wave motion (horizontal across the field) is parallel to the direction of the wheat plant vibration.
15. Reflection (this is called *echolocation*), because the sound is reflected by the prey
17. In principle, any frequency higher than the normal fundamental could be generated by pressing the fingers along the bridge to shorten the string. Lower frequencies cannot be generated because we cannot lengthen the string.
19. If the swing is pushed at a frequency of $f_1/2$, it is pushed only once (in one direction near equilibrium) every other oscillation. It is a smooth action since the swing is pushed in phase with its oscillations, and the amplitude of the motion can build up. Similarly, if it pushed at a frequency of $f_1/3$, it is pushed only once every third oscillation. If it is pushed at a frequency of $2f_1$, it is pushed twice per oscillation, but always at the extreme point in the swing.

CHAPTER 13 EXERCISES

1. $12A$
3. 2.0 Hz
5. Decrease of 2.0 s
7. 0.13 m/s
9. 0.50 m/s
11. (a) The correct choice is (1) $x = 0$, because at $x = 0$ there is no elastic potential energy, so all the energy of the system is kinetic, thus maximum speed. (b) 2.0 m/s
13. 9.26 cm
15. 1.4 m/s

17. (a) 0.38 m (b) 8.5×10^{-3} m
19. (a) 0.140 m (b) 10.0 cm (original position)
21. (a) 1.7 s (b) 0.57 Hz
23. 444 N/m
25. (a) $y = A \cos \omega t$ (b) $y = -A \cos \omega t$
27. (a) 5.0 cm (b) 10 Hz (c) 0.10 s
29. 0.33 Hz
31. $\omega = \sqrt{k/m} \therefore k = \omega^2 m$
$\therefore E = \frac{1}{2}kA^2 = \frac{1}{2}m\omega^2 A^2$
33. (a) 0.188 m (b) 3.00 m/s^2
35. (a) The correct choice is (2) since for constant mass $T \propto \sqrt{1/k}$. If $k \to k/2, T \to T\sqrt{2}$. (b) $T_2 = 3.5$ s
37. (a) 4.8 cm (b) 4.4 cm/s (c) -1.2 cm/s^2
39. The frequency f_1 of the heavy-mass system is $f_1 = \dfrac{1}{2\pi}\sqrt{k_1/m_1}$ and that of the light-mass system is $f_2 = \dfrac{1}{2\pi}\sqrt{k_2/m_2} = \dfrac{1}{2\pi}\sqrt{2k_1/(m_2/2)} = \dfrac{1}{2\pi}\sqrt{4k_1/m_1} = 2\left(\dfrac{1}{2\pi}\sqrt{k_1/m_1}\right) = 2f_1$. Thus $f_2/f_1 = 2$.
41. (a) The correct answer is (1), increase, because $T = 2\pi\sqrt{L/g}$, a smaller value of g means a longer period, T. (b) 4.9 s
43. (a) $y = (-0.10$ m$) \sin(10\pi t/3)$ (b) 27 N/m
45. (a) 0.15 m (b) 0.30 m/s (c) 2.5 kg
47. (a) $y = (0.100$ m$) \cos(6t)$ (b) The mass starts at $y = 0.100$ m with an initial velocity of zero. It will begin moving downward ($-y$-direction) because v will be negative just after $t = 0$. (c) 1.05 s (d) 18.0 N
49. 2.0 Hz
51. 2.59 m
53. 6.00 km
55. (a) Low frequency corresponds to a longer wavelength. Thus AM frequencies are associated with (1) longer wavelengths. (b) AM: $\lambda_{min} = 188$ m, $\lambda_{max} = 545$ m; FM: $\lambda_{min} = 2.78$ m, $\lambda_{max} = 3.41$ m
57. 6.0 km/s
59. (a) 0.20 s (b) 0.40 s
61. 200 Hz
63. 170 Hz
65. 7.5 m
67. 478 N
69. The third harmonic will be set up in the string.
71. (a) The correct answer choice is (2). A's frequency is $f_B = \dfrac{1}{2L}\sqrt{F_B/\mu_B}$ and it can be written in terms of B's:
$f_A = \dfrac{1}{2L}\sqrt{2F_B/(\mu_B/2)} = \dfrac{1}{2L}\sqrt{4F_B/\mu_B} = 2\left(\dfrac{1}{2L}\sqrt{F_B/\mu_B}\right) = 2f_B$. (b) $f_A = 100$ Hz, $f_B = 50$ Hz
73. (a) The answer is (3) shortened, because a shorter string can support a shorter wavelength and therefore a higher frequency given that the wave speed is a constant. (b) 594 Hz
75. $M_1 = 0.22$ kg, $M_2 = 0.055$ kg, $M_3 = 0.024$ kg, $M_4 = 0.14$ kg
77. (a) 2.00 s (b) 1.86 s (c) 2.17 s (d) In free fall, the effective value of g is zero, so the period would be infinite. In other words, it would not swing. (e) 2.00 s

79. (a) 0.31 m/s (b) 1.0 m/s^2 (c) 15 N
81. (a) 1.21 m (b) 0.301 m/s (c) 0.0746 m/s^2

CHAPTER 14 MULTIPLE CHOICE

1. (b)
3. (a)
5. (c)
7. (b)
9. (c)
11. (c)
13. (b)
15. (b)
17. (a)

CHAPTER 14 CONCEPTUAL QUESTIONS

1. Some insects produce sounds with frequencies that are not all in our audible range.
3. They arrive at the same time because sound is not dispersive, i.e., speed does not depend on frequency.
5. Ultrasonic refers to sound having frequencies higher than the limit of human hearing ($f > 20$ kHz), while supersonic refers to speeds greater than the speed of sound.
7. The correct choice is (1) by a factor of 2.
9. Yes. For any intensity *below* the threshold intensity of hearing, β is negative.
11. (a) No, because there is no relative velocity between the observer and the source. (b) An increasing frequency is heard since the source is moving toward the observer, and its speed is increasing.
13. It uses echolocation to measure the cloud location and the Doppler effect to measure their motion (direction and speed).
15. Yes. The air inside the plane is moving with the plane, so the speed of the plane has no effect on what the pilot hears. Sound inside propagates as usual.
17. By pressing on the frets, the player reduces the length of the string that is vibrating. This decreases the wavelength of the standing wave, thereby increasing its frequency and allowing the player to play higher and higher notes. The spacing of successive frets is designed so that the *fractional* (or percent) change in the frequency is the same from one fret to another to preserve the musical intervals of the notes of the scale. As the string gets shorter, the fractional change in frequency remains the same, but the absolute change gets smaller because it is the same fraction of a smaller and smaller length. If the frets were equally spaced, the note *changes* from fret to fret would be different as the string got shorter and shorter.
19. The increased temperature would cause two things to happen: thermal expansion of the pipe and an increase in the speed of sound in the pipe. The pipe would get longer, which would increase the wavelength of the fundamental and other harmonics. A greater wavelength would give *lower* frequencies for the sounds. The increased speed of sound, on the other hand, would result in *higher* frequencies for the sounds. Of these two effects, thermal expansion is usually quite small, so the dominant effect would be due to the increased speed of sound. The net result, then, would be to produce *higher* frequencies.

21. By rubbing on the rim, vibration is set up in the glass and hence in the air. The air above the water behaves like an organ pipe. When the frequency of vibration matches one of the resonant (standing wave) frequencies of the air column, standing waves are set up in this air column, producing the sounds we hear.

CHAPTER 14 EXERCISES

1. (a) 337 m/s **(b)** 343 m/s
3. 1.5×10^3 m
5. If the SI units are correct, then the dimensions are also correct. For a liquid:
$$\sqrt{\frac{N/m^2}{kg/m^3}} = \sqrt{\frac{N \cdot m}{kg}} = \sqrt{\frac{kg \cdot m^2/s^2}{kg}} = \sqrt{\frac{m^2}{s^2}}$$
= m/s. Since Y has the same units as B, the expression for the solid is also dimensionally correct.
7. (a) The correct answer is (1), increases, because the speed of sound increases with temperature **(b)** +0.047 m
9. (a) 7.5×10^{-5} m **(b)** 1.5×10^{-2} m
11. (a) 1.08 s **(b)** 1.04 s
13. 90 m
15. (a) 0.107 m **(b)** 1.43×10^{-4} s **(c)** 4.29×10^{-4} m
17. (a) The correct answer is (1), less than double, because the total time is the sum of the time it takes for the stone to hit the ground (free fall) and the time it takes sound to travel back that distance. While the time for sound *is* proportional to the distance, the time for free fall is *not*. ($t = \sqrt{2d/g}$ from Chapter 2). Thus doubling the distance will only increase the free-fall time by a factor of $\sqrt{2}$. **(b)** 1.0×10^2 m **(c)** 8.7 s
19. 4.5 %
21. (a) The correct answer is (4) 1/9, because I is inversely proportional to the square of R. Tripling R will reduce I by $1/3^2 = 1/9$. **(b)** 1.4 times
23. (a) 3.0 Hz **(b)** Not enough information given
25. (a) 100 dB **(b)** 60 dB **(c)** −30 dB
27. (a) The intensity level (1) increases but will not double. **(b)** 5.0 W: 96 dB; 10 W: 99 dB
29. 24 W/m²
31. 2.0×10^{-5} W/m²
33. 1.01×10^{-3} W/m²
35. (a) The correct choice is (5), none of the preceding. **(b)** Increases by 6 dB
37. (a) 63 dB **(b)** 83 dB **(c)** 113 dB
39. (a) The intensity level is (2) between 40 and 80 dB. **(b)** 43 dB
41. 379 m
43. No, the manager does *not* achieve his goal. The new intensity level is 37 dB.
45. 10.6 m
47. 4 Hz
49. 840 Hz
51. (a) 431 Hz **(b)** 373 Hz
53. 58 km/h
55. Approaching: 755 Hz; receding: 652 Hz
57. (a) The half angle (3) decreases as M increases. **(b)** 42°
59. (a) 100 Hz **(b)** 6 Hz
61. (a) 36.3 kHz **(b)** 37.6 kHz **(c)** Yes

63. (a) It is a closed pipe. **(b)** 0.675 m
65. (a) 3.4 kHz **(b)** 3.4 kHz **(c)** The frequency is lower.
67. For the open pipe, 330 Hz; for the closed pipe, 165 Hz.
69. 264 m/s
71. (a) The mouthpiece is at (2) an antinode. **(b)** 0.655 m **(c)** 0.390 m
73. 0.249 m and 0.251 m
75. (a) $I_1 = 1.59$ W/m²; $I_2 = 0.398$ W/m²; $I_{total} = 1.99$ W/m² **(b)** $\beta_1 = 122$ dB; $\beta_2 = 116$ dB; $\beta_{total} = 123$ dB **(c)** About 15 minutes
77. (a) The correct answer is (1), yes, because the observer will hear beats between the source and the reflection. **(b)** 1.03×10^3 Hz **(c)** 12 Hz
79. (a) The answer is (3) the intensities are the same. **(b)** Both are 0.318 W/m². **(c)** Each speaker is 115 dB and the total is 118 dB.

CHAPTER 15 MULTIPLE CHOICE

1. (c)
3. (c)
5. (d)
7. (a)
9. (c)
11. (c)
13. (a)
15. (d)
17. (a)
19. (b)
21. (c)

CHAPTER 15 CONCEPTUAL QUESTIONS

1. There would be no effect because it is simply an arbitrary sign convention.
3. If an object is positively charged, its mass decreases, because it loses electrons. If an object is negatively charged, its mass increases, because it gains electrons.
5. No, the charges simply change location. There is no gain or loss of electrons.
7. The water would bend the same way as with the balloon. In this case, the positive charges in the rod would attract the negative ends of the water molecules, causing the stream to bend toward the rod.
9. A positive charge can be placed in the middle of the two electrons so that each will experience a repulsive force from the other and an attractive force from the positive charge. With the proper amount of positive charge, the two forces on each electron will cancel.
11. (a) The object is positively charged because the downward repulsion of the nearby positive charge of the dipole is greater than the upward attraction of the more distant negative charge of the dipole. **(b)** The dipole would accelerate downward because the downward attraction on the nearby negative end would be greater than the upward repulsion of the more distant positive end
13. It is determined by the relative density or spacing of the field lines. The closer the lines are, the greater the field magnitude.
15. If a positive charge is at the center of the spherical shell, the electric field is *not* zero

inside. The field lines run radially outward to the inside surface of the shell where they stop at the induced negative charges on this surface. The field lines reappear on the outside shell surface (positively charged) and continue radially outward as if emanating from the point charge at the center. If the charge were negative, the field lines would reverse their directions.
17. (a) Yes, this is possible, for example, when the electric fields are equal in magnitude and opposite in direction at some location. For example, at the midway point in between and along a line joining two charges of the same type and magnitude, the electric field is zero. **(b)** No, this is not possible.
19. At very large distances, the object looks very small—like a point. So the electric field pattern looks like that due to a point charge located at the (point) object.
21. The surface must be spherical.
23. Note that electric field is zero inside the metallic slab. Since charges are mobile, the negative charges are attracted toward the upper portion of the slab, while the positive charges move toward the lower portion. The amount of charge induced on each side of the slab is the same in magnitude as that on each of the plates.

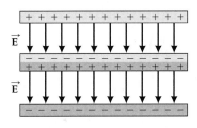

25. The net charges are equal in magnitude but opposite in sign.

CHAPTER 15 EXERCISES

1. -1.6×10^{-13} C
3. (a) 6.40×10^{-19} C **(b)** 2 electrons
5. (a) (1) Positive, because of the conservation of charge. When one object becomes negatively charged, it gains electrons. These electrons must be lost by another object, which becomes positively charged. **(b)** $+4.8 \times 10^{-9}$ C, 2.7×10^{-20} kg **(c)** 2.7×10^{-20} kg
7. 5.15×10^{-11} C
9. (a) 1 **(b)** 1/4 **(c)** 1/2
11. (a) 5.8×10^{-11} N **(b)** zero
13. 2.24 m
15. (a) $x = 0.25$ m **(b)** Nowhere **(c)** $x = -0.94$ m for $\pm q_3$
17. (a) The two forces on the electron add numerically because they are in the same direction. Putting the negative charge on the left and the positive charge on the right, the net force points to the right (+). At 5.0 cm: $F_{net} = 2.7 \times 10^{-18}$ N. At 10.0 cm, the net force is 8.1×10^{-19} N. At 15.0 cm it is 5.8×10^{-19} N. At 20.0 cm it is 8.1×10^{-19} N. At 25.0 cm it is 2.7×10^{-18} N. **(b)** According to the graph (sketched qualitatively below), the electron

feels the least force midway (or at least approximately so) between the two charges.

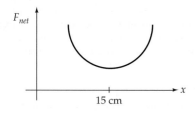

19. (a) 96 N, 39° below positive x-axis **(b)** 61 N, 84° above negative x-axis
21. (a) E is *inversely* proportional to the square of the distance, thus its magnitude is (2) decreased. **(b)** 2.5×10^{-5} N/C
23. 2.9×10^5 N/C
25. 1.2×10^{-7} m away from the charge
27. 1.0×10^{-7} N/C upward, 5.6×10^{-11} N/C downward
29. $\vec{E} = (2.2 \times 10^5 \text{ N/C})\hat{x} + (-4.1 \times 10^3 \text{ N/C})\hat{y}$
31. 5.4×10^6 N/C toward the charge of $-4.0 \ \mu C$
33. 3.8×10^7 N/C in the $+y$-direction
35. (a) 1.5×10^{-5} C/m^2 **(b)** 3.4×10^{-7} C
37. $\vec{E} = (-4.4 \times 10^6 \text{ N/C})\hat{x} + (7.3 \times 10^7 \text{ N/C})\hat{y}$
39. (a) (1) Negative due to induction **(b)** Zero **(c)** $+Q$ **(d)** $-Q$ **(e)** $+Q$
41. (a) Zero **(b)** kQ/r^2 **(c)** Zero **(d)** kQ/r^2

43.
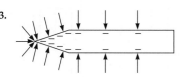

45. -6 lines, or net of 6 lines entering it
47. (a) 10 field lines entering (negative) **(b)** Zero field lines
49. (a) The field is to the right. **(b)** 8.55×10^3 N/C
51. (a) Positive on right plate and negative on left plate **(b)** From right to left **(c)** 1.13×10^{-13} C
53. (a) The forces $\vec{F}_+$ and $\vec{F}_-$ on either end of the dipole create a torque on it, which tends to rotate its dipole moment $\vec{p}$ in a direction parallel to the electric field.

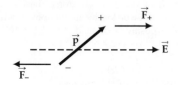

(b) Add the torques due to the two forces, realizing that the torques are equal. Calling d the distance between the charges and θ the angle between $\vec{p}$ and $\vec{E}$, then $\tau = 2F_+(d/2)\sin\theta = qdE\sin\theta = pE\sin\theta$. **(c)** $\vec{F}_{net} = 0$ because the two forces are of equal magnitude but opposite in direction. **(d)** The torque is a maximum when $|\sin\theta| = 1$, which is when $\theta = 90°$ or $270°$. When $\theta = 0°$ or $180°$, $\sin\theta = 0$ and the torque is at its minimum value of zero.
55. (a) 6.78 m/s^2 in the $-x$-direction **(b)** 1.97×10^{-12} m · N counterclockwise

CHAPTER 16 MULTIPLE CHOICE

1. **(d)**
3. **(b)**
5. **(b)**
7. **(a)**
9. **(b)**
11. **(b)**
13. **(e)**
15. **(a)**
17. **(c)**
19. **(a)**
21. **(b)**
23. **(d)**
25. **(c)**
27. **(b)**
29. **(b)**

CHAPTER 16 CONCEPTUAL QUESTIONS

1. (a) Electrical potential is the electrostatic potential energy *per unit charge.* **(b)** No difference
3. Approaching a negative charge means moving toward a region of lower electric potential. Positive charges tend to move toward regions of lower potential, thus losing potential energy and gaining kinetic energy (speeding up).
5. It doesn't accelerate. It is in a region where the potential is constant and so feels no force.
7. The ball would accelerate in the direction from the beach to the ocean (from higher potential energy to lower potential energy).
9. It takes zero work, because there is no change in kinetic or potential energy.
11. (a) Cylindrical **(b)** Near the outer surface **(c)** Near the inner surface
13. (a) 1.60×10^{-13} J **(b)** It would double.
15. The electrostatic potential energy stored in the system is $U_c = Q^2/2C$, with $C = \varepsilon_o/Ad$. With Q fixed, increasing d decreases C, and therefore increases U_c.
17. (a) If $d \to d' = 3d$, then $C \to C' = \varepsilon_o/Ad' = \varepsilon_o/[A(3d)] = C/3$. Since $Q = CV$, with V fixed by connecting it to the battery, then $V' = V$, and $Q' = Q/3$. **(b)** $U'_C = U_C/3$ **(c)** $E' = E/3$
19. (a) The capacitance increases since $C = \kappa C_o > C_o$. **(b)** The potential difference decreases since $V = Q/C = V_o/\kappa < V_o$. **(c)** The electric field decreases because $E = V/d = E_o/\kappa < E_o$.
21. For two capacitors in series, $V_1 = Q/C_1$ and $V_2 = Q/C_2$. If $V_1 = V_2$ then it must be that $C_1 = C_2$. In a parallel connection, $V_1 = V_2$ regardless of the values of C_1 and C_2.
23. (a) Connect them in parallel to get maximum equivalent capacitance. **(b)** Connect them in series to get minimum equivalent capacitance.

CHAPTER 16 EXERCISES

1. 1.0 cm
3. (a) 2.7 μC **(b)** Negative to positive
5. (a) 5.9×10^5 m/s, down **(b)** Lose potential energy

7. (a) (2) 3, because electric potential is inversely proportional to the distance **(b)** 0.90 m **(c)** -6.7 kV
9. (a) Gains 6.2×10^{-19} J **(b)** Loses 6.2×10^{-19} J **(c)** Gains 4.8×10^{-19} J
11. 1.1 J
13. (a) $+0.27$ J **(b)** No
15. -0.72 J
17. (a) 3.1×10^5 V **(b)** 2.1×10^5 V
19. (a) (3) A lower, because electrons have a negative charge; they move toward higher potential regions where they have lower potential energy. **(b)** 4.2×10^7 m/s **(c)** 6.0×10^{-9} s
21. 70 cm
23. 8.3 mm from the positive plate
25. (a) The correct answer is (1) since V is inversely proportional to r. **(b)** $+297$ eV
27. (a) 2.0×10^7 eV **(b)** 2.0×10^4 keV **(c)** 20 MeV **(d)** 2.0×10^{-2} GeV **(e)** 3.2×10^{-12} J
29. 6.2×10^7 m/s (proton), 4.4×10^7 m/s (alpha)
31. (a) 3.5 V, 1.1×10^6 m/s **(b)** 4.1 kV, 3.8×10^7 m/s **(c)** 5.0 kV, 4.2×10^7 m/s
33. (a) $\Delta V = +0.40$ V. This potential difference is positive, which means the electric field is opposite the direction moved. **(b)** Same as (a), except $\Delta V = -0.40$ V because you moved toward the lower potential plate or in the direction of the electric field. **(c)** $\Delta V = 0$ because you stay the same distance from the plates and are moving along an equipotential surface. These results mean that the electric field has no component parallel to the plates, thus it points perpendicularly from positive to negative plate.
35. 2.4×10^{-5} C
37. 0.418 mm
39. (a) 4.54×10^{-9} C **(b)** 2.72×10^{-8} J **(c)** 2.29×10^3 V/m
41. 2.2 V
43. (a) 2.2×10^4 V/m **(b)** 1.1×10^{-5} C **(c)** 5.7×10^{-4} J **(d)** $E = 6.7 \times 10^4$ V/m, $\Delta Q = 0$, $\Delta U_C = -1.7 \times 10^{-3}$ J
45. 3.1×10^{-9} C; 3.7×10^{-8} J
47. (a) 2.4 **(b)** The answer is (2). The stored energy under constant charge conditions is inversely related to the capacitance ($U_C = Q^2/2C \propto 1/C$). Since the capacitance increases with the dielectric insertion ($C = \kappa C_o > C_o$), inserting it increased the capacitance, and thus *decreased* the stored energy. **(c)** -6.3×10^{-5} J
49. (a) 0.24 μF **(b)** 1.0 μF
51. (a) The correct answer is (1). The equivalent capacitances for two capacitors connected in series and parallel are $C_{series} = \dfrac{C_1 C_2}{C_1 + C_2}$ and $C_{parallel} = C_1 + C_2$. The energy stored in a capacitor system is $U = \frac{1}{2}CV^2$. Since $C_{parallel} > C_{series}$, $U_{parallel} > U_{series}$. **(b)** In the series connection, the energy supplied by the battery is 1.5×10^{-5} J. In the parallel connection, it is 7.6×10^{-5} J.
53. (a) (3) $Q/3$, because $Q_{total} = Q_1 + Q_2 + Q_3$. Also $Q_1 = Q_2 = Q_3$ because the capacitors have the same capacitance. Therefore, each capacitor has only 1/3 of the total charge. **(b)** 3.0 μC **(c)** 9.0 μC

55. Max. 6.5 μF; min. 0.67 μF

57. $Q_1 = 2.4 \ \mu$C and $U_1 = 7.2 \ \mu$J. The answers for C_2 are the same as C_1 since it has the same capacitance as C_1 and they are in parallel. $Q_3 = 1.2 \ \mu$C and $U_3 = 3.6 \ \mu$J; $Q_4 = 3.6 \ \mu$C and $U_4 = 11 \ \mu$J

59. **(a)** The acceleration due to gravity can be neglected [see (c)]. The electron's initial kinetic energy is 4.66×10^{-18} J. If the electron did reach the bottom of the tube, the electric field would have to do -1.2×10^{-17} J on it. Since this is more than the initial kinetic energy of the electron, it cannot reach the bottom. **(b)** 0.306 m or 30.6 cm from the bottom of the well **(c)** $F_g = 8.93 \times 10^{-31}$ N upward; $F_e = -2.4 \times 10^{-17}$ N upward. Since $|F_e| \gg F_g$, the gravitational force can be ignored.

61. **(a)** 0.17 μF **(b)** 2.1 μC **(c)** $V_3 = 6.9$ V and $V_1 = V_2 = 12$ V $- 6.9$ V $= 5.1$ V **(d)** $U_1 = 2.0 \ \mu$J; $U_2 = 3.3 \ \mu$J; $U_3 = 7.1 \ \mu$J

63. **(a)** 0.030 eV **(b)** $E \approx \Delta V / \Delta x = 3.0 \times 10^6$ V/m. The potential difference is $\Delta V = V_{in} - V_{out} = +30.0$ mV, so $V_{in} > V_{out}$. Thus the electric field points *outward* across the membrane. **(c)** 4.8×10^{-13} N **(d)** $E \approx \Delta V / \Delta x = 7.0 \times 10^6$ V/m. Now ΔV is negative, thus $V_{out} > V_{in}$, and the field points *inward* across the membrane.

65. (a) 4.65×10^{-11} F **(b)** $E_1 = 2.9 \times 10^3$ V/m and $E_2 = 1.8 \times 10^3$ V/m

67. **(a)** The same charge of 20 μC is on each capacitor since they are in series. **(b)** Letting "1" be the larger capacitor and "2" be the smaller: $V_1 = 3.5$ V and $V_2 = 8.6$ V. **(c)** The *charge* remains the *same* on each capacitor since there is no place for it to go. Therefore, $\Delta Q = 0$ and $V_1 = 3.5$ V (no change), so $\Delta V_1 = 0$. However, the new voltage across 2 is $V_2 = 4.3$ V, thus $\Delta V_2 = -4.3$ V.

CHAPTER 17 MULTIPLE CHOICE

1. **(b)**
3. **(b)**
5. **(c)**
7. **(c)**
9. **(a)**
11. **(c)**
13. **(a)**
15. **(a)**
17. **(b)**
19. **(d)**
21. **(d)**
23. **(b)**

CHAPTER 17 CONCEPTUAL QUESTIONS

1. Although electrode A is negative, it is *less negative* than B and is therefore at a higher potential than B. We say that electrode A is at a positive potential *relative to B*.

3. No. When current is flowing through the battery, the terminal voltage will be *less than* 12 V due to the voltage drop across the battery's internal resistance.

5. **(a)** Upward **(b)** Downward **(c)** Upward

7. Electrons flow from A to B inside the battery, but in the wire the flow is from B to A, thereby completing a closed loop and ensuring continuous current.

9. Write the relationship between voltage and current as $V = (R)I$. If resistance is constant, then the plot will be a straight line of the form $y = mx$, where m is the slope. Thus the slope is the resistance. Therefore, the shallower slope implies lower resistance.

11. **(a)** Same **(b)** One-quarter the current

13. Since $R = \rho \dfrac{L}{A} = \rho \dfrac{L}{\pi(D/2)^2}$, if L is changed to $L/2$, then the denominator must also be reduced by $\frac{1}{2}$ to keep the resistance the same. Therefore, $(D/2)^2$ should become $\frac{1}{2}(D/2)^2$, which means that D should become $D/\sqrt{2}$. In other words, the diameter must be reduced to $1/\sqrt{2}$ or 0.707, or about 71% of its original value.

15. Since $P = V^2/R$, the bulb of higher power has a smaller resistance, which means *thicker* wire if the length is the same. So the wire in the 60-W bulb would be thicker.

17. Current is also affected by the resistance. Rewrite the relationship as $R = V^2/P$ or $R \propto 1/P$ if V is constant. Thus a high-wattage bulb has *less* resistance than a low-wattage bulb.

CHAPTER 17 EXERCISES

1. **(a)** 4.5 V **(b)** 1.5 V
3. **(a)** 24 V **(b)** Two 6.0-V in series, together in parallel with the 12-V
5. **(a)** The answer is (2) the same, because the total voltage of identical batteries in parallel is the same as the voltage of each individual battery, and the total voltage of the batteries in series is the sum of the voltages of each individual battery. Each arrangement has one parallel and one series so they have the same total voltage. **(b)** 3.0 V, 3.0 V
7. 0.25 A
9. **(a)** 0.30 C **(b)** 0.90 J
11. 56 s
13. **(a)** Since both conventional currents are to the left, they add to create a net current to the left, so the answer is (2). **(b)** 1.8 A to the left **(c)** 1.5 A to the left **(d)** 3.3 A to the left
15. **(a)** 11.4 V **(b)** 0.32 Ω
17. 1.0 V
19. **(a)** The answer is (1) a greater diameter, because aluminum has a higher value of resistivity. Its area (and thus its diameter) must be greater, if the length of the wire is the same, to have the same resistance as copper. The relationship $R = \rho L/A$ shows this mathematically. **(b)** 1.29
21. $1.3 \times 10^{-2} \ \Omega$
23. **(a)** 4 **(b)** 4
25. **(a)** 0.13 Ω **(b)** 0.038 Ω
27. **(a)** 4.6 mΩ **(b)** 8.5 mA
29. **(a)** 0.054 m or 5.4 cm **(b)** 0.11 mΩ
31. **(a)** The answer is (1) greater than, because after the stretch, the length L is increased and the cross-sectional area A must therefore decrease (to keep the volume constant), so R increases due to the changes in both factors according to $R = \rho L/A$. **(b)** 1.6
33. **(a)** 7.8 Ω **(b)** 0.77 A **(c)** 16.4 °C
35. 144 Ω
37. 2.0×10^3 W
39. 1.2 Ω

41. **(a)** The answer is (4) because $P_2 = V_2^2/R = (V_1/2)^2/R = V_1^2/(4R) = P_1/4$. **(b)** 5.63 W
43. **(a)** 4.3 kW **(b)** 13 Ω **(c)** Non-ohmic
45. **(a)** 58 Ω **(b)** 86 Ω
47. **(a)** 0.600 kWh **(b)** 9¢ **(c)** 417 days
49. **(a)** 0.15 A **(b)** $1.4 \times 10^{-4} \ \Omega \cdot$ m **(c)** 2.3 W
51. **(a)** 1.1×10^2 J **(b)** 6.8 J
53. **(a)** 22 Ω **(b)** 5.6 A.
55. $R_{120}/R_{60} = 4/3$
57. **(a)** Costs rounded to nearest dollar: central air, $130; blender, $0; dishwasher, $1; microwave oven, $1; refrigerator, $6; stove (oven and burners), $13; color TV, $1. **(b)** The total cost using the rounded costs is $152. The percents, to two significant figures, are central air, 86%; blender, 0%,; dishwasher, 0.66%; microwave oven, 0.66%; refrigerator, 4.0%; stove, 8.6%; and color TV, 0.66%. **(c)** Central air, $I = 41.7$ A, $R = 2.88 \ \Omega$; blender, $I = 6.7$ A, $R = 18 \ \Omega$; dishwasher, $I = 10.0$ A, $R = 12.0 \ \Omega$; microwave oven, $I = 5.2$ A, $R = 33 \ \Omega$; refrigerator, $I = 4.2$ A, $R = 28 \ \Omega$; stove (oven), $I = 37.5$ A, $R = 3.20 \ \Omega$; stove (top burners), $I = 50.0$ A, $R = 2.40 \ \Omega$; and color TV, $I = 0.83$ A, $R = 150 \ \Omega$.
59. **(a)** There are two temperatures: 117 °C and -72.6 °C. **(b)** At the high temperature the ratio is 0.664 and at the lower it is 1.94. **(c)** At the high temperature the ratio is 0.642 and at the lower it is 2.31.
61. **(a)** It is not ohmic. **(b)** At the lower emf: 5.80 Ω; at the higher emf: 3.80 Ω **(c)** At the lower emf it is 29.0:1. At the higher emf it is 19.0:1.
63. **(a)** 0.0833 A or 83.3 mA **(b)** 1.44 kΩ **(c)** Assuming there are 10^8 households in the United States (half have the feature), and the output of a power plant is 10^9 W (1 GW), then about half of a power plant is needed to supply this power.
65. **(a)** 1.6 kΩ **(b)** 72.0 W **(c)** -63.1 W
67. **(a)** 400 A **(b)** $4.5 \times 10^{-3} \ \Omega$ **(c)** 1.8 V **(d)** 250 kV
69. **(a)** $2.82 \times 10^{14} \ \Omega$ **(b)** 3.54×10^{-11} A **(c)** 2.54×10^3 m or 2.54 km

CHAPTER 18 MULTIPLE CHOICE

1. **(b)**
3. **(b)**
5. **(a)**
7. **(c)**
9. **(a)**
11. **(d)**
13. **(b)**
15. **(c)**
17. **(b)**
19. **(b)**
21. **(a)** and **(b)**
23. **(c)**
25. **(b)** and **(d)**

CHAPTER 18 CONCEPTUAL QUESTIONS

1. No, not generally. However, if all resistors are equal, the voltages across them are the same.

3. No, not generally. However, if all resistors are equal, the currents in them are the same.

5. If they are in series, the effective resistance will be closer in value to that of the large resistance because $R_s = R_1 + R_2$. If $R_1 \gg R_2$, then $R_s \approx R_1$. If they are in parallel, the effective resistance will be closer in value to that of the small resistance because $R_p = R_1R_2/(R_1 + R_2)$. If $R_1 \gg R_2$ then $R_p \approx R_1R_2/R_1 = R_2$.

7. (a) The third resistor has the largest current, because the total current through the two other resistors is equal to the current through the third resistor. **(b)** The third resistor also has the largest voltage, because the current through it is the largest and all the resistors have the same resistance value ($V = IR$). **(c)** The third resistor has the largest power output because it has both the largest current and largest voltage.

9. Not necessarily. If two batteries of unequal emfs are connected with opposite polarity in series with a resistor, the larger battery will force current to enter the positive terminal of the smaller battery.

11. The 60-W bulb has a higher resistance than the 100-W bulb. When these are in series, they have the same current. Therefore, the 60-W bulb will have a higher voltage. Thus, the 60-W bulb has more power because $P = IV$.

13. In series, the current is the same in all resistors. Since each resistor's voltage drop is related to its resistance by $V_i = IR_i$, it is clear that the greater the resistance the larger the voltage drop.

15. During the charging of a the capacitor, its charge as a function of time is $Q(t) = Q_o(1 - e^{-t/\tau})$. After one time constant, τ, has elapsed, the charge is $Q(\tau) = Q_o(1 - e^{-1}) \approx 0.632Q_o$, so the time it takes to charge up to $0.25Q_o$ is *less than* one time constant. When discharging, the capacitor's charge is $Q(t) = Q_oe^{-t/\tau}$. After one time constant, the charge is $Q(\tau) = Q_oe^{-1} \approx 0.368Q_o$, so the time to discharge to $0.25Q_o$ is *more than* one time constant.

17. (a) An ammeter has very low resistance, so if it were connected in parallel in a circuit, the circuit current would be very high and its galvanometer could burn out. **(b)** A voltmeter has very high resistance, so if it were connected in series in a circuit, it would read the voltage of the source because it has the highest resistance (most probably) and therefore the most voltage drop among the circuit elements. The circuit current would drop close to zero.

19. An ammeter is used to measure current when it is in series with a circuit element. If it has very small resistance, there will be very little voltage across it, so it will not affect the voltage across the circuit element, nor its current.

21. (a) The voltmeter could be connected across any one of the resistors.

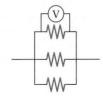

(b) Same as (a)
(c) The voltmeter is connected in parallel with whole series.

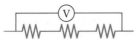

(d) The voltmeter should be connected across just the resistor whose potential difference is to be determined.

23. No, a high voltage can produce high harmful current, even if resistance is high because current is caused by voltage (potential difference).

25. It is safer to jump. If you step off the car one foot at a time, there will be a high voltage between your feet. If you jump, the voltage (i.e., potential *difference*) between your feet is zero because your feet will be at the same potential all the time.

CHAPTER 18 EXERCISES

1. (a) In series, 60 Ω **(b)** In parallel, 5.5 Ω
3. 30 Ω
5. (a) 30 Ω **(b)** 0.30 A **(c)** 1.4 W
7. (a) (1) $R/4$. Each shortened segment has a resistance of $R/2$ because resistance is proportional to length (Chapter 17). Then two $R/2$ resistors in parallel gives $R/4$. **(b)** 3.0 $\mu\Omega$
9. 1.0 A (for all); $V_{8.0} = 8.0$ V; $V_{4.0} = 4.0$ V
11. 2.7 Ω
13. (a) 1.0 A; 0.50 A; 0.50 A **(b)** 20 V; 10 V; 10 V **(c)** 30 W
15. (a) There would be no change in I_1, I_2, and I_4. In the lower branch, the current will be halved since the new resistance is now doubled to 4 Ω. **(b)** $I_1 = I_2 = 3.0$ A and $I_4 = 6.0$ A. Originally, $I_3 = 6.0$ A, but now the new currents are $I_3' = I_5' = 3.0$ A.
17. (a) 0.085 A **(b)** 7.0 W, 2.6 W, 0.24 W, 0.41 W
19. (a) $I_2 = I_1 = 0.67$ A; $I_3 = 1.0$ A; and $I_4 = I_5 = 0.40$ A **(b)** $V_1 = 6.7$ V; $V_2 = 3.3$ V; $V_3 = 10$ V; $V_4 = 2.0$ V; and $V_5 = 8.0$ V **(c)** $P_1 = 4.4$ W; $P_2 = 2.2$ W; $P_3 = 10$ W; $P_4 = 0.80$ W; and $P_5 = 3.2$ W **(d)** $P_{total} = 21$ W
21. (a) Around loop 3 in the opposite direction from the figure, $-V_1 + I_2R_2 + I_1R_1 = 0$, or, multiplying by -1, $+V_1 - I_1R_1 - I_2R_2 = 0$, which is the same as the original result. **(b)** Around loop 1 (reverse), $-V_1 + I_3R_3 + V_2 + I_1R_1 = 0$. After multiplying by -1, it is the same as the equation for loop 1 (forward). Around loop 2 (reverse), $I_2R_2 - V_2 - I_3R_3 = 0$. Again, if we multiply by -1 on both sides, it is the same as the equation for loop 2 (forward).
23. $I_1 = 1.0$ A; $I_2 = I_3 = 0.50$ A
25. $I_1 = 0.33$ A (left); $I_2 = 0.33$ A (right)
27. $I_1 = 3.75$ A (up); $I_2 = 1.25$ A (left); $I_3 = 1.25$ A (right)
29. $I_1 = 0.664$ A (left); $I_2 = 0.786$ A (right); $I_3 = 1.450$ A (up); $I_4 = 0.770$ A (down); $I_5 = 0.016$ A (down); $I_6 = 0.664$ A (right)
31. (a) $V_C = 0$; $V_R = V_o$ **(b)** $V_C = 0.86V_o$; $V_R = 0.14V_o$ **(c)** $V_C = V_o$; $V_R = 0$

33. 0.693τ
35. (a) 0.86 mJ **(b)** 17 V. No, it is more than half of 24 V because the energy storage depends on the square of the voltage; thus, cutting it in half would reduce the energy to one-fourth the original value. **(c)** 480 kΩ **(d)** 35 μA
37. (a) 9.4×10^{-4} C **(b)** $V_C = 24$ V; $V_R = 0$
39. (a) The current is maximum at $t = 0$, when the switch is closed. Its magnitude is $I_o = 2.0 \times 10^{-6}$ A. **(b)** 0.86% **(c)** The maximum charge is 1.7×10^{-6} C. It attains its maximum value as $t \rightarrow \infty$, or in practical terms, when $t \gg \tau$. **(d)** 99% **(e)** 2.0×10^{-6} J
41. (a) (3) A multiplier resistor, because a galvanometer cannot have a large voltage across it; the large voltage has to be across a series resistor (multiplier). **(b)** 7.4 kΩ
43. 50 kΩ
45. (a) The voltmeter is connected in parallel with the 10-Ω resistor. $I = 0.60000$ A **(b)** The potential difference is 6.0000 V because the resistor and voltmeter are connected across the terminals of the battery.
47. (a) The ammeter is in series with R, the voltmeter is connected across this combination of R and the ammeter, and this total combination is connected across the power supply. The ammeter reads the current through R but the voltmeter does not read the voltage across R because there is some voltage drop across the ammeter. See figure. **(b)** Applying Ohm's law to R gives $V_R = RI$. But $V_R = V - IR_A$, so $V - IR_A = RI$. Solving: $R = (V/I) - R_A < V/I$. **(c)** For an ideal ammeter, $R_A = 0$, so $R = V/I$.

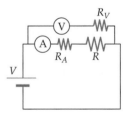

49. (a) $I/I' = 3.3 \times 10^{-4}$ **(b)** The current in the wire is 1200 A. This is certainly big enough to cause the circuit breaker to trip immediately.
51. (a) $I_1 = 2.6$ A (to the right); $I_2 = 1.7$ A (to the left); and $I_3 = 0.867$ A (down) **(b)** $V_1 = 5.1$ V; $V_2 = 6.9$ V; and $V_3 = 6.9$ V **(c)** $P_1 = 13$ W; $P_2 = 12$ W; and $P_3 = 5.9$ W
53. (a) $2.73R$ **(b)** The current in the two resistors closest to points A and B is 0.439 A. The current in the third resistor closest to points A and B is 0.322 A. The current in the fourth and fifth resistors closest to points A and B is 0.117 A. The current in the fourth resistor farthest away from points A and B is 0.0878 A. The current in the last three resistors farthest away from points A and B is 0.0293 A.
55. (a) 4.47 V **(b)** 0.447 A **(c)** $P_{batt} = 0.012$ W and $P_{circuit} = 2.00$ W (167 times more than the battery)
57. (a) The patient and physician are in series across a 120-V power source. **(b)** 58 kΩ
59. (a) 39.3 pF **(b)** 19.7 ns **(c)** Cut it in half to 1.13 mm. **(d)** Increase each side to 14.1 cm. **(e)** 2.00

CHAPTER 19 MULTIPLE CHOICE

1. (a)
3. (c)
5. (d)
7. (a)
9. (d)
11. (b)
13. (d)
15. (b)
17. (c)
19. (b)
21. (b)
23. (d)
25. (b)
27. (a)

CHAPTER 19 CONCEPTUAL QUESTIONS

1. The magnet would attract the unmagnetized iron bar when a pole end is placed at the center of its long side. If the end of the unmagnetized bar were placed at the center of the long side of the magnet, it would not be attracted.

3. (a) The iron filings get farther apart. Thus the magnetic field strength decreases with distance from the middle. (b) The magnetic field points up and down (parallel to the line along which the magnets are aligned). But we cannot tell if it points up or down from looking at the filing pattern alone.

5. Not necessarily, because there still could be a magnetic field. If the magnetic field and the velocity of the charged particle make an angle of either 0° or 180°, there is no magnetic force on the particle

7. (a) The fields should be uniform and of equal magnitude but point in opposite directions. The lower field should point into the paper and the upper field should point out of the paper. (b) The emerging kinetic energy is the same as the initial kinetic energy. Since the magnetic force is perpendicular to the velocity, it changes only the direction of the velocity, not its magnitude, so the kinetic energy does not change.

9. The magnetic force on the electron beam causes the deflection.

11. The electric force is qE and the magnetic force is qvB. Since both depend linearly on the charge, the selected speed, found from equating their magnitudes, is independent of the charge.

13. (a) If the electric field is reduced, the magnetic force will be greater than the electric force. Therefore, the positive charges in the velocity selector will be deflected upward and not enter B_2. (b) If B_1 is reduced, the electric force will be greater than the magnetic force. Then positive charges will be deflected downward and will not enter B_2.

15. It shortens because the coils of the spring attract each other due to the magnetic fields created in the coils. (Parallel wires with current in same direction will attract each other.)

17. Pushing the button in both cases completes the circuit. The current in the wires activates the electromagnet, causing the clapper to be attracted and ring the bell. However, this breaks the armature contact and opens the circuit. Holding the button causes this to repeat, and the bell rings continuously. For the chimes, when the circuit is completed, the electromagnet attracts the core and compresses the spring. Inertia causes it to hit one tone bar, and the spring force then sends the core in the opposite direction to strike the other bar.

19. (a) SI units of $IAB = (A)(m^2)[T] = \left(\frac{C}{s}\right)(m^2)\left[\frac{N}{C \cdot (m/s)}\right] = m \cdot N$ (b) The magnetic moment is out of the page, toward you.

21. The compass points downward so the magnetic field is downward at the center of the loop. The current direction is clockwise according to the right-hand source rule.

23. Not necessarily. The magnetic field in a solenoid depends on the current in it and the number of turns per unit length, not just the number of turns. For example, if the 200 turns is over 0.20 m and the 100 turns is over 0.10 m, then they will have the same turns per unit length and the same magnetic field.

25. The direction of the current should be counterclockwise to cancel the magnetic field of the outer loop. Its current should be smaller than 10 A, because the field created by a loop is *inversely* related to the radius of that loop. With a smaller radius for the inner loop, its current must be less than 10 A.

27. It is to increase the magnetic permeability and magnetic field, because the magnetic field is proportional to the magnetic permeability of the material.

29. Hawaii is slightly north of the equator, so the magnetic field there points northward and mostly parallel to the surface of the Earth but slightly downward. This would be the direction of the remnant magnetism.

31. It will be the north magnetic pole. Right now, the pole near the Earth's geographical North pole is actually a south magnetic pole.

CHAPTER 19 EXERCISES

1. (a) The second magnet has its N end just below the N end of the first one (figure not shown). (b) A third identical horizontal bar magnet should be placed, with its N end to the right of the N end of the first magnet (figure not shown). The N ends of magnet #2 and #3 both repel the N end of magnet #1 and it feels a net force of 2.1 mN at an angle of 45° above the horizontal and to the left.

3. (a) $\vec{B} = (-B_0)\,\hat{x} + (-B_0/2)\,\hat{y}$. $\vec{B}$ is 27° below the $-x$ axis. (b) $\vec{B} = B_0\,\hat{x} + (-B_0/2)\,\hat{y}$ and $\vec{B}$ is 27° below the $+x$-axis.

5. 3.5×10^3 m/s

7. 2.0×10^{-14} T, left, looking in the direction of the velocity

9. (a) 3.8×10^{-18} N (b) 2.7×10^{-18} N (c) Zero (d) Zero

11. (a) 8.6×10^{12} m/s^2 southward (b) 8.65×10^{12} m/s^2, north (c) Same magnitude but opposite direction (d) The acceleration of the electron is about 1830 times that of the proton.

13. (a) 1.8×10^3 V (b) Same voltage, independent of charge

15. 5.3×10^{-4} T

17. (a) 4.8×10^{-26} kg (b) 2.4×10^{-18} J (c) No, work equals zero.

19. (a) (i) To the right (ii) Upward (iii) Into the paper (iv) To the left (v) Into or out of the paper; either would give zero force (b) All forces are 8.3×10^{-4} N except situation (v), which is zero.

21. 1.2 N perpendicular to the plane of $\vec{B}$ and I

23. (a) Zero (b) 4.0 N/m in the $+z$-direction (c) 4.0 N/m in the $-y$-direction (d) 4.0 N/m in the $-z$-direction (e) 4.0 N/m in the $+y$-direction (f) 2.8 N/m in the $+z$-direction

25. (a) 0.400 N/m in the $+z$-direction (b) 0.400 N/m in the $+z$-direction (c) 0.500 N/m in the $-z$-direction

27. 7.5 N upward in the plane of the paper

29. (a) 0.013 m · N (b) Doubling just the field doubles the torque. (c) Doubling just the current doubles the torque. (d) Doubling just the area doubles the torque. (e) You *cannot* double the torque by just increasing the angle.

31. At the center of a circular coil, $B = \mu_0 NI/2r$. To double the field, r must be cut in half. Since the area is proportional to the *square* of the radius, the area would be decreased to one-fourth its initial value.

33. 11 A

35. 0.25 m

37. (a) 2.0×10^{-5} T (b) 9.6 cm from wire 1

39. At the right-hand point, the two fields are both into the plane of the paper, so the net field is 2.9×10^{-6} T. The field at the left-hand point will have the same magnitude (by symmetry), but will point out of the plane of the paper.

41. 1.0×10^{-4} T, away from the observer

43. 4.0 A

45. (a) 8.8×10^{-2} T (b) To the right

47. (a) The correct answer is (1). The force between two wires carrying currents in the same direction is attractive (see Figure 19.26). (b) 170 A (c) At a point midway between the two wires, the field is zero.

49. (a) 3.74×10^{-3} T · m/A (b) 3.0×10^3

51. (a) The proton should be directed at 10° north of east. (b) 9.6×10^6 m/s^2

53. (a) From $F = qvB = \frac{mv^2}{r} \Rightarrow v = \frac{qBr}{m}$, thus $T = \frac{2\pi r}{v} = \frac{2\pi}{qB/m} = \frac{2\pi m}{qB}$, $f = \frac{1}{T} = \frac{qB}{2\pi m}$. Thus T turns out to be independent of both r and v. (b) The cyclotron frequency is 2.8×10^6 Hz. The path radius is 5.69×10^{-3} m.

55. (a) The answer is (2). Let the wires be parallel, one above the other with the upper current (I_1) going to the right. The magnetic field at wire 2 (the lower wire) due to wire 1 points into the paper. By the right-hand rule, the force on the lower wire is away from wire 1, so wire 1 repels wire 2. Similar reasoning shows that wire 2 also repels wire 1. (b) 6.0×10^{-6} T (c) 1.8×10^{-5} N/m

57. (a) The top wire must attract the lower wire so it can stay in equilibrium. For the forces to attract, the currents should be in (1) the same direction. (b) 38 A

59. (a) 0.44 T (b) $I_2 = 7.5$ A in the same direction as I_1 (c) $I_2 = 7.5$ A in the *opposite* direction as I_1

61. 2.3×10^{-13} T down. No, it does not seem likely this would interfere with a magnetic strip on an ATM card, because it is much smaller than the Earth's field in the range of 10^{-5} T.

63. (a) 2.18×10^{-4} T (b) 3.80×10^{-4} T

65. (a) 6.4×10^{-3} T (b) 6.7×10^{-3} T (c) 2.1×10^{-3} T The field strength as a function of x is shown in the sketch. The field is nearly uniform in the region between the two coils.

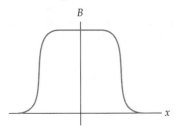

CHAPTER 20 MULTIPLE CHOICE

1. (a) and (b)
3. (d)
5. (d)
7. (a)
9. (a)
11. (c)
13. (b)
15. (a)
17. (d)
19. (f)

CHAPTER 20 CONCEPTUAL QUESTIONS

1. (a) When the bar magnet enters the coil, the needle deflects to one side, and when it leaves the coil, the needle reverses direction. (b) No, because of induced currents according to Lenz's law, it is repelled as it moves toward the loop and attracted as it leaves the loop.

3. Move with the same velocity as the bar magnet, so there is no change in magnetic flux

5. The units of $\mathcal{E}$ are J/C = (N·m)/C. The units of $\Delta\Phi/\Delta t$ are (the N can be ignored since it has no SI units) $(\mathrm{T}\cdot\mathrm{m}^2)/\mathrm{s} = \left[\dfrac{\mathrm{N}}{\mathrm{C}\cdot(\mathrm{m/s})}\right]\cdot(\mathrm{m}^2/\mathrm{s}) = \dfrac{\mathrm{N}\cdot\mathrm{s}\cdot\mathrm{m}^2}{\mathrm{C}\cdot\mathrm{m}\cdot\mathrm{s}} = (\mathrm{N}\cdot\mathrm{m})/\mathrm{C}$, the same as emf.

7. To prevent induced current, the magnetic flux must remain constant. Since the magnitude of the magnetic field has increased, the area must decrease, so the diameter must also decrease.

9. (a) The magnetic flux through the coil is proportional to $\cos\omega t$, while the induced emf is proportional to $\sin\omega t$. The max emf occurs when $\sin\omega t = 1$, which is when $\omega t = \pi/2$. This means that at this time, $\cos\omega t = 0$. (b) As in part (a), the flux is a maximum when $\cos\omega t = 1$, which is when $\omega t = 0$. But at this time, $\sin\omega t = 0$, so the emf is zero. The fundamental general answer is that the flux and emf are always 90° out of phase; when one is at a maximum (magnitude) the other is zero.

11. If the armature is jammed or turns very slowly, there is no back emf and thus there is a large current.

13. It is transmitted at high voltage and thus low current to reduce the Joule heating rate, which depends on the *square* of the current (I^2R).

15. (a) The induced current is clockwise since the induced field points away from you. (b) The induced current is clockwise since the induced field points away from you. (c) The induced current is counterclockwise since the induced field points toward you. (d) The induced current is zero since the magnetic flux does not change.

17. UV radiation causes sunburn and much of the solar radiation in that wavelength range *can* penetrate the clouds. You feel cool because infrared radiation, which is partially responsible for the sensation of "feeling hot," is *absorbed* by the water molecules in the clouds.

19. (a) The car acts as a moving observer. Due to its oncoming motion, it strikes the electromagnetic waves at a higher rate than if it were at rest. Therefore, the reflected waves will have a higher frequency than the waves had when emitted by the radar gun. (b) The frequency is higher (see above) and the wavelength is shorter. The reflected wave speed is the same as the original wave speed since both are electromagnetic waves (light) in the same medium, air, which is close to a vacuum, so that speed would be c.

CHAPTER 20 EXERCISES

1. 32 cm
3. 42° or 138°
5. (a) 1.3×10^{-6} T·m² (b) 3.0 A
7. (a) 1.6 V (b) 0.40 V
9. (a) 0.30 s (b) 0.60 s
11. (a) (1) At the equator, because the velocity of the metal rod is parallel to the magnetic field at the equator (b) 0.20 mV at the pole, zero at the equator
13. (a) 0.60 V (b) The current would be zero because the circuit in not complete. (c) 4.0 A
15. (a) 2.6 V (b) For a complete cycle, the induced emf is zero. (c) The induced emf is a maximum when the flux is changing at its maximum rate, when the magnetic field is zero. This occurs twice per cycle. The minimum emf is zero when the magnetic field is at its maximum, which also occurs twice per cycle.
17. (a) 0.057 V (b) He should use 10 loops.
19. (a) 100 V (b) Zero (c) (100 V) $\sin(120\pi t)$ (d) 1/120 s (e) 200 V
21. (a) 16 Hz (b) The amplitude remains at 24 V.
23. (a) The answer is (3) lower than 44 A (maximum possible current with no back emf is 110 V/2.50 Ω = 44 A). This is because the back emf lowers the effective voltage of the motor; thus, the current is lower than 44 A. (b) 4.00 A
25. (a) 216 V (b) 160 A (c) 8.1 Ω
27. (a) 16 (b) 5.0×10^2 A
29. (a) 24 (b) 2.0 A
31. (a) 17.5 A (b) 15.7 V
33. (a) The answer is (2) non-ideal, because the power in the secondary is lower than that in the primary. (b) 45%

35. (a) 128 kWh (b) $1840
37. (a) 53 W (b) $N_p/N_s = 200$
39. 326 m and 234 m
41. 2.6 s
43. AM: 67 m; FM: 0.77 m
45. Sound waves cause the resistance of the button to change as described. This results in a change in the current, so the sound waves produce electrical pulses. These pulses travel through the phone lines and to a receiver. The receiver has a coil wrapped around a magnet, and the pulses create a varying magnetic field as they pass through the coil, causing the diaphragm to vibrate and thus produce sound waves as the diaphragm vibrates in the air.
47. (a) Since $P_{in} \neq P_{out}$, the transformer is not ideal. (b) 90.9% (c) 120 W
49. (a) The answer is (2). The induced magnetic field must point in the same direction as the external field (out of the page) to oppose the decrease in flux, so the current must be counterclockwise, by the right-hand rule. (b) 10.0 mV (c) 69.1 ms
51. (a) 3.8 m. This is much too large to be practical. (b) 4.2 m (c) A manageable size would have a diameter of about 20 cm, which would require 3.6×10^5 windings.
53. (a) As the coil enters the magnetic field, the induced current is counterclockwise, which we can designate as a negative emf. The rate of change of the flux is due to the rate at which the area of the coil enters the field.

Since the coil moves at constant speed, the amount of area ΔA that *enters* the field in a given time Δt is largest as it just enters the field. Therefore, the induced emf is largest at first and gradually decreases. Once the coil is all in the field, there is no flux change through it so the induced emf is zero. As the coil leaves the field, the induced emf is the same as when it entered except reversed in shape and direction. The graph (sketch) would look like the one below. (b) (1) Zero (2) -3.53 mV (3) Zero (4) $+3.53$ mV (5) Zero

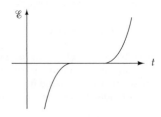

CHAPTER 21 MULTIPLE CHOICE

1. (a)
3. (a)
5. (b)
7. (b)
9. (b)
11. (d)
13. (d)

CHAPTER 21 CONCEPTUAL QUESTIONS

1. The average current is zero due to directional change, so it can be either positive or negative. However, power is delivered regardless of current direction, because power depends on the current squared. Therefore, power does not average out to zero.

3. The current is cut in half because the voltage is halved. Since power is proportional to the *square* of the current, the power drops to 25% of the designed power.

5. The time averaged power is inversely proportional to the resistance, so the power is cut in half. The rms voltage does not change since it depends only on the ac power source. The rms current is cut in half because it is inversely proportional to the resistance.

7. For a capacitor, the *lower the frequency*, the longer the charging time in each cycle. If the frequency is very low (dc), then the charging time is very long, so it acts as an ac open circuit. For an inductor, the *lower the frequency*, the more slowly the current changes in the inductor The more slowly the current changes, the less back emf is induced in the inductor, resulting in less impedance to current.

9. At $t = 0$, $I = 120$ A, or at maximum. The voltage is then zero, because current leads voltage by 90° in a capacitor. When current is maximum, voltage is 1/4 period behind, or at zero. They are out of phase.

11. Capacitive reactance is proportional to the product of the frequency and the capacitance. To keep it constant, the frequency should be halved if the capacitance is doubled.

13. **(a)** f_o is halved. **(b)** f_o is reduced to one-third its value. **(c)** Changing the resistance has no effect on the resonance frequency. **(d)** f_o is halved.

15. The circuit is not at resonance because the inductive reactance and capacitance are not equal. Since the inductive reactance is greater than the capacitive reactance, the driving frequency is *greater than* the resonance frequency.

CHAPTER 21 EXERCISES

1. For a 120-V line, the peak voltage is 170 V. For a 240-V line, the peak voltage is 339 V.

3. 1.2 A

5. **(a)** 10.0 A **(b)** 14.1 A **(c)** 12.0 Ω

7. **(a)** 4.47 A, 6.32 A **(b)** 112 V, 158 V

9. $V = (170 \text{ V}) \sin(119\pi t)$

11. $I_{rms} = 0.333$ A; $I_0 = 0.471$ A; $R = 360$ Ω

13. **(a)** 20 Hz, 0.050 s **(b)** 2.4×10^2 W

15. **(a)** 60 Hz **(b)** 1.4 A **(c)** 1.2×10^2 W
(d) $V = (120 \text{ V}) \sin 380t$
(e) $P = (240 \text{ W}) \sin^2 380t$
(f) $P = (240 \text{ W}) [1 - \cos 2(380t)]/2 = 120 \text{ W} - (120 \text{ W}) \cos 2(380t)$. The average of a sine or cosine function is zero. So $\overline{P} = 120$ W, the same as in part (c).

17. 1.3×10^3 Ω

19. 2.3 A

21. **(a)** −38% **(b)** +60%

23. **(a)** 250 Hz **(b)** 990 Hz

25. **(a)** 90 V **(b)** Voltage leads current by 90°

27. **(a)** 4.42 μF **(b)** 0.10 A

29. **(a)** 1.7×10^2 Ω **(b)** 2.0×10^2 Ω

31. **(a)** 38 Ω; 1.1×10^2 Ω **(b)** 1.1 A

33. **(a)** (3) Negative, because this is a capacitive circuit **(b)** −27°

35. **(a)** (3) In resonance, because $X_L = X_C$, so $Z = R$. **(b)** 72 Ω

37. **(a)** 50 W **(b)** 115 W

39. **(a)** 53 pF **(b)** 31 pF

41. **(a)** 9.0 Ω **(b)** 13 A

43. $(V_{rms})_R = 12$ V; $(V_{rms})_L = 2.7 \times 10^2$ V; $(V_{rms})_C = 2.7 \times 10^2$ V

45. **(a)** (2) Equal to 25 Ω. At resonance, $X_L = X_C$, so $Z = R$. **(b)** 362 Ω

47. **(a)** 0.55 **(b)** 0.30

49. **(a)** 38 Ω **(b)** 63 Ω **(c)** 1.8 A **(d)** Zero **(e)** 37°

51. **(a)** $\phi = 0$ **(b)** 9.47×10^{-6} F **(c)** 0.743 H **(d)** 31.9°

53. **(a)** A "step-up" transformer is needed with a winding ratio of $N_s/N_p = 10$. **(b)** 9.6 Ω **(c)** 60 Hz, 120 V, and 13 A **(d)** 18 A, 170 V, and 3000 W **(e)** 17 V, 180 A, and 3.0 kW

55. **(a)** The answer is (2). The equivalent parallel capacitance would be less. A smaller value of C means a larger value for f_o. **(b)** 119 Hz

CHAPTER 22 MULTIPLE CHOICE

1. **(b)**
3. **(d)**
5. **(b)**
7. **(a)** and **(c)**
9. **(c)**

CHAPTER 22 CONCEPTUAL QUESTIONS

1. The angle of reflection is always equal to the angle of incidence.

3. After a rain, the road surface is wet, with water filling the crevices and turning the road into a relatively smooth surface. The normally diffuse reflection turns into specular reflection.

5. The laser beam has a better chance to hit the fish. The fish appears to the hunter to be at a location different from its true location due to refraction. The laser beam obeys the same law of refraction and retraces the light the hunter sees from the fish. The arrow goes into the water in a near-straight line path and thus passes above the fish.

7. This severed look is because the angle of refraction is different for the air–glass interface than for the water–glass interface. The top portion refracts from air to glass, and the bottom portion refracts from water to glass. This is different from what's in Fig. 22.13b. In that figure, we see the top portion in air directly and the bottom portion in water through refraction from water to air. The angle of refraction made the pencil appear to be bent.

9. Total internal reflection could *not* occur because in this case, medium 2 is more optically dense than medium 1. We know this because the light is bent toward the normal in medium 2.

11. In a prism, there are two refractions and two dispersions because both refractions cause the refracted light to bend downward, therefore doubling the effect or dispersion.

13. No, the light will be further dispersed by the second prism.

15. A glass pane is typically a few millimeters thick, so the distance over which it separates the colors is too small for detection by our eyes. The speeds of each color are different in the glass, but not by much. Dispersion does occur, but it is usually too small for detection under ordinary circumstances.

CHAPTER 22 EXERCISES

1. 60°

3. **(a)** The answer is (2). The angle of reflection is $\theta_r = \theta_i = 90° - \alpha$. **(b)** 57°

5. **(a)** (3) $\tan^{-1}(w/d)$ **(b)** 27°

7. When the mirror rotates through a small angle of θ, the normal will rotate through an angle of θ and the angle of incidence is $35° + \theta$. The angle of reflection is also $35° + \theta$. Since the original angle of reflection is 35°, the reflected ray will rotate through an angle of 2θ. If the mirror rotates in the opposite direction, the angle of reflection will be $35° - \theta$. However, the normal will again rotate through an angle of θ but also in the opposite direction. Thus, the reflected ray still rotates through an angle of 2θ.

9. 90°, any θ_{i_1}

11. 1.41

13. 1.34

15. **(a)** (1) Greater than, because water has a lower index of refraction **(b)** 17°

17. **(a)** The correct answer is (2). For total internal reflection to occur, light must go from a more optically dense medium into a less optically dense one; that is, it must go from a high-n medium into a low-n medium, which is the case for water to air. **(b)** 48.8°

19. 47°

21. 1.55×10^{15} Hz

23. **(a)** (3) Less than, because its index of refraction is higher **(b)** 15/16

25. **(a)** This is caused by refraction of light in the water–air interface. The angle of refraction in air is greater than the angle of incidence in water, so the object immersed in water appears closer to the surface.

27. 66.7%

29. **(a)** (3) Less than, because it is equal to $90° - \theta_1$. $\theta_1 > 45° = \theta_2$ and $n_1 < n_2$ **(b)** 20°

31. Seen for 40° but not for 50°; $\theta_c = 49°$

33. **(a)** This arrangement depends on (3) the indices of refraction of both, because $\theta_c \geq \sin^{-1}(n_2/n_1)$. **(b)** Air: $n_1 \geq 1.41$; water: $n_1 \geq 1.88$

35. 43°

37. **(a)** 25° **(b)** 1.97×10^8 m/s **(c)** 362 nm

39. 11 cm

41. 1.41

43. $n_R = 1.362$; $n_B = 1.371$

45. 1.498

47. **(a)** 21.7° **(b)** 0.22° **(c)** 0.37°

49. **(a)** 49° **(b)** 1.5 **(c)** 1 **(d)** 42°

51. **(a)** (1) More than, because red light will have a smaller index of refraction and thus a higher speed of light than blue light **(b)** 1.3 mm

53. No light leaks into the air for either angle of incidence.

CHAPTER 23 MULTIPLE CHOICE

1. **(c)**
3. **(b)**
5. **(a)**
7. **(a)**
9. **(d)**
11. **(c)**
13. **(b)**
15. **(c)**
17. **(b)**

CHAPTER 23 CONCEPTUAL QUESTIONS

1. No, virtual images cannot be seen on a screen, because no rays intersect at the image.

3. During the day, the reflection is mainly from the silvered back surface. During the night, when the switch is flipped, the reflection comes from the front side. There is a reduction of intensity and glare because the front side reflects only about 5% of the light, which is more than enough to see due to the dark background.

5. When viewed by a driver through a rearview mirror, the right-left reversal of the image formed by a plane mirror will make it read "AMBULANCE."

7. **(a)** A spoon can behave as either a concave or a convex mirror depending on which side you use for reflection. If you use the concave side, you normally see an inverted image. If you use the convex side, you always see an upright image. **(b)** In theory, the answer is yes. If you are very close (inside the focal point) to the spoon on the concave side, an upright image exists. However, it might be difficult for you to see the image in practice, because your eyes might be too close to the image. Eyes cannot see things that are closer than the near point (Chapter 25).

9. **(a)** The image is smaller than the object, and it is possible to "see your full body in 10 cm" in a diverging mirror. **(b)** As the ball swings toward the mirror and approaches the focal point, the image enlarges. An enlarged image appears to be closer to our eyes, and it appears to move toward the observer; therefore, it produces the effect of appearing to "jump" out of the mirror as the ball swings through the focal point.

11. The image is upright, virtual, and twice as large as her face.

13. The object distance should be between the focal length and twice the focal length. In this region, the image is real, inverted, and magnified.

15. The object must be inside the focal point of the lens. This cannot be done with a diverging lens because it always produces a smaller image.

17. +, +; +, ∞; +, −; −−; ∞, −; +, −

19. No. The more general lens maker's equation is $\dfrac{1}{f} = \left[(n/n_m) - 1\right]\left(\dfrac{1}{R_1} + \dfrac{1}{R_2}\right)$, where n_m is the index of refraction of the surrounding material. If $n_m > n$, f is negative, meaning the lens is diverging.

21. Spherical aberration is caused by a spherical lens surface. The rays that pass through the outer edges of the lens are *not* focused to the same place as those that pass through the center of the lens. This causes a fuzzy image.

CHAPTER 23 EXERCISES

1. 5.0 m

3. **(a)** 0.80 m **(b)** 5.0 cm **(c)** +1.0

5. **(a)** The dog's image is 3.0 m behind the mirror. **(b)** They approach each other at 2.0 m/s.

7. **(a)** You see multiple images caused by reflections off two mirrors. **(b)** 3.0 m behind the *north* mirror, 11 m behind the *south* mirror, 5.0 m behind the *south* mirror, 13 m behind the *north* mirror

9. The two triangles (with d_o and d_i as base, respectively) are similar to each other because all three angles of one triangle are the same as those of the other triangle due to the law of reflection. Furthermore, the two triangles share the same height, the common vertical side. Therefore, the two triangles are congruent. Hence $d_o = d_i$.

11. **(a)** See Fig. 23.8 except that the object is beyond C. The image is (1) real, (2) inverted, and (3) reduced. **(b)** −66.7 cm (in front of the mirror), and the lateral magnification is 0.667

13. $f = -33.3$ cm, $M = 0.400$

15. $d_i = -30$ cm; $h_i = 9.0$ cm; and the image is virtual, upright, and magnified

17. The image is inverted but the same size as the object.

19. 3.0 cm, real and inverted

21. **(a)** The correct answer is (1). Since the image is virtual but reduced in size, the mirror must be diverging (convex). **(b)** −10 cm

23. −25 cm (behind the mirror)

25. **(a)** Concave, because only concave mirrors can form real images (formed on a screen) **(b)** 24 cm

27. **(a)** Virtual and upright **(b)** 1.5 m

29. 2.3 cm

31. **(a)** See ray diagram below.

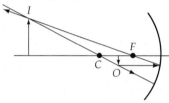

(b) $d_i = 60$ cm, $M = -3.0$, real and inverted

33. **(a)** The mirror is concave since the image is magnified. Only a concave mirror can form a magnified image. **(b)** $f = 18$ cm and $R = 36$ cm

35. **(a)** $d_i = \dfrac{d_o f}{d_o - f} = \dfrac{f}{1 - f/d_o}$

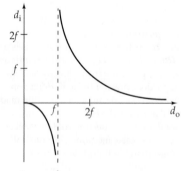

$|M| = \dfrac{d_i}{d_o} = \dfrac{f}{d_o - f}$

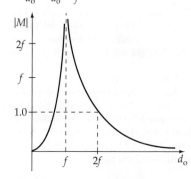

(b) $d_i = \dfrac{d_o(-f)}{d_o + f} = \dfrac{-f}{1 + f/d_o}$

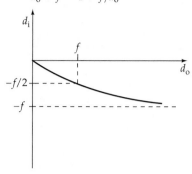

$|M| = \dfrac{d_i}{d_o} = \dfrac{-f}{d_o + f}$

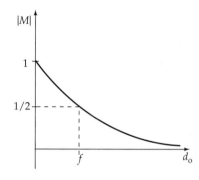

37. 0.69

39. Yes, it is possible. One is a real image and the other is a virtual image. 13 cm; 27 cm

41. $d_i = 12.5$ cm; $M = -0.250$

43. 22 cm, −9.0

45. **(a)** From the ray diagram below, the image is virtual, upright and magnified.

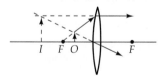

(b) $d_i = -47$ cm and $M = +3.1$

47. **(a)** −6.4 cm; +0.64 (virtual and upright) **(b)** −10.5 cm; +0.42 (virtual and upright)

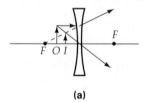

(a)

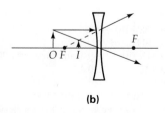

(b)

49. **(a)** 18 cm **(b)** 6.0 cm

51. Since the mirror and lens equations are the same and the definitions of the lateral magnification are also the same, the graphs are exactly the same as those in Exercise 23.35.

53. 0.55 mm

55. (a) 40 cm (b) −1.0

57. (a) From similar triangles, $\dfrac{d_i - f}{f} = -\dfrac{y_i}{y_o}$, where the negative is introduced, because the image is inverted. Also, $-\dfrac{y_i}{y_o} = \dfrac{d_i}{d_o}$. So $\dfrac{d_i - f}{f} = \dfrac{d_i}{d_o}$, or $d_o d_i - d_o f = d_i f$, that is, $d_i f + d_o f = d_o d_i$. Dividing both sides by $d_o d_i f$ gives $\dfrac{1}{d_o} + \dfrac{1}{d_i} = \dfrac{1}{f}$. (b) $M = \dfrac{y_i}{y_o} = -\dfrac{d_i}{d_o}$ from the similar triangles in part (a).

59. −37 cm

61. (a) The lens should be convex since the image is magnified. Only convex lenses can form magnified images. (b) 6.25 cm

63. The image is 18 cm to the left of the eyepiece. It is virtual, inverted, and 92× larger than the object.

65. $M_1 = -h_{i1}/h_{o1}$, $M_2 = -h_{i2}/h_{o2}$, and $M = h_{i2}/h_{o1}$. Since $h_{o2} = h_{i1}$ (the image formed by the first lens is the object for the second lens), $M_1 M_2 = (h_{i1}/h_{o1})(h_{i2}/h_{o2}) = h_{i2}/h_{o1} = M_{\text{total}}$.

67. −25 cm

69. (a) According to the sign convention, the signs are (2) +, −. (b) 27.2 cm

71. −40 cm, concave

73. 29 cm (in air), 84 cm (under water)

75. (a) Since the index of refraction of the lens is greater than that of air, the angle of refraction is less than the angle of incidence at the air–lens interface and greater than the angle of incidence at the lens–air interface. So both refractions refract the incident light toward the axis. (b) For the same reason, the rays refract away from the axis due to the opposite curvatures of the surfaces.

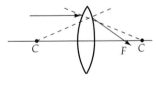

(a)

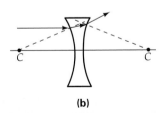

(b)

77. The image formed by the converging lens is at the mirror. This image is the object for the diverging lens. If the mirror is at the focal point of the diverging lens, the rays refracted after the diverging lens will be parallel to the axis. These rays will be reflected back parallel to the axis by the mirror and will form another image at the mirror. This second image is now the object for the converging lens. By reversing the rays, a sharp image is formed on the screen located where the original object is. Therefore,

the distance from the diverging lens to the mirror is the focal length of the diverging lens.

79. 6.00 cm

81. 1.50 D (the lower portion of the lens for near vision), −1.25 D (the upper portion of the lens for distant vision)

CHAPTER 24 MULTIPLE CHOICE

1. (b)
3. (a)
5. (a)
7. (b)
9. (a)
11. (d)
13. (b)
15. (c)

CHAPTER 24 CONCEPTUAL QUESTIONS

1. The maxima angles are $\theta = \sin^{-1}(n\lambda/d)$. As d decreases, θ increases; thus the pattern spreads out.

3. No, this is not a violation of the conservation of energy. Energy is redistributed (moved from the minima to the maxima). *Total* energy is still conserved.

5. It is always dark because of destructive interference due to the 180° phase shift. If there had not been the 180° phase shift, zero thickness would have corresponded to constructive interference.

7. If the slit length is comparable to the width, a second diffraction pattern perpendicular to the first will also be observed.

9. Since $d \sin \theta = n\lambda$, the advantage is a wider diffraction pattern, as d is smaller.

11. (a) Twice (b) Four times (c) None (d) Six times

13. The numbers appear and disappear as the sunglasses are rotated because the light from the numbers on a calculator is polarized.

15. There is no air on the surface of the Moon, and so an astronaut would see a black sky.

CHAPTER 24 EXERCISES

1. The path difference is 1.5λ, so the waves interfere destructively. The path difference is 2λ, so the waves interfere constructively.

3. 0.37°

5. 489 nm

7. (a) $\Delta y = L\lambda/d \propto \lambda$. Thus the distance between the maxima will (2) also decrease if λ decreases. (b) 600 nm (orange-yellow) (c) 0.41 cm

9. (a) 402 nm, violet (b) 3.45 cm

11. (a) The correct answer is (1), increase. Since $\Delta y = L\lambda/d \propto L$, the distance between the maxima will also increase if the distance from the slits to the screen is increased. (b) 0.63 cm (c) 0.94 cm

13. 4.2×10^{-5} m

15. 450 nm

17. (a) Both the light waves reflected off the air–film interface and those reflected off the film–glass interface will suffer a 180° phase change because they are incident on a more optically dense material. (b) 2.4×10^{-5} m (c) Constructive interference

19. 54.3 nm

21. (a) Yes. If $n_{solar} > n_{film}$, the minimum film thickness has to be $\lambda'/2$ for destructive interference because the reflective phase shifts will cancel out. If $n_{solar} < n_{film}$, there will be only one reflective phase shift of 180°, so the minimum film thickness for cancellation will be λ'. (b) 113 nm (c) 196 nm

23. (a) 158.2 nm (b) 316.4 nm

25. 1.51×10^{-6} m

27. 2.0×10^{-4} m

29. (a) 5.4 cm (b) 2.7 cm

31. (a) 4.3 mm (b) Microwave

33. (a) The correct answer is (1). The first angle at which destructive interference occurs is given by $w \sin \theta = \lambda$, so if λ increases, so will θ. This makes the central maximum wider. (b) 3.6 mm (c) 6.2 mm

35. 1.24×10^3 lines/cm

37. 7.1×10^{-10} m

39. $n = 0, \theta_0 = 0$; $n = \pm 1, \theta_1 = \pm 22.31°$; $n = \pm 2, \theta_2 = \pm 49.41°$. There are a total of four *side* maxima, two on each side of a central. If you include the central maximum, there are five.

41. For $n = 1, \theta_1 = \pm 23°$; and for $n = 2$, $\theta_2 = \pm 52°$

43. (a) The correct choice is (1). Maxima occur when $d \sin \theta = n\lambda$. Thus the smaller the wavelength, the smaller the θ. Hence the smaller the wavelength, the closer its maximum will be to the central maximum. (b) $\theta_R = 34.1°$ and $\theta_R = 18.7°$

45. For violet, $\theta_{3v} = \sin^{-1}(3)(400 \text{ nm})/d = \sin^{-1}(1200 \text{ nm})/d$. For yellow-orange, $\theta_{3y} = \sin^{-1}(2)(600 \text{ nm})/d = \sin^{-1}(1200 \text{ nm})/d$. Since $\theta_{3v} = \theta_{2y}$, they overlap.

47. 39.2°

49. (a) The correct answer is (1), also increase, because $\tan \theta_p = n_2/n_1 = n_2$ (since $n_1 = 1$). If n_2 increases, so does θ_p. (b) 58°, 61°

51. 31.7°

53. 40.5°

55. (a) The correct answer is (2). The Brewster angle is given by $\tan \theta_p = n_2/n_1$. Thus the larger the n_1 (the incident medium), the smaller the Brewster angle. Since $n_{water} > n_{air}$, the Brewster angle in water will be less than the Brewster angle in air. (b) In air, $\theta_p = 58.9°$; in water: $\theta_p = 51.3°$

57. In water, the angle of incidence at the water–glass interface must be $\theta_p = \tan^{-1}(1.52/1.33) = 48.8°$. For an air–water interface, $n_1 \sin \theta_1 = n_2 \sin \theta_2$, so $\sin \theta_1 = n_2 \sin \theta_2/n_1 = (1.33) \sin 48.8° > 1$. Since the maximum of $\sin \theta_1$ is 1, the answer is no.

59. (a) The correct choice is (1). The shorter the wavelength, the greater the intensity of the scattered light. Since the other color has less intensity than the 550-nm light, its wavelength must be longer than 550 nm. (b) 822 nm

61. (a) No. The critical angle (for total internal reflection) is not the same as the Brewster angle (for polarized reflection). (b) 33.8°

63. $\tan \theta_p = \tan \theta_1 = n_2/n_1 = \sin \theta_1/\cos \theta_1$, (Eq. 1). From Snell's law, $n_1 \sin \theta_1 = n_2 \sin \theta_2$ (Eq. 2). From (1) and (2), $\cos \theta_1 = \sin \theta_2 =$

cos $(90° − θ_2)$ [since sin x = cos $(90° − x)$].
Thus $θ_1 = 90° − θ_2$, and $θ_1 + θ_2 = 90°$.

65. $n_{max} = 3$, that is, there are, at most, 3 orders of the complete spectrum.

CHAPTER 25 MULTIPLE CHOICE

1. (c)
3. (a)
5. (d)
7. (b)
9. (b)
11. (c)
13. (d)
15. (d)

CHAPTER 25 CONCEPTUAL QUESTIONS

1. Iris, crystalline lens, and retina correspond to the aperture, lens, and film, respectively, of the camera.

3. Yes. The image is smaller for nearsightedness and larger for farsightedness.

5. To correct nearsightedness, a diverging lens is needed to form an image at the far point of an object at infinity. Thus $f = d_i$, where d_i is the distance to the far point. When replacing ordinary glasses with contacts, the distance from the lens to the image is now a few centimeters *larger*, which makes the focal length slightly larger. Since $P = 1/f$, an increase in f means a *decrease* in P, thus contacts are weaker than regular lenses.

7. A short focal length lens has a small radius. The aberration (angle approximation is no longer valid if the object is large compared with the size of the lens) will get more important as the focal length of the lens gets smaller. This limits the magnification to about 3× to 4×.

9. A telescope is supposed to magnify objects, so its angular magnification should be greater than 1. Since the magnification is $m = −f_o/f_e$, the eyepiece should have a shorter focal length than the objective lens. So use the lens with the shorter focal length as the eyepiece and the other lens as the objective.

11. A reflecting telescope employs a concave parabolic mirror instead of a lens. A parabolic mirror does not exhibit spherical aberration and also is free of chromatic aberration.

13. Smaller minimum angle of resolution corresponds to higher resolution because smaller angle of resolution means more details can be resolved.

15. The smaller lens has a lower resolution. The smaller the lens, the greater the minimum angle of resolution and the lower the resolving power.

17. Under red light, red and white appear red; blue appears black. Under green light, only white appears green; both red and blue appear black. With blue light, red appears black; white and blue appear blue.

19. The liquid is dark or colored because it absorbs all colors of light *except* that color. The amount of light absorbed by an object depends on how much material is absorbing the light. Foam has low density and it thus absorbs very little light and a lot is reflected; therefore, foam generally appears as white.

CHAPTER 25 EXERCISES

1. (a) +5.0 D **(b)** −2.0 D

3. (a) The correct answer is (2), a diverging contact lens should be prescribed, because the person is nearsighted. **(b)** −1.1 D

5. −2.00 m, a diverging lens

7. (a) 2.8 D **(b)** The patient can focus clearly on any object beyond 85 cm without the glasses. So she can see any image beyond 85 cm, which would include all objects beyond 25 cm. Therefore, she should leave the glasses on for distant objects.

9. (a) The correct answer is (2). Since he cannot focus on close objects, he is farsighted. **(b)** The correct choice is (1). Since he is farsighted, he needs a converging lens to correct his vision. **(c)** +3.3 D

11. (a) −0.505 D **(b)** −0.500 D

13. 83 cm

15. His far point is 85 cm from his eyes and his near point is 50.0 cm from his eyes.

17. +3.0 D

19. (a) 0.17° **(b)** 0.082°

21. 3.1×

23. 2.5×

25. (a) The correct answer is (2). Since the maximum magnification is given by $m = 1 + (25 \text{ cm}/f)$, a small f gives a large magnification. **(b)** The maximum magnifications are 1.9× with the 28-mm lens and 1.6× with the 40-mm lens.

27. +6.0 D

29. +77 D

31. (a) −340× **(b)** 3900%

33. (a) The correct choice is (2), the one with the shorter focal length, because the total magnification is inversely proportional to the focal length of the objective. **(b)** −280× and −360×

35. 25×

37. (a) Greatest: 1.6 mm/10×; least: 16 mm/5× **(b)** $M_{max} = −930×$; $M_{min} = −42×$

39. (a) −4.0× **(b)** 75 cm

41. 1.00 m and 2.0 cm

43. (a) 13 cm **(b)** +7.0×

45. (a) 60.0 cm and 80.0 cm; 40.0 cm and 90.0 cm **(b)** −75×; −44×

47. 650 nm

49. $1.32 × 10^{−7}$ rad; $θ_{min}$ by Hale is 1.6 times as large

51. (a) The correct choice is (3). For a circular aperture, the minimum angle of resolution is $θ_{min} ∝ λ$. Thus the smaller the wavelength, the smaller the $θ_{min}$ and the finer the detail that can be resolved. Thus blue light (shortest wavelength) gives the finest details. **(b)** $θ_{min} = 7.0 × 10^{−5}$ rad = 0.0040° for 400-nm light and $θ_{min} = 1.2 × 10^{−4}$ rad = 0.0070° for 700-nm light.

53. 17 km

55. (a) $2.20 × 10^{−5}$ rad = 0.00126° **(b)** $1.8 × 10^{−5}$ mm

57. $4.1 × 10^{16}$ km

59. (a) d_o = 7.14 cm **(b)** 3.5×

61. A refracting telescope forms an inverted image. $θ_i ≈ −\tan θ_i = −y_i/f_e$ and $θ_o = y_i/f_o$. Therefore, $m = θ_i/θ_o = (−y_i/f_e)/(y_i/f_o) = −f_o/f_e$.

63. (a) (1) B, (2) A **(b)** −110×, $8.95 × 10^{−7}$ rad

65. (a) 6.3 and 0.25 **(b)** 1/120 s

CHAPTER 26 MULTIPLE CHOICE

1. (d)
3. (b)
5. (b)
7. (a)
9. (a)
11. (c)
13. (a)
15. (b)
17. (b)
19. (a)
21. (a)
23. (d)
25. (c)
27. (c)

CHAPTER 26 CONCEPTUAL QUESTIONS

1. No, she cannot. Newton's laws apply only in *inertial* reference frames. The carousel is a noninertial reference frame because it is spinning (centripetal acceleration).

3. They are exactly the same, because an elevator moving at constant velocity is an inertial reference frame. Thus in both frames, the acceleration is zero.

5. Yes, there is such a reference frame. It would have to move along the x-axis in the direction of A toward B.

7. In frame O, the bullet takes 1 s to hit the target. Light takes $10^{−6}$ s to get to the target (the time for light of speed $3.00 × 10^8$ m/s to travel 300 m). Frame O' would have to travel to the right. The light flash from the gun reaches the target in $10^{−6}$ s. The observer in frame O' would have to cover the 300 m in less than $10^{−6}$ s to intercept the signals at the same time, which means $v > c$. Since $v > c$ is not possible, all observers agree that the gun fires before the bullet hits the target.

9. Rocket B moves faster since its length appears to contract more than that of A. Similarly, you would conclude that the clock in B runs slower since it's moving at a greater speed than A.

11. (a) You are measuring the proper time, because you and the clock are in the same frame of reference (no relative motion). **(b)** Your professor measures the proper length of the spacecraft, because your professor and the spacecraft are in the same frame of reference (no relative motion).

13. No, the acceleration cannot be constant. Since v must be less than c, it cannot accelerate at a constant rate. If it did, eventually v would become more than c.

15. The rest energy of a proton is 938 MeV. Since its kinetic energy is much much less than this, classical physics is adequate. On the other hand, when its kinetic energy is 2000 MeV, large in comparison to its rest energy, relativistic physics is required.

17. Drop the cup with the pole vertical. By the principle of equivalence, the weight of the ball appears to be zero in the downward accelerating reference frame. In that frame, the ball is subject only to the tension force of the stretched rubber band and is pulled inside the cup.

19. Light is bent by the gravity of the black hole. At a certain distance, this light would be able to orbit around the black hole. Therefore, light from the back of your head could go into orbit and come around to strike your eyes.

21. The relative speed between A and B must be less than c. However, you as the third-party observer could observe the distance between them closing at a rate greater than speed of light. (In this case, that would be 1.5c.) But this does not violate the principle of relativity, because no information is being transmitted at that rate.

CHAPTER 26 EXERCISES

1. **(a)** With the wind blowing toward you, the time difference is 0.10 s, less time. **(b)** Now the time difference is 3.58 s − 3.48 s = 0.10 s, more time.

3. **(a)** v_{max} = 55.0 m/s and v_{min} = 45.0 m/s **(b)** 4.0 s

5. The time between adjacent gaps is $t = \Delta\theta/\omega = (2\pi/N)/(2\pi f) = 1/Nf$. So the speed of light is $c = 2L/t = 2L/[1/(Nf)] = 2fNL$.

7. **(a)** The time for light to travel from B to A is 20.0 μs. Since B occurred 25.0 μs before A, B could have caused A. **(b)** B could not have caused A because no signal could travel from B to A in less than 20.0 μs.

9. 23 min

11. **(a)** 14.1 m **(b)** $c/3$

13. 0.998c

15. **(a)** 4.8 y **(b)** 2.1 y **(c)** 4.3 ly **(d)** 1.9 ly

17. **(a)** 5.3 m **(b)** 27 ns **(c)** 36 ns. The difference is due to the fact that the Earth observer sees a shorter length, and therefore, less time to pass the location.

19. **(a)** 0.952c = 2.86 × 10^8 m/s **(b)** 16.4 min

21. **(a)** 2.14 × 10^{-14} m, which is approximately 20 nuclear diameters, hardly noticeable even to people with very sharp vision! **(b)** 6.00 × 10^5 m/s

23. **(a)** 0.985c **(b)** 2.50 MeV **(c)** 1.56 × 10^{-21} kg·m/s

25. 0.45 kg

27. 1.6 × 10^{24} J or about 4.6 × 10^{17} kWh, which is over a *million* times more energy than the United States uses per year for electricity.

29. **(a)** 79 keV **(b)** 79 keV **(c)** 1.6 × 10^{-22} kg·m/s

31. **(a)** 0.96c = 2.9 × 10^8 m/s **(b)** 3.8 × 10^{-10} J **(c)** 1.7 × 10^{-18} kg·m/s

33. **(a)** Liquid water will have (1) more mass than when it is in the form of ice, because energy must be added to convert ice to water. **(b)** Δm = 3.7 × 10^{-12} kg. No, this is not detectable, as it is extremely small.

35. **(a)** The total energy is $E = m\gamma c^2$ and the momentum is $p = m\gamma v$. Therefore, $p^2 c^2 = (m\gamma v)^2 c^2 = m^2\gamma^2 v^2 c^2$ and $(mc^2)^2 = m^2 c^4$. Adding these two quantities and simplifying gives

$$p^2 c^2 + (mc^2)^2 = m^2\gamma^2 v^2 c^2 + m^2 c^4 = m^2 c^2(\gamma^2 v^2 + c^2).$$

In the solution to Exercise 28, it was shown that $v = c\sqrt{1 - \dfrac{1}{\gamma^2}}$. Using this in the previous result proves what was asked for:

$$p^2 c^2 + (mc^2)^2 = m^2 c^2\left[\gamma^2 c^2\left(1 - \frac{1}{\gamma^2}\right) + c^2\right] =$$

$(m\gamma c^2)^2 = E^2$. **(b)** The total energy is $K + mc^2$, so the result in part (a) gives $(K + mc^2)^2 = p^2 c^2 + (mc^2)^2$. Solving for p gives

$$p = \sqrt{\frac{K^2 + 2Kmc^2}{c^2}}$$

$$= \sqrt{\frac{(1000\ \text{MeV})^2 + 2(1000\ \text{MeV})(938.27\ \text{MeV})}{c^2}}$$

= 1700 MeV/c = 9.0 × 10^{-19} kg·m/s.

37. As a black hole, the sun's density would be 1.8 × 10^{19} kg/m^3. The sun's actual density is 1.4 × 10^3 kg/m^3. Thus as a black hole, our sun would be abut 10^{16} times denser than it is now!

39. At 2 Schwarzschild radii, v_{esc} = 2.1 × 10^8 m/s = 0.71c. At twice the sun's present radius, the escape velocity would be 4.4 × 10^5 m/s = 0.0015c.

41. **(a)** 2R **(b)** $\rho/4$

43. 0.43c

45. −0.154c (toward Earth)

47. **(a)** 0.988c to the left **(b)** 0.988c to the right

49. **(a)** K_{rel} = 1.27 × 10^{-14} J and $K_{non-rel}$ = 1.02 × 10^{-14} J. The relativistic kinetic energy is about 25% larger than the nonrelativistic kinetic energy. **(b)** E_{rel} = 9.47 × 10^{-14} J. The nonrelativistic total energy is just the kinetic energy, or $E_{non-rel}$ = 1.02 × 10^{-14} J. **(c)** p_{rel} = 1.58 × 10^{-22} kg·m/s and $p_{non-rel}$ = 1.37 × 10^{-22} kg·m/s **(d)** The classical rest energy is zero. The relativistic rest energy is E_o = 8.20 × 10^{-14} J.

51. **(a)** 2.5 × 10^{10} kWh **(b)** 6 days

53. 1.9 kg

55. **(a)** 5.31m **(b)** 2.31mc^2 **(c)** 0.0382c left

CHAPTER 27 MULTIPLE CHOICE

1. **(b)**
3. **(c)**
5. **(b)**
7. **(a)**
9. **(d)**
11. **(d)**
13. **(d)**
15. **(d)**

CHAPTER 27 CONCEPTUAL QUESTIONS

1. The temperature of a black body is inversely proportional to the wavelength at which the maximum amount of energy is radiated. Therefore, a red star radiates most of its energy at longer wavelengths, while the blue star radiates most of its energy at shorter wavelengths. Therefore, the blue star must be hotter than the red star.

3. $\lambda_{max}T$ = 2.9 × 10^{-3} m·K, thus $(c/f_{max})T$ = 2.9 × 10^{-3} m·K. Thus

$$f_{max} = \frac{c}{2.9 \times 10^{-3}\ \text{m}\cdot\text{K}}T =$$

[1.03 × 10^{11}/(s·K)]T. The graph is a straight line as shown. If T is tripled, f_{max} becomes three times as large.

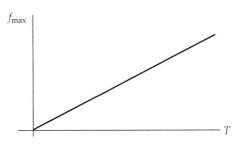

5. It is not possible. The frequency of IR radiation is less than the frequency of UV radiation. Since the energy of a photon is proportional to the light frequency, the IR photon must have less energy than the UV photon.

7. The greater the work function, the more energy it takes to dislodge photoelectrons and thus the less kinetic energy these electrons have if the incident light wavelength is kept constant. A smaller kinetic energy means that a lower stopping potential is needed to stop them. Hence a larger work function results in a lower stopping potential.

9. For each scattering the wavelength shift is on the order of $\Delta\lambda = \lambda_C = 0.00243$ nm. The wavelength change from X-ray to visible light is about 550 nm − 0.01 nm = 550 nm. So it would take about $\dfrac{550\ \text{nm}}{0.00243\ \text{nm/scattering}}$, or 200 000 scatterings, for an X-ray photon to become a visible light photon.

11. Calling the +x-axis the direction of the incident photon and the +y-axis the direction in which the scattered photon goes, the electron would have momentum components in the +x-direction and the −y-direction (due to momentum conservation). Hence it would move off at an angle below the +x-axis, i.e., in the fourth quadrant.

13. It takes less energy to ionize the electron that is in an excited state than one that is in the ground state. The excited state already has more energy.

15. During an optical pumping process, light is used to "pump" electrons from lower energy levels to higher ones. This results in a deviation of population of quantized energy states from its thermal equilibrium distribution.

17. In a *spontaneous* emission, electrons jump from a higher-energy state to a lower-energy state without any external stimulation, and a photon is released in the process. *Stimulated* emission, on the other hand, is an induced process. The electron in the higher-energy orbit can jump to a lower-energy orbit when a photon of energy that equals the difference of the energy between the two orbits is introduced. Once atoms are prepared with enough electrons in the higher-energy state, stimulating photons triggers them to jump down to the lower-energy state. The emitted photons trigger the rest of the electrons and eventually all the electrons will be in the lower-energy state.

CHAPTER 27 EXERCISES

1. 9670 nm
3. 1.06×10^{-5} m, 2.83×10^{13} Hz
5. 690 °C
7. 4800 °C
9. 3.8×10^{19} per m² per s
11. 306 nm in the UV region
13. (a) 1.32×10^{-18} J (b) 8.27 eV
15. 4.0×10^{-19} J or 2.5 eV
17. 354 nm
19. 254 nm
21. (a) 6.7×10^{-34} J·s (b) 2.9×10^{-19} J
23. 4.5 eV
25. (a) 0.625 V (b) 2.33 eV (c) Zero
27. 0.44 eV
29. 180°
31. 54°
33. (a) The correct answer is (2), less than 5.0 keV but not zero. According to conservation of momentum and energy, the electron must recoil so it has some kinetic energy at the expense of the photon energy. Thus the photon has some energy but less than its initial amount. (b) 20 eV
35. (a) 66.0° (b) 3.19×10^6 m/s
37. (a) $E_2 = -3.40$ eV (b) $E_3 = -1.51$ eV
39. $n = 310$
41. (a) 10.2 eV (b) 1.89 eV (c) The first is UV; the second is visible (red).
43. (a) The correct choice is (1). The kinetic energy of the electron is $K = E_\gamma - 13.6$ eV. If $E_\gamma \to E_\gamma' = 2E_\gamma$, then
$K \to K' = E_\gamma' - 13.6 \text{ eV} = 2E_\gamma - 13.6 \text{ eV}$,
and then we have $\dfrac{K'}{K} = \dfrac{2E_\gamma - 13.6 \text{ eV}}{E_\gamma - 13.6 \text{ eV}} =$
$\dfrac{2(E_\gamma - 13.6 \text{ eV}) + 13.6 \text{ eV}}{E_\gamma - 13.6 \text{ eV}} =$
$2 + \dfrac{13.6 \text{ eV}}{E_\gamma - 13.6 \text{ eV}} > 2$. So, K'/K is more than doubled. (b) If frequency is 7.0×10^{15} Hz, then the kinetic energy of the electron is 15.4 eV. If the frequency is 1.40×10^{16} Hz, then the kinetic energy of the electron is 44.4 eV. Clearly, K_2 is more than double K_1.
45. (a) The correct choice is (1) because longer wavelengths are associated with smaller values of ΔE. To get a small ΔE, n_i should be as large as possible and $\Delta n = |n_f - n_i|$ as small as possible. Thus, $n = 5 \to n = 3$ gives the longest wavelength. (b) $\Delta E_{5\to3} = 0.967$ eV; $\Delta E_{6\to2} = 3.02$ eV; $\Delta E_{2\to1} = 10.2$ eV and the corresponding wavelengths are $\lambda_{5\to3} = 1280$ nm; $\lambda_{6\to2} = 410$ nm; $\lambda_{2\to1} = 122$ nm.
47. (a) 2.55 eV (b) Setting $\Delta E =$
$(-13.6 \text{ eV})\left[\dfrac{1}{n_f{}^2} - \dfrac{1}{n_i{}^2}\right] = 2.55$ eV, then it can be seen that only the transition with $n_i = 2$ and $n_f = 4$ yields an energy difference of 2.55 eV.
49. (a) The answer is (1) one. (b) 2 to 3 (c) 1.89 eV and 656 nm (red)
51. $(2.2 \times 10^6 \text{ m/s})/n$
53. (a) For the 2.0-eV state, $\lambda = 620$ nm. For the 4.0-eV state, $\lambda = 310$ nm. (b) The 2.0-eV transition is in the visible range.
55. (a) 3.3 eV (b) 0.35 V (c) 376 nm
57. (a) 1.02 MeV (b) 1.21×10^{-3} nm (c) 1.02 MeV

59. (a) 2.00 MeV (b) about 1 keV
61. 9.13×10^5 m/s

CHAPTER 28 MULTIPLE CHOICE

1. (a)
3. (b)
5. (c)
7. (a)
9. (a)
11. (b)
13. (a)
15. (a)

CHAPTER 28 CONCEPTUAL QUESTIONS

1. Its wavelength is too short compared to everyday dimensions, so we do not observe a wave nature.
3. The wavelength will be shorter, as a higher potential difference yields more linear momentum and the de Broglie wavelength is inversely proportional to momentum.
5. If the proton's charge were decreased, it would attract the electron less strongly, so the electron would not be held as close to the proton. Therefore, the electron would be less likely to be found as close to the proton as it now is, and the radius of the probability cloud would increase.
7. The principal quantum number n provides information on the electron energies as well as orbital radii of the states. The quantum number l is associated with the orbital angular momentum of the electron.
9. The atoms in a given group all have similar outer shells with similar numbers of valence electrons and hence have similar chemical properties. The atoms in a given period all have the same maximum principal quantum number n.
11. According to the uncertainty principle, the product of uncertainty in position and the uncertainty in momentum is on the order of Plank's constant. A bowling ball's large diameter and momentum (mostly due to its large mass) make the uncertainty in them (determined by the extremely small value of Planck's constant) negligible. However, for an electron, with its very small mass, the uncertainty in both its location and momentum cannot be ignored.
13. Linear momentum needs to be conserved so the particles will be moving afterward because the initial photon has momentum. Therefore, the electron–positron pair must have some kinetic energy. The input energy is converted into not only their rest energy but also their kinetic energy.
15. In pair annihilation, two photons are created because momentum conservation requires that the total momentum be zero. The two photons must travel in exactly opposite directions for the momenta to add to zero afterward.

CHAPTER 28 EXERCISES

1. 2.7×10^{-38} m
3. (a) The electron will have (3) a longer de Broglie wavelength due to its smaller mass. The de Broglie wavelength of a particle is inversely proportional to its mass. (b) $\lambda_{\text{electron}} = 7.28 \times 10^{-6}$ m and $\lambda_{\text{proton}} = 3.97 \times 10^{-9}$ m.

5. 1.5×10^4 V
7. (a) The correct choice is (3), a decrease due to the potential difference. The proton gains speed from the potential difference and its de Broglie wavelength is inversely proportional to its speed. (b) −53%
9. $\lambda_2/\lambda_1 = 1/\sqrt{2} \approx 0.71$
11. 8.89×10^{-18} J = 55.6 eV
13. (a) 3.71×10^{-63} m (b) 1.18×10^{72} (c) There would be an increase but it would be undetectable because 1 is negligible compared to 1.18×10^{72}.
15. $\dfrac{|\psi_1|}{|\psi_2|} = \sqrt{2} \approx 1.41$
17. (a) 4.07 MeV, 16.3 MeV, 36.6 MeV (b) 32.5 MeV, gamma ray
19. (a) $\ell = 2, m_\ell = 0, m_s = \pm\frac{1}{2} \to 2$ states
$\ell = 2, m_\ell = +2, m_s = \pm\frac{1}{2} \to 2$ states
$\ell = 2, m_\ell = +1, m_s = \pm\frac{1}{2} \to 2$ states
$\ell = 2, m_\ell = -2, m_s = \pm\frac{1}{2} \to 2$ states
$\ell = 2, m_\ell = -1, m_s = \pm\frac{1}{2} \to 2$ states
So there are 10 states.
(b) $\ell = 3, m_\ell = 0, m_s = \pm\frac{1}{2} \to 2$ states
$\ell = 3, m_\ell = +3, m_s = \pm\frac{1}{2} \to 2$ states
$\ell = 3, m_\ell = +2, m_s = \pm\frac{1}{2} \to 2$ states
$\ell = 3, m_\ell = +1, m_s = \pm\frac{1}{2} \to 2$ states
$\ell = 3, m_\ell = -3, m_s = \pm\frac{1}{2} \to 2$ states
$\ell = 3, m_\ell = -2, m_s = \pm\frac{1}{2} \to 2$ states
$\ell = 3, m_\ell = -1, m_s = \pm\frac{1}{2} \to 2$ states
So there are 14 states.
21. (a) $\ell = 2$ (b) $n = 3$
23.

Na has 11 electrons

$3s$ ——•——

$2s$ ——•—— $2p$ ——•••••——

$1s$ ——•——

(a)

Ar has 18 electrons

$3s$ ——•—— $3p$ ——•••••——

$2s$ ——•—— $2p$ ——•••••——

$1s$ ——•——

(b)

25. (a) $1s^2 2s^2 2p^1$ (b) $1s^2 2s^2 2p^6 3s^2 3p^6 4s^2$ (c) $1s^2 2s^2 2p^6 3s^2 3p^6 3d^{10} 4s^2$ (d) $1s^2 2s^2 2p^6 3s^2 3p^6 3d^{10} 4s^2 4p^6 4d^{10} 5s^2 5p^2$
27. It would have a $1s^3$ configuration, which would be the first closed shell (inert) gas. With three spin orientations, s states can contain three electrons without violating the Pauli exclusion principle.
29. (a) The correct answer is (2) the same, because both have the same uncertainty in momentum. (b) Both 0.21 m

31. 1.1×10^{-12} s

33. 7.0×10^{-7} m

35. **(a)** The correct answer is (1). By the uncertainty principle $(\Delta E)(\Delta t) \geq h/2\pi$. If $\Delta t_A > \Delta t_B$, then $(\Delta E)_A < (\Delta E)_B$. So the width of the spectral line for A will be smaller. **(b)** $\Delta E_B/\Delta E_A = 10^4$

37. 938.27 MeV

39. **(a)** The correct answer is (3). Since $m_p < m_n$, it takes *less* energy to produce a proton–antiproton pair than a neutron–antineutron pair. **(b)** The threshold frequency for proton–antiproton pair production is 4.541×10^{23} Hz. The threshold frequency for neutron–antineutron pair production is slightly higher at 4.547×10^{23} Hz.

41. 12 cm

43. $\theta_{el}/\theta_{light} = 3.15 \times 10^{-4}$

45. $\Delta p \geq 2.48 \times 10^{-20}$ kg·m/s and $\Delta K = 2.11 \times 10^3$ MeV. This is much greater than a few MeV, so the electron would not stay in the nucleus.

CHAPTER 29 MULTIPLE CHOICE

1. **(a)**
3. **(d)**
5. **(b)**
7. **(c)**
9. **(b)**
11. **(d)**
13. **(d)**
15. **(b)**
17. **(a)**
19. **(a)**
21. **(d)**
23. **(b)**

CHAPTER 29 CONCEPTUAL QUESTIONS

1. The minimum distance of approach is larger than the nucleus radius. If the alpha particles got closer than the nucleus radius, it would feel the strong nuclear force and the scattering would have a different pattern since it would *not* be due just to Coulomb force.

3. In Rutherford scattering, the distance of closest approach is inversely proportional to the kinetic energy of the incident particle, thus $r'_{min}/r_{min} = K/K' = 3.0 \text{ MeV}/6.0 \text{ MeV} = 0.500$.

5. Carbon-13 has 13 nucleons. The nitrogen isobar of carbon-13 should have 13 nucleons, so it is nitrogen-13.

7. Nucleon number (the sum of the proton number and the neutron number) is still conserved in the process $n \rightarrow p + e^- + \overline{v}_e$, even though neutron and proton numbers are not.

9. Due to momentum conservation, the decay particles are moving in opposite directions so they carry away some of the kinetic energy.

11. The decay processes can be represented as $^A_Z X_N \xrightarrow{\alpha} {}^{A-4}_{Z-2} Y_{N-2} \xrightarrow{\beta-} {}^{A-4}_{Z-1} Z^*_{N-3} \xrightarrow{\gamma} {}^{A-4}_{Z-1} Z_{N-3}$. The final nucleus has mass number $A - 4$, atomic number $Z - 1$, and neutron number $N - 3$. The final product remains the same regardless of whether alpha or beta decay occurs first; however, the intermediate products will vary.

13. None. It is totally independent of temperature, environment, and chemistry.

15. **(a)** Infinite **(b)** Zero

17. Since the number of original nuclei that remain is $N = N_o e^{-\lambda t}$, the number of those that have decayed is therefore
$N' = N_o - N = N_o(1 - e^{-\lambda t})$.

19. $^{238}_{92}U$ is even-even so tends to be more stable than $^{235}_{92}U$, which is even-odd. Even so, both are unstable because $Z > 82$, but $^{238}_{92}U$ has a longer half-life, that is, it is closer to stability.

21. In both the fusion of very light nuclei and the fission of very heavy nuclei, the average binding energy per nucleon *increases*. The binding energy is the energy released during each of these processes.

23. From Table 29.4, the RBE for X-rays is 1 and 20 for alphas. Equal absorbed doses means a dose 20 times more effective for alphas, thus the effective dose of an alpha particle would be 20 times that of an X-ray.

25. The detectors need to be accurate to measure the gamma ray energy of 0.511 MeV. Also, the two photons arrive in coincidence exactly. The detectors need to be able to identify both photons as originating from the same annihilation event.

27. X-rays and gamma rays are absorbed continuously as they pass through tissue and therefore release their energy gradually over a fairly long range. Particle beams, on the other hand, are charged, and the interaction of these charged particles with the tissue causes them to be stopped suddenly, and hence to release their energy very quickly over a short distance. This energy is therefore mostly deposited to a small portion of tissue (the tumor).

CHAPTER 29 EXERCISES

1. **(a)** 40p, 50n, 40e **(b)** 82p, 126n, 82e

3. $^{41}_{19}K$

5. All uranium nuclei have 92 protons. For U-235: $N = A - Z = 235 - 92 = 143$ neutrons. Since it is neutral, it has 92 electrons. For U-238: $N = 238 - 92 = 146$ neutrons, 92 protons, 92 electrons.

7. **(a)** $R_{He} = 1.9 \times 10^{-15}$ m; $R_{Ne} = 3.3 \times 10^{-15}$ m; $R_{Ar} = 4.1 \times 10^{-15}$ m; $R_{Kr} = 5.3 \times 10^{-15}$ m; $R_{Xe} = 6.1 \times 10^{-15}$ m; $R_{Rn} = 7.3 \times 10^{-15}$ m
(b) $M_{He} = 6.68 \times 10^{-27}$ kg; $M_{Ne} = 3.34 \times 10^{-26}$ kg; $M_{Ar} = 6.68 \times 10^{-26}$ kg; $M_{Kr} = 1.40 \times 10^{-25}$ kg; $M_{Xe} = 2.20 \times 10^{-25}$ kg; 3.71×10^{-25} kg
(c) 2.3×10^{17} kg/m³; yes, the answer surprises because the density is huge compared to everyday densities.

9. **(a)** The correct choice is (2). Tritium, denoted as ^3_1H, has one proton and two neutrons, one more than the stable isotope ^2_1H. Thus, we expect it to undergo β^- decay. **(b)** The decay is $^3_1H \rightarrow {}^3_2He + {}^0_{-1}e$. The daughter nucleus is ^3_2He, which is stable.

11. **(a)** $^{237}_{93}Np \rightarrow {}^{233}_{91}Pa + {}^4_2He$
(b) $^{32}_{15}P \rightarrow {}^{32}_{16}S + {}^0_{-1}e$ **(c)** $^{56}_{27}Co \rightarrow {}^{56}_{26}Fe + {}^0_{+1}e$
(d) $^{56}_{27}Co + {}^0_{-1}e \rightarrow {}^{56}_{26}Fe$ **(e)** $^{42}_{19}K^* \rightarrow {}^{42}_{19}K + \gamma$

13. **(a)** $\alpha - \beta$: $^{209}_{82}Pb + {}^0_{-1}e \rightarrow {}^{209}_{81}Tl$; $^{209}_{81}Tl + {}^4_2He \rightarrow {}^{213}_{83}Bi$.
(b) $\beta - \alpha$: $^{209}_{82}Pb + {}^4_2He \rightarrow {}^{213}_{84}Po$; $^{213}_{84}Po + {}^0_{-1}e \rightarrow {}^{213}_{83}Bi$.

15. **(a)** $^{238}_{92}U \rightarrow {}^{234}_{90}Th + {}^4_2He$
(b) $^{40}_{19}K \rightarrow {}^{40}_{20}Ca + {}^0_{-1}e$
(c) $^{236}_{92}U \rightarrow {}^{131}_{53}I + 3({}^1_0n) + {}^{102}_{39}Y$
(d) $^{23}_{11}Na^* \rightarrow {}^{23}_{11}Na + \gamma$ **(e)** $^{11}_6C + {}^0_{-1}e \rightarrow {}^{11}_5B$

17. The possible decay modes of ^{237}Np are as follows:

$^{237}Ac \xrightarrow{\alpha} {}^{233}Pa \xrightarrow{\beta-} {}^{233}U \xrightarrow{\alpha}$
$^{229}Th \xrightarrow{\alpha} {}^{225}Ra \xrightarrow{\beta-} {}^{225}Ac \xrightarrow{\alpha}$
$^{221}Fr \xrightarrow{\alpha} {}^{217}At \xrightarrow{\alpha} {}^{213}Bi \xrightarrow{\alpha}$
$^{209}Tl \xrightarrow{\beta-} {}^{209}Pb \xrightarrow{\beta-} {}^{209}Bi$

or

$^{237}Ac \xrightarrow{\alpha} {}^{233}Pa \xrightarrow{\beta-} {}^{233}U \xrightarrow{\alpha}$
$^{229}Th \xrightarrow{\alpha} {}^{225}Ra \xrightarrow{\beta-} {}^{225}Ac \xrightarrow{\alpha}$
$^{221}Fr \xrightarrow{\alpha} {}^{217}At \xrightarrow{\alpha} {}^{213}Bi \xrightarrow{\beta-}$
$^{213}Po \xrightarrow{\alpha} {}^{209}Pb \xrightarrow{\beta-} {}^{209}Bi$

19. **(a)** 7.4×10^8 decays/s
(b) 4.4×10^{10} betas/min

21. 2.06×10^{-8} W

23. 28.6 years

25. **(a)** (2) Younger than, because the activity decreases as time passes, so a higher activity indicates a short time, that is, younger
(b) 1.1×10^4 years

27. 3.1%

29. **(a)** 6.75×10^{19} nuclei
(b) 3.58×10^{16} decays/s or 3.58×10^{16} Bq
(c) 2.11×10^{18} nuclei **(d)** 1.12×10^{15} Bq

31. **(a)** 9.1 mg **(b)** 2.3×10^{18} nuclei

33. 7.6×10^6 kg

35. **(a)** $^{17}_8O$ **(b)** $^{42}_{20}Ca$ **(c)** $^{10}_5B$ **(d)** Approximately the same

37. 2.013 553 u

39. **(a)** 92.2 MeV **(b)** 7.68 MeV/nucleon

41. **(a)** Tritium has the higher *total* binding energy. The total binding energy for deuterium is 2.22 MeV. For tritium it is 8.48 MeV. **(b)** Deuterium has the lower *average* binding energy *per nucleon*. For deuterium, $E_b/A = 1.11$ MeV/nucleon. For tritium, $E_b/A = 2.83$ MeV/nucleon.

43. **(a)** 104.7 MeV **(b)** 7.476 MeV/nucleon

45. **(a)** $^{27}_{13}Al \rightarrow {}^{23}_{11}Na + {}^4_2He$ **(b)** Energy must be put *into* the system. This energy is +10.09 MeV.

47. 7.59 MeV/nucleon

49. 36 μCi

51. Dose (in rem) = Σ[Dose (in rad) × RBE] = $(0.5 \text{ rad})(1) + (0.3 \text{ rad})(4) + (0.1 \text{ rad})(20) = 3.7$ rem in 2 months. So yes, the maximum permissible radiation dosage is exceeded.

53. Effective dose = 26.6 rem = 0.266 Sv

55. **(a)** 2.15 g **(b)** $^{215}_{83}Bi \rightarrow {}^{215}_{84}Po + {}^0_{-1}e$; the product nucleus is $^{215}_{84}Po$ **(c)** 3.35×10^{20} nuclei **(d)** 1.80×10^{14} nuclei **(e)** After 10 min: $R_{10} = 1.61 \times 10^{18}$ Bq = 4.36×10^7 Ci; after 1.0 h: $R_{60} = 8.67 \times 10^{11}$ Bq = 23.4 Ci

57. **(a)** 11.5 MeV **(b)** 1.08×10^{-13} m

59. **(a)** $^{223}_{88}Ra \rightarrow {}^{219}_{86}Rn + {}^4_2He$ and $^{219}_{85}At \rightarrow {}^{219}_{86}Rn + {}^0_{-1}e + \overline{v}_e$
(b) $^{219}_{86}Rn \rightarrow {}^{215}_{84}Po + {}^4_2He$.

61. **(a)** *Unstable* because $Z > N$. The likely decay is positron emission (or electron capture). Positron emission would yield nitrogen-15 (as would EC, not shown):
$^{15}_8O_7 \rightarrow {}^{15}_7N_8 + {}^0_{+1}e$. **(b)** *Unstable* because

$N > Z$, which is not indicative of a stable low-A nucleus. The likely decay is β^-: $^8_3\text{Li}_5 \rightarrow ^8_4\text{Be}_4 + ^0_{-1}\text{e}$. **(c)** *Unstable* because $A > 83$. It could decay by alpha or beta decay: $^{222}_{86}\text{Rn}_{136} \rightarrow ^{218}_{84}\text{Po}_{134} + ^4_2\text{He}$ or $^{222}_{86}\text{Rn}_{136} \rightarrow ^{222}_{87}\text{Fr}_{135} + ^0_{-1}\text{e}$. **(d)** *Unstable* because it is a low-A nucleus and has $N > Z$. The most likely decay would be β^-: $^{27}_{12}\text{Mg}_{15} \rightarrow ^{27}_{13}\text{Al}_{14} + ^0_{-1}\text{e}$. **(e)** This isotope is likely to be stable because $N \approx Z$ for a low-A nucleus. **(f)** The reactions are shown with the individual parts above.

63. (a) $R = 198$ decays/s $= 200$ Bq
(b) 8.0×10^6 y

CHAPTER 30 MULTIPLE CHOICE

1. (c)
3. (a)
5. (d)
7. (a)
9. (c)
11. (a)
13. (a)
15. (b)
17. (a)
19. (d)
21. (c)
23. (a)

CHAPTER 30 CONCEPTUAL QUESTIONS

1. When both particles are moving toward each other, the magnitude of the total momentum is only the difference between the magnitudes of the two momenta of the two particles. According to the conservation of momentum, the total momentum after the collision will be less than if one were at rest. In turn, the particles have less kinetic energies, so the reaction requires less incident kinetic energy.

3. ^{25}Mg would have a larger capture cross-section because it has an odd neutron number and thus the likelihood of neutron capture (so as to create a neutron pair in the nucleus) is higher.

5. The hydrogen nuclei in water can capture neutrons to form deuterium and thus remove neutrons from the chain reaction.

7. Most of the uranium in a reactor is ^{238}U. Although this isotope is not fissionable, it can be converted into ^{239}Pu by fast neutron bombardment and absorption. ^{239}Pu *is* fissionable and can be used to construct a bomb.

9. For a sustained viable fusion plant, a large number of fusion reactions and thus a high plasma density are required. For the nuclei to get close enough to fuse, high kinetic energies and hence very high temperatures are also needed. These requirements are "at odds" because very high temperatures tend to make the plasma less dense (expansion). Alternatively, if the higher density is acheived, it would be at a temperature that is not high enough to allow fusion.

11. $p + \bar{\nu}_e \rightarrow n + \beta^+$ and $p + n \rightarrow ^2_1\text{H} + \gamma$. The 2.22 MeV gamma ray is from the formation of deuterium. The two 0.511 MeV gamma

rays result from pair annihilation when the positron meets an electron.

13. The Q value is the energy released in the decay. Some of this energy goes to the beta particle, some to the recoil of the nucleus, and some to the neutrino. Therefore, the electron does not get all of the released energy.

15. The forces created by the virtual exchange particles can be predicted, and these predictions agree with experiment.

17. The range of the strong nuclear force is about 100 times greater than the range of the weak nuclear force. The range of the force varies inversely with the mass of the exchange particle. So the mass of the short-range weak force exchange particle should be greater than the mass of the relatively long-range strong force exchange particle. The ratio of the masses of the exchange particles should be on the order of 100 to 1000 times, because $m_W/m_S = R_S/R_W \approx 10^{-15}$ m/10^{-17} m ≈ 100.

19. The stable leptons are the electron and all three neutrinos (ν_e, ν_μ, ν_τ). Of the hadrons, only the proton is truly stable (and there is even some question about that). The neutron is *much* more stable than the other hadrons, but outside the nucleus it does decay in about 90 s.

21. All hadrons are thought to be composed of quarks and/or antiquarks. Quarks are not believed to be able to exist freely outside the nucleus.

23. The total energy E of a particle is the sum of its rest energy E_o and its kinetic energy K. At *very* high energies, the kinetic energy is so much greater than the rest energy that the particle behaves as though it has no rest energy: $E = E_o + K \approx K$. This is just like a photon, which has no rest energy because it is truly massless.

CHAPTER 30 EXERCISES

1. (a) $^1_0\text{n} + ^{40}_{18}\text{Ar} \rightarrow ^{37}_{16}\text{S} + ^4_2\text{He}$
(b) $^1_0\text{n} + ^{235}_{92}\text{U} \rightarrow ^{98}_{40}\text{Zr} + ^{135}_{52}\text{Te} + 3(^1_0\text{n})$
(c) $^{14}_7\text{N} \, (\alpha, p) \, ^{17}_8\text{O}$ **(d)** $^{13}_6\text{C} \, (^1_1\text{H}, \alpha) \, ^{10}_5\text{B}$
3. (a) $^{22}_{11}\text{Na}$, but since $Q < 0$ it will *not* occur spontaneously. **(b)** $^{222}_{86}\text{Rn}$, and since $Q > 0$ it will occur spontaneously **(c)** $^{12}_6\text{C}$, but since $Q < 0$ it will *not* occur spontaneously
5. (a) The correct answer is (1) positive, because this is a decay (spontaneous) process. **(b)** $+4.28$ MeV
7. 4.36 MeV
9. Exoergic, $Q = +6.50$ MeV
11. 5.38 MeV
13. (a) The correct answer is (1) greater, because $K_{min} = (1 + m_a/M_A)|Q|$. The difference in Q value (three times) is greater than the difference caused by the target mass (15 and 20). **(b)** 3.05
15. 36 collisions
17. (a) $^2_1\text{H} + ^2_1\text{H} \rightarrow ^3_2\text{He} + ^1_0\text{n}$
(b) $^2_1\text{H} + ^3_1\text{H} \rightarrow ^4_2\text{He} + ^1_0\text{n}$ **(c)** For (a) 3.270 MeV and for (b) 17.59 MeV
19. (a) The correct choice is (2). If the daughter does not recoil, the 5.35 MeV is shared by the beta particle and the neutrino. Since neutrino has 2.65 MeV of the full amount of

energy, the beta particle must have less than 5.35 MeV. **(b)** 2.70 MeV **(c)** If the daughter nucleus does not recoil appreciably, the beta particle and neutrino must have equal but opposite momenta, so they must move in opposite directions.
21. 13.4 MeV
23. In a β^+ decay: $^A_Z\text{P} \rightarrow ^A_{Z-1}\text{D} + ^0_{+1}\text{e}$, so $Q = (m_p - m_D - m_e)c^2 = \{(m_p + Zm_e) - [m_D + (Z-1)m_e] - 2m_e\}c^2 = (M_P - M_D - 2m_e)c^2$
25. (a) From Exercise 20, $Q = (M_P - M_D)c^2$. For β^- decay to occur, Q must be positive. Thus $(M_P - M_D)c^2 > 0$ or $M_P > M_D$. **(b)** By the same reasoning as in part (a), $Q > 0$. From Exercise 23, $Q = (M_P - M_D - 2m_e)c^2 > 0$, thus $M_P - M_D - 2m_e > 0$, so $M_P > M_D + 2m_e$.
27. (a) The correct choice is (3) decreases, because if energy increases, the mass increases, and the range is inversely proportional to the mass. **(b)** 1.98×10^{-16} m
29. 4.71×10^{-24} s
31. $v = 0.999999999928c(\approx$ but $<c!)$
33. (a) 0.27 mm **(b)** 3930 MeV
35. (a) The correct answer is (1): A neutron is udd, so the antineutron is $\bar{u}\bar{d}\bar{d}$.
(b) $q(\bar{n}) = q(\bar{u}) + q(\bar{d}) + q(\bar{d}) = (-2e/3) + (e/3) + (e/3) = 0$
37. $R = 6.11 \times 10^{-17}$ decays/s $= 1.7 \times 10^{-27}$ Ci. A source of 1.0 μCi decays at a rate of 3.70×10^4 decays/s, which is *much* greater than the predicted proton decay from 1 liter of water.
39. Depending on the estimate assumptions, the answer is on the order of several μCi.
41. (a) $^{14}_6\text{C} \rightarrow ^{14}_7\text{N} + ^0_{-1}\text{e} + \bar{\nu}_e$ **(b)** The neutrino is an anti-electron neutrino, and the beta particle is an electron (β^-). **(c)** 0.1565 MeV **(d)** $0.498c = 1.49 \times 10^8$ m/s opposite the direction of the neutrino
43. (a) The correct choice is (1) only EC can happen. From Exercise 42, $Q_{EC} = (M_P - M_D)c^2 = (M_{Be} - M_{Li})c^2 = +0.8616$ MeV. Since $Q_{EC} > 0$, this *can* occur. From Exercise 23, $Q_{\beta^+} = (M_P - M_D - 2m_e)c^2 = (M_{Be} - M_{Li} - 2m_e)c^2 = -0.160$ MeV. Since $Q_{\beta^+} < 0$, this *cannot* occur. **(b)** There is no neutrino emitted in EC.
45. 5.4×10^{16} K
47. (a) $\pi^0 \rightarrow 2\gamma$ **(b)** 155 MeV **(c)** 8.01×10^{-15} m **(d)** 5.20×10^{-8} m
49. (a) $\bar{\nu}_e + ^1_1\text{H} \rightarrow ^1_0\text{n} + ^0_{+1}\text{e}$. By charge conservation, the charged particle product must be a positive electron, or a positron.
(b) $Q = -1.8$ MeV, and this reaction is endothermic since $Q < 0$
(c) $^1_0\text{n} + ^1_1\text{H} \rightarrow ^2_1\text{H} + \gamma$ and $^0_{+1}\text{e} + ^0_{-1}\text{e} \rightarrow \gamma + \gamma$
(d) For the neutron–proton process, $E_\gamma = 2.224$ MeV, and for the beta-positron annihilation process, $E_\gamma = 1.022$ MeV
51. 1.0×10^{-22} m, which is approximately 10^{-7} times smaller than the range of the strong force and about 10^{-5} smaller than the range of the weak force. At these energies, the collisions should serve as a good way to probe the details of both of these forces.

APPENDIX I

A: Symbols, Arithmetic Operations, Exponents, and Scientific Notation

Exercises on Symbol Usage

1. Any x between +3 and +8 and any between −3 and −8 (including the end values of each interval)

3. 490 widgets

5. The value of y quadruples (increases by a multiplicative factor of 4).

Exercises on Arithmetic Operations

1. Without parentheses following the mnemonic leads to $3^2 + 4^2/5^2 - \sqrt{4} + 6 = 9 + 16/25 - \sqrt{4} + 6 = 9 + 0.64 - 2 + 6 = 13.64$. To make the required result of +5 uses parentheses as follows: $(3^2 + 4^2)/(5^2) - \sqrt{4} + 6 = (25)/(25) - 2 + 6 = 1 - 2 + 6 = +5$

3. 9

Exercises on Exponents and Exponential Notation

1. 1/2

3. ±5184

Exercises on Scientific Notation (Powers-of-10 Notation)

1. If your weight is 175 lbs, for example, then it is written as 1.75×10^2 lb.

3. **(a)** 1.10×10^2 **(b)** 2.0×10^{-3} **(c)** 3.27×10^2 **(d)** 1.80×10^5

5. 9.0×10^{16}

B: Algebra and Common Algebraic Relationships

Exercises on Algebra

1. $y^2 - 4xy + 4x^2$

3. 0.35 s and 5.8 s, both are physically reasonable.

5. $x = 2$ and $y = -1$

C: Geometric Relationships

Exercises on Geometric Relationships

1. **(a)** on the order of 1.4×10^4 cm^3 **(b)** on the order of 900 in.^{3}

3. 0.16 L

5. 160 cm^3.

D: Trigonometric Relationships

Exercises on Trigonometric Relationships

1. **(a)** 87 m **(b)** 100 m

3. $\sin(30° + 60°) = \sin(30°)\cos(60°) + \cos(30°)\sin(60°) = \frac{1}{2} \cdot \frac{1}{2} + \frac{\sqrt{3}}{2} \cdot \frac{\sqrt{3}}{2} = \frac{1}{4} + \frac{3}{4} = 1$

5. **(a)** 4.6 cm and 10 cm **(b)** 23 cm^2 **(c)** 26 cm

E: Logarithms

Exercises on Logarithms

1. **(a)** 1.30 **(b)** 1.70 **(c)** 3.40 **(d)** 0.48

3. **(a)** 3.00 **(b)** 0.69 **(c)** 4.61 **(d)** 1.10

5. **(a)** $\log 1500 = 3.18$ and $\log(15 \times 100) = \log 15 + \log 100 = 1.18 + 2.00 = 3.18$ **(b)** $\log 6400 = 3.81$ and $\log(64/0.010) = \log 64 - \log 0.010 = 1.81 - (-2.00) = 3.81$ **(c)** $\log 8 = 0.90$ and $\log 2^3 = 3 \log 2 = 3(0.30) = 0.90$

7. 35 min

Photo Credits

Index

for ideal gas, 424–31
irreversible, 419–20
isobaric process, 427–28, 431t
isometric process, 427–28, 431t
isothermal process, 427–28, 431t, 433, 443
reversible, 419, 443
Projectile motion, 80–88
at arbitrary angles, 82–88
horizontal, 80–81
and momentum change, 196
vertical, 53, 55
Projectiles, 67
Promethium, 976, 981, 1003
Propagation, of waves, 456, 468–69, 471, 497, 827–28
Propagation of errors, 18
Propellant gases in aerosol cans, 427
Proper length, 887, 889, 901
Properties. *See* Materials, various.
Proper time interval, 884–85, 889, 901
Proportional limit, 313
Propulsion
airboat fan, 214
jet, 114–15, 208–11
via magnetohydrodynamics, 666–67
Proton(s), 967–68
atomic mass unit, 983
in Bohr theory of hydrogen atom, 921–23
electric charge of, 530–31
electric potential, 563–64, 566
as hadrons, 1020
nuclear stability and, 981–83, 985
in unification theory, 1024
Proton number (Z), 968–69, 970, 973
Proton-proton cycle, 1012
p–T **diagram,** 419
Pulleys, and inertia, 285–86
Pulmonary physics, 109, 340, 376
Pulse, wave, 456, 469
Pulsed induction (PI), 702
Pumping, laser, 927
Pumps
gasoline-powered water, 439
human heart as, 326–27
thermal and heat, 440–42
p–V **diagrams,** 419, 423
adiabatic process, 428
ideal heat cycle, 443
isobaric process, 425, 426
isometric process, 427
isothermal process, 424, 430
P waves (primary), 472. *See also* Compressional waves.
Pythagorean theorem, 25

Q

Quadratic formula, A-3
Quality factor, 988
Quality of tone, 518

Quantative ultrasound, 320
Quantization
of angular momentum, 940
of light, 911, 914–18
Planck's hypothesis and, 913
thermal radiation, 911–13
Quantized charge, 532
Quantum chromodynamics (QCD), 1023, 1024
Quantum electrodynamics (QED), 1017, 1023
Quantum mechanics, 911, 939–64
de Broglie hypothesis, 939–42
Heisenberg uncertainty principle, 955–58, 1016, 1017
particles and antiparticles, 938, 958–59
Pauli exclusion principle, 950–52, 1023
periodic table of elements, 952–55
probability and radioactivity, 971
Schrödinger's wave equation, 944–45, 947
Quantum numbers, 922, 945–52. *See also* Atomic quantum numbers.
magnetic, 946, 947
orbital, 946, 947, 949
principal, 922, 923, 924, 946, 947, 949
spin, 947, 1015
Quantum of energy, 913
Quantum particles, 918–20
Quantum physics, 910–37
Bohr theory, 920–26, 940, 945, 966
Compton effect, 918–20
lasers, 926–31 (*See also* Lasers)
Planck's hypothesis, 913
quantization, 911–13, 914–18, 940
Quantum theory, 897n, 911, 913, 914–16
Quark confinement, 1023
Quark model, 1021–23
Quarks, 531n, 1001, 1021–23, 1024
Quarter-wavelength radio antenna, 727
Quasars, 896–97, 898
Quinine sulfide periodide (herapathite), 828
Q value, 1003–5, 1014

R

rad (radiation absorbed dose), 988
Radar, 491, 510, 513, 718. *See also* Doppler effect.
Radial ray (chief ray), 783
Radian (rad), 224
radians per second (rad/s), 226
radians per second squared (rad/s²), 236

Radiant energy, 405
Radiation. *See also* Radioactivity.
electromagnetic, 714–20 (*See also* Electromagnetic waves)
heat transfer by, 405–9
infrared, 405–6, 610, 718–19, 861–62
nonvisible, for telescopes, 861–62
thermal, 405, 911–13
ultraviolet, 719, 862
Van Allen belts, 684, 685f
Radiation, nuclear, 965
biological effects and medical applications, 987–91
blackbody, 407, 912–13
damage, 974, 987–89
detection of, 986–87, 991–92
domestic and industrial applications, 991–92
Radiation dosage, 988, 990
Radiation penetration, 973–74
Radiation pressure, 716–17
Radio
capacitors and oscillator circuits, 743
quarter-wavelength antenna, 727
reception, 810, 822
resonance frequency and, 742–44
waves, 717–18, 822
Radioactive dating, 978–80
Radioactive decay. *See* Decay, radioactive.
Radioactive half-lives. *See* Half-life.
Radioactive isotopes (nuclides), 969, 970–73, 974, 976, 988–91
Radioactive tracers, 991
Radioactive waste, 1011
Radioactivity, 965, 969–74
alpha decay, 970–71, 976t, 981
beta decay, 971–72, 973, 976t, 979, 981, 1014–15
decay rate, 975–80
electron capture decay, 972, 976t
gamma decay, 972–73
radiation penetration, 973–74
Radio-frequency (RF) radiation, 948
Radioisotopes, 988–91. *See also* Radioactive isotopes.
Radio telescopes, 861, 864–65
Radium, 969, 978
Radius, Schwarzschild, 897–98
Radius, speed, energy for circular orbital motion, 252
Radius of curvature, 782, 791
Radon (Ra), 378, 974
Rainbows, 769
Range, of projectile, 82–87
Rankine scale, 366

Rarefactions of sound waves, 471, 490, 504
Ray(s)
chief, of lens (central ray), 791, 792
chief, of mirror (radial ray), 783
cosmic, 958, 974, 979, 1020
extraordinary, 832
focal, 783, 791, 792
gamma, 720, 970, 974 (*See also* Gamma rays)
light, 752–53, 755, 759
ordinary, 832
parallel, 783, 791, 792
Ray diagrams
for lenses, 791–92, 793, 795, 796
for spherical mirrors, 783–88
Rayleigh, Lord, 835, 862
Rayleigh criterion, 862–65
Rayleigh scattering, 835
RC circuits, 637–41, 737–38
Reactance
capacitive, 733–35, 737, 740, 742, 745
inductive, 735–37, 740
Reaction cross-sections, 1006
Reaction forces, 113–16
Reactions. *See* Nuclear reactions.
Reaction time, 53–54
Real image, 778, 786, 793, 794, 857
Rectangular vector components, 74–75
Rectifier circuit, 702
Red light, 770, 814, 825, 835
Red shift, Doppler, 510
Reference, displacement, 150
Reference circle for vertical motion, 460–62
Reference circle object, 462
Reference frame, 88, 90
absolute, 877–78, 879
inertial, 877 (*See also* Inertial reference frame)
noninertial, 877
Reference points, 156–57
Reflecting telescopes, 859–61
Reflection, 751–52, 753–55
angle of, 753
diffuse (irregular), 753–55
law of, 753, 755, 782
phase shifts and, 815
polarization by, 830–31
of sound, 503–4
specular (regular), 753–55
thin-film interference, 815–16
total internal, 764–67, 769
of waves, 475–76
Reflection gratings, 824
Refracting telescopes, 857–59
Refraction, 751–52, 756–63
angle of, 756, 758, 759, 764
atmospheric, 762–63
defined, 756
dispersion and, 770–71

The Periodic Table of Elements

(See Section 28.3)

Main-group elements

Period	1 / 1A	2 / 2A		Transition elements											13 / 3A	14 / 4A	15 / 5A	16 / 6A	17 / 7A	18 / 8A
			3 / 3B	4 / 4B	5 / 5B	6 / 6B	7 / 7B	8	9 / 8B	10	11 / 1B	12 / 2B							
1	1 **H** 1.00794																			2 **He** 4.00260
2	3 **Li** 6.941	4 **Be** 9.01218												5 **B** 10.811	6 **C** 12.011	7 **N** 14.0067	8 **O** 15.9994	9 **F** 18.9984	10 **Ne** 20.1797
3	11 **Na** 22.9898	12 **Mg** 24.3050												13 **Al** 26.9815	14 **Si** 28.0855	15 **P** 30.9738	16 **S** 32.066	17 **Cl** 35.4527	18 **Ar** 39.948
4	19 **K** 39.0983	20 **Ca** 40.078	21 **Sc** 44.9559	22 **Ti** 47.88	23 **V** 50.9415	24 **Cr** 51.9961	25 **Mn** 54.9381	26 **Fe** 55.847	27 **Co** 58.9332	28 **Ni** 58.693	29 **Cu** 63.546	30 **Zn** 65.39	31 **Ga** 69.723	32 **Ge** 72.61	33 **As** 74.9216	34 **Se** 78.96	35 **Br** 79.904	36 **Kr** 83.80
5	37 **Rb** 85.4678	38 **Sr** 87.62	39 **Y** 88.9059	40 **Zr** 91.224	41 **Nb** 92.9064	42 **Mo** 95.94	43 **Tc** (98)	44 **Ru** 101.07	45 **Rh** 102.906	46 **Pd** 106.42	47 **Ag** 107.868	48 **Cd** 112.411	49 **In** 114.818	50 **Sn** 118.710	51 **Sb** 121.76	52 **Te** 127.60	53 **I** 126.904	54 **Xe** 131.29
6	55 **Cs** 132.905	56 **Ba** 137.327	57 *****La** 138.906	72 **Hf** 178.49	73 **Ta** 180.948	74 **W** 183.84	75 **Re** 186.207	76 **Os** 190.23	77 **Ir** 192.22	78 **Pt** 195.08	79 **Au** 196.967	80 **Hg** 200.59	81 **Tl** 204.383	82 **Pb** 207.2	83 **Bi** 208.980	84 **Po** (209)	85 **At** (210)	86 **Rn** (222)
7	87 **Fr** (223)	88 **Ra** 226.025	89 †**Ac** 227.028	104 **Rf** (261)	105 **Db** (262)	106 **Sg** (263)	107 **Bh** (262)	108 **Hs** (265)	109 **Mt** (266)	110 **Ds** (281)	111 ** (272)	112 ** (285)		114 ** (289)		116 ** (292)		

Metals **Nonmetals** **Noble gases**

*Lanthanide series	58 **Ce** 140.115	59 **Pr** 140.908	60 **Nd** 144.24	61 **Pm** (145)	62 **Sm** 150.36	63 **Eu** 151.965	64 **Gd** 157.25	65 **Tb** 158.925	66 **Dy** 162.50	67 **Ho** 164.930	68 **Er** 167.26	69 **Tm** 168.934	70 **Yb** 173.04	71 **Lu** 174.967
†Actinide series	90 **Th** 232.038	91 **Pa** 231.036	92 **U** 238.029	93 **Np** 237.048	94 **Pu** (244)	95 **Am** (243)	96 **Cm** (247)	97 **Bk** (247)	98 **Cf** (251)	99 **Es** (252)	100 **Fm** (257)	101 **Md** (258)	102 **No** (259)	103 **Lr** (260)

** Not yet named

Notes: (1) Values in parentheses are the mass numbers of the most common or most stable isotopes of radioactive elements. (2) Some elements adjacent to the stair-step line between the metals and nonmetals have a metallic appearance but some nonmetallic properties. These elements are often called metalloids or semimetals. There is no general agreement on just which elements are so designated. Almost every list includes Si, Ge, As, Sb, and Te. Some also include B, At, and/or Po.

Physical Data*

Quantity	Symbol	Approximate Value
Universal gravitational constant	G	$6.67 \times 10^{-11} \, \text{N} \cdot \text{m}^2/\text{kg}^2$
Acceleration due to gravity (generally accepted value on surface of Earth)	g	$9.80 \, \text{m/s}^2 = 980 \, \text{cm/s}^2 = 32.2 \, \text{ft/s}^2$
Speed of light	c	$3.00 \times 10^8 \, \text{m/s} = 3.00 \times 10^{10} \, \text{cm/s} = 1.86 \times 10^5 \, \text{mi/s}$
Boltzmann's constant	k_B	$1.38 \times 10^{-23} \, \text{J/K}$
Avogadro's number	N_A	$6.02 \times 10^{23} \, \text{mol}^{-1}$
Gas constant	$R = N_A k_B$	$8.31 \, \text{J/(mol} \cdot \text{K)} = 1.99 \, \text{cal/(mol} \cdot \text{K)}$
Coulomb's law constant	$k = 1/4\pi\epsilon_o$	$9.00 \times 10^9 \, \text{N} \cdot \text{m}^2/\text{C}^2$
Electron charge	e	$1.60 \times 10^{-19} \, \text{C}$
Permittivity of free space	ϵ_o	$8.85 \times 10^{-12} \, \text{C}^2/(\text{N} \cdot \text{m}^2)$
Permeability of free space	μ_o	$4\pi \times 10^{-7} \, \text{T} \cdot \text{m/A} = 1.26 \times 10^{-6} \, \text{T} \cdot \text{m/A}$
Atomic mass unit	u	$1.66 \times 10^{-27} \, \text{kg} \leftrightarrow 931 \, \text{MeV}$
Planck's constant	h	$6.63 \times 10^{-34} \, \text{J} \cdot \text{s}$
	$\hbar = h/2\pi$	$1.05 \times 10^{-34} \, \text{J} \cdot \text{s}$
Electron mass	m_e	$9.11 \times 10^{-31} \, \text{kg} = 5.49 \times 10^{-4} \, \text{u} \leftrightarrow 0.511 \, \text{MeV}$
Proton mass	m_p	$1.672\,62 \times 10^{-27} \, \text{kg} = 1.007\,276 \, \text{u} \leftrightarrow 938.27 \, \text{MeV}$
Neutron mass	m_n	$1.674\,93 \times 10^{-27} \, \text{kg} \times 1.008\,665 \, \text{u} \leftrightarrow 939.57 \, \text{MeV}$
Bohr radius of hydrogen atom	r_1	$0.053 \, \text{nm}$

*Values from NIST Reference on Constants, Units, and Uncertainty.

Solar System Data*

Equatorial radius of the Earth	$6.378 \times 10^3 \, \text{km} = 3963 \, \text{mi}$
Polar radius of the Earth	$6.357 \times 10^3 \, \text{km} = 3950 \, \text{mi}$
	Average: $6.4 \times 10^3 \, \text{km}$ (for general calculations)
Mass of the Earth	$5.98 \times 10^{24} \, \text{kg}$
Diameter of Moon	$3500 \, \text{km} \approx 2160 \, \text{mi}$
Mass of Moon	$7.4 \times 10^{22} \, \text{kg} \approx \frac{1}{81}$ mass of Earth
Average distance of Moon from the Earth	$3.8 \times 10^5 \, \text{km} = 2.4 \times 10^5 \, \text{mi}$
Diameter of Sun	$1.4 \times 10^6 \, \text{km} \approx 864\,000 \, \text{mi}$
Mass of Sun	$2.0 \times 10^{30} \, \text{kg}$
Average distance of the Earth from Sun	$1.5 \times 10^8 \, \text{km} = 93 \times 10^6 \, \text{mi}$

*See Appendix III for additional planetary data.

Mathematical Symbols

$=$	is equal to
$\neq$	is not equal to
$\approx$	is approximately equal to
$\sim$	about
$\propto$	is proportional to
$>$	is greater than
$\geq$	is greater than or equal to
$\gg$	is much greater than
$<$	is less than
$\leq$	is less than or equal to
$\ll$	is much less than
$\pm$	plus or minus
$\mp$	minus or plus
$\bar{x}$	average value of x
Δx	change in x
$|x|$	absolute value of x
Σ	sum of
∞	infinity

The Greek Alphabet

Alpha	A	α	Nu	N	ν
Beta	B	β	Xi	Ξ	ξ
Gamma	Γ	γ	Omicron	O	o
Delta	Δ	δ	Pi	Π	π
Epsilon	E	ε	Rho	P	ρ
Zeta	Z	ζ	Sigma	Σ	σ
Eta	H	η	Tau	T	τ
Theta	Θ	θ	Upsilon	Y	υ
Iota	I	ι	Phi	Φ	ϕ
Kappa	K	κ	Chi	X	χ
Lambda	Λ	λ	Psi	Ψ	ψ
Mu	M	μ	Omega	Ω	ω